J

Chemistry 12

Chemistry 12

Authors

Hans van Kessel
St. Albert Protestant Schools

Dr. Frank Jenkins
University of Alberta

Lucille Davies
Limestone District School Board

Donald Plumb
The Bishop Strachan School

Maurice Di Giuseppe
Toronto Catholic District School Board

Dr. Oliver Lantz
University of Alberta

Dick Tompkins
Edmonton Public School Board

Program Consultant

Maurice Di Giuseppe
Toronto Catholic District School Board

THOMSON
NELSON

Australia   Canada   Mexico   Singapore   Spain   United Kingdom   United States

Nelson Chemistry 12

Authors
Hans van Kessel, Dr. Frank Jenkins,
Lucille Davies, Donald Plumb,
Maurice Di Giuseppe, Dr. Oliver Lantz,
Dick Tompkins

Contributing Writer
Milan Sanader

Director of Publishing
David Steele

Publisher
Kevin Martindale

Program Manager
Colin Bisset

Developmental Editors
Julia Lee
George E. Huff

Research
Keith Lennox

On-line Quiz Editor
Karim Dharssi

Senior Managing Editor
Nicola Balfour

Senior Production Editor
Rosalyn Steiner

Copy Editor
Ruth Peckover

Proofreader
Gilda Mekler

Production Coordinator
Sharon Latta Paterson

Editorial Assistant
Matthew Roberts

Creative Director
Angela Cluer

Art Director
Ken Phipps

Art Management
Suzanne Peden

Illustrators
Andrew Breithaupt
Steven Corrigan
Deborah Crowle
Irma Ikonen
Dave Mazierski
Dave McKay
Peter Papayanakis
Ken Phipps
Marie Price
Katherine Strain

Interior Design
Kyle Gell and Allan Moon

Cover Design
Ken Phipps

Cover Image
Tek Image/Science Photo Library

Composition
Marnie Benedict
Susan Calverley
Zenaida Diores
Krista Donnelly
Tammy Gay
Nelson Gonzalez
Janet Zanette

Design Team
Anne Bradley
Peter Papayanakis
Katherine Strain

Photo Shoot Coordinator
Julie Greener

Photo Research and Permissions
Linda Tanaka

Printer
Transcontinental Printing Inc.

National Library of Canada Cataloguing in Publication Data

Main entry under title:
Nelson chemistry 12

Includes index.
ISBN 0-17-625986-4

1. Chemistry.
I. Jenkins, Frank, 1944–

QD33.N442 2002   540
C2002-900872-7

Acknowledgments
Nelson and the authors of *Nelson Chemistry 12* thank the staff and students of Mary Ward Catholic Secondary School and The Bishop Strachan School for the use of their facilities, and for the grace and generosity of their help.

Reviewers

Advisory Panel

Carl Twiddy
Formerly of York Region District School Board, ON

Doug De La Matter
Formerly of Renfrew District School Board, ON

Milan Sanader
Dufferin-Peel Catholic District School Board, ON

Patricia Thomas
Ottawa-Carleton District School Board, ON

Accuracy Reviewer

Prof. Carey Bissonnette
Department of Chemistry
University of Waterloo

Safety Reviewer

Ian Mackellar
STAO Safety Committee

Technology Reviewer

Patricia Thomas
Ottawa-Carleton District School Board, ON

Teacher Reviewers

Thomas S.H. Baxter
Lakehead District School Board, ON

Peter Bloch
Toronto District School Board, ON

Richard Christensen
Peel District School Board, ON

Katy Farrow
Thames Valley District School Board, ON

Brenda Forbes
York Region District School Board, ON

Ann Harrison
Niagara Catholic District School Board, ON

Robin Howard
Ottawa-Carleton Catholic District School Board, ON

Lisa MacLachlan
Dufferin-Peel Catholic District School Board, ON

Robert Nalepa
Halifax Regional School Board, NS

Al Orlando
Nipissing–Parry Sound Catholic District School Board, ON

Mike Penrose
Peel District School Board, ON

Marc James Robillard
Lambton Kent District School Board, ON

Paul E. St.Louis
Renfrew County District School Board, ON

CONTENTS

Unit 3
Energy Changes and
Rates of Reaction

▶ Unit 4
Chemical Systems and Equilibrium

Unit 5
Electrochemistry

Appendixes

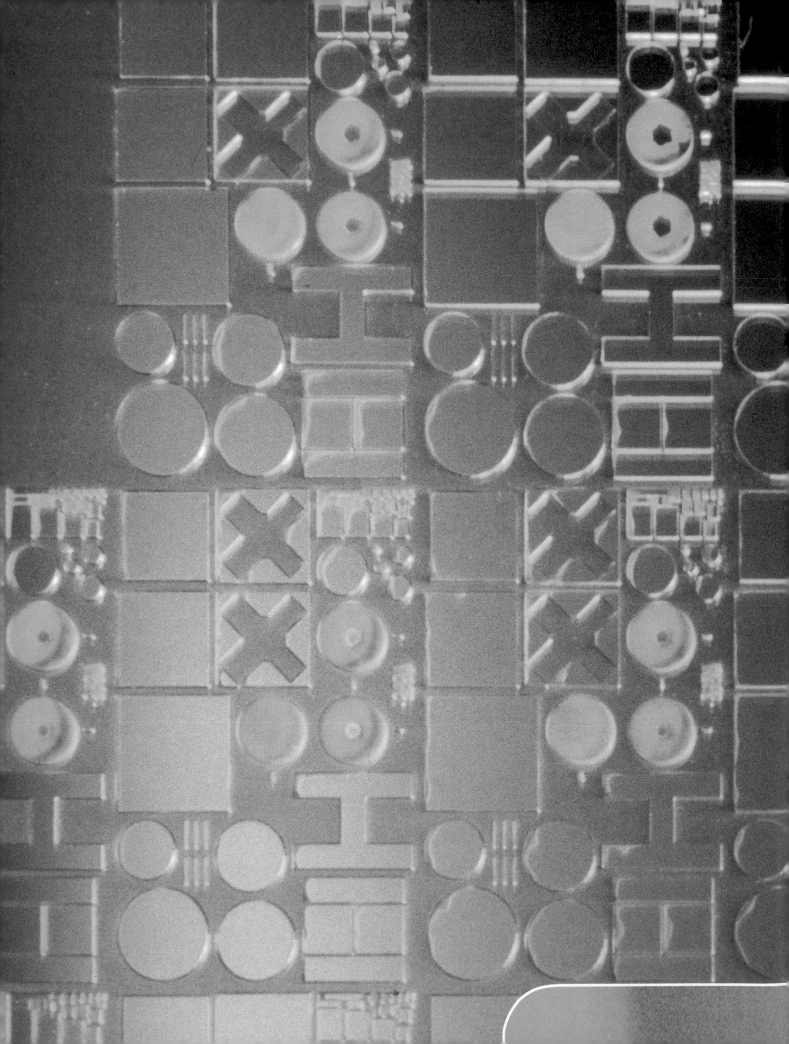

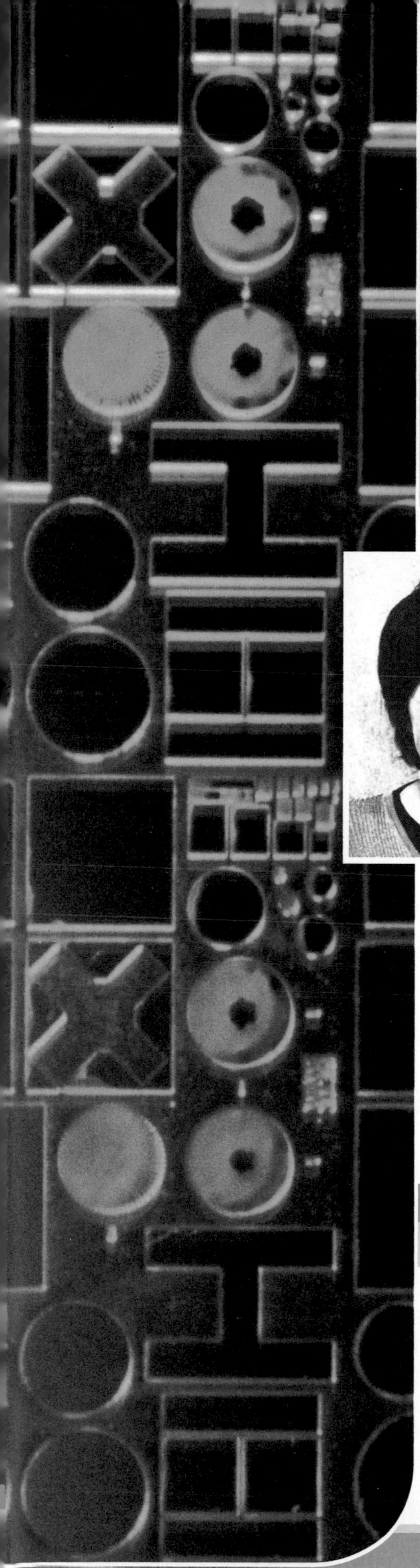

Organic Chemistry

Eugenia Kumacheva
Associate Professor
University of Toronto

"By clever synthesis, organic chemists obtain new molecules with fascinating architectures, compositions, and functions. My research group studies polymers (long-chain molecules with many repeating units) that possess fluorescent, non-linear optical, and electroactive properties. In particular, we are interested in nanostructured materials made from very small polymer particles. For example, we work on synthesizing polymers for high-density optical data storage. One of the materials designed and created in our laboratory is often pictured as a piece of new plastic about the size of a cube of sugar on which one can store the entire Canadian National Library collection. Other polymers can change their transparency when illuminated with high-intensity light. The coatings and films made from such polymers can be used to protect pilots' eyes from damaging laser light and in optical networks in telecommunication. New synthetic polymers have found a variety of exciting applications, and their use in materials science will grow even more rapidly in the future."

▶ Overall Expectations

In this unit, you will be able to

- demonstrate an understanding of the structure of various organic compounds, and of chemical reactions involving these compounds;
- investigate various organic compounds through research and experimentation, predict the products of organic reactions, and name and represent the structures of organic compounds using the IUPAC system and molecular models; and
- evaluate the impact of organic compounds on our standard of living and the environment.

▶ **Prerequisites**

Concepts

- IUPAC nomenclature of simple aliphatic hydrocarbons, including cyclic compounds
- structural and geometric isomers
- characteristic physical properties and chemical reactions of saturated and unsaturated hydrocarbons
- electronegativity and polar bonds
- chemical bonding, including ionic bonds, covalent bonds, hydrogen bonds, van der Waals forces
- formation of solutions involving polar and nonpolar substances

Understanding Concepts

1. Write the IUPAC name for each of the following compounds.

   (a)
   $$CH_3 - \overset{\displaystyle CH_2 - CH_3}{\underset{\displaystyle CH_3}{\overset{|}{\underset{|}{C}}} } - CH_2 - CH_3$$

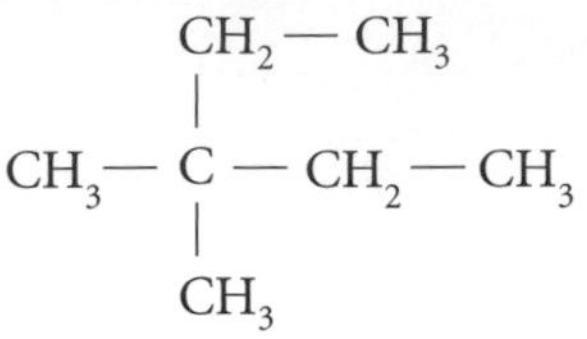

   (b)
   $$CH_2 - CH_2 - CH_3$$

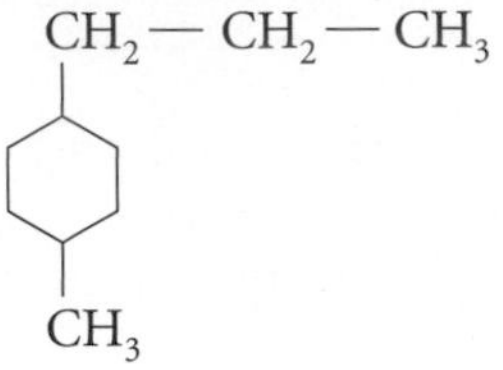

2. Draw structures of the following compounds.
   (a) pentane
   (b) 2,2-dimethylheptane
   (c) 4-ethyl-1-methylcyclohexane
   (d) 5-methyl-1-hexene
   (e) 1-butyne

3. Write a balanced chemical equation to show the complete combustion of heptane, a component of gasoline.

4. Which of the following are structural isomers?

   (a)

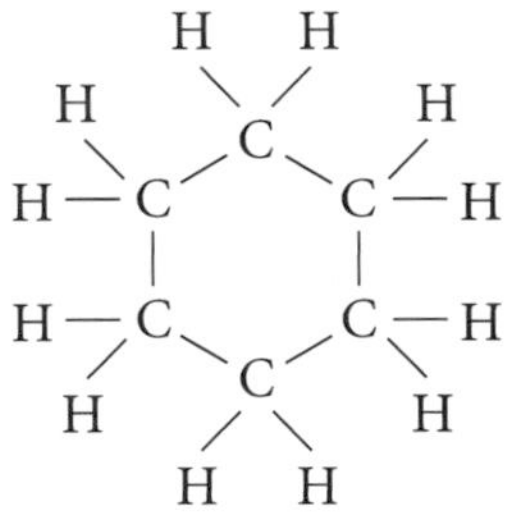

   (b)

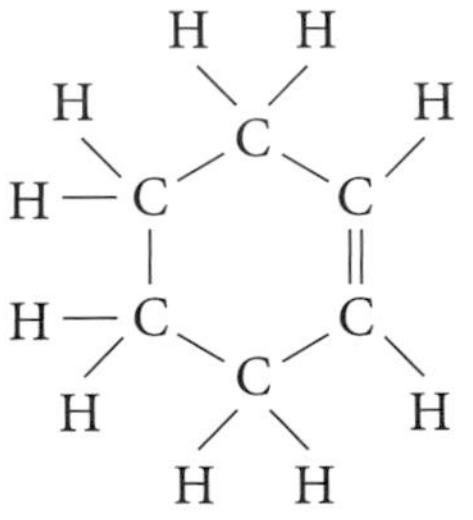

   (c)
   $$H - C \equiv C - \overset{H}{\underset{H}{\overset{|}{\underset{|}{C}}}} - \overset{H}{\underset{H}{\overset{|}{\underset{|}{C}}}} - \overset{H}{\underset{H}{\overset{|}{\underset{|}{C}}}} - \overset{H}{\underset{H}{\overset{|}{\underset{|}{C}}}} - H$$

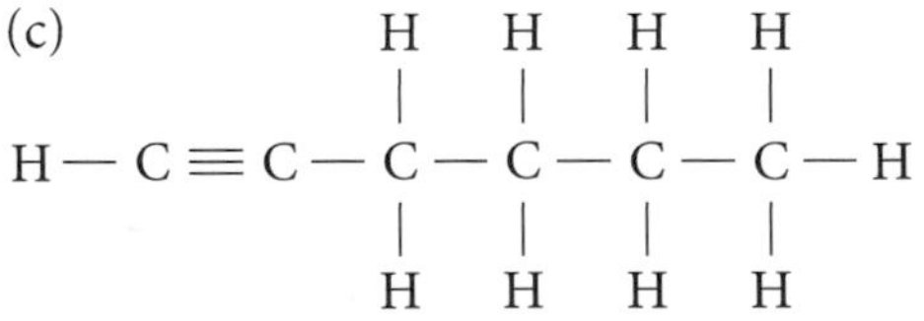

   (d)
   $$CH_3C \overset{\displaystyle CH_3}{\overset{|}{}} = \overset{\displaystyle CH_3}{\overset{|}{}} CCH_3$$

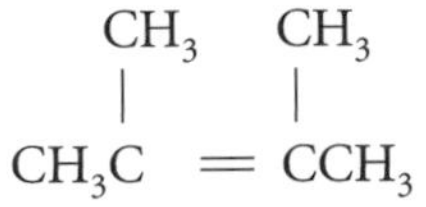

5. Predict the relative boiling points of the following two compounds.
   (a) $CH_3CH_2CH_2CH_2CH_3$

   pentane

   (b) 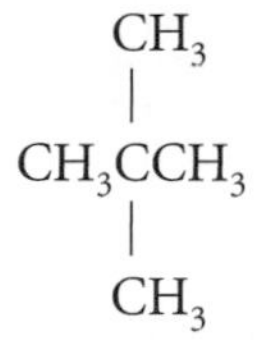
   $$CH_3$$
   $$|$$
   $$CH_3CCH_3$$
   $$|$$
   $$CH_3$$

   2,2-dimethylpropane

6. Predict the relative solubilities of the following compounds in water.

   (a) 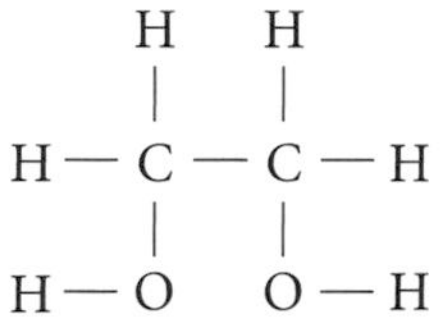
   $$\begin{array}{c} H \quad\ H \\ |\qquad | \\ H-C-C-H \\ |\qquad | \\ H-O \quad\ O-H \end{array}$$

   (b)

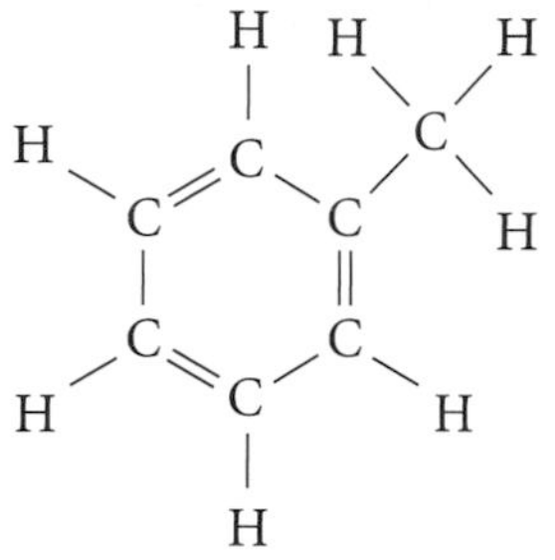

7. Write the following elements in order of increasing electronegativity: carbon, chlorine, hydrogen, nitrogen, oxygen, sulfur.

8. For each of the following compounds, describe the intramolecular bond types and the intermolecular forces.
   (a) CH_4
   (b) H_2O
   (c) NH_3

Applying Inquiry Skills

9. Three liquids are tested with aqueous bromine (**Figure 1**). Samples of the solutions are also vaporized and their boiling points determined. The evidence is shown in **Table 1**.

   Table 1

Compound	Liquid 1	Liquid 2	Liquid 3
$Br_{2(aq)}$ test	no change	turns colourless	no change
boiling point (°C)	36	39	−12

   Which of the liquids is pentane, 2-methylbutane, and 2-methyl-2-butene?

   liquid 1   liquid 2

   Figure 1

Safety and Technical Skills

10. List the safety precautions needed in the handling, storage, and disposal of
    (a) concentrated sulfuric acid;
    (b) flammable liquids, e.g., ethanol.

Organic Compounds

In this chapter, you will be able to

- classify organic compounds by identifying their functional groups, by name, by structural formula, and by building molecular models;

- use the IUPAC system to name and write structural diagrams for different classes of organic compounds, and identify some nonsystematic names for common organic compounds;

- relate some physical properties of the classes of organic compounds to their functional groups;

- describe and predict characteristic chemical reactions of different classes of organic compounds, and classify the chemical reactions by type;

- design the synthesis of organic compounds from simpler compounds, by predicting the products of organic reactions;

- carry out laboratory procedures to synthesize organic compounds;

- evaluate the use of the term "organic" in everyday language and in scientific terminology;

- describe the variety and importance of organic compounds in our lives, and evaluate the impact of organic materials on our standard of living and the environment.

In a supermarket or in a pharmacy, the term "organic" is used to describe products that are grown entirely through natural biological processes, without the use of synthetic materials. "Organic" fruits and vegetables are not treated with synthetic fertilizers or pesticides; "organic" chickens or cows are raised from organically grown feed, without the use of antibiotics. The growing "organic" market, despite higher prices over "conventionally grown" foods, indicates that some consumers believe that molecules made by a living plant or animal are different from, and indeed better than, those made in a laboratory.

In the early 18th century, the term "organic" had similar origins in chemistry. At that time, most chemists believed that compounds produced by living systems could not be made by any laboratory procedure. Scientists coined the chemical term "organic" to distinguish between compounds obtained from living organisms and those obtained from mineral sources.

In 1828, a German chemist, Friedrich Wöhler, obtained urea from the reaction of two inorganic compounds, potassium cyanate and ammonium chloride. Since then, many other organic compounds have been prepared from inorganic materials.

Organic chemistry today is the study of compounds in which carbon is the principal element. Animals, plants, and fossil fuels contain a remarkable variety of carbon compounds. What is it about the carbon atom that allows it to form such a variety of compounds, a variety that allows the diversity we see in living organisms? The answer lies in the fact that carbon atoms can form four bonds. Carbon atoms have another special property: They can bond together to form chains, rings, spheres, sheets, and tubes of almost any size and can form combinations of single, double, and triple covalent bonds. This versatility allows the formation of a huge variety of very large organic molecules.

In this chapter, we will examine the characteristic physical properties of families of organic molecules, and relate these properties to the elements within the molecule and the bonds that hold them together. We will also look at the chemical reactions that transform one organic molecule into another. Finally, we will see how these single transformations can be carried out in sequence to synthesize a desired product, starting with simple compounds.

REFLECT on your learning

1. Much of the research in organic chemistry is focused on a search for new or improved products. Suppose that you wish to develop a new stain remover, or a more effective drug, or a better-tasting soft drink. What should be the properties of the ingredients of your chosen product?

2. In the field of biology, complex systems have been developed to classify and name the countless different living organisms. Suggest an effective method of classifying and naming the vast range of organic compounds that exist.

3. From your knowledge of intramolecular and intermolecular attractions, describe features in the molecular structure of a compound that would account for its solubility and its melting and boiling points.

4. What does "organic" mean? Give as many definitions as you can.

TRY THIS activity

How Do Fire-Eaters Do That?

Have you ever wondered how some street performers can extinguish a flaming torch by "swallowing" the fire, without burning themselves? Here is an activity that might help you answer the puzzle of "how do they do that?"

Materials: 2 large glass beakers or jars; 2-propanol (rubbing alcohol); water; table salt; tongs; paper; safety lighter or match

2-propanol is highly flammable. Ensure that containers of the alcohol are sealed and stored far from any open flame.

- In a large glass beaker or jar, mix together equal volumes of 2-propanol and water, to a total of about 100 mL.

- Dissolve a small amount of NaCl (about 0.5 g) in the solution, to add colour to the flame that will be observed.

- Using tongs, dip a piece of paper about 5 cm × 5 cm into the solution until it is well soaked. Take the paper out and hold it over the jar for a few seconds until it stops dripping.

- Dispose of the alcohol solution by flushing it down the sink (or as directed by your teacher), and fill another beaker or jar with water as a precautionary measure to extinguish any flames if necessary.

- Still holding the soaked paper with tongs, ignite it using the lighter or match.

(a) From your observations, suggest a reason why "fire-eaters" do not suffer severe burns from their performance.

Organic Compounds **7**

Figure 1
The design and synthesis of new materials with specific properties, like the plastic in this artificial ski run, is a key focus of the chemical industry.

organic family a group of organic compounds with common structural features that impart characteristic physical properties and reactivity

functional group a structural arrangement of atoms that imparts particular characteristics to the molecule

With the huge number of organic substances, we would have great difficulty memorizing the properties of each compound. Fortunately, the compounds fall into **organic families** according to particular combinations of atoms in each molecule. The physical properties and reactivity of the compounds are related to these recognizable combinations, called **functional groups**. These functional groups determine whether the molecules are readily soluble in polar or non-polar solvents, whether they have high or low melting and boiling points, and whether they readily react with other molecules.

So, if we can recognize and understand the influence of each functional group, we will be able to predict the properties of any organic compound. If we can predict their properties, we can then design molecules to serve particular purposes, and devise methods to make these desired molecules.

In this chapter, we will discuss each organic family by relating its properties to the functional groups it contains. Moreover, we will focus on how one organic family can be synthesized from another; that is, we will learn about the reaction pathways that allow one functional group to be transformed into another. By the end of the chapter, we will have developed a summary flow chart of organic reactions, and we will be able to plan synthetic pathways to and from many different organic molecules. After all, designing the synthesis of new molecules, ranging from high-tech fabrics to "designer drugs," is one of the most important aspects of modern organic chemistry (**Figure 1**).

Before discussing each organic family, let's take a look at what makes up the functional groups. Although there are many different functional groups, they essentially consist of only three main components, one or more of which may be present in each functional group. Understanding the properties of these three components will make it easy to understand and predict the general properties of the organic families to which they belong (**Figure 2**):

- carbon–carbon multiple bonds, $-C=C-$ or $-C\equiv C-$
- single bonds between a carbon atom and a more electronegative atom, e.g., $-C-O-$, $-C-N-$, or $-C-Cl$
- carbon atom double-bonded to an oxygen atom, $-C=O$

(a)

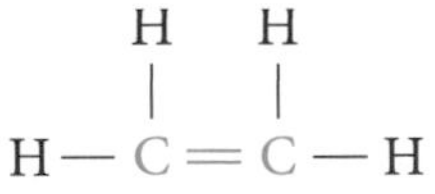

$$H - \underset{\underset{H}{|}}{C} = \underset{\underset{H}{|}}{C} - H$$

ethene (an alkene)

(b)

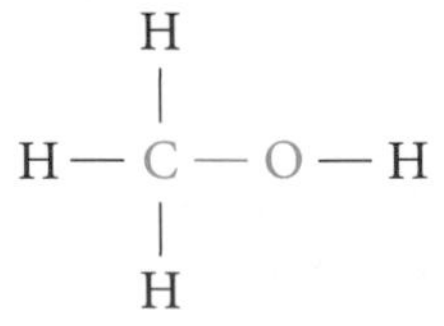

$$H - \underset{\underset{H}{|}}{\overset{\overset{H}{|}}{C}} - O - H$$

methanol (an alcohol)

(c)

$$H - \underset{}{\overset{\overset{H}{|}}{C}} = O$$

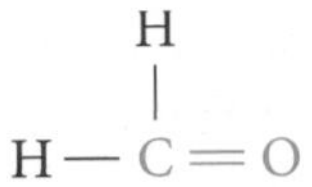

methanal (an aldehyde)

Figure 2
Examples of the three main components of functional groups:
(a) A double bond between two carbon atoms
(b) A single bond between carbon and a more electronegative atom (e.g., oxygen)
(c) A double bond between carbon and oxygen

Carbon–Carbon Multiple Bonds

When a C atom is single-bonded to another C atom, the bond is a strong covalent bond that is difficult to break. Thus, the sites in organic molecules that contain C—C bonds are not reactive. However, double or triple bonds between C atoms are more reactive. The second and third bonds formed in a multiple bond are not as strong as the first bond and are more readily broken. This allows carbon–carbon multiple bonds to be sites for reactions in which more atoms are added to the C atoms. The distinction between single and multiple bonds is not always clear-cut. For example, the reactivity of the six-carbon ring structure found in benzene indicates that there may be a type of bond intermediate between a single and a double bond. This theory is supported by measured bond lengths. You will learn more about the strengths of single and multiple bonds in Chapter 4.

Single Bonds Between Carbon and More Electronegative Atoms

Whenever a C atom is bonded to a more electronegative atom, the bond between the atoms is polar; that is, the electrons are held more closely to the more electronegative atom. This results in the C atom having a partial positive charge and the O, N, or halogen atom having a partial negative charge. Any increase in polarity of a molecule also increases intermolecular attractions, such as van der Waals forces. As more force is required to separate the molecules, the melting points and boiling points also increase (**Figure 3**).

Table 1 Electronegativities of Common Elements

Element	Electro-negativity
H	2.1
C	2.5
N	3.0
O	3.5

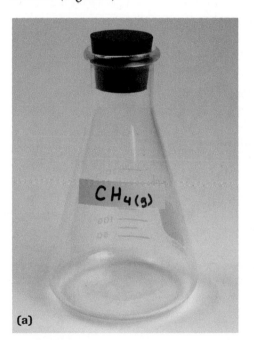

(a)

(b)

Figure 3
(a) Nonpolar substances, with weak forces of attraction among the molecules, evaporate easily. In fact, they are often gases at room temperature.
(b) Polar substances, with strong forces of attraction among the molecules, require considerable energy to evaporate.

If the O or N atoms are in turn bonded to an H atom, an —OH or —NH group is formed, with special properties. The presence of an —OH group enables an organic molecule to form hydrogen bonds with other —OH groups. The formation of these hydrogen bonds not only further increases intermolecular attractions, it also enables these molecules to mix readily with polar solutes and solvents. You may recall the saying "like dissolves like." The solubility of organic compounds is affected by nonpolar components and polar components within the molecule. Since N is only slightly less electronegative than O, the effect of an N—H bond is similar to that of an O—H bond: —NH groups also participate in hydrogen bonding.

Double Bonded Carbon and Oxygen

The third main component of functional groups consists of a C atom double-bonded to an O atom. The double covalent bond between C and O requires that *four* electrons be shared between the atoms, all four being more strongly attracted to the O atom. This makes the C=O bond strongly polarized, with the accompanying effects of raising boiling and melting points, and increasing solubility in polar solvents.

 **SUMMARY** *Three Main Components of Functional Groups*

Multiple bonds between C atoms

−C=C−   Unlike single C−C bonds, double and triple bonds allow atoms
−C≡C−   to be added to the chain.

C atom bonded to a more electronegative atom (O, N, halogen)

C−O   Unequal sharing of electrons results in polar bonds,
C−N   increasing intermolecular attraction, and raising boiling and
C−Cl, C−Br, C−F melting points.

C−OH or   These groups enable hydrogen bonding, increasing solubility
C−NH−   in polar substances.

C atom double-bonded to an O atom

C=O   The resulting polar bond increases boiling point and melting point.

▶ **Practice**

Understanding Concepts

1. Explain the meaning of the term "functional group."

2. Are double and triple bonds between C atoms more reactive or less reactive than single bonds? Explain.

3. Would a substance composed of more polar molecules have a higher or lower boiling point than a substance composed of less polar molecules? Explain.

4. Describe the three main components of functional groups in organic molecules.

▶ **Section 1.1 Questions**

Understanding Concepts

1. What is the effect of the presence of an −OH group or an −NH group on
   (a) the melting and boiling points of the molecule? Explain.
   (b) the solubility of the molecule in polar solvents? Explain.

2. Identify all components of functional groups in the following structural diagrams. Predict the solubility of each substance in water.
   (a) CH_3-O-H
   (b) $CH_3CH=CHCH_3$
   (c) $CH_3CH=O$
   (d) $CH_3CH_2C=O$
       $\quad\quad\quad |$
       $\quad\quad\quad OH$

3. The compounds water, ammonia, and methane are formed when an oxygen atom, a nitrogen atom, and a carbon atom each bonds with hydrogen atoms.
   (a) Write a formula for each of the three compounds.
   (b) Predict, with reference to electronegativities and intermolecular forces, the solubility of each of the compounds in the others.
   (c) Of the three compounds, identify which are found or produced by living organisms, and classify each compound as organic or inorganic. Justify your answer.

We will begin our study of organic families with a review of **hydrocarbons**, many of which contain multiple bonds between carbon atoms, a functional group with characteristic properties.

Fossil fuels (**Figure 1**) contain mainly hydrocarbons: simple molecules of hydrogen and carbon that are the result of the breakdown of living organisms from long ago. These compounds include the natural gas that is piped to our homes, the propane in tanks for barbecues, and the gasoline for our cars. Hydrocarbons are classified by the kinds of carbon−carbon bonds in their molecules. In **alkanes**, all carbons are bonded to other atoms by single bonds, resulting in the maximum number of hydrogen atoms bonded to each carbon atom. These molecules are thus called *saturated hydrocarbons*. **Alkenes** are hydrocarbons that contain one or more carbon−carbon double bonds, and **alkynes** contain one or more carbon–carbon triple bonds. These two groups are called *unsaturated hydrocarbons* because they contain fewer than the maximum possible number of hydrogen atoms. Because alkenes and alkynes have multiple bonds, they react in characteristic ways. The multiple bond is the functional group of these two chemical families.

In all of these hydrocarbons, the carbon−carbon backbone may form a straight chain, one or more branched chains, or a **cyclic** (ring) structure (**Table 1**). All of these molecules are included in a group called **aliphatic hydrocarbons**.

A hydrocarbon branch that is attached to the main structure of the molecule is called an **alkyl group**. When meth*ane* is attached to the main chain of a molecule, it is called a meth*yl* group, $-CH_3$. An eth*yl* group is CH_3CH_2, the branch formed when eth*ane* links to another chain.

Figure 1
Crude oil is made up of a variety of potentially useful hydrocarbons.

hydrocarbon an organic compound that contains only carbon and hydrogen atoms in its molecular structure

alkane a hydrocarbon with only single bonds between carbon atoms

alkene a hydrocarbon that contains at least one carbon−carbon double bond; general formula, C_nH_{2n}

alkyne a hydrocarbon that contains at least one carbon−carbon triple bond; general formula, C_nH_{2n-2}

cyclic hydrocarbon a hydrocarbon whose molecules have a closed ring structure

aliphatic hydrocarbon a compound that has a structure based on straight or branched chains or rings of carbon atoms; does not include aromatic compounds such as benzene

alkyl group a hydrocarbon group derived from an alkane by the removal of a hydrogen atom; often a substitution group or branch on an organic molecule

Table 1 Examples of Hydrocarbons

Hydrocarbon group	Example	Formula	Spacefill diagram	Bond and angles diagram
Aliphatic				
alkane	ethane	CH_3CH_3		
	cyclohexane	C_6H_{12}		
alkene	ethene	CH_2CH_2		120°
alkyne	ethyne	CHCH		
Aromatic				
	benzene	C_6H_6		

aromatic hydrocarbon a compound with a structure based on benzene: a ring of six carbon atoms

IUPAC International Union of Pure and Applied Chemistry; the organization that establishes the conventions used by chemists

Figure 2
Benzene, C_6H_6, is colourless, flammable, toxic, and carcinogenic, and has a pleasant odour. Its melting point is 5.5°C and its boiling point 80.1°C. It is widely used in the manufacture of plastics, dyes, synthetic rubber, and drugs.

A fourth group of hydrocarbons with characteristic properties and structures is called the **aromatic hydrocarbons**. The simplest aromatic hydrocarbon is benzene; all other members of this family are derivatives of benzene. The formula for benzene is C_6H_6, and the six carbon atoms form a unique ring structure. Unlike cyclohexane, C_6H_{12}, the benzene ring has a planar (flat) structure, and is unsaturated (**Table 1**). As we will learn later in this chapter and in Chapter 10, the bonds in the benzene ring have properties intermediate between single bonds and double bonds; the common structural diagram for benzene shows a hexagon with an inscribed circle, symbolizing the presence of double bonds in unspecified locations within the six-carbon ring (**Figure 2**). The unique structure and properties of compounds containing benzene rings have prompted their classification as a broad organic family of their own. Named historically for the pleasant aromas of compounds such as oil of wintergreen, aromatic compounds include all organic molecules that contain the benzene ring. All other hydrocarbons and their oxygen or nitrogen derivatives that are not aromatic are called aliphatic compounds.

Nomenclature of Hydrocarbons

Because there are so many organic compounds, a systematic method of naming them is essential. In this book, we will use the **IUPAC** system of nomenclature, with additional nonsystematic names that you may encounter in common usage. It is especially important to have a good grasp of the nomenclature of hydrocarbons, as the names of many organic molecules are based on those of hydrocarbon parent molecules.

Alkanes

All alkanes are named with the suffix *-ane*. The prefix in the name indicates the number of carbon atoms in the *longest straight chain* in the molecule (**Table 2**). Thus a 5-C straight-chained alkane would be named pentane.

Any alkyl branches in the carbon chain are named with the prefix for the branch, followed by the suffix *-yl*. Thus, a branch that contains a 2-C chain is called an ethyl group. The name of a branched alkane must also indicate the point of attachment of the branch. This is accomplished by assigning numbers to each C atom of the parent alkane, and pointing out the location of the branch chain by the numeral of the C atom where the branching occurs. The naming system always uses the lowest numbers possible to denote a position on the chain. Finally, all numerals are separated by commas; numerals and letters are separated by hyphens; and names of branches and parent chains are not separated.

Table 2 Alkanes and Related Alkyl Groups

Prefix	IUPAC name	Formula	Alkyl group	Alkyl formula
meth-	methane	$CH_{4(g)}$	methyl-	$-CH_3$
eth-	ethane	$C_2H_{6(g)}$	ethyl-	$-C_2H_5$
prop-	propane	$C_3H_{8(g)}$	propyl-	$-C_3H_7$
but-	butane	$C_4H_{10(g)}$	butyl-	$-C_4H_9$
pent-	pentane	$C_5H_{12(l)}$	pentyl-	$-C_5H_{11}$
hex-	hexane	$C_6H_{14(l)}$	hexyl-	$-C_6H_{13}$
hept-	heptane	$C_7H_{16(l)}$	heptyl-	$-C_7H_{15}$
oct-	octane	$C_8H_{18(l)}$	octyl-	$-C_8H_{17}$
non-	nonane	$C_9H_{20(l)}$	nonyl-	$-C_9H_{19}$
dec-	decane	$C_{10}H_{22(l)}$	decyl-	$-C_{10}H_{21}$

We will take a special look at naming propyl groups and butyl groups. When alkyl groups have three or more C atoms, they may be attached to a parent chain either at their end C atom, or at one of the middle C atoms. For example, **Figure 4** shows two points of attachment for a propyl group. The two arrangements are structural **isomers** of each other, and are commonly known by their nonsystematic names. The prefix *n-* (normal) refers to a straight-chain alkyl group, the point of attachment being at an end C atom. The isomer of the *n*-propyl group is the isopropyl group. **Figure 5** shows the common names for isomers of the butyl group; in this book, we will not concern ourselves with isomers of alkyl groups greater than 4 C atoms.

(a) CH₃ — CH₂ — CH₂ — CH₂ —

   n-butyl (normal butyl)

(b) CH₃ — CH — CH₃
                |
               CH₂
                |

   isobutyl

(c) CH₃ — CH — CH₂ — CH₃
                |

   s-butyl (secondary butyl)

(d)   CH₃
                |
       CH₃ — C — CH₃
                |

   t-butyl (tertiary butyl)

Figure 5
Four isomers of the butyl group

(a)
CH₃ — CH₂ — CH₂ —

n-propyl (normal propyl)

(b)
CH₃ — CH — CH₃
            |

isopropyl

Figure 4
Two isomers of the propyl group. The coloured bond indicates where the group is attached to the larger molecule.

isomer a compound with the same molecular formula as another compound, but a different molecular structure

---

Naming Alkanes

SAMPLE problem ◀

1. Write the IUPAC name for the chemical with the following structural diagram.

        CH₃   CH₃
         | |
        CH₂   CH₂   CH₃
         | | |
CH₃ — CH₂ — CH — CH — CH — CH₃

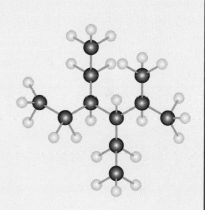

First, identify the longest carbon chain. Note that you may have to count along what appear to be branches in the structural diagram to make sure you truly have the longest chain. In this case, the longest carbon chain is 6 C long. So the parent alkane is *hexane*. Next, number the C atoms as shown.

        CH₃   CH₃
       6 | |
        CH₂   CH₂   CH₃
       5 | | | 1b
CH₃ — CH₂ — CH — CH — CH — CH₃
  6   5   4   3   2   1a

In this case, there are several possible six-carbon chains. Choose the one that gives the lowest possible total of numbers identifying the location of the branches. Usually it is best to start numbering with the end carbon that is closest to a branch. In this case, the first branch is on C 2. Notice that it makes no difference whether we choose C 1a or C 1b to be the actual C 1.

Name each branch and identify its location on the parent chain. In this example, there is a methyl group on C 2 and an ethyl group on each of C 3 and C 4. Thus the branches are 2-methyl, 3-ethyl, and 4-ethyl.

To check that you've got the lowest total, try naming the structure from the other ends of the chain. If we had counted from either of the C 6 ends, we would arrive at 3-ethyl, 4-ethyl, and 5-methyl—a set of numbers with a higher total.

When the same alkyl group (e.g., ethyl) appears more than once, they are grouped as di-, tri-, tetra-, etc. In this compound, the two ethyl groups are combined as 3,4-diethyl.

Finally, write the complete IUPAC name, following this format: (number indicating location)-(branch name)(parent chain). In this book, when more than one branch is present, the branches are listed in alphabetical order. (Note that other sources may list the branches in order of complexity.) Alphabetically, ethyl comes before methyl. So the name begins with the ethyl groups, followed by the methyl group, and ends with the parent alkane. Watch the use of commas and hyphens, and note that no punctuation is used between the alkane name and the alkyl group that precedes it.

The IUPAC name for this compound is 3,4-diethyl-2-methylhexane.

2.   Write the IUPAC name for the following hydrocarbon.

$$CH_3 - CH_2 - CH_3$$
$$|$$
$$CH_3 - CH_2 - CH - CH_2 - CH_2 - CH_2 - CH_2 - CH_3$$

First, identify the longest carbon chain: 8 C atoms. So the molecule is an octane.
Next, number the C atoms as shown.

$$CH_3 - CH_2 - CH_3$$
$$|$$
$$\underset{1}{CH_3} - \underset{2}{CH_2} - \underset{3}{CH} - \underset{4}{CH_2} - \underset{5}{CH_2} - \underset{6}{CH_2} - \underset{7}{CH_2} - \underset{8}{CH_3}$$

If we start counting at C 1, the branch group attached to C 3 contains 3 C atoms, so it is a propyl group. However, the propyl group is attached to the parent chain at its middle C atom, not at an end C atom. This arrangement of the propyl group is called isopropyl (**Figure 4(b)**).

One possible name for this compound is 3-isopropyloctane.

However, a different name results if we number this hydrocarbon from the top branch.

$$\overset{1}{CH_3} - \overset{2}{CH} - CH_3$$
$$|$$
$$CH_3 - CH_2 - \underset{3}{CH} - \underset{4}{CH_2} - \underset{5}{CH_2} - \underset{6}{CH_2} - \underset{7}{CH_2} - \underset{8}{CH_3}$$

This shows a methyl group on C 2 and an ethyl group on C 3, giving the name 3-ethyl-2-methyloctane. Where more than one name is correct, we use the one that includes the lowest possible numerals.

The correct name of this compound is 3-ethyl-2-methyloctane.

3.   Draw a structural diagram for 1,3-dimethylcyclopentane.

The parent alkane is cyclopentane. Start by drawing a ring of 5 C atoms single-bonded to each other, in the shape of a pentagon.
Next, number the C atoms in the ring, starting anywhere in the ring.
Then attach a methyl group to C 1 and another to C 3.
Finally, add H atoms to the C atoms to complete the bonding and the diagram.

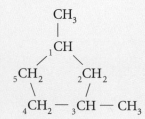

Example

Write the IUPAC name for the following hydrocarbon.

$$CH_3—CH_2—CH—\overset{\overset{\displaystyle CH_3}{|}}{\underset{\underset{\underset{\underset{\displaystyle CH_3}{|}}{\displaystyle CH_2}}{|}}{C}}—\overset{}{\underset{\underset{\displaystyle CH_3}{|}}{CH}}—CH_3$$

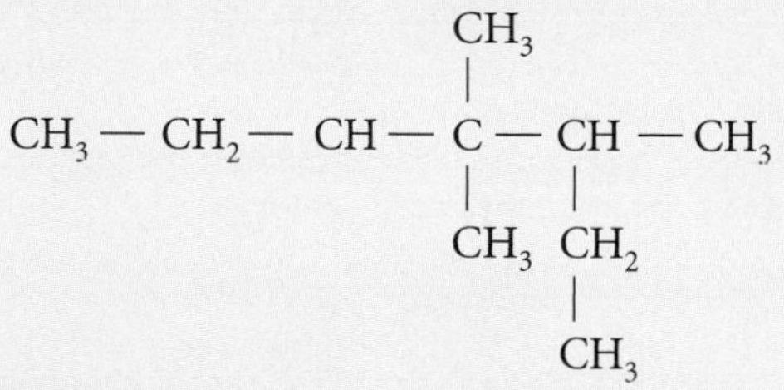

Solution

This alkane is 3,4,4-trimethylheptane.

SUMMARY   *Naming Branched Alkanes*

Step 1   Identify the longest carbon chain; note that structural diagrams can be deceiving—the longest chain may travel through one or more "branches" in the diagram.

Step 2   Number the carbon atoms, starting with the end that is closest to the branch(es).

Step 3   Name each branch and identify its location on the parent chain by the number of the carbon at the point of attachment. Note that the name with the lowest numerals for the branches is preferred. (This may require restarting your count from the other end of the longest chain.)

Step 4   Write the complete IUPAC name, following this format: (number of location)-(branch name)(parent chain).

Step 5   When more than one branch is present, the branches are listed either in alphabetical order or in order of complexity; in this book, we will follow the alphabetical order.

Note: When naming cyclic hydrocarbons, the carbon atoms that form the ring structure form the parent chain; the prefix *cyclo-* is added to the parent hydrocarbon name, and the naming of substituted groups is the same as for noncyclic compounds.

▶ Practice

Understanding Concepts

1. Write IUPAC names for the following hydrocarbons.

(a)

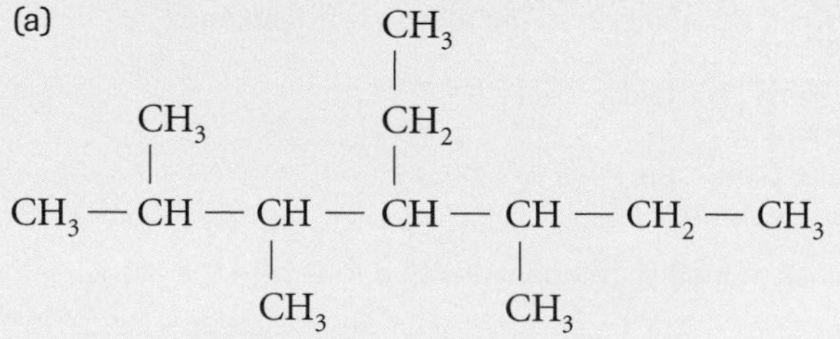

$$CH_3—\overset{\overset{\displaystyle CH_3}{|}}{CH}—CH—\overset{\overset{\displaystyle CH_2}{|}}{CH}—\overset{\overset{\displaystyle CH_3}{|}}{\underset{\underset{\displaystyle CH_3}{|}}{CH}}—CH_2—CH_3$$

with CH_3 below the third carbon

4-ethyl-2,3,5-trimethylheptane

LEARNING *TIP*

The structure of an organic molecule can be represented in many different ways: some representations give three-dimensional detail; others are simplified to show only the carbon backbone and functional groups. The following structural diagrams all show the same molecule—pentanoic acid—but in slightly different ways.

(a)

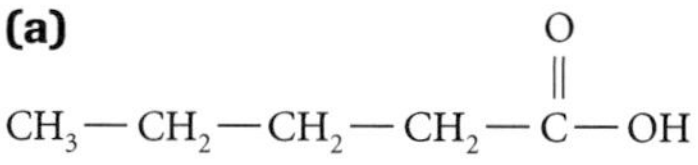

$$CH_3—CH_2—CH_2—CH_2—\overset{\overset{\displaystyle O}{\|}}{C}—OH$$

(b)

(c)

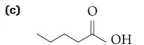

(d)

(b) 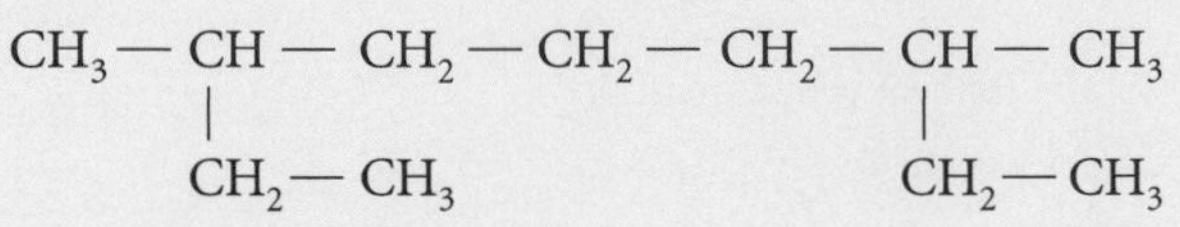 $CH_3 — CH — CH_2 — CH_2 — CH_2 — CH — CH_3$

$\qquad\quad\ |\qquad\qquad\qquad\qquad\qquad\ \ |$

$\qquad\quad CH_2 — CH_3\qquad\qquad\qquad CH_2 — CH_3$

(c) 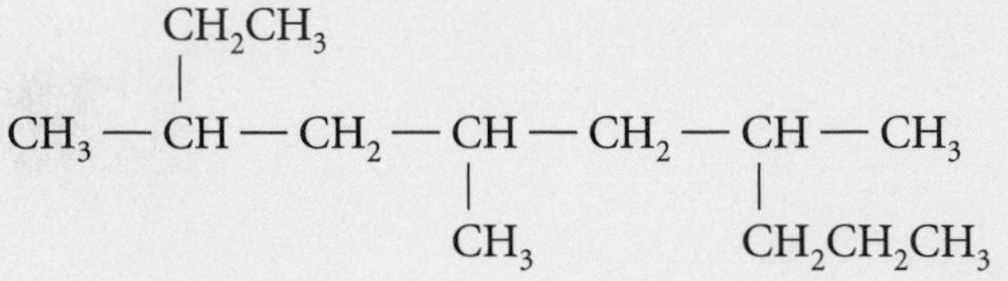 $\qquad\qquad\qquad CH_2CH_3$

$\qquad\qquad\qquad\qquad |$

$CH_3 — CH — CH_2 — CH — CH_2 — CH — CH_3$

$\qquad\qquad\qquad\qquad\quad\ |\qquad\qquad\qquad\ |$

$\qquad\qquad\qquad\qquad\ CH_3\qquad\quad CH_2CH_2CH_3$

(d) 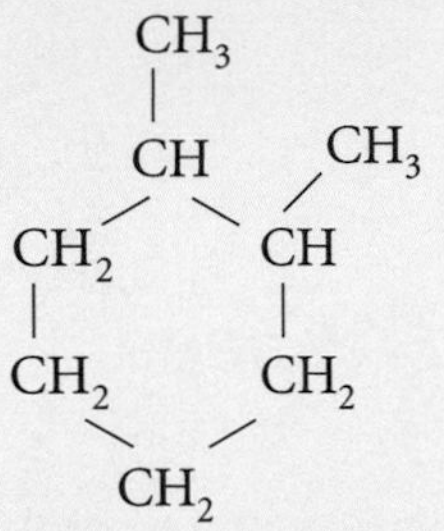
$\qquad\qquad\quad CH_3$

$\qquad\qquad\qquad |$

$\qquad\qquad\quad\ CH\qquad CH_3$

$\qquad\qquad\ CH_2\qquad CH$

$\qquad\qquad\ \ |\qquad\qquad\ |$

$\qquad\qquad\ CH_2\qquad\ CH_2$

$\qquad\qquad\qquad\ CH_2$

2. Draw a structural diagram for each of the following hydrocarbons:
   (a) 3,3,5-trimethyloctane
   (b) 3,4-dimethyl-4-ethylheptane
   (c) 2-methyl-4-isopropylnonane
   (d) cyclobutane
   (e) 1,1-diethylcyclohexane

Alkenes and Alkynes

The general rules for naming alkenes and alkynes are similar to those for alkanes, using the alkyl prefixes and ending with *-ene* or *-yne* respectively.

▶ **SAMPLE** problem | **Naming Alkenes and Alkynes**

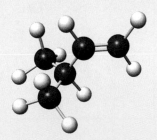

1. **Write the IUPAC name for the hydrocarbon whose structural diagram and ball-and-stick model are shown.**

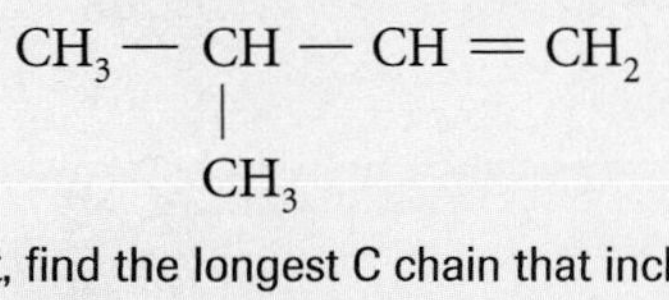

$CH_3 — CH — CH = CH_2$

$\qquad\qquad\ |$

$\qquad\quad CH_3$

First, find the longest C chain that includes the multiple bond. In this case, it is 4 C long, so the alkene is a butene.
   Number the C atoms, beginning with the end closest to the double bond.
   The double bond is between C 1 and C 2, so the alkene is a 1-butene.

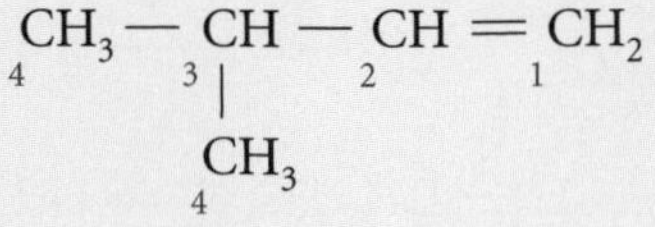

$CH_3 — CH — CH = CH_2$

$\ _4\qquad _3\ |\qquad _2\qquad\ _1$

$\qquad\quad CH_3$

$\qquad\qquad\ _4$

   Next, identify any branches: A methyl group is attached to C 3, so the branch is 3-methyl.
   Finally, write the name, following the conventions for hyphenation and punctuation. Since a number precedes the word butene, hyphens are inserted and the alkene is 3-methyl-1-butene.

2. *Draw a structural diagram for 2-methyl-1,3-pentadiene.*

First, draw and number a 5 C chain for the pentadiene.

$$C - C - C - C - C$$
1   2   3   4   5

Now insert the double bonds. The name "diene" tells us that there are two double bonds, one starting at C 1 and another starting at C 3.

$$C = C - C = C - C$$
1   2   3   4   5

Draw a methyl group attached to C atom 2.

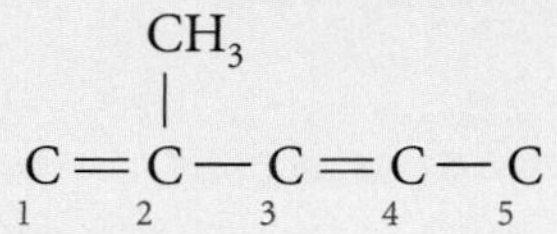

$$CH_3$$
|
$$C = C - C = C - C$$
1   2   3   4   5

Finally, write in the remaining H atoms.

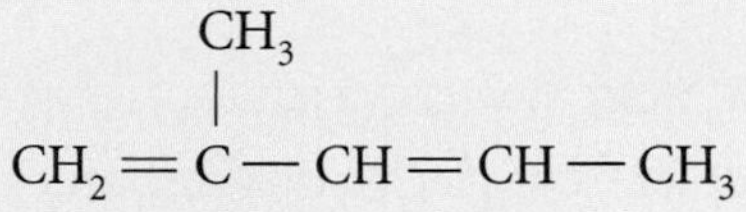

$$CH_3$$
|
$$CH_2 = C - CH = CH - CH_3$$

3. *Write the IUPAC name for the compound whose structural diagram and ball-and-stick model are shown.*

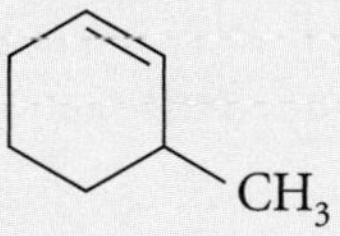

First, identify the ring structure, which contains 6 C atoms with one double bond. The parent alkene is therefore cyclohexene.

Next, number the C atoms beginning with one of the C atoms in the double bond. The numbering system should result in the attached group having the lowest possible number, which places the methyl group at C 3.

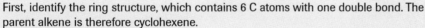

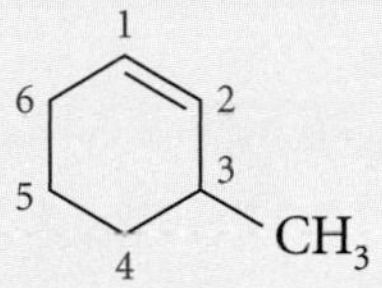

The IUPAC name for this compound is 3-methylcyclohexene.

Example 1

Draw a structural diagram for 3,3-dimethyl-1-butyne.

Solution

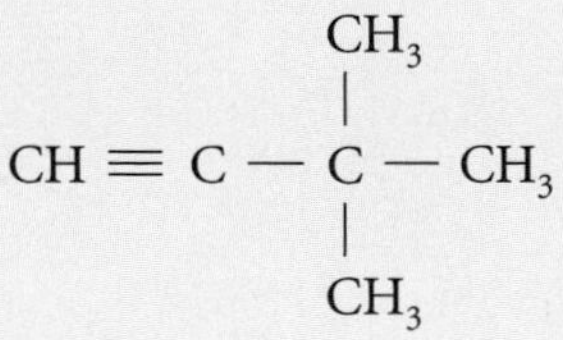

$$CH_3$$
|
$$CH \equiv C - C - CH_3$$
|
$$CH_3$$

Example 2

Write the IUPAC name for the following compound.

$$CH_2 = CH - C = CH - CH_2 - CH_3$$
$$CH_3 - CH - CH_3$$

Solution

The compound is 3-isopropyl-1,3-hexadiene.

SUMMARY Naming Alkenes and Alkynes

Step 1. The parent chain must be an alkene or alkyne, and thus must contain the multiple bond.

Step 2. When numbering the C atoms in the parent chain, begin with the end closest to the multiple bond.

Step 3. The location of the multiple bond is indicated by the number of the C atom that begins the multiple bond; for example, if a double bond is between the second and third C atoms of a pentene, it is named 2-pentene.

Step 4. The presence and location of multiple double bonds or triple bonds is indicated by the prefixes *di-*, *tri-*, etc.; for example, an octene with double bonds at the second, fourth, and sixth C atoms is named 2,4,6-octatriene.

▶ Practice

Understanding Concepts

3. Explain why no number is used in the names ethene and propene.

4. Write the IUPAC name and the common name for the compound in **Figure 6**.

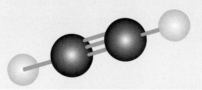

Figure 6
When this compound combusts, it transfers enough heat to melt most metals.

5. Write IUPAC names for the compounds with the following structural diagrams:

(a)
$$CH_3 - C \equiv C - CH - CH - CH_2$$
with CH_3 branches and $CH_2 - CH_3$

(b)

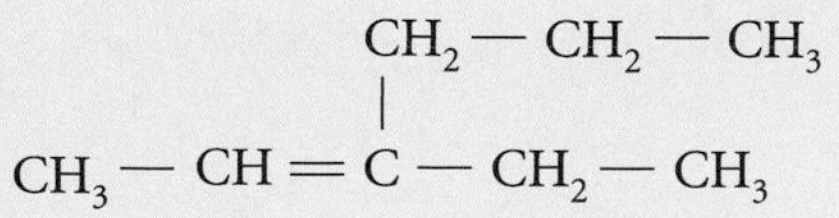

$$CH_3 - CH = C - CH_2 - CH_3$$
with branch $CH_2 - CH_2 - CH_3$

(c) $CH_3CH = CH - CH_2 - CH = CH - CH_2 - CH = CH_2$

(d) $CH_2 = CH - CH = CH - CH - CH_3$
with branch $CH_3 - CH_2 - CH_2$

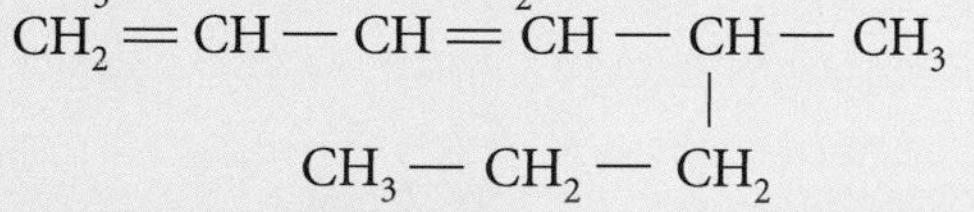

(e) CH_3— (cyclohexene ring with CH_3)

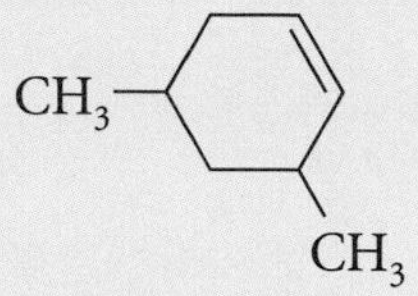

6. Draw structural diagrams for each of the following compounds:
   (a) 2-methyl-5-ethyl-2-heptene
   (b) 1,3,5-hexatriene
   (c) 3,4-dimethylcyclohexene
   (d) 1-butyne
   (e) 4-methyl-2-pentyne

Aromatic Hydrocarbons

In naming simple aromatic compounds, we usually consider the benzene ring to be the parent molecule, with alkyl groups named as branches attached to the benzene. For example, if a methyl group is attached to a benzene ring, the molecule is called methylbenzene (**Figure 7**). Since the 6 C atoms of benzene are in a ring, with no beginning or end, we do not need to include a number when naming aromatic compounds that contain only one additional group.

When two or more groups are attached to the benzene ring, we do need to use a numbering system to indicate the locations of the groups. We always number the C atoms so that we have the lowest possible numbers for the points of attachment. Numbering may proceed either clockwise or counterclockwise. As shown in the examples in **Figure 8**, we start numbering with one of the attached ethyl groups, then proceed in the direction that is closest to the next ethyl group.

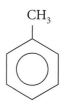

Figure 7
Methylbenzene, commonly called toluene, is a colourless liquid that is insoluble in water, but will dissolve in alcohol and other organic fluids. It is used as a solvent in glues and lacquers and is toxic to humans. Toluene reacts with nitric acid to produce the explosive trinitrotoluene (TNT).

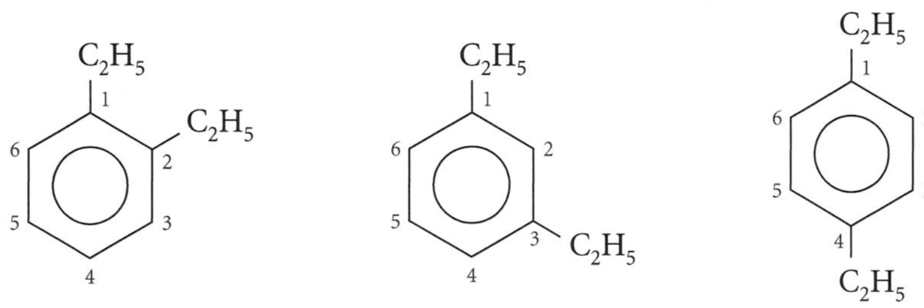

(a) 1,2-diethylbenzene

(b) 1,3-diethylbenzene

(c) 1,4-diethylbenzene

Figure 8
Three isomers of diethylbenzene

▶ **TRY THIS** *activity*   **Building Hydrocarbons**

Materials: molecular model kits.

- From a molecular model kit, obtain 6 carbons and 14 hydrogens.
  (a) Build, draw, and name as many hydrocar-bons as you can, using all 20 pieces in each model.
  (b) Put away 2 hydrogen atoms, and build and name as many different structures as pos-sible, using all of the remaining atoms.
  (c) Repeat (b) as many times as possible, until you can no longer construct any hydro-carbon molecules.
  (d) Which of the compounds are isomers?

For some aromatic molecules where the attached group is not easily named, it is more convenient to consider the benzene ring as a branch rather than as the parent molecule. When the benzene ring is the attached branch, $-C_6H_5$, it is called a *phenyl* group. For example, the compound shown in **Figure 9** is named 2-phenylbutane. According to the naming system for branched alkanes (see page 13), it may also be called *s*-butylbenzene.

Either naming system for aromatic compounds is acceptable; the object is to choose the more con-venient method for the compound in question.

$$CH_3 - CH - CH_2 - CH_3$$

Figure 9
2-phenylbutane, or *s*-butylbenzene

▶ **SAMPLE** problem   **Naming Aromatic Hydrocarbons**

Draw the structural diagram for 3-ethyl-1-methylbenzene.

First, draw the benzene ring, then add a methyl group to any C atom of the ring; this C atom automatically becomes C 1 in the numbering system. Finally, add an ethyl group to C 3, which can be clockwise or counterclockwise from C 1.

$$CH_3$$

$$CH_2CH_3$$

Example

Write IUPAC names for the following aromatic hydrocarbons.

(a)
$$CH_2 - CH_3$$
$$CH_3$$
$$CH_3$$

(b)
$$CH_2CH_2CH_3$$
$$CH_2 = CH - CH_2 - CH - CH_2 - CH_3$$

Solution

(a) 1-ethyl-2,4-dimethylbenzene

(b) 4-phenyl-3-propyl-1-hexene

SUMMARY *Naming Aromatic Hydrocarbons*

1. If an alkyl group is attached to a benzene ring, the compound is named as an alkylbenzene. Alternatively, the benzene ring may be considered as a branch of a large molecule; in this case, the benzene ring is called a phenyl group.

2. If more than one alkyl group is attached to a benzene ring, the groups are numbered using the lowest numbers possible, starting with one of the added groups.

▶ **Practice**

7. Write IUPAC names for the following hydrocarbons.

(a) 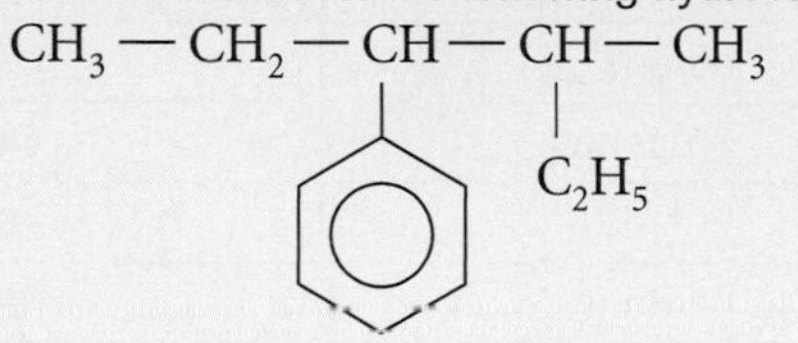 $CH_3 - CH_2 - CH - CH - CH_3$

(b) 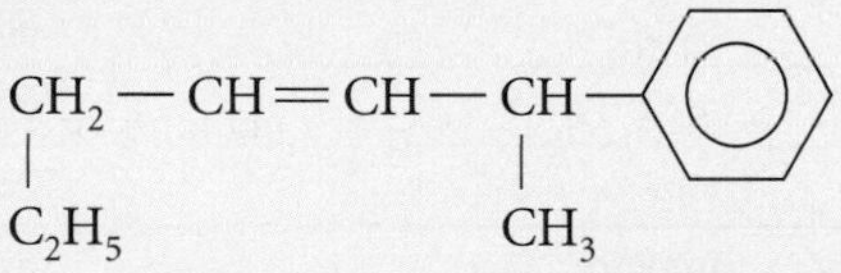 $CH_2 - CH = CH - CH -$ phenyl

(c) 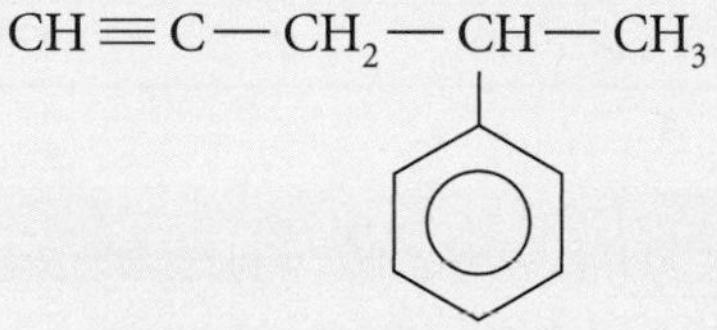 $CH \equiv C - CH_2 - CH - CH_3$

(d) 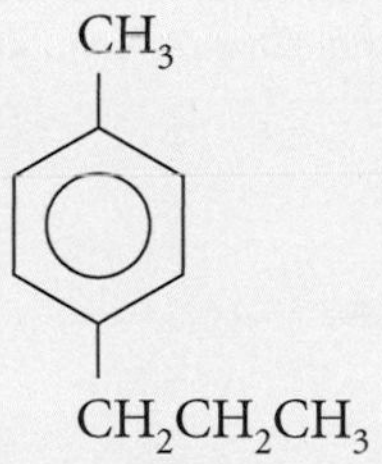 CH_3 ... $CH_2CH_2CH_3$

8. Draw structural diagrams for the following hydrocarbons:

(a) 1,2,4-trimethylbenzene

(b) 1-ethyl-2-methylbenzene

(c) 3-phenylpentane

(d) *o*-diethylbenzene

(e) *p*-ethylmethylbenzene

Physical Properties of Hydrocarbons

Figure 10
The nonpolar hydrocarbons in gasoline are insoluble in water and remain in a separate phase.

Since hydrocarbons contain only C and H atoms, two elements with very similar electronegativities, bonds between C and H are relatively nonpolar. The main intermolecular interaction in hydrocarbons is van der Waals forces: the attraction of the electrons of one molecule for the nuclei of another molecule. Since these intermolecular forces are weak, the molecules are readily separated. The low boiling points and melting points of the smaller molecules are due to the fact that small molecules have fewer electrons and weaker van der Waals forces, compared with large molecules (**Table 3**). These differences in boiling points of the components of petroleum enable the separation of these compounds in a process called **fractional distillation**. Hydrocarbons, being largely nonpolar, generally have very low solubility in polar solvents such as water, which is why gasoline remains separate from water (**Figure 10**). This property of hydrocarbons makes them good solvents for other nonpolar molecules.

fractional distillation the separation of components of petroleum by distillation, using differences in boiling points; also called fractionation

Table 3 Boiling Points of the First 10 Straight Alkanes

Formula	Name	b.p. (°C)
$CH_{4(g)}$	methane	−161
$C_2H_{6(g)}$	ethane	−89
$C_3H_{8(g)}$	propane	−44
$C_4H_{10(g)}$	butane	−0.5
$C_5H_{12(l)}$	pentane	36
$C_6H_{14(l)}$	hexane	68
$C_7H_{16(l)}$	heptane	98
$C_8H_{18(l)}$	octane	125
$C_9H_{20(l)}$	nonane	151
$C_{10}H_{22(l)}$	decane	174

▶ Section 1.2 Questions

Understanding Concepts

1. Draw a structural diagram for each hydrocarbon:
   (a) 2-methyloctane
   (b) 2-methyl-3-isopropylnonane
   (c) methylcyclopentane
   (d) 3-hexyne
   (e) 3-methyl-1,5-heptadiene
   (f) 1,2,4-trimethylbenzene
   (g) 4-s-butyloctane
   (h) 2-phenylpropane
   (i) 3-methyl-2-pentene
   (j) n-propylbenzene
   (k) p-diethylbenzene
   (l) 1, 3-dimethylcyclohexane

2. For each of the following names, determine if it is a correct name for an organic compound. Give reasons for your answer, including a correct name.
   (a) 2-dimethylhexane
   (b) 3-methyl-1-pentyne
   (c) 2,4-dimethylheptene
   (d) 3,3-ethylpentane
   (e) 3,4-dimethylhexane
   (f) 3,3-dimethylcyclohexene
   (g) 2-ethyl-2-methylpropane
   (h) 2,2-dimethyl-1-butene
   (i) 1-methyl-2-ethylpentane
   (j) 2-methylbenzene
   (k) 1,5-dimethylbenzene
   (l) 3,3-dimethylbutane

3. Write correct IUPAC names for the following structures.

(a) 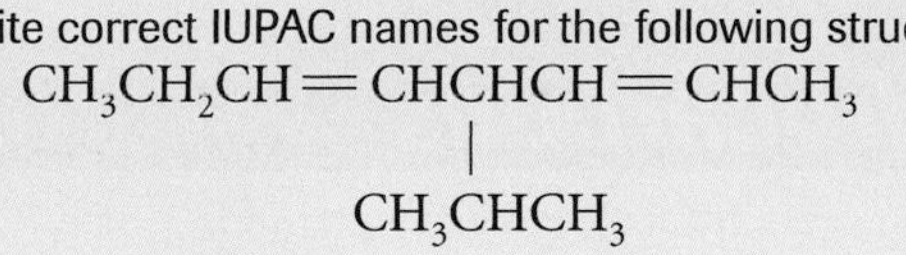 $CH_3CH_2CH = CHCHCH = CHCH_3$
$$|$$
$$CH_3CHCH_3$$

(b) 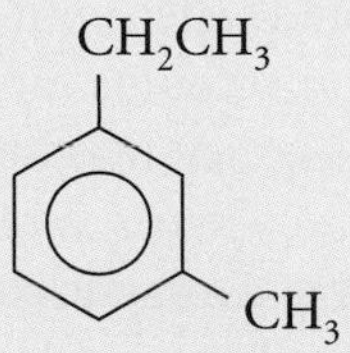 CH_2CH_3

CH_3

(c) 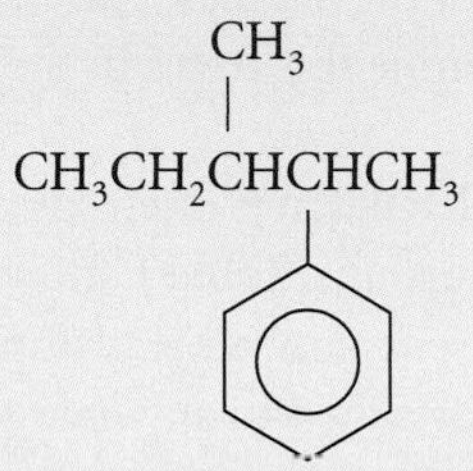 CH_3
$$|$$
$CH_3CH_2CHCHCH_3$

(d) 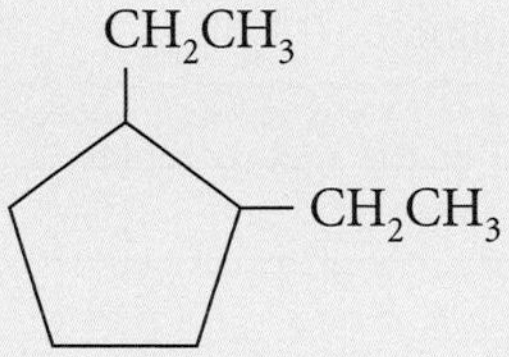 CH_2CH_3

— CH_2CH_3

(e) 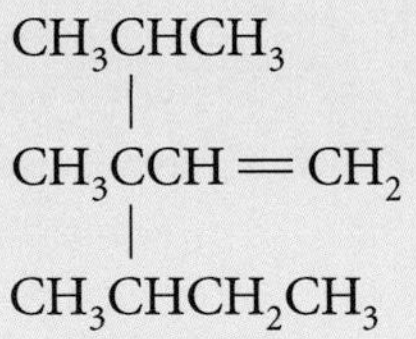 CH_3CHCH_3
$$|$$
$$CH_3CCH = CH_2$$
$$|$$
$$CH_3CHCH_2CH_3$$

4. Draw a structural diagram for each of the following compounds, and write the IUPAC name for each:

(a) ethylene

(b) propylene

(c) acetylene

(d) toluene, the toxic solvent used in many glues

(e) the *o*-, *m*-, and *p*- isomers of xylene (dimethylbenzene), used in the synthesis of other organic compounds such as dyes

Making Connections

5. (a) Use the information in **Table 3** to plot a graph showing the relationship between the number of carbon atoms and the boiling points of the alkanes. Describe and propose an explanation for the relationship you discover.

(b) Research a use for each of the first 10 alkanes, and suggest why each is appropriate for this use.

GO www.science.nelson.com

Figure 1
Hydrocarbons are found as solids, liquids, and gases, all of which burn to produce carbon dioxide and water, and large amounts of light and heat energy.

LAB EXERCISE 1.3.1

Preparation of Ethyne (p. 84)
How close does the actual yield come to the theoretical yield in the reaction between calcium carbide and water?

---

combustion reaction the reaction of a substance with oxygen, producing oxides and energy

substitution reaction a reaction in which a hydrogen atom is replaced by another atom or group of atoms; reaction of alkanes or aromatics with halogens to produce organic halides and hydrogen halides

alkyl halide an alkane in which one or more of the hydrogen atoms have been replaced with a halogen atom as a result of a substitution reaction

All hydrocarbons readily burn in air to give carbon dioxide and water, with the release of large amounts of energy (**Figure 1**); this chemical reaction accounts for the extensive use of hydrocarbons as fuel for our homes, cars, and jet engines. In other chemical reactions, alkanes are generally less reactive than alkenes and alkynes, a result of the presence of more reactive double and triple bonds in the latter. Aromatic compounds, with their benzene rings, are generally more reactive than the alkanes, and less reactive than the alkenes and alkynes. In this section, we will examine this trend in the chemical reactivity of hydrocarbons.

When we are representing reactions involving large molecules, it is often simpler to use a form of shorthand to represent the various functional groups. **Table 1** shows some of the commonly used symbols. For example, $R-\emptyset$ represents any alkyl group attached to a benzene ring, and $R-X$ represents any alkyl group attached to any halogen atom.

Table 1 Examples of Symbols Representing Functional Groups

Group	Symbol
alkyl group	R, R′, R″, etc. (R, R-prime, R-double prime)
halogen atom	X
phenyl group	Ø

Reactions of Alkanes

The characteristic reactions of saturated and unsaturated hydrocarbons can be explained by the types of carbon−carbon bonds in saturated and unsaturated hydrocarbons. Single covalent bonds between carbon atoms are relatively difficult to break, and thus alkanes are rather unreactive. They do undergo **combustion reactions** if ignited in air, making them useful fuels. Indeed, all hydrocarbons are capable of combustion to produce carbon dioxide and water. The reaction of propane gas, commonly used in gas barbecues, is shown below:

$$C_3H_{8(g)} + 5\ O_{2(g)} \rightarrow 3\ CO_{2(g)} + 4\ H_2O_{(g)}$$

While the C−C bonds in alkanes are difficult to break, the hydrogen atoms may be *substituted* by a halogen atom in a **substitution reaction** with F_2, Cl_2, or Br_2. Reactions with F_2 are vigorous, but Cl_2 and Br_2 require heat or ultraviolet light to first dissociate the halogen molecules before the reaction will proceed. In each case, the product formed is a halogenated alkane; as the halogen atom(s) act as a functional group, halogenated alkanes are also referred to as an organic family called **alkyl halides**.

In the reaction of ethane with bromine, the orange colour of the bromine slowly disappears, and the presence of $HBr_{(g)}$ is indicated by a colour change of moist litmus paper from blue to red. A balanced equation for the reaction is shown below.

$$\underset{\displaystyle H\ \ \ H}{\overset{\displaystyle H\ \ \ H}{H-C-C-H_{(g)}}} + Br_{2(l)} \xrightarrow{\text{heat or UV light}} \underset{\displaystyle H\ \ \ H}{\overset{\displaystyle H\ \ \ Br}{H-C-C-H_{(l)}}} + HBr_{(l)} \quad \text{(substitution reaction)}$$

bromoethane,
(ethyl bromide)

As the reaction proceeds, the concentration of bromoethane increases and bromine reacts with it again, leading to the substitution of another of its hydrogen atoms, forming 1,2-dibromoethane.

$$H-\underset{\underset{H}{|}}{\overset{\overset{H}{|}}{C}}-\underset{\underset{H}{|}}{\overset{\overset{Br}{|}}{C}}-H_{(g)} + Br_{2(g)} \xrightarrow{\text{heat or UV light}} H-\underset{\underset{H}{|}}{\overset{\overset{Br}{|}}{C}}-\underset{\underset{H}{|}}{\overset{\overset{Br}{|}}{C}}-H_{(l)} + HBr_{(g)} \quad \text{(substitution reaction)}$$

1,2-dibromoethane

Additional bromines may be added, resulting in a mixture of brominated products that (because of differences in physical properties) can be separated by procedures such as distillation.

Reactions of Alkenes and Alkynes

Alkenes and alkynes exhibit much greater chemical reactivity than alkanes. For example, the reaction of these unsaturated hydrocarbons with bromine is fast, and will occur at room temperature (**Figure 2**). (Recall that the bromination of alkanes requires heat or UV light.) This increased reactivity is attributed to the presence of the double and triple bonds. This characteristic reaction of alkenes and alkynes is called an **addition reaction** as atoms are *added* to the molecule with *no loss of hydrogen atoms.*

Alkenes and alkynes can undergo addition reactions not only with halogens, but also with hydrogen, hydrogen halides, and water, given the appropriate conditions. Examples of these reactions are shown below.

Halogenation (with Br₂ or Cl₂)

$$H-\underset{\underset{H}{|}}{\overset{\overset{H}{|}}{C}}=\underset{\underset{H}{|}}{\overset{\overset{H}{|}}{C}}-H + Br_{2(g)} \xrightarrow{\text{room temperature}} H-\underset{\underset{H}{|}}{\overset{\overset{Br}{|}}{C}}-\underset{\underset{H}{|}}{\overset{\overset{Br}{|}}{C}}-H \quad \text{(addition reaction)}$$

ethene   1,2-dibromoethane

Hydrogenation (with H₂)

$$H-C\equiv C-H + 2H_{2(g)} \xrightarrow[\text{heat, pressure}]{\text{catalyst}} H-\underset{\underset{H}{|}}{\overset{\overset{H}{|}}{C}}-\underset{\underset{H}{|}}{\overset{\overset{H}{|}}{C}}-H \quad \text{(addition reaction)}$$

ethyne   ethane

Hydrohalogenation (with hydrogen halides)

$$H-\underset{\underset{H}{|}}{\overset{}{C}}=CH-CH_3 + HBr_{(g)} \xrightarrow{\text{room temperature}} H-\underset{\underset{H}{|}}{\overset{\overset{H}{|}}{C}}-\overset{\overset{Br}{|}}{CH}-CH_3 \quad \text{(addition reaction)}$$

propene   2-bromopropane

Figure 2
The reaction of cyclohexene and bromine water, $Br_{2(aq)}$, is rapid, forming a layer of brominated cyclohexane (clear).

addition reaction a reaction of alkenes and alkynes in which a molecule, such as hydrogen or a halogen, is added to a double or triple bond

Table 2 Prefixes for Functional Groups

Functional group	Prefix
–F	fluoro
–Cl	chloro
–Br	bromo
–I	iodo
–OH	hydroxy
–NO₂	nitro
–NH₂	amino

DID YOU *KNOW* ?

Margarine
Vegetable oils consist of molecules with long hydrocarbon chains containing many double bonds; these oils are called "polyunsaturated." The oils are "hardened" by undergoing hydrogenation reactions to produce more saturated molecules, similar to those in animal fats such as lard.

Organic Compounds **25**

Hydration (with H_2O)

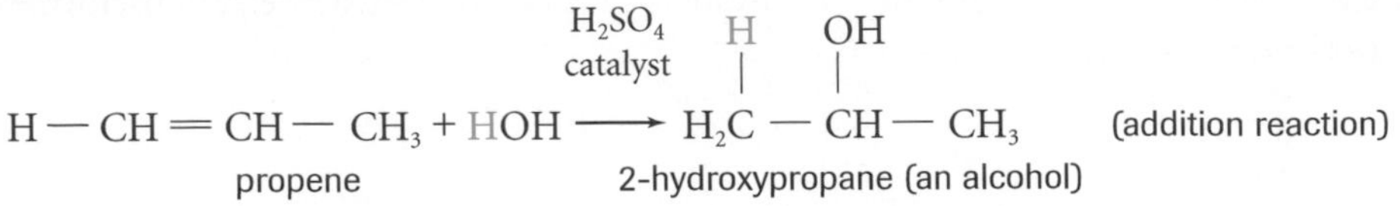

$$H-CH=CH-CH_3 + HOH \xrightarrow[\text{catalyst}]{H_2SO_4} H_2C-\overset{\overset{\displaystyle H}{|}}{C}H-\overset{\overset{\displaystyle OH}{|}}{C}H_3 \quad \text{(addition reaction)}$$

propene   2-hydroxypropane (an alcohol)

Markovnikov's Rule

When molecules such as H_2, consisting of two *identical* atoms, are added to a double bond, only one possible product is formed; in other words, addition of identical atoms to either side of the double bond results in identical products.

When molecules of *nonidentical* atoms are added, however, two *different* products are theoretically possible. For example, when HBr is added to propene, the H may add to C atom 1, or it may add to C 2; two different products are possible, as shown below.

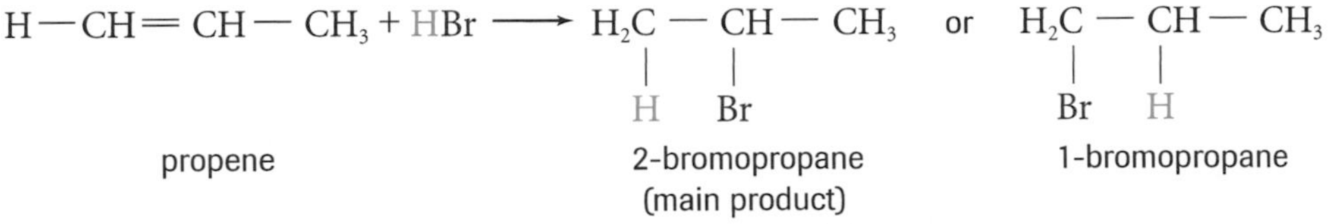

$$H-CH=CH-CH_3 + HBr \longrightarrow H_2C-\overset{\overset{\displaystyle H}{|}}{C}H-\overset{\overset{\displaystyle Br}{|}}{C}H_3 \quad \text{or} \quad H_2C-\overset{\overset{\displaystyle Br}{|}}{C}H-\overset{\overset{\displaystyle H}{|}}{C}H_3$$

propene   2-bromopropane (main product)   1-bromopropane

Experiments show that, in fact, only one main product is formed. The product can be predicted by a rule known as Markovnikov's rule, first stated by Russian chemist V. V. Markovnikov (1838–1904).

> **Markovnikov's Rule**
>
> When a hydrogen halide or water is added to an alkene or alkyne, the hydrogen atom bonds to the carbon atom within the double bond that *already has more hydrogen atoms*. This rule may be remembered simply as "the rich get richer."

As illustrated in the reaction of propene above, the first C atom has two attached H atoms, while the second C atom has only one attached H atom. Therefore, the "rich" C1 atom "gets richer" by gaining the additional H atom; the Br atom attaches to the middle C atom. The main product formed in this reaction is 2-bromopropane.

▶ **SAMPLE** problem | **Predicting Products of Addition Reactions**

What compound will be produced when water reacts with 2-methyl-1-pentene?

First, write the structural formula for 2-methyl-1-pentene.

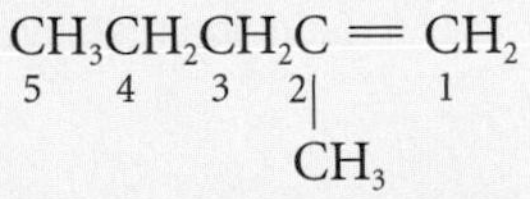

$$\underset{5}{C}H_3\underset{4}{C}H_2\underset{3}{C}H_2\underset{2}{\overset{\overset{\displaystyle |}{}}{C}}=\underset{1}{C}H_2$$
$$\underset{}{C}H_3$$

Next, identify the C atom within the double bond that has more H atoms attached. Since carbon 1 has two H atoms attached, and carbon 2 has no H atoms attached, the H atom in the HOH adds to carbon 1, and the OH group adds to carbon 2.

We can now predict the product of the reaction.

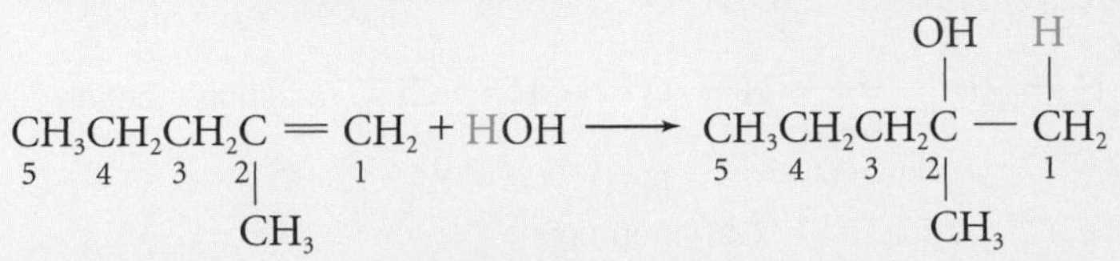

$$\underset{5}{C}H_3\underset{4}{C}H_2\underset{3}{C}H_2\underset{2}{C}=\underset{1}{C}H_2 + HOH \longrightarrow \underset{5}{C}H_3\underset{4}{C}H_2\underset{3}{C}H_2\underset{2}{\overset{\overset{\displaystyle OH}{|}}{C}}-\underset{1}{\overset{\overset{\displaystyle H}{|}}{C}}H_2$$
$$CH_3 \qquad\qquad\qquad CH_3$$

The compound produced is 2-hydroxy-2-methylpentane.

Example

Draw structural diagrams to represent an addition reaction of an alkene to produce 2-chlorobutane.

Solution

$$H_2C=CHCH_2CH_3 + HCl \rightarrow H_2C \overset{\overset{\displaystyle H}{|}}{-} \overset{\overset{\displaystyle Cl}{|}}{CHCH_2CH_3}$$

1-butene

> ### ▶ Practice

Understanding Concepts

1. What compounds will be produced in the following addition reactions?

   (a)
   $$CH_3CH=\underset{\underset{\displaystyle CH_2CH_3}{|}}{C}CH_2CH_3 + H_2 \xrightarrow[500°C]{Pt\ catalyst}$$

   (b)
   $$CH_3CH=\underset{\underset{\displaystyle CH_3}{|}}{C}CH_2CH_3 + HBr \longrightarrow$$

   (c)
   $$CH_3CH_2\underset{\underset{\displaystyle CH_2CH_3}{|}}{C}HCH=CH_2 + H_2O \xrightarrow{H_2SO_4}$$

   (d) 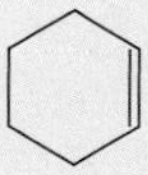
   ⬡ $+ Cl_2 \longrightarrow$

Synthesis: Choosing Where to Start

Addition reactions are important reactions that are often used in the synthesis of complex organic molecules. Careful selection of an alkene as starting material allows us to strategically place functional groups such as a halogen or a hydroxyl group ($-OH$) in desired positions on a carbon chain. As we will see later in this chapter, the products of these addition reactions can in turn take part in further reactions to synthesize other organic compounds, such as vinegars and fragrances.

> ### ▶ Practice

Understanding Concepts

2. Explain the phrase "the rich get richer" as it applies to Markovnikov's rule.

3. Draw structural diagrams to represent addition reactions to produce each of the following compounds:
   (a) 2,3-dichlorohexane
   (b) 2-bromobutane
   (c) 2-hydroxy-3-methylpentane
   (d) 3-hydroxy-3-methylpentane

Reactions of Aromatic Hydrocarbons

Since aromatic hydrocarbons are unsaturated, one might expect that they would readily undergo addition reactions, as alkenes do. Experiments show, however, that the benzene ring does not undergo addition reactions except under extreme conditions of temperature and pressure. They are less reactive than alkenes.

Aromatic hydrocarbons do undergo substitution reactions, however, as alkanes do. In fact, the hydrogen atoms in the benzene ring are more easily replaced than those in alkanes. When benzene is reacted with bromine in the presence of a catalyst, bromobenzene is produced.

Overall, the reactivity of aromatic hydrocarbons appears to be intermediate between that of alkanes and alkenes.

(a) cyclohexane + Br_2 → bromocyclohexane

cyclohexane + Br_2 $\xrightarrow{\text{heat UV}}$ bromocyclohexane + HBr (substitution reaction)

(b) benzene + Br_2 → bromobenzene

benzene + Br_2 $\xrightarrow{\text{FeBr}_3}$ bromobenzene + HBr (substitution reaction; addition does not occur)

(c) cyclohexene + Br_2 → bromocyclohexane

cyclohexene + Br_2 $\xrightarrow[\text{temperature}]{\text{room}}$ 1,2-dibromocyclohexane (addition reaction)

Further reaction of bromobenzene with Br_2 results in the substitution of another Br on the ring. In theory, this second Br atom may substitute for an H atom on any of the other C atoms, resulting in three possible isomers of dibromobenzene.

1, 2-dibromobenzene   and/or   1, 3-dibromobenzene   and/or   1, 4-dibromobenzene

In practice, the 1,3 isomer appears to be favoured.

The relatively low reactivity of aromatic hydrocarbons indicates that the benzene structure is particularly stable. It seems that the bonds in a benzene ring are unlike the double or triple bonds in alkenes or alkynes. In 1865, the German architect and chemist Friedrich August Kekulé (1829–1896) proposed a cyclic structure for benzene, C_6H_6. With 6 C atoms in the ring, and one H atom on each C atom, it appears that there might be 3 double bonds within the ring, each alternating with a single bond. As carbon−carbon double bonds are shorter than single bonds, we would predict that the bonds in the benzene ring would be of different lengths. Experimental evidence, however, shows otherwise. The technique of X-ray diffraction indicates that all the C−C bonds in benzene are identical in length and in strength (intermediate between that of single and double bonds). Therefore, rather than having 3 double bonds and 3 single bonds, an acceptable model for benzene would require that the valence electrons be shared *equally* among all 6 C atoms, making 6 identical bonds. A model of benzene is shown in **Figure 3**. In this model, the 18 valence electrons are shared equally, in a *delocalized* arrangement; there is no specific location for the shared electrons, and all bond strengths are intermediate between that of single and double bonds. This explains why benzene rings do not undergo addition reactions as double bonds do, and why they do undergo substitution reactions as single bonds do, and do so more readily. In Chapter 4, you will examine in more detail the unique bonding that is present in the benzene ring.

In another substitution reaction, benzene reacts with nitric acid in the presence of H_2SO_4 to form nitrobenzene. Benzene also reacts with alkyl halides (R−X) in the presence of an aluminum halide catalyst (AlX_3); the alkyl group attaches to the benzene ring, displacing an H atom on the ring. These products can undergo further reactions, enabling the design and synthesis of aromatic compounds with desired groups attached in specific positions.

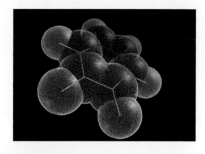

Figure 3
Kekulé wrote the following diary entry about a dream he had, in which he gained a clue to the structure of benzene: "Again the atoms gambolled before my eyes. This time the smaller groups kept modestly to the background. My mind's eyes, rendered more acute by repeated visions of a similar kind, could now distinguish larger structures, of various shapes; long rows, sometimes more closely fitted together; all twining and twisting in snakelike motion. But look! What was that? One of the snakes grabbed its own tail, and the form whirled mockingly before my eyes. As if struck by lightning I awoke; ... I spent the rest of the night in working out the consequences of the hypothesis.... If we learn to dream we shall perhaps discover the truth."

$$\text{benzene} + HNO_3 \xrightarrow{H_2SO_4} \text{nitrobenzene (}NO_2\text{)} + H_2O \quad \text{(substitution reaction)}$$

nitrobenzene

$$\text{benzene} + CH_3CH_2Cl \xrightarrow{AlCl_3} \text{ethylbenzene (}CH_2CH_3\text{)} + HCl \quad \text{(substitution reaction)}$$

ethylbenzene

Predicting Reactions of Aromatic Hydrocarbons   *SAMPLE* problem ◀

Predict the product or products formed when benzene is reacted with 2-chlorobutane, in the presence of a catalyst ($AlCl_3$). Draw structural diagrams of the reactants and products.

The methyl group of chloromethane substitutes for one of the H atoms on the benzene ring, forming methylbenzene and releasing the chloride to react with the displaced hydrogen.

$$\text{benzene} + Cl - \overset{\displaystyle H}{\underset{\displaystyle H}{C}} - H \longrightarrow \text{methylbenzene (}CH_3\text{)} + HCl$$

The products formed are methylbenzene (toluene) and hydrogen chloride.   ▶

Example

Draw balanced chemical equations (including structural diagrams) to represent a series of reactions that would take place to synthesize ethylbenzene from benzene and ethene. Classify each reaction.

Solution

Reaction 1: Halogenation (by addition) of ethene by hydrogen chloride

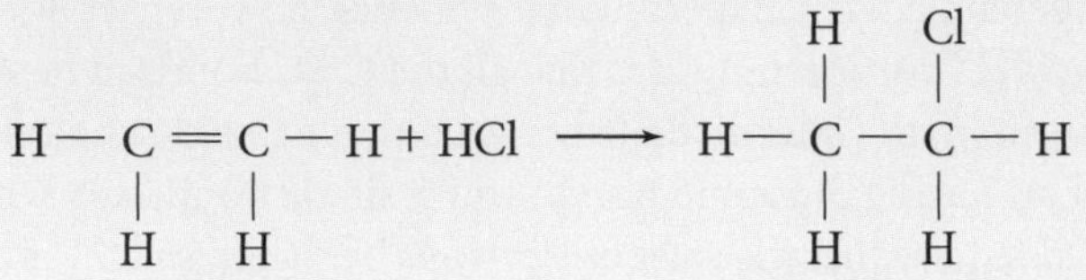

Reaction 2: Halogenation (by substitution) of benzene by chloroethane

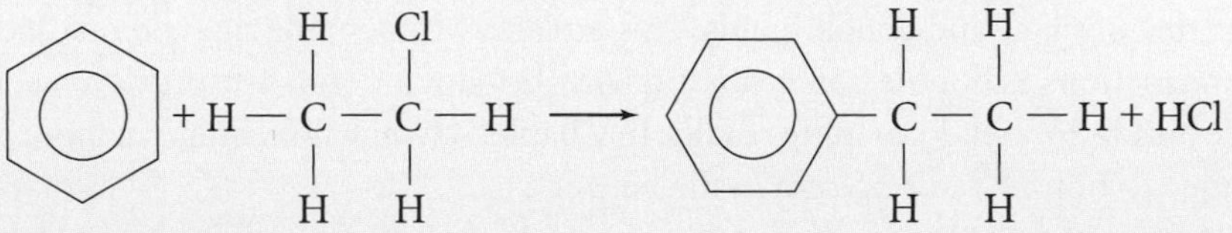

▶ Practice

Understanding Concepts

4. Predict the product or products formed in each of the following reactions:

    (a) + $Cl_2 \longrightarrow$

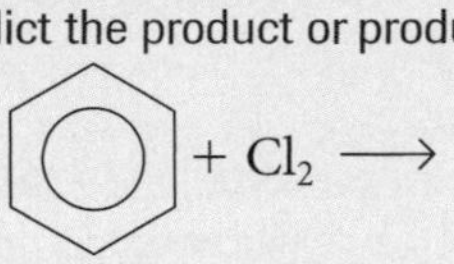

    (b) + $HNO_3 \xrightarrow{H_2SO_4}$

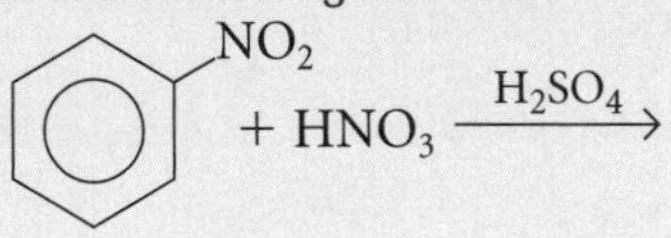

5. Propose a reaction series that would produce 2-phenylbutane, starting with benzene and 1-butene as reactants.

6. Which of the terms "addition," "substitution," or "halogenation" describes the reaction between benzene and bromine? Explain.

7. Describe the bonding structure in benzene, and explain the experimental evidence in support of this structure.

SUMMARY   Reactions of Hydrocarbons

- All hydrocarbons undergo combustion reactions with oxygen to produce carbon dioxide and water.

Alkanes

- Primarily undergo **substitution** reactions, with heat or UV light:
  with halogens or hydrogen halides: halogenation
  with nitric acid

Alkenes and Alkynes

- Primarily undergo **addition** reactions:
  with H_2: hydrogenation
  with halogens or hydrogen halides: halogenation
  with water: hydration

Aromatics

- Primarily undergo **substitution** reactions:
  with X_2: halogenation, $\emptyset-X$
  with HNO_3: nitration, $\emptyset-NO_2$
  with RX: alkylation, $\emptyset-R$

- Do *not* undergo addition reactions.

▶ **Section 1.3** *Questions*

Understanding Concepts

1. Write a balanced equation for each of the following types of reactions of acetylene:
   (a) addition
   (b) hydrogenation
   (c) halogenation
   (d) hydration

2. Classify each of the following reactions as one of the following types: addition, substitution, hydrogenation, halogenation, or combustion. Write the names and the structures for all reactants and products.
   (a) methyl-2-butene + hydrogen →
   (b) ethyne + Cl_2 →
   (c) $CH_3-C\equiv C-CH_3 + H_2$ (excess) ⟶
   (d)
   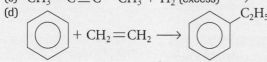
   (e)
   $$CH_3-CH=\overset{\overset{\displaystyle C_2H_5}{|}}{C}-CH-CH_3 + O=O \longrightarrow$$
   $$\underset{\displaystyle CH_3}{|}$$

3. Classify and write structural formula equations for the following organic reactions:
   (a) 3-hexene + water $\overset{H_2SO_4}{\longrightarrow}$
   (b) 2-butene + hydrogen → butane
   (c) 4,4-dimethyl-2-pentyne + hydrogen → 2,2-dimethylpentane
   (d) methylbenzene + oxygen → carbon dioxide + water
   (e) 2-butene → 3-methylpentane

Applying Inquiry Skills

4. To make each of the following products, select the reactants and describe the experimental conditions needed.
   (a) 2-hydroxypropane
   (b) 1, 3-dichlorocyclohexane from cyclohexane
   (c) 2-methyl-2-hydroxypentane from an alkene
   (d) chlorobenzene

Making Connections

5. If a certain volume of propane gas at SATP were completely combusted in oxygen, would the volume of gaseous product formed be greater or smaller than that of the reactant? By how much?

6. From your knowledge of intermolecular attractions, which of these organic compounds—2-chlorononane, 2-hydroxynonane, or nonane—would be the most effective solvent for removing oil stains? Give reasons for your answer.

7. TNT is an explosive with a colourful history (**Figure 5**). Research and report on who discovered it, and its development, synthesis, uses, and misuses.

Figure 5

GO www.science.nelson.com

1.4 Organic Halides

organic halide a compound of carbon and hydrogen in which one or more hydrogen atoms have been replaced by halogen atoms

Organic halides are a group of compounds that includes many common products such as Freons (chlorofluorocarbons, CFCs) used in refrigerators and air conditioners, and Teflon (polytetrafluoroethene), the nonstick coating used in cookware and labware.

While we use some organic halides in our everyday lives, many others are toxic and some are also carcinogenic, so their benefits must be balanced against potential hazards. Two such compounds, the insecticide DDT (dichlorodiphenyltrichloroethane) and the PCBs (polychlorinated biphenyls) used in electrical transformers, have been banned because of public concern about toxicity.

In Section 1.3 you learned that when H atoms in an alkane are replaced by halogen atoms, the resulting organic halide is more specifically referred to as an alkyl halide.

Naming Organic Halides

When naming organic halides, consider the halogen atom as an attachment on the parent hydrocarbon. The halogen name is shortened to fluoro-, chloro-, bromo-, or iodo-. For example, the structure shown below is 1,2-dichloroethane, indicating an ethane molecule substituted with a chlorine atom on carbon 1 and a chlorine atom on carbon 2.

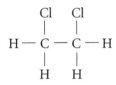

1,2-dichloroethane

▶ **SAMPLE** problem   **Drawing and Naming Organic Halides**

Draw a structural diagram for 2,2,5-tribromo-5-methylhexane.

First, draw and number the parent alkane chain, the hexane:

$$C - C - C - C - C - C$$
$$1 \quad 2 \quad 3 \quad 4 \quad 5 \quad 6$$

Next, add two Br atoms to carbon 2, one Br atom to carbon 5, and a methyl group to carbon 5.

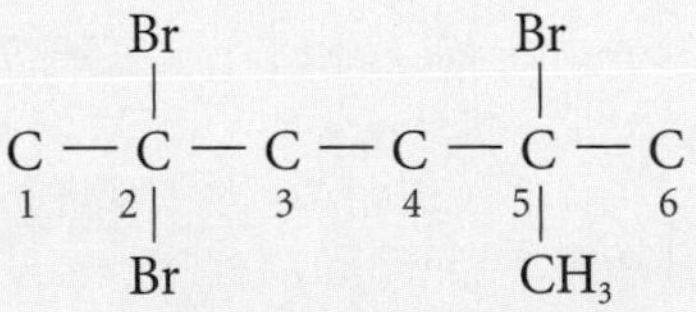

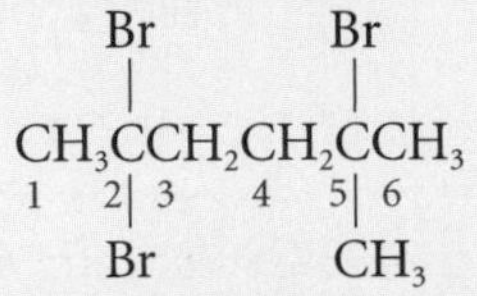

$$CH_3CCH_2CH_2CCH_3$$

Finally, complete the bonding by adding H atoms to the C atoms.

Example 1

Write the IUPAC name for $CH_3CH_2CH_2CH(Cl)CH_2CH(Br)CH_3$.

Solution

This compound is 2-bromo-4-chloroheptane.

Example 2

Draw a structural diagram of 1,2-dichlorobenzene.

Solution

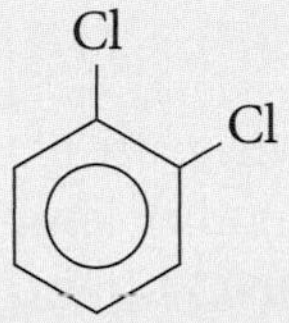

> ▶ **Practice**

Understanding Concepts

1. Draw structural diagrams for each of the following alkyl halides:
   (a) 1,2-dichloroethane (solvent for rubber)
   (b) tetrafluoroethene (used in the manufacture of Teflon)
   (c) 1,2-dichloro-1,1,2,2-tetrafluoroethane (refrigerant)
   (d) 1,4-dichlorobenzene (moth repellent)

2. Write IUPAC names for each of the formulas given.
   (a) CHI_3 (antiseptic)
   (b) $CH_2 = C - CH_2Cl$
              |
            CH_3
         (insecticide)
   (c) CH_2Cl_2 (paint remover)
   (d) $CH_2Br - CHBr - CH_2Br$ (soil fumigant)

Properties of Organic Halides

The presence of the halogen atom on a hydrocarbon chain or ring renders the molecule more polar. This is because halogens are more electronegative than C and H atoms, and so carbon–halogen bonds are more polar than C–H bonds. The increased polarity of alkyl halides increases the strength of the intermolecular forces. Thus alkyl halides have higher boiling points than the corresponding hydrocarbons. Because "like dissolves like," the increased polarity also makes them more soluble in polar solvents than hydrocarbons of similar size.

When organic halides are formed from halogenation of hydrocarbons, the product obtained is often a mixture of halogenated compounds. These compounds may contain 1, 2, 3, or more halogens per molecule, reflecting intermediate compounds that can be further halogenated. The molecules that contain more halogen atoms are usually more polar than the less halogenated molecules, and thus have higher boiling points (**Table 1**). This difference in boiling points conveniently enables us to separate the components of a mixture by procedures such as fractional distillation.

Table 1 Boiling Points of Some Hydrocarbons and Corresponding Organic Halides

Hydrocarbon	Boiling point (°C)	Alkyl halide	Boiling point (°C)
CH_4	−164	CH_3Cl	−24
C_2H_6	−89	C_2H_5Cl	12
C_3H_8	−42	C_3H_7Cl	46
C_4H_{10}	−0.5	C_4H_9Cl	78

The Cost of Air Conditioning

The cost of a new car with air conditioning includes the price of the unit plus an additional "air-conditioner" tax. On top of that, there is another, less obvious, cost: possible environmental damage. Let us take a look at how organic chemistry can be used to solve some problems, and how sometimes new problems are created along the way.

In the late 1800s, refrigerators were cooled using toxic gases such as ammonia, methyl chloride, and sulfur dioxide. When several fatal accidents occurred in the 1920s as a result of leaked coolant, the search began for a safer refrigerant. In 1930 the DuPont company manufactured Freon, a chlorofluorocarbon, $CF_2Cl_{2(g)}$, also called CFC-12. (Industrial chemists sometimes name Freons using a non-SI system.) As it was inert, it was considered very safe and its use spread to aerosol sprays, paints, and many other applications.

In the 1970s, large "ozone holes" were detected in the upper atmosphere, particularly over the polar regions. It appears that although CFCs are inert in the lower atmosphere, they are reactive in the upper atmosphere. In the presence of UV light, CFC molecules—including Freon—decompose, releasing highly reactive chlorine atoms.

The chlorine destroys the ozone molecules in the stratosphere, leaving us unprotected from harmful UV radiation (**Figure 1**). You may have learned about these reactions in a previous chemistry course.

Automobile air conditioners use over one-third of the total amount of Freon in Canada, and about 10% of this total is released into the atmosphere each year. Hence the search is on again to find a new chemical to meet the demand for inexpensive air-conditioning systems, and to minimize environmental damage. Two types of chemicals have been developed as substitute refrigerants: the hydrochlorofluorocarbons (HCFCs), and the hydrofluorocarbons (HFCs). These molecules differ from CFCs in that they contain hydrogen atoms in addition to carbon and halogen atoms. The H atoms react with hydroxyl groups in the atmosphere, decomposing the molecules. Since the HCFCs and HFCs readily decompose, they have less time to cause damage to the ozone layer. Note that HCFCs still contain chlorine, the culprit in ozone depletion; HFCs contain no chlorine and so are the preferred substitute for CFCs. These molecules are more readily decomposed in the lower atmosphere and thus have less time to cause damage. However, they do release carbon dioxide, a major greenhouse gas, upon decomposition.

The most commonly used coolant now is HFC-134a. Since 1995, it has been used in all new automobile air conditioners. In 2001, the Ontario government introduced legislation requiring that all old units, when refilling is needed, be adapted to use one of the new alternative refrigerants.

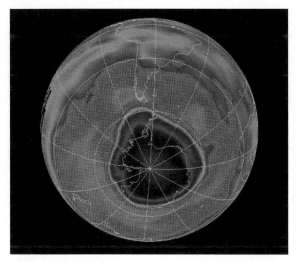

Figure 1
An ozone "hole" (blue) forms over the Antarctic every spring (September and October).

> **Practice**

Understanding Concepts

3. Create a flow chart outlining the effects of an accidental leak of refrigerant from a car's air conditioner. Include chemical equations wherever possible.

4. Draw a time line showing the use and effects of various refrigerants over the last 150 years.

5. (a) Write chemical equations predicting the decomposition of HCFCs and HFCs.
   (b) Why might HCFCs and HFCs decompose more quickly than CFCs?
   (c) Why might this make them less damaging than CFCs?

> **EXPLORE** an issue

Role Play: Can We Afford Air Conditioning?

Decision-Making Skills

- Define the Issue
- Analyze the Issue
- Identify Alternatives
- Defend the Position
- Research
- Evaluate

When a car manufacturer is planning to develop a new model, all aspects of the vehicle are reconsidered. Government regulations prohibit the manufacture of new vehicles with air-conditioning units that use CFCs. Alternative coolants have been developed, and now most manufacturers use HFC-134a. However, HFCs are greenhouses gases and so, if released, are likely to be contributors to global warming. Imagine that a committee is set up to decide whether the next new model should have air conditioning using HFC-134a, or no air-conditioning unit at all. Committee members include: a union representative for the production-line workers; the local MP; an environmentalist; a reporter from a drivers' magazine; a physician; a representative from the Canadian Automobile Association; an advertising executive; shareholders in the car company.

(a) Costs can be measured in many ways: financial, social, environmental, political, etc. Choose one way of

assessing cost and collect and sort information to help you decide whether the costs of automobile air conditioners are justified.

(b) Select a role for yourself—someone who would be concerned about the kinds of costs that you have researched. Consider how this person might feel about the issue of air conditioning.

(c) Role-play the meeting, with everyone taking a turn to put forward his/her position on whether the new car model should have air conditioning.

(d) After the "meeting," discuss and summarize the most important points made. If possible, come to a consensus about the issue.

GO www.science.nelson.com

Preparing Organic Halides

Alkyl halides are produced in halogenation reactions with hydrocarbons, as we learned in Section 1.3. Alkenes and alkynes readily add halogens or hydrogen halides to their double and triple bonds. Recall also that Markovnikov's rule of "the rich get richer" applies when hydrogen halides are reactants, and must be considered in designing the synthesis of specific alkyl halides. These alkyl halides can then be transformed into other organic compounds.

An example of the halogenation of an alkyne is shown below for a review of the reactions that produce alkyl halides. These reactions readily take place at room temperature.

(a)

$$H-C\equiv C-H + Br-Br \longrightarrow H-\underset{|}{\overset{Br}{C}}=\underset{|}{\overset{Br}{C}}-H$$

ethyne + bromine ⟶ 1,2-dibromoethene

(b)

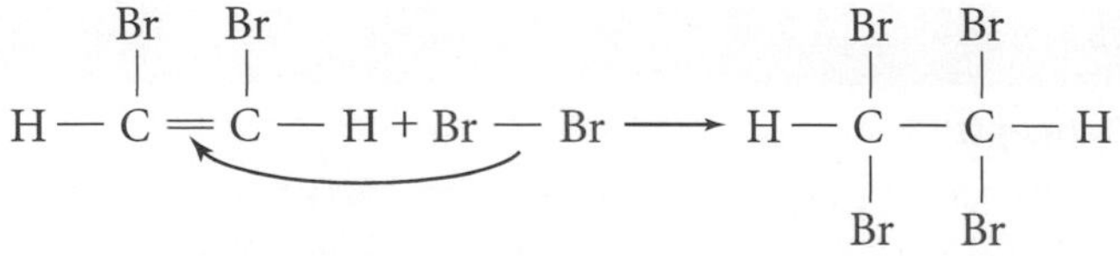

$$\text{1,2-dibromoethene} + \text{bromine} \longrightarrow \text{1,1,2,2-tetrabromoethane}$$

If we wanted to produce a halide of a benzene ring, we would need to arrange a substitution reaction with a halogen. The following example illustrates the chlorination of benzene in the presence of a catalyst. Further substitution can occur in the benzene ring until all hydrogen atoms are replaced by halogen atoms.

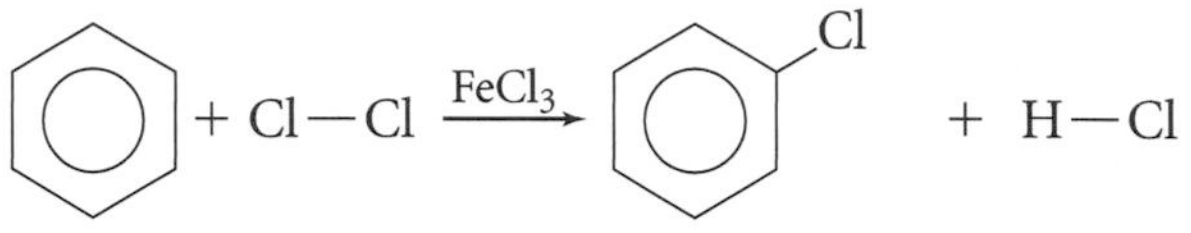

$$\text{benzene} + \text{chlorine} \xrightarrow{FeCl_3} \text{chlorobenzene} + \text{hydrogen chloride}$$

Preparing Alkenes from Alkyl Halides: Elimination Reactions

Alkyl halides can eliminate a hydrogen and a halide ion from adjacent carbon atoms, forming a double bond in their place, thereby becoming an alkene. The presence of a hydroxide ion is required, as shown in the example below. This type of reaction, in which atoms or ions are removed from a molecule, is called an **elimination reaction**. Elimination reactions of alkyl halides are the most commonly used method of preparing alkenes.

elimination reaction a type of organic reaction that results in the loss of a small molecule from a larger molecule; e.g., the removal of H_2 from an alkane

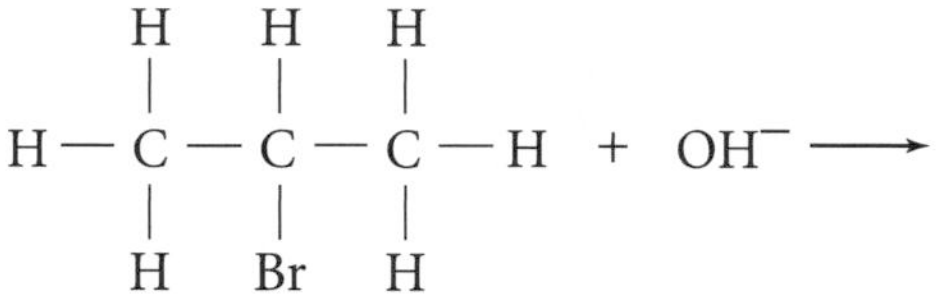

2-bromopropane + hydroxide ion

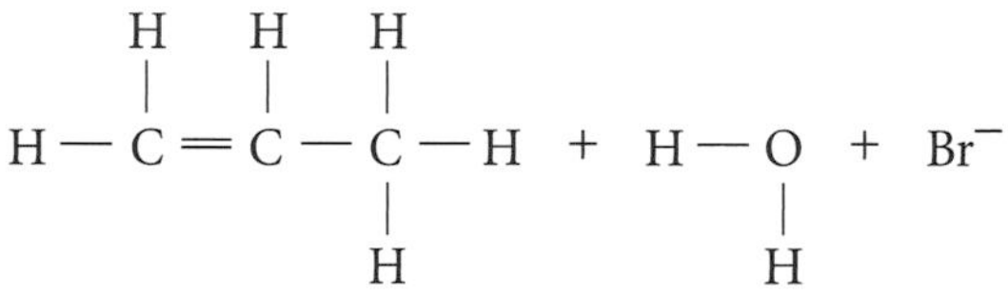

propene + water + bromide ion

SUMMARY   *Organic Halides*

Functional group: R–X

Preparation:

• alkenes and alkynes → organic halides
  addition reactions with halogens or hydrogen halides

• alkanes and aromatics → organic halides
  substitution reactions with halogens or hydrogen halides

Pathway to other groups:

• alkyl halides → alkenes
  elimination reactions, removing hydrogen and halide ions

▶ *Practice*

Understanding Concepts

6. Classify the following as substitution or addition reactions. Predict all possible products for the initial reaction only. Complete the word equation and the structural diagram equation in each case. You need not balance the equations.
   (a) trichloromethane + chlorine →
   (b) propene + bromine →
   (c) ethylene + hydrogen iodide →
   (d) ethane + chlorine →
   (e) Cl—C≡C—Cl + F—F (excess) ⟶
   (f)
$$\begin{array}{cccc} & H & H & H & H \\ & | & | & | & | \\ H-C &=& C - C - C - H + H - Cl \longrightarrow \\ & & & | & | \\ & & & H & H \end{array}$$

   (g)

$$+ \; Cl-Cl \longrightarrow$$

Extension

7. Why are some organic halides toxic while others are not? And why are some organisms affected more than others? Use the Internet to find out, using the following key words in your search: bioaccumulation; fat soluble; food chain. Report on your findings in a short article for a popular science magazine or web site.

 www.science.nelson.com

▶ *Section 1.4* *Questions*

Understanding Concepts

1. Draw structural diagrams to represent the elimination reaction of 2-chloropentane to form an alkene. Include reactants, reaction conditions, and all possible products and their IUPAC names.

2. Classify and write structural formula equations for the following organic reactions:
   (a) propane + chlorine →
         1-chloropropane + 2-chloropropane + hydrogen chloride
   (b) propene + bromine → 1,2-dibromopropane
   (c) benzene + iodine → iodobenzene + hydrogen iodide

Applying Inquiry Skills

3. The synthesis of an organic compound typically involves a series of reactions, for example, some substitutions and some additions.
   (a) Plan a reaction beginning with a hydrocarbon to prepare 1,1,2-trichloroethane.
   (b) What experimental complications might arise in attempting the reactions suggested in part (a)?

Making Connections

4. Research examples of the use of organic chemistry to address health, safety, or environmental problems, and write a report or present one such case study. Examples of topics include: leaded and unleaded gasoline, use of solvents in dry cleaning, use of aerosol propellants, and use of pesticides and fertilizers.

5. Why was mustard gas such an effective weapon, both during World War 1 and more recently? Research its properties and effects, and what defences have been developed against it.

 www.science.nelson.com

6. Shortly after the connection was made between the "hole" in the ozone layer and the release of chlorofluorocarbons, many manufacturers stopped using CFCs as propellants in aerosol cans.
   (a) Research what alternatives were developed, and the effectiveness of each in the marketplace. Are the alternatives still in use? Have any of them been found to cause problems?
   (b) Design a product (one that must be sprayed under pressure) and its packaging. Plan a marketing strategy that highlights the way in which your product is sprayed from the container.

 www.science.nelson.com

alcohol an organic compound characterized by the presence of a hydroxyl functional group; R—OH

hydroxyl group an —OH functional group characteristic of alcohols

Figure 1
Molecular models and general formulas of **(a)** water, H—O—H, **(b)** the simplest alcohol, CH_3—OH, methanol and **(c)** the simplest ether, H_3C—O—CH_3, methoxymethane (dimethyl ether)

Alcohols and ethers are structurally similar in that they are essentially water molecules with substituted alkyl groups. In **alcohols**, one of the two H atoms in H_2O is replaced by an alkyl group; in ethers, both H atoms are replaced by alkyl groups. The molecular models in **Figure 1** show water, the simplest alcohol, and the simplest ether. The properties of these compounds are related to the effects of the polar **hydroxyl groups** (—OH) and the nonpolar alkyl groups.

(a)   **(b)**   **(c)**

H — O — H   R — O — H   R — O — R

Alcohols

The "alcohol" in wine and beer is more correctly called ethanol. It is formed by yeast, a fungus that derives its energy from breaking down sugars, producing carbon dioxide and ethanol as waste products. Once the concentration of ethanol reaches a critical level, the yeast cannot survive and fermentation ceases. The alcohol content in wine is therefore limited to about 13% (26 proof).

Other alcohols that are produced by living organisms include cholesterol and retinol, commonly known as vitamin A.

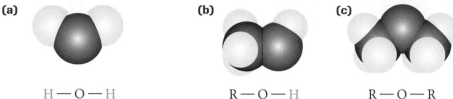

retinol
(vitamin A)

DID YOU KNOW ?

Alcohol Toxicity
It may be argued that all chemicals are toxic, to widely varying degrees. Some substances, such as methanol, are toxic in very small amounts, while others, such as NaCl, are generally harmless in moderate quantities. Toxicity is expressed by an LD_{50} rating, found in Material Safety Data Sheets (MSDS). It is the quantity of a substance, in grams per kilogram of body weight, that researchers estimate would be a lethal dose for 50% of a particular species exposed to that quantity of the substance. The LD_{50} values for several alcohols in human beings are shown below.

Alcohol	LD_{50} (g/kg body weight)
methanol	0.07
ethanol	13.7
1-propanol	1.87
2-propanol (rubbing alcohol)	5.8
glycerol (glycerine)	31.5
ethylene glycol (car antifreeze)	<1.45
propylene glycol (plumber's antifreeze)	30

While ethanol is not as toxic as other alcohols, it is recognized as a central nervous system depressant and a narcotic poison. Ethanol can be purchased in alcoholic beverages, the only "safe" form to consume. The ethanol commonly used in science laboratories is not intended for drinking and is purposely mixed with methanol, benzene, or other toxic materials in order to make it unpalatable.

Naming Alcohols

In the IUPAC system of naming alcohols, the —OH functional group is named -*ol*, and is added to the prefix of the parent alkane. As before, the parent alkane is the longest carbon chain to which an —OH group is attached. For example, the simplest alcohol, with one —OH group attached to methane, is named "methan*ol*." It is highly toxic and ingesting even small quantities can lead to blindness and death. The alcohol with two carbon atoms is ethan*ol*, the active ingredient in alcoholic beverages. It is an important synthetic organic chemical, used also as a solvent in lacquers, varnishes, perfumes, and flavourings, and is a raw material in the synthesis of other organic compounds.

When an alcohol contains more than two C atoms, or more than two —OH groups, we add a numbering system to identify the location of the —OH group(s). This is necessary because different isomers of the polyalcohol have entirely different properties. The location of the —OH group is indicated by a number corresponding to the C atom bearing the hydroxyl group. For example, there are two isomers of propanol, C_3H_7OH: 1-propanol is used as a solvent for lacquers and waxes, as a brake fluid, and in the manufacture of propanoic acid; 2-propanol (commonly called *iso*propanol, *i*-propanol, or isopropyl alcohol) is sold as rubbing alcohol and is used to manufacture oils, gums, and acetone. Both isomers of propanol are toxic to humans if taken internally.

$$CH_3 - CH_2 - CH_2 - OH$$
1–propanol

$$CH_3 - CH - CH_3$$ with OH attached
2–propanol

primary alcohol an alcohol in which the hydroxyl functional group is attached to a carbon which is itself attached to only one other carbon atom

secondary alcohol an alcohol in which the hydroxyl functional group is attached to a carbon which is itself attached to two other carbon atoms

tertiary alcohol an alcohol in which the hydroxyl functional group is attached to a carbon which is itself attached to three other carbon atoms

1°, 2°, and 3° Alcohols

Alcohols are subclassified according to the type of carbon to which the —OH group is attached. Since C atoms form four bonds, the C atom bearing the —OH group can be attached to a further 1, 2, or 3 alkyl groups, the resulting alcohols classifed as **primary, secondary,** and **tertiary alcohols,** respectively (1°, 2°, and 3°). Thus, 1-butanol is a primary alcohol, 2-butanol is a secondary alcohol, and 2-methyl-2-propanol is a tertiary alcohol. This classification is important for predicting the reactions each alcohol will undergo, because the reactions and products are determined by the availability of H atoms or alkyl groups in key positions. It is therefore useful in the selection of starting materials for a multistep reaction sequence to synthesize a final product.

$$CH_3 - CH_2 - CH_2 - C - H$$ (with H above and OH below)
1-butanol,
a 1° alcohol

$$CH_3 - CH_2 - C - CH_3$$ (with H above and OH below)
2-butanol,
a 2° alcohol

$$CH_3 - C - CH_3$$ (with CH_3 above and OH below)
2-methyl-2-propanol,
a 3° alcohol

INVESTIGATION 1.5.1

Comparison of Three Isomers of Butanol (p. 84)
How does the molecular structure of an organic molecule affect its properties? To find out, explore the physical and chemical properties of three isomers of butanol.

Polyalcohols

Alcohols that contain more than one hydroxyl group are called **polyalcohols**; the suffixes -*diol* and -*triol* are added to the entire alkane name to indicate two and three —OH groups, respectively. The antifreeze used in car radiators is 1,2-ethanediol, commonly called ethylene glycol. It is a liquid that is soluble in water and has a slightly sweet taste; caution is required when storing or disposing of car antifreeze as spills tend to attract animals who enjoy the taste, but who may suffer from its toxic effects.

Another common polyalcohol is 1,2,3-propanetriol, commonly called glycerol or glycerine. Like ethylene glycol, it is also a sweet-tasting syrupy liquid, and is soluble in water. However, unlike ethylene glycol, glycerol is nontoxic. Its abundance of —OH groups makes it a good participant in hydrogen bonding with water, a property that makes it a valuable ingredient in skin moisturizers, hand lotions, and lipsticks, and in foods such as chocolates. As you will see in Chapter 2, glycerol is a key component in the molecular structure of many fats and oils. You will also see in Chapter 2 that sugar molecules such as glucose and sucrose consist of carbon chains with many attached —OH groups, in addition to other functional groups.

polyalcohol an alcohol that contains more than one hydroxyl functional group

cyclic alcohol an alcohol that contains an alicyclic ring

aromatic alcohol an alcohol that contains a benzene ring

The hydroxyl groups in an alcohol can also be considered an added group to a parent hydrocarbon chain; the prefix for the hydroxyl group is *hydroxy-*. Thus 1,2,3-propanetriol is also named 1,2,3-trihydroxypropane.

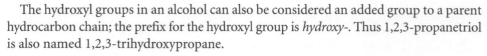

1,2-ethanediol (ethylene glycol)

1,2,3-propanetriol (glycerol)

Cyclic Alcohols

Many cyclic compounds have attached −OH groups, and are classified as **cyclic alcohols**; many large molecules are known by their common names which often end in *-ol*. Menthol, for example, is an alcohol derived from the oil of the peppermint plant. Its molecule consists of a cyclohexane ring with attached methyl, isopropyl, and hydroxyl groups. It is a white solid with a characteristic odour, used as a flavouring and in skin lotions and throat lozenges.

A more complex cyclic alcohol is cholesterol (**Figure 2**), a compound that is biochemically significant due to its effect on the cardiovascular system. The structure of cholesterol shows that it is a large molecule of which the polar −OH group forms only a small part, making it largely insoluble in water.

Aromatic compounds can also have attached −OH groups, forming the **aromatic alcohols**. The simplest aromatic alcohol is a benzene ring with one attached −OH group; it is named hydroxybenzene, also called phenol. Phenol is a colourless solid that is slightly soluble in water; the effect of the polar −OH group is apparent when compared with benzene, a liquid with a lower melting point than phenol, and which is insoluble in water. Phenol is used in the industrial preparation of many plastics, drugs, dyes, and weedkillers.

When naming cyclic or aromatic alcohols, the −OH (hydroxyl) groups may be considered as groups attached to the parent ring. Thus, phenol is named hydroxybenzene. A benzene ring with two hydroxyl groups adjacent to each other is named 1,2-dihydroxybenzene.

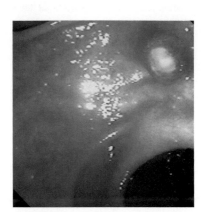

Figure 2
Cholesterol is only slightly soluble in water (0.26 g per 100 mL of water). Excessive amounts will precipitate from solution. In the gallbladder, crystals of cholesterol form gallstones. This endoscope photo shows a gallstone (yellow) blocking the duct from the gall bladder to the small intestine.

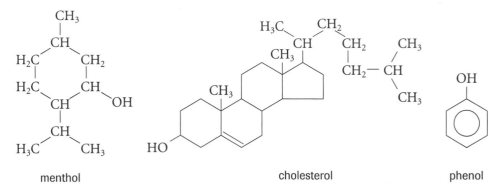

menthol

cholesterol

phenol

▶ **SAMPLE** problem | **Naming and Drawing Alcohols**

1. Name the following alcohol and indicate whether it is a primary, secondary, or tertiary alcohol.

First, identify the longest C chain. Since it is five Cs long, the alcohol is a pentanol.

Next, look at where the hydroxyl groups are attached. An −OH group is attached to the second C atom, so the alcohol is a 2-pentanol.

Look to see where any other group(s) are attached. A methyl group is attached to the second C atom, so the alcohol is 2-methyl-2-pentanol.

Since the second C atom, to which the OH is attached, is attached to three alkyl groups, the alcohol is a tertiary alcohol.

2. Draw a structural diagram for 1,3-butanediol.

First, write the C skeleton for the "parent" molecule, butane.
  Next, attach an —OH group to the first and third C atoms.
  Finally, complete the remaining C bonds with H atoms.

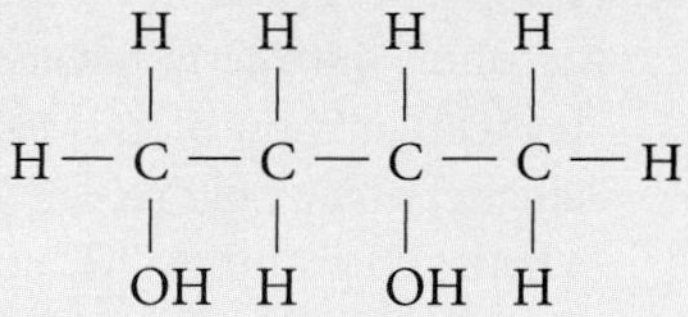

$$H-\overset{\overset{\displaystyle H}{|}}{C}-\overset{\overset{\displaystyle H}{|}}{\underset{\underset{\displaystyle OH}{|}}{C}}-\overset{\overset{\displaystyle H}{|}}{\underset{\underset{\displaystyle H}{|}}{C}}-\overset{\overset{\displaystyle H}{|}}{\underset{\underset{\displaystyle OH}{|}}{C}}-\overset{\overset{\displaystyle H}{|}}{\underset{\underset{\displaystyle H}{|}}{C}}-H$$

Example 1

Draw a structural diagram for 3-ethyl-2-pentanol and indicate whether it is a primary, secondary, or tertiary alcohol.

Solution

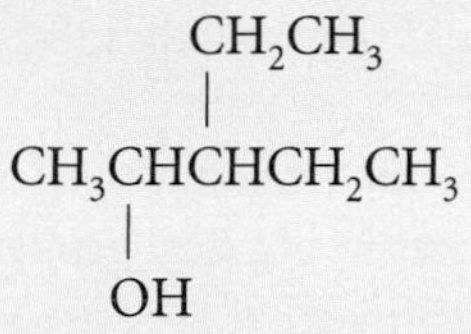

$$\overset{\overset{\displaystyle CH_2CH_3}{|}}{CH_3CHCHCH_2CH_3}$$
$$\underset{\underset{\displaystyle OH}{|}}{}$$

This is a secondary alcohol.

Example 2

Name the following alcohol.

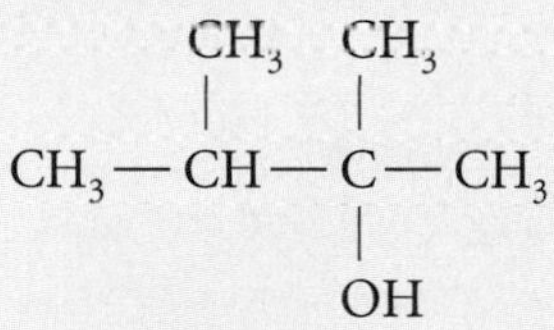

$$CH_3-\overset{\overset{\displaystyle CH_3}{|}}{CH}-\overset{\overset{\displaystyle CH_3}{|}}{\underset{\underset{\displaystyle OH}{|}}{C}}-CH_3$$

Solution

The alcohol is 2,3-dimethyl-2-butanol.

▶ Practice

Understanding Concepts

1. Write IUPAC names for the following compounds.

(a) 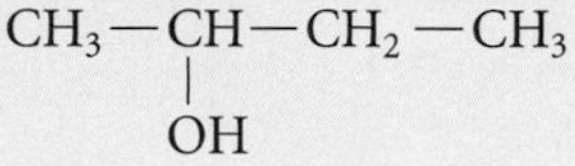 $CH_3-\overset{\overset{\displaystyle }{}}{CH}-CH_2-CH_3$
$\qquad\quad \underset{\underset{\displaystyle OH}{|}}{}$

(b) 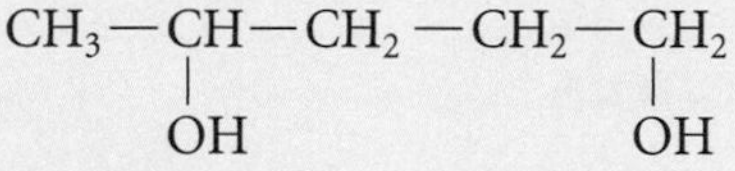 $CH_3-\overset{\overset{\displaystyle }{}}{CH}-CH_2-CH_2-\overset{\overset{\displaystyle }{}}{CH_2}$
$\qquad\quad \underset{\underset{\displaystyle OH}{|}}{}\qquad\qquad\qquad \underset{\underset{\displaystyle OH}{|}}{}$

(c) 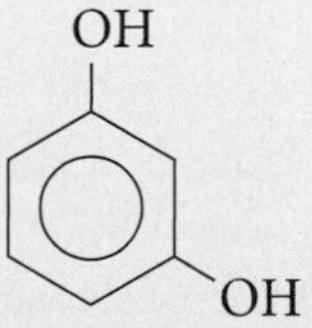 OH

OH

2. Draw a structural diagram for
(a) 3-methyl-1-butanol
(b) 1,2-propanediol
(c) glycerol
(d) phenol

3. Draw structural diagrams showing:
(a) an isomer of butanol that is a secondary alcohol
(b) all the pentanols that are isomers

Table 1 Boiling Points for Some Short-Chain Alcohols

Name	Formula	Boiling point (°C)
methanol	CH_3OH	65
ethanol	C_2H_5OH	78
1-propanol	C_3H_7OH	97
1-butanol	C_4H_9OH	117

INVESTIGATION 1.5.2

Trends in Properties of Alcohols (p. 86)

Is there a link between the molecular size of alcohols and their properties? Predict a trend, and then see if your evidence supports your prediction.

DID YOU KNOW ?

Fill up with Methanol

Alcohols have many uses, one of the more recent being a fuel for motor vehicles. The problem with methanol as a fuel for cars is its hydroxyl group. This functional group makes it less volatile than the hydrocarbons that make up gasoline, and the low volatility makes it difficult to ignite. In our cold Canadian winters, there is little methanol vapour in the engine and an electrical spark is insufficient to start the car. Canadian scientists are investigating a variety of dual ignition systems, one of which is a plasma jet igniter that is 100 times more energetic than conventional ignition systems.

hydration reaction a reaction that results in the addition of a water molecule

Properties of Alcohols

Alcohols have certain characteristic properties, including boiling points that are much higher than those of their parent alkanes. For example, ethanol boils at 78°C, compared with ethane, which boils at −89°C (**Table 1**). This property can be explained by the presence of a hydroxyl group, −OH, attached to a hydrocarbon chain. This functional group not only makes alcohol molecules polar, it also gives them the capacity to form hydrogen bonds.

Furthermore, simple alcohols are much more soluble in polar solvents such as water than are their parent alkanes. This can also be explained by the presence of the polar O−H bond.

In long-chain alcohols, the hydrocarbon portion of the molecule is nonpolar, making larger alcohols good solvents for nonpolar molecular compounds as well. Thus, alcohols are frequently used as solvents in organic reactions because they will dissolve both polar and nonpolar compounds.

When one of the H atoms in water is replaced by an alkyl group, the resulting alcohol, R−O−H, is less polar than water, with accompanying differences in physical properties. We will see later that when both H atoms in water are replaced by alkyl groups, we get another group of organic compounds named ethers, R−O−R. Perhaps you can predict now what their physical properties will be.

▶ Practice

Understanding Concepts

4. Explain briefly why methanol has a higher boiling point than methane.

5. Arrange the following compounds in order of increasing boiling point, and give reasons for your answer.
   (a) butane
   (b) 1-butanol
   (c) octane
   (d) 1-octanol

Making Connections

6. Glycerol is more viscous than water, and can lower the freezing point of water; when added to biological samples, it helps to keep the tissues from freezing, thereby reducing damage. From your knowledge of the molecular structure of glycerol, suggest reasons to account for these properties of glycerol.

Reactions Involving Alcohols

Preparing Alcohols: Hydration Reactions

If you recall the reactions of alkenes, the double bonds readily undergo addition reactions. If we start with an alkene, we can introduce the −OH functional group by adding HOH, water. Indeed, many alcohols are prepared industrially by addition reactions of water to unsaturated hydrocarbons. For example, 2-butanol is formed by the reaction between water and butene, using sulfuric acid as a catalyst. Since the overall result is the addition of a water molecule, this type of addition reaction is also referred to as a **hydration reaction**. This reaction follows Markovnikov's rule: The hydrogen attaches to the carbon atom that already has more hydrogen atoms; the −OH group attaches to the other carbon atom in the double bond.

$$CH_3CH_2CH{=}CH_2 + HOH \xrightarrow{\text{acid}} CH_3CH_2{-}\underset{\underset{OH}{|}}{CH}{-}\underset{\underset{H}{|}}{CH_2}$$

$$\text{1-butene} \qquad \text{water} \qquad\qquad \text{2-butanol}$$

The simplest alcohol, methanol, is sometimes called wood alcohol because it was once made by heating wood shavings in the absence of air. Methanol is toxic to humans. Drinking even small amounts of it or inhaling the vapour for prolonged periods can lead to blindness or death. The modern method of preparing methanol combines carbon monoxide and hydrogen at high temperature and pressure in the presence of a catalyst.

$CO_{(g)} + 2 H_{2(g)} \rightarrow CH_3OH_{(l)}$

As discussed earlier, ethanol, $C_2H_5OH_{(l)}$, can be prepared by the fermentation of sugars, using a yeast culture. This process occurs in the absence of oxygen, and produces energy for growth of the yeast. While ethanol is considered one of the least toxic alcohols, a combination of ethanol and driving is more deadly than many other chemical compounds.

$C_6H_{12}O_{6(s)} \rightarrow 2 CO_{2(g)} + 2 C_2H_5OH_{(l)}$

Combustion of Alcohols

Like hydrocarbons, alcohols undergo complete combustion to produce only carbon dioxide and water.

$2 C_3H_7OH_{(l)} + 9 O_{2(g)} \rightarrow 8 H_2O_{(g)} + 6 CO_{2(g)}$
propanol   oxygen   water   carbon dioxide

From Alcohols to Alkenes: Elimination Reactions

The addition reaction between an alkene and water can be made to proceed in reverse. Under certain conditions alcohols decompose to produce alkenes and water. This type of reaction is catalyzed by concentrated sulfuric acid, which removes a hydrogen atom and a hydroxyl group from neighbouring C atoms, leaving a C=C double bond. Since this *elimination* reaction results in the removal of water, this type of reaction is also called a **dehydration reaction**. It is essentially the reverse of the hydration reaction in the preparation of alcohols. Alkenes are similarly formed in elimination reactions of alkyl halides, as we discussed earlier in section 1.4.

dehydration reaction a reaction that results in the removal of water

$$CH_3CH-CH_2 \xrightarrow[\text{catalyst}]{\text{conc. } H_2SO_4} CH_3CH{=}CH_2 + HOH$$
      | |
     OH   H
    propanol   propene   water

Reactions of Alcohols   *SAMPLE problem* ◀

1. Draw structural diagrams to represent the addition reaction of propene to form an alcohol.

To form an alcohol, water is added to the double bond of propene. Markovnikov's rule predicts that the H is added to the atom of the double bond that is richer in H atoms, and the OH group to the other C atom.

$$\begin{array}{ccc} & H & H \\ & | & | \\ CH_3-C{=}C-H + H_2O \longrightarrow CH_3- & C-C-H \\ & | & | \\ & OH & H \end{array}$$

2. **Write balanced equations and name the reactants and products for:**
   (a) the dehydration reaction of ethanol
   (b) the complete combustion of 1-propanol

(a) A dehydration reaction is the elimination of H_2O from a molecule. When H_2O is eliminated from an alcohol, the corresponding alkene is formed.

$$CH_3CH_2OH \xrightarrow{H_2SO_4} CH_2{=}CH_2 + H_2O$$
$$\text{ethanol} \qquad\quad \text{ethene}$$

(b) The only products formed in the complete combustion of an alcohol are carbon dioxide and water.

$$CH_3CH_2CH_2OH + \frac{9}{2}O_2 \longrightarrow 3\,CO_2 + 4\,H_2O$$

Example

Draw structural diagrams and write IUPAC names to represent the formation of 2-methyl-2-butanol.

Solution

$$
\underset{\text{2-methyl-1-butene}}{CH_3CH_2\overset{\displaystyle CH_3}{\overset{\displaystyle |}{C}}{=}CH_2 + H_2O} \longrightarrow \underset{\text{2-methyl-2-butanol}}{CH_3CH_2\overset{\displaystyle CH_3}{\underset{\displaystyle OH}{\overset{\displaystyle |}{\underset{\displaystyle |}{C}}}}CH_3}
$$

▶ **Practice**

Understanding Concepts

7. Alcohols can be made by addition reactions.
   (a) Draw structural diagrams to represent the reaction
      2-butene + water → 2-butanol
   (b) Write a word equation, with IUPAC names, for the reaction

$$CH_2{=}CH_2 + H{-}O{-}Cl \longrightarrow \underset{\substack{OH \quad Cl}}{CH_2{-}CH_2}$$
$$\underset{\substack{\text{hydrogen}\\\text{hypochlorite}}}{}$$

8. Elimination reactions of alcohols are generally slow, and require an acid catalyst and heating.
   (a) Draw structural diagrams to represent the reaction
      1-propanol → propene + water
   (b) Write a word equation, with IUPAC names, for the dehydration reaction (in the presence of concentrated H_2SO_4)
      $CH_3{-}CH_2{-}CH_2{-}CH_2{-}OH \rightarrow$

9. Only a few of the simpler alcohols are used in combustion reactions. Alcohol–gasoline mixtures, known as gasohol, are the most common examples. Write a balanced chemical equation, using molecular formulas, for the complete combustion of the following alcohols:
   (a) ethanol (in gasohol)
   (b) 2-propanol (rubbing alcohol)

SUMMARY *Alcohols*

Functional group: –OH, hydroxyl group

Preparation:

- alkenes → alcohols
  addition reaction with water: hydration

$$R-\underset{\underset{H}{|}}{\overset{\overset{H}{|}}{C}}=\underset{\underset{H}{|}}{\overset{\overset{H}{|}}{C}}-R' + H-OH \longrightarrow R-\underset{\underset{H}{|}}{\overset{\overset{H}{|}}{C}}-\underset{\underset{OH}{|}}{\overset{\overset{H}{|}}{C}}-R'$$

Pathways to other groups:

- alcohols → alkenes
  elimination reaction: dehydration

$$R-\underset{\underset{H}{|}}{\overset{\overset{H}{|}}{C}}-\underset{\underset{OH}{|}}{\overset{\overset{H}{|}}{C}}-R' \xrightarrow{\text{conc. }H_2SO_4} R-\underset{\underset{H}{|}}{\overset{\overset{H}{|}}{C}}=\underset{\underset{H}{|}}{\overset{\overset{H}{|}}{C}}-R' + HOH$$

- alcohols → aldehydes → carboxylic acids (see Sections 1.6, 1.7)
  controlled oxidation reaction

▶ Practice

Making Connections

10. (a) Research the reactions, along with the necessary conditions, that take place in a methanol-burning car engine.
    (b) What are the advantages and disadvantages of methanol as a fuel, over more conventional gasoline or diesel fuels?
    (c) Give arguments for and against the implementation of a law requiring that all cars in Canada should be adapted to burn methanol.

GO www.science.nelson.com

 INVESTIGATION 1.5.2

Trends in Properties of Alcohols (p. 86)
Is there a link between the molecular size and the physical properties of alcohols? Investigate three 1° alcohols to find out.

Ethers

Physicians and chemists have long experimented with the use of chemical substances for eliminating pain during medical procedures. Many different compounds have been tried over the years. Dinitrogen monoxide (nitrous oxide, N_2O) was first tested in 1844 when a Boston dentist demonstrated a tooth extraction using the chemical as an anesthetic. Unhappily for the patient, the tooth was pulled before the "laughing gas" had taken effect and the public was not impressed.

At about the same time, tests were also done with an organic compound that proved effective. This compound is ethoxyethane, more commonly called diethyl ether, or simply, ether; it is a volatile and highly flammable liquid ($CH_3CH_2OCH_2CH_{3(l)}$). You may have seen old movies in which a handkerchief dabbed in ether, held over the face, would render a victim unconscious in seconds. With better anesthetics available, diethyl ether is now used mainly as a solvent for fats and oils.

Properties of Ethers

ether an organic compound with two alkyl groups (the same or different) attached to an oxygen atom

The structure of **ethers** is similar to that of water, $H-O-H$, and alcohol, $R-O-H$. In an ether, the oxygen atom is bonded to two alkyl groups. The two alkyl groups may be identical ($R-O-R$) or different ($R'-O-R''$).

Since ethers do not contain any $-OH$ groups, they cannot form hydrogen bonds to themselves, as alcohols can. However, their polar $C-O$ bonds and the V-shape of the $C-O-C$ group do make ether molecules more polar than hydrocarbons. Thus, the intermolecular attractions between ether molecules are stronger than those in hydrocarbons and weaker than those in alcohols. Accordingly, the boiling points of ethers are slightly higher than those of analogous hydrocarbons, but lower than those of analogous alcohols (**Table 2**). Note that H_2O, with two $-OH$ groups per molecule, has the highest boiling point in this series.

Table 2 Boiling Points of Analogous Compounds

Compound	Structure	Boiling point (°C)
ethane	CH_3-CH_3	−89
methoxymethane (dimethyl ether)	CH_3-O-CH_3	−23
ethanol	CH_3-CH_2-O-H	78.5
water	$H-O-H$	100

Ethers are good solvents for organic reactions because they mix readily with both polar and nonpolar substances. Their $C-O$ bonds make them more polar than hydrocarbons and thus ethers are more miscible with polar substances than are hydrocarbons. Meanwhile, their alkyl groups allow them to mix readily with nonpolar substances. In addition, the single covalent $C-O$ bonds in ethers are difficult to break, making ethers quite unreactive, another property of a good solvent.

Naming Ethers

Ethers are named by adding *oxy* to the prefix of the *smaller* hydrocarbon group and joining it to the alkane name of the *larger* hydrocarbon group. Hence, the IUPAC name for $CH_3-O-C_2H_5$ is methoxyethane (*not* ethoxymethane). You may also encounter names for ethers derived from the two alkyl groups, followed by the term ether; methoxyethane would thus be methyl ethyl ether. When the two alkyl groups are the same, the prefix *di-* is used; for example, ethoxyethane is *di*ethyl ether.

▶ SAMPLE problem | *Naming and Drawing Ethers*

Draw a structural diagram for 1-ethoxypropane.

First, draw the C skeleton for the propane molecule, then add the ethoxy group, CH_3CH_2-O- to the first carbon. Finally, add the remaining H atoms:

$$CH_3CH_2-O-CH_2CH_2CH_3$$

▶ Practice

Making Connections

11. Write IUPAC names for the following compounds:
    (a) $CH_3-CH_2-CH_2-O-CH_3$
    (b) $CH_3-CH_2-O-CH_2-CH_2-CH_2-CH_3$ or $CH_3-CH_2-O-(CH_2)_3-CH_3$

Preparing Ethers from Alcohols: Condensation Reactions

Ethers are formed when two alcohol molecules react. A molecule of water is eliminated and the remaining portions of the two alcohol molecules combine to form an ether. This type of reaction, in which two molecules interact to form a larger molecule with a loss of a small molecule such as water, is called a **condensation reaction**. As illustrated in the equation below, two molecules of methanol interact to produce methoxymethane and water; as with many dehydration reactions, concentrated sulfuric acid is needed as a catalyst.

condensation reaction a reaction in which two molecules combine to form a larger product, with the elimination of a small molecule such as water or an alcohol

$$CH_3OH_{(l)} + CH_3OH_{(l)} \xrightarrow{H_2SO_4} CH_3OCH_{3(l)} + HOH_{(l)} \qquad \text{(condensation reaction)}$$

methanol   methanol   methoxymethane   water

SUMMARY

Condensation Reactions of Alcohols to Ethers

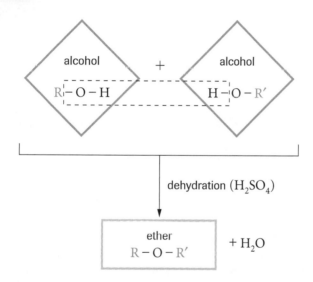

Reactions Involving Ethers

SAMPLE problem ◄

Write a balanced equation to show the formation of an ether from 1-propanol. Name the ether formed and the type of reaction.

   Two 1-propanol molecules can react together, in a condensation reaction, with concentrated sulfuric acid as a catalyst.

$$CH_3CH_2CH_2OH + CH_3CH_2CH_2OH \xrightarrow{H_2SO_4} CH_3CH_2CH_2OCH_2CH_2CH_3 + H_2O$$

1-*n*-propoxypropane

The ether formed is 1-*n*-propoxypropane; since water is eliminated, it is also a dehydration reaction.

Functional group: R–O–R′

Preparation:

- alcohols → ethers + water
  Condensation reaction, eliminating H_2O; dehydration

$$R-OH + R'-OH \rightarrow R-O-R' + H-OH$$

▶ Practice

Understanding Concepts

12. Like many organic compounds, alcohols and ethers undergo complete combustion reactions to produce carbon dioxide and water. Select one alcohol and one ether, and write stuctural diagrams for their complete combustion.

13. The major disadvantages of using ethoxyethane as an anesthetic are its irritating effects on the respiratory system and the occurrence of post-anesthetic nausea and vomiting. For this reason, it has been largely replaced by methoxypropane, which is relatively free of side effects.
  (a) Draw structural formulas of ethoxyethane and methoxypropane, and determine if they are isomers.
  (b) Write an equation to show the formation of ethoxyethane from ethanol.

▶ Section 1.5 Questions

Understanding Concepts

1. Write structural formulas and IUPAC names for all saturated alcohols with five carbon atoms and one hydroxyl group.

2. Explain why the propane that is used as fuel in a barbecue is a gas at room temperature, but 2-propanol used as rubbing alcohol is a liquid at room temperature.

3. Draw the structures and write the IUPAC names of the two alkenes that are formed when 2-hexanol undergoes a condensation reaction in the presence of an acid catalyst.

4. Write an equation using structural diagrams to show the production of each of the following alcohols from an appropriate alkene:
  (a) 2-butanol   (b) 2-methyl-2-propanol

5. Draw the structures and write the IUPAC names of all the ethers that are isomers of 2-butanol.

6. Classify and write structural formula equations for the following organic reactions.
  (a) ethene + water → ethanol
  (b) 2-butanol → 1-butene + 2-butene + water
  (c) ethoxyethane + oxygen →
  (d) ethene + hypochlorous acid ($HOCl_{(aq)}$) →
      2-chloroethanol
  (e) methanol + oxygen →

7. For each of the following pairs of compounds, select the one that has the higher boiling point. Give reasons for your answer.
  (a) ethylene glycol or glycerol
  (b) water or methoxymethane
  (c) methanol or propanol
  (d) methoxyethane or propanol

Applying Inquiry Skills

8. Ethylene glycol, although toxic, has occasionally, and illegally, been added to wines to enhance the sweet flavour. Propose an experimental design, including the procedure and the equipment that you would use, to remove the ethylene glycol and purify the contaminated wine. Include in your answer any safety precautions needed.

Making Connections

9. Alcohols have gained increased popularity as an additive to gasoline, as a fuel for automobiles. "Gasohols" may contain up to 10% methanol and ethanol, and are considered more environmentally friendly than gasoline alone.
  (a) Write balanced chemical reactions for the complete combustion of methanol and ethanol.
  (b) Although methanol is less expensive to produce, ethanol is blended with methanol in gasoline. This is because methanol does not mix well with gasoline, and ethanol is used as a co-solvent. Explain, with reference to molecular structure, why ethanol is more soluble than methanol in gasoline (which is mostly octane).
  (c) Methanol and ethanol are considered to be more environmentally friendly than gasoline. Research, and explain why.
  (d) When small amounts of water are present in the gasoline in the gas lines of a car, the water may freeze and block gasoline flow. Explain how using a gasohol would affect this problem.

 www.science.nelson.com

A key to the survival of a species is the ability of its members to communicate with each other, often over great distances. Insects use chemical signals to share information about location of food or of danger, and to attract potential mates. You may have seen a line of ants following a chemical trail left by other ants who have found food or water. Many of the molecules used belong to a group of organic compounds called **ketones**. For example, some ants use a simple ketone called 2-heptanone (**Figure 1(a)**) to warn of danger. A group of more complex ketones, called *pheromones*, are used by many insects as sex attracters. One example, 9-ketodecenoic acid (**Figure 1(b)**), is used by a honeybee queen when she takes flight to establish a new hive.

ketone an organic compound characterized by the presence of a carbonyl group bonded to two carbon atoms

aldehyde an organic compound characterized by a terminal carbonyl functional group; that is, a carbonyl group bonded to at least one H atom

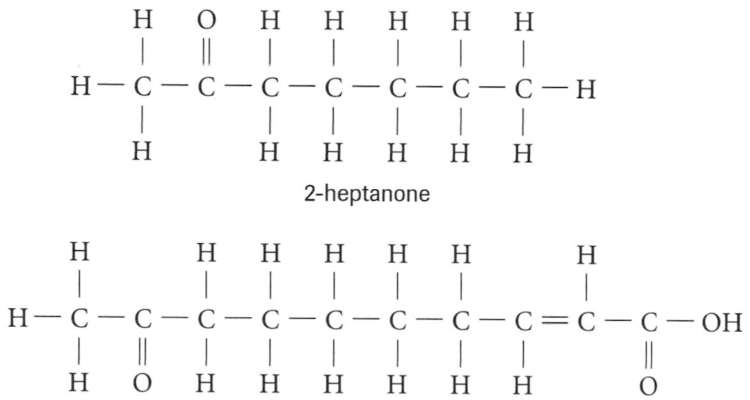

(a)

2-heptanone

(b)

9-keto-2-decenoic acid

Figure 1

These pheromones are very specific to the species and are powerful in their effect. Foresters make use of the potency of pheromones to attract and trap pest insects such as the gypsy moth (**Figure 2**). The trapping programs are designed to control specific insect populations by inhibiting breeding—an environmentally benign alternative to spraying nonspecific insecticides.

Closely related to the ketone family is the **aldehyde** family of organic compounds. These compounds also seem to be detectable over long distances, by our sense of smell. The smaller aldehydes have strong, unpleasant odours. Formaldehyde is the simplest aldehyde and is a colourless gas at room temperature; in aqueous solution, it is used as an antiseptic and a disinfectant. The next simplest aldehyde is acetaldehyde, which is a colourless liquid used in the synthesis of resins and dyes, and also as a preservative. Both formaldehyde and acetaldehyde can form solid trimers; that is, three molecules can join together into a single large molecule. The trimer of formaldehyde is used to fumigate rooms against pests, and the trimer of acetaldehyde is used as a hypnotic drug.

Aldehydes of higher molecular weight have pleasant flowery odours and are often found in the essential oils of plants. These oils are used for their fragrance in perfumes and aromatherapy products. The oil of bitter almond, for example, is benzaldehyde, the simplest aldehyde of benzene.

The functional group in ketones is the **carbonyl group**, consisting of a carbon atom joined with a double covalent bond to an oxygen atom. In ketones, the carbonyl group is attached to two alkyl groups and no H atoms. In aldehydes, the carbonyl group is attached to at least one H atom; it is also attached to either another H atom or an alkyl group. In other words, in an aldehyde, the carbonyl group always occurs at the end of a carbon chain; in a ketone, the carbonyl group occurs in the interior of a carbon chain.

Figure 2
Many ketones are volatile compounds that can travel long distances, making them good carriers of chemical signals for insects. Ants, bees, and moths like the gypsy moths shown here, produce and detect minute quantities of ketones to communicate the presence of food or water or the availability of a mate.

carbonyl group a functional group containing a carbon atom joined with a double covalent bond to an oxygen atom; C=O

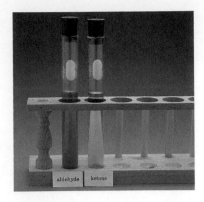

Figure 3
In a diagnostic test, Fehling's solution distinguishes aldehydes from ketones. An aldehyde converts the blue copper(II) ion in the Fehling's solution to a red precipitate of copper(I) oxide. A ketone does not react with Fehling's solution, so the solution remains blue.

This difference in the position of the carbonyl group affects the chemical reactivity of the molecule, and is used in a test to distinguish aldehydes from ketones (**Figure 3**).

(a) carbonyl group   (b) an aldehyde   (c) two ketones

$$\begin{array}{cccc} O & O & O & O \\ \| & \| & \| & \| \\ -C- & R-C-H & R-C-R \quad \text{or} & R-C-R' \end{array}$$

Later in this chapter, we will look at compounds in which the carbonyl group is attached to other groups containing oxygen or nitrogen atoms.

Naming Aldehydes and Ketones

The IUPAC names for aldehydes are formed by taking the parent alkane name, dropping the final -*e*, and adding the suffix -*al*. The simplest aldehyde consists of a carbonyl group with no attached alkyl group; its formula is HCHO. It has only one C atom; thus the parent alkane is methane and the aldehyde is methan*al*, although it is more often known by its common name—formaldehyde. The next simplest aldehyde is a carbonyl group with a methyl group attached; this two-carbon aldehyde is called ethanal, also known as acetaldehyde.

Ketones are named by replacing the -*e* ending of the name of the corresponding alkane with -*one*. The simplest ketone is propanone, CH_3COCH_3, commonly known as acetone. If, in a ketone, the carbon chain containing the carbonyl group has five or more carbon atoms, it is necessary to use a numerical prefix to specify the location of the carbonyl group. For example, in 2-pentanone, the carbonyl group is the second carbon atom in the carbon chain.

$$\begin{array}{cccc} O & O & O & O \\ \| & \| & \| & \| \\ H-C-H & CH_3-C-H & CH_3-C-CH_3 & CH_3-C-CH_2-CH_3 \end{array}$$

methanal   ethanal   propanone   butanone
(formaldehyde)   (acetaldehyde)   (acetone)

▶ **SAMPLE** problem   **Drawing and Naming Aldehydes and Ketones**

1. **Write the IUPAC name for the following compound.**

$$\begin{array}{c} O \\ \| \\ CH_3CH_2CH_2-C-H \end{array}$$

The carbonyl group at the end of the chain indicates that the compound is an aldehyde, so will have the suffix -*al*.
    There are four C atoms in the molecule, so the corresponding alkane is butane. The compound is, therefore, butanal.

2. **Draw a structural diagram for 3-hexanone.**

The ending -*one* indicates that the compound is a ketone, so will have its carbonyl group in the interior of a carbon chain.
    The prefix *hexan-* indicates the corresponding alkane is hexane and so has 6 carbons, with the carbonyl group being in the third position.

$$\begin{array}{c} O \\ \| \\ CH_3CH_2CCH_2CH_2CH_3 \end{array}$$

Example

Draw structural diagrams and write IUPAC names for an aldehyde and a ketone, each containing three C atoms.

Solution

$$CH_3CH_2 - \overset{\displaystyle O}{\overset{\|}{C}} - H$$

propanal

$$CH_3\overset{\displaystyle O}{\overset{\|}{C}}CH_3$$

propanone

(a)

> ### ▶ Practice

Understanding Concepts

1. Draw structural diagrams for each of the following compounds:
   (a) ethanal
   (b) 2-hexanone
   (c) pentanal
   (d) benzaldehyde (**Figure 4**)

2. Write IUPAC names for
   (a) all possible heptanones.
   (b) all possible heptanals.

3. Write the IUPAC names for the following compounds:
   (a)
   $$CH_3CH_2CH_2CH_2\overset{\displaystyle O}{\overset{\|}{C}}H$$
   (b)
   $$CH_3CH_2CH_2\overset{\displaystyle O}{\overset{\|}{C}}CH_2CH_3$$
   (c)
   $$\overset{\displaystyle O}{\overset{\|}{H}}CH$$

(b)

Figure 4
Many essential oils contain aldehydes, which contribute their pleasant fragrances. Benzaldehyde is called oil of bitter almond; it is formed by grinding almonds or apricot pits and boiling them in water. In this process, the poisonous gas hydrogen cyanide is also produced.

Properties of Aldehydes and Ketones

Aldehydes and ketones have lower boiling points than analogous alcohols (**Table 1**), and are less soluble in water than alcohols; this is to be expected as they do not contain $-OH$ groups and so do not participate in hydrogen bonding. However, the carbonyl group is a strongly polar group due to the four shared electrons in the double $C=O$ bond. Thus, aldehydes and ketones are more soluble in water than are hydrocarbons. The ability of these compounds to mix with both polar and nonpolar substances makes them good solvents (**Figure 5**).

Table 1 Boiling Points of Analogous Compounds

Compound	Structure	Boiling point (°C)
ethanol	CH_3CH_2OH	78
ethanal	CH_3CHO	21
1-propanol	$CH_3CH_2CH_2OH$	97
propanal	CH_3CH_2CHO	49
propanone	CH_3COCH_3	56
1-butanol	$CH_3CH_2CH_2CH_2OH$	118
butanal	$CH_3CH_2CH_2CHO$	75
butanone	$CH_3CH_2COCH_3$	80

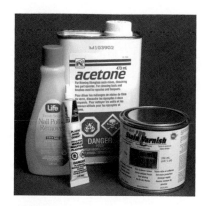

Figure 5
Propanone (acetone) is an effective organic solvent found in many nail polish removers, plastic cements, resins, and varnishes. Like many ketones, acetone is both volatile and flammable and should be used only in well-ventilated areas.

Finding chemicals that have desired odours is the subject of much research in the perfume industry. It appears that our olfactory receptors respond to the shape of the molecules that we smell rather than to their chemical composition. For example, the molecules shown below all smell like camphor, a component of many perfumes. All three molecules are "bowl" shaped, a geometrical shape that fits in our receptor sites for camphor-like odours.

Cl   Cl
| |
Cl — C — C — Cl
| |
Cl   Cl

hexachloroethane

cyclooctane

CH₃   CH₃
\   /
C

CH₂   CH
CH₂ — C   CH₂
CH₃   C
‖
O

oxidation reaction a chemical transformation involving a loss of electrons; historically used in organic chemistry to describe any reaction involving the addition of oxygen atoms or the loss of hydrogen atoms

---

▶ Practice

Understanding Concepts

4. Write the IUPAC name for the following compounds:
   (a) acetone
   (b) formaldehyde
   (c) acetaldehyde

5. Arrange the following compounds in increasing order of predicted boiling points. Give reasons for your answer.

   (a)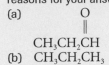
   $$\overset{\displaystyle O}{\overset{\|}{CH_3CH_2CH}}$$
   (b) $CH_3CH_2CH_3$
   (c) $CH_3CH_2CH_2OH$

Preparing Aldehydes and Ketones from Alcohols: Oxidation Reactions

Historically, the term "oxidation" was used to describe any reaction involving oxygen. The term has since been broadened to include all chemical processes that involve a loss of electrons. These processes are always accompanied by a reaction partner that undergoes "reduction", or a gain of electrons. This system of tracking electrons is useful in describing the changes in the types of bonds or the number of bonds in a chemical transformation. This further helps us to understand and select the type of reagent to use for a particular reaction.

Aldehydes and ketones can be prepared by the *controlled* oxidation of alcohols. In organic chemistry, the term **oxidation reaction** generally implies a gain of oxygen or a loss of hydrogen. (The term "oxidation" in fact encompasses many other chemical processes, and you will learn more about them in Chapter 9.) When alcohols are burned in oxygen, complete oxidation occurs and carbon dioxide and water are the only products formed. However, the conditions of oxidation reactions can be controlled to form other products. In these reactions, oxygen atoms are supplied by compounds called *oxidizing agents*. Some examples of oxidizing agents are hydrogen peroxide, H_2O_2; potassium dichromate, $K_2Cr_2O_7$; and potassium permanganate, $KMnO_4$. In the following equations, (O) will be used to indicate the reactive oxygen atom supplied by an oxidizing agent; the involvement of the oxidizing agents is not important in discussions here.

Let us examine the transformation of an alcohol to an aldehyde or ketone.

$$R-OH + (O) \longrightarrow \overset{\displaystyle O}{\overset{\|}{R-C-H}} \text{ or } \overset{\displaystyle O}{\overset{\|}{R-C-R}} + H_2O$$

alcohol   aldehyde or ketone

Essentially, the reactive (O) atom removes two H atoms, one from the −OH group, and one from the adjacent C atom, resulting in a C=O group; a water molecule is also produced.

- When a *primary* alcohol is oxidized, an H atom remains on the C atom, and an aldehyde is produced.

$$CH_3-\overset{\displaystyle OH}{\underset{\displaystyle H}{\overset{|}{\underset{|}{C}}}}-H + (O) \longrightarrow CH_3-\overset{\displaystyle O}{\overset{\|}{C}}-H + HOH \quad \text{(oxidation reaction)}$$

ethanol (1° alcohol)   ethanal

- When a *secondary* alcohol is similarly oxidized, the carbonyl group formed is necessarily attached to two alkyl groups, forming a ketone.

$$CH_3-\overset{\displaystyle OH}{\underset{\displaystyle H}{C}}-CH_3 + (O) \longrightarrow CH_3-\overset{\displaystyle O}{C}-CH_3 + HOH \quad \text{(oxidation reaction)}$$

2-propanol   propanone
(2° alcohol)

- Tertiary alcohols do not undergo this type of oxidation; no H atom is available on the central C atom.

$$CH_3-\overset{\displaystyle OH}{\underset{\displaystyle CH_3}{C}}-CH_3 + (O) \longrightarrow \text{not readily oxidized} \quad \text{(no reaction)}$$

2-methyl-2-propanol
(3° alcohol)

SUMMARY *Oxidation of Alcohols → Aldehydes and Ketones*

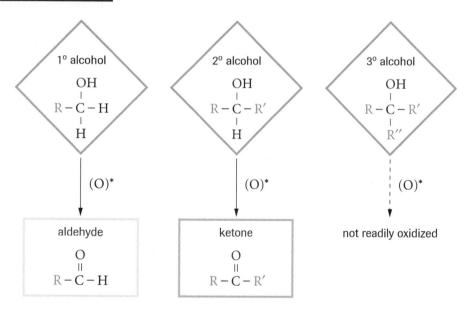

*(O) indicates controlled oxidation with $KMnO_4$ or $Cr_2O_7^{2-}$, in H_2SO_4

DID YOU KNOW?

Did You Know? Steroids

Steroids are unsaturated compounds based on a structure of four rings of carbon atoms. The best-known and most abundant steroid is cholesterol. Other steroids are ketones, including the male and female sex hormones, testoster*one* and progester*one*, and anti-inflammatory agents such as cortis*one*. Oral contraceptives include two synthetic steroids. Some athletes use anabolic steroids to enhance muscle development and physical performance, but such use may cause permanent damage.

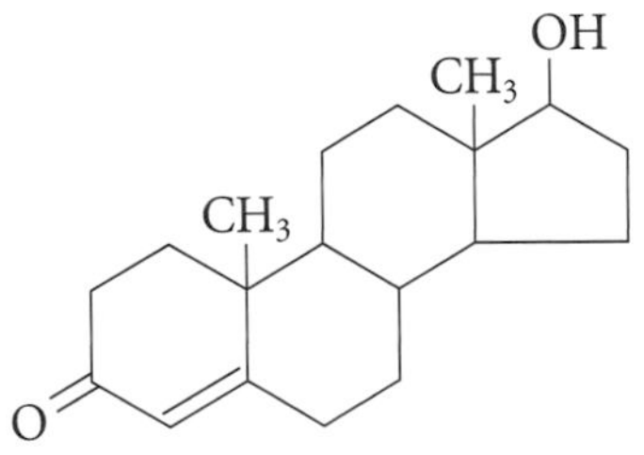

testosterone

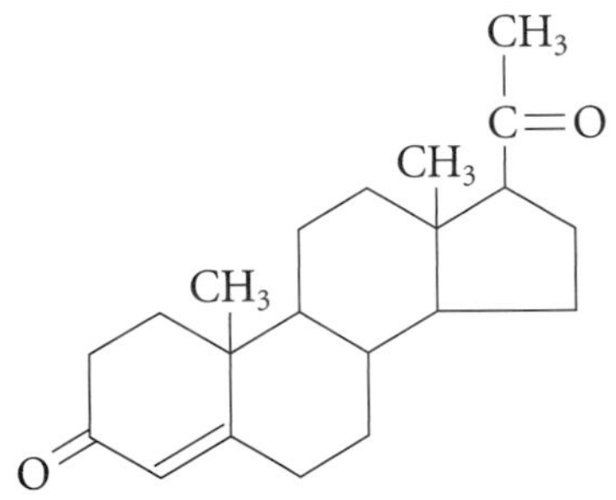

progesterone

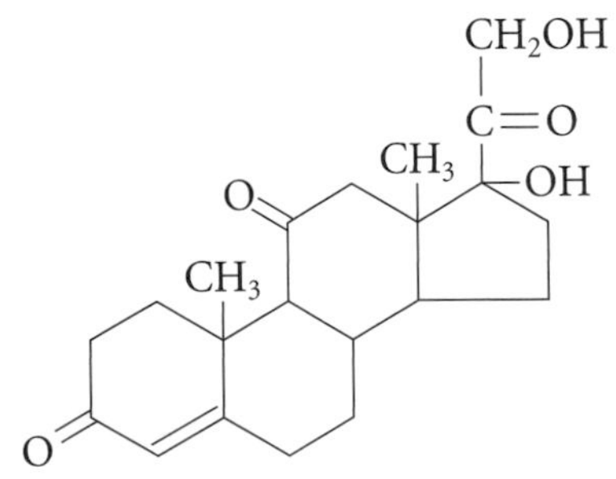

cortisone

▸ TRY THIS activity — How Many Can You Build?

Materials: molecular model kit

- From a molecular model kit, obtain 2 carbons, 6 hydrogens, and 1 oxygen. Build two different structures that use all 9 atoms.
  (a) Name these compounds and their functional groups.
- Obtain one additional carbon atom and build as many different structures as possible, using all 10 atoms.
  (b) Name these compounds.
  (c) Which of these compounds are isomers?

From Aldehydes and Ketones to Alcohols: Hydrogenation Reactions

The C=O double bond in carbonyl groups can undergo an addition reaction with hydrogen, although not with other reactants. High temperatures and pressures and the presence of a catalyst are needed for this hydrogenation reaction. When the H atoms are added to the carbonyl group, an —OH group results, producing an alcohol. This is, in effect, a reversal of the controlled oxidation of alcohols, discussed above. For example, ethanal forms ethanol and propanone forms 2-propanol. As you can see, because of the type of groups attached to the carbonyl C atom, aldehydes always produce primary alcohols, and ketones always produce secondary alcohols.

$$\underset{\text{ethanal}}{CH_3-\overset{\overset{\displaystyle O}{\|}}{C}-H} + H-H \xrightarrow[\text{heat, pressure}]{\text{catalyst}} \underset{\substack{\text{ethanol}\\(1^\circ\text{ alcohol})}}{CH_3-\underset{\underset{\displaystyle H}{|}}{\overset{\overset{\displaystyle OH}{|}}{C}}-H} \quad \text{(hydrogenation reaction)}$$

$$\underset{\text{propanone}}{CH_3-\overset{\overset{\displaystyle O}{\|}}{C}-CH_3} + H-H \xrightarrow[\text{heat, pressure}]{\text{catalyst}} \underset{\substack{\text{2-propanol}\\(2^\circ\text{ alcohol})}}{CH_3-\underset{\underset{\displaystyle H}{|}}{\overset{\overset{\displaystyle OH}{|}}{C}}-CH_3} \quad \text{(hydrogenation reaction)}$$

The difference in the position of the carbonyl group in aldehydes and ketones accounts for their difference in behaviour with oxidizing agents. As we will see in the next section when we discuss organic acids, aldehydes can be oxidized to form organic acids; ketones (because they have no hydrogen atoms on the carbonyl carbon) cannot. The synthesis pathway from alcohols to aldehydes and then to acids is an important tool in the preparation of many organic substances.

SUMMARY   *Aldehydes and Ketones*

Functional group: $>C=O$, carbonyl group

- aldehydes:

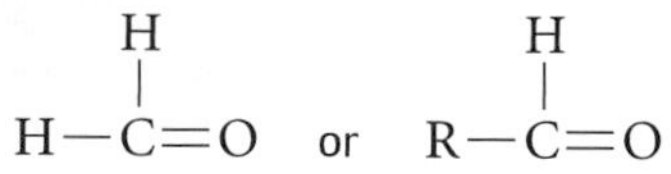

$$\overset{\overset{\displaystyle H}{|}}{H-C}=O \quad \text{or} \quad \overset{\overset{\displaystyle H}{|}}{R-C}=O$$

- ketones:

$$\overset{\overset{\displaystyle R}{|}}{R-C}=O \quad \text{or} \quad \overset{\overset{\displaystyle R'}{|}}{R-C}=O$$

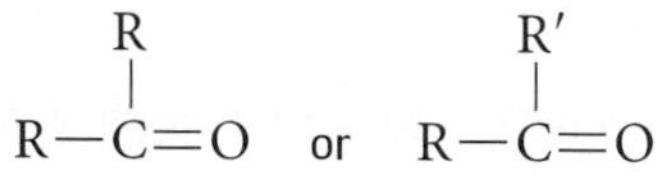

Preparation:

- primary alcohols → aldehydes
  controlled oxidation reactions

$$R-\overset{\overset{\displaystyle H}{|}}{\underset{\underset{\displaystyle H}{|}}{C}}-O-H + (O) \longrightarrow \overset{\overset{\displaystyle H}{|}}{R-C}=O + HOH$$

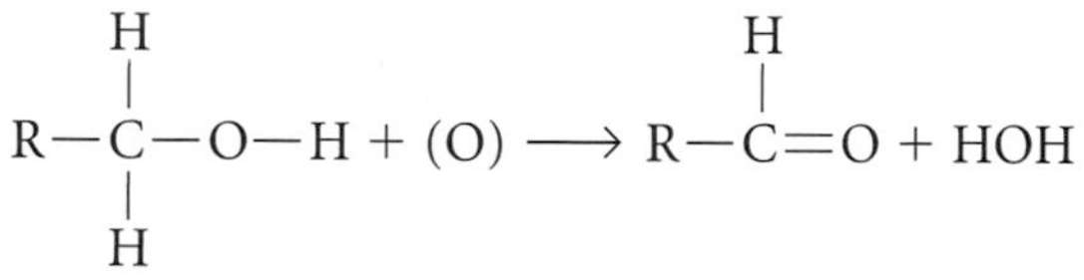

- secondary alcohols → ketones
  controlled oxidation reactions

$$R-\overset{\overset{\displaystyle R'}{|}}{\underset{\underset{\displaystyle H}{|}}{C}}-O-H + (O) \longrightarrow \overset{\overset{\displaystyle R'}{|}}{R-C}=O + HOH$$

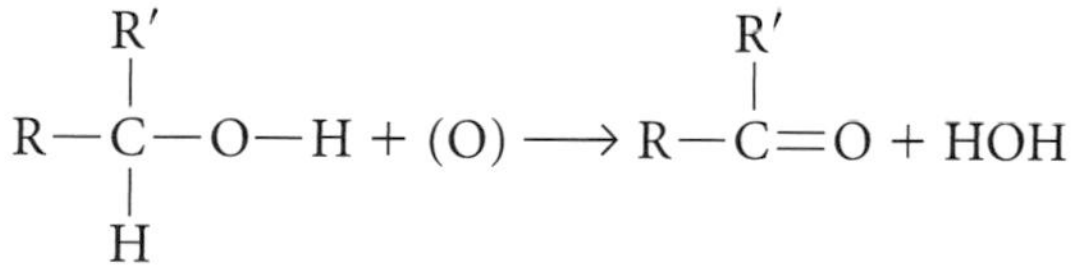

Pathways to other groups:

- aldehydes → primary alcohols
  addition reaction with hydrogen: hydrogenation

$$\overset{\overset{\displaystyle H}{|}}{R-C}=O + H_2 \longrightarrow R-\overset{\overset{\displaystyle H}{|}}{\underset{\underset{\displaystyle H}{|}}{C}}-OH$$

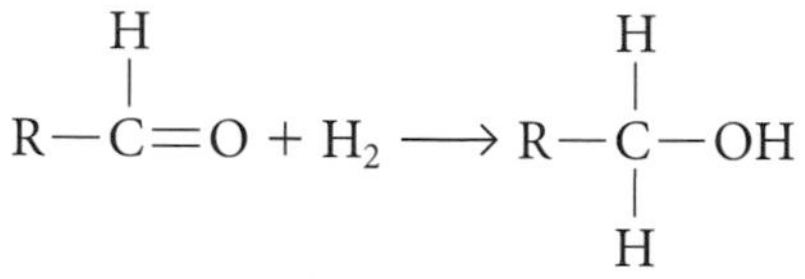

- ketones → secondary alcohols
  addition reaction with hydrogen: hydrogenation

$$\overset{\overset{\displaystyle R'}{|}}{R-C}=O + H_2 \longrightarrow R-\overset{\overset{\displaystyle R'}{|}}{\underset{\underset{\displaystyle H}{|}}{C}}-OH$$

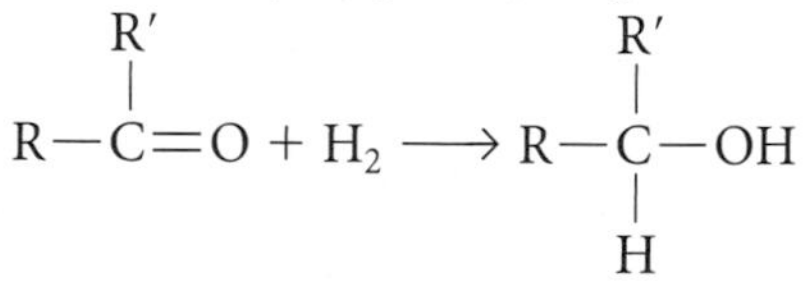

Reactions of Aldehydes and Ketones   **SAMPLE** problem ◀

Draw structural formulas and write IUPAC names to represent the controlled oxidation of an alcohol to form butanone.

First, draw the structural formula for butanone.

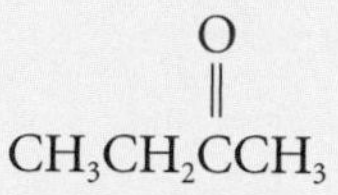

$$\overset{\overset{\displaystyle O}{\|}}{CH_3CH_2CCH_3}$$
butanone

As the oxygen atom of the carbonyl group is attached to the second C atom, the alcohol to be oxidized must be 2-butanol, so draw the structural formula for 2-butanol.

$$\underset{\text{2-butanol}}{\overset{\displaystyle\text{OH}}{\underset{\displaystyle\text{H}}{\text{CH}_3\text{CH}_2\text{CCH}_3}}} + (\text{O}) \longrightarrow \underset{\text{butanone}}{\overset{\displaystyle\text{O}}{\text{CH}_3\text{CH}_2\text{CCH}_3}}$$

The (O) represents controlled oxidation using an oxidizing agent such as KM_nO_4.

Example
Draw structural diagrams and write IUPAC names to illustrate the hydrogenation of formaldehyde.

Solution

$$\underset{\substack{\text{methanal}\\ \text{(formaldehyde)}}}{\overset{\displaystyle\text{O}}{\text{H} - \text{C} - \text{H}}} + \text{H}_2 \longrightarrow \underset{\text{methanol}}{\overset{\displaystyle\text{OH}}{\underset{\displaystyle\text{H}}{\text{H} - \text{C} - \text{H}}}}$$

▶ Practice

Understanding Concepts

6. Draw structural diagrams and write IUPAC names to illustrate the controlled oxidation of the following alcohols. Is the product an aldehyde or a ketone?
   (a) 2-pentanol
   (b) 1-hexanol

7. Predict the relative solubility of the following compounds in water, listing the compounds in increasing order of solubility. Give reasons for your answer.
   (a) $\underset{\displaystyle\text{O}}{\overset{}{\text{CH}_3\text{CCH}_2\text{CH}_3}}$ (b) $CH_3CH_2CH_2CH_2OH$ (c) $CH_3CH_2CH_2CH_3$

8. Briefly explain the meaning of the term "oxidation."

Applying Inquiry Skills

9. Design an experimental procedure to prepare an alcohol, starting with acetone. Describe the main steps in the procedure, list experimental conditions needed, and draw structural diagrams and write IUPAC names to represent the reaction used.

▸ *Section 1.6 Questions*

Understanding Concepts

1. Write an equation for a reaction involving an aldehyde to illustrate a hydrogenation reaction. Write IUPAC names for all reactants and products.

2. Explain why no numeral is needed as a prefix in the naming of butanal and butanone.

3. Draw structural diagrams and write IUPAC names for the product(s) formed when 1-propanol undergoes the following reactions:
   (a) controlled oxidation with $Na_2Cr_2O_7$
   (b) complete combustion

4. Consider the two compounds, $CH_3CH_2CH_2CH_2CHO$ and $CH_3CH_2OCH_2CH_2CH_3$. Giving reasons for your choice, select the compound that:
   (a) will evaporate at a lower temperature;
   (b) has higher solubility in a nonpolar solvent;
   (c) can undergo an addition reaction with hydrogen.

5. Use a molecular model kit to build models for the following compounds, and write the IUPAC name for each:
   (a) an aldehyde with four carbon atoms
   (b) two isomers of a ketone with five carbon atoms
   (c) an aldehyde that is an isomer of acetone

Applying Inquiry Skills

6. Suppose that you are given three alcohols: a primary alcohol, a secondary alcohol, and a tertiary alcohol. Design an experimental procedure that you could carry out with commonly available materials and equipment that would identify the tertiary alcohol. Describe the main steps in the procedure and explain your experimental design.

7. Design an experimental procedure for the synthesis of butanone from an alkene. Identify the starting alkene of your choice, describe the steps in the procedure, and include the experimental conditions needed. Your answer should also contain any precautions required in the handling and disposal of the materials.

Making Connections

8. Many organic compounds have been in everyday use for many years, and are commonly known by nonsystematic names. Make a list of common names of organic compounds found in solvents, cleaners, and other household items. Conduct research using electronic or print sources to find out the chemical names of five of these compounds, identify the functional groups that are present, and discuss the useful properties that these functional groups may impart to the compound.

 www.science.nelson.com

9. The smell of formaldehyde was once common in the hallways of high schools, as it was used as a preservative of biological specimens. This use has largely been discontinued.
   (a) What is the IUPAC name and structure of formaldehyde?
   (b) Why was its use as a preservative curtailed, and what substances are being used in its place?

 www.science.nelson.com

10. In cases of severe diabetes, a patient's tissues cannot use glucose, and, instead, the body breaks down fat for its energy. The fats are broken down in the liver and muscles, producing several compounds called "ketone bodies," one of which is acetone.
   (a) The acetone produced in this process is carried in the blood and urine. Explain why acetone is soluble in these aqueous solutions.
   (b) When fats are the main source of energy production, there is overproduction of ketone bodies, leading to a condition called ketosis. A patient with untreated diabetes may have a blood concentration of acetone of 20 mg/100 mL. Convert this concentration to mol/L.
   (c) Acetone is volatile and is exhaled with the breath. Suggest a reason why, like untreated diabetic patients, people who are severely starved or dieting may also have a smell of acetone on their breath—a diagnostic symptom of ketosis.
   (d) Other ketone bodies lower blood pH, causing a condition called acidosis, which can lead to coma and death. Research the symptoms and effects of ketosis and acidosis and how these conditions may be avoided.

 www.science.nelson.com

1.7 Carboxylic Acids and Esters

Organic acids are characterized by the presence of a carboxyl functional group, $-COOH$, and hence are called carboxylic acids. As with inorganic acids, carboxylic acids can react with compounds containing $-OH$ groups to form an organic "salt" called an ester.

Carboxylic Acids

When wine is opened and left in contact with air for a period of time, it will likely turn sour. The alcohol in the wine has turned into vinegar. Grocery stores sell wine vinegars for cooking or for salad dressings. The chemical reaction in this souring process is the *oxidation* of ethanol, and the vinegar produced belongs to a family of organic compounds called **carboxylic acids**.

Carboxylic acids are generally weak acids and are found in citrus fruits, crab apples, rhubarb, and other foods characterized by a sour, tangy taste. Sour milk and yogurt contain lactic acid, produced by a bacteria culture. If you have ever felt your muscles ache after prolonged exertion, you have experienced the effect of lactic acid in your muscles. The lactic acid is produced when the supply of oxygen cannot keep up with the demand during extended exercise. The gamey taste of meat from animals killed after a long hunt is due to the high concentration of lactic acid in the muscles.

$$CH_3 - \overset{\overset{\displaystyle OH}{|}}{\underset{\underset{\displaystyle H}{|}}{C}} - \overset{\overset{\displaystyle OH}{|}}{C} = O$$

lactic acid

Carboxylic acids also have distinctive odours that can be used to advantage in law enforcement (**Figure 1**).

Figure 1
Tracking dogs, with their acute sense of smell, are trained to identify odours in police work. As carboxylic acids have distinctive odours, the dogs may follow the characteristic blend of carboxylic acids in a person's sweat. Trained dogs are also used to seek out illegal drug laboratories by the odour of acetic acid. Acetic acid is formed as a byproduct when morphine, collected from opium poppies, is treated to produce heroin.

carboxylic acid one of a family of organic compounds that is characterized by the presence of a carboxyl group; $-COOH$

TRY THIS activity   ***Making Flavoured Vinegar***

If vinegar (a carboxylic acid) can be made from ethanol (the alcohol resulting from the fermentation of sugars), it should be possible to make a nice flavoured vinegar at home starting with apples, and a good supply of oxygen for oxidation.
Materials: apples; blender or food processor; sieve or cheesecloth; glass or plastic jars with lids, flavouring (e.g., ginger, garlic, raspberries)

• Wash and chop the apples. Purée them, peel included, in the blender.

• Pour the pulp into a sieve or a bowl lined with cheesecloth. Strain out most of the pulp.

• Pour the juice into glass or plastic jars. Replace the lids loosely to maintain a good oxygen supply. Keep at room temperature and out of direct sunlight.

• Stir well each day to increase oxygen access. The yeast normally found in the fruit will start the fermentation process,

and vinegar should be produced in three to four weeks, identifiable by smell.
(a) Write a series of chemical equations showing the reactions that produce vinegar from ethanol.
(b) Describe a test to confirm that an acid is present in the solution.

• Filter the vinegar through a coffee filter to remove any sediment, then pasteurize it by heating the filled jars (loosely lidded) in a pan of hot water until the vinegar is between 60°C and 70°C.
(c) What is the purpose of the preceding step? Why does it work?

• To add flavour, tie flavourings such as ginger, garlic, or raspberries in a small cheesecloth bag and suspend in the vinegar for several days. Enjoy the final oxidation product on a salad.

Naming Carboxylic Acids

The functional group of carboxylic acids is the **carboxyl group**, written in formulas as $-COOH$. This functional group combines two other functional groups already familiar to us: the hydroxyl ($-OH$) group in alcohols, and the carbonyl ($-C=O$) group in aldehydes and ketones.

The IUPAC name for a carboxylic acid is formed by taking the name of the alkane or alkene with the same number of carbon atoms as the longest chain in the acid. Remember to count the C atom in the carboxyl group in the total number of the parent chain. The -*e* ending of the alkane name is replaced with the suffix -*oic*, followed by the word "acid."

The simplest carboxylic acid is methanoic acid, HCOOH, commonly called formic acid; the name is derived from the Latin word *formica* which means "ant," the first source of this acid (**Figure 2**). Methanoic acid is used in removing hair from hides and in coagulating and recycling rubber.

$$H - \overset{\overset{\displaystyle O}{\|}}{C} - O - H$$

formic acid

Ethanoic acid, commonly called acetic acid, is the compound that makes vinegar taste sour. This acid is used extensively in the textile dyeing process and as a solvent for other organic compounds.

The simplest aromatic acid is phenylmethanoic acid, better known by its common name, benzoic acid. Benzoic acid is largely used to produce sodium benzoate, a common preservative in foods and beverages.

Some acids contain multiple carboxyl groups. For example, oxalic acid, which is found naturally in spinach and in the leaves of rhubarb, consists of two carboxyl groups bonded to each other; it is used in commercial rust removers and in copper and brass cleaners. Tartaric acid occurs in grapes; it is often used in recipes that require a solid edible acid to react with baking soda as a leavening agent. Citric acid is responsible for the sour taste of citrus fruits. Vitamin C, or ascorbic acid, found in many fruits and vegetables, is a cyclic acid. A familiar aromatic acid is acetylsalicylic acid (ASA), the active ingredient in Aspirin; you may have experienced its sour taste when swallowing a tablet.

When naming acids with multiple carboxyl groups, the suffix -*dioic acid* is used for acids with a carboxyl group at each end of the parent chain. The compound $HOOC-CH_2-COOH$ is named propanedioic acid; the carboxyl C atoms are counted in the parent chain. When more carboxyl groups are present, all COOH groups may be named as substituents on the parent chain; in this case, the parent chain does not include the carboxyl C atoms. An example is citric acid, shown below; it is named as a tricarboxylic acid of propane.

carboxyl group a functional group consisting of a hydroxyl group attached to the C atom of a carbonyl group; $-COOH$

Figure 2
Most ants and ant larvae are edible and are considered quite delicious. They have a vinegary taste because they contain methanoic acid, HCOOH, commonly called formic acid. In some countries, large ants are squeezed directly over a salad to add the tangy ant juice as a dressing.

DID YOU KNOW ?

Water-Soluble Vitamins
With its many polar hydroxyl groups, vitamin C ascorbic acid is highly water-soluble. The water-soluble vitamins are not stored in the body; rather, they are readily excreted in the urine. It is therefore important that we include these vitamins in our daily diet. However, although taking too much vitamin C is not dangerous, taking excessive amounts is truly sending money down the drain.

oxalic acid

tartaric acid

citric acid

ascorbic acid (Vitamin C)

acetylsalicylic acid (ASA)

Naming and Drawing Carboxylic Acids

Write the IUPAC name and the structural formula for propenoic acid.

The prefix *propen-* indicates that the acid contains three C atoms with one double bond; the end C atom is in the carboxyl group. Since the carboxyl C atom can only form one more single bond with its neighbouring C atom, the double bond is between carbon 2 and carbon 3.

The structural formula for propenoic acid is

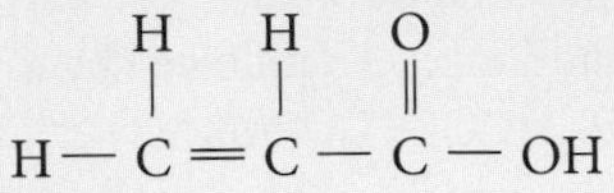

$$H-\underset{\underset{H}{|}}{C}=\underset{\underset{H}{|}}{C}-\overset{\overset{O}{\|}}{C}-OH$$

Example

What is the IUPAC name for this carboxylic acid?

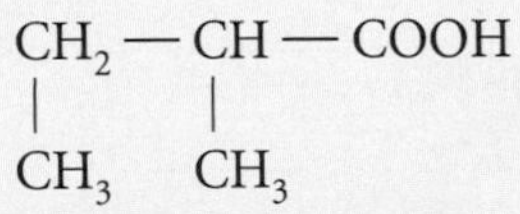

$$\underset{\underset{CH_3}{|}}{CH_2}-\underset{\underset{CH_3}{|}}{CH}-COOH$$

Solution

The structure represents 2-methylbutanoic acid.

▶ Practice

Understanding Concepts

1. Draw a structural diagram for each of the following compounds:
   (a) octanoic acid
   (b) benzoic acid
   (c) 2-methylbutanoic acid

2. Give IUPAC and, if applicable, common names for these molecules:
   (a)
   $$H-\overset{\overset{O}{\|}}{C}-OH$$

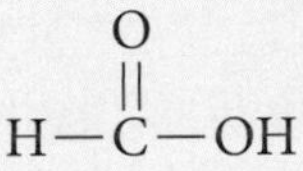

   (b)
   $$CH_3-CH_2-\underset{\underset{CH_2CH_3}{|}}{CH}-CH_2-\overset{\overset{O}{\|}}{C}-OH$$

   (c)
   $$CH_3-CH_2-CH_2-\underset{\underset{CH_2CH_3}{|}}{\overset{\overset{CH_3CH_2}{|}}{CH}}-\overset{\overset{O}{\|}}{CH}-\overset{\overset{O}{\|}}{C}-OH$$

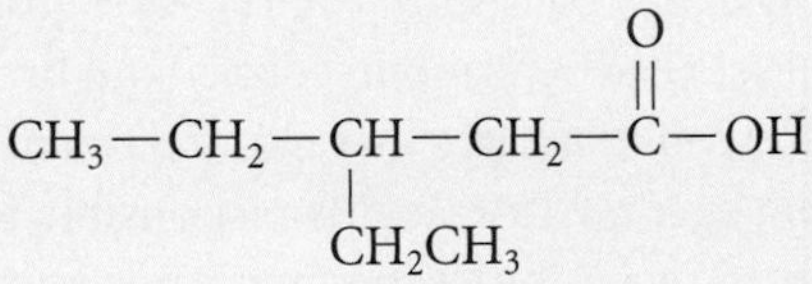

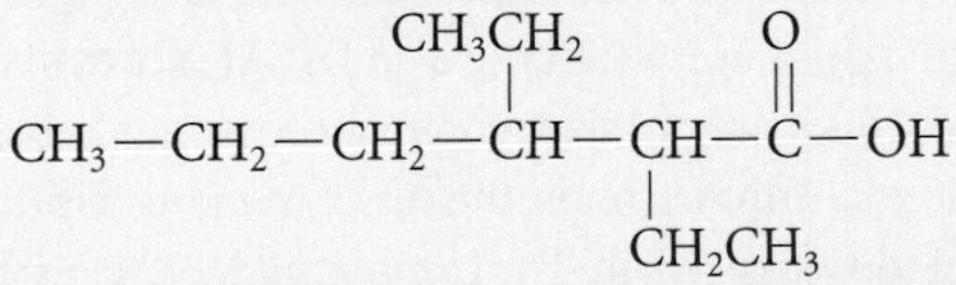

Properties of Carboxylic Acids

The carboxyl group is often written in condensed form as $-COOH$. However, the two oxygen atoms are not bonded to each other. In fact, the carboxyl group consists of a hydroxyl group ($-OH$) attached to the C atom of a carbonyl group ($-C=O$).

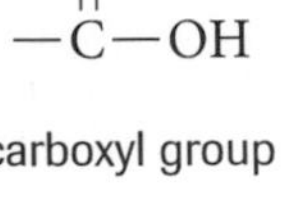

$$-\overset{\overset{O}{\|}}{C}-OH$$

carboxyl group

As one would predict from the presence of both carbonyl ($-CO$) and hydroxyl groups ($-OH$), the molecules of carboxylic acids are polar and form hydrogen bonds with each other and with water molecules. These acids exhibit similar solubility behaviour to that of alcohols; that is, the smaller members (one to four carbon atoms) of the acid series are soluble in water, whereas larger ones are relatively insoluble. Carboxylic acids have the properties of acids: a litmus test can distinguish these compounds from other hydrocarbon derivatives. They also react with organic "bases" in neutralization reactions to form organic "salts," as we will see later, in Chapter 2.

The melting points of carboxylic acids, as **Table 1** shows, are higher than those of their corresponding hydrocarbons (**Figure 3**). We can explain this by the increased intermolecular attractions of the polar carboxyl functional groups. This explanation is supported by the significantly higher melting points of analogous acids with an abundance of carboxyl groups. 🧪

Table 1 Melting Points of Some Carboxylic Acids and Their Parent Alkanes

Number of C atoms	Number of COOH groups	Compound	Melting point (°C)
1	0	methane	−182
1	1	methanoic acid	8
2	0	ethane	−183
2	1	ethanoic acid	17
2	2	oxalic acid	189
4	0	butane	−138
4	1	*n*-butanoic acid	-8
4	2	tartaric acid	206
6	0	hexane	−95
6	1	hexanoic acid	13
6	3	citric acid	153

Preparing Carboxylic Acids

When an *alcohol* is mildly oxidized, an *aldehyde* is produced. Further controlled oxidation of the aldehyde results in the formation of a *carboxylic acid*, containing a carboxyl group.

The general oxidation pathway in this process is from alcohol to aldehyde to carboxylic acid; the functional group in the parent molecule changes from the hydroxyl group to the carbonyl group, then to the carboxyl group. As you can see in the example below, the difference between the carbonyl group and the carboxyl group is one additional O atom, present in the $-OH$ group. In the case of ethanol, the aldehyde formed is ethanal (acetaldehyde), which is further oxidized to ethanoic acid, commonly known as acetic acid.

$$\underset{\text{ethan}\textit{ol}}{CH_3CH_2\overset{\overset{\displaystyle OH}{|}}{}} + (O) \longrightarrow \underset{\substack{\text{ethan}\textit{al} \\ \text{(acetaldehyde)}}}{CH_3\overset{\overset{\displaystyle O}{||}}{C}-H} + H_2O$$

$$\underset{\text{ethan}\textit{al}}{CH_3\overset{\overset{\displaystyle O}{||}}{C}-H} + (O) \longrightarrow \underset{\substack{\text{ethan}\textit{oic acid} \\ \text{(acetic acid)}}}{CH_3\overset{\overset{\displaystyle O}{||}}{C}-OH}$$

Cockroaches can swim happily in sulfuric acid because they have an unreactive outer layer that consists of hydrocarbons. Human skin contains functional groups that react with acids, which is why we can be badly injured by contact with concentrated strong acids.

Figure 3
Oxalic acid, found in rhubarb, differs from vinegar in that its molecule contains an additional carboxyl group. This increased polarity explains why oxalic acid is a solid while vinegar is a liquid at the same temperature.

⬛ INVESTIGATION 1.7.1

Properties of Carboxylic Acids (p. 87)
What gives a carboxylic acid its own unique properties? Compare many properties of a large and a small carboxylic acid, and draw your own conclusions.

The active oxygen, (O), in these reactions is supplied by an oxidizing agent (one of many compounds that itself becomes reduced). The clever selection of an oxidizing agent that changes colour as this reaction proceeds is the basis of the breathalyzer test for alcohol. In this system, the (O) is supplied by the chromate ion in its Cr^{6+} oxidation state—an ion with an orange colour in aqueous solution. When a measured volume of air containing ethanol passes through the breathalyzer tube, the ethanol is oxidized to acetaldehyde and then to acetic acid. The oxidation process is accompanied by a reduction of the chromate ion to its Cr^{3+} oxidation state—an ion with a green colour in aqueous solution. The extent of the green colour down the breathalyzer tube provides a measure of the concentration of alcohol in the breath.

$$CH_3CH_2OH + (Cr^{6+}) \rightarrow CH_3COOH + Cr^{3+}$$
ethanol   (orange)   acetic acid (green)

SUMMARY | *Oxidation of Aldehydes to Carboxylic Acids*

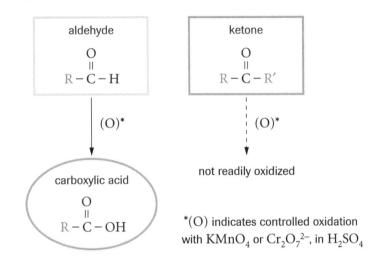

*(O) indicates controlled oxidation with $KMnO_4$ or $Cr_2O_7^{2-}$, in H_2SO_4

▶ SAMPLE problem | *Formation of Carboxylic Acids*

Write an equation to show the controlled oxidation of an aldehyde to form butanoic acid.

First, write the structural formula for butanoic acid.

$CH_3CH_2CH_2COOH$

The aldehyde required must have the same number of C atoms, so must be butanal.

$CH_3CH_2CH_2CHO$

The reaction equation is therefore

$CH_3CH_2CH_2CHO + (O) \rightarrow CH_3CH_2CH_2COOH$

Example

Write a series of equations to show the reactions needed to produce methanoic acid from methanol. Write IUPAC names and nonsystematic names for the organic compounds in the reactions.

Solution

$$CH_3OH + (O) \longrightarrow H-\overset{\overset{\displaystyle O}{\|}}{C}-H + H_2O$$

methanol   methanal
                    (formaldehyde)

$$H-\overset{\overset{\displaystyle O}{\|}}{C}-H + (O) \longrightarrow H-\overset{\overset{\displaystyle O}{\|}}{C}-OH$$

methanal   methanoic acid
(formaldehyde)   (formic acid)

▶ Practice

Understanding Concepts

3. Draw a structural diagram and write the IUPAC name of an alcohol that can be used in the synthesis of oxalic (ethanedioic) acid.

4. The labels have fallen off three bottles. Bottle A contains a gas, bottle B contains a liquid, and bottle C contains a solid. The labels indicate that the compounds have the same number of carbon atoms, one being an alkane, one an alcohol, and the other a carboxylic acid. Suggest the identity of the contents of each bottle, and give reasons for your answer.

5. Write a series of chemical equations to illustrate the synthesis of a carboxylic acid from the controlled oxidation of 1-propanol.

6. Name and draw a general structure for the functional group in a carboxylic acid. Explain the effect of the components of this functional group on the molecule.

7. When a bottle of wine is left open to the air for a period of time, the wine often loses its alcoholic content and starts to taste sour. Write a series of equations to illustrate the reactions.

Applying Inquiry Skills

8. Suppose that you are given three colourless liquids whose identities are unknown. You are told that one is an aldehyde, one a ketone, and the other a carboxylic acid. What physical and chemical properties would you examine in order to identify each compound? Give reasons for your strategy.

Making Connections

9. Think about common products in the home and identify several that contain alcohols other than ethanol. Do these alcohols also turn sour over time? Explain, and illustrate your answer with structural diagrams.

10. Some cosmetic facial creams contain an ingredient manufacturers call "alpha hydroxy," which is designed to remove wrinkles. These compounds are carboxylic acids that contain a hydroxyl group attached to the C atom adjacent to the carboxyl group.
    (a) Explain why "alpha hydroxy" is an incorrect name for any compound. ▶

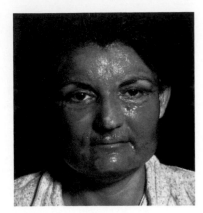

Figure 4
Cosmetic treatment with carboxylic acids to remove surface skin may lead to irritation or sun sensitivity. This patient is in the third day of a skin-peeling treatment.

ester an organic compound characterized by the presence of a carbonyl group bonded to an oxygen atom

esterification a condensation reaction in which a carboxylic acid and an alcohol combine to produce an ester and water

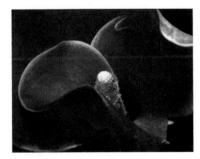

Figure 5
The rich scent of the lily is at least partially due to the esters produced in the flower.

(b) The alpha hydroxy acids in cosmetics may include glycolic acid, lactic acid, malic acid, and citric acid. Research and draw structural diagrams for these compounds.

 GO www.science.nelson.com

(c) In order to be absorbed through the skin, a substance must have both polar and nonpolar components. Explain why these alpha hydroxy acids are readily absorbed through the skin.

(d) Over-the-counter creams may contain up to 10% alpha hydroxy acids, while 25-30% concentrations are used by cosmetologists for "chemical peels" (**Figure 4**). In each case, the ingredients cause the surface of the skin to peel, revealing younger-looking skin. Suggest reasons why physicians recommend daily use of sun protection to accompany the use of these facial creams.

From Carboxylic Acids to Organic "Salts": Esterification

Carboxylic acids react as other acids do, in neutralization reactions, for example. A carboxylic acid can react with an alcohol, forming an **ester** and water. In this reaction, the alcohol acts as an organic base and the ester formed may be considered an organic salt. This condensation reaction is known as **esterification**. As we will see later in the chapter, carboxylic acids also react with organic bases other than alcohols to form important biological compounds.

The general reaction between a carboxylic acid and an alcohol is represented below. An acid catalyst, such as sulfuric acid, and heat are generally required. It is interesting to note that, by tracking the oxygen atoms using isotopes, it has been found that the acid contributes the $-OH$ group to form the water molecule in the reaction.

$$\underset{\text{acid}}{\overset{O}{\underset{\|}{RC}}-OH} + \underset{\text{alcohol}}{R'OH} \xrightarrow[\text{heat}]{\text{conc. } H_2SO_4} \underset{\text{ester}}{\overset{O}{\underset{\|}{RC}}-O-R'} + \underset{\text{water}}{HOH}$$

Esters

Esters occur naturally in many plants (**Figure 5**) and are responsible for the odours of fruits and flowers. Synthetic esters are often added as flavourings to processed foods, and as scents to cosmetics and perfumes. **Table 2** shows the main esters used to create certain artificial flavours.

Table 2 The Odours of Selected Esters

Odour	Name	Formula
apple	methyl butanoate	$CH_3CH_2CH_2COOCH_3$
apricot	pentyl butanoate	$CH_3CH_2CH_2COOCH_2CH_2CH_2CH_3$
banana	3-methylbutyl ethanoate	$CH_3COOCH_2CH_2\overset{\overset{\displaystyle CH_3}{\|}}{C}HCH_3$
cherry	ethyl benzoate	$C_6H_5COOC_2H_5$
orange	octyl ethanoate	$CH_3COOCH_2CH_2CH_2CH_2CH_2CH_2CH_3$

Table 2 The Odours of Selected Esters (continued)

Odour	Name	Formula
pineapple	ethyl butanoate	$CH_3CH_2CH_2COOCH_2CH_3$
red grape	ethyl heptanoate	$CH_3CH_2CH_2CH_2CH_2CH_2COOCH_2CH_3$
rum	ethyl methanoate	$HCOOCH_2CH_3$
wintergreen	methyl salicylate	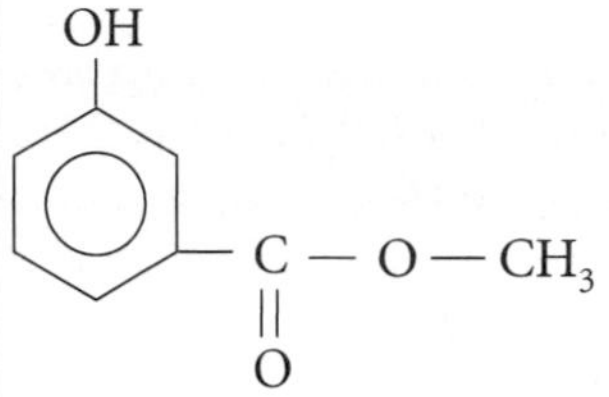 (structural diagram of methyl salicylate with OH group on benzene ring and $C-O-CH_3$ with double-bonded O)

> **LEARNING TIP**
>
> Esters are organic salts, and are named in a similar way to inorganic salts:
> sodium hydroxide + nitric acid →
> sodium nitrate + water
> methanol + butanoic acid →
> methyl butanoate + water

Naming and Preparing Esters

As we learned earlier, esters are organic "salts" formed from the reaction of a carboxylic acid and an alcohol. Consequently, the name of an ester has two parts. The first part is the name of the alkyl group from the alcohol used in the esterification reaction. The second part comes from the acid. The ending of the acid name is changed from -*oic* acid to -*oate*. For example, in the reaction of ethanol and butanoic acid, the ester formed is ethyl butan*oate*, an ester with a banana odour.

> **LEARNING TIP**
>
> Note that the names of carboxylic acids and esters are written as *two separate words* (e.g., propanoic acid, ethyl butanoate), unlike the single names of most other organic compounds (e.g., ethoxybutane, 2-methyl-3-pentanol).

$$CH_3CH_2CH_2\overset{O}{\overset{||}{C}}-OH + CH_3CH_2OH \longrightarrow CH_3CH_2CH_2\overset{O}{\overset{||}{C}}-O-CH_2CH_3 + HOH$$

butanoic acid	ethanol	ethyl butanoate	water
acid	alcohol	ester	

The functional group for an ester is a carboxyl group in which the H atom is substituted by an alkyl group: −COOR. The general structural formula for an ester is shown below.

$$(H\ or)R-\overset{O}{\overset{||}{C}}-O-R'$$

The general formula of an ester is written as RCOOR′. When read from left to right, RCO− comes from the carboxylic acid, and −OR′ comes from the alcohol. Hence, $CH_3COOCH_2CH_3$ is propyl ethanoate. Note that, for an ester, the acid is the first part of its formula as drawn, but is the second part of its name.

$$CH_3CH_2CH_2COOH + CH_3OH \rightarrow CH_3CH_2CH_2COOCH_3 + HOH$$

butanoic acid   methanol methyl butanoate

Reactions Involving Carboxylic Acids and Esters   **SAMPLE** problem ◀

1. Draw a structural diagram and write the IUPAC name for the ester formed in the reaction between propanol and benzoic acid.

To name the ester:
- the first part of the name comes from the alcohol—propyl, and
- the second part of the name comes from the acid—benzoate, so
- the IUPAC name of the ester is propyl benzoate.

To draw the structure:
- draw structural diagrams of the reactants and complete the condensation reaction.

▶

$$\underset{\text{benzoic acid}}{\begin{array}{c} \text{O} \\ \| \\ \text{C}-\text{OH} \\ \end{array}} + \underset{\text{1-propanol}}{\text{HO}-\text{CH}_2\text{CH}_2\text{CH}_3} \xrightarrow{\text{H}_2\text{SO}_4} \underset{\text{propyl benzoate}}{\begin{array}{c} \text{O} \\ \| \\ \text{C}-\text{O}-\text{CH}_2\text{CH}_2\text{CH}_3 \\ \end{array}} + \underset{\text{water}}{\text{H}_2\text{O}}$$

2. **Write a condensed structural diagram equation for the esterification reaction to produce the ester CH₃CH₂CH₂COOCH₂CH₃. Write IUPAC names for each reactant and product.**

First, identify the acid (four carbons—butanoic acid) and the alcohol (two carbons—ethanol) that may be used in the synthesis of the ester. Then draw structures and include the conditions in the chemical equation.

$$\underset{\text{butanoic acid}}{\text{CH}_3\text{CH}_2\text{CH}_2\text{COOH}} + \underset{\text{ethanol}}{\text{HOCH}_2\text{CH}_3} \xrightarrow{\text{H}_2\text{SO}_4} \underset{\text{ethyl butanoate}}{\text{CH}_3\text{CH}_2\text{CH}_2\text{COOCH}_2\text{CH}_3} + \underset{\text{water}}{\text{H}_2\text{O}}$$

Example

Name the ester CH_3COOCH_3 and the acid and alcohol from which it can be prepared.

Solution

The ester is methyl ethanoate, and it can be prepared from methanol and ethanoic acid.

▶ Practice

Understanding Concepts

11. Write complete structural diagram equations and word equations for the formation of the following esters. Refer to **Table 2** and identify the odour of each ester formed.
    (a) ethyl methanoate
    (b) ethyl benzoate
    (c) methyl butanoate
    (d) 3-methylbutyl ethanoate

12. Name the following esters, and the acids and alcohols from which they could be prepared.
    (a) $CH_3CH_2COOCH_2CH_3$
    (b) $CH_3CH_2CH_2COOCH_3$
    (c) $HCOOCH_2CH_2CH_3$
    (d) $CH_3COOCH_2CH_2CH_3$

Properties of Esters

The functional group of an ester is similar to the carboxyl group of an acid. What it lacks in comparison to an acid is its $-OH$ group; the hydroxyl group is replaced by an $-OR$ group. With the loss of the polar $-OH$ group, esters are less polar, and therefore are less soluble in water, and have lower melting and boiling points than their parent acids. Moreover, the acidity of the carboxylic acids is due to the H atom on their $-OH$ group, and so esters, having no $-OH$ groups, are not acidic.

$$\underset{\text{carboxylic acid}}{\begin{array}{c} \text{O} \\ \| \\ \text{R}\overset{}{\text{C}}-\text{OH} \\ \end{array}} \qquad \underset{\text{ester}}{\begin{array}{c} \text{O} \\ \| \\ \text{R}\overset{}{\text{C}}-\text{OR}' \\ \end{array}}$$

It is the low-molecular-mass esters that we can detect by scent, because they are gases at room temperature. The larger, heavier esters more commonly occur as waxy solids.

Reactions of Esters: Hydrolysis

When esters are treated with an acid or a base, a reversal of esterification occurs; that is, the ester is split into its acid and alcohol components. This type of reaction is called **hydrolysis**. In the general example shown below, the reaction is carried out in a basic solution, and the products are the sodium salt of the carboxylic acid and the alcohol.

$$\underset{\text{ester}}{\overset{O}{\underset{\|}{RC}}-O-R' + Na^+ + OH^-} \longrightarrow \underset{\text{acid}}{\overset{O}{\underset{\|}{RC}}-O^- + Na^+} + \underset{\text{alcohol}}{R'OH}$$

As we shall see in more detail in the next chapter, fats and oils are esters of long-chain acids (**Figure 6**). When these esters are heated with a strong base such as sodium hydroxide (NaOH), a hydrolysis reaction occurs. The sodium salts of the acids that result are what we call soap. This soap-making reaction is called **saponification**, from the Latin word for soap, *sapon*.

When certain reactants are used, esters can be formed repeatedly and joined together to form long chains. These large molecules of repeating units are called polymers, and when the repeating units are esters, the polymer is the familiar polyester. We will learn more about these and other polymers in the next chapter.

Figure 6
Edible oils such as vegetable oils are liquid glycerol esters of unsaturated fatty acids. Fats such as shortening are solid glycerol esters of saturated fatty acids. Adding hydrogen to the double bonds of the unsaturated oil converts the oil to a saturated fat. Most saturated fats are solids at room temperature.

 **SUMMARY** *Carboxylic Acids and Esters*

hydrolysis a reaction in which a bond is broken by the addition of the components of water, with the formation of two or more products

saponification a reaction in which an ester is hydrolyzed

Functional groups:

- carboxylic acid: −COOH carboxyl group

$$\overset{O}{\underset{\|}{-C}}-OH$$

- ester: −COOR alkylated carboxyl group

$$\overset{O}{\underset{\|}{-C}}-OR$$

Preparation:

- alcohol + (O) → aldehyde + (O) → carboxylic acid
  oxidation reaction; add (O)

- carboxylic acid + alcohol → ester + H₂O
  condensation reaction

Pathway to other compounds:

- ester + NaOH → sodium salt of acid + alcohol
  hydrolysis; saponification

ACTIVITY 1.7.2

Synthesis of Esters (p. 89)
What do esters really smell like? Make some from alcohols and carboxylic acids, and find out for yourself!

▶ *Practice*

Understanding Concepts

13. In what way is the functional group of an ester different from that of a carboxylic acid? How does this difference account for any differences in properties?

14. Describe the experimental conditions in the hydrolysis of ethyl formate. Write a balanced equation for the reaction, and name the product(s). ▶

▶ *Section 1.7 Questions*

Understanding Concepts

1. Write IUPAC names for the following compounds:
   (a) $CH_3CH_2COOCH_2CH_2CH_3$
   (b)
   $$\overset{\displaystyle O}{\overset{\|}{CH_3CH_2C}}OCH_2\overset{\displaystyle CH_3}{\overset{|}{CH}}CH_2CH_2CH_3$$
   (c)
   $$CH_3\overset{}{\underset{\underset{\displaystyle Br}{|}}{CH}}\overset{\displaystyle O}{\overset{\|}{C}}OH$$
   (d) acetic acid
   (e) benzoic acid

2. Draw structural diagrams for each of the following compounds:
   (a) methanoic acid
   (b) the product of controlled oxidation of propanal
   (c) the acid formed from saponification of butyl ethanoate
   (d) the ester that is produced in the esterification of 1-propanol and formic acid
   (e) the ester that is produced in the esterification of phenol and vinegar

3. Draw the structures of the compounds formed by condensation reactions between the following reactants, and write IUPAC names for each product.
   (a) formic acid and 2-butanol
   (b) acetic acid and 1-propanol
   (c) benzoic acid and methanol

4. Name the carboxylic acid and the alcohol that may be used to produce each of the following compounds:
   (a)
   $$\overset{\displaystyle O}{\overset{\|}{CH_3CH_2C}}OCH_2CH_2CH_2CH_2CH_3$$
   (b)
   $$CH_3CH_2CH_2O\overset{}{\underset{\underset{\displaystyle OCH_2CH_3}{\|\,|}}{CCHCH}}CH_2CH_2CH_3$$
   (c)

   $\quad$ COOCH_3

Applying Inquiry Skills

5. Describe an experimental procedure to carry out the saponification of propyl butanoate. Explain the evidence that will indicate that the reaction has been completed.

6. In the laboratory synthesis of an ester, what procedure can be used to recover the ester from the other components in the reaction mixture? Explain the strategy behind this procedure.

Making Connections

7. From what you have learned about controlled oxidations in chemical reactions, describe some controlled oxidation reactions that occur in our everyday lives. In what situations are controlled oxidations ideal, and in what situations are "uncontrolled" oxidations ideal?

8. Working with a partner or a small group, brainstorm and list several occupations that require a knowledge of alcohols, carboxylic acids, or esters. Research one of these careers and write a brief report on the main strengths and qualities needed, academic training, and job opportunities in the field.

 GO ▸ | www.science.nelson.com

9. Tannic acid, originally obtained from the wood and bark of certain trees, has for centuries been used to "tan" leather (**Figure 8**).
   (a) Give the chemical formula for tannic acid.
   (b) What effect does tannic acid have on animal hides? Explain your answer with reference to the chemical reactions that take place.

GO ▸ | www.science.nelson.com

Figure 8
Tanneries are notorious for the bad smells they produce, as a result of the chemical reactions between the animal hides and the chemicals used to process them.

So far, we have been discussing compounds containing only C, H, O, and halogen atoms. Many of the compounds that are naturally produced by living organisms also contain nitrogen. In fact, many of these nitrogenous organic compounds such as proteins and DNA have essential biological functions. In this section, we will examine the nitrogenous organic families of **amines** and **amides**.

Amines can be thought of as ammonia (NH_3) with one, two, or all three of its hydrogens substituted by alkyl groups; these are classified as primary (1°), secondary (2°), or tertiary (3°) amines, respectively. The alkyl groups in an amine may be identical or different.

amine an ammonia molecule in which one or more H atoms are substituted by alkyl or aromatic groups

amide an organic compound characterized by the presence of a carbonyl functional group (C=O) bonded to a nitrogen atom

$$
\begin{array}{cccc}
H & H & R' & R' \\
| & | & | & | \\
H-N-H & R-N-H & R-N-H & R-N-R'' \\
\text{ammonia} & \text{primary amine (1°)} & \text{secondary amine (2°)} & \text{tertiary amine (3°)}
\end{array}
$$

Amines are organic bases, and can react with carboxylic acids to form nitrogenous organic "salts," called amides. Amide functional groups (−CON, **Figure 1**) occur in proteins, the large molecules formed in all living organisms.

$$
\begin{array}{c}
\quad\;\; O \quad R'' \\
\quad\;\; || \quad\; | \\
R-C-N-R'
\end{array}
$$

Figure 1
General structure of an amide. The R groups may be the same or different, or they may be replaced by H atoms.

Amines

When organisms decompose, large and complex molecules such as proteins are broken down to simpler organic compounds called amines. As with many compounds of nitrogen, such as ammonia, NH_3, amines often have an unpleasant odour. For example, the smell of rotting fish is due to a mixture of amines (**Figure 2**). The putrid odour of decomposing animal tissue is caused by amines appropriately called putrescine and cadaverine, produced by bacteria. Spermine, an amine with its own distinctive aroma, can be isolated from semen.

$$H_2N-CH_2-CH_2-CH_2-CH_2-NH_2$$

putrescine

$$H_2N-CH_2-CH_2-CH_2-CH_2-CH_2-NH_2$$

cadaverine

$$H_2N-(CH_2)_3-NH-(CH_2)_4-NH-(CH_2)_3-NH_2$$

spermine

Naming Amines

Amines can be named in either of two ways:

- as a nitrogen derivative of an alkane (IUPAC system); e.g., CH_3NH_2 would be aminomethane; or
- as an alkyl derivative of ammonia; e.g., CH_3NH_2 would be methylamine.

$$
\begin{array}{c}
\quad\quad\; H \\
\quad\quad\; | \\
H-C-N-H \\
\quad\; | \quad\; | \\
\quad\; H \quad H
\end{array}
$$

aminomethane (methylamine)

$$
\begin{array}{c}
\quad\quad\; CH_3 \\
\quad\quad\; | \\
CH_3-N-H
\end{array}
$$

dimethylamine

Figure 2
Low-molecular-weight amines are partly responsible for the characteristic "fishy" smell. Lemon juice is often provided in restaurants to neutralize the taste of these amines, which are weak bases.

Let us look at some examples of amines and how they are named, first with the IUPAC system. Consider each molecule as an alkane; the NH_2 group is called an *amino* group. Thus, the structure in **Figure 3(a)** is a butane with an amino group on C atom 1; it is named 1-aminobutane. The structure in **Figure 3(b)** is a hexane with an amino group on C atom 3; it is thus 3-aminohexane.

(a)

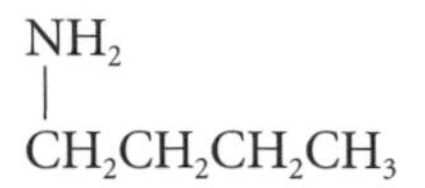

$$NH_2$$
$$|$$
$$CH_2CH_2CH_2CH_3$$

1-aminobutane (butylamine); a 1° amine

(b)

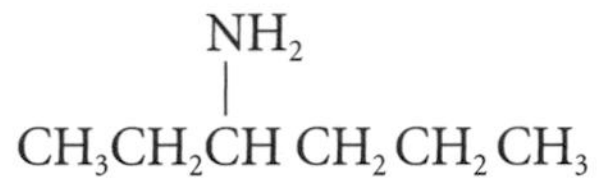

$$NH_2$$
$$|$$
$$CH_3CH_2CH\ CH_2\ CH_2\ CH_3$$

3-aminohexane, a 1° amine

(c)

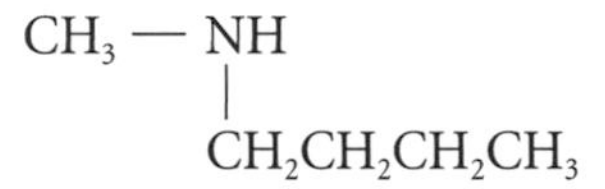

$$CH_3 - NH$$
$$|$$
$$CH_2CH_2CH_2CH_3$$

N-methyl-1-aminobutane (*n*-butylmethylamine); a 2° amine

(d)

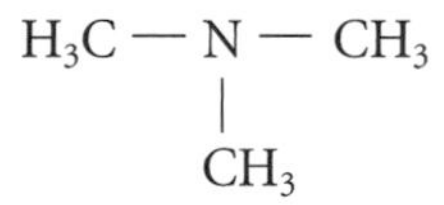

$$H_3C - N - CH_3$$
$$|$$
$$CH_3$$

N,N-dimethylaminomethane
(trimethylamine), a 3° amine

Figure 3
Structures of amines

Many compounds contain more than one amino group. The molecule cadaverine, for example, is a 5-carbon chain with an amino group at each end: $NH_2CH_2CH_2CH_2CH_2CH_2NH_2$. Molecules with 2 amino groups are called diamines, and the IUPAC name for cadaverine is 1,5-diaminopentane.

The IUPAC names for 2° and 3° amines include the *N*- prefix to denote the substituted groups on the N atom of the amino group. The IUPAC names for the structures in **Figures 3(c)** and **3(d)** are given.

There is a convenient alternative system of naming amines, in which the names imply an alkyl derivative of ammonia. The structural diagram in **Figure 3(a)** shows an ammonia molecule with one of its H atoms substituted by a butyl group. Its alternative name is butylamine. The structure in **Figure 3(c)** has two alkyl groups on the N atom: a methyl group and a butyl group. It is named *n*-butylmethylamine. **Figure 3(d)** shows an ammonia molecule with all its H atoms substituted by methyl groups; it is trimethylamine. Note that, generally, the alkyl groups are listed alphabetically in the amine name.

We mentioned earlier that amines with one, two, or three alkyl groups attached to the central nitrogen atom are referred to, respectively, as primary, secondary, and tertiary amines. Note that this designation is different from that of alcohols where the attachments of the *carbon* atom are indicated.

LEARNING TIP

Be careful not to confuse the names of these two nitrogenous groups:
$-NH_2$ is an amino group, and
$-NO_2$ is a nitro group.

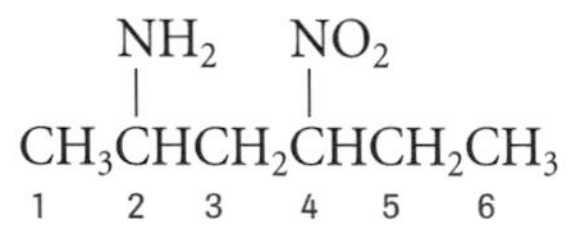

$$NH_2\quad NO_2$$
$$|\qquad |$$
$$CH_3CHCH_2CHCH_2CH_3$$
$$\ 1\ \ \ \ 2\ \ \ \ \ 3\ \ \ \ 4\ \ \ \ \ 5\ \ \ \ 6$$

2-amino-4-nitrohexane

1. **Write two names for the following structure:**

$$CH_3 - N - CH_3$$
$$\overset{|}{\underset{1 \quad 2 \quad 3}{CH_3CHCH_3}}$$

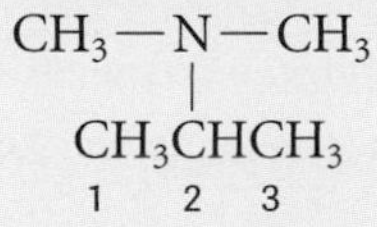

Using the IUPAC system, name the compound as a substituted alkane. The longest hydro-carbon chain is propane. The amino group is on C atom 2 of the propane, so it is 2-amino-propane. The amino group has two methyl groups attached, so the amino group is *N,N*-dimethyl.

The IUPAC name for this compound is *N,N*-dimethyl-2-aminopropane.

Using the alternative system, name the compound as a substituted amine. The three alkyl groups attached to the N atom are methyl, methyl, and isopropyl (*i*-propyl). The alternative name for this compound is dimethyl-*i*-propylamine.

2. **Draw structural diagrams of (a) a 1° amine, (b) a 2° amine, and (c) a 3° amine, each with 3 C atoms in the molecule. Write two names for each amine.**

(a) In a primary amine, the N atom is attached to only 1 alkyl group and 2 H atoms. Since there are 3 C atoms, a single alkyl group may be *n*-propyl or *i*-propyl. So there are two possible 1° amines for this formula.

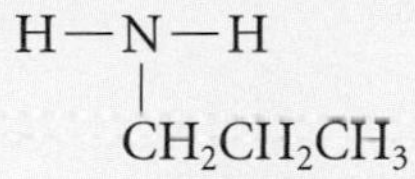

$$H - N - H$$
$$\overset{|}{CH_2CH_2CH_3}$$

1-aminopropane
n-propylamine

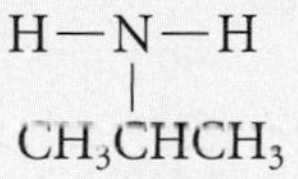

$$H - N - H$$
$$\overset{|}{CH_3CHCH_3}$$

2-aminopropane
i-propylamine

(b) In a secondary amine, the N atom is attached to 2 alkyl groups and only 1 H atom. Since there are 3 C atoms, one alkyl group must be methyl and the other ethyl. There is only one possible 2° amine for this formula:

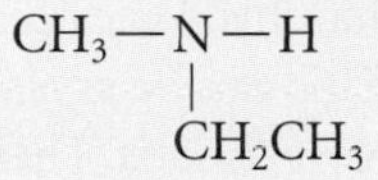

$$CH_3 - N - H$$
$$\overset{|}{CH_2CH_3}$$

N-methylaminoethane
ethylmethylamine

(c) In a tertiary amine, the N atom is attached to 3 alkyl groups and no H atoms. Since there are 3 C atoms, each alkyl group must be a methyl group. There is only one pos-sible 3° amine for this formula.

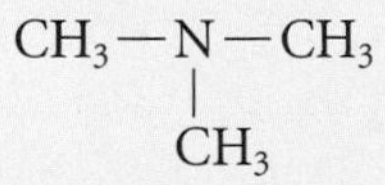

$$CH_3 - N - CH_3$$
$$\overset{|}{CH_3}$$

N,N-dimethylaminomethane
trimethylamine

Example

Write the IUPAC name for the following structure:

$$CH_3NCH_3$$
$$\overset{|}{CH_3CHCH_2CHCH_2CH_3}$$
$$\overset{|}{Cl}$$

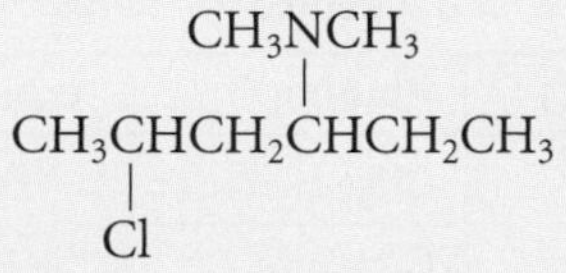

Solution

The IUPAC name for this structure is 2-chloro-*N,N*-dimethyl-4-aminohexane.

Properties of Amines

Amines have higher boiling points and melting points than hydrocarbons of similar size, and the smaller amines are readily soluble in water. This can be explained by two types of polar bonds in amines: the N—C bonds and any N—H bonds. These bonds are polar because N is more electronegative than either C or H. These polar bonds increase intermolecular forces of attraction, and therefore, higher temperatures are required to melt or to vaporize amines. The series of amines below (all of which have characteristic fishy odours) illustrates the effect of hydrogen bonding on boiling point.

$$H-\underset{\underset{H}{|}}{N}-H \qquad CH_3-\underset{\underset{H}{|}}{N}-H \qquad CH_3-\underset{\underset{H}{|}}{N}-CH_3 \qquad CH_3-\underset{\underset{CH_3}{|}}{N}-CH_3$$

b.p. –33°C   b.p. –6°C   b.p. 8°C   b.p. 3°C

Where N—H bonds are present, hydrogen bonding also occurs with water molecules, accounting for the high solubility of amines in water. It is worth noting that since N—H bonds are less polar than O—H bonds, amines boil at lower temperatures than do alcohols of similar size (**Table 1**).

Table 1 Boiling Points of Analogous Hydrocarbons, Amines, and Alcohols

Hydrocarbon	b.p. (°C)	Amine	b.p. (°C)	Alcohol	b.p. (°C)
CH_3CH_3	–89	CH_3NH_2	–6	CH_3OH	65
$C_2H_5CH_3$	–42	$C_2H_5NH_2$	16	C_2H_5OH	78
$C_3H_7CH_3$	–0.5	$C_3H_7NH_2$	48	C_3H_7OH	97
$C_4H_9CH_3$	36	$C_4H_9NH_2$	78	C_4H_9OH	117

Preparing Amines

Amines can be prepared by the reaction of ammonia, which is a weak base, with an alkyl halide. For example, the reaction of ammonia with ethyl iodide (iodoethane) yields ethylamine.

$$CH_3CH_2-I + H-\underset{\underset{H}{|}}{N}-H \longrightarrow CH_3CH_2-\underset{\underset{H}{|}}{N}-H + HI$$

ethyl iodide   ammonia   ethylamine
1° amine

The ethylamine formed is a primary amine that can also react with alkyl halides—in this case, the ethyl iodide already present in the reaction mixture. Thus, the secondary amine diethylamine is formed.

$$CH_3CH_2-I + CH_3CH_2-\underset{\underset{H}{|}}{N}-H \longrightarrow CH_3CH_2-\underset{\underset{H}{|}}{N}-CH_2CH_3 + HI$$

ethyl iodide   ethylamine   diethylamine
1° amine   2° amine

This last product can also react with the ethyl iodide to produce the tertiary amine, triethylamine.

$$CH_3CH_2-I + CH_3CH_2-\underset{\underset{H}{|}}{N}-CH_2CH_3 \longrightarrow CH_3CH_2-\underset{\underset{CH_2CH_3}{|}}{N}-CH_2CH_3 + HI$$

ethyl iodide   diethylamine   triethylamine
2° amine   3° amine

The final product is a mixture of primary, secondary, and tertiary amines. The components can be separated by fractional distillation by virtue of their different boiling points; however, this is generally not a useful method for the synthesis of primary amines. More specific synthesis methods are available, and you will learn about them in more advanced chemistry courses.

SUMMARY ***Synthesizing Amines from Alkyl Halides***

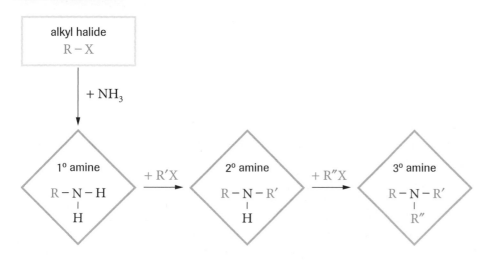

Amides

Amides are structurally similar to esters, with a N atom replacing the O atom in the chain of an ester. The amide functional group consists of a carbonyl group directly attached to an N atom. Compare this group to the ester functional group, shown below. The amide linkage is of major importance in biological systems as it forms the backbone of all protein molecules. In proteins, the amide bonds are called peptide bonds, and it is the forming and breaking of these bonds that gives specificity to the proteins and their functions. We will learn more about proteins in the next chapter.

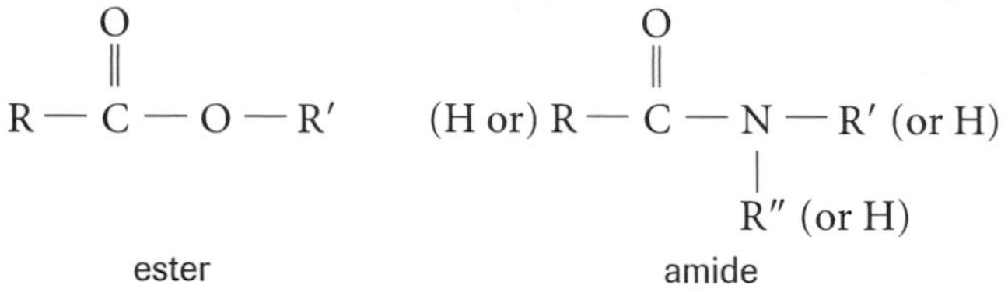

$$
\underset{\text{ester}}{R-\overset{\overset{\displaystyle O}{\|}}{C}-O-R'} \qquad \underset{\text{amide}}{(H \text{ or}) \; R-\underset{\underset{\displaystyle R'' \,(\text{or } H)}{|}}{\overset{\overset{\displaystyle O}{\|}}{C}}-N-R' \,(\text{or } H)}
$$

Naming and Preparing Amides

We have learned that carboxylic acids react with alcohols (organic bases) to produce esters (organic salts). Similarly, carboxylic acids react with ammonia or with 1° and 2° amines (also organic bases) to produce amides (another type of organic salt). Both are condensation reactions, and water is formed. The amide functional group consists of a carbonyl group bonded to a nitrogen atom. The nitrogen makes two more bonds: to H atoms or alkyl groups.

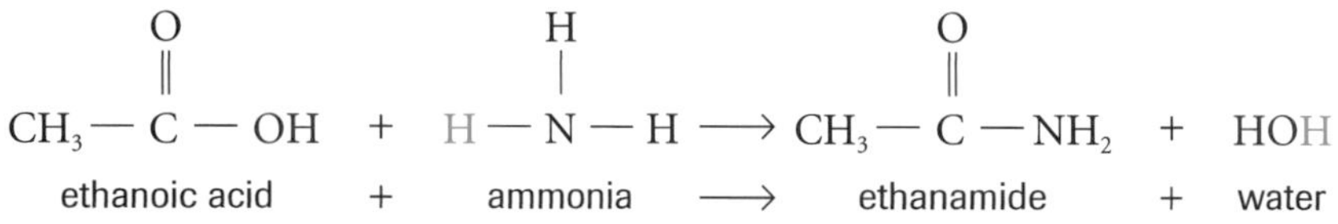

$$
\underset{\substack{\text{ethanoic acid} \\ +}}{CH_3-\overset{\overset{\displaystyle O}{\|}}{C}-OH} \;+\; \underset{\substack{\text{ammonia} \\ \longrightarrow}}{H-\overset{\overset{\displaystyle H}{|}}{N}-H} \longrightarrow \underset{\text{ethanamide}}{CH_3-\overset{\overset{\displaystyle O}{\|}}{C}-NH_2} \;+\; \underset{\text{water}}{HOH}
$$

The equation above illustrates the reaction of ethanoic acid with ammonia, with the elimination of a molecule of water. As an H atom is needed on the amine for this reaction to occur, tertiary amines do not undergo this type of reaction. The equations below

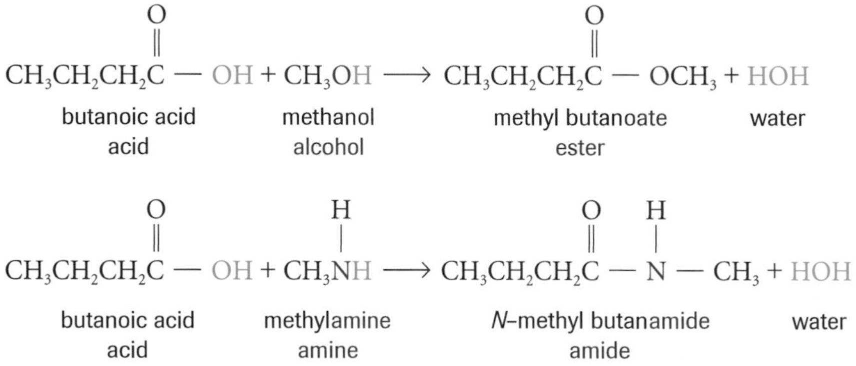

$$
\underset{\substack{\text{butanoic acid} \\ \text{acid}}}{CH_3CH_2CH_2C-OH} + \underset{\substack{\text{methanol} \\ \text{alcohol}}}{CH_3OH} \longrightarrow \underset{\substack{\text{methyl butanoate} \\ \text{ester}}}{CH_3CH_2CH_2\overset{\overset{\displaystyle O}{\|}}{C}-OCH_3} + \underset{\text{water}}{HOH}
$$

$$
\underset{\substack{\text{butanoic acid} \\ \text{acid}}}{CH_3CH_2CH_2\overset{\overset{\displaystyle O}{\|}}{C}-OH} + \underset{\substack{\text{methylamine} \\ \text{amine}}}{CH_3N\overset{\overset{\displaystyle H}{|}}{H}} \longrightarrow \underset{\substack{\textit{N}-\text{methyl butanamide} \\ \text{amide}}}{CH_3CH_2CH_2\overset{\overset{\displaystyle O}{\|}}{C}-\overset{\overset{\displaystyle H}{|}}{N}-CH_3} + \underset{\text{water}}{HOH}
$$

LEARNING TIP

IUPAC name	Common name
ethanoic acid	acetic acid
ethanoate	acetate
ethanamide	acetamide

illustrate the similarity between the synthesis of an ester from an alcohol, and the synthesis of an amide from a primary amine. Both are condensation reactions.

Naming amides is again similar to naming esters: While esters end in -oate, amides end in -amide. The name of an amide can be considered to be in two parts: The first part is derived from the amine. In the example above, the amine is methylamine, so the amide's name begins with methyl. The second part is derived from the acid. In the example, the acid is butanoic acid; the -oic of the acid name is dropped and replaced with -amide—in this case, it is butanamide. The complete name of the amide is then methyl butanamide. As with esters, the name of an amide is separated into two words.

When one or more alkyl groups are attached to the N atom in the amide linkage, the italicized uppercase letter *N* is used to clarify the location of the group. This is similar to the naming of amines, discussed earlier. The structural diagrams and names of two amides are shown below as examples.

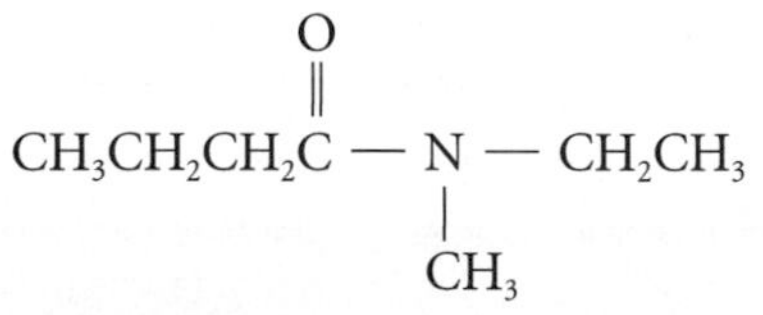

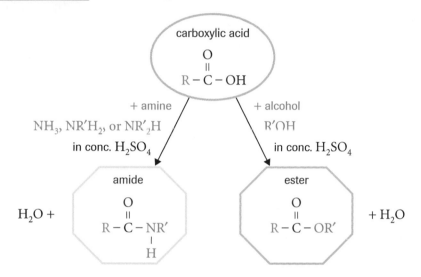

N-ethyl-*N*-methyl butanamide 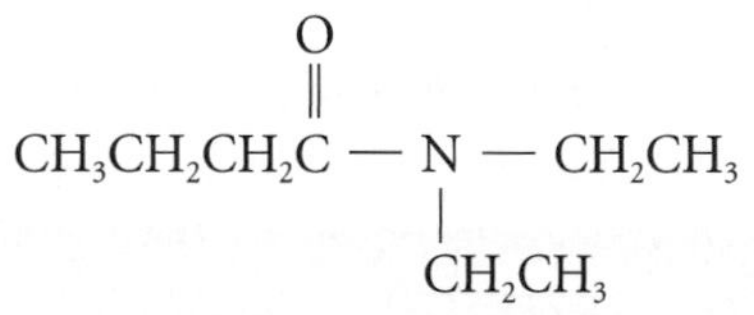   *N,N*-diethyl butanamide

SUMMARY — Condensation Reactions of Carboxylic Acids

carboxylic acid

$$\text{R} - \overset{\overset{\displaystyle O}{\|}}{\text{C}} - \text{OH}$$

+ amine   + alcohol
NH_3, $NR'H_2$, or NR'_2H $R'OH$
in conc. H_2SO_4 in conc. H_2SO_4

amide

$H_2O +$

$$\text{R} - \overset{\overset{\displaystyle O}{\|}}{\underset{\underset{\displaystyle H}{|}}{\text{C}}} - \text{NR'}$$

ester

$$\text{R} - \overset{\overset{\displaystyle O}{\|}}{\text{C}} - \text{OR'}$$ $+ H_2O$

Drawing and Naming Amides

SAMPLE problem ◀

Name the following structure:

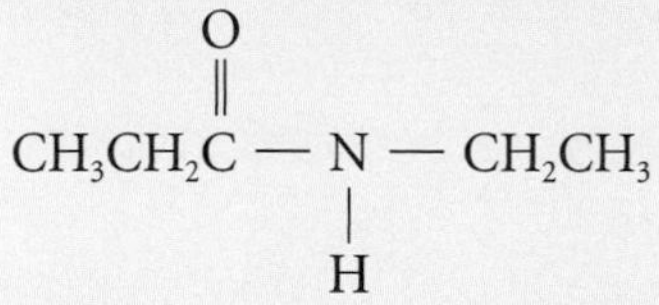

$$\text{CH}_3\text{CH}_2\text{C} - \overset{\overset{\displaystyle O}{\|}}{\underset{\underset{\displaystyle H}{|}}{\text{N}}} - \text{CH}_2\text{CH}_3$$

We can see by the CON functional group that this is an amide, so the last part of the structure's name will be *-amide*.

The alkyl group attached to the N has 2 atoms, so it will be an *N*-ethyl amide of the carboxylic acid.

The carbonyl group is on the end carbon atom of the carboxylic acid. We can see that the acid has 3 C atoms, including the carbonyl group; so the acid is propanoic acid. The amide is a propanamide.

The IUPAC name is *N*-ethyl propanamide.

Example
Draw structures of
(a) *N*-isopropyl ethanamide
(b) *N,N*-diethyl acetamide

▶

Solution

(a)

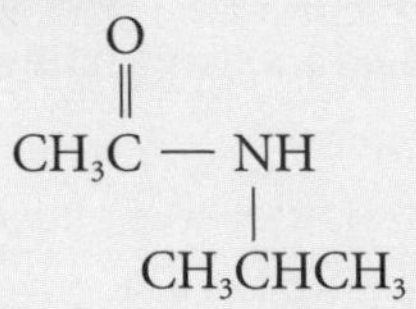

$$CH_3C \overset{\overset{\displaystyle O}{\|}}{-} NH$$
$$|$$
$$CH_3CHCH_3$$

(b)

$$CH_3C \overset{\overset{\displaystyle O}{\|}}{-} N - CH_2CH_3$$
$$|$$
$$CH_2CH_3$$

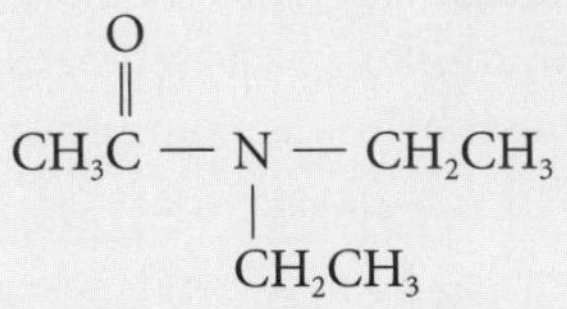

► **Practice**

Understanding Concepts

4. Write the IUPAC name for each of the following compounds:

   (a)

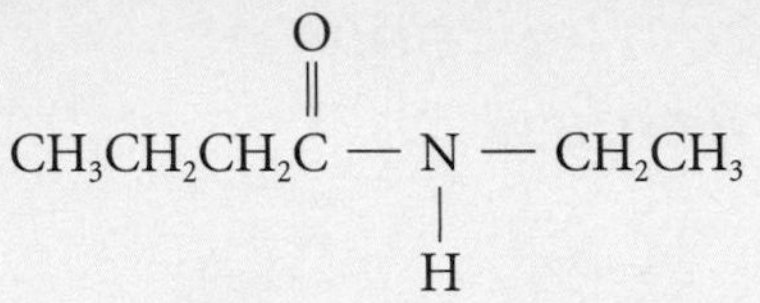

   $$CH_3CH_2CH_2C \overset{\overset{\displaystyle O}{\|}}{-} N - CH_2CH_3$$
   $$|$$
   $$H$$

   (b)

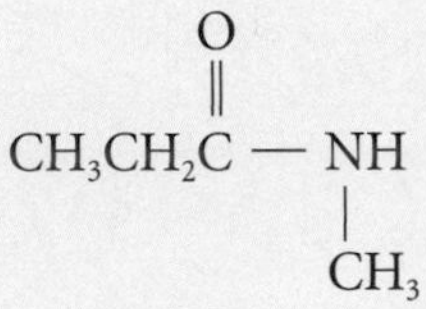

   $$CH_3CH_2C \overset{\overset{\displaystyle O}{\|}}{-} NH$$
   $$|$$
   $$CH_3$$

   (c)

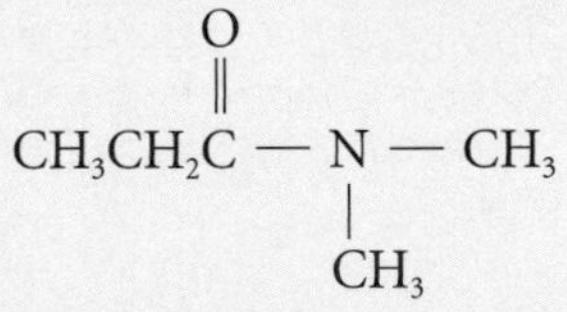

   $$CH_3CH_2C \overset{\overset{\displaystyle O}{\|}}{-} N - CH_3$$
   $$|$$
   $$CH_3$$

   (d) $CH_3(CH_2)_3CONCH_3$
   $$|$$
   $$CH_2CH_3$$

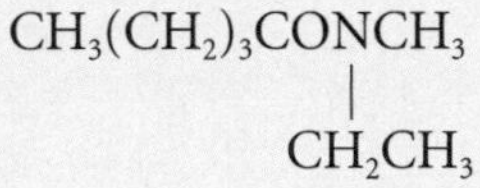

5. Draw structures for the following amides:
   (a) *N,N*-dimethyl hexanamide
   (b) *N*-methyl acetamide
   (c) hexanamide
   (d) *N*-isopropyl-*N*-methyl butanamide

6. Classify each of the following compounds as amines or amides, and write the IUPAC name for each:
   (a) $CH_3CH_2CH_2NH_2$
   (b) $CH_3NHCH_2CH_3$
   (c)

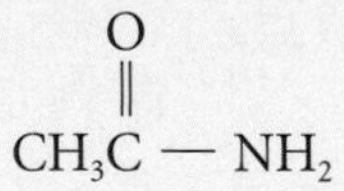

   $$CH_3C \overset{\overset{\displaystyle O}{\|}}{-} NH_2$$

Properties of Amides

Amides are weak bases, and are generally insoluble in water. However, the lower-molecular-weight amides are slightly soluble in water because of the hydrogen bonding taking place between the amides' polar N—H bonds and the water molecules.

Amides whose N atoms are bonded to two H atoms have higher melting points and boiling points than amides that have more attached alkyl groups. This can also be explained by increased hydrogen bonding, requiring higher temperatures for vaporization.

Reactions of Amides

Like esters, amides can be hydrolyzed in acidic or basic conditions to produce a carboxylic acid and an amine. This hydrolysis reaction is essentially the reverse of the formation reactions of amides that we have just discussed.

The hydrolysis reaction of amides proceeds more slowly than that of esters, however. Since amide linkages are the bonds that hold smaller units together to make proteins, the resistance of amide linkages to hydrolysis is an important factor in the stability of proteins.

Amine and Amide Reactions | **SAMPLE** problem ◀

Draw a structural diagram of the amide formed from the reaction between 3-methylbutanoic acid and ethylmethylamine. Name the amide formed.

First, draw the structural diagram of each reactant. Since the structure of an amide begins with the acid component, write the acid structure first, then the amine structure with the amine group first.

$$\underset{\substack{\text{3-methylbutanoic acid}}}{\underset{\substack{4 \quad 3 \quad 2 \quad 1}}{CH_3CHCH_2C}}\overset{\overset{\displaystyle CH_3}{|}}{\underset{}{}}\overset{\overset{\displaystyle O}{\parallel}}{\underset{}{}} - OH \qquad \underset{\substack{\text{ethylmethylamine}}}{HN}\overset{\overset{\displaystyle CH_3}{|}}{\underset{}{}} - CH_2CH_3$$

An OH group is eliminated from the carboxyl group of the acid and an H atom from the amine group of the alcohol. A water molecule is eliminated and a bond forms between the C atom of the carboxyl group and the N atom of the amine group.

$$CH_3CHCH_2C \overset{\overset{\displaystyle CH_3}{|}}{\underset{}{}}\;\overset{\overset{\displaystyle O}{\parallel}}{\underset{}{}} - N \overset{\overset{\displaystyle CH_3}{|}}{\underset{}{}} - CH_2CH_3$$

The name of the amide formed is *N*-ethyl-*N*,3-dimethyl butanamide.

Example

Write a structural formula for
(a)   6-aminohexanoic acid;
(b)   the product formed when two molecules of 6-aminohexanoic acid react to form an amide linkage.

Solution

(a) $H_2N{-}CH_2CH_2CH_2CH_2CH_2\overset{\overset{\displaystyle O}{\parallel}}{C}OH$

(b) $H_2N{-}CH_2CH_2CH_2CH_2CH_2\overset{\overset{\displaystyle }{\underset{\overset{\parallel}{O}}{C}}}N{-}HCH_2CH_2CH_2CH_2CH_2\overset{\overset{\displaystyle }{\underset{\overset{\parallel}{O}}{C}}}{-}OH + H_2O$

▶ Practice

Understanding Concepts

7. Draw structural diagrams and write IUPAC names for the carboxylic acid and amine which may be used to produce the following compound.

$$CH_3C\overset{\overset{\displaystyle O}{\parallel}}{\underset{}{}} - \underset{\underset{\displaystyle CH_3}{|}}{N} - CH_3$$

SUMMARY **Amines and Amides**

Functional Groups:

- amines:

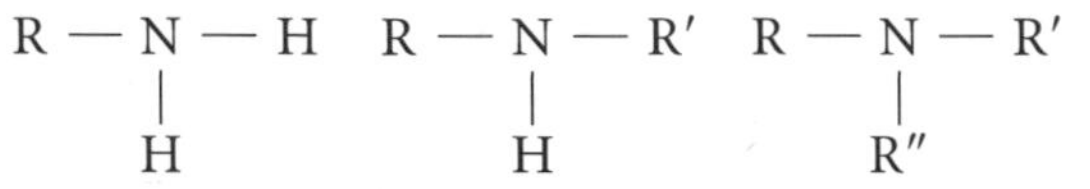

$$R - N - H \qquad R - N - R' \qquad R - N - R'$$
$$\qquad | \qquad\qquad\quad | \qquad\qquad\quad |$$
$$\qquad H \qquad\qquad\quad H \qquad\qquad\quad R''$$

- amides:

$$O$$
$$\|$$
$$\text{or } RC - N - R' \text{ or } H$$
$$\qquad\qquad |$$
$$\qquad\quad H \text{ or } R''$$

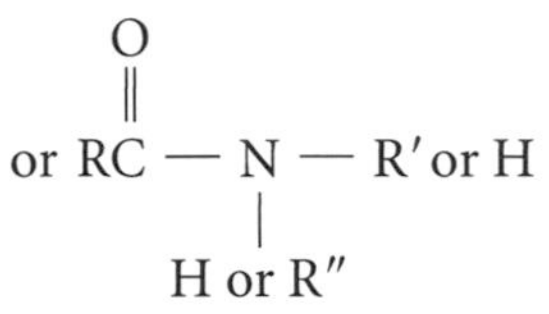

Preparation:

- amines:

$$RX + NH_3 \rightarrow \text{amine} + HX$$
$$RX + R_2NH \rightarrow \text{amine} + HX$$

- amides:

carboxylic acid + amine → amide + H_2O

Reactions:

$$\text{amine} + \text{carboxylic acid} \xrightarrow{\text{heat}} \text{amide} + H_2O \quad \text{(condensation reaction)}$$

$$\text{amide} + H_2O \xrightarrow{\text{acid or base}} \text{carboxylic acid} + \text{amine} \quad \text{(hydrolysis reaction)}$$

▶ **Practice**

Understanding Concepts

8. Explain why the formation of an amide from a carboxylic acid and an amine is a condensation reaction.

9. Proteins are built from many smaller molecules, each of which contains an amine group and a carboxylic acid group; these small molecules are called amino acids.
   (a) Draw structural diagrams for the amino acids glycine (2-aminoethanoic acid) and alanine (2-aminopropanoic acid).
   (b) Draw structural diagrams for two possible products of the condensation reaction between glycine and alanine.

10. Explain why amines generally have lower boiling points than alcohols of comparable molar mass.

11. Write a series of equations to represent the formation of N-methyl ethanamide from methane, ethanol, and inorganic compounds of your choice.

▶ **Section 1.8** *Questions*

Understanding Concepts

1. Write an equation to represent the formation of an amide linkage between propanoic acid and diethylamine.

2. Look at the following pairs of compounds and arrange each pair in order of increasing solubility in nonpolar solvents. Give reasons for your answer.

(a) an alcohol and an amine of similar molecular mass
(b) a primary amine and a tertiary amine of similar molecular mass
(c) a hydrocarbon and a tertiary amine of similar molecular mass

    (d) a primary amine of low molecular mass and one of high molecular mass

3. Draw structural formulas for three isomers of C_3H_9N, and classify them as primary, secondary, or tertiary amines. Write IUPAC names for each isomer.

4. Write structural formula equations to represent the formation of the following amides:
    (a) methanamide
    (b) propanamide

5. Write IUPAC names for the following compounds:

(a)
$$\begin{array}{c} O \\ \parallel \\ CH_3CH_2C - NH_2 \end{array}$$

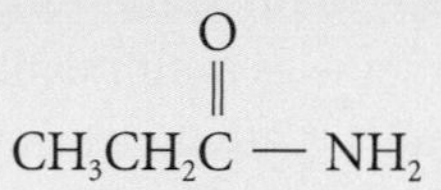

(b)
$$\begin{array}{c} CH_3CH_2CH_2 - N - CH_3 \\ | \\ CH_3 \end{array}$$

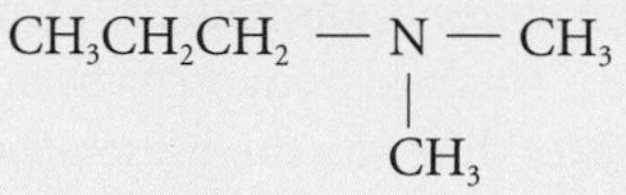

(c)
$$\begin{array}{c} O \\ \parallel \\ CH_3CH_2C - N - CH_2CH_3 \\ | \\ CH_2CH_3 \end{array}$$

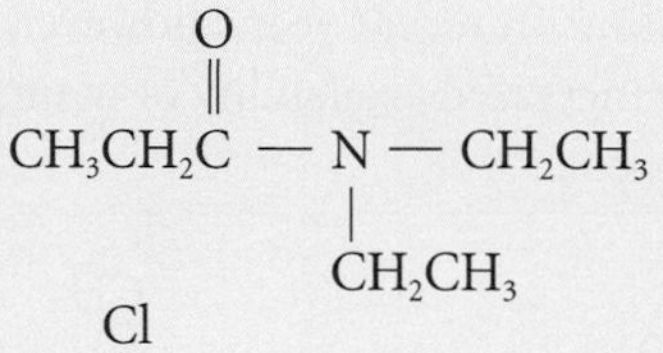

(d)
$$\begin{array}{c} Cl \\ | \\ CH_3CH_2CH_2CHCH_2CH_2CH_3 \\ | \\ NH_2 \end{array}$$

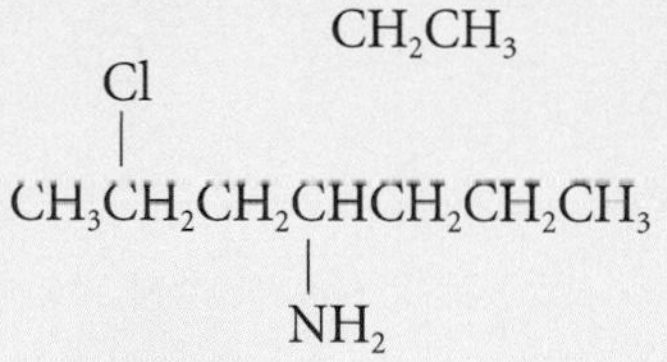

(e)
$$\begin{array}{c} NH_2 \\ | \\ CH_3CH_2CH_2CHCH_2CH = CH_2 \\ | \\ NH_2 \end{array}$$

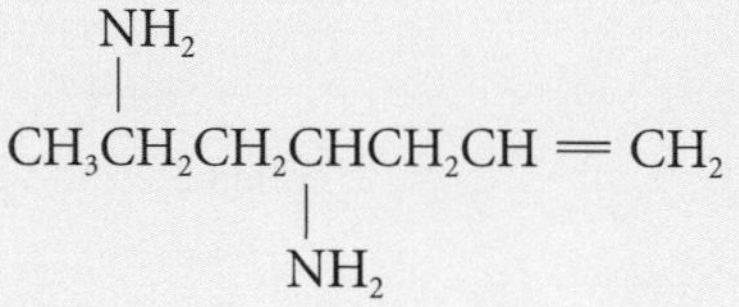

(f)
$$\begin{array}{c} NH_2 \\ | \\ CH_2C = O \\ | \\ OH \end{array}$$

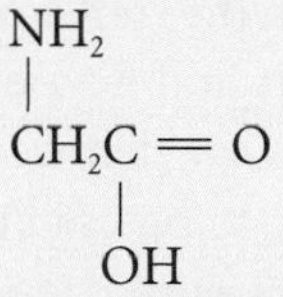

6. In this chapter, you have encountered many different chemical names for a range of chemical families.
    (a) Summarize all the families, functional groups, and naming convention(s) in a chart.
    (b) Comment on whether you feel that this is a logical naming system, and suggest any improvements that would, in your opinion, make the system simpler to use.

7. What features in the molecular structure of a compound affect its solubility and its melting and boiling points? Illustrate your answer with several examples, including structural diagrams.

Applying Inquiry Skills

8. In a reaction of ammonia with an ethyl bromide, a mixture of products is formed. Describe the procedure that can be used to separate the components of the mixture, and explain the theory behind the procedure.

Making Connections

9. In some cuisines, fish recipes include lemon garnish or a vinegar sauce such as sweet-and-sour sauce. Suggest a reason why these common culinary technique might be used. Support your answer with chemical information.

10. Protein molecules in all living systems are long molecules made up of small units called amino acids. From what you have learned in this chapter,
    (a) deduce the structural features of amino acids; and
    (b) predict some physical and chemical properties of amino acids.

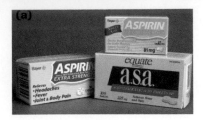

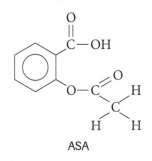

ASA

urea

Figure 1

(a) More than ten million kilograms of Aspirin, or acetylsalicylic acid (ASA), are produced in North America annually. Because of this compound's ability to reduce pain and inflammation, it is the most widely used medicine in our society.

(b) Urea, the organic compound first synthesized by Wöhler in 1828, is produced in even larger quantities than ASA. Urea is used primarily as plant fertilizer and as an animal feed additive.

There are many reasons why chemists create new organic substances. They may be synthesized as part of research or to demonstrate a new type of reaction. Others are synthesized if a compound is needed with specific chemical and physical properties. Large amounts of some synthetic compounds are routinely produced industrially (**Figure 1**).

Chemical engineers can design and create synthetic replicas of naturally produced compounds by adding or removing key functional groups to or from available molecules. For example, there is a large market demand for caffeine as a stimulant for use in non-coffee beverages such as soft drinks and in solid foods. Caffeine is a cyclic compound containing nitrogen, and is extracted from coffee beans and tea leaves. To meet the large demand, a compound called theobromine is obtained from cocoa fruits and modified chemically to produce caffeine; the structures differ by the presence of a single methyl group, which is added to the theobromine. In recent years, however, the popularity of decaffeinated coffee has resulted in the increased availability of natural caffeine, removed from coffee in the decaffeination process.

theobromine   caffeine

Throughout our study of organic families, we have focused on the transformation of functional groups. Let us take our knowledge of these reactions and devise a method of preparing a more complex molecule from simpler ones. Let's say that we have been assigned the task of preparing an ester with a pineapple flavour, to be used in candies. We have available a number of short-chain alkanes and alkenes as starting materials. What substance will we make, which starting materials should we use, and what steps should we take in the synthesis?

Pineapple Flavouring: Ethyl Butanoate

When planning a synthesis pathway for any substance, we have a couple of options. One approach is to look at the available compounds and examine the transformations they can undergo. For example, what products can we obtain from short-chain alkanes and alkenes? We can then look at the possible products of these transformations and proceed further. Because there are many possible reactions and products, however, this approach presents several pathways to consider, many of which are ultimately unproductive.

Another approach is to start with the compound we want to prepare—ethyl butanoate—and think "backwards." Ethyl butanoate is, like all synthetic esters, derived from an acid and an alcohol. Having identified the acid and the alcohol, we then identify a reaction that would produce each of them from our choice of short-chain alkanes and alkenes. This approach limits the possible reactions and is a more efficient method of selecting a synthesis pathway.

$$
\underset{\text{ethyl butanoate}}{CH_3CH_2CH_2\overset{\displaystyle O}{\overset{\displaystyle \|}{C}}-O-CH_2CH_3}
$$

Let us apply the "backwards" approach to the synthesis of ethyl butanoate, the ester with a pineapple flavour. As the name and structure show, this ester is formed from a reaction between ethanol and butanoic acid. The steps in our multi-step procedure are shown in the sample problem below. Of course, there is more than one possible synthesis pathway for any product; when selecting a procedure, industrial chemists always pay attention to the availability and cost of starting materials, as well as safety and environmental concerns.

Designing Synthesis Pathways   *SAMPLE* problem ◀

Write a series of equations to illustrate the synthesis of ethyl butanoate from an alkene and an alcohol from simpler molecules.

First, plan a pathway for the production of ethyl butanoate.

To make the "ethyl" portion of the ester, we will need to synthesize the necessary alcohol, ethanol. Ethanol can be prepared by the addition reaction of water to ethene, a hydration reaction. We will select ethene and water as the starting materials for this alcohol.

Next, we will need to prepare butanoic acid. Recall that acids can be prepared from the controlled oxidation of aldehydes, which, in turn, are prepared from the controlled oxidation of alcohols. We need to determine which alcohol to use for this sequence. 1-butanol will produce butanal, which will produce butanoic acid. However, being a primary alcohol, 1-butanol is difficult to prepare from the reactions we have studied so far. The addition reaction of H_2O to either 1-butene or 2-butene would produce predominantly 2-butanol, as predicted by Markovnikov's rule.

$$
CH_2{=}CHCH_2CH_3 + H_2O \xrightarrow{H_2SO_4} \underset{\displaystyle \overset{|}{OH}}{CH_3CHCH_2CH_3}
$$

$$
CH_2CH{=}CHCH_3 + H_2O \xrightarrow{H_2SO_4} \underset{\displaystyle \overset{|}{OH}}{CH_3CHCH_2CH_3}
$$

Other reagents and reactions are available for preparing primary alcohols, as you may learn in future chemistry courses. For this synthesis, we will use 1-butanol as our starting material to prepare butanoic acid. Our pathway is now complete.

ethyl butanoate (an ester)

ethanol (alcohol)   +   butanoic acid (carboxylic acid)

ethene (alkene) + water   butanal (aldehyde) + (O)

1-butanol (alcohol) + (O)

The equations are

1. $\underset{\text{1-butanol}}{CH_3CH_2CH_2CH_2OH} + (O) \rightarrow \underset{\text{butanal}}{CH_3CH_2CH_2CHO} + H_2O$

2. $\underset{\text{butanal}}{CH_3CH_2CH_2CHO} + (O) \rightarrow \underset{\text{butanoic acid}}{CH_3CH_2CH_2COOH}$

---

3. $CH_2{=}CH_2 + H_2O \rightarrow CH_3CH_2OH$
   ethene   ethanol

4. $CH_3CH_2CH_2COOH + CH_3CH_2OH \rightarrow CH_3CH_2CH_2COOCH_2CH_3$
   butanoic acid   ethanol   ethyl butanoate

Example

Write a series of equations for the synthesis of propanone, starting with an alkene.

Solution

1. $CH_3CH{=}CH_2 + H_2O \longrightarrow CH_3CH(OH)CH_3$
   propene   2-propanol

2.
$$CH_3CH(OH)CH_3 + (O) \longrightarrow CH_3\overset{\displaystyle O}{\overset{||}{C}}CH_3$$
   2-propanol   propanone

ACTIVITY 1.9.1

Building Molecular Models to Illustrate Reactions (p. 90)
What do molecules really look like, and how does one substance turn into another? This activity helps you visualize, using molecular models, how reactions happen.

ACTIVITY 1.9.2

Preparation of an Ester— Aspirin (p. 90)
The reaction between salicylic acid and acetic anhydride produces a well-known pharmaceutical product.

▶ Practice

Understanding Concepts

1. A food additive that is named as a single ingredient is often a mixture of many compounds. For example, a typical "artificial strawberry flavour" found in a milkshake may contain up to 50 different compounds. A few of these are listed below. For each, write a structural formula and an equation or a series of equations for a method of synthesis from the suggested reactants.
   (a) 2-pentyl butanoate from pentene and butanal
   (b) phenyl 2-methylpropanoate from phenol and an appropriate alcohol
   (c) 4-heptanone from an alcohol
   (d) ethyl ethanoate from ethene

▶ Section 1.9 Questions

Understanding Concepts

1. For each product, write a structural formula and an equation or a series of equations for a method of synthesis from other compounds.
   (a) pentyl ethanoate from ethene and an alcohol
   (b) phenyl ethanoate from an alkene and an alcohol
   (c) 3-octanone from a simpler compound
   (d) methyl benzoate from two alcohols
   (e) sodium salt of butanoic acid from an ester
   (f) trimethylamine from ammonia and alkanes
   (g) N-ethylethanamide from an alkane and ammonia

2. The smell of freshly cut grass can be simulated by the addition of hexanal to substances such as air-fresheners. Describe how hexanal can be made from an alcohol.

Applying Inquiry Skills

3. In artificial apple flavour, the main ingredient is ethyl 2-methylbutanoate.
   (a) Draw a flow chart to show the synthesis of this compound, starting with simpler compounds.
   (b) Describe the steps in the procedure for this synthesis. Include experimental conditions and safety precautions in the handling and disposal of materials.

Making Connections

4. The flavour of almonds is obtained from phenyl methanal, also called benzaldehyde. "Natural" almond flavour is extracted from the pit of peaches and apricots, and may contains traces of the poison hydrogen cyanide.
   (a) Write a structural diagram for benzaldehyde.
   (b) Write equations for a series of reactions in the synthesis of benzaldehyde from an ester.

(c) Research and describe the toxicity of the reactants listed in (b), and compare the toxicity to that of hydrogen cyanide. In your opinion, do your findings substantiate consumer perception that "natural" products are always healthier than their "artificial" counterparts?

 www.science.nelson.com

5. Quinine is a fairly effective treatment for malaria, a mainly tropical disease that kills or incapacitates millions of people every year. Quinine has a bitter taste and was mixed with lemon or lime to make it palatable; it was later included as a component of a beverage called tonic water.
   (a) Tonic water contains approximately 20 mg quinine per 375-mL can; the dosage required for the treatment of malaria is approximately six 300-mg tablets per day. In your opinion, would drinking tonic water be an effective approach to the treatment of malaria? Explain.
   (b) Refer to the structural diagram of quinine (**Figure 2**), and explain why quinine is soluble in water.
   (c) Research the effectiveness of quinine as an antimalarial agent from the early 1800s to the present.

 www.science.nelson.com

6. If you were told that a neighbour was starting a new career in "organic products," why might there be some confusion about what the new job involves?

7. Look at the product labels on foods, pharmaceuticals, and dietary supplements in a supermarket or drugstore. In particular, look for the terms "organic," "natural," and "chemical." What is the purpose of using these terms on the packaging? Do they always mean the same thing? Do they have a positive or negative connotation? Write a short opinion piece on the use of these three terms, and what these terms mean to consumers.

Figure 2
Structure of quinine

SUMMARY · *Flow Chart of Organic Reactions*

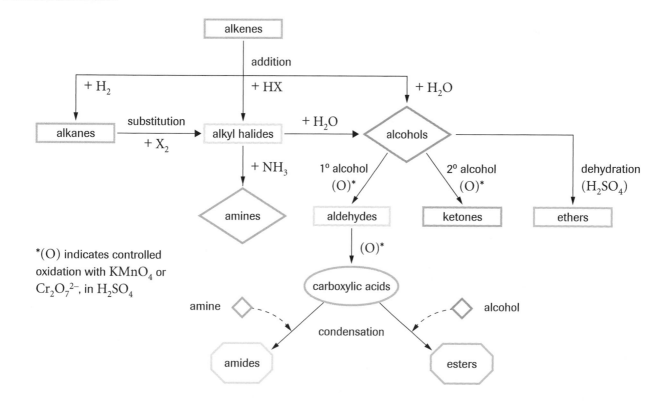

*(O) indicates controlled oxidation with $KMnO_4$ or $Cr_2O_7^{2-}$, in H_2SO_4

✎ LAB EXERCISE 1.3.1

Preparation of Ethyne

Ethyne is the simplest and most widely used alkyne. It can be produced from the reaction of calcium carbide with water. Calcium hydroxide is the only other product.

$$CaC_{2(s)} + 2H_2O_{(l)} \rightarrow C_2H_{2(g)} + Ca(OH)_{2(aq)}$$

Knowing the quantity of reactants and the balanced chemical equation for a reaction, we can predict the theoretical yield of a product. How close does the actual yield come to this theoretical yield? The difference in these two calculations can be used to assess the purity of the reactants.

Purpose
The purpose of this lab exercise is to determine the yield of ethyne from a synthesis procedure.

Question
What mass of ethyne is produced from a known mass of calcium carbide, and what is the purity of the calcium carbide?

Prediction
(a) Predict the mass of ethyne and calcium hydroxide produced from the hydration reaction of 6.78 g of calcium carbide.

Inquiry Skills

○ Questioning	○ Planning	● Analyzing
○ Hypothesizing	○ Conducting	● Evaluating
● Predicting	○ Recording	○ Communicating

Experimental Design
A known mass of calcium carbide is reacted with an excess of water. The amount of calcium hydroxide formed is determined by titration of the entire volume of $Ca(OH)_{2(aq)}$ with 1.00 mol/L HCl. The amount of ethyne produced is then calculated stoichiometrically.

Evidence
mass of CaC_2 reacted: 6.78 g
volume of 1.00 mol/L $HCl_{(aq)}$ added to neutralize the $Ca(OH)_2$ solution: 100.0 mL

Analysis
(b) Write a balanced chemical equation for the neutralization reaction between $Ca(OH)_{2(aq)}$ and $HCl_{(aq)}$.

(c) Calculate the amount of $Ca(OH)_{2(aq)}$ actually produced in the reaction of $CaC_{2(aq)}$ and water.

(d) Calculate the actual yield of ethyne, in grams.

(e) Calculate the percent purity of the CaC_2.

Evaluation
(f) What assumptions did you make in your calculations?

🧪 INVESTIGATION 1.5.1

Comparison of Three Isomers of Butanol

The reactivity of alcohols can be accounted for by their molecular structure—particularly by the attachment of their hydroxyl functional group. The isomers of butanol are used as examples of 1°, 2°, and 3° alcohols to examine this relationship.

Purpose
The purpose of this investigation is to test our theories of how the molecular structure of an organic molecule affects its properties. To do this, we will determine and compare the chemical properties of three isomers of butanol.

Inquiry Skills

○ Questioning	○ Planning	● Analyzing
○ Hypothesizing	● Conducting	● Evaluating
● Predicting	● Recording	● Communicating

Question
Does each alcohol undergo halogenation and controlled oxidation?

Prediction
(a) Make predictions about your observations, with reasons.

INVESTIGATION 1.5.1 *continued*

Experimental Design

Each of the three isomers of butanol is mixed with concentrated $HCl_{(aq)}$. The presence of an alkyl halide product is indicated by cloudiness of the mixture, as the halides are only slightly soluble in water. Each alcohol is also mixed with dilute $KMnO_{4(aq)}$ solution, which provides conditions for controlled oxidation. Any colour change of the permanganate solution indicates that an oxidation reaction has taken place.

Materials

lab apron
eye protection
gloves
1-butanol (pure)
2-butanol (pure)
2-methyl-2-propanol (pure)
concentrated $HCl_{(aq)}$ (12 mol/L)
3 test tubes
test-tube rack
4 eye droppers
$KMnO_{4(aq)}$ solution (0.01 mol/L)
10-mL graduated cylinder

 Concentrated hydrochloric acid is corrosive and the vapour is very irritating to the respiratory system. Avoid contact with skin, eyes, clothing, and the lab bench. Wear eye protection and a laboratory apron.

 All three alcohols are highly flammable. Do not use near an open flame.

Procedure

1. Place 3 test tubes in a test-tube rack. Using a clean eye dropper for each alcohol, place 2 drops of 1-butanol in the first tube; in the second, place 2 drops of 2-butanol; and in the third, place 2 drops of 2-methyl-2-propanol.

2. Carry the test-tube rack and tubes to the fume hood, and use a clean pipet to add 10 drops of concentrated $HCl_{(eq)}$ to each of the three test tubes. Shake the mix-

ture very gently and carefully. Return to your lab bench with the test-tube rack and tubes.

3. Allow the tubes to stand for 1 min and observe for evidence of cloudiness.

4. Follow your teacher's instructions for the disposal of the contents of the test tubes and for cleaning the test tubes.

5. Set up three test tubes as described in step 1, this time using 4 drops of each alcohol.

6. To each tube, carefully add 2 mL of 0.01 mol/L $KMnO_{4(aq)}$ solution. Shake the mixture carefully.

7. Allow the tubes to stand for 5 min, with occasional shaking. Observe and record the colour of the solution in each tube.

Evidence

(b) Prepare a table in which to record the observed properties of the three alcohols. Include structural diagrams for each alcohol.

Analysis

(c) Answer the Question.

Evaluation

(d) Evaluate the theory that you used to make your predictions.

Synthesis

(e) How can the results of this investigation be accounted for in terms of intermolecular forces?

(f) Write a structural diagram equation to represent the reaction between each alcohol and $HCl_{(aq)}$. Where no reaction occurred, write the starting materials and the words "no reaction."

(g) Write a structural diagram equation to represent the controlled oxidation of each alcohol in $KMnO_{4(aq)}$ solution. Where no reaction occurred, write the starting materials and the words "no reaction."

(h) Summarize in a few sentences the halogenation and controlled oxidation reactions of 1°, 2°, and 3° alcohols.

INVESTIGATION 1.5.2

Trends in Properties of Alcohols

We might expect to see a trend in properties within a chemical family, such as alcohols. Is there a link, in the first four primary alcohols, between molecular size and physical properties? In this investigation we will first use our knowledge of intermolecular forces of the hydrocarbon components and the hydroxyl functional group to predict trends. We will then test our predictions experimentally.

Purpose
The purpose of this investigation is to create a theoretical model of 1° alcohols, that will explain their properties.

Question
What are the melting points, boiling points, solubilities, and acidity of the three 1° alcohols?

Experimental Design
Reference sources are used to look up the melting point and boiling point of several alcohols. The solubility of each alcohol is determined by mixing with a nonpolar solvent (cyclohexane) and with water, and observing miscibility. The acidity of each alcohol in aqueous solution is tested with litmus paper.

Materials
lab apron
blue and pink litmus paper
eye protection
3 test tubes
ethanol
test-tube rack
1-propanol
test-tube holder
1-butanol
graduated cylinder (10 mL)
cyclohexane

 All three alcohols and cylohexane are highly flammable. Do not use near an open flame.

Procedure
1. Place 1 mL of each alcohol in a separate test tube. To each alcohol, add 1 mL of cyclohexane. Observe and record any evidence of miscibility of the two liquids.

2. Follow your teacher's instructions for the disposal of the contents of the test tubes and for cleaning the test tubes.

3. Repeat step 1, using water in place of cyclohexane.

4. Before disposing of the contents of the test tubes, add a small piece of blue and red litmus paper to each mixture. Record the colour of the litmus paper.

Evidence
(a) Prepare a table with the following headings: alcohol name, structural diagram, melting point, boiling point, solubility in cyclohexane, solubility in water, colour with litmus.

(b) Draw in the structural diagrams of each alcohol.

(c) Using a print or electronic reference source, find the melting point and boiling point of each of the alcohols listed, and record the information in your table.

 www.science.nelson.com

Analysis
(d) Your completed table provides answers to the Question. Summarize the trends in properties that you discovered.

Synthesis
(e) Use your answers in (d) to develop a model of the primary alcohols. Represent your model in some way, and hypothesize how its features affect its properties.

🧪 INVESTIGATION 1.7.1

Properties of Carboxylic Acids

A carboxylic acid is identified by the presence of a carboxyl group. The physical properties and reactivity of carboxylic acids are accounted for by the combination of their polar functional group and their nonpolar hydrocarbon components. We will look at some the following properties of carboxylic acids in this investigation: melting and boiling points; solubility; acidity; reaction with $KMnO_{4(aq)}$; and reaction with $NaHCO_{3(aq)}$. The carboxylic acids we will be using are ethanoic (acetic) acid and octadecanoic (stearic) acid (**Figure 1**).

Figure 1
Stearic acid is used to harden soaps, particularly those made with vegetable oils, that otherwise tend to be very soft.

Purpose

The purpose of this investigation is to test our theoretical prediction of some chemical properties of carboxylic acids.

Question

(a) Devise a question that this investigation will allow you to answer.

Prediction

(b) Use your theoretical knowledge of the structure and functional groups of carboxylic acids to predict answers to the Question. (In some cases, you will only be able to compare the acids and give relative answers.)

Experimental Design

Several properties of two carboxylic acids of different molecular size are determined and compared. The melting points and boiling points of acetic acid and stearic acid are obtained from reference resources. The solubility of each acid in polar and nonpolar solvents is determined by mixing each acid with water and with oil. The reactions, if any, of each acid with $NaHCO_{3(aq)}$, a base, and with $KMnO_{4(aq)}$, an oxidizing agent, are observed and compared.

Materials

ethanoic acid (glacial acetic acid)
dilute ethanoic acid (8% acetic acid, vinegar)
octadecanoic acid (stearic acid, solid)
water
vegetable oil
pH meter or universal indicator
$KMnO_{4(aq)}$ (0.01 mol/L aqueous solution)
$NaHCO_{3(aq)}$ (saturated aqueous solution)
2 test tubes
test-tube holder
test-tube rack
pipet and bulb
eye dropper
graduated cylinder (10 mL)

 Glacial acetic acid is corrosive. Avoid contact with skin and eyes. It is also volatile, so you must be careful to avoid inhalation. Wear eye protection and a laboratory apron.

Procedure

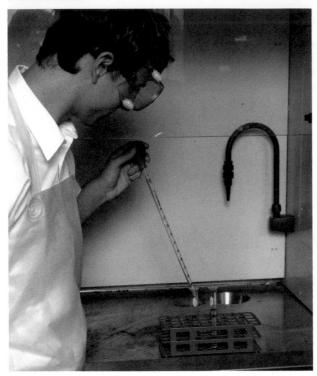

Figure 2

1. Add 5 mL of water to one test tube and 5 mL of oil to another test tube. In the fume hood, using an eye dropper, add one drop of glacial acetic acid to each tube (**Figure 2**). Shake each tube very carefully to mix. Make and record observations on the miscibility of the contents of each tube.

2. Still in the fume hood, add a drop of pH indicator to each of the test tubes in step 1, or use a pH meter to measure the pH. Record the results.

3. Follow your teacher's instructions to dispose of the contents of each test tube, and clean the test tubes.

4. Repeat steps 1, 2, and 3, using a small amount of solid stearic acid (enough to cover the tip of a toothpick). These steps do not need to be performed in the fume hood.

5. Place about 2 mL of saturated $NaHCO_{3(aq)}$ solution in each of two test tubes. Add 2 mL of dilute acetic acid to one tube, and a small amount of solid stearic acid to the other tube. Shake the tubes gently to mix and observe for formation of bubbles. Record your observations.

6. Place about 2 mL of $KMnO_{4(aq)}$ solution in each of two test tubes. Add 2 mL of dilute acetic acid to one tube, and a small amount of solid stearic acid to the other tube. Shake the tubes gently to mix and observe for any change in colour. Record your observations.

Evidence

(c) Prepare a table to record structural diagrams, molar mass, and the properties under investigation for acetic acid and for stearic acid. Complete the information in the table for acid name, structural diagram, and molar mass.

(d) Using print or electronic resources, obtain the melting point and boiling point for acetic acid and stearic acid, and record them in the table.

 www.science.nelson.com

Analysis

(e) Your table of evidence should contain your answer to the Question. Compare the melting points and boiling points of acetic acid and stearic acid. Account for the differences in these properties in terms of the molecular structure and intermolecular forces of each acid.

(f) Compare the solubilities of acetic acid and stearic acid in water and in oil. Account for any differences in their solubilities in terms of molecular structure and intermolecular forces.

(g) Do acetic acid and stearic acid appear to be organic acids in this investigation? Explain why or why not, with reference to experimental reactants and conditions.

(h) Write an equation to illustrate any reaction between the acids and $NaHCO_{3(aq)}$.

(i) Do acetic acid and stearic acid undergo controlled oxidation reactions? Explain why or why not. Draw a structural diagram equation to illustrate your answer.

Evaluation

(j) Did the Experimental Design enable you to collect appropriate evidence?

(k) Compare the answer obtained in your Analysis to your Prediction. Account for any differences.

(l) Did your theoretical model of carboxylic acids enable you to correctly predict the chemical properties of these acids? Give reasons for your answer.

ACTIVITY 1.7.2

Synthesis of Esters

Many esters are found naturally in fruits, and are responsible for some of their pleasant odours. Synthetic esters are produced from condensation reactions between alcohols and carboxylic acids. In this activity you will synthesize, and detect the odours of, several esters.

Materials

lab apron
eye protection
ethanol
2-propanol
1-pentanol
glacial acetic acid
graduated cylinder
2 mL concentrated $H_2SO_{4(aq)}$
3 test tubes

test-tube rack
wax pencils
test-tube holder
500-mL beaker
hot plate
pipet and bulb
evaporating or petri dish
balance

 Both acids are corrosive. Avoid contact with skin and eyes. Glacial acetic acid is also volatile, so you must be careful to avoid inhalation. Wear eye protection and a laboratory apron.

 All three alcohols and cylohexane are highly flammable. Do not use near an open flame.

• Prepare a hot-water bath by half filling a 500-mL beaker with water and heating it carefully on a hot plate until it comes to a gentle boil.

• Number three test tubes in a test tube rack. Place 1 mL of an alcohol in each test tube, as indicated in **Table 1**.

• In the fume hood, use a pipet to add carefully the glacial acetic acid and concentrated $H_2SO_{4(aq)}$ to each tube, as indicated in **Table 1**. Gently shake each tube to mix.

• Carefully return the test tubes, to your lab bench, and place all three of them in the hot-water bath. Be careful not to point the test tubes at anybody. After about 5 min of heating, remove the test tubes from the heat and put them back in the rack.

• Pour the contents of the first test tube into an evaporating dish half filled with cold water, and identify the odour of the ester as instructed by your teacher (**Figure 3**). Repeat for each ester.

(a) For each of the three reactions, identify the odours of the esters.

(b) Draw structural diagram equations to represent each of the three esterification reactions in this investigation. Write the IUPAC name of each reactant and product.

(c) What was the function of the concentrated sulfuric acid in these reactions?

(d) What evidence is there that the carboxylic acid used in this investigation is soluble or insoluble in aqueous solution? Explain this evidence in terms of molecular structure of the acid.

(e) What evidence is there that the esters synthesized in this investigation are soluble or insoluble in aqueous solution? Explain this evidence in terms of molecular structure of the esters.

Figure 3

Table 1 Contents of Test Tubes for Synthesizing Esters

Contents	Tube #1	Tube #2	Tube #3
Alcohol (1 mL)	ethanol	2-propanol	1-pentanol
Acid (1 mL or 1 g)	glacial acetic acid	glacial acetic acid	glacial acetic acid
Catalyst (0.5 mL)	conc. $H_2SO_{4(aq)}$	conc. $H_2SO_{4(aq)}$	conc. $H_2SO_{4(aq)}$

Building Molecular Models to Illustrate Reactions

In this activity, you will build molecular models of organic compounds and use them to illustrate a variety of chemical reactions. This should help you to see the differences in molecular structure between reactants and products. Your teacher may assign you to work in pairs or small groups.

Materials
molecular model kit

Part 1: Plan Your Own Reactions

(a) Copy **Table 2** and fill in the reactant and product columns with the chemical names and formulas of each compound as you complete the molecular models for each reaction. Supply the name of any missing reaction type.

- For the first reaction indicated in **Table 2**, build a molecular model of a reactant (alkane) of your choice, and its product.
- Compare the models of the reactants and products, and examine how the structure changes as a result of the reaction.
- Disassemble the models and repeat the process for the next reaction. Continue for all the reactions listed.

(b) Classify each of the named substances as aliphatic, cyclic, or aromatic.

Part 2: Modelling the Synthesis of Acetic Acid

(c) Draw a flow chart to show the synthesis of acetic acid, starting with ethene.

- Build a molecular model of ethene and make changes to the model to illustrate the reactions in the flow chart, until you finally make a model of acetic acid.

Table 2 Reactions to be Modelled

Reaction #	Reactant	Reaction type	Product
1	alkane	substitution	
2	alkene	addition	
3	alcohol		aldehyde
4			ether
5	aldehyde	controlled oxidation	
6			chlorobenzene
7		controlled oxidation	ketone
8		esterification	
9			amide
10	an aromatic compound		aromatic halide

Preparation of an Ester—Aspirin

The analgesic properties of willow bark led chemists to the isolation of its active ingredient, acetylsalicylic acid, commonly called Aspirin or ASA. ASA is synthesized from the reaction between salicylic acid and acetic anhydride, the structures of which are shown. In this activity, you will prepare Aspirin and calculate the percentage yield.

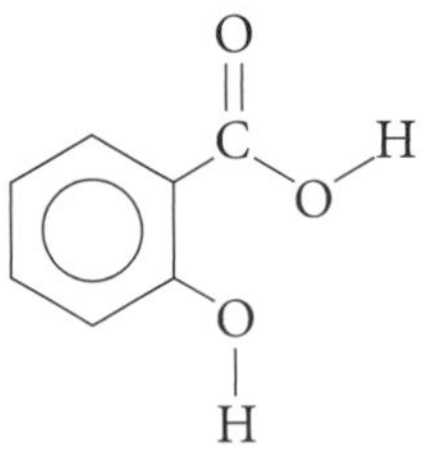

salicylic acid

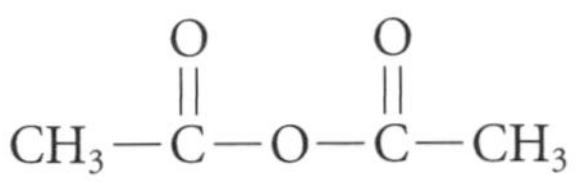

acetic anhydride

ACTIVITY 1.9.2 *continued*

Materials

lab apron	20 g crushed ice
eye protection	ice water
4 g salicylic acid	ice bath (large container
8 mL acetic anhydride	of crushed ice and water)
4 drops concentrated $H_2SO_{4(aq)}$	filter funnel
250-mL conical flask	ring stand with ring
600-mL beaker	clamp
eye dropper	filter paper
glass stirring rod	balance
hot plate	

Acetic anhydride is a severe eye irritant and must only be handled in a fume hood.

Concentrated sulfuric acid is corrosive. Wear eye protection and a laboratory apron.

Chemicals used or produced in a chemistry lab must *never* be consumed; they may contain toxic impurities.

- Prepare a hot-water bath by heating 300-mL of water in a 600 mL beaker on a hot plate until boiling.

- Obtain approximately 4 g of salicylic acid and determine its mass to the nearest 0.01 g. Transfer the sample to a 250-mL conical flask.

- In a fume hood, add to the conical flask 8.0 mL of acetic anhydride, and stir with a glass stirring rod until all the solid has dissolved.

- Remain in the fume hood and, using an eye dropper, carefully add 4 drops of concentrated $H_2SO_{4(aq)}$. Stir the mixture.

- Return to the lab bench and place the conical flask in the hot-water bath for 15 min, to allow the reaction to proceed.

- Remove the conical flask from the hot-water bath and add to it about 20 g of crushed ice and about 20 mL of ice water. Place the flask in the ice bath and stir for about 10 min (**Figure 4**).

Figure 4

- Determine the mass of a piece of filter paper to the nearest 0.01 g, and set up the filtration apparatus. Carefully filter the contents of the conical flask and wash the crystals with 10 mL of ice water.

- Allow the filter paper and crystals to dry completely. Determine the mass of the dry filter paper and crystals.

(a) Calculate the percentage yield of the ASA obtained. Comment on the laboratory procedure, and on your lab skills.

(b) Explain why ASA tablets have a sour taste.

(c) The acidic character of ASA may be used to quantify the amount of pure ASA in a sample. From your knowledge of acids and bases, suggest a procedure that may be performed in a school laboratory to test the purity of the ASA sample that you prepared. Include a description of the apparatus needed and how the results may be interpreted, as well as any safety precautions and emergency laboratory procedures.

Key Expectations

Throughout this chapter, you have had the opportunity to do the following:

- Distinguish among the different classes of organic compounds, including alcohols (1.5), aldehydes (1.6), ketones (1.6), carboxylic acids (1.7), esters (1.7), ethers (1.5), amines (1.8), and amides (1.8), by name and by structural formula.

- Describe some physical properties of the classes of organic compounds in terms of solubility in different solvents, molecular polarity, odour, and melting and boiling points. (all sections)

- Describe different types of organic reactions, such as substitution (1.3, 1.4, 1.8), addition (1.4), elimination (1.4, 1.5), oxidation (1.5, 1.6, 1.7), esterification (1.7), and hydrolysis (1.7, 1.8), and predict and correctly name their products. (1.3, 1.4, 1.5, 1.6, 1.7, 1.8)

- Use appropriate scientific vocabulary to communicate ideas related to organic chemistry. (all sections)

- Use the IUPAC system to name and write appropriate structures for the different classes of organic compounds. (all sections)

- Build molecular models of a variety of aliphatic, cyclic, and aromatic organic compounds. (1.2, 1.6, 1.9)

- Identify some nonsystematic names for organic compounds. (1.2, 1.3, 1.5, 1.6, 1.7)

- Carry out laboratory procedures to synthesize organic compounds. (1.7, 1.9)

- Present informed opinions on the validity of the use of the terms "organic," "natural," and "chemical" in the promotion of consumer goods. (1.1, 1.9)

- Describe the variety and importance of organic compounds in our lives. (all sections)

- Analyze the risks and benefits of the development and application of synthetic products. (1.4, 1.6)

- Provide examples of the use of organic chemistry to improve technical solutions to existing or newly identified health, safety, and environmental problems. (1.4, 1.5, 1.9)

Key Terms

addition reaction
alcohol
aldehyde
aliphatic hydrocarbon
alkane
alkene
alkyl group
alkyl halide
alkyne
amide
amine
aromatic alcohol
aromatic hydrocarbon
carbonyl group
carboxylic acid
carboxyl group
combustion reaction
condensation reaction
cyclic alcohol
cyclic hydrocarbon
dehydration reaction
elimination reaction
ester
esterification
ether
fractional distillation
functional group
hydration reaction
hydrocarbon
hydrolysis
hydroxyl group
isomer
IUPAC
ketone
Markovnikov's rule
organic family
organic halide
oxidation reaction
polyalcohol
primary alcohol
saponification
secondary alcohol
substitution reaction
tertiary alcohol

Key Symbols and Equations

Table 1 Families of Organic Compounds

Family name	General formula	Example	
alkanes	$$-\overset{\textstyle\mid}{\underset{\textstyle\mid}{C}}-\overset{\textstyle\mid}{\underset{\textstyle\mid}{C}}-$$	propane	$CH_3 - CH_2 - CH_3$
alkenes	$$-\overset{\textstyle\mid}{C}=\overset{\textstyle\mid}{C}-$$	propene (propylene)	$CH_2 = CH - CH_3$
alkynes	$-C \equiv C-$	propyne	$CH \equiv C - CH_3$
aromatics		methyl benzene (phenyl methane, toluene)	CH_3
organic halides	$R - X$	1-chloropropane	$CH_3 - CH_2 - CH_2 - Cl$
alcohols	$R - OH$	propanol	$CH_3 - CH_2 - CH_2 - OH$
ethers	$R - O - R'$	methoxyethane (ethyl methyl ether)	$CH_3 - O - CH_2 - CH_3$
aldehydes	$$R[H] - \overset{\textstyle O}{\overset{\textstyle \|}{C}} - H$$	propanal	$$CH_3 - CH_2 - \overset{\textstyle O}{\overset{\textstyle \|}{C}} - H$$
ketones	$$R - \overset{\textstyle O}{\overset{\textstyle \|}{C}} - R'$$	propanone (acetone)	$$CH_3 - \overset{\textstyle O}{\overset{\textstyle \|}{C}} - CH_3$$
carboxylic acids	$$R[H] - \overset{\textstyle O}{\overset{\textstyle \|}{C}} - OH$$	propanoic acid	$$CH_3 - CH_2 - \overset{\textstyle O}{\overset{\textstyle \|}{C}} - OH$$
esters	$$R[H] - \overset{\textstyle O}{\overset{\textstyle \|}{C}} - O - R'$$	methyl ethanoate (methyl acetate)	$$CH_3 - \overset{\textstyle O}{\overset{\textstyle \|}{C}} - O - CH_3$$
amines	$$R - \overset{\textstyle R'[H]}{\overset{\textstyle \mid}{N}} - R''[H]$$	1-aminopropane (*n*-propylamine)	$$CH_3 - CH_2 - CH_2 - \overset{\textstyle H}{\overset{\textstyle \mid}{N}} - H$$
amides	$$R[H] - \overset{\textstyle O}{\overset{\textstyle \|}{C}} - \overset{\textstyle R''[H]}{\overset{\textstyle \mid}{N}} - R'[H]$$	propanamide	$$CH_3 - CH_2 - \overset{\textstyle O}{\overset{\textstyle \|}{C}} - \overset{\textstyle H}{\overset{\textstyle \mid}{N}} - H$$

Problems You Can Solve

- Write the IUPAC name for an alkane or an aromatic hydrocarbon, given the structural formula, or draw the structural formula, given the name. (1.2)

- Write the IUPAC name for alkenes and alkynes, given the structural formulas, or draw the structural formulas, given the names. (1.2)

- Given the reactants or products of an addition reaction (of a hydrocarbon), predict the products or reactants. (1.3)

- Given the reactants or products of an addition reaction (of an aromatic hydrocarbon), predict the products or reactants. (1.3)

- Write the IUPAC name for an organic halide, given the structural formula, or draw the structural formula, given the name. (1.4)

- Write the IUPAC name for an alcohol, given the structural formula, or draw the structural formula, given the name; identify the alcohol as 1°, 2°, or 3°. (1.5)

- Write equations or draw structural diagrams to represent the reactions of alcohols. (1.5)

- Write either the IUPAC name or the structural formula for an ether, given the other. (1.5)

- Write equations or draw structural diagrams to represent the formation of ethers. (1.5)

- Write the IUPAC names for aldehydes and ketones, given the structural formula, or draw the structural formula given the name. (1.6)

- Write equations and draw structural formulas to represent reactions, given a reaction type and the name of an aldehyde or ketone as a reactant or a product. (1.6)

- Given one of the IUPAC name, the common name, or the structural formula of a carboxylic acid, provide the other two. (1.7)

- Given either the reactant or product of a specified reaction involving a carboxylic acid, write the reaction equation and structural formula. (1.7)

- Draw structural diagrams and write IUPAC names for the esters formed in a reaction, given either the reactants or the products. (1.7)

- Write the names of amines, given structural diagrams, or vice versa; specify whether the amines are 1°, 2°, or 3°. (1.8)

- Write the names of amides, given structural diagrams, or vice versa. (1.8)

- Draw structural diagrams and write equations to represent reactions involving amines and amides, given either the reactants or products and the type of reaction. (1.8)

- Select a series of reactions to get from starting materials to a desired end product. (1.9)

▶ *MAKE* a summary

On one or more large sheets of paper, construct a table like **Table 1**. Complete the table with sample reaction equations and the IUPAC name and structural formula of at least one specific example for each family.

Table 1 Summary of Organic Families and Reaction Types

Family	Substitution	Addition	Elimination	Oxidation	Condensation	Hydrolysis
alkanes						
alkenes						
alkynes						
organic halides						
1° alcohols						
2° alcohols						
3° alcohols						
ethers						
aldehydes						
ketones						
carboxylic acids						
esters						
amines						
amides						

Identify each of the following statements as true, false, or incomplete. If the statement is false or incomplete, rewrite it as a true statement.

1. An ester is formed when the hydroxyl group of an alcohol and the hydrogen atom of the carboxyl group of an acid are eliminated, and water is released.

2. Benzene is generally more reactive than alkanes and less reactive than alkenes because of the bonding in its aliphatic ring.

3. Aldehydes and ketones differ in the location of the carbonyl group and can be isomers of each other.

4. In an amide, the nitrogen atom is connected to at least one carbon atom, while in an amine, the nitrogen is connected to at least one hydrogen atom.

5. When 1-pentanol and 3-pentanol are each oxidized in a controlled way, they produce pentanal and propoxyethane, respectively.

Identify the letter that corresponds to the best answer to each of the following questions.

6. Which of the following compounds may be a structural isomer of 2-butanone?
   (a) $CH_3CH_2CH_2CH_2OH$
   (b) $CH_3CH_2CH_2CHO$
   (c) $CH_3CH_2OCH_2CH_3$
   (d) $CH_3CH_2CH_2COOH$
   (e) $CH_3CH{=}CHCH_3$

7. What is the major product of the reaction between 1-pentene and HCl?
   (a) 1-chloropentane   (d) 1-pentachlorine
   (b) 2-chloropentane   (e) 2-pentachlorine
   (c) 1,2-dichloropentane

8. The ester represented by the structural formula

$$CH_3CH_2 - \overset{\overset{\displaystyle O}{\|}}{C} - O - CH_2CH_3$$

   can by synthesized from
   (a) ethanol and propanoic acid
   (b) ethanol and ethanoic acid
   (c) methanol and propanoic acid
   (d) propanol and ethanoic acid
   (e) propanol and propanoic acid

9. What type of organic compound is represented by the following general formula?

$$R - \overset{\overset{\displaystyle O}{\|}}{C} - R$$

(a) an aldehyde   (d) an ester
(b) an amide   (e) a ketone
(c) a carboxylic acid

10. The reaction between a carboxylic acid and an amine is an example of this type of reaction:
    (a) addition   (d) hydrogenation
    (b) condensation   (e) hydrolysis
    (c) esterification

11. Acetaldehyde can be synthesized from the controlled oxidation of:
    (a) acetic acid   (d) methanol
    (b) acetone   (e) 2-propanol
    (c) ethanol

12. The compound whose structure is shown below

$$CH_3CH_2\overset{\overset{\displaystyle CH_3}{|}}{\underset{\underset{\displaystyle OH}{|}}{C}}CH_2CH_2CH_3$$

    (a) can be oxidized to an aldehyde
    (b) can be oxidized to a ketone
    (c) can be synthesized from the addition reaction of an alkane
    (d) is a secondary alcohol
    (e) is a tertiary alcohol

13. The compound whose structure is shown below

$$CH_3CH_2\overset{\overset{\displaystyle CH_3}{|}}{C}HCH_2\overset{\overset{\displaystyle O}{\|}}{C} - OH$$

    (a) can be oxidized to an aldehyde
    (b) can be synthesized by the oxidation of a ketone
    (c) can undergo, a condensation reaction with methyl amine
    (d) has a lower boiling point than 3-methylpentane
    (e) is less soluble than 3-methylpentane in water

14. Which of the following reaction types occurs when benzene reacts with bromine?
    (a) addition   (d) polymerization
    (b) elimination   (e) substitution
    (c) hydration

15. Which of the following statements is true of the compounds (i) $CH_3CH_2OCH_3$ and (ii) CH_3COOCH_3?
    (a) i and ii are both hydrocarbons.
    (b) i and ii each contains a carbonyl group.
    (c) i is less polar than ii.
    (d) i is less soluble than ii in nonpolar solvents.
    (e) Neither i nor ii contains any double bonds.

Understanding Concepts

1. Write balanced chemical equations, with structural diagrams, to show each of the following reactions. Explain in general terms any differences in the three reactions.
   (a) One mole of Cl_2 reacts with one mole of 2-hexene.
   (b) One mole of Cl_2 reacts with one mole of cyclohexene.
   (c) One mole of Cl_2 reacts with one mole of benzene. (1.4)

2. The following compounds have comparable molar mass. Arrange the compounds in order of increasing boiling points and give reasons for your answer.

   CH_3COOH CH_3CH_2CHO

        A   B

   $CH_3CH_2CH_2CH_3$ $CH_3CH_2CH_2OH$

        C   D   (1.6)

3. Write the structural formula for each of the following compounds:
   (a) A secondary alcohol with the formula $C_4H_{10}O$
   (b) A tertiary alcohol with the formula $C_4H_{10}O$
   (c) An ether with the formula $C_4H_{10}O$
   (d) A ketone with the formula C_4H_8O
   (e) An aromatic compound with the formula C_7H_8
   (f) An alkene with the formula C_6H_{10}
   (g) An aldehyde with the formula C_4H_8O
   (h) A carboxylic acid with the formula $C_2H_4O_2$
   (i) An ester with the formula $C_2H_4O_2$ (1.7)

4. Analysis of an unknown organic compound gives the empirical formula $C_5H_{12}O$. It is slightly soluble in water. When this compound is oxidized in a controlled way with $KMnO_4$, it is converted into a compound of empirical formula $C_5H_{10}O$, which has the properties of a ketone. Draw diagrams of the possible structure(s) of the unknown compound.   (1.7)

5. Name the family to which each of the following compounds belongs and write the IUPAC name for each:
   (a) $CH_2 = CHCH_2CH_3$
   (b) $CH_3 - O - CH_2CH_3$
   (c) $CH_3CH_2CHCH_2OH$
                  |
                 $CH3$
   (d) $CH_3C \equiv CH$
   (e) $CH_3CH_2CH_2COOCH_3$
   (f)   O
               ‖
       CH_3CH_2CH

   (g)   O
            ‖
       $CH_3CCH_2CH_3$
   (h) $CH_3CH_2CH_2CH_2NH_2$
   (i) $CH_2 = CHCHCH_2CH_3$
                   | |
               OHOH
   (j)   O
            ‖
      $CH_3CH_2CH_2CONHCH_2CH_3$ (1.8)

6. Many organic compounds have more than one functional group in a molecule. Copy the following structural diagrams. Circle and label the functional groups: hydroxyl, carboxyl, carbonyl, ester, amine, and amide. Suggest either a source or a use for each of these substances.

   (a) 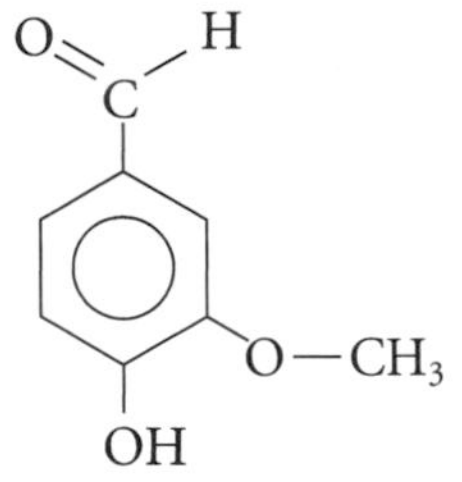
   (b) 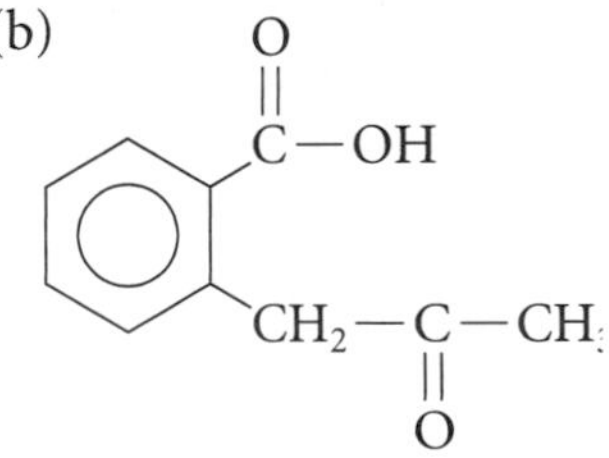
   (c) 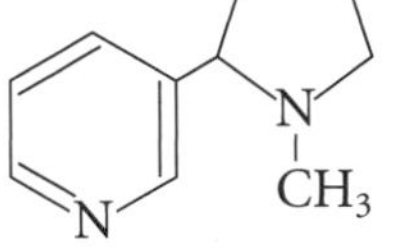
   (d) $H_2N - \overset{\overset{O}{\|}}{C} - NH_2$

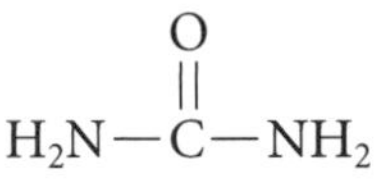

   (e) 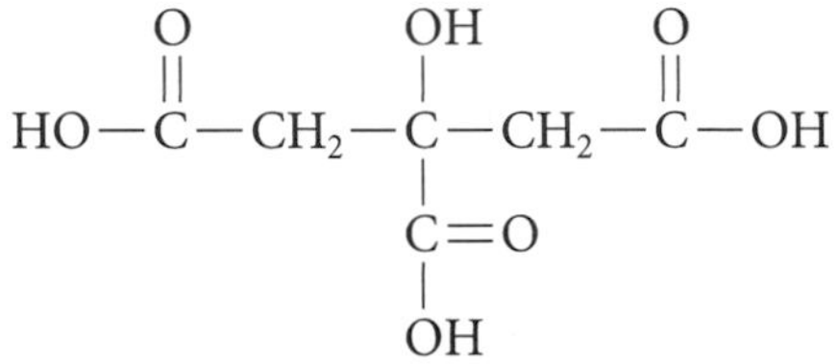
   $$HO - \overset{\overset{O}{\|}}{C} - CH_2 - \overset{\overset{OH}{|}}{\underset{\underset{\underset{OH}{|}}{\overset{|}{C=O}}}{C}} - CH_2 - \overset{\overset{O}{\|}}{C} - OH$$
                               (1.8)

7. Name and classify the organic compounds and write a complete structural diagram equation for each of the following reactions. Where possible, classify the reactions.
   (a) $C_2H_6 + Br_2 \rightarrow C_2H_5Br + HBr$
   (b) $C_3H_6 + Cl_2 \rightarrow C_3H_6Cl_2$
   (c) $C_6H_6 + I_2 \rightarrow C_6H_5I + HI$
   (d) $CH_3CH_2CH_2CH_2Cl + OH^- \rightarrow$
        $CH_3CH_2CHCH_2 + H_2O + Cl^-$
   (e) $C_3H_7COOH + CH_3OH \rightarrow$
        $C_3H_7COOCH_3 + HOH$
   (f) $C_2H_5OH \rightarrow C_2H_4 + H_2O$
   (g) $C_6H_5CH_3 + O_2 \rightarrow CO_2 + H_2O$
   (h) $C_2H_5OH \rightarrow CH_3CHO \rightarrow CH_3COOH$
   (i) $CH_3CHOHCH_3 \rightarrow CH_3COCH_3$
   (j) $NH_3 + C_4H_9COOH \rightarrow C_4H_9CONH_2 + H_2O$
   (k) $NH_3 + CH_4 \rightarrow CH_3NH_2 + H_2$ (1.8)

8. Write a series of equations to illustrate each of the following reactions:
   (a) a substitution reaction of propane
   (b) a halogenation reaction of benzene
   (c) the complete combustion of ethanol
   (d) an elimination reaction of 2-butanol
   (e) the controlled oxidation of butanal
   (f) the preparation of 2-pentanone from an alcohol
   (g) the preparation of hexyl ethanoate from an acid and an alcohol
   (h) the hydrolysis of methyl pentanoate
   (i) the controlled oxidation of 1-propanol
   (j) an addition reaction of an alkene to produce an alcohol
   (k) a condensation reaction of an amine   (1.8)

9. Write equations to show the synthesis pathway for ethyl 3-hydroxybutanoate, the flavouring of marshmallows. An alkene and an alcohol of your choice are the starting materials.   (1.9)

10. Write balanced chemical equations to represent the synthesis of:
    (a) propanone from a hydrocarbon
    (b) sodium ethanoate from an ester
    (c) propanal from an alcohol
    (d) propanoic acid from 1-propanol
    (e) 1,2-dichloroethane from ethene
    (f) *N,N*-dimethylethanamide from an alkane, an alkene, a halogen, and ammonia   (1.9)

Applying Inquiry Skills

11. In an experimental synthesis of 1,1-dichloroethane from ethene and chlorine, the following evidence was obtained:

    mass of ethene = 2.00 kg

    mass of 1,1-dichloroethane = 6.14 kg

    (a) Write an equation for the synthesis reaction.
    (b) Calculate the percent yield of 1,1-dichloroethane.
    (c) Suggest some reasons why the actual yield may be different from the theoretical yield.   (1.4)

12. Design a procedure to separate a mixture of alcohols containing methanol, ethanol, and 1-pentanol. Explain, with reference to intermolecular forces, why this separation method is effective in this situation.   (1.5)

13. Describe a procedure to synthesize the ester ethyl ethanoate, starting from ethene. Include in your answer details of the conditions and safety precautions required for the procedure.   (1.7)

14. When esters are prepared in the reaction of an alcohol with a carboxylic acid, the product formed can often be separated from the reactants by cooling the reaction mixture. Is the solid (formed when the reaction mixture is cooled) the alcohol, the acid, or the ester? Explain your answer with reference to the molecular structure of each reactant and product.   (1.7)

Making Connections

15. Use print or electronic resources to obtain the molecular structure of glucose, glycerol, and ethylene glycol. All three compounds have a sweet taste.
    (a) Predict the relative melting points and boiling points of rubbing alcohol, ethylene glycol, glycerol, and glucose. Give reasons for your answer.
    (b) Predict the solubility of each of these compounds in water, and in gasoline. Give reasons for your answer.
    (c) Ethylene glycol is toxic and is used as an antifreeze in automobile radiators. Suggest a reason why car antifreeze must be stored safely and spills must be cleaned up.
    (d) Do the structures of these three compounds support the hypothesis that taste receptors respond to functional groups in the compounds tasted? Explain.   (1.5)

 www.science.nelson.com

16. The distinction between "natural" and "synthetic" products is usually based on the source of the product, whether it is made by living organisms or by a laboratory procedure. Sometimes, the product is in fact the same, but the distinction is made in the way the product is processed. For example, when bananas are dissolved in a solvent and the flavouring extracted, the pentyl ethanoate obtained is labelled "natural flavour." When pentyl ethanoate is synthesized by esterification of ethanoic acid and pentanol, it is labelled "artificial flavour."
    (a) Write an equation for the synthesis of pentyl ethanoate.
    (b) In your opinion, what criteria should be used to distinguish a "natural" product from a "synthetic" product?
    (c) Research the differences in the source and processing methods of vanilla flavouring. Write a report on your findings.   (1.9)

 www.science.nelson.com

chapter

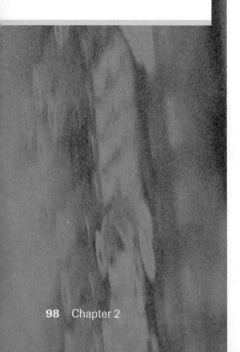

2

Polymers—Plastics, Nylons, and Food

▶ In this chapter, you will be able to

- demonstrate an understanding of the formation of polymers through addition reactions and condensation reactions;

- predict and correctly name the products of polymerization reactions and describe some of the properties of these products, e.g., plastics and nylons;

- make some polymers in the laboratory;

- describe a variety of large organic molecules including proteins, carbohydrates, nucleic acids and fats, and explain their importance to living organisms;

- supply examples of the use of polymers to provide technical solutions to health, safety, and environmental problems.

If you take a look around you, you will likely find that you are surrounded by plastic products of many shapes and sizes. They may include pens, buttons, buckles, and parts of your shoes, chair, and lamp. Your lunch may be wrapped in a plastic film, and you may have a drink from a plastic cup or bottle. In fact, plastic containers hold anything from margarine and shampoo to paint thinners and sulfuric acid. There are plastic components in your calculator, telephone, computer, sporting equipment and even the building in which you live. What are plastics and what makes them such desirable and versatile materials?

Plastics belong to a group of substances called polymers: large molecules made by linking together many smaller molecules, much like paper clips in a long chain. Different types of small molecules form links in different ways, by either addition or condensation reactions. The types of small units and linkages can be manipulated to produce materials with desired properties such as strength, flexibility, high or low density, transparency, and chemical stability. As consumer needs change, new polymers are designed and manufactured.

Plastics are synthetic polymers, but many natural polymers have similar properties recognized since early times. Amber from tree sap, and tortoise shell, for example, can be processed and fashioned into jewellery or ornaments. Rubber and cotton are plant polymers, and wool and silk are animal polymers that have been shaped and spun into useful forms. In fact, our own cells manufacture several types of polymers, molecules so large that they offer us the variety that makes us the unique individuals that we are.

Throughout this chapter, you will notice that we commonly use many different systems for naming complex organic compounds. In many cases, common names for compounds—frequently related to their origins—were used long before their structures (and hence their IUPAC names) were discovered. For example, many of us are more familiar with the common name vitamin C or ascorbic acid than the IUPAC name L-3-ketothreohexuronic acid lactone, and the name neoprene is much more widely used than poly-2-chlorobutadiene. Even for quite simple organic compounds, the common names (e.g., acetylene and acetone) are so familiar that they are more frequently used than the IUPAC names (ethyne and propanone). You will become familiar with both the common and the systematic names for many organic compounds.

💡 REFLECT on your learning ▼

1. Using a paper clip as an analogy, discuss what you think is a structural requirement for a molecule to be part of a long polymer chain.

2. Describe some properties that are characteristic of plastics.

3. Plastics are made mainly from petroleum products. From your knowledge of organic compounds, describe the types of bonding you would expect to find within plastic molecules, and between the long polymer molecules.

4. Discuss whether the characteristic properties of plastics can be explained by the intramolecular and intermolecular bonds that you described.

5. From what you have learned in previous courses, list and describe the functions of any large molecules synthesized by biological systems.

▶ *TRY THIS* activity   *It's a Plastic World*

You can probably name half a dozen plastic things in the next minute, but walking around and using your eyes will give you an even better idea of the vast number of plastic products we use every day.

- Take a walk through your home and make a list of at least 20 different plastic products, selecting from a wide variety of functions.
- For each product, gather the following information and make a summary in a table with suitable column headings.
  (a) Describe the function of the product.
  (b) Describe the properties of the plastic that make it suitable for its function, e.g., transparency, strength, electrical insulation, etc.
  (c) Note the recycling number code, if any, on the product (**Figure 1**).

(d) Briefly describe what materials could be used to make this product if plastics were not available, and discuss the advantages or disadvantages of using plastics in this product.

Figure 1

Figure 1
How many addition polymers have you used today? Teflon, Styrofoam, and polyethylene are all produced from addition reactions of monomers containing double bonds.

Polyesters are made up of many ester molecules linked end to end, and polyethylene (IUPAC polyethene) is made by connecting many ethene molecules. The small subunits are called **monomers**, and the long chains of many monomers (generally 10 or more) are called **polymers**, derived from the Greek word *polumeres*, meaning "having many parts." The chemical process by which monomers are joined to form polymers is called **polymerization**. The monomers in any polymer might be identical, or they may occur in a repeating pattern.

The properties of a polymer (**Figure 1**) are determined by the properties of its monomers, which may include functional groups such as alkyl groups, aromatic groups, alcohols, amines, and esters. The monomers of an organic polymer may be linked together by carbon–carbon bonds, carbon–oxygen bonds, or carbon–nitrogen bonds.

In this section, we will look at **addition polymers**, which result from the addition reactions of monomers containing unsaturated carbon–carbon bonds. In Section 2.2, we will look at condensation polymers, resulting from reactions between other functional groups.

Polyethylene: a Polymer of Ethene

Let us first consider the synthesis of a very simple addition polymer, polyethene (commonly called polyethylene); it is used for insulating electrical wires and for making plastic containers. You may recall that alkenes undergo addition reactions in which substances such as hydrogen halides, chlorine, bromine, or hydrogen add to the carbon–carbon double bond. Under certain conditions, alkenes can also undergo addition reactions with other alkenes; the double bond in each alkene monomer is transformed into a single bond, freeing up an unbonded electron pair that then forms a single carbon–carbon bond with another monomer. Let us look at the addition reaction of ethene molecules with other ethene molecules.

$$\underset{\text{ethene}}{C=C} + \underset{\text{ethene}}{C=C} + \underset{\text{ethene}}{C=C} + \longrightarrow \underset{\text{polyethene (polyethylene)}}{-C-C-C-C-C-C-} \text{ or } -\left[C-C\right]_n-$$

monomer a molecule of relatively low molar mass that is linked with other similar molecules to form a polymer

polymer a molecule of large molar mass that consists of many repeating subunits called monomers

polymerization the process of linking monomer units into a polymer

addition polymer a polymer formed when monomer units are linked through addition reactions; all atoms present in the monomer are retained in the polymer

The polymerization reaction may continue until thousands of ethene molecules have joined the chain. The structure of a polymer can be written in condensed form, the repeating unit being bracketed and a subscript "n" denoting the number of repeating units, which may be into the thousands.

Other Addition Polymers: Carpets, Raincoats, and Insulated Cups

There are probably hundreds of different industrial polymers, all with different properties, and formed from different reactants.

Polypropene

Propene also undergoes addition polymerization, producing polypropene, commonly called polypropylene (**Figure 2**). You may have used polypropylene rope, or walked on polypropylene carpet.

The chemical structures at the top:

$$
\begin{array}{ccccccccc}
& H & H & H & H & H & H \\
& | & | & | & | & | & | \\
& C{=}C & + & C{=}C & + & C{=}C & + \\
& | & | & | & | & | & | \\
& H & CH_3 & H & CH_3 & H & CH_3
\end{array}
\longrightarrow
\begin{array}{cccccc}
& H & H & H & H & H & H \\
& | & | & | & | & | & | \\
-& C & - & C & - & C & - & C & - & C & - & C & - \\
& | & | & | & | & | & | \\
& H & CH_3 & H & CH_3 & H & CH_3
\end{array}
\quad \text{or} \quad
-\left[\begin{array}{cc}
H & H \\
| & | \\
C & - C \\
| & | \\
H & CH_3
\end{array}\right]_n-
$$

propene   monomers   polypropene (polypropylene)

The polymerization reaction in the formation of polypropene is very similar to that of polyethene. The propene molecule can be considered as an ethene molecule with the substitution of a methyl group in place of a hydrogen atom. The polymer formed contains a long carbon chain with methyl groups attached to every other carbon atom in the chain.

Figure 2
When the double bonds in propene molecules undergo addition reactions with other propene molecules, the polymer formed is structurally strong, as evidenced by the load entrusted to polypropene ropes.

Polyvinyl Chloride

Ethene molecules with other substituted groups produce other polymers. For example, polyvinyl chloride, commonly known as PVC, is an addition polymer of chloroethene. A common name for chloroethene is vinyl chloride. PVC is used as insulation for electrical wires and as a coating on fabrics used for raincoats and upholstery materials.

$$
\begin{array}{cccccc}
| & | & | & | & | & | \\
C{=}C & + & C{=}C & + & C{=}C & + \\
| & | & | & | & | & | \\
Cl & & Cl & & Cl &
\end{array}
\longrightarrow
\begin{array}{cccccc}
| & | & | & | & | & | \\
-C & - & C & - & C & - & C & - & C & - & C & - \\
| & | & | & | & | & | \\
Cl & & Cl & & Cl &
\end{array}
$$

vinyl chloride monomers   polyvinyl chloride (PVC)

$$
\text{or} \quad -\left[\begin{array}{cc}
| & | \\
C & - C \\
| & | \\
Cl &
\end{array}\right]_n-
$$

Polystyrene

When a benzene ring is attached to an ethene molecule, the molecule is vinyl benzene, commonly called styrene. An addition polymer of styrene is polystyrene.

$$
\begin{array}{cccccc}
| & | & | & | & | & | \\
C{=}C & + & C{=}C & + & C{=}C & + \\
\end{array}
\longrightarrow
\begin{array}{cccccc}
| & | & | & | & | & | \\
-C & - & C & - & C & - & C & - & C & - & C & -
\end{array}
$$

styrene monomers   polystyrene

$$
\text{or} \quad -\left[\begin{array}{cc}
| & | \\
-C & - C \\
| &
\end{array}\right]_n-
$$

Addition Polymerization

Some unsaturated hydrocarbon groups have special names; for example, the ethene group is sometimes called a vinyl group. Many synthetic products commonly called "vinyl" or plastics are addition polymers of vinyl monomers with a variety of substituted groups.

$$
\begin{array}{cc}
H & H \\
| & | \\
C & = C \\
| & | \\
H & H
\end{array}
$$

vinyl

$$
\begin{array}{cc}
H & Cl \\
| & | \\
C & = C \\
| & | \\
H & Cl
\end{array}
$$

monomer of Saran wrap
(with vinyl chloride)

$$
\begin{array}{cc}
H & H \\
| & | \\
C & = C \\
| & | \\
H & CN
\end{array}
$$

monomer of acrylic

$$
\begin{array}{cc}
H & COOCH_3 \\
| & | \\
C & = C \\
| & | \\
H & CN
\end{array}
$$

monomer of instant glue

Draw a structural diagram of three repeating units of the addition polymer of 2-butene.

First, draw structural diagrams of three molecules of 2-butene, $CH_3CH{=}CHCH_3$. For clarity, draw the double-bonded C atoms and add other atoms above or below the double bond; it is not important whether the methyl groups and H atoms are written above or below the double bond because, after the addition reactions, only single bonds remain and there is free rotation about the single bonds.

$$
\begin{array}{cccccc}
CH_3 & H & \quad & CH_3 & H & \quad & CH_3 & H \\
| & | & & | & | & & | & | \\
C & = C & & C & = C & & C & = C \\
| & | & & | & | & & | & | \\
H & CH_3 & & H & CH_3 & & H & CH_3
\end{array}
$$

Next, the double bonds participate in addition reactions, linking together the double-bonded C atoms with single bonds.

$$
\begin{array}{c}
\quad CH_3 \quad H \quad\ CH_3 \quad H \quad\ CH_3 \quad H \\
\qquad | \qquad\ | \qquad | \qquad\ | \qquad | \qquad\ | \\
-\ C\ -\ C\ -\ C\ -\ C\ -\ C\ -\ C\ - \\
\qquad | \qquad\ | \qquad | \qquad\ | \qquad | \qquad\ | \\
\quad H \quad CH_3 \quad H \quad CH_3 \quad H \quad CH_3
\end{array}
$$

polymer of 2-butene

Example

Draw and name a structural diagram of the monomer of the polymer PVA, used in hair sprays and styling gels.

$$
\begin{array}{c}
OH \quad H \quad\ OH \quad H \quad\ OH \quad H \quad\ OH \\
| \qquad | \qquad | \qquad | \qquad | \qquad | \qquad | \\
-\ C\ -\ C\ -\ C\ -\ C\ -\ C\ -\ C\ -\ C\ - \\
| \qquad | \qquad | \qquad | \qquad | \qquad | \qquad | \\
H \quad\ H \quad\ H \quad\ H \quad\ H \quad\ H \quad\ H
\end{array}
$$

Solution

The monomer could be

$$
\begin{array}{cc}
| & | \\
C & = C \\
| & | \\
& OH
\end{array}
$$

vinyl alcohol (hydroxyethene)

In practice, poly(vinyl alcohol) is made from vinyl acetate to form poly(vinyl acetate), which is then hydrolyzed to poly(vinyl alcohol).

▶ Practice

Understanding Concepts

1. Draw a structural diagram of three repeating units of a polymer of 1-butene.

2. Draw a structural diagram of the monomer of the following polymer.

$$
\begin{array}{c}
\quad H \quad\ F \quad\ H \quad\ F \quad\ H \quad\ F \\
\quad | \qquad | \qquad | \qquad | \qquad | \qquad | \\
-\ C\ -\ C\ -\ C\ -\ C\ -\ C\ -\ C\ - \\
\quad | \qquad | \qquad | \qquad | \qquad | \qquad | \\
\quad H \quad CH_3 \quad H \quad CH_3 \quad H \quad CH_3
\end{array}
$$

3. Draw a structural diagram of three repeating units of a polymer of vinyl fluoride.

Models of Monomers

Materials: molecular model kit

- Build four or more models of ethene.
- Link the models together to form an addition polymer.
  (a) Name the polymer you have made.
- Repeat the first two steps, replacing ethene with propene.
  (b) Name this polymer.
- Design and construct a monomer that is capable of linking within a main polymer chain, and also linking across to a neighbouring polymer chain. Demonstrate your monomer's ability to form bonds between chains.
  (c) Name your monomer and its polymer.
- From what you have learned about condensation reactions, design and build a model of two monomers that may be used in a condensation polymerization reaction. Show how the monomers may be linked together to form a polymer.

The Addition Polymerization Process

There are three stages in the synthesis of addition polymers: initiation, propagation, and termination. In the first stage, an initiating molecule (such as a peroxide) with an unpaired electron forms a bond to one of the carbon atoms in the double bond of a molecule. The electrons shift in the newly bonded molecule, which now has an unpaired electron at the other end of its original double bond. This unpaired electron is now available to form another covalent bond with another atom or group. The addition reaction continues and the chain "propagates." The chain growth is terminated when any two unpaired electron ends combine, forming a covalent bond that links two growing chains together.

Properties of Plastics

Polymers of substituted ethene (or vinyl) monomers are generally categorized as **plastics**. Many of them are very familiar to us, so let us consider their characteristic properties in the context of their structure. Plastics are often used for containers for chemicals, solvents, and foods because they are chemically unreactive. We can explain this stability by the transformation of carbon–carbon double bonds into less reactive single bonds; in effect, the *unsaturated* alkene monomers have been transformed into less reactive *saturated* carbon skeletons of alkanes. The strong bonds within the molecules make the chemical structure very stable.

Plastics are also generally flexible and mouldable solids or viscous liquids. The forces of attraction between the long polymer molecules are largely van der Waals attractions (**Figure 3**), with some electrostatic attractions due to substituted groups. Because the molecules may have carbon chains many thousands of atoms long, the number of these

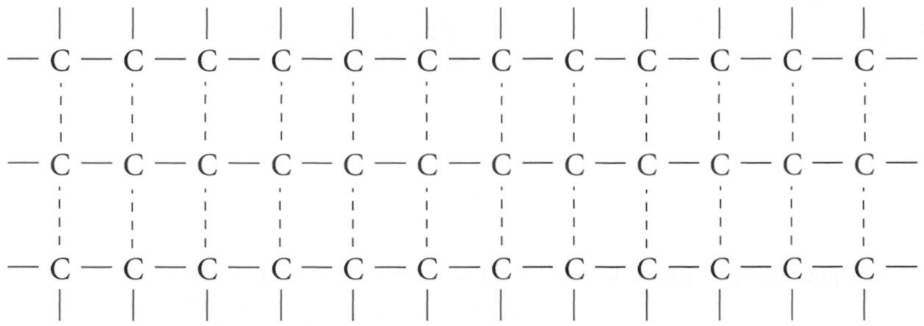

plastic a synthetic substance that can be moulded (often under heat and pressure) and that then retains its given shape

Figure 3
Polymers gain strength from numerous van der Waals attractions between chains; however, these weak forces allow chains to slide over each other, making plastics flexible and stretchable.

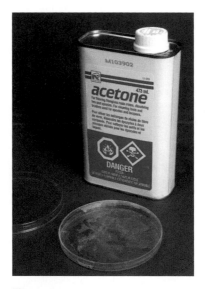

INVESTIGATION 2.1.1

Identification of Plastics (p. 138)
Have you ever wondered what those little triangular symbols on the bottom of plastic bottles and tubs are for? They are part of a code, which this investigation will help you unravel.

attractions is very large. These van der Waals attractions, although numerous, are individually weak forces that allow the polymer chains to slide along each other, rendering them flexible and stretchable. Heating increases the flexibility of plastics because molecular motion is increased, disrupting the structure imposed by the relatively weak van der Waals forces. Thus, plastics can be softened and moulded by heating.

The Effect of Substituted Groups on Polymer Properties

Let us look at a few other addition polymers and the effect of substituted groups on the properties of the polymer. Teflon is the common name for an addition polymer with nonstick properties that are much desired in cookware. The monomer used to synthesize Teflon is the simple molecule tetrafluoroethene, $F_2C=CF_2$, an ethene molecule in which all hydrogen atoms are replaced with fluorine atoms. The absence of carbon–hydrogen bonds and the presence of the very strong carbon–fluorine bonds make Teflon highly unreactive with almost all reagents. It is this unreactivity that allows it to be in contact with foods at high temperatures without "sticking."

tetrafluoroethene   polytetrafluoroethene (Teflon)

Plexiglas is the common name for the plastic used as a lightweight replacement for glass (**Figure 4**). This plastic is produced from the addition polymerization of an alkene monomer containing a carboxymethyl group, $-COOCH_3$. This functional group is responsible for its optical properties, including its transparency. Although Plexiglas may be as clear and smooth as glass and can substitute for glass in many functions, it is easily damaged by organic solvents such as the acetone in nail polish remover. This is due to the carbonyl group in its monomer, which makes the polymer readily soluble in other carbonyl compounds (**Figure 5**).

(a) plexiglass monomer   **(b)** acetone (propanone)   **(c)** ethyl acetate (ethyl ethanoate)

Figure 4
Plexiglas has many properties of glass, but, unlike glass, it is easily marred by organic solvents.

Figure 5
The Plexiglas monomer contains a carbonyl group, accounting for its high solubility in organic solvents with carbonyl groups, such as acetone and ethyl acetate.

Strengthening Polymers with Crosslinking

The ethene and propene monomers we have been discussing contain one double bond. Let us now take a look at monomers with two double bonds. What possibilities for polymerization are available when a monomer can form linkages in two directions?

Alkenes with two double bonds are called dienes, and 1,3-butadienes are the monomers in several common polymers whose names often end in "-ene." Neoprene, well known for its use in wet suits, is an addition polymer whose properties are similar to those of natural rubber. Natural rubber is an addition polymer of 2-methylbutadiene (**Figure 6**).

The monomer used to synthesize neoprene is 2-chlorobutadiene; the presence of the highly electronegative chlorine atom renders neoprene more polar and therefore less miscible with hydrocarbons. This gives neoprene the advantage of being more resistant to substances such as oils and gasoline than natural rubber is.

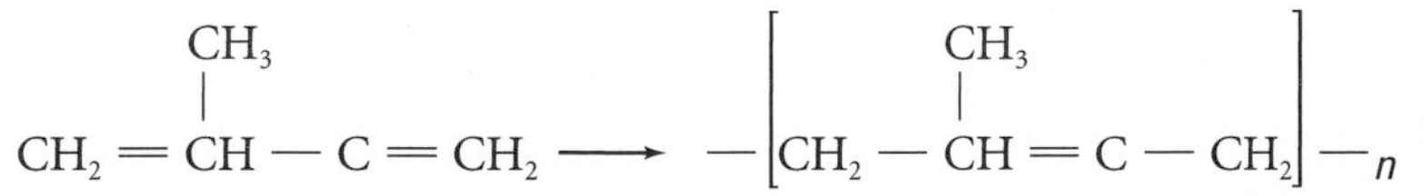

$$CH_2 = CH - \underset{\underset{CH_3}{|}}{C} = CH_2 \longrightarrow - \left[CH_2 - CH = \underset{\underset{CH_3}{|}}{C} - CH_2 \right] -_n$$

2-methyl-1,3-butadiene (isoprene),   rubber (natural and synthetic)
rubber monomer

$$CH_2 = \underset{\underset{Cl}{|}}{C} - CH = CH_2 \longrightarrow - \left[CH_2 - \underset{\underset{Cl}{|}}{C} = CH - CH_2 \right] -_n$$

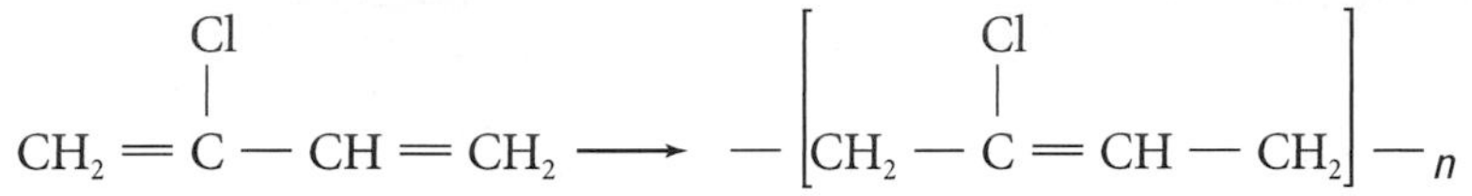

2-chloro-1,3-butadiene (chloroprene)   neoprene
neoprene monomer

Figure 6
Neoprene is more resistant to dissolving in oils than is rubber, due to increased polarity conferred by its Cl atoms.

Diene monomers are able to add to other molecules in two locations, which is analogous to having two hands to grab onto nearby objects. These dienes can be incorporated into two separate polymer chains at the same time, joining them together by forming bridges between the chains. These bridges, called "crosslinks," may be formed intermittently along the long polymer chains. Of course, the more crosslinks that form, the stronger the attraction holding the chains to each other. These links between polymer molecules are covalent bonds—much stronger than the van der Waals attractions that are otherwise the main forces holding the chains together. Consequently, crosslinked polymers form much stronger materials. You can demonstrate this with chains of paperclips (**Figure 7**).

The degree of crosslinking depends on how many of the monomers are dienes. The diene is not necessarily the main ingredient of the polymer. Selected crosslinking monomers can be added to other monomer mixtures. This is an effective way of controlling the degree of rigidity in the polymer synthesized. A more rigid product requires a higher concentration of the crosslinking monomer. For example, polyethylene is produced with varying degrees of crosslinking (which is generally directly related to the density), to be used for garbage bags, fencing, netting wrap for trees....

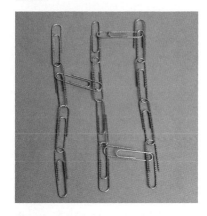

Figure 7
Strong crosslinks make for a more rigid structure.

An example of a diene used for crosslinking is *p*-divinylbenzene (1,4-diethenylbenzene). Each of the two vinyl groups contains a double bond, and each double bond can be incorporated in a separate polymer chain. In polystyrene polymer shown below, the main monomer is styrene; the styrene monomers are crosslinked by *p*-divinylbenzene. The crosslinking monomer holds the chains together with strong covalent bonds. The relative concentrations of the two types of monomers determine the degree of crosslinking and hence the degree of rigidity of the polystyrene produced.

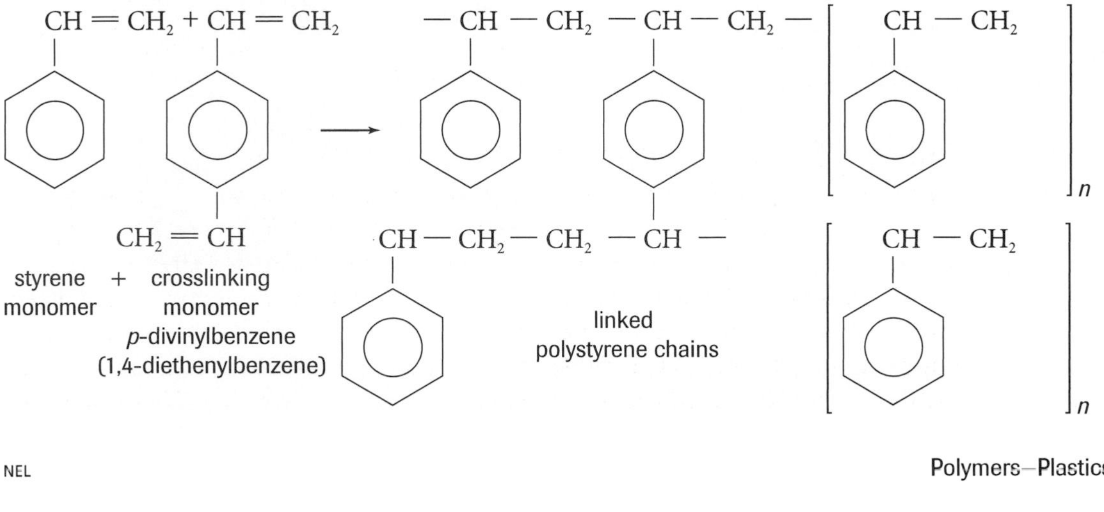

styrene + crosslinking
monomer   monomer
p-divinylbenzene
(1,4-diethenylbenzene)

linked
polystyrene chains

Figure 8
A ride on bumpy roads was made more comfortable thanks to sulfur–sulfur bonds forming crosslinks between the polymer molecules in natural rubber.

Inorganic crosslinking agents can also be used. Sulfur, for example, can form two covalent bonds and is used to harden latex rubber from the rubber tree. This reaction of natural rubber was an accidental discovery that made Charles Goodyear a very successful tire manufacturer. Natural rubber is made from resin produced by the rubber tree, *Hevea brasiliensis*. As we discussed earlier, the monomer 2-methylbutadiene forms polymer chains that remain partially unsaturated, resulting in a natural rubber that is soft and chemically reactive. Goodyear found that when sulfur is added to rubber and the mixture heated, the new product was harder and less reactive. This process was named "vulcanization," after Vulcan, the Roman god of fire. The sulfur atoms form crosslinks by adding to the double bonds of two different polymer molecules. When the rubber is stretched, the crosslinks hold the chains together and return them to their original position, imparting an elasticity to the rubber tire, for a comfortable, shock-free car ride (**Figure 8**).

$$CH = CH_2 + \text{sulfur} \longrightarrow$$
$$|$$
$$CH_3$$

$$— CH(CH_3) — CH — CH(CH_3) — CH — CH(CH_3) —$$
$$\qquad\qquad\qquad | \qquad\qquad\qquad |$$
$$\qquad\qquad\qquad S \qquad\qquad\qquad S$$
$$\qquad\qquad\qquad | \qquad\qquad\qquad |$$
$$— CH(CH_3) — CH — CH(CH_3) — CH — CH(CH_3) —$$

rubber   + crosslinking   linked rubber polymer chains
monomer   agent

ACTIVITY 2.1.1

Making Guar Gum Slime (p. 140)
The huge guar gum molecule, when crosslinked with borax, becomes a gooey substance with weird properties.

For manufacturers of plastic products, the heat resistance of plastics determines how they can be processed, as well as their potential uses. Some plastics soften when heated, then take on a new shape upon cooling—they are readily moulded. These polymers belong to the general group called *thermoplastics*. Such polymers are not crosslinked and so are held together only by the weak van der Waals forces. Plastics that are highly crosslinked are not softened by heat because their polymer chains are linked together by strong covalent bonds; these are called *thermoset* polymers.

▶ TRY THIS activity

Skewering Balloons

The polymers in synthetic rubber are strong, stretchy, and flexible. Just how much can you push them around? Try this!

Materials: balloon; thin knitting needle or long bamboo skewer

- Inflate a large round balloon. Release a little of the air until the balloon is not taut, and tie a knot at the neck.
- Place the tip of the needle or skewer at the thick part of the balloon, directly opposite the knot. Slowly rotate the needle or skewer and push it into the balloon. Continue pushing the needle through the balloon and out through the thick area around the knot.
  (a) Explain your observations. **Figure 9** may help!

Figure 9
How do chopsticks pass through a pile of noodles?

▶ Practice

Understanding Concepts

4. What functional group(s), if any, must be present in a monomer that undergoes an addition polymerization reaction?

5. Addition polymers may be produced from two different monomers, called co-monomers. Saran, the polymer used in a brand of food wrap, is made from the monomer vinyl chloride and 1,1-dichloroethene. Draw structural diagrams for each monomer, and for three repeating units of the polymer, with alternating co-monomers.

▶ Section 2.1 Questions

Understanding Concepts

1. (a) Describe the intramolecular and intermolecular forces of attraction between long addition polymer chains.
   (b) Explain why these polymers are more useful, as materials, than their monomers.
   (c) Explain why these polymers are chemically more stable than their monomers.

2. Chlorotrifluoroethene is a monomer that forms an addition polymer.
   (a) Draw a structural diagram for three repeating units of this polymer.
   (b) Predict the properties of this polymer in terms of solubility in organic solvents, rigidity, and resistance to heating.

3. What monomer could be used to produce each of these polymers?

(a)
$$-CH-CH-CH-CH-CH-CH-$$
with CH_3 substituent on each CH

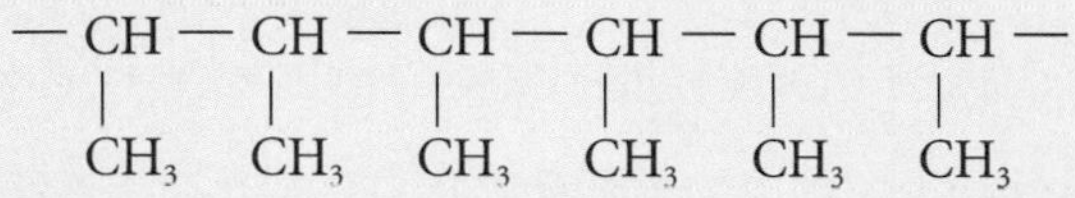

(b)
$$-C-C-C-C-C-C-$$
with F above each C and CH_3, Cl, CH_3, Cl, CH_3, Cl below

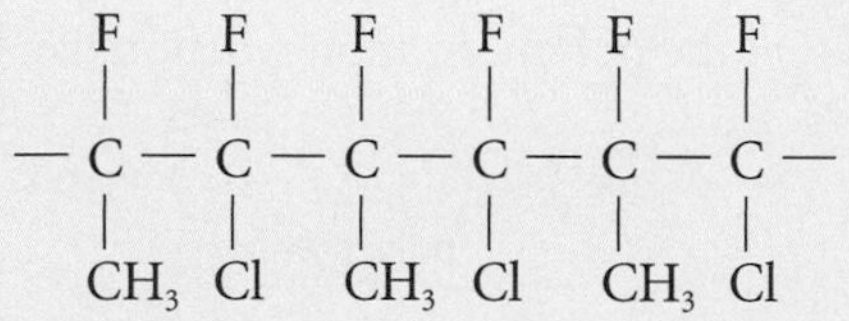

4. Which is more soluble in an organic solvent such as acetone, a polymer whose monomer contains a methyl group or one that contains a carbonyl group? Explain.

5. Describe the structural features necessary in a monomer that is added for crosslinking polymer chains. Illustrate your answer with a structural diagram of an example.

6. What characteristics must a molecule have, to be part of
   (a) a condensation polymer?
   (b) an addittition polymer?

7. (a) What are the typical properties of a plastic?
   (b) What types of bonding would you expect to find within and between the long polymer molecules?
   (c) Explain the properties of plastics by referring to their bonding.

Applying Inquiry Skills

8. Describe a simple experimental procedure to determine whether a sample of an unknown plastic contains crosslinked polymers. Explain how the observations are interpreted.

Making Connections

9. Find out from your community recycling facility what types of plastic products are accepted for recycling in your area. If there are some plastics that are not accepted, find out the reason. Summarize your findings in a well-organized table.

10. When household waste is deposited in a landfill site, some of it decomposes and the products of biodegradation may seep into the ground, contaminating water supplies. Working with a partner,
    (a) brainstorm and make a list of properties that would be ideal for a plastic liner for the landfill site, to contain the potentially toxic seepage;
    (b) describe the general structural features of a polymer that would provide the properties you listed.

11. Natural rubber is made from resin produced by the rubber tree, *Hevea brasiliensis*. Research the commercial production and use of natural rubber, and the circumstances that stimulated the development of synthetic rubber. Write a brief report on your findings.

 GO www.science.nelson.com

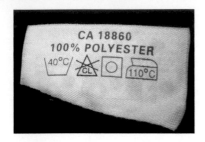

Figure 1
Polyester is a familiar term to anyone buying clothing these days. It can be made into a shiny, slippery fabric that is very easy to care for.

dimer a molecule made up of two monomers

condensation polymer a polymer formed when monomer units are linked through condensation reactions; a small molecule is formed as a byproduct

polyester a polymer formed by condensation reactions resulting in ester linkages between monomers

As we saw in the last section, monomers that contain carbon–carbon double bonds link together by addition reactions. What about molecules that do not contain double bonds? Can they link together to form long polymer chains?

If a molecule has one functional group, it can react and link with one other molecule to form a **dimer**. To form a chain, a molecule must link to one other molecule at each end; that is, it needs to have two reactive functional groups, one at each end of the molecule. For example, carboxylic acids react with alcohols to form esters, and with amines to form amides; these are both condensation reactions. When monomers join end to end in ester or amide linkages, we get polyesters and polyamides. Many of these synthetic **condensation polymers** are widely used for fabrics and fibres (**Figure 1**).

Polyesters from Carboxylic Acids and Alcohols

When a carboxylic acid reacts with an alcohol in an esterification reaction, a water molecule is eliminated and a single ester molecule is formed, as you learned in Chapter 1. The two reactant molecules are linked together into a single ester molecule. This condensation reaction can be repeated to form not just one ester molecule, but many esters joined in a long chain, a **polyester**. This is accomplished using a dicarboxylic acid (an acid with a carboxyl group at each end of the molecule), and a diol (an alcohol with a hydroxyl group at each end of the molecule). Ester linkages can then be formed end to end between alternating acid molecules and alcohol molecules.

$$\boxed{HO}OCCH_2CH_2CO\boxed{OH} + \boxed{H}OCH_2CH_2O\boxed{H} + \boxed{HO}OCCH_2CH_2CO\boxed{OH} + \longrightarrow$$
$$\text{di-ethanoic acid} \qquad \text{1,2-ethanediol}$$

$$— OCCH_2CH_2COOCH_2CH_2OOCCH_2CH_2CO — + \text{water}$$

If we were to depict the acid with the symbol $\triangle\text{--}\triangle$, the alcohol with $\circ\text{--}\circ$, and the ester linkage with $\boxed{\circ\triangle}$, we could represent the polymerization reaction like this:

$$\triangle\text{--}\triangle + \circ\text{--}\circ + \triangle\text{--}\triangle + \circ\text{--}\circ \longrightarrow \triangle\text{--}\boxed{\triangle\circ}\text{--}\boxed{\circ\triangle}\text{--}\boxed{\triangle\circ}\text{--}\circ\text{-} + \text{water}$$
$$\text{polyester}$$

The two functional groups on each monomer do not have to be identical; for example, the monomer may contain a carboxyl group at one end and a hydroxyl group at the other end. Ester linkages can be formed between the carboxyl group and the hydroxyl group of adjacent identical monomers.

Again, if we replace the formula for 3-hydroxypropanoic acid with the symbol $\triangle\text{-}\circ$, we can represent the reaction as shown below.

$$\triangle\text{--}\circ + \triangle\text{--}\circ + \triangle\text{--}\circ + \triangle\text{--}\circ \longrightarrow \triangle\text{--}\boxed{\circ\triangle}\text{--}\boxed{\circ\triangle}\text{--}\boxed{\circ\triangle}\text{--}\circ + \text{water}$$
$$\text{polyester}$$

LEARNING TIP

A numbering system is often used to name polyamides and polyesters. Nylon 6,10, for example, denotes the polyamide made from a 6-carbon diamine and a 10-carbon dicarboxylic acid. Polyester 3,16 represents the polyester made from a 3-carbon diol and a 16-carbon dicarboxylic acid.

We encounter many polyesters, one of the most familiar being Dacron, used in clothing fabrics. The monomers used in the production of Dacron are the dicarboxylic acid *p*-phthalic acid (1,4-benzene dicarboxylic acid), and the diol ethylene glycol (1,2-ethanediol). The resulting polymer molecules are very long and contain the polar carbonyl groups at regular intervals. The attractive forces between these polar groups hold the polymer chains together, giving these polyesters considerable strength.

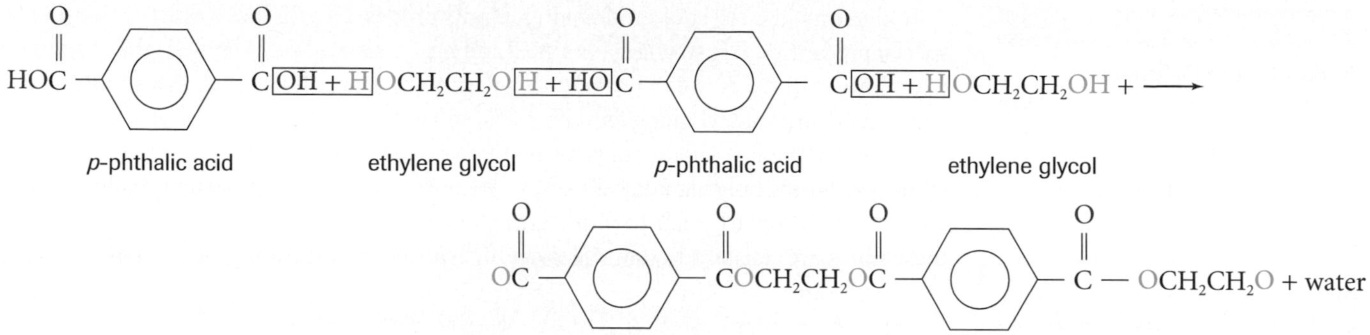

p-phthalic acid   ethylene glycol   p-phthalic acid   ethylene glycol

polyester (Dacron)

Polyamides from Carboxylic Acids and Amines

There are other types of condensation polymers, some of which have familiar names. **Polyamides**, for example, are polymers consisting of many amides. Analogous to esters, amides are formed from the condensation reaction between a carboxylic acid and an amine, with the removal of a water molecule. Thus, the monomers used must each contain two functional groups: carboxyl groups, amine groups, or one of each.

Let us look at the production of nylon 6,6 as an example of a synthetic polyamide. It is formed from adipic acid (a dicarboxylic acid—hexanedioic acid), and 1,6-diaminohexane (a diamine). The name nylon 6,6 refers to the number of carbon atoms in the monomers: The first "6" refers to 1,6-diaminohexane, which contains 6 carbon atoms, and the second "6" refers to adipic acid, which also contains 6 carbon atoms.

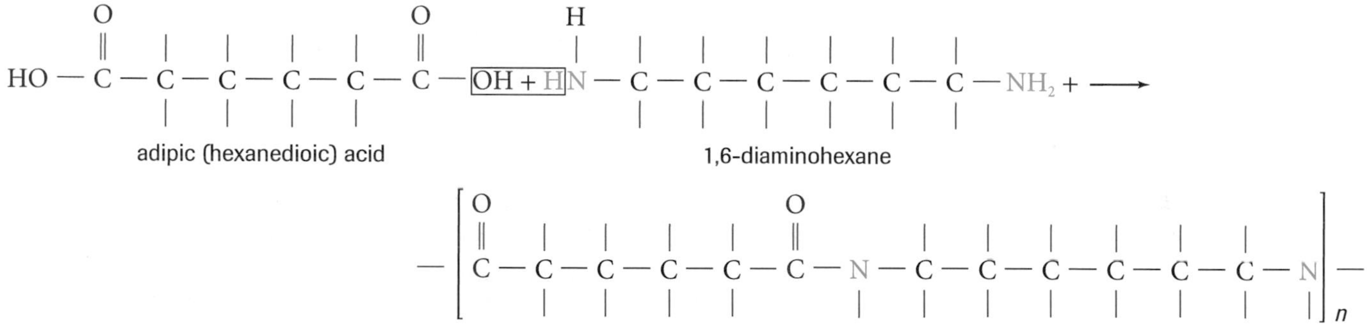

adipic (hexanedioic) acid   1,6-diaminohexane

nylon 6,6

polyamide a polymer formed by condensation reactions resulting in amide linkages between monomers

Nylon was synthesized as a substitute for silk, a natural polyamide whose structure nylon mimics. Nylon production was speeded up by the onset of the Second World War when it was used to make parachutes, ropes, cords for aircraft tires, and even shoelaces for army boots. It is the amide groups that make nylon such a strong fibre. When spun, the long polymer chains line up parallel to each other and amide groups form hydrogen bonds with carbonyl groups on adjacent chains.

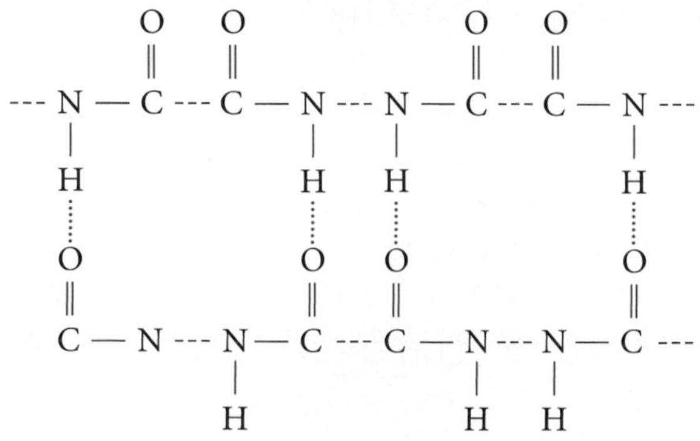

DID YOU KNOW ?

Nylon
Nylon was designed in 1935 by Wallace Carothers, a chemist who worked for Du Pont; the name *nylon* is a contraction of New York and London.

Polymers—Plastics, Nylons, and Food **109**

INVESTIGATION 2.2.1

**Preparation of Polyester—
A Condensation Polymer (p. 141)**
Use the theoretical knowledge you
have gained to predict the structure
of a thermoset polymer, and then
make it in the lab.

DID YOU *KNOW* ?

Pulling Fibres
When a polymer is to be made into
a fibre, the polymer is first heated
and melted. The molten polymer is
then placed in a pressurized container and forced through a small
hole, producing a long strand, which
is then stretched. The process,
called extrusion, causes the polymer
chains to orient themselves length-
wise along the direction of the
stretch. Interchain bonds (either
hydrogen bonds or crosslinks) are
formed between the chains, giving
the fibres added strength.

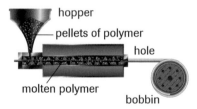

To illustrate the effect of crosslinking in polyamides, let's discuss a polymer with very special properties. It is stronger than steel and heat resistant, yet is lightweight enough to wear. This material is called Kevlar and is used to make products such as aircraft, sports equipment, protective clothing for firefighters, and bulletproof vests for police officers. What gives Kevlar these special properties? The polymer chains form a strong network of hydrogen bonds holding adjacent chains together in a sheet-like structure. The sheets are in turn stacked together to form extraordinarily strong fibres. When woven together, these fibres are resistant to damage, even that caused by a speeding bullet.

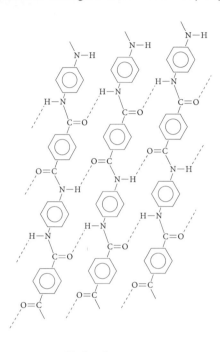

Kevlar sheet

▶ **SAMPLE** problem | **Drawing & Naming Condensation Reactants & Products**

Draw a structural diagram and write the names of the monomers used to synthesize nylon 6,8.

The first number in the name nylon 6,8 indicates that the diamine monomer has 6 carbon
atoms. Draw its structure by adding an amino group to each end of a hexane molecule.

$$H_2N - CH_2CH_2CH_2CH_2CH_2CH_2 - NH_2$$

1, 6-diaminohexane

The second number in the name nylon 6,8 indicates that the dicarboxylic acid has 8
carbon atoms. Draw its structure by adding a carboxyl group to each end of a hexane
molecule, for a total of 8 carbons in the molecule.

$$HOOC - CH_2CH_2CH_2CH_2CH_2CH_2 - COOH$$

Example
Draw two repeating units of the polyamide formed from the monomer 4-amino-butanoic acid.

Solution

$$-\underset{H}{\overset{}{N}} - CH_2CH_2CH_2\overset{O}{\overset{\|}{C}} - \underset{H}{\overset{}{N}} - CH_2CH_2CH_2\overset{O}{\overset{\|}{C}} -$$

▶

> ▶ *Practice*

Understanding Concepts

1. Draw a structural diagram to show a repeating unit of a condensation polymer formed from the following compounds.

$$\underset{\text{HOCCH}_2\text{CH}_2\text{COH}}{\overset{\text{O} \quad\quad\; \text{O}}{\overset{||\quad\quad\;\; ||}{}}} \quad \text{and} \quad \text{HOCH}_2\text{CH}_2\text{CH}_2\text{CH}_2\text{OH}$$

2. What functional group(s), if any, must be present in a monomer of a condensation polymer?

3. Describe the type of chemical bonding within a polyamide chain, and between adjacent polyamide chains.

Keeping Baby Dry with Polymers

From the time your grandparents were babies to the time you were born, diapers have been made entirely of polymers. Cotton cloth diapers were, and still are, made of cellulose, a natural polymer. Nowadays, disposable diapers made mainly of synthetic polymers are a popular choice. Which is better for the baby? How do they affect our environment?

The typical disposable diaper has many components, mostly synthetic plastics, that are designed with properties particularly desirable for its function.

- Polyethylene film: The outer surface is impermeable to liquids, to prevent leakage. It is treated with heat and pressure to appear cloth-like for consumer appeal.

- Hot melts: Different types of glue are used to hold components together. Some glues are designed to bond elastic materials.

- Polypropylene sheet: The inner surface at the leg cuffs is designed to be impermeable to liquids and soft to the touch. The main inner surface is designed to be porous, to allow liquids to flow through and be absorbed by the bulk of the diaper.

- Polyurethane, rubber, and Lycra: Any or all of these stretchy substances may be used in the leg cuffs and the waistband.

- Cellulose: Basically processed wood pulp, this natural polymer is obtained from pine trees. It forms the fluffy filling of a diaper, absorbing liquids into the capillaries between the fibres.

- Polymethylacrylate: This crystalline polymer of sodium methylacrylate absorbs water through osmosis and hydrogen bonding. The presence of sodium ions in the polymer chains draws water that is held between the chains, forming crosslinks that result in the formation of a gel (**Figure 2**).

Manufacturers claim that grains of sodium polymethylacrylate can absorb up to 400 times their own mass in water. If sodium ions are present in the liquid they act as contaminants, reducing absorbency because the attraction of water to the polymer chains is diminished. Urine always contains sodium ions, so the absorbency of diapers for urine is actually less than the advertised maximum.

Proponents of natural products argue that disposable diapers pose a long- term threat to our environment, filling waste disposal sites with non-biodegradable plastics for future centuries. The industry has developed some new "biodegradable" materials—a

DID YOU KNOW ?

Paintball: A Canadian Invention
The sport of paintball was invented in Windsor, Ontario. Paintballs were first used to mark cattle for slaughter and trees for harvesting, using oil-based paints in a gelatin shell. When recreational paintball use demanded a water-based paint, the water-soluble gelatin shell was modified by adjusting the ratio of the synthetic and natural polymers used.

combination of cellulose and synthetic polymers, for example. Some of these materials have proven too unstable to be practical, while others appear to biodegrade in several years, which gives the diapers a reasonable shelf life. The energy and raw materials needed to manufacture the huge quantity of disposable diapers used is also of concern.

Diaper manufacturers and their supporters argue, however, that using the natural alternative—cloth diapers—consumes comparable amounts of energy in the laundry. In addition, the detergents used in the cleaning process are themselves non-biodegradable synthetic compounds made from petroleum products.

There is rarely a single, simple solution to such problems. The "best" solution depends on many factors, including changing views and priorities of society and individuals.

Figure 2
Polymethylacrylate

dry

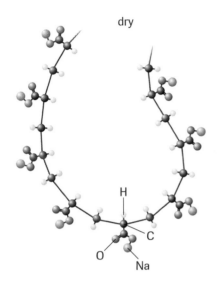

in water

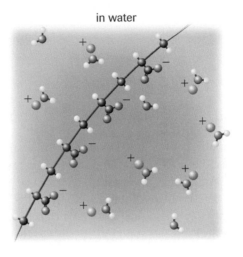

 TRY THIS activity   **_Diaper Dissection_**

What is it in disposable diapers that keeps a baby dry? Perform a dissection and find out (**Figure 3**).

Figure 3

Materials: eye protection; thin ("ultra") disposable diaper; sharp scissors; plastic grocery bag; 3 paper cups or beakers, water; table salt; sugar; calcium chloride

- Cut a 2 cm × 2 cm sample of each of the following components: outer back sheet, inside top sheet (surface that touches the baby's skin), adhesive tape, elastic material around leg cuffs, fluffy filling. Examine the properties of each of these samples. (They are all polymers, synthetic or natural.)

- Cut down the centre of the inside surface. Remove the fluffy filling and put it in a plastic grocery bag. Pull the filling apart into smaller pieces and tie a knot to close the bag. Shake the bag to dislodge the absorbent crystals within.

- Open and tilt the bag to collect the dislodged crystals in a corner of the bag. Cut a small hole in that corner, and empty the crystals into a paper cup or beaker.

🤚 **Do not ingest the crystals or touch your face or eyes after coming in contact with them; the crystals cause irritation and dehydration.**

- Add about 100 mL of water to the crystals and observe. After a few minutes, divide the material formed into 3 equal portions in 3 cups. To one portion add about 2 mL of table salt; to the second, add 2 mL sugar; and to the third add 2 mL calcium chloride.
(a) Note and explain any changes.

> ▶ **Practice**

Applying Inquiry Skills

4. Design an investigation to answer the following question. By how much does the absorption rate of sodium polymethylacrylate decrease if there are sodium or potassium ions in the liquid absorbed?

Making Connections

5. From what you have learned about organic reactions, suggest the type of reaction and conditions needed to break down polyamides into their monomers. Write an equation to illustrate your answer.

6. The first nylon product that was introduced to the public, in 1937, was a nylon toothbrush called Dr. West's Miracle-Tuft toothbrush. Earlier toothbrush bristles were made of hair from animals such as boar. From your knowledge of the properties of nylon, suggest some advantages and drawbacks of nylon toothbrushes compared with their natural counterparts.

> ▶ **Section 2.2** *Questions*

Understanding Concepts

1. Explain the difference in the structure of a polyester and a polyamide, and give an example of each.

2. Draw the structure and write the name of the monomers that make up the polyamide nylon 5,10.

3. Oxalic acid is a dicarboxylic acid found in rhubarb and spinach. Its structure is shown below.

$$\underset{\text{HOC}-\text{COH}}{\overset{\text{O} \quad \text{O}}{\overset{\|}{}\overset{\|}{}}}$$

   Draw three repeating units of the condensation polymer made from oxalic acid and ethanediol.

4. Nylon 6, used for making strong ropes, is a condensation polymer of only one type of monomer, 6-aminohexanoic acid. Draw a structural diagram of the monomer and a repeating unit of the polymer.

5. Draw structural diagrams and write the names of the monomers used in the synthesis of the following polyamide.

$$-\left[\begin{array}{c}\overset{\text{O}}{\overset{\|}{\text{C}}}-\overset{|}{\underset{|}{\text{C}}}-\overset{|}{\underset{|}{\text{C}}}-\overset{|}{\underset{|}{\text{C}}}-\overset{|}{\underset{|}{\text{C}}}-\overset{\text{O}}{\overset{\|}{\text{C}}}-\overset{|}{\underset{|}{\text{N}}}-\overset{|}{\underset{|}{\text{C}}}-\overset{|}{\underset{|}{\text{C}}}-\overset{|}{\underset{|}{\text{C}}}-\overset{|}{\underset{|}{\text{C}}}-\overset{|}{\underset{|}{\text{N}}}\end{array}\right]_n-$$

6. Describe the role of each of the following types of chemical bonding in a polyamide:
   (a) covalent bonds
   (b) amide bonds
   (c) hydrogen bonds

Applying Inquiry Skills

7. Suppose that two new polymers have been designed and synthesized for use as potting material for plants.
   (a) List and discuss the properties of an ideal polymer to be used to hold and supply water and nutrients for a plant over an extended period of time.
   (b) Design an experiment to test and compare the two polymers for the properties you listed. Write a brief description of the procedures followed, and possible interpretations of experimental results.

Making Connections

8. The "superabsorbency" of sodium polymethylacrylate is ideally suited to its use in baby diapers and other hygiene products. Suggest other applications for which this polymer would be useful.

9. When we purchase a product, we may wish to consider not only the source of its components, but also the requirements for its use and maintenance. In your opinion, how valid is the use of the terms "organic," "natural," and "chemical" in the promotion of consumer goods?

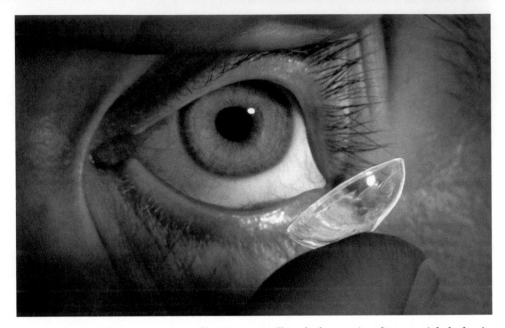

Figure 1
The explosive development of polymer technology has thrown up one product that makes you shudder and blink when you first hear of it—a large object that sits on the surface of your eyeball.

Have you ever felt the irritation of having a small eyelash or a tiny dust particle lodge in your eye? Now imagine placing a circular plastic disk, nearly a centimetre in diameter, onto the cornea of each eye (**Figure 1**). You may be one of the many contact lens wearers who do this daily, and who is hardly aware of the intrusion except for the fact that you can see much better (**Figure 2**). Since the 1960s, when contact lenses became widely available, there have been many new developments in the materials used, and the process is continuing—patent lawyers submit many new patent applications each month, but only a few materials pass the rigorous tests of health agencies and manufacturers.

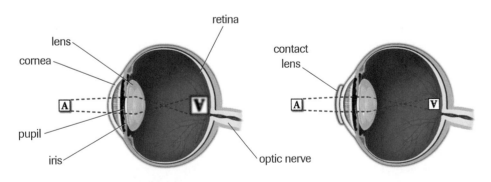

Figure 2
If the natural lens cannot bring the images of distant objects into focus on the retina, a contact lens can solve the problem.

> ### ▶ Practice

Making Connections

1. Brainstorm with a partner or in a small discussion group, and make a list of all the desired properties of a material that is used to manufacture contact lenses.

2. From what you have learned in this unit, suggest some types of molecules that may provide some of the properties that you listed in the last question.

Hard Contact Lenses

The first contact lenses, made in 1887, were manufactured from glass and were designed to cover the entire eye. These were replaced by plastic lenses in 1938, and over the next 10 years the shape of the lens was changed to cover only the cornea. The material used in these hard lenses was polymethylmethacrylate (PMMA), a plastic that is firm but uncomfortable to wear. The experience of contact lens wearers and further understanding about eye physiology led to concerns about PMMA, the most alarming of which was the finding in the mid-1970s that these lenses did not allow much oxygen to reach the eye. The cornea does not contain blood vessels and all of its oxygen is obtained from a film of tears on its surface. In the absence of oxygen, the cornea swells and may develop microcysts, becoming more susceptible to infection. The old PMMA hard lenses are now obsolete and have been replaced by soft contact lenses and rigid gas-permeable lenses.

Soft Contact Lenses

More comfortable soft contact lenses were introduced in 1971. They are made of a material known as polyHEMA, a polymer of monomers of 2-hydroxyethylmethylacrylate (HEMA).

$$H_2C\!=\!C\!-\!\underset{\underset{O}{\|}}{C}\!-\!O\!-\!CH_2\!-\!CH_2OH$$

with CH_3 on the central carbon

HEMA

In the dry form, polyHEMA is glassy and hard, but the polymer swells with water to form a hydrogel, rendering it soft and flexible. As with many polymers, crosslinking between polymer strands in the hydrogel gives the lens elasticity, a property that provides a comfortable fit for the lens wearer. With elasticity comes deformation, so the lens material must be able to recover from the numerous daily deformations caused by eyelid action; otherwise, the visual performance of the lens would rapidly deteriorate.

The water content of the polymer material is a key factor to consider in its use as a contact lens. The water in the soft contact lens supplies much-needed oxygen to the cornea. The gas permeability of a hydrogel is dependent on its water content; the higher the water content, the more oxygen becomes available to the cornea.

With increasing water content, however, the refractive index of the material decreases; that is, the degree of bending of light decreases. This lowers the corrective effect of the lens, requiring thicker lenses to achieve the prescription requirements.

> ▶ **Practice**

Understanding Concepts

3. PolyHEMA is formed by an addition polymerization reaction. Draw the structure of a polyHEMA strand showing three linked monomers.

4. Explain how crosslinking provides elastic properties to the contact lens material.

Making Connections

5. As is often the situation with the development of new materials, some properties of a material are beneficial in one respect but disadvantageous in another. Describe two situations in the use of hydrogels for contact lenses where a compromise may be needed.

Rigid Gas-Permeable Lenses

While soft lenses were being developed, a race also began for the best material for a hard lens that is gas permeable. These new lenses were introduced in 1978 and are commonly called rigid gas permeables, or RGPs. The first material used was cellulose acetate butyrate, a polymer that showed improved gas permeability, but at the cost of low stability and increased risk of protein and lipid deposits on the lens surface. Later generations of materials included the addition of silicone and other groups to the polymers, with the introduction of fluorosilicone acrylate materials in 1987.

Manufacturers of RGPs claim that since their lenses do not contain water, they are easier to handle, are not prone to buildup of deposits, and last longer than the soft lenses. The demand for the future is for a lens that is gas permeable, rigid, and easy to maintain, and that can be worn for long periods of time. Research is ongoing in this field and the world of polymer chemistry offers endless possibilities.

▶ Practice

Understanding Concepts

6. Discuss the validity of the notion that "the world of polymer chemistry offers endless possibilities."

Making Connections

7. Speculate on future developments in the field of contact lenses. What properties might be desirable for a lens? Suggest how these properties could be achieved.

▶ Section 2.3 Questions

Understanding Concepts

1. Compare the features, good or bad, of each of the three types of contact lenses: hard lenses, soft lenses, and rigid gas-permeable lenses. Use the comparison to illustrate how organic chemistry has contributed to making improvements in the field of vision correction and eye care.

Making Connections

2. Research and provide examples of the use of organic chemistry to improve technical solutions in the medical field. Write a report on one of the examples you found. Include in your report a description of the problem to be solved, existing technical solutions, and the role of organic chemistry in the improved solution. A few examples are provided as a starting point for your research.

- Drug delivery systems, e.g., nicotine patches, estrogen patches, gel capsules for timed release
- Artificial flexible joints
- Medical textiles, e.g., adhesives
- Medical equipment, e.g., materials for angioplasty
- Polymers as UV blockers

 www.science.nelson.com

Living systems produce a large variety of molecules whose properties fulfill many different functions in the organism. Some substances need to be mobile and water soluble in order to be transported throughout the system. Other substances serve a structural function, providing support or motion; so the properties of strength, flexibility, and insolubility in water are desirable in these molecules.

Many of the substances that make up living organisms are large molecules of high molecular mass, and are polymers of functionally similar subunits. We will be looking at four groups of these biological **macromolecules**: proteins, carbohydrates, nucleic acids, and fats and lipids.

Proteins make up about half of the dry mass of our bodies. Our muscles, skin, cartilage, tendons, and nails are all made up of protein molecules. We also produce thousands of different enzymes, all proteins, to catalyze specific reactions, and many other protein molecules such as hemoglobin and some hormones. Although these proteins appear so diverse in function and in structure, they are made from the same set of monomers: a group of 20 molecules called **amino acids**.

Amino Acids

Just as their name suggests, amino acids contain two functional groups—an amine group and a carboxylic acid group—attached to a central carbon atom. The central carbon atom completes its bonding with a hydrogen atom and a substituted group, shown in the margin as R.

Each of the 20 natural amino acids has a different R group. The simplest amino acid, glycine, has a hydrogen atom for its R group. Some of the R groups are acidic, others basic; some are polar and others nonpolar. Each protein molecule is a polymer of these amino acids, linked together in a sequence that is specific to the protein, each monomer contributing its characteristics to the overall molecule.

Let us consider the diversity possible in this construction. Using the alphabet as an analogy, consider the number of words that exist with an alphabet of 26 letters—enough to fill a dictionary. Now imagine the number of words that we can make up if each word may be hundreds of letters long, and any sequence of letters is permissible. This limitless number of combinations of amino acids affords the diversity of structure and properties that we find in the proteins of living organisms.

Chiral Molecules

Any molecule containing a carbon with four different attached groups is capable of existing as two different isomers that (like our two hands) are mirror images of each other. These are known as **chiral** molecules (**Figure 1**). All the amino acids, with the exception of glycine, can therefore exist in two different configurations: L and D. In fact, natural amino acids appear in only one configuration, designated by convention as "L."

macromolecule a large molecule composed of several subunits

amino acid a compound in which an amino group and a carboxyl group are attached to the same carbon atom

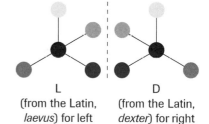

amino acid

chiral able to exist in two forms that are mirror images of each other

L
(from the Latin, *laevus*) for left

D
(from the Latin, *dexter*) for right

amino acid

Figure 1
Chiral molecules

> ▶ **TRY THIS** activity

Making Chiral Molecules

Any molecule that contains a carbon atom bonded to four different atoms or groups is a chiral molecule. Prove it to yourself!
Materials: molecular model kit; mirror

- Join four different atoms or groups to a single carbon atom. Hold the model next to the mirror and observe the reflection.
  (a) What is the correct name for your molecule?

- Now take a duplicate set of atoms and create a model of the image you can see in the mirror. Compare it to the original.
  (b) What is the correct name for your second molecule?
  (c) Would you expect the two molecules to have the same properties? Explain.

peptide bond the bond formed when the amine group of one amino acid reacts with the acid group of the next

polypeptide a polymer made up of amino acids joined together with peptide bonds

dipeptide two amino acids joined together with a peptide bond

Polypeptides from Amino Acids

The difference between chiral isomers is dramatic in biological systems. For example, in the late 1950s the L-isomer of the drug thalidomide was found to be an effective treatment for "morning sickness" in pregnant women. However, the drug that was sold contained a mixture of both isomers. The other isomer turned out to cause "errors" in fetus development and suppress natural abortions. As a result, the use of the drug led to a significant increase in physical deformities among newborns. Synthetic processes often produce a mixture of the two possible isomers, and the pharmaceutical industry has to take great care to market only the isomer with the desired effect.

A list of the names and condensed structural formulas of the 20 amino acids is shown in **Figure 2**. Let us look at how these amino acids are linked together. In the previous section, we discussed the formation of polyamides; the same reaction occurs here. The amine group of one amino acid reacts with the acid group of the next amino acid in the sequence, with the elimination of a water molecule. The bond formed in this reaction between amino acids is given a special name—a **peptide bond**. These biological polyamides are accordingly called **polypeptides**.

How is a polypeptide different from a protein? Proteins may consist of several polypeptide chains, and may also have other components, such as the "heme unit" in hemoglobin.

Protein molecules are long and flexible, and can form bonds and links with themselves or with other protein molecules.

The diagram below shows two amino acids reacting to form a **dipeptide**.

$$
\begin{array}{ccc}
\underset{\text{amino acid 1}}{\text{H}_2\text{N}-\overset{\overset{\displaystyle H}{|}}{\underset{\underset{\displaystyle R'}{|}}{C}}-\overset{\overset{\displaystyle O}{\|}}{C}-\boxed{\text{OH}+\text{H}}} & \underset{\text{amino acid 2}}{\text{N}-\overset{\overset{\displaystyle H}{|}}{\underset{\underset{\displaystyle R''}{|}}{C}}-\overset{\overset{\displaystyle O}{\|}}{C}-\text{OH}} \longrightarrow & \underset{\text{dipeptide}}{\text{H}_2\text{N}-\overset{\overset{\displaystyle H}{|}}{\underset{\underset{\displaystyle R'}{|}}{C}}-\overset{\overset{\displaystyle O}{\|}}{C}-\overset{\overset{\displaystyle H}{|}}{\underset{\underset{\displaystyle H}{|}}{N}}-\overset{\overset{\displaystyle H}{|}}{\underset{\underset{\displaystyle R''}{|}}{C}}-\overset{\overset{\displaystyle O}{\|}}{C}-\text{OH}+\text{H}_2\text{O}}
\end{array}
$$

amino acid 1   amino acid 2   dipeptide

$$
\begin{array}{ccc}
\underset{\text{glycine}}{\text{H}_2\text{N}-\overset{\overset{\displaystyle H}{|}}{\underset{\underset{\displaystyle H}{|}}{C}}-\overset{\overset{\displaystyle O}{\|}}{C}-\boxed{\text{OH}+\text{H}}} & \underset{\text{alanine}}{\text{N}-\overset{\overset{\displaystyle H}{|}}{\underset{\underset{\displaystyle CH_3}{|}}{C}}-\overset{\overset{\displaystyle O}{\|}}{C}-\text{OH}} \longrightarrow & \underset{\substack{\text{glycyl alanine}\\\text{(gly-ala)}}}{\text{H}_2\text{N}-\overset{\overset{\displaystyle H}{|}}{\underset{\underset{\displaystyle H}{|}}{C}}-\overset{\overset{\displaystyle O}{\|}}{C}-\overset{\overset{\displaystyle H}{|}}{\underset{\underset{\displaystyle H}{|}}{N}}-\overset{\overset{\displaystyle H}{|}}{\underset{\underset{\displaystyle CH_3}{|}}{C}}-\overset{\overset{\displaystyle O}{\|}}{C}-\text{OH}+\text{HOH}}
\end{array}
$$

glycine   alanine   glycyl alanine (gly-ala)

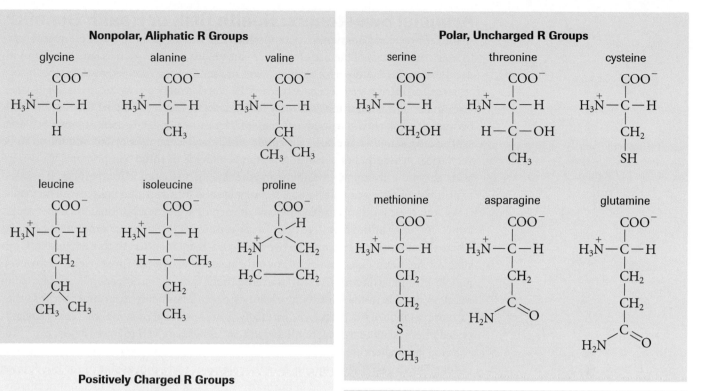

Nonpolar, Aliphatic R Groups

glycine   alanine   valine

leucine   isoleucine   proline

Polar, Uncharged R Groups

serine   threonine   cysteine

methionine   asparagine   glutamine

Positively Charged R Groups

lysine   arginine   histidine

Aromatic R Groups

phenylalanine   tyrosine   tryptophan

Negatively Charged R Groups

aspartate   glutamate

Figure 2
The 20 amino acids used by living things

sodium cyclamate

saccharine

aspartame

Artificial Sweeteners: Health Risk or Health Benefit?

Artificial sweeteners are synthetic compounds that often bear little chemical resemblance to sugar. Supporters of the use of sugar substitutes list benefits such as reduction of tooth decay, better food choices for diabetics, not to mention the fight against obesity.

Several artificial sweeteners have reached the North American market in the last decades. Cyclamates and saccharin have been widely used, the former being banned in the 1970s because of a link with increased incidence of liver cancer. In 1965, chemist James Schlatter synthesized a methyl ester of a dipeptide which he accidentally tasted and found to be very sweet. This dipeptide, called aspartame, is made of three components: the amino acids aspartic acid and phenylalanine, and the alcohol methanol. The aspartic acid is linked to phenylalanine by an amide bond. Phenylalanine also forms an ester linkage with methanol.

As with most food substitutes, there is some controversy surrounding the use of aspartame. Enzymes in the digestive tract break it down into its three components. This is where the controversy begins. The two amino acids are identical to those found in foods such as meat and eggs, and thus are not cause for alarm for most people. However, people with phenylketonuria (PKU, a rare genetic disease that is generally diagnosed in infancy) must be warned of the presence of phenylalanine in aspartame, because they cannot metabolize phenylalanine from any source, natural or synthetic. The unmetabolized phenylalanine can accumulate in their bodies and lead to brain damage.

Some researchers are concerned about the consumption of aspartame by the general public. The danger, they contend, is in the release of methanol as one of the breakdown products. Methanol is oxidized in the body by the same enzymes that metabolize ethanol, but the products of methanol oxidization are much more lethal. The first reaction produces formaldehyde. It forms crosslinks between polypeptide chains, changing the structure of proteins, in turn causing drastic changes in protein function. Formaldehyde is oxidized further to formic acid, the acid responsible for the pain associated with ant bites. This acid disrupts the function of mitochondria, the energy production mechanism of the cell. When these breakdown products are formed in the cells of the retina and the optic nerve they cause irreparable damage. Blindness is one of the first detectable symptoms of methanol poisoning.

Manufacturers of aspartame claim that the amounts of methanol produced from consuming normal amounts of the sweetener are so small that any harmful effects are negligible. Indeed, methanol itself is naturally produced in many fruits and vegetables, and is also present in low concentrations in wine and beer, up to 300 mg/L.

Is aspartame indeed dangerous? If so, why is it so readily available? Aspartame has been approved for use in more than 90 countries.

▶ Practice

Understanding Concepts

1. (a) Draw the structure of aspartame and calculate its molar mass.
   (b) Calculate the percentage by mass of aspartame attributable to methanol.

2. (a) Outline the series of products resulting from the breakdown of aspartame.
   (b) Write two equations to show the steps in the controlled oxidation of methanol by cellular enzymes.

3. (a) Each package of NutraSweet contains 35 mg of aspartame, and delivers sweetness equivalent to 10 g (2 teaspoons) of sugar. A can of diet soft drink contains approximately 200 mg of aspartame. Calculate the mass of methanol produced from the ingestion of one can of diet soft drink.
   (b) The toxicity of a substance is usually measured by the LD_{50} rating (the dosage that would kill 50% of the test subjects) and is expressed in g/kg of body weight. The LD_{50} for methanol is 0.07 g/kg. Calculate the lethal dose for a 70-kg individual.
   (c) How many cans of diet pop would a person have to drink to ingest this lethal dose?

Take a Stand: Will That Be "Regular" or "Diet"?

You have probably seen its name on cans of "diet" soft drinks. You might also have tasted its flavour, which is not quite the same as that of sugar. Aspartame, sold as NutraSweet and Equal, is probably the most widely used sugar substitute in North America. More than 200 studies have been done on the safety of the use of aspartame, some claiming that it produces side effects ranging from headaches to brain tumours, others claiming that there is no evidence of any harmful effects.

(a) Research some of these studies, writing brief summaries of the arguments for and against the use of aspartame or another food substitute or additive (e.g., artificial flavour, food colour).

(b) In a small group discuss the factors that you would consider in reviewing any such scientific study to determine the validity of the results.

(c) Analyze the risks and benefits of the use of your chosen food substitute or additive. Write a report on the results of your analysis, and make a recommendation on its use, with supporting arguments.

 GO www.science.nelson.com

Protein Structure

While some proteins provide the structure and support in a living system, others are transported throughout the organism, serving as carriers of oxygen or regulators of biological processes. If all proteins are long strands of polypeptides, how do they show such variation in function? The answer can be found in the sequence of amino acids, which is determined genetically and is unique to each protein. The complex structure of giant protein molecules (**Figure 3**) all depends on the electronic attractions and repulsions that develop between the functional groups on the component amino acids.

Primary Structure of Proteins

As you know, all proteins are polypeptides: long strings of amino acids arranged in a very specific order. This is the primary structure of the protein. It can be altered by changes (mutations) in the DNA. A change in primary structure can cause a "ripple effect," interfering with all the other levels of structure, and so possibly making the protein completely useless for its "intended" task.

primary structure the sequence of the monomers in a polymer chain; in polypeptides and proteins, it is the sequence of amino acid subunits

secondary structure the three-dimensional organization of segments of a polymer chain, such as alpha-helices and pleated-sheet structures

alpha-helix a right-handed spiraling structure held by intramolecular hydrogen bonding between groups along a polymer chain

pleated-sheet conformation a folded sheetlike structure held by intramolecular or intermolecular hydrogen bonding between polymer chains

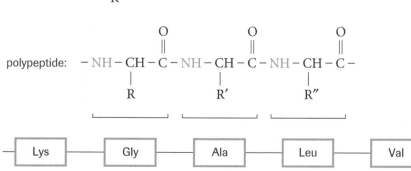

Primary structure, 1°
The sequence of amino acids in the polypeptide chain determines which protein is created.

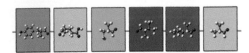

Secondary structure, 2°
Polar and nonpolar amino acids at different locations within the long polypeptide chain interact with each other, forming coils or pleated sheets. The interactions may be van der Waals forces, hydrogen bonding, or other attractions.

Alpha-helix This coiled secondary structure results from hydrogen bonds between the amine group in a peptide linkage and the carbonyl group of an amino acid further along on the same chain. The R groups of the amino acids protrude outward from the coil.

Pleated-sheet (or beta-pleated sheet) This folded secondary structure results from the zigzag shape of the backbone of the polypeptide chain forming a series of pleated sheets. Hydrogen bonds form between the amine groups and the carbonyl groups on adjacent pleated sheets.

Tertiary structure, 3°
Proteins may have helical sections and pleated-sheet sections within the same molecule. These sections attract each other, within the molecule, folding a long, twisted ribbon into a specific shape. Proteins such as enzymes, hemoglobin, and hormones tend to have a spherical or globular tertiary structure, to travel easily through narrow vessels.

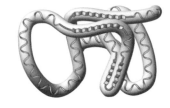

Figure 3
Proteins are very complicated compounds, with up to four levels of organization giving each protein a unique physical shape with unique physical characteristics.

Quaternary structure, 4°
Some proteins are complexes formed from two or more protein subunits, joined by van der Waals forces and hydrogen bonding between protein subunits. For example, hemoglobin has four subunits held together in a roughly tetrahedral arrangement.

Identifying Fibres by Odour

Fibre artists, such as weavers and felters, need to know the composition of the fibres or textiles they are thinking of using. Are they cellulose fibres (e.g., linen, cotton, hemp), protein fibres (e.g., silk, wool, fur, mohair), or synthetics (e.g., nylon or polyester)? Sometimes the fibres aren't labelled, so the artists use a burn test to narrow down their identification. Wool, hair, and fur smell of sulfur when exposed to flame, and don't burn well. Cellulose fibres have a "burning wood" smell and burn very readily. Synthetic fibres have an acrid smell.
Materials: eye protection; small pieces (1 cm × 1 cm) of several fabrics, including wool, cotton, polyester, and a few strands of hair or fur; a similar sample of unknown composition; laboratory burner; metal tweezers or forceps; fume hood or well-ventilated room

- Take a small piece of several known fabrics.
  (a) Classify the fibres as protein, cellulose, or synthetic.
- Over a nonflammable surface, burn each sample by passing it slowly through the flame of the laboratory burner. Carefully smell the odour by wafting the smoke toward your nose.
- Repeat the test with the unknown fabric sample.
  (b) Classify the unknown sample as protein, cellulose, or synthetic.

Secondary Structure of Proteins

The alpha-helical secondary structure accounts for the strength of fibrous proteins such as alpha-keratin in hair and nails (**Figure 4**), and collagen in tendons and cartilage. In these proteins, alpha-helical chains are coiled in groups of three, into a "rope" structure; these ropes are further bundled into thicker fibres. The strength of these fibres is increased by crosslinking between polypeptide chains, provided by disulfide (–S–S–) bonds.

X-ray analysis has revealed that the pleated-sheet secondary structure is indeed found in fibroin, the protein in silk, and in a similar protein in spider webs. These proteins are rich in Ala and Gly, the amino acids with the two smallest R groups, allowing the pleated sheets to be closely packed into layers. The pleated-sheet structure affords the protein added strength—the strength of silk and spider webs is well respected—so pleated-sheet proteins are used in products ranging from parachutes to the cross hairs of rifles.

Figure 4
Keratin is structurally strong and smooth because of the way the molecules pack together and form crosslinkages.

Tertiary Structure of Proteins

The tertiary structure of proteins is highly specific and is closely related to its role in the organism. The winter flounder (**Figure 5**), a fish found off the coast of Newfoundland, is able to lower its own freezing point sufficiently to survive the winter without migrating to warmer climes. (The formation of ice crystals in tissue fluids causes irreparable damage to cell membranes.) Canadian scientists have found that these fish, when stimulated with the appropriate trigger, produce an "antifreeze protein" whose three-dimensional structure fits well into the surface of a developing ice crystal, inhibiting further ice growth. Similar antifreeze mechanisms exist in other organisms, such as insects, each using different structural features of the proteins and ice surfaces.

tertiary structure a description of the three-dimensional folding of the alpha-helices and pleated-sheet structures of polypeptide chains

Quaternary Structure of Proteins

Quaternary structure, in which several protein subunits join together, is found in many proteins that serve a regulatory function, such as insulin, and the catalytic enzyme kinase. One of the best-known protein complexes is hemoglobin, which has four protein subunits. Interactions between subunits permit responses to changes in concentrations of the substance regulated, such as oxygen.

Figure 5
The winter flounder produces its own antifreeze protein.

Denaturation of Proteins

When the bonds responsible for the secondary and tertiary protein structures are broken, the protein loses its three-dimensional structure; this process is called *denaturation*. When fish is cooked, for example, no covalent bonds are broken by the mild heating, but the weaker forces of attraction such as van der Waals forces and hydrogen bonds are disrupted. Changing the pH affects electrostatic forces and disrupts hydrogen bonding, as witnessed in the curdling of milk in vinegar or orange juice. Organic solvents such as formaldehyde and acetone interact with the nonpolar components of the amino acids; these solvents denature the proteins in the specimens they are used to preserve. In all cases, even mild denaturation of a protein is accompanied by severe loss of function.

▶ **Practice**

Understanding Concepts

4. Are proteins addition polymers or condensation polymers? Explain.

5. How do chiral molecules differ from each other?

6. Draw a structural diagram of the linkage between amino acids in a peptide chain.

7. Differentiate between the primary, secondary, tertiary, and quaternary structure of proteins. Sketch a simple diagram of each structure to illustrate your answer.

8. Give one example of a fibrous protein and one example of a globular protein. For each, describe its function in the organism and how its structure serves its function.

▶ **Section 2.4 Questions**

Understanding Concepts

1. Explain why amino acids, with the exception of glycine, can occur in more than one chiral form.

2. Explain how it is possible to make millions of different proteins from only 20 amino acid monomers.

3. Describe the type of protein structure that gives fibrous proteins such as collagen their exceptional strength.

4. Explain why an alteration in the primary structure of a protein could result in a change in its tertiary structure.

5. What is meant by the quaternary structure of proteins? Give an example to illustrate your answer.

6. For each of the following types of chemical bonding, describe an example of its occurrence in protein molecules and its effect on the protein's structure:
   (a) covalent bonds   (c) van der Waals forces
   (b) hydrogen bonds   (d) disulfide bonds

Applying Inquiry Skills

7. In an experiment on the effects of artificial sweeteners on health, the sweetener saccharin was fed to lab rats. The experimenters reported an increase of 50% in the incidence of liver cancer in the saccharin-fed rats. You are asked to evaluate the experimental results with respect to any health risk of saccharin to humans.
   (a) List the missing information that you would require about the experimental design and conditions.
   (b) Describe essential experimental controls that must be incorporated.
   (c) Suggest any circumstances that might render the results of the experiment inconclusive.

Making Connections

8. When fresh vegetables are prepared for storage in the freezer, they are dipped momentarily in boiling water. This procedure, called blanching, stops the vegetables from further ripening through enzyme action. Give an explanation at the molecular level for the success of this technique.

9. Research the secondary and tertiary structures of one of the following proteins:
   (a) fibrinogen, the protein involved in blood clotting
   (b) collagen, a connective tissue
   (d) cytochrome c, used in electron transport
   (c) myoglobin, an oxygen-binding protein
   (d) myosin, a muscle protein

   Present your findings in a report that includes a description of the secondary and tertiary structures that make the protein ideally suited to its function.

    www.science.nelson.com

10. Thalidomide, so harmful when administered as a mix of L and D configurations, has been banned in many countries. However, it is a very versatile and inexpensive drug when the configurations are isolated and used selectively.
    (a) Research the current use of thalidomide (if any), and other pharmaceuticals (if any) that are used in its place.
    (b) Compare the costs of using thalidomide with the costs of developing alternative drugs or the costs of having no drugs available.
    (c) Write a brief report addressing the question, "Should thalidomide continue to be banned?" for use in a popular science magazine.

     www.science.nelson.com

Starch and cellulose are natural polymers, belonging in a group of compounds called carbohydrates. Carbohydrates include sugars (which are monosaccharides or disaccharides) as well as starches and cellulose (which are polysaccharides composed of sugar monomers). Carbohydrates are very important components in our food, being a major source of energy for many of us. Plants are the source of most of the carbohydrates we eat: sugar from sugar cane or beets; starch from potatoes or wheat; and cellulose in bran and vegetables. The differences among these forms of carbohydrates are those of molecular size and shape, sugars being the smallest units and starches and cellulose being polymers of the simple sugar, glucose.

Monosaccharide Sugars

All **carbohydrates** have the empirical formula $C_x(H_2O)_y$, hence their name, which is derived from "hydrated carbon." Like other compounds of carbon, hydrogen, and oxygen, carbohydrates undergo complete combustion to produce carbon dioxide and water.

Glucose, the sugar formed in the process of photosynthesis, has the formula $C_6H_{12}O_6$, or $C_6(H_2O)_6$; however, the molecules are not actually in the form of carbon and water. The atoms in simple sugar molecules are arranged as a carbon backbone, often six carbons long, with functional groups attached to each carbon atom. In some simple sugars, the first carbon atom is a part of a carbonyl group (C=O), forming an aldehyde. In other sugars, it is the second carbon atom that is a part of a C=O group, forming a ketone. The other carbon atoms in the molecule each hold a hydroxyl group.

Two simple sugars, glucose and fructose, are shown below. Glucose is the sugar molecule most widely produced by plants, and is the monomer that makes up all the larger carbohydrates. Glucose belongs to the group of sugars called **aldoses**, due to its aldehyde group. Fructose, with its ketone group, is a **ketose**. It is the sugar found widely in fruits, so is sometimes called fruit sugar, and is a key sugar in honey. Both glucose and fructose are single units of sugar and are called **monosaccharides**.

glucose   fructose

The presence of both –OH groups and C=O groups on a flexible backbone also provides opportunity for these groups on the same molecule to react with each other. When a bond is formed in this way, like a snake grabbing its own tail, a ring structure is formed. Essentially, the C=O group reacts with the –OH group to form an oxygen link, resulting in either a five-membered ring or a six-membered ring. When aqueous solutions of glucose are analyzed, it is found that more than 99.9% of glucose molecules are in ring form.

When these rings are formed, a very important structural "decision" is made. In the open-ended linear sugar molecule, the single covalent bonds in the carbon backbone allow for free rotation of any attached atoms or groups. Once a ring structure is formed, there is no longer free rotation about the ring. The ring itself is not planar, but is in a "chair" conformation (**Figure 1**). Functional groups may be fixed in a position "above the ring"

carbohydrate a compound of carbon, hydrogen, and oxygen, with a general formula $C_x(H_2O)_y$

aldose a sugar molecule with an aldehyde functional group at C 1

ketose a sugar molecule with a ketone functional group, usually at C 2

monosaccharide a carbohydrate consisting of a single sugar unit

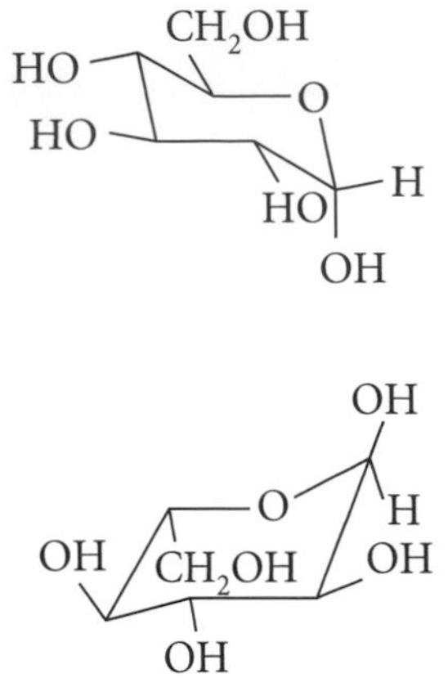

Figure 1
The "chair" conformation of the glucose molecule: functional groups are "fixed" above or below the ring.

disaccharide a carbohydrate consisting of two monosaccharides

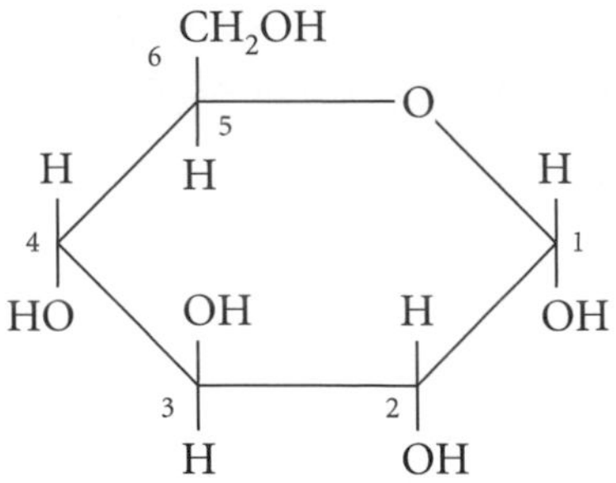

glucose

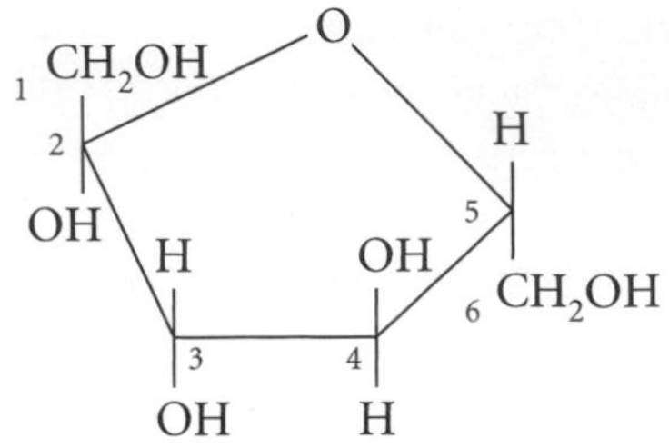

fructose

or "below the ring". The orientation of the functional groups determines the orientation of any further bonds the molecule forms with other molecules. As we will see later, this orientation affects the shape, and hence the properties, of the polymers of these sugars.

Disaccharide Sugars

When a glucose molecule is joined to a fructose molecule, the dimer formed is a **disaccharide** called sucrose, the common sugar we use in our coffee or on our cereal. All disaccharides, as their name implies, are made of two simple sugar molecules. Lactose, the sugar found in milk, is not as sweet as sucrose. It is also a disaccharide: a dimer of glucose and galactose (which is another isomer of glucose).

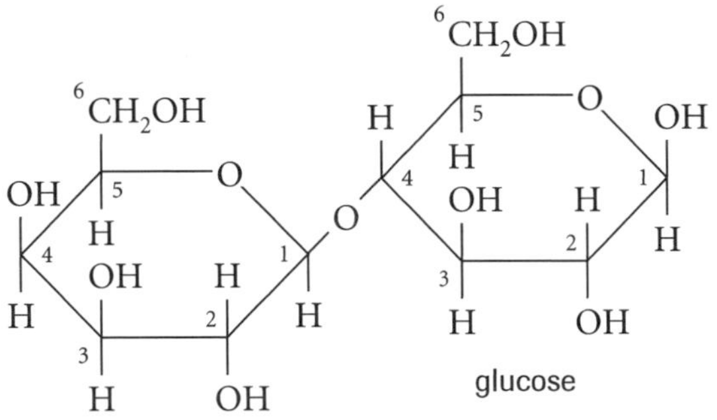

galactose glucose

DID YOU KNOW?

Lactose Intolerance
Nearly all babies thrive on milk, but many grow up to become lactose intolerant. These children and adults have stopped producing the enzyme lactase, which breaks down the milk sugar lactose. Undigested lactose accumulates in the small intestine and, by osmosis, causes an influx of fluid with accompanying consequences, such as diarrhea. The occurrence of lactose intolerance in human populations varies greatly. It appears that the ability to continue production of lactase beyond infancy has developed in some human populations since dairy farming began, about 10 000 years ago.

When disaccharides are ingested, specific enzymes are required to break the bonds between the monomers. This is a hydrolysis reaction in which the disaccharide is broken down to its component monosaccharides. For example, the enzyme lactase is required in the hydrolysis of lactose, to separate the glucose from the galactose before further digestion can occur. People who lack this enzyme cannot break down this sugar, and are thus "lactose intolerant."

While many low-molecular-mass organic compounds are gases or volatile liquids at room temperature, sugars are solids with relatively high melting points. You may have heated table sugar in a pan on a stove, to melt it into a thick, sticky liquid, perfect for gluing gingerbread houses or for making peanut brittle. Or you may have melted marshmallows over a campfire. Sugars are also readily soluble in water, as the number of sweet drinks in our refrigerators attests. These two properties—high melting point and solubility in water—are accounted for by the abundance of –OH groups in sugar molecules, allowing for hydrogen bonding both with other sugar molecules and with water molecules (**Table 1**).

Table 1 Physical Properties of Common Sugars

Sugar	Melting point (°C)	Solubility
glucose	150	extremely soluble (91 g/100 mL at 25°C)
fructose	103–105	very soluble
sucrose	185–186	very soluble
maltose	102	soluble

Starch for Energy; Cellulose for Support

Starchy foods such as rice, wheat, corn, and potatoes provide us with readily available energy. They are also the main method of energy storage for the plants that produce them, as seeds or tubers. **Starches** are polymers of glucose, in either branched or unbranched chains; they are thus **polysaccharides**. Animals also produce a starch-like substance, called **glycogen**, that performs an energy storage function. Glycogen is stored in the muscles as a ready source of energy, and also in the liver, where it helps to regulate the blood glucose level.

We have, in our digestive tracts, very specific enzymes: one that breaks down starch and another that breaks down glycogen. However, the human digestive system does not have an enzyme to break down the other polymer of glucose: **cellulose**. Cellulose is a straight-chain, rigid polysaccharide with glucose–glucose linkages different from those in starch or glycogen. It provides structure and support for plants, some of which tower tens of metres in height. Wood is mainly cellulose; cotton fibres and hemp fibres are also cellulose. Indeed, it is because cellulose is indigestible that whole grains, fruits, and vegetables are good sources of dietary fibre. Herbivores such as cattle, rabbits, termites, and giraffes rely on some friendly help to do their digesting: They have specially developed stomachs and intestines that house enzyme-producing bacteria or protozoa to aid in the breakdown of cellulose.

It is the different glucose–glucose linkages that make cellulose different from starch or glycogen. Recall that, when glucose forms a ring structure, the functional groups attached to the ring are fixed in a certain orientation above or below the ring (**Figure 1**). Our enzymes are specific to the orientation of the functional groups, and cannot break the glucose–glucose linkages found in cellulose.

In starch and glycogen, glucose monomers are added at angles that lead to a helical structure, which is maintained by hydrogen bonds between –OH groups on the same polymer chain (**Figure 2(a)**). The single chains are sufficiently small to be soluble in water. Thus, starch and glycogen molecules are both mobile and soluble—important properties in their role as readily available energy storage for the organism.

In cellulose, glucose monomers are added to produce linear polymer chains that can align side by side, favouring interchain hydrogen bonding (**Figure 2(b)**). These interchain links produce a rigid structure of layered sheets of cellulose. This bulky and inflexible structure not only imparts exceptional strength to cellulose, it also renders it insoluble in water. It is, of course, essential for plants that their main building material does not readily dissolve in water.

(a) starch

(b) cellulose

DID YOU KNOW ?

The Centre of the Chocolate
Sucrose, a disaccharide, is slightly sweeter than glucose but only half as sweet as fructose. The enzyme sucrase, also called invertase, can break sucrose down into glucose and fructose: a mixture that is sweeter and more soluble than the original sucrose. The centres of some chocolates are made by shaping a solid centre of sucrose and invertase, and coating it with chocolate. Before long, the enzyme transforms the sucrose centre into the sweet syrupy mixture of glucose and fructose.

polysaccharide a polymer composed of monosaccharide monomers

starch a polysaccharide of glucose; produced by plants for energy storage

glycogen a polysaccharide of glucose; produced by animals for energy storage

cellulose a polysaccharide of glucose; produced by plants as a structural material

Figure 2
The difference in linkages between glucose monomers gives very different three-dimensional structures.
(a) In starch the polymer takes on a tightly coiled helical structure
(b) In cellulose, the linked monomers can rotate, allowing formation of straight fibres.

Prehistoric Polymer

Contrary to popular belief, amber is not a fossil but is a natural polymer formed by the crosslinking of the resin molecules produced by some plants; hence its hard, plastic-like properties. Small insects, lizards, and even frogs have been trapped in the sticky resin (a viscous liquid, generally composed of mixtures of organic acids and esters) that gradually penetrates their bodies and replaces the water in their tissues.

▶ Practice

Understanding Concepts

1. Identify the functional groups present in a molecule of glucose and in a molecule of fructose.

2. Describe several functions of polysaccharides and how these functions are served by their molecular structures,
   (a) in animals
   (b) in plants.

3. Compare the following pairs of compounds, referring to their structure and properties:
   (a) sugars and starch
   (b) starch and cellulose

4. (a) Draw a structural diagram of the most common configuration of a glucose molecule.
   (b) Why does glucose exist in two different forms?

5. Explain in terms of molecular structure why sugars have a relatively high melting point compared with hydrocarbons of comparable size.

6. Discuss why starch molecules are helical and cellulose molecules are linear, given that they are both polymers of glucose. Draw a simple sketch to illustrate your answer.

▶ Section 2.5 Questions

Understanding Concepts

1. Give an example of each of the following:
   (a) aldose
   (c) monosaccharide
   (b) ketose
   (d) disaccharide

2. Are polysaccharides addition polymers or condensation polymers? Explain.

3. Write a chemical equation, using condensed formulas, to show
   (a) the hydrolysis of sucrose;
   (b) the complete combustion of sucrose.

4. Some sugars are referred to as "reducing sugars," indicating that they will undergo oxidation under suitable conditions. Give an example of a sugar containing an aldehyde functional group and an example of a sugar with a ketone functional group. Which of these two sugars is a reducing sugar?

5. Many organic compounds that we use in the home are gases or volatile liquids, e.g., propane for cooking, rubbing alcohol, and paint remover. However, table sugar is a solid organic compound that is stable enough to store for long periods of time. Identify the functional group(s) in each of the compounds named above, and give reasons for the differences in their physical state and properties.

6. When a starchy food such as boiled potato is chewed in the mouth for a long time, the potato begins to taste sweet, even though no sugar is added.
   (a) Explain why potatoes taste sweet to us after chewing.
   (b) Would grass, which is mostly cellulose, taste sweet after chewing?

7. Starch and cellulose have the same caloric value when burned, but very different food values when eaten by humans. Explain.

8. Explain why the sugars in a maple tree dissolve in the sap but the wood in the tree trunk doesn't.

Applying Inquiry Skills

9. From what you have learned about the reactions of aldehydes and ketones, design a test to distinguish an aldose sugar such as glucose from a ketose sugar such as fructose. Include your Experimental Design, a list of the Materials you will need, and a Prediction of the observations you would expect to make for each sugar.

Making Connections

10. Many consumer products are available in natural or synthetic materials: paper or plastic shopping bags, wood or plastic lawn furniture, cotton or polyester clothing. Choose one consumer product and discuss the advantages and disadvantages of the natural and synthetic alternatives, with particular reference to structure and properties of the material used as it relates to the function of the product.

11. Chitin is the main component of the exoskeleton of insects, crabs, lobsters, and other arthropods. It is structurally similar to cellulose, with the sole difference of the substitution of one of the hydroxyl groups on the glucose ring with an acetylated amino group:

$$CH_3 - \overset{\overset{\displaystyle O}{\|}}{C} - NH -$$

   (a) Predict the effect of the acetylated amino group on the attractive forces in the chitin molecule.
   (b) Predict the physical properties of chitin.
   (c) Discuss the suitability of chitin as a protective covering for insects, crabs, and lobsters.

DNA, the well-known abbreviation for **deoxyribonucleic acid**, is vital for cell function and reproduction. Its task is to direct the synthesis of proteins. As you learned in Section 2.4, each protein is a polymer of amino acids in a unique sequence. The role of DNA, therefore, is to put amino acids in the desired sequence *before* peptide bonds form between each pair of amino acids.

There are 20 different amino acids that are incorporated into proteins; so the DNA molecules must be able to "write code" for at least 20 amino acids. The coding system of DNA is created from only four different **nucleotides** that are the monomers of the large DNA polymer.

Another type of nucleic acid, **ribonucleic acid (RNA)**, is similar to DNA and also serves important roles in protein synthesis. Messenger RNAs carry the genetic code from the DNA to the sites of protein synthesis in the cell, and transfer RNAs are relatively small molecules that bring the amino acids to the site where they are aligned in proper sequence.

deoxyribonucleic acid (DNA) a polynucleotide that carries genetic information; the cellular instructions for making proteins

ribonucleic acid a polynucleotide involved as an intermediary in protein synthesis

nucleotide a monomer of DNA, consisting of a ribose sugar, a phosphate group, and one of four possible nitrogenous bases

DNA Structure

monomers:
- four nucleotides (adenine, thymine, guanine, or cytosine), differing only in their nitrogenous bases

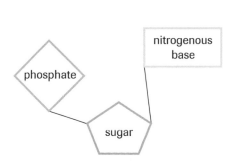

could be:

A

T

G

C

polymer:
- double helix with backbones of alternating sugar and phosphate groups
- opposite strands held together by hydrogen bonding between pairs of nitrogenous bases (A–T and G–C)

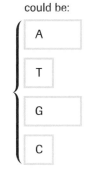

hydrogen bonds

adenine (A)

thymine (T)

guanine (G)

cytosine (C)

Figure 1
The structure of DNA

double helix the coiled structure of two complementary, antiparallel DNA chains

There are three main parts to each nucleotide: a phosphate group, a 5-carbon sugar called ribose (**Figure 2(a)**), and a nitrogenous base, which contains an amino group. The four different nucleotides differ only in the composition of the bases: adenine, thymine, cytosine, and guanine (**Figure 1**). In DNA, the ribose is "deoxy," meaning that it is "lacking an oxygen," a result of an –OH group being replaced by an H atom (**Figure 2(b)**).

(**a**) Ribose (in RNA)   (**b**) Deoxyribose (in DNA)

Figure 2
The sugar part of monomers of RNA and DNA

These nucleotide monomers are linked to form polynucleotides—called nucleic acids—by condensation reactions involving hydroxyl groups of the phosphate on one monomer and the ribose of another monomer. The resulting ribose–phosphate–ribose–phosphate backbone is staggered rather than linear, giving the polymer chain a helical shape. Attached to this spiral backbone are the different bases with their amino groups.

As it turns out, not only is deoxyribonucleic acid helical in structure, it in fact occurs as two strands coiled together in a **double helix** (**Figure 3**). The ribose–phosphate backbone of the two DNA chains is held together by pairs of amine bases, one from each DNA chain. As shown in **Figure 1**, the bases contain –NH groups and C=O groups, between which hydrogen bonds can form. The helical structure of the long DNA strands allows the genetic material to be flexible and easily stored in the nucleus of the cell.

The double-stranded arrangement is also an ingenious design for cell replication, ensuring that exact copies of DNA can be made. Each strand in the DNA double helix is an exact "opposite" match of the other strand. When needed, each of the two strands makes its own new "opposite" partner, and so two new double-stranded DNA pairs are produced. How is one strand an exact "opposite" of the other? The maximum hydrogen bonding between DNA strands occurs only when adenine is paired with thymine, and when cytosine is paired with guanine; that is, only A–T pairs and C–G pairs are formed.

Figure 3
(**a**) The two DNA chains are held together by hydrogen bonding between the nitrogenous bases on each chain, A pairing with T, and C pairing with G.
(**b**) The double-helical structure of a DNA molecule

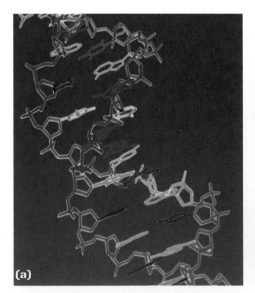
(a)

(b)

This restriction is dictated by the shape and size of the amine side chains (**Figure 3**). Thus, if one DNA strand has a sequence of AACC, its "opposite" partner has to be TTGG.

The deduction of this double helix structure was made in 1953 and won James Watson, Francis Crick, and Maurice Wilkins the 1962 Nobel Prize in Physiology and Medicine. Rosalind Franklin was credited for her contribution of X-ray diffraction work in the study.

Denaturation of DNA

Just as heat and changes in pH can denature globular proteins, they can also denature or "melt" double-helical DNA. Hydrogen bonds between the paired bases are disrupted, causing the two strands in the double helix to unwind and separate. This "melting" process is reversible on cooling, however, and if the two separated strands are partially attached, or come into contact, the remaining portions of the strands quickly "zip" together into the double-helical structure once again.

Some changes to DNA structure *mutations* occur spontaneously, others are accelerated by chemical agents. For example, the nitrogen bases of the nucleotides can lose an amino group; this reaction is accelerated by the presence of nitrites and nitrates, often found in small amounts as preservatives in meats such as hot dogs and sausages. High-energy radiation such as gamma rays, beta rays, X rays, and UV light can also cause chemical changes in DNA. For example, UV radiation can lead to the joining of two adjacent nitrogen bases, introducing a bend or a kink in the DNA chain.

Since the DNA sequence dictates the amino acid sequence in proteins synthesized by the cell, even a minor error can cause the wrong protein or no protein to be synthesized, leading to major malfunctions in an organism.

> ### ▶ *Practice*

Understanding Concepts

1. What do the letters DNA stand for, and what is its main function in an organism?

2. Describe the three main components of a monomer of a nucleic acid.

3. What type of linkage joins the nucleotides
   (a) within a single DNA strand?
   (b) between two single DNA strands?

4. Write a balanced chemical equation for the condensation reaction between deoxyribose and phosphoric acid.

5. In what ways does the double-helical structure of DNA serve its function as a carrier of genetic information in a cell?

6. (a) Describe three causes of chemical alterations to DNA.
   (b) Explain briefly why a minor alteration in a DNA sequence can cause a change in cell function.

Understanding Concepts

1. (a) Name the four nucleotides that make up deoxyribonucleic acid.
   (b) In what ways are these four nucleotides similar, and in what way are they different?

2. What is RNA and how is it similar to or different from DNA?

3. Discuss the advantage of a helical structure over a linear fibrous structure, in view of DNA's cellular function.

4. Discuss the role of hydrogen bonding in the structure of
   (a) single-stranded DNA
   (b) double-stranded DNA

5. Alterations in DNA structure, however minor, can have serious consequences for the organism. Explain what the consequences are and why they result.

6. An experiment was performed in which DNA molecules were extracted and completely hydrolyzed into their component nucleotides. The nucleotides were analyzed and it was found that the number of adenine bases was the same as the number of thymine bases, and the number of cytosine bases was the same as the number of guanine bases. What conclusions might you draw about the structure of DNA, based on the evidence obtained?

Applying Inquiry Skills

7. There are only four nucleotides in DNA to code for 20 amino acids in proteins. Each amino acid must have its own unique DNA code. Design and perform a numerical test to find out the minimum number of nucleotides that would be needed in order to assign a different code to each of the 20 amino acids. Show calculations to support your answer.

Making Connections

8. In this chapter, we have discussed the helical structure of three natural polymers: proteins, carbohydrates, and nucleic acids. For the three polymers,
   (a) compare the functional groups on their monomers;
   (b) compare the types of interchain interactions; and
   (c) compare the properties and function of the polymers.

9. Canadian biochemist Dr. Michael Smith won a Nobel Prize in Chemistry in 1993 for an ingenious technique he developed called "site-specific mutagenesis." Simply, he "knocked out" specific sites in a DNA sequence, and observed the effects on protein and cell function. Using the analogy of an unlabelled electrical fuse box or circuit-breaker box in a home, describe how Dr. Smith's technique can be used to decipher the circuitry and wiring in the home.

10. List and describe as many ways as possible in which you can reduce the risk of mutation of DNA in your skin cells from UV exposure.

Fats and Oils 2.7

We are familiar with many different fats and oils produced by living systems for energy storage: corn oil and vegetable shortening from plants; butter from cows' milk; and lard from pork or beef fat (**Figure 1**). Fats are usually solid at ordinary room temperatures and oils are liquid; they both belong to a class of organic compounds called lipids that includes other substances such as steroids and waxes.

Chemically, fats and oils are **triglycerides**: esters formed between the alcohol glycerol and long-chained carboxylic acids called **fatty acids**. Glycerol is a 3-carbon alcohol with three hydroxyl groups (**Figure 2**). The three fatty acids that are attached to each molecule of glycerol may be identical (simple triglyceride) or may be different (mixed triglyceride); most fats and oils consist of mixed triglycerides.

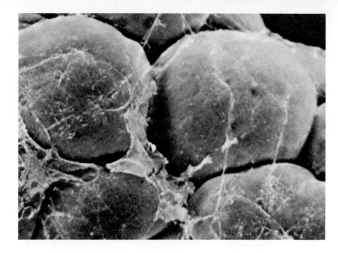

Figure 1
In the aqueous environment of living cells, fats and oils usually exist as droplets. In some animals, they are stored in specialized fat cells called adipocytes (shown here). In many types of plants, such as corn, peanut, and olive, they are stored as oil droplets.

Fatty Acids

The hydrocarbon chains of fatty acids may be 4 to 36 carbons long (**Table 1**). These long carbon chains are generally unbranched, and may be saturated or unsaturated. Except for the carboxylic acid group, fatty acids are very similar to hydrocarbons, and burn as efficiently. When fatty acids are "burned" in the cell, the amount of energy produced is equivalent to that of burning fossil fuels, and is much greater than the energy released from an equal mass of carbohydrates. Lipids, which can be metabolized into fatty acids, are an efficient form of energy storage.

triglyceride an ester of three fatty acids and a glycerol molecule

fatty acid a long-chain carboxylic acid

Table 1 Formula and Source of Some Fatty Acids

Name	Formula	Source
butanoic acid	$CH_3(CH_2)_2COOH$	butter
lauric acid	$CH_3(CH_2)_{10}COOH$	coconuts
myristic acid	$CH_3(CH_2)_{12}COOH$	butter
palmitic acid	$CH_3(CH_2)_{14}COOH$	lard, tallow, palm, and olive oils
stearic acid	$CH_3(CH_2)_{16}COOH$	lard, tallow, palm, and olive oils
oleic acid	$CH_3(CH_2)_7CH{=}CH(CH_2)_7COOH$	corn oil
linoleic acid	$CH_3(CH_2)_4CH{=}CHCH_2CH{=}CH(CH_2)_7COOH$	vegetable oils

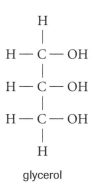

glycerol

Figure 2
Structure of glycerol, 1,2,3-propanetriol

Triglycerides

An example of a simple triglyceride, one with three identical fatty acids, is palmitin. It is found in most fats and oils: palm oil, olive oil, lard, and butter. The fatty acid in palmitin is palmitic acid, $CH_3(CH_2)_{14}COOH$. Palmitin is the tripalmitate ester of glycerol. Another simple triglyceride found in palm oil and olive oil is stearin. The fatty acid that composes this oil is stearic acid, $CH_3(CH_2)_{16}COOH$.

You may recall from our study of esters in Section 1.7 that esters can be hydrolyzed, or broken down into their alcohols and acids. When triglycerides are hydrolyzed with sodium hydroxide, we get glycerol and the sodium salt of the fatty acids. These salts of fatty acids are what we call soap, and the process is called soap making, or **saponification**. Palmitin and stearin from palm oil and olive oil are often used to make soap.

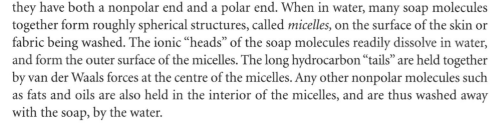

$$CH_3(CH_2)_{14}COO — CH_2$$
$$|$$
$$CH_3(CH_2)_{14}COO — CH + 3\ NaOH \xrightarrow{\text{heat}} 3\ CH_3(CH_2)_{14}COONa + CH_2(OH) — CH(OH) — CH_2OH$$
$$|$$
$$CH_3(CH_2)_{14}COO — CH_2$$

palmitin   sodium palmitate   glycerol
(triglyceride)   (soap: Na$^+$ salt of fatty acid)

ACTIVITY 2.7.1

Making Soap (p. 143)
Now that you know the chemistry, you can make soap from kitchen fats and oils.

Soap molecules are effective in washing fats and oils from fabrics or from skin because they have both a nonpolar end and a polar end. When in water, many soap molecules together form roughly spherical structures, called *micelles,* on the surface of the skin or fabric being washed. The ionic "heads" of the soap molecules readily dissolve in water, and form the outer surface of the micelles. The long hydrocarbon "tails" are held together by van der Waals forces at the centre of the micelles. Any other nonpolar molecules such as fats and oils are also held in the interior of the micelles, and are thus washed away with the soap, by the water.

$$CH_3CH_2CH_2CH_2CH_2CH_2CH_2CH_2CH_2CH_2CH_2CH_2CH_2CH_2CH_2COONa$$

non-polar "tail"   polar "head"

Structure and Properties of Fats and Oils

Fats and oils are largely insoluble in water. For example, an oil and vinegar salad dressing needs to be shaken to mix the two liquids, as the oil does not dissolve in the aqueous vinegar. Butter for a recipe can be measured by immersing the solid fat in water to determine its volume, without losing any in solution. The immiscibility of fats and oils in water is due to the nonpolar nature of the triglyceride molecules. All the polar C=O groups and OH groups are bound in ester linkages, and the extending fatty acids contain long hydrocarbon chains that are nonpolar.

The hydrocarbon chains in fatty acids also affect the physical state of the fat or oil. The shape of the fatty acids dictates how closely the fat and oil molecules can be packed together. This, in turn, affects their melting point (**Table 2**).

DID YOU *KNOW* **?**

Rancid Butter
One of the fatty acids incorporated in the fat of butter is butanoic acid, $CH_3CH_2CH_2COOH$, commonly called butyric acid. It is a foul-smelling liquid at room temperatures, and its presence as a free fatty acid in rancid butter accounts for the distinctive odour.

Table 2   Melting Points of Some Saturated and Unsaturated Fatty Acids

Name	Formula	# Carbons	m.p. (°C)
caproic acid	$CH_3(CH_2)_3COOH$	6	−3
lauric acid	$CH_3(CH_2)_{10}COOH$	12	44
myristic acid	$CH_3(CH_2)_{12}COOH$	14	58
palmitic acid	$CH_3(CH_2)_{14}COOH$	16	63
stearic acid	$CH_3(CH_2)_{16}COOH$	18	70
oleic acid	$CH_3(CH_2)_7CH{=}CH(CH_2)_7COOH$	18	4
linoleic acid	$CH_3(CH_2)_4CH{=}CHCH_2CH—CH(CH_2)_7COOH$	18	−5
linolenic acid	$CH_3CH_2CH{=}CHCH_2CH{=}CHCH_2CH{=}CH(CH_2)_7COOH$	18	−12

If *saturated* hydrocarbon chains are present, the chain can rotate freely about the single C–C bonds. The carbon backbone of each chain has a flexible zig-zag shape, allowing the chains to pack together tightly, with maximum van der Waals interaction between molecules. More thermal energy is therefore needed to separate the saturated fatty acids, leaving these triglycerides solids at 25°C.

However, *unsaturated* hydrocarbon chains *cannot* rotate about their double bonds (**Figure 3**). This restriction introduces one or more "bends" in the molecule, a result of a *cis* configuration at the double bonds (**Figure 4**). These bent fatty acids cannot pack together as tightly, reducing the strength of their van der Waals attractions; triglycerides with these unsaturated fatty acids are generally oils.

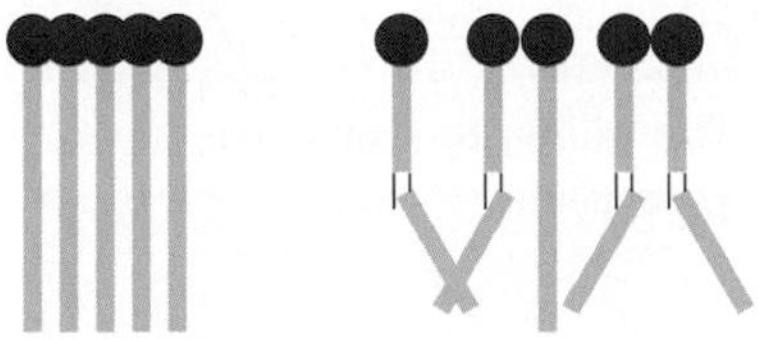

saturated   unsaturated
fatty acids   fatty acids

Figure 3
Saturated fatty acids can pack together much more closely than unsaturated molecules, allowing more bonding.

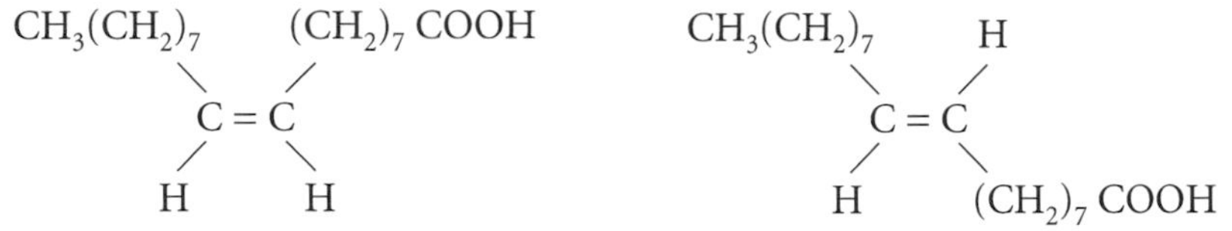

$$CH_3(CH_2)_7 \qquad (CH_2)_7\ COOH$$
$$\backslash \qquad /$$
$$C = C$$
$$/ \qquad \backslash$$
$$H \qquad H$$

cis-oleic acid

$$CH_3(CH_2)_7 \qquad H$$
$$\backslash \qquad /$$
$$C = C$$
$$/ \qquad \backslash$$
$$H \qquad (CH_2)_7\ COOH$$

trans-oleic acid

Figure 4
Geometric isomers differ in the position of attached groups relative to a double bond. In a *cis* isomer, two attached groups are on the same side of the double bond; in a *trans* isomer, they are across from each other.

Vegetable oils contain more unsaturated fatty acids than do animal fats, and are said to be polyunsaturated compounds. When oils such as corn oil or canola oil are made into margarine, the oils are hydrogenated to convert the double bonds into single bonds. The increase in saturation of the hydrocarbon chains leads to a decrease of bending of the hydrocarbon chains, and thus closer packing, converting the liquid oil into a solid. Whether saturated fats made from vegetable oils are better for our health than naturally saturated fats from animal sources is yet unclear.

The high caloric value of fats and oils makes them a good energy source, although their poor solubility in water makes them less readily convertible to energy than carbohydrates. Many animals rely on their fat stores to survive long periods of food deprivation. Bears hibernate for up to seven months each year, oxidizing their stored fat for heat and metabolic processes. The cellular "burning" of fats and oils also produces CO_2 and large amounts of water, which the animal uses to replace lost moisture. Camels are able to make long journeys through the desert using the energy and water released from the stored fat in their humps. Another advantage of lipids is their low density, compared to that of carbohydrates or proteins—a lighter load for a migratory bird to carry.

DID YOU KNOW ?

Waxes—Another Lipid

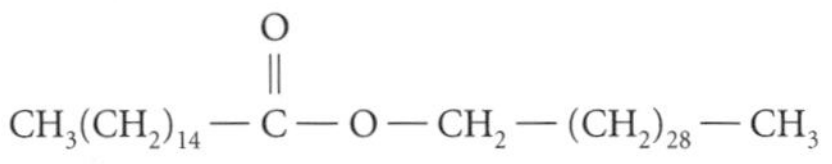

$$CH_3(CH_2)_{14}-\overset{\overset{\displaystyle O}{\|}}{C}-O-CH_2-(CH_2)_{28}-CH_3$$

Beeswax is a type of lipid made by bees to create their combs, in which their larvae grow and honey is stored. Waxes are another form of lipid made by both plants and animals. Being nonpolar, waxes are water-repellent. Beeswax is formed from a long-chain alcohol and a long-chain acid, and may be over 40 carbon atoms long. The long lipid molecules are closely packed, with many interchain forces, so waxes have relatively high melting points (60°C to 100°C).

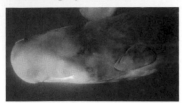

Sperm whales dive 1 to 3 km into cold water to feed on squid. To facilitate descent and ascent, the whale regulates the buoyancy of its body. Its head is almost entirely filled with oil. As the whale dives down, its body cools from 37°C to about 31°C, at which temperature the oil solidifies, becoming more dense. This helps the whale stay at the bottom of the ocean, with its prey. As the whale ascends, it warms up, melting the oil to its more buoyant state.

▸ Practice

Understanding Concepts

1. Draw a structural diagram for the simple triglyceride of oleic acid, $CH_3(CH_2)_7CH{=}CH(CH_2)_7COOH$, the fatty acid found in corn oil.

2. Given the physical properties of corn oil, would you expect the fatty acid components to be saturated or unsaturated? What process may be necessary to convert corn oil into margarine?

3. Write a balanced chemical equation for the saponification of a simple triglyceride of stearic acid.

4. Explain, with the aid of a sketch, why the presence of double bonds in fatty acids tends to lower the melting points of their triglycerides.

5. Suggest reasons why fats and oils are an efficient form of energy storage for living systems.

Applying Inquiry Skills

6. Describe the general conditions required for making soap, given vegetable oil or animal fat as a starting material. List the main reactants and experimental conditions, and safety precautions needed.

▸ *Section 2.7 Questions*

Understanding Concepts

1. Describe three different functions that fats and oils serve in living organisms.

2. Write a balanced chemical equation for the esterification of glycerol with lauric acid, $CH_3(CH_2)_{10}COOH$, the fatty acid found in coconuts.

3. There are claims that some oils in fish, when consumed, can lower blood cholesterol levels. Some of these oils are called omega-3 fatty acids, referring to the presence of a carbon–carbon double bond at the third C atom from the hydrocarbon end. Draw a structural diagram for an omega-3 fatty acid that contains 16 carbon atoms.

4. Distinguish between the terms in the following pairs:
   (a) glycerol and triglyceride
   (b) fatty acids and fats
   (c) fats and oils
   (d) lipids and fats
   (e) saponification and esterification

5. Write a balanced equation to represent the saponification of a triglyceride of myristic acid, $CH_3(CH_2)_{12}COOH$, a fatty acid found in butter.

6. Describe how intramolecular and intermolecular forces act in fats and oils. How do these forces affect their melting points?

7. Explain, with reference to molecular structure, why fats and oils are insoluble in water even though their components, glycerol and fatty acids, contain polar functional groups.

8. (a) Draw structural diagrams to illustrate an example of a saponification reaction.
   (b) Explain why soap molecules are soluble in water as well as in fats and oils.

9. Summarize your learning in this chapter by creating a table of at least 12 polymers synthesized by biological systems. For each polymer, provide the name, list the characteristic functional groups, describe the formation reactions, and describe the function.

Applying Inquiry Skills

10. From what you know of reactions of alkenes, design a test for identifying a fat or an oil as saturated or unsaturated. Describe briefly the materials needed, the procedural steps, and an interpretation of possible results.

11. From your knowledge of reactions of alkenes, suggest a laboratory method to "harden" a vegetable oil into a margarine. Write a chemical equation to illustrate your answer.

Making Connections

12. Commercial drain-cleaners often contain sodium hydroxide. Explain how this ingredient may help to clear a grease-clogged drainpipe.

13. Research the following aspects of linseed oil, and write a report to present your findings:
    (a) chemical composition
    (b) properties
    (c) common uses
    (d) safety precautions in its use and disposal
    (e) classification as natural, organic, or synthetic

 www.science.nelson.com

14. Research to find out why olive oil is considered to be a better dietary choice than coconut oil. Summarize your findings in a pamphlet to be distributed at a cooking class for people recovering from heart attacks.

Organic chemistry is a highly creative science in which chemists can propose, design, and synthesize new molecules for specific purposes. Organic chemists are employed in industry, in academic institutions, and in government, where they contribute to research and development in numerous areas of basic science and technological applications. Most areas of health science, including medicine, nursing, and biotechnology, require a background in organic chemistry.

Pharmaceutical or Polymer Chemist

In industries such as plastics and pharmaceuticals, organic chemists have the exciting challenge of designing and synthesizing new molecules with desired properties. For example, they may determine the structure of a natural antibiotic or antitumour agent, and then modify it to enhance its activity or to decrease undesired side effects. Plastics engineers may develop new materials or new parts for equipment or manufacturing processes. Polymer chemists manipulate the molecular structure of polymers used as adhesives and films, for example.

Environmental Chemist

Organic chemists are involved in many areas of environmental chemistry, such as developing environmentally friendly products to replace potentially hazardous materials, or designing strategies for recycling and waste minimization. Environmental chemists work closely with biologists and toxicologists to evaluate new methods for storage and cleanup of hazardous waste.

▶ **Practice**

Making Connections

1. Select one of the careers described above, or another career in organic chemistry that interests you, and research details of the position. Write a short report to present your findings, including examples of typical projects, features of the career that you find attractive, and training and education requirements.

Computer Analyst in Bioinformatics

Information technology is increasingly applied to handling and interpreting the vast amounts of data collected on large biological polymers. This field is called bioinformatics, and involves the use of computers to store, retrieve, and analyze information such as the sequences of nucleic acids and amino acids. Organic chemists or biochemists can determine sequences of small segments of a molecule and assemble them to obtain a complete sequence. Computer programs produce three-dimensional models of the sequences, allowing predictions of structure and function. Data analysis matches sequences from different organisms to reveal similarities and differences in composition and so, possibly, evolutionary relationships.

Patent Lawyer

Organic chemists are often involved in the research and development of new molecules that may have important properties and applications in industry or in medical fields. These scientists seek to obtain proprietary protection for their work and for the compounds that they have designed. Industries and academic institutions often employ patent lawyers with a background in organic chemistry to represent their interests.

X-Ray Crystallographer

The main task of an X-ray crystallographer is to determine the three-dimensional structure of molecules, from relatively simple organic compounds to incredibly complicated proteins, DNA strands, and even viruses. The crystallographer does this by aiming X rays at the molecules, and (with the help of computer-imaging technology) analyzing the patterns produced as the molecules diffract the X rays. This career requires extensive education and training, usually including a Ph.D. in a university with access to the costly hardware needed. It's also a very specialized field, with a relatively limited range of employment: generally, pharmaceutical companies and university research labs.

⚗ INVESTIGATION 2.1.1

Identification of Plastics

Inquiry Skills

○ Questioning	○ Planning	● Analyzing
○ Hypothesizing	● Conducting	● Evaluating
○ Predicting	● Recording	● Communicating

In recent times, the massive quantities of plastics we throw away have spurred communities to implement recycling programs in an effort to reduce the amount of waste going to landfill sites. As different types of plastics are made of different components, effective recycling requires a systematic identification of each type of polymer. To aid this identification, a resin identification coding system was established in 1988 by the Society of the Plastics Industry, Inc. (SPI) (**Figure 1**). **Table 1** gives the properties and end products of resins identified by their SPI codes.

Figure 1
The triangular symbols on plastic containers allow us to identify the kind of plastic from which they are made.

We will differentiate among several unknown samples of plastics by their density, flame colour, solubility in acetone, and resistance to heating; then we will use the results to identify the plastics and their SPI codes.

Purpose
The purpose of this investigation is to use the properties of plastics to identify unknown samples and their SPI codes.

Question
What is the SPI resin code for each of the six unknown plastic samples?

Experimental Design
The different composition and structures of the plastics allow us to differentiate them by their properties. The samples are placed in liquids of different densities to separate them by flotation. Some samples are identified by their solubility in acetone. Other samples are identified by their resistance to heating. A flame test with a hot copper wire reveals plastics that contain chlorine atoms. The chlorine reacts with the copper wire on heating to produce copper chloride, which colours the flame green.

Materials
lab apron
eye protection
1 cm × 1 cm samples (unknown) of each of the 6 categories of plastics, each cut into an identifiable shape
water
60 g 2-propanol (70%)
corn oil (e.g., Mazola)
50 mL acetone
three 250-mL beakers
100-mL beaker
glass stirring rod
copper wire, 15 cm
cork or rubber stopper
tongs
paper towel
hot plate
lab burner

 Acetone and 2-propanol are highly flammable liquids and must be kept well away from open flames.

Procedure
Part 1   Testing for Density

1. Obtain one sample of each of the 6 plastic materials.

2. Place all 6 samples in a 250-mL beaker containing 100 mL of water and stir with a stirring rod. Allow the samples to settle. Use tongs to separate the samples that float from the samples that sink, and dry each sample with paper towel.

3. Prepare an alcohol solution by weighing out 60 g of 2-propanol (rubbing alcohol) in a 250-mL beaker and adding water to make a total of 100 g. Mix well.

4. Take any samples that float in water and place them in the alcohol solution. Stir and allow the samples to settle. Use tongs to separate the samples that float from the ones that sink, and dry each sample with paper towel.

5. Take any samples that float in the alcohol solution and place them in a 250-mL beaker containing 100 mL of corn oil. Stir and allow the samples to settle for a few minutes. Note any samples that float and any that sink.

Table 1 Codes on Everyday Plastics

SPI resin code	Structure of monomer	Density (g/cm³)	Properties	End products
1 **PETE** polyethylene terephthalate		1.38–1.39	Transparent, strong, impermeable to gas and to oils, softens at approximately 100°C	Bottles for carbonated drinks, containers for peanut butter and salad dressings
2 **HDPE** high density polyethylene		0.95–0.97	Naturally milky white in colour, strong and tough, readily moulded, resistant to chemicals, permeable to gas	Containers for milk, water, and juice, grocery bags, toys, liquid detergent bottles
3 **PVC** vinyl (polyvinyl chloride)		1.16–1.35	Transparent, stable over long time, not flammable, tough, electrical insulator	Construction pipe and siding, carpet backing and window-frames, wire and cable insulation, floor coverings, medical tubing
4 **LDPE** low density polyethylene		0.92–0.94	Transparent, tough, and flexible, low melting point, electrical insulator	Dry cleaning bags, grocery bags, wire and cable insulation, flexible containers and lids
5 **PP** polypropylene		0.90–0.91	Excellent chemical resistance, strong, low density, high melting point	Ketchup bottles, yogurt and margarine containers, medicine bottles
6 **PS** polystyrene		1.05–1.07	Transparent, hard and brittle, poor barrier to oxygen and water vapour, low melting point, may be in rigid or foam form, softens in acetone	Cases for compact discs, knives, spoons, forks, trays; cups, grocery store meat trays, fast-food sandwich containers

Part 2   Testing for Flame Colour

6. Take the samples that sank in water, and test each for flame colour in a fume hood, using a 15-cm length of copper wire attached to a cork or rubber stopper. Holding the cork, heat the free end of the copper wire in a lab burner flame until the wire glows. Touch the hot end of the copper wire to a sample so that a small amount melts and attaches to the wire. Heat the melted sample that is attached to the copper wire in a flame. Record the colour of the flame.

Part 3   Testing with Acetone

7. Ensure that all open flames are extingushed. Of the samples tested in step 6, obtain a fresh sample of any material that did not burn with a green flame. Use tongs to test the samples for softness. Place each of these fresh samples in a 100-mL beaker containing 50 mL of nail polish remover. Watch the samples for a few minutes and note any colour changes. Remove the samples with tongs and test each sample for increased softness.

Part 4   Testing for Resistance to Melting

8. Heat a 250-mL beaker half filled with water on a hot plate until the water comes to a rolling boil. Place any sample that remained unchanged in step 7 into the boiling water and keep the water at a boil for a few minutes. Note any change in shape and softness in the sample.

Evidence

(a) Make a table in which to record the observations you expect from each test.

Analysis

(b) From the information given in **Table 1** and your own observations, identify and give possible SPI codes for each of the 6 samples tested.

Evaluation

(c) Obtain the actual SPI resin codes for each sample from your teacher, and evaluate the reliability of your results. Suggest any changes to the procedure that would improve the reliability.

Synthesis

(d) From your analysis and your understanding of the properties of the plastics tested, design an effective process for a recycling plant to separate plastic products that are collected. Draw a flow chart for such a process and briefly describe the steps involved, any problems you can foresee, and safety precautions required.

 Activity 2.1.1

Making Guar Gum Slime

Guar gum, a vegetable gum derived from the guar plant, has a molar mass of about 220 000–250 000 g/mol. It is used as a protective colloid, a stabilizer, thickener, and film-forming agent for cheese, in salad dressing, ice cream, and soups; as a binding and disintegrating agent in tablet formulations; and in suspensions, emulsions, lotions, creams, and toothpastes. In short, it is a very useful polymer.

In this activity, you will make "slime" by creating a reversible crosslinked gel made from guar gum. The crosslinking is accomplished by adding sodium borate, $Na_2B_4O_7 \cdot 10\ H_2O$, commonly called borax.

Materials

guar gum—food grade from health food stores
water
saturated sodium borate (borax) solution
vinegar
graduated cylinder or measuring spoons
balance or measuring spoons
Popsicle stick or glass rod, for stirring
glass or disposable cup, or beaker
food colouring
sealable plastic bags for storing slime
small funnel and funnel support

Guar gum, if not food grade, is not safe to taste.

Sodium borate is moderately toxic in quantities of more than 1 g/1 kg of body weight. Wash any borax from hands with water. Wash hands after handling the slime.

Slime may stain or mar clothing, upholstery, or wood surfaces. Clean up spills immediately by wetting with vinegar, followed by soapy water.

- Measure 80 mL of water into the cup.
- Add one to two drops of food colouring if desired.
- Measure 0.5 g of guar gum (1/8 tsp). Add it to the water and stir until dissolved. Continue stirring until the mixture thickens (approximately 1 or 2 min).
- Add 15 mL (3 tsp) of saturated borax solution, and stir. The mixture will gel in 1–2 min. Let the slime sit in the cup for a few minutes to gel completely. (To store the slime for more than a few minutes, put it into a plastic bag and seal the top.)

(a) Describe what happens to the slime when you pull it slowly.

(b) Describe what happens to the slime when you pull it sharply.

- Put some slime on a smooth, hard surface and hit it with your hand.

(c) Describe what happens to the slime.

- Place a funnel on a funnel support. Put some slime into the funnel. Push it through the funnel.

(d) Describe what happens as the slime comes out of the hole.

- Take about 20 mL of the slime and add about 5 mL of vinegar.

(e) Describe any changes in the properties of the slime.

(f) Using your knowledge of the structure and bonding of polymers, explain each of your observations.

- Dispose of the slime as directed by your teacher.

Investigation 2.2.1

Preparation of Polyester— a Condensation Polymer

Inquiry Skills		
○ Questioning	○ Planning	● Analyzing
○ Hypothesizing	● Conducting	● Evaluating
● Predicting	● Recording	● Communicating

Condensation polymers are produced by reactions between functional groups on monomers, with the elimination of a small molecule such as water.

The reaction between an alcohol with two or more hydroxyl groups (such as glycerol) and an acid with two or more carboxyl groups produces a polyester known as an alkyd resin, commonly used in making paints and enamels. One of these is the Glyptal resin.

Glyptal is a thermoset polymer, meaning that it solidifies or "sets" irreversibly when heated. The heating causes a crosslinking reaction between polymer molecules. (This is the effect seen when proteins such as egg whites (**Figure 2**), another thermoset polymer, are cooked.) In Glyptal resin, phthalic anhydride forms crosslinks with other glycerol molecules, holding them together in a polymer.

Purpose

The purpose of this investigation is to test the principle of combining monomers to produce a thermoset polymer: glyptal.

Figure 2
Egg white, when cooked, changes its structure irreversibly.

Question

What are the properties of the polymer produced from the reaction of orthophthalic acid and glycerol?

orthophthalic acid   glycerol

Prediction

(a) Write the equation for the reaction you predict will occur.

(b) Predict the properties that you will observe, if the polymerization reaction takes place.

Experimental Design

The component monomers, glycerol and orthophthalic acid (in the form of phthalic anhydride, are mixed and heated together to form the thermoset polymer, Glyptal. The reaction is an esterification reaction, and any water produced is removed by boiling the mixture. The product is then tested for solubility in a nonpolar solvent.

Materials

lab apron
eye protection
gloves
ethylene glycol (1,2-ethandiol), 2 g
glycerol, 2 g
phthalic anhydride powder, 3 g
solvent (paint thinner or nail polish remover), 5 mL
two 100-mL beakers
glass stirring rod
beaker tongs
watch glass to fit beaker
hot plate
small metal container (e.g., aluminum pie dish)

 Phthalic anhydride is toxic and a skin irritant. Handle with care, and wear gloves, eye protection, and a lab apron.

The solvent is flammable, so must be kept well away from any open flame.

Procedure

1. Place 2 g of glycerol and 3 g of phthalic anhydride in a 100-mL beaker and mix with a glass stirring rod.

2. Heat gently on a hot plate, while stirring, to dissolve all the solid.

3. Cover with a watch glass and continue heating gently until the mixture boils, and then boil for five minutes.

4. Carefully pour the solution into a metal container. Let the plastic cool completely at room temperature.

5. Observe and record the properties of the plastic formed.

6. Place about 5 mL of the solvent supplied in a beaker and try to dissolve a portion of the plastic in the solvent.

7. Allow the solvent containing any dissolved plastic to evaporate and observe any residue formed.

8. Dispose of materials according to your teacher's instructions.

Analysis

(c) Describe the properties of the product that you observed. Do they match your prediction?

(d) What makes this product suitable for use in household paints?

Evaluation

(e) Evaluate the concept that heating two component monomers together will produce a polymer with the desired properties.

Synthesis

(f) Suppose this experimental design were repeated using 1,2-ethanediol in place of glycerol. Predict the reaction and properties of its product. Explain why your predictions are different from those made in (a) and (b), illustrating your answer with molecular structures.

⌕ **ACTIVITY 2.7.1**

Making Soap

Fats and oils are triglycerides of glycerol and fatty acids. The ester linkages in the triglycerides are broken in the presence of a strong base, such as NaOH, and heat. This reaction is called saponification and the sodium salt of the fatty acids produced is called soap. The other product formed is glycerol.

Materials

lab apron; eye protection; fats (lard or vegetable shortening); oils (cooking oils such as corn oil, canola oil, olive oil); NaOH pellets; ethanol; vinegar; $NaCl_{(s)}$; distilled water; food colouring and perfume (optional); 250-mL beaker; two 100-mL beakers; forceps; glass stirring rods; beaker tongs; filter funnel and paper; ring stand and ring clamp; hot plate; balance

 Ethanol is flammable. Ensure that there are no open flames.

 Sodium hydroxide pellets are extremely corrosive to eyes and skin. They must be handled with forceps. If NaOH comes into contact with skin, rinse with copious amounts of cold water. If it is splashed in the eyes, flush with water at an eyewash station for at least 10 minutes, then get medical attention.

Procedure

- Put on a lab apron and eye protection.
- Set up a 100-mL beaker and label it Beaker A. (See **Table 2** for a summary of the contents of Beaker A and other beakers used in this activity.) Using forceps, add 18 pellets of solid NaOH to Beaker A. Do not allow the pellets to touch your skin. Add 10 mL of distilled water to the NaOH pellets and stir with a glass rod to dissolve. Set this beaker aside.
- Set up a 250-mL beaker and label it Beaker B. Add 15 g of a fat, such as lard or shortening, or an oil such as corn oil or olive oil to Beaker B. Add 15 mL of ethanol to the fat or oil and warm the mixture very gently on a hot plate, stirring with a glass rod to dissolve.
- Pour the contents of Beaker A into Beaker B and heat the mixture gently on the hot plate (low setting). Stir the mixture continuously for at least 20 min. If the mixture bubbles or splatters, use tongs to remove the beaker from the hot plate. Then return the beaker to the hot plate when the mixture has cooled slightly.
- When the reaction is complete, the mixture should thicken and have the appearance and consistency of a

creamy pudding. Remove the beaker from the heat and allow to cool. You may add a drop or two of food colouring to the mixture at this stage to colour the soap.

- Set up another 100-mL beaker and label it Beaker C. Add 4 g of NaCl and 20 mL of cold distilled water. Stir to dissolve.
- Add the cold salt solution from Beaker C to the soap mixture in Beaker B. This should cause the soap to precipitate from the solution.
- Add 10 mL of vinegar to the mixture to neutralize any excess NaOH. Pour off any liquid into the sink.
- Add 10 mL of distilled water to wash the excess vinegar off the soap. Pour off any liquid into the sink.
- Set up a filter funnel and filter paper. Pour the soap mixture into the funnel, taking care not to puncture the filter paper. If desired, you may add a few drops of perfume or scent to the soap at this stage.
- The soap remains in the filter funnel and can be left in the filter paper to dry, or it can be taken out of the filter paper and shaped and left to dry on a paper towel.

Table 2 Contents of Beakers A, B, and C

Beaker A (100 mL)	Beaker B (250 mL)	Beaker C (100 mL)
18 pellets of $NaOH_{(s)}$	15 g of fat or oil	4 g NaCl
10 mL distilled water	15 mL of ethanol	20 mL cold distilled water
stir to dissolve	warm gently on hot plate to dissolve	stir to dissolve

Analysis

(a) Why is a saponification reaction considered to be the reverse of an esterification reaction?

(b) Compare the soaps made by other students in the class. Is there any difference in hardness of the soaps made from different fats and oils? Explain your answer.

(c) Are soap molecules polar or nonpolar? Draw a sketch of a soap molecule to illustrate your answer.

(d) From your knowledge of fatty acids, what are some possible fatty acids present in the fat or oil you used? Draw structural diagrams for a possible triglyceride in the fat or oil used to make your soap, and for a possible soap molecule in your soap.

(e) What substances may be in the filtrate after the soap was filtered out?

Key Expectations

Throughout this chapter, you have had the opportunity to do the following:

- Describe some physical properties of the classes of organic compounds in terms of solubility in different solvents, molecular polarity, odour, and melting and boiling points. (2.1, 2.5)
- Demonstrate an understanding of the processes of addition and condensation polymerization. (2.1, 2.2, 2.4, 2.5)
- Describe a variety of organic compounds present in living organisms, and explain their importance to those organisms. (2.4, 2.5, 2.6, 2.7)
- Use appropriate scientific vocabulary to communicate ideas related to organic chemistry. (all sections)
- Predict and correctly name (using IUPAC and nonsystematic names) the products of organic reactions. (all sections)
- Carry out laboratory procedures to synthesize organic compounds. (2.1, 2.2, 2.7)
- Present informed opinions on the validity of the use of the terms "organic," "natural," and "chemical" in the promotion of consumer goods. (2.1, 2.2)
- Describe the variety and importance of organic compounds in our lives. (all sections)
- Analyze the risks and benefits of the development and application of synthetic products. (2.1, 2.2, 2.3)
- Provide examples of the use of organic chemistry to improve technical solutions to existing or newly identified health, safety, and environmental problems. (2.1, 2.2, 2.3, 2.4)

Key Terms

addition polymer
aldose
alpha-helix
amino acid
carbohydrate
cellulose
chiral
condensation polymer
deoxyribonucleic acid (DNA)
dimer
dipeptide
disaccharide
double helix
fatty acid
glycogen
ketose
macromolecules
monomer
monosaccharide
nucleotide
peptide bond
plastic
pleated-sheet conformation
polyamide
polyester
polymer
polymerization
polypeptide
polysaccharide
primary structure
quaternary structure
ribonucleic acid (RNA)
saponification
secondary structure
starch
tertiary structure
triglyceride

Problems You Can Solve

- Name and draw the structures of the monomers and polymers in addition polymerization reactions.
- Predict the polymers that will form by addition reactions of given monomers.
- Name and draw the structures of the monomers and polymers in condensation polymerization reactions.

▶ **MAKE** a summary

- Make a table with the following column headings: Polymer; Monomer; Example; Structure; Properties
- Under the Polymer heading, enter these row headings: synthetic polyesters; synthetic polyamides; proteins; nucleic acids; and carbohydrates.
- Complete the table.

Identify each of the following statements as true, false, or incomplete. If the statement is false or incomplete, rewrite it as a true statement.

1. Polymers such as nylon are formed from monomer subunits that are identical.

2. The vulcanization of rubber involves sulfur in forming crosslinkages.

3. Thermoplastics can be softened by heat and moulded into a new shape.

4. Amino acids contain a hydrocarbon chain with an amino group at one end and a carboxyl group at the other end.

5. Starch, sucrose, and cellulose are polysaccharides of the monosaccharide glucose.

Identify the letter that corresponds to the best answer to each of the following questions.

6. Polyethylene is
   (a) an addition polymer of ethyl ethanoate.
   (b) a condensation polymer of ethene.
   (c) a condensation polymer of ethyl ethanoate.
   (d) a saturated hydrocarbon.
   (e) an unsaturated hydrocarbon.

7. A monomer of a condensation polymer
   (a) contains an amino group.
   (b) contains a carboxyl group.
   (c) contains a double bond.
   (d) reacts in a condensation reaction.
   (e) reacts to form an ester.

8. The polymer produced from the polymerization of HOOCCOOH and $HOCH_2CH_2OH$ is
   (a) $[-OOCCOCH_2CH_2O-]_n$
   (b) $[-OCH_2CH_2OOCCO-]_n$
   (c) $[-OOCOOCH_2CH_2O-]_n$
   (d) $[-OCCH_2OOCCH_2O-]_n$
   (e) $[-COOHCH_2CH_2CO-]_n$

9. Carbohydrates
   (a) are polymers of glucose.
   (b) are polymers of sucrose.
   (c) are polymers that make up enzymes.
   (d) contain only carbon atoms and water molecules.
   (e) contain only carbon, hydrogen, and oxygen atoms.

10. Amino acids
    (a) are the monomers of nucleic acids.
    (b) are the monomers of fatty acids.
    (c) are the monomers of proteins.
    (d) polymerize by forming ester linkages.
    (e) undergo condensation reactions to eliminate ammonia molecules.

11. The process by which triglycerides are broken down in the presence of NaOH to glycerol and the sodium salts of fatty acids is called
    (a) condensation.   (d) saponification.
    (b) elimination.   (e) substitution.
    (c) hydrogenation.

12. Deoxyribonucleic acid
    (a) acts as energy storage for the cell.
    (b) determines the primary structure of proteins.
    (c) forms peptide bonds with the amine group of amino acids.
    (d) is a polymer of the monomers adenine, lysine, glycine, and cysteine.
    (e) provides structure and support in the nucleus of the cell.

13. Fats are different from oils in that they
    (a) are more soluble in aqueous solvents.
    (b) contain fatty acid chains that are more closely packed together.
    (c) contain more carbon–carbon double bonds.
    (d) have lower melting points.
    (e) undergo condensation reactions with glycerol.

14. The secondary and tertiary structures of proteins
    (a) are a result of hydrogen bonding between the ribose–phosphate backbone of the polymer chains.
    (b) are determined by the primary structure, the sequence of nucleic acids in the protein.
    (c) indicate the position of the hydroxyl groups in the carbon chain of the polymer.
    (d) result from hydrogen bonding between the adenine and thymine bases and the cytosine and guanine bases.
    (e) result in the protein being globular, as in some enzymes, or fibrous, as in muscle fibres.

15. There is an enormous variety of proteins because
    (a) DNA consists of at least 20 amino acids which code for proteins.
    (b) DNA consists of 4 different amino acids which code for proteins.
    (c) proteins are made up of 20 different amino acids that can combine to form millions of sequences.
    (d) proteins are made up of 4 different amino acids in coded sequence.
    (e) proteins are synthesized from groups of three amino acids for each nucleotide.

NEL   An interactive version of the quiz is available online.
GO   www.science.nelson.com

Polymers—Plastics, Nylons, and Food   **145**

Understanding Concepts

1. Orlon is the name of a synthetic material used to make fabric and clothing. It is an addition polymer. A portion of its structure is shown below.

$$-CH_2-CH-CH_2-CH-CH_2-CH-$$
$$\qquad\quad |\qquad\qquad\quad |\qquad\qquad\quad |$$
$$\qquad\quad CN\qquad\qquad CN\qquad\qquad CN$$

Draw a structural diagram of its monomer. (2.1)

2. Lactic acid (2-hydroxy-propanoic acid), produced by bacterial culture in yogurt, contains a hydroxyl group and a carboxyl group in each molecule. Draw a structure of two repeating units of a condensation polymer of lactic acid molecules. (2.2)

3. Crosslinking between polymer strands contributes to the elastic properties of a polymer.
    (a) Explain briefly why crosslinking increases the elasticity of a polymer.
    (b) What structural features of a monomer are needed for crosslinking to occur between polymer chains? (2.2)

4. For each of the polymers whose structures are shown below, identify any possible bonding interactions within the same chain, and between chains. Describe the type of structure that results from the identified interactions.

    (a)
    $$\left[-CH_2CH(CH_2)_m\overset{\displaystyle O}{\overset{\displaystyle \|}{C}}O-\right]_n$$
    $$\qquad\quad |$$
    $$\qquad\quad OH$$

    (b)
    $$\left[-CH_2CH(CH_2)_m-\right]_n$$
    $$\qquad\qquad |$$
    $$\qquad\qquad CH=CH_2$$

    (c)
    $$\left[-\overset{\displaystyle O}{\overset{\displaystyle \|}{N}C(CH)_m}-\right]_n$$
    $$\qquad |$$
    $$\qquad H$$

5. Use molecular model kits to build models of the following compounds and use them to show the elimination of a small molecule in a condensation polymerization reaction. (2.2)

    (a)
    $$\qquad\quad\overset{\displaystyle O}{\overset{\displaystyle \|}{}}\;\overset{\displaystyle O}{\overset{\displaystyle \|}{}}$$
    $$HO-C-C-OH\quad\text{and}\quad H_2NCH_2CH_2NH_2$$

    (b)
    $$\qquad\qquad\quad\overset{\displaystyle O}{\overset{\displaystyle \|}{}}$$
    $$H_2N-CH_2-C-OH$$

6. (a) Use molecular model kits to build a molecule of 1,4-benzenedioic acid.

    (b) Predict whether this compound can be a monomer of an addition polymer or a condensation polymer. Explain. (2.2)

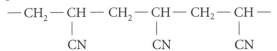

7. The terms "primary," "secondary," "tertiary," and "quaternary" are used to describe protein structure.
    (a) Draw a sketch to illustrate each type of structure of a protein molecule.
    (b) Explain how each type of structure plays an important role in the function of proteins in an organism. (2.4)

8. Describe the importance of each of the following polymers of glucose to an organism, and relate the function of each to its molecular structure:
    (a) starch   (c) glycogen   (2.5)
    (b) cellulose

9. Write a balanced chemical equation to illustrate each of the following reactions of sucrose:
    (a) complete combustion   (b) hydrolysis   (2.5)

10. Name and describe the general structural features of the monomers of each of the following biological polymers. Include a structural diagram of each.
    (a) a protein   (b) DNA   (2.6)

11. Identify the functional groups that are involved in the formation of each of the following polymers, and name the type of reaction that occurs.
    (a) starch from its monomers
    (b) fats and oils from their components   (2.7)

12. Products such as canola oil are advertised as "polyunsaturated." Are there any double bonds in the molecules of polyunsaturated oils? Explain.   (2.7)

13. Write a balanced chemical equation for the esterification of glycerol with myristic acid, $CH_3(CH_2)_{12}COOH$, a fatty acid found in butter. (2.7)

14. (a) Write structural diagrams of the products of complete hydrolysis of the following triglyceride.

    $$CH_3(CH_2)_7CH=CH(CH_2)_7COOCH_2$$
    $$\qquad\qquad\qquad\qquad\qquad\qquad\quad |$$
    $$CH_3(CH_2)_{10}COOCH$$
    $$\qquad\qquad\qquad\qquad\qquad\quad |$$
    $$CH_3(CH_2)_4CH=CHCH_2CH=CH(CH_2)_7COOCH_2$$

    (b) Predict the physical state of the triglyceride above at ordinary room temperatures. Illustrate your answer with a diagram.   (2.7)

15. Explain, with reference to molecular structure, why
    (a) the caloric value of 100 g of lard is greater than the caloric value of 100 g of table sugar.
    (b) a substantial amount of energy is released from burning 1 kg of paper, but we would not derive any energy from eating the same mass of paper. (2.7)

Applying Inquiry Skills

16. Describe the physical or chemical properties that you would investigate to help you to identify a polymer whose structure includes each of the following features. Give reasons for your answer.
    (a) crosslinking between polymer chains
    (b) presence of carbonyl groups on the polymer chain
    (c) a carbon backbone saturated with halogen atoms (2.2)

17. DNA can be extracted from onion cells by the following procedure.

    • Coarsely chopped onion is covered with 100 mL of an aqueous solution containing 10 mL of liquid detergent and 1.5 g of table salt.
    (a) Explain why this step dissolves the fatty molecules in the cell membranes.

    • The mixture is heated in a hot-water bath at 55–60°C for 10 min, while gently crushing the onion with a spoon.
    (b) Explain how this separates the DNA strands.

    • The mixture is then cooled in an ice bath for 5 min.
    (c) Suggest a purpose for this step.

    • The mixture is filtered through four layers of cheesecloth to remove protein and lipid. Ice-cold ethanol is added to the filtrate to visibly precipitate the DNA, which can be spooled onto a glass rod.
    (d) Explain why DNA is insoluble in alcohol. (2.6)

18. Plant oils from different sources often have different degrees of saturation in their fatty acids. Consider the two cooking oils, olive oil and coconut oil, and design a procedure to compare relative saturation of their fatty acids. Describe steps in the procedure, the materials used, and an interpretation of possible results. (2.7)

Making Connections

19. Give five examples of different synthetic polymers that are important in our lives. For each example, identify the monomer and the type of polymerization reaction involved. (2.2)

20. Many synthetic products arise from studying the structure of biological substances, which are then copied in the lab. Give several examples of natural substances for which a synthetic equivalent has been developed, and comment on the advantages and disadvantages of each. Consider a variety of points of view. (2.2)

21. When we purchase groceries, we are sometimes offered a choice of paper or plastic bags.
    (a) Which material is stronger? Explain your answer.
    (b) Describe and explain the different rates of decomposition of these two materials.
    (c) In your opinion, is each of these materials correctly labelled as organic, natural, or chemical? Justify your answer. (2.5)

22. Linoleum is a term that describes a natural material used as a floor covering, although common usage of the term includes synthetic versions made of vinyl polymers. "Real" linoleum was originally made from wood flour, cork, limestone dust, rosin from pine trees, and linseed oil. The mixture was coloured with organic pigments and baked onto a jute backing.
    (a) According to your assessment of the ingredients listed above, would real linoleum be classed as a polymer? Is it organic, natural, or chemical? Give reasons for your answers.
    (b) Manufacturers caution that high pH cleaning agents should not be used with real linoleum, but that it is resistant to acid. Give chemical reasons for this information.
    (c) Manufacturers of real linoleum claim that their product is environmentally friendly because it is biodegradable; when no longer needed, it can be shredded and turned into compost. What are the advantages and disadvantages of this property?
    (d) Suppose that you are choosing between real linoleum and a vinyl material for your kitchen floor, and cost is comparable. Describe the factors that you would consider, and give reasons for your choice. (2.7)

23. Monosodium glutamate (MSG) is a derivative of glutamic acid. It is used widely as a flavour enhancer in canned soups and meats, sauces, and potato chips.
    (a) Draw a structural diagram for MSG.
    (b) Could this molecule have two different conformations? Explain.
    (c) Research and report on any concerns about the use of MSG in foods. (2.4)

 www.science.nelson.com

Chemistry in a Bathtub

It is time for a well-deserved relaxing bath in the tub! Have you ever tried one of those scented bath bombs that add fizz and oils to your bath water (**Figure 1**)? Well, your task as an organic chemist is to apply your knowledge and skills to making one of these bath bombs, using available materials and equipment in the school laboratory. Of course, as with any consumer product, you will also need to provide a detailed product information report, listing and describing all the ingredients and packaging of your product.

Your Task

You are given a procedure for making a bath bomb, and all the ingredients will be provided except for the fragrance, which you will need to synthesize. The task is divided into three parts.

1 Research, Plan, and Synthesize an Ester

Experimental Design
(a) Research the composition of esters with known odours and select one that can be synthesized from the available chemicals and that would be appropriate for a bath bomb.

(b) Create a plan for synthesizing the ester, using appropriate terminology, molecular structures, and flow charts.

Materials
(c) Generate a Materials list. You will have available a list of chemicals that you may use as starting materials. You may use common laboratory equipment for the synthesis.

Procedure
(d) Write a Procedure for the synthesis. Include safety precautions and procedures for disposal of materials. Hand the Procedure in to your teacher for approval.

1. After your Procedure has been approved, carry it out.

2. Hand in a sample of your ester to the teacher for evaluation.

2 *Make the Bath Bomb*

Materials

citric acid, 40 mL

sodium hydrogen carbonate, 125 mL

vegetable oil, 60 mL (e.g., olive oil,
   coconut oil, corn oil)

stirring rod

cornstarch, 40 mL

ester (fragrance), several drops

600-mL beaker

100-mL beaker

250-mL graduated cylinder

Procedure

1. In the large beaker, mix together 40 mL of citric acid, 40 mL of cornstarch, and 125 mL of sodium hydrogen carbonate.

2. In the small beaker, mix 60 mL of vegetable oil and a few drops of the ester for fragrance.

3. Pour the contents of the small beaker into the large beaker and mix well with a stirring rod.

4. Form the mixture into a ball or other shape of your choice. Set the bath bomb on a sheet of wax paper and allow to dry completely (approximately 24–48 h).

5. Package the bath bomb in materials of your choice (e.g., tissue paper, polyester or silk ribbon, Styrofoam pellets), and store the entire package in clear plastic wrap or a clear plastic bag.

3 *Prepare a Product Information Package*

(e) Prepare an attractive consumer product information package, either in print pamphlet format, or as a computer presentation or web page. The product information package must contain the following features:

- A list of all materials used, including all starting materials, those in the product, and all packaging materials.

- Chemical names for each non-polymer compound, using the IUPAC system. Include common nonsystematic names for chemicals, where suitable.

- Identification of each chemical by organic family; e.g., alcohol, carboxylic acid, condensation polymer, etc.

- A discussion of the physical properties of each compound.

- A discussion of the relationship between the structure and properties of several representative compounds used, including 3 non-polymers, a natural polymer, and a synthetic polymer.

- An explanation of the chemical reaction that will occur when the bath bomb is immersed in water, with a discussion of any safety concerns in its use.

- A label for the product identifying its organic, natural, or chemical nature, and a justification for your choice.

Identify each of the following statements as true, false, or incomplete. If the statement is false or incomplete, rewrite it as a true statement.

1. Carbonyl groups are present in alcohols, ethers, aldehydes, ketones, and esters.
2. When a primary alcohol is mildly oxidized, an aldehyde is produced; when a secondary alcohol is mildly oxidized, a ketone is produced.
3. The formation of an alcohol when an alkene reacts with water in the presence of an acid is an example of a hydrolysis reaction.
4. Benzene reacts readily with bromine in addition reactions at its double bonds.
5. When methanol and vinegar are allowed to react, ethyl methanoate and water are produced from the esterification reaction.
6. Diethylether is a structural isomer of 2-butanol, and hexanal is a structural isomer of 3-hexanone.
7. 1,2-dibromoethane can be produced from the substitution reaction of bromine with ethene.
8. Polybutene is formed from addition reaction of butene monomers, and the polymer chain consists of carbon atoms single bonded to each other, with ethyl groups attached to each carbon atom in the chain.
9. Condensation polymers such as polystyrene and polypropylene may have physical properties such as flexibility and strength as a result of the degree of crosslinkages present in the polymer.
10. In proteins, the amino acid monomers are held together by amide linkages between the amino group of one amino acid to the carboxyl group of the adjacent amino acid.

Identify the letter that corresponds to the best answer to each of the following questions.

11. What is the name of the compound whose structure is shown below?

H
CH3CH2 — N — CH2CH2CH3

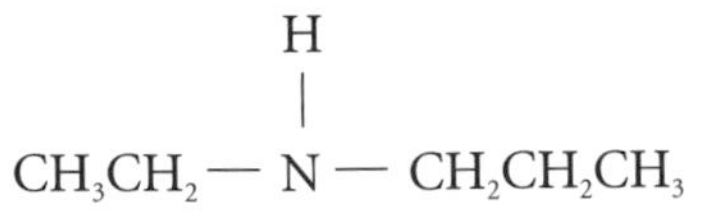

(a) 2-aminopentane
(b) 2-nitropentane
(c) ethylpropylamide
(d) ethylpropylamine
(e) pentylamino acid

12. What is the name of the compound whose structure is shown below?

CH2CH3
CH2CH3
CH3CHCHCH2CHCHCH3
Br
NH2

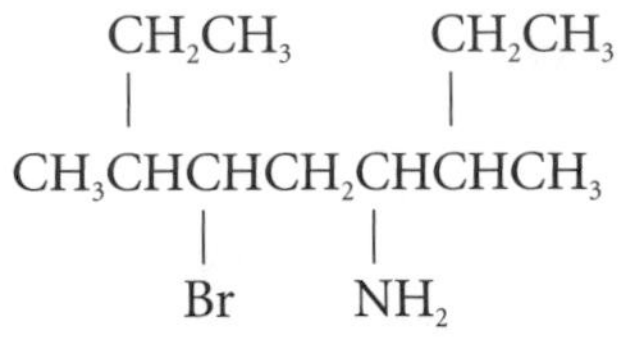

(a) 3-amino-5-bromo-2,2-diethylheptane
(b) 3-amino-5-bromo-2,6-diethylheptane
(c) 3-bromo-5-amino-2,6-diethylheptane
(d) 4-amino-6-bromo-3,7-diethyloctane
(e) 4-amino-6-bromo-3,7-dimethylnonane

13. A compound belonging to the organic family of aromatic compounds is
(a) an aldehyde or a ketone with a distinctive odour.
(b) a compound containing a benzene ring in its structure.
(c) a compound containing a cyclic structure.
(d) a compound containing a cyclic structure with a double bond.
(e) an ester with a pleasant aroma.

14. Which of the following is NOT a correct description of the structure shown:

CH3

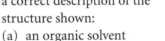

(a) an organic solvent
(b) methylbenzene
(c) phenylmethane
(d) toluene
(e) vanillin

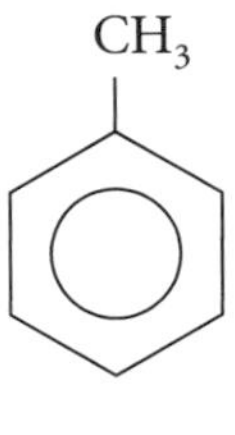

15. What is the compound whose structure is shown?

O
CH3(CH2)7CH=CH(CH2)7C—OCH2
O
CH3(CH2)7CH=CH(CH2)7C—OCH
O
CH3(CH2)7CH=CH(CH2)7C—OCH2

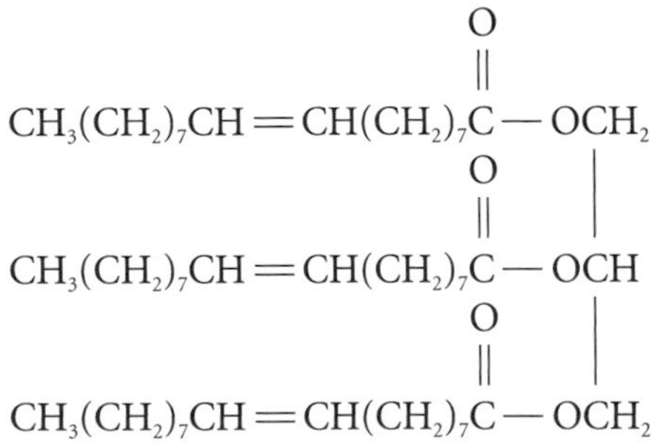

(a) a nucleotide
(b) a product of a saponification reaction
(c) a saturated fatty acid
(d) a triglyceride
(e) an unsaturated fatty acid

16. The formula of acetone is
(a) CH_3CHO
(b) CH_3OCH_3
(c) CH_3COCH_3
(d) CH_3COOCH_3
(e) CH_3CH_2COOH

An interactive version of the quiz is available online.
GO www.science.nelson.com

17. Of the following compounds, which has the highest solubility in water?

    (a) $CH_3CH_2CH=CHCH_3$

    (b) $CH_3CH_2\overset{\overset{\displaystyle O}{\|}}{C}\underset{\underset{\displaystyle CH_3}{|}}{C}HCH_3$

    (c) $CH_3CH_2CH_2CH_2\overset{}{\underset{\underset{\displaystyle O}{\|}}{C}}-OH$

    (d) $CH_3CH_2\underset{\underset{\displaystyle OH}{|}}{C}HCH_2CH_3$

    (e) $CH_3CH_2CH_2\overset{}{\underset{\underset{\displaystyle O}{\|}}{C}}-OCH_3$

18. The formula for methyl acetate is:

    (a) $CH_3CH_2COCH_3$ (d) CH_3COOCH_3
    (b) $CH_3COOCH_2CH_3$ (e) $CH_3CHOHCH_3$
    (c) $CH_3CH_2OOCH_2CH_3$

19. When 1-butene undergoes a hydration reaction, the product is

    (a) 1-butanal   (d) 2-butanol
    (b) 1-butanol   (e) 2-butanone
    (c) 2-butanal

20. The amide represented by the structural formula

    $CH_3-\overset{\overset{\displaystyle O}{\|}}{C}-\underset{\underset{\displaystyle CH_3CHCH_3}{|}}{N}-CH_2CH_3$

    can be synthesized from
    (a) acetic acid and a secondary amine
    (b) ethanoic acid and a primary amine
    (c) ethanoic acid and a tertiary amine
    (d) formic acid and a secondary amine
    (e) methanoic acid and a primary amine

21. Of the following compounds — aldehyde, amide, carboxylic acid, ester, ketone — all of analogous size, the compound with the highest boiling point is the
    (a) aldehyde   (d) ester
    (b) amide   (e) ketone
    (c) carboxylic acid

22. Which of the following compounds is an ether?

    (a) $CH_3\overset{\overset{\displaystyle O}{\|}}{C}CH_2CH_3$

    (b) $CH_3CH_2CH_2\overset{\overset{\displaystyle O}{\|}}{C}H$

    (c) $CH_3CH(OH)CH_3$
    (d) $CH_3CH_2CH_2OCH_3$
    (e) $CH_3CH_2COOCH_3$

23. A monomer of an addition polymer must
    (a) contain a carboxyl group
    (b) contain a double bond at each end of the monomer
    (c) contain at least one double bond
    (d) react to form crosslinkages
    (e) react to form an ester

24. The polymer produced from the polymerization of H_2NCH_2COOH is
    (a) $[-COONHCH_2CH_2COO-]_n$
    (b) $[-HNCH_2CONHCH_2CO-]_n$
    (c) $[-HNCH_2COOCCH_2NH-]_n$
    (d) $[-HNCOCH_2OCNH-]_n$
    (e) $[-OOCH_2NHOOCH_2-]_n$

25. Deoxyribonucleic acid, DNA, is a polymer made up of monomers
    (a) linked in the polymer chain by amide linkages
    (b) linked in the polymer chain by hydrogen bonds
    (c) of glucose units
    (d) of nucleic acids
    (e) of nucleotides

26. The compound whose formula is
    $CH_3(CH_2)_4CH=CHCH_2CH=CH(CH_2)_7COOH$
    (a) is a fat at ordinary room temperatures
    (b) is an oil at ordinary room temperatures
    (c) will undergo a hydrolysis reaction to form an oil
    (d) will undergo an addition reaction to form a polysaccharide with glycerol
    (e) will undergo an esterification reaction to form a triglyceride with glycerol

Understanding Concepts

1. For each of the following compounds, identify the organic family to which it belongs:
   (a) 1-propanol
   (b) CH_3CH_2COOH
   (c) hexanal
   (d) $CH_3CH_2OCH_2CH_2CH_3$
   (e) CH_3NH_2
   (f) 2-pentanone
   (g) propyl ethanoate
   (h) $CH_3CH_2CONHCH_3$
   (i) $CH_3CH(CH_3)CCH_2CH_3$
   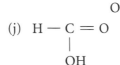

   (j) H — C = O
       |
       OH

2. Name the functional group(s) in each of the following compounds:
   (a) 2-hexanone
   (b) 2-methylpentanal
   (c) 1,3-pentandiol
   (d) 1,3-butadiene
   (e) butanamide
   (f) propoxybutane
   (g) ethyl ethanoate   (1.8)

3. Draw a structural diagram for each of the following organic compounds:
   (a) 2-ethyl-4-methyl-2-pentanol
   (b) 1,2-ethandiol
   (c) 1,3-dimethylbenzene
   (d) 1,2-dichloropropane
   (e) 2-methylbutanal
   (f) 3-hexanone
   (g) 1-ethoxypropane
   (h) 2-aminoethanoic acid
   (i) 2,2-dichloropropane
   (j) cyclohexanol   (1.8)

4. Give the IUPAC name and draw the structural formula for the compounds with the following nonsystematic names:
   (a) toluene
   (b) acetone
   (c) acetic acid
   (d) formaldehyde
   (e) glycerol
   (f) diethyl ether   (1.8)

5. Identify the functional groups in each of the following molecules:
   (a) testosterone (hormone)

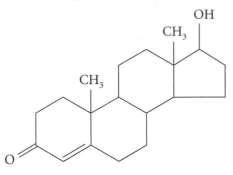

   (b) ibuprofen (pain reliever)

   $C_{12}H_{16}COOH$

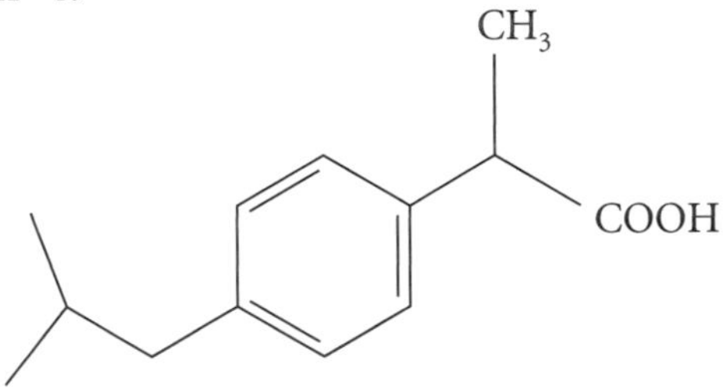

   (c) amphetamine (stimulant)

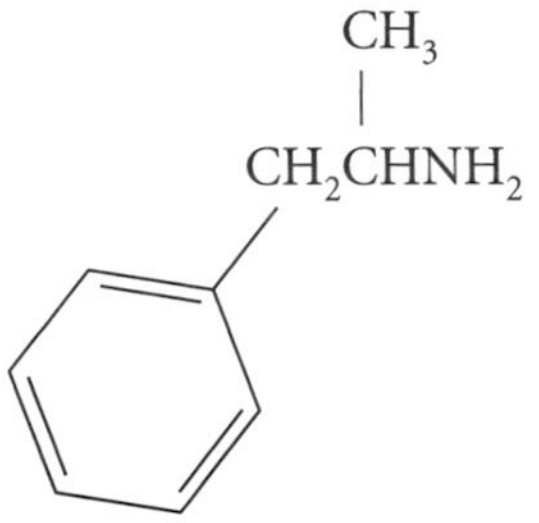

       CH_3
       |
   CH_2CHNH_2

                                                           (1.8)

6. Predict the relative boiling points of the following pairs of compounds, and arrange the two compounds of each pair in order of increasing boiling point. Give reasons for your answers.
   (a) C_2H_5—O—C_2H_5 and C_2H_5—C—C_2H_5
                                            ||
                                            O

   (b)   O   O
           || ||
       CH_3CH and CH_3COH
   (c) CH_3CH_2OH and $CH_3CH_2CH_2CH_2CH_2OH$

7. Predict the relative solubility of the following pairs of compounds in aqueous solvent, and arrange the two compounds in each pair in order of increasing solubility. Give reasons for your answers.

 (a)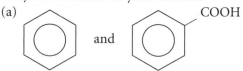

 (b) CH_3COOH and CH_3COOCH_3
 (c) 2-butanol and 2-butanone (1.7)

8. Write a structural diagram equation to represent a reaction for the synthesis of each of the following compounds, and categorize the type of reaction involved.
 (a) ethene from ethanol
 (b) ethoxyethane from ethanol
 (c) propanal from an alcohol
 (d) a secondary pentanol from an alkene
 (e) acetic acid from an alcohol
 (f) methoxymethane
 (g) ethyl formate (1.7)

9. Draw structural diagrams and write IUPAC names for
 (a) three ketones with the molecular formula $C_5H_{10}O$
 (b) two esters with the formula $C_3H_6O_2$
 (c) a primary, a secondary, and a tertiary amine, with the formula $C_5H_{13}N$ (1.8)

10. Write structural formulas for the products formed in each of the following reactions, and categorize the type of reaction involved.
 (a)
$$CH_3CCH_3 + H_2 \xrightarrow[\text{catalyst}]{\text{heat, pressure}}$$
where the carbonyl is shown as $\overset{\text{O}}{\underset{\|}{}}$

 (b) $CH_3CH_2CHO + (O) \rightarrow$
 (c) $CH_3CH=CH_2 + HCl \rightarrow$
 (d) $HOCH_2CH_2OH + HOOC\text{–}\emptyset\text{–}COOH + HOCH_2CH_2OH + HOOC\text{–}\emptyset\text{–}COOH + \text{etc.}$
 (e) $CH_3CH=CH_2 + CH_3CH=CH_2 + \text{etc.} \rightarrow$
 (f) $CH_3COOCH_3 + H_2O \xrightarrow{\text{heat}}$
 (g) $\emptyset + Br_2 \rightarrow$ (1.8)

11. Draw structural diagrams and write IUPAC names for the monomers that make up the polymers whose structures are shown below. Identify each as an addition polymer or a condensation polymer.

 (a) $-\overset{\|}{\underset{O}{C}}-O-CH_2-CH_2-\overset{\|}{\underset{O}{C}}-O-CH_2-CH_2-\overset{\|}{\underset{O}{C}}-O-$

 (b) $-CH-CH_2-CH-CH_2-CH-CH_2-$ with CH_3 branches on the CH carbons

 (c) $-CH-CH-CH_2-O-CH-CH-CH_2-O-$ with O double bonds ($\overset{\|}{\underset{O}{}}$) on the first two carbons of each repeat

 (d) $-\overset{O}{\overset{\|}{C}}-\phi-\overset{O}{\overset{\|}{C}}-NH-\phi-NH-\overset{O}{\overset{\|}{C}}-\phi-\overset{O}{\overset{\|}{C}}-NH-\phi-NH-$

 (e) $-C-C-C-C-C-C-C-C-$ with F and Cl substituents:
$$\begin{array}{cccccccc} F & F & F & F & F & F & F & F \\ | & | & | & | & | & | & | & | \\ C & C & C & C & C & C & C & C \\ | & | & | & | & | & | & | & | \\ Cl & Cl & Cl & Cl & Cl & Cl & Cl & Cl \end{array}$$

12. Using molecular model kits, build molecular models of each of the following compounds:
 (a) glycerol
 (b) 1,3-dihydroxybenzene
 (c) a primary, a secondary, and a tertiary alcohol of molecular formula $C_4H_{10}O$
 (d) the product of the controlled oxidation of ethanal
 (e) a section of a polymer chain of polypropylene (2.1)

13. The monomer used to maufacture some types of "instant glue" is methylcyanoacrylate.

$$CH_2=\overset{\displaystyle |}{\underset{\displaystyle COOCH_3}{C}}-CN$$

Draw a structure of two repeating units of the addition polymer of this monomer. (2.1)

14. The search for new materials with desired properties most often begins in the field of organic chemistry. Explain why the potential for synthesizing new materials is higher in organic chemistry than in inorganic chemistry, and give an example of the use of organic chemistry to help solve an existing problem in health, safety, or the environment. (2.3)

15. For each of the following polymers of glucose, explain its importance to the organism and how its structure is related to its function:
 (a) starch
 (b) glycogen
 (c) cellulose (2.5)

16. Consider a breakfast cereal that is produced in a factory from genetically modified corn that is grown without the use of synthetic fertilizers or pesticides. In your opinion, which of the following terms is an accurate description of the product: organic, natural, or chemical? Give reasons for your answer. (2.6)

17. Identify the functional groups in a monomer of a nucleic acid polymer, and describe the type of polymerization reaction that occurs when the monomers are joined together. (2.6)

18. Describe, with the help of diagrams, the primary, secondary, and tertiary structure of proteins, and explain the role of DNA in protein synthesis. (2.6)

19. Explain, referring to the structure and properties of triglycerides, why fats and oils are better energy storage molecules than carbohydrates. (2.7)

20. Margarine can be produced by hydrogenation of "polyunsaturated" oils such as corn oil. Explain why an increase in the degree of saturation of an oil increases its melting point. (2.7)

Applying Inquiry Skills

21. The labels have fallen off three bottles of organic compounds. These labels indicate that the contents of the bottles are 1-propanol, propanal, and propanoic acid. Design a series of measurements or chemical reactions to determine the identity of each unlabelled compound. Write a detailed design for your investigation, including a Prediction of the results of each measurement or reaction for each compound; the Hypothesis that would explain the Prediction; and a Procedure for each measurement or reaction (with safety precautions for handling and disposal of all chemicals used). (1.7)

22. Design a procedure to synthesize propanone, starting with propane. Describe each step of the procedure, including any necessary experimental conditions and safety precautions. (1.9)

Making Connections

23. The compound *p*-aminobenzoic acid (PABA) is the active ingredient in some sunscreen lotions.
    (a) Draw a structural diagram for PABA.
    (b) Predict some properties of PABA such as solubility, melting point, and chemical reactivity.
    (c) Research, using electronic or print resources, the properties of PABA, explanations of its role in protection against UV radiation, and any possible hazards in its use to the wearer or to the environment. (1.8)

 www.science.nelson.com

24. Research the use of chlorinated hydrocarbons as solvents in the removal of contaminants from microchips in the manufacture of computers. Write a report discussing
    (a) any hazards in the use of chlorinated hydrocarbons; and
    (b) other polymers that have been used by the computer industry as cleaning agents for microchips. (2.2)

 www.science.nelson.com

25. There is ongoing research in the development of polymeric materials for use in dentistry and related fields, such as cranial and facial restoration. Polymers used to make dental impressions, for example, require certain properties, including low solubility, high tensile strength, and a high softening temperature.

    Research current practice and recent advances in the use of organic polymers in the dental industry and the dental profession. Write a report on your findings, covering the following topics.

    • The desired properties of dental polymers and the shortfalls of available materials

    • The key structural features of the monomers in current and prospective dental polymers

    • The rate and degree of the formation of crosslinks between polymer chains, and properties of the structural matrix formed

    • The role of hydrogen bonding on the physical and mechanical properties of the polymer

    • The pros and cons of different methods of initiating the polymerization process (2.2)

 www.science.nelson.com

26. Various chromatographic methods are used for the separation of mixtures into their individual components. The mixture is adsorbed onto a stationary support such as a gel, and the compounds are separated by differences in polarity and molecular size. Research one of the following aspects of chromatography, and write a report to present your findings.

- The selective use of solvents in liquid chromatography to separate compounds with differences in solubility in the solvent

- The structure and properties of proteins and nucleic acids that allow their purification through electrophoresis

- The structure and properties of synthetic polymers, such as polyacrylamide, used as stationary support in gel electrophoresis

- The features of a gas chromatography system used in the analysis of trace amounts of organic substances such as illicit drugs or pesticides   (2.6)

   www.science.nelson.com

27. Give examples of 10 different organic compounds that are significant in your life. For each example, indicate if the substance is natural or synthetic, whether it is a polymer, and how you use it.   (2.7)

Extension

28. What is the theoretical yield of methyl propanoate, prepared from 20.0 g of methanol and 40.0 g of propanoic acid? Calculate the percent yield if 35.2 g of the ester is obtained.

29. A compound of carbon, hydrogen, and oxygen is a colourless liquid at ordinary room temperatures, with a minty, sweet odour. It is soluble in alcohols, ethers, benzene, acetone, and oils. Aqueous bromine does not undergo a colour change with this compound. A 3.00-g sample of this compound is a gas at 100.0°C and 100.0 kPa, and occupies 1.29 L. When this sample is burned in excess oxygen, 7.35 g of carbon dioxide and 3.01 g of water are formed.
    (a) Determine the molecular formula of this compound.
    (b) Draw a possible structural formula for this compound.
    (c) Use the evidence provided to justify your answer in (b).

Exploring

30. Polymers can be made from other substances, besides organic monomers. One of the more familiar is silicone. Research its structure, manufacture, properties, and uses. Conclude with an example of a situation in which the use of silicone has provided some benefits and some drawbacks.

     www.science.nelson.com

Structure and Properties

Geoffrey A. Ozin
Professor, Materials Chemistry Research Group
University of Toronto

"I have always been fascinated by the ability to fashion, through synthetic chemistry, the structure and properties of solid-state materials. When I began my career, materials were made by a 'trial-and-error' approach. Now, materials chemistry of a different kind is emerging—materials are devised that self-assemble by design rather than happenstance. Of course, the ultimate dream is a molecular electronics world made of nanometre-scale molecules and assemblies that perform the function of today's semiconductor devices. Right now we are just learning how to get the molecular world to "kiss" macroscopic wires, a difficult love affair that must be made to work. To make headway in the field of nanoscience, one has to cross boundaries between the disciplines of chemistry, physics, biology, materials science, and engineering (what I call panoscience). I enjoy this challenge and have always been driven by the desire to discover new things in interesting ways and in new fields. Nanochemistry was clearly the undisputed career choice for me."

▶ Overall Expectations

In this unit, you will be able to

- demonstrate an understanding of quantum mechanical theory, and explain how types of chemical bonding account for the properties of ionic, molecular, covalent network, and metallic substances;

- investigate and compare the properties of solids and liquids, and use bonding theory to predict the shape of simple molecules;

- describe products and technologies whose development has depended on understanding molecular structure, and technologies that have advanced atomic and molecular theory.

► **Prerequisites**

Concepts

- Use definitions of ionic and molecular compounds to classify compounds.
- Use the Bohr model and the periodic table to describe the atoms of elements.
- Draw orbit diagrams for the first 20 elements on the periodic table.
- Use the Bohr model and the periodic table to explain and predict the formation of monatomic ions of the representative elements.
- Use bonding theory to explain the formation of ionic and molecular compounds.

Skills

- Work safely in the laboratory.
- Write lab reports for investigations.

Safety and Technical Skills

1. Before using an electrical device, such as a high-voltage spark-producing device, what safety precautions should you take?

2. To handle a hazardous chemical safely, what is required
   (a) for general personal protection from the chemical?
   (b) immediately if your skin comes in contact with the chemical?
   (c) to determine disposal procedures for the chemical?

Knowledge and Understanding

3. Each of the following statements applies only to elements or only to compounds.
   i.   cannot be decomposed into simpler substances by chemical means
   ii. composed of two or more kinds of atoms
   iii. can be decomposed into simpler substances using heat or electricity
   iv. composed of only one kind of atom
   (a) Which statements apply only to elements?
   (b) Which statements apply only to compounds?
   (c) Which statements are empirical?
   (d) Which statements are theoretical?

4. Describe the atomic models presented by each of the following:
   (a) J. J. Thomson
   (b) Ernest Rutherford
   (c) John Dalton

5. Provide the labels, (a) to (e), for **Figure 1**. In addition to the name, provide the international symbol for the three subatomic particles.

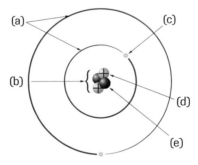

(a) (b) (c) (d) (e)

Figure 1
An atom in the 1930s

6. Use the periodic table and atomic theory to complete **Table 1**.

Table 1 Components of Atoms and Ions

Atom/Ion	Number of protons	Number of electrons	Net charge
hydrogen atom			
sodium atom			
chlorine atom			
hydrogen ion			
sodium ion			
chloride ion			

7. Draw orbit (Bohr) diagrams for the following atoms:
   (a) nitrogen, N
   (b) calcium, Ca
   (c) chlorine, Cl

8. Compounds can be classified as ionic and molecular. Copy and complete **Table 2**, indicating the properties of these compounds.

Table 2 Properties of Ionic and Molecular Compounds

Class of compound	Classes of elements involved	Properties		
		Melting point (high/low)	State at SATP (s, l, g)	Electrolytes (yes/no)
ionic				
molecular				

9. Classify compounds with the following properties as ionic, molecular, or either. Explain each classification.
   (a) high solubility in water; aqueous solution conducts electricity
   (b) solid at SATP; low solubility in water
   (c) solid at SATP; low melting point; aqueous solution does not conduct electricity

10. According to atomic and bonding theories, atoms react by rearranging their electrons.
    (a) Silicon tetrafluoride is used to produce ultra-pure silicon wafers (**Figure 2**). Draw an electron orbit diagram for each atom or ion in the following word equation for the formation of silicon tetrafluoride. (Include coefficients where necessary to balance the equation.)
    silicon atom   +   fluorine atom   →   silicon tetrafluoride
    (b) Calcium fluoride occurs naturally as fluorite (pure compound) and as fluorspar (mineral, **Figure 3**). Calcium fluoride is the principal source of the element fluorine, and is also used in a wide variety of applications such as metal smelting, certain cements, and paint pigments. Draw an electron orbit diagram for each atom or ion in the following word equation for the formation of calcium fluoride. (Include coefficients where necessary to balance the equation.)
    calcium atom   +   fluorine atom   →   calcium fluoride
    (c) What is the difference in the electron rearrangement in (a) compared with (b)?
    (d) What property of atoms is used to explain why electrons are sometimes shared and why they are sometimes transferred between atoms? Describe this property using a "tug of war" analogy.

Inquiry and Communication

11. What is the difference between a scientific law and a scientific theory?

12. In the work of scientists, what generally comes first, laws or theories?

Figure 2
In 1972, Intel's 8008 computer processor had 3500 transistors. In 2000, the Pentium 4 processor had 42 million transistors on a silicon chip.

Figure 3
Fluorspar is a mineral that can be found in many countries of the world. It takes a variety of colours and has different properties, depending on contaminants. The variety found near Madoc, Ontario, is unusual in that it can be used to make optical lenses and prisms.

chapter

3

Atomic Theories

In this chapter, you will be able to

- describe and explain the evidence for the Rutherford and Bohr atomic models;
- describe the origins of quantum theory and the contributions of Planck and Einstein;
- state the four quantum numbers and recognize the evidence that led to these numbers;
- list the characteristics of *s*, *p*, *d*, and *f* blocks of elements;
- write electron configurations and use them to explain the properties of elements;
- describe the development of the wave model of quantum mechanics and the contributions of de Broglie, Schrödinger, and Heisenberg;
- describe the shapes (probability densities) of *s* and *p* orbitals;
- use appropriate scientific vocabulary when describing experiments and theories;
- conduct experiments and simulations related to the development of the Rutherford, Bohr, and quantum mechanical atomic models;
- describe some applications of atomic theories in analytical chemistry and medical diagnosis;
- describe advances in Canadian research on atomic theory.

The stories of the development of atomic theories from Dalton's model to the quantum mechanical model of the atom are fascinating accounts of achievements of the human mind and the human spirit. These stories, perhaps more than any other set of stories in the sciences, illustrate how scientific theories are created, tested, and then used by scientists. The stories usually start with some empirical result in the laboratory that cannot be explained by existing theories, or may even contradict them. Scientists live for this kind of challenge. Hypotheses are then repeatedly created and tested until a hypothesis is found that can survive the testing. Successful hypotheses are able to describe and explain past evidence and, most importantly, predict future evidence. The scientists who create the hypotheses and the scientists who devise the most severe tests of the hypotheses they can imagine (experimental designs) experience excitement and joy not unlike that of a gold medal winner at the Olympics.

Chemists strive to understand the nature of matter through a combination of gathering and analyzing evidence and creating theoretical concepts to explain the evidence. This demands teamwork among experimental and theoretical chemists. Experimental chemists often work on the scale of our macroscopic world, using observable quantities and equipment. However, the theoretical explanations come in the form of particles and forces that we cannot see. The creativity of the empirical chemist in the laboratory is complemented by the creativity of the theoretical chemist in the mind's eye — and perhaps both will win the Nobel Prize for Chemistry and earn praise for themselves, their families, their research facilities, and their countries.

REFLECT on your learning

1. (a) How did our understanding of atomic structure evolve in the 20th century? You may want to sketch a series of atomic models.
   (b) What are some significant differences between the current quantum mechanical model and all previous atomic models?

2. Why are scientific concepts revised and/or replaced?

3. (a) How has quantum mechanics changed the way we describe electrons in atoms?
   (b) How has our understanding of the periodic table changed with quantum mechanics?

4. (a) Use your knowledge of atomic structure to explain a couple of technologies in common use.
   (b) Are all technological products and processes related to our modern understanding of atomic structure good? Provide examples.

▶ *TRY THIS* activity *Molecules and Light*

In this activity you observe fluorescence in which molecules absorb light and then release light, and also chemiluminescence in which light is produced as a result of a chemical reaction. An example of a chemiluminescent system is that used by fireflies to emit flashes of light as signals.

Materials: eye protection; 600-mL beaker; sodium fluorescein; lab spatula; 18 × 150 mm test tube; 0.02 mol/L $NaOH_{(aq)}$; glow stick

 0.02 mol/L sodium hydroxide is an irritant. Wear eye protection.

- Add about 500 mL of 0.02 mol/L NaOH to a 600-mL beaker.

- Observe the solid sodium fluorescein, and add a small scoop (about 5 g) to the water in the large beaker. Initially, just sprinkle the solid on the surface using a lab spatula and observe. Stir to thoroughly dissolve the compound — enough should be dissolved to give a definite orange colour to the solution.

- Fill a test tube two-thirds full of the solution and hold the solution up between your eye and a window (or another bright white light source). What you see is mainly the light transmitted by the solution.

- Now turn your back to the light and observe the solution colour by reflected light. What you see is mainly the light emitted by the fluorescein.

  (a) Compare the colour of the light transmitted with the colour of the light emitted by the fluorescein solution.

(b) If the fluorescein solution did not emit light when illuminated, what colour would it appear to be when viewed by reflected light?

- Darken the room and follow the package instructions to illuminate a commercial glow stick.

  (c) Based on the instructions, speculate on what happens inside the glow stick.

  (d) Compare and contrast fluorescence and chemiluminescence.

  (e) How do think these effects relate to molecular structure?

- When you are finished this unit, you can return to these examples and speculate about how the energy is stored in the molecules.

- Dispose of the fluorescein solutions down the sink and wash your hands.

sodium fluorescein

The history of atomic theories is full of success and failure stories for hundreds of chemists. In textbooks such as this one, only the success of a few is documented. However, the success of these chemists was often facilitated by both the success and failure of many others.

Recall that by the use of deductive logic the Greeks (for example, Democritus) in about 300 B.C. hypothesized that matter cut into smaller and smaller pieces would eventually reach what they called the atom — literally meaning indivisible. This idea was reintroduced over two thousand years later by an English chemist/schoolteacher named John Dalton in 1805. He re-created the modern theory of atoms to explain three important scientific laws — the laws of definite composition, multiple proportions, and conservation of mass. The success of Dalton's theory of the atom was that it could explain all three of these laws and much more. Dalton's theory was that the smallest piece of matter was an atom that was indivisible, and that an atom was different from one element to another. All atoms of a particular element were thought to be exactly the same. Dalton's model of the atom was that of a featureless sphere — by analogy, a billiard ball (**Figure 1**). Dalton's atomic theory lasted for about a century, although it came under increasing criticism during the latter part of the 1800s.

Figure 1
In Dalton's atomic model, an atom is a solid sphere, similar to a billiard ball. This simple model is still used today to represent the arrangement of atoms in molecules.

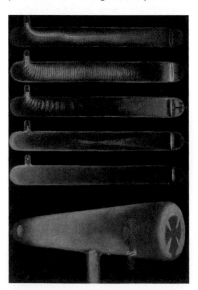

SUMMARY	Creating the Dalton Atomic Theory (1805)

Table 1

Key experimental work	Theoretical explanation	Atomic theory
Law of definite composition: elements combine in a characteristic mass ratio	Each atom has a particular combining capacity.	Matter is composed of indestructible, indivisible atoms, which are identical for one element, but different from other elements.
Law of multiple proportions: there may be more than one mass ratio	Some atoms have more than one combining capacity.	
Law of conservation of mass: total mass remains	Atoms are neither created nor destroyed constant in a chemical reaction.	

The Thomson Atomic Model

The experimental studies of Svante Arrhenius and Michael Faraday with electricity and chemical solutions and of William Crookes with electricity and vacuum tubes suggested that electric charges were components of matter. J. J. Thomson's quantitative experiments with cathode rays resulted in the discovery of the electron, whose charge was later measured by Robert Millikan. The Thomson model of the atom (1897) was a hypothesis that the atom was composed of electrons (negative particles) embedded in a positively charged sphere (**Figure 2(a)**). Thomson's research group at Cambridge University in England used mathematics to predict the uniform three-dimensional distribution of the electrons throughout the atom. The Thomson model of the atom is often communicated by using the analogy of a raisin bun, with the raisins depicting the electrons and the bun being the positive material of his atom (**Figure 2(b)**).

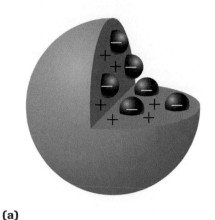

(a)

(b)

INVESTIGATION 3.1.1

The Nature of Cathode Rays (p. 209)

The discovery of cathode rays led to a revision of the Dalton atomic model. What are their properties?

Figure 2
(a) In Thomson's atomic model, the atom is a positive sphere with embedded electrons.
(b) This model can be compared to a raisin bun, in which the raisins represent the negative electrons and the bun represents the region of positive charge.

SUMMARY

Creating the Thomson Atomic Theory (1897)

Table 2

Key experimental work	Theoretical explanation	Atomic theory
Arrhenius: the electrical nature of chemical solutions	Atoms may gain or lose electrons to form ions in solution.	Matter is composed of atoms that contain electrons (negative particles) embedded in a positive material. The kind of element is characterized by the number of electrons in the atom.
Faraday: quantitative work with electricity and solutions	Particular atoms and ions gain or lose a specific number of electrons.	
Crookes: qualitative studies of cathode rays	Electricity is composed of negatively charged particles.	
Thomson: quantitative studies of cathode rays	Electrons are a component of all matter.	
Millikan: charged oil drop experiment	Electrons have a specific fixed electric charge.	

The Rutherford Atomic Theory

One of Thomson's students, Ernest Rutherford (**Figure 3**), eventually showed that some parts of the Thomson atomic theory were not correct. Rutherford developed an expertise with nuclear radiation during the nine years he spent at McGill University in Montreal. He worked with and classified nuclear radiation as alpha (α), beta (β), and gamma (γ) — helium nuclei, electrons, and high-energy electromagnetic radiation from the nucleus, respectively. Working with his team of graduate students, at Manchester in England, he devised an experiment to test the Thomson model of the atom. They used radium as a source of alpha radiation, which was directed at a thin film of gold. The prediction, based on the Thomson model, was that the alpha particles should be deflected little, if at all. When some of the alpha particles were deflected at large angles and even backwards

DID YOU KNOW ?

Rutherford Quotes
- "You know it is about as incredible as if you fired a 15-inch (350-mm) shell at a piece of tissue paper and it came back and hit you."
- "Now I know what the atom looks like." 1911
- The electrons occupy most of the space in the atom, "like a few flies in a cathedral."
- The notion that nuclear energy could be controlled is "moonshine." 1933

Figure 3
Rutherford's work with radioactive materials at McGill helped prepare him for his challenge to Thomson's atomic theory.

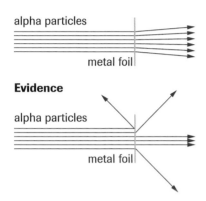

ACTIVITY 3.1.1

Rutherford's Gold Foil Experiment (p. 210)
Rutherford's famous experiment involved shooting "atomic bullets" at an extremely thin sheet of gold. You can simulate his experiment.

Prediction

alpha particles

metal foil

Evidence

alpha particles

metal foil

Figure 4
Rutherford's experimental observations were dramatically different from what he had expected based on the Thomson model.

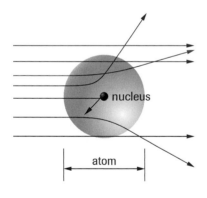

atom

Figure 5
To explain his results, Rutherford suggested that an atom consisted mostly of empy space, explaining why most of the alpha particles passed nearly straight through the gold foil.

proton ($^1_0$p or p$^+$) a positively charged subatomic particle found in the nucleus of atoms

from the foil, the prediction was shown to be false, and the Thomson model judged unacceptable (**Figure 4**). Rutherford's nuclear model of the atom was then created to explain the evidence gathered in this scattering experiment. Rutherford's analysis showed that all of the positive charge in the atom had to be in a very small volume compared to the size of the atom. Only then could he explain the results of the experiment (**Figure 5**). He also had to hypothesize the existence of a nuclear (attractive) force, to explain how so much positive charge could occupy such a small volume. The nuclear force of attraction had to be much stronger than the electrostatic force repelling the positive charges in the nucleus. Even though these theoretical ideas seemed far-fetched, they explained the experimental evidence. Rutherford's explanation of the evidence gradually gained widespread acceptance in the scientific community.

SUMMARY *Creating the Rutherford Atomic Theory (1911)*

Table 3

Key experimental work	Theoretical explanation	Atomic theory
Rutherford: A few positive alpha particles are deflected at large angles when fired at a gold foil.	The positive charge in the atom must be concentrated in a very small volume of the atom.	An atom is composed of a very tiny nucleus, which contains positive charges and most of the mass of the atom. Very small negative electrons occupy most of the volume of the atom.
Most materials are very stable and do not fly apart (break down).	A very strong nuclear force holds the positive charges within the nucleus.	
Rutherford: Most alpha particles pass straight through gold foil.	Most of the atom is empty space.	

Protons, Isotopes, and Neutrons

The Thomson model of the atom (1897) included electrons as particles, but did not describe the positive charge as particles; recall the raisins (electrons) in a bun (positive charge) analogy. The Rutherford model of the atom (1911) included electrons orbiting a positively charged nucleus. There may have been a hypothesis about the nucleus being composed of positively charged particles, but it was not until 1914 that evidence was gathered to support such a hypothesis. Rutherford, Thomson, and associates studied positive rays in a cathode ray tube and found that the smallest positive charge possible was from ionized hydrogen gas. Rutherford reasoned that this was the fundamental particle of positive charge and he named it the **proton**, meaning first. (Again Rutherford showed his genius by being able to direct the empirical work and then interpret the evidence theoretically.) By bending the hydrogen-gas positive rays in a magnetic field they were able to determine the charge and mass of the hypothetical proton. The proton was shown to have a charge equal to but opposite to that of the electron and a mass 1836 times that of an electron. All of this work was done in gas discharge tubes that evolved into the version of the mass spectrometer (**Figure 6**) developed by Francis Aston during the period 1919–1925.

Evidence from radioactivity and mass spectrometer investigations falsified Dalton's theory that all atoms of a particular element were identical. The evidence indicated that

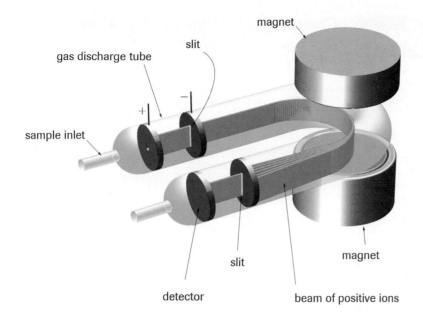

Figure 6
A mass spectrometer is used to determine the masses of ionized particles by measuring the deflection of these particles as they pass through the field of a strong magnet.

there were, for example, atoms of sodium with different masses. These atoms of different mass were named **isotopes**, although their existence could not yet be explained.

Later, James Chadwick, working with Rutherford, was bombarding elements with alpha particles to calculate the masses of nuclei. When the masses of the nuclei were compared to the sum of the masses of the protons for the elements, they did not agree. An initial hypothesis was that about half of the mass of the nucleus was made up of proton–electron (neutral) pairs. However, in 1932 Chadwick completed some careful experimental work involving radiation effects caused by alpha particle bombardment. He reasoned that the only logical and consistent theory that could explain these results involved the existence of a neutral particle in the nucleus. According to Chadwick, the nucleus would contain positively charged protons and neutral particles, called **neutrons**. The different radioactive and mass properties of isotopes could now be explained by the different nuclear stability and different masses of the atom caused by different numbers of neutrons in the nuclei of atoms of a particular element.

isotope ($_Z^A X$) a variety of atoms of an element; atoms of this variety have the same number of protons as all atoms of the element, but a different number of neutrons

neutron ($_0^1 n$ or n) a neutral (uncharged) subatomic particle present in the nucleus of atoms

SUMMARY   *Rutherford Model*

- An atom is made up of an equal number of negatively charged electrons and postively charged protons.
- Most of the mass of the atom and all of its positive charge is contained in a tiny core region called the nucleus.
- The nucleus contains protons and neutrons that have approximately the same mass.
- The number of protons is called the atomic number (Z).
- The total number of protons and neutrons is called the mass number (A).

 **SUMMARY** **Creating the Concepts of Protons, Isotopes, and Neutrons**

Table 4

Key experimental work	Theoretical explanation	Atomic theory
Rutherford (1914): The lowest charge on an ionized gas particle is from the hydrogen ion	The smallest particle of positive charge is the proton.	Atoms are composed of protons, neutrons, and electrons. Atoms of the same element have the same number of protons and electrons, but may have a varying number of neutrons (isotopes of the element).
Soddy (1913): Radioactive decay suggests different atoms of the same element	Isotopes of an element have a fixed number of protons but varying stability and mass.	
Aston (1919): Mass spectrometer work indicates different masses for some atoms of the same element	The nucleus contains neutral particles called neutrons.	
Radiation is produced by bombarding elements with alpha particles.		

▶ *Section 3.1 Questions*

Understanding Concepts

1. Summarize, using labelled diagrams, the evolution of atomic theory from the Dalton to the Rutherford model.

2. Present the experimental evidence that led to the Rutherford model.

3. How did Rutherford infer that the nucleus was
   (a) very small (compared to the size of the atom)?
   (b) positively charged?

4. (a) State the experimental evidence that was used in the discovery of the proton.
   (b) Write a description of a proton.

5. (a) State the experimental evidence that was used in the discovery of the neutron.
   (b) Describe the nature of the neutron.

Applying Inquiry Skills

6. What is meant by a "black box" and why is this an appropriate analogy for the study of atomic structure?

7. Theories are often created by scientists to explain scientific laws and experimental results. To some people it seems strange to say that theories come after laws. Compare the scientific and common uses of the term "theory."

8. What is the ultimate authority in scientific work (what kind of knowledge is most trusted)?

Making Connections

9. State some recent examples of stories in the news media that mention or refer to atoms.

10. Describe some contributions Canadian scientists and/or scientists working in Canadian laboratories made to the advancement of knowledge about the nature of matter.

 www.science.nelson.com

Extension

11. Rutherford's idea that atoms are mostly empty space is retained in all subsequent atomic theories. How can solids then be "solid"? In other words, how can your chair support you? Why doesn't your pencil go right through the atoms that make up your desk?

12. When you look around you, the matter you observe can be said to be made from electrons, protons, and neutrons. Modern scientific theories tell us something a little different about the composition of matter. For example, today protons are not considered to be fundamental particles; i.e., they are now believed to be composed of still smaller particles. According to current nuclear theory, what is the composition of a proton? Which Canadian scientist received a share of the Nobel Prize for his empirical work in verifying this hypothesis of sub-subatomic particles?

Case Study:
A Canadian Nuclear Scientist 3.2

Who was Ernest Rutherford's first graduate student at McGill University in Montreal? Who was the first woman to receive a master's degree in physics from McGill? Who was referred to by Rutherford as, next to Marie Curie, the best woman experimental physicist of her time? Who was likely the only scientist to work in the laboratories of Ernest Rutherford, J. J. Thomson, *and* Marie Curie? Who provided the evidence upon which Rutherford based many of his theories? The answer to all these questions is Canadian Harriet Brooks (**Figure 1**).

Harriet Brooks was born on July 2, 1876 in Exeter in southwest Ontario, about 48 km from London. After moving around Ontario and Quebec, her family finally settled in Montreal, just in time for Brooks and her sister to enter McGill University. She was extraordinarily successful — the only student in her year to graduate with first-rank honours marks in both mathematics and natural philosophy (physics).

At the time, Rutherford was researching radioactivity at McGill (for the nine years from 1898 to 1907). As the head of the laboratory, Rutherford had his choice of the best graduate students with whom to work, and Brooks was the first of them. After graduation Brooks was invited by Rutherford to join his research team, working in what has been described as the best laboratory in North America, with equipment second to none in the world (**Figure 2**). She learned the research process from Rutherford and earned her master's degree in 1901.

The work Brooks and other McGill graduate students contributed helped Rutherford earn the Nobel Prize, although it was not given until one year after he left McGill. Rutherford was one of those rare scientists who excelled in both empirical and theoretical work. However, Brooks did a lot of the empirical work for Rutherford, and he often provided the theoretical interpretation.

Brooks studied the reactivity of radium and found an "emanation," which we now know was radon from the radioactive (alpha particle) decay of radium.

$$^{226}_{88}\text{Ra} \rightarrow ^{222}_{86}\text{Rn} + ^{4}_{2}\alpha$$

At that time, everyone (including Becquerel and the Curies) believed that an element remained the same when it emitted radiation. Brooks gathered evidence to falsify that concept. She used diffusion of the emanated gas to determine the molar mass of what we now know as radon. The alchemists' dream of the transformation of elements was real.

In the process, she gathered evidence that led to Rutherford's theoretical interpretation that radiation resulted in the recoil (action–reaction) of the radiating nucleus. Often the recoiling nucleus was ejected from the radioactive sample. Brooks pioneered the use of this recoil effect to capture and identify decay products.

Her third significant contribution was important evidence that Rutherford interpreted as a series of radioactive transformations. Rutherford and Frederick Soddy were to receive the accolades for this theory, but it was Brooks who gathered the initial evidence for this theoretical effect. For example:

$$^{226}_{88}\text{Ra} \rightarrow ^{222}_{86}\text{Rn} \rightarrow ^{218}_{84}\text{Po} \rightarrow ^{214}_{82}\text{Pb}$$

(emitting an alpha particle in each step)

Figure 1
Harriet Brooks broke ground as a nuclear scientist and as a woman in science.

Figure 2
Although the Rutherford laboratory was well equipped, often the researchers made their own equipment.

Remember that atomic, nuclear, and radioactive decay theories were in their infancy at that time. The importance of Brooks's work was not evident until much later — after theories to explain her evidence became acceptable (**Table 1**). Based upon what she accomplished in such a short career, one has to wonder what she might have done in a full career as a teacher and researcher.

Table 1 Brooks's Place in the History of Radioactivity

1896	Becquerel discovers radioactivity (of uranium)
1898	Radium, polonium, and thorium are identified as radioactive (by the Curies)
1898	Brooks earns B.S. (first in her class) from McGill University and starts graduate work with Rutherford
1900	Rutherford's team identifies radioactive emissions as alpha, beta, and gamma
1901	Brooks identifies radon as a product of the radioactivity of radium
1901	Brooks earns M.A. from McGill University
1901–02	Brooks starts Ph.D. in Pennsylvania
1902–03	Brooks earns fellowship to work with J. J. Thomson in Cambridge, England
1903–04	Brooks returns to work with Rutherford at McGill and publishes further evidence for radon as a radioactive decay product
1904–06	Brooks works at Barnard College, New York
1906–07	Brooks works with Marie Curie in Paris, France
1907	Brooks marries and discontinues her research life
1908	Rutherford wins Chemistry Nobel Prize for Canadian research
1911	Rutherford scattering experiment and the nuclear atom
1913	Soddy invents isotopes to explain his experimental results
1913	Bohr publishes his atomic interpretation of the periodic table and line spectra
1914	Rutherford coins the word "proton"
1916	Lewis theorizes that valence electrons are shared to form covalent bonds
1932	Chadwick creates the neutron to explain the mass of the alpha particle

▶ Section 3.2 Questions

Making Connections

1. Complete a case study like the one in this section on the important work done by one of the following contributors to modern atomic theory. Provide summaries of his/her evidence or arguments and indicate the placement and the importance of his/her work in the ongoing story of atomic theory.
   - John Dalton and the laws of definite composition, multiple proportions, and conservation of mass in chemical reactions.
   - Michael Faraday, his electrolysis experiments, and his explanations.
   - Svante Arrhenius's experiments on conductivity and freezing-point depression and his explanation.
   - William Crookes's cathode ray experiments and the explanation of the evidence.
   - J. J. Thomson's charge-to-mass ratio experiments with cathode rays and his explanation.
   - Robert Millikan's experiment to determine the charge on electrons.
   - Antoine Becquerel's discovery of radioactivity.
   - Marie Curie's discovery of radium.
   - Ernest Rutherford's alpha-particle scattering experiment.
   - Ernest Rutherford's and Frederick Soddy's research on radioactive series and their explanation of the evidence.
   - Henry Moseley's X-ray studies.
   - James Chadwick's alpha-particle experiments with light elements, and the hypothesis that explained the results.

Origins of Quantum Theory *3.3*

Max Planck (1858–1947) is credited with starting the quantum revolution with a surprising interpretation of the experimental results obtained from the study of the light emitted by hot objects, started by his university teacher, Gustav Kirchhoff (**Figure 1**). Kirchhoff was interested in the light emitted by blackbodies. The term "blackbody" is used to describe an ideal, perfectly black object that does not reflect any light, and emits various forms of light (electromagnetic radiation) as a result of its temperature.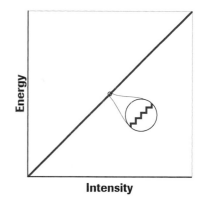

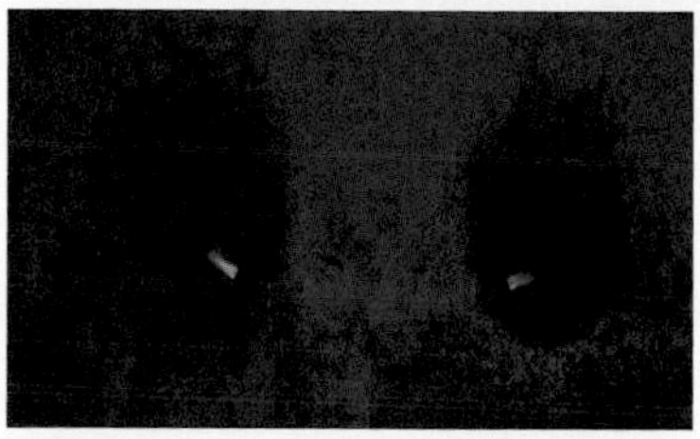

Planck's Quantum Hypothesis

As a solid is heated to higher and higher temperatures, it begins to glow. Initially, it appears red and then becomes white when the temperature increases. Recall that white light is a combination of all colours, so the light emitted by the hotter object must now be accompanied by, for example, blue light. The changes in the colours and the corresponding spectra do not depend on the composition of the solid.

If electronic instruments are used to measure the intensity (brightness) of the different colours observed in the spectrum of the emitted light, a typical bell-shaped curve is obtained.

For many years, scientists struggled to explain the curves shown in **Figure 2**. Some were able to create an equation to explain the intensity curve at one end or the other, but not to explain the overall curve obtained from experiments. In 1900 Planck developed a mathematical equation to explain the whole curve, by using a radical hypothesis. Planck saw that he could obtain agreement between theory and experiment by hypothesizing that the energies of the oscillating atoms in the heated solid were multiples of a small quantity of energy; in other words, energy is not continuous. Planck was reluctant to pursue this line of reasoning, and so it was Albert Einstein who later pointed out that the inevitable conclusion of Planck's hypothesis is that the light emitted by a hot solid is also quantized — it comes in "bursts," not a continuous stream of energy (**Figure 3**). One little burst or packet of energy is known as a **quantum** of energy.

This is like dealing with money — the smallest quantity of money is the penny and any quantity of money can be expressed in terms of pennies; e.g., $1.00 is 100 pennies. Of course, there are other coins. The $1.00 can be made up of two quarters, three dimes, three nickels, and five pennies. We can apply this thought to light. You could think of the coins representing the energy of the light quanta — the penny is infrared, the nickel is red, the dime is blue, and the quarter is ultraviolet radiation. Heat (without colour) would then be emitted

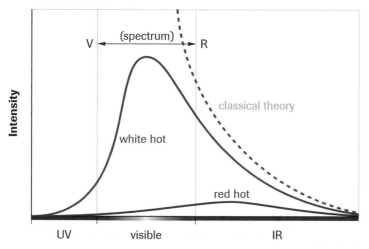

Figure 1
Kirchhoff and other experimenters studied the light given off by heated objects, such as this red-hot furnace.

⦿ **ACTIVITY 3.3.1**

Hot Solids (p. 210)
What kind of light is given off when a solid is heated so that it becomes "white hot"?

quantum a small discrete, indivisible quantity (plural, quanta); a quantum of light energy is called a photon

Figure 3
Scientists used to think that as the intensity or brightness of light changes, the total energy increases continuously, like going up the slope of a smooth hill. As a consequence of Planck's work, Einstein suggested that the slope is actually a staircase with tiny steps, where each step is a quantum of energy.

Figure 2
The solid lines show the intensity of the colours of light emitted by a red-hot wire and a white-hot wire. Notice how the curve becomes higher and shifts toward the higher-energy UV as the temperature increases. The dotted line represents the predicted curve for a white-hot object, according to the existing classical theory before Planck.

Photon Energy
The energy, *E*, of a photon of light is the product of Planck's constant, *h*, and the frequency, *f*, of the light. If you are a Star Trek fan, you will recognize that the creators of this popular series borrowed the photon term to invent a "photon torpedo" that fires bursts or quanta of light energy at enemy ships. An interesting idea, but not practical.

Figure 4
Max Planck was himself puzzled by the "lumps" of light energy. He preferred to think that the energy was quantized for delivery only, just like butter, which is delivered to stores only in specific sizes, even though it could exist in blocks of any size.

Figure 5
The electromagnetic spectrum, originally predicted by Maxwell, includes all forms of electromagnetic radiation from very short wavelength gamma (γ) rays to ordinary visible light to very long wavelength radio waves.

as pennies only, red-hot radiation would include nickels, white-hot radiation would add dimes, and blue-hot would likely include many more dimes and some quarters. An interpretation of the evidence from heating a solid is that a sequence of quanta emissions from IR to red to blue to UV occurs — pennies, to nickels, to dimes, to quarters, by analogy.

A logical interpretation is that as the temperature is increased, the proportion of each larger quantum becomes greater. The colour of a heated object is due to a complex combination of the number and kind of quanta.

Although Planck (**Figure 4**) was not happy with his own hypothesis, he did what he had to do in order to get agreement with the ultimate authority in science — the evidence gathered in the laboratory. Planck thus started a trend that helped to explain other experimental results (for example, the photoelectric effect) that previously could not be explained by classical theory.

> ### Practice

Understanding Concepts

1. The recommended procedure for lighting a laboratory burner is to close the air inlet, light the burner, and then gradually open the air inlet. What is the initial colour of the flame with the air inlet closed? What is the final colour with sufficient air? Which is the hotter flame?

2. How would observations of a star allow astronomers to obtain the temperature of the star?

3. Draw staircase diagrams (like **Figure 3**) to show the difference between low-energy red light quanta versus higher-energy violet light quanta.

4. Liquids and solids, when heated, produce continuous spectra. What kind of spectrum is produced by a heated gas?

The Photoelectric Effect

The nature of light has been the subject of considerable debate for centuries. Greek philosophers around 300 B.C. believed light was a stream of particles. In the late 17th century, experiments led the Dutch scientist Christiaan Huygens to propose that light can best be explained as a wave. Not everyone agreed. The famous English scientist, Isaac Newton, bitterly opposed this view and continued to try to explain the properties of light in terms of minute particles or "corpuscles." However, mounting evidence from experiments with, for example, reflection, refraction, and diffraction clearly favoured the wave hypothesis over the particle view.

In the mid-19th century, James Maxwell produced a brilliant theory explaining the known properties of light, electricity, and magnetism. He proposed that light is an electromagnetic wave composed of electric and magnetic fields that can exert forces on charged particles. This electromagnetic-wave theory, known as the classical theory of light, eventually became widely accepted when new experiments supported this view. Most scientists thought this was the end of the debate about the nature of light — light is (definitely) an electromagnetic *wave* consisting of a continuous series of wavelengths (**Figure 5**).

Electromagnetic Spectrum

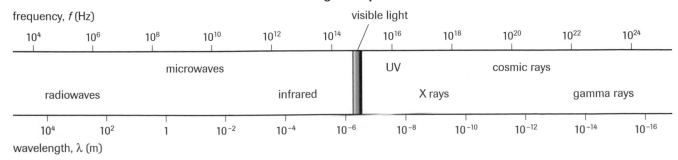

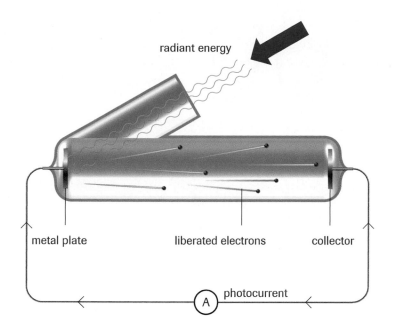

radiant energy

metal plate   liberated electrons   collector

(A) photocurrent

INVESTIGATION 3.3.1

The Photoelectric Effect (p. 209)
The photoelectric effect has had important modern applications such as solar cells and X-ray imaging. You can investigate it using an electroscope.

Figure 6
In the photoelectric effect, light shining on a metal liberates electrons from the metal surface. The ammeter (A) records the electric current (the number of electrons per second) in the circuit.

The photoelectric effect is one of the key experiments and stories leading to quantum theory. Heinrich Hertz discovered the photoelectric effect by accident in 1887. It involves the effect of electromagnetic radiation or light on substances, particularly certain metals. Hertz studied this effect qualitatively but had no explanation for it.

Although Heinrich Hertz described his discovery of the **photoelectric effect** (**Figure 6**) as minor, it was to have a major contribution in changing the accepted, classical theory of light. According to the classical theory, the brightness (intensity) of the light shone on the metal would determine the kinetic energy of the liberated electrons; the brighter the light, the greater the energy of the electrons ejected. This prediction was shown to be false. Further experimental work showed that the frequency (colour/energy) of the light was the most important characteristic of the light in producing the effect. Classical theory was therefore unacceptable for explaining the photoelectric effect. 🧪▮

Albert Einstein was awarded the Nobel Prize in 1905 for using Planck's idea of a quantum of energy to explain the photoelectric effect. He reasoned that light consisted of a stream of energy packets or quanta—later called **photons**. A photon of red light contains less energy than a photon of UV light (**Figure 7**). Einstein suggested that the ejection of an electron from the metal surface could be explained in terms of a photon–electron collision. The energy of the photon is transferred to the electron. Some of this energy is used by the electron to break free from the atom and the rest is left over as kinetic energy of the ejected electron. The electron cannot break free from the atom unless a certain minimum quantity of energy is absorbed from a single photon.

An electron held in an atom by electrostatic forces is like a marble trapped statically in a bowl. If you bang the bowl (with incrementally larger bumps), the marble can move higher from rest in the bowl, but may still be trapped. A certain, minimum quantity of potential energy is required by the marble to escape from the bowl (**Figure 8**). This explains why the energy of the electrons produced by the photoelectric effect is independent of light intensity. If one electron absorbs one photon, then the photon energy (related only to the type of light) needs to be great enough for the electron to be able to escape. No electrons are detected at low photon energies because the energy of the single photon captured was insufficient for the electron to escape the metal. This quantum explanation worked, where no classical explanation could. Quantum theory

photoelectric effect the release of electrons from a substance due to light striking the surface of a metal

photon a quantum of light energy

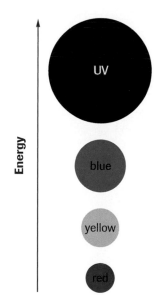

Figure 7
Each photon of light has a different energy, represented by the relative sizes of the circles.

Figure 8

(a) Using a bowl analogy, different atoms would be represented with bowls of different depths.

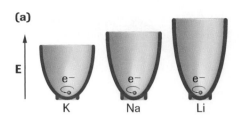

(a)

K   Na   Li

(b) For most atoms, the energy of a red photon is not great enough to boost the electron (marble) out of the atom (bowl). The electron can absorb the energy but is still stuck in the atom. This process simply results in the heating of the sample.

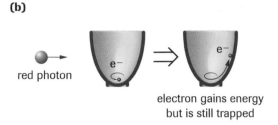

(b)

red photon

electron gains energy but is still trapped

(c) A higher-energy photon, such as a UV photon, has more than enough energy to boost the electron out of many atoms.

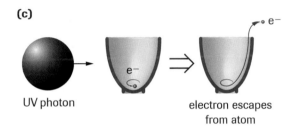

(c)

e−

UV photon

electron escapes from atom

received a huge boost in popularity for explaining this and other laboratory effects at the atomic and subatomic levels.

Quantum theory is heralded as one of the major scientific achievements of the 20th century. There were results from many scientific experiments that could not be explained by classical chemistry and physics, but these experimental results could be explained by quantum theory.

Two of the experiments leading to quantum theory are summarized below, but there were many more that could only be explained using quantum theory.

 **SUMMARY**   *Creating Quantum Theory*

Table 1

Key experimental work	Theoretical explanation	Quantum theory
Kirchhoff (1859): blackbody radiation	Planck (1900): The energy from a blackbody is quantized; i.e., restricted to whole number multiples of certain energy	Electromagnetic energy is not infinitely subdivisible; energy exists as packets or quanta, called photons. A photon is a small packet of energy corresponding to a specific frequency of light ($E = hf$).
Hertz (1887): the photo-electric effect	Einstein (1905): The size of a quantum of electromagnetic energy depends directly on its frequency; one photon of energy ejects one electron	

▶ *Section 3.3* **Questions**

Understanding Concepts

1. State the two important experimental observations that established the quantum theory of light.

2. Although Einstein received the Nobel Prize for his explanation of the photoelectric effect, should Max Planck be considered the father of quantum theory?

3. Write a brief description of the photoelectric effect experiment.

4. Distinguish between the terms "quantum" and "photon."

Applying Inquiry Skills

5. What effect does the type or colour of light have on the release of electrons from a sodium metal surface?
   (a) Write a brief experimental design to answer this question, based on **Figure 6**. Be sure to identify all variables.
   (b) Would you expect all colours of light to release electrons from the sodium metal? Justify your answer, in general terms, using the idea of photons.

Extension

6. Einstein won the Nobel Prize in 1921 for explaining the photoelectric effect in 1905. Einstein calculated the energy of an incoming photon from the Planck equation

   $$E = hf$$

where E is energy in joules (J), h is Planck's constant (6.6×10^{-34} J/Hz), and f is the frequency in hertz (Hz) of light shining on the metal.
   (a) If the minimum frequency of light required to have an electron escape from sodium is 5.5×10^{14} Hz, calculate the energy of photons of this frequency.
   (b) What is the minimum energy of the quantum leap that an electron makes to escape the sodium atom as a photoelectron?

7. Ultraviolet (UV) light that causes tanning and burning of the skin has a higher energy per photon than infrared (IR) light from a heat lamp.
   (a) Use the Planck equation from the previous question to calculate the energy of a 1.5×10^{15} Hz UV photon and a 3.3×10^{14} Hz IR photon.
   (b) Compare the energy of the UV and IR photons, as a ratio.
   (c) From your knowledge of the electromagnetic spectrum, how does the energy of visible-light photons and X-ray photons compare with the energy of UV and IR photons?

The development of modern atomic theory involved some key experiments and many hypotheses that attempted to explain empirical results. Along the way, some ideas were never completed and some hypotheses were never accepted. For example, Thomson and his students worked long and hard to explain the number and arrangement of electrons and relate this to the periodic table and the spectra of the elements. Their attempts did not even come close to agreeing with the evidence that existed. Eventually, this work was abandoned.

Rutherford's model of a nuclear atom was a significant advance in the overall understanding of the atom, but it did little to solve the problem of the electrons that frustrated Thomson. In fact, the Rutherford model created some new difficulties. Before looking at Bohr's atomic theory, let us look at some of the problems Bohr was to solve—the stability of the Rutherford atom and the explanation of atomic spectra.

The Big Problem with the Rutherford Model

Rutherford and other scientists had guessed that the electrons move around the nucleus as planets orbit the Sun or moths flutter around a light bulb. This seemed like a logical idea. Planets are attracted to the Sun by gravity but maintain their orbit because they are moving. The same could be said for negatively charged electrons orbiting a positively charged nucleus. However, it was already well established, both experimentally and conceptually, that accelerating charges continuously produce some type of light (electromagnetic radiation). As you may have learned in your study of physics, bodies are accelerating when they change speed and/or direction. An electron travelling in a circular orbit is constantly changing its direction, and is, therefore, accelerating. According to classical theory, the orbiting electron should emit photons of electromagnetic radiation, losing energy in the process, and so spiral in toward the nucleus and collapse the atom (**Figure 1**). This prediction from classical theory of what happens in an atom is obviously not correct. Materials we see around us are very stable and so the atoms that compose them must be very stable and not in immediate danger of collapse. Even the most vigorous supporter of the existing science did not believe that, if atoms contained electrons, the electrons would be motionless and not accelerating.

Atomic Spectra

Robert Bunsen and Gustav Kirchhoff worked together to invent the spectroscope (**Figure 2**). The spectroscope forms the basis of an analytic method called **spectroscopy**, a method first reported to the scientific community by Bunsen and Kirchhoff in 1859. They studied the spectra of chemicals, especially elements, heated in a Bunsen burner flame, and the spectrum of the Sun. What they discovered was that an element not only produced a characteristic flame colour but, on closer examination through a spectroscope, also produced a **bright-line spectrum** that was characteristic of the element (**Figure 2**). The spectra of known elements were quickly catalogued and when a new spectrum was found, the spectrum was used as evidence of a new element. The elements cesium and rubidium were discovered within a year of the invention of spectroscopy. Once you quantitatively know the line spectrum of an element, it can be used as an analytic technique to identify an unknown element — a powerful technique that goes beyond the flame tests you have used previously.

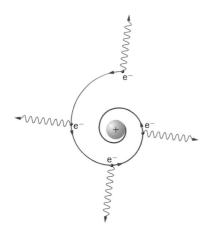

Figure 1
According to existing scientific knowledge at the time of the Rutherford atom, an orbiting electron should continuously emit electromagnetic radiation, lose energy, and collapse the atom. The evidence is to the contrary.

spectroscopy a technique for analyzing spectra; the spectra may be visible light, infrared, ultraviolet, X-ray, and other types

bright-line spectrum a series of bright lines of light produced or emitted by a gas excited by, for example, heat or electricity

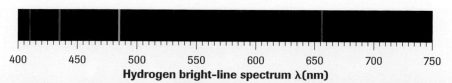

Hydrogen bright-line spectrum λ(nm)

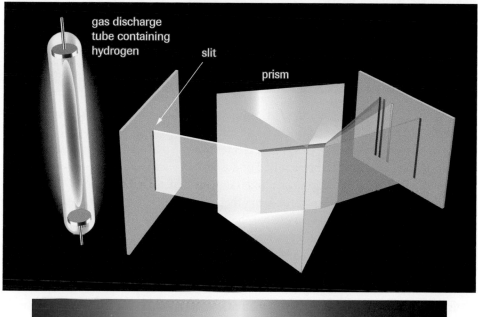

Visible spectrum λ (nm)

Figure 2
Light from a flame test, or any other source of light, is passed through slits to form a narrow beam. This beam is split into its components by the prism to produce a series of coloured lines. This kind of spectroscope was invented by Bunsen and Kirchhoff. The visible region of the hydrogen spectrum includes four coloured lines at the wavelengths shown by the scale. (1 nm = 10^{-9} m)

absorption spectrum a series of dark lines (i.e., missing parts) of a continuous spectrum; produced by placing a gas between the continuous spectrum source and the observer; also known as a dark-line spectrum

As early as 1814, **absorption** or dark-line **spectra** (**Figure 3**) were investigated qualitatively and quantitatively by Joseph von Fraunhofer. Kirchhoff, among others, was able to show in the 1860s that dark lines in an element's spectrum were in the same position as the bright lines in the spectrum of the same element. This provided a powerful tool to determine the composition of gases far away in the universe. When light passes through a gas, for example, the atmosphere around the Sun, some light is absorbed by the atoms present in the gas. This refuted the statement by the French philosopher Auguste Comte who said in 1835 that the composition of the stars was an example of something that scientists could never know.

Spectroscopes may separate the light by using a prism (**Figure 2**) or a diffraction grating. The most modern, compact, and inexpensive school spectrometers use a diffraction grating. 🔘▮

Bohr's Model of the Atom

In his mid-20s, Niels Bohr went to Cambridge University in England to join the group working under the famous J. J. Thomson. At this time, Thomson's group was attempting, quite unsuccessfully, to explain electrons in atoms and atomic spectra. Bohr suggested that tinkering with this model would never work, and some revolutionary change was required. Bohr's hunch was that a new model required using the new quantum theory of light developed by Planck and Einstein. Thomson did not like these revolutionary ideas, especially from a young man fresh out of university in Denmark. There were many heated arguments and Bohr decided to abandon Thomson's group in Cambridge and go to the University of Manchester to work with Rutherford, one of Thomson's former students.

🔘 **ACTIVITY 3.4.1**

Line Spectra (p. 212)
Use a spectroscope to "dissect" light into its components.

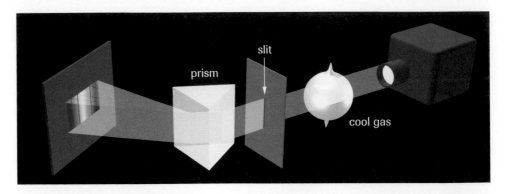

Figure 3
If you start with a complete colour spectrum of all possible colours, then pass this light through a gas and analyze what is left, you get a dark-line spectrum; in other words, the complete spectrum with some lines missing.

ACTIVITY 3.4.2

The Hydrogen Line Spectrum and the Bohr Theory (p. 213)
This computer simulation helps illustrate the electron transitions that produce bright lines in a spectrum.

stationary state a stable energy state of an atomic system that does not involve any emission of radiation

transition the jump of an electron from one stationary state to another

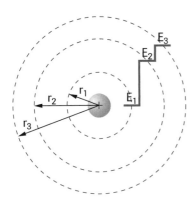

Figure 4
In the Bohr model of the atom, electrons orbit the nucleus, as the planets orbit the Sun. However, only certain orbits are allowed, and an electron in each orbit has a specific energy.

This turned out to be a much better environment for Bohr to develop his still vague ideas about the quantum theory and electrons in atoms.

The bright- and dark-line spectra of the elements mean that only certain quanta of light (certain photon energies) can be emitted or absorbed by an atom. Bohr reasoned that if the light released or absorbed from an atom was quantized, then the energy of the electron inside the atom must also be quantized. In other words, an electron can only have certain energies, just as the gearbox in a car can only have certain gears — first, second, third,…. The simplest arrangement would be a planetary model with each electron orbit at a fixed distance and with a fixed energy (**Figure 4**). In this way, the energy of the electron was quantized; in other words, the electrons could not have any energy, only certain allowed energies. To avoid the problem of the Rutherford model, Bohr boldly stated that these were special energy states (called **stationary states**), and the existing rules did not apply inside an atom.

Bohr's First Postulate
Electrons do not radiate energy as they orbit the nucleus. Each orbit corresponds to a state of constant energy (called a stationary state).

Like many other scientists of the time, Bohr was familiar with the long history of atomic spectra and shared the general feeling that the electrons were somehow responsible for producing the light observed in the line spectra. But no one knew how or why. The visible hydrogen spectrum (**Figure 2**) was the simplest spectrum and corresponded to the smallest and simplest atom. This was obviously the place to start. Although Bohr was familiar with atomic spectra, he did not know about the mathematical analysis of the hydrogen spectrum by Jacob Balmer, a teacher at a girls' school in Switzerland. According to a common story, someone showed Bohr the formula. "As soon as I saw Balmer's equation, the whole thing was immediately clear to me" (Niels Bohr).

Without going into the mathematical detail, what was clear to Bohr was that electrons "jump" from one orbit and energy level to another. This is called an electron **transition**. A transition from a higher energy state to a lower energy state means that the electron loses energy and this energy is released as a photon of light, explaining a bright line in a bright-line spectrum (**Figure 5a**). When some energy is absorbed, for example from a photon of light, the electron undergoes a transition from a lower energy state to a higher one, explaining a dark line in an absorption spectrum (**Figure 5b**). This was the crucial idea that Bohr was seeking. 🔍▮

Bohr's Second Postulate
Electrons can change their energy only by undergoing a transition from one stationary state to another.

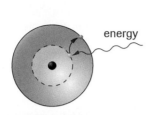
(a) An electron gains a quantum of energy.

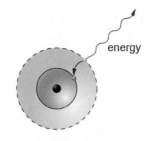
(b) An electron loses a quantum of energy.

Figure 5
In both types of transitions, the energy of the photon must match the difference in energies of the two electron states.
(a) The energy of a photon is absorbed by the electron to move it from a lower to a higher energy state.
(b) When the electron returns from a higher to a lower energy state, a photon is released.

Bohr recognized the need to use Planck's quantum theory to explain the spectral evidence from several decades of experiments. Fifty-four years after line spectra were first observed by Bunsen and Kirchhoff, and twenty-eight years after line spectra were described quantitatively by Balmer, an acceptable theoretical description was created by Bohr. The Bohr theory was another significant step forward in the evolution of modern atomic theories, which started with Dalton.

The Successes and Failure of the Bohr Model

Most importantly for our everyday work in chemistry, the Bohr model of the atom is able to offer a reasonable explanation of Mendeleev's periodic law and its representation in the periodic table. According to the Bohr model, periods in the periodic table result from the filling of electron energy levels in the atom; e.g., atoms in Period 3 have electrons in three energy levels. A period comes to an end when the maximum number of electrons is reached for the outer level. The maximum number of electrons in each energy level is given by the number of elements in each period of the periodic table; i.e., 2, 8, 8, 18, etc. You may also recall that the last digit of the group number in the periodic table provides the number of electrons in the valence (outer) energy level. Although Bohr did his calculations as if electrons were in circular orbits, the most important property of the electrons was their energy, not their motion. Energy-level diagrams for Bohr atoms are presented in the sample problem below. These diagrams have the same procedure and rationale as the orbit diagrams that you have drawn in past years. Since the emphasis here is on the energy of the electron, rather than the motion or position of the electron, orbits are not used.

DID YOU KNOW ?

Analogy for Electron Transitions
In an automobile, the transmission shifts the gears from lower to higher gears such as first to second, or downshifts from higher to lower gears. The gears are fixed, for example, first, second, third. You cannot shift to "$2\frac{1}{2}$." Similarly, electron energies in the Bohr model are fixed and electron transitions can only be up or down between specific energy levels. A satellite is not a good analogy for electron energy levels, because a satellite can be in any orbit, so any change in its energy is continuous, not in jumps.

Bohr Energy-Level Diagram

SAMPLE *problem* ◄

Use the Bohr theory and the periodic table to draw energy-level diagrams for the phosphorus atom.

First, we need to refer to the periodic table to find the position of phosphorus. Use your finger or eye to move through the periodic table from the top left along each period until you get to the element phosphorus. Starting with period 1, your finger must pass through 2 elements, indicating that there is the maximum of 2 electrons in energy level 1. Moving on to period 2, your finger moves through the full 8 elements, indicating 8 electrons in energy level 2. Finally, moving on to period 3, your finger moves 5 positions to phosphorus, indicating 5 electrons in energy level 3 for this element.

The position of 2, 8, and 5 elements per period for phosphorus tells you that there are 2, 8, and 5 electrons per energy level for this atom. The information about phosphorus atoms in the periodic table can be interpreted as follows:

atomic number, 15:   15 protons and 15 electrons (for the atom)
period number, 3:   electrons in 3 energy levels
group number, 15:   5 valence electrons (the last digit of the group number)

▶

To draw the energy-level diagram, work from the bottom up:

Sixth, the 3rd energy level,	5 e⁻	(from group 15)
Fifth, the 2nd energy level,	8 e⁻	(from eight elements in period 2)
Fourth, the 1st energy level,	2 e⁻	(from two elements in period 1)
Third, the protons:	15 p⁺	(from the atomic number)
Second, the symbol:	P	(uppercase symbol from the table)
First, the name of the atom:	phosphorus	(lowercase name)

Although the energy levels in these diagrams are (for convenience) shown as equal distance apart, we must understand that this is contrary to the evidence. Line spectra evidence indicates that the energy levels are increasingly closer together at higher energy levels.

Example

Use the Bohr theory and the periodic table to draw energy-level diagrams for hydrogen, carbon, and sulfur atoms.

Solution

		6 e⁻
	4 e⁻	8 e⁻
1 e⁻	2 e⁻	2 e⁻
1 p⁺	6 p⁺	16 p⁺
H	C	S
hydrogen atom	carbon atom	sulfur atom

▶ **Practice**

Understanding Concepts

1. Draw energy-level diagrams for each of the following:
   (a) an atom of boron
   (b) an atom of aluminum
   (c) an atom of helium

Not only was Bohr able to explain the visible spectrum for hydrogen, he was also able to successfully predict the infrared and ultraviolet spectra for hydrogen. For a theory to be able to explain past observations is good; for it to be able to predict some future observations is very good. Unfortunately, Bohr's theory was not excellent, because it works very well only for the spectrum for hydrogen atoms (or ions with only one electron). The calculations of spectral lines using Bohr's theory for any atom or ion containing more than one electron did not agree with the empirical results. In fact, the discrepancy became worse as the number of electrons increased. Nevertheless, Bohr's theory was a great success because it was the start of a new approach — including the new quantum ideas in a model of the atom (**Figure 6**).

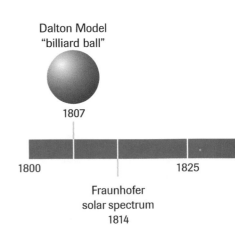

Dalton Model
"billiard ball"

1807

1800   1825

Fraunhofer
solar spectrum
1814

SUMMARY	*Creating the Bohr Atomic Theory (1913)*

Table 1

Key Experimental evidence	Theoretical explanation	Bohr's atomic theory
Mendeleev (1869–1872): There is a periodicity of the physical and chemical properties of the elements.	A new period begins in the periodic table when a new energy level of electrons is started in the atom.	• Electrons travel in the atom in circular orbits with quantized energy—energy is restricted to only certain discrete quantities. • There is a maximum number of electrons allowed in each orbit. • Electrons "jump" to a higher level when a photon is absorbed. A photon is emitted when the electron "drops" to a lower level.
Mendeleev (1872): There are two elements in the first period and eight elements in the second period of the periodic table.	There are two electrons maximum in the first electron energy level and eight in the next level.	
Kirchhoff, Bunsen (1859), Johann Balmer (1885): Emission and absorption line spectra, and not continuous spectra, exist for gaseous elements.	Since the energy of light absorbed and emitted is quantized, the energy of electrons in atoms is quantized.	

Figure 6
In a little over a hundred years, the idea of an atom has changed from the original indivisible sphere of Dalton to a particle with several components and an internal organization.

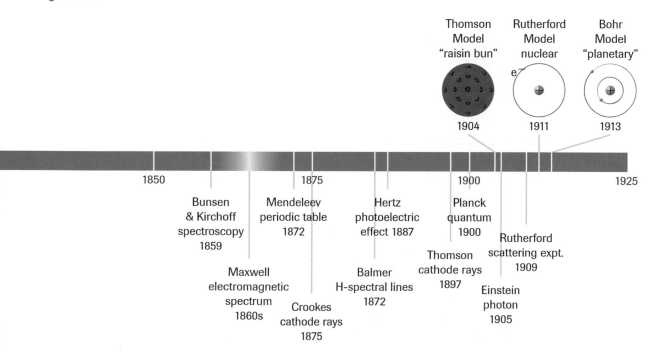

Understanding Concepts

1. What was the main achievement of the Rutherford model? What was the main problem with this model?

2. State Bohr's solution to the problem with the Rutherford atomic model.

3. When creating his new atomic theory, Bohr used one important new idea (theory) and primarily one important experimental area of study. Identify each of these.

4. (a) What is the empirical distinction between emission and absorption spectra?
   (b) In general terms, how did Niels Bohr explain each of these spectra?

5. Niels Bohr and Ernest Rutherford both worked at the same university at approximately the same time. In this text, their work has been largely separated for simplicity, but scientists often refer to the "Bohr–Rutherford" model. How did the accomplishments of Rutherford and Bohr complement each other?

6. Draw an energy-level diagram for each of the following:
   (a) fluorine atom
   (b) neon atom
   (c) sodium atom

7. What do the atomic, period, and group numbers contribute in energy-level diagrams?

8. State two or more reasons why Bohr's theory was considered a success.

9. Identify one significant problem with the Bohr theory.

Applying Inquiry Skills

10. Element 118 was reported to have been discovered in 1999. However, as of July 2001 no one, including the original researchers, has been able to replicate the experiments. Using your present knowledge, you can make predictions about this element. Predict the properties of element 118 based on the periodic law and the Bohr theory of the atom.

Making Connections

11. Read as much as you can from Bohr's original paper about the periodic table. List the content presented in Bohr's writing that you recognize. Approximately how much of the content is beyond your understanding at this time?

 www.science.nelson.com

12. Use your knowledge from this section to determine if a sample of table salt ($NaCl_{(s)}$) contains some potassium chloride.

Extension

13. In 1885 Balmer created an equation that described the visible light spectrum for hydrogen. This evolved to become the Rydberg equation presented below. Bohr used Balmer's work as an insight into the structure of the hydrogen atom.

$$\frac{1}{\lambda} = R_H \left(\frac{1}{n_f^2} - \frac{1}{n_i^2} \right)$$

$$R_H = 1.10 \times 10^7 \text{ /m}$$

   (a) The visible portion of the hydrogen spectrum is called the Balmer series. The visible light photons emitted from the hydrogen atom all involve electron transitions from higher (excited) energy levels down to the $n_f = 2$ level. Calculate the wavelength of the light emitted from the quantum leap of an electron from the $n_i = 4$ level to the $n_f = 2$ level.
   (b) Use the wave equation, $\lambda = c/f$, to calculate the frequency of the light emitted ($c = 3.00 \times 10^8$ m/s).
   (c) Use the Planck equation, $E = hf$, to calculate the energy of the electron transition and, therefore, the difference in energy between the $n_i = 4$ and the $n_f = 2$ levels ($h = 6.63 \times 10^{-34}$ J/Hz).
   (d) Repeat (a) through (c) for the $n_i = 3$ to the $n_f = 2$ electron transition.
   (e) Draw an energy-level diagram for hydrogen showing the $n_i = 3$ and 4 to the $n_f = 2$ transitions. Add the energy difference values to the diagram. From these values, what is the energy difference between $n = 4$ and $n = 3$?

14. Using your knowledge of the history of atomic theories from Dalton to Bohr, state what you think will happen next in the historical story. Provide some general comments without concerning yourself with any details.

The Bohr atomic theory, which included Planck's revolutionary idea of the quantum of energy, started a flurry of activity in the scientific community. The success of the Bohr theory in terms of explaining the line spectrum for hydrogen and the periodic table caused excitement. Not surprisingly, there was much more to come both experimentally and theoretically. Most of the new evidence came from the same area of study that had inspired Bohr — atomic spectra. For each observation, the theory had to be evaluated and revised. This is a common pattern in science, where new empirical knowledge requires revisions to the existing theory.

The Principal Quantum Number, *n*

The integer, *n*, that Bohr used to label the orbits and energies describes a main *shell* of electrons, and is referred to today as the **principal quantum number**. Although, as you will see later, we no longer use the planetary model for the electron position and motion, this quantum number invented by Bohr is still used to designate the main energy levels of electrons (**Figure 1**). Bohr's theory used only one quantum number, which is the main reason that it worked well for hydrogen but not for other atoms.

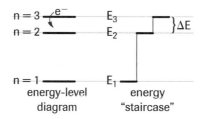

Figure 1
Bohr's principal energy levels, designated by *n*, are like unequal steps on an energy staircase. If an electron "falls" from a higher energy level such as $n = 3$ to a lower energy level, $n = 2$, the difference between the steps is released as a photon of light.

> The principal quantum number relates primarily to the main energy of an electron. $n = 1, 2, 3, 4, \ldots$

The Secondary Quantum Number, *l*

The success of the Bohr theory in explaining line spectra prompted many scientists to investigate line spectra in more detail. One obvious place to start was to use existing observations that still needed to be explained. For example, Albert Michelson in 1891 had found that the main lines of the bright-line spectrum for hydrogen were actually composed of more than one line. Although difficult to see, these lines were natural and remained unexplained for decades. Arnold Sommerfeld (1915) boldly employed elliptical orbits to extend the Bohr theory and successfully explain this line-splitting (**Figure 2**). He introduced the **secondary quantum number**, *l*, to describe additional electron energy sublevels, or *subshells*, that formed part of a main energy level. Using the analogy of a staircase for an energy level, this means that one of Bohr's main energy "steps" is actually a group of several little "steps" (**Figure 3**).

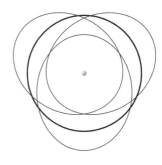

Figure 2
The original Bohr orbit for $n = 2$ is circular (red line). Sommerfeld's revision is three, slightly elliptical orbits for $n = 2$ (blue lines).

> The secondary quantum number relates primarily to the shape of the electron orbit. The number of values for *l* equals the volume of the principal quantum number.

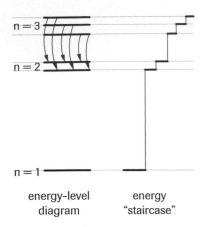

energy-level diagram   energy "staircase"

Figure 3
Sommerfeld's model adds some closely spaced energy levels to all of the Bohr main levels, except the first. Notice that the single electron transition in Figure 1 becomes multiple transitions that are very close together in energy.

(Historically, each value of l was given a letter designation related to details of the spectra that were observed.) Notice from **Figure 3** that the number of sublevels equals the value of the principal quantum number; e.g., if $n = 3$, then there are 3 sublevels, $l = 0, 1, 2$. Using existing observations and some new ideas, Sommerfeld improved the Bohr model so that a better agreement could be obtained between the laboratory evidence and atomic theory.

The secondary quantum number, l, eventually relates energy levels to the shape of the electron orbitals (covered in Section 3.7). It will also help to explain the regions of the periodic table.

> ▶ *Practice*

Understanding Concepts

1. What is the similarity in the type of observations used by Bohr and Sommerfeld?

2. What is the difference in the electron orbits proposed by Bohr and those of Sommerfeld?

3. Complete **Table 1**.

Table 1 Sommerfeld's Electron Energy Sublevels

Primary energy level	Principal quantum number, n	Possible secondary quantum numbers, l	Number of sublevels per primary level
1	1	0	1
2			
3			
4			

4. Using your answers in **Table 1**, for any principal quantum number, n, state what is the highest possible value of l.

5. Write a general rule that can be used to predict all possible values, from lowest to highest, of the secondary quantum number for any value of the principal quantum number.

The Magnetic Quantum Number, m_l

The scientific work of analyzing atomic spectra was still not complete. If a gas discharge tube is placed near a strong magnet, some single lines split into new lines that were not initially present. This observation was first made by Pieter Zeeman in 1897 and is called the normal Zeeman effect. He observed, for example, triplets where only one line existed without the magnetic field (**Figure 4**). The Zeeman effect was explained using another quantum number, the **magnetic quantum number**, m_l, added by Arnold Sommerfeld and Peter Debye (1916). Their explanation was that orbits could exist at various angles. The idea is that if orbits are oriented in space in different planes, the energies of the orbits are different when the atom is near a strong magnet.

As with the secondary quantum number, this new quantum number has restrictions. The number of lines observed in the spectra is used to determine that, for each value of l, m_l can vary from $-l$ to $+l$ (**Table 2**). Each of the non-zero values of the secondary quantum number, l, corresponds to some orientation in space. For example, if $l = 1$, then m_l can be -1, 0, or $+1$, suggesting that there are three orbits that have the same energy and shape, but differ only in their orientation in space.

> The magnetic quantum number, m_l, relates primarily to the direction of the electron orbit. The number of values for m_l is the number of independent orientations of orbits that are possible.

Table 2 Values for the Magnetic Quantum Number

Value of l 0 to n-1	Values of m_l $-l$ to $+l$
0	0
1	-1, 0, +1
2	-2, -1, 0, +1, +2
3	-3, -2, -1, 0, +1, +2, +3

The Spin Quantum Number, m_s

The final (fourth) quantum number was added to account for two kinds of evidence—some additional spectral line-splitting in a magnetic field and different kinds of magnetism. Once again we see the familiar pattern — new evidence requires a revision to the existing atomic model.

The magnetism that you are most familiar with is known as *ferromagnetism* and is most commonly associated with substances containing iron, cobalt, and nickel metals. This type of magnetism is relatively strong and has been known for thousands of years. *Paramagnetism* is another kind of magnetism of substances and is recognized as a relatively weak attraction to a strong magnet. Paramagnetism refers to the magnetism of individual atoms; ferromagnetism is due to the magnetism of a collection of atoms.

There are many substances, elements and compounds, that are known to be paramagnetic. There was no acceptable explanation for this effect until, in 1925, a student of Sommerfeld and of Bohr, Wolfgang Pauli, suggested that each electron spins on its axis. He suggested not only that a spinning electron is like a tiny magnet, but that it could have only two spins. Using an analogy, we can say an electron is like a spinning top, which can spin either clockwise or counterclockwise (**Figure 5**). For an electron, the two spins are equal in magnitude but opposite in direction, and these are the only choices; i.e., the spin is quantized to two and only two values. This fourth quantum number is called the **spin quantum number**, m_s, and is given values of either $+1/2$ or $-1/2$. Qualitatively, we refer to the spin as either clockwise or counterclockwise or as up or down.

You may know that bar magnets are stored in pairs arranged opposite to each other. This arrangement will not attract iron objects and keeps the magnets from losing their strength. Similarly, an opposite pair of electron spins is a stable arrangement and produces no magnetism of the substance. However, a single, unpaired electron, like a single bar magnet, shows magnetism and can be affected by a magnetic field. In Section 3.7, you will learn a method to predict this property. 🧪▌

> The spin quantum number, m_s, relates to a property of an electron that can best be described as its spin. The spin quantum number can only be $+1/2$ or $-1/2$ for any electron.

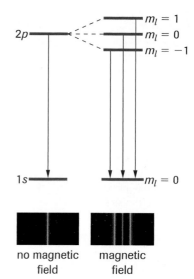

no magnetic field   magnetic field present

Figure 4
The triplets in spectra produced in a magnetic field are explained by creating a third quantum number corresponding to some new energy levels with angular orbits.

Figure 5
Spinning tops are a familiar children's toy. The top can be twisted to spin in one direction or another. This is a common analogy to understand electron spin, although scientists do not think that the electron is actually spinning.

🧪 **INVESTIGATION 3.5.1**

Paramagnetism (p. 214)
In this investigation, you can experimentally determine which substances are paramagnetic (have unpaired electrons).

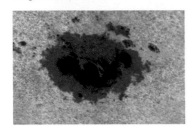
In the earliest days of quantum mechanics that we are describing in this section, an electron was thought to orbit the nucleus in a well-defined orbit. These orbits had different sizes, shapes, orientations, and energies. Today, the concept of a well-defined orbit has been abandoned and the meaning and interpretation of quantum numbers has also changed, as you will see in Section 3.7.

SUMMARY *Creating the Four Quantum Numbers*

Table 3

Key experimental work	Theoretical explanation	Quantum theory
low-resolution line spectra	principal quantum number, n	All electrons in all atoms can be described by four quantum numbers.
high-resolution line spectra	secondary quantum number, l	
spectra in magnetic field	magnetic quantum number, m_l	
ferro- and paramagnetism	spin quantum number, m_s	

Four numbers are required to describe the energy of an electron in an atom (**Table 4**). All values are quantized—restricted to a few discrete values.

Table 4 Summary of Quantum Numbers

Principal quantum number, n: the main electron energy levels or shells (n)	Secondary quantum number, l: the electron sublevels or subshells (0 to $n-1$)	Magnetic quantum number, m_l: the orientation of a sublevel ($-l$ to $+l$)	Spin quantum number, m_s: the electron spin ($+1/2$ or $-1/2$)
1	0	0	$+1/2, -1/2$
2	0	0	$+1/2, -1/2$
	1	$-1, 0, +1$	$+1/2, -1/2$
3	0	0	$+1/2, -1/2$
	1	$-1, 0, +1$	$+1/2, -1/2$
	2	$-2, -1, 0, +1, +2$	$+1/2, -1/2$

▶ Section 3.5 Questions

Understanding Concepts

1. What is the main kind of evidence used to develop the description of electrons in terms of quantum numbers?

2. Briefly, what is the theoretical description of electrons in atoms provided by each of the four quantum numbers?

3. Each value of the secondary quantum number is used to determine the possible values of the magnetic quantum number.
   (a) How many possible values of m_l are there for $l = 0, 1, 2$, and 3?
   (b) What pattern do you notice in these numbers?
   (c) Using your answer to (b), predict the number of possible values of m_l for $l = 4$.

4. Theoretical knowledge in science develops from a need to explain what is observed. What is the fourth quantum number and why is it necessary?

5. Using **Table 4** as a guide, complete the next section of this table using $n = 4$.

6. How many quantum numbers does it take to fully describe an electron in an atom? Provide an example, listing labels and values of each quantum number.

7. If every electron must have a unique set of four quantum numbers, how many different electrons (sets of four quantum numbers) can there be for each principal quantum number from $n = 1$ to $n = 3$?

8. In the development of scientific knowledge, which comes first – empirical or theoretical knowledge? Justify your answer by providing two examples from this section.

In the few years following the announcement of the Bohr theory, a series of revisions to this model occurred. Bohr's single quantum number (n) was expanded to a total of four quantum numbers (n, l, m_l, m_s). These quantum numbers were necessary to explain a variety of evidence associated with spectral lines and magnetism. In addition, these same quantum numbers also greatly improved the understanding of the periodic table and chemical bonding. You will recall that the atomic theory you used previously allowed only a limited description of electrons in atoms up to atomic number 20, calcium. In this section, you will see that the four quantum numbers improve the theoretical description to include all atoms on the periodic table and they improve the explanation of chemical properties.

Section 3.5 provided the empirical and theoretical background to quantum numbers. The main thing that you need to understand is that there are four quantized values that describe an electron in an atom. Quantized means that the values are restricted to certain discrete values — the values are not on a continuum like distance during a trip. There are quantum leaps between the values. **Table 4** in Section 3.5 illustrates the values to which the quantum numbers are restricted.

The advantage of quantized values is that they add some order to our description of the electrons in an atom. In this section, the picture of the atom is based upon the evidence and concepts from Section 3.5, but the picture is presented much more qualitatively. For example, as you shall see, the secondary quantum number values of 0, 1, 2, and 3 are presented as s, p, d, and f designations to represent the shape of the orbitals (**Table 1**). What is truly amazing about the picture of the atom that is coming in this section is that the energy description in Section 3.5 fits perfectly with both the arrangement of electrons and the structure of the periodic table. The unity of these concepts is a triumph of scientific achievement that is unparalleled in the past or present.

Figure 1
Orbitals are like "electron clouds." This computer-generated image shows a $3d$ orbital of the hydrogen atom, which has four symmetrical lobes (in this image, two blue and two red-orange), with the nucleus at the centre. The bands in the lobes show different probability levels: the probability of finding an electron decreases while moving away from the nucleus. This is quite a different image from the Bohr electron orbits.

orbital a region of space around the nucleus where an electron is likely to be found

Table 1 Values and Letters for the Secondary Quantum Number

value of l	0	1	2	3
letter designation*	s	p	d	f
name designation	**s**harp	**p**rincipal	**d**iffuse	**f**undamental

* This is the primary method of communicating values of l later in this section.

Electron Orbitals

Although the Bohr theory and subsequent revisions were based on the idea of an electron travelling in some kind of orbit or path, a more modern view is that of an electron **orbital**. A simple description of an electron orbital is that it defines a region (volume) of space where an electron may be found. **Figure 1** and **Table 2** present some of the differences between the concepts of orbit and orbital.

At this stage in your chemistry education, the four quantum numbers apply equally well to electron orbits (paths) or electron orbitals (clouds). A summary of what is coming is presented in **Table 3**.

Table 2 Orbits and Orbitals

Orbits	Orbitals
2-D path	3-D region in space
fixed distance from nucleus	variable distance from nucleus
circular or elliptical path	no path; varied shape of region
$2n^2$ electrons per orbit	2 electrons per orbital

Table 3 Energy Levels, Orbitals, and Shells

Principal energy level	Energy sublevel	Energy in magnetic field	Additional energy differences
n shell	l orbital shape subshell	m_l orbital orientation	m_s electron spin

The first two quantum numbers (n and l) describe electrons that have different energies under normal circumstances in multi-electron atoms. The last two quantum numbers (m_l, m_s) describe electrons that have different energies only under special conditions, such as the presence of a strong magnetic field. In this text, we will consider only the first two quantum numbers, which deal with energy differences for normal circumstances. As we move from focusing on the energy of the electrons to focusing on their position in space, the language will change from using main (principal) energy level to **shell**, and from energy sublevel to **subshell**. The terms can be taken as being equivalent, although the contexts of energy and space can be used to decide when they are primarily used.

Rather than a complete mathematical description of energy levels using quantum numbers, it is common for chemists to use the number for the main energy level and a letter designation for the energy sublevel (**Table 4**). For example, a $1s$ orbital, a $2p$ orbital, a $3d$, or a $4f$ orbital in that energy sub-level can be specified. This $1s$ symbol is simpler than communicating $n = 1$, $l = 0$, and $2p$ is simpler than $n = 2$, $l = 1$. Notice that this orbital description includes both the principal quantum number and the secondary quantum number; e.g., $5s$, $2p$, $3d$, or $4f$. Including the third quantum number, m_l, requires another designation, for example $2p_x$, $2p_y$, and $2p_z$.

> **shell** main energy level; the shell number is given by the principal quantum number, n; for the representative elements the shell number also corresponds to the period number on the periodic table for the s and p subshells
>
> **subshell** orbitals of different shapes and energies, as given by the secondary quantum number, l; the subshells are most often referred to as s, p, d, and f

Table 4 Classification of Energy Sublevels (Subshells)

Value of l	Sublevel symbol	Number of orbitals
0	s	1
1	p	3
2	d	5
3	f	7

Although the s-p-d-f designation for orbitals is introduced here, the shape of these orbitals is not presented until Section 3.7. The emphasis in this section is on more precise energy-level diagrams and their relationship to the periodic table and the properties of the elements.

Creating Energy-Level Diagrams

Our interpretation of atomic spectra is that electrons in an atom have different energies. The fine structure of the atomic spectra indicates energy sublevels. The designation of these energy levels has been by quantum number. Now we are going to use energy-level diagrams to indicate which orbital energy levels are occupied by electrons for a particular atom or ion.

These energy-level diagrams show the relative energies of electrons in various orbitals under normal conditions. Note that the previous energy-level diagrams that you have drawn included only the principal quantum number, n. Now you are going to extend these diagrams to include all four quantum numbers.

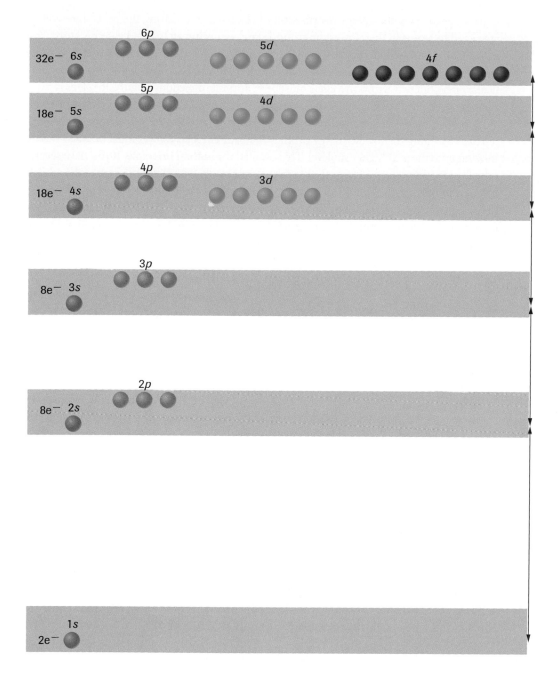

In **Figure 2**, you see that as the atoms become larger and the main energy levels become closer together, some sublevels start to overlap in energy. This figure summarizes the experimental information from many sources to produce the correct order of energies. A circle is used to represent an electron orbital within an energy sublevel. Notice that the energy of an electron increases with an increasing value of the principal quantum number, n. For a given value of n, the sublevels increase in energy, in order, $s<p<d<f$. Note also that the restrictions on the quantum numbers (**Table 4**) require that there can be only one s orbital, three p orbitals, five d orbitals, and seven f orbitals. Completing this diagram for a particular atom provides important clues about chemical properties and patterns in the periodic table. We will now look at some rules for completing an orbital energy-level diagram and then later use these diagrams to explain some properties of the elements and the arrangement of the periodic table.

Figure 2
Diagram of relative energies of electrons in various orbitals. Each orbital (circle) can potentially contain up to two electrons.

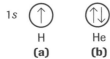

Figure 3
Energy-level diagrams for
(a) hydrogen and **(b)** helium atoms

Figure 4
Energy-level diagrams for lithium,
carbon, and fluorine atoms. Notice
that all of the 2*p* orbitals at the same
energy are shown, even though
some are empty.

Pauli exclusion principle no two
electrons in an atom can have the
same four quantum numbers; no
two electrons in the same atomic
orbital can have the same spin; only
two electrons with opposite spins
can occupy any one orbital

aufbau principle "aufbau" is
German for building up; each electron is added to the lowest energy
orbital available in an atom or ion

Hund's rule one electron occupies
each of several orbitals at the same
energy before a second electron
can occupy the same orbital

In order to show the energy distribution of electrons in an atom, the procedure will be restricted to atoms in their lowest or ground state, assuming an isolated gaseous atom. You show an electron in an orbital by drawing an arrow, pointed up or down to represent the electron spin (**Figure 3a**). It does not matter if you point the arrow up or down in any particular circle, but two arrows in a circle must be in opposite directions (**Figure 3b**). This is really a statement of the **Pauli exclusion principle**, which requires that no two electrons in an atom have the same four quantum numbers. Electrons (arrows) are placed into the orbitals (circles) by filling the lowest energy orbitals first. An energy sublevel must be filled before moving onto the next higher sublevel. This is called the **aufbau principle**. If you have several orbitals at the same energy (e.g., *p*, *d*, or *f* orbitals), one electron is placed into each of the orbitals before a second electron is added. In other words, spread out the electrons as much as possible horizontally before doubling up any pair of electrons. This rule is called **Hund's rule**. You follow this procedure until the number of electrons placed in the energy-level diagram for the atom is equal to the atomic number for the element (**Figure 4**).

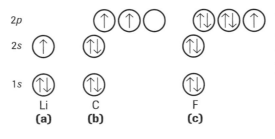

According to these rules, when electrons are added to the second ($n = 2$) energy level, there are *s* and *p* sublevels to fill with electrons (**Figure 2**). The lower energy *s* sublevel is filled before the *p* sublevel is filled. According to Hund's rule, one electron must go into each of the *p* orbitals before a second electron is used for pairing (**Figure 4**).

There are several ways of memorizing and understanding the order in which the energy levels are filled without having the complete chart shown in **Figure 2**. One method is to use a pattern like the one shown in **Figure 5**. In this aufbau diagram, all of the orbitals with the same principal quantum number are listed horizontally. You can follow the diagonal arrows starting with the 1s orbital to add the required number of electrons.

An alternate procedure for determining the order in which energy levels are filled comes from the arrangement of elements in the periodic table. As you move across the periodic table, each atom has one more electron (and proton) than the previous atom. Because the electrons are added sequentially to the lowest energy orbital available (aufbau principle), the elements can be classified by the sublevel currently being filled (**Figure 6**). To obtain the correct order of orbitals for any atom, start at hydrogen and move from left to right across the periodic table, filling the orbitals as shown in **Figure 6**. Check to see that this gives exactly the same order as shown in **Figure 5**.

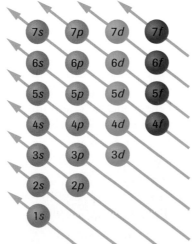

Figure 5
In this aufbau diagram, start at
the bottom (1*s*) and add electrons in the order shown by the
diagonal arrows. You work your
way from the bottom left corner
to the top right corner.

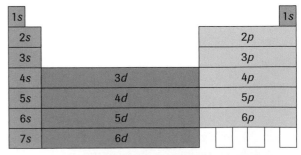

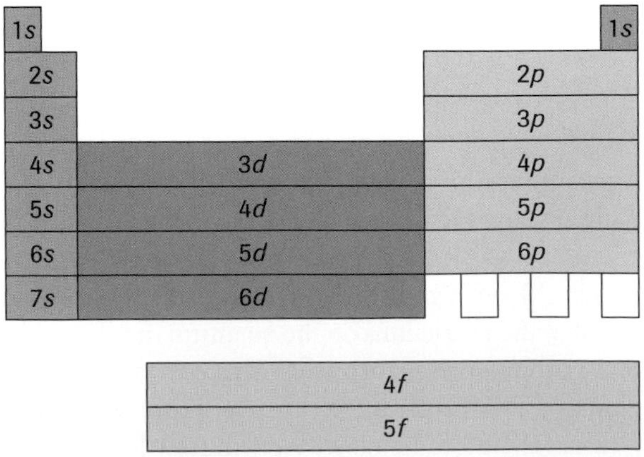

Figure 6
Classification of elements by the sublevels
that are being filled

Drawing Energy-Level Diagrams for Atoms

Draw the electron energy-level diagram for an oxygen atom.

Since oxygen (O) has an atomic number of 8, there are 8 electrons to be placed in energy levels. As the element is in period 2, there are electrons in the first two main energy levels.

- Using either the aufbau diagram (**Figure 5**) or the periodic table (**Figure 6**), we can see that the first two electrons will occupy the 1s orbital.

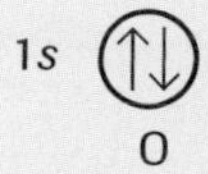

1s (↑↓)

O

- The next two electrons will occupy the 2s orbital.

2s (↑↓)

1s (↑↓)

O

- The next three electrons are placed singly in each of the 2p orbitals.

2p (↑↓)(↑)(↑)

2p (↑↓)(↑)(↑)
2s (↑↓)

- The last (eighth) electron must be paired with one of the electrons in the 2p orbitals. It does not matter into which of the three p orbitals this last electron is placed. The final diagram is drawn as shown.

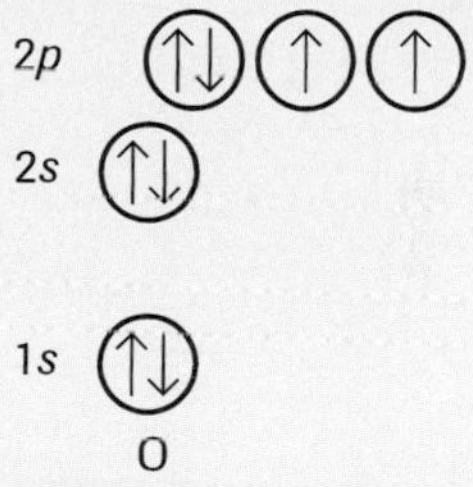

1s (↑↓)

O

Note that this energy-level diagram is not drawn to scale. The "actual" gap in energy is much larger between the 1s and 2s levels than between the 2s and 2p levels.

Example

Draw the energy-level diagram for an iron atom.

Solution

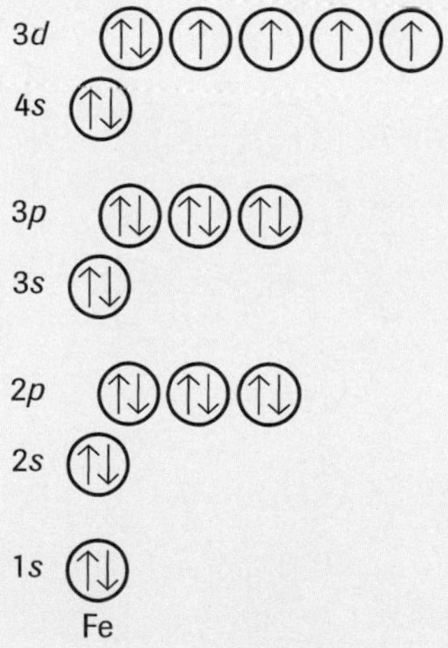

3d (↑↓)(↑)(↑)(↑)(↑)
4s (↑↓)

3p (↑↓)(↑↓)(↑↓)
3s (↑↓)

2p (↑↓)(↑↓)(↑↓)
2s (↑↓)

1s (↑↓)

Fe

Creating Energy-Level Diagrams for Anions

The energy-level diagrams for anions, or negatively charged ions, are done using the same method as for atoms. The only difference is that you need to *add the extra electrons corresponding to the ion charge to the total number of electrons before proceeding to distribute the electrons into orbitals*. This is shown in the following sample problem.

Drawing Energy-Level Diagrams for Anions

Draw the energy-level diagram for the sulfide ion.

Sulfur has an atomic number of 16 and is in period 3. A sulfide ion has a charge of 2–, which means that it has two more electrons than a neutral atom. Therefore, we have 18 electrons to distribute in three principal energy levels.

- Using either the aufbau diagram (**Figure 5**) or the periodic table (**Figure 6**), we can see that the first two electrons will occupy the 1s orbital.
- The next two electrons will occupy the 2s orbital, and six more electrons will complete the 2p orbitals.
- The next two electrons fill the 3s orbital, which leaves the final six electrons to completely fill the 3p orbitals. Notice that all orbitals are now completely filled with the 18 electrons.

Creating Energy-Level Diagrams for Cations

For cations, positively charged ions, the procedure for constructing energy-level diagrams is slightly different than for anions. *You must draw the energy-level diagram for the corresponding neutral atom first, and then remove the number of electrons (corresponding to the ion charge) from the orbitals with the highest principal quantum number, n.* The electrons removed might not be the highest-energy electrons. However, in general, this produces the correct arrangement of energy levels based on experimental evidence.

Drawing Energy-Level Diagrams for Cations

Draw the energy-level diagram for the zinc ion.

First, we need to draw the diagram for the zinc atom (atomic number 30). Using either the aufbau diagram (**Figure 5**) or the periodic table (**Figure 6**), we can see that the 30 electrons are distributed as follows:

- The first two electrons will occupy the 1s orbital.
- The next two electrons will occupy the 2s orbital, and six more electrons complete the 2p orbitals.
- The next two electrons fill the 3s orbital, and six more electrons complete the 3p orbitals.
- The next two electrons fill the 4s orbital and the final 10 electrons fill the 3d orbitals.
- The zinc ion, Zn^{2+}, has a two positive charge, and therefore has two fewer electrons than the zinc atom. Remove the two electrons from the orbital with the highest *n* — the 4s orbital in this example.

SUMMARY Electron Energy-Level Diagrams

Electrons are added into energy levels and sublevels for an atom or ion by the following set of rules. Remembering the names for the rules is not nearly as important as being able to apply the rules. These rules were created to explain the spectral and periodic-table evidence for the elements.

- Start adding electrons into the lowest energy level (1*s*) and build up from the bottom until the limit on the number of electrons for the particle is reached — the aufbau principle.
- For anions, add extra electrons to the number for the atom. For cations, do the neutral atom first, then subtract the required number of electrons from the orbitals with the highest principal quantum number, *n*.
- No two electrons can have the same four quantum numbers; if an electron is in the same orbital with another electron, it must have opposite spin — the Pauli exclusion principle.
- No two electrons can be put into the same orbital of equal energy until one electron has been put into each of the equal-energy orbitals — Hund's rule.

This process is made simpler by labelling the sections of the periodic table and then creating the energy levels and electron configurations in the order dictated by the periodic table.

DID YOU KNOW?

Gerhard Herzberg

Spectroscopy was an essential tool in developing quantum numbers and electron energy levels. A key figure in the development of modern spectroscopy was Gerhard Herzberg (1904–1999). His research in spectroscopy at the University of Saskatchewan and the National Research Council in Ottawa earned him an international reputation and the Nobel Prize in chemistry (1971).

Practice

Understanding Concepts

1. State the names of the three main rules/principles used to construct an energy-level diagram. Briefly describe each of these in your own words.
2. How can the periodic table be used to help complete energy-level diagrams?
3. Complete electron energy-level diagrams for the
   (a) phosphorus atom
   (b) potassium atom
   (c) manganese atom
   (d) nitride ion
   (e) bromide ion
   (f) cadmium ion
4. (a) Complete electron energy-level diagrams for a potassium ion and a chloride ion.
   (b) Which noble gas atom has the same electron energy-level diagram as these ions?

Extension

5. If the historical letter designations were not used for the sublevels, what would be the label for the following orbitals, using only quantum numbers: 1*s*, 2*s*, 2*p*, 3*d*?

Electron Configuration

Electron energy-level diagrams are a better way of visualizing the energies of the electrons in an atom than quantum numbers, but they are rather cumbersome to draw. We are now going to look at a third way to convey this information. **Electron configurations** provide the same information as the energy-level diagrams, but in a more concise format. An electron configuration is a listing of the number and kinds of electrons in order of increasing energy, written in a single line; e.g., Li: $1s^2 2s^1$. The order, from left to right, is the order of increasing energy of the orbitals. The symbol includes both the type of orbital and the number of electrons (**Figure 7**).

For example, if you were to look back at the energy-level diagrams shown previously for the oxygen atom, the sulfide ion, and the iron atom, then you could write the electron configuration from the diagram by listing the orbitals from lowest to highest energy.

oxygen atom, O: $1s^2 2s^2 2p^4$

sulfide ion, S^{2-}: $1s^2 2s^2 2p^6 3s^2 3p^6$

iron atom, Fe: $1s^2 2s^2 2p^6 3s^2 3p^6 4s^2 3d^6$

Note that some of the information is lost when going from the energy-level diagram to the electron configuration, but the efficiency of the communication is much improved by using an electron configuration. Fortunately, there is a method for writing electron configurations that does not require drawing an energy-level diagram first. Let us look at this procedure.

electron configuration a method for communicating the location and number of electrons in electron energy levels; e.g., Mg: $1s^2 2s^2 2p^6 3s^2$

principal quantum number

$$3p^5$$ ← number of electrons in orbital(s)

↑ orbital

Figure 7
Example of electron configuration

Writing Electron Configurations

1. Write the electron configuration for the chlorine atom.

First, locate chlorine on the periodic table. Starting at the top left of the table, follow with your finger through the sections of the periodic table (in order of atomic number), listing off the filled orbitals and then the final orbital.

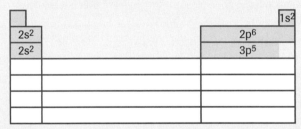

				1s²
2s²			2p⁶	
2s²			3p⁵	

You now have the electron configuration for chlorine: $1s^2$, $2s^2$, $2p^6$, $3s^2$, and $3p^5$.

2. Identify the element whose atoms have the following electron configuration:

$1s^2 2s^2 2p^6 3s^2 3p^6 4s^2 3d^{10} 4p^6 5s^2 4d^{10} 5p^6 6s^2 4f^{14} 5d^{10} 6p^4$

Notice that the highest *n* is 6, so you can go quickly to the higher periods in the table to identify the element. The highest *s* and *p* orbitals always tell you the period number, and in this case the electron configuration finishes with $6s^2 4f^{14} 5d^{10} 6p^4$: the element must be in period 6. Going across period 6 through the two *s*-elements, the 14 *f*-elements, and the 10 *d*-elements, you come to the fourth element in the *p*-section of the periodic table.

The fourth element in the $6p$ region of the periodic table is polonium, Po (**Figure 8**).

Example 1
Write the electron configuration for the tin atom and the tin(II) ion.

Solution
Sn: $1s^2 2s^2 2p^6 3s^2 3p^6 4s^2 3d^{10} 4p^6 5s^2 4d^{10} 5p^2$

Sn^{2+}: $1s^2 2s^2 2p^6 3s^2 3p^6 4s^2 3d^{10} 4p^6 5s^2 4d^{10}$

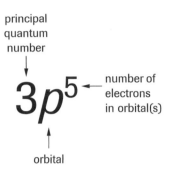

Figure 8
Polonium is a very rare, radioactive natural element found in small quantities in uranium ores. Polonium is also synthesized in gram quantities by bombarding Bi-209 with neutrons. The energy released from the radioactive decay of Po-210, the most common isotope, is very large (100 W/g) — a half gram of the isotope will spontaneously heat up to 500°C. Surprisingly, Po-210 has several uses, including as a thermoelectric power source for satellites.

Example 2

Identify the atoms that have the following electron configurations:
(a) $1s^2\,2s^2\,2p^6\,3s^2\,3p^6\,4s^2\,3d^{10}\,4p^5$
(b) $1s^2\,2s^2\,2p^6\,3s^2\,3p^6\,4s^2\,3d^{10}\,4p^6\,5s^2\,4d^5$

Solution

(a) bromine atom, Br
(b) technetium atom, Tc

Shorthand Form of Electron Configurations

There is an internationally accepted shortcut for writing electron configurations. The core electrons of an atom are expressed by using a symbol to represent all of the electrons of the preceding noble gas. Just the remaining electrons beyond the noble gas are shown in the electron configuration. This reflects the stability of the noble gases and the theory that only the electrons beyond the noble gas (the outer shell electrons) are chemically important for explaining chemical properties. Let's rewrite the full electron configurations for the chlorine and tin atoms into this shorthand format.

Cl: $1s^2\,2s^2\,2p^6\,3s^2\,3p^5$ becomes Cl: [Ne] $3s^2\,3p^5$
Sn: $1s^2\,2s^2\,2p^6\,3s^2\,3p^6\,4s^2\,3d^{10}\,4p^6\,5s^2\,4d^{10}\,5p^2$ becomes Sn: [Kr] $5s^2\,4d^{10}\,5p^2$

Writing Shorthand Electron Configurations

SAMPLE problem ◀

Write the shorthand electron configuration for the strontium atom.

Follow the same procedure as before, but start with the noble gas immediately preceding the strontium atom, which is krypton. Then continue adding orbitals and electrons until you obtain the required number of electrons for a strontium atom (two beyond krypton).
Sr: [Kr] $5s^2$

Example

Write the shorthand electron configuration for the lead atom and the lead(II) ion.

Solution

Pb: [Xe] $6s^2\,4f^{14}\,5d^{10}\,6p^2$
Pb^{2+}: [Xe] $6s^2\,4f^{14}\,5d^{10}$

LEARNING TIP

Electron Configurations for Cations
Recall that energy-level diagrams for cations are done by first doing the energy level for the neutral atom, and then subtracting electrons from the highest principal quantum number, n. Notice in this Example that electrons are removed from the $n = 6$ orbital. The $6p$ electrons are removed before the $6s$ electrons.

SUMMARY | **Procedure for Writing an Electron Configuration**

Step 1 Determine the position of the element in the periodic table and the total number of electrons in the atom or simple ion.

Step 2 Start assigning electrons in increasing order of main energy levels and sublevels (using the aufbau diagram, **Figure 5**, or the periodic table, **Figure 6**).

Step 3 Continue assigning electrons by filling each sublevel before going to the next sublevel, until all of the electrons are assigned.

- For anions, add the extra electrons to the total number in the atom.
- For cations, write the electron configuration for the neutral atom first and then remove the required number of electrons from the highest principal quantum number, n.

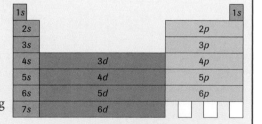

Understanding Concepts

6. Identify the elements whose atoms have the following electron configurations:
   (a) $1s^2\, 2s^2$
   (b) $1s^2\, 2s^2\, 2p^5$
   (c) $1s^2\, 2s^2\, 2p^6\, 3s^1$
   (d) $1s^2\, 2s^2\, 2p^6\, 3s^2\, 3p^4$

7. Write full electron configurations for each of the Period 3 elements.

8. (a) Write shorthand electron configurations for each of the halogens.
   (b) Describe how the halogen configurations are similar. Does this general pattern apply to other families?

9. Write the full electron configurations for a fluoride ion and a sodium ion.

10. A fluoride ion, neon atom, and sodium ion are theoretically described as isoelectronic. State the meaning of this term.

11. Write the shorthand electron configurations for the common ion of the first three members of Group 12.

Explaining the Periodic Table

The modern view of the atom based on the four quantum numbers was developed using experimental studies of atomic spectra and the experimentally determined arrangement of elements in the periodic table. It is no coincidence that the maximum number of electrons in the s, p, d, and f orbitals (**Table 5**) corresponds exactly to the number of columns of elements in the s, p, d, and f blocks in the periodic table (**Figure 6**). This by itself is a significant accomplishment that the original Bohr model could not adequately explain.

representative elements the metals and nonmetals in the main blocks, Groups 1-2, 13-18, in the periodic table; in other words, the s and p blocks

transition elements the metals in Groups 3-12; elements filling d orbitals with electrons

Table 5   Electron Subshells and the Periodic Table

Period	# of elements	Electron distribution groups: 1-2 orbitals: s	13-18 p	3-12 d	– f
Period 1	2	2			
Period 2	8	2	6		
Period 3	18	2	6	10	
Period 4-5	18	2	6	10	
Period 6-7	32	2	6	10	14

Groups or families in the periodic table were originally created by Mendeleev to reflect the similar properties of elements in a particular group. The noble gas family, Group 18, is a group of gases that are generally nonreactive. The electron configurations for noble gas atoms show that each of them has a filled ns^2np^6 outer shell of electrons (**Table 6**). The original idea from the Bohr theory — filled energy levels as stable (nonreactive) arrangements — still holds, but is more precisely defined. Similar outer shell or valence electron configurations also apply to most families, in particular, the **representative elements**.

Similarly, the **transition elements** can now be explained by our new theory as elements that are filling the d energy sublevel with electrons. The transition elements are sometimes referred to as the d block of elements. The 5 d orbitals can accommodate 10 electrons, and there are 10 elements in each transition-metal period (**Table 6**).

LEARNING TIP

The lanthanides are also called the rare earths, and the elements after uranium (the highest-atomic-number naturally occurring element) are called transuranium elements.

Table 6 Explaining the Periodic Table

Sublevel	Elements	Orbitals	Electrons	Series of elements
s and p	$2 + 6 = 8$	$1 + 3 = 4$	$2 + 6 = 8$	representative
d	10	5	10	transition
f	14	7	14	lanthanides and actinides

Using the same test of the theory on the **lanthanides** and the **actinides**, we can explain these series of elements as filling an f energy level. The f block of elements is 14 elements wide, as expected by filling 7 f orbitals with 14 electrons. The success of the quantum-number and s-p-d-f theories in explaining the long-established periodic table led to these approaches being widely accepted in the scientific community.

lanthanides and **actinides** the 14 metals in each of periods 6 and 7 that range in atomic number from 57-70 and 89-102, respectively; the elements filling the f block

Explaining Ion Charges

Previously, we could not explain transition-metal ions and multiple ions formed by heavy representative metals. Now many of these can be explained, although some require a more detailed theory beyond this textbook. For example, you know that zinc forms a 2+ ion. The electron configuration for a zinc atom:

Zn: $[Ar] 4s^2 3d^{10}$

shows 12 outer electrons. If another atom or ion removes the two $4s$ electrons (the ones with the highest n) this would leave zinc with filled $3d$ orbitals — a relatively stable state, like those of atoms with filled sub-shells:

Zn^{2+}: $[Ar] 3d^{10}$

(Note that it is unlikely that zinc would give up 10 electrons to leave filled $4s$ orbitals.)

Another example that illustrates the explanatory power of this approach is the formation of either 2+ or 4+ ions by lead. The electron configuration for a lead atom:

Pb: $[Xe] 6s^2 4f^{14} 5d^{10} 6p^2$

shows filled $4f$, $5d$, $6s$ orbitals, and a partially filled $6p$ orbital. The lead atom could lose the two $6p$ electrons to form a 2+ ion or lose four electrons from the $6s$ and $6p$ orbitals to form a 4+ ion. (From the energy-level diagram (**Figure 2**), you can see that all of these outer electrons are very similar in energy and it is easier to remove fewer electrons than large numbers such as 10 and 14.) Again, our new theory passes the test of being able to initially explain what we could not explain previously. Let's put it to another test.

Explaining Magnetism

To create an explanation for magnetism, let's start with the evidence of ferromagnetic (strongly magnetic) elements and write their electron configurations (**Table 7**).

Table 7 Ferromagnetic Elements and Their Electron Configurations

Ferromagnetic element	Electron configuration	d-Orbital filling	Pairing of d electrons
iron	$[Ar] 4s^2 3d^6$	⇅ ↑ ↑ ↑ ↑	1 pair; 4 unpaired
cobalt	$[Ar] 4s^2 3d^7$	⇅ ⇅ ↑ ↑ ↑	2 pairs; 3 unpaired
nickel	$[Ar] 4s^2 3d^8$	⇅ ⇅ ⇅ ↑ ↑	3 pairs; 2 unpaired

Based on the magnetism associated with electron spin and the presence of several unpaired electrons, an initial explanation is that the unpaired electrons cause the magnetism. However, ruthenium, rhodium, and palladium, immediately below iron, cobalt, and nickel in the

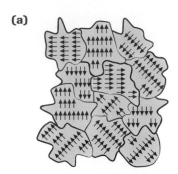

(a)

unmagnetized

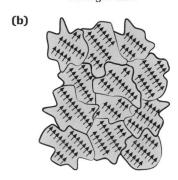

(b)

magnetized

Figure 9
The theory explaining ferromagnetism in iron is that in unmagnetized iron **(a)** the domains of atomic magnets are randomly oriented. In magnetized iron **(b)** the domains are lined up to form a "permanent" magnet.

 LAB EXERCISE 3.6.1

Quantitative Paramagnetism (p. 215)
Paramagnetism is believed to be related to unpaired electrons. This lab exercise explores this relation.

Figure 10
The stability of half-filled and filled subshells is used to explain the anomalous electron configurations of chromium and copper.

periodic table (i.e., in the same groups) are only paramagnetic (weakly magnetic) and are not ferromagnetic. The presence of several unpaired electrons may account for some magnetism, but not for the strong ferromagnetism. The eventual explanation for this anomaly is that iron, cobalt, and nickel (as smaller, closely packed atoms) are able to orient themselves in a magnetic field. The theory is that each atom acts like a little magnet. These atoms influence each other to form groups (called *domains*) in which all of the atoms are oriented with their north poles in the same direction. If most of the domains are then oriented in the same direction by an external magnetic field of, for example, a strong bar magnet, the ferromagnetic metal becomes a "permanent" magnet. However, the magnet is only permanent until dropped or heated or subjected to some other procedure that allows the domains to become randomly oriented again. (**Figure 9**). *Ferromagnetism is a based on the properties of a collection of atoms, rather than just one atom.*

Paramagnetism is also explained as being due to unpaired electrons within substances where domains do not form. In other words, paramagnetism is based on the magnetism of individual atoms. Again, the theory of electron configurations is able to at least partially explain an important property of some chemicals. In this case, a full description of each electron, including its spin, is involved in the explanation.

Anomalous Electron Configurations

Electron configurations can be determined experimentally from a variety of sophisticated experimental designs. Using the rules created above, let's test our ability to accurately predict the electron configurations of the atoms in the $3d$ block of elements. First, the predictions:

Sc: [Ar] $4s^2 3d^1$ Fe: [Ar] $4s^2 3d^6$
Ti: [Ar] $4s^2 3d^2$ Co: [Ar] $4s^2 3d^7$
V: [Ar] $4s^2 3d^3$ Ni: [Ar] $4s^2 3d^8$
Cr: [Ar] $4s^2 3d^4$ Cu: [Ar] $4s^2 3d^9$
Mn: [Ar] $4s^2 3d^5$ Zn: [Ar] $4s^2 3d^{10}$

Then the evidence:
Sc: [Ar] $4s^2 3d^1$ Fe: [Ar] $4s^2 3d^6$
Ti: [Ar] $4s^2 3d^2$ Co: [Ar] $4s^2 3d^7$
V: [Ar] $4s^2 3d^3$ Ni: [Ar] $4s^2 3d^8$
Cr: [Ar] $4s^1 3d^5$ Cu: [Ar] $4s^1 3d^{10}$
Mn: [Ar] $4s^2 3d^5$ Zn: [Ar] $4s^2 3d^{10}$

Overall, the configurations based on experimental evidence agree with the predictions, with two exceptions — chromium and copper. A slight revision of the rules for writing electron configurations seems to be required.

The evidence suggests that half-filled and filled subshells are more stable (lower energy) than unfilled subshells. This appears to be more important for d orbitals compared to s orbitals. In the case of chromium, an s electron is promoted to the d subshell to create two half-filled subshells; i.e., [Ar] $4s^2 3d^4$ becomes [Ar] $4s^1 3d^5$. In the case of copper, an s electron is promoted to the d subshell to create a half-filled s subshell and a filled d subshell; i.e., [Ar] $4s^2 3d^9$ becomes [Ar] $4s^1 3d^{10}$ (**Figure 10**). The justification is that the overall energy state of the atom is lower after the promotion of the electron. Apparently, this is the lowest possible energy state for chromium and copper atoms.

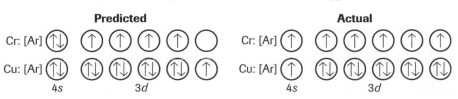

▶ *Section 3.6* *Questions*

Understanding Concepts

1. Determine the maximum number of electrons with a principal quantum number
   (a) 1   (c) 3
   (b) 2   (d) 4

2. Copy and complete **Table 8**.

   Table 8 Orbitals and Electrons in *s, p, d,* and *f* Sublevels

Sublevel symbol	Value of *l*	Number of orbitals	Max. # of electrons
(a) *s*	0		
(b) *p*	1		
(c) *d*	2		
(d) *f*	3		

3. State the aufbau principle and describe two methods that can be used to employ this principle.

4. If four electrons are to be placed into a *p* subshell, describe the procedure, including the appropriate rules.

5. (a) Draw electron energy-level diagrams for beryllium, magnesium, and calcium atoms.
   (b) What is the similarity in these diagrams?

6. The last electron represented in an electron configuration is related to the position of the element in the periodic table. For each of the following sections of the periodic table, indicate the sublevel (*s,p,d,f*) of the last electron:
   (a) Groups 1 and 2
   (b) Groups 3 to 12 (transition metals)
   (c) Groups 13 to 18
   (d) lanthanides and actinides

7. (a) When the halogens form ionic compounds, what is the ion charge of the halide ions?
   (b) Explain this similarity, using electron configurations.

8. The sodium ion and the neon atom are isoelectronic; i.e., have the same electron configuration.
   (a) Write the electron configurations for the sodium ion and the neon atom.
   (b) Describe and explain the similarities and differences in properties of these two chemical entities.

9. Use electron configurations to explain the common ion charges for antimony; i.e., Sb^{3+} and Sb^{5+}.

10. Predict the electron configuration for the gallium ion, Ga^{3+}. Provide your reasoning.

11. Evidence indicates that copper is paramagnetic, but zinc is not. Explain the evidence.

12. Predict the electron configuration of a gold atom. Provide your reasoning.

13. Use electron configurations to explain the
    (a) 3+ charge on the scandium ion
    (b) 1+ charge on a silver ion
    (c) 3+ and 2+ charges on iron(III) and iron(II) ions
    (d) 1+ and 3+ charges on the Tl^{1+} and Tl^{3+} ions

14. Carbon, silicon, and germanium all form four bonds. Explain this property, using electron configurations.

Applying Inquiry Skills

15. The ingenious Stern-Gerlach experiment of 1921 is famous for providing early evidence of quantized electron spin. The experimental design called for a beam of gaseous silver atoms from an oven to be sent through a nonuniform magnetic field. There were two possible results: one predicted by classical and one by quantum theory (**Figure 11**).
    (a) Which of the expected results is likely the classical prediction and which is likely the quantum theory prediction? Explain your choice.
    (b) Use quantum theory, the Pauli exclusion principle, and the electron configuration of silver to explain the results of the Stern-Gerlach experiment.

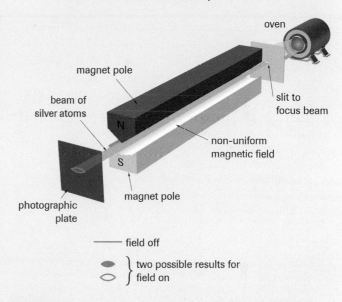

Figure 11
The Stern–Gerlach experiment expected two possible results: one predicted from classical magnetic theory in which all orientations of electron spin are possible and the other predicted from quantum theory in which only two orientations are possible.

Making Connections

16. Prior to 1968 Canadian dimes were made from silver rather than nickel. A change was made because the value of the silver in the dime had become greater than ten cents and the dimes were being shipped out of the country to be melted down.
    (a) Why were the dimes shipped out of Canada before being melted?
    (b) If you had a box full of Canadian dimes and you wanted to efficiently separate the silver from the nickel ones, what empirical properties of silver and nickel learned in this section could assist you in completing your task?

(c) Use theoretical concepts learned in this section to explain your separation technique.

17. Electron spin resonance (ESR) is an analytical technique that is based on the spin of an electron. State some examples of the uses of ESR in at least two different areas.

 www.science.nelson.com

18. Magnetic Resonance Imaging (MRI) is increasingly in demand for medical diagnosis (**Figure 12**).
   (a) How is this technique similar to and different from electron spin techniques?
   (b) Provide some examples of the usefulness of MRI results.
   (c) What political issue is associated with MRI use?

 www.science.nelson.com

Figure 12
MRI was developed using the quantum mechanical model of the atom.

After many revisions, the quantum theory of the atom produced many improvements in the understanding of different electron energy states within an atom. This was very useful in explaining properties such as atomic spectra and some periodic trends. However, these advances did not address some fundamental questions such as, "What is the electron doing inside the atom?" and "Where does the electron spend its time inside the atom?" Scientists generally knew that a planetary model of various orbits was not correct because an atom should collapse, according to known physical laws. We know that this is not true; atoms are generally very stable. Bohr's solution to this problem was to state that an electron orbit is somehow stable and doesn't obey the classical laws of physics. Even Bohr knew that this was not a satisfactory answer because it does not offer any explanation of the behaviour of the electron.

The solution to this dilemma of electron behaviour came surprisingly in 1923 from a young graduate student, Louis de Broglie (**Figure 1**). By 1923, the idea of a photon as a quantum of energy was generally accepted. This meant that light appeared to have a dual nature—sometimes it behaved like a continuous electromagnetic wave and sometimes it behaved like a particle (photon). De Broglie's insight was essentially to reverse this statement—if a wave can behave like a particle, then a particle should also be able to behave like a wave. Of course, he had to justify this hypothesis and he did this by using a number of formulas and concepts from the work of Max Planck and Albert Einstein. At first, this novel idea by de Broglie was scorned by many respected and established scientists. However, like all initial hypotheses in science, the value of the idea must be determined by experimental evidence. This happened a few years later. Clinton Davisson accidentally discovered evidence for the wave properties of an electron, although he did not realize what he had done at first. Shortly after Davisson's report, G. P. Thomson (son of J. J. Thomson) independently demonstrated the wave properties of an electron. Davisson and Thomson shared the Nobel prize in 1937 for their experimental confirmation of de Broglie's hypothesis.

When Erwin Schrödinger heard of de Broglie's electron wave, it immediately occurred to him that this idea could be used to solve the problem of electron behaviour inside an atom. Schrödinger and others created the physics to describe electrons behaving like waves inside an atom. Schrödinger's proposed wave mechanics was firmly based on the existing quantum concepts, and for this reason is usually referred to as **quantum mechanics**. According to Schrödinger, the electron can only have certain (quantized) energies because of the requirement for only whole numbers of wavelengths for the electron wave. This is illustrated in **Figure 2**.

Figure 1
Louis de Broglie obtained his first university degree in history. After serving in the French army, he became interested in science and returned to university to study physics. De Broglie was a connoisseur of classical music and he used his knowledge of basic tones and overtones as inspiration for electron waves. In spite of the fact that his hypothesis was ridiculed by some, he graduated with a doctorate in physics in 1924.

quantum mechanics the current theory of atomic structure based on wave properties of electrons; also known as wave mechanics

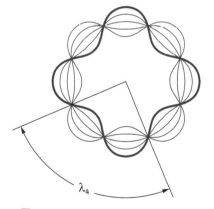

Figure 2
Schrödinger envisaged electrons as stable circular waves around the nucleus.

Electron Orbitals

Schrödinger's quantum or wave mechanics provided a complete mathematical framework that automatically included all four quantum numbers and produced the energies of all electron orbitals. But what does it tell us about where the electrons are and what they are doing? A significant problem in trying to answer these questions is the difficulty in picturing

ACTIVITY 3.7.1

Modelling Standing Electron Waves (216)
Standing waves on a string are interesting, but standing waves on a circular loop are very cool.

a particle as a wave. This seems contrary to our experience and we really have no picture to comfort us. A way around this problem is to still retain our picture of an electron as a particle, but one whose location we can only specify as a statistical probability; i.e., what are the odds of finding the electron at this location? This probabilistic approach was shown to be necessary as a result of the work by Werner Heisenberg, a student of Bohr and Sommerfeld. Heisenberg realized that to measure any particle, we essentially have to "touch" it. For ordinary-sized objects this is not a problem, but for very tiny, subatomic particles we find out where they are and their speed by sending photons out to collide with them. When the photon comes back into our instruments, we can make interpretations about the particle. However, the process of hitting a subatomic particle with a photon means that the particle is no longer where it was and it has also changed its speed. The very act of measuring changes what we are measuring. This is the essence of the famous **Heisenberg's uncertainty principle**, in which he showed mathematically that there are definite limits to our ability to know both where a particle is and its speed. Because it is impossible to know exactly where an electron is, we are stuck with describing the likelihood or probability of an electron being found in a certain location.

We do not know what electrons are doing in the atom — circles, ellipses, figure eights, the mambo.... In fact, quantum mechanics does not include any description of how an electron goes from one point to another, if it does this at all. In terms of a location, all that we know is the probability of finding the electron in a particular position around the nucleus of an atom. (This is somewhat like knowing that the caretaker is somewhere in the school doing something, but we do not know where or what.) Since we can never know what the electrons are doing, scientists use the term *orbital* (rather than orbit) to describe the region in space where electrons may be found.

Fortunately, the wave equations from quantum mechanics can be manipulated to produce a three-dimensional probability distribution of the electron in an orbital specified by the quantum numbers. This is known as an **electron probability density** and can be represented in a variety of ways. The electron probability density for a 1s orbital of the hydrogen atom (**Figure 3**) shows a spherical shape with the greatest probability of finding an electron at r_{max}. Interestingly, this distance is the same as the one calculated by Bohr for the radius of the first circular orbit. Notice, however, that the interpretation of the electron has changed substantially. The probability densities of other orbitals, such as *p* and *d* orbitals, can also be calculated; some of these are presented in **Figure 4**. Looking at these diagrams, you can see why orbitals are often called "electron clouds."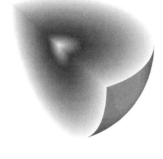

Heisenberg uncertainty principle it is impossible to simultaneously know exact position and speed of a particle

electron probability density a mathematical or graphical representation of the chance of finding an electron in a given space

DID YOU KNOW?

Crazy Enough?
Niels Bohr had this to say to Heisenberg and Pauli when reporting his colleagues' response to their theory: "We are all agreed that your theory is crazy. The question that divides us is whether it is crazy enough to have a chance of being correct. My own feeling is that it is not crazy enough."

ACTIVITY 3.7.2

Simulation of Electron Orbitals (p. 217)
Computers are necessary to do calculations in quantum mechanics and can quickly provide electron probability densities.

Figure 3
A 1s orbital. The concentration of dots near each point provides a measure of the probability of finding the electron at that point. The more probable the location, the more dots per unit volume. The same information, only in 2-D (like a slice into the sphere), is shown by the graph.

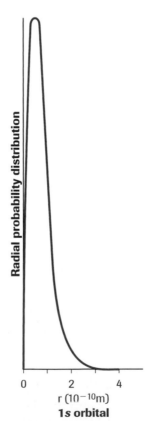

Radial probability distribution

r (10⁻¹⁰m)
1s orbital

(a)

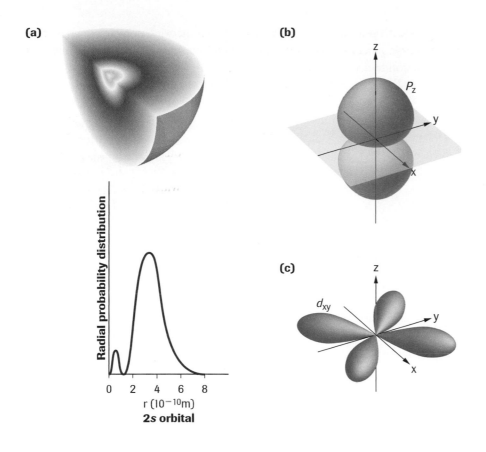

(b)

(c)

(d)

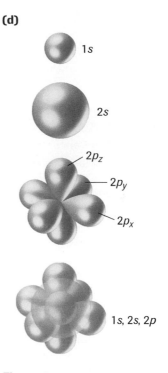

Figure 4
Some of the electron clouds representing the electron probability density.
(a) In the cross-section, the darker the shading, the higher the probability of finding the electron.
(b) A $2p_z$ orbital
(c) A d_{xy} orbital
(d) A superposition of $1s$, $2s$, and $2p$ orbitals

Problems with Quantum Mechanics

As is the case with all theories, there are areas of quantum mechanics that are not well understood. There is evidence of quantum phenomena that are not explainable with our current concepts. As chemists examine larger and more complicated molecules, the analysis of the structure becomes mathematically very complex. Also, Werner Heisenberg pointed out in his uncertainty principle that there is a limit to how precise we can make any measurement. Many scientists (Einstein among them) have found this concept disturbing, since it means that many rules thought to be true in our normal world may not be true in the subatomic world, including the basic concept of cause and effect.

Technology requires that something works, not necessarily that we understand why it works. A typical technology used without complete understanding is superconductivity. In 1911 Heike Kamerlingh-Onnes first demonstrated, using liquid helium as a coolant, that the electrical resistance of mercury metal suddenly decreased to zero at 4.2 K. Since then, science has discovered many superconducting materials, and technology has found them to be incredibly useful. A good example is the coil system that produces the magnetic field for MRIs. If the electromagnets were not superconducting, the current used could not be nearly as great, and the magnetic field would not be nearly strong enough to work.

A Nobel prize was awarded for an "electron-pairing" superconductor quantum theory in 1972, to John Bardeen, William Cooper, and John Schrieffer (B, C, S). However, there are still many aspects of superconductivity that are not explained by the BCS theory. The most notable point is that their theory sets an upper temperature limit for superconductivity of 23 K, and recently, materials have been found that are superconducting at temperatures that are much higher (**Figure 5**). Some superconductors work at temperatures over 150 K, and oddly enough, recently produced superconducting substances are not even conductors at room temperature. Why they should become superconducting

Figure 5
Levitation of magnets suspended above a superconducting ceramic material at the temperature of boiling nitrogen (77 K) has become a common science demonstration in high schools and universities.

as the temperature drops is the subject of several current theories, none of which is complete enough to have general acceptance by the scientific community. What does seem certain is that superconductivity is a quantum effect, and it seems to support the argument that at subatomic levels, we don't necesarily know what the rules are.

▶ **Section 3.7** *Questions*

Understanding Concepts

1. Briefly state the main contribution of each of the following scientists to the development of quantum mechanics:
   (a) de Broglie
   (b) Schrödinger
   (c) Heisenberg

2. What is an electron orbital and how is it different from an orbit?

3. State two general characteristics of any orbital provided by the quantum mechanics atomic model.

4. What information about an electron is not provided by the quantum mechanics theory?

5. Using diagrams and words, describe the shapes of the 1*s*, 2*s*, and three 2*p* orbitals.

Making Connections

6. Statistics are used in many situations to describe past events and predict future ones. List some examples of the use of statistics. How is this relevant to quantum mechanics?

7. When the police use a radar gun to measure a car's speed, photons are fired at the car. The photons hit the car and bounce back to the radar gun. If you got a speeding ticket, could you use Heisenberg's uncertainty principle in your defence? Explain briefly.

8. Dr. Richard Bader and his research group at McMaster University are well known for their work on atomic and molecular structure. Find out the nature of their work and give a brief, general description of how it relates to quantum mechanics.

 www.science.nelson.com

9. There are many present and projected technological applications for superconductivity. Research these applications and make a list of at least four, with a brief description of each.

 www.science.nelson.com

10. Research for the highest temperature at which superconductivity has been achieved. What substance is used for this highest temperature?

 www.science.nelson.com

▼ *Applications of Quantum Mechanics* **3.8**

The 20th century saw an increase in human knowledge and understanding of nature far greater than in all earlier recorded human history. To a large extent this is true because knowledge builds on knowledge—the more we learn, the more we discover there is to be learned. Communication within the scientific community is the key factor—it can be argued that the human knowledge "explosion" of the last few centuries occurred largely because of the sharing of information among scientists and technologists. This kind of communication has two significant benefits. It assures that knowledge acquired by an individual does not get lost by accident, or die with the discoverer. Even more importantly, it ensures that many minds look at any new concept or phenomenon. The result is that any new idea will generate a multitude of others, which in turn will generate still more in a continuous, ongoing process.

Some discoveries are (with the benefit of hindsight) considered exceptionally important. Such discoveries are pivotal because they affect many areas of knowledge and change many long-established ways of thinking. The work of Isaac Newton in physics and of Gregor Mendel in biology are typical examples of revolutionary science immediately providing new and better descriptions and explanations, and leading quickly to many other important advances in their areas of science.

The quantum mechanics model of matter will likely always be regarded as one of the most important scientific advances of the 20th century. Albert Einstein's explanation of the photoelectric effect by assuming the quantization of energy won him the Nobel prize. This was revolutionary science at its finest. Max Planck's suggestion that energy was not continuous had seemed at the time to be nonsense—just a way of dealing with light emission evidence to make calculations come out correct. Einstein's greatness lay in his ability to suspend common assumptions and preconceived notions, and to ask "what if...." If, for example, energy really exists in discrete "packages," then not only the photoelectric effect, but also a whole host of other evidence becomes explainable.

A critical concept arises from the idea suggested by Louis de Broglie that led to the wave mechanics model of Erwin Schrödinger. If the electron has characteristics of both a wave and a particle, then all matter must have this wave-particle duality. Since chemical change is essentially interactions of electrons of substances, a better understanding of the nature of the electron and of atomic structure has also had some far-reaching effects on the understanding of chemical bonding and on the chemical technologies of our society.

Laser Technology

The laser is one of the few technologies that is actually applied science. Science was used to create the technology. Most often, the science is only "applied" to explain the technology after the fact. In the case of the laser, the creative thinking required a knowledge of quantum theory. To even try to get the laser technology started required the inventor to have a good working knowledge of quantum theory, in particular electron quantum leaps in an atom.

The first visible light device operating on a laser principle was developed in 1960 by Theodore Maiman. The acronym for Light Amplification by Stimulated Emission of Radiation (LASER) has become a word used extensively in English as the device has become extremely common in a wide variety of applications. Actually, the original device produced microwave radiation and was developed in 1953 by Charles Townes using ammonia molecules. Townes's demonstration actually verified a prediction about

DID YOU *KNOW* ?

Semiconductors
The fundamental substances in transistors, microchips, thermistors, and light-emitting diodes are semiconductors, which work because of a quantum effect. Electrons unable to move through full outer energy levels of atoms become able to move freely from atom to atom when shifted to a higher energy level.

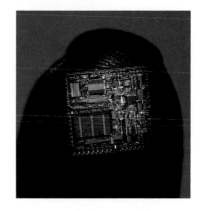

DID YOU *KNOW* ?

John Polanyi (1929–)
John Polanyi is a professor of chemistry at the University of Toronto. He is internationally recognized for his brilliant work on the dynamics of chemical reactions and chemical lasers. Polanyi was co-winner of the 1986 Nobel Prize for Chemistry, for his work on infrared chemiluminescence.

stimulated emission of photons made by Einstein some three decades earlier. The term "laser" is now routinely used for devices that produce infrared and ultraviolet light, as well as those that produce light visible to humans. The laser has become far more useful than was ever initially imagined. The device was created as a scientific demonstration of a theory, but as was the case with X-rays, the scientific and engineering communities quickly found a multitude of uses for the technology.

Lasers produce a light beam with unique characteristics. The light is completely monochromatic, meaning it is all precisely one wavelength (or colour). The beam is made up of coherent waves, meaning that each photon in line follows the previous one precisely, acting like one continuous wave. As well, in a laser beam all of these waves are quite precisely parallel, so the beam does not spread apart (diverge) significantly as it travels. Each of these light beam characteristics turns out to be immensely useful and important technologically.

Lasers are everywhere in our society. Lasers read information in CD and DVD players, transfer images in laser printers and copiers, and scan bar codes on merchandise at store checkout counters (**Figure 1**). Medical lasers are used for microsurgery to illuminate fibre-optic tubes for viewing areas inside the body, and to reshape corneas and reattach dislodged retinas in optical surgery (**Figure 2**). Low-power diode laser pointers presentations are commonly sold in stationery stores.

The operation of a laser can only be explained using the principles of quantum mechanics. Einstein predicted as early as 1917 that a rise in energy level of an electron in an atom or molecule requires a specific photon to be absorbed. If a photon of exactly that energy were to hit an atom, already in its excited state, the electron would be "stimulated" to drop back down to a lower level. That drop would emit another photon of exactly the same energy, and moving in the same direction as the first. Therefore, if the electrons of many atoms could be moved to a higher state and held there temporarily, a single photon could cause a kind of chain reaction. One photon hitting one atom would release two photons, which would hit two atoms releasing four, then eight, and so on until a huge number of photons would emerge. All of these photons would be exactly the same wavelength (colour), would be arranged as coherent (in phase) waves, and would be moving in the same direction. This creation of a beam of a huge number of photons from just one initial photon is the reason for the "amplification" part of the original LASER acronym.

Figure 1
Merchandise bar code scanners use laser beams to read product information. These lasers are very low power and harmless.

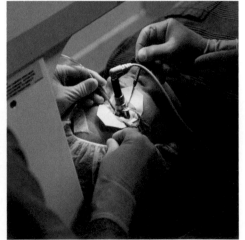

Figure 2
Laser heat can be used to vaporize tissue to reshape human corneas to permanently correct vision distortion problems.

TRY THIS activity *Bar Code Scanners*

Never look directly into a laser beam of any kind, any more than you would stare directly at the Sun, and for the same reason. The light is relatively bright so one should only look at it briefly, and only by reflection. Use only the laser provided, not your own personal laser.

Materials: diode laser pointer

• Obtain a diode laser pointer from your teacher, and in a darkened classroom scan the beam slowly across any area that has varying colours. The spines of different books on a bookshelf will work very well, or you can prepare a sheet of white

paper with broad vertical stripes of different-coloured magic markers. Alternatively, you can use strips of different colours of construction paper or even an article of clothing that has bright and dark stripes.

• Squint your eyes as you observe the reflection of the spot of light. Try to concentrate only on the apparent brightness of the reflection, and how the brightness changes from one material to another.
  (a) Which colour reflects the light most brightly? Which, least brightly?
  (b) Explain how this activity is similar to the way bar code scanners work.

Quantum Analysis and Diagnosis Technologies

Quantum mechanics eventually explained the bright-line spectrum from flame tests that Bunsen and Kirchhoff had discovered and employed more than 50 years earlier. The particular line spectrum of colours of light emitted or absorbed by substances is unique to a substance and can therefore be used to identify that substance. The broad field of study based on this fact is *spectroscopy* (literally, looking at a spectrum). Its technological applications are seemingly endless. An immediate (and now famous) application was the discovery of the element cesium in 1860 and rubidium in 1862. This was done by spectroscopic identification of colour patterns not belonging to any known element.

Spectrophotometers are devices that measure light photons electronically, allowing detection of wavelengths both far longer and far shorter than the human eye can see (**Figure 3**). These devices can be used for both qualitative and quantitative analysis. These instruments can also measure light of much lower intensity (fewer photons) than the eye can detect, which means they can be made to have incredible sensitivity, routinely measuring to precisions of parts per billion or better.

(a) Absorption spectrometer

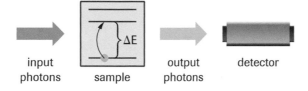

(b) Emission spectrometer

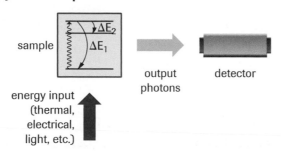

Figure 3
(a) In the absorption spectrophotometer, the detector determines which photons are absorbed by the atoms/ions or molecules in the sample.
(b) In the emission process, electrons in the sample are first excited to higher empty orbitals, and the detector determines the energies of the photons released when the electrons return to their initial, ground states.
In both processes, the energy changes (ΔE) are characteristic of the atoms/ions or molecules present. The greater the energy absorbed or released, the greater the number of specific species present.

Figure 4
Athletes are routinely tested at competitions for the presence of prohibited substances.

By measuring the characteristics of the light emitted or absorbed, such an instrument can both identify (from the spectrum) and also measure the quantity (from the intensity of light) of any substance targeted in the sample. Not surprisingly, research and commercial chemical analysis depend heavily on spectroscopy, and there are particularly extensive diagnostic applications in the field of medicine and medical research.

The human body contains an incredible variety of substances, some of them present in quantities so low that until recently they were not even detectable. Modern spectroscopy allows very precise detection and measurement of substances in human tissue or fluids. This technology can be used to monitor athletes for the use of performance-enhancing substances (**Figure 4**). More importantly, medical patients can be tested for substances that indicate the onset or progress of disease. Advancing technology means that ever-smaller traces of more substances can be detected. Advancing science means that doctors have more diagnostic tests than ever to use in their fight against disease, as research identifies more compounds that are linked to disease mechanisms.

The earliest device to let doctors "see" inside the body without surgery was the X-ray machine. Electrons are accelerated to very high velocities and then collided with atoms in a target material, such as tungsten. Some electrons in the target atoms jump to a much higher energy level. The quantum leap down is very large and produces extremely high-energy photons, called X rays. In a simplified way, this is like the reverse of the photoelectric effect. Many X-ray photons will pass completely through the body, while some of them are blocked or absorbed. (Note that X rays damage tissue and therefore there is an annual limit for X-ray exposure. Lead aprons are often used, for example in dental X rays, to limit the exposure.)

By exposing film to the X-ray photons that pass through the body, a two-dimensional shadow image of the internal body structure can be obtained. Combining the X-ray machine with digital computer technology led to the development of computerized axial tomography (CT), the so-called "cat scan." This produces a three-dimensional image of body structures, a huge advance for diagnosticians.

Perhaps the most revolutionary device based on quantum mechanics to be used by the medical community is the magnetic resonance imaging (MRI) unit. This device uses superconducting electromagnets to create extremely powerful magnetic fields. These magnetic fields together with microwaves are used to detect "spin" changes of hydrogen nuclei of water molecules. This effect is very sensitive to the local environment, for example, in various kinds of tissues. Tissues that are similar enough to appear identical in X rays can usually be distinguished in an MRI image. This can give doctors the ability to distinguish cancerous tissue from normal tissue, or to obtain more precise detail about critical soft-tissue areas such as the brain and spinal cord.

▶ *Section 3.8 Questions*

Understanding Concepts

1. State three important special characteristics of laser light.

2. Briefly describe the role energy levels play in the operation of a laser.

3. Describe some applications of the principles of quantum mechanics in medical diagnosis.

Making Connections

4. Use the Internet to write a short report on X-ray crystallography, describing how X rays can be used to give information on structures of solids at the atomic level.

 www.science.nelson.com

The field of spectroscopy involves three main technologies. Infrared spectroscopy is primarily concerned with determining molecular structure. Nuclear magnetic resonance spectroscopy analyzes for the presence of atoms by detecting changes in nuclear spin. This is called Magnetic Resonance Imaging when applied to hydrogen nuclei in the human body. Finally, mass spectroscopy analyzes ionized molecular fragments to help determine the structure and composition of substances.

Medical MRI Technologist

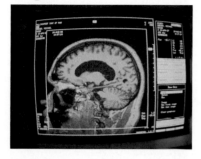

Technologists require specialized training to operate MRI units. MRI units use intense magnetic fields and very large current flows, as well as extremely cold (cryogenic) temperatures in their electromagnetic coils. Precise imaging of soft body tissues, without the tissue damage that high X-ray doses would cause, make this the diagnostic tool of choice for doctors. Most medical facilities have a waiting list for MRI use, and the future employment opportunities for MRI technicians appear excellent.

Pharmaceutical Research Spectroscopist

The pharmaceutical industry is huge and still growing. Infrared spectroscopy is one tool used to examine the structure of compounds that may prove to be useful pharmaceuticals. Protein and peptide analysis (see Chapter 2) may also be done to provide information about the effects of trial substances on living organisms. Scientists and technologists in this area are largely concerned with molecular structures and their effects on subsequent reactions among organic substances.

Petrochemical and Plastics Spectroscopists

Structural analysis is critical to understanding reactions controlling the production of petrochemicals and to controlling and customizing the characteristics of the growing number of commercial plastic materials. Infrared spectroscopy is used extensively in these areas, both for research and to provide information for production control. Technologists need to be familiar with organic chemistry and the effects of pressure and temperature on rates of reaction to comprehend the role played by molecular structure in these industries.

Mass Spectrometer Research Technologist

Technicians using mass spectrometers sometimes provide analyses for pure scientific research. At other times they may be collecting information for drug analysis in law enforcement or for precise geological identifications for the mining or petroleum industries. An understanding of physics and mathematics is part of the knowledge required by this technology.

▶ **Practice**

Making Connections

1. Choose one of the food and beverage industry, biotechnology, clinical research, environmental research, or the pharmaceutical industry, and use Internet information to write a short report on a career in that area associated with spectroscopy. Include training and education requirements, approximate salary expectations, and employment prospects.

 www.science.nelson.com

Development of Quantum Theory

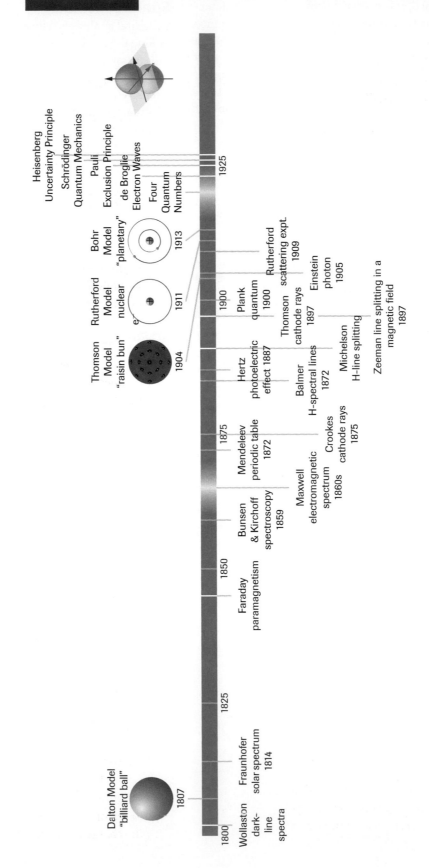

🧪 INVESTIGATION 3.1.1

The Nature of Cathode Rays

When cathode rays were first discovered in about 1860, their nature was a mystery. One hypothesis was that they were a form of electromagnetic radiation, like visible light or ultraviolet light. An obvious test would be to check some properties of cathode rays and compare them to the known properties of light. The purpose of this experiment is to test the hypothesis relating cathode rays to light.

Question

What effect do electric charges and magnets have on the direction of motion of a cathode ray?

Experimental Design

A cathode ray is tested separately with a bar magnet and with two charged, parallel plates (independent variables). In each case, the dependent variable is the deflection of the cathode ray. Controlled variables include the charged plates and the magnet. As a control, the same tests are repeated with a thin beam of light, such as a laser.

Materials

cathode-ray tube
power supply with variable voltage
connecting wires
bar magnet
2 lab stands
2 clamps
laser light source (e.g., laser pointer)

 **Serious shock hazard may result from the use of a high-voltage supply. Unplug or turn off power supply when connecting and disconnecting wires.
Do not shine a laser beam into anyone's eyes.**

Procedure

1. Clamp the cathode tube on a lab stand so that it is horizontal.

2. Connect the cathode-ray tube to the power supply. Check all connections before plugging the power supply into the electrical outlet. Turn on the power supply.

3. Once the cathode ray is visible (either inside the tube or on the screen at the end of the tube), bring a bar magnet near the tube and note any effect on the cathode ray.

4. Connect the two parallel plates to the same or different power supply and slowly increase the potential difference (voltage) on the plates. Note the effect on the cathode ray. (The positively charged plate is at the red connection; the negatively charged plate at the black connection.)

5. Turn off the power supply and disconnect the cathode-ray tube.

6. Clamp the laser light source horizontally on a lab stand so that the end of the beam is visible on a wall or screen.

7. Hold the bar magnet near the beam and note any effect by observing the dot on the screen.

8. Aim the laser beam so that it shines between the parallel plates inside the cathode-ray tube and the end of the beam is again visible.

9. Connect the parallel plates as in step 4 and note any effect of the charged plates on the laser beam.

Evidence

(a) Create a table to record your observations.

Analysis

(b) Based on your evidence, does it appear that cathode rays are like electromagnetic radiation such as visible light? Justify your answer.

Evaluation

(c) Are there any obvious flaws in this experiment? Suggest some improvements in the Experimental Design, Materials, and/or Procedure to improve the quality or quantity of the evidence.

(d) Evaluate the hypothesis that cathode rays are a form of electromagnetic radiation.

Synthesis

(e) Use your knowledge about attraction and repulsion of electric charges. What does the bending of the cathode ray when passing near electrically charged plates suggest about the composition of the cathode rays?

(f) Based on the evidence collected, what is the sign of the charge of the particles in a cathode ray?

ACTIVITY 3.1.1

Rutherford's Gold Foil Experiment

The purpose of this activity is to illustrate Rutherford's famous alpha-particle-scattering experiment, using a computer simulation.

Materials
Nelson Chemistry 12 CD; PC

- Start the "Rutherford" simulation from the Nelson Chemistry 12 CD on your computer.
- Follow the instructions and choose the initial settings suggested by your teacher.
- Record general and specific observations while the experiment is running.
- If requested, change the settings and run the experiment again.

(a) According to the Thomson model of the atom, what result is predicted for a stream of alpha particles striking a layer of gold atoms?

(b) Summarize the main evidence from Rutherford's experiment.

(c) Compare the relative numbers of alpha particles that travelled relatively undeflected to the number that recoiled. What does this suggest about the relative size of the nucleus?

(d) Evaluate the prediction from the Thomson model and the Thomson model itself.

(e) Use your modern knowledge of the components of an atom. How would the results of Rutherford's experiment with an aluminum foil be similar to his experiment with a gold foil? How would the results be different?

ACTIVITY 3.3.1

Hot Solids

All hot solids, liquids, and gases produce some form of light, occasionally in the visible region but often in other invisible regions. The purpose of this activity is to study the light produced by a hot solid.

Materials

Variac (variable power supply)	diffraction grating
clear lamp with vertical filament	(e.g., 6000–7500
overhead projector	lines/cm)
2 pieces of heavy paper	tape

- Set up a clear incandescent lamp with a vertical filament. Plug the cord into a Variac that is plugged into an electrical outlet. Set the Variac at its lowest setting.

(a) Switch on the Variac and turn off all lights. Observe the filament as the voltage is slowly increased from 0 to about 110 V and describe the change in colour.

(b) People sometimes describe hot objects using terms such as "red hot" or "white hot." Which do you think is hotter?

(c) State two objects in your home which may be red hot at certain times.

- An overhead projector bulb produces a very white, bright light when the filament is white hot. Set up the projector to shine its light on a white wall or board. Two straight-edge pieces of heavy paper are placed about 5 mm apart on the horizontal stage of the projector so that they block all of the light except for a thin strip visible on the wall or board (**Figure 1**).

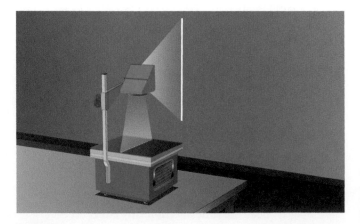

Figure 1

- A diffraction grating is used to separate the light into its various components. Hang the grating in the centre of the final lens using a piece of plastic tape. Shut off all room lights and observe.

(d) Sketch and label the main colours in the visible spectrum that you see on the right-hand side of the central white line. This spectrum is known as a continuous spectrum. (The same spectrum, but reversed, appears on the left side.)

(e) If your eyes were constructed differently, you might be able to see beyond the blue/violet end and beyond the red end of the spectrum. What are these regions called? Label them on your diagram.

(f) Which do you think is more dangerous to you—being exposed to a red lamp or an ultraviolet lamp? Justify your choice.

(g) Assuming the danger is related to the energy of different types of light, label your spectrum from (d) with an arrow going from low to high energy.

INVESTIGATION 3.3.1

The Photoelectric Effect

An electroscope is a device that is used to detect and measure electric charges.

Purpose
The purpose of Investigation 3.3.1 is to create an initial theoretical understanding of the photoelectric effect.

Question
What effect does light have on a negatively charged metal plate?

Experimental Design
An electroscope with a zinc plate is charged negatively using a vinyl strip rubbed with paper towel, and then observed when white light and ultraviolet light are each shone onto the zinc. The control is the charged electroscope under normal room lighting.

Materials
electroscope (with a flat top)
vinyl plastic strip
paper towel
zinc plate (about 3–5 cm sized square)
steel wool
lamp with 100-W bulb
ultraviolet light source

Inquiry Skills

○ Questioning	○ Planning	● Analyzing
○ Hypothesizing	● Conducting	● Evaluating
○ Predicting	● Recording	● Communicating

Procedure

1. Set the electroscope on a flat surface and make certain the vane or leaves can move freely.

2. Rub the vinyl strip several times with a piece of paper towel to negatively charge the strip.

3. Touch the vinyl strip one or more times to the top of the electroscope until the vane or leaves move and remain in their new position. (If nothing happens, have your teacher check your electroscope.)

4. Observe the electroscope for about one minute.

5. Touch the electroscope with your finger to discharge or neutralize it. Note the change in the electroscope.

6. Scrub the zinc plate with the steel wool until the plate is shiny. Place the zinc plate on the top of the electroscope.

7. Charge the electroscope again (steps 2–3).

8. Plug in the lamp with the 100-W bulb, turn it on, and shine the light onto the zinc plate from a distance of about 10 cm.

9. Observe the electroscope for about 1 min.

10. If necessary, charge the electroscope to the same angle as before. Repeat steps 8 and 9 using the ultraviolet light.

Evidence

(a) Create a table to record your observations.

Analysis

(b) What effect did a bright white light from a normal lamp have on the negatively charged zinc plate? Answer relative to the control — the charged electroscope sitting for 1 min under room-light conditions.

(c) What effect, compared to the control, did the ultraviolet light have on the charged zinc plate?

(d) If the intensity (brightness) of the light is responsible for the effect, which light should work better? How certain are you of your answer? Provide your reasoning.

(e) Explain, in terms of electrons and protons, the existence of a negatively charged zinc plate.

(f) Did the electroscope become more or less charged when illuminated with ultraviolet light compared to visible light?

(g) Using your answers to (e) and (f), suggest an initial explanation for the effect of the ultraviolet light on a negatively charged zinc plate.

Evaluation

(h) What other test could be done with the electroscope and zinc plate to make the interpretation in (g) more certain?

(i) What other general improvement to the materials could be made to improve the quality of the evidence?

(j) How certain are you about your answer to the question in part (g)? Provide reasons.

Synthesis

(k) Ordinary glass absorbs ultraviolet light and does not allow it to pass through. Predict the results of this experiment if a glass plate were placed between the ultraviolet light and the charged zinc plate.

(l) Would direct sunlight through an open or a closed window discharge the electroscope? Provide your reasoning.

ACTIVITY 3.4.1

Line Spectra

All substances absorb or emit some part of the electromagnetic spectrum. Some substances absorb some of the visible spectrum, while other substances may absorb in the ultraviolet region or other regions. The spectrum produced is called a dark-line or absorption spectrum. Substances can also emit light in different parts of the electromagnetic spectrum. Under certain conditions, the light emitted appears as bright lines. The spectrum produced is called a bright-line spectrum. The purpose of this activity is to illustrate the two main types of line spectra—bright-line and absorption.

Materials

overhead projector
2 pieces of heavy paper
diffraction grating (600 lines/mm)
2 large beakers
flat glass plate to cover beaker
pure water
tiny crystals of potassium permanganate and of iodine
spectrum tube power supply
hydrogen gas tube
spectroscope

 Serious shock hazard may result from the use of a high-voltage supply.

• An overhead projector is set up to shine its light on a white wall or board. Two straight-edged pieces of heavy paper are placed about 5 mm apart on the horizontal stage of the projector so that they block all of the light except for a thin strip.

• A diffraction grating (containing about 6000–7500 lines/cm) is hung in the centre of the vertical projecting lens using a piece of plastic tape.

ACTIVITY 3.4.1 continued

- Place a large clean beaker on the slit formed by the two pieces of paper. Shut off the room lights and observe the continuous visible spectrum.

- Now fill the beaker with some pure water and observe the spectrum again.

  (a) Do either glass or water change the colours in the visible spectrum? Do these substances absorb visible light?

- Add a few crystals of potassium permanganate to the water and stir. Observe the visible spectrum.

  (b) What effect does aqueous potassium permanganate have on the visible spectrum?

- Warm some solid iodine in a beaker in a fume hood or by some other safe method. Cover the beaker with a flat glass plate. Place the beaker containing the iodine vapour on the overhead projector and observe the spectrum.

  (c) Compare the spectrum obtained to the one for the potassium permanganate solution in step 5. Can gases also absorb electromagnetic radiation?

- Observe the spectrum of sunlight recorded on Earth's surface (**Figure 2**).

  (d) What evidence is there that some light is being absorbed? Suggest some possible gases that might be responsible.

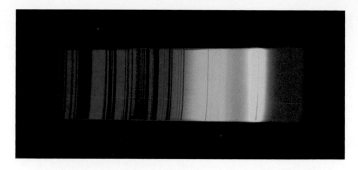

Figure 2
The spectrum of sunlight recorded on the surface of Earth is a dark-line (absorption) spectrum due to gases in the atmospheres of Earth and the Sun absorbing specific parts of the sunlight.

- Set up a hydrogen gas tube in a gas discharge power supply. Switch on the power and turn off all lights. Observe the spectrum with a handheld spectroscope. Draw and label a diagram of this spectrum. If time permits, observe the spectra of other gases.

  (e) Under what conditions do gases produce light that is in the visible region of the electromagnetic spectrum?

  (f) In this case, is the spectrum produced a bright-line or continuous spectrum?

  (g) How are line spectra used in chemical analysis?

ACTIVITY 3.4.2

The Hydrogen Line Spectrum and the Bohr Theory

The bright-line or emission spectrum of hydrogen (**Figure 3**) has been known since the mid-1800s. The position or wavelength of each of the coloured lines in the visible region was precisely measured by the Swedish scientist, Anders Ångström in 1862. (A unit of length was named after Ångström, 1 Å = 0.1 nm, exactly.) The purpose of this activity is to use a computer simulation to demonstrate some of the main concepts of the Bohr theory and to relate these to the hydrogen emission spectrum.

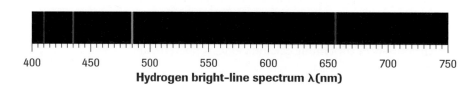

Hydrogen bright-line spectrum λ(nm)

Figure 3

Materials

Nelson Chemistry 12 CD, PC

(a) Using **Figure 3**, estimate the wavelength of each of the four lines in the visible region of the hydrogen spectrum. These visible lines belong to the group of lines known as the Balmer series.

- Start the "Bohr" simulation from the Nelson Chemistry 12 CD on your computer. Under the "Series" menu, select "Balmer."

- Set the "New State" at 3 and click the "Photon" button.

- With the electron now in $n_i = 3$, set the "New State" to 2.

(b) Note and record the wavelength of the light to the nearest nanometre. Click the "Photon" button.

(c) Is some light (a photon) absorbed or released in this transition?

(d) To which line in the Balmer series does this transition correspond?

- Repeat the simulation using the following settings:

$n_i = 4, n_f = 2$
$n_i = 5, n_f = 2$
$n_i = 6, n_f = 2$

- Answer questions (b) to (d) for each of these transitions.

(e) How do your answers from (a) using **Figure 3** compare with your answers to (d) using the computer simulation? Is this surprising? Explain briefly.

(f) If some light (a photon) is absorbed by an electron, what happens to the electron? Try this with the simulation program.

(g) How does the wavelength of light corresponding to the transition from $n_i = 3$ to $n_f = 2$ compare with $n_i = 2$ to $n_f = 3$? Explain briefly why this is necessary, according to the Bohr theory.

(h) An electron cannot undergo a transition from $n_i = 1$, to $n_f = 2.5$. According to the Bohr theory, why is this not possible?

🧪 INVESTIGATION 3.5.1

Paramagnetism

Inquiry Skills

○ Questioning	● Planning	● Analyzing
○ Hypothesizing	● Conducting	● Evaluating
○ Predicting	● Recording	● Communicating

Paramagnetism was first investigated and named by Michael Faraday in the mid-1800s. At this time, before the discovery of the electron, there was no theoretical explanation of the cause of paramagnetism. According to modern atomic theory, paramagnetism is believed to be caused by the presence of unpaired electrons in an atom or ion.

Purpose

The scientific purpose of this investigation is to determine experimentally which substances are paramagnetic.

Question

Which substances containing calcium, zinc, copper(II), and manganese(II) ions are paramagnetic?

Experimental Design

Test tubes containing the sulfates of each of the ions are suspended by threads from a support. Evidence for any attraction of each test tube toward a strong magnet is observed.

(a) Identify the independent, dependent, and controlled variables.

Materials

eye protection	strong magnet (e.g.,
4 small test tubes	neodymium
stirring rod	a few grams of the solids:
thread	calcium sulfate
laboratory stand	zinc sulfate
clamp	copper(II) sulfate
horizontal bar	manganese(II) sulfate

Procedure

(b) Write a complete procedure for this experiment. Include safety precautions with respect to handling and disposal of the chemicals used. Have the procedure checked by your teacher before you proceed.

Analysis

(c) Answer the Question based on the evidence collected.

⚗ INVESTIGATION 3.5.1 *continued*

Evaluation

(d) Are there any flaws or possible improvements in the Experimental Design? Describe briefly.

(e) Suggest some improvements to the Materials and Procedure.

(f) How certain are you about the evidence obtained? Include possible sources of error or uncertainty.

📝 LAB EXERCISE 3.6.1

Quantitative Paramagnetism

Inquiry Skills

○ Questioning	○ Planning	● Analyzing
○ Hypothesizing	○ Conducting	● Evaluating
○ Predicting	○ Recording	● Communicating

In Investigation 3.5.1, you obtained some preliminary evidence for a possible connection between unpaired electrons (as determined by the electron configuration) and paramagnetism. The purpose of this lab exercise is to test this hypothesis with a quantitative experiment.

Question

What effect does the number of unpaired electrons have on the strength of the paramagnetism of metal salts?

Prediction/Hypothesis

(a) Write a prediction and provide your reasoning based on electron configurations.

Experimental Design

A sensitive electronic balance is used to measure the attraction between a powerful magnet and a test tube containing a metal salt. The balance is tared (zeroed) before the test tube is lowered (**Figure 4**). The mass reading is taken just before contact of the test tube with the magnet. Several ionic compounds containing different metal ions are individually tested using the same mass of each compound.

(b) Identify the independent, dependent, and controlled variables.

Evidence

Table 1: Change in Mass in a Strong Magnetic Field

Ionic compound	Mass reading, Δm (g)
$CaSO_{4(s)}$	0.00
$Al_2(SO_4)_{3(s)}$	0.00
$CuCl_{(s)}$	0.00
$CuSO_4 \cdot 5H_2O_{(s)}$	−0.09
$NiSO_4 \cdot 7H_2O_{(s)}$	−0.22
$CoCl_2 \cdot 6H_2O_{(s)}$	−0.47
$FeSO_4 \cdot 7H_2O_{(s)}$	−0.51
$MnSO_4 \cdot H_2O_{(s)}$	−1.26
$FeCl_3 \cdot 6H_2O_{(s)}$	−0.95

mass of each compound in test tube = 3.00 g

Analysis

(c) What is the significance of a zero-mass reading for some substances and negative-mass readings for other substances?

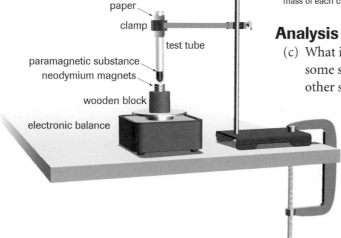

paper
clamp
test tube
paramagnetic substance
neodymium magnets
wooden block
electronic balance

Figure 4
A strong magnet or magnets (such as neodymium magnets) and a paramagnetic substance attract each other. This means that the magnet and block are slightly lifted toward the fixed test tube.

(d) How does this change in mass relate to the paramagnetic strength of the substance? (Each of the compounds has a different molar mass and therefore a different amount in moles in the controlled mass of 3.00 g. In order to make a valid comparison, you need to know the change in mass per mole of the substance.)

(e) Create a table with headings ionic compound, molar mass, number of moles. Create and complete another table with the following headings: metal ion, electron configuration, number of unpaired electrons, mass decrease per mole.

(f) Plot a graph of the number of unpaired electrons (x-axis) and mass decrease per mole (y-axis). Draw a best-fit line.

(g) Answer the Question asked at the beginning of this investigation.

Evaluation

(h) Evaluate the Experimental Design. Are there any obvious flaws? Any improvements?

(i) Suggest some improvements to the materials and procedure that would improve the quality and quantity of the evidence collected.

(j) How confident are you with the experimental answer to the question?

(k) Evaluate the Prediction (verified, falsified, or inconclusive). State your reasons.

(l) Does the hypothesis appear to be acceptable based on your evaluation of the prediction?

ACTIVITY 3.7.1

Modelling Standing Electron Waves

A mechanical model of Schrödinger's standing waves associated with electrons can be made using a thin, stiff, loop of wire which is vibrated with a variable frequency mechanical oscillator. The mechanical oscillator is like a heavy-duty speaker cone with a rod attached to its centre. As the cone and rod move up and down, whatever is attached to the rod oscillates up and down. Vibrating one point in the loop sets up waves in the wire. This is like holding the edge of a long spring, oscillating one end back and forth, and generating waves that move along the spring. When returning waves meet they interfere with each other, either constructively (increasing the amplitude) or destructively (decreasing the amplitude). Standing waves are a special case of wave interference that results in apparently stationary nodes (zero amplitude points) and antinodes (maximum amplitude points).

- Secure the oscillator on a sturdy stand. Attach the plug containing the loop of wire and adjust so that the plane of the loop is horizontal.

- Set the frequency to its lowest setting. Plug in the oscillator and turn it on.

- Slowly increase the frequency and observe the results.

- Continue increasing the frequency until no further observations are possible because the nodes and antinodes are no longer visible.

- Slowly decrease the frequency back down to its lowest setting and view the changes in reverse order.

- Repeat this procedure, if necessary, to complete your observations.

  (a) Describe, in general, the appearance of the nodes and antinodes.

  (b) Do all frequencies produce standing wave patterns? Discuss briefly.

  (c) List the number of antinodes from the lowest possible to as many as you were able to observe.

  (d) How does this physical model relate to the wave mechanics model of the atom? What are some limitations of this model?

ACTIVITY 3.7.2

Simulation of Electron Orbitals

Materials
Nelson Chemistry 12 CD, PC

- Start the SIRs program on the Nelson Chemistry 12 CD.
- Select "SIR Orbital." Read and then click on the SIR Orbital title screen.
- You should now see two screens that will allow comparisons of various $n = 1$ and $n = 2$ orbitals. The "Bohr" choice is representative only and does not present corresponding Bohr orbits. It serves as a general reminder of the view of an electron in an atom, according to the Bohr theory.
- There are a number of options available.

"Overwrite"—By selecting this option, you can overlay several diagrams onto one on the screen.

"Replace"—When this is selected, the existed diagram is erased and replaced by a new diagram.

"Quickplot"—This produces a summary of many calculations of the location of an electron at an instant in time. This final diagram represents the electron probability density for an electron of energy corresponding to the selected orbital.

"Flashplot"—Selecting this option initiates individual calculations performed by your computer. Each flashing "star" (which leaves a yellow dot) indicates the instantaneous location for an electron of energy corresponding to the selected orbital. This illustrates how the final probability density (as seen in Quickplot) is generated.

"Symbol"—Chemists find it convenient to reduce the probability density diagram to a figure that contains at least a 90% probability of the electron being located within the bounds of the figure. These kinds of diagrams are very useful for describing chemical bonding.

Note: Orbitals are three dimensional. The SIR program eliminates one dimension (z) to produce a two-dimensional diagram (like a slice or cross section of the full 3-D view).

- In the "Flashplot" choices, select "1s" and let this run until it is finished.
- At the bottom right of the right window, select "Classic (Bohr) Orbit."
  (a) Compare the views of the electron according to the quantum mechanics theory (left window) and the Bohr theory (right window).
- Select "End" for the Classic (Bohr) Orbit.
- Click "End." Select the "Overwrite" and "Figure" boxes, then click "1s" in the left window.
- In the right window, use "Replace" and "Flashplot" for the 2s orbital. When the simulation is finished, select "Overwrite" and "Figure" for 2s.
  (b) Compare the electron probability density for the 1s and 2s orbitals.
  (c) How do the sizes of these orbitals compare?
  (d) How does the energy of an electron in these orbitals compare?
- Using the functions of this program (e.g., "Quickplot" and "Symbol") create representations of 2s and 2px and then 2px and 2py.
  (e) Compare the 2s and 2p orbitals and the 2px and 2py orbitals.
  (f) Which p orbital is missing? What would be the relative size, shape, and orientation of this orbital?
- In the left window, select the "Overwrite" and "Symbol" boxes, then click "1s".
  (g) Assuming full 1s and 2s orbitals, and half-filled 2px and 2py, what atom would this represent?
  (h) In 3-D, what would be the appearance of this combined electron probability density?
- The view of the complete atom is not very useful. Now overwrite this view using the "Symbol" for each of the four orbitals.
  (i) Sketch this diagram and label each of the four orbitals.

Key Expectations

- Explain the experimental observations and inferences made by Rutherford and Bohr in developing the planetary model of the hydrogen atom. (3.1, 3.2, 3.3, 3.4)

- Describe the contributions of Planck, Bohr, Sommerfeld, de Broglie, Einstein, Schrödinger, and Heisenberg to the development of the quantum mechanical model. (3.3, 3.4, 3.5, 3.6, 3.7)

- Describe the quantum mechanical model of the atom. (3.5, 3.6, 3.7)

- Use appropriate scientific vocabulary to communicate ideas related to atomic structure sections. (all)

- Write electron configurations for elements in the periodic table, using the Pauli exclusion principle and Hund's rule. (3.5, 3.6)

- Describe some applications of principles relating to atomic structure in analytical chemistry and medical diagnosis. (3.8)

- Describe advances in Canadian research on atomic theory. (3.2, 3.7, 3.8)

Key Terms

absorption spectrum

actinides

aufbau principle

bright-line spectrum

electron configuration

electron probability density

Heisenberg uncertainty principle

Hund's rule

isotope

lanthanides

magnetic quantum number, m_l

neutron

orbital

Pauli exclusion principle

photoelectric effect

photon

principal quantum number, n

proton

quantum

quantum mechanics

representative elements

secondary quantum number

shell

spectroscopy

spin quantum number, m_s

stationary state

subshell

transition

transition elements

Key Symbols

- $E, n, l, m_l, m_s, s, p, d, f$

Problems You Can Solve

- Determine possible values of the four quantum numbers. (3.4)

- Write electron configurations for atoms of elements in the periodic table. (3.5).

▶ **MAKE** a summary

- The following chart is intended to summarize the key experimental work that directly led to major steps in the evolution of atomic theories. Copy and complete this chart.

Atomic theory	Key experimental work	Contribution to theory
Rutherford		
Bohr		
Quantum Mechanics (including initial development of quantum numbers)		

- Many scientists contributed to the development of the quantum mechanical model of the atom. For each of the following scientists, state one significant contribution.

  Planck, Bohr, Sommerfeld, de Broglie, Einstein, Schrödinger, Heisenberg

- (a) Sketch an outline of the periodic table and label the *s, p, d,* and *f* blocks of elements.
  (b) What is the empirical justification for this labelling?
  (c) What is the theoretical justification for this labelling?

Identify each of the following statements as true, false, or incomplete. If the statement is false or incomplete, rewrite it as a true statement.

1. The region in space where an electron is most likely to be found is called an energy level.

2. Electron configurations are often condensed by writing them using the previous noble gas core as a starting point. In this system, $[Ar] 3d^3 4s^2$ would represent calcium.

3. The f sublevel is thought to have five orbitals.

4. Orbital diagrams generally include the region of space in which the electron may be found most of the time.

5. For some alpha particles to be reflected backward by a gold nucleus in Rutherford's experiment, the gold nucleus had to be both very massive and strongly positive.

6. Rutherford knew the nucleus had to be very small because most alpha particles were deflected when fired through a layer of gold atoms.

7. Electrons shifting to higher levels, according to Bohr, would account for emission spectra.

8. Elements with atomic electron configurations ending in np^5, where n is an integer from 2 to 6, are called the halogens.

9. Photon is the term used to refer to a quantum of electromagnetic energy.

10. The serious shortcoming of Bohr's theory was failure to predict spectra for atoms other than hydrogen.

11. The Pauli exclusion principle states that two electrons may not occupy the same energy level.

Identify the letter that corresponds to the best answer to each of the following questions.

12. Rutherford's classic experiment produced evidence for a nuclear atom model when atoms in a thin metal foil scattered a beam of
    (a) cathode rays.   (d) electrons.
    (b) alpha particles.   (e) protons.
    (c) X rays.

13. Max Planck's mathematical explanation of blackbody radiation required that he assume that, for atoms
    (a) most of the mass is in a tiny part of the volume.
    (b) electrons orbit the nucleus as planets orbit a star.
    (c) electrons have several different energy levels.
    (d) the energy of the vibrating atoms is quantized.
    (e) all of the positive charge is located in the nucleus.

14. Niels Bohr assumed that when a photon is released from an atom to produce a bright line in the spectrum,
    (a) an electron has dropped from a higher energy level to a lower one.
    (b) the atom must have returned to its ground state.
    (c) an electron has been converted into emitted energy.
    (d) the electron has both wave and particle properties.
    (e) the energy of the atom has increased one quantum.

15. If the ground-state electron configuration of an atom is $[Ne] 3s^2 3p^4$, the atom is
    (a) magnesium.   (d) argon.
    (b) silicon.   (e) selenium.
    (c) sulfur.

16. Which of the following will **not** have an electron configuration ending with $3s^2 3p^6$?
    (a) chloride ion   (d) calcium ion
    (b) sulfide ion   (e) potassium ion
    (c) aluminum ion

17. Which of the following is **not** used to determine an electron configuration for an atom?
    (a) Hund's rule
    (b) Heisenberg's uncertainty principle
    (c) Pauli's exclusion principle
    (d) the aufbau principle
    (e) the periodic table

18. Which of the following statements is *false*, based upon your knowledge of electron configurations?
    (a) Iron is ferromagnetic; copper is paramagnetic.
    (b) The sodium ion is formed by the sodium atom losing one s electron.
    (c) The silver atom has an electron promoted from the $5s$ to a $4d$ orbital.
    (d) The manganese atom has one electron in each $3d$ orbital.
    (e) The tin(IV) ion is formed by the tin atom losing two p electrons and two d electrons.

19. The contribution of Erwin Schrödinger to the quantum mechanical atomic model was
    (a) a theoretical prediction that particles should exhibit wave properties.
    (b) a theoretical principle that precision of measurement has an ultimate limit.
    (c) an explanation of the photoelectric effect utilizing an energy quantum.
    (d) a mathematical description that treats electrons as standing waves.
    (e) experimental verification of the quantization of charge.

Understanding Concepts

1. Scientific theories are usually developed to explain the results of experiments. Describe the evidence that the following scientists used to develop their atomic models. Include the main interpretation of the evidence.
   (a) Rutherford
   (b) Bohr   (3.4)

2. When a theory is not able to explain reliable observations, it is often revised or replaced. The Rutherford and Bohr atomic models represent stages in the development of atomic theory. Describe the problems with each of these models.   (3.4)

3. State a similarity and a difference between the terms "orbit" and "orbital." Which atomic models that you have studied would use each of these terms?   (3.5)

4. What was the main kind of experimental work used to develop the concepts of quantum mechanics? (3.5)

5. The quantum mechanical model of the atom involves several theoretical concepts. Describe each of the following concepts:
   (a) quantum
   (b) orbital
   (c) electron probability density
   (d) photon   (3.7)

6. The Pauli exclusion principle states that no two electrons in an atom can have the same set of four quantum numbers. Draw an energy-level diagram for the ground-state oxygen atom and label the features that provide the following information.
   (a) the main/principal energy level
   (b) the energy sublevel (subshell)
   (c) the orientation of an orbital
   (d) the spin of the electron   (3.6)

7. What evidence indicates that electrons have two directions of spin?   (3.6)

8. Draw an outline of the periodic table and label the sublevels (subshells) being filled in each part of the table.   (3.6)

9. According to quantum mechanics, how does the position of an element on the periodic table relate to its properties?   (3.6)

10. Complete energy-level diagrams for potassium ions and sulfide ions. Which noble gas atom has the same diagrams as these ions?   (3.6)

11. (a) What are some similarities in the chemical properties of alkali metals?

    (b) How is this explained theoretically, using concepts in this chapter?   (3.6)

12. Write a complete ground-state electron configuration for each of the following atoms or ions:
    (a) Mg
    (b) S^{2-}
    (c) K^{+}
    (d) Rb
    (e) Au   (3.6)

13. Write the shorthand electron configuration for each of the following atoms or ions:
    (a) yttrium
    (b) antimony
    (c) barium ion   (3.6)

14. Paramagnetic substances are attracted by a magnet. Indicate which of the following elements are paramagnetic. Justify your answer.
    (a) aluminum
    (b) beryllium
    (c) titanium
    (d) mercury   (3.6)

15. Identify the following atoms or ions from their electron configurations:
    (a) W: $1s^2 2s^2 2p^6 3s^2 3p^6 4s^2 3d^{10} 4p^3$
    (b) X^+: $1s^2 2s^2 2p^6 3s^2 3p^6 4s^2 3d^{10} 4p^6$
    (c) Y^-: $1s^2 2s^2 2p^6 3s^2 3p^6 4s^2 3d^{10} 4p^6 5s^2 4d^{10} 5p^6$
    (d) Z: $1s^2 2s^2 2p^6 3s^2 3p^6 4s^2 3d^{10} 4p^6 5s^2 4d^{10} 5p^6 6s^2 4f^{11}$ (3.6)

16. Calculate the maximum number of electrons with the following principal quantum numbers, n:
    (a) 1
    (b) 2
    (c) 3
    (d) 4   (3.6)

17. Sketch the shape of a $2p_x$ orbital. How is this orbital the same as and different from the $2p_y$ and $2p_z$?   (3.7)

18. The quantum mechanical model of the atom has been called "the greatest collective work of science in the 20th century," because so many individuals contributed to its development. Briefly describe the contributions of each of the following scientists:
    (a) Max Planck
    (b) Louis de Broglie
    (c) Albert Einstein
    (d) Werner Heisenberg
    (e) Erwin Schrödinger   (3.7)

19. A scientific concept can be tested by its ability to describe and explain evidence gathered by scientists.

Use the concepts created in this chapter to describe and/or explain the following observations.

(a) The very reactive metal sodium (it even reacts with water) reacts with chlorine (a reactive poisonous gas) to produce inert sodium chloride (table salt that we eat).

(b) The flame test for lithium produces a red flame while that of sodium is yellow.

(c) Sodium chloride and silver chloride have similar empirical formulas, $NaCl_{(s)}$ and $AgCl_{(s)}$.

(d) The empirically determined formulas for the chlorides of tin are $SnCl_{2(s)}$ and $SnCl_{4(s)}$. (3.7)

Applying Inquiry Skills

20. Using what you have learned in this chapter, state why the evaluation of evidence is such an important part of the scientific process. (3.7)

21. An unknown substance appears on the surface of a city's water reservoir (**Figure 1**). What are some experimental techniques that could be used to help identify the substance? (3.7)

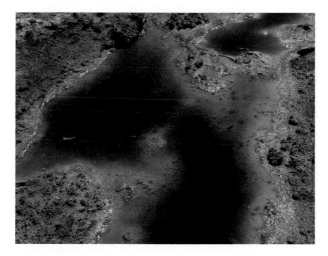

Figure 1

22. Critique the following experimental designs. Suggest better designs to meet each purpose.

(a) A gaseous element is identified from a discharge tube by observing the visible and infrared spectrum through a hand-held spectroscope.

(b) A mixture is identified by conducting a flame test.

(c) The presence of iron in iron-fortified breakfast cereal is tested by taping a strong magnet to the outside of a half-full cereal box and shaking the box.

(d) The paramagnetism of the element calcium is tested by determining the effect of a magnet on a saturated solution of calcium sulfate in a small test tube suspended by a thread. (3.6)

23. Critique the following analogies, physical models, or simulations.

(a) Climbing a staircase is used as an analogy for the transition of electrons to different energy levels in an atom.

(b) A computer simulation for plotting the $1s$ orbital of the hydrogen atom is used to test the quantum/wave mechanical model of the atom. (3.7)

Making Connections

24. Medical diagnosis has benefited substantially from advances in our understanding of the atom.

(a) State three or more examples.

(b) Provide some positive and negative arguments, from at least three perspectives, about government support of fundamental research. (3.8)

25. Ernest Rutherford and Frederick Soddy collaborated in researching radioactivity at McGill University (1900–02). Their empirical work completely transformed the understanding of radioactivity, and earned each of them a Nobel Prize.

(a) Research their Nobel Prizes and report on the year of the award, the subject area, and specific contributions cited.

(b) Describe the effects of their discoveries on our society. (3.8)

 www.science.nelson.com

chapter

4

Chemical Bonding

▶ **In this chapter, you will be able to**

- relate the periodic table and atomic theory to chemical bonding theory;

- use chemical bonding theory to explain and predict properties of elements and compounds;

- use VSEPR theory or orbital theory, electronegativity, and bond polarity to explain and predict the polarity of molecules;

- use intermolecular-force theories to explain and predict physical properties of elements and compounds;

- describe scientific and technological advances that are related to properties of chemicals and to chemical bonding.

The study of the elements and of atoms leads naturally into how they bond to one another to form compounds. As you complete this chapter, you are following approximately the same sequence as chemists in the early 1900s. As atomic theory was being refined by some chemists, other chemists who were interested in how atoms bonded took over. Theories of how atoms combine evolved quickly then and continue to do so today. The initial theories were restricted first to simple ionic and organic compounds, and became more sophisticated as chemists tried to create theories that would describe, explain, and predict the structure and properties of more and more complex molecules.

Although the number of natural substances is relatively constant, we are still discovering more and more of these substances. It is the lifetime job of many chemists and biologists to identify and describe naturally occurring chemicals. Their quest is not only to understand but also to find substances that can be useful.

Chemists need to understand the structure and bonding of molecules in order to reproduce substances already present and to create new synthetic substances. For example, nylon resembles proteins, synthetic fertilizers are related to natural fertilizers, and artificial flavours mimic natural flavours.

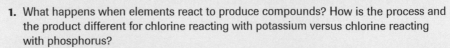

❡ *REFLECT* on your learning

1. What happens when elements react to produce compounds? How is the process and the product different for chlorine reacting with potassium versus chlorine reacting with phosphorus?

2. How do we explain and predict the bonding and shapes of molecules such as water, methane, ammonia, and ethene?

3. Explain the structure and properties of liquids such as $C_5H_{12(l)}$ and $CH_3OH_{(l)}$, and solids such as $Na_{(s)}$, $NaCl_{(s)}$, $CO_{2(s)}$, and $C_{(s)}$ (diamond).

> ▶ **TRY THIS** activity   *Properties and Forces*

Every property of a substance should be predictable once we have a complete understanding of the interactions between atoms and molecules. To obtain this understanding we need to observe carefully and develop explanations. Record your observations for the following activity and see what explanations flow from them.

Materials: 2 small, flat-bottomed drinking glasses or beakers; 2 small ceramic bread plates (china or stoneware or glass); some dishwashing liquid; and some synthetic 75–90 W gear oil

- Place the drinking glass on the bread plate, and press down firmly while trying to move the glass in a small horizontal circle.
  (a) Both the glass and the plate are very smooth. Does this mean they slide over each other easily?

- Add dishwashing liquid to the plate until it is about 2 mm deep, and try the first step again.
  (b) Dishwashing liquid makes your fingers feel slippery. Does dishwashing liquid actually make the contact surface between the glass and plate slippery?

- Add synthetic gear oil to the second plate until it is about 2 mm deep, and try the first step again.
  (c) The gear oil makes your fingers feel slippery. Does oil actually make the contact surface between the glass and plate slippery?

- Lift the glass on both plates vertically, and note the tendency of the plate to stick.
  (d) Which liquid seems to be a more effective adhesive?

- In a sink, run warm water over the plate with the dishwashing liquid, and over the plate with the gear oil.
  (e) What evidence do you have for these liquids about the strengths of the attractive intermolecular forces between their molecules (cohesion), and between their molecules and the water molecules (adhesion)?

- Add some dishwashing liquid to the plate with the gear oil, mix the liquid contents with your fingers, and run some warm water over the plate.
  (f) Describe and explain your observations.
  (g) Both these liquids are designed to have specific properties of adhesion and cohesion. What can you say, based on your observations, about the properties of dishwashing detergent and gear oil?

 Wash your hands thoroughly after completing this activity.

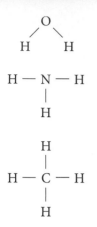

Figure 1
Kekulé structures, now known as structural diagrams, for water, ammonia, and methane explain their empirically determined chemical formulas.

Figure 2
Light is first polarized by passing through polarized lenses like those in sunglasses. When the polarized light passes through most transparent substances, nothing unusual is noticed. However, certain substances dramatically change the light to produce some beautiful effects. This effect can be explained only by 3-D versions of structural diagrams.

ionic bonding the electrostatic attraction between positive and negative ions in the crystal lattice of a salt

covalent bonding the sharing of valence electrons between atomic nuclei within a molecule or complex ion

Bonding is one of the most theoretical concepts in chemistry. First of all, we are lucky if we can even "see" an atom with an electron microscope. Since we are unable to "see" a chemical bond, our theoretical picture of bonds is based on a strong experimental and logically consistent case for the nature of chemical bonding.

The Dalton atom story starts before the Mendeleev periodic table with Edward Frankland (1852) stating that each element has a fixed valence that determines its bonding capacity. Friedrich Kekulé (1858) extended the idea to illustrating a bond as a dash between bonding atoms; i.e., what we now call a structural diagram (**Figure 1**).

Sixteen years after Kekulé created structural diagrams, Jacobus van't Hoff and Joseph Le Bel independently extended these structures to three dimensions (3-D). They revised the theory in order to explain the ability of certain substances to change light as it passes through a sample of the substance (optical activity, **Figure 2**). Recall that all of this work was done by working only with the Dalton atom—no nucleus or bonding electrons. As yet there was no explanation for the bonds that were being represented in the diagrams.

That story starts with Richard Abegg, a German chemist, in 1904. He suggested that the stability of the "inert" (noble) gases was due to the number of outermost electrons in the atom. He looked at the periodic table and theorized that a chlorine atom had one less electron than the stable electron structure of argon, and was likely to gain one electron to form a stable, unreactive chloride ion. Likewise, he suggested that a sodium atom had one more electron than needed for stability. The sodium atom should, according to his theory, lose one electron to form a stable sodium ion, as indicated below. The ions would, in turn, be held together by electrostatic charge (positive charge attracts negative charge), resulting in **ionic bonding** in a crystal of table salt.

$$Na + Cl \rightarrow Na^+ + Cl^- \rightarrow NaCl$$
sodium + chlorine → sodium chloride

In 1916 Gilbert Lewis, an American chemist, used the evidence of many known chemical formulas, the concept of valence, the octet rule, and the electron-shell model of the atom to explain chemical bonding. Lewis's work produced the first clear understanding of chemical bonding, especially **covalent bonding**.

> The key ideas of the Lewis theory of bonding are:
>
> - Atoms and ions are stable if they have a noble gas-like electron structure; i.e., a stable octet of electrons.
> - Electrons are most stable when they are paired.
> - Atoms form chemical bonds to achieve a stable octet of electrons.
> - A stable octet may be achieved by an exchange of electrons between metal and nonmetal atoms.
> - A stable octet of electrons may be achieved by the sharing of electrons between nonmetal atoms.
> - The sharing of electrons results in a covalent bond.

Lewis structures were created before the development of quantum mechanics and they are still used today to communicate the bonding for a wide range of substances. To help you review what you have learned previously, the rules for drawing Lewis structures of atoms and molecules are presented in the following summary.

SUMMARY *Rules for Drawing Lewis Structures*

Step 1 Use the last digit of the group number from the periodic table to determine the number of valence electrons for each atom.

Step 2 Place one electron on each of the four sides of an imaginary rectangle enclosing the central atom before pairing any electrons.

Step 3 If there are more than four valence electrons, pair up the electrons as required to place all of the valence electrons.

Step 4 Use the unpaired electrons to bond additional atoms with unpaired electrons to the central atom until a stable octet is obtained.

There are, of course, exceptions to the rules for writing Lewis structures. The atoms of hydrogen through boron do not achieve an octet of electrons when they combine to form molecules. The octet rule applies only to the carbon atom and beyond. For atoms beyond the second period, exceptions are frequent. Other exceptions to the octet rule will appear later in this chapter.

Let's review some simple examples of Lewis structures—the element chlorine, the ionic compound sodium chloride, and the molecular compound ammonia (**Figure 3**).

Chemists in Lewis's time knew chlorine was diatomic. The Lewis structure explains the diatomic character of chlorine and shows the explanatory power of Lewis structures. Had Lewis structures not been able to explain the diatomic character of chlorine, the theory would have been revised until it could explain the evidence, or it would been discarded.

The transfer of an electron from the sodium atom to the chlorine atom during their reaction results in a stable octet of electrons for both atoms, with all electrons paired. Their success in explaining the chemical formulas of ionic compounds helped Lewis structures gain acceptance.

The Lewis symbol for the nitrogen (central) atom indicates a pair of electrons and three unpaired (bonding) electrons. According to Lewis theory, a nitrogen atom forms three bonds with other atoms in order to achieve a stable octet with all electrons paired.

(a)

(b)

(c)

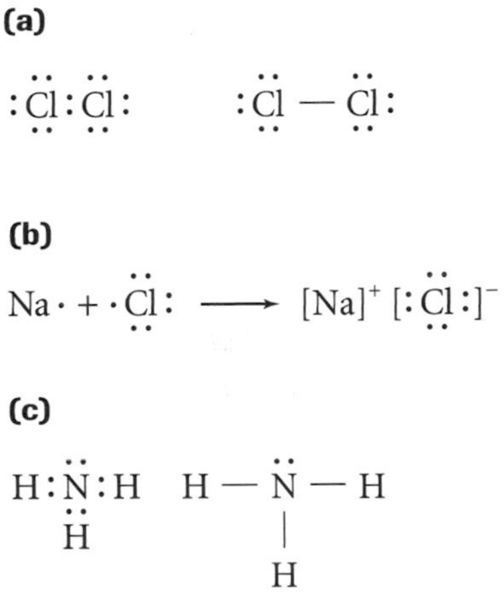

Figure 3
Lewis structures for
(a) $Cl_{2(g)}$,
(b) $NaCl_{(s)}$,
(c) $NH_{3(g)}$.

Drawing Lewis Structures   SAMPLE problem ◄

Draw a Lewis structure and a structural diagram for each of the following known chemical formulas:

(a) $HF_{(g)}$ *(b) $F_{2(g)}$*   *(c) $OF_{2(g)}$*   *(d) $O_2F_{2(g)}$*

Lewis Structural Diagrams:

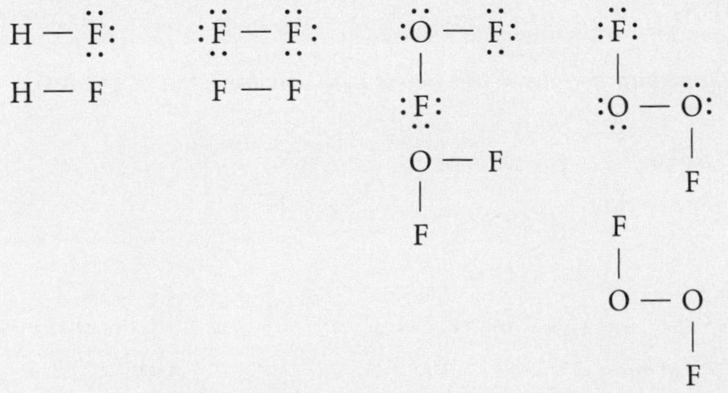

Lewis Structures and Quantum Mechanics

Because Lewis was doing his work at the same time as quantum mechanics was first being developed, it is not surprising that there is a connection between developments in atomic theory and bonding theory. In a Lewis structure, the four sides of the rectangle of electrons around the core or kernel of the atom (beyond the previous noble gas) correspond to the four orbitals of the *s* and *p* energy sublevels. Sommerfeld added these energy sublevels to account for spectral lines in 1915 and Lewis presented his electron accounting theory in 1916. It is hard to know how much Lewis was influenced by the work of Sommerfeld, but there are definite parallels in the theoretical work with atoms and molecules. The unity of chemical theory that was being reached is illustrated in **Table 1**.

Table 1 The Unity of the Atomic and Bonding Theories

Element	Magnesium, Mg	Nitrogen, N	Sulfur, S
Valence	2	3	2
Lewis symbol	·Mg·	·N̈·	:S̈·
Energy-level diagram	$3s$ (↑↓) $2p$ (↑↓)(↑↓)(↑↓) $2s$ (↑↓) $1s$ (↑↓) Mg	$2p$ (↑)(↑)(↑) $2s$ (↑↓) $1s$ (↑↓) N	$3p$ (↑↓)(↑)(↑) $3s$ (↑↓) $2p$ (↑↓)(↑↓)(↑↓) $2s$ (↑↓) $1s$ (↑↓) S
Electron configuration	$1s^2\,2s^2\,2p^6\,3s^2$ or [Ne] $3s^2$	$1s^2\,2s^2\,2p^3$ or [He] $2s^2\,2p^3$	$1s^2\,2s^2\,2p^6\,3s^2\,3p^4$ or [Ne] $3s^2\,3p^4$

The rules of quantum mechanics clarify the arbitrary nature of electron pairs and octets in the initial Lewis structures. The octet comes from the maximum of two electrons in the *s* energy-sublevel and six electrons in the *p* sublevel. In terms of orbitals, this means two electrons in the *s* orbital and and six electrons in the *p* orbitals. Hund's rule from quantum mechanics also requires that one electron be put into each orbital before putting a second electron into an orbital of the same energy.

▶ **SAMPLE** problem | **Writing Electron Configurations**

Write the electron configuration and draw the Lewis symbols for each of the following atoms:

(a) hydrogen　　**(b) boron**　　**(c) silicon**

(a) hydrogen　　$1s^1$　　　　　　　H·

(b) boron　　　$1s^2\,2s^2\,2p^1$　　　　·B̈·

(c) silicon　　　$1s^2\,2s^2\,2p^6\,3s^2\,3p^2$　　·S̈i·

▶ *Practice*

Understanding Concepts

1. Use the following experimental formulas to determine the valence of each element in the formula:
   (a) $MgCl_{2(s)}$
   (b) $CH_{4(g)}$
   (c) $H_2O_{(l)}$
   (d) $H_2S_{(g)}$
   (e) $NH_{3(g)}$

2. Place the following chemistry concepts in the order that they were created:
   (a) Lewis structures
   (b) empirical formulas
   (c) Dalton atom
   (d) Kekulé structures
   (e) Schrödinger quantum mechanics

3. Write the electron configuration and draw the Lewis symbols for each of the following atoms:
   (a) aluminum
   (b) chlorine
   (c) calcium
   (d) germanium

4. Draw Lewis symbols for the atoms with the following electron configurations:
   (a) $1s^2\,2s^2\,2p^4$
   (b) $1s^2\,2s^2\,2p^6\,3s^2\,3p^3$
   (c) $[Ar]\,4s^2\,3d^{10}\,4p^5$
   (d) $[Kr]\,5s^1$

5. Draw Lewis structures and structural diagrams for the following molecules:
   (a) CH_4
   (b) H_2O
   (c) CO_2
   (d) H_2S
   (e) NH_3

Applying Inquiry Skills

6. List the criteria that must be met by a new scientific concept, such as the Lewis theory, before it is accepted by the scientific community.

7. Chemical-bonding theories developed from the knowledge of chemical formulas. Describe a general design of an experiment that could be done to determine a chemical formula.

Extensions

8. Describe the difference in sophistication between the Kekulé structures created in 1858 and the Lewis structures created in 1916.

9. Lewis structures were derived from quantum theory. Write a short paragraph presenting pro and con arguments for this statement.

Extending the Lewis Theory of Bonding

Many simple molecules, like the ones you have seen, were explained by the original, main ideas of the Lewis theory of bonding. However, some molecules and, in particular, polyatomic ions, could not easily be explained. One of the first scientists to extend the Lewis theory was Nevil Sidgwick, who showed that the Lewis structures of many other molecules can work if we do not require that each atom contribute one electron to a shared pair in a covalent bond. In other words, one atom could contribute both electrons that are shared. Sidgwick also recognized that an octet of electrons around an atom may be desirable, but is not necessary in all molecules and polyatomic ions.

The pattern is a familiar one in the development of scientific theories. We know an experimental result (such as a chemical formula) and the theory evolves to successfully explain this result. Consider the nitrate ion, NO_3^-, which is common in nature.

DID YOU KNOW ?

Irving Langmuir
Irving Langmuir was an excellent speaker and often lectured on the Lewis theory. He did such a good job in popularizing the theory that it started to be known as the Lewis-Langmuir theory. When Lewis became somewhat annoyed by this development, Langmuir backed off and returned to other research.

What is the Lewis structure for the nitrate ion, NO_3^-?

First, we need to guess the arrangement of the atoms. Fortunately, there are some guidelines that have been developed.

- The central atom is usually the least electronegative atom and generally appears first in the formula; e.g., N in NO_3^-.

- Oxygen atoms surround the central atom in oxyanions.

- Hydrogen is bonded to an oxygen in oxyacids.

- If possible, choose the most symmetrical arrangement.

Using these guidelines, we guess the arrangement for a nitrate ion to be

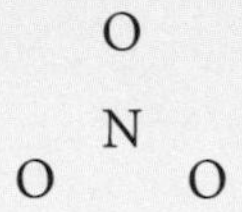

O

N

O   O

Now we need to count the total number of valence electrons. Add the valence electrons for each atom and then add additional electrons for a negative ion or subtract electrons for a positive ion. The number of electrons added or subtracted corresponds to the size of the charge on the polyatomic ion. For NO_3^-, we have $5e^-$ for the nitrogen, three times $6e^-$ for the oxygen atoms plus $1e^-$ for the single negative charge of the nitrate ion.

total valence electrons $= 5 + 3(6) + 1 = 24e^-$

Start allocating these electrons by placing them, in pairs, between the central atom, N, and each of the surrounding O atoms. This uses $6e^-$ out of the total of $24e^-$.

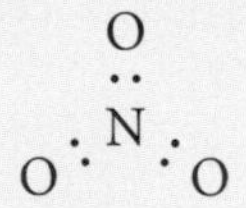

Now complete the octets of the surrounding atoms. (Remember that hydrogen is an exception and its valence shell is complete with only two electrons.) We now have used all of the $24e^-$.

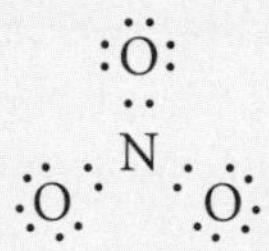

Check the central N atom to see if the octet rule is met. If not, try moving lone pairs of electrons from the surrounding atoms to form double or triple covalent bonds. Whenever possible, you should satisfy the octet rule for all atoms in the structure. For the nitrate ion, we can move one lone pair of electrons from an oxygen atom to form one double bond. This will complete the octet of electrons for the N atom.

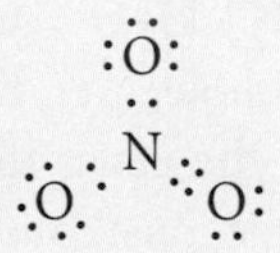

Finally, draw the Lewis structure, showing the covalent bonds with a short line. For polyatomic ions, add a large square bracket with the ion charge.

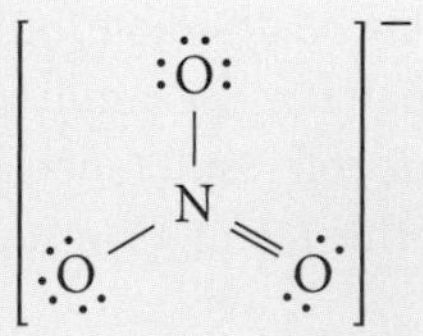

Using this procedure, you can now draw Lewis structures for many more molecules and most polyatomic ions. This procedure extends the Lewis theory of bonding and makes it much more useful.

SUMMARY	*Procedure for Drawing Lewis Structures*

Step 1 Arrange atoms symmetrically around the central atom (usually listed first in the formula, not usually oxygen and never hydrogen).

Step 2 Count the number of valence electrons of all atoms. For polyatomic ions, add electrons corresponding to the negative charge, and subtract electrons corresponding to the positive charge on the ion.

Step 3 Place a bonding pair of electrons between the central atom and each of the surrounding atoms.

Step 4 Complete the octets of the surrounding atoms using lone pairs of electrons. (Remember hydrogen is an exception.) Any remaining electrons go on the central atom.

Step 5 If the central atom does not have an octet, move lone pairs from the surrounding atoms to form double or triple bonds until the central atom has a complete octet.

Step 6 Draw the Lewis structure and enclose polyatomic ions within square brackets showing the ion charge.

> **LEARNING TIP**
>
> **Octet Rule and Violations**
> When drawing Lewis structures you should always try to obtain an octet of electrons around each atom. For most molecules and ions you will encounter, this will be possible. However, do not be surprised if an occasional example violates the octet rule around a central atom. This may happen when the central atom has *d* orbitals at a relatively low energy available for electrons.

Example 1

Draw a Lewis structure for the ammonium ion, NH_4^+.

Solution

total valence electrons $= 5 + 4(1) - 1 = 8e^-$

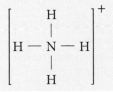

Example 2

Draw Lewis structures for the sulfur trioxide molecule, SO_3.

Solution

total valence electrons $= 6 + 3(6) = 24e^-$

▶ Practice

Understanding Concepts

10. Draw Lewis structures for each of the following molecules or polyatomic ions.
    - (a) ClO_4^-
    - (b) CO_3^{2-}
    - (c) CN^-
    - (d) H_3O^+
    - (e) HCO_3^-
    - (f) HNO_3
    - (g) NO^+

Applying Inquiry Skills

11. Hydrogen chloride and ammonia gases are mixed in a flask to form a white solid product.
    - (a) Write a balanced chemical equation for the reaction of ammonia and hydrogen chloride gases to form ammonium chloride.
    - (b) Rewrite the equation using Lewis structures.
    - (c) Suggest at least two diagnostic tests that could be done to analyze the product.

Extension

12. Draw Lewis structures for a molecule of each of the following substances.
    - (a) $ClF_{3(g)}$
    - (b) $SCl_{4(l)}$
    - (c) $PF_{5(g)}$

Table 2

Chemist	Contribution to bonding theory
Edward Frankland (1852) English	Explained bonding as due to a fixed valence for each element that determined the number of atoms with which it bonded—the theory of valence or combining capacity.
Friedrich Kekulé (1858) German	Kekulé structures represented the bonding of atoms in a molecule using a dash between bonded atoms in a 2-D diagram, even explaining organic isomers.
Dmitri Mendeleev (1869) Russian	Used the theory of valence to help create the periodic law and its representation as the periodic table, even predicting the combining capacity of yet undiscovered elements.
Jacobus van't Hoff (1874) Joseph Le Bel (1874) Dutch & French	Extended Kekulé structures to represent molecules in 3-D, explaining isomers that affect polarized light (optical isomers).
Richard Abegg (1904) German	Explained bonding as due to a transfer of electrons and the subsequent attractive force between positive and negative ions, restricted to simple electrolytes.
Gilbert Lewis (1916) American	Explained the bonding of simple organic compounds as due to the sharing of a pair of electrons between nonmetal atoms to achieve an "inert-gas"-like electron structure.
Irving Langmuir (1920) American	Independently created a theory of electron-pair bonds and described both covalent and ionic bonds.
Nevil Sidgwick (1920s) English	Extended the Lewis theory of electron-pair bonds beyond organic chemicals to explain the chemical formula of polyatomic ions.

▶ Section 4.1 Questions

Understanding Concepts

1. Place the following chemistry concepts in the order that they were created by chemists and briefly explain how one concept led to the next:
   (a) Bohr atom
   (b) empirical formulas
   (c) Dalton atom
   (d) Lewis structures

2. Write the electron configuration and draw the Lewis symbol for an atom of each of the following elements:
   (a) beryllium
   (b) phosphorus
   (c) magnesium
   (d) oxygen

3. Draw Lewis symbols for the atoms with the following electron configurations:
   (a) $1s^2 2s^2 2p^5$
   (b) $1s^2 2s^2 2p^6 3s^2 3p^1$
   (c) [Ar] $4s^2 3d^{10} 4p^5$
   (d) [Kr] $5s^2$

4. Draw Lewis structures and structural diagrams for the following molecules or ions:
   (a) H_2S
   (b) CCl_4
   (c) NCl_3
   (d) SO_4^{2-}
   (e) HSO_4^-
   (f) NO_2^-
   (g) CO
   (h) PO_4^{3-}

Extensions

5. (a) Write a balanced chemical equation for the reaction of boron trichloride and ammonia gases to form a single molecular product.
   (b) Rewrite the equation using Lewis structures.

6. One of the compounds that chemists initially had difficulty explaining was ammonium chloride.
   (a) Use Lewis structures to represent this compound.
   (b) Describe in your own words how the initial difficulty was resolved.

The mathematical equation developed by Schrödinger describes the standing wave for an electron and is used to construct a particular volume of space (orbital), one for each quantum state of the electron, in which the probability of finding an electron is high. The various orbitals are usually visualized as electron clouds and have characteristic shapes and orientations; e.g., $1s$, $2s$, and $2p$ (Figures 3, 4 in Section 3.7). An atom has only one nucleus to consider, but a molecule has two or more nuclei and more electrons to take into account. The molecule is more complex than the atom, and the application of quantum mechanics to molecules is still an area of active research.

In order to deal with the complexity of several atoms in a molecule, some simplifications are required. One approach is to start with individual atoms, with their orbitals, and then build the molecule using atomic orbitals to form covalent bonds. This approach is commonly known as the **valence bond theory** and was developed primarily by Linus Pauling (**Figure 1**). According to this theory, a covalent bond is formed when two orbitals overlap (share the same space) to produce a new combined orbital containing two electrons of opposite spin. This arrangement results in a decrease in the energy of the atoms forming the bond. For example, the $1s$ orbitals of two hydrogen atoms overlap to form the single covalent bond of a hydrogen molecule (**Figure 2**). The two shared electrons spend most of their time between the two hydrogen nuclei.

Figure 1
Linus Pauling was a long-time friend of Gilbert Lewis. He dedicated a famous textbook, *The Nature of the Chemical Bond*, to Lewis. Pauling is one of only four two-time winners of a Nobel Prize—his in two different fields, Chemistry (1954) and Peace (1962).

Figure 2
The formation of a single covalent bond in a hydrogen molecule by the overlap of two $1s$ orbitals of individual hydrogen atoms. This represents a new, lower-energy state of the two atoms.

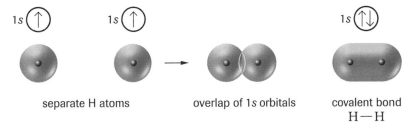

separate H atoms   overlap of $1s$ orbitals   covalent bond
H—H

Notice that the new combined orbital contains a pair of electrons of opposite spin, just like a filled atomic orbital. Any two half-filled orbitals can overlap in the same way. Consider the hydrogen fluoride, HF, molecule. The hydrogen $1s$ orbital is believed to overlap with the half-filled $2p$ orbital of the fluorine atom to form a covalent bond (**Figure 3**).

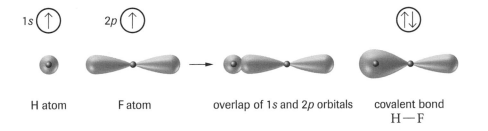

H atom   F atom   overlap of $1s$ and $2p$ orbitals   covalent bond
H—F

Figure 3
A hydrogen atom has only one occupied orbital, the $1s$ orbital. For simplicity, only the one half-filled p orbital of the fluorine atom is shown. The final combined, bonding orbital contains a pair of electrons and the fluorine atom now has a complete octet.

valence bond theory atomic orbitals or hybrid orbitals overlap to form a new orbital containing a pair of electrons of opposite spin

So far we have considered only diatomic molecules, but this approach can also be used for larger molecules. An oxygen atom has two half-filled p orbitals.

O atom:
$1s^2$ $2s^2$ $2p^4$

It is reasonable to propose that the $1s$ orbitals of two hydrogen atoms overlap with the two half-filled $2p$ orbitals of the oxygen atom to produce a stable, lower-energy state (**Figure 4**). Two covalent bonds are created by two sets of combined s-p orbitals. However, the proposition leads us to predict that the angle between the two bonds will be 90°, no matter which two p orbitals are used. The measured angle for the H—O—H bond is about 105°—indicating either a serious problem with this valence bond approach or that some other factor needs to be considered. As you will see, some additional manipulation of atomic orbitals and a consideration of the repulsions between pairs of electrons improves the explanation of observed angles.

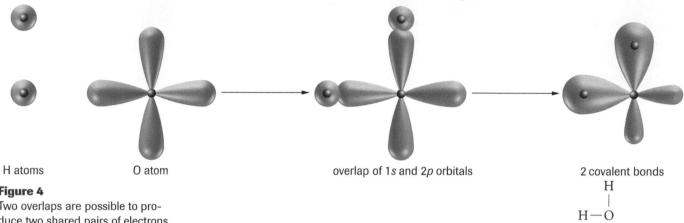

H atoms   O atom   overlap of $1s$ and $2p$ orbitals   2 covalent bonds

$$\begin{array}{c} H \\ | \\ H{-}O \end{array}$$

Figure 4
Two overlaps are possible to produce two shared pairs of electrons forming two covalent bonds. As before, the oxygen atom completes its octet of electrons.

SUMMARY   *Valence Bond Theory*

- A half-filled orbital in one atom can overlap with another half-filled orbital of a second atom to form a new, bonding orbital.
- The new, bonding orbital from the overlap of atomic orbitals contains a pair of electrons of opposite spin.
- The total number of electrons in the bonding orbital must be two.
- When atoms bond, they arrange themselves in space to achieve the maximum overlap of their half-filled orbitals. Maximum overlap produces a bonding orbital of lowest energy.

▶ Practice

Understanding Concepts

1. Use the valence bond theory to explain why two helium atoms do not bond to form a diatomic molecule.

2. In terms of electrons, how is a bonding orbital formed from the overlap of two atomic orbitals?

3. Consider **Figure 5**, showing a proposed overlap of the half-filled atomic orbitals of three atoms. Explain why this overlap is not possible.

$1s$ $2p$ $1s$

Figure 5

4. For each of the following molecules, draw a diagram showing the overlap of atomic orbitals to form the covalent bond. For simplicity, show and label only the half-filled orbitals being used for the covalent bond.
   (a) HCl   (b) Cl_2 (c) ClF

Applying Inquiry Skills

5. (a) Draw a diagram showing the overlap of atomic orbitals to form the covalent bonds in a hydrogen sulfide molecule. For simplicity, show and label only the half-filled orbitals being used for the covalent bonds.
   (b) Predict the angle for the H−S−H bonds. Justify your answer.
   (c) Experimental measurements show that the angle is 92°. Evaluate the prediction and valence bond theory for this molecule.

Making Connections

6. Why is an understanding of a covalent bond useful? Provide several reasons, together with some general or specific examples.

Extension

7. Research and write a brief report outlining the basic principles of molecular orbital theory.

 www.science.nelson.com

Hybrid Orbitals

Two problems left over from the Lewis bonding theory were its inability to explain (1) the four equal bonds represented by the four pairs of electrons in a carbon compound like methane, $CH_{4(g)}$, and (2) the existence of double and triple bonds. Pauling and others were able to explain the evidence that, for example, carbon forms four bonds of equal length distributed in a tetrahedron. They first created the idea of electron promotion from an s to an empty p orbital—justifying this on the basis of getting back more energy from the bonding than was put in to promote the electron to a slightly higher energy level (**Figure 6 (a)**). However, experimental evidence indicated that the electron orbitals were equivalent in shape and energy—there isn't one that is more "s" than the others. As a result, the four bonds for carbon in molecules such as methane are explained by **hybridization** to four identical hybrid sp^3 atomic orbitals. The two 2s electrons and the two 2p electrons form four sp^3 **hybrid orbitals** with one bonding electron in each. This explains the bonding capacity of four for carbon (**Figures 6 (b, c)**). Note that these orbitals are hybridized only when bonding occurs to form a molecule; they do not exist in an isolated atom.

The Pauling theory is that these sp^3 hybrid orbitals are spontaneously formed as these orbitals overlap with, say, the 1s orbitals of the hydrogen atom (**Figure 7**). This results in a pair of electrons in each orbital and a noble–gas-like structure around each atom.

There were additional anomalies to explain. How do you explain the covalent bonding in the empirically determined formulas for boron and beryllium, e.g., $BF_{3(g)}$ and $BeH_{2(s)}$? Pauling suggested that there were a whole series of hybridizations that could occur (**Table 1**).

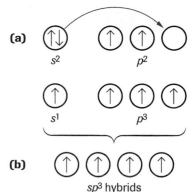

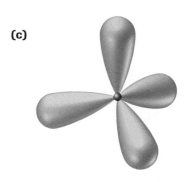

Figure 6
(a) An s electron is promoted to an empty p orbital in a carbon atom.
(b) The four orbitals are combined to produce four hybrid sp^3 orbitals.
(c) Each sp^3 orbital is equivalent in energy and shape. Electron repulsion requires that the orbitals are as far apart as possible—pointing to the corners of a regular tetrahedron.

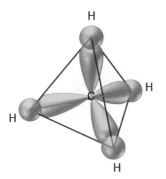

Figure 7
The bonding in a methane, $CH_{4(g)}$, molecule is described by the valence bond theory as being due to the overlap of the four identical sp^3 hybrid orbitals of carbon with the single 1s orbital of four hydrogen atoms.

Period 2 Outlaws

It is tempting to think that the lighter elements should be simpler than the heavier elements, and they are in many ways. However, many of the elements in period 2 do not follow the patterns in the periodic table. Period 2 elements are more often exceptions to the rules. Beryllium is a good example. We would expect it to be a typical metal belonging to Group 2. In fact, beryllium forms compounds that are always molecular in properties rather than ionic.

hybridization a theoretical process involving the combination of atomic orbitals to create a new set of orbitals that take part in covalent bonding

hybrid orbital an atomic orbital obtained by combining at least two different orbitals

Table 1 Forms of Hybridization

Initial atomic orbitals	Changes in orbital configuration	Hybrid orbitals of central atom	Example
s, p	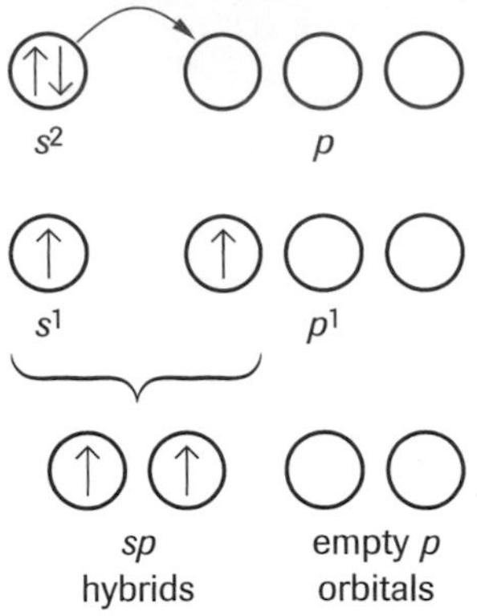 s^2 → p; s^1, p^1; two *sp* hybrids, empty *p* orbitals	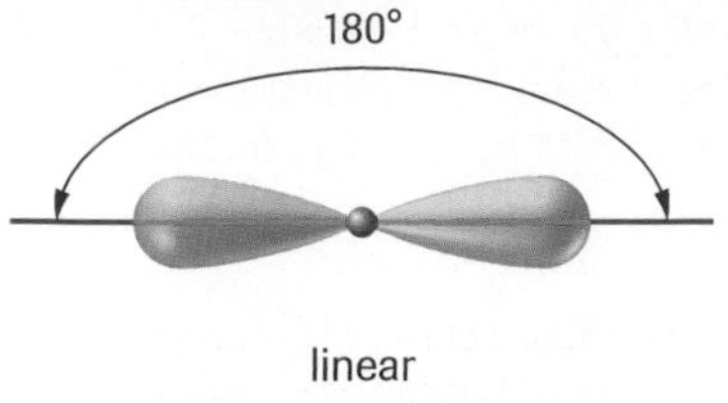 180° linear; two *sp* hybrid orbitals, linear, 180°	BeH_2
s, p, p	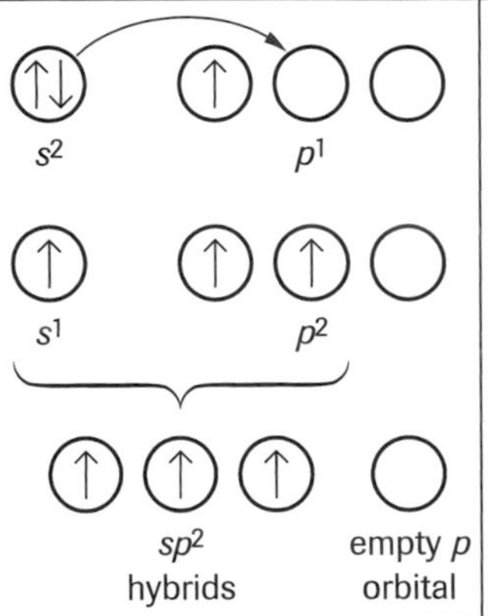 s^2, p^1; s^1, p^2; three *sp²* hybrids, empty *p* orbital	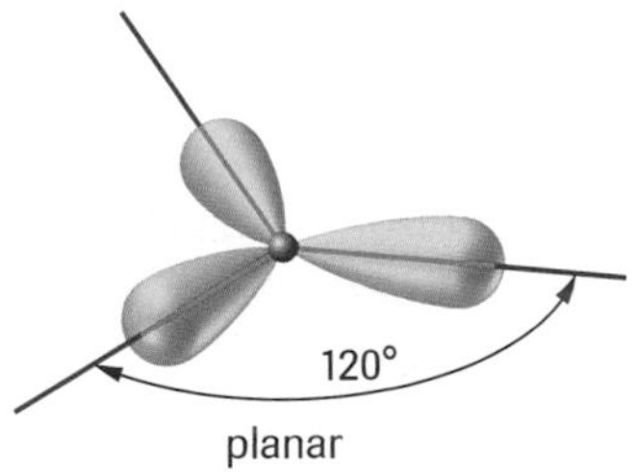 120° planar; three *sp²* orbitals, trigonal planar, 120°	BCl_3
s, p, p, p	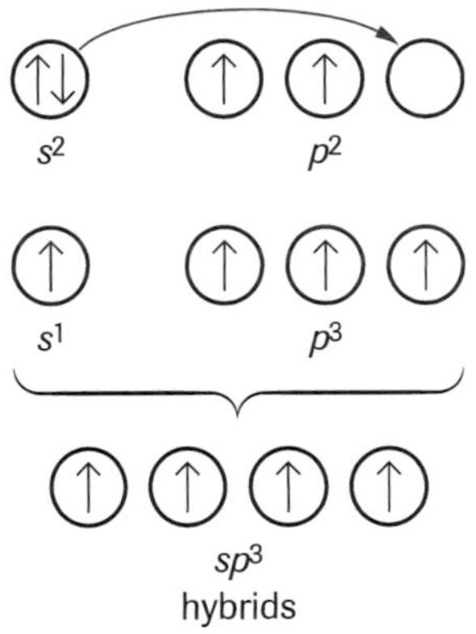 s^2, p^2; s^1, p^3; four *sp³* hybrids	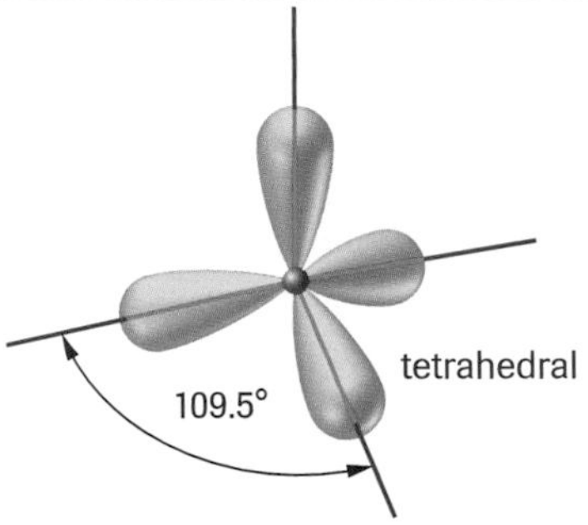 109.5° tetrahedral; four *sp³* orbitals, tetrahedral, 109.5°	CH_4

Notes:

1. The number of hybrid orbitals can be readily obtained from the designation; e.g., sp^3 means s^1p^3, which means $1 + 3 = 4$ orbitals.

2. The empty boxes for the *p* orbitals mean an unfilled or empty orbital for all of the examples given. For other examples you will see later, these empty *p* orbitals may be occupied.

LEARNING *TIP*

Labels for Hybrid Orbitals

The label sp^3 means that the hybrid orbitals are formed from one *s* orbital and three *p* orbitals to give a total of four hybrid orbitals $(1 + 3)$. The "1" superscript in s^1p^3 is assumed and not written. Other hybrid orbitals you will see can be interpreted in the same way.

▶ **SAMPLE** *problem*   ***Explaining the Bonding in a Molecule***

What are the bonding orbitals and the structure of the BF_3 molecule?

First, by looking at the molecule you know that boron has a valence (bonding capacity) of three and fluorine one. You should write the ground-state electron configuration of the boron atom and see how this configuration can be expanded by promoting an electron to another level (usually s to p) to obtain three half-filled orbitals.

B: $1s^2\, 2s^2\, 2p^1$ promoted to $1s^2\, 2s^1\, 2p^1\, 2p^1$

By promoting an s electron into an empty p orbital, there are one s orbital and two p orbitals available for hybridization—a total of three identical sp^2 hybrids arranged trigonally in a plane (**Table 1**).

A fluorine atom forms only one bond, which must be from its single, half-filled p orbital.

F: $1s^2\, 2s^2\, 2p_x^{\,2}\, 2p_y^{\,2}\, 2p_z^{\,1}$

Therefore, the three covalent bonds in the BF_3 molecule are formed from the end-to-end overlap of three sp^2 orbitals of the boron atom with three p orbitals of the fluorine atoms, and the structure is trigonal planar.

Example

Provide the ground-state and the promoted-state electron configurations for beryllium, and then describe the bonding and structure of a BeH_2 molecule.

Solution

Be: $1s^2\, 2s^2$ promoted to $1s^2\, 2s^1\, 2p^1$

sp hybridization provides two bonding orbitals that each overlap with the $1s$ orbitals of two hydrogen atoms to form a linear molecule.

▶ Practice

Understanding Concepts

8. What atomic orbital or orbitals are available for bonding for each of the following atoms?
   (a) H
   (b) F
   (c) S
   (d) Br

9. How many electrons are present in an orbital formed from the overlap of two different atomic orbitals?

10. Provide ground-state and promoted-state electron configurations for each of the following atoms and indicate the type of hybridization involved when each atom forms a compound:
    (a) carbon in CH_4
    (b) boron in BH_3
    (c) beryllium in BeH_2

11. Draw and label the hybrid orbitals of the central atom, and then add the overlapping orbitals forming the covalent bonds for each of the following molecules. (For simplicity, ignore filled orbitals that are not part of the bonding.)
    (a) $BeCl_2$
    (b) CF_4
    (c) BBr_3
    (d) $CHCl_3$
    (e) B_2F_4

12. If the most stable form of a molecule is the lowest energy state, then how did Pauling justify promoting electrons to higher energy levels?

13. Do hybridized orbitals exist in isolated atoms? Why or why not?

14. Why did Pauling create the concept of hybridized orbitals?

Applying Inquiry Skills

15. Empirical scientific knowledge generally appears before theoretical knowledge. Provide two examples.

16. To determine the hybridization of atomic orbitals in a molecule, state the two pieces of experimental evidence that are required.

H H
H:C:C:H
H H

H H
H:C::C:H

H:C:::C:H

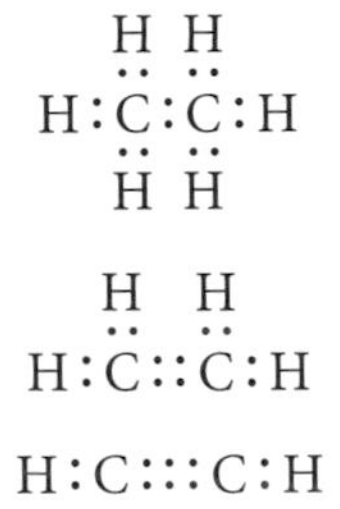

Figure 8
Counting all of the shared electrons around each carbon atom gives a total of eight in each case.

sigma (σ) bond a bond created by the end-to-end overlap of atomic orbitals

pi (π) bond a bond created by the side-by-side (or parallel) overlap of atomic orbitals, usually p orbitals

Double and Triple Covalent Bonds

The problem of explaining double and triple bonding had existed since the valence of carbon was determined experimentally a hundred years before. For example, it was known that there were three substances that could be formed between carbon and hydrogen, each containing two carbon atoms in their chemical formulas—$C_2H_{6(g)}$, $C_2H_{4(g)}$, and $C_2H_{2(g)}$. How could these formulas be explained theoretically? Lewis suggested that between the carbon atoms there must be a sharing of one, two, and three electron pairs in order to obtain a stable octet around the carbon atoms (**Figure 8**).

However, even though the Lewis structures follow the octet rule and the pairing of electrons, they do not explain how the electrons are shared. How is it possible that electrons in what we would predict as being sp^3 hybrid orbitals could overlap not once, but twice or three times with just one other atom? The answer, according to valence bond theory, is that two kinds of orbital overlap are possible. One kind you have already seen, the end-to-end overlap of s orbitals, p orbitals, hybrid orbitals, or some pair of these orbitals. In the valence bond theory, this type of overlap produces a **sigma (σ) bond** (**Figure 9**). Think of sigma bonds as the usual single covalent bonds that you are used to drawing in structural diagrams.

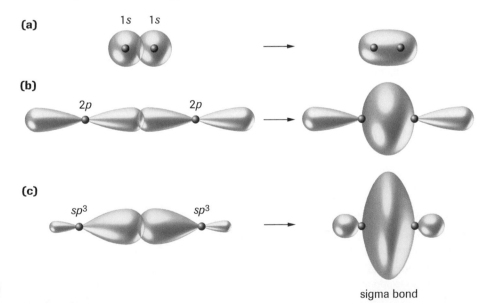

Figure 9
Sigma bonds form with the overlap of **(a)** s orbitals; **(b)** p orbitals; and **(c)** hybrid orbitals.

According to the valence bond theory, two orbitals can overlap side by side to form a **pi (π) bond** (**Figure 10**). Pi bonds are the second and third lines in the structural diagrams for double and triple covalent bonds.

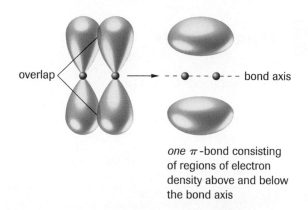

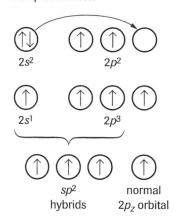

Figure 10
P orbitals form with the side-by-side overlap of orbitals.

overlap — bond axis

one π-bond consisting of regions of electron density above and below the bond axis

Figure 11
Instead of mixing all four orbitals, valence bond theory suggests that only three are mixed to form sp^2 hybrid orbitals and an unhybridized *p* orbital for a carbon atom.

Double Bonds

The carbon atom is the most common central atom in molecules with double and triple covalent bonds. We have already seen that the orbitals of a carbon atom can be hybridized to form four sp^3 hybrid orbitals. This would be the standard explanation for any carbon atom with four single bonds to other atoms. *The key new idea is a partial hybridization of the available orbitals leaving one or two p orbitals with single unpaired electrons.* For example, suppose that after promoting an electron in carbon's 2s orbital to a 2p orbital, we form three sp^2 hybrid orbitals leaving one *p* orbital with a single electron (**Figure 11**). We will still have four orbitals to form bonds but three of these are hybrids and one is a "normal" *p* orbital (**Figure 12**). In a molecule like C_2H_4, the three hybrid orbitals are used to form sigma bonds between the carbon atoms and to the hydrogen atoms (**Figure 13 (a)**). The half-filled *p* orbitals on each carbon are believed to overlap sideways (**Figure 13 (b)**) to form a pi bond. Notice that the pi bond is a region of electron density appearing above and below the sigma bond directly joining the two carbon atoms (**Figure 13(c)**). A pi bond is a combined orbital containing a pair of electrons of opposite spin, just as you have seen previously with the overlap of other orbitals. The additional shared pair of electrons in the pi bond provides greater attraction to the two carbon nuclei, which explains why the double covalent bond is shorter and stronger than a single bond.

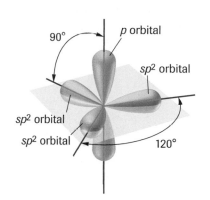

Figure 12
For this carbon atom, the sp^2 hybrids are planar at 120° to each other and the *p* orbital is at right angles to the plane of the hybrid orbitals.

Figure 13
(a) The sigma bonds for a C_2H_4 molecule use the sp^2 hybrid orbitals.
(b) The two half-filled *p* orbitals of the adjacent carbon atoms overlap sideways.
(c) The complete bonding orbitals for a C_2H_4 molecule.

(a)

(b)

(c)

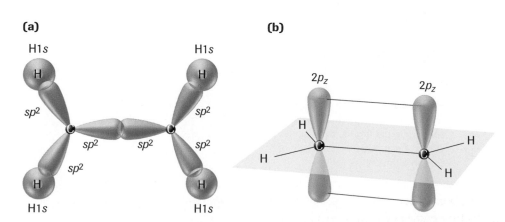

SUMMARY *Double Bonds*

- Sigma (σ) bonds are covalent bonds formed from the end-to-end overlap of atomic orbitals.

- Pi (π) bonds are covalent bonds formed from the side-to-side overlap of atomic orbitals.

- A double covalent bond contains a σ and a π bond.

▶ Practice

Understanding Concepts

18. When are π bonds formed?

19. Why was the concept of the π bond created by scientists?

20. Provide an explanation of the experimental formula and trigonal-planar structure of a molecule in each of the following substances:
    (a) $C_2Cl_{4(l)}$ (b) $H_2CO_{(g)}$ (c) $CO_{2(g)}$

21. Describe and explain the structure of a molecule of propene, $C_3H_{6(g)}$.

Applying Inquiry Skills

22. Jacobus van't Hoff drew structural diagrams for ethene by showing the sideways overlap of two tetrahedral bonds. Draw what van't Hoff may have drawn in his time.

23. Propadiene (also known as allene) is an unstable, colourless gas with the formula $C_3H_{4(g)}$. Experimental evidence shows that the carbon backbone of the molecule is linear.
    (a) Draw a structural diagram for a molecule of propadiene.
    (b) Use hybridization to explain the bonding.
    (c) Predict the shape around the end carbon atoms.

Extension

24. One of the most exciting new molecules discovered in the last two decades is buckminsterfullerene, $C_{60(s)}$, commonly known as the buckyball. X-ray crystallography has shown this molecule is a sphere made up of hexagons and pentagons of carbon atoms, like a soccer ball (**Figure 14**).
    (a) How many carbon atoms are joined to each other at each junction?
    (b) Suggest an explanation for the bonding around each carbon atom. You need only consider a portion of this molecule.
    (c) Experimental evidence shows that the bonds in each hexagon and pentagon ring are identical. Evaluate your answer to (b).

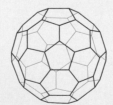

Figure 14
A buckyball is a spherical molecule containing pentagons of carbon atoms surrounded by hexagons of carbon atoms.

Triple Bonds

The ethyne (acetylene) molecule has been determined in the laboratory to be a linear molecule with the formula $C_2H_{2(g)}$. A tetrahedral sp^3 hybridization of the carbon atomic orbitals does not seem likely here. As you have seen with the explanation for double covalent bonds, a partial hybridization produced a successful explanation for ethene so the same approach will be used again.

Focusing on the carbon atom, **Figure 15** describes the ground, promoted, and hybridized states of the carbon atoms that form ethyne. To explain the bonding demanded by the formula, C_2H_2, two carbon atoms bond by overlapping one of their sp hybrid orbitals, and the s orbital of the two hydrogen atoms overlap with the other two available sp hybrid orbitals (**Figure 16 (a)**). According to the valence bond theory, the unpaired electrons in the two p orbitals of the two adjacent carbon atoms share electrons by forming two pi (π) bonds (**Figure 16 (b)**). In this view of carbon–carbon triple bonds, the carbons are bonded by one sigma (σ) bond and two pi (π) bonds (**Figure 16 (c)**). Note that the two identical sp hybrid orbitals oriented at 180° contribute to determine the 3-D orientation about each of the central carbon atoms. The result is a linear molecule for ethyne.

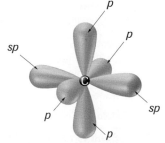

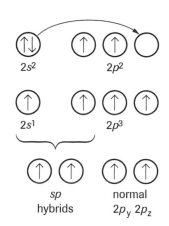

Figure 15
Instead of mixing all four orbitals, valence bond theory suggests that only two are mixed to form sp hybrid orbitals and two unhybridized p orbitals for a carbon atom.

(a)

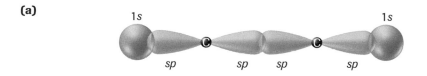

(b) **(c)**

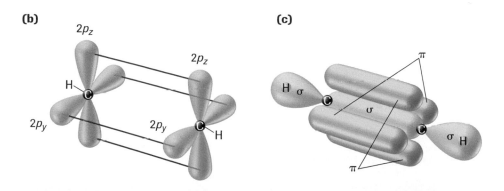

Figure 16
(a) The sigma bonds for a C_2H_2 molecule use the sp hybrid orbitals.
(b) The two pairs of half-filled p orbitals of the adjacent carbon atoms overlap sideways.
(c) The complete bonding orbitals for a C_2H_2 molecule.

▶ **Practice**

Understanding Concepts

25. Why did scientists create the concept of triple bonds?
26. Provide an explanation of the experimental formula and linear structure of
   (a) $C_2F_{2(g)}$
   (b) $HCN_{(g)}$
27. Describe and explain the structure of a molecule of propyne, $C_3H_{4(g)}$.
28. Carbon atoms can form single, double, and triple covalent bonds, but there is no experimental evidence for the formation of a quadruple covalent bond. Explain why quadruple bonds are not likely for a carbon atom.

Applying Inquiry Skills

29. Test your hypothesis concerning the relative lengths and strengths of single and multiple bonds.

▶

Question

What are the relative lengths and strengths of single, double, and triple bonds?

Hypothesis

(a) Which bonds would you hypothesize to be longest and strongest, single, double, or triple bonds? Provide your reasoning.

Evidence

Table 2 Average Bond Strengths and Bond Energies

Bond type	Bond length (pm)	Bond energy (kJ/mol)
C−O	143	351
C=O	121	745
C−C	154	348
C=C	134	615
C≡C	120	812
C−N	143	276
C=N	138	615
C≡N	116	891

Note: These are average bond lengths and bond energies from many molecules.

Analysis

(b) Answer the Question, based upon the Evidence provided in **Table 2**.

Evaluation

(c) Evaluate your Hypothesis and your reasoning.

Making Connections

30. Information about atoms and molecules is often obtained from the interaction of substances with various forms of electromagnetic radiation. (Recall the bright-line and dark-line spectra of atoms.) For molecular substances, the infrared region of the electromagnetic spectrum is particularly useful for obtaining information about covalent bonds. Research and write a brief report answering the following questions.
    (a) What characteristic of the atoms in a molecule is related to infrared (IR) radiation?
    (b) How are different covalent bonds distinguished in an IR spectrum? Why is the analysis technique useful in studying and identifying molecules?
    (c) Other than research chemists, who uses this technique?

  www.science.nelson.com

---

SUMMARY   *Valence Bond Theory*

- Covalent bonds form when atomic or hybrid orbitals with one electron overlap to share electrons.

- Bonding occurs with the highest energy (valence shell) electrons.

- Normally, the *s* and *p* orbitals overlap with each other to form bonds between atoms.

- Sometimes *s* and *p* orbitals of one atom hybridize to form identical hybrid orbitals that are used to form bonds with other atoms.

- sp^3, sp^2, and sp hybrid orbitals are formed from one s orbital and three, two, and one p orbital, respectively.
- There are four sp^3 hybrid orbitals, three sp^2 orbitals, or two sp orbitals when hybridization occurs during bond formation.
- The orientations of sp^3, sp^2, and sp hybrid orbitals are tetrahedral (109.5°), trigonal planar (120°), and linear (180°), respectively.
- End-to-end overlap of orbitals (hybrid or not) is called a sigma (σ) bond.
- Single covalent bonds are sigma (σ) bonds.
- Side-by-side overlap of unhybridized p orbitals is called a pi (π) bond.
- Double bonds have one pi (π) bond, while triple bonds have two pi (π) bonds.
- sp^2 and sp hybrid orbital bonding is usually accompanied by pi bond formation to form double or triple bonds.

Case Study   *The Strange Case of Benzene*   ▼

Benzene is a compound whose structure challenged scientists for decades. Use the following questions to guide your study of this special case. The answers to the first few questions have been provided to get you started.

(a) Who discovered benzene and determined its molecular formula?
Answer: Benzene and its formula were discovered in 1825 by the famous English scientist Michael Faraday.

(b) Who created the first acceptable structural diagram for benzene?
Answer: The German architect and chemist Friedrich Kekulé proposed a cyclical structure for benzene in 1865.

(c) What are some of the physical and chemical properties of benzene?
Answer:

- Benzene is a nonpolar liquid that freezes at 5.5°C and boils at 80.1°C.
- X-ray diffraction indicates that all carbon−carbon bonds are the same length.
- Evidence from chemical reactions indicates that all carbons in its structure are identical, and that each carbon is bonded to one hydrogen atom.
- The molecule appears to be relatively stable. It is more like an alkane in reactivity than an unsaturated compound.

(d) What are the natural and technological sources of benzene?

(e) Describe some of the applications of benzene.

(f) What other structures and explanations were proposed between discovery and now, and by whom?

(g) What is the latest theory for the bonding found in benzene?

Figure 1
Dr. Ronald Gillespie co-created VSEPR theory in 1957. He moved from England to McMaster University in Hamilton, Ontario, the following year. His work in molecular geometry and in chemistry education is renowned.

The shape of molecules has long been investigated through crystallography, using microscopes and polarimeters in the late 1800s and X-ray and other spectrographic techniques since the early 1900s. One of the most important applications of molecular shape research is the study of enzymes. Enzymes are large proteins that are highly specific in what they will react with. There are about three thousand enzymes in an average living cell and each one carries out (catalyzes) a specific reaction. There is no room for error without affecting the normal functioning of the cell; different molecular shapes help to ensure that all processes occur properly. Despite extensive knowledge of existing enzymes, the structure of these proteins is so complex that it is still effectively impossible to predict the shape an enzyme will take, even though the sequence of its constituent amino acids is known. The study of molecular shapes, particularly of complex biological molecules, is still a dynamic field.

The Arrival of VSEPR

The valence bond theory created and popularized by Linus Pauling in the late 1930s successfully explained many of the atomic orientations in molecules and ions, including tetrahedral, trigonal planar, and linear orientations. Pauling's main empirical work was with the X-ray analysis of crystals. The valence bond theory of bonding, for which he is primarily responsible, was created to explain what he "saw" in the laboratory. Pauling extended the work of his friend and colleague, Gilbert Lewis, who is famous for creating electron dot structures.

When you studied the valence bond theory, including hybrid orbitals and sigma and pi bonds, in Section 4.2, you became aware of the complexity of that approach. However, it was not until 1957 that Australian Ronald Nyholm and Englishman Ron Gillespie (**Figure 1**) created a much simpler theory for describing, explaining, and predicting the stereochemistry of chemical elements and compounds. The theory that they created is more effective for predicting the shape of molecules.

> **TRY THIS** activity **Electrostatic Repulsion Model**

The electrostatic repulsion of electron pairs around a central atom in a molecule can be modelled using balloons.
Materials: safety glasses, 9 balloons, string

- Blow up the balloons and tie them off.
- Tie two balloons very close together and place them on a table.
  (a) What is the orientation (e.g., angle) of two balloons?
- Repeat with three and then four balloons.
  (b) What is the orientation of three balloons?
  (c) What is the orientation of four balloons?
- Reuse or recycle the balloons as directed.
  (d) What are the pros and cons for using balloons for a physical model of the repulsion of electron pairs about a central atom in a molecule?

The name of the Nyholm-Gillespie theory is the valence-shell-electron-pair-repulsion theory, or **VSEPR** (pronounced "vesper") theory. The theory is based on the electrical repulsion of bonded and unbonded electron pairs in a molecule or polyatomic ion. The number of electron pairs can be counted by adding the number of bonded atoms plus the number of lone pairs of electrons (**Figure 3**). Once the counting is done, we can predict the 3-D distribution about the **central atom** by arranging all pairs of electrons as far apart as possible.

VSEPR Valence **S**hell **E**lectron **P**air **R**epulsion; pairs of electrons in the valence shell of an atom stay as far apart as possible to minimize the repulsion of their negative charges

central atom the atom or atoms in a molecule that has or have the most bonding electrons; form the most bonds

SUMMARY   *VSEPR Theory*

- Only the valence shell electrons of the central atom(s) are important for molecular shape.
- Valence shell electrons are paired or will be paired in a molecule or polyatomic ion.
- Bonded pairs of electrons and lone pairs of electrons are treated approximately equally.
- Valence shell electron pairs repel each other electrostatically.
- The molecular shape is determined by the positions of the electron pairs when they are a maximum distance apart (with the lowest repulsion possible).

Using the VSEPR Theory

What is the shape of the hydrogen compounds of period 2: $BeH_{2(s)}$, $BH_{3(g)}$, $CH_{4(g)}$, $NH_{3(g)}$, $H_2O_{(l)}$, and $HF_{(g)}$?

First, we draw Lewis structures of each of the molecules and then consider the arrangement of all pairs of electrons. *The key idea is that all pairs of electrons repel each other and try to get as far from each other as possible.*

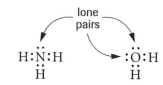

Figure 3
Both the ammonia molecule and the water molecule have four pairs of electrons surrounding the central atom. Some of these are bonding pairs and some are lone pairs.

Lewis structure	Bond pairs	Lone pairs	Total pairs	General formula	Electron pair arrangement	Molecular geometry
H:Be:H	2	0	2	AX_2	linear	H —— Be —— H linear

The Lewis structure indicates that BeH_2 has two bonds and no lone pairs of electrons. The total number of pairs of electrons around the central atom (Be) is two. These electron pairs repel each other. The farthest the electrons can get away from each other is 180°—a *linear* orientation.

Lewis structure	Bond pairs	Lone pairs	Total pairs	General formula	Electron pair arrangement	Molecular geometry
H H:B:H	3	0	3	AX_3	trigonal planar	H | B H H trigonal planar

BH_3 has three bonds, which means three pairs of electrons around the central atom, B. The three pairs of electrons repel one another to form a plane of bonds at 120° to each other. This arrangement or geometry is called *trigonal planar*.

Lewis structure	Bond pairs	Lone pairs	Total pairs	General formula	Electron pair arrangement	Molecular geometry
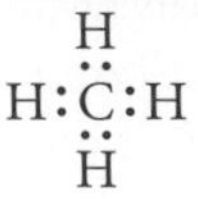 H:C:H (with H above and below)	4	0	4	AX_4	tetrahedral	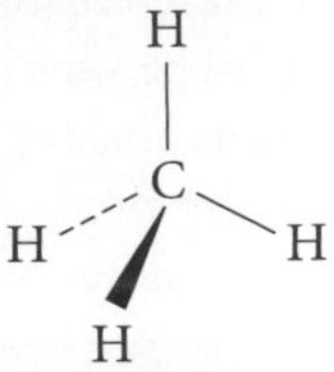 tetrahedral

Lewis theory indicates that CH_4 has four bonds or four pairs of electrons repelling each other around the central atom, C. Experimental work and mathematics both agree that a *tetrahedral* arrangement minimizes the repulsion. Tetrahedral bonds point toward the corners of an equilateral pyramid at an angle of 109.5° to each other.

Lewis structure	Bond pairs	Lone pairs	Total pairs	General formula	Electron pair arrangement	Molecular geometry
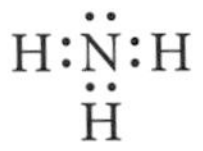 H:N:H (with H below)	3	1	4	AX_3E	tetrahedral	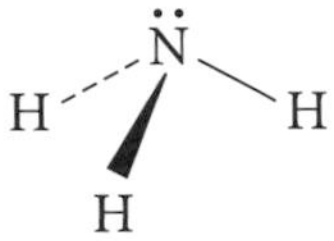 pyramidal

The Lewis structure shows that NH_3 has three bonding pairs (X in general formula) and one lone pair (E in general formula) of electrons. The four groups of electrons should repel each other to form a tetrahedral arrangement of the electron pairs just like methane, CH_4. The molecular geometry is always based on the atoms present and therefore, if we ignore the lone pair, the shape of the ammonia molecule is like a pyramid (called *pyramidal*). We would expect the angle between the atoms, H—N—H to be 109.5°, which is the angle for an ideal pyramid. However, in ammonia, the atoms form a pyramidal arrangement with an angle of 107.3°. This small difference is believed to occur because there is slightly stronger repulsion between the lone pair of electrons and the bonding pairs than between the bonding pairs. This causes the bonding pairs to be pushed closer together.

Lewis structure	Bond pairs	Lone pairs	Total pairs	General formula	Electron pair arrangement	Molecular geometry
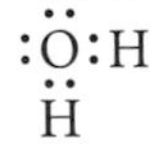 :O:H (with H below)	2	2	4	AX_2E_2	tetrahedral	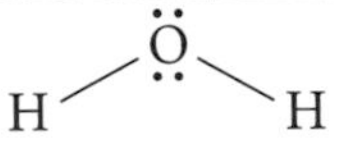 V-shaped

According to the Lewis structure, the water molecule has two bonding pairs and two lone pairs of electrons. The four pairs of electrons repel each other to produce a tetrahedral orientation. The geometry of the water molecule is called *V-shaped* with an angle of 104.5°. Notice that this angle is again less than the ideal angle of 109.5° for a tetrahedral arrangement of electron pairs. The slightly stronger repulsion between the lone pairs of electrons and the lone pairs and the bonding pairs is thought to force the bonding electron pairs closer together.

Lewis structure	Bond pairs	Lone pairs	Total pairs	General formula	Electron pair arrangement	Molecular geometry
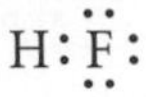 H:F:	1	3	4	AXE_3	tetrahedral	H — F linear

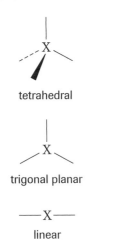

Based upon the Lewis theory of bonding, the hydrogen fluoride molecule has one bonding pair and three lone pairs of electrons. The four electron pairs repel to create a tetrahedral arrangement for the electrons. This has little effect on the geometry of the hydrogen fluoride molecule, which is linear at 180°—as all diatomic molecules are.

SUMMARY   *Shapes of Molecules*

VSEPR theory explains and predicts the geometry of molecules by counting pairs of electrons that repel each other to minimize repulsion. The process for predicting the shape of a molecule is summarized below.

Step 1   Draw the Lewis structure for the molecule, including the electron pairs around the central atom.

Step 2   Count the total number of bonding pairs (bonded atoms) and lone pairs of electrons around the central atom.

Step 3   Refer to **Table 1** or Appendix C3 and use the number of pairs of electrons to predict the shape of the molecule.

Table 1   Using VSEPR Theory to Predict Molecular Shape

General formula*	Bond pairs	Lone pairs	Total pairs	Molecular shape Geometry**	Shape diagram	Examples
AX_2	2	0	2	linear (linear)	X — A — X	CO_2, CS_2
AX_3	3	0	3	trigonal planar (trigonal planar)		BF_3, BH_3
AX_4	4	0	4	tetrahedral (tetrahedral)		CH_4, SiH_4
AX_3E	3	1	4	trigonal pyramidal (tetrahedral)		NH_3, PCl_3
AX_2E_2	2	2	4	V-shaped (tetrahedral)		H_2O, OCl_2
AXE_3	1	3	4	linear (tetrahedral)	A — X	HCl, BrF

*A is the central atom; X is another atom; E is a lone pair of electrons.
**The electron pair arrangement is in parenthesis.

Shape of a Polyatomic Ion

SAMPLE problem

Use the Lewis structure and VSEPR theory to predict the shape of a sulfate ion, $SO_4{}^{2-}$.

Determining the shape of a polyatomic ion is no different than determining the shape of a molecule. Again, you first obtain the Lewis structure of the ion, as shown in Section 4.1. For the sulfate ion, the central sulfur atom is surrounded by four oxygen atoms. There is a total of 32e⁻. An acceptable Lewis structure is shown.

$$6 + 4(6) + 2 = 32e^-$$

DID YOU KNOW ?

Molecular Shape-Shifting
Knowing a molecule's shape is useful, but knowing how the shape changes during a chemical reaction is invaluable. For example, the process by which HIV latches onto its cellular host is believed to depend on a molecular shape change. A new technique of ultrafast X-ray diffraction now allows scientists to observe how a molecule changes shape as it reacts. This technique uses very short X-ray and laser pulses to determine shapes on a time scale of tens of picoseconds.

LEARNING TIP

For more molecular shapes see Appendix C3.

Notice that you have four pairs of electrons around the central sulfur atom. This corresponds to the AX$_4$ category and therefore, the ion has a tetrahedral shape.

$$\left[\begin{array}{c} \overset{O}{\underset{O}{\overset{|}{\underset{|}{S}}}} \end{array} \right]^{2-}$$

SAMPLE problem | Shapes of Molecules with Two Central Atoms

Use the Lewis structure and VSEPR theory to predict the geometry of the B$_2$F$_4$ molecule. Provide your reasoning.

If a molecule has more than one central atom, such as two boron atoms in this example, consider the shape around each atom first, using the same procedure as molecules with one central atom. Then combine these individual shapes to describe or draw the overall geometry of the molecule. As in previous examples, draw the Lewis structure first to determine the number of pairs of electrons.

Three pairs of electrons around a boron atom means an AX$_3$ general case, and hence a trigonal planar arrangement around each central boron atom. This arrangement of electrons creates the minimum repulsion of electron pairs.

Diboron tetrafluoride is composed of molecules with trigonal planar shape around each central boron atom, producing an overall planar molecule as shown to the right.

$$:\overset{..}{F}: \quad :\overset{..}{F}:$$
$$:\overset{..}{F}:\overset{..}{B} \quad : \quad \overset{..}{B}:\overset{..}{F}:$$

▶ Practice

Understanding Concepts

1. Explain how the words that the VSEPR acronym represents communicate the main ideas of this theory.

2. Use VSEPR theory to predict the geometry of a molecule of each of the following substances. Draw a diagram showing the shape of each molecule.
   - (a) BeI$_{2(s)}$
   - (b) PF$_{3(g)}$
   - (c) H$_2$S$_{(g)}$
   - (d) BBr$_{3(g)}$
   - (e) SiBr$_{4(l)}$
   - (f) HCl$_{(g)}$

3. Use VSEPR theory to determine the shape of each of the following polyatomic ions:
   - (a) PO$_4{}^{3-}$
   - (b) IO$_3{}^{-}$

4. Cubane is a hydrocarbon with the formula, C$_8$H$_8$. It has a cubic shape, as its name implies, with a carbon atom at each corner of the cube. This molecule is very unstable and some researchers have been seriously injured when crystals of the compound exploded while being scooped out of a bottle. Not surprisingly, it has some uses as an explosive.
   - (a) According to VSEPR theory, what should be the shape around each carbon atom? Why?
   - (b) If we assume an ideal cubic shape, what would be the bond angles around the carbon?
   - (c) Explain how your answers to (a) and (b) suggest why this molecule is so unstable.

Applying Inquiry Skills

5. Where did the evidence come from that led to the creation of VSEPR theory?

6. Locate two or more VSEPR Web sites and compare them. Which do you prefer? List two or more criteria for evaluating the sites and indicate how each site did based upon each criterion.

 www.science.nelson.com

Making Connections

7. Enzymes make up the largest and most highly specialized class of protein molecules. Describe briefly how their three-dimensional structure influences their function. How does the "lock-and-key" analogy relate to molecular shapes and the highly specific nature of enzyme reactions?

 www.science.nelson.com

8. What are optical isomers? Describe the role that molecular shape plays in classifying optical isomers.

 www.science.nelson.com

Extension

9. The VSEPR theory can be extended to five and six electron pairs to explain several other shapes of molecules, as shown in **Table 2**.
   Several other shapes are possible if one or more of the total number of electron pairs is a lone pair.
   (a) Draw Lewis structures for PCl_5 and SF_6.
   (b) Draw the Lewis structure for ClF_3. If the two lone pairs are in the trigonal plane, predict the molecular shape.
   (c) Draw the Lewis structure for ICl_4^-. If the two lone pairs are above and below the plane of the atoms, predict the molecular shape.

Table 2 Expanded VSEPR Theory to Predict Molecular Shape

General formula	Bond pairs	Lone pairs	Total pairs	Molecular shape		Example
				Geometry	Shape diagram	
AX_5	5	0	5	trigonal bipyramidal (trigonal bipyramidal)		PCl_5
AX_6	6	0	6	octahedral (octahedral)		SF_6

Multiple Bonding in VSEPR Models

Evidence of multiple bonding can be obtained from, for example, the reaction rate of hydrocarbons; e.g., the fast reaction of alkenes and alkynes with bromine or potassium permanganate, compared to the slower reaction rate with alkanes. Further evidence indicates that the multiple bonds are shorter and stronger than single bonds between the same kind of atoms. Evidence from crystallography (e.g., the X-ray analysis of crystals) indicates that these multiple bonds can be treated like single bonds for describing, explaining, and predicting the shape of a molecule. This has implications for using VSEPR theory for molecules containing multiple bonds. Let's look at some examples.

SAMPLE problem — VSEPR and Double Covalent Bonds

Ethene (ethylene, $C_2H_{4(g)}$) is the simplest hydrocarbon with multiple bonding. Crystallography indicates that the orientation around the central carbon atoms is trigonal planar. Is VSEPR theory able to explain the empirically determined shape of this molecule?

The first step in testing the ability of VSEPR theory to explain the shape of ethene is to draw a Lewis structure of the molecule.

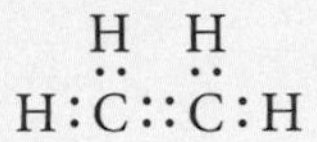

The second step is to count the number of "pairs" of electrons around the central atoms (the carbon atoms). In the case of multiple bonding such as a double bond, the double bond contains two pairs of electrons. The crystallographic evidence indicates a trigonal planar arrangement. The only way that VSEPR theory can accommodate this evidence is to count a multiple bond as a single group of electrons. In other words, you are counting the number of bonded atoms. There are three bonded atoms or sets of bonding electrons around each of the central carbon atoms (AX_3). These electron groups repel each other to obtain minimum repulsion and, thus, a minimum energy state. The result, according to VSEPR theory, is a trigonal planar orientation—three atoms on a plane at 120°.

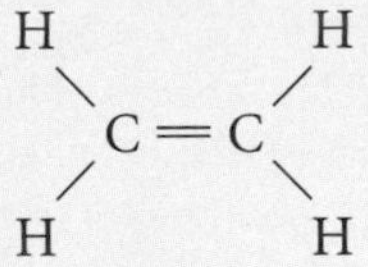

VSEPR theory passes the test by being able to explain the trigonal planar shape of ethene. Now let's see if VSEPR theory can pass another test by predicting the stereochemistry of the ethyne molecule.

DID YOU KNOW?

Smelling Molecular Shapes

Limonene, $C_{10}H_{16}$, is a naturally occurring compound that can exist in two different isomeric forms. One form is present in orange rinds and has a pleasant orange odour. The other isomer is identical in composition but different in shape. This isomer smells like pine or turpentine and is found in lemons and pine needles.

Example 1

Predict the shape and draw the diagram of the ethyne (acetylene, $C_2H_{2(g)}$) molecule.

Solution

H:C:::C:H

H — C ≡ C — H

The shape of the ethyne molecule is linear.

LEARNING TIP

Treat double and triple covalent bonds as one group of electrons when using VSEPR theory to predict shapes of molecules containing these bonds.

Example 2

Predict the shape and draw the diagram for a nitrite ion, NO_2^-.

Solution

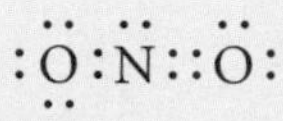

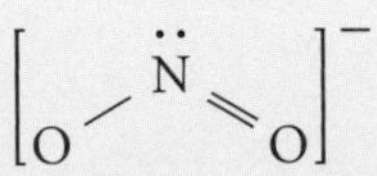

The shape of the nitrite ion is V-shaped.

ACTIVITY 4.3.1

Shapes of Molecules (p. 277)

Molecular models are important tools for understanding molecular shapes.

▶ *Practice*

Understanding Concepts

10. In order to make the rules of VSEPR theory work, how must multiple (double and triple) bonds be treated?

11. Use Lewis structures and VSEPR theory to predict the shapes of the following molecules:
    (a) $CO_{2(g)}$, carbon dioxide (dry ice)
    (b) $HCN_{(g)}$, hydrogen cyanide (odour of bitter almonds)
    (c) $C_3H_{6(g)}$, propene (monomer for polypropylene)
    (d) $C_3H_{4(g)}$, propyne
    (e) $H_2CO_{(g)}$, methanal (formaldehyde)
    (f) $CO_{(g)}$, carbon monoxide (deadly gas)

Applying Inquiry Skills

12. Is VSEPR a successful scientific theory? Defend your answer.

Making Connections

13. Astronomers have detected an amazing variety of molecules in interstellar space.
    (a) One interesting molecule is cyanodiacetylene, HC_5N. Draw a structural diagram for this molecule and predict its shape.
    (b) How do astronomers detect molecules in space?

DID YOU KNOW ?

Theories
Just as scientists have special definitions for words that apply in the context of science (e.g., the definition of "work" in physics), philosophers who study the nature of chemistry have special definitions for terms such as "theory." To them a theory is not a hypothesis. To philosophers of chemistry a theory uses the unobservable (such as electrons and bonds) to explain observables (such as chemical and physical properties).

As a student you have probably found words that are used in different contexts in and out of the science classroom, such as "salt" and "decomposition." What are some other words that you use differently in other contexts?

Molecular Geometry Research

A Canadian researcher doing important work in molecular geometry is Dr. Richard Bader (**Figure 4**), Professor Emeritus at McMaster University. Dr. Bader's work includes the theoretical determination of electron density maps for small molecules, giving a visible interpretation of the molecular shapes and bonding within molecules (**Figure 5**). Note that the shape shown is consistent with the type of structure you have previously used to represent ethene. Viewed this way, single and double bonds are not really single or double structures—they are just different concentrations of electron density. Similarly, the unique carbon–carbon bonds in benzene are part of a total molecular electron orbital structure, resulting in a particular electron density in the region around the ring.

Dr. Bader's work builds on previous bonding theories that have a critical limitation. "We have in chemistry an understanding based on a classification scheme that is both powerful and at the same time, because of its empirical nature, limited." Dr. Bader applies quantum mechanics theory to determine the atomic structures of molecules and crystals. To the extent that this theory is supported by empirical evidence, it allows development of new theory, and may eventually lead to the ability to use computer models to accurately explain and predict forces, structures, and properties that currently can only be observed and measured.

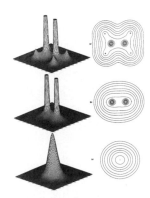

Figure 4
Dr. Richard Bader

Figure 5
A representation of the electron density for an ethene molecule.

Take a Stand: Linus Pauling and the Vitamin C Controversy

Linus Pauling became interested in chemistry at a young age because a friend had a chemistry set. Pauling graduated from Oregon State University in 1922 and obtained his Ph.D. from the California Institute of Technology in 1925. After a year in Europe studying with Sommerfeld, he became a chemistry professor at the California Institute of Technology in 1927 and remained there throughout his academic career.

Pauling's scientific fame came from his theory of chemical bonding, including the ideas of a shared pair of electrons, polar covalent bonds, electronegativity, and resonance structures. These ideas revolutionized thinking about molecular structure. For this work Pauling was awarded the Nobel Prize in chemistry in 1954. He continued his study of molecular structure and was one of the first to suggest helical structures of proteins and the relationship between disease and abnormal molecular structure.

After the Second World War, Pauling used his fame as a Nobel Prize winner to vigorously fight the nuclear arms race of the United States and the Soviet Union. For his outspoken leadership against nuclear testing he was awarded the Nobel Peace Prize in 1962, becoming one of a very few people who have won two Nobel Prizes.

Pauling's fame as a scientist and as a social activist meant that he could easily command media attention whenever he spoke. When he announced in 1970 that large doses (mega-doses) of vitamin C could prevent the common cold, and other illnesses as well, many people paid close attention. However, not everyone agreed that vitamin C is as useful in megadoses as Pauling claimed. There is still a huge interest in Pauling's suggestion, in spite of no clear scientific evidence supporting his claim and some scientific studies that dispute it.

(a) Briefly describe some claims being made today for the beneficial use of large doses of vitamin C.

(b) Briefly describe some objections and criticisms of the claimed benefits.

In small groups, discuss the following questions and obtain a consensus within the group. Report on your conclusions.

(c) Pauling and other proponents of the benefits of mega-doses of vitamin C, and the doctors and scientists opposed to this view have all claimed that science is on their side. Anyone can have an opinion or belief, but science requires more. What are the requirements for a claim to be scientifically valid? List some criteria.

(d) To what extent do you think Pauling's fame influenced public and scientific opinion about the benefits of vitamin C? Suppose someone with no scientific training and unknown to the public made this claim, would anyone notice or consider it seriously? If a Nobel Prize winner makes a claim disputed by other lesser-known scientists, how do we decide what to believe?

(e) The vitamin C controversy is not the first time a famous scientist has made a claim that is disputed by most of the scientific community. What are the repercussions for a scientist who goes against the rest of the scientific community? Who usually "wins"? Is the practice and work of science completely objective?

 www.science.nelson.com

▶ Section 4.3 Questions

Understanding Concepts

1. Use Lewis structures and VSEPR theory to predict the molecular shape of the following molecules. Include a 3-D representation of each molecule.
   (a) $H_2S_{(g)}$, hydrogen sulfide (poisonous gas)
   (b) $BBr_{3(l)}$, boron tribromide (density of 2.7 g/mL)
   (c) $PCl_{3(l)}$, phosphorus trichloride
   (d) $SiBr_{4(l)}$, silicon tetrabromide
   (e) $BeI_{2(s)}$, beryllium iodide (soluble in $CS_{2(l)}$)

2. Use Lewis structures and VSEPR theory to predict the molecular shape around the central atom(s) of each of the following molecules. Provide a 3-D representation of each molecule.
   (a) $CS_{2(l)}$, carbon disulfide (solvent)
   (b) $HCOOH_{(g)}$, acetic acid (vinegar)
   (c) $N_2H_{4(l)}$, hydrazine (toxic; explosive)
   (d) $H_2O_{2(l)}$, hydrogen peroxide (disinfectant)
   (e) $CH_3CCCH_{3(l)}$, 2-butyne (reacts rapidly with bromine)

3. Draw the Lewis structure and describe the shape of each of the following ions:
   (a) IO_4^- (b) SO_3^{2-} (c) ClO_2^-

Making Connections

4. Briefly describe Dr. Bader's contribution to our understanding of molecules.

5. Search the Internet for information on the current workplace and position of Dr. Ronald Gillespie, the co-creator of VSEPR theory. What degrees does he hold? What are some of the awards that he has won? What is his major topic of research?

6. Some scientists argue that taste has developed as a protective mechanism. Many poisonous molecules taste bitter and ones that are useful to us have a more pleasant, often sweet, taste. Write a brief summary about the relation of taste to molecular structure.

 www.science.nelson.com

Chemists believe that molecules are made up of charged particles (electrons and nuclei). A polar molecule is one in which the charge is not distributed symmetrically among the atoms making up the molecule. The existence of polar molecules can be demonstrated by running a stream of any liquid past a charged rod (**Figure 1**). When repeated with a large number of pure liquids, this experiment produces a set of empirical rules for predicting whether a molecule is polar or not (**Table 1**).

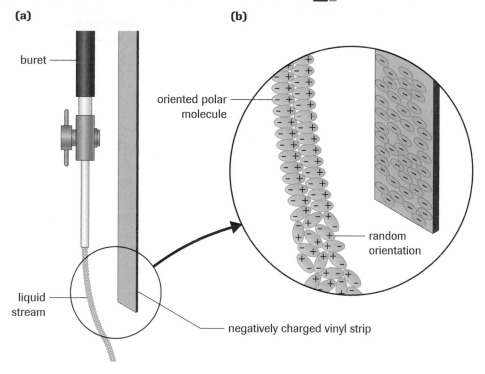

(a)

(b)

buret

oriented polar molecule

random orientation

liquid stream

negatively charged vinyl strip

INVESTIGATION 4.4.1

Testing for Polar Molecules (p. 277)
Test the rules for polar and nonpolar molecules presented in **Table 1** using the effect of electric charges on a liquid.

Figure 1
(a) Testing a liquid with a charged strip provides evidence for the existence of polar molecules in a substance.
(b) In a liquid, molecules are able to move in a limited way. Polar molecules in a liquid become oriented so that their positive poles are closer to a negatively charged material. Near a positively charged material they become oriented in the opposite direction. Polar molecules are thus attracted by either kind of charge.

Table 1   Empirical Rules for Polar and Nonpolar Molecules

	Type	Description of molecule	Examples
Polar	AB	diatomic with different atoms	$HCl_{(g)}$, $CO_{(g)}$
	N_xA_y	containing nitrogen and other atoms	$NH_{3(g)}$, $NF_{3(g)}$
	O_xA_y	containing oxygen and other atoms	$H_2O_{(l)}$, $OCl_{2(g)}$
	$C_xA_yB_z$	containing carbon and two other kinds of atoms	$CHCl_{3(l)}$, $C_2H_5OH_{(l)}$
Nonpolar	A_x	all elements	$Cl_{2(g)}$, $N_{2(g)}$
	C_xA_y	containing carbon and only one other kind of atom	$CO_{2(g)}$, $CH_{4(g)}$

DID YOU KNOW ?

Evidence for Polar Bonds
The energy required to break a bond can be determined experimentally. The energy required to break the H−F bond is considerably greater than either H−H or F−F bonds. Pauling realized that an unequal sharing of the electron pair produced a polar bond that enhanced the bonding between the atoms.

Electronegativity and Polarity of Bonds

Linus Pauling realized that while two atoms can share one or more pairs of electrons, there is nothing that requires them to share those electron pairs equally. He saw the need for a theory to explain and predict the polarity of molecules and so combined properties such as bond energies with valence bond theory to create a new property of atoms called

polar bond a polar bond results from a difference in electronegativity between the bonding atoms; one end of the bond is, at least partially, positive and the other end is equally negative

Figure 2
Electronegativity of the elements increases as you move up the periodic table and to the right. Fluorine has the highest electronegativity of all atoms.

nonpolar bond a nonpolar bond results from a zero difference in electronegativity between the bonded atoms; a covalent bond with equal sharing of bonding electrons

ionic bond a bond in which the bonding pair of electrons is mostly with one atom/ion

Figure 3
Since the transfer of an electron in a polar covalent bond is not complete, only a partially negative ($\delta-$) and a partially positive ($\delta+$) charge appear on the bond.

covalent bond or **nonpolar bond** a bond in which the bonding electrons are shared equally between atoms

polar covalent bond a bond in which electrons are shared somewhat unequally

electronegativity. Because Pauling's calculations produced differences in the attraction of nuclei for a shared pair of electrons in a covalent bond, he arbitrarily assigned values so that fluorine, the most electronegative atom, had a value of about 4.0. All other atoms have a lower electronegativity and, therefore, a lower attraction for electrons when bonded. Electronegativity increases when moving to the right and up the periodic table toward fluorine (**Figure 2**).

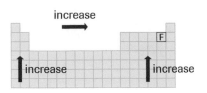

Pauling used the difference in electronegativity (i.e., the difference in attraction for the pair of electrons in a bond) to explain the polarity of a chemical bond. The greater the difference in electronegativity, the more polar the bond. The smaller the difference, the more non-polar the bond. A very **polar bond** is an **ionic bond**. A **nonpolar bond** is a **covalent bond**. A somewhat polar bond is a **polar covalent bond**.

According to Pauling, a polar covalent bond (or simply a polar bond) results when two different kinds of atoms (usually nonmetals) form a bond. The bond is covalent because the electrons are being shared. The bond is polar covalent because the sharing of electrons is unequal. This means that the *electrons spend more of their time closer to one atomic nucleus than the other*. The end of the bond where the negatively charged electrons spend more time is labelled as being partially negative ($\delta-$). The end of the bond that is partially positive is labelled $\delta+$ (**Figure 3**).

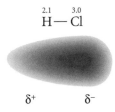

Pauling liked to think of chemical bonds as being different in degree rather than different in kind. According to him, all chemical bonds involved a sharing of electrons, even ionic bonds (**Figure 4**). The degree of sharing depends upon the difference in electronegativities of the bonded atoms.

A general rule is that when the difference in electronegativity exceeds 1.7, the percent ionic character exceeds 50%.

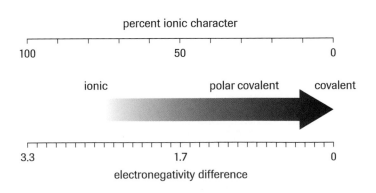

Figure 4
The greater the difference in electronegativity between bonding atoms, the greater the percent ionic character of the bond.

Classifying Bonds

Label the following atoms and bonds with electronegativity and bond polarity, and classify the bond:

(a) H–H   **(b) P–Cl**   **(c) Na–Br**

From the periodic table, assign each atom an electronegativity.

(a) H – H
     2.1   2.1
The electronegativity difference is 0.0, indicating a nonpolar covalent bond.

(b) $\delta+$ $\delta-$
      P – Cl
     2.1   3.0
The electronegativity difference is 0.9, indicating a polar covalent bond.

(c)   +   –
      Na – Br
      0.9   2.8
The electronegativity difference is 1.9, indicating an ionic bond.

DID YOU KNOW ?

Poles and Polar

In general, the term "pole" refers to one or the other of two opposite ends of something; e.g., North and South Poles of Earth or a magnet, or the positive and negative charges on two ends of an object such as a molecule. When we say something is polar we mean it has two opposite ends.

▶ Practice

Understanding Concepts

1. Draw the following bonds, label the electronegativities, and label the charges (if any) on the ends of the bond. Classify the bond as ionic, polar covalent, or nonpolar covalent:
   (a) H–Cl
   (b) C–H
   (c) N–O
   (d) I–Br
   (e) Mg–S
   (f) P–H

2. Using electronegativity as a guide, classify the following bonds as ionic, polar covalent, or nonpolar covalent:
   (a) the bond in $HBr_{(g)}$
   (b) the bond in $LiF_{(s)}$
   (c) the C–C bond in propane, $C_3H_{8(g)}$

3. List and order the bonds in the following substances according to increasing bond polarity. Provide your reasoning.
   (a) $H_2O_{(l)}$, $H_{2(g)}$, $CH_{4(g)}$, $HF_{(g)}$, $NH_{3(g)}$, $LiH_{(s)}$, $BeH_{2(s)}$
   (b) $PCl_{3(l)}$, $LiI_{(s)}$, $I_{2(s)}$, $ICl_{(s)}$, $RbF_{(s)}$, $AlCl_{3(s)}$
   (c) $CH_3OH_{(l)}$
   (d) $CHFCl_{2(g)}$

Applying Inquiry Skills

4. Values for the Pauling electronegativities have changed over time. Why would these values change? Are the new values the "true" values?

Extension

5. There are other electronegativity scales created by different chemists. Research and compare the scales created by Pauling, Mulliken, and Allred-Rochow. For example, what properties did they use when calculating their electronegativity scales?

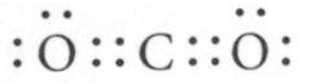

$$\ddot{\mathrm{O}} :: \mathrm{C} :: \ddot{\mathrm{O}}$$

Figure 5
The central carbon atom has no lone pairs and two groups or sets of electrons (remember that multiple bonds count as one group of electrons). The least repulsion of two groups of electrons is a linear arrangement.

bond dipole the electronegativity difference of two bonded atoms represented by an arrow pointing from the lower ($\delta+$) to the higher ($\delta-$) electronegativity

nonpolar molecule a molecule that has either nonpolar bonds or polar bonds whose bond dipoles cancel to zero

polar molecule a molecule that has polar bonds with dipoles that do not cancel to zero

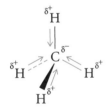

Figure 6
Notice how all of the bond dipoles point into the central carbon atom. There are no positive and negative ends on the outer part of the methane molecule.

Polar Molecules

The existence of polar bonds in a molecule does not necessarily mean that you have a polar molecule. For example, carbon dioxide is considered to be a nonpolar molecule, although each of the C=O bonds is a polar bond. To resolve this apparent contradiction, we need to look at this molecule more closely. Based on the Lewis structure and the rules of VSEPR, carbon dioxide is a linear molecule (**Figure 5**). Using electronegativities, we can predict the polarity of each of the bonds. It is customary to show the bond polarity as an arrow, pointing from the positive ($\delta+$) to the negative ($\delta-$) end of the bond. This arrow represents the **bond dipole**.

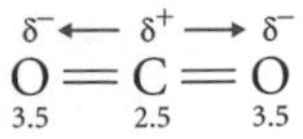

$$\underset{3.5}{\mathrm{O}} \overset{\delta^-}{=} \underset{2.5}{\mathrm{C}} \overset{\delta^+}{=} \underset{3.5}{\mathrm{O}} \overset{\delta^-}{}$$

These arrows are vectors and when added together produce a zero total. In other words, the bond dipoles cancel to produce no polarity for the complete molecule, or a **nonpolar molecule.**

Let's try this procedure again with another small molecule. As you know, water is a polar substance and the O−H bonds in water are polar bonds. The Lewis structure and VSEPR rules predict a V-shaped molecule, shown here with its bond dipoles.

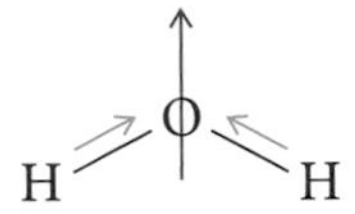

In this case, the bond dipoles (vectors) do not cancel. Instead, they add together to produce a non-zero molecular dipole (shown in red). The water molecule has an overall polarity and that it is why it is a **polar molecule**. Note that the water molecule has a partially negative end near the oxygen atom and a partially positive end at the two hydrogen atoms. Although the "ends" of the water molecule are not initially obvious, the V shape produces two oppositely charged regions on the outside of the molecule. This explains why a stream of water is attracted to a positively charged strip or rod.

From the two examples, carbon dioxide and water, you can see that the shape of the molecule is as important as the bond polarity.

> Both the shape of the molecule and the polarity of the bonds are necessary to determine if a molecule is polar or nonpolar.

Methane is a nonpolar substance and its C−H bonds are polar. Does the same explanation we used for carbon dioxide apply to methane? The shape diagram with bond dipoles (**Figure 6**) shows that the outer part of the molecule is uniformly positive and therefore the molecule has no ends that are charged differently. A nearby molecule would "see" the same charge from all sides of the methane molecule. This is true because the CH_4 molecule is symmetrical. In fact, all symmetrical molecules, such as CCl_4 and BF_3, are nonpolar for the same reason.

> In all symmetrical molecules, the sum of the bond dipoles is zero and the molecule is nonpolar.

The theory created by combining the concepts of covalent bonds, electronegativity, bond polarity, and VSEPR logically and consistently explains the polar or nonpolar nature of molecules. We are now ready to put this combination of concepts to a further test—to predict the polarity of a molecule.

Predicting the Polarity of a Molecule

Predict the polarity of the ammonia, NH_3, molecule, including your reasoning.

First, draw the Lewis structure.

$$H\!:\!\overset{\cdot\cdot}{\underset{H}{N}}\!:\!H$$

Based on the Lewis structure, draw the shape diagram.

Add the electronegativities of the atoms, from the periodic table, and assign $\delta+$ and $\delta-$ to the bonds.

Draw in the bond dipoles.

The ammonia molecule is polar because it has polar bonds that do not cancel to zero. The electron pairs are in a tetrahedral arrangement, but one of these pairs is a lone pair and three are bonding pairs. Therefore, the bond dipoles do not cancel.

SUMMARY | *Theoretical Prediction of Molecular Polarity*

To use molecular shape and bond polarity to determine the polarity of a molecule, complete these steps.

Step 1 Draw a Lewis structure for the molecule.

Step 2 Use the number of electron pairs and VSEPR rules to determine the shape around each central atom.

Step 3 Use electronegativities to determine the polarity of each bond.

Step 4 Add the bond dipole vectors to determine if the final result is zero (nonpolar molecule) or nonzero (polar molecule).

▶ Practice

Understanding Concepts

6. Predict the shape of the following molecules. Provide Lewis and shape structures.
   (a) silicon tetrabromide, $SiBr_{4(l)}$
   (b) nitrogen trichloride, $NCl_{3(l)}$
   (c) beryllium fluoride, $BeF_{2(s)}$
   (d) sulfur dichloride, $SCl_{2(l)}$

7. Predict the bond polarity for the following bonds. Use a diagram that includes the partial negative and positive charges and direction of the bond dipole.
   (a) $C\equiv N$ in hydrogen cyanide
   (b) $N=O$ in nitrogen dioxide
   (c) $P-S$ in $P(SCN)_{3(s)}$
   (d) $C-C$ in $C_8H_{18(l)}$

8. Predict the polarity of the following molecules. Include a shape diagram, bond dipoles, and the final resultant dipole (if nonzero) of the molecule.
   (a) boron trifluoride, $BF_{3(g)}$
   (b) oxygen difluoride, $OF_{2(g)}$
   (c) carbon tetraiodide, $CI_{4(s)}$
   (d) phosphorus trichloride, $PCl_{3(l)}$

9. Use the empirical rules from **Table 1** to predict the polarity of an octane, $C_8H_{18(l)}$, molecule. Explain your answer without drawing the molecule.

10. Why is $N_2H_{4(l)}$ nonpolar?

Applying Inquiry Skills

11. Predict the polarity of hydrogen sulfide, $H_2S_{(g)}$, a toxic gas with a rotten-egg odour. Design an experiment to test your prediction.

▶ *Section 4.4* *Questions*

Understanding Concepts

1. Scientific concepts are tested by their ability to explain current observations and predict future observations. To this end, explain why the following molecules are polar or nonpolar, as indicated by the results of the diagnostic tests.
   (a) beryllium bromide, $BeBr_{2(s)}$; nonpolar
   (b) nitrogen trifluoride, $NF_{3(g)}$; polar
   (c) methanol, $CH_3OH_{(l)}$; polar
   (d) hydrogen peroxide, $H_2O_{2(l)}$; nonpolar
   (e) ethylene glycol, $C_2H_4(OH)_{2(l)}$; nonpolar

2. Predict the polarity of the following molecules. Include shape diagrams and bond dipoles in your reasoning for your prediction.
   (a) dichlorofluoroethane, $CHFCl_{2(g)}$; a refrigerant (a CFC)
   (b) ethene, $C_2H_{4(g)}$; monomer of polyethylene
   (c) chloroethane, $C_2H_5Cl_{(g)}$
   (d) methylamine, $CH_3NH_{2(g)}$
   (e) ethanol, $C_2H_5OH_{(l)}$; beverage alcohol
   (f) diboron tetrafluoride, $B_2F_{4(g)}$

3. Polar substances are used in a capacitor—a device for storing electrical energy. For example, a capacitor may store enough electrical energy to allow you to change the battery in your calculator without losing what you have stored in the memory.

   (a) Based upon polarity alone, which of water or pentane, $C_5H_{12(l)}$, is a good candidate for use in a capacitor? Provide your reasoning.
   (b) What are some other considerations for choosing the liquid inside a capacitor?

Applying Inquiry Skills

4. Geometric isomers are substances with the same molecular formula but a different molecular geometry. One type that you have seen is the *cis* (same side) and *trans* (diagonally opposite) forms of a substituted alkene; for example, *cis*-1,2-dichloroethene and *trans*-1,2-dichloroethene, with the same formula, $C_2H_2Cl_{2(l)}$.

   Predict the polarity of each molecule, including your reasoning. Design an experiment to distinguish these two isomers.

Making Connections

5. Various consumer products and books exist to help people remove stains from clothing, carpets, etc. Discuss how a knowledge of polar and nonpolar substances is related to the removal of stains.

There are many physical properties that demonstrate the existence of **intermolecular forces**. For example, you would not want a raincoat made from untreated cotton. Cotton is very good at absorbing water and easily gets wet. The molecules that make up cotton can form many intermolecular attractions with water molecules. On the other hand, rubber or plastic materials do not absorb water because there is little intermolecular attraction between water molecules and the molecules of the rubber or plastic. A simple property like "wetting" depends to a large extent on intermolecular forces (**Figure 1**). Rubber or plastic may not be desirable materials to wear, but you could still use cotton if it is treated with a water repellent—a coating that has little attraction to water molecules. The development of water repellents requires a good knowledge of intermolecular forces.

Some bugs, like water spiders, walk on water and trees move water up large distances from the ground to the tops of the trees (**Figure 2**). Surface tension and capillary action are directly related to intermolecular attractions between molecules. In this section, we will look at these and other properties in terms of various intermolecular forces.

In 1873 Johannes van der Waals suggested that the deviations from the ideal gas law arose because the molecules of a gas have a small but definite volume and the molecules exert forces on each other. These forces are often simply referred to as van der Waals forces. It is now known that in many substances van der Waals forces are actually a combination of many types of intermolecular forces including, for example, *dipole–dipole forces* and *London forces*. Later, the concept of *hydrogen bonding* was created to explain anomalous (unexpected) properties of certain liquids and solids. In general, intermolecular forces are considerably weaker than the covalent bonds inside a molecule.

> Intermolecular forces are much weaker than covalent bonds. As an approximate comparison, if covalent bonds are assigned a strength of about 100, then intermolecular forces are generally 0.001 to 15.

The evidence for this comparison comes primarily from experiments that measure bond energies. For example, it takes much less energy to boil water (breaking intermolecular bonds) than it does to decompose water (breaking covalent bonds).

$$H_2O_{(l)} \rightarrow H_2O_{(g)} \qquad\qquad 41 \text{ kJ/mol}$$
$$H_2O_{(l)} \rightarrow H_{2(g)} + \frac{1}{2}O_{2(g)} \qquad 242 \text{ kJ/mol}$$

Dipole–Dipole Force

In the last section, you learned how to test a stream of liquid to see whether the molecules of the liquid are polar. You also learned how to predict whether a molecule was polar or nonpolar. (Recall that polar molecules have dipoles—oppositely charged ends.) Attraction between dipoles is called the **dipole–dipole force** and is thought to be due to a simultaneous attraction between a dipole and surrounding dipoles (**Figure 3**).

intermolecular force the force of attraction and repulsion between molecules

Figure 1
Cotton (left) will absorb a lot more water than polyester, because the molecules in cotton are better able to attract and hold water molecules.

Figure 2
Water is transported in thin, hollow tubes in the trunk and branches of a tree. This is accomplished by several processes, including capillary action. It is also crucial that water molecules attract each other to maintain a continuous column of water.

DID YOU KNOW ?

Other van der Waals Forces
Scientists today recognize that there are other varieties of van der Waals forces, such as dipole-induced dipole (part of a group of forces called induction forces). Dipole–dipole forces are now considered to be a special case of multipolar forces that also include quadrupole (4 pole) effects.

dipole–dipole force a force of attraction between polar molecules

(a)

(b)

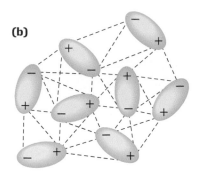

attraction - - - - - - -
repulsion - - - - - - -

Figure 3
(a) Oppositely charged ends of polar molecules attract.
(b) In a liquid, polar molecules can move and rotate to maximize attractions and minimize repulsions. The net effect is a simultaneous attraction of dipoles.

London force the simultaneous attraction of an electron by nuclei within a molecule and by nuclei in adjacent molecules

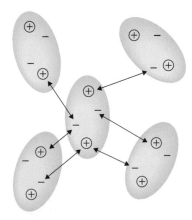

Figure 4
London force is an intermolecular attraction between all molecules. In this figure, only the attractions are shown.

In the past you studied the effect of molecular polarity on solubility. The empirical generalization from that study was that "like dissolves like"; i.e., polar solutes dissolve in polar solvents and nonpolar solutes dissolve in nonpolar solvents. The most important polar solvent is, of course, water. Now we continue the study of intermolecular forces and their effect on several other physical properties of substances, such as boiling point, rate of evaporation, and surface tension. We will interpret these properties using intermolecular forces, but there are other factors that also affect these properties. To minimize this problem, we will try to compare simple, similar substances and do only qualitative comparisons. Before we try to tackle this problem we need to continue our look at kinds of intermolecular forces.

London Force

After repeated failures to find any pattern in physical properties like the boiling points of polar substances, Fritz London suggested that the van der Waals force was actually two forces—the dipole–dipole force and what we now call the **London force**. It was natural that London would describe the force of attraction between molecules (the intermolecular force) in the same way that he described the force of attraction within molecules (the intramolecular force). In both cases, London considered the electrostatic forces between protons and electrons. The difference is that for intramolecular forces the protons and electrons are in the same molecule, while for intermolecular forces the protons and electrons are in different molecules. London's analysis was actually more complicated than this, but for our purposes we only need one basic idea—an atomic nucleus not only attracts the electrons in its own molecule but also those in neighbouring molecules (**Figure 4**). London force is also called dispersion force or London dispersion force. This intermolecular force exists between all molecules. Logically, the strength of the London force depends on the number of electrons (and protons) in a molecule. The greater the number of electrons to be attracted to neighbouring nuclei, the stronger the resulting London force should be.

Using Dipole–Dipole and London Forces to Predict Boiling Points

Let's take a look at the boiling points of group 14 hydrogen compounds (**Table 1**). We would expect these molecules to be nonpolar, based on their four equivalent bonds and their tetrahedral shape.

In **Table 1**, you can see that as the number of electrons in the molecule increases (from 10 to 54), the boiling point increases (from −164°C to −52°C). The evidence presented in Table 1 supports the London theory, and provides a generalization for explaining and predicting the relative strength of London forces among molecules.

Table 1 The Boiling Points of Group 14 Hydrogen Compounds

Compound (at SATP)	Electrons	Boiling point (°C)
$CH_{4(g)}$	10	−164
$SiH_{4(g)}$	18	−112
$GeH_{4(g)}$	36	−89
$SnH_{4(g)}$	54	−52

Predicting Boiling Points

1. **Use London force theory to predict which of these alkanes has the highest boiling point—methane (CH_4), ethane (C_2H_6), propane (C_3H_8), or butane (C_4H_{10}).**

According to intermolecular-force theory, butane should have the highest boiling point. The reasoning behind this prediction is that all of these molecules are nonpolar, but butane has the most attractive London force, because it has the greatest number of electrons in its molecules.

This prediction is borne out by the evidence (**Table 2**).

Molecules with the same number of electrons (called **isoelectronic**) are predicted to have the same or nearly the same strengths for the London force of intermolecular attraction. Isoelectronic molecules help us to study intermolecular forces. For example, if one of two isoelectronic substances is polar and the other is nonpolar, then the polar molecule should have a higher boiling point, as shown in the following problem.

2. **Consider the two isoelectronic substances, bromine (Br_2) and iodine monochloride (ICl). Based upon your knowledge of intermolecular forces, explain the difference in their boiling points (bromine, 59°C; iodine monochloride 97°C).**

Both bromine and iodine monochloride have 70 electrons per molecule (**Table 3**). Therefore, the strength of the London forces between molecules of each should be the same. Bromine is nonpolar and therefore has only London forces between its molecules. Iodine monochloride is polar, which means it has an extra dipole−dipole force between its molecules, in addition to London forces. This extra attraction among ICl molecules produces a higher boiling point.

Table 2 Boiling Points of Alkanes

Alkane	Boiling point (°C)
methane	−162
ethane	−89
propane	−42
butane	−0.5

isoelectronic having the same number of electrons per atom, ion, or molecule

Table 3 Isoelectronic Substances

Substance	Electrons	Boiling point (°C)
$Br_{2(l)}$	70	59
$ICl_{(l)}$	70	97

This is the same London who worked with Sommerfeld and who Pauling indicated had made "the greatest single contribution to the chemist's conception of valence" since Lewis created the concept of the shared pair of electrons. The use of quantum mechanics to describe the covalent bond in the hydrogen molecule is the contribution referred to by Pauling.

SUMMARY | Predicting with Dipole–Dipole and London Forces

- Isoelectronic molecules have approximately the same strength of the London force.
- If all other factors are equal, then
  - the more polar the molecule, the stronger the dipole−dipole force and therefore, the higher the boiling point.
  - the greater the number of electrons per molecule, the stronger the London force and therefore, the higher the boiling point.
- You can explain and predict the relative boiling points of two substances if:
  - the London force is the same, but the dipole-dipole force is different.

— the dipole—dipole force is the same, but the London force is different.

— the influence of both the London force and the dipole—dipole force are in the same direction; e.g., both are tending to increase the boiling point of one of the chemicals.

- You cannot explain and predict with any certainty the relative boiling points of two chemicals if:

— one of the substances has a stronger dipole—dipole force and the other substance has a stronger London force.

> ### Practice

Understanding Concepts

1. Using London forces and dipole—dipole forces, state the kind of intermolecular force(s) present between molecules of the following substances:
   (a) water (solvent)
   (b) carbon dioxide (dry ice)
   (c) ethane (in natural gas)
   (d) ethanol (beverage alcohol)
   (e) ammonia (cleaning agent)
   (f) iodine (disinfectant)

2. Which of the following pure substances has stronger dipole—dipole forces than the other? Provide your reasoning.
   (a) hydrogen chloride or hydrogen fluoride
   (b) chloromethane or iodomethane
   (c) nitrogen tribromide or ammonia
   (d) water or hydrogen sulfide

3. Based upon London force theory, which of the following pure substances has the stronger London forces? Provide your reasoning.
   (a) methane or ethane
   (b) oxygen or nitrogen
   (c) sulfur dioxide or nitrogen dioxide
   (d) methane or ammonia

4. Based upon dipole—dipole and London forces, predict which substance in the following pairs has the higher boiling point. Provide your reasoning.
   (a) beryllium difluoride or oxygen difluoride
   (b) chloromethane or ethane

5. Why is it difficult to predict whether NF_3 or Cl_2O has the higher boiling point?

Applying Inquiry Skills

6. A common method in science is to gather or obtain experimental information and look for patterns. This is common when an area of study is relatively new and few generalizations exist. Analyze the information in **Table 4** to produce some possible patterns and then interpret as many patterns as possible using your knowledge of molecules and intermolecular forces.

7. Write an experimental design to test the ability of the theory and rules for the dipole—dipole force or the London force to predict the trend in melting points of several related substances. What are some possible complications with this proposed experiment?

Extension

8. Using a chemical reference, look up the boiling points for the substances in questions 4 and 5. Evaluate your predictions.

⟋⟋ LAB EXERCISE 4.5.1

Boiling Points and Intermolecular Forces (p. 278)
This lab exercise shows some successes and some failures of the London force and dipole—dipole theories to predict and explain boiling points.

Table 4 Boiling Points of Hydrocarbons

Compound	Formula	Boiling point (°C)
ethane	C_2H_6	−89
ethene	C_2H_4	−104
ethyne	C_2H_2	−84
propane	C_3H_8	−42
propene	C_3H_6	−47
propyne	C_3H_4	−23
butane	C_4H_{10}	−0.5
1-butene	C_4H_0	−6
1-butyne	C_4H_6	8
pentane	C_5H_{12}	36
1-pentene	C_5H_{10}	30
1-pentyne	C_5H_8	40
hexane	C_6H_{14}	69
1-hexene	C_6H_{12}	63
1-hexyne	C_6H_{10}	71

Hydrogen Bonding

The unexpectedly high boiling points of hydrogen compounds of nitrogen (ammonia), oxygen (water), and fluorine (hydrogen fluoride) compared to those of hydrogen compounds of other elements in the same groups is evidence that some other effect in addition to dipole–dipole and London forces exists. Chemists have found that this behaviour is generalized to compounds where a hydrogen atom is bonded to a highly electronegative atom with a lone pair of electrons, such as nitrogen, oxygen, or fluorine (**Figure 5**). The explanation is that of **hydrogen bonding**. This process is not unlike a covalent bond, but in this case a proton is being shared between two pairs of electrons, rather than a pair of electrons being shared between protons.

The hydrogen bond was an extension of Lewis theory in 1920. The special properties of water and some other hydrogen-containing compounds required a new explanation. Maurice Huggins, a graduate student of Lewis, and two colleagues (Wendell Latimer and Worth Rodebush) devised the idea of a bond where hydrogen could be shared between some atoms like nitrogen, oxygen, and fluorine in two different molecules. This requires one of the two atoms to have a lone pair of electrons. Lewis referred to this as "a most important addition to my theory."

Additional evidence for hydrogen bonding can be obtained by looking at energy changes associated with the formation of hydrogen bonds. Recall that endothermic and exothermic reactions are explained by the difference between the energy absorbed to break bonds in the reactants and the energy released when new bonds in the products are formed. For example, in the exothermic formation of water from its elements, more energy is released in forming the new O−H bonds than is required to break H−H and O=O bonds. In a sample of glycerol, you would expect some hydrogen bonding between glycerol molecules. However, these molecules are rather bulky and this limits the number of possible hydrogen bonds. If water is mixed with glycerol, additional

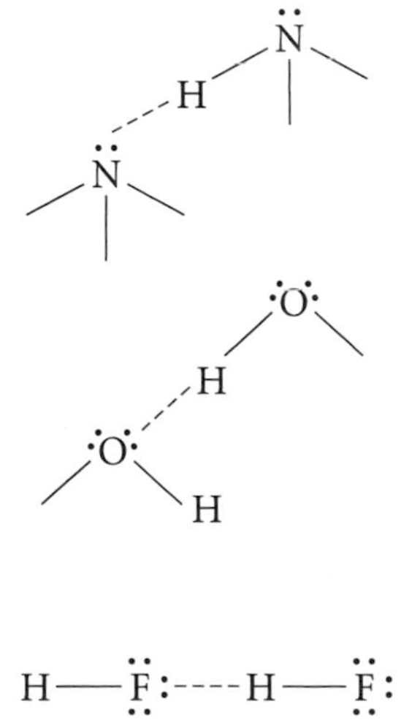

hydrogen bonding the attraction of hydrogen atoms bonded to N, O, or F atoms to a lone pair of electrons of N, O, or F atoms in adjacent molecules

Figure 5
A hydrogen bond (--) occurs when a hydrogen atom bonded to a strongly electronegative atom is attracted to a lone pair of electrons in an adjacent molecule.

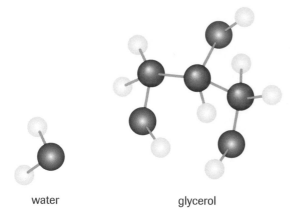

water   glycerol

hydrogen bonds should be possible. The small size of the water molecule should make it possible for water to form many hydrogen bonds with the glycerol molecules. Experimentally, you find that mixing water with glycerol is an exothermic process. 🧪

The theory of hydrogen bonding is necessary to explain the functions of biologically important molecules. Recall that proteins are polymers of amino acids, and that amino acids have $-NH_2$ and $-COOH$ functional groups, both of which fulfill the conditions for hydrogen bonding. Similarly, the double helix of the DNA molecule owes its unique structure largely to hydrogen bonding. The central bonds that hold the double helix together are not covalent (**Figure 6**). If the helix were held together by covalent bonds, the DNA molecule would not be able to unravel and replicate.

Figure 6
According to Francis Crick, co-discoverer with James D. Watson of the DNA structure, "If you want to understand function, study structure." Hydrogen bonding (blue dashes) explains the shape and function of the DNA molecule. The interior of the double helix is cross-linked by hydrogen bonds.

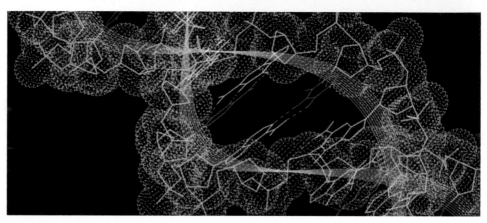

🧪 **INVESTIGATION 4.5.1**

Hydrogen Bonding (p. 278)
Mixing liquids provides evidence for hydrogen bonding.

Figure 7
The weight of the water boatman is not enough to overcome the intermolecular forces between the water molecules. This would be like you walking on a trampoline. The fabric of the trampoline is strong enough to support your weight.

Other Physical Properties of Liquids

Liquids have a variety of physical properties that can be explained by intermolecular forces. As you have seen, boiling point is a property often used in the discussion of intermolecular forces. Comparing boiling points provides a relatively simple comparison of intermolecular forces in liquids, if we assume that the gases produced have essentially no intermolecular forces between their molecules. What about some other properties of liquids, such as surface tension, shape of a meniscus, volatility, and ability to "wet" other substances? Surface tension is pretty important for water insects (**Figure 7**). The surface tension on a liquid is like an elastic skin. Molecules within a liquid are attracted by molecules on all sides, but molecules right at the surface are only attracted downward and sideways (**Figure 8**). This means that the liquid tends to stay together. Not surprisingly, substances containing molecules with stronger intermolecular forces have higher surface tensions. Water is a good example—it has one of the highest surface tensions of all liquids.

The shape of the meniscus of a liquid and capillary action in a narrow tube are both thought to be due to intermolecular forces. In both cases, two intermolecular attractions need to be considered—the attraction between like molecules (called cohesion) and the attraction between unlike molecules (called adhesion). Both cohesion and adhesion are intermolecular attractions. If you compare two very different liquids such as water and mercury (**Figure 9**), water rises in a narrow tube, but mercury does not. The adhesion between the water and the glass is thought to be greater than the cohesion between the water molecules. In a sense, the water is pulled up the tube by the intermolecular forces between the water molecules and the glass. Notice that this also produces a concave (curved downward) meniscus. For mercury, it is the opposite. The cohesion between the mercury atoms is greater than the adhesion of mercury to glass. Mercury atoms tend to stay together, which also explains the convex (curved upward) meniscus.

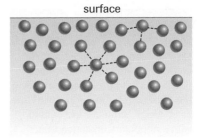

surface

Figure 8
The intermolecular forces on a molecule inside a liquid are relatively balanced. The forces on a molecule right at the surface are not balanced—the net pull is toward the centre.

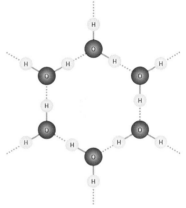

Figure 10
In ice and snowflakes, hydrogen bonds between water molecules result in an open hexagonal structure instead of a more compact structure with a higher density. The dashed lines represents the hydrogen bonds.

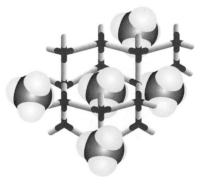

Figure 11
The very high pressures and low temperatures produce ice with methane trapped inside. The source of the methane is believed to be the decay of organic sediment deposited over a long time.

> ▶ **TRY THIS** activity   *Floating Pins*
>
> **Materials:** beaker or glass; water; several other different liquids; dishwashing detergent; straight pin; tweezers; toothpick
>
> • Make sure the straight pin is clean and dry.
> • Using clean tweezers, carefully place the pin in a horizontal position on the surface of each liquid, one at a time. Wash and dry the pin between tests.
>   (a) What happens for each liquid? Why?
> • Using tweezers, carefully place the pin vertically into the surface of the water. Try both ends of the pin.
>   (b) What happens this time? Why do you think the result is different than before?
> • Place the pin horizontally onto the surface of water. Using a toothpick, add a small quantity of dish detergent to the water surface away from the pin.
>   (c) Describe and explain what happens.

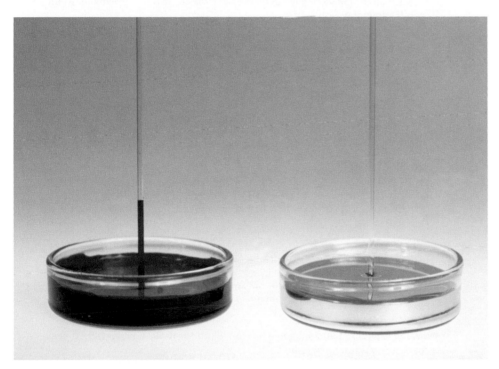

Figure 9
Capillary action is the movement of a liquid up a narrow tube. For water, capillary action is very noticeable; for mercury, it is nonexistent. (The water is coloured to make it more visible.)

We tend to think of water as a "normal" substance because it is so common and familiar to us. However, compared with other substances, water has some unusual properties. For example, the lower density of the solid form (ice) compared to the liquid form (water) is uncommon for a pure substance. Experimentally, the structure of ice is an open hexagonal network (**Figure 10**). Hydrogen bonding and the shape of the water molecule are believed to be responsible for this arrangement. Scientists have discovered that atoms and molecules can be trapped inside this hexagonal cage of water molecules. One of the more interesting and potentially valuable discoveries is the presence of large quantities of ice containing methane molecules on the ocean floor in the Arctic and around the globe (**Figure 11**). These deposits could be a vast new resource of natural gas in the future.

Magic Sand

Ordinary sand is composed of silicates. Water can form intermolecular bonds with the oxygen atoms in the silicates and therefore "wet" the sand. Magic sand is sand coated with a nonpolar substance that cannot form intermolecular bonds with water molecules. Magic sand cannot be wetted. It forms shapes underwater and is completely dry when it is removed.

Figure 12

Figure 13

Understanding Concepts

9. For each of the following molecular compounds, hydrogen bonds contribute to the attraction between molecules. Draw a Lewis diagram using a dashed line to represent a hydrogen bond between two molecules of the substance.
   (a) hydrogen peroxide, $H_2O_{2(l)}$ (disinfectant)
   (b) hydrogen fluoride, $HF_{(l)}$ (aqueous solution etches glass)
   (c) ethanol, $C_2H_5OH_{(l)}$ (beverage alcohol)
   (d) ammonia, $NH_{3(l)}$ (anhydrous ammonia for fertilizer)

10. (a) Refer to or construct a graph of the evidence from Lab Exercise 4.5.1. Extrapolate the group 15 and 16 lines to estimate the boiling points of water and ammonia if they followed the trend of the rest of their family members.
    (b) Approximately how many degrees higher are the actual boiling points for water and ammonia compared to your estimate in (a)?
    (c) Explain why the actual boiling points are significantly higher for both water and ammonia.
    (d) Propose an explanation why the difference from (b) is much greater for water than for ammonia.

11. Water beads on the surface of a freshly waxed car hood. Use your knowledge of intermolecular forces to explain this observation.

12. A lava lamp is a mixture of two liquids with a light bulb at the bottom to provide heat and light (**Figure 12**). What interpretations can you make about the liquids, intermolecular forces, and the operation of the lamp?

Applying Inquiry Skills

13. To gather evidence for the existence of hydrogen bonding in a series of chemicals, what variables must be controlled?

14. (a) Design an experiment to determine the volatility (rate of evaporation) of several liquids. Be sure to include variables.
    (b) Suggest some liquids to be used in this experiment. Predict the results. Explain your reasoning.

15. **Question**
   What is the solubility (high or low) of ammonia in water?

   Prediction/Hypothesis
   (a) Write a prediction, including your reasoning.

   Experimental Design
   Some water is squirted into ammonia gas in a Florence flask. Another tube is available for drawing water up into the flask.

   Evidence
   The water starts to move slowly up the tube and then suddenly flows into the upper flask like a fountain (**Figure 13**).

   Analysis
   (b) Answer the question, based upon the evidence gathered.

   Evaluation
   (c) Assuming that you have confidence in the evidence presented, evaluate the prediction and the reasoning used to make the prediction.

Making Connections

16. Wetting agents are very important in agriculture and other industries. What are wetting agents? Where are they used and for what purposes? Briefly explain how the function of wetting agents relates to the principles of intermolecular forces.

 www.science.nelson.com

17. In 1966 Soviet scientists claimed to have discovered a new form of water, called poly-water. The story of polywater is an interesting example of how people, including scientists, want to believe in a new, exciting discovery even if the evidence is incomplete. Write a brief report about polywater, including how it is supposedly formed, some of its claimed properties, the explanation in terms of intermolecular forces, and the final evaluation of the evidence (specifically, the flaws).

GO www.science.nelson.com

SUMMARY   *Intermolecular Forces*

- Intermolecular forces, like all bonds, are electrostatic—they involve the attraction of positive and negative charges.

- In this section we considered three intermolecular forces—London, dipole–dipole, and hydrogen bonding.

- All molecules attract each other through the London force—the simultaneous attraction of electrons and nuclei in adjacent molecules.

- Dipole–dipole force exists between polar molecules—the simultaneous attraction of a dipole of one molecule for adjacent dipoles.

- Hydrogen bonding exists when hydrogen atoms are bonded to highly electronegative atoms like N, O, and F—the hydrogen is simultaneously attracted to a pair of electrons on the N, O, or F atom of an adjacent molecule.

- Intermolecular forces affect the melting point, boiling point, capillary action, surface tension, volatility, and solubility of substances.

Current Research—Intermolecular Forces

In almost any area of science today, the experimental work runs parallel to the theoretical work and there is constant interplay between the two areas. In Canada there are several theorists whose research teams examine the forces between atoms and molecules to increase our understanding of physical and chemical properties. One such individual is Dr. Robert LeRoy (**Figure 14**), currently working in theoretical chemical physics at the University of Waterloo.

Dr. LeRoy's interest is intermolecular forces. He uses quantum mechanics and computer models to define and analyze the basic forces between atoms and molecules. Early in his career, Dr. LeRoy developed a technique for mathematically defining a radius of a small molecule, now known as the LeRoy radius. This established a boundary. Within the boundary, intramolecular bonding is important, and beyond the boundary, intermolecular forces predominate. In his work, the study of atomic and molecular spectra (called spectroscopy) plays a crucial role. Measurements from spectroscopy help theoreticians develop better models and theories for explaining molecular structure. Computer programs that Dr. LeRoy has developed for the purpose of converting experimental evidence to information on forces, shape, and structure are free, and are now routinely used around the world.

It is important not to assume that forces and structures are well established. Our knowledge of bonding and structure becomes more and more scanty and unreliable for larger structures. A huge amount of research remains to be done if we are ever to be able to describe bonding and structure very accurately for even microscopic amounts of

Figure 14
Dr. Robert J. LeRoy and his research team study intermolecular forces and the behaviour of small molecules and molecular clusters; develop methods to simulate and analyze the decomposition of small molecules; and create computer models to simulate and predict molecular properties. Visit Dr. LeRoy at http://leroy.uwaterloo.ca.

complex substances. Dr. LeRoy states "... except for the simplest systems, our knowledge of (interactions between molecules) is fairly primitive...." A classic example is our understanding of the structure and activity of proteins—the stuff of life. We know the composition of many proteins quite precisely and the structure can be experimentally determined, but the structure of these large molecules depends on how bonding folds and shapes the chains and branches. How a protein behaves and what it does depends specifically on its precise shape and structure, and that is something scientists often state is "not well understood."

▶ Section 4.5 Questions

Understanding Concepts

1. All molecular compounds may have London, dipole–dipole, and hydrogen-bonding intermolecular forces affecting their physical and chemical properties. Indicate which intermolecular forces contribute to the attraction between molecules in each of the following classes of organic compounds:
   (a) hydrocarbon; e.g., pentane, $C_5H_{12(l)}$ (in gasoline)
   (b) alcohol; e.g., 2-propanol, $CH_3CHOHCH_{3(l)}$ (rubbing alcohol)
   (c) ether; e.g., dimethylether, $CH_3OCH_{3(g)}$ (polymerization catalyst)
   (d) carboxylic acid; e.g., acetic acid, $CH_3COOH_{(l)}$ (in vinegar)
   (e) ester; e.g., ethylbenzoate, $C_6H_5COOC_2H_{5(l)}$ (cherry flavour)
   (f) amine; e.g., dimethylamine, $CH_3NHCH_{3(g)}$ (depilatory agent)
   (g) amide; e.g., ethanamide, $CH_3CONH_{2(s)}$ (lacquers)
   (h) aldehyde; e.g., methanal, $HCHO_{(g)}$ (corrosion inhibitor)
   (i) ketone; e.g., acetone, $(CH_3)_2CO_{(l)}$ (varnish solvent)

2. Use Lewis structures and hydrogen bonds to explain the very high solubility of ammonia in water.

3. Predict the solubility of the following organic compounds in water as low (negligible), medium, or high. Provide your reasoning.
   (a) 2-chloropropane, $C_3H_7Cl_{(l)}$ (solvent)
   (b) 1-propanol, $C_3H_7OH_{(l)}$ (brake fluids)
   (c) propanone, $(CH_3)_2CO_{(l)}$ (cleaning precision equipment)
   (d) propane, $C_3H_{8(g)}$ (gas barbecue fuel)

4. For each of the following pairs of chemicals, which one is predicted to have the stronger intermolecular attraction? Provide your reasoning.
   (a) chlorine or bromine
   (b) fluorine or hydrogen chloride
   (c) methane or ammonia
   (d) water or hydrogen sulfide
   (e) silicon tetrahydride or methane
   (f) chloromethane or ethanol

5. Which liquid, propane (C_3H_8) or ethanol (C_2H_5OH), would have the greater surface tension? Justify your answer.

6. In cold climates, outside water pipes, such as underground sprinkler systems, need to have the water removed before it freezes. What might happen if water freezes in the pipes? Explain your answer.

7. A glass can be filled slightly above the brim with water without the water running down the outside. Explain why the water does not overflow even though some of it is above the glass rim.

8. Explain briefly what the "LeRoy radius" of a molecule represents.

Applying Inquiry Skills

9. Design an experiment to determine whether or not hydrogen bonding has an effect on the surface tension of a liquid. Clearly indicate the variables in this experiment.

10. Critique the following experimental design.
    The relative strength of intermolecular forces in a variety of liquids is determined by measuring the height to which the liquids rise in a variety of capillary tubes.

Making Connections

11. Some vitamins are water soluble (e.g., B series and C), while some are fat soluble (e.g., A, D, E, and K).
    (a) What can you infer about the polarity of these chemicals?
    (b) Find and draw the structure of at least one of the water-soluble and one of the fat-soluble vitamins.
    (c) When taking vitamins naturally or as supplements, what dietary requirements are necessary to make sure that the vitamins are used by the body?
    (d) More of a vitamin is not necessarily better. Why can you take a large quantity of vitamin C with no harm (other than the cost), but an excess of vitamin E can be dangerous?

 www.science.nelson.com

12. Many of the new materials that are being invented for specific purposes show an understanding of structure and bonding. One candidate that has been suggested as a future product is commonly known as the "fuzzyball," $C_{60}F_{60(s)}$. What is the structure of this molecule? What use is proposed for this substance? Explain this use in terms of intermolecular forces.

 www.science.nelson.com

13. People who wear contact lenses know that there are hard and soft contact lenses. The polymers used in each type of lens are specifically chosen for their properties. What is the property that largely determines whether the lens is a hard or soft lens? Write a brief explanation using your knowledge of intermolecular forces.

 www.science.nelson.com

14. Plastic cling wrap is widely used in our society. Why does it cling well to smooth glass and ceramics, but not to metals? Describe the controversial social issue associated with the use of this plastic wrap. How are intermolecular forces involved in starting the process that leads to this controversy?

 www.science.nelson.com

Extensions

15. The London force is affected by more than just the number of electrons. What other variable(s) affect(s) the strength of the London force?

16. (a) Draw a bar graph with the temperature in kelvin on the vertical axis and the three isoelectronic compounds listed below on the horizontal axis. For each compound draw a vertical bar from 0 K to its boiling point: propane ($-42°C$), fluoroethane ($-38°C$), and ethanol ($78°C$).

(b) Divide each of the three bar graphs into the approximate component for the intermolecular force involved. (Assume that the London force is the same for each chemical and that the dipole–dipole force is the same for the two polar molecules.)

(c) Based upon the proportional components for the three possible intermolecular forces, order the relative strength of these forces.

Figure 1
Different solids behave very differently under mechanical stress.

All solids, including elements and compounds, have a definite shape and volume, are virtually incompressible, and do not flow readily. However, there are many specific properties such as hardness, melting point, mechanical characteristics, and conductivity that vary considerably for different solids. If you hit a piece of copper with a hammer, you can easily change its shape. If you do the same thing to a lump of sulfur, you crush it. A block of paraffin wax when hit with a hammer may break and will deform (**Figure 1**). Why do these solids behave differently?

In both elements and compounds, the structure and properties of the solid are related to the forces between the particles. Although all forces are electrostatic in nature, the forces vary in strength. What we observe are the different properties of substances and we classify them into different categories (**Table 1**). To explain the properties of each category, we use our knowledge of chemical bonding.

Table 1 Classifying Solids

Class of substance	Elements combined	Examples
ionic	metal + nonmetal	$NaCl_{(s)}$, $CaCO_{3(s)}$
metallic	metal(s)	$Cu_{(s)}$, $CuZn_{3(s)}$
molecular	nonmetal(s)	$I_{2(s)}$, $H_2O_{(s)}$, $CO_{2(s)}$
covalent network	metalloids/carbon	$C_{(s)}$, $SiC_{(s)}$, $SiO_{2(s)}$

crystal lattice a regular, repeating pattern of atoms, ions, or molecules in a crystal

Ionic Crystals

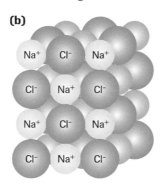

Figure 2
From a cubic crystal of table salt **(a)** and from X-ray analysis, scientists infer the 3-D arrangement for sodium chloride **(b)**. In this cubic crystal, each ion is surrounded by six ions of opposite charge.

Ionic compounds in their pure solid form are described as a 3-D arrangement of ions in a crystal structure. The arrangement of ions within the **crystal lattice** (**Figure 2(b)**) can be inferred from the crystal shape (**Figure 2(a)**) and from X-ray diffraction experiments. The variation of crystalline structures is not a topic here, but the variety of crystal shapes suggests that there is an equally wide variety of internal structures for ionic compounds.

Ionic compounds are relatively hard but brittle solids at SATP, conducting electricity in the liquid state but not in the solid state, forming conducting solutions in water, and having high melting points. These properties are interpreted to mean that ionic bonds are strong (evidence of hardness and melting points of the solid) and directional (evidence of brittleness of the solid) and that the lattice is composed of ions (evidence of electrical conductivity). *Ionic bonding* is defined theoretically as the simultaneous attraction of an ion by the surrounding ions of opposite charge. The full charge on the ions provides a greater force of attraction than do the partial charges (i.e., $\delta+$ and $\delta-$) on polar molecules. In general, ionic bonding is much stronger than all intermolecular forces. For example, calcium phosphate, $Ca_3(PO_4)_{2(s)}$, in tooth enamel (ionic bonds) is much harder than ice, $H_2O_{(s)}$, (hydrogen bonding).

> The properties of ionic crystals are explained by a 3-D arrangement of positive and negative ions held together by strong, directional ionic bonds.

Metallic Crystals

Metals are shiny, silvery, flexible solids with good electrical and thermal conductivity. The hardness varies from soft to hard (e.g., lead to chromium) and the melting points from low to high (e.g., mercury to tungsten). Further evidence from the analysis of X-ray diffraction patterns shows that all metals have a continuous and very compact crystalline structure (**Figure 3**). With few exceptions, all metals have closely packed structures.

An acceptable theory for metals must explain the characteristic metallic properties, provide testable predictions, and be as simple as possible.

Figure 3
Metal crystals are small, and usually difficult to see. Zinc-plated or galvanized metal objects often have large flat crystals of zinc metal that are very obvious.

According to current theory, the properties of metals are the result of the bonding between fixed, positive nuclei and loosely held, mobile valence electrons. This attraction is not localized or directed between specific atoms, as occurs with ionic crystals. Instead, the electrons act like a negative "glue" surrounding the positive nuclei. As illustrated in **Figure 4**, valence electrons are believed to occupy the spaces between the positive centres (nuclei). This simple model, known as the *electron sea* model, incorporates the ideas of

- low ionization energy of metal atoms to explain loosely held electrons
- empty valence orbitals to explain electron mobility
- electrostatic attractions of positive centres and the negatively charged electron "sea" to explain the strong, nondirectional bonding

Figure 4 shows a cross-section of the crystal structure of a metal. Each circled positive charge represents the nucleus and inner electrons of a metal atom. The shaded area surrounding the circled positive charges represents the mobile sea of electrons. The electron sea model is used to explain the empirical properties of metals (**Table 2**).

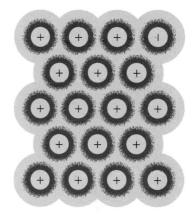

Figure 4
In this model of metallic bonding, each positive charge represents the nucleus and inner electrons of a metal atom, surrounded by a mobile "sea" of valence electrons.

Table 2 Explaining the Properties of Metals

Property	Explanation
shiny, silvery	valence electrons absorb and re-emit the energy from all wavelengths of visible and near-visible light
flexible	nondirectional bonds mean that the planes of atoms can slide over each other while remaining bonded
electrical conductivity	valence electrons can freely move throughout the metal; a battery can force additional electrons onto one end of a metal sample and remove other electrons from the other end
hard solids	electron sea surrounding all positive centres produces strong bonding
crystalline	electrons provide the "electrostatic glue" holding the atomic centres together producing structures that are continuous and closely packed

The properties of metallic crystals are explained by a 3-D arrangement of metal cations held together by strong, nondirectional bonds created by a "sea" of mobile electrons.

Metallic Bonding Analogy
The next time you have a Rice Krispie square, look at it carefully and play with it. The marshmallow is the glue that binds the rice together. If you push on the square, you can easily deform it, without breaking it. The marshmallow is like the "electron sea" in a metal; the rice represents the nuclei. The mechanical properties of a rice krispie square are somewhat similar to those of a metal.

Molecular Crystals

Molecular solids may be elements such as iodine and sulfur or compounds such as ice or carbon dioxide. The molecular substances, other than the waxy solids (large hydrocarbons) and giant polymers (such as plastics), are crystals that have relatively low melting points, are not very hard, and are nonconductors of electricity in their pure form as well as in solution. From X-ray analysis, molecular crystals have a crystal lattice like ionic compounds, but the arrangement may be more complicated (**Figure 5**). In general, the molecules are packed as close together as their size and shape allows (**Figure 6**).

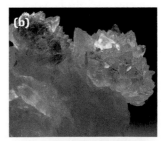

Figure 5
A model of an iodine crystal based on X-ray analysis shows a regular arrangement of iodine molecules.

 indicates an I_2 molecule

The properties of molecular crystals can be explained by their structure and the intermolecular forces that hold them together. London, dipole–dipole, and hydrogen bonding forces are not very strong compared with ionic or covalent bonds. This would explain why molecular crystals have relatively low melting points and a general lack of hardness. Because individual particles are neutral molecules, they cannot conduct an electric current even when the molecules are free to move in the molten state.

> The properties of molecular crystals are explained by a 3-D arrangement of neutral molecules held together by relatively weak intermolecular forces.

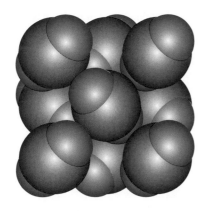

Figure 6
Solid carbon dioxide, or dry ice, also has a crystal structure containing individual carbon dioxide molecules.

Covalent Network Crystals

Most people recognize diamonds in either jewellery or cutting tools, and quartz as gemstones (**Figure 7**) and in various grinding materials, including emery sandpapers. These substances are among the hardest materials on Earth and belong to a group known as covalent network crystals. These substances are very hard, brittle, have very high melting points, are insoluble, and are nonconductors of electricity. Covalent network crystals are usually much harder and have much higher melting points than ionic and molecular crystals. They are described as brittle because they don't bend under pressure, but they

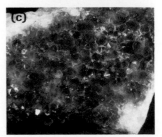

Figure 7
Amethyst **(a)**, rose quartz **(b)**, and citrine **(c)** are all variations of quartz, which is $SiO_{2(s)}$.

are so hard that they seldom break. Diamond ($C_{(s)}$) is the classic example of a covalent crystal. It is so hard that it can be used to make drill bits for drilling through the hardest rock on Earth (**Figure 8**). Another example is silicon carbide ($SiC_{(s)}$)—used for grinding stones to sharpen axes and other metal tools. Carbide-tipped saw blades are steel blades coated with silicon carbide.

The shape and X-ray diffraction analysis of diamond shows that the carbon atoms are in a large tetrahedral network with each carbon covalently bonded to four other carbon atoms (**Figure 8**). Each diamond is a crystal and can be described as a single macromolecule with a chemical formula of $C_{(s)}$. The network of covalent bonds leads to a common name for these covalent crystals as **covalent network**. This name helps to differentiate between the covalent bonds within molecules and polyatomic ions and the covalent bonds within covalent network crystals. Most covalent networks involve the elements and compounds of carbon and silicon.

Crystalline quartz is a covalent network of $SiO_{2(s)}$ (**Figure 9(a)**). Glass shares the same chemical formula as quartz but lacks the long-range, regular crystalline structure of quartz (**Figure 9(b)**). Purposely, glass is cooled to a rigid state in such a way that it will not crystallize.

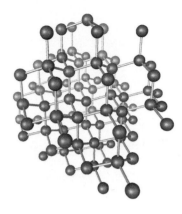

Figure 8
In diamond, each carbon atom has four single covalent bonds to each of four other carbon atoms. As you know from VSEPR theory, four pairs of electrons lead to a tetrahedral shape around each carbon atom.

covalent network a 3-D arrangement of covalent bonds between atoms that extends throughout the crystal

(a)   **(b)**

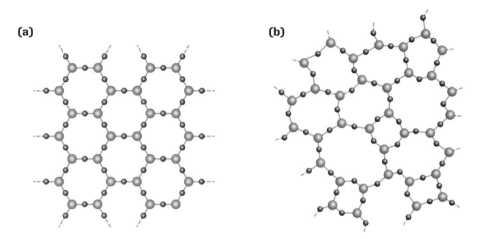

Figure 9
(a) Quartz in its crystalline form has a 3-D network of covalently bonded silicon and oxygen atoms.
(b) Glass is not crystalline because it does not have an extended order; it is more disordered than ordered.

The properties of hardness and high melting point provide the evidence that the overall bonding in the large macromolecule of a covalent network is very strong—stronger than most ionic bonding and intermolecular bonding. Although an individual carbon−carbon bond in diamond is not much different in strength from any other single carbon−carbon covalent bond, it is the interlocking structure that is thought to be responsible for the strength of the material. This is similar to the strength of a steel girder and the greater strength of a bridge built from a three-dimensional arrangement of many steel girders. The final structure is stronger than any individual component. This means that individual atoms are not easily displaced and that is why the sample is very hard. In order to melt a covalent network crystal, many covalent bonds need to be broken, which requires considerable energy, so the melting points are very high. Electrons in covalent network crystals are held either within the atoms or in the covalent bonds. In either case, they are not free to move through the network. This explains why these substances are nonconductors of electricity.

DID YOU *KNOW* **?**

Mohs Hardness Scale
A common method used to measure hardness is the Mohs scale, based on how well a solid resists scratching by another substance. The scale goes from 1 (talc) to 10 (diamond). Any substance will scratch any other substance lower on the scale. One disadvantage of this scale is that it is not linear. Diamond (10) is much harder than corundum (9), but apatite (5), a calcium phosphate mineral, is only slightly harder than fluorite (4), a calcium fluoride mineral.

> The properties of network covalent crystals are explained by a 3-D arrangement of atoms held together by strong, directional covalent bonds.

Other Covalent Networks of Carbon

Carbon is an extremely versatile atom in terms of its bonding and structures. More than any other atom, carbon can bond to itself to form a variety of pure carbon substances. It can form 3-D tetrahedral arrangements (diamond), layers of sheets (graphite), large spherical molecules (buckyballs), and long, thin tubes (carbon nanotubes) (**Figure 10**). Graphite is unlike covalent crystals in that it conducts electricity, but it is still hard and has a high melting point. Graphite also acts as a lubricant. All of these properties, plus the X-ray diffraction of the crystals, indicate that the structure for graphite is hexagonal sheets of sp^2 hybridized carbon atoms. Within these planar sheets the bonding is a covalent network and therefore strong, but between the sheets the bonding is relatively weak—due to London forces. The lubricating property of graphite arises as the covalent network planes slide over one another while maintaining the weak intermolecular attractions. The electrical conductivity arises through formation of π bonds by the unhybridized p orbitals. These π bonds extend over the entire sheet, and electrons within them are free to move from one end of the sheet to the other.

(a) 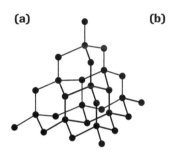 **(b)** 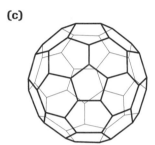 **(c)** 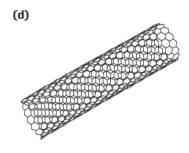 **(d)**

Figure 10
Models of the many forms of pure carbon:
(a) diamond
(b) graphite
(c) buckyball
(d) carbon nanotubes

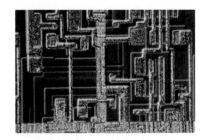

Figure 11
Semiconductors in transistors are covalent crystals that have been purposely manipulated by doping them with atoms that have more or fewer electrons than the atoms in the main crystal.

⚗ **INVESTIGATION 4.6.1**

Classifying Mystery Solids (p. 279)
Properties of various solids are used to determine the type of solid.

Semiconductors

The last five decades have seen an electronic technological revolution driven by the discovery of the transistor—a solid-state "sandwich" of crystalline semiconductors. Semiconductor material used in transistors is usually pure crystalline silicon or germanium with a tiny quantity (e.g., 5 ppm) of either a group 13 or 15 element added to the crystal in a process called doping. The purpose of this doping is to control the electrical properties of the covalent crystal to produce the conductive properties desired. Transistors are the working components of almost everything electronic (**Figure 11**).

In an atom of a semiconductor, the highest energy levels may be thought of as being full of electrons that are unable to move from atom to atom. Normally, this would make the substance a nonconductor, like glass or quartz. In a semiconductor, however, electrons require only a small amount of energy to jump to the next higher energy level, which is empty. Once in this level, they may move to another atom easily (**Figure 12**). Semiconductors can be manipulated chemically by adding small quantities of other atoms to the crystals to make them behave in specific ways. Semiconductors are an example of a chemical curiosity where research into atomic structure has turned out to be amazingly useful and important. Power supplies for many satellites, and for the International Space Station (**Figure 13**), come from solar cells that are semiconductors arranged to convert sunlight directly to electricity. Other arrangements convert heat to electricity, or electricity to heat, or electricity to light—all without moving parts in a small, solid device. Obviously, improving the understanding of semiconductor structure was of great value to our society. ⚗▮

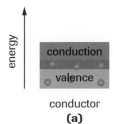

conductor
(a)

insulator
(b)

semiconductor
(c)

(a) In a conductor, these orbitals and the valence orbitals are at or about the same energy and electrons can be easily transported throughout the solid.

(b) In insulators, there is a large energy gap between empty orbitals and the valence orbitals. Electrons cannot easily get to these orbitals and insulators do not conduct electricity.

(c) In semiconductors, there is a relatively small energy gap between the valence orbitals and the empty orbitals that extend throughout the crystal. Thermal energy can easily promote some electrons into the empty orbitals to provide conductivity.

Figure 12
All atoms and molecules have empty orbitals. In large macromolecules, partially filled or empty orbitals extend throughout the solid.

Figure 13
The huge solar panels that power this space station are multiple solid-state devices that use semiconductors to change light energy to electric current.

SUMMARY ***Properties of Ionic, Metallic, Molecular, and Covalent Network Crystals***

Table 3

Crystal	Particles	Force/Bond	Properties	Examples
Ionic	ions (+, −)	ionic	hard; brittle; high melting point; liquid and solution conducts	$NaCl_{(s)}$, $Na_3PO_{4(s)}$, $CuSO_4 \cdot 5H_2O_{(s)}$
Metallic	cations	metallic	soft to very hard; solid and liquid conducts; ductile; malleable; lustrous	$Pb_{(s)}$, $Fe_{(s)}$, $Cu_{(s)}$, $Al_{(s)}$
Molecular	molecules	London dipole–dipole hydrogen	soft; low melting point; nonconducting solid, liquid, and solution	$Ne_{(g)}$, $H_2O_{(l)}$, $HCl_{(g)}$, $CO_{2(g)}$, $CH_{4(g)}$, $I_{2(s)}$
Covalent Network	atoms	covalent	very hard; very high melting point; nonconducting	$C_{(s)}$, $SiC_{(s)}$, $SiO_{2(s)}$

▶ **Practice**

Understanding Concepts

1. In terms of chemical bonds, what are some factors that determine the hardness of a solid?

2. Identify the main type of bonding and the type of solid for each of the following:
(a) SiO_2 (c) CH_4 (e) Cr
(b) Na_2S (d) C   (f) CaO

3. How does the melting point of a solid relate to the type of particles and forces present?

4. Explain why metals are generally malleable, ductile, and flexible.

5. State the similarities and differences in the properties of each of the following pairs of substances. In terms of the particles and forces present, briefly explain each answer.
(a) $Al_{(s)}$ and $Al_2O_{3(s)}$ (b) $CO_{2(s)}$ and $SiC_{(s)}$ ▶

6. To cleave or split a crystal you tap a sharp knife on the crystal surface with a small hammer.
   (a) Why is the angle of the blade on the crystal important to cleanly split the crystal?
   (b) If you wanted to cleave a sodium chloride crystal, where and at what angle would you place the knife blade?
   (c) Speculate about what would happen if you tried to cleave a crystal in the wrong location or at the wrong angle.
   (d) State one application of this technique.

7. Match the solids, $NaBr_{(s)}$, $V_{(s)}$, $P_2O_{5(s)}$, and $SiO_{2(s)}$, to the property listed below.
   (a) high melting point, conducts electricity
   (b) low melting point, soft
   (c) high melting point, soluble in water
   (d) very high melting point, nonconductor

Applying Inquiry Skills

8. Metals are generally good conductors of heat and electricity. Is there a relationship between a metal's ability to conduct heat and its ability to conduct electricity?
   (a) Predict the answer to this question. Include your reasoning.
   (b) Design an experiment to test your prediction and reasoning using common examples of metals.

Making Connections

9. Suggest some reasons why graphite may be better than oil in lubricating moving parts of a machine.

10. Nitinol is known as the "metal with a memory." It is named after the alloy and place where it was accidentally discovered: "**Ni**ckel **ti**tanium **n**aval **o**rdinance **l**aboratory" This discovery has revolutionized manufacturing and medicine in the form of many products that can "sense" and respond to changes. Research and write a brief report about Nitinol including its composition, a brief description of how it works, and some existing or proposed technological applications.

 www.science.nelson.com

11. The synthetic material moissanite (silicon carbide) looks like diamond and is used to simulate diamonds in jewellery.
    (a) Compare the physical properties of moissanite and diamond.
    (b) Do these properties suggest a method to distinguish between a real diamond and a simulated diamond like moissanite? Describe briefly.
    (c) What test do jewellers use to distinguish between these materials? Describe the principle used and the distinction made.

 www.science.nelson.com

Extension

12. If graphite did not conduct electricity, describe how you would change its model to explain this, but still explain its lubricating properties.

The 20th century saw an explosive development of electronic devices, mostly due to the development of the transistor and the science–technology cycle we commonly simplify as "miniaturization." There now exists a whole range of consumer goods, such as personal computers, cellular phones, and digital cameras, that are in many ways changing the way our society operates. The quest for smaller and smaller working devices and electrical circuitry has now taken us to the realm of nanotechnology—devices built on an atomic scale. Science visionaries foresee a time when invisibly tiny machines move through human arteries to clear blockages or hunt down cancerous cells. Nanomachines might also be made to build substances and structures molecule by molecule, producing stronger and better materials than exist today. This type of advance involves a whole network of careers with many people in different, but related, fields cooperating to further our knowledge.

Research Project/Team Director

Scientists who lead teams doing research into the nature of molecular structure and interactions are generally experimentalists or theoreticians. The distinction is often blurry, but generally a theoretical researcher is more concerned with understanding the fundamental nature of intermolecular forces and bonding, while an experimental researcher is more concerned with trying to obtain new results. People heading research teams generally hold Ph.D. degrees in their subject areas, and have usually done considerable postdoctoral work as well. Their research projects are generally funded by government, university, and/or commercial grants.

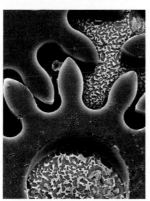

 Practice

Making Connections

1. Choose a university and use its web site to obtain biographical information on one of its research scientists who is involved in research of intermolecular forces and structures. Report on the scientist's training and education qualifications, and briefly summarize the nature of his/her current research project.

GO   www.science.nelson.com

Software Engineer— Scanning Probe Microscopy

The new field of scanning probe microscopy is expanding rapidly. The ability to examine product surfaces and to create new materials at an atomic level seems to hold the same kind of promise as the initial development of the 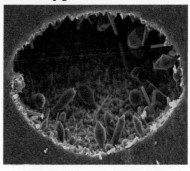 laser—no one can imagine yet how many uses and applications there might be. This career requires at least one degree in engineering, with specialization in computer software production and in atomic structure. Software routines that allow interpretation of evidence obtained by force probes are essential to create molecular "pictures."

Semiconductor Applications Engineer

Semiconductors work on quantum mechanical principles, involving atomic-level electron transfer phenomena. Much atomic-level research is devoted to the creation of smaller and more efficient semiconducting devices, and of course, semiconductors are involved at every level in every type of equipment used to perform such research. Finding new applications for existing  technology is also a large part of this career description. This type of career requires an advanced degree in engineering, with specialization in the physics of electronics and the chemistry of semiconductors.

Atomic-Force Microscope Research Technologist

Technicians using atomic-force microscopes are operating state-of-the-art equipment that uses pure quantum mechanical effects to provide information, sometimes for pure scientific research and at other times for configuring nanotechnology. A good background in atomic theory, physics, and mathematics is necessary for understanding the characteristics of atomic-force probes, which scan a surface one atom at a time with quantum electron "tunnelling" effects.

Understanding Concepts

1. Describe and explain the different electrical conductivity properties of ionic substances under different conditions.

2. In terms of particles and forces present, what determines whether a substance conducts electricity?

3. Calcium oxide (m.p. 2700°C) and sodium chloride (m.p. 801°C) have the same crystal structure and the ions are about the same distance apart in each crystal. Explain the significant difference in their melting points.

4. Compare and contrast the bonding forces in carbon dioxide (dry ice) and silicon dioxide (quartz). How does this explain the difference in their properties?

5. Why do most metals have a relatively high density?

6. State the order of strength of intermolecular, ionic, covalent network, and metallic bonding. Defend your answer.

7. If the zipper on your jacket does not slide easily, how could using your pencil help? Describe what you would do and explain why this would work.

8. Compare diamond and graphite, using the following categories: appearance, hardness, electrical conductivity, crystal structure, and uses.

Applying Inquiry Skills

9. Use the evidence in **Table 4** to classify the type of substance and type of bonding for each unknown chloride listed.

Table 4

Substance	Melting point (°C)	Boiling point (°C)	Solubility in water	Solubility in benzene
XCl_a	750	1250	high	very low
YCl_b	−25	92	very low	high

Making Connections

10. Describe two areas where research into the structure of molecules has caused a dramatic improvement in materials available to society in general.

11. Clays and ceramics are substances closely related to silica, $SiO_{2(s)}$. List some properties of clay and ceramics. How does the structure and bonding change when clay is fired (strongly heated) to produce a ceramic?

 www.science.nelson.com

12. Research and report on the properties, applications, structure, and bonding in boron nitride, $BN_{(s)}$.

 www.science.nelson.com

13. Experts agree that we are reaching the physical limit of how many transistors can be put onto a computer chip of a given size. Some scientists are already looking at a new generation of biological computers. Research one proposed biological computer. Describe some similarities and differences between this proposed computer and present computers. What are some of the promises of this new technology?

 www.science.nelson.com

ACTIVITY 4.3.1

Shapes of Molecules

Chemists use molecular models to study the shapes of molecules. Since the models are built to reflect the theory, they can be used to test your understanding of the theory. The purpose of this activity is to use VSEPR theory to predict the stereochemistry of some common molecules.

Materials: molecular models kit; a legend of colour codes for models of atoms

(a) Use VSEPR theory to predict the shape of the following molecules: CCl_4, C_2H_4, C_2F_2, NCl_3, OF_2, and NH_2OH.
  • Assemble molecular models for the chemicals listed in the question.

(b) Sketch a 3-D diagram of each molecule assembled and classify its shape.

(c) Evaluate the predictions that you made. How does your understanding of VSEPR theory have to be revised?

INVESTIGATION 4.4.1

Testing for Polar Molecules

In the winter when the humidity is relatively low, it is not unusual to acquire an electric charge by walking across a carpet in stocking feet or by pulling off a sweater.

A nonmetal rod or strip can also easily acquire a positive or negative charge by friction. As a diagnostic test, if you bring a charged rod or plastic strip near a liquid stream, and the stream is attracted to the rod or strip, then the molecules in the liquid are polar.

Purpose

The purpose of this investigation is to test the empirical rules provided in **Table 1,** Section 4.4.

Question

Which of the liquids provided have polar molecules?

Prediction

(a) Use the empirical rules in **Table 1**, Section 4.4 to predict whether the liquids provided contain polar molecules.

Experimental Design

A thin stream of each liquid is tested by holding a positively or negatively charged rod or plastic strip near the liquid (**Figure 1**, Section 4.4).

 Check the MSDS for all liquids used and follow appropriate safety precautions.

Inquiry Skills

○ Questioning	○ Planning	● Analyzing
○ Hypothesizing	● Conducting	● Evaluating
● Predicting	● Recording	● Communicating

Materials

lab apron	buret funnel
safety glasses	(if buret is used)
medicine dropper or	400-mL beaker
50-mL buret	acetate strip (marked +)
clamp and stand	vinyl strip (marked −)
(if buret is used)	paper towel
	various liquids

Procedure

1. Fill the dropper/buret with one of the liquids.

2. Rub the acetate strip back and forth several times with a paper towel.

3. Allow drops or a thin stream of the liquid to pour into the waste beaker.

4. Hold the charged acetate strip close to the liquid stream and observe any effect or none.

5. Repeat steps 1 through 4 with the charged vinyl strip.

6. Move to the next station and repeat steps 1 through 5 using the next liquid provided.

Evidence

(b) Prepare and complete a table for your observations.

Analysis

(c) Answer the Question.

Evaluation

(d) Evaluate the experimental design, materials, procedure, and skills employed.

(e) Evaluate your prediction and the empirical rules used to make the prediction.

LAB EXERCISE 4.5.1

Boiling Points and Intermolecular Forces

Inquiry Skills

○ Questioning   ○ Planning   ● Analyzing
● Hypothesizing   ○ Conducting   ● Evaluating
● Predicting   ○ Recording   ● Communicating

In all liquids, intermolecular forces are important, but these forces become negligible in the gas state for the conditions at which liquids boil. Therefore, we are looking at a situation where intermolecular forces must be overcome by adding energy, but no new bonds are formed. The temperature at which a liquid boils reflects the strength of the intermolecular forces present among the molecules. Higher temperatures mean more energy has been added and the intermolecular forces must have been stronger.

Purpose

The purpose of this lab exercise is to test the theory and rules for London and dipole−dipole forces.

Question

What is the trend in boiling points of the hydrogen compounds of elements in groups 14-17?

Hypothesis/Prediction

(a) Based upon dipole−dipole and London forces, write a prediction for the trend in boiling points within and between groups. Your prediction could include a general sketch of a graph of boiling point versus number of electrons per molecule. Provide your reasoning.

Analysis

(b) Complete a graph of the evidence by plotting boiling point versus number of electrons per molecule.

(c) Answer the Question.

Evidence

Table 1   Boiling Points of the Hydrogen Compounds of Elements in Groups 14–17

Group	Hydrogen compound	Boiling point (°C)
14	$CH_{4(g)}$	−162
	$SiH_{4(g)}$	−112
	$GeH_{4(g)}$	−89
	$SnH_{4(g)}$	−52
15	$NH_{3(g)}$	−33
	$PH_{3(g)}$	−87
	$AsH_{3(g)}$	−55
	$SbH_{3(g)}$	−17
16	$H_2O_{(l)}$	100
	$H_2S_{(g)}$	−61
	$H_2Se_{(g)}$	−42
	$H_2Te_{(g)}$	−2
17	$HF_{(g)}$	20
	$HCl_{(g)}$	−85
	$HBr_{(g)}$	−67
	$HI_{(g)}$	−36

Evaluation

(d) Assuming that the evidence is valid, evaluate the Prediction and the concept of intermolecular forces used to make the prediction.

(e) Are there any anomalies (unexpected evidence) in the evidence presented? Suggest an explanation.

⚗ INVESTIGATION 4.5.1

Hydrogen Bonding

Inquiry Skills

○ Questioning	● Planning	● Analyzing
● Hypothesizing	● Conducting	● Evaluating
● Predicting	● Recording	● Communicating

Exothermic and endothermic physical and chemical processes are explained by comparing the energy required to break bonds and the energy released when bonds are formed. The net effect of bonds broken and bonds formed is either an exothermic (energy released) or an endothermic (energy absorbed) change. Thermochemical changes can be physical (phase changes) or chemical (chemical changes)—or solution formation, which does not seem to fit the physical–chemical classification.

Purpose

The purpose of this investigation is to test the concept of hydrogen bonding.

Question

How does the temperature change for the mixing of ethanol with water compare with the mixing of glycerol in water?

Hypothesis/Prediction

(a) Write a prediction, complete with reasoning, based upon the concept of hydrogen bonding.

Experimental Design

Equal volumes of ethanol and water, and glycerol and water are mixed and the change in temperature is recorded.

(b) Identify the variables.

Materials

lab apron	two 10-mL graduated cylinders
eye protection	nested pair of polystyrene cups
distilled water	cup lid with centre hole
ethanol	two thermometers
glycerol	250-mL beaker (for support)

Procedure

(c) Write a complete procedure including disposal instructions.

Analysis

(d) Answer the Question.

Evaluation

(e) Evaluate the evidence by judging the experimental design, materials, and procedure. Note any flaws and improvements.

(f) How certain are you about the evidence obtained? Justify your answer, including possible experimental errors and uncertainties.

(g) If you think that the evidence is of suitable quality, evaluate the prediction and reasoning used. If not, discuss your reasons.

⚗ INVESTIGATION 4.6.1

Classifying Mystery Solids

Inquiry Skills

○ Questioning	● Planning	● Analyzing
○ Hypothesizing	● Conducting	● Evaluating
○ Predicting	● Recording	● Communicating

When analyzing an unknown solid, physical properties can help to quickly narrow down the possibilities to a particular class of solids—ionic, metallic, molecular, and covalent network.

Question

To what class of solids do the four mystery solids belong?

Experimental Design

(a) Write a general plan to answer the question.

Materials

(b) Using your design and commonly available materials, prepare a list of the materials to be used.

Procedure

(c) List all steps in the appropriate order to answer the question. Include safety precautions.

Analysis

(d) Identify the class of each of the solids.

Evaluation

(e) Discuss the quality of your evidence and how certain you are about the answer obtained. Suggest some improvements to increase the certainty of the classification.

Key Expectations

- Explain how the Valence-Shell-Electron-Pair-Repulsion (VSEPR) model can be used to predict molecular shape. (4.3)
- Predict molecular shape for simple molecules and ions, using the VSEPR model. (4.3)
- Predict the polarity of various substances, using molecular shape and electronegativity values of the elements of the substances. (4.4)
- Explain how the properties of a solid and liquid depend on the nature of the particles present and the types of forces between them. (4.4, 4.5, 4.6)
- Predict the type of solid (ionic, molecular, covalent network, or metallic) formed by a substance, and describe its properties. (4.6)
- Conduct experiments to observe and analyze the physical properties of different substances, and to determine the type of bonding that contributes to the attraction between molecules. (4.4, 4.5, 4.6)
- Describe some specialized new materials that have been created on the basis of the findings of research on the structure of matter, chemical bonding, and other properties of matter. (4.6)
- Describe advances in Canadian research on atomic and molecular theory. (4.6)
- Use appropriate scientific vocabulary to communicate ideas related to structure and bonding. (all sections)

Key Terms

bond dipole	ionic bonding
central atom	isoelectronic
covalent bond	London force
covalent bonding	nonpolar bond
covalent network	nonpolar molecule
crystal lattice	polar bond
dipole–dipole force	polar covalent bond
hydrogen bonding	polar molecule
intermolecular force	VSEPR
ionic bond	

EXTENSIONS

hybrid orbital	sigma (σ) bond
hybridization	valence bond theory
pi (π) bond	

Key Symbols

- $\sigma, \pi, \delta^+, \delta^-$

Problems You Can Solve

1. Predict the shape of simple molecules, using VSEPR theory. (4.4)
2. Predict the type of solid formed by a substance. (4.6)
3. Explain the properties of substances, based upon intermolecular forces and bonding. (4.5, 4.6)

▶ MAKE a summary

- Intra- and intermolecular forces can be explained in a unified way by describing the central particle that is simultaneously attracted (electrostatically) to the surrounding particles. Complete the following table.

Force or bond	Central particle	Surrounding particles
covalent		
covalent network		
dipole–dipole		
hydrogen		
ionic		
London		
metallic		

- Each class of substance has a characteristic set of properties. Complete the following table using relative descriptions such as negligible, low, medium, and high. (Indicate n/a if not applicable.)

Substance	Hardness	Melting point	Electrical conductivity		
			Solid	Liquid	Solution
molecular					
ionic					
covalent network					
metallic					

Identify each of the following statements as true, false, or incomplete. If the statement is false or incomplete, rewrite it as a true statement.

1. The shape of molecules of the rocket fuel hydrazine, $N_2H_{4(l)}$, is predicted by VSEPR theory to be trigonal planar around each nitrogen.

2. Diborane gas, $B_2H_{6(g)}$, is used to dope semiconductors. However, a Lewis structure cannot be drawn without modifying the theory; nor is a VSEPR diagram possible even though the compound exists, and is well known.

3. A central atom with two bonded atoms and two unshared electron pairs has a linear arrangement of its electron pairs.

4. Ionic substances are network solids, with a special type of metallic bonding.

5. Hydrogen bonding is possible whenever the molecule contains hydrogen atoms as well as N, O, and F atoms.

6. A molecule with a pyramidal shape and polar bonds will be nonpolar.

7. Larger atoms, like sulfur, can bond as central atoms in more ways than smaller atoms, like oxygen, because they have more complex electron structures.

8. Of the molecules HCl, HBr, and HI, the HI should have the highest boiling point.

9. The end of a soap molecule that attracts and dissolves oily dirt must be polar.

10. Covalent network solids generally have high melting points compared with molecular crystals.

Identify the letter that corresponds to the best answer to each of the following questions.

11. The Lewis model of the atom emphasizes the concept of
    (a) atoms gaining or losing electrons to become ions.
    (b) orbital hybridization.
    (c) electron energy level changes.
    (d) electron orbital overlap.
    (e) the stable octet of electrons.

12. The Lewis symbol for an oxide ion would show dots to represent
    ca) 2 electrons.
    (b) 8 electrons.
    (c) 10 electrons.
    (d) 18 electrons.
    (e) 32 electrons.

13. A Lewis symbol for an atom with a configuration of $1s^2\ 2s^2\ 2p^3$ would show
    (a) 1 unpaired electron and 3 electron pairs.
    (b) 2 unpaired electrons and 2 electron pairs.
    (c) 1 unpaired electron and 2 electron pairs.
    (d) 3 unpaired electrons and 1 electron pair.
    (e) 3 unpaired electrons and 2 electron pairs.

14. A Lewis structure for the molecule NCl_3 would show
    (a) 13 electron pairs.
    (b) 10 electron pairs.
    (c) 8 electron pairs.
    (d) 4 electron pairs.
    (e) 3 electron pairs.

15. X-ray diffraction evidence about the structure of compounds in crystals led to development of
    (a) structural models.
    (b) Lewis models.
    (c) VSEPR theory.
    (d) the octet rule.
    (e) energy level theory.

16. Which of the following molecules has a trigonal planar shape?
    (a) NH_3 (d) H_2O
    (b) CO_2 (e) BBr_3
    (c) PCl_3

17. Which of the following covalent bonds is the most polar?
    (a) $N-O$ (d) $H-Cl$
    (b) $C-H$ (e) $C-Cl$
    (c) $O-H$

18. The property that is best explained by intermolecular forces is
    (a) surface tension of a liquid.
    (b) electrical conductivity of a metal.
    (c) hardness of a covalent network solid.
    (d) melting point of an ionic solid.
    (e) the colour of copper.

19. Metallic bonding depends on
    (a) high electronegativity.
    (b) delocalized electrons.
    (c) polar covalent bonds.
    (d) electrical conductivity.
    (e) a full valence orbitals.

20. A molecule of a substance with physical properties primarily determined by London forces would be
    (a) SiC (d) PCl_3
    (b) KCl (e) H_2O_2
    (c) Na_3P

NEL An interactive version of the quiz is available online.
GO www.science.nelson.com

Chemical Bonding **281**

1. Draw Lewis symbols for atoms of the following elements and predict their bonding capacity:
   (a) calcium
   (b) chlorine
   (c) phosphorus
   (d) silicon
   (e) sulfur   (4.1)

2. Describe the requirements for valence electrons and orbitals in order for a covalent bond to form between two approaching atoms.   (4.2)

3. According to atomic theory, how many lone electron pairs are on the central atom in molecules of the following substances?
   (a) $HF_{(g)}$
   (b) $NH_{3(g)}$
   (c) $H_2O_{(l)}$
   (d) $CCl_{4(l)}$
   (e) $PCl_{3(l)}$ (4.2)

4. Draw Lewis symbols for atoms with the following electron configurations:
   (a) $1s^2\, 2s^2\, 2p^5$
   (b) $1s^2\, 2s^2\, 2p^6\, 3s^2\, 3p^3$
   (c) $1s^2\, 2s^2\, 2p^6\, 3s^2\, 3p^6\, 4s^1$
   (d) $[Ar]\, 4s^2\, 3d^{10}\, 4p^4$
   (e) $[Kr]\, 5s^2$ (4.2)

5. The American chemist G. N. Lewis suggested that atoms react in order to achieve a more stable electron configuration. Describe the electron configuration that gives an atom maximum stability.   (4.2)

6. Compounds of metals and carbon are used in engineering because of their extreme hardness and strength. The carbon in these metallic carbides behaves as the C^{4-} ion.
   (a) Write the electron configuration for a carbide ion.
   (b) Calculate the number of protons, electrons, and neutrons in a carbide ion, $^{12}_{6}C^{4-}$, and state their relative position in the ion.   (4.2)

7. The theory of hybridization of atomic orbitals was developed to explain molecular geometry. Sketch and name the shape of each of the following hybrid orbitals of a carbon atom in a compound:
   (a) sp (b) sp^2 (c) sp^3 (4.2)

8. Identify the types of hybrid orbitals found in molecules of the following substances:
   (a) $CCl_{4(l)}$
   (b) $BH_{3(g)}$
   (c) $BeI_{2(s)}$
   (d) $SiH_{4(g)}$ (4.2)

9. What is the difference between a sigma bond and a pi bond?   (4.2)

10. Indicate the number of sigma and the number of pi bonds in each of the following molecules:
    (a) $H_2O_{(l)}$
    (b) $C_2H_{2(g)}$
    (c) $C_2H_{4(g)}$
    (d) $C_2H_{6(g)}$ (4.2)

11. Describe any changes in the hybridization of the nitrogen and boron atoms in the following reaction.
    $$BF_{3(g)} + NH_{3(g)} \rightarrow F_3B - NH_{3(s)} \qquad (4.2)$$

12. The VSEPR model includes several concepts related to atomic theory. Explain the following concepts:
    (a) valence shell   (c) lone pair
    (b) bonding pair   (d) electron pair repulsion   (4.3)

13. Outline the steps involved in predicting the shape of a molecule using the VSEPR model.   (4.3)

14. Using VSEPR theory, predict the shape around each central atom in a molecule of each of the following substances:
    (a) $HI_{(g)}$
    (b) $BF_{3(g)}$
    (c) $SiCl_{4(l)}$
    (d) $CH_{4(g)}$
    (e) $HCN_{(g)}$
    (f) $OCl_{2(g)}$
    (g) $NH_4^+{}_{(aq)}$
    (h) $H_2O_{2(l)}$ (4.3)

15. (a) Draw Lewis and shape diagrams for ammonia, methane, and water molecules.
    (b) Use these diagrams to explain why the molecular bond angles decrease in the order, methane > ammonia > water.   (4.3)

16. Carbon dioxide is used by green plants in the process of photosynthesis and is also a greenhouse gas produced by fossil fuel combustion.
    (a) Draw a Lewis structure for carbon dioxide.
    (b) Name the shape of a carbon dioxide molecule and give its bond angle.
    (c) Using appropriate bonding theories, predict and explain the polarity of carbon dioxide.   (4.4)

17. The polarity of a molecule is determined by bond polarity and molecular shape.
    (a) Compare the polarity of the bonds $N-Cl$ and $C-Cl$.
    (b) Predict whether the molecules, $NCl_{3(l)}$ and $CCl_{4(l)}$, are polar or nonpolar. Explain your predictions.   (4.4)

18. Use appropriate bonding theory to explain the following experimental observations:
    (a) BeH_2 is nonpolar; H_2S is polar.
    (b) BH_3 is planar; NH_3 is pyramidal.
    (c) LiH has a melting point of 688°C; that of HF is −83°C.   (4.5)

19. Use the theory of intermolecular bonding to explain the sequence of boiling points in the following alkyl bromides: $CH_3Br_{(g)}$ (4°C), $C_2H_5Br_{(l)}$ (38°C), and $C_3H_7Br_{(l)}$ (71°C).   (4.5)

20. Name the intermolecular forces present in the following compounds and account for the difference in

their boiling points: $CH_{4(g)}$ (−164°C), $NH_{3(g)}$ (−33°C), and $BF_{3(g)}$ (−100°C). (4.5)

21. Ionic compounds and metals have different physical properties because of the different forces involved. For example, while sodium chloride and nickel have nearly identical molar masses, their melting points, conductivity, and solubility in water are quite different.
    (a) Explain the large difference in melting point between sodium chloride (801°C) and nickel metal (1453°C).
    (b) Predict the electrical conductivity of each of these substances in the solid state, and provide a theoretical explanation for your prediction.
    (c) Predict the solubility in water of each substance, and provide a theoretical explanation for your prediction. (4.6)

22. Name the forces acting between particles in each of the following substances:
    (a) hexane, $C_6H_{14(l)}$
    (b) 1-butanol, $C_4H_9OH_{(l)}$
    (c) ethylamine, $C_2H_5NH_{2(l)}$
    (d) chloroethane, $C_2H_5Cl_{(l)}$
    (e) calcium carbonate, $CaCO_{3(s)}$
    (f) diamond, $C_{n(s)}$ (4.6)

Applying Inquiry Skills

23. An investigation is to be done to see how well intermolecular force concepts can predict differences in solubility.

    Question
    What is the order from lowest to highest solubility in water for: pentane, $C_5H_{12(l)}$, 1-butanol, $C_4H_9OH_{(l)}$, diethyl ether, $(C_2H_5)_2O_{(l)}$, butanoic acid, $C_3H_7COOH_{(l)}$?

    Prediction
    (a) Predict the answer to the question, including your reasoning for each substance.

    Experimental Design
    (b) Design an experiment to answer the question. Include a brief plan and variables.

    Materials
    (c) Prepare a list of materials.

    Procedure
    (d) Write a numbered list of steps, including disposal instructions. (4.5)

24. Hydrocarbons can be oxidized step by step through a series of compounds until they are converted to carbon dioxide and water, e.g., methane (CH_4), methanol (CH_3OH), methanal (CH_2O), methanoic acid ($HCOOH$), carbon dioxide (CO_2). For each compound in this series draw a structural diagram, and then describe the molecular shape. (4.5)

25. Compare the particles and forces in the following pairs of solids:
    (a) metallic and covalent network
    (b) covalent network and molecular
    (c) molecular and ionic (4.6)

Making Connections

26. Methylamine, CH_3NH_2, is one of the compounds responsible for the unpleasant odour of decomposing fish.
    (a) Draw Lewis and structural diagrams for methylamine.
    (b) Use VSEPR theory to predict the shape around the carbon and nitrogen atoms in methylamine.
    (c) Methylamine and ethane have similar molar masses. Explain why the boiling point of methylamine is −6°C while that of ethane is −89°C.
    (d) Since amines are bases they react readily with acids. Use structural diagrams to rewrite the following equation for the reaction of methylamine with acetic acid:

    $$CH_3NH_{2(aq)} + CH_3COOH_{(aq)} \rightarrow$$
    $$CH_3NH_3{}^+_{(aq)} + CH_3COO^-_{(aq)}$$

    (e) Explain how vinegar and lemon juice can be used to reduce the odour of fish. (4.3)

27. What material is used in the outer skin of a stealth bomber (**Figure 1**)? Describe how the structure and properties of this material relate to its function.

    Figure 1 (4.6)

     www.science.nelson.com

Extension

28. Chlorine is a very reactive element that forms stable compounds with most other elements. For each of the following chlorine compounds, draw Lewis and structural diagrams, and then predict the polarity of the molecules:
    (a) NCl_3 (c) PCl_5
    (b) $SiCl_4$ (d) SCl_6

A Study of an Element

To gain as much understanding of an element as possible one must know its properties. In the first chapter of this unit, you studied concepts describing the internal structure of the atom of the elements. In the second chapter, you studied the intra- and intermolecular bonding of the elements. Chemistry, more than any other science, produces useful products and processes. Some of the technological uses of the elements have been mentioned in this unit, but many more exist. When completing the task below, you have an opportunity to learn about one element in detail—and then share your information with others.

Task

Choose an element to research and report about. Your report should be on a different element from that of any other student, so you may not get your first choice. The information provided on the element should cover as many of the concepts presented in this unit as possible. Act as if you are an (ethical) salesperson for the element. Sell its usefulness, while also presenting its negative side, to research, consumer, commercial, and/or industrial users. You can choose the type of communication that best suits your purpose, abilities, and interests. Although the emphases in your report should be scientific and technological, try to go beyond these to include economic, ecological, legal, ethical, social, and aesthetic perspectives.

▶ Criteria

Your completed task will be assessed according to the following criteria:

Process

- Use a variety of sources for your research.

Product

- Prepare a report in written or electronic format that follows the stated guidelines.
- Demonstrate an understanding of the concepts developed in this unit.
- Use terms, symbols, equations, and SI metric units correctly.

Figure 1
Gold, iodine, and titanium are only three of about 114 elements.

Report Guidelines

Your report or sales pitch should include the following:

- a brief history of the element
- a complete description of the element from a scientific perspective
  —the isolated atom
  —the atom in the element
  —the atom in its compounds
- the physical properties of the element
  —an explanation of its physical properties
  —uncertainties that you have about these explanations
- the chemical properties of the element
  —an explanation of its chemical properties
  —uncertainties that you have about these explanations
- technological applications of the element
- other perspectives on the element, including MSD sheet and WHMIS symbol, if applicable
- careers associated with the element
- a bibliography of the literature used to create your report

Identify each of the following statements as true, false, or incomplete. If the statement is false or incomplete, rewrite it as a true statement.

1. The term "orbital" refers to the path or trajectory an electron follows as it orbits a nucleus.

2. The configuration [Ne] $2s^2 2p^1$ represents an aluminum atom in its lowest energy state. *F*

3. Rutherford knew that alpha particles were small and massive, and when moving fast should act much as bullets do when striking a target. He expected them to punch through his foil target and be slowed enough to let him determine the density of the atoms in the foil.

4. Light passed through a flame may have certain frequencies absorbed, because ions in the gas have electrons jump from lower energy levels to higher energy levels.

5. The ground state electron configuration for all alkali metals shows that the highest energy electrons are in a p sub level. *F*

6. There are thought to be seven d energy sublevels. *10 F*

7. Spectra from atoms larger than hydrogen do not follow simple "rules" because when an atom has multiple electrons they repel each other and interfere with each other's orbital.

8. Schrödinger became famous by predicting that the particles called electrons might behave like waves under certain conditions, and then demonstrating this experimentally.

9. The aufbau principle states that when electron configurations are written, the lower energy levels must be filled before the higher levels.

10. VSEPR theory predicts that a sulfate ion, SO_4^{2-}, should be tetrahedral in shape. *F*

11. VSEPR theory predicts that a central atom with three bonded atoms and one lone pair of electrons should have a trigonal planar shape. *F*

12. A hydrogen bond is a particularly strong intermolecular bond existing between a hydrogen on one molecule and a lone pair of electrons on another molecule.

13. VSEPR and Lewis theories are not complete enough to explain the structure and shape of the molecules in gaseous uranium hexafluoride, $UF_{6(g)}$, which is used in uranium nuclear fuel-enriching processes.

14. VSEPR and Lewis theories are not complete enough to explain the structure and shape of the molecules in gaseous silane, $SiH_{4(g)}$, which is used as a doping agent in the manufacture of semiconductors for solid-state devices.

15. A molecule with tetrahedral shape and all bonds equally polar will be nonpolar, overall. *T*

16. A three-atom molecule with linear shape and two identical atoms attached to the central atom will always be nonpolar. *T*

17. Metallic bonding involves 3-D structures with vacant valence orbitals and mobile valence electrons. *T*

18. Ionic bonding involves 3-D structures with vacant valence orbitals and mobile valence electrons. *F*

19. Silver normally forms a 1+ ion, indicating that normally only one electron occupies its highest energy level. *T*

Identify the letter that corresponds to the best answer to each of the following questions.

20. The atomic structure that did not follow directly from Rutherford's experiments is the idea of the
    (a) electron.
    (c) neutron.
    (e) "empty" atom.
    (b) proton.
    (d) nucleus.

21. Observing a frequency of light emitted by a hot gas will also allow prediction of a frequency that this same gas will absorb, when cool, according to theory advanced by
    (a) Rutherford.
    (c) Planck.
    (e) Chadwick.
    (b) Bohr.
    (d) Heisenberg.

22. The concept of atomic structure contributed by Niels Bohr is that
    (a) atoms can absorb and release only specific frequencies of electromagnetic radiation.
    (b) protons are extremely close together in a tiny part of the atomic volume.
    (c) electrons can have only certain specific different levels of energy.
    (d) electrons orbit a nucleus like tiny planets orbiting a star.
    (e) uncharged particles exist in the nucleus.

23. The biggest flaw in Bohr's theory was that it
    (a) did not apply to atoms larger than hydrogen.
    (b) predicted electrons would slow and spiral into the nucleus.
    (c) did not explain blackbody radiation.
    (d) predicted that protons in nuclei would repel and fly apart.
    (e) ignore the structure of the nucleus.

24. An energy-level diagram for fluorine would show the highest level of energy for
    (a) 7 electrons.   (c) 10 electrons.   (e) 19 electrons
    (b) 9 electrons.   (d) 18 electrons.

25. The major differences in electron energy levels are described by the
    (a) principal quantum number, n.
    (b) secondary quantum number, l.
    (c) magnetic quantum number, m_l.
    (d) spin quantum number, m_s.
    (e) exclusion principle.

26. The concept that electrons are oriented along different axes in 3-dimensional space is described by the
    (a) principal quantum number, n.
    (b) secondary quantum number, l.
    (c) magnetic quantum number, m_l.
    (d) spin quantum number, m_s.
    (e) electron configuration.

27. The evidence that all substances are attracted or repelled by a magnetic field is described by the
    (a) principal quantum number, n.
    (b) secondary quantum number, l.
    (c) magnetic quantum number, m_l.
    (d) spin quantum number, m_s.
    (e) electron configuration.

28. The electron configuration of a chlorine atom in its lowest energy state is
    (a) $1s^2\,2s^2\,2p^6\,3s^2\,3p^6$.   (d) $s^2\,2s^2\,2p^6$.
    (b) $1s^2\,2s^2\,2p^6\,3s^2\,3p^5$.   (e) $1s^2\,2s^2\,2p^5$.
    (c) $1s^2\,2s^2\,2p^6\,3s^1$.

29. The electron configuration of a calcium ion in its lowest energy state is
    (a) $1s^2\,2s^2\,2p^6\,3s^2\,3p^6\,4s^2$.
    (b) $1s^2\,2s^2\,2p^6\,3s^2\,3p^6\,4s^1$.
    (c) $1s^2\,2s^2\,2p^6\,3s^2\,3p^4\,4s^2$.
    (d) $1s^2\,2s^2\,2p^6\,3s^2\,3p^6$.
    (e) $1s^2\,2s^2\,2p^6\,3s^1\,3p^5$.

30. The electron configuration that could be a fluoride ion in an "excited" energy state is
    (a) $1s^2\,2s^2\,2p^5$.   (d) $1s^2\,2s^2\,2p^5\,3s^1 4s^1$.
    (b) $1s^2\,2s^2\,2p^4\,3s^1 4s^1$.   (e) $1s^2\,2s^2\,2p^6\,3s^1 4p^1$.
    (c) $1s^2\,2s^2\,2p^6$.

31. The idea of special stability due to the presence of a stable octet of electrons is central to
    (a) Kekulé line diagrams.
    (b) Bohr atomic structure.
    (c) VSEPR molecular shape prediction.
    (d) Pauling hybrid orbitals.
    (e) Lewis dot diagrams.

32. The Lewis symbol of a calcium ion would show dots to represent
    (a) 0 electrons.   (c) 10 electrons.   (e) 12 electrons.
    (b) 2 electrons.   (d) 18 electrons.

33. The Lewis symbol of a magnesium atom would show dots to represent
    (a) 0 electrons.   (c) 8 electrons.   (e) 12 electrons.
    (b) 2 electrons.   (d) 10 electrons.

34. A Lewis symbol for an atom with a most stable electron configuration of $1s^2\,2s^2\,2p^6\,3s^2\,3p^6\,4s^2$ would show
    (a) 2 unpaired electrons and 2 electron pairs.
    (b) 1 unpaired electron and 3 electron pairs.
    (c) 2 unpaired electrons.
    (d) 1 electron pair.
    (e) 1 electron.

35. A Lewis symbol for a negative ion with a configuration of $1s^2\,2s^2\,2p^6$ would show
    (a) 3 unpaired electrons and 2 electron pairs.
    (b) 1 unpaired electron and 3 electron pairs.
    (c) 2 unpaired electrons and 3 electron pairs.
    (d) no electrons, either single or paired.
    (e) 4 electron pairs.

36. A Lewis structure for the sulfur dichloride molecule, SCl_2, would show
    (a) 3 electron pairs.   (d) 12 electron pairs.
    (b) 4 electron pairs.   (e) 18 electron pairs.
    (c) 10 electron pairs.

37. A Lewis structure for the hydroxide ion, OH^-, would show
    (a) 3 electron pairs.   (d) 10 electron pairs.
    (b) 4 electron pairs.   (e) 12 electron pairs.
    (c) 7 electron pairs.

38. The molecule in the following list that has a linear shape, according to VSEPR theory, is
    (a) H_2O.   (c) CO_2.   (e) CH_3COOH.
    (b) OF_2.   (d) H_2O_2.

39. The hydrogen bonding of large molecules is a very important area of study in biochemistry. A hydrogen bond can only form at a location on a large molecule where a hydrogen atom is bonded either to an oxygen atom, or to
    (a) a nitrogen atom.   (d) a sulfur atom.
    (b) a chlorine atom.   (e) another hydrogen atom.
    (c) a fluorine atom.

40. The atoms of hard, brittle substances with high melting points are essentially all joined in a network of
    (a) ceramic bonds.   (d) covalent bonds.
    (b) coordinate bonds.   (e) hydrogen bonds.
    (c) metallic bonds.

NEL   An interactive version of the quiz is available online.
GO www.science.nelson.com

Structure and Properties   287

Understanding Concepts

1. In Rutherford's classic experiment, it was found that most of the alpha particles in a directed beam passed through a metal foil essentially unaffected, while a very few of them were quite significantly deflected—some, almost straight backward. Explain what each part of this evidence indicates about the structure of the layers of atoms within the metal foil. (3.1)

2. Briefly outline the experimentation that led to the discovery of each of these subatomic particles:
   (a) the electron
   (b) the proton
   (c) the neutron (3.1)

3. Atoms of an element may differ in mass, and sometimes also in radioactivity.
   (a) Describe how this is explained as a result of structure within the atom.
   (b) State the term applied to such atoms. (3.1)

4. In 1900, classical theory suggested that warm substances radiating electromagnetic energy (like chemistry students) should emit mostly very short wavelengths. Thus, that theory predicts that people will radiate mostly ultraviolet light—but evidence shows they emit about 0.10 kJ/s, almost all of it as very long-wave infrared energy. Explain what Max Planck suggested as a way to deal with this evidence, which conflicted so dramatically with accepted theory. (3.3)

5. Rutherford suggested that electrons be thought of as orbiting a nucleus, like little planets orbiting a star. Explain why he assumed electrons could not be stationary—that is, why they somehow had to be moving around a nucleus. (3.3)

6. Bohr knew that according to electromagnetic theory, evidence clearly indicated that electrons could not really be travelling in circular orbits (or elliptical orbits, as Sommerfeld suggested) around a nucleus. Describe the evidence that would be observed if this were, in fact, the way electrons behave. (3.4)

7. Bohr had to ignore classical electromagnetic theory to make his own theory consistent. State Bohr's First Postulate, the first example of this break with tradition. (3.4)

8. The theory that electron energy change is quantized, that is, can occur only in specific amounts, is central to the development of Bohr's theory. State Bohr's Second Postulate, which establishes this concept. (3.4)

9. Draw an electron orbital energy-level diagram for each of the following simple atoms or ions:
   (a) nitrogen atom
   (b) sulfide ion
   (c) potassium ion
   (d) beryllium atom
   (e) zirconium atom (3.6)

10. Technetium metal (element 43) does not exist in nature because it has no isotope that is not highly radioactive. The metal is created in nuclear reactors as a fission byproduct of uranium radioactive decay, and can be obtained from spent nuclear reactor fuel. Based on an electron orbital energy-level diagram for technetium, predict whether it is attracted by a magnet, and explain what theory enables you to use the electron configuration to make this prediction. (3.6)

11. When water is poured into a glass, the bottom of the glass fills first, and when allowed to stand, the water surface becomes level. State which of these phenomena is similar to the application of the aufbau principle for electron configurations of atoms, and which is similar to Hund's Rule. (3.6)

12. The so-called transition elements, in Groups 3 to 12 of the periodic table (the "B" group), have chemical and physical properties that do not vary with the same simple periodicity as do those of "A" group elements. Explain why this is so, in terms of electron orbital configuration. (3.6)

13. Write a complete electron configuration for each of the following entities:
   (a) titanium atom
   (b) technetium atom
   (c) iron(III) ion
   (d) bromide ion
   (e) selenide ion (3.6)

14. Write a shorthand electron configuration for each of the following entities:
   (a) zirconium atom
   (b) mercury atom
   (c) radium atom
   (d) iodide ion
   (e) uranium(VI) ion (3.6)

15. State the maximum possible number of orbitals, and of electrons, in an f sublevel. (3.6)

16. Identify the following atoms or ions from their electron configurations:
   (a) atom : $1s^2\ 2s^2\ 2p^6\ 3s^2\ 3p^6\ 4s^2\ 3d^{10}\ 4p^5$
   (b) 1^+ ion : $1s^2\ 2s^2\ 2p^6\ 3s^2\ 3p^6\ 4s^2\ 3d^{10}\ 4p^6\ 4d^{10}$

(c) 4^+ ion : (Xe) $4f^{14}$

(d) atom : (Kr) $5s^2 4d^{10} 5p^1$

(e) 2^- ion $1s^2 2s^2 2p^6 3s^2 3p^6$ (3.6)

17. State the valence orbital and valence electron conditions that must exist on *each* atom, in order for an ionic bond to form between two approaching neutral atoms. (4.2)

18. State the primary factor controlling the packing together of ions (formation of the crystal lattice) in solid ionic compounds. What other factor(s) might affect the structure of the lattice? (4.2)

19. Describe the structural conditions that must apply on *each* molecule in order for a single hydrogen bond to form between two approaching neutral molecules. (4.2)

20. Draw Lewis symbols for atoms with the following electron configurations:
    (a) $1s^2 2s^2 2p^6 3s^2 3p^6 4s^2 3d^{10} 4p^4$
    (b) $1s^2 2s^2 2p^6 3s^2 3p^6 4s^2 3d^{10} 4p^6 5s^2 4d^{10} 5p^2$
    (c) [Ar] $4s^2$
    (d) [Kr] $5s^2 4d^{10} 5p^1$
    (e) [Xe] $6s^2$ (4.2)

21. Describe the physical and chemical properties of elements with electron configurations ending in $\#s^1$, where $\#$ is any integer from 2 to 6. (4.2)

22. Describe the physical and chemical properties of elements with electron configurations ending in $\#p^6$, where $\#$ is any integer from 2 to 6. (4.2)

23. Describe the chemical properties of elements with electron configurations ending in $\#p^5$, where $\#$ is any integer from 2 to 6, and explain why the physical properties cannot be generalized. (4.2)

24. Write out the words represented by the acronym VSEPR. (4.3)

25. Use VSEPR theory to predict the shape around the central atom(s) of the following molecules:
    (a) H_2O_2
    (b) SiF_4
    (c) NI_3
    (d) H_2S
    (e) CS_2 (4.3)

26. State which of the molecules SiF_4 and NF_3 should have smaller bond angles, and how VSEPR theory explains this. (4.3)

27. For the common substance found in household white vinegar—acetic acid, $CH_3COOH_{(aq)}$, do the following:
    (a) Draw the Lewis structure.
    (b) Use VSEPR theory to predict the shape around the three atoms that act as central atoms.

(c) Draw a 3-D representation of the molecular shape.

(d) Predict the polarity of each bond in the molecule, and whether the molecule will be polar overall.

(e) Predict the predominant type of intermolecular bonding between acetic acid molecules in pure liquid state. (4.5)

28. The molecules H_2S and F_2 are isoelectronic. Explain what this means and what type of intermolecular bonding force may be predicted approximately for isoelectronic substances. (4.5)

29. Predict which of the substances $H_2S_{(g)}$ and $F_{2(g)}$ will have a higher boiling point; what this means in terms of the intermolecular forces present; and how VSEPR theory and electronegativity tables allow this prediction to be made. (4.6)

Applying Inquiry Skills

30. Write a brief experimental plan to distinguish the following pairs of substances. Identify the property to be tested and include a brief explanation of the principles involved in each test.
    (a) $He_{(g)}$ and $Ne_{(g)}$
    (b) $MnCl_{2(s)}$ and $ZnCl_{2(s)}$
    (c) $Zn_{(s)}$ and $I_{2(s)}$
    (d) $CaCO_{3(s)}$ and $SiO_{2(s)}$ (3.1)

31. Theories are valued for how well they explain and predict. Bohr's first and second postulates established the conditions necessary for his theory, but they were arbitrary statements, explaining nothing about themselves. State how de Broglie's concept of an electron considered as a standing wave *explained* Bohr's first and second postulates. (3.7)

32. Explain how de Broglie's theory removed the concept problem caused by the lack of observed electromagnetic energy radiation from any electron in any stable atom, which would naturally be expected from any *moving* negative particle. (3.7)

33. In an experiment to study a group of solids, each solid is rubbed across the surface of each other solid to see if a scratch mark occurs.
    (a) Identify the independent, dependent, and controlled variables.
    (b) What property of a solid is being tested?
    (c) How does this property depend on the nature of the particles present and the types of forces between them?

(d) What type of solid would not be able to be scratched by any other solids? Give an example. (4.6)

34. As part of an inquiry skill test, a student was required to use her knowledge of London forces to predict the boiling point of the compound $SiBr_4$, given boiling points of the other silicon halides. Complete the Analysis and Evaluation of her report.

Evidence

The boiling point of $SiF_{4(g)}$ is –85°C, the boiling point of $SiCl_{4(l)}$ is 58°C, and the boiling point of $SiI_{4(s)}$ is 290°C. The physical states shown in each case are for the compound at SATP conditions.

Analysis

(a) Plot a graph of boiling point versus the total molecular electron count, and use it to make a prediction.

Evaluation

(b) Look up an accepted value for the boiling point of $SiBr_4$, and calculate the accuracy of your prediction as a % difference. (4.6)

Making Connections

35. In steel mills and foundries, a device called an optical or wire pyrometer is often used to measure the approximate temperature of steel that is hot enough to glow by its own emitted light. Research and report on the operating principle of this type of pyrometer and refer to Planck's theory of light emission in your explanation. (3.3)

 www.science.nelson.com

36. In DNA, the pairing of nitrogen bases depends on hydrogen bonding between two molecular structures that must be precisely the right size and shape to bring the hydrogen atoms and lone pairs of electrons together. Research and report on the names of the nitrogen bases that bond together in a DNA double helix structure; and on how many hydrogen bonds are formed between the two different base pairings. (4.5)

37. Describe a carbon nanotube and some of its properties. How does the bonding within the nanotube molecule compare with other forms of pure carbon? What have scientists been able to do with these tubes? List some potential applications of these nanotubes. (4.6)

 www.science.nelson.com

38. Kevlar is a specialized synthetic material that has some remarkable properties. List some of these properties and describe some different applications. What makes Kevlar so strong? Include information about molecular structure, shape, and bonding in your answer. (4.6)

 www.science.nelson.com

39. Diamond is the hardest material known and it is used in cutting tools for any other substance, including very hard rocks like granite. And yet, diamonds are cut and polished. How is this possible? Describe the procedure and materials used. (4.6)

 www.science.nelson.com

Extensions

40. In metal refining, a device called an immersion thermocouple is often used to measure the temperature of metals in liquid state. Research and report on the operating principle of this device and refer to quantum theory effects in your discussion of the thermocouple principle.

41. Light-emitting diodes are becoming quite common as light sources in applications where low power consumption, very long life, and simple construction are required. LEDs have been used for appliance displays for years. High-level brake lights on automobiles are now often made of multiple arrays of small LEDs. Research and report on the interaction of semiconductors and electrons that can cause monochromatic light to be released by these kinds of diodes.

42. Most metallic crystals reflect (absorb and re-emit) all colours (wavelengths) of visible light well—so well, in fact, that thin films of silver or aluminum make excellent mirrors. Gold, obviously, does not reflect all colours equally well. Research and report on the appearance of very thin films of gold under transmitted and under reflected light conditions.

43. Use the relationship $c = f\lambda$, where c is the speed of light, 3.00×10^8 m/s, and λ is the wavelength, to calculate the frequencies of the longest (700 nm—red) and shortest (400 nm—blue) wavelengths of visible light.

44. Use the relationship $\Delta E = hf$, where h is Planck's constant, 6.63×10^{-34} J•s, and f is the frequency, to calculate the energy of photons of the highest and lowest frequencies of light visible to humans.

45. The relationships among frequency, wavelength, and photon energy for all electromagnetic radiation, including that portion of the spectrum we call "light," is described mathematically by $c = f\lambda$, and $\Delta E = hf$. Complete the following statement: The shorter the wavelength, the _____ the frequency, and the _____ the photon energy, for electromagnetic radiation. Use this relationship to explain why X rays are highly dangerous to living things, and radio waves are not.

46. Concern over a decrease in concentration of the ozone in the stratosphere is based on the fact that ozone absorbs some of the Sun's ultraviolet light, before it reaches Earth's surface. The so-called ozone "layer" is very diffuse—the ozone present would make a layer only 3 mm thick if it were collected together at SATP. Ultraviolet light is arbitrarily divided into categories by wavelength: UVA is 320–400 nm, UVB is 280–320 nm, and UVC is 200–280 nm. The ozone layer passes all UVA and blocks all UVC—so any human exposure concern is about changes in levels of UVB. Calculate the energy of 300-nm wavelength ultraviolet photons from the relationship $\Delta E = hf$, and compare this energy to that of 600-nm wavelength of light, which is a visible colour we see as orange.

47. Photoluminescence such as that exhibited by basic fluorescein solution (see the Try This Activity at the beginning of this Unit) is based on molecules absorbing the energy of photons and then releasing energy, after a time delay, as photons that fall into the human visible range. Fluorescein, when illuminated with white light, is observed to emit green photons. Based on your knowledge of the process by which atoms absorb and emit electromagnetic energy, predict whether basic fluorescein solution would fluoresce if illuminated with
    (a) red light.
    (b) ultraviolet light.
    (c) infrared light.
    (d) blue light.

48. In the fourth row of the periodic table, chromium and copper electron configurations seem to violate Hund's rule, indicating that a half-filled or full d orbital energy level is especially stable. Assume that this "subrule" holds true. What will be the predicted electron configurations of molybdenum and silver?

49. Paired electrons in any orbital are very slightly repelled by a magnetic field—a property known as diamagnetism. All atoms are diamagnetic, but if they have enough unpaired electrons, the stronger magnetic attraction called paramagnetism predominates, and they are slightly attracted. Five elements show a very much stronger magnetic attraction—strong enough to be significant under ordinary conditions. We commonly call such materials "magnetic," but the correct term is ferromagnetic—meaning similar to the magnetism shown by the period 4, Group VIIIB elements, iron, cobalt, and nickel (**Figure 1**). Ferromagnetism is a quantum effect, involving what scientists call *domains*. Research and report how electron spin and domain conditions cause ferromagnetic elements to have very much stronger magnetic effects than other elements, which two other elements show ferromagnetism, and why strong heating weakens magnetic effects in solids.

50. VSEPR theory describes three possible shapes for molecules when there are six electron pairs around a central atom. Refer to the Web or any university-level chemistry resource to draw the structures showing these general shapes, and predict the shape of molecules of gaseous uranium hexafluoride, $UF_{6(g)}$, which exists at very high temperatures and is used in processing enriched nuclear fuels. The electron configuration for uranium is $[Rn]\ 7s^2\ 5f^3\ 6d^1$.

Figure 1
Extremely strong "neodymium" permanent magnets have recently become available, and are now sold and used for all kinds of science and workshop applications. These magnets are actually intermetallic compounds, like $Nd_2Fe_{14}B$—which is interesting, because neither neodymium nor boron is ferromagnetic in elemental form.

Energy Changes and Rates of Reaction

Paul Berti
Professor
McMaster University

"I hated high school. I was bored out of my wits and nearly dropped out more than once. Instead, I stuck with it and got the necessary educational background, and I am glad I did, because discovery is an incredible rush. Doing research means studying things that are not understood by anyone in the world. Doing science is active. For example, we've all seen the headlines in newspapers and on TV about the increasing threat of antibiotic-resistant bacteria. We are working to invent new kinds of antibiotics. Our main targets are enzymes that bacteria need to survive, but that don't exist in humans or other animals. If we understand how these enzymes work in detail, we can design inhibitors that will be antibiotics without being toxic to humans. Like all catalysts, enzymes lower the activation energy of reactions by binding to and stabilizing the transition state. The challenge is to use chemical techniques to study something in detail that exists for one ten-trillionth of a second."

▶ Overall Expectations

In this unit, you will be able to

- demonstrate an understanding of energy transformations and kinetics of physical and chemical changes;
- gather and analyze experimental evidence using calculations and graphs to determine energy changes and rates of reaction;
- show how chemical technologies and processes depend on the energetics and rates of chemical reactions.

ARE YOU READY?

Concepts

- endothermic and exothermic chemical reactions
- potential energy, kinetic energy, and conversions of energy from one form to another
- heat terms as part of chemical reactions, including combustion of hydrocarbons
- factors affecting the rate of a chemical reaction

Skills

- using calorimetry to determine the amount of heat exchanged during a reaction
- calculation of quantity of heat using $q = mc\Delta T$
- expressing rate of reaction in terms of quantity of reactant consumed or product produced per unit time.

Knowledge and Understanding

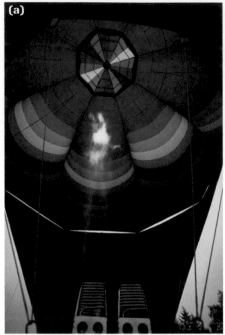

(a)

(b)

Figure 1
There are different changes of energy in these examples.

(d)

(c)

1. This unit is about energy in chemical systems. What type of energy is converted to thermal energy in each of the photos in **Figure 1**?

2. Chemical energy is described as a form of potential energy. Why?

3. (a) Explain the difference between exothermic and endothermic changes, with an example of each.
   (b) Identify each of the following as an exothermic or an endothermic change:
      (i) a campfire burns;
      (ii) ice melts;
      (iii) frost forms on a window;
      (iv) when two chemicals are mixed together, the container becomes very cold.

4. When a piece of chalk is placed in dilute hydrochloric acid, the chalk slowly dissolves and a colourless gas is produced. Suggest as many ways as possible that this chemical change can be made to go faster.

5. An experiment is performed to measure the heat produced when a chemical is dissolved in water in a calorimeter (**Figure 2**). The formula for heat transferred in this chemical change is $q = mc\Delta T$.
   (a) What quantity does each symbol represent?
   (b) Name a common unit for each of the quantities.
   (c) Using the evidence in **Table 1**, calculate how much heat, measured in joules, is transferred during the experiment.

Table 1 Calorimetry Evidence

mass of water	200.0 g
specific heat capacity of water	4.18 J/(g·°C)
initial temperature of water	30°C
final temperature of water	45°C

Inquiry and Communication

6. When a hydrocarbon fuel is burned in excess oxygen, it reacts completely (**Figure 3**). Write balanced thermochemical equations to represent the combustion of
   (a) methane ($CH_{4(g)}$) (molar heat of combustion = 890 kJ/mol);
   (b) ethane ($C_2H_{6(g)}$) (molar heat of combustion = 1560 kJ/mol);
   (c) propane ($C_3H_{8(g)}$) (molar heat of combustion = 2220 kJ/mol);
   (d) butane ($C_4H_{10(g)}$) (molar heat of combustion = 2858 kJ/mol).

7. Rewrite each of the equations in the previous question so that it represents one mole of fuel burning. You may need to use fractional coefficients.

Mathematical Skills

8. When solid sodium hydrogen carbonate is added to aqueous sulfuric acid, carbon dioxide gas is produced. The unbalanced chemical equation is:

   $$NaHCO_{3(s)} + H_2SO_{4(aq)} \rightarrow Na_2SO_{4(aq)} + H_2O_{(l)} + CO_{2(g)}$$
   (a) Balance the equation.
   (b) When 10 mol of the solid is added, the reaction takes 4 min. What is the average rate of consumption of sodium hydrogen carbonate (in mol $NaHCO_3$/min)?
   (c) What amount (in moles) of carbon dioxide gas will be produced if all 10 mol of sodium hydrogen carbonate solid is consumed?
   (d) What is the average rate of production of carbon dioxide (in mol CO_2/min) if 10 mol of sodium hydrogen carbonate is consumed in 4 min?

Technical Skills and Safety

9. You will be performing activities that involve combustion in this unit. Suggest four safety procedures that should be followed to prevent or deal with a fire in the laboratory.

Making Connections

10. The rate of a chemical reaction is affected by a number of different factors. Imagine that you were about to assemble materials to build and start a campfire. How could you apply what you know about factors affecting rate to make the campfire burn more quickly?

Figure 2
A simple calorimeter

Figure 3
Lighters burn butane to produce carbon dioxide, water vapour, and thermal energy.

chapter

5

Thermochemistry

▶ **In this chapter, you will be able to**

- compare the energy changes resulting from physical, chemical, and nuclear changes;

- represent such energy changes using thermochemical equations and potential energy diagrams;

- determine enthalpies of reaction both experimentally and by calculation from Hess's law and standard enthalpies of formation;

- compare conventional and alternative sources of energy;

- recognize examples of technologies that depend on exothermic and endothermic changes.

Energy transformations are the basis for all of the activities that make up our lives. When we breathe or walk or ride a bicycle, we use the chemical process of respiration and a whole series of complex metabolic reactions to convert the chemical energy in food into mechanical energy. Our home furnaces burn fossil fuels such as wood, coal, oil, or natural gas to produce heat that keeps us comfortable in this northern climate. Plants in forests and our fields take in sunlight and change the solar energy into chemical potential energy — stored carbohydrates — that may be further processed by animals and by ourselves — consider the many transformations necessary to cook corn, for example.

You are already familiar with many energy changes, both chemical and physical. As a tennis player sprints for the ball, glucose molecules in muscle cells react to form carbon dioxide, water, and energy. At the same time a physical change — the evaporation of perspiration — consumes energy and helps the player maintain a constant body temperature.

Many energy changes have significant effects on our way of life and our future. Transportation, whether in cars or by mass transit, depends to a large extent on the combustion of fossil fuels. This combustion produces energy of motion but also releases carbon dioxide, with its possible links to the greenhouse effect, and other pollutants. Even electrically powered vehicles use energy that is generated in part in nuclear or fossil fuel-burning power plants.

Energy is a major factor in decision making on our planet. Most sources of energy are finite, and using each has its advantages and disadvantages. The control and use of our present energy resources and the decisions that we make for the future will continue to have far-reaching environmental, economic, social, and political effects for many years.

💡 ▎ **REFLECT** on your learning ▼

1. Consider the following changes: ice melting, water evaporating, water vapour condensing, photosynthesis, respiration, and combustion of gasoline. Classify these changes as absorbing or releasing thermal energy.

2. Based on your current understanding of energy, how is electrical power produced in Ontario? What are the sources of energy that produce this power?

3. How is nuclear power different from hydroelectric power? How is it similar?

▶ *TRY THIS* activity   *Burning Food*

Have you heard of fat-burning exercises? Now you are going to not only burn fat, but also measure how much energy is released in the process.

Engineers who design furnaces to heat homes and nutritionists who calculate the energy value of different foods need to analyze the energy-producing ability of different fuels. In experiments in which heat is absorbed by water, they use the formula:

  heat = (mass of water) × (temperature change of water)
  × 4.18 J/(g°C)

Materials: eye protection; centigram balance; pecan or other nut; paper clip; small tin can, open at one end and punctured under the rim on opposite sides; pencil; thermometer; measuring cup; matches

Figure 1

- Measure the mass of the nut or assume that an average pecan weighs about 0.5 g.

- Bend a paper clip so that it forms a stand that will support a nut above the lab bench (**Figure 1**).

- Place 50 mL of tap water in the tin can. Measure the temperature of the water.

- Suspend the can of water above the nut by putting the pencil through the holes under the rim of the can.

- Light the nut and allow the reaction to continue until the nut stops burning. Measure the final temperature of the water.

(a) Calculate how much energy was absorbed by the water.
(b) Where did this energy come from?
(c) Calculate the amount of heat produced per gram of fuel (nut) burned.
(d) Compare this combustion reaction to the reaction that would happen if you were to eat the pecan instead of burning it. Possible areas of comparison could include: reactants and products, total energy production, energy storage, efficiency of energy production, and so on.
(e) What were some sources of experimental error? How would you improve this experiment?

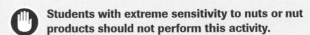

  Students with extreme sensitivity to nuts or nut products should not perform this activity.

What happens when matter undergoes change? Clearly, new substances or states are produced, but energy changes also occur. If chemistry is the study of matter and its transformations, then **thermochemistry** is the study of the energy changes that accompany these transformations. Changes that occur in matter may be classified as physical, chemical, or nuclear, depending on whether a change has occurred in the arrangements of the molecules, their electronic structure, or the nuclei of the atoms involved (**Figure 1**). Whether ice melts, iron rusts, or an isotope used in medical therapy undergoes radioactive decay, changes occur in the energy of chemical substances.

thermochemistry the study of the energy changes that accompany physical or chemical changes in matter

(a)

(b)

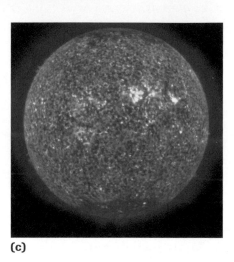

(c)

Figure 1
Hydrogen may undergo a physical, chemical, or nuclear change.
(a) Physical: Hydrogen boils at $-252°C$ (or only about $20°C$ above absolute zero):
$$H_{2(l)} \rightarrow H_{2(g)}$$
(b) Chemical: Hydrogen is burned as fuel in the space shuttle's main engines:
$$2\,H_{2(g)} + O_{2(g)} \rightarrow 2\,H_2O_{(l)}$$
(c) Nuclear: Hydrogen undergoes nuclear fusion in the Sun, producing helium:
$$H + H \rightarrow He$$

thermal energy energy available from a substance as a result of the motion of its molecules

chemical system a set of reactants and products under study, usually represented by a chemical equation

surroundings all matter around the system that is capable of absorbing or releasing thermal energy

Heat and Energy Changes

Both physical and chemical changes are involved in the operation of an oxyacetylene torch to weld metals together. A chemical reaction, which involves ethyne (or acetylene) and oxygen as reactants, produces carbon dioxide gas, water vapour, and considerable energy. This energy is released to the surroundings as **thermal energy**, a form of kinetic energy that results from the motion of molecules. The result is a physical change — the melting of the metal — when the increased vibration of metal particles causes them to break out of their ordered solid pattern. When you are studying such transfers of energy, it is important to distinguish between the substances undergoing a change, called the **chemical system**, and the system's environment, called the **surroundings**. A system is often represented by a chemical equation. For the burning of ethyne, the equation is:

$$2\,C_2H_{2(g)} + 5\,O_{2(g)} \rightarrow 4\,CO_{2(g)} + 2\,H_2O_{(g)} + \text{ energy}$$

The surroundings in this reaction would include anything that could absorb the thermal energy that has been released, such as metal parts, the air, and the welder's protective clothing.

When the reaction occurs, **heat**, q, is transferred between substances. (An object possesses thermal energy but cannot possess heat.) When heat transfers between a system and its surroundings, measurements of the temperature of the surroundings are used to classify the change as **exothermic** or **endothermic** (**Figure 2**).

The acetylene torch reaction is clearly an exothermic reaction because heat flows into the surroundings. Chemical potential energy in the system is converted to heat energy, which is transferred to the surroundings and used to increase the thermal energy of the molecules of metal and air. Since the molecules in the surroundings have greater kinetic energy, the **temperature** of the surroundings increases measurably.

Chemical systems may be further classified. A chemical reaction that produces a gas in a solution in a beaker is described as an **open system**, since both energy and matter can flow into or out of the system. The surroundings include the beaker itself, the surface on which the beaker sits, and the air around the beaker. In the same way, most explosive reactions are considered to be open systems because it is so difficult to contain the energy and matter produced. **Figure 3** shows an open system. Most calculations of energy changes involve systems in which careful measurements of mass and temperature changes are made (**Figure 4**). These are considered to be **isolated systems** for the purpose of calculation. However, it is impossible to completely prevent energy from entering or leaving any system. In reality, the contents of a calorimeter, or of any container that prevents movement of matter, form a **closed system**.

Exothermic   Endothermic

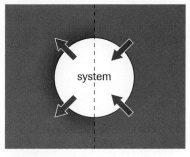

Figure 2
In exothermic changes, energy is released from the system, usually causing an increase in the temperature of the surroundings. In endothermic changes, energy is absorbed by the system, usually causing a decrease in the temperature of the surroundings.

heat amount of energy transferred between substances

exothermic releasing thermal energy as heat flows out of the system

endothermic absorbing thermal energy as heat flows into the system

temperature average kinetic energy of the particles in a sample of matter

open system one in which both matter and energy can move in or out

isolated system an ideal system in which neither matter nor energy can move in or out

closed system one in which energy can move in or out, but not matter

Figure 3
A burning marshmallow is an example of an open system. Gases and energy are free to flow out of the system.

Figure 4
A bomb calorimeter is a device in which a fuel is burned inside an insulated container to obtain accurate measurements of heat transfer during chemical reactions. Because neither mass nor energy can escape, the chemical system is described as isolated.

Understanding Concepts

1. Identify each of the following as a physical, chemical, or nuclear change, with reasons for your choice:
   (a) a gas barbecue operating
   (b) an ice cube melting in someone's hand
   (c) white gas burning in a camping lantern
   (d) wax melting on a hot stove
   (e) zinc metal added to an acid solution in a beaker
   (f) ice applied to an athletic injury

2. Identify the system and surroundings in each of the examples in the previous question.

3. Identify the following as examples of open or isolated systems and explain your identification:
   (a) gasoline burning in an automobile engine
   (b) snow melting on a lawn in the spring
   (c) a candle burning on a restaurant table
   (d) the addition of baking soda to vinegar in a beaker
   (e) a gas barbecue operating

4. A thimbleful of water at 100°C has a higher temperature than a swimming pool full of water at 20°C, but the pool has more thermal energy than the thimble. Explain.

5. Identify each of the following as an exothermic or endothermic reaction:
   (a) hydrogen undergoes nuclear fusion in the Sun to produce helium atoms;
   (b) the butane in a lighter burns;
   (c) the metal on a safety sprinkler on the ceiling of an office melts when a flame is brought near it.

Making Connections

6. (a) List five changes that you might encounter outside your school laboratory. Create a table to classify each change as physical, chemical, or nuclear; endothermic or exothermic; and occurring in an open or an isolated system.
   (b) What are the most commonly encountered types of chemical reactions in terms of energy flow?

7. The energy content of foods is sometimes stated in "calories" rather than the SI unit of joules. Physical activity is described as "burning calories". Research the answers to the following questions:
   (a) What are the relationships among a calorie, a Calorie, and a joule?
   (b) Are calories actually burned? Why is this terminology used?
   (c) What laboratory methods are used to determine the energy content of foods?

 www.science.nelson.com

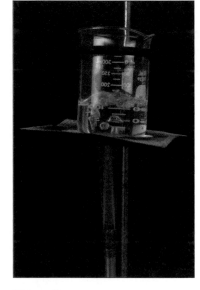

Figure 5
The fuel in the burner releases heat energy that is absorbed by the surroundings, which include the beaker, water, and air.

calorimetry the technological process of measuring energy changes in a chemical system

Measuring Energy Changes: Calorimetry

When methane reacts with oxygen in a lab burner, enough heat is transferred to the surroundings to increase the temperature and even to cause a change of state (**Figure 5**). How is this amount of heat measured? The experimental technique is called **calorimetry** and it depends on careful measurements of masses and temperature changes. When a fuel like methane burns, heat is transferred from the chemical system into the surroundings (which include the water in the beaker). If more heat is transferred, the observed temperature rise in the water is greater. Similarly, given the same amount of heat, a small amount of water will undergo a greater increase in temperature than a large amount of water. Finally, different substances vary in their ability to absorb amounts of heat.

These three factors — mass (m), temperature change (ΔT), and type of substance — are combined in an equation to represent the quantity of heat (q) transferred:

$$q = mc\Delta T$$

where c is the **specific heat capacity**, the quantity of heat required to raise the temperature of a unit mass (e.g., one gram) of a substance by one degree Celsius or one kelvin. For example, the specific heat capacity of water is 4.18 J/(g•°C). (Recall that the SI unit for energy is the joule, J.) Specific heat capacities vary from substance to substance, and even for different states of the same substance (**Table 1**).

As the equation indicates, the quantity of heat, q, that flows varies directly with the quantity of substance (mass m), the specific heat capacity, c, and the temperature change, ΔT.

Cancelling units in your calculations will help ensure that you have applied the formula correctly. In this book, quantities of heat transferred are calculated as absolute values by subtracting the lower temperature from the higher temperature.

specific heat capacity quantity of heat required to raise the temperature of a unit mass of a substance 1°C or 1K

Table 1 Specific Heat Capacities of Substances

Substance	Specific heat capacity, c
ice	2.01 J/(g•°C)
water	4.18 J/(g•°C)
steam	2.01 J/(g•°C)
aluminum	0.900 J/(g•°C)
iron	0.444 J/(g•°C)
methanol	2.918 J/(g•°C)

Calculating Quantity of Heat   SAMPLE problem ◀

When 600 mL of water in an electric kettle is heated from 20°C to 85°C to make a cup of tea, how much heat flows into the water?

First, use the density formula to calculate the mass of water.

$m = dV$

$\quad = 1.00$ g/mL $\times$ 600 mL

$\quad = 600$ g

Use the heat formula, $q = mc\Delta T$, to calculate the quantity of heat transferred.

$q = ?$

$m = 600$ g

$c = 4.18$ J/(g•°C) (from **Table 1**)

$\Delta T = 85°C - 20°C = 65°C$

$q = mc\Delta T$

$\quad = 600$ g $\times \dfrac{4.18 \text{J}}{(\text{g•°C})} \times 65°C$

$\quad = 1.63 \times 10^5$ J or 163 kJ.

163 kJ of heat flows into the water.

Example

What would the final temperature be if 250.0 J of heat were transferred into 10.0 g of methanol initially at 20.0°C?

Solution

$m = 10.0$ g

$c = 2.918$ J/(g•°C)

$T_1 = 20.0$ °C

$\Delta T = T_2 - T_1 = T_2 - 20.0°C$

$q = 250$ J

$$q = mc\Delta T$$
$$\Delta T = \frac{q}{mc}$$
$$= \frac{250 \text{ J}}{10.0 \text{ g} \times 2.918 \text{ J/(g} \cdot °C)}$$
$$= 8.57°C$$
$$T_2 - 20°C = 8.57°C$$
$$T_2 = 20.0 + 8.57$$
$$T_2 = 28.6°C$$

The final temperature of the methanol is 28.6°C.

▶ *Practice*

Understanding Concepts

Answers

9. 506 kJ

10. 38 g

11. 20°C

12. (a) 15 M

    (b) $77

13. (a) 1.0×10^4 kJ

    (b) $55

8. If the same amount of heat were added to individual 1-g samples of water, methanol, and aluminum, which substance would undergo the greatest temperature change? Explain.

9. There is 1.50 kg of water in a kettle. Calculate the quantity of heat that flows into the water when it is heated from 18.0°C to 98.7°C.

10. On a mountaineering expedition, a climber heats water from 0°C to 50°C. Calculate the mass of water that could be warmed by the addition of 8.00 kJ of heat.

11. Aqueous ethylene glycol is commonly used in car radiators as an antifreeze and coolant. A 50% ethylene glycol solution in a radiator has a specific heat capacity of 3.5 J/(g•°C). What temperature change would be observed in a solution of 4 kg of ethylene glycol if it absorbs 250 kJ of heat?

12. Solar energy can preheat cold water for domestic hot-water tanks.
    (a) What quantity of heat is obtained from solar energy if 100 kg of water is pre-heated from 10°C to 45°C?
    (b) If natural gas costs 0.351¢/MJ, calculate the money saved if the volume of water in part (a) is heated 1500 times per year.

13. The solar-heated water in the previous question might be heated to the final temperature in a natural gas water heater.
    (a) What quantity of heat flows into 100 L (100 kg) of water heated from 45°C to 70°C?
    (b) At 0.351¢/MJ, what is the cost of heating 100 kg of water by this amount, 1500 times per year?

$\Delta T = 50°C$

$m = ?$

$q = 8000$ J.

$c = 4.18$ J/(g·°C)

$8000 = m \cdot 4.18 \times 50.$

$m = 38 g.$

Heat Transfer and Enthalpy Change

Chemical systems have many different forms of energy, both kinetic and potential. These include the kinetic energies of

- moving electrons within atoms;
- the vibration of atoms connected by chemical bonds; and
- the rotation and translation of molecules that are made up of these atoms.

More importantly, they also include

- the nuclear potential energy of protons and neutrons in atomic nuclei; and
- the electronic potential energy of atoms connected by chemical bonds.

Researchers have not yet found a way to measure the sum of all these kinetic and potential energies of a system. For this reason chemists usually study the **enthalpy change**, or the energy absorbed from or released to the surroundings when a system changes from reactants to products.

An enthalpy change is given the symbol ΔH, pronounced "delta H," and can be determined from the energy changes of the surroundings. A useful assumption that will be applied in more detail later in this chapter is that the enthalpy change of the system equals the quantity of heat that flows from the system to its surroundings, or from the surroundings to the system (**Figure 6**). This assumption applies as long as there is no significant production of gas, which is the case in most reactions you will encounter.

This idea is consistent with the law of conservation of energy — energy may be converted from one form to another, or transferred from one set of molecules to another, but the total energy of the system and its surroundings remains the same.

$$\Delta H_{\text{system}} = \pm \left| q_{\text{surroundings}} \right|$$

For example, consider the reaction that occurs when zinc metal is added to hydrochloric acid in a flask:

$$Zn_{(s)} + 2\,HCl_{(aq)} \rightarrow H_{2(g)} + ZnCl_{2(aq)}$$

Some of the chemical potential energy in the system is converted initially to increased kinetic energy of the products. Eventually, through collisions, this kinetic energy is transferred to particles in the surroundings. The enthalpy change in the system is equal to the heat released to the surroundings. We can observe this transfer of energy, and can measure it by recording the increase in temperature of the surroundings (which include the solvent water molecules, the flask, and the air around the flask). Our calculations of the heat released will involve the masses of the various substances as well as their temperature change and specific heat capacities.

In order to control variables and allow comparisons, energy changes in chemical systems are measured at standard conditions of temperature and pressure, such as SATP, before and after the reaction. Under these conditions, the enthalpy change of a chemical

enthalpy change (ΔH) the difference in enthalpies of reactants and products during a change

INVESTIGATION 5.1.1

Medical Cold Packs (p. 347)
Can you identify the active chemical in a medical cold pack?

DID YOU KNOW ?

Setting Hard
The setting of concrete is quite exothermic, and the rate at which it sets or cures determines the hardness of the concrete. If the concrete sets too quickly (for example, if the heat of reaction is not dissipated quickly enough into the air), the concrete may expand and crack.

Changes in Kinetic and Potential Energy

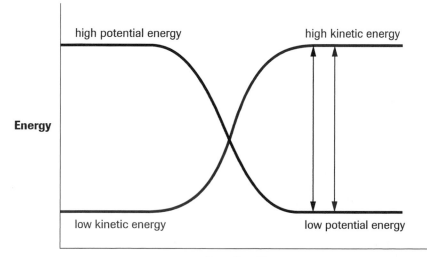

Reaction Progress

Figure 6
In this example of an exothermic change, the change in potential energy of the system (ΔH) equals the change in kinetic energy of the surroundings (q). This is consistent with the law of conservation of energy.

physical change a change in the form of a substance, in which no chemical bonds are broken

chemical change a change in the chemical bonds between atoms, resulting in the rearrangement of atoms into new substances

nuclear change a change in the protons or neutrons in an atom, resulting in the formation of new atoms

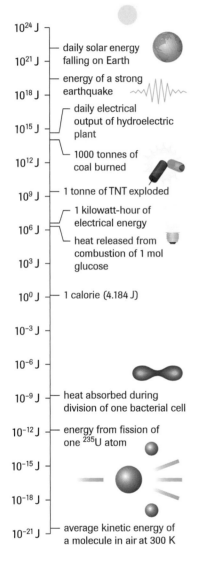

Figure 7
Log scale of the enthalpy changes resulting from a variety of physical, chemical, and nuclear changes

system is the change in the chemical potential energy of the system because the kinetic energies of the system's molecules stay constant (for our purposes at this stage).

We can observe enthalpy changes during phase changes, chemical reactions, or nuclear reactions. Although the magnitudes of the enthalpy changes that accompany these events vary considerably (**Table 2** and **Figure 7**), the basic concepts of enthalpy change and heat transfer apply. Notice how much more energy is produced in a nuclear change than in a chemical change, and in a chemical change than in a physical change.

Table 2 Types of Enthalpy Changes

Physical changes
• Energy is used to overcome or allow intermolecular forces to act.
• Fundamental particles remain unchanged at the molecular level.
• Temperature remains constant during changes of state (e.g., water vapour sublimes to form frost: $H_2O_{(g)} \rightarrow H_2O_{(s)}$ + heat).
• Temperature changes during dissolving of pure solutes (e.g., potassium chloride dissolves: $KCl_{(s)}$ + heat $\rightarrow KCl_{(aq)}$).
• Typical enthalpy changes are in the range $\Delta H = 10^0 - 10^2$ kJ/mol.

Chemical changes
• Energy changes overcome the electronic structure and chemical bonds within the particles (atoms or ions).
• New substances with new chemical bonding are formed (e.g., combustion of propane in a barbecue: $C_3H_{8(g)} + 5\ O_{2(g)} \rightarrow 3\ CO_{2(g)} + 4\ H_2O_{(g)}$ + heat); (e.g., calcium reacts with water: $Ca_{(s)} + 2\ H_2O_{(l)} \rightarrow H_{2(g)} + Ca(OH)_{2(aq)}$ + heat).
• Typical enthalpy changes are in the range $\Delta H = 10^2 - 10^4$ kJ/mol.

Nuclear changes
• Energy changes overcome the forces between protons and neutrons in nuclei.
• New atoms, with different numbers of protons or neutrons, are formed (e.g., nuclear decay of uranium-238: $^{238}_{92}U \rightarrow\ ^{4}_{2}He +\ ^{234}_{90}Th$ + heat).
• Typical enthalpy changes are in the range $\Delta H = 10^{10} - 10^{12}$ kJ/mol. The magnitude of the energy change is a consequence of Einstein's equation (**Figure 8**).

▶ *Practice*

Understanding Concepts

14. Explain how ΔH_{system} and $q_{surroundings}$ are different and how they are similar.

15. How do enthalpy changes of physical, chemical, and nuclear changes compare?

Applying Inquiry Skills

16. Design an experiment to determine the identity of an unknown metal, clearly describing the set of observations that you would make and the calculations that you would perform (including units), given the following information:
- The metal is zinc, magnesium, or aluminum, all of which are shiny, silvery metals.
- These metals react when placed in dilute acid solution.
- Dilute acid has the same density and specific heat capacity as water.
- A Chemical Handbook provides values for the heat (in J) released per unit mass (g) of metal reacting in acid.

Figure 8
In Einstein's famous equation, large amounts of energy, E, are produced when a small amount of mass, m, is destroyed because c, the speed of light, is such a large value $(3.0 \times 10^8$ m/s).

> **Section 5.1** *Questions*

Understanding Concepts

1. For the three states of matter (solid, liquid, and gas), there are six possible changes of state. Which changes of state are exothermic? Which are endothermic?

2. What three factors are involved in calculations of the amount of heat absorbed or released in a chemical reaction?

3. Identify each of the following as a physical, chemical, or nuclear change, giving reasons for your choice:
   (a) gasoline burning in a car engine
   (b) water evaporating from a lake
   (c) uranium fuel encased in concrete in a reactor

4. Identify the chemical system and the surroundings in each of the examples in question 3.

5. Identify each of the examples in question 3 as an open or an isolated system. Explain your classifications.

6. Describe the chemical system in each of the examples in question 3. Compare the relative amounts of energy per mole that would be transferred in each of the changes of state (to the nearest power of ten).

Making Connections

7. The bomb calorimeter is a commonly used laboratory apparatus. Research and write a brief report describing the applications of this technology.

 www.science.nelson.com

8. Hot packs and cold packs use chemical reactions to produce or absorb energy. Write a brief report describing the chemical systems used in these products and their usefulness.

 www.science.nelson.com

5.2 Molar Enthalpies

When we write an equation to represent changes in matter, the chemical symbols may represent individual particles but usually they represent numbers of moles of particles. Thus, the thermochemical equation

$$1\,H_{2(g)} + \frac{1}{2}\,O_{2(g)} \rightarrow 1\,H_2O_{(g)} + 241.8\text{ kJ}$$

molar enthalpy, ΔH_x the enthalpy change associated with a physical, chemical, or nuclear change involving one mole of a substance

represents the combustion reaction of 1 mol of hydrogen with 0.5 mol of oxygen to form 1 mol of water vapour. The enthalpy change per mole of a substance undergoing a change is called the **molar enthalpy** and is represented by the symbol ΔH_x, where x is a letter or a combination of letters to indicate the type of change that is occurring. Thus, the molar enthalpy of combustion of hydrogen is

$$\Delta H_{comb} = -241.8\text{ kJ/mol}$$

Note the negative sign in the value of ΔH. Changes in matter may be either endothermic or exothermic. The following sign convention has been adopted.

- Enthalpy changes for exothermic reactions are given a negative sign.
- Enthalpy changes for endothermic reactions are given a positive sign.

Stating the molar enthalpy is a convenient way of describing the energy changes involved in a variety of physical and chemical changes involving 1 mol of a particular reactant or product. **Table 1** shows some examples.

Table 1 Some Molar Enthalpies of Reaction (ΔH_x)

Type of molar enthalpy	Example of change
solution (ΔH_{sol})	$NaBr_{(s)} \rightarrow Na^+_{(aq)} + Br^-_{(aq)}$
combustion (ΔH_{comb})	$CH_{4(g)} + 2\,O_{2(g)} \rightarrow CO_{2(g)} + H_2O_{(l)}$
vaporization (ΔH_{vap})	$CH_3OH_{(l)} \rightarrow CH_3OH_{(g)}$
freezing (ΔH_{fr})	$H_2O_{(l)} \rightarrow H_2O_{(s)}$
neutralization (ΔH_{neut})*	$2\,NaOH_{(aq)} + H_2SO_{4(aq)} \rightarrow 2\,Na_2SO_{4(aq)} + 2\,H_2O_{(l)}$
neutralization (ΔH_{neut})*	$NaOH_{(aq)} + 1/2\,H_2SO_{4(aq)} \rightarrow 1/2\,Na_2SO_{4(aq)} + H_2O_{(l)}$
formation (ΔH_f)**	$C_{(s)} + 2\,H_{2(g)} + 1/2\,O_{2(g)} \rightarrow CH_3OH_{(l)}$

* Enthalpy of neutralization can be expressed per mole of either base or acid consumed.

** Molar enthalpy of formation will be discussed in more detail in Section 5.5.

We can express the molar enthalpy of a physical change, such as the vaporization of water, as follows:

$$H_2O_{(l)} + 40.8\text{ kJ} \rightarrow H_2O_{(g)}$$

What we may think of as the change in potential energy in the system, the molar enthalpy of vaporization for water, is

$$\Delta H_{vap} = 40.8\text{ kJ/mol}$$

Molar enthalpy values are obtained empirically and are listed in reference books in tables such as **Table 2**.

Table 2 Molar Enthalpies for Changes in State of Selected Substances

Chemical Name	Formula	Molar enthalpy of fusion (kJ/mol)	Molar enthalpy of vaporization (kJ/mol)
sodium	Na	2.6	101
chlorine	Cl_2	6.40	20.4
sodium chloride	NaCl	28	171
water	H_2O	6.03	40.8
ammonia	NH_3	–	1.37
freon-12	CCl_2F_2	–	34.99
methanol	CH_3OH	–	39.23
ethylene glycol	$C_2H_4(OH)_2$	–	58.8

The amount of energy involved in a change (the enthalpy change ΔH, expressed in kJ) depends on the quantity of matter undergoing that change. This is logical: twice the mass of ice will require twice the amount of energy to melt. To calculate an enthalpy change ΔH for some amount of substance other than a mole, you need to obtain the molar enthalpy value ΔH_x from a reference source, and then use the formula $\Delta H = n\Delta H_x$. Note the cancellation of units in the following problem.

Using Molar Enthalpies in Heat Calculations

A common refrigerant (Freon-12, molar mass 120.91 g/mol) is alternately vaporized in tubes inside a refrigerator, absorbing heat, and condensed in tubes outside the refrigerator, releasing heat. This results in energy being transferred from the inside to the outside of the refrigerator. The molar enthalpy of vaporization for the refrigerant is 34.99 kJ/mol. If 500.0 g of the refrigerant is vaporized, what is the expected enthalpy change ΔH?

ΔH_{vap} = 34.99 kJ/mol

ΔH = ?

First, find the amount of refrigerant, n, in moles. From the problem statement,

$M_{refrigerant}$ = 120.91 g/mol, and $m_{refrigerant}$ = 500.0 g, so

$n_{refrigerant} = 500.0 \ \cancel{g} \times \dfrac{1 \ mol}{120.91 \ \cancel{g}}$

$= 4.35 \ mol$

Then calculate the enthalpy change, ΔH.

$\Delta H = n\Delta H_{vap}$

$= 4.35 \ \cancel{mol} \times \dfrac{34.99 \ kJ}{1 \ \cancel{mol}}$

$\Delta H = 144.7 \ kJ$

Because the refrigerant vaporizes by absorbing heat, the enthalpy change is positive.

> **LEARNING TIP**
>
> In calculations involving molar enthalpies, we assume that the number of moles indicated in the chemical equation is exact (has infinite certainty) and so does not affect the number of significant digits in the answer.

Example

What amount of ethylene glycol would vaporize while absorbing 200.0 kJ of heat?

Solution

$\Delta H = 200.0$ kJ

$\Delta H_{vap} = 58.8$ kJ/mol (from **Table 2**)

$n = ?$

$$\Delta H = n\Delta H_{vap}$$

$$n = \frac{\Delta H}{\Delta H_{vap}}$$

$$= \frac{200.0 \text{ kJ}}{58.8 \text{ kJ/mol}}$$

$$n = 3.40 \text{ mol}$$

The amount of ethylene glycol that would vaporize is 3.40 mol.

▶ **Practice**

Understanding Concepts

Answers

1. 227 kJ

2. 474 kJ

3. -3.4×10^5 kJ

1. Calculate the enthalpy change ΔH for the vaporization of 100.0 g of water at 100.0°C (**Table 2**).

2. Ethylene glycol is used in automobile coolant systems because its aqueous solutions lower the freezing point of the coolant liquid and prevent freezing of the system during Canadian winters. What is the enthalpy change needed to completely vaporize 500.0 g of ethylene glycol? (See **Table 2**)

3. Under certain atmospheric conditions, the temperature of the surrounding air rises as a snowfall begins, because energy is released to the atmosphere as water changes to snow. What is the enthalpy change ΔH for the freezing of 1.00 t of water at 0.0°C to 1.00 t of snow at 0.0°C? (Recall that 1 t = 1000 kg.)

Calorimetry of Physical Changes

So far, you have been provided with values for molar enthalpies. How are these values obtained? Studying energy changes requires an isolated system, that is, one in which neither matter nor energy can move in or out. Carefully designed experiments and precise measurements are also needed. Two nested disposable polystyrene cups are a fairly effective calorimeter for making such measurements (**Figure 1**).

When we investigate energy changes we base our analysis on the law of conservation of energy: the total energy change of the chemical system is equal to the total energy change of the surroundings.

$$\Delta H_{system} = \pm |q_{surroundings}|$$

There are three simplifying assumptions often used in calorimetry:

- no heat is transferred between the calorimeter and the outside environment;

- any heat absorbed or released by the calorimeter materials, such as the container, is negligible; and

- a dilute aqueous solution is assumed to have a density and specific heat capacity equal to that of pure water (1.00 g/mL and 4.18 J/g·°C or 4.18 kJ/kg · °C).

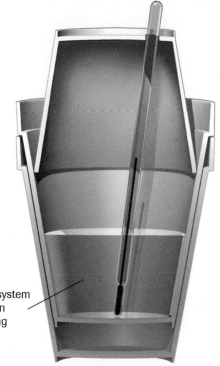

Chemical system dissolved in surrounding water

Figure 1
A simple laboratory calorimeter consists of an insulated container made of two nested polystyrene cups, a measured quantity of water, and a thermometer. The chemical system is placed in or dissolved in the water of the calorimeter. A third cup, with a hole punched in the bottom, can be inverted and used as a lid. Energy transfers between the chemical system and the surrounding water are monitored by measuring changes in the temperature of the water.

Using Calorimetry to Find Molar Enthalpies

In a calorimetry experiment, 7.46 g of potassium chloride is dissolved in 100.0 mL (100.0 g) of water at an initial temperature of 24.1°C. The final temperature of the solution is 20.0°C. What is the molar enthalpy of solution of potassium chloride?

First, calculate the amount of potassium chloride.

$$m_{KCl} = 7.46 \text{ g}$$

$$M_{KCl} = 74.6 \text{ g/mol}$$

$$n_{KCl} = 7.46 \text{ g} \times \frac{1 \text{ mol}}{74.6 \text{ g}}$$

$$n_{KCl} = 0.100 \text{ mol KCl}$$

The next step is to recognize the law of conservation of energy.

$$\Delta H = q$$

(KCl dissolving)   (calorimeter water)

By combining this with the mathematical formulas used earlier in this chapter,

$$\Delta H = n\Delta H_{sol} \text{ and } q = mc\Delta T,$$

we can derive a formula to determine the enthalpy change of potassium chloride dissolving to form a solution:

$$n\Delta H_{sol} = mc\Delta T$$

Assuming that the dilute solution has the same physical properties as pure water, we can now find the molar enthalpy of solution by rearranging this new equation to isolate the quantity we wish to solve for and substituting the given information and the appropriate constants. Note that the mass quantity we are considering is the mass of water in the solution.

$$m_{water} = 100.0 \text{ g}$$

$$c_{water} = 4.18 \text{ J/(g·°C)}$$

$$\Delta T = 24.1°C - 20.0°C = 4.1°C$$

KEY EQUATION

$$n\Delta H_{sol} = mc\Delta T$$

where n and ΔH_{sol} refer to the solute, and m, c, and ΔT refer to the solvent (assuming the solution has the same physical properties as the solvent — as it will if it is dilute).

▶

$$n\Delta H_{sol} = mc\Delta T$$

$$\Delta H_{sol(KCl)} = \frac{mc\Delta T}{n}$$

$$= \frac{100.0 \text{ g} \times 4.18 \text{ J/g·°C} \times 4.1°C}{0.100 \text{ mol}}$$

$$\Delta H_{sol(KCl)} = 1.7 \times 10^4 \text{ J/mol or 17 kJ/mol}$$

The enthalpy change for each mole of potassium chloride that dissolves is 17 kJ/mol. Because the reaction is endothermic, the molar enthalpy of solution for potassium chloride is reported as +17 kJ/mol. Note that the certainty of the final answer (two significant digits) is determined by the certainty of the temperature change, 4.1°C.

Example

What mass of lithium chloride must have dissolved if the temperature of 200.0 g of water increased by 6.0°C? The molar enthalpy of solution of lithium chloride is −37 kJ/mol.

Solution

$$m_{water} = 200.0 \text{ g}$$

$$\Delta T = 6.0°C$$

$$\Delta H_{sol} = 37 \text{ kJ/mol}$$

$$c_{water} = 4.18 \text{ kJ/kg·°C}$$

$$M_{LiCl} = 42.4 \text{ g/mol}$$

$$m_{LiCl} = ?$$

$$n\Delta H_{sol} = q_{water}$$

$$n\Delta H_{sol} = mc\Delta T$$

$$n_{LiCl} = \frac{mc\Delta T}{\Delta H_{sol}}$$

$$n_{LiCl} = \frac{0.2000 \text{ kg} \times \frac{4.18 \text{ kJ}}{\text{kg·°C}} \times 6.0°C}{37 \text{ kJ/mol}}$$

$$= 0.14 \text{ mol}$$

$$m_{LiCl} = 0.14 \text{ mol} \times \frac{42.4 \text{ g}}{1 \text{ mol}}$$

$$m_{LiCl} = 5.7 \text{ g}$$

The mass of lithium chloride required to raise the temperature of the water 6.0°C is 5.7 g.

▶ Practice

Understanding Concepts

Answers

4. +13.9 kJ/mol

5. −5.54 kJ/mol

4. In a chemistry experiment to investigate the properties of a fertilizer, 10.0 g of urea, $NH_2CONH_{2(s)}$, is dissolved in 150 mL of water in a simple calorimeter. A temperature change from 20.4°C to 16.7°C is measured. Calculate the molar enthalpy of solution for the fertilizer urea.

5. A 10.0-g sample of liquid gallium metal, at its melting point, is added to 50.0 g of water in a polystyrene calorimeter. The temperature of the water changes from 24.0°C to 27.8°C as the gallium solidifies. Calculate the molar enthalpy of solidification for gallium.

Calorimetry of Chemical Changes

Chemical reactions that occur in aqueous solutions can also be studied using a polystyrene calorimeter. The chemical system usually involves aqueous reactant solutions that are considered to be equivalent to water. The assumptions and formulas applied are identical to those used in the analysis of energy changes during state changes and dissolving.

When aqueous solutions of acids and bases react, they undergo a neutralization reaction. For example, potassium hydroxide and hydrobromic acid solutions react to form water and aqueous potassium bromide:

$$KOH_{(aq)} + HBr_{(aq)} \rightarrow H_2O_{(l)} + KBr_{(aq)}$$

The molar enthalpy of reaction for systems such as this is sometimes called the heat of neutralization, or enthalpy of neutralization. 🧪▌

▶ *Practice*

Understanding Concepts

6. List three assumptions made in student investigations involving simple calorimeters.

7. The energy involved in the process $H_2O_{(g)} \rightarrow H_2O_{(l)}$ could be described as a molar enthalpy of condensation. Describe the type of molar enthalpy that would be associated with each of the following reactions:

   (a) $Br_{2(l)} \rightarrow Br_{2(g)}$

   (b) $CO_{2(g)} \rightarrow CO_{2(s)}$

   (c) $LiBr_{(s)} \rightarrow Li^+_{(aq)} + Br^-_{(aq)}$

   (d) $C_3H_{8(g)} + 5 O_{2(g)} \rightarrow 3 CO_{2(g)} + 4 H_2O_{(l)}$

   (e) $NaOH_{(aq)} + HCl_{(aq)} \rightarrow 2 NaCl_{(aq)} + H_2O_{(l)}$

Applying Inquiry Skills

8. In a calorimetry experiment in which you are measuring mass and temperature using equipment available to you in your school lab, which measurements limit the certainty of the experimental result? Explain.

9. (a) A laboratory technician adds 43.1 mL of concentrated, 11.6 mol/L hydrochloric acid to water to form 500.0 mL of dilute solution. The temperature of the solution changes from 19.2°C to 21.8°C. Calculate the molar enthalpy of dilution of hydrochloric acid.

   (b) What effect would there be on the calculated value for the molar enthalpy of dilution if the technician accidentally used too much water so that the total volume was actually more than 500.0 mL? Explain.

   (c) The dissolving of an acid in water is a very exothermic process. Dilute acid solutions should always be made by adding acid to water. Explain why adding water to acid is very dangerous.

10. In a laboratory investigation into the reaction

    $$Ba(NO_3)_{2(s)} + K_2SO_{4(aq)} \rightarrow BaSO_{4(s)} + 2 KNO_{3(aq)}$$

    a researcher adds a 261-g sample of barium nitrate to 2.0 L of potassium sulfate solution in a polystyrene calorimeter.

    #### Evidence

    As the barium nitrate dissolves, a precipitate is immediately formed.

    $T_1 = 26.0°C$

    $T_2 = 29.1°C$

    #### Analysis

    (a) Calculate the molar enthalpy of reaction of barium nitrate.

🧪 **INVESTIGATION 5.2.1**

Molar Enthalpy of a Chemical Change (p. 348)
How are molar enthalpies determined? Use the equations and generalizations you've learned to determine a value for the molar enthalpy of neutralization of sodium hydroxide by sulfuric acid.

Answers

9. (a) -10.9 kJ/mol

10. (a) -26 kJ/mol

Understanding Concepts

1. If the molar enthalpy of combustion of ethane is -1.56 MJ/mol, how much heat is produced in the burning of
   (a) 5.0 mol of ethane?
   (b) 40.0 g of ethane?

2. The molar enthalpy of solution of ammonium chloride is $+14.8$ kJ/mol. What would be the final temperature of a solution in which 40.0 g of ammonium chloride is added to 200.0 mL of water, initially at 25°C?

3. The molar enthalpy of combustion of decane ($C_{10}H_{22}$) is -6.78 MJ/mol. What mass of decane would have to be burned in order to raise the temperature of 500.0 mL of water from 20.0°C to 55.0°C?

4. During sunny days, chemicals can store solar energy in homes for later release. Certain hydrated salts dissolve in their water of hydration when heated and release heat when they solidify. For example, Glauber's salt, $Na_2SO_4 \cdot 10\ H_2O_{(s)}$, solidifies at 32°C, releasing 78.0 kJ/mol of salt. What is the enthalpy change for the solidification of 1.00 kg of Glauber's salt used to supply energy to a home (**Figure 2**)?

Figure 2
Glauber's salt is an ideal medium for storing solar energy during the day and releasing it at night. Its melting point is convenient, at 32°C, and it has a high enthalpy of fusion, so a lot of energy can be stored by a small mass of salt. Tubes filled with this salt are part of the heat system in a solar-heated home.

Applying Inquiry Skills

5. In a laboratory investigation into the neutralization reaction
   $$HNO_{3(aq)} + KOH_{(s)} \rightarrow KNO_{3(aq)} + H_2O_{(l)}$$
   a researcher adds solid potassium hydroxide to nitric acid solution in a polystyrene calorimeter.

   Evaluation
   mass KOH = 5.2 g
   volume of nitric acid solution = 200 mL
   $T_1 = 21.0$°C
   $T_2 = 28.1$°C

   Analysis
   (a) Calculate the molar enthapy of neutralization of potassium hydroxide.

6. A student noticed that chewing fast-energy dextrose tablets made her mouth feel cold. Design an investigation, including a Question, Hypothesis, Experimental Design, Materials list, and Procedure, to find out whether there really is a temperature change.

Making Connections

7. The propane refrigerator seems to be a contradiction in terms: the exothermic combustion of a hydrocarbon is used to cool food.
   (a) Find out how this device functions and what changes in matter occur in its operation.
   (b) Calculate enthalpy changes expected in a typical example.

 www.science.nelson.com

Representing Enthalpy Changes 5.3

How do scientists communicate to each other the size of enthalpy changes and determine whether they are endothermic or exothermic? Combustion reactions are often spectacular and are obviously exothermic. However, it is usually not obvious whether a chemical change will absorb or release energy, so, when we are discussing thermochemical reactions, we must indicate this information clearly. The equations we use to do this are called themochemical equations.

You have already seen that the value of an enthalpy change, ΔH, depends on the quantity of a substance that undergoes a change. For example, one mole of hydrogen as it burns has an enthalpy change of −285.8 kJ, and the enthalpy change for two moles of hydrogen is twice that: −571.6 kJ. You have also learned that a sign convention identifies reactions as endothermic or exothermic:

• endothermic enthalpy changes are reported as positive values; and

• exothermic enthalpy changes are reported as negative values.

When water decomposes, the system gains energy from the surroundings and so the molar enthalpy is reported as a positive quantity to indicate an endothermic change:

$$H_2O_{(l)} \rightarrow H_{2(g)} + \frac{1}{2}O_{2(g)} \quad \Delta H_{decomp} = +285.8 \text{ kJ/mol } H_2O$$

The law of conservation of energy implies that the reverse process (combustion of hydrogen) has an equal and opposite energy change.

$$H_{2(g)} + \frac{1}{2}O_{2(g)} \rightarrow H_2O_{(l)} \quad \Delta H_{comb} = -285.8 \text{ kJ/mol } H_2$$

The sign convention represents the change from the perspective of the chemical system itself, not from that of the surroundings. An increase in the temperature of the surroundings implies a decrease in the enthalpy of the chemical system, because the change was exothermic.

Most information about energy changes, for example, the enthalpy change that accompanies the burning of methanol (**Figure 1**), comes from the experimental technique of calorimetry. We can communicate the energy changes, obtained from these empirical studies, in four different ways. Three use thermochemical equations and one uses a diagram:

• by including an energy value as a term in the thermochemical equation

  e.g., $CH_3OH_{(l)} + \frac{3}{2}O_{2(g)} \rightarrow CO_{2(g)} + 2H_2O_{(g)} + 726 \text{ kJ}$

• by writing a chemical equation and stating its enthalpy change

  e.g., $CH_3OH_{(l)} + \frac{3}{2}O_{2(g)} \rightarrow CO_{2(g)} + 2H_2O_{(g)} \quad \Delta H = -726 \text{ kJ}$

• by stating the molar enthalpy of a specific reaction

  e.g., $\Delta H_{combustion}$ or $\Delta H_c = -726 \text{ kJ/mol } CH_3OH$

• by drawing a chemical potential energy diagram (**Figure 2**)

All four of these methods of expressing energy changes are equivalent and are described in more detail as follows.

Figure 1
Hydrocarbons such as acetone burn with a readily visible flame. The flame produced by combusting methanol (right) is difficult to see, and so more dangerous.

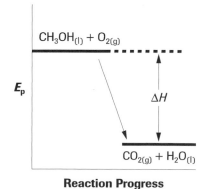

Potential Energy Diagram for an Exothermic Reaction

$CH_3OH_{(l)} + O_{2(g)}$

E_p

ΔH

$CO_{2(g)} + H_2O_{(l)}$

Reaction Progress

Figure 2

Method 1:
Thermochemical Equations with Energy Terms

You are already familiar, from your grade 11 Chemistry course, with the first way to describe the enthalpy change in a chemical reaction: include it as a term in a thermochemical equation. If a reaction is endothermic, it requires a certain quantity of energy to be supplied to the reactants. This energy (like the reactants) is "consumed" as the reaction progresses and is listed along with the reactants.

For example, in the electrolysis of water, energy is absorbed. For our purposes, SATP conditions are usually assumed for all equations.

$$H_2O_{(l)} + 285.8 \text{ kJ} \rightarrow H_{2(g)} + \frac{1}{2}O_{2(g)}$$

If a reaction is exothermic, energy is released as the reaction proceeds (**Figure 3**) and is listed along with the products. For example, magnesium burns in oxygen as follows:

$$Mg_{(s)} + \frac{1}{2}O_{2(g)} \rightarrow MgO_{(s)} + 601.6 \text{ kJ}$$

Figure 3
Combustion reactions are the most familiar exothermic reactions. The searing heat produced by a burning building is a formidable obstacle facing firefighters.

▶ **SAMPLE** problem | **Writing Thermochemical Equations with Energy Terms**

Write a thermochemical equation to represent the exothermic reaction that occurs when two moles of butane burn in excess oxygen gas. The molar enthalpy of combustion of butane is –2871 kJ/mol.

First, write the equation for the combustion of butane:

$$2 C_4H_{10(g)} + 13 O_{2(g)} \rightarrow 8 CO_{2(g)} + 10 H_2O_{(l)}$$

Then obtain the amount of butane, n, from the balanced equation. In this case, $n = 2$ mol. From the problem, $\Delta H_c = -2871$ kJ/mol,

$$\Delta H = n\Delta H_c$$

$$= 2 \text{ mol} \times \frac{-2871 \text{ kJ}}{1 \text{ mol}}$$

$$\Delta H = -5742 \text{ kJ}$$

The reaction is exothermic, so the energy term must be a product. Report the enthalpy change for the reaction by writing it as a product in the thermochemical equation, as follows:

$$2 C_4H_{10(g)} + 13 O_{2(g)} \rightarrow 8 CO_{2(g)} + 10 H_2O_{(l)} + 5742 \text{ kJ}$$

Write a thermochemical equation to represent the dissolving of one mole of silver nitrate in water. The molar enthalpy of solution is + 22.6 kJ/mol.

Solution

$$AgNO_{3(s)} + 22.6 \text{ kJ} \rightarrow Ag^+_{(aq)} + NO^-_{3(aq)}$$

Method 2: Thermochemical Equations with ΔH Values

A second way to describe the enthalpy change in a reaction is to write a balanced chemical equation and then the ΔH value beside it, making sure that ΔH is given the correct sign. Thus, the production of methanol from carbon monoxide and hydrogen could be written as:

$$CO_{(g)} + 2 H_{2(g)} \rightarrow CH_3OH_{(l)} \qquad \Delta H = -128.6 \text{ kJ}$$

Note that the units for the enthalpy change are kilojoules (not kJ/mol), because the enthalpy change applies to the reactants and products *as written*, with the numbers of moles of reactants and products given in the equation. The same equation could be written as:

$$\frac{1}{2} CO_{(g)} + H_{2(g)} \rightarrow \frac{1}{2} CH_3OH_{(l)} \qquad \Delta H = -64.3 \text{ kJ}$$

Writing Thermochemical Equations with ΔH Values

SAMPLE problem ◀

Sulfur dioxide and oxygen react to form sulfur trioxide (Figure 4). The molar enthalpy for the combustion of sulfur dioxide, ΔH_{comb}, in this reaction is -98.9 kJ/mol SO_2. What is the enthalpy change for this reaction?

First, write the balanced chemical equation:

$$2 SO_{2(g)} + O_{2(g)} \rightarrow 2 SO_{3(g)}$$

Then obtain the amount of sulfur dioxide, n, from the balanced equation and use

$$\Delta H = n\Delta H_c$$

$$n = 2 \text{ mol and } \Delta H_c = -98.9 \text{ kJ/mol, so}$$

$$\Delta H = 2 \text{ mol} \times \frac{-98.9 \text{ kJ}}{1 \text{ mol}}$$

$$= -197.8 \text{ kJ}$$

The enthalpy change and the reaction are

$$2 SO_{2(g)} + O_{2(g)} \rightarrow 2 SO_{3(g)} \qquad \Delta H = -197.8 \text{ kJ}$$

Figure 4
Most sulfuric acid is produced in plants like this by the contact process, which includes two exothermic combustion reactions. Sulfur reacts with oxygen, forming sulfur dioxide; sulfur dioxide, in contact with a catalyst, reacts with oxygen, forming sulfur trioxide.

Write a thermochemical equation, including a ΔH value, to represent the exothermic reaction between xenon gas and fluorine gas to produce solid xenon tetrafluoride, given that the reaction produces 251 kJ per mol of Xe reacted.

$$Xe_{(g)} + 2 F_{2(g)} \rightarrow XeF_{4(s)} \qquad\qquad \Delta H = -251 \text{ kJ}$$

As previously described, the enthalpy change ΔH depends on the chemical equation as written. Therefore, if the balanced equation for the reaction is written differently, the enthalpy change should be reported differently. For example,

$$SO_{2(g)} + \frac{1}{2} O_{2(g)} \rightarrow SO_{3(g)} \qquad\qquad \Delta H = -98.9 \text{ kJ}$$

Both this thermochemical equation and the one in the sample problem above agree with the empirically determined molar enthalpy for sulfur dioxide in this reaction.

$$\Delta H_c = \frac{-197.8 \text{ kJ}}{2 \text{ mol}}$$

$$= \frac{-98.9 \text{ kJ}}{1 \text{ mol}}$$

$$= -98.9 \text{ kJ/mol } SO_2$$

The enthalpy changes for most reactions must be accompanied by a balanced chemical equation that includes the state of matter of each substance.

Method 3: Molar Enthalpies of Reaction

molar enthalpy of reaction, ΔH_x the energy change associated with the reaction of one mole of a substance (also called molar enthalpy change)

As you have seen in the previous section, molar enthalpies are convenient ways of describing the energy changes involved in a variety of physical and chemical changes. In each case, one mole of a particular reactant or product is specified. For example, the enthalpy change involved in the dissolving of one mole of solute is called the molar enthalpy of solution and can be symbolized by ΔH_{sol}. In **Table 1**, the substance under consideration in each reaction is highlighted in red.

A molar enthalpy that is determined when the initial and final conditions of the chemical system are at SATP is called a **standard molar enthalpy of reaction**. The symbol ΔH_x° distinguishes standard molar enthalpies from molar enthalpies, ΔH_x, which are measured at other conditions of temperature and pressure. Standard molar enthalpies allow chemists to create tables to compare enthalpy values, as you will see in the next two sections.

standard molar enthalpy of reaction, ΔH_x° the energy change associated with the reaction of one mole of a substance at 100 kPa and a specified temperature (usually 25°C)

Table 1 Some Molar Enthalpies of Reaction

Type of molar enthalpy	Example of change
solution (ΔH_{sol})	$NaBr_{(s)} \rightarrow Na^+_{(aq)} + Br^-_{(aq)}$
combustion (ΔH_{comb})	$CH_{4(g)} + 2 O_{2(g)} \rightarrow CO_{2(g)} + H_2O_{(l)}$
vaporization (ΔH_{vap})	$CH_3OH_{(l)} \rightarrow CH_3OH_{(g)}$
freezing (ΔH_{fr})	$H_2O_{(l)} \rightarrow H_2O_{(s)}$
neutralization (ΔH_{neut})*	$2 NaOH_{(aq)} + H_2SO_{4(aq)} \rightarrow 2 Na_2SO_{4(aq)} + 2 H_2O_{(l)}$
neutralization (ΔH_{neut})*	$NaOH_{(aq)} + 1/2 H_2SO_{4(aq)} \rightarrow 1/2 Na_2SO_{4(aq)} + H_2O_{(l)}$
formation (ΔH_f)**	$C_{(s)} + 2 H_{2(g)} + 1/2 O_{2(g)} \rightarrow CH_3OH_{(l)}$

*   Enthalpy of neutralization can be expressed per mole of either base or acid consumed.

**   Molar enthalpy of formation will be discussed in more detail in Section 5.5.

LEARNING TIP

For the purposes of this textbook, tabulated values will be standard values at 25°C, so that molar enthalpies will be assumed to be standard molar enthalpies. For example, the values for ΔH_c and ΔH_c° will be equivalent.

For an exothermic reaction, the standard molar enthalpy is measured by taking into account all the energy required to change the reaction system from SATP, in order to initiate the reaction, *and* all the energy released following the reaction, as the products are cooled to SATP. For example, the standard molar enthalpy of combustion of methanol (**Figure 5**) is

$$\Delta H_c^\circ = -726 \text{ kJ/mol CH}_3\text{OH}$$

This quantity takes into account the energy input to initiate the reaction, the burning of 1 mol of methanol in oxygen to produce 1 mol $CO_{2(g)}$ and 2 mol $H_2O_{(g)}$, then the energy released as the products are cooled to SATP.

Molar enthalpies can be used to describe reactions other than combustion, as long as the reaction is clearly described. For example, methanol is produced industrially by the high-pressure reaction of carbon monoxide and hydrogen gases.

$$CO_{(g)} + 2 H_{2(g)} \rightarrow CH_3OH_{(l)}$$

Chemists have determined the standard molar enthalpy of reaction for methanol in this reaction, ΔH_r°, to be –128.6 kJ/mol CH_3OH. To describe the reaction fully, we would write the thermochemical equation

$$CO_{(g)} + 2 H_{2(g)} \rightarrow CH_3OH_{(l)} \qquad \Delta H_r^\circ = -128.6 \text{ kJ/mol CH}_3\text{OH}$$

The symbol for the molar enthalpy of reaction uses the subscript "r" to refer to the reaction under consideration, with the stated number of moles of reactants and products. Since two moles of hydrogen are consumed as 128.6 kJ of heat are produced, the standard molar enthalpy of reaction in terms of hydrogen could be described as half the above value, or -64.3 kJ/mol H_2.

Figure 5
Methanol burns more completely than gasoline, producing lower levels of some pollutants. The technology of methanol-burning vehicles was originally developed for racing cars because methanol burns faster than gasoline. However, its energy content is lower so it takes twice as much methanol as gasoline to drive a given distance.

LEARNING *TIP*

The combustion of fuels is always exothermic: heat is released to the surroundings. Enthalpies of combustion are often called heats of combustion and given as absolute values. For example,
$\Delta H_{comb(methanol)} = 726$ kJ/mol.

Describing Molar Enthalpies of Reaction   *SAMPLE* problem ◄

Write an equation whose energy change is the molar enthalpy of combustion of propanol (C_3H_7OH).

Hydrocarbons such as propanol undergo combustion in air by reacting with oxygen gas to produce carbon dioxide gas and water. Since SATP is assumed unless further information is provided, water is produced in liquid form.

Since it is a molar enthalpy, we must write the equation for 1 mol of C_3H_7OH, which requires a fractional coefficient in front of oxygen gas.

The equation is

$$C_3H_7OH_{(g)} + \frac{9}{2} O_{2(g)} \rightarrow 3 CO_{2(g)} + 4 H_2O_{(l)}$$

Example
Write an equation whose enthalpy change is the molar enthalpy of reaction of calcium with hydrochloric acid to produce hydrogen gas and calcium chloride solution.

Solution

$$Ca_{(s)} + 2 HCl_{(aq)} \rightarrow H_{2(g)} + CaCl_{2(aq)}$$

Method 4: Potential Energy Diagrams

Chemists sometimes explain observed energy changes in chemical reactions in terms of chemical potential energy. This stored energy is related to the relative positions of particles and the strengths of the bonds between them. Potential energy is stored or released as the positions of the particles change, just as it is when a spring is stretched and then released. As bonds break and re-form and the positions of atoms are altered, changes occur in potential energy. As you have seen before, the potential energy change in the system is equivalent to the heat transferred to or from the surroundings.

We can visually communicate this energy transferred by using a **potential energy diagram**. In this theoretical description, the energy transferred during a change is represented as changes in the chemical potential energy of the particles as bonds are broken or formed.

The vertical axis on the diagram represents the potential energy of the system. Since the reactants are written on the left and the products on the right, the horizontal axis is sometimes called a reaction coordinate or reaction progress. In an exothermic change (**Figure 6(a)**), the products have less potential energy than the reactants: energy is released to the surroundings as the products form. In an endothermic change (**Figure 6(b)**), the products have more potential energy than the reactants: energy is absorbed from the surroundings. Neither of the axes is numbered; only the numerical change in potential energy (enthalpy change, ΔH) of the system is shown in the diagrams.

Potential energy diagrams can be used to describe a wide variety of chemical changes as shown in **Figure 7**. 🧪▮

potential energy diagram a graphical representation of the energy transferred during a physical or chemical change

🧪 **INVESTIGATION 5.3.1**

Combustion of Alcohols (p. 349)
Do different alcohols produce different quantities of heat when they combust? How do their molar enthalpies compare?

Figure 6
(a) During an exothermic reaction, the enthalpy of the system decreases and heat flows into the surroundings. We observe a temperature increase in the surroundings.
(b) During an endothermic reaction, heat flows from the surroundings into the chemical system. We observe a temperature decrease in the surroundings. This corresponds to an increase in the enthalpy of the chemical system.

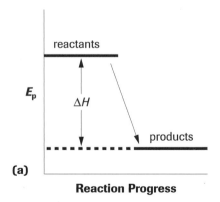

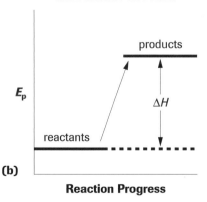

Figure 7
(a) The reaction in which one mole of magnesium oxide is formed from its elements is exothermic, so the reactants must have a higher potential energy than the product.
(b) The reaction in which water decomposes to form hydrogen and oxygen gases is endothermic, so the reactant (water) must have a lower potential energy than the products (hydrogen and oxygen).

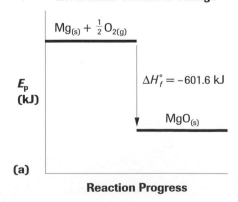

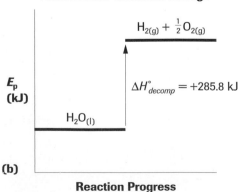

Communicating Enthalpy Changes

Figure 8 uses the chemical reactions for photosynthesis and respiration to summarize the four methods of communicating the molar enthalpy or change in enthalpy of a chemical reaction. Each method has advantages and disadvantages. To best communicate energy changes in chemical reactions, you should learn all four methods.

1. $C_6H_{12}O_{6(s)} + 6\,O_{2(g)} \longrightarrow 6\,CO_{2(g)} + 6\,H_2O_{(l)} + 2802.7\ kJ$

2. $C_6H_{12}O_{6(s)} + 6\,O_{2(g)} \longrightarrow 6\,CO_{2(g)} + 6\,H_2O_{(l)} \quad \Delta H = -2802.7\ kJ$

3. Molar enthalpy for cellular respiration:
   $\Delta H_{respiration} = -2802.7\ kJ/mol\ glucose$

4. Potential energy diagram for **cellular respiration:**

1. $6\,CO_{2(g)} + 6\,H_2O_{(l)} + 2802.7\ kJ \longrightarrow C_6H_{12}O_{6(s)} + 6\,O_{2(g)}$

2. $6\,CO_{2(g)} + 6\,H_2O_{(l)} \longrightarrow C_6H_{12}O_{6(s)} + 6\,O_{2(g)} \quad \Delta H = +2802.7\ kJ$

3. Molar enthalpy for photosynthesis:
   $\Delta H_{photosynthesis} = +2802.7\ kJ/mol\ glucose$

4. Potential energy diagram for **photosynthesis:**

Cellular Respiration of Glucose

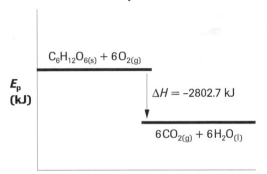

Photosynthesis

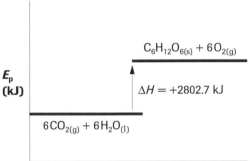

Figure 8
Energy is transformed in cellular respiration and in photosynthesis. Cellular respiration, a series of exothermic reactions, is the breakdown of foodstuffs, such as glucose, that takes place within cells. Photosynthesis, a series of endothermic reactions, is the process by which green plants use light energy to make glucose from carbon dioxide and water.

▶ **Practice**

Understanding Concepts

1. Communicate the enthalpy change by using the four methods described in this section for each of the following chemical reactions. Assume standard conditions (SATP) for the measurements of initial and final states.
   (a) The formation of acetylene (ethyne, C_2H_2) fuel from solid carbon and gaseous hydrogen ($\Delta H° = +228\ kJ/mol$ acetylene)
   (b) The simple decomposition of aluminum oxide powder ($\Delta H° = +1676\ kJ/mol$ aluminum oxide)
   (c) The complete combustion of pure carbon fuel ($\Delta H° = -393.5\ kJ/mol\ CO_2$)

2. For each of the following balanced chemical equations and enthalpy changes, write the symbol and calculate the molar enthalpy of combustion for the substance that reacts with oxygen.
   (a) $2\,H_{2(g)} + O_{2(g)} \rightarrow 2\,H_2O_{(g)}$ $\Delta H° = -483.6\ kJ$
   (b) $4\,NH_{3(g)} + 7\,O_{2(g)} \rightarrow 4\,NO_{2(g)} + 6\,H_2O_{(g)} + 1134.4\ kJ$
   (c) $2\,N_{2(g)} + O_{2(g)} + 163.2\ kJ \rightarrow 2\,N_2O_{(g)}$
   (d) $3\,Fe_{(s)} + 2\,O_{2(g)} \rightarrow Fe_3O_{4(s)}$ $\Delta H° = -1118.4\ kJ$

3. The neutralization of a strong acid and a strong base is an exothermic process.

   $H_2SO_{4(aq)} + 2\,NaOH_{(aq)} \rightarrow Na_2SO_{4(aq)} + 2\,H_2O_{(l)} + 114\ kJ$

   (a) What is the enthalpy change for this reaction?
   (b) Write this thermochemical equation, using the $\Delta H_x°$ to produce $H_2O_{(g)}$ notation.
   (c) Calculate the molar enthalpy of neutralization in kJ/mol sulfuric acid.
   (d) Calculate the molar enthalpy of neutralization in kJ/mol sodium hydroxide.

Answers
2. (a) $-241.8\ kJ/mol\ H_2$
   (b) $-283.6\ kJ/mol\ NH_3$
   (c) $+81.6\ kJ/mol\ N_2$
   (d) $-372.8\ kJ/mol\ Fe$
3. (c) $-114\ kJ/mol\ H_2SO_4$
   (d) $-57\ kJ/mol\ NaOH$

4. The standard molar enthalpy of combustion for hydrogen to produce $H_2O_{(g)}$ is –241.8 kJ/mol. The standard molar enthalpy of decomposition for water vapour is 241.8 kJ/mol.

(a) Write both chemical equations as thermochemical equations with a $\Delta H°$ value.

(b) How does the enthalpy change for the combustion of hydrogen compare with the enthalpy change for the simple decomposition of water vapour? Suggest a generalization to include all pairs of chemical equations that are the reverse of one another.

5. Classify the reactions in **Figure 9** as endothermic or exothermic. Explain your classification.

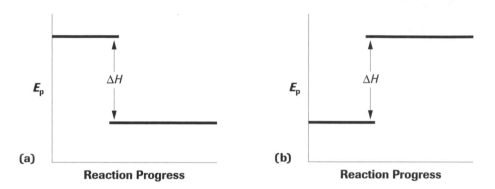

(a) E_p ΔH **(b)** E_p ΔH

Figure 9   Reaction Progress   Reaction Progress

▶ *Section 5.3* Questions

Understanding Concepts

1. Draw a potential energy diagram with appropriately labelled axes to represent

(a) the exothermic combustion of octane ($\Delta H° = -5.47$ MJ)

(b) the endothermic formation of diborane (B_2H_6) from its elements ($\Delta H° = +36$ kJ)

2. Translate each of the molar enthalpies given below into a balanced thermochemical equation, including the enthalpy change, ΔH.

(a) The enthalpy change for the reaction in which solid magnesium hydroxide is formed from its elements at SATP is -925 kJ/mol.

(b) The standard molar enthalpy of combustion for pentane, C_5H_{12}, is -2018 kJ/mol.

(c) The standard molar enthalpy of simple decomposition, $\Delta H°_{decomp}$, for nickel(II) oxide to its elements is 240 kJ/mol.

3. For each of the following reactions, write a thermochemical equation including the energy as a term in the equation.

(a) Butane obtained from natural gas is used as a fuel in lighters (**Figure 10**). The standard molar enthalpy of combustion for butane is -2.86 MJ/mol.

(b) Carbon exists in two different forms, graphite and diamond, which have very different crystal forms. The molar enthalpy of transition of graphite to diamond is 2 kJ.

(c) Ethanol, obtained from the fermentation of corn and other plant products, can be added to gasoline to act as a cheaper alternative. The standard molar enthalpy of combustion for ethanol is -1.28 MJ/mol.

Figure 10
Butane is the fuel used in lighters.

Applying Inquiry Skills

4. A calorimeter is used to determine the enthalpy change involved in the combustion of eicosane ($C_{20}H_{42}$), a solid hydrocarbon found in candle wax. Complete the Analysis and Evaluation sections of the investigation report.

Experimental Design

A candle is placed under a copper can containing water, and a sample of candle wax (eicosane) is burned such that the heat from the burning is transferred to the calorimeter.

Evidence

Table 2 Observations When Burning Candle Wax

Quantity	Measurement
mass of water, m	200.0 g
specific heat capacity of copper, c_{copper}	0.385 J/(g•°C)
mass of copper can, m_{copper}	50.0 g
initial temperature of calorimeter, T_1	21.0°C
final temperature of calorimeter, T_2	76.0°C
initial mass of candle wax, $m_{wax,1}$	8.567 g
final mass of candle wax, $m_{wax,2}$	7.357 g

Analysis

(a) Calculate the molar enthalpy of combustion of eicosane.
(b) Was the reaction exothermic or endothermic? Explain.
(c) Write two thermochemical equations to represent the combustion of eicosane: using an energy term in the equation, and using a ΔH value.

Evaluation

(d) If the accepted value for the molar enthalpy of combustion of eicosane is -13.3 MJ, calculate the percentage error of this procedure.

5.4 Hess's Law of Additivity of Reaction Enthalpies

Figure 1
Some reactions are too slow to be studied experimentally.

Calorimetry is an accurate technique for determining enthalpy changes, but how do chemists deal with chemical systems that cannot be analyzed using this technique? For example, the rusting of iron (**Figure 1**) is extremely slow and, therefore, the resulting temperature change would be too small to be measured using a conventional calorimeter. Similarly, the reaction to produce carbon monoxide from its elements is impossible to measure with a calorimeter because the combustion of carbon produces both carbon dioxide and carbon monoxide simultaneously.

Chemists have devised a number of methods to deal with this problem. These methods are based on the principle that net (or overall) changes in some properties of a system are independent of the way the system changes from the initial state to the final state. An analogy for this concept is shown in **Figure 2**. The net vertical distance that the bricks rise is the same whether they go up in one stage or in two stages. The same principle applies to enthalpy changes: If a set of reactions occurs in different steps but the initial reactants and final products are the same, the overall enthalpy change is the same (**Figure 3**).

Figure 2
In this illustration, bricks on a construction site are being moved from the ground up to the second floor, but there are two different paths that the bricks could follow. In one path, they could move from the ground up to the third floor and then be carried down to the second floor. In the other path, they would be carried up to the second floor in a single step. In both cases, the overall change in position — one floor up — is the same.

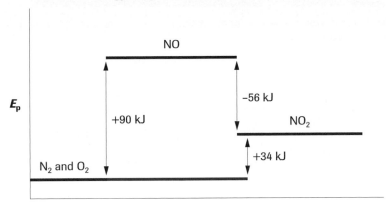

Potential Energy Diagram Showing Additive Enthalpy Changes

Figure 3
In this potential energy diagram, nitrogen gas and oxygen gas combine to form nitrogen dioxide, but there are two different paths to reach the products. In one path, nitrogen (N_2) and oxygen (O_2) gases react to form nitrogen monoxide (NO), a reaction for which $\Delta H = +90$ kJ. Then, nitrogen monoxide and more oxygen react to form nitrogen dioxide (NO_2) gas, a reaction for which $\Delta H = -56$ kJ. In the other path, nitrogen (N_2) and oxygen (O_2) gases react directly to form nitrogen dioxide (NO_2) gas. In both cases, the overall enthalpy change, $\Delta H = +34$ kJ, is the same.

Predicting ΔH Using Hess's Law

Based on experimental measurements of enthalpy changes, the Swiss chemist G. H. Hess suggested that there is a mathematical relationship among a series of reactions leading from a set of reactants to a set of products. This generalization has been tested in many experiments and is now accepted as the law of additivity of reaction enthalpies, also known as

Hess's Law

The value of the ΔH for any reaction that can be written in steps equals the sum of the values of ΔH for each of the individual steps.

Another way to state Hess's law is: If two or more equations with known enthalpy changes can be added together to form a new "target" equation, then their enthalpy changes may be similarly added together to yield the enthalpy change of the target equation.

Hess's law can also be written as an equation using the uppercase Greek letter $\sum$ (pronounced "sigma") to mean "the sum of."

$$\Delta H_{target} = \Delta H_1 + \Delta H_2 + \Delta H_3 + \ldots$$

or

$$\Delta H_{target} = \sum \Delta H_{known}$$

Hess's discovery allowed chemists to determine the enthalpy change of a reaction without direct calorimetry, using two familiar rules for chemical equations and enthalpy changes:

- If a chemical equation is reversed, then the sign of ΔH changes.

- If the coefficients of a chemical equation are altered by multiplying or dividing by a constant factor, then the ΔH is altered in the same way.

Using Hess's Law to Find ΔH

1. *What is the enthalpy change for the formation of two moles of nitrogen monoxide from its elements?*

$$N_{2(g)} + O_{2(g)} \rightarrow 2\,NO_{(g)} \qquad\qquad \Delta H° = ?$$

This reaction, which may be called the *target equation* to distinguish it clearly from the other equations, is difficult to study calorimetrically since the combustion of nitrogen produces nitrogen dioxide as well as nitrogen monoxide. However, we can measure the enthalpy of complete combustion in excess oxygen (to nitrogen dioxide) for both nitrogen and nitrogen monoxide by calorimetry. Consider the following two known reference equations:

$$(1)\ \tfrac{1}{2}N_{2(g)} + O_{2(g)} \rightarrow NO_{2(g)} \qquad\qquad \Delta H°_1 = +34\ kJ$$

$$(2)\ NO_{(g)} + \tfrac{1}{2}O_{2(g)} \rightarrow NO_{2(g)} \qquad\qquad \Delta H°_2 = -56\ kJ$$

If we work with these two equations, which may be called *known equations*, and then add them together, we obtain the chemical equation for the formation of nitrogen monoxide.

The first term in the target equation for the formation of nitrogen monoxide is one mole of nitrogen gas. We therefore need to double equation (1) so that $N_{2(g)}$ will appear on the reactant side when we add the equations. However, from equation (2) we want 2 mol of $NO_{(g)}$ to appear as a product, so we must reverse equation (2) and double each of its terms (including the enthalpy change). Effectively, we have multiplied known equation (1) by +2, and multiplied known equation (2) by –2.

$$2 \times (1):\ N_{2(g)} + 2\,O_{2(g)} \rightarrow 2\,NO_{2(g)} \qquad\qquad \Delta H° = 2(+34)\ kJ$$

$$-2 \times (2):\ 2\,NO_{2(g)} \rightarrow 2\,NO_{(g)} + O_{2(g)} \qquad\qquad \Delta H° = -2(-56)\ kJ$$

Note that the sign of the enthalpy change in equation (2) will change, since the equation has been reversed.

Now add the reactants, products, and enthalpy changes to obtain a net reaction equation. Note that $2\,NO_{2(g)}$ can be cancelled because it appears on both sides of the net equation. Similarly, $O_{2(g)}$ can be cancelled from each side of the equation, yielding the target equation:

$$N_{2(g)} + \cancel{2}\,O_{2(g)} + 2\,\cancel{NO_{2(g)}} \rightarrow 2\,\cancel{NO_{2(g)}} + 2\,NO_{(g)} + \cancel{O_{2(g)}}$$

or $\qquad\qquad N_{2(g)} + O_{2(g)} \rightarrow 2\,NO_{(g)}$

Now we can apply Hess's law: *If* the known equations can be added together to form the target equation, *then* their enthalpy changes can be added together.

$$\Delta H° = (2 \times 34)\ kJ + (-2 \times (-56))\ kJ$$

$$= +68\ kJ + 112\ kJ$$

$$\Delta H° = +180\ kJ$$

The enthalpy change for the formation of two moles of nitrogen monoxide from its elements is 180 kJ.

When manipulating the known equations, you should check the target equation and plan ahead to ensure that the substances end up on the correct sides and in the correct amounts.

LEARNING TIP

Generally, a subscript on a ΔH value indicates a molar enthalpy value, expressed in kJ/mol. The "known" equations in Hess's law problems are exceptions. The subscript is used as a convenience in distinguishing equations from each other and the units of the ΔH_n values are kilojoules.

2. **What is the enthalpy change for the formation of one mole of butane (C_4H_{10}) gas from its elements? The reaction is:**

$$4 \, C_{(s)} + 5 \, H_{2(g)} \rightarrow C_4H_{10(g)} \qquad \Delta H° = \, ?$$

The following known equations, determined by calorimetry, are provided:

(1) $C_4H_{10(g)} + \frac{13}{2} O_{2(g)} \rightarrow 4 \, CO_{2(g)} + 5 \, H_2O_{(g)}$ $\qquad \Delta H°_1 = -2657.4 \text{ kJ}$

(2) $C_{(s)} + O_{2(g)} \qquad \rightarrow CO_{2(g)}$ $\qquad \Delta H°_2 = -393.5 \text{ kJ}$

(3) $2 \, H_{2(g)} + O_{2(g)} \qquad \rightarrow 2 \, H_2O_{(g)}$ $\qquad \Delta H°_3 = -483.6 \text{ kJ}$

Reversing known equation (1), which will require multiplying its ΔH by -1, will make $C_4H_{10(g)}$ a product; multiplying known equation (2) by 4 will provide the required amount of $C_{(s)}$ reactant; and multiplying known equation (3) by 5/2 will provide the required amount of $H_{2(g)}$ reactant. Cancellation when the equations are added will determine whether the required amount of $O_{2(g)}$ remains.

$-1 \times$ (1): $\quad 4 \, CO_{2(g)} + 5 \, H_2O_{(g)} \qquad\qquad \rightarrow C_4H_{10(g)} + \frac{13}{2} O_{2(g)}$

$$\Delta H° = -1(-657.4) \text{ kJ}$$

$4 \times$ (2): $\quad 4 \, C_{(s)} + 4 \, O_{2(g)} \qquad\qquad \rightarrow 4 \, CO_{2(g)} \quad \Delta H° = 4(-393.5) \text{ kJ}$

$\frac{5}{2} \times$ (3): $\quad 5 \, H_{2(g)} + \frac{5}{2} O_{2(g)} \qquad\qquad \rightarrow 5 \, H_2O_{(g)} \quad \Delta H° = \frac{5}{2}(-483.6) \text{ kJ}$

$$4 \, \cancel{CO}_{2(g)} + 5 \, \cancel{H_2O}_{(g)} + 4 \, C_{(s)} + \frac{13}{2} \, \cancel{O}_{2(g)} + 5 \, H_{2(g)} \rightarrow C_4H_{10(g)} + \frac{13}{2} \, \cancel{O}_{2(g)}$$

$$+ \, 4 \, \cancel{CO}_{2(g)} + 5 \, \cancel{H_2O}_{(g)}$$

or $\qquad 4 \, C_{(s)} + 5 \, H_{2(g)} \qquad\qquad \rightarrow C_4H_{10(g)}$

If the known equations can be added together to form the target equation, *then* their enthalpy changes can be added together. In this case,

$$\Delta H°_{total} = (+2657.4) + (-1574.0) + (-1209.0) \text{ kJ}$$
$$\Delta H°_{total} = -125.6 \text{ kJ}$$

The enthalpy change for the formation of one mole of butane is -125.6 kJ.

Example
Determine the enthalpy change involved in the formation of two moles of liquid propanol.

$$6 \, C_{(s)} + 8 \, H_{2(g)} + O_{2(g)} \rightarrow 2 \, C_3H_7OH_{(l)}$$

The standard enthalpies of combustion of propanol, carbon, and hydrogen gas at SATP are -2008, -394, and -286 kJ/mol, respectively.

Solution
The known equations are

(1) $C_3H_7OH_{(l)} + \frac{9}{2} O_{2(g)} \rightarrow 3 \, CO_{2(g)} + 4 \, H_2O_{(l)}$ $\qquad \Delta H°_1 = -2008 \text{ kJ}$

(2) $C_{(s)} + O_{2(g)} \qquad \rightarrow CO_{2(g)}$ $\qquad \Delta H°_2 = -394 \text{ kJ}$

(3) $H_{2(g)} + \frac{1}{2} O_{2(g)} \qquad \rightarrow H_2O_{(l)}$ $\qquad \Delta H°_3 = -286 \text{ kJ}$

$$-2 \times (1): \quad 6\,CO_{2(g)} + 8\,H_2O_{(l)} \rightarrow 2\,C_3H_7OH_{(l)} + 9\,O_{2(g)} \quad \Delta H° = -2(-2008\ kJ)$$

$$6 \times (2): \quad 6\,C_{(s)} + 6\,O_{2(g)} \rightarrow 6\,CO_{2(g)} \quad\quad\quad\quad\quad\quad \Delta H° = 6(-394\ kJ)$$

$$8 \times (3): \quad 8\,H_{2(g)} + 4\,O_{2(g)} \rightarrow 8\,H_2O_{(l)} \quad\quad\quad\quad\quad\quad \Delta H° = 8(-286\ kJ)$$

$$\overline{6\,C_{(s)} + 8\,H_{2(g)} + O_{2(g)} \quad\rightarrow 2\,C_3H_7OH_{(l)} \quad\quad\quad\quad\quad\quad \Delta H° = -636\ kJ}$$

The enthalpy change involved in the formation of propanol is -636 kJ.

▶ Practice

Understanding Concepts

Answers

1. -851.5 kJ
2. 131.3 kJ
3. -524.8 kJ

Figure 4
Electric power generating stations that use coal as a fuel are only 30% to 40% efficient. Coal gasification and combustion of the coal gas provide one alternative to burning coal. Efficiency is improved by using both a combustion turbine and a steam turbine to produce electricity.

1. The enthalpy changes for the formation of aluminum oxide and iron(III) oxide from their elements are:

(1) $2\,Al_{(s)} + \frac{3}{2}\,O_{2(g)} \rightarrow Al_2O_{3(s)}$ $\quad\quad\quad\quad\quad\quad\quad \Delta H_1° = -1675.7$ kJ

(2) $2\,Fe_{(s)} + \frac{3}{2}\,O_{2(g)} \rightarrow Fe_2O_{3(s)}$ $\quad\quad\quad\quad\quad\quad\quad \Delta H_2° = -824.2$ kJ

Calculate the enthalpy change for the following target reaction.

$Fe_2O_{3(s)} + 2\,Al_{(s)} \rightarrow Al_2O_{3(s)} + 2\,Fe_{(s)}$ $\quad\quad\quad \Delta H° = ?$

2. Coal gasification converts coal into a combustible mixture of carbon monoxide and hydrogen, called coal gas (**Figure 4**), in a gasifier.

$H_2O_{(g)} + C_{(s)} \rightarrow CO_{(g)} + H_{2(g)}$ $\quad\quad\quad\quad\quad \Delta H° = ?$

Calculate the standard enthalpy change for this reaction from the following chemical equations and enthalpy changes.

(1) $2\,C_{(s)} + O_{2(g)} \rightarrow 2\,CO_{(g)}$ $\quad\quad\quad\quad\quad\quad \Delta H_1° = -221.0$ kJ

(2) $2\,H_{2(g)} + O_{2(g)} \rightarrow 2\,H_2O_{(g)}$ $\quad\quad\quad\quad\quad\quad \Delta H_2° = -483.6$ kJ

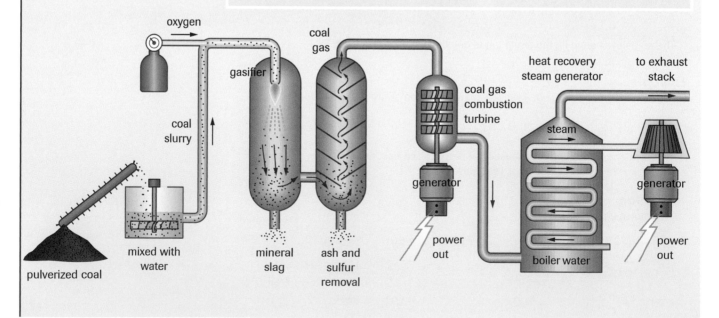

3. The coal gas described in the previous question can be used as a fuel, for example, in a combustion turbine.

$$CO(g) + H_{2(g)} + O_{2(g)} \rightarrow CO_{2(g)} + H_2O_{(g)} \qquad \Delta H° = ?$$

Predict the change in enthalpy for this combustion reaction from the following information.

(1) $2\,C_{(s)} + O_{2(g)} \rightarrow 2\,CO_{(g)}$ $\Delta H_1° = -221.0$ kJ

(2) $C_{(s)} + O_{2(g)} \rightarrow CO_{2(g)}$ $\Delta H_2° = -393.5$ kJ

(3) $2\,H_{2(g)} + O_{2(g)} \rightarrow 2\,H_2O_{(g)}$ $\Delta H_3° = -483.6$ kJ

INVESTIGATION 5.4.1

Hess's Law (p. 351)
Use calorimetry to determine your own "known" equations and use them to calculate the molar enthalpy of combustion of magnesium.

Multistep Energy Calculations

In practice, energy calculations rarely involve only a single-step calculation of heat or enthalpy change. Several energy calculations might be required, involving a combination of energy change definitions such as

- heat flows, $q = mc\Delta T$
- enthalpy changes, $\Delta H = n\Delta H_r$
- Hess's law, $\Delta H_{target} = \sum \Delta H_{known}$

In these multi-step problems, ΔH is often found by using standard molar enthalpies or Hess's law and then equated to the transfer of heat, q. As shown in the following sample problem, if we know the enthalpy change of a reaction and the quantity of reactant or product, we can predict how much energy will be absorbed or released.

Solving Multistep Enthalpy Problems

SAMPLE problem ◄

In the Solvay process for the production of sodium carbonate (or washing soda), one step is the endothermic decomposition of sodium hydrogen carbonate:

$$2\,NaHCO_{3(s)} + 129.2\,kJ \rightarrow Na_2CO_{3(s)} + CO_{2(g)} + H_2O_{(g)}$$

What quantity of chemical energy, ΔH, is required to decompose 100.0 kg of sodium hydrogen carbonate?

First, calculate the energy absorbed per mole of $NaHCO_3$, that is, the molar enthalpy of reaction with respect to sodium hydrogen carbonate.

$$\Delta H = n\Delta H_r$$

$$\Delta H_r = \frac{\Delta H}{n}$$

$$= \frac{129.2\,kJ}{2\,mol}$$

$$\Delta H_r = 64.6\,kJ/mol$$

This means that 64.6 kJ of energy is required for every mole of $NaHCO_3$ decomposed. Converting 100.0 kg to an amount in moles and multiplying by the molar enthalpy will give us the required enthalpy change, ΔH, for the equation.

$$n_{NaHCO_3} = 100.0 \text{ kg} \times \frac{1 \text{ mol}}{84.01 \text{ g}}$$

$$= 1.190 \text{ kmol}$$

$$\Delta H = n\Delta H_r$$

$$= 1.190 \text{ kmol} \times \frac{64.6 \text{ kJ}}{1 \text{ mol}}$$

$$\Delta H = 76.9 \text{ MJ}$$

The decomposition of 100 kg of sodium hydrogen carbonate requires 76.9 MJ of energy.

Example

How much energy can be obtained from the roasting of 50.0 kg of zinc sulfide ore?

$$ZnS_{(s)} + \frac{3}{2} O_{2(g)} \rightarrow ZnO_{(s)} + SO_{2(g)}$$

You are given the following thermochemical equations.

$$(1) \; ZnO_{(s)} \qquad\qquad \rightarrow Zn_{(s)} + \frac{1}{2} O_{2(g)} \qquad\qquad \Delta H_1^\circ = 350.5 \text{ kJ}$$

$$(2) \; S_{(s)} + O_{2(g)} \qquad \rightarrow SO_{2(g)} \qquad\qquad\qquad \Delta H_2^\circ = -296.8 \text{ kJ}$$

$$(3) \; ZnS_{(s)} \qquad\qquad \rightarrow Zn_{(s)} + S_{(s)} \qquad\qquad \Delta H_3^\circ = 206.0 \text{ kJ}$$

Solution

$M_{ZnS} = 97.44 \text{ g/mol}$

$$-1 \times (1): \quad Zn_{(s)} + \frac{1}{2} O_{2(g)} \rightarrow ZnO_{(s)} \qquad\qquad \Delta H^\circ = -1(350.5 \text{ kJ})$$

$$1 \times (2): \quad\; S_{(s)} + O_{2(g)} \quad\; \rightarrow SO_{2(g)} \qquad\qquad\quad \Delta H^\circ = 1(-296.8 \text{ kJ})$$

$$1 \times (3): \quad\; ZnS_{(s)} \qquad\quad \rightarrow Zn_{(s)} + S_{(s)} \qquad\qquad \Delta H^\circ = 1(206.0 \text{ kJ})$$

$$ZnS_{(s)} + \frac{3}{2} O_{2(g)} \quad \rightarrow ZnO_{(s)} + SO_{2(g)} \qquad \Delta H^\circ = -441.3 \text{ kJ}$$

According to Hess's law, 441.3 kJ of energy can be obtained from the roasting of 1 mol of ZnS for which reaction $\Delta H_r^\circ = -441.3$ kJ/mol.

$$n_{ZnS} = 50.0 \text{ kg} \times \frac{1 \text{ mol}}{97.44 \text{ g}}$$

$$= 513 \text{ mol}$$

$$\Delta H^\circ = n_{ZnS}\Delta H_r^\circ$$

$$= 513 \text{ mol} \times \frac{(-441.3 \text{ kJ})}{1 \text{ mol}}$$

$$\Delta H^\circ = -2.26 \times 10^5 \text{ kJ or } -226 \text{ MJ}$$

According to Hess's law, 226 MJ of energy can be obtained from the roasting of 50.0 kg of zinc sulfide ore.

> **Practice**

Understanding Concepts

4. Ethyne gas may be reduced by reaction with hydrogen gas to form ethane gas in the following reduction reaction:

$$C_2H_{2(g)} + 2 H_{2(g)} \rightarrow C_2H_{6(g)}$$

Predict the enthalpy change for the reduction of 200 g of ethyne, using the following information.

(1) $C_2H_{2(g)} + \dfrac{5}{2} O_{2(g)} \rightarrow 2 CO_{2(g)} + H_2O_{(l)}$ $\Delta H_1^\circ = -1299$ kJ

(2) $H_2 + \dfrac{1}{2} O_{2(g)} \rightarrow H_2O_{(l)}$ $\Delta H_2^\circ = -286$ kJ

(3) $C_2H_{6(g)} + \dfrac{7}{2} O_{2(g)} \rightarrow 2 CO_{2(g)} + 3 H_2O_{(l)}$ $\Delta H_3^\circ = -1560$ kJ

5. As an alternative to combustion of coal gas described earlier in this section, coal gas can undergo a process called methanation.

$$3 H_{2(g)} + CO_{(g)} \rightarrow CH_{4(g)} + H_2O_{(g)} \qquad \Delta H = ?$$

Determine the enthalpy change involved in the reaction of 300 g of carbon monoxide in this methanation reaction, using the following reference equations and enthalpy changes.

(1) $2 H_{2(g)} + O_{2(g)} \rightarrow 2 H_2O_{(g)}$ $\Delta H_1^\circ = -483.6$ kJ

(2) $2 C_{(s)} + O_{2(g)} \rightarrow 2 CO_{(g)}$ $\Delta H_2^\circ - -221.0$ kJ

(3) $CH_{4(g)} + 2 O_{2(g)} \rightarrow CO_{2(g)} + 2 H_2O_{(g)}$ $\Delta H_3^\circ = -802.7$ kJ

(4) $C_{(s)} + O_{2(g)} \rightarrow CO_{2(g)}$ $\Delta H_4^\circ = -393.5$ kJ

Answers

4. -2.39 MJ

5. -2.20 MJ

Understanding Concepts

1. (a) Write three balanced thermochemical equations to represent the combustions of one mole each of octane, hydrogen, and carbon, given that their molar enthalpies of combustion are, respectively, -5.47 MJ, -285.8 kJ, and -393.5 kJ/mol.

(b) Use Hess's law to predict the enthalpy change for the formation of octane from its elements.

$$8\, C_{(s)} + 9\, H_{2(g)} \rightarrow C_8H_{18(l)} \qquad \Delta H = ?$$

2. Predict the enthalpy change for the reaction

$$HCl_{(g)} + NaNO_{2(s)} \rightarrow HNO_{2\,(g)} + NaCl_{(s)}$$

using the following information:

(1) $2\, NaCl + H_2O \rightarrow 2\, HCl + Na_2O$ $\qquad \Delta H_1^\circ = 507$ kJ

(2) $NO + NO_2 + Na_2O \rightarrow 2\, NaNO_2$ $\qquad \Delta H_2^\circ = -427$ kJ

(3) $NO + NO_2 \rightarrow N_2O + O_2$ $\qquad \Delta H_3^\circ = -43$ kJ

(4) $2\, HNO_2 \rightarrow N_2O + O_2 + H_2O$ $\quad \Delta H_4^\circ = 34$ kJ

3. Bacteria sour wines and beers by converting ethanol (C_2H_5OH) into acetic acid (CH_3COOH). The reaction is

$$C_2H_5OH + O_2 \rightarrow CH_3COOH + H_2O$$

The molar enthalpies of combustion of ethanol and acetic acid are, respectively, -1367 kJ/mol and -875 kJ/mol. Write thermochemical equations for the combustions, and use Hess's law to determine the enthalpy change for the conversion of ethanol to acetic acid.

Applying Inquiry Skills

4. A series of calorimetric experiments is perfomed to test Hess's law. Complete the Experimental Design, Prediction, and Analysis sections of the investigation report.

Question

Can Hess's law be verified experimentally by combining enthalpies of reaction?

Experimental Design

Three calorimetry experiments are performed, with the choice of chemical systems such that the enthalpy changes of two of the reactions should equal the enthalpy change of the third. The three thermochemical reactions are:

(1) $HBr_{(aq)} + KOH_{(aq)} \rightarrow H_2O_{(l)} + KBr_{(aq)}$ $\quad \Delta H_1 = ?$ kJ

(2) $KOH(s) \rightarrow KOH_{(aq)}$ $\qquad \Delta H_2 = ?$ kJ

(3) $KOH(s) + HBr_{(aq)} \rightarrow H_2O_{(l)} + KBr_{(aq)}$ $\quad \Delta H_3 = ?$ kJ

(a) Use Hess's law to show how two of the thermochemical equations can be added together to yield the third thermochemical equation.

Evidence

See **Table 1**.

Analysis

(b) Use the experimental values to calculate the enthalpy change in each system.

Evaluation

(c) Calculate a percentage error in the experiment, given that the ΔH for one equation should equal exactly the sum of the other two.

Table 1 Observations for Hess's Law Investigation

Observation	Experiment 1	Experiment 2	Experiment 3
quantity of reactant 1	100.0 mL of 1.00 mol/L $KOH_{(aq)}$	5.61 g $KOH_{(s)}$	5.61 g $KOH_{(s)}$
quantity of reactant 2	100.0 mL of 1.00 mol/L $HBr_{(aq)}$	N/A	200.0 mL of 0.50 mol/L $HBr_{(aq)}$
initial temperature	20.0°C	20.0°C	20.0°C
final temperature	22.5°C	24.1°C	26.7°C

Calorimetry and Hess's law are two ways of determining enthalpies of reaction. A third method uses tabulated enthalpy changes (**standard enthalpies of formation**) for a special set of reactions called formation reactions, in which compounds are formed from their elements. For example, the formation reaction and standard enthalpy of formation for carbon dioxide are:

$$C_{(s)} + O_{2(g)} \rightarrow CO_{2(g)} \qquad \Delta H_f^\circ = -393.5 \text{ kJ/mol}$$

Both the elements on the left side of the equation are in their standard states — their most stable form at SATP (25°C and 100 kPa). Note also that the units of standard enthalpies of formation are kJ/mol because they are always stated for that quantity of substance.

standard enthalpy of formation
the quantity of energy associated with the formation of one mole of a substance from its elements in their standard states

LEARNING TIP

Although the name "standard enthalpies of formation" does not include the word *molar*, it is always a quantity of energy per mole.

Writing Formation Equations

Formation equations are always written for one mole of a particular product, which may be in any state or form, but the reactant elements must be in their standard states. For example, the standard states of most metals are monatomic solids ($Mg_{(s)}$, $Ca_{(s)}$, $Fe_{(s)}$, $Au_{(s)}$, $Na_{(s)}$), some nonmetals are diatomic gases ($N_{2(g)}$, $O_{2(g)}$, $H_{2(g)}$), and the halogen family shows a variety of states ($F_{2(g)}$, $Cl_{2(g)}$, $Br_{2(l)}$, $I_{2(s)}$). The periodic table at the back of this text identifies the states of elements.

SUMMARY | *Writing Formation Equations*

Step 1: Write one mole of product in the state that has been specified.

Step 2: Write the reactant elements in their standard states.

Step 3: Choose equation coefficients for the reactants to give a balanced equation yielding one mole of product.

Writing Formation Equations

SAMPLE problem

Write the formation equation for liquid ethanol.

Start with one mole of product.

$$C_2H_5OH_{(l)}$$

Write the reactant elements in their standard states.

$$C_{(s)} + H_{2(g)} + O_{2(g)} \rightarrow C_2H_5OH_{(l)}$$

Balance the equation to yield one mole of product.

$$2\,C_{(s)} + 3\,H_{2(g)} + \tfrac{1}{2}\,O_{2(g)} \rightarrow C_2H_5OH_{(l)}$$

LEARNING TIP

The standard state of most elements in the periodic table is solid. There are five common gaseous elements at SATP that form compounds readily: H_2, O_2, N_2, F_2, and Cl_2. There are only two liquid elements at SATP: Hg and Br_2.

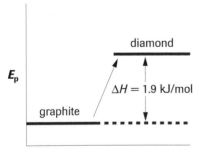

Example

What is the formation equation for liquid carbonic acid?

Solution

$$H_{2(g)} + C_{(s)} + \frac{3}{2} O_{2(g)} \rightarrow H_2CO_{3(l)}$$

> ### ▶ Practice

Understanding Concepts

1. Write formation equations for the compounds:
   (a) benzene (C_6H_6), used as a solvent
   (b) potassium bromate, used in commercial bread dough
   (c) glucose ($C_6H_{12}O_6$), found in soft drinks
   (d) magnesium hydroxide, found in antacids

Using Standard Enthalpies of Formation

Consider the equation for the formation of hydrogen gas:

$$H_{2(g)} \rightarrow H_{2(g)}$$

The product and reactant are the same, so there is no change in the enthalpy of the system. This observation can be generalized to all elements in their standard states:

> ### ΔH_f° for Elements
>
> The standard enthalpy of formation of an element already in its standard state is zero.

Thus, the standard enthalpies of formation of, for example, $Fe_{(s)}$, $O_{2(g)}$, and $Br_{2(l)}$ are all zero.

Standard molar enthalpies of formation give us a means of comparing the stabilities of substances. For example, the element carbon exists in two solid forms at SATP: diamond, used in jewellery and mining drill bits; and graphite, the black substance used in pencil "leads" and composite plastics. Graphite is the more stable form of carbon at SATP, so

$$\Delta H_{f(graphite)}^\circ = 0 \text{ kJ/mol}$$

Diamond is slightly less stable at SATP and has a greater potential energy than graphite (**Figure 1**). For the formation of diamond,

$$C_{(graphite)} \rightarrow C_{(diamond)} \qquad \Delta H_{f(diamond)}^\circ = +1.9 \text{ kJ/mol}$$

You have seen that Hess's law may be applied to a set of known equations to find an unknown enthalpy change (Section 5.4). We can apply this problem-solving method to predict the energy changes for many reactions.

Relative Potential Energies of Graphite and Diamond

diamond

E_p

$\Delta H = 1.9$ kJ/mol

graphite

Reaction Progress

Figure 1
Graphite is the more stable form of carbon. The formation of diamond requires an increase in potential energy.

Using Enthalpies of Formation to Find ΔH

What is the thermochemical equation for the reaction of lime (calcium oxide) and water?

We can express the target equation as

$$CaO_{(s)} + H_2O_{(l)} \rightarrow Ca(OH)_{2(s)} \qquad \Delta H = ?$$

Consider the following set of formation reactions for each compound in the equation, each with its corresponding standard enthalpy of formation. (Tables of standard enthalpies of formation are readily available, including a sample reference in Appendix C4 in this text.) Use these with Hess's law to find the enthalpy change ΔH for the target equation.

$$(1)\ Ca_{(s)} + \frac{1}{2}O_{2(g)} \qquad\qquad \rightarrow CaO_{(s)} \qquad \Delta H_1 = -634.9\ \text{kJ/mol}$$

$$(2)\ H_{2(g)} + \frac{1}{2}O_{2(g)} \qquad\qquad \rightarrow H_2O_{(l)} \qquad \Delta H_2 = -285.8\ \text{kJ/mol}$$

$$(3)\ Ca_{(s)} + H_{2(g)} + O_{2(g)} \qquad \rightarrow Ca(OH)_{2(s)} \qquad \Delta H_3 = -986.1\ \text{kJ/mol}$$

Manipulating and adding the equations according to Hess's law results in a sum for the thermochemical equation

$$-1 \times (1):\ Ca_{(s)} + \frac{1}{2}O_{2(g)} \qquad \rightarrow CaO_{(s)} \qquad -1 \times \Delta H_1$$

$$-1 \times (2):\ H_{2(g)} + \frac{1}{2}O_{2(g)} \qquad \rightarrow H_2O_{(l)} \qquad -1 \times \Delta H_2$$

$$1 \times (3):\ Ca_{(s)} + H_{2(g)} + O_{2(g)} \quad \rightarrow Ca(OH)_{2(s)} \qquad 1 \times \Delta H_3$$

$$CaO_{(s)} + H_2O_{(l)} \qquad \rightarrow Ca(OH)_{2(s)} \qquad \Delta H = \Delta H_3 + (-\Delta H_2) + (-\Delta H_1)$$

Notice that the enthalpy change for the target equation equals the enthalpy of formation for the products (calcium hydroxide) minus the enthalpies of formation of the reactants (calcium oxide and water). This observation can be generalized to any chemical equation: The enthalpy change for any given equation equals the sum of the enthalpies of formation of the products minus the sum of the enthalpies of formation of the reactants, or, symbolically

$$\Delta H = \sum n\Delta H^{\circ}_{f(products)} - \sum n\Delta H^{\circ}_{f(reactants)}$$

where n represents the amount (in moles) of each particular product or reactant. Substitute our known molar enthalpies of formation into this equation:

$$\Delta H = n\Delta H^{\circ}_{f(Ca(OH)_2(s))} - (n\Delta H^{\circ}_{f(CaO_{(s)})} + n\Delta H^{\circ}_{f(H_2O)_{(l)}})$$

$$= -(1\ \text{mol} \times \frac{986.1\ \text{kJ}}{1\ \text{mol}}) - ((1\ \text{mol} \times \frac{-634.9\ \text{kJ}}{1\ \text{mol}}) + (1\ \text{mol} \times \frac{-285.8\ \text{kJ}}{1\ \text{mol}}))$$

$$\Delta H = -65.4\ \text{kJ}$$

The thermochemical equation for the slaking of lime with water is

$$CaO_{(s)} + H_2O_{(l)} \rightarrow Ca(OH)_{2(s)} \qquad \Delta H = -65.4\ \text{kJ}$$

KEY EQUATION

$$\Delta H = \sum n\Delta H^{\circ}_{f(products)} - \sum n\Delta H^{\circ}_{f(reactants)}$$

Example 1

The main component in natural gas used in home heating or laboratory burners is methane. What is the molar enthalpy of combustion of methane fuel?

Solution

$$CH_{4(g)} + 2\,O_{2(g)} \rightarrow CO_{2(g)} + 2\,H_2O_{(l)}$$

$$\Delta H = \sum n\Delta H^{\circ}_{f(products)} - \sum n\Delta H^{\circ}_{f(reactants)}$$

$$= (1\text{ mol} \times \frac{-393.5\text{ kJ}}{1\text{ mol}}) + 2\text{ mol} \times \frac{-285.8\text{kJ}}{1\text{ mol}}) - (1\text{ mol} \times \frac{-74.4\text{ kJ}}{1\text{ mol}} + 2\text{ mol} \times \frac{0\text{ kJ}}{1\text{ mol}})$$

$$= -965.1\text{ kJ} - (-74.4\text{ kJ})$$

$$\Delta H = -890.7\text{ kJ}$$

$$\Delta H_c = \frac{\Delta H}{n} = \frac{-890.7\text{ kJ}}{1\text{ mol CH}_4}$$

$$= -890.7\text{ kJ/mol CH}_4$$

The molar enthalpy of combustion of methane fuel is –890.7 kJ/mol.

LEARNING TIP

Enthalpies of combustion of hydrocarbons generally assume production of $CO_{2(g)}$ and $H_2O_{(l)}$, the states of these compounds at SATP.

A variation on the application of standard enthalpies of formation to thermochemical equations is a problem in which the enthalpy change of reaction is provided, and you are asked to find one of the ΔH°_f values.

Example 2

The standard enthalpy of combustion of benzene ($C_6H_{6(l)}$) to carbon dioxide and liquid water is –3273 kJ/mol. What is the standard enthalpy of formation of benzene, given the tabulated values for carbon dioxide and liquid water (Appendix C4)?

Solution

$$C_6H_{6(g)} + \frac{15}{2}\,O_{2(g)} \rightarrow 6\,CO_{2(g)} + 3\,H_2O_{(l)}$$

$$\Delta H = \sum n\Delta H^{\circ}_{f(products)} - \sum n\Delta H^{\circ}_{f(reactants)}$$

$$-3273\text{ kJ} = (6\text{ mol} \times \frac{-393.5\text{ kJ}}{1\text{ mol}} + 3\text{ mol} \times \frac{-285.8\text{ kJ}}{1\text{ mol}})$$

$$- \left(1\text{ mol} \times \Delta H^{\circ}_{f(benzene)} + \frac{15}{2}\text{ mol} \times \frac{0\text{ kJ}}{1\text{ mol}}\right)$$

$$-3273\text{ kJ} = -3217.5\text{ kJ} - 1\text{ mol} \times \Delta H^{\circ}_{f(benzene)}$$

$$\Delta H^{\circ}_{f(benzene)} = \frac{-3217.5\text{ kJ} + 3273\text{ kJ}}{1\text{ mol}}$$

$$\Delta H^{\circ}_{f(benzene)} = 56\text{ kJ/mol}$$

The standard enthalpy of formation of benzene is +56 kJ/mol.

LEARNING TIP

Note the importance of using the standard enthalpy of formation appropriate to the state of a substance. The standard enthalpy of formation of $H_2O_{(g)}$ (–241.8 kJ/mol) is different from that of $H_2O_{(l)}$ (–285.8 kJ/mol).

> ▶ **Practice**

Understanding Concepts

2. Use standard enthalpies of formation to calculate:
   (a) the molar enthalpy of combustion for pentane to produce carbon dioxide gas and liquid water;
   (b) the enthalpy change that accompanies the reaction between solid iron(III) oxide and carbon monoxide gas to produce solid iron metal and carbon dioxide gas.

3. The standard enthalpy of combustion of liquid cyclohexane to carbon dioxide and liquid water is –3824 kJ/mol. What is the standard enthalpy of formation of cyclohexane ($C_6H_{12(l)}$)?

4. Methane, the major component of natural gas, is used as a source of hydrogen gas to produce ammonia. Ammonia is used as a fertilizer and a refrigerant, and is used to manufacture fertilizers, plastics, cleaning agents, and prescription drugs. The following questions refer to some of the chemical reactions of these processes. For each of these equations, use standard enthalpies of formation to calculate ΔH:
   (a) The first step in the production of ammonia is the reaction of methane with steam, using a nickel catalyst.

   $$CH_{4(g)} + H_2O_{(g)} \xrightarrow{Ni} CO_{(g)} + 3\ H_{2(g)}$$

   (b) The second step of this process is the further reaction of water with carbon monoxide to produce more hydrogen. Both iron and zinc–copper catalysts are used.

   $$CO_{(g)} + H_2O_{(g)} \xrightarrow{Fe,\,Zn\,-\,Cu} CO_{2(g)} + H_{2(g)}$$

   (c) After the carbon dioxide gas is removed by dissolving it in water, the hydrogen reacts with nitrogen in the air to form ammonia.
   $$N_{2(g)} + 3\ H_{2(g)} \rightarrow 2\ NH_{3(g)}$$

5. Nitric acid, required in the production of nitrate fertilizers, is produced from ammonia by the Ostwald process. Use standard enthalpies of formation to calculate the enthalpy changes in each of the following systems.

   (a) $4\ NH_{3(g)} + 5\ O_{2(g)} \rightarrow 4\ NO_{(g)} + 6\ H_2O_{(g)}$

   (b) $2\ NO_{(g)} + O_{2(g)} \rightarrow 2\ NO_{2(g)}$

   (c) $3\ NO_{2(g)} + H_2O_{(l)} \rightarrow 2\ HNO_{3(l)} + NO_{(g)}$

Making Connections

6. Energy is used in the manufacture of fertilizers, to grow crops. We then extract food energy from these crops.
   (a) Trace the energy path through the various steps in this process.
   (b) If more thermal energy is put into the process than food energy is gained from the process, should we abandon the practice of manufacturing fertilizers? Discuss.

Answers

2. (a) −3509 kJ/mol
   (b) −24.8 kJ
3. −252 kJ/mol
4. (a) 205.7 kJ
   (b) −41.2 kJ
   (c) −91.8 kJ
5. (a) −906.4 kJ
   (b) −114.2 kJ
   (c) −71.8 kJ

Multistep Energy Calculations Using Standard Enthalpies of Formation

Chemical engineers frequently need to do calculations in which they determine the heats produced by an internal-combustion engine. Such calculations involve bringing together many of the problem-solving skills that you are developing. In Section 3.4 you learned how to solve multistep problems where enthalpy changes, heats transferred, and masses of reactants were interrelated. Two key relationships that you applied were

1. enthalpy change in the system = heat transferred to/from the surroundings

   $\Delta H = q$

   and

2. $\Delta H = n\Delta H_r$

The following sample problem illustrates a situation involving both a multistep problem approach and the technique of referring to standard enthalpies of formation introduced earlier in this section.

Solving Multistep Problems Using Standard Enthalpies of Formation

When octane burns in an automobile engine, heat is released to the air and to the metal in the car engine, but a significant portion is absorbed by the liquid in the cooling system—an aqueous solution of ethylene glycol. What mass of octane is completely burned to cause the heating of 20.0 kg of aqueous ethylene glycol automobile coolant from 10°C to 70°C? The specific heat capacity of the aqueous ethylene glycol is 3.5 J/(g •°C). Assume water is produced as a gas and that all the heat flows into the coolant.

First, write and balance the combustion equation for octane. To simplify later calculations, write the equation so there is 1 mol of octane:

$$C_8H_{18(g)} + \frac{25}{2} O_{2(g)} \rightarrow 8\ CO_{2(g)} + 9\ H_2O_{(g)}$$

Use a reference (such as Appendix C4) to find the standard enthalpies of formation for the products and reactants:

$\Delta H^\circ_{f\ (CO_{2(g)})} = -393.5$ kJ/mol $\qquad\qquad \Delta H^\circ_{f\ (C_8H_{18(g)})} = -250.1$ kJ/mol

$\Delta H^\circ_{f\ (H_2O_{(g)})} = -241.8$ kJ/mol $\qquad\qquad \Delta H^\circ_{f\ (O_{2(g)})} = 0.0$ kJ/mol

Next, use the coefficients from the combustion equation and the standard molar enthalpies to calculate the molar enthalpy of combustion of octane:

$$\Delta H = \sum n\Delta H^\circ_{f(products)} - \sum n\Delta H^\circ_{f(reactants)}$$

$$= (8\ mol \times \Delta H^\circ_{f\ (CO_{2(g)})} + 9\ mol \times \Delta H^\circ_{f\ (H_2O_{(g)})})$$

$$- \left(2\ mol \times \Delta H^\circ_{f\ (C_8H_{18(g)})} + \frac{25}{2}\ mol \times \Delta H^\circ_{f\ (O_{2(g)})}\right)$$

$$\Delta H = \left(8\ mol \times \frac{-393.5\ kJ}{1\ mol} + 9\ mol \times \frac{-241.8\ kJ}{1\ mol}\right)$$

$$- \left(1\ mol \times \frac{-250.1\ kJ}{1\ mol} + \frac{25}{2}\ mol \times \frac{0\ kJ}{1\ mol}\right)$$

$$\Delta H = -5074.1\ kJ$$

Note that if you check tabulated values, you will find a listing of −5471 kJ/mol for octane, because the standard reaction is for $H_2O_{(l)}$, not $H_2O_{(g)}$.

Next, we assume that the enthalpy change in the reaction in the engine equals the heat flow into the aqueous ethylene glycol coolant, or

$$\Delta H_{octane} = q_{coolant}$$

$$n\Delta H_c = mc\Delta T$$

Rearrange to solve for the required amount of octane, n_{octane}:

$$n_{octane} = \frac{mc\Delta T}{\Delta H_c}$$

Substitute the givens from the problem and the calculated value of ΔHc:

$$m_{coolant} = 20.0 \text{ kg}$$
$$\Delta T = 70°C - (-10°C) = 80°C$$
$$c_{coolant} = 3.5 \text{ J}/(g•°C)$$

$$n_{octane} = \frac{20.0 \text{ kg} \times \frac{3.5 \text{ J}}{g•°C} \times 80°C}{\frac{5074.1 \text{ kJ}}{1 \text{ mol}}}$$

$$= 1.1 \text{ mol}$$

To find the required mass of octane, convert the amount into mass, using the molar mass of octane (114.26 g/mol):

$$m_{octane} = 1.1 \text{ mol} \times \frac{114.26 \text{ g}}{1 \text{ mol}}$$

$$m_{octane} = 0.13 \text{ kg}$$

If you are well practised in this technique, you may want to do the final steps in one calculation. Since

$$n_{octane} = \frac{m_{octane}}{M_{octane}}$$

we can rewrite

$$n_{octane} = \frac{mc\Delta T}{\Delta H_c}$$

as

$$m_{octane} = \frac{m_{coolant} \, c_{coolant} \, \Delta T M_{octane}}{\Delta H_{c(octane)}}$$

$$= \frac{20.0 \text{ kg} \times \frac{3.5 \text{ J}}{g•°C} \times 80°C \times \frac{114.26 \text{ g}}{1 \text{ mol}}}{\frac{5074.1 \text{ kJ}}{1 \text{ mol}}}$$

$$m_{octane} = 0.13 \text{ kg}$$

According to the molar enthalpy of formation method and the law of conservation of energy, the mass of octane required is 0.13 kg.

Example

One way to heat water in a home or cottage is to burn propane. If 3.20 g of propane burns, what temperature change will be observed if all of the heat from combustion transfers into 4.0 kg of water?

Solution

$$C_3H_{8(g)} + 5\,O_{2(g)} \rightarrow 3\,CO_{2(g)} + 4\,H_2O_{(l)}$$

$$\Delta H^\circ_{f\,(CO_{2(g)})} = -393.5\ \text{kJ/mol}$$

$$\Delta H^\circ_{f\,(H_2O_{(l)})} = -285.8\ \text{kJ/mol}$$

$$\Delta H^\circ_{f\,(C_3H_{8(g)})} = -104.7\ \text{kJ/mol}$$

$$\Delta H^\circ_{f\,(O_{2(g)})} = 0.0\ \text{kJ/mol}$$

$$m_{\text{propane}} = 3.20\ \text{g}$$

$$m_{H_2O} = 4.0\ \text{kg}$$

$$c_{H_2O} = 4.18\ \text{J/(g}\cdot{}^\circ\text{C)}$$

The enthalpy of change for the reaction is:

$$\Delta H = \sum n\Delta H^\circ_{f(\text{products})} - \sum n\Delta H^\circ_{f(\text{reactants})}$$

$$\Delta H = \left(3\ \text{mol} \times \frac{-393.5\ \text{kJ}}{1\ \text{mol}} + 4\ \text{mol} \times \frac{-285.8\ \text{kJ}}{1\ \text{mol}}\right)$$
$$- \left(1\ \text{mol} \times \frac{-104.7\ \text{kJ}}{1\ \text{mol}} + 5\ \text{mol} \times \frac{0.0\ \text{kJ}}{1\ \text{mol}}\right)$$

$$\Delta H = -2219\ \text{kJ}$$

Therefore, the molar enthalpy of combustion of propane is

$$\Delta H_{c(\text{propane})} = 2219\ \text{kJ/mol}$$

$$\Delta H_{c(\text{propane})} = q_{\text{water}}$$

$$n\Delta H_c = mc\Delta T$$

$$\Delta T = \frac{n\Delta H_c}{mc}$$

$$= \frac{m_{\text{propane}}\Delta H_c}{M_{\text{propane}}m_{\text{water}}c}$$

$$= \frac{3.20\ \text{g} \times \dfrac{2219\ \text{kJ}}{1\ \text{mol}}}{\dfrac{44.11\ \text{g}}{1\ \text{mol}} \times 4.0\ \text{kg} \times \dfrac{4.18\ \text{J}}{\text{g}\cdot{}^\circ\text{C}}}$$

$$\Delta T = 9.6\,^\circ\text{C}$$

The temperature change of the water is 9.6°C.

Figure 2
The production of canola crops are dependent on the use of fertilizers such as ammonium nitrate.

▶ Practice

Making Connections

7. Ammonium nitrate fertilizer is produced by the reaction of ammonia with nitric acid:

$$NH_{3(g)} + HNO_{3(l)} \rightarrow NH_4NO_{3(s)}$$

Ammonium nitrate is one of the most important fertilizers for increasing crop yields (**Figure 2**).
(a) Use standard enthalpies of formation to calculate the standard enthalpy change of the reaction used to produce ammonium nitrate.

(b) Sketch a potential energy diagram for the reaction of ammonia and nitric acid.

(c) Calculate the heat that would be produced or absorbed in the production of 50 t of ammonium nitrate.

8. Coal is a major energy source for electricity, of which industry is the largest user (**Figure 3**). Anthracite coal is a high-molar-mass carbon compound with a composition of about 95% carbon by mass. A typical simplest-ratio formula for anthracite coal is $C_{52}H_{16}O_{(s)}$. The standard enthalpy of formation of anthracite can be estimated at -396.4 kJ/mol. What is the quantity of energy available from burning 100.0 kg of anthracite coal in a thermal electric power plant, according to the following equation?

$$2 C_{52}H_{16}O_{(s)} + 111 O_{2(g)} \rightarrow 104 CO_{2(g)} + 16 H_2O_{(g)}$$

9. Alternative transportation fuels include methanol and hydrogen.
   (a) Use standard enthalpies of formation to calculate the energy produced per mole of methanol burned.
   (b) Use standard enthalpies of formation to calculate the energy produced per mole of hydrogen burned.
   (c) In terms of energy content, how do these two alternative fuels compare with gasoline, which is mostly octane? The molar enthalpy of combustion of octane is -5.07 MJ/mol.
   (d) What factors other than energy content are important when comparing different automobile fuels? Include several perspectives.

Making Connections

10. In a typical household, about one-quarter of the energy consumed is used to heat water.
    (a) What mass of methane undergoing complete combustion is required to heat 100.0 kg of water from 5.0°C to 70.0°C in a gas water heater?
    (b) How might we heat water more efficiently?
    (c) What alternative energy resources are available for heating water?

Figure 3
Coal is an important current source of electrical energy in Ontario but its use has serious environmental consequences.

Answers

7. (a) -145.6 kJ
   (c) -9.10×10^4 MJ
8. 3.34×10^3 MJ
9. (a) -726 kJ/mol
   (b) -285.8 kJ/mol
10. (a) .48 kg

▶ *Section 5.5* *Questions*

Understanding Concepts

1. Write balanced equations for the formation of the following:
   (a) acetylene gas (C_2H_2), used in welding
   (b) creatine ($C_4H_9N_3O_2$), used as a food supplement
   (c) potassium iodide (KI), used as a salt substitute
   (d) iron(II) sulfate ($FeSO_4$), used as a diet supplement

2. Use standard enthalpies of formation to calculate the enthalpy changes in each of the following equations:
   (a) Magnesium carbonate decomposes when strongly heated.
   (b) Ethene burns in air.
   (c) Sucrose ($C_{12}H_{22}O_{11(s)}$) decomposes to carbon and water vapour when concentrated sulfuric acid is poured onto it (**Figure 4**).

3. When octane is strongly heated with hydrogen gas in the presence of a suitable catalyst, it "cracks," forming a mixture of hydrocarbons. A typical cracking reaction yields 1 mol of methane, 2 mol of ethane, and 1 mol of propane from each mole of octane.

(a) Write the balanced equation for this reaction.
(b) Use standard enthalpies of formation to calculate the enthalpy change associated with the cracking of one mole of octane.

Figure 4
The decomposition of sucrose produces black carbon and water.

Applying Inquiry Skills

4. A sample of acetone (C_3H_6O) is burned in an insulated calorimeter to produce carbon dioxide gas and liquid water.

Question

What is the molar enthalpy of combustion of acetone?

Experimental Design

A sample of liquid acetone is burned such that the heat from the burning is transferred into an aluminum calorimeter and its water contents.

Prediction

(a) Use tabulated standard enthalpies of formation to calculate a theoretical value for the molar enthalpy of combustion of acetone.

Evidence

Table 1 Observations on the Combustion of Acetone

Quantity	Measurement
mass of water	100.0 g
specific heat capacity of aluminum	0.91 J/(g • °C)
mass of aluminum can	50.0 g
initial temperature of calorimeter	20.0°C
final temperature of calorimeter	25.0°C
mass of acetone burned	0.092 g

Analysis

(b) Calculate the molar enthalpy of combustion of acetone, using the experimental evidence.

Evaluation

(c) Calculate the percentage error by comparing the predicted and experimental values.

(d) Does the evidence support the law of conservation of energy?

(e) If heat is lost to the surroundings instead of being transferred only into the aluminum can and water, will the experimental molar enthalpy be higher or lower? Explain.

Extension

5. Remote controlled model boats are powered by burning methanol or a racing mixture of 80% methanol and 20% nitromethane by mass (**Figure 5**). The standard enthalpies of formation of liquid methanol and nitromethane are, respectively, –238.6 kJ/mol and –74.78 kJ/mol.

(a) Draw structural diagrams for all reactants and products.

(b) Calculate the percentage change in energy output achieved by using a racing mixture.

(c) Considering the various desirable characteristics of a fuel, why do you suppose such a racing mixture is used?

Figure 5

Energy is central to our way of life. Everything we do during the day involves energy changes. Earlier in this chapter, we talked about three types of changes in matter (physical, chemical, and nuclear) and the energy changes (net gain or net loss) that accompany them. Our consumption of energy also involves these types of changes, but to varying degrees. Freon gas vaporizes in the refrigerator coil inside a freezer, the physical change absorbs energy. When methane burns in a natural gas oven, energy is released to the surroundings. A smoke detector on the ceiling emits tiny quantities of energetic radioactive particles. Most of these changes involve conversion of potential energy into some form of kinetic energy.

Our household and industrial energy needs are, by and large, met by two energy sources: electricity, and the burning of fossil fuels such as gasoline and natural gas. The majority of our household technologies, from the microwave oven in the kitchen to the alarm clock in the bedroom to the television in the living room, use electricity. The production of electricity for consumer use is a multi-billion-dollar industry and, in Ontario, this electricity comes mainly from three sources: hydroelectric power, nuclear power, and the burning of fossil fuels. All of these sources involve some system falling from higher potential energy to lower potential energy (**Figure 1**), releasing a form of kinetic energy in the process.

The major energy sources that we use in Ontario are all in some sense finite: There are limited numbers of geographical regions where one can harness the flow of water, and both fossil fuels and nuclear fuel are "used up" as they generate energy. Each of these resources also has significant advantages and disadvantages. Hydroelectric projects often involve diversion of rivers, displacement of wildlife and other environmental effects, and huge initial capital expenses, but the result is "clean" energy powered by the Sun (as the Sun provides the energy to allow water to evaporate) and producing comparatively little pollution. Fossil fuels are relatively cheap and available, but they are a finite resource that cannot last and their use damages the environment, contributing to acid rain and the greenhouse effect. Nuclear power can generate huge amounts of energy from a tiny amount of fuel, but reactors are very expensive and there are concerns about the safety and disposal of nuclear waste products. As you will see in the following pages, the huge potential benefits and risks involved in the use of nuclear power make it a central issue in the ongoing energy debate.

Power from Nuclear Fission

Nuclear power stations have much in common with conventional power stations fired by fossil fuels (**Figure 2**). In both, heat is used to boil water and the resulting steam drives a turbine. The spinning turbines, in turn, drive generators that produce electricity. In a conventional power station, combustion of natural gas, oil, or coal supplies the necessary heat; in a nuclear power station, nuclear fission provides the heat.

The most widely used nuclear reaction is the fission or splitting of uranium into two smaller nuclei. Nuclear fission reactions such as those shown below for uranium provide the energy for nuclear power generating stations; the enthalpy changes for the reactions are in the order of 10^{10} kJ/mol of uranium. Nuclear reactors use the energy that is

Radiodecay of Uranium

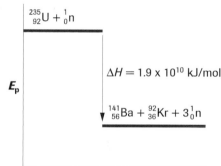

$$^{235}_{92}U + {}^{1}_{0}n$$

$$\Delta H = 1.9 \times 10^{10} \text{ kJ/mol}$$

$$^{141}_{56}Ba + {}^{92}_{36}Kr + 3{}^{1}_{0}n$$

E_p

Reaction Progress

Figure 1
Water falling at Niagara and nuclear fission in a reactor are both examples of systems in which the amount of stored potential energy decreases and is converted into kinetic energy.

Figure 2

Many energy sources can generate electricity.

(a) In a hydroelectric power station, water collected behind a dam is released through a pipe to the turbine.

(b) Water in a boiler is heated in one of several ways:
- chemical energy from the combustion of fossil fuels in a thermal electric power plant;
- nuclear energy from the fission of uranium in a nuclear power plant;
- direct radiant solar energy reflected from many mirrors onto the boiler in a solar power plant; and
- geothermal energy from the interior of the Earth in a geothermal power plant.

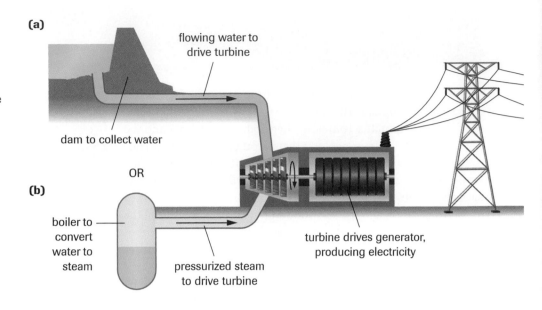

(a)
flowing water to drive turbine

dam to collect water

OR

(b)
boiler to convert water to steam

pressurized steam to drive turbine

turbine drives generator, producing electricity

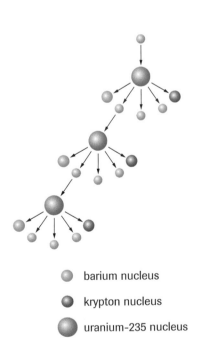

barium nucleus

krypton nucleus

uranium-235 nucleus

neutron

Figure 3

In the fission of U-235, one neutron is absorbed, the nucleus splits, and three more neutrons are produced.

released when uranium-235 undergoes nuclear fission to produce any of several possible radioisotopes. One of the nuclear reactions that occurs is

$$^{235}_{92}U + ^{1}_{0}n \rightarrow ^{92}_{36}Kr + ^{141}_{56}Ba + 3 ^{1}_{0}n + energy \qquad \Delta H = 1.9 \times 10^{10} \text{ kJ/mol}$$

This process is initiated by relatively slow-moving neutrons hitting a uranium nucleus (**Figure 3**), releasing large amounts of energy. Moreover, more neutrons are generated, which are able to collide with yet more uranium nuclei to generate even more heat energy. If there is only a small amount of uranium, most of the neutrons produced just fly out of the fuel without causing further fission. However, if enough uranium nuclei are available—a situation known as critical mass—the neutrons collide with more uranium nuclei before they leave the fissionable material. The result is an explosive chain reaction with the sudden release of massive amounts of energy. The complete chain reaction and loss of energy can take place in less than one microsecond (1×10^{-6} s).

Uranium is a very concentrated energy source. For example, when placed in a CANDU reactor, a fuel bundle 50 cm long and 10 cm in diameter, with a mass of 22 kg, can produce as much energy as 400 t of coal or 2000 barrels of oil. At present, approximately 16% of the world's electricity is generated by nuclear power stations like the ones shown in **Figure 4**.

Canadian nuclear reactors (known as CANDU reactors) use natural uranium containing about 0.7% uranium-235 and 99.3% uranium-238. The energy is produced by the fission of uranium-235 inside the reactor. Because of the vast quantities of energy released in fission, it has great potential as a commercial power source.

There are strong arguments both for and against nuclear power. Advocates of nuclear energy point out that nuclear power

- has low uranium fuel costs, including transportation;
- causes very little air pollution, such as greenhouse and acid gases; and
- reduces our dependence on fossil fuels for electricity generation, allowing those materials to be used for other purposes.

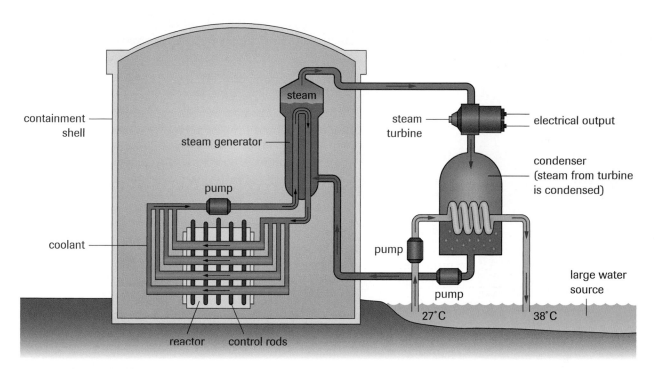

Figure 4
In nuclear reactors, the energy released in nuclear fission is absorbed by a primary coolant, which then transfers the energy to a secondary coolant. The energy from the secondary coolant is used in a turbine or engine to generate electrical energy. A typical reactor consists of five components: fuel, moderator, coolant, control rods, and shielding. All of these become radioactive, to some degree, and pose a disposal challenge.

Opponents of nuclear power are concerned about

- the possible release of radioactive materials in a reactor malfunction;
- the difficulty of disposing of the highly toxic radioactive wastes;
- the large capital costs of building nuclear reactors and then decommissioning them at the end of their relatively short lifetime;
- unknown health effects of long-term low-level exposure to radiation; and
- thermal pollution from cooling water.

In the 1950s, people had high expectations of endless, inexpensive nuclear energy. Few would have predicted that concerns about reactor safety and radioactive wastes would severely dampen the enthusiasm for nuclear energy. But public attitudes, already changing as the public concentrated on the risks of nuclear generation, changed decisively after the nuclear accident at Chernobyl in Ukraine (**Figure 5**), where a serious accident

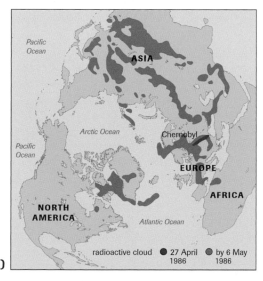

(a)

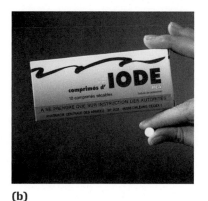

(b)

Figure 5
The most serious nuclear accident in history happened at Chernobyl, Ukraine.
(a) Radioactive material, carried by the wind, spread for thousands of kilometres.
(b) The element iodine reaches a maximum concentration in the thyroid gland. To reduce the effects of radioactive iodine from fallout, children in several countries were given tablets containing non-radioactive iodine. This iodine would concentrate in the thyroid, so any ingested radioiodine would not stay in the body but be excreted.

Figure 6
Large tanks of water provide safe short-term storage of nuclear waste.

DID YOU KNOW ?

Potential Energy
When a kilogram of water (1 L) flows over Niagara Falls, it loses roughly 1 kJ of gravitational potential energy, some fraction of which may be converted into hydroelectricity. When a kilogram of gasoline burns, it releases about 4×10^4 kJ of energy. When a kilogram of uranium in a nuclear reactor undergoes fission, it releases about 1×10^{11} kJ of energy.

in a nuclear reactor on April 28, 1986, spewed a deadly, steam-driven cloud of radioactive plutonium, cesium, and uranium dioxide into the atmosphere. A nuclear reactor does not explode like a nuclear bomb; rather, the steam pressure can build up, resulting in an explosion more like a dynamite explosion.

In 1979, a reactor malfunctioned at Three Mile Island in Pennsylvania, but the containment structure worked well. There have been no major nuclear accidents in Canada, but no new nuclear generators have been built here or in the United States since 1986.

Less dramatic than a reactor malfunction, but also serious, is the continuing problem of radioactive waste disposal (**Figure 6**). Scientists and engineers are working to devise a safe and economically feasible method of disposing of radioactive waste, but the public remains skeptical. Burial in arid regions or in granite layers in mine shafts—to avoid contaminating ground water—are two possibilities, although both of these "stable" situations could change over geological time. Chemists have been developing suitable materials for encasing radioactive substances to prevent their escape into the environment. Lead–iron–phosphate glass is a promising material, since the nuclear waste can be chemically incorporated into a stable glass and then buried in the safest possible place.

Practice

Understanding Concepts

1. (a) What are the original sources of energy for most of Ontario's electricity?
   (b) Outline the similarities and differences of power generation from these various sources.

2. Nuclear reactors in the United States and Europe use different systems than the CANDU system. Research and make a chart to summarize the similarities and differences among these systems.

 www.science.nelson.com

Take a Stand: Energy Options

Energy use is an unavoidable part of our everyday life. As citizens, we have to make decisions about the sources of energy we use. Our choices affect the future of our country and the world. Some of the options include:

- hydroelectric power
- conventional fossil fuel power
- nuclear power
- alternative energy generation (non-traditional fuels; wind, geothermal, and solar power)
- "soft energy paths" (conservation, alternatives to electricity use, changes in lifestyle to use less energy rather than finding ways to generate more power)

What are the advantages and disadvantages of each of the above options? What direction should we go in the future? There are many aspects of power generation to consider, such as:

- efficiency, in terms of energy output per dollar spent
- efficiency, in terms of energy output per gram of fuel used
- capital cost of technology
- environmental impact

Decision-Making Skills

○ Define the Issue ● Identify Alternatives ● Research
● Analyze the Issue ● Defend the Position ● Evaluate

(a) Separate into small groups and choose one of the above power options for study.

(b) As a group, research the advantages and disadvantages of the option that you have chosen, considering as many different aspects of your power source as possible, such as technological innovations that exist or are proposed to make the option more attractive; the practicality of this option, including efficiency, initial capital, and ongoing operation costs; public attitudes and effects on habits; environmental impact; and comparisons of Canada to the rest of the world with respect to this option.

(c) Summarize your research, including explanations of terminology as required to make major points understandable.

(d) Present your research to the class. Compare your points with those of other groups. Be prepared to assess your efforts as well as those of your classmates.

(e) Within your original group, write a consensus statement that summarizes how your group feels about each energy option.

(f) Write a report, intended to influence government leaders, expressing the direction that your group feels the provincial or federal government should take on energy policy.

Nuclear Fusion

The nuclear reactions that occur in the Sun are important to us because they supply the energy that sustains life on Earth. There are many different nuclear reactions taking place in the Sun, as in other stars. In one of the main reactions, four hydrogen atoms fuse to produce one helium atom.

$$4\,{}^{1}_{1}\text{H} + 2\,{}^{0}_{-1}\text{e} \rightarrow {}^{4}_{2}\text{He}$$

Scientists and engineers think that using a similar reaction, the fusion of two isotopes of hydrogen, is a promising possibility for the development of commercial nuclear fusion reactors on Earth (**Figure 7**). In this reaction, a helium atom (${}^{4}_{2}\text{He}$), a neutron (${}^{1}_{0}\text{n}$), and a large quantity of energy are produced.

$${}^{2}_{1}\text{H} + {}^{3}_{1}\text{H} \rightarrow {}^{4}_{2}\text{He} + {}^{1}_{0}\text{n} \qquad \Delta H = -1.70 \times 10^{9}\ \text{kJ}$$

This means that 1.7×10^{9} kJ of energy is released for every mole of helium produced. The large energies involved in fusion reactions make them a promising area for research on power generation, but the technology for controlling the process is less well developed than that for fission reactions.

Figure 7
This tokomak, an experimental fusion reactor, is in use at Princeton University in the United States.

Answers

3. (a) 8.1×10^{10} kJ

 (b) 4.3×10^{11} kJ

Understanding Concepts

3. (a) Canadian (CANDU) nuclear reactors produce energy by nuclear fission of uranium-235:

$$^{235}_{92}U + ^{1}_{0}n \rightarrow ^{141}_{56}Ba + ^{92}_{36}Kr + 3\,^{1}_{0}n$$

If the molar enthalpy of fission for uranium-235 is 1.9×10^{10} kJ/mol, how much energy can be obtained from the fission of 1.00 kg (4.26 mol) of pure uranium-235?

 (b) In the Sun, hydrogen atoms undergo nuclear fusion to produce He and 1.7×10^{9} kJ/mol He. How much energy can be obtained by producing 1.00 kg of helium?

 (c) Explain why, although the molar enthalpies of reaction of U-235 (fission) and He (fusion) are similar, their energy outputs per kilogram are so different.

Making Connections

4. Current nuclear power generation uses nuclear fission reactions. Why are fission reactions used instead of fusion reactions? Report on what progress has been made in making nuclear fusion practical as an energy source.

 www.science.nelson.com

Understanding Concepts

1. Compare, with examples, the energy produced from nuclear fusion and nuclear fission reactions.

Making Connections

2. Research and report on atomic energy in Ontario. Include descriptions of:
 (a) locations of nuclear facilities;
 (b) amount of power generated; and
 (c) advantages and disadvantages of this energy source.

 www.science.nelson.com

3. Coal is often described as the "dirtiest" fossil fuel used for power generation. Name other fossil fuels that can be burned to generate power and write a short report discussing their advantages and disadvantages, or list the advantages and disadvantages in a table.

 www.science.nelson.com

4. Both geothermal and solar energy have been suggested as clean and efficient alternatives to fossil fuels and atomic energy. Research and write a brief report on the practicality of one of these energy sources.

 www.science.nelson.com

5. Hydrogen power is described by some as the ideal alternative to fossil fuel combustion because the only product is water. Others argue that "dirty" energy sources are used to produce hydrogen fuel, so that the pollution is just produced somewhere else. Find out how hydrogen fuel is obtained and report on the advantages and disadvantages of this alternative energy source.

 www.science.nelson.com

6. Every year several groups in Ontario organize for Canadian families to host children from the Chernobyl area of Ukraine. Research and report on the effects of the Chernobyl nuclear disaster, and how these Ontario communities are trying to improve life for a few survivors of the Chernobyl disaster.

 www.science.nelson.com

⚗ INVESTIGATION 5.1.1

Medical Cold Packs

Inquiry Skills

○ Questioning	● Planning	● Analyzing
○ Hypothesizing	● Conducting	● Evaluating
○ Predicting	● Recording	● Communicating

Figure 1
A simple laboratory calorimeter consists of an insulated container made of three nested polystyrene cups, a measured quantity of water, and a thermometer. The chemical system is placed in or dissolved in the water of the calorimeter. Energy transfers between the chemical system and the surrounding water are monitored by measuring changes in the temperature of the water.

We can study physical changes that involve liquids and aqueous solutions, using a polystyrene calorimeter like the one shown in **Figure 1**. Such materials can be applied to practical thermochemical systems. For example, when athletes are injured, they may immediately hold an "instant cold pack" against the injury. The medical cold pack operates on the principle that certain salts dissolve endothermically in water. The amount of heat per unit mass involved in the dissolving of a compound is a characteristic property of that substance. It is called the enthalpy of solution. **Table 1** lists some compounds that might be possible candidates for a cold pack because they absorb energy when they dissolve.

Table 1   Enthalpies of Solution for Compounds in a
Medical Cold Pack

Salt	Enthalpy of solution
ammonium chloride, NH_4Cl	0.277 kJ/g
potassium nitrate, KNO_3	0.345 kJ/g
ammonium nitrate, NH_4NO_3	0.321 kJ/g
sodium acetate trihydrate, $NaC_2H_3O_2 \cdot 3\ H_2O$	0.144 kJ/g
potassium chloride, KCl	0.231 kJ/g

Purpose

The purpose of this investigation is to use calorimetry to determine the enthalpy of solution of an unknown salt, and then to use that value to identify the salt in a medical cold pack.

Question

What is the identity of the unknown salt?

Experimental Design

You will be given a solid sample of about 10 g of a salt typically used in medical cold packs and listed in **Table 1**.

(a) Design an experiment to determine the heat transferred when a measured mass of the salt is dissolved in water.

You will then apply calorimetric calculations to discover the identity of the salt.

(b) Write a detailed description of the calculations that you will use to determine the identity of the salt. Show all formulas and units that you will use. You will find it convenient to assign symbols (e.g., m_1, m_2, T_1, etc.) to the measurements that you expect to make so that your calculations are clear.

Materials

lab apron
eye protection
centigram or analytical balance
8–10 g of an unknown salt (from the list in **Table 1**)
water
thermometer
3 Styrofoam cups
100-mL graduated cylinder

Procedure

(c) Write your Procedure as a series of numbered steps, clearly identifying the masses, volumes, and specific equipment to be used (see Materials list), necessary safety and disposal considerations, and a table in which to record your observations.

1. Have your Experimental Design, calculations, and Procedure approved by your teacher before performing your experiment.

Analysis

(d) Determine, by calculation, the identity of the unknown salt.

Evaluation

(e) Calculate a percentage difference between your experimental value and the accepted value for enthalpy of solution of the identified salt.

(f) How confident are you in your identification? Suggest three sources of experimental error in this experiment.

(g) If some heat were transferred to the air or Styrofoam cups, would your calculated enthalpy of solution of the unknown salt be too high or too low? Explain.

(h) If some salt were accidentally spilled as it was transferred from the balance to the Styrofoam cup, would your calculated enthalpy of the unknown salt be too high or too low? Explain.

INVESTIGATION 5.2.1

Molar Enthalpy of a Chemical Change

Inquiry Skills

○ Questioning	○ Planning	● Analyzing
○ Hypothesizing	● Conducting	● Evaluating
○ Predicting	● Recording	● Communicating

When aqueous solutions of acids and bases react in a calorimeter, the solutions may act as both system and surroundings. The acid and base form the system as they react to form water and a dissolved salt. For example, sodium hydroxide and sulfuric acid react to form sodium sulfate and water:

$$NaOH_{(aq)} + \frac{1}{2}H_2SO_{4\,(aq)} \rightarrow H_2O_{(l)} + \frac{1}{2}Na_2SO_{4(aq)} + heat$$

The reactant solutions are mostly water containing dissolved and dispersed acid and base particles. Thus, a solution of an acid or a base may be regarded for *calorimetric* purposes to have the same specific heat capacity as water, at least to the degree of experimental accuracy that applies to simple calorimeters. For example, when 100 mL of a dilute acid solution reacts exothermically with 150 mL of a solution of a dilute base, the acid and base may be thought to react and release heat to a total of 250 mL of water. Assuming that the solutions have the same density as water (1.00 g/mL) also makes calculations simpler.

Purpose

The purpose of this investigation is to use calorimetry to obtain an empirical value for the molar enthalpy of neutralization of sodium hydroxide by sulfuric acid.

Question

What is the molar enthalpy of neutralization (ΔH_{neut}) of sodium hydroxide with sulfuric acid?

Experimental Design

A measured volume of sodium hydroxide solution of known concentration will be combined with excess sulfuric acid solution of known concentration. (The sodium hydroxide will be the limiting reagent.) Calorimetric measurements will be made to determine the heat produced by the reaction.

(a) Demonstrate by calculation that the quantity of acid identified in the Procedure will be enough to completely consume the base, and that the base will therefore be the limiting reagent for calculation purposes.

Materials

lab apron
eye protection
1.0 mol/L sodium hydroxide solution
1.0 mol/L sulfuric acid solution
thermometer
polystyrene calorimeter
two 100-mL graduated cylinders

 CAUTION: 1.0 mol/L sodium hydroxide solution is corrosive, and 1.0 mol/L sulfuric acid solution is an irritant. Wear eye protection and a lab apron. Clean up any spills immediately. At these dilutions, the chemicals may be disposed of down the drain with lots of water.

Procedure

1. Add 50 mL of 1.0 mol/L sodium hydroxide solution to a polystyrene calorimeter. Measure and record its temperature.

⚗ INVESTIGATION 5.2.1 *continued*

2. Measure and record the temperature of a 30-mL sample of 1.0 mol/L sulfuric acid solution in a graduated cylinder.

3. Carefully add the acid to the base, stirring slowly with the thermometer. Measure and record the maximum temperature obtained.

Analysis

(b) Calculate

   (i)   the masses of acid and base solutions;

   (ii)   the temperature changes in the acid and base solutions;

   (iii) the total heat absorbed by the calorimetric liquids (acid and base);

   (iv) the amount of base (in moles) that reacted; and

   (v)   the molar enthalpy of neutralization of the sodium hydroxide.

Evaluation

(c) Calculate a percentage difference by comparing your experimental values and the accepted values for enthalpy of neutralization with respect to sodium hydroxide (-56 kJ/mol).

(d) Suggest any sources of experimental error in this investigation. Evaluate the experimental design, and your skill in carrying it out.

Synthesis

(e) What would have been the effect on the calculated enthalpy of reaction if you had used

   (i) 100 mL of sulfuric acid solution? Explain.

   (ii) 20 mL of sulfuric acid solution? Explain.

⚗ INVESTIGATION 5.3.1

Inquiry Skills

○ Questioning	○ Planning	● Analyzing
○ Hypothesizing	● Conducting	● Evaluating
● Predicting	● Recording	● Communicating

Combustion of Alcohols

When alcohols burn they produce heat. Do different alcohols produce different quantities of heat—do their molar enthalpies of combustion differ? In this investigation, you will link your study of thermochemistry to your knowledge of the molecular structure of alcohols to investigate energy relationships.

Purpose

The purpose of this investigation is to use calorimetry to determine the molar enthalpy change in the combustion of each of a series of alcohols.

Question

How do the enthalpies of combustion change as the alcohol molecules become larger (i.e., ethanol to butanol)?

Prediction

(a) Predict what should happen to the enthalpies of combustion per mole of alcohol as the alcohol molecules become larger.

Experimental Design

You will burn measured masses of a series of alcohols, and calculate the amount of each alcohol burned. Assume that the energy produced is transferred to a measured volume of water, the temperature change of which is calculated. The enthalpy change involved in the combustion of one mole of each alcohol can then be calculated and compared.

Materials

lab apron
eye protection
centigram or analytical balance
alcohol burners containing ethanol, propanol, and butanol
thermometer
small tin can, open at one end, and cut under the rim on opposite sides
stirring rod
ring stand with iron ring
large tin can, open at both ends, with vent openings cut around the rim at the base
100-mL graduated cylinder

Figure 2
Apparatus for burning alcohols

⚠ **Alcohols are highly flammable. Do not attempt to refuel the burners. If refuelling is necessary, ask your teacher to do so.**

Do not move the burners after they are lit.

Procedure

1. Obtain an alcohol burner containing one of the alcohols.

2. Put on your eye protection.

3. Place 100 mL of cold water in a small can suspended over an alcohol burner surrounded by a larger can (**Figure 2**).

4. Measure the mass of the alcohol burner and the temperature of the water.

5. Burn the alcohol such that the heat from the reaction is transferred as efficiently as possible into the water.

6. Cease heating when the temperature has risen about 20°C.

7. Re-weigh the alcohol burner and remaining alcohol.

8. Repeat steps 3 to 7 for the remaining alcohols.

Analysis

(b) Assume that 100 mL of water has a mass of 100 g. For each of the alcohols, calculate and tabulate:

  (i) the heat absorbed by the water in the calorimeter, which equals the heat produced by alcohol burning;

  (ii) the heat produced per gram of alcohol;

  (iii) the amount (in moles) of alcohol burned;

  (iv) the heat produced per mole of alcohol.

(c) Write a thermochemical equation to represent the burning of 1 mol of each of the alcohols, including your experimental value for ΔH_{comb}.

(d) Use the four methods described in Section 3.3 to represent the experimentally determined enthalpy change for the burning of 2 mol of ethanol.

(e) Use your representations to answer the Question.

Evaluation

(f) What are the sources of experimental error in this experiment?

(g) Accepted values for combustion of the three alcohols are: 1369 kJ/mol, 2008 kJ/mol, and 3318 kJ/mol. Calculate a percentage difference between your experimental value and the accepted values. Based on this calculation, was the Experimental Design an acceptable method to answer the Question?

Synthesis

(h) If you were in the business of buying, transporting, and storing alcohols for use as home-heating fuels, which of these alcohols would you choose to work with? Explain.

INVESTIGATION 5.4.1

Hess's Law

Inquiry Skills

○ Questioning   ○ Planning   ● Analyzing
○ Hypothesizing   ● Conducting   ● Evaluating
● Predicting   ● Recording   ● Communicating

Figure 3
The reaction of magnesium in air is very exothermic.

The combustion of magnesium is very rapid and exothermic (**Figure 3**), and is represented by the equation

$$Mg_{(s)} + \frac{1}{2} O_{2(g)} \rightarrow MgO_{(s)}$$

It is possible to observe and measure a series of reactions that enable us, with the use of Hess's law, to determine the enthalpy change for this reaction.

(1) Magnesium reacts in acid to form hydrogen gas and a salt:

$$Mg_{(s)} + 2\,HCl_{(aq)} \rightarrow H_{2(g)} + MgCl_{2(aq)} \qquad \Delta H_1 = ?\ kJ$$

(2) Magnesium oxide reacts in acid to form water and a salt:

$$MgO_{(s)} + 2\,HCl_{(aq)} \rightarrow H_2O_{(l)} + MgCl_{2(aq)} \qquad \Delta H_2 = ?\ kJ$$

The values ΔH_1 and ΔH_2 can be determined empirically using a simple calorimeter. The enthalpy of reaction for a third reaction can be found in a reference table.

(3) Hydrogen and oxygen gases react to form water:

$$H_{2(g)} + \frac{1}{2} O_{2(g)} \rightarrow H_2O_{(l)} \qquad \Delta H_3 = -285.8\ kJ$$

Purpose

The purpose of this investigation is to use Hess's law to determine the molar enthalpy of combustion of magnesium, using calorimetry.

Question

What is the molar enthalpy of combustion, ΔH_c, of magnesium?

Experimental Design

Measured masses of magnesium and magnesium oxide will be added to measured volumes of hydrochloric acid solution of known concentration. The temperature changes will be determined. Calculated enthalpies of reaction will be combined, using Hess's law, to determine the enthalpy of combustion of magnesium.

Prediction

(a) Show how the three known equations and their enthalpies of reaction may be combined, using Hess's law, to yield the target equation and its enthalpy of combustion.

Materials

lab apron
eye protection
steel wool
centigram or milligram balance
thermometer
polystyrene calorimeter
100-mL graduated cylinder
scoopula
10- to 15-cm strip of magnesium ribbon
magnesium oxide powder
1.00 mol/L hydrochloric acid

 Hydrochloric acid is corrosive. Eye protection and a lab apron should be worn. All spills should be cleaned up quickly and any skin that has come into contact with acid should be immediately and thoroughly rinsed with cold water.

 Magnesium ribbon is highly flammable and should be kept far from any source of ignition.

Procedure

1. Measure 100.0 mL of 1.00 mol/L hydrochloric acid into a polystyrene cup. Measure the initial temperature of the acid solution to the nearest 0.2°C.

2. Polish a length of magnesium ribbon with steel wool. Determine the mass (to ±0.01 g) of approximately 0.5 g of magnesium metal. Add the solid to the solution, stir it, and record the maximum temperature that the solution attains.

3. Dispose of the products as directed by your teacher, and rinse and dry the equipment.

4. Repeat the first three steps, using approximately 1 g of magnesium oxide powder measured to ±0.01 g.

Analysis

(b) Were the changes exothermic or endothermic? Explain.

(c) For the first reaction, calculate the enthalpy change per mole of magnesium.

(d) Write a thermochemical equation for the reaction of magnesium in acid, including your experimental value.

(e) Repeat steps (c) and (d) for magnesium oxide.

(f) Using Hess's law, the values you have found experimentally, and the given value for the enthalpy change for the formation of water from its elements, determine the molar enthalpy of combustion of magnesium.

(g) Which of the measured values limited the precision of your value? Explain.

Evaluation

(h) Explain why (and how) your calculated enthalpies of reaction would be inaccurate if

(i) some heat were transferred to the air or Styrofoam cup;

(ii) the surface of the magnesium ribbon had a coating of MgO.

(i) Suggest some other possible sources of experimental error in this investigation.

(j) The accepted value for the molar enthalpy of combustion of magnesium is –601.6 kJ/mol. Calculate a percentage difference by comparing your experimental values and the accepted values. Comment on your confidence in the evidence.

(k) Based on your evaluation of the Experimental Design and the evidence, is Hess's law an acceptable method to calculate enthalpies of reaction?

Synthesis

(l) Suggest an experimental technique that could be used to determine the enthalpy of combustion of magnesium directly.

LAB EXERCISE 5.5.1

Testing Enthalpies of Formation

Inquiry Skills

○ Questioning	○ Planning	● Analyzing
○ Hypothesizing	○ Conducting	● Evaluating
● Predicting	○ Recording	● Communicating

Calorimetry is the basic experimental tool used to determine enthalpies of reaction. When carefully obtained calorimetric results are used to find an enthalpy of reaction, the calculations should be consistent with results obtained using standard enthalpies of formation.

Question

What is the molar enthalpy of combustion of methanol?

Prediction

(a) Use the given values for standard enthalpy of formation to calculate the molar enthalpy of combustion of methanol. (Assume that the products of the reaction are gaseous carbon dioxide and liquid water only.)

Experimental Design

A known mass of methanol is burned in a calibrated bomb calorimeter. The enthalpy of combustion is also calculated using standard enthalpies of formation. The two values are compared to test the standard enthalpies.

Materials

methanol
bomb calorimeter
thermometer

Evidence

Table 2 Observations for Burning Methanol

Quantity	Measurement
mass of methanol reacted	4.38 g
heat capacity of bomb calorimeter	10.9 kJ/C°
initial temperature of calorimeter	20.4°C
final temperature of calorimeter	27.9°C

Analysis

(b) Use the experimental values to calculate, to the appropriate number of significant figures, the molar enthalpy of combustion of methanol.

Evaluation

(c) Calculate the percentage difference between the experimental and predicted values. Does this experiment support the standard enthalpies?

Key Expectations

- Determine enthalpy of reaction using a calorimeter, and use the evidence obtained to calculate the enthalpy change for a reaction. (5.1, 5.2, 5.3, 5.4)

- Compare the energy changes resulting from physical change, chemical reactions, and nuclear reactions (fission and fusion). (5.1, 5.6)

- Describe examples of technologies that depend on exothermic or endothermic changes. (5.1, 5.6)

- Analyze simple potential energy diagrams of chemical reactions. (5.3)

- Write thermochemical equations, expressing the energy change as a ΔH value or as a heat term in the equation. (5.3)

- Explain Hess's law, using examples. (5.4)

- Apply Hess's law to solve problems, including problems that involve evidence obtained through experimentation. (5.4, 5.5)

- Calculate enthalpies of reaction, using tabulated standard enthalpies of formation. (5.5)

- Compare conventional and alternative sources of energy with respect to efficiency and environmental impact. (5.6)

- Use appropriate scientific vocabulary to communicate ideas related to the energetics of chemical reactions. (all sections)

Key Terms

calorimetry	nuclear change
chemical change	open system
chemical system	physical change
endothermic	potential energy diagram
enthalpy	
enthalpy change	specific heat capacity
exothermic	standard enthalpy of formation
heat	
Hess's law	standard molar enthalpy of reaction
isolated system	
law of additivity of reaction enthalpies	surroundings
	temperature
molar enthalpy	thermal energy
molar enthalpy of reaction	thermochemistry

Key Symbols and Equations

- $q = mc\Delta T$

- $\Delta H_{system} = \pm |q_{surroundings}|$

- $\Delta H = n\Delta H_r$

- $\Delta H_{target} = \sum \Delta H_{known}$ (Hess's law)

- ΔH_f°

- $\Delta H^\circ = \sum n\Delta H_{f(products)}^\circ - \sum n\Delta H_{f(reactants)}^\circ$

Problems You Can Solve

- How much heat is transferred to a known mass of matter, for a given temperature change? (5.1)

- What is the enthalpy change for a change in state, given the mass and molar enthalpy? (5.2)

- What is the molar enthalpy of a change taking place in a calorimeter, given the mass of the system and the solution, and the temperature change? (5.2)

- What is the thermochemical equation (including the energy term) for a given reaction? (5.3)

- What is the thermochemical equation (including the enthalpy change, ΔH) for a given reaction? (5.3)

- What is the equation for a molar enthalpy of reaction? (5.3)

- What is the enthalpy change for a target reaction, given enthalpy changes for other known reactions and using Hess's law? (5.4)

- What is the enthalpy change for the reaction of a given quantity of a substance, given *either* the energy involved for the reaction of a different amount of that substance *or* the energy involved in a series of related known reactions? (5.4)

- What is the formation equation for a substance? (5.5)

- What is the enthalpy change of a target reaction, given the enthalpies of formation of various known reactions? (5.5)

- What mass of a substance is involved in a reaction, *or* what temperature change results from a reaction, given the enthalpies of formation of various known reactions? (5.5)

▶ **MAKE** a summary

Make a concept map to summarize what you have learned in this chapter. Start with the phrase "Energy Changes" in the centre, and try to include as many as possible of the Key Terms and Key Symbols and Equations listed above.

Identify each of the following statements as true, false, or incomplete. If the statement is false or incomplete, rewrite it as a true statement.

1. Nuclear changes generally absorb more energy than chemical changes.

2. In exothermic reactions, the reactants have more kinetic energy than the products.

3. On a potential energy diagram, the horizontal axis may be called reaction progress.

4. In endothermic reactions, the heat term is written on the right side of the equation.

5. In an isolated system, neither matter nor energy may enter or leave the system.

6. In calorimetry, the assumption is made that the enthalpy change of the system equals the heat transferred to the surroundings.

7. The burning of gasoline is an example of an endothermic physical change.

8. A formation reaction has elements in their standard states as reactants.

9. Specific heat capacity is the amount of heat required to change a given mass through 1°C.

10. The ΔH value for an exothermic reaction is negative.

Identify the letter that corresponds to the best answer to each of the five following questions.

11. Which of the following would involve the largest production of heat per mole?
    (a) the burning of gasoline
    (b) the evaporation of water
    (c) the fission of uranium
    (d) the freezing of water
    (e) the rusting of iron

12. Which of the following would involve the greatest absorption of heat per mole?
    (a) the burning of gasoline
    (b) the evaporation of water
    (c) the fission of uranium
    (d) the freezing of water
    (e) the rusting of iron

13. The evaporation of methanol (molar mass 32.0 g/mol) involves absorption of 1.18 kJ/g. What is the molar enthalpy of vaporization of methanol?
    (a) 0.0369 kJ/mol   (d) 32.0 kJ/mol
    (b) 1.18 kJ/mol   (e) 37.8 kJ/mol
    (c) 3.78 kJ/mol

14. When solid ammonium chloride is added to water, the solution feels cold to the hand. Which statement best describes the observation?
    (a) $NH_4Cl_{(s)} \rightarrow NH_4Cl_{(aq)} + 34$ kJ
    (b) The reaction is exothermic.
    (c) $NH_4Cl_{(s)} \rightarrow NH_4Cl_{(aq)}$ $\Delta H = 34$ kJ
    (d) The system releases heat, so it feels colder.
    (e) The boiling point is increased, so it feels colder.

15. Which of the following statements is *not* true?
    (a) ΔH is the difference in enthalpy between reactants and products.
    (b) ΔH may be written as part of the equation.
    (c) ΔH is negative for an endothermic reaction.
    (d) A reaction consumes heat if ΔH is positive.
    (e) ΔH equals the heat transferred to or from the surroundings.

16. Referring to the reaction below, how much heat is released if 120 g of potassium metal reacts?
    $$2\,K_{(s)} + 2\,H_2O_{(l)} \rightarrow 160\text{ kJ} + H_{2(g)} + 2\,KOH_{(aq)}$$
    (a) 0.0 kJ since no temperature change is indicated
    (b) 52.1 kJ   (d) 491 kJ
    (c) 246 kJ   (e) 280 KJ

17. The standard heat of formation of solid ammonium nitrate, NH_4NO_3, is -330 kJ/mol . The equation that represents this process is:
    (a) $N_{2(g)} + 4\,H_{2(g)} + 3\,O_{2(g)} \rightarrow NH_4NO_{3(s)} + 330$ kJ
    (b) $2\,N_{(g)} + 4\,H_{(g)} + 3\,O_{(g)} \rightarrow NH_4NO_{3(s)} + 330$ kJ
    (c) $NH_{4(g)}^+ + NO_{3(g)}^- \rightarrow NH_4NO_{3(s)} + 330$ kJ
    (d) $\frac{1}{2}N_{2(g)} + 2\,H_{2(g)} + \frac{3}{2}O_{2(g)} \rightarrow NH_4NO_{3(s)} + 330$ kJ
    (e) $N_{2(g)} + 2\,H_{2(g)} + \frac{3}{2}O_{2(g)} \rightarrow NH_4NO_{3(s)} + 330$ KJ

18. Consider the following two thermochemical equations:
    $$N_{2(g)} + \frac{5}{2}O_{2(g)} \rightarrow N_2O_{5(s)} \qquad \Delta H = x \text{ kJ}$$
    $$N_{2(g)} + \frac{5}{2}O_{2(g)} \rightarrow N_2O_{5(g)} \qquad \Delta H = y \text{ kJ}$$
    The enthalpy change, in kJ, for the sublimation of one mole of N_2O_5 solid to gas would be represented by the quantity
    (a) $x + y$ (d) $-x - y$
    (b) $x - y$ (e) xy
    (c) $y - x$

Understanding Concepts

1. Make a chart to summarize the energy changes involved in physical, chemical, and nuclear changes. For each type of change
   (a) describe the change in matter;
   (b) give a range of energies in kJ/mol; and
   (c) give two examples. (5.1)

2. Bricks in a fireplace will absorb heat and release it long after the fire has gone out. A student conducted an experiment to determine the specific heat capacity of a brick. Based on the evidence obtained in this experiment, 16 kJ of energy was transferred to a 938-g brick as the temperature of the brick changed from 19.5°C to 35.0°C. Calculate the specific heat capacity of the brick. (5.1)

3. What basic assumptions are made in calorimetry experiments that involve reacting solutions? (5.2)

4. A 2.0-kg copper kettle (specific heat capacity 0.385 J/g•°C) contains 0.500 kg of water at 20.0°C. How much heat is needed to raise the temperature of the kettle and its contents to 80.0°C? (5.2)

5. Propane gas is used to heat a tank of water. If the tank contains 200.0 L of water, what mass of propane will be required to raise its temperature from 20.0°C to 65.0°C? (ΔH_{comb} for propane = –2220 kJ/mol) (5.2)

6. Draw a labelled potential energy diagram to represent the exothermic combustion of nonane (C_9H_{20}) in oxygen. (5.3)

7. (a) Describe four ways in which enthalpy changes for a reaction may be represented.
   (b) Use the four methods to represent the dissolving of ammonium chloride (ΔH_{sol} = +14 kJ/mol). (5.3)

8. If one mole of water absorbs 44 kJ of heat as it changes state from liquid to gas, write thermochemical equations, with appropriate energy terms, to represent
   (a) the evaporation of one mole of water;
   (b) the condensation of one mole of water vapour; and
   (c) the evaporation of two moles of water. (5.3)

9. (a) Write the equation for the formation of liquid acetone (C_3H_6O).
   (b) Write thermochemical equations for the combustion of one mole each of hydrogen, carbon, and acetone, given that their molar enthalpies of combustion are –285.8 kJ/mol, –393.5 kJ/mol, and –1784 kJ/mol, respectively.

   (c) Use Hess's law and the three combustion reactions to calculate the enthalpy of formation of acetone. (5.4)

10. When glucose is allowed to ferment, ethanol and carbon dioxide are produced. Given that the enthalpies of combustion of glucose and ethanol are –2813 kJ/mol and –1369 kJ/mol respectively, use Hess's law to calculate the enthalpy change when 0.500 kg of glucose is allowed to ferment. (5.4)

11. The molar enthalpy of combustion of natural gas is –802 kJ/mol. Assuming 100% efficiency and assuming that natural gas consists only of methane, what is the minimum mass of natural gas that must be burned in a laboratory burner at SATP to heat 3.77 L of water from 16.8°C to 98.6°C? (5.4)

12. Pyruvic acid is a molecule involved as an intermediate in metabolic reactions such as cellular respiration. Pyruvic acid ($CH_3COCOOH$) is converted into acetic acid and carbon monoxide in the reaction

    $$CH_3COCOOH \rightarrow CH_3COOH + CO$$

    If the molar enthalpies of combustion of these substances are, respectively, –1275 kJ/mol, –875.3 kJ/mol, and –282.7 kJ/mol, use Hess's law to calculate the enthalpy change for the given reaction. (5.4)

13. Ammonia forms the basis of a large fertilizer industry. Laboratory research has shown that nitrogen from the air reacts with water, using sunlight and a catalyst to produce ammonia and oxygen. This research, if technologically feasible on a large scale, may lower the cost of producing ammonia fertilizer.
    (a) Determine the enthalpy change of the reaction, using a chemical equation balanced with whole-number coefficients.
    (b) Calculate the quantity of solar energy needed to produce 1.00 kg of ammonia.
    (c) If 3.60 MJ of solar energy is available per square metre each day, what area of solar collectors would provide the energy to produce 1.00 kg of ammonia in one day?
    (d) What assumption is implied in the previous calculation? (5.4)

14. When sodium bicarbonate ($NaHCO_3$) is heated, it decomposes into sodium carbonate (Na_2CO_3), water vapour, and carbon dioxide. If the standard enthalpies of formation of sodium bicarbonate and sodium carbonate are –947.7 kJ/mol and –1131 kJ/mol respectively, calculate the enthalpy change for the reaction. (5.5)

15. When benzoic acid is burned, the enthalpy of combustion is -3223.6 kJ/mol. Use this information and tabulated values of the standard enthalpies of formation of liquid water and carbon dioxide to calculate the standard enthalpy of formation of benzoic acid. (5.5)

Applying Inquiry Skills

16. A series of calorimetric experiments is performed to test Hess's law. Enthalpy changes for two known reactions are determined experimentally:

(1) $CaCO_{3(s)} + 2\,HCl_{(aq)} \rightarrow CO_{2(g)} + H_2O_{(l)} + CaCl_{2(aq)}$

$$\Delta H_1 = ?\ kJ$$

(2) $CaO_{(s)} + 2\,HCl_{(aq)} \rightarrow H_2O_{(l)} + CaCl_{2(aq)}$

$$\Delta H_2 = ?\ kJ$$

The target equation is

$$CaCO_{3(s)} \rightarrow CO_{2(g)} + CaO_{(s)} \quad \Delta H = ?\ kJ$$

Experimental Design

Two calorimetry experiments are performed, with the choice of chemical systems such that the enthalpy changes of these two known reactions are equal to the enthalpy change of the unknown target equation.

(a) Use Hess's law to show how the two known equations may be added together to yield the target equation.

Procedure

(b) Describe the series of procedural steps that you would follow to produce the following observations.

Evidence

Table 1 Observations for Calorimetry Investigation

Observation	Experiment 1	Experiment 2
mass of reactant 1	4.2 g $CaCO_{3(s)}$	4.7 g CaO
mass of cup	3.0 g	3.1 g
mass of cup and acid	173.2 g	158.6 g
initial temperature of acid	29.0°C	29.0°C
final temperature of mixture	31.0°C	36.0°C

Analysis

(c) Use the evidence to calculate the enthalpy change in each system. (Assume that the specific heat capacity of the acid solution is 4.18 J/g•°C).

(d) Calculate the enthalpy change for the target equation.

(e) Explain the effect that you would expect on the calculated ΔH for reaction (1) if:
   (i) some of the calcium carbonate remained on the weighing paper as it was added to the acid;
   (ii) some heat was lost to the air. (5.4)

17. (a) A pure liquid is suspected to be ethanol. Using the energy concepts from this chapter, list as many experimental designs as possible to confirm or refute the suspected identity of the liquid.

(b) Describe some other experimental designs that could be used to determine if the unknown liquid is ethanol. (5.5)

Making Connections

18. (a) Calculate the enthalpy change for
   • the condensing of 1.00 mol of steam to water at 100°C.
   • the formation of 1.00 mol of water from its elements.
   • the formation of 1.00 mol of helium-4 in a hydrogen fusion reaction.

(b) If the preceding enthalpy changes were represented on a graph with a scale of 1 cm per 100 kJ, calculate the distance for each enthalpy change in parts (a), (b), and (c).

(c) Develop an analogy for the relative amounts of energy released in part (a). (5.5)

19. Canadians use more energy per capita than almost any other country in the world.

(a) List some factors that contribute to this level of consumption.

(b) Compare Canada's energy consumption with that of another country. At the same time, compare the factors listed in (a).

(c) Choose one source of energy used in Canada, and research the efficiency and environmental impact of our use of this type of energy.

(d) Write a short opinion piece on whether it is morally appropriate for Canada to have the present level of energy consumption. (5.6)

GO www.science.nelson.com

chapter

6

Chemical Kinetics

▶ In this chapter, you will be able to

- represent and analyze rates of reaction graphically and mathematically;

- use collision theory and both potential and kinetic energy diagrams to explain factors that affect rate of reaction;

- determine rate and order of reaction by calculation and experimentally, and relate these variables to reaction mechanism;

- recognize and explain practical examples of slow reactions, fast reactions, and catalyzed reactions;

- describe complex reactions as a series of steps, with energy changes associated with each step.

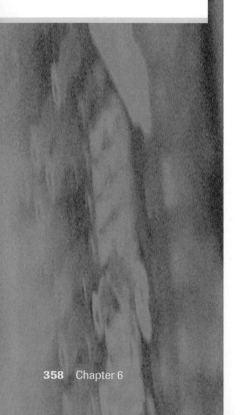

A fireworks display is a spectacular and rapid series of chemical changes—the slow burning of various timed fuses followed by rapid explosive reactions. The technicians who design fireworks use several techniques to control the rates of these reactions. They choose specific chemical compounds and manipulate their concentrations; they mix and pack the chemicals in certain ways; they consider reaction temperatures; and they sometimes use catalysts. Much of their knowledge of the rates of these reactions has been gathered by trial and error: They try different combinations of substances and conditions to see what works best, and learn from experience.

Chemists take a more systematic approach to understanding and predicting the speeds at which chemical reactions occur. We can use theories of molecular behaviour effectively to explain experimental observations in the study of rates of reaction, or chemical kinetics. Rates of reaction also help us deduce how reactions happen "step by step" at the molecular level.

Chemical kinetics is an area of chemistry that crosses over into many other areas of science and engineering. Biologists are interested in the rates of metabolic reactions, as well as in the progress of reactions involved in growth and bone regeneration. Automotive engineers want to increase the rate at which noxious pollutants are oxidized in car exhaust systems and to decrease the rate of rusting of car bodies. Agriculture specialists are concerned with the progress of reactions involved in spoilage and decay and with the ripening rate of many foodstuffs. The study and understanding of rates of chemical reactions has implications for many areas of our lives.

💡 REFLECT on your learning ▼

1. (a) List as many ways as you can think of to make a chemical reaction go faster.
   (b) Give a practical example of each of the ways described.

2. Consider the chemical reaction that happens in an automobile engine when hydrocarbon molecules (mostly octane) combine with oxygen molecules to produce carbon dioxide and water vapour. Describe the bonding that exists within the reactant and product molecules, using structural diagrams as appropriate. Now imagine that you have an extraordinary microscope that enables you to look at individual molecules. Describe what you see as the chemical reaction occurs. Describe the positions and movements of the molecules and the changes in their bonds as specifically as possible.

3. Suggest why some chemical reactions occur slowly while others occur quickly.

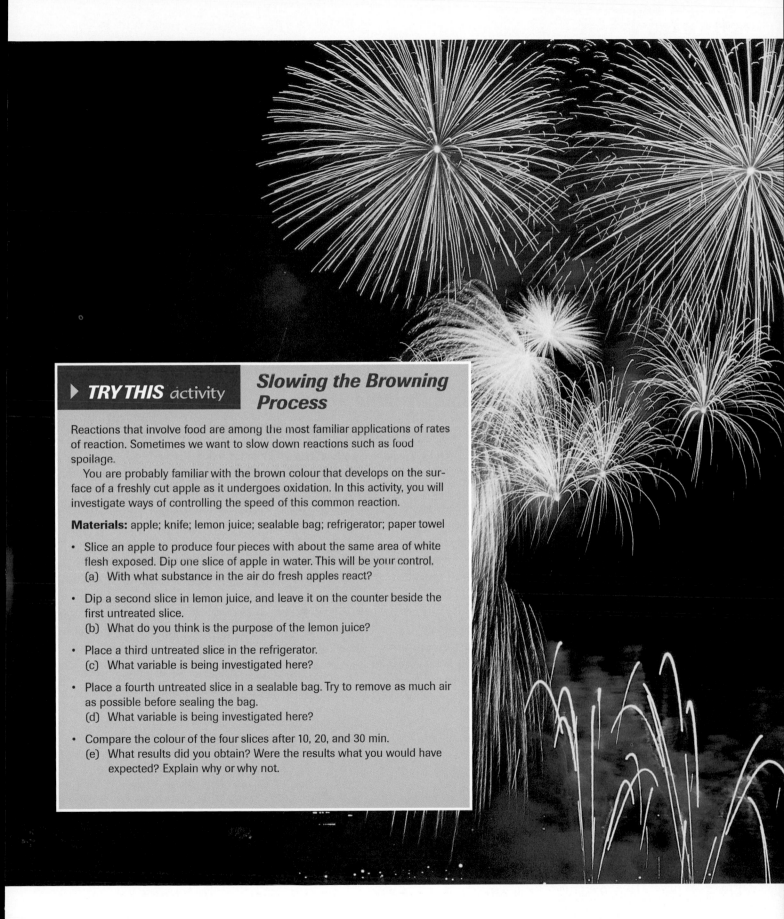

> ► **TRY THIS** activity

Slowing the Browning Process

Reactions that involve food are among the most familiar applications of rates of reaction. Sometimes we want to slow down reactions such as food spoilage.

You are probably familiar with the brown colour that develops on the surface of a freshly cut apple as it undergoes oxidation. In this activity, you will investigate ways of controlling the speed of this common reaction.

Materials: apple; knife; lemon juice; sealable bag; refrigerator; paper towel

- Slice an apple to produce four pieces with about the same area of white flesh exposed. Dip one slice of apple in water. This will be your control.
  (a) With what substance in the air do fresh apples react?

- Dip a second slice in lemon juice, and leave it on the counter beside the first untreated slice.
  (b) What do you think is the purpose of the lemon juice?

- Place a third untreated slice in the refrigerator.
  (c) What variable is being investigated here?

- Place a fourth untreated slice in a sealable bag. Try to remove as much air as possible before sealing the bag.
  (d) What variable is being investigated here?

- Compare the colour of the four slices after 10, 20, and 30 min.
  (e) What results did you obtain? Were the results what you would have expected? Explain why or why not.

chemical kinetics the area of chemistry that deals with rates of reactions

Figure 1
Industrial chemists and chemical engineers design processes to speed up the reaction that forms the sulfuric acid in this reaction vessel, but slow the reaction that corrodes the vessel itself.

Figure 2
For centuries, people have observed that the mineral pyrolusite (manganese dioxide) speeds up the decomposition of hydrogen peroxide. However, the details of why pyrolusite increases the rate are still not well understood.

rate of reaction the speed at which a chemical change occurs, generally expressed as change in concentration per unit time

From a practical perspective, **chemical kinetics** is the study of ways to make chemical reactions go faster or slower. This control of the speed of reactions is crucial in the chemical industry, where starting materials are expensive and any improvement in reaction yield can have major economic effects (**Figure 1**). Most pharmaceutical drugs could not be manufactured without application of knowledge about chemical kinetics.

From a scientific perspective, chemical kinetics starts with laboratory research of reaction times and changes in concentrations of reactants and products. Chemical kinetics also includes the development of theories and models to explain and predict observed rates of reaction. In the past few years, technological advances in instrumentation and computers have greatly aided researchers in the field of reaction rates, especially for very high-speed reactions such as explosions.

Chemical kinetics is an area of chemistry that still presents good research opportunities for future chemists. Many common reactions are understood in terms of *what* happens when reactants are put together, but not *how* or *why* they behave the way they do (**Figure 2**).

Describing Reaction Rates

How can we find out how fast a reaction is progressing? A **rate of reaction** (or reaction rate) is usually obtained by measuring the rate at which a product is formed or the rate at which a reactant is consumed over a series of time intervals. Properties such as mass, colour, conductivity, volume, and pressure can be measured, depending on the particular reaction.

We express rates of reaction mathematically in terms of a change in property of a reactant or product per unit of time. This process is similar to our expression of the average speed of a vehicle on a trip. If a car travels from Montreal to Toronto (roughly 700 km) in 7 h, the average speed for the trip is about 100 km/h. In the same way, if 700 mmol of a product is produced in 7 min, we can express the average rate of reaction as 100 mmol/min.

A variety of units are used to express rate, such as mol/min or mL/s. To allow easy comparison of many reaction types, we often express reaction rates as "change in concentration per unit of time" — as mol/(L·s), for example. Symbolically, where r is the average reaction rate, Δc is the change in concentration, and Δt is the elapsed time, the expression

$$\text{average reaction rate} = \frac{\text{change in concentration}}{\text{elapsed time}}$$

becomes

$$r = \frac{\Delta c}{\Delta t}$$

▶ SAMPLE problem | Finding a Rate of Reaction

What is the overall rate of production of nitrogen dioxide in the system

$$2\,N_{2(g)} + O_{2(g)} \rightarrow 2\,NO_{2(g)}$$

if the concentration of nitrogen dioxide changes from 0.32mol/L to 0.80 mol/L in 3 min?

First, calculate the change in concentration of nitrogen dioxide:

$$\Delta c = \Delta[NO_2]$$
$$= (0.80 - 0.32) \text{ mol/L}$$
$$\Delta c = 0.48 \text{ mol/L}$$

Then, apply the formula for rate, being careful to note the units:

$$r_{NO_2} = \frac{\Delta c}{\Delta t}$$
$$= \frac{0.48 \text{ mol/L}}{3 \text{ min}}$$
$$r_{NO_2} = 0.16 \text{ mol/(L·min)}$$

The overall rate of production of $NO_{2(g)}$ for the system given is 0.16 mol/(L·min).

Example

What is the average rate of production of ammonia for the system, between 1.0 and 4.0 min,

$$N_{2(g)} + 3\,H_{2(g)} \rightarrow 2\,NH_{3(g)}$$

if the concentration of ammonia is 3.5 mol/L after 1.0 min and 6.2 mol/L after 4.0 min?

Solution

$$\Delta c = \Delta[NH_3]$$
$$= (6.2 - 3.5) \text{ mol/L}$$
$$\Delta c = 2.7 \text{ mol/L}$$

$$\Delta t = (4.0 - 1.0) \text{ min}$$
$$\Delta t = 3.0 \text{ min}$$

$$r_{NH_3} = \frac{\Delta c}{\Delta t}$$
$$= \frac{2.7 \text{ mol/L}}{3.0 \text{ min}}$$
$$r_{NH_3} = 0.90 \text{ mol/(L·min)}$$

The rate of production of $NH_{3(g)}$ over the given time interval is 0.90 mol/(L·min).

▶ Practice

Understanding Concepts

1. In the reaction between solid phosphorus and oxygen gas to form $P_4O_{10(s)}$, what is the overall rate of reaction if the concentration of oxygen gas changes from 0.200 mol/L to 0.000 mol/L over a 40 s period?

2. One way to remove rust stains from ceramic bathroom fixtures is to apply a solution of oxalate ion, as shown in the following equation.

$$C_2O_{4(aq)}^{2-} + 2\,Fe_{(aq)}^{3+} \rightarrow 2\,CO_{2(g)} + 2\,Fe_{(aq)}^{2+}$$

What is the average rate of reaction for this system over the specified period if the concentration of oxalate ion changes from 0.80 mol/L at 3.0 min to 0.20 mol/L at 8.0 min?

Answers

1. 0.005 mol/(L·s)

2. 0.12 mol/(L·min)

LEARNING *TIP*

In gas reactions, the partial pressure of the reactants is a convenient measure of their concentration.

average rate of reaction the speed at which a reaction proceeds over a period of time (often measured as change in concentration of a reactant or product over time)

Figure 3
During the progress of a reaction, the concentration of the reactant decreases continuously. However, the rate of the reaction is not constant. The average rate of reaction over a time interval is the absolute value of the slope of the secant (a line drawn between two points on a curve) for that time interval.

instantaneous rate of reaction the speed at which a reaction is proceeding at a particular point in time

LEARNING *TIP*

The initial rate is a particular example of an instantaneous rate of reaction. It is the absolute value of the slope of the tangent at the instant the reactants are mixed (i.e., at $t = 0$).

Figure 4
The instantaneous rate of reaction of a particular reactant during the progress of a reaction can be obtained at times A and B from the slopes of the tangents.

Typically for a reaction such as

$$CH_{4(g)} + Cl_{2(g)} \rightarrow CH_3Cl_{(g)} + HCl_{(g)}$$

the concentration of the reactant $Cl_{2(g)}$ would fall during the progress of the reaction. A graph of its concentration plotted against time would appear as a curve with a negative slope, as shown in **Figure 3**. On this graph, the **average rate of reaction** (or rate of consumption of Cl_2) over a time period is the absolute value of the slope of the secant (a line between two points on a curve) drawn between the two points.

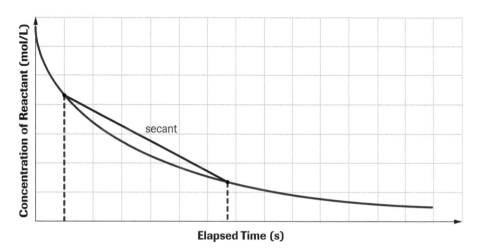

Finding Average Rate of Reaction

It is also possible to determine the reaction rate at any particular point in time: The **instantaneous rate of reaction** (or rate of consumption of Cl_2) is the slope of the curve at that point. **Figure 4** shows lines drawn, at a tangent to the curve, at two different times during the progress of the reaction. Later in the reaction the slope is less steep, indicating that the rate is slower.

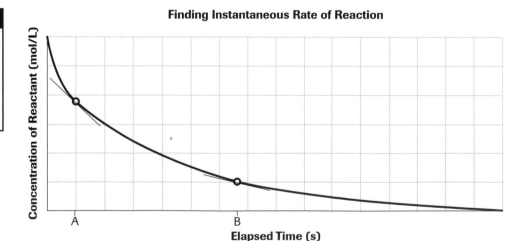

Finding Instantaneous Rate of Reaction

Evidence shows that, for most reactions, the concentration changes are more rapid near the beginning of the reaction and the rate decreases with the time elapsed (**Figure 5**).

If we plot data for concentration of a product (such as HCl or CH$_3$Cl) against time, the result is a rising curve with a steadily decreasing positive slope (Figure 5). Plotting the concentration of the reactant gives a falling curve with a steadily decreasing negative slope. When we measure reaction rates from a graph, we often describe them using a notation that represents the negative slope of the reactant graph and the positive slope of the product graph, and allows the numerical value of the rate to be an absolute value. For example, the rate of reaction for the methane–chlorine system above could be represented by any of the following expressions:

$$-\frac{\Delta[CH_4]}{\Delta t} = x \text{ mol/(L·s)} \quad \text{or} \quad -\frac{\Delta[Cl_2]}{\Delta t} = x \text{ mol/(L·s)}$$

$$\text{or} \quad +\frac{\Delta[CH_3Cl]}{\Delta t} = x \text{ mol/(L·s)} \quad \text{or} \quad +\frac{\Delta[HCl]}{\Delta t} = x \text{ mol/(L·s)}$$

Note that a negative ($-$) sign indicates a rate of consumption of reactant and positive ($+$) sign indicates a rate of production of product. The numerical value x can therefore be an absolute value.

We should specify the substance whose concentration is changing when we describe rates of reaction, because the concentration of the reactant may not change in a 1:1 ratio with the concentration of the product. The following Sample Problem illustrates this.

Concentration of a Product Changing over Time

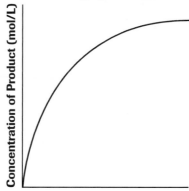

Reaction Progress

Figure 5
The concentration of a product, such as HCl, during the progress of a reaction increases continuously. The slope, $\Delta[HCl]/\Delta t$, is greatest in the early stages of the reaction.

Calculating Rates of Reaction   *SAMPLE* problem ◀

Consider the reaction of iodate, iodide, and hydrogen ions to yield iodine and water.

$$1\ IO_{3(aq)}^- + 5\ I_{(aq)}^- + 6\ H_{(aq)}^+ \rightarrow 3\ I_{2(aq)} + 3\ H_2O_{(l)}$$

What are the rates of reaction with respect to the various reactants and products? The rate of reaction with respect to iodate ions (rate of consumption of IO_3^-) is determined experimentally to be 3.0×10^{-5} mol/(L·s).

As you can see in the equation, for every 1 mol of iodate ions (IO_3^-) consumed, 5 mol of iodide ions (I^-) and 6 mol of hydrogen ions are used up. As a result, the rates of consumption of the three reactant ions will be quite different.

We can express the rate of reaction of iodate ions as

$$-\frac{\Delta[IO_3^-]}{\Delta t} = 3.0 \times 10^{-5} \text{ mol/(L·s)}$$

We can therefore express the rate of the same reaction for the other reactants, taking into account that they are consumed or produced in different amounts, using the mole ratio as a conversion factor:

$$-\frac{\Delta[I^-]}{\Delta t} = \frac{5 \text{ mol } I^-}{1 \text{ mol } IO_3^-} \times 3.0 \times 10^{-5} \text{ mol/(L·s)}$$

$$= 1.5 \times 10^{-4} \text{ mol/(L·s)}$$

You can check that you have the mole ratio correct, rather than inverted, by considering how much of the reactant is being consumed. In this case, because iodine ions are being consumed more rapidly than iodate ions, you would expect the change in concentration to be greater for the iodide ions than for the iodate ions.
Similarly,

$$-\frac{\Delta[H^+]}{\Delta t} = \frac{6 \text{ mol } H^+}{1 \text{ mol } IO_3^-} \times 3.0 \times 10^{-5} \text{ mol/(L·s)}$$

$$= 1.8 \times 10^{-4} \text{ mol/(L·s)}$$

▶

We can also express the rate of the reaction in terms of the products:

$$+\frac{\Delta[I_2]}{\Delta t} = \frac{3 \text{ mol } I_2}{1 \text{ mol } IO_3^-} \times 3.0 \times 10^{-5} \text{ mol/(L·s)}$$

$$= 9.0 \times 10^{-5} \text{ mol/(L·s)}$$

and, since equal numbers of moles of each product are formed,

$$+\frac{\Delta[H_2O]}{\Delta t} = 9.0 \times 10^{-5} \text{ mol/(L·s)}$$

Note that all of these expressions describe the same system reacting at the same time. Chemists need to measure or calculate the rate of reaction for only one convenient reactant or product because we can use the balanced equation to deduce the other rates. However, we should state which substance we are considering.

Example

When nitrogen and hydrogen gases react to produce ammonia gas, the reaction is

$$N_{2(g)} + 3\,H_{2(g)} \rightarrow 2\,NH_{3(g)}$$

What is the rate of consumption of each of the reactants when the rate of production of ammonia is 4.0×10^{-3} mol/(L·s)?

Solution

$$-\frac{\Delta[N_2]}{\Delta t} = \frac{1 \text{ mol } N_2}{2 \text{ mol } NH_3} \times 4.0 \times 10^{-3} \text{ mol/(L·s)}$$

$$= 2.0 \times 10^{-3} \text{ mol/(L·s)}$$

$$-\frac{\Delta[H_2]}{\Delta t} = \frac{3 \text{ mol } N_2}{2 \text{ mol } NH_3} \times 4.0 \times 10^{-3} \text{ mol/(L·s)}$$

$$= 6.0 \times 10^{-3} \text{ mol/(L·s)}$$

▶ **Practice**

Understanding Concepts

Answers

3. 0.20 mol $Fe^{2+}_{(aq)}$/(L·min)

   0.32 mol $H^+_{(aq)}$/(L·min)

   0.040 mol $Mn^{2+}_{(aq)}$/(L·min)

   0.20 mol $Fe^{3+}_{(aq)}$/(L·min)

   0.16 mol $H_2O_{(l)}$/(L·min)

4. (a) 25 mmol $O_{2(g)}$/(L·s)

   (b) 30 mmol $H_2O_{(g)}$/(L·s)

6. (a) 0.16 g N_aHCO_3/s

   (b) 95 mg H_2SO_4/s

   (c) 0.97 mmol H_2SO_4/s

   (d) 0.19 mmol CO_2/s

3. The reaction among permanganate, iron(II), and hydrogen ions occurs in aqueous solution as follows:

   $$MnO_{4(aq)}^- + 5\,Fe^{2+}_{(aq)} + 8\,H^+_{(aq)} \rightarrow Mn^{2+}_{(aq)} + 5\,Fe^{3+}_{(aq)} + 4\,H_2O_{(l)}$$

   Given that the rate of this reaction is 4.0×10^{-2} mol/L $MnO_{4(aq)}^-$ consumed per minute, calculate and express the rate of reaction with respect to each of the other reactants or products in the equation.

4. Under certain conditions, ammonia and oxygen can react as shown by the equation

   $$4\,NH_{3(g)} + 5\,O_{2(g)} \rightarrow 4\,NO_{(g)} + 6\,H_2O_{(g)}$$

   When the instantaneous rate of consumption of ammonia is 2.0×10^{-2} mol/(L·s), what will be the instantaneous rate
   (a) of consumption of oxygen?
   (b) of formation of water vapour?

5. Nitric oxide and oxygen gases react to form nitrogen dioxide gas in the reaction

   $$2\,NO + O_2 \rightarrow 2\,NO_2$$

   (a) Write a mathematical expression to represent the rate of production of nitrogen dioxide: 4.0×10^{-3} mol/(L·s).
   (b) Write expressions to represent the rates of consumption of nitric oxide and oxygen gases, and calculate the numerical value of each rate.

6. A 3.25-g sample of sodium hydrogen carbonate reacts completely with sulfuric acid in a time of 20 s. Express the average rate of reaction as
   (a) mass of $NaHCO_3$ consumed per second
   (b) mass of H_2SO_4 consumed per second
   (c) amount of H_2SO_4 consumed per second
   (d) amount of CO_2 produced per second

Measuring Reaction Rates

Chemists have many methods to choose from when measuring reaction rates. The method they choose depends on the kinds of substances and type of reaction. Ideally, the experimenter takes direct measurements of a reactant or product without disturbing the progress of the reaction itself. This is possible for some reactions, including those that produce a gas, produce more or fewer charged entities, or change colour.

Figure 6
The rate of production of a low-solubility gaseous product can be measured by volume of water displaced.

Reactions That Produce a Gas

In the reaction of zinc and hydrochloric acid in aqueous solution, the investigator can collect the gas and measure its volume and/or pressure as the reaction proceeds (**Figure 6**). The faster the reaction, the greater the change in volume or pressure in the same time interval.

Reactions That Involve Ions

For some reactions that occur in solution, the conductivity of the solution changes as the reaction proceeds. For example, the hydrolysis of alkyl halides in aqueous solution starts with neutral molecules as reactants but produces charged ions. As the reaction proceeds and more ions form, the conductivity of the solution increases (**Figure 7**). This conductivity can be measured and plotted graphically as a function of time.

$$(CH_3)_3CCl_{(aq)} + H_2O_{(l)} \rightarrow (CH_3)_3COH_{(aq)} + H^+_{(aq)} + Cl^-_{(aq)}$$

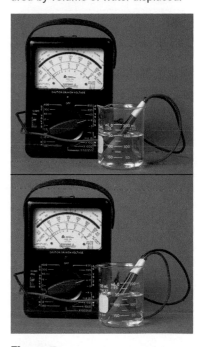

Figure 7
The rate of production of an ionic product can be measured by conductivity.

Reactions That Change Colour

For some reactions, especially those in solution, we can measure the intensity (strength of colour) of a coloured reactant or product. A spectrophotometer is used for precise measurements of colour intensity. This technique can also be used to measure accurately wavelengths of light that are invisible to the human eye (**Figure 8**). The following chemical equation represents a reaction for which we can use colour to measure the concentration of a reactant or product.

$$ClO^-_{(aq)} + I^-_{(aq)} \rightarrow IO^-_{(aq)} + Cl^-_{(aq)}$$
colourless   colourless   yellow   colourless

For this reaction, the yellow colour of the aqueous hypoiodite ions ($IO^-_{(aq)}$) appears initially, then becomes more intense as the reaction progresses and more hypoiodite ions form.

▶ *Practice*

Understanding Concepts

7. State three examples of properties, directly related to reactants or products, that could be used to measure a reaction rate.

8. What property would be appropriate to measure rate in each of the following reactions?

   (a) $MnO_4^-{}_{(aq)} + 5\ Fe^{2+}_{(aq)} + 8\ H^+_{(aq)} \rightarrow Mn^{2+}_{(aq)} + 5\ Fe^{3+}_{(aq)} + 4\ H_2O_{(l)}$

   (b) $Zn_{(s)} + H_2SO_{4(aq)} \rightarrow H_{2(g)} + ZnSO_{4(aq)}$ ▶

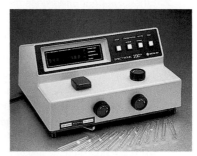

Figure 8
Solutions of different colour intensity absorb proportionately different amounts of light, so the rate of production of a coloured product can be measured by changes in light absorbency.

LAB EXERCISE 6.1.1

Determining a Rate of Reaction (p. 401)

Graph and interpret evidence collected from the reaction of calcium carbonate with hydrochloric acid.

Table 1 Concentration of Product

Time (s)	Concentration (mol/L)
0	0
10.0	0.23
20.0	0.40
30.0	0.52
40.0	0.60
50.0	0.66
60.0	0.70
70.0	0.73
80.0	0.75
90.0	0.76
100.0	0.76

9. Rates of reaction are generally fastest at the beginning of a reaction. Explain why this is so.

10. Sketch the graph of **Figure 5** and add curves to represent the molar concentrations of methane, $CH_{4(g)}$, and chlorine, $Cl_{2(g)}$, as the reaction progresses. Assume that the initial concentration of methane is slightly more than the initial concentration of chlorine.

11. A kinetics experiment is performed in which the concentration of a product is measured every 10 s. **Table 1** shows the evidence obtained.
    (a) Plot a properly labelled graph of concentration vs. time.

    Use the graph to estimate:
    (b) the average rate of formation of product during the first 60.0 s;
    (c) the average rate of formation of product between $t = 20.0$ and $t = 60.0$ s;
    (d) the instantaneous rate of formation of product at $t = 20.0$ s.

▶ *Section 6.1* *Questions*

Understanding Concepts

1. In a combustion experiment, 8.0 mol of methane gas reacts completely in a 2.00 L container containing excess oxygen gas in 3.2 s.
   (a) Express the average rate of consumption of the methane in units of mol/(L·s).
   (b) Express the average rate of consumption of the oxygen gas in units of mol/(L·s).
   (c) Express the average rate of production of carbon dioxide gas in units of mol/(L·s).
   (d) Express the average rate of production of water vapour in units of mol/(L·s).

2. Ethanal vapour undergoes thermal decomposition in the reaction

   $$C_2H_4O_{(g)} \rightarrow CH_{4(g)} + CO_{(g)}$$

   (a) Use the data in **Table 2** to plot a graph of the concentration of ethanal vapour vs. time.
   (b) From the graph, determine the average rate of reaction between the times $t = 0$ s and $t = 420$ s.
   (c) From the graph, determine the average rate of reaction between the times $t = 420$ s and $t = 1250$ s.
   (d) Describe qualitatively what happens to the rate of reaction over time and suggest a reason why this occurs.
   (e) Predict how long it will take for the concentration of ethanal to fall from 0.045 mol/L to 0.090 mol/L.
   (f) Determine the instantaneous rates of decomposition of ethanal when its concentrations are 0.20 mol/L and 0.10 mol/L.

Table 2 Concentration of Ethanal During Thermal Decomposition

$[C_2H_4O_{(g)}]$ (mol/L)	Time (s)
0.360	0
0.290	100
0.250	185
0.200	270
0.180	420
0.150	575
0.130	730
0.110	950
0.090	1250
0.080	1440

Making Connections

3. (a) List as many ways as you can think of to make a chemical reaction go faster.
   (b) Give a practical example of each of the ways described.
   (c) How do your answers differ from those you gave to Reflect on your Learning, question 1, at the beginning of the chapter?

Factors Affecting Reaction Rate 6.2

Many different physical and chemical processes are involved in food preparation, and a number of factors control how long these processes take. For example, a piece of meat takes longer to cook than a vegetable of similar mass. If we want potatoes to cook more quickly, we chop them into smaller pieces. And the easiest way to speed up the cooking process is to provide more heat. Recipes in cookbooks are designed to produce results in specific times. Many of the same factors that control cooking processes also affect chemical reactions in general.

There are five basic factors that control the speed of a chemical reaction.

The Five Factors Affecting Rate

- chemical nature of reactants
- concentration of reactants
- temperature
- presence of a catalyst
- surface area

In this section we will consider each of these factors in turn.

Chemical Nature of Reactants

Gold and silver have been used over the centuries for jewellery and precious objects because they are slow to react in air. By contrast, sodium and potassium are so reactive that they are never found naturally in their elemental state. The type of reaction that the metals might undergo with atmospheric oxygen is similar, but their rates of reaction are quite different (**Figure 1**).

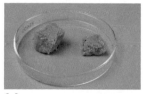

(a) **(b)** **(c)**

Figure 1
Sodium (a), silver (b), and gold (c) react at different rates with gases in the atmosphere – an example of the nature of reactants affecting reaction rate.

As you know, similar elements (such as those in the same group in the periodic table) tend to react similarly, but at different rates. The chemical nature of the reactant affects the reaction rate. For example, the metals zinc, iron, and lead all react with hydrochloric acid to produce hydrogen gas as one of the products. Even when all other conditions (such as temperature, initial concentration, amount, and physical shape) are the same for these reactions, the rates are quite different. It has become traditional to speak of the activity series for common metals based on exactly this comparison: the rate of reaction with common acids, such as sulfuric and hydrochloric acids. Historically, a concern with metals has always been their rate of corrosion, and scientists have searched for methods to reduce the corrosion rate.

In homogeneous chemical systems, such as reactions in aqueous solution, most reactions of monatomic ions (e.g., Ag^{+} and Cl^{-}) are extremely fast. Reactions of molecular substances are often much slower. For example, glucose molecules ($C_6H_{12}O_6$) and iron(II) ions (Fe^{2+}) both react with purple permanganate ions (MnO_4^{-}) in a highly visible way:

DID YOU KNOW?

Elemental, Naturally!
The first metals to be discovered were the least reactive: Gold, silver, and copper can be found in their "native" state as elements. Metals such as lead and iron were the next to be discovered because they could be produced easily by heating some of their compounds. Elements such as sodium and potassium are so reactive that they are found in nature only as compounds such as sodium chloride and potassium nitrate.

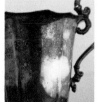

Figure 2
The different reaction rates of glucose molecules and of iron(II) ions with acidified permanganate ions: at initial mixing (left); after 5 s (centre); and after 100 s (right).

When the colour disappears, all of the permanganate ions have reacted. The reaction rate of covalently bonded glucose molecules with permanganate ions is very slow compared with the rate of reaction with monatomic iron(II) ions, as shown in **Figure 2**.

Concentration

Concentrated hydrochloric acid and pure acetic acid are dangerous reagents that can cause serious burns if allowed to contact skin. Nonetheless, human stomach acid is dilute hydrochloric acid, and the vinegar we put on fish and chips is a solution of acetic acid. These dilute acids are still capable of the same reactions, but their reaction rates are so slow that their effects are much reduced.

Experiments suggest that if the initial concentration of a reactant is increased, then the reaction rate generally increases. For example, when a metal is added to solutions of different concentrations of acid, a higher initial concentration of acid noticeably increases the rate of gas production. This is such a common effect that everyday language routinely uses the word "concentrated" to imply faster and more effective, as we see in advertising for cleaning compounds and fabric softeners. For the same reason, warning labels about smoking or open flames are posted near tanks of oxygen because any combustion reaction could be dangerously accelerated by concentrated oxygen gas. Similarly, fireworks contain a mixture of chemicals, one of which is usually a substance that produces a high enough initial concentration of oxygen to promote an explosive reaction.

Temperature

When the ingredients for a cake are mixed together in a bowl at room temperature, they do not appear to change. Sometimes, a recipe even suggests that the ingredients first be placed in the refrigerator to slow any changes that may occur. However, putting the mixture in an oven quickly produces the results that the cook wants: The baking soda reacts with the liquid to produce bubbles of carbon dioxide, making the cake rise. Cooking, whether in the oven, the frying pan, or the microwave, uses increased temperature to make changes in food happen more quickly.

We know from experience that when the temperature of the system increases then the reaction rate generally increases: paint dries faster and glue sets faster (really the completion of a series of reactions) at higher temperatures. Chemists have long known that as a rule of thumb, around SATP, a 10°C rise in temperature often doubles or triples the rate of a chemical reaction.

Lowering the temperature of the system can decrease the rate of reaction, which is useful for food storage. Reptiles (**Figure 3**) are more active in warm sunlight than at night because their metabolic rates are slowed by low temperatures. Humans apply cold substances to burned skin to slow — and thereby minimize the effect of — unwanted physiological reactions.

Figure 3
Snakes, like this Massasauga rattlesnake, have metabolic rates that increase and decrease in response to the amount of thermal energy they absorb from their surroundings.

Presence of a Catalyst

Soda crackers are mostly starch, large carbohydrate molecules made up of smaller sugar molecules bonded together. Starch does not taste sweet, but if you chew crackers for several minutes you can taste the sweet sugar molecules in your mouth. The role that amylase, a molecule found in saliva, plays in the conversion of starch into smaller sugar molecules is an example of a process called catalysis.

Catalysis refers to the effect on reactions of a **catalyst**. The chemical composition and amount of a catalyst are identical at the start and at the end of a catalysis reaction. In green plants, for example, the process of photosynthesis can take place only in the presence of the catalyst chlorophyll (**Figure 4**). Most catalysts accelerate reactions quite significantly, even when present in very tiny amounts compared with the amount of reactants present.

The action of catalysts remains an area of some mystery to chemists, and effective catalysts for reactions have almost all been discovered by trial and error. Chemists learned early that finely divided (powdered) metals catalyze many reactions. Perhaps the most common consumer example of catalysis today is the use of finely divided platinum and palladium in catalytic converters in car exhaust systems (**Figure 5**). These catalysts speed the combustion of the exhaust gases so that a higher proportion of the exhaust products will be relatively harmless, completely oxidized substances. Highly toxic CO gas, for example is oxidized to CO_2 before being emitted to the atmosphere. Although carbon dioxide is less toxic, it remains a serious threat to the environment as a greenhouse gas.

catalyst a substance that alters the rate of a chemical reaction without itself being permanently changed

Figure 4
Green plants contain chlorophyll, which acts as a catalyst during photosynthesis to help convert carbon dioxide and water into glucose and oxygen.

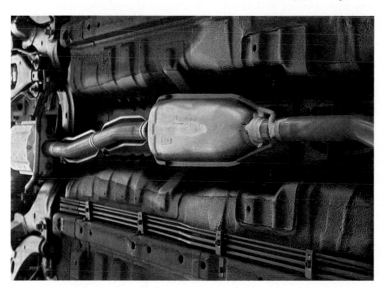

Figure 5
An automobile catalytic converter increases the rate of oxidation of exhaust gases: $NO_{(g)}$ is converted into $N_{2(g)}$ and $O_{2(g)}$; $CO_{(g)}$ is converted into $CO_{2(g)}$; and unburned hydrocarbons are converted into $CO_{2(g)}$ and $H_2O_{(g)}$.

Enzymes in human body processes are complex molecular substances (proteins) that act as catalysts. A great number of human physiological reactions are actually controlled by the amount of enzyme present. The amylase found in saliva is an example of an enzyme. Lactase is a digestive enzyme found in all infants that speeds up the breakdown of lactose or milk sugar. Adults frequently have much lower concentrations of this enzyme. A person with low levels of lactase has difficulty in metabolizing milk and milk products. If lactase is added to the products, they can be consumed by people with "lactose intolerance." Enzymes are also of great importance for catalyzing specific reactions in the food, beverage, cleaner, and pharmaceutical industries. Enzymes are often very sensitive to temperature and pH, so these conditions are critical for enzyme-assisted reactions.

The success of any chemical industry depends on the control of chemical reactions: The operators select conditions that will maximize the yield of the desired product and minimize the production of unwanted substances. For many industrial processes, the

enzyme a molecular substance (protein) in living cells that controls the rate of a specific biochemical reaction

use of catalysts (**Table 1**) means the difference between success and failure by making the reaction rate fast enough to be profitable, but slow enough that it is not dangerous.

Table 1 Examples of Industrial Catalysts

Industrial Product	Catalyst
sulfuric acid	vanadium(V) oxide
ammonia	magnetic iron oxide
acetaldehyde	vanadium(V) oxide
margarine	nickel
nitrogen dioxide	platinum
polyethylene	titanium(IV) oxide
polyester	nickel(II) oxide
methanol	chromium(VI) oxide

Surface Area

Powdered sugar dissolves more quickly in water than do sugar cubes. Similarly, lighting a block of wood with a match is next to impossible, but if the same wood block is whittled into fine shavings, the wood lights easily and burns rapidly. If the same mass of wood is made into very fine sawdust and blown into the air, it might burn so rapidly that it becomes an explosion hazard.

Where a reaction system is heterogeneous (e.g., a solid in a liquid, such as a piece of reactive metal in an acid solution), the reaction occurs at the interface of two different phases present. The amount of exposed surface area, where the two reacting phases are in contact, affects the reaction rate. When a metal reacts in acid, the reaction takes place only where the metal surface is in contact with the acid solution. A reaction with finely divided iron is much faster than one with a solid piece of iron, even if the mass of iron is the same, because the small pieces have a much greater overall surface area. In general, the reaction rate increases proportionally with the increase in surface area.

Both systems and reactions can be referred to as heterogeneous. Heterogeneous reactions are very common, including any reaction of a gas with a solid or a liquid and any reaction of a solid with a solution. 🧪▮

🧪 **INVESTIGATION 6.2.1**

Chemical Kinetics and Factors Affecting Rate (p. 402)
Can you design an investigation to look at the effects of various factors on the rate of a chosen reaction?

SUMMARY *Factors Affecting Reaction Rates*

- The rate of any reaction depends on the chemical nature of the substances reacting.
- An increase in reactant concentration increases the rate of a reaction.
- An increase in temperature increases the rate of a reaction.
- A catalyst increases the rate of a reaction, without itself being consumed.
- In heterogeneous systems, an increase in reactant surface area increases the rate of a reaction.

> **Practice**

Understanding Concepts

1. Identify five different factors that are likely to affect the rate of a reaction. Give a practical example of each.

2. Which of the five factors that affect the rate of reactions applies only to heterogeneous systems? Give an example of such a system.

3. What would happen to the rate of a reaction if the temperature were raised from 20°C to 40°C? Explain qualitatively and make a quantitative prediction.

4. A match can be applied to a lump of coal with little effect. However, the ignition of coal dust has caused many fatal mining explosions. Explain.

5. Signs warn about the dangers of having sparks or open flames near oxygen tanks or near flammable fuels. Which of the five factors that affect reaction rate are involved in each of these warnings?

Making Connections

6. Enzymes in your body are generally present in extremely small quantities, but any substances that affect your enzymes are almost always very toxic and dangerous. Explain why this should be so, referring to reaction rates in your explanation.

7. Many boxed, dry cereals contain BHT (butylated hydroxytoluene, also named 2,6-di-tert-butyl-*p*-cresol) in the packaging material. Use the Internet or other resources to discover and report on the effect of BHT on reaction rates.

 www.science.nelson.com

> **Section 6.2 Questions**

Understanding Concepts

1. In each of the following examples, identify the factor that affects the rate of the reaction described:
   (a) Gold and copper are both used in jewellery, but copper bracelets will turn green over time.
   (b) Milk kept in a refrigerator will keep for a week or more, but milk left out on the counter will quickly turn sour.
   (c) Papain is a food additive that is sometimes added to meat to make it more tender.
   (d) The dust in grain silos has been known to explode, whereas kernels of grain are almost nonflammable.
   (e) Vinegar is safe to add to food and consume, but pure acetic acid will burn skin on contact.

2. You have learned in past studies that strong acids, such as hydrochloric acid, are completely ionized whereas weak acids, like carbonic acid (H_2CO_3), are only partially ionized.

Explain why a solution of 1 mol/L hydrochloric acid is more dangerous than a solution of 1 mol/L carbonic acid.

Making Connections

3. List the characteristics of fuel for starting a campfire, giving reasons why each characteristic is desirable.

4. Heterogeneous catalysts are generally preferred in industrial reactions. Explain why this is so, with reference to several industrial examples not mentioned in this text.

 www.science.nelson.com

5. How do catalytic converters work? Research and write a brief report on the materials and design of these devices.

 www.science.nelson.com

When chemists and engineers make decisions about operating conditions in a chemical industry, they need to know the quantitative effects of various factors on rate of reaction. Increasing the temperature or initial concentration of reactants may make a chemical reaction occur more quickly, but by how much?

There is a mathematical relationship—a pattern—between reaction rate and the various factors that affect it. This pattern cannot be predicted initially from theory, but must be determined empirically — from analysis of experimental evidence. The balanced chemical equation is just a starting point for examining the relationship. If we look at the reaction of nitric oxide and oxygen gas

$$2\,NO_{(g)} + O_{2(g)} \rightarrow 2\,NO_{2(g)}$$

it seems logical that the rate depends on the concentration of the reactants. But, does nitric oxide concentration have a greater effect on rate than oxygen concentration? Only experimental evidence, not theory, can answer the question.

Empirical Determination of Rate Laws

Experimental evidence suggests that the rate of a reaction is exponentially proportional to the product of the initial concentrations of the reactants. This generalization has been verified for many chemical reactions and is now known as the rate law. Imagine a chemical reaction equation in which a and b are the coefficients used in balancing the equation, and X and Y are formulas for reactant molecules:

$$a\,X + b\,Y \rightarrow (\text{products})$$

> **The Rate Law**
>
> The rate, r, will always be proportional to the product of the initial concentrations of the reactants, where these concentrations are raised to some exponential values. This can be expressed as
>
> $r \propto [X]^m[Y]^n$

Note that m and n, the exponents that describe the relationship between rate and initial concentration, can only be determined empirically: They can have any real number value, including fractions or zero, and do not have to equal the coefficients (a and b) in the balanced equation.

This relationship can be rewritten as a

rate law equation the relationship among rate, the rate constant, the initial concentrations of reactants, and the orders of reaction with respect to the reactants; also called rate equation or rate law

> **Rate Law Equation**
> (also called a rate equation or rate law)
>
> $r = k[X]^m[Y]^n$

rate constant the proportionality constant in the rate law equation

where k is a **rate constant**, determined empirically, that is valid only for a specific reaction at a specific temperature.

For example, the empirically determined rate law equation for the reaction between nitrogen dioxide and fluorine gas

$$2\,NO_2 + F_2 \rightarrow 2\,NO_2F$$

is

$$r = k[NO_2]^1[F_2]^1$$

As we have already stated, the exponents do not have to be the same as the coefficients in the balanced equation. In this reaction, the rate depends equally on the initial concentrations of each reactant ($m = 1$ and $n = 1$), despite the fact that their reaction coefficients are different.

The exponents in the rate law describe the mathematical dependence of rate on initial concentration and are called the individual **orders of reaction**. For the nitrogen dioxide–fluorine system, we say that

order of reaction the exponent value that describes the initial concentration dependence of a particular reactant

- the order of reaction with respect to NO_2 is 1;
- the order of reaction with respect to F_2 is 1;
- the overall order of reaction is 2 $(1 + 1)$.

Note that the **overall order of reaction** is the sum of the individual orders of reaction for each reactant.

overall order of reaction the sum of the exponents in the rate law equation

If we were to perform a series of experiments with different initial concentrations of the same reactants, we could use the orders of reaction to predict the reaction rates. For example, consider the theoretical equation

$$2X + 2Y + 3Z \rightarrow \text{products}$$

You are told that experimental evidence gives the following rate law equation:

$$r = k\,[X]^1[Y]^2[Z]^0$$

Because $r \propto [X]^1$,

- if the initial concentration of X is doubled, the rate will double (multiply by 2^1).
- if the initial concentration of X is multiplied by 3, the rate will multiply by 3 (3^1).

Because $r \propto [Y]^2$,

- if the initial concentration of Y is doubled, the rate will multiply by 4 (2^2).
- if the initial concentration of Y is multiplied by 3, the rate will multiply by 9 (3^2).

Because $r \propto [Z]^0$, the rate does not depend on $[Z]$, and

- if the initial concentration of Z is doubled, the rate will multiply by 1 (2^0), and so will be unchanged.
- if the initial concentration of Z is tripled, the rate will multiply by 1 (3^0), and so will be unchanged.

The overall order of this reaction is 3 $(1 + 2 + 0)$.
Since the rate does not depend on the initial concentration of Z, for this reaction
$r = k\,[X]^1[Y]^2[Z]^0$ becomes $r = k\,[X]^1[Y]^2$.

LEARNING TIP

The rate law is most accurate only for the initial concentrations at a specified temperature. In most reactions, the concentrations and temperature change as soon as the reaction begins.

How Rates Depend on Concentration Change and Order of Reaction

Table 1

Concentration Change	Order of Reaction			
	0	**1**	**2**	**3**
$\times$ **1**	$1^0 = 1$	$1^1 = 1$	$1^2 = 1$	$1^3 = 1$
$\times$ **2 (doubling)**	$2^0 = 1$	$2^1 = 2$	$2^2 = 4$	$2^3 = 8$
$\times$ **3 (tripling)**	$3^0 = 1$	$3^1 = 3$	$3^2 = 9$	$3^3 = 27$

SAMPLE problem

Figure 1
Butadiene is a monomer that forms the polymer polybutadiene, used in car tires.

LEARNING TIP

Rate depends on initial reactant concentrations raised to various exponents:

- if rate depends on [reactant]^0 (i.e., does not depend on this reactant), then doubling initial concentration has no effect on rate;
- if rate depends on [reactant]^1, then doubling initial concentration doubles rate; and
- if rate depends on [reactant]^2, then doubling initial concentration quadruples rate.

Using a Rate Equation to Predict Rate

The decomposition of dinitrogen pentoxide,

$$2\,N_2O_5 \rightarrow 2\,NO_2 + O_2$$

is first order with respect to N_2O_5. If the initial rate of consumption is 2.1×10^{-4} mol/(L·s) when the initial concentration of N_2O_5 is 0.40 mol/L, predict what the rate would be if another experiment were performed in which the initial concentration of N_2O_5 were 0.80 mol/L.

First, write the rate law equation for the system:

$$r = k[N_2O_5]^1$$

Since the rate is described as first order with respect to $[N_2O_5]$, any change in the $[N_2O_5]$ will have the same effect on the rate.

The $[N_2O_5]$ is doubled from 0.40 to 0.80 mol/L, so the rate of consumption will double from 2.1×10^{-4} to 4.2×10^{-4} mol/(L·s).

Example

The dimerization reaction of 1,3–butadiene (C_4H_6) (**Figure1**).

$$2\,C_4H_6 \rightarrow C_8H_{12}$$

is second order with respect to C_4H_6. If the initial rate of reaction were 32 mmol C_4H_6/(L·min) at a given initial concentration of C_4H_6, what would be the initial rate of reaction if the initial concentration of C_4H_6 were doubled?

Solution

The rate equation is

$$r = k[C_4H_6]^2$$

If the initial concentration is doubled (multiplied by 2), the initial rate will be multiplied by 2^2, or 4.

The new rate is

4×32 mmol C_4H_6/(L·min), or 0.13 mol C_4H_6/(L·min).

The initial rate of the second reaction would be 0.13 mol C_4H_6/(L·min).

Rate law equations must be determined empirically, by measuring initial rates of reaction for systems in which many trials are performed with different initial concentrations of reactants.

Finding a Rate Equation

When aqueous bromate and bisulfite ions react to produce bromine, the overall equation is

$$2\ BrO_{3(aq)}^{-} + 5\ HSO_{3(aq)}^{-} \rightarrow Br_{2(g)} + 5\ SO_{4(aq)}^{2-} + H_2O_{(l)} + 3\ H_{(aq)}^{+}$$

Consider the series of experiments recorded in Table 2, in which initial reactant concentrations are varied and rates are compared. From the evidence provided, determine a rate equation.

Table 2 Initial Concentrations of Reactants and Rate of Product Production

Trial	Initial $[BrO_{3(aq)}^{-}]$ (mmol/L)	Initial $[HSO_{3(aq)}^{-}]$ (mmol/L)	Initial rate of $Br_{2(g)}$ production (mmol/(L·s))
1	4.0	6.0	1.60
2	2.0	6.0	0.80
3	2.0	3.0	0.20

The reaction rate, r, is proportional to the initial concentrations of bromate ions and of bisulfite ions.

$$r = k\,[BrO_{3(aq)}^{-}]^{m}[HSO_{3(aq)}^{-}]^{n}$$

where m, n, and k are to be determined.

A key to solving problems of this type is to look for pairs of data in which the initial concentration of only one reactant changes.

To find m, look at the data from Trials 1 and 2, because the initial concentration of bromate changed while the initial concentration of bisulfite remained constant. Comparing these trials shows that when the initial concentration of $BrO_{3(aq)}^{-}$ is doubled (from 2.0 to 4.0 mol/L), the rate changes by a factor of 1.60/0.80, or 2. This is in direct proportion to the change in initial concentration of $BrO_{3(aq)}^{-}$: As $[BrO_{3(aq)}^{-}]$ doubles, the rate doubles. The exponent m in the rate equation is therefore 1; thus, the order of reaction with respect to $BrO_{3(aq)}^{-}$ is 1.

To find n, look at the data from Trials 2 and 3, where the initial concentration of bisulfite changed while the initial concentration of bromate remained constant. Comparing these trials shows that when the concentration of $HSO_{3(aq)}^{-}$ is doubled from 3.0 to 6.0 mol/L, the rate changes by a factor of 0.80/0.20, or 4. This is a direct square proportion to the change in concentration of bisulfite. Since $2^2 = 4$, the exponent n in the rate equation is 2, and the order of reaction with respect to bisulfite ions is 2.

To find the rate constant k, enter the values from Trial 1 (or any of the trials) into the rate equation, with the concentrations expressed in mol/L. For example, if we use the data from Trial 1 (converted to mol/L from mmol/L),

$$r = k\,[BrO_{3(aq)}^{-}]^{1}[HSO_{3(aq)}^{-}]^{2}$$
$$0.00160\ \text{mol/(L·s)} = k \times (0.0040\ \text{mol/L})^{1} \times (0.0060\ \text{mol/L})^{2}$$

We can then solve for k:

$$k = \frac{0.00160\ \text{mol/(L·s)}}{0.0040\ \text{mol/L} \times (0.0060\ \text{mol/L})^{2}}$$

$$k = 1.1 \times 10^{4}\ \text{L}^{2}/(\text{mol}^{2}\text{·s})$$

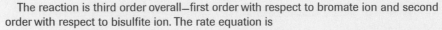

The reaction is third order overall—first order with respect to bromate ion and second order with respect to bisulfite ion. The rate equation is

$$r = k[BrO_{3\,(aq)}^-][HSO_{3\,(aq)}^-]^2$$

where

$$k = 1.1 \times 10^4 \; L^2/(mol^2 \cdot s)$$

Example

Mixing an acidic solution containing iodate, $IO_{3\,(aq)}^-$, ions with another solution containing $I_{(aq)}^-$ ions begins a reaction that proceeds, in several reaction steps, to finally produce molecular iodine as one of the products (**Figure 2**).

$$IO_{3\,(aq)}^- + 5\,I_{(aq)}^- + 6\,H_{(aq)}^+ \rightarrow 3\,I_{2(aq)} + 3\,H_2O_{(l)}$$

The data in **Table 3** are obtained for rate of production of iodine (I_2):

Table 3 Concentrations of Reactants and Rate of Product Production

Trial	Initial $[IO_3^-]$ (mmol/L)	Initial $[I^-]$ (mmol/L)	Initial $[H^+]$ (mmol/L)	Rate of production of iodine (mmol/(L·s))
1	0.10	0.10	0.10	$r_1 = 5.0 \times 10^{-4}$
2	0.20	0.10	0.10	$r_2 = 1.0 \times 10^{-3}$
3	0.10	0.30	0.10	$r_3 = 1.5 \times 10^{-3}$
4	0.10	0.30	0.20	$r_4 = 6.0 \times 10^{-3}$

(a) What is the rate equation for this reaction?
(b) What will the rate of reaction be when $[IO_3^-] = 0.20$ mmol/L, $[I^-] = 0.40$ mmol/L, and $[H^+] = 0.10$ mmol/L?

Solution

(a) We assume that the rate equation follows the pattern

$$r = k\,[IO_3^-]^a\,[I^-]^b\,[H^+]^c.$$

Comparing Trials 1 and 2, we find that only $[IO_3^-]$ changes:

$$\frac{rate_2}{rate_1} = \frac{1.0 \times 10^{-3}}{5.0 \times 10^{-4}} = 2.0$$

As the $[IO_3^-]$ doubles, the rate doubles, so rate $\alpha\,[IO_3^-]^1$. The rate is first order with respect to the initial concentration of iodate ions.
Comparing Trials 1 and 3, we find that only $[I^-]$ changes:

$$\frac{rate_3}{rate_1} = \frac{1.5 \times 10^{-3}}{5.0 \times 10^{-4}} = 3.0$$

As the $[I^-]$ triples, the rate triples, so rate $\alpha\,[I^-]^1$. The rate is first order with respect to the initial concentration of iodide ions.
Comparing Trials 3 and 4, we find that only $[H^+]$ changes:

$$\frac{rate_4}{rate_3} = \frac{6.0 \times 10^{-3}}{1.5 \times 10^{-3}} = 4.0$$

As the $[H^+]$ doubles, the rate quadruples, so rate $\alpha\,[H^+]^2$. The rate is second order with respect to the initial concentration of hydrogen ions.
The rate equation for this reaction is

$$r = k\,[IO_3^-]^1\,[I^-]^1\,[H^+]^2$$

or

$$r = k\,[IO_3^-][I^-][H^+]^2.$$

Figure 2
Starch turns blue-black in the presence of iodine. Biologists often use this reaction as a test for starch, whereas chemists use it as a test for iodine.

Using the data from Trial 1 to find k, we see that

$$k = \frac{r_1}{[IO_3^-]^1 [I^-]^1 [H^+]^2}$$

$$= \frac{5.0 \times 10^{-4} \text{ mmol}/(L \cdot s)}{(0.10 \text{ mmol}/L)(0.10 \text{ mmol/L})(0.10 \text{ mmol/L})^2}$$

$$= 5.0 \text{ L}^3/(\text{mmol}^3 \cdot s)$$

The rate constant for the equation is $5.0 \text{ L}^3/(\text{mmol}^3 \cdot s)$.

(b) Substituting the given concentrations, we see that

$$r = k[IO_3^-]^1 [I^-]^1 [H^+]^2$$

$$= 5.0 \text{ L}^3/(\text{mmol}^3 \cdot s) \times (0.20 \text{ mmol}/L) \times (0.40 \text{ mmol}/L) \times (0.10 \text{ mmol/L})^2$$

$$= 4.0 \times 10^{-3} \text{ mmol}/(L \cdot s)$$

The rate at the new set of conditions is $4.0 \times 10^{-3} \text{ mmol}/(L \cdot s)$.

▶ **Practice**

Understanding Concepts

1. Explain, with an example, the difference between order of reaction and overall order of reaction.

2. The decomposition of $SO_2Cl_{2(g)}$

$$SO_2Cl_{2(g)} \rightarrow SO_{2(g)} + Cl_{2(g)}$$

is known to be first order with respect to SO_2Cl_2. If the initial rate of reaction is $3.5 \times 10^{-3} \text{ mol } SO_2Cl_2/(L \cdot s)$ when the initial concentration of SO_2Cl_2 is 0.25 mol/L, what would the rate be if another experiment were performed in which the initial concentration of SO_2Cl_2 were 0.50 mol/L?

3. For a reaction where the rate equation is $r = k[NH_{4(aq)}^+][NO_{2(aq)}^-]$,
    (a) calculate k at temperature T_1, if the rate, r, is 2.40×10^{-7} mol/(L·s) when $[NH_{4(aq)}^+]$ is 0.200 mol/L and $[NO_{2(aq)}^-]$ is 0.00500 mol/L.
    (b) calculate r at temperature T_2, if the rate constant, k, is 3.20×10^{-4} L/(mol·s) when $[NH_{4(aq)}^+]$ is 0.100 mol/L and $[NO_{2(aq)}^-]$ is 0.0150 mol/L.

4. A series of experiments is performed for the system

$$2A + 3B + C \rightarrow D + 2E.$$

    • When the initial concentration of A is doubled, the rate increases by a factor of 4.
    • When the initial concentration of B is doubled, the rate is doubled.
    • When the initial concentration of C is doubled, there is no effect on rate.

    (a) What is the order of reaction with respect to each of the reactants?
    (b) Write an expression for the rate equation.

5. State the effect on the value of a reaction rate constant, k, in a given rate equation when
    (a) the temperature of the reaction is increased;
    (b) the initial concentration of any reactant is decreased.

6. The experimental observations in **Table 4** are obtained for the reaction

$$2A + B + 2C \rightarrow 3X$$

    (a) What is the order of reaction with respect to each of the reactants?
    (b) Write an expression for the rate equation.
    (c) Calculate a value for the rate constant.
    (d) Calculate the rate of production of X when $[A] = [B] = [C] = 0.40$ mol/L.

Answers

3. (a) 0.24 mL/(mol·s)
    (b) 0.48 μL/(mol·s)

6. (c) 0.3 L²/(mol²·s)
    (d) 19 mmol/(L·s)

Table 4 Observations on the Rate of Production of X

Trial	Initial [A] (mol/L)	Initial [B] (mol/L)	Initial [C] (mol/L)	Rate of production of X (mol/(L·s))
1	0.10	0.10	0.10	3.0×10^{-4}
2	0.20	0.10	0.10	1.2×10^{-3}
3	0.10	0.30	0.10	3.0×10^{-4}
4	0.20	0.10	0.20	2.4×10^{-3}

Relating Reaction Rate to Time

So far, we have discussed the relationship between rate and initial concentration. However, it can be inconvenient to determine the initial rate of a reaction in a school laboratory. It is much easier to measure the time that elapses before a certain point in the reaction (such as a visible change) is reached. As you know, the average rate is inversely related to the elapsed time:

$$r_{av} \propto \frac{1}{\Delta t}$$

Therefore, if the rate of a reaction in which some reactant A is consumed is

$r_{av} \propto [A]^n$, then

$$\frac{1}{\Delta t} \propto [A]^n$$

(Remember that [A] is the initial concentration of the reactant, not the concentration that changes as the reaction proceeds.)

We often find it easier to recognize relationships if we can see them on graphs. For example, we can use graphs to help us recognize the order of reaction with respect to a particular reactant. Plotting experimental data as shown in **Figure 3**, and looking for a straight line (indicating a direct relationship) will determine the value of n.

INVESTIGATION 6.3.1

The Iodine Clock Reaction (p. 403)
A dramatic colour change allows you to investigate the effect of concentration on rate of reaction.

Figure 3
When a series of kinetics experiments is performed on a given system, the rates of reaction (1/time) are measured for different initial concentrations of a reactant. When the evidence is graphed, you may see one or more of these results.
(a) In this plot, $r \propto [A]^0$. The reaction is zeroth order with respect to [A].
(b) In this plot, $r \propto [A]^1$. The reaction is first order with respect to [A].
(c) In this plot, $r \propto [A]^n$, where n is greater than 1.
(d) In this plot, $r \propto [A]^2$. The reaction is second order with respect to [A].

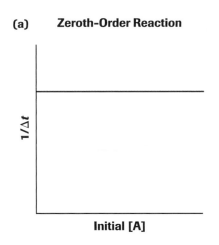

(a) **Zeroth-Order Reaction**

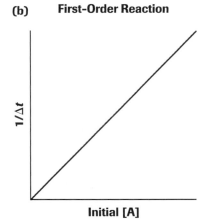

(b) **First-Order Reaction**

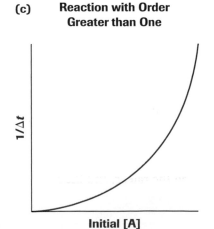

(c) **Reaction with Order Greater than One**

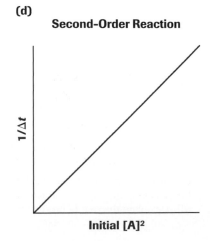

(d) **Second-Order Reaction**

Chemical Kinetics and Half-Life

You have seen that most reactions occur most quickly at the beginning, and then slow down as the concentrations of reactants decrease and fewer collisions occur that can lead to product. The rate of reaction of many reactions is a first-order process:

rate α [reactant]^1

Such first-order processes may occur in both chemical and nuclear changes. For example, nuclear decay — the change that occurs as a radioactive isotope breaks down into smaller isotopes — is a first-order process. This means that it occurs at a predictable rate in which the order of reaction with respect to the radioactive isotope is 1: If the initial concentration of reactant doubles, the rate of reaction also doubles. For example, when uranium-238 undergoes radioactive decay

$$^{238}_{92}U \rightarrow {}^4_2He + {}^{234}_{90}Th$$

the rate equation is

$r = k[\text{U-238}]^1$

and the concentration–time graphs that describe the kinetics of the change are typical first-order curves (**Figure 4**): The slope of the straight line (i.e., the rate) is directly proportional to the concentration of the reactant.

A consequence of this first-order process is that the quantity of radioisotope remaining in a sample follows a predictable pattern. The **half-life** is the time required for half of the sample to react.

Consider the example of a sample of 100 g of a substance with a half-life of 12 min. Note the mass of the original isotope remaining after each half-life, as shown in **Table 5**.

Table 5 Radioactive Decay of an Isotope

t (min)	Half-lives	Mass of original isotope (g)	Mass of products (g)
0	0	100.0	0.0
12	1	50.0	50.0
24	2	25.0	75.0
36	3	12.5	87.5

Different isotopes have different half-lives. For example, the half-life of uranium-235 used in nuclear reactors is 710 million years, and the half-life of carbon-14 used in dating archaeological artifacts is 5730 a (**Figure 5**), but the half-life of thallium-200 used in medical diagnosis is about one day. The half-life of this isotope is so short that it is made artificially in hospitals and used immediately in the patient, with very little residual radioactivity after a few days.

The concept of half-life may be applied to chemical, as well as nuclear, systems. For example, caffeine is a stimulant found in many foods, including coffee, tea, cola, and chocolate. Its physiological effects depend on its concentration in the bloodstream, and this concentration decreases steadily over time, according to first-order kinetics. Similarly, so-called "heavy metals" such as lead and cadmium may accumulate in the body. Cadmium has adverse physiological effects and is most toxic when it is inhaled: A major source is cigarette smoke. The concept of half-life and rate of reaction can help us predict how much of such substances will remain in the body after a certain number of half-lives, or calculate how many half-lives must pass before a certain proportion of the original substance has been eliminated from the body.

Reaction kinetics yields a simple relationship between the rate, r, and the half-life, $t_{1/2}$. For a first-order reaction of some reactant A,

$$-\frac{\Delta[A]}{\Delta t} = r = k[A]$$

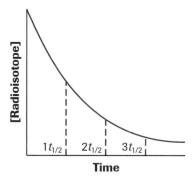

Concentration of Radioisotope Changing over Time

Figure 4
The decay or reaction of a reactant (such as carbon-14) shows first-order dependence in this plot of concentration vs. time. After each half-life, the concentration of the reactant is halved.

half-life the time for half of the nuclei in a radioactive sample to decay, or for half the amount of a reactant to be used up (in a first-order reaction)

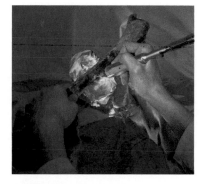

Figure 5
When the reindeer this bone belongs to died, the quantity of carbon-14 in its tissues was fixed. Radioactive decay over the centuries reduced the quantity of C-14 in a predictable first-order reaction, allowing chemists to determine the number of half-lives and, therefore, the number of years since death.

This equation can be manipulated using integral calculus to yield the expression

$$\ln\left(\frac{[A]_0}{[A]}\right) = kt$$

where $[A]$ is the concentration at any given time and $[A]_0$ is the initial concentration. After one half-life, $t_{1/2}$, the concentration of A is half what it was at the beginning:

$$[A] = \frac{1}{2} \times [A]_0$$

The previous expression then simplifies to

$$\ln 2 = kt_{\frac{1}{2}} \quad \text{or}$$

> **Equation Relating Half-Life and the Rate Constant**
>
> $$kt_{\frac{1}{2}} = 0.693$$

This equation can be applied *only* to a reaction based on first-order kinetics.

▶ SAMPLE problem | *Calculating with Half-Lives*

The radioisotope lead-212 has a half-life of 10.6 h. What is the rate constant for this isotope?

First, write the rate equation:

$$r = k[\text{Pb-212}]^1$$

Then, apply the equation that relates half-life and the rate constant:

$$kt_{1/2} = 0.693$$

$$k = \frac{0.693}{t_{1/2}}$$

$$= \frac{0.693}{10.6 \text{ h}}$$

$$k = 0.0654 \text{ h}^{-1}$$

The rate constant for the decay of lead-212 is 0.0654 h^{-1}.

Example

If the mass of an antibiotic in a patient is 2.464 g, what mass of antibiotic will remain after 6.0 h, if the half-life is 2.0 h, and no further drug is taken?

Solution

Number of half-lives in 6.0 h $= \dfrac{6.0 \text{ h}}{2.0 \text{ h}} = 3.0$ half-lives

After each half-life, the mass of the antibiotic will be halved.
The mass remaining will be

$$m = 2.464 \text{ g} \times \frac{1}{2} \times \frac{1}{2} \times \frac{1}{2} \text{ or } 2.464 \text{ g} \times \left(\frac{1}{2}\right)^3$$

$$m = 0.31 \text{ g}$$

The mass of original antibiotic remaining will be 0.31 g.

▶ Practice

Understanding Concepts

7. A radioisotope has a half-life of 24 a and an initial mass of 0.084 g.
  (a) What mass of radioisotope will remain after
    (i) 72 a?
    (ii) 192 a?
  (b) Approximately how many years will have passed if only 10% of the radioisotope remains?

8. The specific rate constant for a first-order reaction is 2.34×10^{-3} s^{-1}.
  (a) What is the half-life of the reaction?
  (b) How long will it take for the concentration of the reactant to reach 12.5% of its original value?

Making Connections

9. Patients are kept isolated for several days after receiving some types of radiation therapy. Comment on the types of radioisotopes that are suitable for radiation therapy, and why this isolation is enforced.

Answers

7. (a) (i) 10.5 mg
       (ii) 0.33 mg

   (b) 80 a

8. (a) 296 s

   (b) 888 s

▶ Section 6.3 Questions

Understanding Concepts

1. Explain what is meant by the terms "first-order reaction" and "second-order reaction," including an example of each.

2. For a reaction where the rate law equation is $r = k[Cl_{2(g)}][NO_{(g)}]^2$,
  (a) what is the order of the reaction with respect to each of the reactants?
  (b) what would the effect on rate be if the initial concentration of $Cl_{2(g)}$ were doubled?
  (c) what would the effect on rate be if the initial concentration of $NO_{(g)}$ were tripled?
  (d) calculate k at a temperature where the rate, r, is 0.0242 mol/(L·s) when $[Cl_{2(g)}]$ is 0.20 mol/L and $[NO_{(g)}]$ is 0.20 mol/L.
  (e) calculate r at a temperature where the rate constant, k, is 3.00 when $[Cl_{2(g)}]$ is 0.44 mol/L and $[NO_{(g)}]$ is 0.025 mol/L.

3. When antibiotics are administered to fight bacterial infections, one concern is to maintain the concentration of the antibiotic at a minimum level. Antibiotics also decompose naturally when stored by pharmacists and drug manufacturers. The rate law equation for the decomposition of a certain antibiotic is known to be first order with respect to the antibiotic concentration.
  (a) Write the rate law expression for this process.
  (b) If the rate constant for the decomposition of the antibiotic stored in water at 20°C is 1.40 a^{-1}, what is the half-life of the antibiotic?
  (c) If a bottle contains 20 g of antibiotic, roughly what mass of the antibiotic would remain after 2.0 years?

4. The half-life of uranium-234 is 2.3×10^5 a. If 10.0 g of this isotope were to be placed in underground storage for 1.84×10^6 a, how much of the original isotope would remain?

Applying Inquiry Skills

5. A "clock" experiment is performed to investigate the rate of reaction of hydrogen peroxide, hydrogen ions, and iodide ions:

$$H_2O_{2(aq)} + 2H^+_{(aq)} + 2I^-_{(aq)} \rightarrow I_{2(aq)} + 2H_2O_{(g)}$$

Question

What is the order of reaction with respect to initial hydrogen peroxide concentration?

Experimental Design

A series of solutions is prepared in which the only variable is the initial concentration of hydrogen peroxide. Constant quantities of other reactant solutions are mixed with these solutions, so that the time from mixing to formation of a blue-black product can be timed. The evidence is analyzed graphically.

Evidence

Table 6 Time to Colour Change at Different Concentrations of Hydrogen Peroxide

$[H_2O_2]$ (mol/L)	Time (s)
0.0042	71
0.0034	91
0.0026	115
0.0018	167
0.0010	301

Analysis

   (a) Use the experimental values in **Table 6** to determine the order of reaction with respect to hydrogen peroxide.

6. Design an experiment that would enable you to study the effect of changing temperature, using the iodine clock reaction. Write a Question, Prediction, Experimental Design, Materials list, and Procedure (with safety and disposal precautions).

Making Connections

7. Radiocarbon dating is a common way of determining the age of archaeological specimens, but the carbon-14 isotope is only one of a number of radioisotopes used. Research and report on the application of carbon-14 and other isotopes to such dating techniques.

 www.science.nelson.com

8. Nuclear waste storage is a controversial issue. Research the Internet to answer the following questions:

   (a) What kinds of nuclear wastes are produced by CANDU reactors, what are their half-lives, and what are the products of their radioactive decay?

   (b) Where are these wastes currently stored?

   (c) What are some proposed long-term solutions to the problem of nuclear waste?

   (d) Imagine that you are a member of a mining community. A local abandoned mine has been proposed as a site for storage of nuclear waste. Write a letter to your MP supporting or opposing the proposal. Back up your opinion with arguments based on specific information.

 www.science.nelson.com

Experimental observations and rate law equations describe what happens to the rate of reaction when various factors such as initial concentration are changed. But how do reactions actually occur? Why do some reactions happen quickly and others slowly? For example, why does potassium react so much more quickly than calcium, in water? Why can butane gas from a lighter mix with air with no visible reaction until a spark is provided?

Rates of reaction can be explained with **collision theory**. Molecules are held together with chemical bonds. According to collision theory, chemical reactions can occur only if energy is provided to break those bonds; the source of that energy is the kinetic energy of the molecules.

Concepts of the Collision Theory

- A chemical system consists of *particles* (atoms, ions, or molecules) that are in constant random motion at various speeds. The average kinetic energy of the particles is proportional to the temperature of the sample. **Figure 1** shows the distribution of kinetic energies among particles in a sample at two different temperatures.

- A chemical reaction must involve *collisions of particles* with each other or the walls of the container.

- An *effective collision* is one that has sufficient energy and correct orientation (alignment or positioning) of the colliding particles so that bonds can be broken and new bonds formed.

- *Ineffective collisions* involve particles that rebound from the collision, essentially unchanged in nature.

- The rate of a given reaction depends on the *frequency* of collisions and the *fraction* of those collisions that are effective.

Kinetic Energy Distribution at Two Temperatures

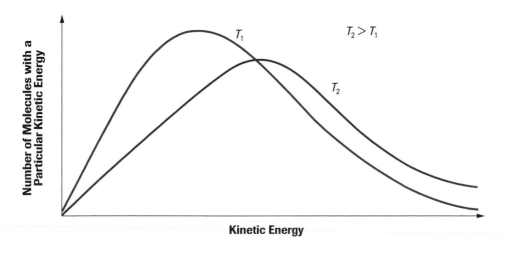

Figure 1
Temperature is a measure of the average kinetic energy of the particles. This graph shows how the distribution of kinetic energies changes when a substance is heated or cooled. At any temperature there are some particles with low kinetic energy and some with high kinetic energy. The higher the temperature, the more particles there are with higher kinetic energies.

Figure 2
The reaction between nitrogen and oxygen in air to produce yellow-brown nitrogen dioxide ($N_2 + 2O_2 \rightarrow 2NO_2$) is a very slow reaction under normal conditions, but occurs more quickly in the high temperatures of automobile engines. The result can be a yellow haze of toxic gas.

An Analogy for Activation Energy

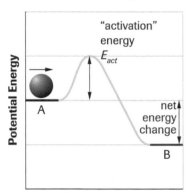

Figure 3
On a trip from A to B there is a net decrease in overall (net) energy, but there must be an initial increase in potential energy (activation energy) for the trip to be possible.

activation energy the minimum increase in potential energy of a system required for molecules to react

The effects of these factors can be explained in terms of the last statement, which can be expressed mathematically as

rate = frequency of collisions × fraction of collisions that are effective

For example, consider a system in which the collision frequency is 1000 collisions/s and the fraction of effective collisions is 1/100 (meaning that only 1 collision of every 100 results in reaction, and the other 99 collisions involve the molecules simply rebounding without changing).

The reaction rate, expressed in terms of number of reactions/s, is

$$\text{reactions per second} = \frac{1000 \text{ collisions}}{1 \text{ s}} \times \frac{1 \text{ reaction}}{100 \text{ collisions}}$$
$$= 10 \text{ reactions/s}$$

What happens if the collision frequency is increased to 5000 collisions/s?

$$\text{reactions per second} = \frac{5000 \text{ collisions}}{1 \text{ s}} \times \frac{1 \text{ reaction}}{100 \text{ collisions}}$$
$$= 50 \text{ reactions/s}$$

The rate becomes 50 reactions/s.

What happens if the fraction of effective collisions is increased to 1/20 from 1/100?

$$\text{reactions per second} = \frac{1000 \text{ collisions}}{1 \text{ s}} \times \frac{1 \text{ reaction}}{20 \text{ collisions}}$$
$$= 50 \text{ reactions/s}$$

The rate again becomes 50 reactions/s.

Clearly, increasing either the collision frequency or the fraction of effective collisions will increase the rate.

Activation Energy

Consider an empty soft drink can. In the nitrogen–oxygen air mixture in the can, an estimated 10^{30} molecular collisions occur every second. This is an absolutely enormous number: one thousand billion billion billion events per second. When nitrogen and oxygen react, one of the products they form is a colourful, very toxic gas called nitrogen dioxide, $NO_{2(g)}$ (**Figure 2**). However, our air remains relatively colourless so, although nitrogen and oxygen can react at normal conditions, they must do so very slowly. How can we use collision theory to explain this observation?

If each collision produced a reaction when reacting substances were combined, the rate of any reaction would be extremely rapid, appearing essentially instantaneous. In the pop can, the nitrogen and oxygen molecules would react completely to form nitrogen dioxide in about 5×10^{-9} s (five-billionths of a second). Since the actual rate is too slow to be measurable, we conclude that normally only an extremely tiny fraction of the collisions between oxygen and nitrogen molecules actually produce the new substances. The collision frequency is high but the fraction of effective collisions is small.

The theoretical explanation for the empirical evidence involves the concept of **activation energy** — the minimum energy with which particles must collide before they can rearrange in structure, resulting in an effective collision. The concept of activation energy, E_{act}, can be illustrated by an analogy with gravitational potential energy. Consider a billiard ball rolling (friction-free) on a smooth track shaped as shown in **Figure 3**. The ball leaves point A moving toward the right. As it rises on the uphill portion of the track it slows down: Kinetic energy converts to potential energy. The ball can only successfully overcome the rise of the track and proceed to point B if it has enough initial speed (kinetic energy). We could call this situation an effective trip. The minimum kinetic energy required is analogous to the activation energy for a reaction. If the ball does not have enough kinetic energy it will not reach the top of the track and will just roll back to point A. This is analogous to two molecules colliding without enough energy to rearrange their bonds: They just rebound. Since the activation energy sometimes seems to prevent reaction, it is often called an activation energy barrier.

Note that a ball that returns to point A will have the same potential and kinetic energy it began with, but a ball that makes it to point B will have a different combination of energies: It will have less potential energy because it is at a lower point on the diagram, but it will have more kinetic energy because it will be moving faster. This billiard-ball example is an analogy for the energy change taking place during an exothermic reaction, in which energy is released to the surroundings. In kinetic theory terms, this means that the energy is released to any other nearby molecules. These other molecules then move faster, collide with more energy, and become more likely to react. The reaction, once begun, is self-sustaining as long as enough molecules remain to make collisions likely. For example, the energy released when wood burns allows the reaction to proceed unaided by external sources of energy. Exothermic reactions often drive themselves in this way once begun, as shown in **Figure 4**.

The concept of activation energy fits well with the fundamental idea from collision theory that rate is proportional to collision frequency and the proportion of effective collisions:

rate = frequency of collisions × fraction of collisions that are effective

Increasing the concentration or the surface area of the reactants increases the collision frequency — more collisions occur per unit time. Changing the nature of the reactants or using a catalyst changes the size of the activation energy barrier, making it easier or more difficult for molecules to react. This has the effect of changing the fraction of effective collisions. Increasing temperature has a particularly dramatic effect on the reaction rate because it increases both collision frequency *and* the fraction of effective collisions. Since the molecules are moving more quickly, not only do they collide more often but they also collide with more energy, on average, making it more likely that bonds will break.

Get in Line!
Some collisions appear to be energetic enough for reaction to occur but are still not effective because the molecules do not collide with the correct alignment. The smaller and simpler in structure the molecules, the more likely it is that a collision will be effective.

Figure 4
Once a fire has begun, the exothermic combustion of wood is self-sustaining and can destroy large areas of forest.

SUMMARY	***Factors Affecting Rate and Collision Theory***

Each of the five factors that affect rate increases either collision frequency or the fraction of collisions that are effective (or both) to increase rate of reaction.

rate = collision frequency × fraction effective

concentration	nature of reactant
surface area	catalyst
temperature	temperature

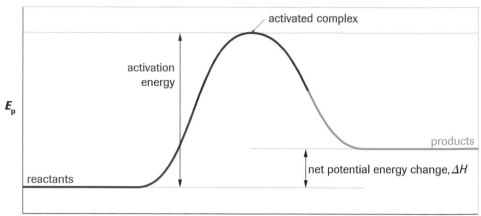

Potential Energy Changes During an Endothermic Reaction

Figure 5

Over the progress of an effective collision between molecules in the gas phase, the potential energy increases to a maximum at the point of closest approach, then decreases to a final value higher than the initial energy (as the reaction is endothermic). The potential energy gain of the molecules comes from conversion of kinetic energy. The overall reaction would lower the temperature of the system and surroundings.

activated complex an unstable chemical species containing partially broken and partially formed bonds representing the maximum potential energy point in the change; also known as *transition state*

Consider the reaction of hydrogen and iodine molecules, a single-step reaction at high temperatures, plotted as potential energy of the molecules versus progress of the reaction (**Figure 5**).

$$H_{2(g)} + I_{2(g)} \rightarrow 2\,HI_{(g)}$$

We can discuss the progress of this reaction in terms of molecular collisions, by moving from left to right along the plot shown in Figure 5. Along the flat region to the left, the molecules are moving toward each other, but are still distant from each other. As the molecules approach more closely, they are affected by repulsion forces and begin to slow down, as some of their kinetic energy is changed to potential energy (stored as a repelling electric field between them). If the molecules have enough kinetic energy, they can approach closely enough for their bond structures to rearrange to form an **activated complex**. The activated complex is an unstable molecule with a particular geometry. It is unstable because it possesses the maximum potential energy possible.

When the reacting system reaches the activated complex stage it may reverse to reactants, or it may continue to form product molecules. In either case, repulsion forces push the molecules apart, converting potential energy to kinetic energy. Overall, there are potential energy changes as bonds are broken and formed and products are formed. If the energy difference is measured at constant pressure (the usual situation for a reaction open to the atmosphere), it is called the enthalpy change or enthalpy of reaction, ΔH, which is $+53$ kJ/mol for the endothermic formation of hydrogen iodide.

When the products of a reaction have higher potential energy than the reactants, they will have lower kinetic energy (temperature). In their subsequent collisions with other molecules in the system and in the surroundings, they will tend to decrease the speed of molecules they collide with, resulting in a drop in the temperature of the system. This is why endothermic reactions have the effect of cooling their surroundings.

In other types of reactions, the final potential energy may be lower than the initial potential energy, meaning that the reaction is exothermic. In such cases the enthalpy of the system decreases, so ΔH has a negative value. For example, the thermochemical equation for the production of carbon dioxide and nitric oxide from carbon monoxide and nitrogen dioxide is

$$CO_{(g)} + NO_{2(g)} \rightarrow CO_{2(g)} + NO_{(g)} \qquad\qquad \Delta H = -227\,kJ$$

During such reactions the temperature of the surroundings tends to rise, as heat is released to the surroundings in the progress of the reaction (**Figure 6**).

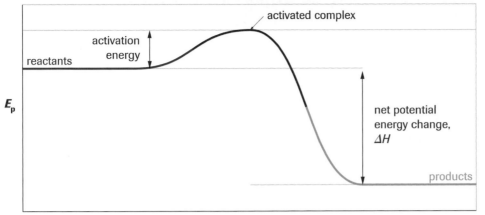

Potential Energy Changes During an Exothermic Reaction

Reaction Progress

Figure 6
Over the progress of this exothermic reaction, the potential energy (or enthalpy) increases to a maximum as the activated complex forms, then decreases to a final value lower than the initial energy. The potential energy lost by the molecules is converted to kinetic energy. The overall reaction would raise the temperature of the surroundings.

▶ *Practice*

Understanding Concepts

1. (a) In your notebook sketch the graph shown in **Figure 7**, and add labels for the axes.
  (b) What does each curve represent?
  (c) What type of reaction is occurring in terms of energy flow to or from the surroundings?
  (d) What does each number (i, ii, iii) represent?

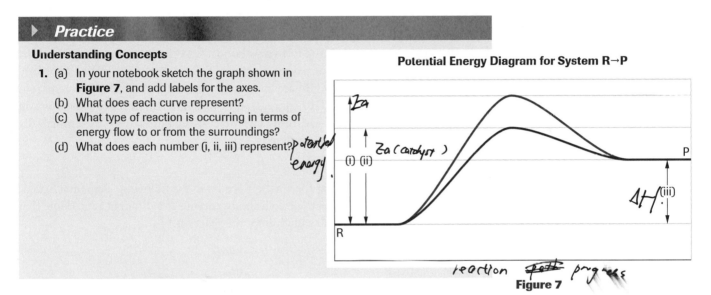

Potential Energy Diagram for System R→P

Figure 7

Reaction Mechanisms

How can you have a "collision" involving one particle? Consider the reaction in which hydrogen peroxide decomposes to water and oxygen gas. How do some molecules obtain enough energy to decompose? Some reaction mechanisms involve a step in which a single molecule apparently hits container walls or any other particle in order to convert enough energy from kinetic to potential for the molecule to decompose. Still other molecules absorb light energy in a reaction step to break a bond, resulting in two or more atoms, such as the following reaction:

$$Cl_2 + light \rightarrow 2\,Cl\cdot$$

Common sense, and calculations, indicates that collisions of three particles simultaneously must be much less frequent than two-particle collisions and that any collision involving four or more particles is extremely unlikely indeed. Imagine a circle of students tossing Velcro-covered Ping-Pong balls toward the centre of the circle. The chances of any two Ping-Pong balls colliding and adhering in the air is small, and the probability of three being in the same place at the same time is much smaller. Think how incredibly unlikely it would be for four balls to collide simultaneously.

Scientists believe that most chemical reactions actually occur as a sequence of **elementary steps**. This overall sequence is called the **reaction mechanism.**

elementary step a step in a reaction mechanism that only involves one-, two-, or three-particle collisions

reaction mechanism a series of elementary steps that makes up an overall reaction

rate-determining step the slowest step in a reaction mechanism

reaction intermediates molecules formed as short-lived products in reaction mechanisms

Oxidation of HBr$_{(g)}$

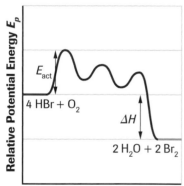

Figure 8
Over the progress of this reaction, the potential energy increase necessary to reach the first activated complex stage is the greatest increase required, so this is the rate-determining (slowest) step. Energy released as kinetic energy past this point is sufficient to quickly carry the reaction mechanism to completion.

An automobile assembly line is a reasonable analogy to a reaction mechanism, because a car is not built in a single, concerted step, but rather in a sequence of steps. Imagine a car being assembled in a plant by a series of workers. One assembles the chassis, another adds wheels, another the seats, and so on. The worker who controls the overall rate of production of cars will not necessarily be the first or the last but, rather, the slowest worker. This worker could be called the "rate-determining worker." If the slowest worker is at the beginning of the line, other workers will wait for cars to arrive at their station; if the slowest worker is at the end, partially assembled cars will stack up waiting to be finished. Adding workers—increasing concentration—at the fast steps will have no effect on car production. However, increasing the "concentration" of workers at the slowest step should increase the rate of production.

A chemical example of a reaction mechanism is the oxidation of hydrogen bromide, which is rapid between 400°C and 600°C. This reaction has been studied extensively because all substances are simple molecules and in the gas phase.

$$4\,HBr_{(g)} + O_{2(g)} \rightarrow 2\,H_2O_{(g)} + 2\,Br_{2(g)}$$

It is highly unlikely that this reaction would occur in a single step, because a total of five reactant molecules would have to collide simultaneously with the proper alignment and sufficient energy to break and form new bonds. Experimental evidence shows that increasing the concentration of oxygen increases the reaction rate, just as we would expect. Since four molecules of HBr are involved for every molecule of O_2, it seems logical to expect that a change in HBr concentration would have a much greater effect on the rate, but measurement shows this is not the case. Quantitatively, the empirically determined rate equation for this system is

$$r = k[HBr][O_2]$$

We explain by theorizing that the reaction occurs in the following elementary steps, each of which involves a two-particle collision occurring at a different rate. (Note that the steps sum to give the overall equation for the reaction.)

$HBr_{(g)} + O_{2(g)}$	$\rightarrow HOOBr_{(g)}$	(slow)
$HOOBr_{(g)} + HBr_{(g)}$	$\rightarrow 2\,HOBr_{(g)}$	(fast)
$2\,\{HOBr_{(g)} + HBr_{(g)}$	$\rightarrow H_2O_{(g)} + Br_{2(g)}\}$	(fast)
$4\,HBr_{(g)} + O_{2(g)}$	$\rightarrow 2\,H_2O_{(g)} + 2\,Br_{2(g)}$	

The theoretical interpretation is that the first elementary step is relatively slow because it has a fairly high activation energy. The rate of the overall reaction is basically controlled by this step, just as the slowest worker determined the overall rate at which cars were produced. The second step cannot use HOOBr any faster than the first step can produce it, so the rate of the reaction overall is the same as the rate of the slowest step — in this case, the first. The slowest reaction step in any reaction mechanism is called the **rate-determining step**. Substances such as HOBr and HOOBr — which are formed during the reaction but immediately react again and are not present when the reaction is complete — are called intermediate products or **reaction intermediates**. On a potential energy diagram like **Figure 8**, unstable activated complexes exist at the "peaks" and slightly more stable reaction intermediates exist at the small "valleys" within a mechanism.

How do chemists determine a mechanism for a reaction? The first step is to perform experiments and thereby determine a rate equation. There is a direct correlation between the exponents in the rate equation and the equation coefficients in the rate-determining step in the mechanism.

For example, in the hydrogen bromide–oxygen system, the empirically determined rate equation has the exponent 1 on each of the concentrations of reactants.

$r = k[HBr]^1[O_2]^1$

Thus, the rate depends on the concentration of one molecule of each of the reactants. The reaction coefficients of the rate-determining step are also 1 for each of the molecules.

$1\ HBr + 1\ O_2 \rightarrow$ reaction intermediate

In general, if the empirically determined rate equation is

$r \propto$ [molecule X]^m[molecule Y]^n

then the rate-determining step in the mechanism must be

$m\ X + n\ Y \rightarrow$ products or reaction intermediates

Reaction mechanisms are only "best guesses" at the behaviour of molecules, but there are three rules that must be followed in proposing a mechanism:

- each step must be elementary, involving no more than three reactant (and more usually only one or two) molecules;
- the slowest or rate-determining step must be consistent with the rate equation; and
- the elementary steps must add up to the overall equation.

It is often possible to create two or more mechanisms, each of which could account for the empirically derived rate equation.

Finding the Rate-Determining Step — SAMPLE problem

Consider the decomposition of dinitrogen pentoxide

$2\ N_2O_{5(g)} \rightarrow 2\ N_2O_{4(g)} + O_{2(g)}$

(a) What would the rate equation be if the reaction occurred in a single step?

(a) The exponent in the rate equation would be the same as the coefficient on the reactant. Therefore, the rate law equation would be

$r = k[N_2O_5]^2$

(b) The actual experimentally derived rate equation is

$r = k\,[N_2O_5]^1$

What is the rate-determining step?

(b) Because the coefficient on the reactant must be the same as the exponent in the rate equation, the rate-determining step must be

$1\ N_2O_{5(g)} \rightarrow$ some product or reaction intermediate

(c) Suggest a possible mechanism and indicate the slowest step.

(c) The only step that we are sure of is the rate-determining step; the others are guesses. The following is a possibility.

$N_2O_5 \rightarrow N_2O_4 + O$ (slow)
$O + N_2O_5 \rightarrow N_2O_4 + O_2$ (fast)

The rate-determining step is the slow step, which could occur at any point in the mechanism. Each step in the mechanism involves two or fewer reactant molecules, and the steps sum up to the overall equation.

Example

Consider the overall reaction involving three elements as reactants, and a compound as the product:

$$X + 2Y + 2Z \rightarrow XY_2Z_2$$

When a series of reactions is performed with different initial concentrations of reactants, the results are as follows:

- doubling the concentration of X has no effect on the overall rate
- doubling the concentration of Y multiplies the overall rate by 4
- doubling the concentration of Z doubles the overall rate

State
- (a) the rate law equation for this system;
- (b) the rate-determining step;
- (c) a possible mechanism, indicating the slow step; and
- (d) a possible reaction intermediate in your mechanism.

Solution

(a) From the empirical information provided,

$r \alpha [X]^0$; $r \alpha [Y]^2$; and $r \alpha [Z]^1$, giving a rate equation of

$r = k[Y]^2[Z]^1$

(b) Therefore, the rate-determining step must be

$2Y + 1Z \rightarrow$ some product(s)

(c) Any mechanism consistent with the above rules is acceptable. For example, one possibility is

$X + Z$	$\rightarrow XZ$	(fast)
$2Y + Z$	$\rightarrow Y_2Z$	(slow)
$XZ + Y_2Z$	$\rightarrow XY_2Z_2$	(fast)

$$X + 2Y + 2Z \rightarrow XY_2Z_2$$

(d)   Two possible reaction intermediates are XZ and Y_2Z

⟋⟋ LAB EXERCISE 6.4.1

The Sulfur Clock (p. 405)
Analyze the evidence – how long it takes for the "X" to disappear – to find the order of the thiosulfate reaction.

▸ Practice

Understanding Concepts

2. Consider the overall reaction in which two elements combine to form a compound:

$$A_2 + 2B \rightarrow 2AB$$

When a series of reactions is performed with different initial concentrations of reactants, the results are as shown in **Table 1**:

Table 1 Formation of AB

Trial	Initial [A₂]	Initial [B]	Initial rate of production of AB
1	0.10	0.10	3×10^{-4}
2	0.20	0.10	6×10^{-4}
3	0.10	0.20	3×10^{-4}

State (a) the rate law equation for this system;

(b) the rate-determining step; and

(c) a possible mechanism, indicating the slowest step.

▶ *Section 6.4 Questions*

Understanding Concepts

1. What is the difference between an elementary step and a rate-determining step?

2. Consider the following mechanism, in which A, B, and E may be elements or compounds, and C, D, and F are compounds:

   (1) 2A + B → C
   (2) C → D
   (3) D + E → F

   $2A + B + E → F$

   (a) What is the overall equation?
   (b) Which step is most likely to be the rate-determining step? Explain.

3. (a) What is the (overall) activation energy for the following reaction in the potential energy diagram in **Figure 9**?

   reactants → products

   (b) What is the reaction enthalpy (ΔH) for the reaction?
   (c) What is the rate-determining step for the reaction?
   (d) Is the reaction exothermic or endothermic?
   (e) Which letters represent activated complexes?
   (f) Which letters represent reaction intermediates?

A Multistep Reaction

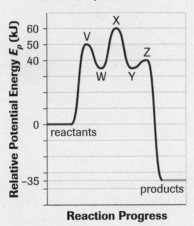

Figure 9
Potential energy diagram

Making Connections

4. (a) Consider the chemical reaction that happens in an automobile engine when hydrocarbon molecules (mostly octane) combine with oxygen molecules to produce carbon dioxide and water vapour. Imagine that you have an extraordinary microscope that enables you to look at individual molecules. Describe what you see as the chemical reaction occurs. Make sure that you describe the positions and movements of the molecules as specifically as possible.

   (b) Compare your answer with what you gave for Reflect on your Learning, question 2, at the beginning of this chapter. How has your understanding changed?

5. Ozone is a molecule that is helpful in the upper atmosphere but harmful at ground level. Research on the Internet to answer the following questions:
   (a) Why is ground-level ozone a problem?
   (b) What are the mechanisms of the reactions that lead to its production?
   (c) What is the connection between the production of ozone and kinetics?

    GO   www.science.nelson.com

6. John Polanyi, a scientist at the University of Toronto, won the 1986 Nobel Prize for his work in chemical kinetics. Imagine that you are a newspaper reporter describing Polanyi's work in an article written the day after he received the prize. Include in your article a description of his area of research and the experiments that he performed.

   GO   www.science.nelson.com

When pyrotechnists design fireworks they mix the chosen chemicals together very carefully, and assemble the various parts so that explosive chemical reactions will happen in the right sequence and at the expected rates. How can we explain the rates at which reactions happen? In Sections 6.2 and 6.4, you learned how factors such as temperature and initial concentration affect rate of reaction. Can we apply collision theory to explain the effects of these factors?

As you know, the rate at which a reaction occurs depends on two criteria:

• the frequency of collisions, and
• the fraction of those collisions that are effective.

In this section we will explore how the frequency of collisions and the fraction of effective collisions depend on factors such as temperature and chemical nature of reactant.

Theoretical Effect of Chemical Nature of Reactant

When nickel metal is added to hydrochloric acid, the reaction is slow. However, magnesium reacts quickly in the same acid (**Figure 1**). To explain these empirical observations, we need to consider factors such as the atomic structure of the reactants and the nature of their bonds, as well as the type of reaction occurring.

Think about the molecules in the air about you. There are billions and billions of them moving rapidly and colliding in every cubic centimetre. However, the molecules are not all moving at the same speed: Some are moving quickly and some slowly, but most are in some mid-range of kinetic energy. This distribution of kinetic energies is called a Maxwell-Boltzmann distribution, and has been experimentally found to fit the pattern in **Figure 2**. Note the axes: the vertical axis represents the number of molecules with a particular kinetic energy, and the horizontal axis represents the different energies. An analogy might be a distribution of heights of students on a Grade 8 field trip to an amusement park (**Figure 3**): Most students are in the mid-range, with fewer students at high and low heights.

In a chemical reaction, an effective collision requires a minimum energy—energy of collision—which is converted into potential energy (the activation energy) as the activated complex is formed. This minimum energy is called the **threshold energy**.

(a)

(b)

Figure 1
The different rates of reaction of
(a) magnesium and **(b)** nickel with hydrochloric acid

threshold energy the minimum kinetic energy required to convert kinetic energy to activation energy during the formation of the activated complex

Figure 2
At a given temperature, the molecules with enough energy to create a successful collision are represented by the area enclosed under the graph line and to the right of the (dashed) minimum energy level. Note the very large increase in the number of these molecules when the threshold energy is decreased.

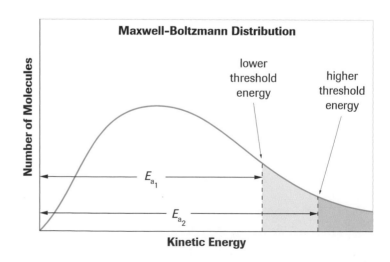

The chemical nature of reactants affects the threshold energy (and the fraction of collisions that are effective) in two possible ways.

(1) Some molecules have bonds that are relatively weak and small activation energy barriers, so the threshold energy is relatively low and a large fraction of molecules is capable of colliding effectively. In the student field trip analogy, this is equivalent to the students choosing a different ride with a lower "threshold height" so that a larger fraction of students can go on the ride. Other molecules have strong bonds and high activation energy barriers, so most collisions are ineffective.

(2) A second factor is what is sometimes called collision geometry — some reactions involve complicated molecular substances or complex ions that are often less reactive because more bonds have to be broken and the molecules have to collide in the correct orientation relative to each other for a reaction to occur.

In conclusion, reactions that occur quickly, such as the reaction between magnesium and acid, have lower activation energies than those that are slow to occur, such as nickel reacting in acid.

Theoretical Effect of Concentration and Surface Area

When zinc is added to concentrated sulfuric acid, the reaction occurs much more quickly than in dilute acid. Similarly, a flammable liquid burns much more quickly when the surface area of the fuel that is exposed to air is increased. Concentration and surface area both affect the collision frequency.

Concentration

If the initial concentration of a reactant is increased, the reaction rate generally increases. A higher concentration of a reactant means a greater number of particles per unit volume, which are more likely to collide as they move randomly within a fixed space. If twice as many particles are present, there should be twice the probability of an effective collision. Therefore, for elementary reactions, the rate of reaction is generally directly proportional to the concentration of a reactant.

Surface Area

This factor applies only to heterogeneous reactions, such as those where a gas reacts with a solid or a solid with a liquid. Surface area affects collision frequency because reactants can collide only at the surface where the substances are in contact. The number

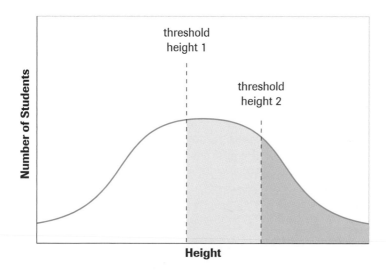

Figure 3
In a student height distribution graph, there is a range of heights with most heights concentrated in the mid-range. The "threshold height" is the minimum allowable height, determined by the amusement park, for a student to be allowed on a ride. If the "threshold height" is lowered, a larger fraction of students will be able to go on the ride.

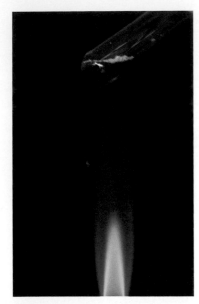

Figure 4
Heating green copper(II) carbonate produces black copper(II) oxide. Heating causes the decomposition of many carbonates into solid oxides and carbon dioxide gas.

of particles per square millimetre of surface of a solid is fixed. However, the area of surface exposed for a given quantity depends on how finely divided the sample of solid is. We make use of this in cooking by finely grinding pepper to add flavour and by using icing sugar (finely powdered) and ground coffee to increase the rate at which they dissolve. Dividing a solid into finer and finer pieces has a limit — when you reach the elementary particles of which the solid is composed. Sugar cannot be divided more finely than into its individual molecules. Dissolving divides a solid or liquid solute into the theoretical maximum number of separate particles, creating the maximum possible surface area. This is why so many reactions occur more quickly in solution, including nearly all of the reactions of human physiology.

Theoretical Effect of Temperature

At room temperature, copper(II) carbonate is stable but, when heated (**Figure 4**), it rapidly decomposes to copper(II) oxide. The reactants have not changed; only the temperature has changed. An increase in temperature has a dramatic effect on rate of reaction, because temperature affects both the collision frequency and the fraction of collisions that are effective.

Theoretically, temperature is believed to be a measure of the average kinetic energy of the particles in a sample. Experimental evidence shows that a relatively small increase in temperature seems to have a very large effect on reaction rate. An increase of about 10°C will often double or triple the rate of a reaction. When you consider that a rise from 27°C to 37°C represents an absolute increase from 300 K to 310 K, you can see that a 3% increase in temperature seems to cause a 100% increase in reaction rate.

The explanation for the temperature effect is that increasing temperature causes molecules to collide both more often and with more force on average, making any individual collision more likely to be effective. The concept of activation energy is a significant part of the explanation: For a given activation energy, E_a, a much larger fraction of molecules has the required kinetic energy at a higher temperature than at a lower temperature. A temperature rise that is a small increase in overall energy might cause a very large increase in the number of particles that have energy exceeding the activation energy (**Figure 5**). In the student field trip analogy used previously, an increase in temperature corresponds to a new group of students. This new group has the same number of students, but they are two years older and have a greater height on average. The height distribution is shifted to taller heights, so a larger fraction of the students are able to go on the ride.

Figure 5
Experiment shows that when the temperature increases from T_1 to T_2, the shape of the Maxwell-Boltzmann distribution curve flattens and shifts to the right. Note the very large increase in the fraction of molecules able to react at the higher system temperature.

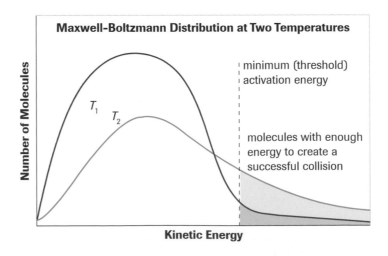

Maxwell-Boltzmann Distribution at Two Temperatures

minimum (threshold) activation energy

molecules with enough energy to create a successful collision

T_1 T_2

Number of Molecules

Kinetic Energy

Theoretical Effect of Catalysis

When margarine is manufactured in the food industry, vegetable oils that are too runny to spread on bread must be made solid. The process of hydrogenation adds hydrogen to some of the double bonds in unsaturated fats. The product molecules pack more closely and have stronger intermolecular interactions and a higher melting point. The reaction is slow under normal conditions but is catalyzed by use of nickel metal.

Theoretically, catalysts accelerate a reaction by providing an alternative lower energy pathway from reactants to products. That is, a catalyst allows the reaction to occur by a different mechanism, inserting different intermediate steps, but resulting in the same products overall. If the new pathway (mechanism) has a

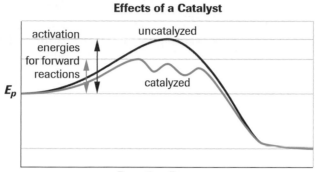

Effects of a Catalyst

activation energies for forward reactions

uncatalyzed

catalyzed

E_p

Reaction Progress

lower activation energy, a greater fraction of molecules possesses the minimum required energy and the reaction rate increases (**Figure 6**).

The actual mechanism by which catalysis occurs is not well understood for most reactions, and discovering acceptable catalysts has traditionally been a hit-and-miss process. A few catalyzed reactions, studied in detail, suggest some general mechanism changes are involved. For example, the action of platinum as a catalyst in the reaction of hydrogen and oxygen gases to produce water has a practical application in fuel cell technology, as shown in **Figure 7**. The reaction of these gases is slow under normal conditions, but the platinum catalyzes the process so that the hydrogen and oxygen can combine at a significant rate. Because the platinum is a solid and the reactants are gases, platinum is called a **heterogeneous catalyst** for this reaction. The nickel catalyst described above is also a heterogeneous catalyst.

The reaction between aqueous tartrate ions and hydrogen peroxide is an example of the action of a **homogeneous catalyst**: cobalt(II) ions. At room temperature this reaction is very slow, with no noticeable activity. When a solution containing cobalt ions is added, the pink solution reacts to form a green intermediate (aqueous cobalt(III) ion), which then further reacts to regenerate the cobalt(II) catalyst (**Figure 8**). Even though the cobalt(II) ions react in the first step of the mechanism, they are a catalyst because they are regenerated in the final step. A catalyst speeds up a reaction without being consumed.

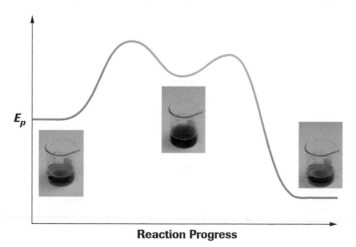

E_p

Reaction Progress

ACTIVITY 6.5.1

Catalysts in Industry and Biochemical Systems (p. 405)
Identify similarities and differences among industrial catalysts and enzymes.

Figure 6
The reaction shown here proceeds by a three-step mechanism when a catalyst is present, but nonetheless proceeds much faster than by the one-step uncatalyzed mechanism. The catalyzed mechanism has a lower activation energy, so more collisions are successful.

Figure 7
The Ballard fuel cell consists of two electrodes, each coated on one side with a thin layer of platinum catalyst. Hydrogen fuel dissociates into free electrons and positive hydrogen ions in the presence of the platinum catalyst, and the protons migrate through a membrane electrolyte to a second electrode where they combine with oxygen from air.

heterogeneous catalyst a catalyst in a reaction in which the reactants and the catalyst are in different physical states

homogeneous catalyst a catalyst in a reaction in which the reactants and the catalyst are in the same physical state

Figure 8
The reaction between colourless tartrate ions and colourless hydrogen peroxide is catalyzed by Co^{2+} ions:
$$C_4H_4O_6^{2-}{}_{(aq)} + 5\,H_2O_{2(aq)} \rightarrow$$
$$6\,H_2O_{(l)} + 4\,CO_{2(g)} + 2\,OH^-_{(aq)}$$
The reactants and pink Co^{2+} react to form an intermediate and green Co^{3+}, which further reacts to form colourless products and pink Co^{2+}. The catalyst reacts but is regenerated at the end.

Chemical Kinetics   **395**

SUMMARY **Explaining Reaction Rates**

- Particles require a minimum activation energy and correct alignment for a collision to be effective. The collision must provide sufficient energy to cause the breaking and forming of bonds, producing new particles.

- Many reactions occur as a sequence of elementary steps that make up the overall reaction mechanism.

- The rate of any reaction depends on the nature of the chemical substances reacting, because both the strength of bond(s) to be broken and the location of the bond(s) in the particle structure affect the likelihood that any given collision is effective.

- An increase in initial reactant concentration or in reactant surface area increases the rate of a reaction because the total number of collisions possible per unit time is increased proportionately.

- A rise in temperature increases the rate of a reaction for two reasons: the total number of collisions possible per unit time is increased slightly; and, more importantly, the fraction of collisions that are sufficiently energetic to be effective is increased dramatically.

- A catalyst increases the rate of a reaction by providing an alternative pathway, with lower activation energy, to the same product formation. A much larger fraction of collisions is effective following the changed reaction mechanism. Catalysts are involved in the reaction mechanism at some point, but are regenerated before the reaction is complete.

▶ **Practice**

Understanding Concepts

1. Which of the five factors that affect rate of reaction do so by
   (a) increasing the collision frequency?
   (b) increasing the fraction of collisions that are effective?

2. The reaction of hydrogen and oxygen is exothermic and self-sustaining.
   (a) Write the equation for this reaction, and provide a reason why it is not likely that the reaction occurs as a single step.
   (b) This reaction is catalyzed by platinum metal, which provides a surface on which hydrogen gas splits to form Pt–H units that react readily with oxygen molecules. Suggest a possible mechanism for this process, given that a catalyst must be regenerated in any change.

3. Identify each of the following as examples of the action of homogeneous or heterogeneous catalysts:
   (a) Rhodium and platinum metals are used in an automobile catalytic converter to convert exhaust gases into safer gases.
   (b) Gaseous chlorofluorocarbons (CFCs) have been shown to catalyze the breakdown of ozone in the upper atmosphere.
   (c) Aqueous sulfuric acid catalyzes the decomposition of aqueous formic acid to carbon monoxide and water.
   (d) Powdered $TiCl_4$ is used in the formation of polyethylene polymer from gaseous ethylene.

4. Use collision theory to explain each of the following observations.
   (a) Permanganate ion (MnO_4^-) reacts much more quickly with iron(II) ions (Fe^{2+}) than with oxalate ions ($C_2O_4^{2-}$).
   (b) When heated in a flame, steel wool burns but a steel nail just glows.
   (c) Liquid nitroglycerin is a dangerous explosive, but people with heart conditions take nitroglycerin tablets.

Making Connections

5. The reaction of hydrogen with chlorine at room temperature is so slow as to be unde-tectable if the container is completely dark, but is explosively fast if sunlight is allowed to fall on the reactants. The following reaction mechanism has been suggested for this reaction:

$$Cl_{2(g)} + \text{light energy} \rightarrow Cl_{(g)} + Cl_{(g)}$$
$$Cl_{(g)} + H_{2(g)} \rightarrow HCl_{(g)} + H_{(g)}$$
$$H_{(g)} + Cl_{2(g)} \rightarrow HCl_{(g)} + Cl_{(g)}$$
$$Cl_{(g)} + Cl_{(g)} \rightarrow Cl_{2(g)}$$

   (a) Write the overall reaction equation.
   (b) Identify the reaction intermediates.
   (c) Compare the activation energy for the collision of molecular chlorine with molec-ular hydrogen to the activation energy for the collision of atomic chlorine with molecular hydrogen. Which reaction must have the greater activation energy, and what evidence can be used to support your argument?

6. "Platinum should be described as a precious metal, not because of its use in jewellery but because of its use as a catalyst." Do you agree or disagree with the statement? Back up your opinion with references to specific applications of platinum.

 www.science.nelson.com

7. List at least four different general methods of food preservation. Write a few sentences about each, explaining the chemical theory behind its effectiveness.

▶ EXPLORE an issue

Debate: Food Preservation

For thousands of years people have tried to preserve food so that it will still be available to them when the source of fresh food has dwindled. Inuit hunters stashed butchered carcasses close to the permafrost, under piles of rocks, to keep the meat frozen until times of shortage. Newfoundland fishers dried their cod on fish flakes to make it last through the winter, while the indigenous Mi'qmaq gathered and dried wild berries. The Haida Gwai, on Canada's west coast, smoked their abundant seasonal catch of salmon, and Aboriginal peoples on the prairies dried buffalo meat into high-protein pemmican. All of these traditional ways of preserving food used readily available materials and cir-cumstances: brisk, dry winds; smoke from wood fires; or the long, cold northern winters.

In the last hundred years, however, we have modified these preserving methods and developed many new ones: We still freeze food to keep it fresh, but now we have indoor freezers. Some of our foods are dried, but generally with the help of com-mercial dehydrators. And salting and smoking are more often used for flavouring than for preserving these days — consider bacon and smoked salmon. More commonly, our foods are cooked at very high temperatures and vacuum-packed in cans or plastic packages, or prepared with chemical preservatives, or irradiated with gamma rays, or genetically modified to be more resistant to rotting....

Decision-Making Skills

● Define the Issue ● Analyze the Issue ● Research
● Defend the Position ● Identify Alternatives ● Evaluate

Which of these many ways of preserving food is best for our health? For our planet? For our wallets? For our community?

• In small groups, select at least five different ways of pre-serving food. Choose some traditional and some more recent techniques. Discuss what you know about each of these methods, looking at them from several different perspectives.

• As a class, decide on a proposition to debate, on the subject of food preservation. Take a preliminary vote: How many people in the class support the proposition?

• Your class will then be divided into one team supporting the proposition, and another opposing it.
   (a) Carry out research and assemble your research into evi-dence to back up your arguments.

• Debate your proposition, following the rules of debating, and conclude with a vote.
   (b) Did the voting results change from the preliminary vote to the final one? What were the most convincing arguments? Which arguments seemed least effective?

 www.science.nelson.com

Biochemist

Biochemists study the chemicals consumed and produced by living organisms during growth, development, and metabolism. Often the reactions are isolated from the organisms, and carried out under controlled conditions. Organic compounds, including enzymes, fats, carbohydrates, and DNA, are often the focus of the research. The rates of the reactions, and what affects these rates, may be a crucial focus of study. Although usually based in a lab, biochemists are employed in many fields: medicine, agriculture, nutrition, materials development (biotechnology), forensics, water purification, and even waste management.

Pharmacologist

Pharmacologists are specialized chemists, pursuing laboratory research into the actions of drugs and chemicals on the human (or animal) body. Their work leads to the development and testing of new drugs to fight both serious and minor illnesses. The stages of drug testing include first in vitro (in a test tube), then in vivo (in live animals, finally including humans). Pharmacologists may work for any of a variety of employers: medical schools, universities, the pharmaceutical industry, private research laboratories, and government testing laboratories.

▸ Practice

Making Connections

8. Choose a career that interests you, either one of those featured or another that involves rates of reactions, and report on the following:
   (a) training required;
   (b) working conditions; and
   (c) current employment opportunities.

 www.science.nelson.com

Industrial Chemist

An industrial chemist studies the makeup and behaviour of chemicals, the way they react with each other and how they can be used in industry. It is truly an applied science involving scientific research and technological development to create an end product such as cosmetics, pharmaceuticals, food additives or preservatives, or new materials such as plastics or computer chips.

Industrial chemists may work with many other people, including nutritionists, medical researchers, or engineers, to make sure that the new product is safe.

Food Scientist

Food scientists provide their technological know-how to the food industry, using their training and experience to convert raw foods into quality products quickly, efficiently, and with a minimum of waste. Research and development of new processed or preserved foods begins on a small scale, and then must be scaled up to mass production, which may require new machinery or processes. The food scientist must carefully consider ingredients, preparation, temperature, humidity, and packaging. Food scientists may also test and monitor commercially produced foods to determine nutrient levels and ensure safety and freedom from contamination.

The Arrhenius Equation

The rate law equation

$r = k[A]^n[B]^m$

clearly describes quantitative rate dependence with respect to concentrations of reactants — specifically, reactants that are involved in the rate-determining step of the reaction mechanism. But how do we explain the large quantitative effect of temperature and catalysis? The rate constant, k, incorporates the quantitative effects of temperature, nature of reactant, and catalysts as described by the Arrhenius equation for k.

> **The Arrhenius Equation**
>
> $k = Ae^{-E_a/RT}$
>
> where E_a is the activation energy (in J)
> A is a constant related to the geometry of the molecules
> R is the gas constant (8.31 J/(mol·K))
> T is the temperature (in K)

Note that changes in both the activation energy and temperature have exponential effects on the value of k and therefore, the rate of reaction. Mathematical calculations show how a relatively small change in either the temperature or the activation energy has a very large effect on the numerical value of k and, hence, the rate of reaction.

For example, consider a typical reaction in which the activation energy is 150.0 kJ/mol and the temperature is 27.0°C, or 300.0 K. The effects on rate can be measured by calculating the exponential factor $(-\frac{E_a}{RT})$ for each set of conditions.

For the given set of starting conditions,

$$\frac{-E_a}{RT} = \frac{-150.0 \text{ kJ/mol}}{8.31 \text{ J/mol·K} \times 300 \text{ K}}$$

$$= -60.2$$

$$e^{-60.2} = 7.2 \times 10^{-27}$$

If the temperature is increased from 27°C to 37°C, or 310 K, the exponential factor becomes

$$\frac{-E_a}{RT} = \frac{-150.0 \text{ kJ/mol}}{8.31 \text{ J/mol·K} \times 310 \text{ K}}$$

$$= -58.2$$

$$e^{-58.2} = 5.3 \times 10^{-26}$$

An increase in temperature of 10°C has multiplied the exponential factor and therefore, the rate, more than seven times ($5.3 \times 10^{-26}/7.2 \times 10^{-27}$).

If a catalyst is used that reduces the activation energy from 150 kJ to 130 kJ at 300 K, the exponential factor becomes

$$\frac{-E_a}{RT} = \frac{-130.0 \text{ kJ/mol}}{8.31 \text{ J/mol·K} \times 300 \text{ K}}$$

$$= -52.1$$

$$e^{-52.1} = 2.4 \times 10^{-23}$$

The use of a catalyst has multiplied the rate more than 3000 times ($2.4 \times 10^{-23}/7.2 \times 10^{-27}$).

Thus, changes in temperature, changes in reactants, and the use of catalysts that affect the activation energy all have dramatic effects on rate.

Understanding Concepts

Answers

9. (a) approx. × 4

 (b) approx. × 3600

9. Use the Arrhenius equation (with the realistic values $A = 10^{27}$ L/mol·s, and $E_a = 200$ kJ/mol) to calculate
 (a) the change in the rate constant if the temperature of a system is raised from 20°C to 25°C, and
 (b) the change produced by the use of a catalyst that lowers the activation energy by 10%.

> **Section 6.5** Questions

Understanding Concepts

1. Draw a potential energy diagram for an endothermic elementary reaction. On the same diagram draw a reasonable curve to represent the same reaction catalyzed. Summarize the effects of a catalyst by labelling
 (a) the ΔH for the overall reaction, catalyzed or uncatalyzed;
 (b) the E_a for the reaction uncatalyzed;
 (c) the E_a for the reaction catalyzed.

2. Draw a kinetic energy distribution diagram with labelled curves for lower (T_1) and higher (T_2) temperatures. On the diagram draw lines to represent threshold energies for catalyzed (E_{cat}) and uncatalyzed (E_{uncat}) reactions. Summarize the effects of temperature and a catalyst by shading and labelling areas to represent the fraction of molecules able to react
 (a) at a lower temperature uncatalyzed;
 (b) at a higher temperature uncatalyzed;
 (c) at a lower temperature catalyzed;
 (d) at a higher temperature catalyzed.

Making Connections

3. Both nitric oxide (NO) and chlorine (Cl) atoms generated by the decomposition of CFCs catalyze the decomposition of ozone in the stratosphere.
 (a) Why is the decomposition of ozone in the stratosphere a problem?

 (b) When NO reacts with ozone in the rate-determining step, the activation energy E_a is about 12 kJ/mol but the activation energy for the reaction of Cl with ozone in a separate mechanism is about 2 kJ/mol. Which of nitric oxide and atomic chlorine is the more effective catalyst? Explain.

4. Create a table with two columns, one headed Fast Reactions and the other Slow Reactions. In each column list at least 10 everyday reactions that we want to control. Give reasons for your categorization of each example (e.g., you might write "iron rusting" in the Slow Reactions column, because we generally want iron to maintain its strength as a metal). In your table, include methods that can be used to control the rate of each of these reactions.

5. Enzymes have application in both body chemistry and industrial reactions. Write a brief report describing the function of an enzyme (not yet mentioned in this chapter) in the human body or in industry.

 www.science.nelson.com

6. (a) Suggest why some chemical reactions occur slowly while others occur quickly.
 (b) How does your answer differ from the one that you gave for Reflect on your Learning, question 3, at the beginning of this chapter? Explain how your new answer shows a change in your understanding.

LAB EXERCISE 6.1.1

Determining a Rate of Reaction

Inquiry Skills

○ Questioning	○ Planning	● Analyzing
○ Hypothesizing	○ Conducting	● Evaluating
● Predicting	○ Recording	● Communicating

The reaction between calcium carbonate (found naturally as chalk or limestone) and acid produces carbon dioxide gas.

$$CaCO_{3(s)} + 2\,HCl_{(aq)} \rightarrow CO_{2(g)} + CaCl_{2(aq)} + H_2O_{(l)}$$

This reaction is used by geologists to confirm the presence of limestone in a mineral sample (**Figure 1**). As you know, the rate of any reaction is determined by measuring the change in quantity or concentration of some reactant or product over a series of time intervals. In this experiment, the change in concentration of hydrochloric acid (in mol/L) and time (in min) are measured. The rates of reaction are, therefore, expressed as

$$-\frac{\Delta[HCl]}{\Delta t} = x\,mol/(L\cdot min)$$

Figure 1
Limestone is a mineral that can be detected by adding concentrated hydrochloric acid to a sample.

Question

What is the rate of reaction of calcium carbonate with acid over various time intervals, and at specific times in the reaction?

Prediction

(a) Sketch a graph of concentration vs. time to show how you would expect the rate of reaction to change as the reaction proceeds.

Experimental Design

The concentration of hydrochloric acid is determined (by calculation from measured solution pH values) at a series of times as the acid reacts with calcium carbonate solid. A graph of concentration vs. time is plotted and analyzed to determine rates of reaction at various times.

Materials

calcium carbonate chips
1.90 mol/L hydrochloric acid
pH meter
beaker

Evidence

Table 1 Concentration of Hydrochloric Acid Remaining

[HCl] (mol/L)	Time (min)
1.90	0.0
1.40	1.0
1.10	2.0
0.90	3.0
0.80	4.0
0.75	5.0
0.72	6.0

Analysis

(b) Plot a graph of [HCl] vs. time.

(c) Determine the average rate of consumption of HCl over the time interval from

    (i) 0 to 2 min   (ii) 3 to 5 min

(d) Determine the instantaneous rate of consumption of HCl at

    (i) 1 min   (ii) 4 min

(e) Communicate the change in reaction rate in words.

Evaluation

(f) Can you detect any flaws in this Experimental Design?

(g) Compare your Prediction to your answer in (e). Account for any differences.

Synthesis

(h) State your experimental rate of consumption of hydrochloric acid at 1 min. Use the appropriate notation to express the rate of consumption of $CaCO_3$ and the rate of production of CO_2 at the same time.

(i) How could conductivity and gas measurements be used to measure the rate of reaction?

INVESTIGATION 6.2.1

Chemical Kinetics and Factors Affecting Rate

Inquiry Skills

- ● Questioning
- ○ Hypothesizing
- ● Predicting
- ● Planning
- ● Conducting
- ● Recording
- ● Analyzing
- ● Evaluating
- ● Communicating

You have learned about five factors that affect rate of reaction. In this investigation you will test as many of these factors as possible with a given chemical system. It is difficult to test all of these factors with just one chemical system, but many systems show clear changes in rates when nature of reactant, temperature, initial concentration, or surface area is varied. Catalysts (such as aqueous copper(II) sulfate and solid manganese dioxide) affect some chemical reactions, but not others. Some possible choices of reactions to test the five factors include the following:

- metals reacting with acids

  e.g., $X_{(s)} + H_2SO_{4(aq)} \rightarrow XSO_{4(aq)} + H_{2(g)}$

  (possible catalyst: aqueous $CuSO_4$)

- carbonate or bicarbonate salts reacting with acids

  e.g., $XCO_{3(s)} + 2HCl_{(aq)} \rightarrow XCl_{2(aq)} + H_2O_{(g)} + CO_{2(g)}$

- the decomposition of hydrogen peroxide

  $H_2O_{2(aq)} \rightarrow H_2O_{(l)} + O_{2(g)}$

Purpose

The purpose of this investigation is to design an experiment to test the factors that affect rate of reaction.

Question

(a) Write an appropriate question that you will attempt to answer in this experiment.

Prediction

(b) After designing your experiment and having it approved, but before carrying it out, predict an answer to your Question.

Experimental Design

(c) Choose a chemical system and design an experiment to investigate as many factors as practical that might affect the rate of reaction of the system. Clearly describe the set of conditions that will be used as the experimental control. Decide on the units you will use to calculate the rates of reaction (for example, amount of reactant/min or mol/L of product).

Materials

lab apron   eye protection

(d) List any necessary chemicals and equipment.

Procedure

(e) Write a step-by-step procedure. Include any safety precautions that are needed, based on what you learn about the reactants and products from safety information sheets.

1. Have your procedure approved by your teacher before carrying it out.

Evidence

(f) Design a table for recording observations, making sure that you can obtain and record both qualitative and quantitative observations.

Analysis

(g) Analyze your quantitative evidence both mathematically and graphically, and determine a rate of reaction for each of the experimental trials.

(h) Answer your Question.

Evaluation

(i) Evaluate your Experimental Design and Procedure. Suggest how they might be improved, if necessary.

(j) Evaluate your Evidence and the Prediction that you made before starting the investigation.

⚗ INVESTIGATION 6.3.1

The Iodine Clock Reaction

Inquiry Skills

○ Questioning	○ Planning	● Analyzing
○ Hypothesizing	● Conducting	● Evaluating
● Predicting	● Recording	● Communicating

The rate of a chemical reaction depends on many physical and chemical factors, including temperature, the chemical nature of the reacting species, and the initial concentration of the reactants. In the following experiment, all factors except one are held constant.

The iodine clock reaction is the classic experiment used to investigate rates of reaction. It is called a clock reaction because at first, when two colourless solutions are mixed, no reaction appears to occur. Then, at a specific time, the mixture suddenly changes colour. The reaction under study combines aqueous iodate ($IO_{3(aq)}^-$) and bisulfite ($HSO_{3(aq)}^-$) ions and involves three reactions in sequence.

In the first reaction, aqueous iodate ions are reduced to iodide ions.

$$IO_{3(aq)}^- + 3\,HSO_{3(aq)}^- \rightarrow 3\,SO_{4(aq)}^{2-} + I_{(aq)}^- + 3\,H_{(aq)}^+$$

In the second reaction, the iodide ion is changed to molecular iodine.

$$6\,H_{(aq)}^+ + 1\,IO_{3(aq)}^- + 5\,I_{(aq)}^- \rightarrow 3\,I_{2(aq)} + 3\,H_2O_{(l)}$$

In the final step, the iodine reacts with starch suspended in solution to form a blue-black complex.

$$I_{2(aq)} + starch \rightarrow blue\text{-}black\ complex$$

The third reaction is extremely rapid in comparison to the first two, and serves to indicate the time of reaction for those two reactions. The shorter the time required, the greater is the rate of reaction. Because the times of reaction are related to the rates of reaction, we will be able to make rate–concentration comparisons.

Graphical analysis is an efficient way to determine order of reaction for iodate, a, where rate $\alpha\ [IO_{3(aq)}^-]^a$. **Figure 3** in section 6.3 shows graphs that are characteristic of zeroth order ($n = 0$), first order ($n = 1$), and second order ($n = 2$) reactions.

Purpose

The purpose of this investigation is to gather and analyze experimental observations to determine the rate dependence of a reactant in a chemical system.

Question

What is the order of reaction with respect to the initial concentration of iodate ions in the iodine clock reaction?

Prediction

(a) Predict qualitatively what will happen to the time of reaction as the initial concentration of iodate ions is increased. What will happen to the rate of reaction?

Experimental Design

A series of solutions will be prepared in which the only variable is the initial concentration of iodate ions. Equal amounts of starch, sodium bisulfite, and hydrochloric acid will be mixed with each of these solutions, so that the time from mixing to formation of a blue-black product can be timed. The evidence will be analyzed graphically.

(b) Calculate the initial concentration of the iodate solution in each of the wells in microtray A, based on dilution from the concentration of the stock iodate solution.

Materials

lab apron
eye protection
0.020 mol/L potassium iodate solution (Solution A)
0.00100 mol/L sodium bisulfite/hydrochloric acid/starch
   solution (Solution B)
distilled water
3 plastic micropipets, labelled A, B, and H_2O
three 100-mL beakers, labelled A, B, and H_2O
2 large-well microtrays or spot plates, labelled A and B
stopwatch

Procedure

1. Place the two microtrays on clean sheets of white paper. For microtray A, using the appropriate micropipets, place 1 drop of Solution A in Well 1, 2 drops of Solution A in Well 2, 3 drops in Well 3, and so on up to 10 drops of Solution A in Well 10.

2. Also in microtray A, place 9 drops of water in Well 1, 8 drops of water in Well 2, 7 drops in Well 3, and so on down to 1 drop of water in Well 9. There are now 10 drops of solution in each well (**Figure 2**).

3. For microtray B, using the appropriate micropipet, put 10 drops of Solution B in each of the first 10 wells.

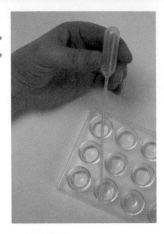

Figure 2

4. With stopwatch ready, and using the water micropipet, transfer the contents of Plate A Well 10 to Plate B Well 10. In doing so, insert the tip of the micropipet below the surface of the liquid in Plate B to ensure that the solutions mix thoroughly and keep stirring (**Figure 3**). Start timing at the moment the micropipet is squeezed and stop when the colour first appears. Record your observations.

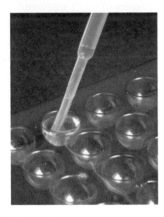

Figure 3

5. Rinse the water micropipet you just used at least twice with water, making sure that no water remains in the micropipet each time.

6. Repeat Steps 4 and 5 for each of the other pairs of wells, recording your observations.

7. Dispose of solutions down the drain with lots of running water.

Analysis

(c) Calculate the initial concentration of iodate solution in each of the wells at the instant of mixing with an equal volume of Solution B.

(d) Make a general statement to summarize your qualitative observations.

(e) Make appropriate evidence tables and plot graphs to identify the order of the reaction with respect to concentration of iodate ions.

(f) Write an expression to answer the Question.

Evaluation

(g) What other variables, apart from initial bisulfite ion concentration, were controlled in this investigation?

(h) Evaluate the Experimental Design, your lab skills, and the Evidence. Suggest any ways in which this experiment could be improved.

(i) If you are confident in your Evidence, evaluate the Prediction that you made before starting the investigation.

Synthesis

(j) Why was it important to add specific volumes of water to the wells in microtray A?

LAB EXERCISE 6.4.1

The Sulfur Clock

Inquiry Skills

○ Questioning   ○ Planning   ● Analyzing
○ Hypothesizing   ○ Conducting   ● Evaluating
○ Predicting   ○ Recording   ○ Communicating

Like the iodine clock reaction in Investigation 6.3.1, the sulfur clock reaction occurs in stages, the final step of which causes a visible change. The overall reaction is between thiosulfate ions and acid to form elemental sulfur.

$$S_2O_3{}^{2-}{}_{(aq)} + 2\,H^+{}_{(aq)} \rightarrow S_{(s)} + H_2SO_{3(aq)}$$

As particles of solid sulfur form, the solution becomes first cloudy and finally opaque. If a visible mark such as an "X" is written on a piece of filter paper and placed under the reaction beaker containing clear starting materials, the rates of reaction under different conditions can be compared by measuring the amount of time required for the "X" to become invisible in the increasingly cloudy solution.

Question

(i) What is the order of reaction in the sulfur clock system with respect to thiosulfate ions?

LAB EXERCISE 6.4.1 *continued*

(ii) How many thiosulfate ions are involved in the rate-determining step?

Experimental Design

The rates of reaction under different conditions are tested by adding a fixed concentration of acid to a series of solutions of different concentrations of thiosulfate ion. The time is measured from the instant of mixing to the point at which an "X" marked under the reaction beaker becomes invisible.

Materials

laboratory apron
eye protection
six 100-mL beakers
2 volumetric pipettes

0.160 mol/L sodium thiosulfate
  pentahydrate solution
2.0 mol/L hydrochloric
  acid solution

Evidence

Table 2 Initial Concentration & Reaction Time Data for Sulfur Clock

Trial	Initial $[S_2O_3^{2-}{}_{(aq)}]$ (mmol/L)	Initial $[H^+]$ (mol/L)	Time (s)
1	0.10	0.050	83
2	0.20	0.050	44
3	0.30	0.050	32
4	0.40	0.050	23
5	0.50	0.050	18

Analysis

(a) Make a general statement to summarize the qualitative observations in the investigation.

(b) Make appropriate data tables and plot graphs to answer Question (i).

(c) Write a mathematical expression that shows how the rate of consumption of $S_2O_3^{2-}{}_{(aq)}$ ions depends on $S_2O_3^{2-}{}_{(aq)}$.

(d) Answer Question (ii).

Evaluation

(e) Why is it important to use the same "X" marked on the same piece of paper for all of the trials?

(f) What measurement could be expected to cause the greatest experimental uncertainty?

(g) Evaluate the Experimental Design. Suggest three ways in which it could be improved.

Synthesis

(h) Create a possible mechanism for this reaction, given your experimental results.

(i) Design an experiment to determine the order of this reaction with respect to hydrogen ions.

ACTIVITY 6.5.1

Catalysts in Industry and Biochemical Systems

Catalysts can generally be classified as either industrial or biological. In industry, catalysts can be the key to the rapid and economical production of a wide range of materials. Catalysts are used in the production of almost all industrial chemicals, including nitric acid, sulfuric acid, and ammonia. They are also used to make plastic polymers such as polyethylene and polypropylene.

Biological catalysts, or enzymes, control thousands of chemical reactions in living things, including metabolic processes such as digestion, growth, and building of cells, and all reactions involving transformation of energy.

In this activity you will gather and organize information on catalysts from print and electronic sources.

- Using the Internet, find at least 10 examples of enzymes and industrial catalysts.

(a) For each of the enzymes or catalysts, record
  - the reaction that is catalyzed;
  - how the catalyst was discovered;
  - where and how the catalyst or enzyme acts;
  - for an enzyme, physiological implications of its presence or deficiency, whether such a condition exists, and, if so, how it is currently treated;
  - for an industrial catalyst, economic implications of its use.

(b) Which of the industrial catalysts has the greatest effect on your own life? Explain.

(c) Summarize your research.

 www.science.nelson.com

Key Expectations

- Describe, with the aid of a graph, the rate of reaction as a function of the change of concentration of a reactant or product with respect to time. (6.1)
- Explain, using collision theory and potential energy diagrams, how factors such as temperature, surface area, nature of reactants, catalysts, and concentration control the rate of chemical reactions. (6.2, 6.5)
- Express the rate of reaction as a rate law equation. (6.3)
- Explain the concept of half-life for a reaction. (6.3)
- Determine a rate of reaction experimentally, and measure the effect on rate of temperature, initial concentration, and catalysis. (6.3)
- Analyze simple potential energy diagrams of chemical reactions. (6.4)
- Demonstrate understanding that most reactions occur as a series of elementary steps in a reaction mechanism. (6.4, 6.5)
- Describe the use of catalysts in industry and in biochemical systems on the basis of information gathered from print and electronic sources. (6.5)
- Use appropriate scientific vocabulary to communicate ideas related to the energetics of chemical reactions. (all sections)
- Describe examples of slow chemical reactions, rapid reactions, and reactions whose rates can be controlled. (all sections)

Key Terms

activated complex

activation energy

average rate of reaction

catalyst

chemical kinetics

collision theory

elementary step

enzyme

half-life

heterogeneous catalyst

homogeneous catalyst

instantaneous rate of reaction

order of reaction

overall order of reaction

rate constant

rate-determining step

rate law equation

rate of reaction

reaction intermediates

reaction mechanism

threshold energy

Key Symbols and Equations

- $r = \dfrac{\Delta c}{\Delta t}$ or $-\dfrac{\Delta[\text{reactant}]}{\Delta t}$ or $+\dfrac{\Delta[\text{product}]}{\Delta t}$

- $r = k\,[X]^m[Y]^n$

- $k\,t_{1/2} = 0.693$

Problems You Can Solve

- What is the overall or average rate of reaction, given change in concentration over time? (6.1)
- What is the rate of reaction with respect to a specific participating substance, given the rate with respect to another substance in the reaction? (6.1)
- What is the initial rate of a reaction, given a rate law equation and information about how the concentration is changed? (6.3)
- What is the rate equation for a reaction, given experimental observations of the initial concentrations of reactants, and initial rates of production of products? (6.3)
- What is the rate constant for a radioisotope undergoing decay, given its half-life and using the rate equation that relates half-life to the rate constant? (6.3)
- What is a possible mechanism for an overall reaction, given experimental rate data? (6.4)

▶ *MAKE* a summary

Create a concept map to summarize what you have learned in this chapter. Start with the phrase "Chemical Kinetics" in the centre and try to include as many as possible of the Key Terms and Key Symbols and Equations listed here.

Identify each of the following 10 statements as true, false, or incomplete. If the statement is false or incomplete, rewrite it as a true statement.

1. The molecular species that exists at a maximum of potential energy is called the activation energy.

2. The instantaneous rate of reaction is determined graphically from the slope of a tangent.

3. Elementary steps in reaction mechanisms generally involve collisions of three or four molecules.

4. The elementary steps in a mechanism must add up to the overall equation.

5. If one molecule is involved in the rate-determining step, the reaction is called first order.

6. The enthalpy change is smaller for a catalyzed chemical reaction.

7. An enzyme is a biological catalyst.

8. The threshold energy is a minimum kinetic energy required for the activated complex to be formed.

9. The rate-determining step in a mechanism is the fastest step.

10. A homogeneous catalyst is one in which the catalyst and the reactants are in different phase.

Identify the letter that corresponds to the best answer to each of the five following questions.

11. In the chemical reaction
$$CH_{4(g)} + 2 O_{2(g)} \rightarrow CO_{2(g)} + 2 H_2O_{(g)}$$
the rate of consumption of oxygen gas is observed to be 4 mol/(L·min). What is the rate of production of carbon dioxide?
(a) 1 mol/(L·min)   (d) 8 mol/(L·min)
(b) 2 mol/(L·min)   (e) 16 mol/(L·min)
(c) 4 mol/(L·min)

12. Which of the following factors that affect rate of reaction applies only to heterogeneous systems?
(a) chemical nature of reactants   (d) catalysis
(b) concentration   (e) surface area
(c) temperature

13. What is the overall order of reaction for the elementary system A + 2 B → products?
(a) 0   (d) 3
(b) 1   (e) 5
(c) 2

14. If the initial rate of reaction is observed to increase by a factor of nine when the concentration of a reactant is tripled, what is the order of reaction with respect to that reactant?
(a) 0
(b) 1
(c) 2
(d) 3
(e) cannot be determined from this information

15. In **Figure 1**, the activation energy for the uncatalyzed reaction is
(a) A   (d) D
(b) B   (e) A or C
(c) C

16. In **Figure 1**, the enthalpy change for the catalyzed reaction is
(a) A   (d) D
(b) B   (e) A or C
(c) C

Potential Energy Diagram of Catalyzed and Uncatalyzed Pathways

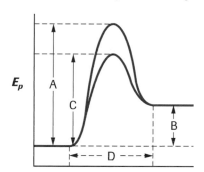

Figure 1

17. When zinc, iron, and lead are reacted with hydrochloric acid, the variable most easily measured in order to determine the effect of the chemical nature of the reactants on reaction rate is
(a) temperature   (d) volume of gas
(b) colour   (e) concentration
(c) surface area

18. What would the units of the rate constant, k, be in a second-order equation if rate was measured in mol/(L·s) and all concentrations in mol/L?
(a) s^{-1} (d) $L^3/(mol^3 \cdot s)$
(b) $L/(mol \cdot s)$ (e) $mol^2/(L^2 \cdot s)$
(c) $L^2/(mol^2 \cdot s)$

Chapter 6 REVIEW

Understanding Concepts

1. What quantitative measurement(s) would be appropriate in order to determine a rate of reaction in each of the following reactions:
   (a) a gas is produced;
   (b) molecular substances form soluble ions;
   (c) colourless ions react to form purple ions. (6.1)

2. List four factors that affect the reaction rate of a homogeneous reaction. (6.2)

3. A cube of solid reactant with sides of 1.00 cm^2 is submerged in a liquid and reacts to form a gas product at an initial rate of 20 mL/s. The solid-liquid interface is 6.00 cm^2 of surface area. If you sliced this cube (like a block of cheese) into 10 slices, and then replaced it in the liquid, predict the initial rate of reaction. (6.2)

4. Use the following chemical reaction equation to predict the effect of the listed changes on the rate of the reaction of zinc metal in hydrochloric acid.

   $$2 HCl_{(aq)} + Zn_{(s)} \rightarrow H_{2(g)} + ZnCl_{2(aq)} + \text{heat energy}$$

   (a) The concentration of hydrochloric acid is increased.
   (b) The reaction mixture is cooled.
   (c) Finely ground zinc is used instead of large chunks of zinc.
   (d) A solution of copper(II) sulfate is used as a catalyst. (6.2)

5. At 25°C a catalyzed solution of formic acid produces 44.2 mL of carbon monoxide gas in 30.0 s.
   (a) Calculate the rate of reaction with respect to $CO_{(g)}$ production.
   (b) What can you state about how long you would expect the production of the same volume to take
      (i) at 30°C?
      (ii) without the catalyst? (6.2)

6. Chlorine dioxide and hydroxide ions react to form chlorate ions, chlorite ions, and water.

   $$2 ClO_{2(aq)} + 2 OH^-_{(aq)} \rightarrow ClO_3^-_{(aq)} + ClO_2^-_{(aq)} + H_2O_{(l)}$$

   The reaction is found to be second order with respect to chlorine dioxide and first order with respect to hydroxide ions.
   (a) Write a rate equation for the reaction.
   (b) What is the overall order of reaction?
   (c) What would you expect the effect on rate to be of doubling the concentration of chlorine dioxide?
   (d) What would you expect the effect on rate to be of doubling the concentration of hydroxide ions? (6.3)

7. Nitric oxide, $NO_{(g)}$, reacts with chlorine gas, $Cl_{2(g)}$, in the reaction

   $$2 NO_{(g)} + Cl_{2(g)} \rightarrow 2 NOCl_{(g)}$$

   Initial rates of reaction are determined for various combinations of initial concentrations of reactants and recorded in **Table 1**.

Table 1 Observations on Rates of NOCl Production

Trial	Initial [NO] (mol/L)	Initial II [Cl$_2$] (mol/L)	Rate of production of NOCl (mol/(L·s))
1	0.10	0.10	1.8×10^{-2}
2	0.10	0.20	3.6×10^{-2}
3	0.20	0.20	1.43×10^{-1}

   (a) What is the rate law equation for the reaction?
   (b) What is the rate-determining step?
   (c) Calculate a value for the rate constant, including units.
   (d) Calculate the expected rate of reaction if the initial concentrations of NO and Cl$_2$ gases were 0.30 and 0.40 mol/L, respectively. (6.3)

8. (a) Explain what is meant by the term "half-life."
   (b) If the half-life of a radioisotope is 3.5 a, what percentage of the original isotope remains after 14.0 a? (6.3)

9. Sketch a potential energy diagram for the endothermic formation reaction of nitrogen and oxygen to produce nitrogen dioxide. Using appropriate symbols, label the activation energy and enthalpy change on the diagram. (6.4)

10. Draw a sketch, roughly to scale, of the potential energy diagram for a system in which $E_a = +80$ kJ and $\Delta H = -20$ kJ. Label the axes, reactants, products, the activation energy, the activated complex, and the enthalpy change. (6.4)

11. (a) How many particles are generally involved in an elementary step in a reaction mechanism?
    (b) Using a collision model, explain why it is unlikely that larger numbers of particles will be involved in an elementary step. (6.4)

12. (a) Explain how the following statement is not quite accurate: "A catalyst is a substance that speeds up a chemical reaction without itself reacting."
    (b) Explain the difference between homogeneous and heterogeneous catalysts, and provide an example of each. (6.5)

13. Consider the reaction of $S_2O_{8\,(aq)}^{2-}$ ions with $I_{(aq)}^-$ ions, for which the following mechanism has been suggested. Assume that the overall reaction is slightly exothermic.

$$Cu_{(aq)}^{2+} + I_{(aq)}^- \rightarrow Cu_{(aq)}^+ + I_{(aq)} \qquad \text{(fast)}$$

$$2\,I_{(aq)} \rightarrow I_{2(aq)} \qquad \text{(fast)}$$

$$Cu_{(aq)}^+ + S_2O_{8\,(aq)}^{2-} \rightarrow CuSO_{4\,(aq)}^+ + SO_{4\,(aq)}^{2-} \qquad \text{(slow)}$$

$$Cu_{(aq)}^+ + CuSO_{4\,(aq)}^+ \rightarrow 2\,Cu_{(aq)}^{2+} + SO_{4\,(aq)}^{2-} \qquad \text{(fast)}$$

(a) Identify the catalyst and the reaction intermediates in this reaction.
(b) Identify the reactants and products, and write the overall reaction equation.
(c) Explain what effect increasing $[I_{(aq)}^-]$ would have on the overall rate.
(d) Explain the effect of increasing $[S_2O_{8\,(aq)}^{2-}]$. (6.5)

Applying Inquiry Skills

14. An investigation is performed in which the concentration of nitrogen dioxide reacting is measured as a function of time.

Evidence

Table 2 Changing $[NO_2]$ over Time

$[NO_2]$ (mol/L)	Time (s)
0.500	0
0.445	12
0.380	30
0.340	45
0.250	90
0.175	180

Analysis

(a) Plot a graph of $[NO_2]$ vs. time.
(b) Determine the average rate of reaction of NO_2 between 10 and 60 s.
(c) Determine the instantaneous rate when
   (i) $[NO_2] = 0.46$ mol/L
   (ii) $[NO_2] = 0.23$ mol/L (6.1)
(d) As the $[NO_2]$ halved, in the previous question, what was the effect on the instantaneous rate of reaction? (6.2)
(e) What does the result in the previous section suggest about the order of reaction with respect to $[NO_2]$? (6.3)

15. An investigation is carried out in which Evidence is collected for the following hypothetical chemical reaction:

$$W + 2X + 2Y \rightarrow Z$$

(W, X, and Y could be either elements or compounds, but Z is a compound.)

Evidence

Table 3 Observations on the Rate of Production of Z

Test	Initial [W] (mol/L)	Initial [X] (mol/L)	Initial [Y] (mol/L)	Rate of production of Z (mmol/(L·s))
1	0.10	0.10	0.10	12
2	0.20	0.10	0.10	12
3	0.10	0.20	0.10	24
4	0.10	0.10	0.20	48

Analysis

(a) What is the rate law equation for the reaction?
(b) What is the rate-determining step? (6.3)
(c) What is a possible mechanism, including the slow step? (6.4)

Making Connections

16. The complex protein hemoglobin, Hbn, is the key molecule involved in oxygen transport between the lungs and the cells. Carbon monoxide, CO, is a highly poisonous gas because it can bind to the hemoglobin molecule and prevent it from carrying oxygen. The rate of this binding reaction can be represented by the equation

$$CO + Hbn \rightarrow HbnCO$$

Consider the experimental observations of the rate of binding of hemoglobin as a function of concentrations of reactants (**Table 4**).

(a) What are the orders of reaction with respect to each of the reactants?
(b) What is the overall order?
(c) Write the rate equation.
(d) If oxygen gas were to follow a similar rate equation in binding to hemoglobin under normal circumstances, which rate constant k would you expect to be larger—that of the oxygen or the carbon monoxide reaction? (6.3)

Table 4 Observations on Rate of Formation of Carboxyhemoglobin

Trial	Initial [CO] (mmol/L)	Initial [Hbn] (mmol/L)	Initial rate (mmol/(L·s))
1	5.0×10^{-4}	1.34×10^{-3}	3.12×10^{-4}
2	5.0×10^{-4}	2.68×10^{-3}	6.24×10^{-4}
3	1.5×10^{-3}	2.68×10^{-3}	1.872×10^{-3}

PERFORMANCE TASK

Energy and Rates Analysis of Chemical Reactions

Magnesium is one of the more active metals in the activity series. When stored on a shelf, it reacts with oxygen in the air to form a coating of magnesium oxide (**Figure 1**). Added to acid, magnesium readily reacts to form hydrogen gas and an aqueous solution of a salt (**Figure 2**).

Figure 1
Magnesium reacts readily in air to produce a coating of magnesium oxide.

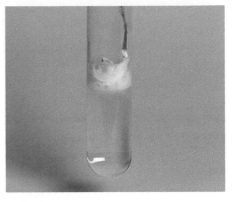

Figure 2
Magnesium reacts readily in acid to produce hydrogen gas.

magnesium$_{(s)}$ + hydrochloric acid$_{(aq)}$ → hydrogen$_{(g)}$ + magnesium chloride$_{(aq)}$ + heat

magnesium$_{(s)}$ + sulfuric acid$_{(aq)}$ → hydrogen$_{(g)}$ + magnesium sulfate$_{(aq)}$ + heat

magnesium$_{(s)}$ + acetic acid$_{(aq)}$ → hydrogen$_{(g)}$ + magnesium acetate$_{(aq)}$ + heat

These reactions are exothermic and proceed at measurable rates.

Your Task

You will quantitatively analyze the enthalpy changes and rates of reaction associated with the reactions of magnesium in different acids. In order to do so, you will design and perform a series of controlled calorimetry experiments to determine the molar enthalpies of reaction of magnesium in different acids.

You will also design and perform a series of controlled experiments to determine the rates of reaction of magnesium in the different acids. In the latter series, it will be important to decide what quantity you will measure—the rate of consumption of a particular reactant or the rate of production of a particular product—in order to calculate a rate of reaction for each system. The reactions will be performed in several separate series of trials in which many variables will be carefully controlled.

Questions

(i) How does the molar enthalpy of reaction of magnesium vary with different acids, namely, hydrochloric, sulfuric, and acetic acids?

(ii) How does the rate of reaction of magnesium vary with these acids?

Prediction/Hypothesis

(a) Write balanced chemical equations for the reactions of magnesium with various acids.

(b) Predict what you would expect to observe for the three acids, explaining why you made those predictions.

Experimental Design

(c) Design experiments that will allow you to answer the Questions. Outline the variables you will measure and any controls you will need. Include sample observation tables.

Materials

magnesium ribbon
1 mol/L hydrochloric acid
1 mol/L sulfuric acid
1 mol/L acetic acid

(d) Complete the Materials list. The equipment you select should be commonly available.

Procedure

(e) Write a detailed Procedure, including safety precautions and disposal considerations.

1. With your teacher's permission, carry out your Procedure.

Evidence

(f) Create sample observation tables before beginning your experiment.

Analysis

(g) Analyze the Evidence you obtained to answer the Questions.

Evaluation

(h) Evaluate your Evidence and use it to evaluate your Predictions.

Synthesis

(i) For any Predictions that you judged unacceptable, suggest a hypothesis that would explain the variance.

(j) In your experiments, you measured the rate of reaction with respect to the consumption or production of a particular substance. How might you have redesigned your experiments to measure rate with respect to some other participant in the reaction?

(k) The reactions of magnesium and acid are exothermic. How might this have affected the rates of reaction and your rate determinations? Briefly outline an Experimental Design that would control for the exothermic nature of the reactions.

(l) The magnesium–acid reactions use readily available materials to release thermal energy. Discuss whether one or more of these reactions might be appropriate in the design of a consumer product, such as a handwarmer for skiers.

Identify each of the following 10 statements as true, false, or incomplete. If the statement is false or incomplete, rewrite it as a true statement.

1. A physical change usually involves a greater enthalpy change than does a chemical change.

2. In a potential energy diagram, the activated complex is at a maximum potential energy for the system.

3. The potential energy of the products is greater than the potential energy of the reactants in an exothermic change.

4. The standard enthalpy of formation of a substance is measured at 25°C and 100 kPa.

5. An exothermic reaction absorbs heat from the surroundings.

6. The overall order of reaction is the sum of the exponents in the rate law equation.

7. The activation energy is lower in a catalyzed reaction than in the same reaction without a catalyst.

8. All of a radioisotope will have decayed after two half-lives.

9. In an endothermic reaction, both the potential and kinetic energies of the chemical system increase.

10. Elementary steps usually involve one- or two-body collisions.

Identify the letter that corresponds to the best answer to each of the following questions.

11. Which of the following statements does *not* apply to the term "enthalpy"?
    (a) It is symbolized by the letter H.
    (b) It is the same for all substances.
    (c) It changes during a chemical reaction.
    (d) It increases during the formation of some substances.
    (e) It is described as potential energy.

12. Which of the following statements is false concerning ΔH?
    (a) It is the difference between the potential energies of reactants and products.
    (b) It represents a change in potential energy.
    (c) If it is negative, it represents an endothermic reaction.
    (d) If the reaction consumes heat, it is positive.
    (e) It may be written as part of the equation.

13. When solid ammonium nitrate is added to water, the solution feels cold to the hand. Which statement best describes this observation?

(a) Heat is released from the system, so it feels colder.
(b) The reaction is exothermic.
(c) The potential energy of the surroundings has increased.
(d) $NH_4NO_{3(aq)} \rightarrow NH_4NO_{3(s)} + 31 \text{ kJ}$
(e) $NH_4NO_{3(s)} \rightarrow NH_4NO_{3(aq)} \quad \Delta H = +31 \text{ kJ}$

14. The reaction that would release the least energy is:
    (a) condensation of a mole of water vapour
    (b) combustion of a mole of hydrogen
    (c) breaking the chlorine–chlorine bonds in a mole of $Cl_{2(g)}$
    (d) nuclear fission in a mole of uranium
    (e) melting a mole of ice

15. A student dissolves some sodium hydroxide in water in a process represented by the equation

    $NaOH_{(s)} \rightarrow Na^+_{(aq)} + OH^-_{(aq)}$

    The following temperatures are recorded:
    initial temperature of water = 19.0°C
    final temperature of resulting solution = 27.0°C

    Based on these observations, which of the following statements is true?
    (a) The dissolving of NaOH is endothermic.
    (b) The kinetic energy of $NaOH_{(s)}$ is higher than the kinetic energy of the ions.
    (c) The reaction has a positive ΔH value.
    (d) Heat is absorbed from the surroundings.
    (e) The potential energy of $NaOH_{(s)}$ is higher than the potential energy of the ions.

16. Refer to the reaction below. How much heat is released if 95.0 g of sodium metal reacts?

    $2 \text{ Na}_{(s)} + 2 \text{ H}_2O_{(l)} \rightarrow H_{2(g)} + 2 \text{ NaOH}_{(aq)} + 150 \text{ kJ}$

    (a) 155 kJ   (d) 620 kJ
    (b) 246 kJ   (e) 734 kJ
    (c) 310 kJ

17. Which of the following representations of enthalpy changes in chemical reactions is inconsistent with the rest?
    (a) $CH_3OH_{(l)} + O_{2(g)} \rightarrow CO_{2(g)} + 2 H_2O_{(g)}$
        $\Delta H = -638.0 \text{ kJ}$
    (b) $CH_3OH_{(l)} + O_{2(g)} \rightarrow CO_{2(g)} + 2 H_2O_{(g)} - 638.0 \text{ kJ}$
    (c) $2 CH_3OH_{(l)} + 2 O_{2(g)} \rightarrow 2 CO_{2(g)} + 4 H_2O_{(g)}$
        $\Delta H = -1276.0 \text{ kJ}$
    (d) $\Delta H_c = -638.0 \text{ kJ/mol } CH_3OH$
    (e) $CH_3OH_{(l)} + O_{2(g)} - 638.0 \text{ kJ} \rightarrow CO_{2(g)} + 2 H_2O_{(g)}$

18. The term "chlorine" is used by people to mean a number of possible types of matter containing chlorine atoms, such as $Cl_{2(g)}$, $Cl_{(g)}$, and $Cl^-_{(aq)}$. Which of these forms of "chlorine" would have a zero standard enthalpy of formation?

(a) $Cl_{2(g)}$, $Cl_{(g)}$, and $Cl^-_{(aq)}$
(b) $Cl_{(g)}$ only
(c) $Cl^-_{(aq)}$ only
(d) $Cl_{2(g)}$ only
(e) $Cl_{2(g)}$ or $Cl_{(g)}$

19. Calculate the molar enthalpy of reaction per mole of carbon in the following reaction:

$3\ C_{(s)}\ +\ 2\ Fe_2O_{3(s)}\ +\ 466\ kJ\ \rightarrow\ 4\ Fe_{(s)}\ +\ 3\ CO_{2(g)}$

(a) $+466$ kJ/mol C
(b) $+155$ kJ/mol C
(c) $+117$ kJ/mol C
(d) -466 kJ/mol C
(e) -155 kJ/mol C

20. In which of the following combinations of reactants would surface area most affect rate of reaction?
(a) sodium chloride and silver nitrate solutions
(b) hydrogen and chlorine gases
(c) glucose and potassium permanganate solutions
(d) sodium hydroxide and hydrochloric acid solutions
(e) copper metal and aqueous nitric acid

21. Which of the following statements is false?
(a) A catalyst increases the rate of a chemical reaction.
(b) A catalyst is consumed in a chemical reaction.
(c) A catalyst can be either homogeneous or heterogeneous.
(d) A catalyst lowers the activation energy barrier for a reaction.
(e) A catalyst has no effect on the enthalpy change in the reaction.

22. At about 300K, an increase in temperature of 10°C roughly
(a) increases the rate of reaction by a factor of 2
(b) decreases the rate of reaction by a factor of 2
(c) increases the rate of reaction by a factor of 10
(d) increases the rate of reaction by a factor of 100
(e) increases the rate of reaction by a factor of 1/3

23. The value of k in the rate law equation for a chemical reaction is affected most by
(a) increasing the concentration of a reactant
(b) finely dividing particles of a reactant
(c) lowering the reaction temperature
(d) decreasing the partial pressure of a gaseous reactant
(e) removing the product as it is formed

24. For a hypothetical reaction with a rate law equation of rate $= k[X]^m[Y]^n$, the rate of the reaction over time will probably decrease due to
(a) a decrease in the value of k
(b) a decrease in the values of m and n
(c) a decrease in the values of $[X]$ and $[Y]$
(d) all of the above
(e) the rate does not decrease

25. For the rate law equation rate $= k[N_2O_{5(g)}]$, the initial rate of the reaction, r, is 2.20×10^{-4} mol/(L·s) when $[N_2O_{5(g)}]$ is 0.140 mol/L. The value of k for this reaction at this temperature is
(a) 2.34×10^{-4}/s
(b) 3.52×10^{-3}/s
(c) 3.60×10^{-2}/s
(d) 6.36×10^{-1}/s
(e) 1.57×10^{-3}/s

26. The rate of a chemical reaction would be expected to increase the most if
(a) the activation energy doubled
(b) the concentration of reactants doubled
(c) the activation energy were halved
(d) the surface area were doubled
(e) the temperature were halved

27. If 1.0 g of copper reacts with excess nitric acid in 45 s, the average rate of reaction is
(a) $\dfrac{1.0}{45}$ mol/s
(b) $\dfrac{1.0}{4.5 \times 63.55}$ mol/s
(c) $\dfrac{1.0 \times 63.55}{45}$ mol/s
(d) $\dfrac{45 \times 63.55}{1.0}$ mol/s
(e) $45 \times 1.0 \times 63.55$ mol/s

28. Which one of the following statements is *not* a significant part of the collision–reaction theory?
(a) motion of molecules
(b) orientation of molecules on collision
(c) kinetic energy of the molecules on collision
(d) enthalpy change of the reaction
(e) temperature dependence of molecule speed

29. Which of the following statements about an activated complex is *not* correct?
(a) The potential energy of the activated complex is greater than that for the reactants.
(b) The activated complex is more complex than the reactants.
(c) The activated complex is more stable than the reactants.
(d) The activated complex has a high potential energy.
(e) The potential energy of the activated complex is greater than that of the products.

30. Which of the following statements is true for a reaction mechanism?
(a) Any catalyst is consumed.
(b) Collisions involving three or more particles are common.
(c) All steps have the same activation energy.
(d) The overall rate depends on the slowest step.
(e) The total activation energy is the same for catalyzed and uncatalyzed mechanisms.

Understanding Concepts

1. Calculate the amount of heat in J and kJ that is required to heat 1.5 kg of water from 20°C to 75°C. (5.1)

2. The molar enthalpy of vaporization of chlorine is +20.7 kJ/mol. Calculate the enthalpy change during the vaporization of 2.25 kg of chlorine. (5.2)

3. In a student lab, 60.0 mL of 0.700 mol/L sodium hydroxide solution was neutralized with 40.0 mL of excess sulfuric acid solution. The temperature increased by 5.6°C.
   (a) Calculate the molar enthalpy of neutralization for sodium hydroxide.
   (b) What assumptions have you made? (5.2)

4. The standard molar enthalpy of formation for vinyl chloride, $C_2H_3Cl_{(g)}$, is +37.3 kJ/mol. Express this information in thermochemical equations as
   (a) a heat term;
   (b) a ΔH value. (5.3)

5. The standard enthalpy of formation of sulfur dioxide is −296.8 kJ/mol.
   (a) Write a thermochemical equation for the formation reaction.
   (b) Sketch a potential energy diagram for the reaction, labelling axes, enthalpy of reactants, enthalpy of products, and ΔH.
   (c) If 9.63 g of sulfur dioxide is formed under standard conditions, what quantity of heat is released? (5.4)

6. Nitromethane is a rapid-burning fuel often used in dragsters where rate, not energy yield, is important.

$$4\,CH_3NO_{2(g)} + 3\,O_{2(g)} \rightarrow 4\,CO_{2(g)} + 2\,N_{2(g)} + 6\,H_2O_{(g)}$$

Use Hess's law and the known thermochemical equations given below to calculate the enthalpy change for the combustion of one mole of nitromethane.

$$C_{(s)} + O_{2(g)} \rightarrow CO_{2(g)} \qquad \Delta H = -393.5 \text{ kJ}$$
$$2\,H_{2(g)} + O_{2(g)} \rightarrow 2\,H_2O_{(g)} \qquad \Delta H = -483.6 \text{ kJ}$$
$$2\,C_{(s)} + 3\,H_{2(g)} + 2\,O_{2(g)} + N_{2(g)} \rightarrow 2\,CH_3NO_{2(g)}$$
$$\Delta H = -226.2 \text{ kJ} \qquad (5.4)$$

7. (a) Write an equation for the combustion of one mole of pentane, $C_5H_{12(l)}$, to form carbon dioxide gas and liquid water.
   (b) Given the standard enthalpies of formation in Appendix C4 and the information that the standard enthalpy of formation of pentane is

−146 kJ/mol, calculate the enthalpy change associated with the combustion of 1 mol of pentane.
   (c) How much heat would be released in the combustion of 20 g of pentane? (5.5)

8. Suggest any three physical properties that may change during a reaction and that may be used to measure the rate of a reaction. (6.1)

9. **Table 1** refers to the reaction between carbon monoxide and nitrogen dioxide:

$$CO_{(g)} + NO_{2(g)} \rightarrow CO_{2(g)} + NO_{(g)}$$

   (a) Predict the missing concentration values.
   (b) If the initial concentration of the $NO_{2(g)}$ was 0.250 mol/L, what will be its concentration after 80 s? (6.1)

Table 1 Concentration of Carbon Monoxide and Carbon Dioxide

Time (s)	$[CO_{(g)}]$ (mol/L)	$[CO_{2(g)}]$ (mol/L)
0	0.100	—
20	0.050	0.050
40	0.033	—
60	0.026	0.074
80	0.020	0.080
100	—	0.083

10. Fire departments warn people about leaving newspapers in large piles in basements. Why would these newspapers be more of a fire hazard than the same quantity of wood? (6.2)

11. Observations were made (**Table 2**) during the decomposition of a compound:

$$2\,X_2O_{5(g)} \rightarrow 4\,XO_{2(g)} + O_{2(g)}$$

   (a) Using the same concentration and time axes, draw a graph to show:
       (i) $[X_2O_{5(g)}]$ vs. time
       (ii) $[O_{2(g)}]$ vs. time
   (b) Calculate the values to fill the blanks in **Table 2**.

Table 2 Concentration of Reactant and Products During Decomposition

Time (h)	$[X_2O_{5(g)}]$ (mol/L)	$[XO_{2(g)}]$ (mol/L)	$[O_{2(g)}]$ (mol/L)
0.0	1.20	0	0
2.0	0.80		0.20
4.0	0.55		0.325
7.0	0.30		0.45
12.0	0.10		0.55

(c) Calculate the overall rates of consumption (or production) in the first 12 h of:
 (i) $X_2O_{5(g)}$
 (ii) $O_{2(g)}$
 (iii) $XO_{2(g)}$
(d) Determine the instantaneous rates of consumption of X_2O_5 at 2.0 h and 7.0 h.
(e) Describe and explain the observed trend in rate of consumption of X_2O_5. (6.2)

12. Aluminum metal is used in many familiar objects, from frying pans to screen doorframes and jet aircraft. However, the bottle of aluminum powder in the chemistry laboratory carries a warning that the contents are potentially dangerously combustible. Explain these observations. (6.2)

13. From a kinetics study of the following reaction,

$$ClO^-_{(aq)} + I^-_{(aq)} \rightarrow Cl^-_{(aq)} + IO^-_{(aq)}$$

the rate was found to be first order with respect to each of the reactants. Predict what would happen to the initial rate, r, as each of the following changes are made.
(a) The initial $[ClO^-_{(aq)}]$ is doubled.
(b) The initial $[I^-_{(aq)}]$ is halved.
(c) The same initial numbers of moles of reactants were placed in a container of half the volume. (6.3)

14. The combustion of propane is represented by the equation

$$C_3H_{8(g)} + 5\,O_{2(g)} \rightarrow 3\,CO_{2(g)} + 4\,H_2O_{(g)}$$

(a) Explain whether you would expect this reaction to happen in a single step or in a series of steps.
(b) If the rate of consumption of propane gas is

$$-\frac{\Delta[C_3H_8]}{\Delta t} = 4 \times 10^{-2} \text{ mol/(L·s)},$$

write expressions and numerical values to represent the rate of reaction with respect to oxygen and carbon dioxide gases. (6.4)

15. Hydrogen iodide and oxygen react together as shown by the equation:

$$4\,HI_{(g)} + O_{2(g)} \rightarrow 2\,I_{2(g)} + 2\,H_2O_{(g)}$$

The observations shown in **Table 3** are obtained when initial concentrations of reactants are varied:

Table 3 Rate Evidence for Consumption of Hydrogen Iodide

Initial [HI] (mol/L)	Initial [O_2] (mol/L)	Initial rate (mol/(L·s))
0.010	0.010	0.0042
0.010	0.020	0.0084
0.020	0.020	0.0168

(a) What is the order of this reaction with respect to the hydrogen iodide?
(b) What is the order of this reaction with respect to the oxygen?
(c) What is the overall order of this reaction?
(d) Write the rate equation for this reaction.
(e) Determine the specific rate constant for this reaction, including units.
(f) How many molecules are involved in the rate-determining step?
(g) The overall reaction clearly proceeds in several steps. Even in the absence of rate data, why would you predict that the reaction would not proceed in a single step? (6.4)

16. Nitrogen monoxide reacts with hydrogen gas to produce nitrogen and water vapour. The mechanism is believed to be:

Step 1 $2\,NO_{(g)} \rightarrow N_2O_{2(g)}$

Step 2 $N_2O_{2(g)} + H_{2(g)} \rightarrow N_2O_{(g)} + H_2O_{(g)}$

Step 3 $N_2O_{(g)} + H_{2(g)} \rightarrow N_{2(g)} + H_2O_{(g)}$

(a) Write the overall balanced equation for this process.
(b) Identify the reaction intermediates.
(c) Write the rate equation for this reaction, given the information that Step 1 is the slow step. (6.4)

17. Sketch a Maxwell-Boltzmann graph showing the distribution of molecular kinetic energies for a sample of gas at temperatures T_1 and T_2, where T_2 is the higher temperature. Label the axes. (6.5)

18. Which one of the following reactions would you expect to be faster at room temperature? Explain your answer briefly.
(a) $Pb^{2+}_{(aq)} + 2\,Cl^-_{(aq)} \rightarrow PbCl_{2(s)}$
(b) $Pb_{(s)} + Cl_{2(g)} \rightarrow PbCl_{2(s)}$ (6.5)

19. Diamond and graphite are different forms of the same element, carbon. Under room conditions, the enthalpy change from diamond to graphite is negative (-1.9 kJ/mol), suggesting that diamond should spontaneously change into graphite, releasing energy as it does so. Yet no observable reaction takes place. In contrast, white phosphorus is so dangerously reactive that it will ignite and burn if exposed to air, forming phosphorus oxide and releasing considerable heat. It has been used in incendiary bombs.
(a) Explain these observations in terms of energy.
(b) Sketch potential energy diagrams for the reactions diamond $\rightarrow$ graphite and phosphorus + oxygen $\rightarrow$ phosphorus oxide. In each case label the enthalpy change and the activation energy. (6.5)

Applying Inquiry Skills

20. A calorimetry investigation involves the use of a copper flame calorimeter.

    Question

    What is the enthalpy of combustion of propanal, C_3H_6O?

    Evidence

Mass of calorimeter	305 g
Mass of water in calorimeter	255 g
Mass of propanal burned	1.01 g
Temperature increase of calorimeter and contents	28.8°C
Specific heat capacity of copper	0.385 J/(g•°C)

    Analysis

    (a) Analyze the Evidence and answer the Question. (5.2)

21. The following investigation was conducted.

    Question

    What is the molar enthalpy of neutralization for hydrochloric acid?

    Experimental Design

    Solid sodium hydroxide is dissolved in a measured quantity of hydrochloric acid solution in a Styrofoam laboratory calorimeter.

    Evidence

mass of $NaOH_{(s)}$	3.40 g
volume of $HCl_{(aq)}$	100.0 mL
concentration of $HCl_{(aq)}$	0.850 mol/L
initial temperature of $HCl_{(aq)}$	14.5°C
final temperature of $HCl_{(aq)}$	35.6°C

    Analysis

    (a) Analyze the Evidence and answer the Question.

    Evaluation

    (b) Evaluate the Experimental Design. (5.3)

22. The following investigation was conducted.

    Question

    What is the molar enthalpy of formation of butane?

    Experimental Design

    The enthalpies of combustion of butane, carbon, and hydrogen are used with Hess's law to determine the molar enthalpy of formation of butane.

    Evidence

    $$C_4H_{10(g)} + \frac{13}{2} O_{2(g)} \rightarrow 4\,CO_{2(g)} + 5\,H_2O_{(g)}$$
    $$\Delta H_c = -2657.3 \text{ kJ}$$

    $$C_{(s)} + O_{2(g)} \rightarrow CO_{2(g)} \qquad \Delta H_c = -393.5 \text{ kJ}$$
    $$H_{2(g)} + 1/2\,O_{2(g)} \rightarrow H_2O_{(g)} \qquad \Delta H_c = -241.8 \text{ kJ}$$

Analysis

(a) Analyze the Evidence and answer the Question. (5.5)

23. (a) Outline an appropriate Experimental Design for the following investigation.

    Question

    What is the rate law for the following hypothetical chemical reaction?

    $$A + C + 2\,D \rightarrow 2\,S + T$$

    Evidence

    Table 4 Observations on Rate of Reaction and Formation

Initial rate of formation of S (mol/L·s)	Initial [A] (mol/L)	Initial [B] (mol/L)	Initial [C] (mol/L)
6.0	0.15	0.20	0.20
12.0	0.30	0.20	0.20
24.0	0.30	0.40	0.20
24.0	0.30	0.40	0.40

    Analysis

    (b) Analyze the Evidence and answer the Question. (6.3)

Making Connections

24. "Nuclear reactors should be used instead of burning fossil fuels to supply electrical energy." Do you agree or disagree with this statement? Justify your answer, from economic and environmental perspectives. (6.6)

25. "The concept of half-life can apply to more than radioactive decay." Research this assertion, and explain it by making reference to at least two examples each of radioactive half-life and chemical half-life. (6.3)

26. Use collision theory to provide an explanation of the following observations:
    (a) Magnesium metal reacts much more rapidly in concentrated $HNO_{3(aq)}$ than in dilute $HNO_{3(aq)}$.
    (b) Paints and stains often have instructions that they should not be applied below 10°C (**Figure 2**).
    (c) Food spoils more rapidly on a counter than in a refrigerator.
    (d) A natural gas furnace requires a pilot light or electronic igniter in order to operate.
    (e) Dust in grain elevators has been blamed for several violent explosions, which have completely demolished the elevators.

KEY FACTS

USES
- Wood siding
- Wood trim
- Pre-finished aluminum, vinyl and galvanized siding
- Primed metal

THINNER (if needed)
- Water

PERFORMANCE
- Resists mildew
- Excellent adhesion
- Durable
- Resists cracking, peeling and blistering
- Excellent colour retention
- Easy clean up
- Fade resistant
- Easy soil removal

DRYING TIME
- 8 hours recoat

COVERAGE
- 1 litre covers approx. 12 sq. m. or 130 sq. ft.

APPLICATOR
- Brush
- Roller
- Sprayer

CLEAN UP
- Soap and water

SURFACE PREPARATION - Surface must be clean, dry, and free from contaminants. Remove all loose peeling paint. Sand to smooth edges. Sand glossy areas and sand exposed wood to a new wood appearance.
UNPAINTED WOOD - Shellac knots and sap streaks. Prime entire area with CIL 979 Primer. Woods prone to staining should be coated with CIL 999 Alkyd Primer.
PAINTED WOOD - Prime bare areas with CIL 979 Primer.
METAL - Prime with appropriate Metal Primer.
PAINTING - Stir thoroughly. Apply one or two coats as required. Apply only if temperature is above 10°C.
MILDEW RESISTANT - Treated with fungicide to protect the paint itself from the growth of mildew fungi.

DISPOSAL: REMOVE LID AND ALLOW CONTENTS TO HARDEN COMPLETELY BEFORE CONTACTING YOUR LOCAL MUNICIPALITY FOR DISPOSAL DETAILS. PLEASE ENSURE OPEN CONTAINER IS KEPT OUT OF REACH OF CHILDREN AND ANIMALS.

Figure 2

(f) People who have contact lenses sometimes use a sterilizing kit containing hydrogen peroxide and a platinum-coated disk. Hydrogen peroxide decomposes to release bubbles of oxygen gas only when the disk is placed in the solution.

(g) Baking powder, containing a mixture of a solid acid and a solid base, does not react when dry but reacts rapidly when dissolved in water. (6.4)

Extension

27. When octane is burned completely in a well-tuned engine, the products are gaseous carbon dioxide and liquid water. A poorly tuned car engine may produce equimolar amounts of carbon monoxide and carbon dioxide, and water is produced in the gaseous state.
    (a) Write balanced equations for the reactions that occur in both the efficient and non-efficient engines.
    (b) Calculate the energy produced in each of the reactions, using standard heats of formation.
    (c) Determine what percentage of available energy is being wasted in the non-efficient engines.
    (d) The engine block and radiator of the car have a mass of 200 kg and a specific heat capacity of 0.50 J/(g·°C). The cooling system contains 15.0 kg of water. The operating temperature of the engine is 95°C, but when the engine is started up on a cold winter morning, the air temperature is −15°C. If the engine has been tuned to make it run efficiently, calculate the mass of octane that must be burned to raise the temperature of the engine to operating temperature.

(e) The car is completely warmed up after travelling 2 km. If it uses fuel at a rate of 0.150 L/km when fully warmed up, what is the rate of consumption during the first 2 km? The density of octane is 0.800 kg/L. What assumptions must you make in order to solve the problem with the data provided? (5.5)

28. The concept of half-lives can be applied to biological, as well as chemical, systems. There is some concern that antibiotics, fed to livestock to speed their growth by killing disease organisms, are excreted in the animals' urine, thereby contaminating ground water. This rate of excretion appears to be a first-order reaction. Research this issue, and find out how long animals keep 50% of the administered antibiotics in their bodies. (6.3)

 www.science.nelson.com

29. Papain is an enzyme derived from tropical papaya fruit. It acts as a catalyst to break down the peptide bonds in proteins. Research to find how this enzyme is being marketed for consumer use, and comment on the potential effectiveness of these uses. (6.5)

 www.science.nelson.com

30. The hydrolysis of dimethyl ether, an exothermic reaction, is represented by the equation

$$CH_3-O-CH_{3(aq)} + H_2O_{(l)} \rightarrow 2\ CH_3OH_{(aq)}$$

For the reaction to proceed, acid ($H^+_{(aq)}$) is required. The experimentally determined rate law equation is

$$r = k\,[CH_3-O-CH_3]^1[H^+_{(aq)}]^1$$

(a) Draw structural diagrams for the reactants and products.
(b) What is the function of the acid?
(c) What reactant species must be involved in the rate-determining step?
(d) Design a simple two- or three-step mechanism, consistent with the above information. You may find it useful to consider CH_3OH^+ as an intermediate. Clearly identify the slowest step in your mechanism.
(e) Draw a reasonable potential energy diagram, consistent with your mechanism, with clearly labelled axes, reactants, intermediates, and products. (6.5)

Chemical Systems and Equilibrium

Miriam Diamond
Associate Professor
Department of Geography
University of Toronto

"In our lab we study the movement of chemical contaminants in the environment. To simplify the task, we assume that the environment behaves as a defined chemical system. We develop mathematical equations that quantify chemical movement, which is controlled by forces such as rainfall and soil erosion and the passive movement of chemicals as they seek to achieve an equilibrium distribution within a phase (such as water) or between phases (from water to air). One novel aspect of our research is discovering that all surfaces are coated with a thin layer of organic material. This film 'buffers' air concentrations— if air concentrations decrease, gases dissolved in the film can volatilize and, conversely, when air concentrations are high, the chemicals partition into the film. This process influences concentrations of contaminants inside our houses and outside. We have developed methods that allow us to sample the concentrations of trace metals in the water surrounding the sediments of lakes and to sample the thin film on surfaces of the inside and outside of buildings (we clean windows!)."

▶ Overall Expectations

In this unit, you will be able to

- demonstrate an understanding of the concept of equilibrium, Le Châtelier's principle, and solution equilibria;
- investigate the behaviour of different equilibrium systems, and solve problems involving the law of chemical equilibrium;
- explain the importance of chemical equilibrium in various systems, including ecological, biological, and technological systems.

ARE YOU READY?

▶ **Prerequisites**

Concepts

- states of matter
- concentration of solutions
- solubility of solutes
- collision–reaction theory
- reaction rate
- molar enthalpy of reaction
- Brønsted-Lowry acid–base theory

Skills

- Write and balance molecular and ionic equations.
- Calculate percent ionization.
- Solve algebraic equations for one unknown.
- Perform acid–base titrations.

Knowledge and Understanding

1. **Figure 1** shows a series of diagrams of a crystal of sucrose, $C_{11}H_{22}O_{11(s)}$, suspended in an aqueous solution of sucrose, $C_{11}H_{22}O_{11(aq)}$, over a period of 11 days. The system was maintained at SATP throughout.

day 1   day 3   day 5   day 7   day 11

Figure 1
A sucrose crystal suspended in a sucrose solution for 11 days

   (a) List the states of matter you see in the first jar (Day 1).
   (b) Why is the system considered closed for the first 7 d and open for the last 4 d?
   (c) Is the solution saturated or unsaturated from Day 1 to Day 5? Explain.
   (d) Why did the crystal stay the same size from Day 5 to Day 7?
   (e) Is the solution saturated or unsaturated from Day 8 to Day 11? Explain.

2. Magnesium metal reacts with hydrochloric acid to produce magnesium chloride, $MgCl_{2(aq)}$, and hydrogen gas.
   (a) Write a balanced chemical equation for this reaction.
   (b) What amount of magnesium chloride will be produced when 250 mL of 0.8 mol/L $HCl_{(aq)}$ reacts with excess magnesium?
   (c) Calculate the concentration of $MgCl_{2(aq)}$ in the resulting solution. (Assume that the volume remains constant.)

3. (a) Distinguish between a strong base and a weak base, and provide an example of each.
   (b) Define the terms "acid" and "base" according to Brønsted-Lowry acid–base theory.
   (c) Classify each entity in the following equation as a Brønsted-Lowry acid or base, and identify conjugate acid–base pairs.

$$HPO_4^{2-}{}_{(aq)} + HSO_4^{-}{}_{(aq)} \rightarrow H_2PO_4^{-}{}_{(aq)} + SO_4^{2-}{}_{(aq)}$$

4. Use standard molar enthalpies of formation to calculate the molar enthalpy of combustion of cyclopropane, $C_3H_{6(g)}$ (assuming all gaseous entities).

Inquiry and Communication

5. (a) Describe an experimental design you could use to determine the solubility of $Ca(OH)_{2(s)}$ in water at 60°C. (Include safety procedures.)
   (b) If you determine that the solubility of $Ca(OH)_{2(s)}$ is 0.10 g/100 mL at 60°C, calculate the molar concentration of $OH^-_{(aq)}$ in 1.0 L of saturated $Ca(OH)_{2(aq)}$ at this temperature.
   (c) What assumption regarding the solute/solvent interaction must you make to justify your calculation?

6. (a) Write balanced chemical, total ionic, and net ionic equations for the reaction between $NaOH_{(aq)}$ and $HCl_{(aq)}$.
   (b) Classify the reaction between $NaOH_{(aq)}$ and $HCl_{(aq)}$.
   (c) The apparatus in **Figure 2** is used to carry out the above reaction. Name the container that holds the $NaOH_{(aq)}$ solution.
   (d) Name the laboratory procedure that the apparatus in **Figure 2** is used to perform.
   (e) What is missing in order to adequately carry out the procedure?
   (f) Using the apparatus illustrated in **Figure 2**, describe the steps you would take to complete the procedure (include safety precautions).
   (g) Using the quantity of $HCl_{(aq)}$ and concentrations of $HCl_{(aq)}$ and $NaOH_{(aq)}$ shown, predict the volume of $NaOH_{(aq)}$ that would be used to complete the procedure.
   (h) Based on your answer to (g), predict the pH of the solution in the receiving flask at the end of the procedure.
   (i) Describe a quantitative test you could perform on the solution in the receiving flask to test your prediction in (h).
   (j) Assuming your prediction is validated by the test you suggested in (i), calculate the $[H^+_{(aq)}]$ of the solution.

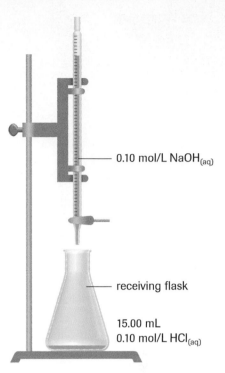

0.10 mol/L $NaOH_{(aq)}$

receiving flask

15.00 mL
0.10 mol/L $HCl_{(aq)}$

Figure 2

Mathematical Skills

7. Solve for x.

   (a) $\dfrac{(0.020)(0.030)}{(0.10)x} = 2.3 \times 10^{-4}$

   (b) $\dfrac{(2x)^2}{(x - 0.20)^2} = 49$

   (c) $\dfrac{(3x)^2}{(x - 1.0)(x - 3.0)} = 2.0$

Making Connections

8. As a result of several complaints from angry citizens regarding acidic-tasting tapwater, Soursville's water commissioner ordered the town's utilities inspector to measure the water's pH monthly, for a period of one year. The following chart records the pH measurements that were taken from January to December.

month	J	F	M	A	M	J	J	A	S	O	N	D
pH	6.1	6.3	5.9	6.1	5.8	5.7	5.3	4.8	5.0	5.5	5.8	6.0

Based on these results, the commissioner recommended that no action be taken since the average pH was 5.7. When asked to justify her decision, she said, "People eat salad dressing at a pH of approximately 2.9 and drink soft drinks at a pH of approximately 3.4 all the time. Surely tapwater with an average pH of 5.7 is no cause for concern." Comment on the commissioner's point of view.

chapter

7 Chemical Systems in Equilibrium

In this chapter, you will be able to

- demonstrate an understanding of dynamic equilibrium in a variety of situations, and use the law of chemical equilibrium to solve problems;
- apply Le Châtelier's Principle to predict how various factors affect a chemical system at equilibrium;
- carry out experiments to determine equilibrium constants;
- explain how equilibrium principles may be applied to optimize the production of industrial chemicals;
- describe the behaviour of ionic solutes in solutions that are unsaturated, saturated, and supersaturated;
- calculate the molar solubility of a pure substance in water or in a solution of a common ion, and predict the formation of precipitates by using the solubility product constant;
- identify the entropy changes associated with chemical and physical processes and the tendency of reactions to achieve minimum energy and maximum entropy;
- identify the effects of solubility on biological systems.

Have you ever felt as though you were at rest while walking up the "down" escalator, or as you kept pace with a speeding treadmill? If so, you have experienced a state of *dynamic equilibrium*—a situation where at least one property remains constant while opposing processes occur at the same rate. On the escalator or the treadmill, your body's position is the property that remains constant, while your legs and the platform move in opposite directions at the same speed.

Consider a swimming pool. Water jets on the sides of the pool continuously inject warm, clean water into the pool, but the water level remains constant. For this to occur, water must be leaving the pool through drains at the same rate as it is entering through the jets. The water in the pool is in a state of dynamic equilibrium. Water molecules are entering and leaving the pool at the same rate while the water level remains constant.

There are many examples of dynamic equilibrium in the world of chemistry. Although they are not exactly the same as the situations described above, there are many similarities. Chemical systems at equilibrium all display at least one constant property. For example, the colour of a reaction mixture could remain constant, giving the impression that nothing is happening. This can be deceiving. In the pool example, close inspection reveals competing activities occurring at the same rate: water entering and water leaving. What competing activities could be occurring in a chemical reaction at equilibrium?

REFLECT on your learning

1. Two children are perfectly balanced on a teeter-totter. Is this a case of dynamic equilibrium? Explain.
2. Describe two household examples of dynamic equilibrium.
3. The molar solubility of $PbI_{2(s)}$ in pure water is 1.3×10^{-3} mol/L. In water that contains 0.1 mol/L $NaI_{(aq)}$, the solubility of $PbI_{2(s)}$ is 7.9×10^{-7} mol/L. Provide reasons for the difference in the two solubilities.
4. (a) What does the double arrow in the following equation mean?

$$2\,CO_{(g)} + O_{2(g)} \rightleftharpoons 2\,CO_{2(g)} \qquad \Delta H^{\circ} = -566.1 \text{ kJ}$$

   (b) If the above reaction occurs in a closed container, would heating the container increase or decrease the production of $CO_{2(g)}$? Explain.
5. Will the following reaction have a tendency to occur spontaneously as written (at SATP)? Explain.

$$4\,C_3H_5(NO_3)_{3(l)} \rightarrow 12\,CO_{2(g)} + 10\,H_2O_{(g)} + 6\,N_{2(g)} + O_{2(g)}$$

▶ TRY THIS activity   *Shakin' the Blues*

When a nugget of mossy zinc is added to a dilute solution of hydrochloric acid in an open test tube, a violent reaction occurs with lots of gas and heat given off. The zinc continues to react and when it is completely consumed, the visible signs of reaction come to an end. After seeing many reactions like this in your science studies, you may have come to think that all chemical reactions go one way: from reactants to products. But do they always?

Materials: lab apron; eye protection; 400-mL flask and stopper; 250 mL water; 5.0 g potassium hydroxide ($KOH_{(s)}$); 3.0 g glucose or dextrose; 2% methylene blue; stirring rod

- Pour 250 mL of water into the flask.
- Add 6 drops of methylene blue and all of the potassium hydroxide and glucose to the flask.
- Stir the mixture with the stirring rod until the solids have dissolved.
- Stopper the flask and set it on the bench. Observe the colour of the solution.
- Shake the solution vigorously and note any changes (**Figure 1**).
- Set the flask on the table and leave it standing until another change is noticed.
- Repeat the previous two steps many times. Make observations each time.

(a) Describe the reaction in the flask in relation to the discussion at the top of this activity.
(b) What evidence do you have to substantiate your answer to question (a)?
(c) Predict whether the colour changes will continue forever.
- Test your prediction over a reasonable period of time.
(d) Evaluate your prediction.

 Potassium hydroxide is poisonous and corrosive.

 Keep away from skin and eyes. Wear eye protection.

Figure 1
How many times does this reaction happen?

Figure 1
A carbonated beverage in a closed bottle (left) displays constant macroscopic properties. When the cap is removed, the pressure on the system is reduced and carbon dioxide bubbles come out of the solution.

closed system a system that may exchange energy but not matter with its surroundings

dynamic equilibrium a balance between forward and reverse processes occurring at the same rate

forward reaction in an equilibrium equation, the left-to-right reaction

reverse reaction in an equilibrium equation, the right-to-left reaction

solubility equilibrium a dynamic equilibrium between a solute and a solvent in a saturated solution in a closed system

phase equilibrium a dynamic equilibrium between different physical states of a pure substance in a closed system

chemical reaction equilibrium a dynamic equilibrium between reactants and products of a chemical reaction in a closed system

Scientists describe the state of a chemical system in terms of properties such as temperature, pressure, chemical composition, and amounts of substances present. In order to do that, the system must be accessible and definable. Chemical changes in a system are easier to study when the system is separated from its surroundings by a definite physical boundary. This arrangement allows us to control the variables and ensure that, while energy may be transferred, matter cannot enter or leave the system. Such a system is called a **closed system**. An aqueous solution in a test tube or a beaker can be considered a closed system, as long as no gas is used or produced in the reaction. Closed systems involving gases must be closed on all sides.

One example of a closed system is a bottle of carbonated soft drink. Nothing appears to change, until the bottle is opened (**Figure 1**).

If you slowly twist the cap off the bottle, you will hear gas escaping even before bubbles begin to form, indicating that the space between the liquid and the cap is filled with a gas under pressure—gas that escapes when you loosen the cap. The bubbles that form quickly after opening indicate that there is carbon dioxide dissolved in the solution, and that the solution in the closed bottle contained more CO_2 molecules than could remain dissolved at the temperature and pressure of the room. After you replace the cap, you can assume that there is some carbon dioxide in the space between the cap and the liquid, and within the liquid. A quick shake reinforces this assumption.

If you focus on the gas/liquid interface within the bottle (**Figure 1**), you will notice that several macroscopic properties remain constant. The liquid continues to look clear, colourless, and bubble-free, and the gas above the liquid continues to appear colourless and transparent. The level of the liquid does not change. These macroscopic properties (those we can see) remain constant. At the molecular level, however, it is not unreasonable to imagine carbon dioxide molecules leaving the dissolved state and entering the gas state, and other carbon dioxide molecules in the gas state colliding with the solvent's surface and dissolving. Since macroscopic properties remain constant, these opposing changes must be occurring at equal rates. A closed system such as this is in **dynamic equilibrium**, a state where a balance is achieved by opposing processes occurring at the same rate. We can symbolize this equilibrium with an equation containing a forward ($\rightarrow$) and a reverse ($\leftarrow$) arrow, usually combined into a single symbol as follows:

$$CO_{2(g)} \rightleftharpoons CO_{2(aq)}$$

When an equation is written with double arrows to show that the change occurs both ways, the left-to-right change is called the **forward reaction**, and the right-to-left change is called the **reverse reaction**. You will learn more about reversible reactions shortly.

In the soft-drink bottle system, the equilibrium involves changes in solubility between a solute, $CO_{2(aq)}$, and a solvent, $H_2O_{(l)}$. This is called **solubility equilibrium**. A saturated aqueous solution of iodine is another example of a solubility equilibrium. In a saturated solution, the concentration of the dissolved solute is constant. According to the dynamic equilibrium concept, the rate of the dissolving process is equal to the rate of the crystallizing process.

$$I_{2(s)} \rightleftharpoons I_{2(aq)}$$

Other types of equilibria include **phase equilibrium**, where two or more states of a pure substance are in dynamic equilibrium, such as ice over a lake, and **chemical reaction equilibrium**, where the reactants and products of a chemical reaction are in dynamic equilibrium.

TRY THIS activity	**The Coin Exchange:** **Establishing Dynamic Equilibrium**

Reversible chemical reactions occurring in closed systems become mixtures of reactants and products in dynamic equilibrium. In this activity, you and a partner will simulate the formation of equilibrium by exchanging coins in fixed ratios.

Materials: 100 pennies or gaming chips

• Divide 100 pennies or chips randomly between you and a partner and record the amount each has in a table similar to **Table 1**.

• Your teacher will assign you and your partner a "transfer rate."

• Pass coins or chips to one another according to your assigned transfer rates. The amount one partner transfers to the other must equal the total possessed by that partner multiplied by that person's transfer rate (rounding fractional values). Record the amount of money you have after each transaction in the columns headed "Partner A's amount" and "Partner B's amount" in your table.

• After each transaction, calculate the ratio of the amounts possessed by each partner and record this in the Amount A/Amount B column in your table.

• Continue exchanging coins or chips until a noticeable pattern develops.

(a) How did the amount of money held by each player change over the course of the transactions?

(b) Compare your team's results with those of other teams that used different transfer rates or started with a different distribution of coins. Suggest reasons for any differences and similarities.

(c) Compare the ratio of the number of coins held by you and your partner at the end of the exchanges to the ratio of your transfer rates. Account for the results.

(d) Describe the quantitative changes that occur over the course of this activity.

Table 1 Coin Exchange Data Table

Exchange No.	Partner A's amount (rate = 0.2)	Partner B's amount (rate = 0.4)	Amount A/Amount B
0	62¢	38¢	1.6
1	65¢	35¢	1.9
2			

Solubility Equilibrium

Most substances dissolve in a solvent to a certain extent, then appear to stop dissolving. If the solution constitutes a closed system, then the observable properties of the solution become constant. If the solvent is water and the solute is copper sulfate, the solution turns an increasingly more intense blue as salt dissolves, until the salt reaches its solubility limit. From then on, the colour remains the same. A copper sulfate solution will also conduct electricity. Its ability to conduct increases as more and more solute dissolves, then levels off when **dissolution** ends.

According to the kinetic molecular theory, particles are always moving and collisions are always occurring in a system, even if no changes are observed. For example, the initial dissolution of sodium chloride in water is thought to be the result of collisions among water molecules and ions that make up the crystals (**Figure 2(a)**). Once ions have entered the dissolved state, collisions between water molecules and the remaining crystal continue. However, dissolved ions will now also collide with the crystal. Whenever molecules of water collide with the crystal, ions may break from the crystal and enter the dissolved state. When dissolved ions collide with the crystal, they may form ionic bonds and crystallize out of solution (**Figure 2(b)**). Early in the dissolving process, the number of ions entering the dissolved state far exceeds the number that crystallize. Nearing equilibrium, the rates of dissolution and crystallization approach one another. At equilibrium, water molecules and ions still collide with the crystal surface, but the rate of dissolution now equals the rate of crystallization (**Figure 2(c)**).

dissolution the process of dissolving

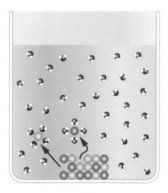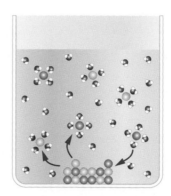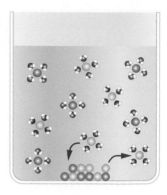

Figure 2

(a) When the solute is first added, many more ions dissociate from the crystal than crystallize onto it.

(b) As more ions come into solution, more ions also crystallize.

(c) At solubility equilibrium, solute ions dissolve and crystallize at the same rate.

If both dissolving and crystallizing processes take place at the same rate, no observable changes would occur in either the concentration of the ions in solution or in the quantity of solid present. The system is in a state of dynamic equilibrium.

Chemists have collected evidence of the dynamic nature of solubility equilibrium: They added a few crystals of radioactive iodine to a saturated aqueous solution of normal iodine. To a similar second sample of normal iodine solution, they added a few millilitres of a saturated solution of radioactive iodine (**Figure 3**).

The radioactive iodine emits radiation that can be detected by a Geiger counter. After a few minutes, the solution and the solid in both samples clearly show increased radioactivity. Assuming the radioactive iodine molecules are chemically identical to normal iodine, the experimental evidence supports the idea that iodine solid is dissolving at the same time as iodine molecules in the solution add back onto the solid surface. If this were not the case, in the first sample, where solid iodine was added, only the solid would show traces of radioactivity, and in the second sample, where iodine solution was added, only the solution would be radioactive.

▶ **TRY THIS** activity **Digesting a Precipitate**

Chemists usually allow precipitates to sit for a while in contact with the solution from which they were precipitated before filtering them. They call this process "digesting the precipitate." This procedure improves the purity of the precipitate. (If the precipitate forms quickly, impurities can be dragged down and trapped in the precipitate.) Digesting also results in the formation of larger crystals that separate from the liquid more effectively during filtration.

The dynamic nature of solubility equilibrium can be demonstrated by "digesting" a saturated aqueous solution of table salt containing excess crystals.

Materials: water; coarse pickling salt, $NaCl_{(s)}$; glass jar with lid; clock or watch

- In a jar, make a saturated solution of pickling salt. (A small amount of crystals will remain at the bottom of the jar.)

- Ensure that the lid is firmly in place, then shake the jar and record the time it takes for the contents to settle so that the solution is clear.

- Repeat this process once a day for two weeks. Note the relative size of the crystals each day.
  (a) What happened to the settling time over the course of one week?
  (b) How did the size of the crystals change?
  (c) What evidence do you have that dynamic equilibrium was established in this system?

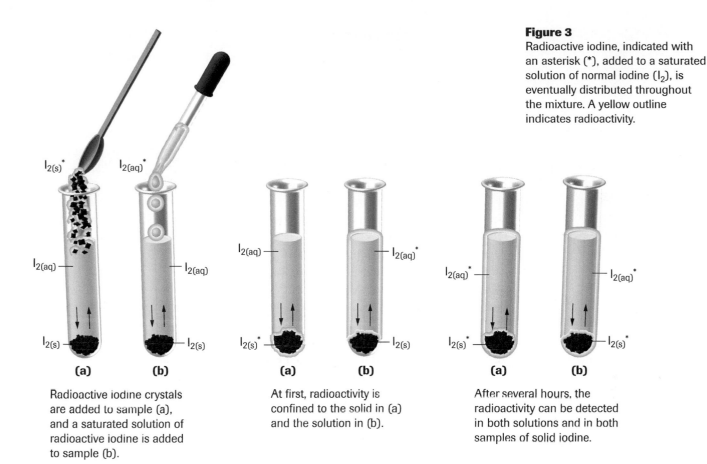

Figure 3
Radioactive iodine, indicated with
an asterisk (*), added to a saturated
solution of normal iodine (I_2), is
eventually distributed throughout
the mixture. A yellow outline
indicates radioactivity.

(a)   **(b)**

Radioactive iodine crystals
are added to sample (a),
and a saturated solution of
radioactive iodine is added
to sample (b).

(a)   **(b)**

At first, radioactivity is
confined to the solid in (a)
and the solution in (b).

(a)   **(b)**

After several hours, the
radioactivity can be detected
in both solutions and in both
samples of solid iodine.

A mixture exhibiting solubility equilibrium must contain both dissolved and undissolved solute at the same time. This state can be established by starting with a solute and adding it to a solvent. Consider adding calcium sulfate to water in a large enough quantity that not all of it will dissolve. We say we have added excess solute. We can write a dissociation equation to represent the equilibrium that is established:

$$CaSO_{4(s)} \rightleftharpoons Ca^{2+}_{(aq)} + SO^{2-}_{4(aq)}$$

Now suppose we have two solutions, one containing a very high concentration of calcium ions and the other containing a very high concentration of sulfate ions. When the two solutions are mixed, the initial rate at which $Ca^{2+}_{(aq)}$ and $SO^{2-}_{4(aq)}$ ions combine to form solid crystals is much greater than the rate at which those crystals dissolve, so we observe precipitation of $CaSO_{4(s)}$. Precipitation continues until the rates become equal and a dynamic equilibrium is established.

The important point is that, at equilibrium, the rates of the forward and reverse reactions are equal. How the equilibrium is established is not relevant. Once equilibrium is established, it is impossible to tell whether we started with separate solutions of $Ca^{2+}_{(aq)}$ and $SO^{2-}_{4(aq)}$, or with excess $CaSO_{4(s)}$ added to pure water.

LEARNING TIP

Sometimes an equals sign (=) is used in place of double arrows in an equilibrium equation.

Figure 4
A liquid placed in a sealed container, like this toy, evaporates until the vapour pressure inside the container becomes constant. A dynamic equilibrium is established at the point where the rate of evaporation is equal to the rate of condensation.

Phase Equilibrium

In a closed system, a phase change may establish an equilibrium (**Figure 4**), such as the evaporation/condensation equilibrium.

$$H_2O_{(l)} \rightleftharpoons H_2O_{(g)}$$

We can use the kinetic molecular theory to explain the establishment of this equilibrium. We assume that when a liquid is placed in a closed container, initially only evaporation occurs: Some molecules in the liquid gain enough energy in collisions to leave the surface of the liquid phase and move into the space above the liquid to start a second phase, the gas phase. As the number of molecules in the gas phase increases, however, increasingly more of them collide with the liquid surface and lose enough energy to join the condensed phase. In time, the rates at which molecules are evaporating and condensing become equal, so that, while both processes are still occurring, no further changes are observed. The amount (and thus, the pressure) of the substance in the gas phase remains constant. Note that the tendency of any liquid to evaporate increases at higher temperatures, so the concentration (and thus, the pressure) of the vapour is greater if the equilibrium is established at a higher temperature. (In an open container no equilibrium can be established because molecules that leave the surface of the liquid escape from the system and do not return to the liquid phase.)

The other common phase equilibrium is solid/liquid phase equilibrium. This equilibrium can normally be established only at the melting/freezing point. If crushed ice is placed in warm water, for example, initially the ice melts much more rapidly than the water freezes, so you observe melting as the net change. As the temperature of the water drops, the rate of melting decreases and the rate of freezing increases, until at a temperature of 0°C the rates become equal—unless more heat is added to the system, the amount of ice in the mixture will remain constant. This phenomenon has useful applications: Chemists know that in a well-stirred ice/water mixture at 101.3 kPa, the temperature must be precisely 0°C, so an ice/water slush can be used to control temperatures for experiments. This situation is illustrated by the following equilibrium equation:

$$H_2O_{(s)} \rightleftharpoons H_2O_{(l)} \qquad t = 0°C$$

> ▶ *Practice*

Understanding Concepts

1. Why does a wet towel dry out if hung over the back of a chair, but not if left in a plastic bag on the seat of the chair?

2. Use the concept of dynamic equilibrium to describe a saturated solution containing excess solute.

3. Why does stirring help to dissolve a solute?

Making Connections

4. How does the construction of a clothes dryer prevent a state of equilibrium from developing while drying clothes?

5. Why might we want to reseal opened bottles of carbonated soft drinks?

Chemical Reaction Equilibrium

Chemical reaction equilibria are more complex than phase or solubility equilibria, due to the large variety of possible chemical reactions and the greater number of substances involved.

quantitative reaction a reaction in which virtually all of the limiting reagent is consumed

You already know that some chemical reactions proceed to completion, meaning that all of the reactants are converted to products in proportions dictated by a balanced chemical equation. Such reactions are called **quantitative reactions**. For example, millions of tonnes of calcium oxide (lime) are produced annually by heating limestone (mostly calcium carbonate) in open kilns. The calcium oxide is used to produce cement, mortar, and plaster for the building industry (**Figure 5**). When heated strongly in an open system, calcium carbonate, $CaCO_{3(s)}$, decomposes into calcium oxide, $CaO_{(s)}$, and carbon dioxide, $CO_{2(g)}$, according to the following equation

$$CaCO_{3(s)} \xrightarrow{\text{heat}} CaO_{(s)} + CO_{2(g)}$$

In an open system such as this, the carbon dioxide gas escapes, preventing the reverse reaction from occurring. The system cannot reach equilibrium. This reaction is quantitative because we assume that, if an amount of calcium carbonate is heated, it will completely decompose into calcium oxide and carbon dioxide. At the end of the process, the reaction vessel contains only calcium oxide. All of the reactants have been consumed. The theoretical amount of product can be calculated using the molar ratios in the balanced chemical equation.

However, if we confine this reaction to a closed container, we find that both reactants and products are present after the reaction appears to have stopped. In a closed container, the reaction is no longer quantitative. This apparent anomaly can be explained by considering the effect of the reverse reaction.

$$CaCO_{3(s)} \longleftarrow CaO_{(s)} + CO_{2(g)}$$

The final state of a chemical system at equilibrium can be explained as a competition between collisions of reactants to form products and collisions of products to form reactants. The equilibrium equation is

$$CaCO_{3(s)} \rightleftharpoons CaO_{(s)} + CO_{2(g)}$$

This competition requires that the system be closed so that reactants and products cannot escape from the reaction container. We assume that any closed reacting system with constant macroscopic properties is in a state of dynamic equilibrium.

Allowing the reaction to reach equilibrium limits the amount of product produced. The industrial process of lime production never allows an equilibrium to be established. The reaction is carried out in a kiln that is open to the environment. The lime and carbon dioxide are continually removed as fresh reactant is fed into the system. This is called *continuous processing*.

Another industrial process, *batch processing*, is done in a closed container: Fixed amounts of each reactant are put into a closed vessel, and the products are removed when the reaction is complete. (Making popcorn in a bag in the microwave is a form of batch processing.) 🧪▌

Figure 5
A large rotary lime kiln. Raw limestone flows continuously into the large tubes, where the reaction occurs. This kiln is attached to a pulp mill in Port Alberni, B.C. The large building in the background is a heat exchanger, which uses waste heat from the kiln to generate electricity.

🧪 **INVESTIGATION 7.1.1**

Discovering the Extent of a Chemical Reaction (p. 513)
A quantitative reaction is one in which the limiting reagent is completely consumed. Is it valid to assume that all reactions are quantitative?

Another relatively simple reaction to study is the decomposition of dinitrogen tetroxide, $N_2O_{4(g)}$, into nitrogen dioxide, $NO_{2(g)}$. When $N_2O_{4(g)}$ is placed in a closed container, it decomposes into $NO_{2(g)}$ according to the equation

$$N_2O_{4(g)} \rightarrow 2\,NO_{2(g)}$$

This reaction occurs relatively quickly at first; the concentration of $N_2O_{4(g)}$ (a colourless gas) decreases as the concentration of $NO_{2(g)}$ (an orange-brown gas) increases. As time advances, the concentrations change more slowly until, eventually, equilibrium is reached (**Figure 6(a)**). At equilibrium, the concentrations of $N_2O_{4(g)}$ and $NO_{2(g)}$ remain constant (**Figure 6(b)**).

Figure 6
(a) $N_2O_{4(g)}$ in equilibrium with $NO_{2(g)}$
(b) When $N_2O_{4(g)}$ is introduced into a closed container its concentration decreases rapidly (red line) as the concentration of $NO_{2(g)}$ increases (blue line). However, soon the concentrations of both substances change more slowly, until equilibrium is reached (dotted vertical line). From this point on, the concentrations no longer change with time.

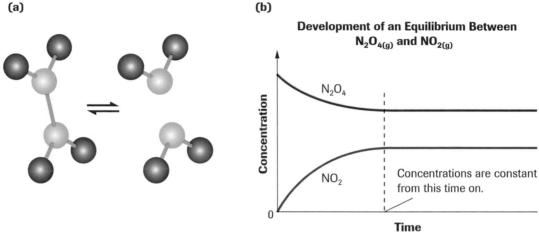

The $N_2O_{4(g)}/NO_{2(g)}$ equilibrium illustrated in **Figure 6** was created by placing $N_2O_{4(g)}$ into a closed container at SATP. Would equilibrium have developed if $NO_{2(g)}$ were the starting material? Consider two experiments, illustrated in **Figure 7**.

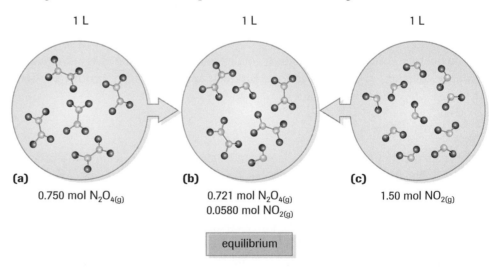

Figure 7
The same dynamic equilibrium composition is reached whether we start from pure $N_2O_{4(g)}$, pure $NO_{2(g)}$, or a mixture of the two, provided that environment, system and total mass remain the same.

In the first experiment (**Figure 7(a)**), 0.750 mol $N_2O_{4(g)}$ is placed in a 1.0-L rigid container. Since there is no $NO_{2(g)}$, the reaction will proceed to the right, reducing the concentration of $N_2O_{4(g)}$ and increasing the concentration of $NO_{2(g)}$. When equilibrium is reached, the concentration of $N_2O_{4(g)}$ is found to be 0.721 mol/L and the concentration of $NO_{2(g)}$ has increased to 0.058 mol/L (**Figure 7(b)**). In the second experiment we place 1.50 mol of $NO_{2(g)}$ in a 1.0-L container (**Figure 7(c)**) and allow the reaction to occur in the reverse direction. Once the system reaches equilibrium, the concentrations of the two compounds are measured. The concentration of $N_2O_{4(g)}$ is found to be 0.721 mol/L and the concentration of $NO_{2(g)}$ is 0.058 mol/L—exactly the same as in the first experiment (**Table 2**). The same equilibrium composition has been achieved whether the system begins with $N_2O_{4(g)}$ only or $NO_{2(g)}$ only. Experiments such as this have led scientists to the following generalization for **reversible reactions** in closed systems:

reversible reaction a reaction that can achieve equilibrium in the forward or reverse direction

> For a given overall system composition, the same equilibrium concentrations are reached whether equilibrium is approached in the forward or the reverse direction.

Table 2 Observations on $N_2O_{4(g)}$–$NO_{2(g)}$ Equilibrium

	Initial concentrations (mol/L)		Final concentrations (mol/L)	
	$N_2O_{4(g)}$	$NO_{2(g)}$	$N_2O_{4(g)}$	$NO_{2(g)}$
Experiment 1	0.75	0	0.721	0.058
Experiment 2	0	1.50	0.721	0.058

Percent Reaction at Chemical Equilibrium

Chemists have also studied the reaction of hydrogen gas and iodine gas extensively, because the molecules are relatively simple and the reaction takes place entirely in the gas phase. When hydrogen and iodine are mixed, the reaction proceeds rapidly at first. The initial dark-purple colour of the iodine vapour fades, then becomes constant (**Figure 8**).

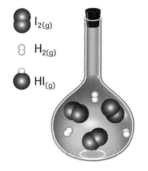 $I_{2(g)}$

$H_{2(g)}$

$HI_{(g)}$

Figure 8
Initially, hydrogen and iodine are added to the system. The colour of iodine vapour is the only easily observable property.

The reaction is

$$H_{2(g)} + I_{2(g)} \rightleftharpoons 2HI_{(g)}$$

Table 3 contains evidence from three experiments involving the hydrogen–iodine system—one in which hydrogen and iodine are mixed; one in which hydrogen, iodine, and hydrogen iodide are mixed; and one in which only hydrogen iodide is present initially.

At a temperature of 448°C, the system quickly reaches an observable equilibrium each time. Chemists use measurements such as those in **Table 3** to describe a state of equilibrium in two ways—in terms of percent reaction and in terms of an equilibrium constant.

Table 3 The Hydrogen–Iodine System at 448°C

System	Initial system concentrations (mmol/L)			Equilibrium system concentrations (mmol/L)		
	$H_{2(g)}$	$I_{2(g)}$	$HI_{(g)}$	$H_{2(g)}$	$I_{2(g)}$	$HI_{(g)}$
1	1.00	1.00	0	0.22	0.22	1.56
2	0.50	0.50	1.70	0.30	0.30	2.10
3	0	0	3.20	0.35	0.35	2.50

percent reaction the yield of product measured at equilibrium compared with the maximum possible yield of product

Analysis of the concentration values of system 1 in Table 3 shows that at 448°C it reaches an equilibrium when 78% of the maximum possible yield (the theoretical yield) is formed. The theoretical yield can be calculated stoichiometrically, as if the reaction were quantitative and forward. In this example, we would expect 2.00 mmol/L of $HI_{(g)}$ product, based on the mole ratios of the chemical equation and the concentrations we started with. The actual yield was 1.56 mmol/L, allowing us to calculate a **percent reaction**:

$$\text{percent reaction} = \frac{\text{actual product yield}}{\text{theoretical product yield}} \times 100\%$$

$$= \frac{1.56 \text{ mmol/L}}{2.00 \text{ mmol/L}} \times 100\%$$

$$= 78.0\%$$

Table 4 summarizes the percent reaction for all three systems in **Table 3**. Notice that the percent reaction is constant at 78%. Percent reaction provides an easily understood way to communicate relative amounts of chemicals present in equilibrium systems. Remember that it always refers to the amount of *product* formed in the stated reaction.

To communicate the extent of a reaction, the percent reaction is usually written above the equilibrium arrows in a chemical equation. The following equation describes the position of the hydrogen–iodine equilibrium system.

$$H_{2(g)} + I_{2(g)} \overset{78\%}{\rightleftharpoons} HI_{(g)}$$

Table 4 Percent Reaction of the Hydrogen–Iodine System at 448°C

System	Equilibrium [HI]* (mmol/L)	Maximum possible [HI]* (mmol/L)	Percent reaction (%)
1	1.56	2.00	78.0
2	2.10	2.70	77.8
3	2.50	3.20	78.1

*Square brackets [ ] indicate molar concentration.

All chemical reactions are now considered to be reversible. The question is not whether reactions go in both directions, but to what extent they go one way or the other. Reactions fall loosely into three categories.

- Reactions that favour reactants very strongly—the percent reaction is much less than 1%. In these reactions, mixing reactants has no observable result.

- Reactions that favour products very strongly, where the percent reaction is more than 99%. These reactions are observed to be complete (quantitative). These reactions are generally written with a single arrow to indicate that the effect of the reverse reaction is negligible.

- Reactions that achieve noticeable equilibrium conditions—the percent reaction lies somewhere between 1% and 99%. In these cases, significant amounts of both reactants and products are always present in a mixture in a closed system. If the percent reaction is less than 50%, reactants are favoured; if greater than 50%, products are favoured.

Table 5 shows how percent reaction is used to classify equilibrium systems and how the classification is communicated in reaction equations.

Table 5 Classes of Chemical Reactions at Equilibrium

Description of equilibrium	Position of equilibrium
no reaction (NR)	<1% $\rightleftharpoons$
reactants favoured	<50% $\rightleftharpoons$
products favoured	>50% $\rightleftharpoons$
quantitative	>99% $\rightleftharpoons$ or $\rightarrow$

As you learned in your previous chemistry courses, stoichiometric calculations are straightforward when reactions proceed to completion. However, when reversible reactions achieve equilibrium before all of the reactants become products, the stoichiometry requires a little more thought. An ICE table is a convenient way to organize the information needed to solve a stoichiometric problem involving an equilibrium system. ICE stands for *Initial*, *Change*, and *Equilibrium*, and refers to the values that are included in the table. For systems composed of aqueous solutions or gases, *I* means *initial* concentrations of reactants and products (before reaction), *C* stands for the *change* in the concentrations of reactants and products between the start and the point at which equilibrium is achieved, and *E* stands for the concentrations of reactants and products at *equilibrium*.

In the following sample problem, we will use an ICE table to calculate equilibrium concentrations.

DID YOU KNOW ?

How Napoleon Helped Discover Reversible Reactions
Napoleon recruited French chemist J. Berthellot to accompany him on an expedition to Egypt in 1798. While there, Berthellot noticed deposits of sodium carbonate, $Na_2CO_{3(s)}$, around the edges of some salt lakes. Already familiar with the reaction

$$Na_2CO_{3(aq)} + CaCl_{2(aq)} \rightarrow$$
$$CaCO_{3(s)} + 2\,NaCl_{(aq)}$$

which proceeds to completion in the laboratory, he realized that the Na_2CO_3 must have been formed by the reverse reaction. He speculated that excess $NaCl_{(aq)}$ in the water and the slow evaporation of water at the shore drove the reaction backward.

Calculating Concentrations at Equilibrium

SAMPLE problem ◀

1. **Consider the following equation for the formation of hydrogen fluoride from its elements at SATP:**

$$H_{2(g)} + F_{2(g)} \rightleftharpoons 2\,HF_{(g)}$$

If the reaction begins with 1.00 mol/L concentrations of $H_{2(g)}$ and $F_{2(g)}$ and no $HF_{(g)}$, calculate the concentrations of $H_{2(g)}$ and $HF_{(g)}$ at equilibrium if the equilibrium concentration of $F_{2(g)}$ is measured to be 0.24 mol/L.

(Notice that, in this chapter, the concentrations of gases are given in moles per litre. A concentration of 1.00 mol/L of $H_{2(g)}$ means that there is one mole of $H_{2(g)}$ per litre of space occupied. For example, a 1.00-L vessel containing 1.00 mol/L $H_{2(g)}$ and 1.00 mol/L $F_{2(g)}$ contains 1.00 mol $H_{2(g)}$ and 1.00 mol $F_{2(g)}$.)

First, list the information given in the question.

$[H_{2(g)}]_{initial} = 1.00$ mol/L

$[F_{2(g)}]_{initial} = 1.00$ mol/L

$[HF_{(g)}]_{initial} = 0.00$ mol/L

$[F_{2(g)}]_{equilibrium} = 0.24$ mol/L

We can now set up an ICE table (**Table 6**) to help organize the information. (Notice that an ICE table has the balanced equation for the process written in the first row.) Start by inserting the known values for initial concentrations into the table, directly below the corresponding entity in the balanced equation.

Table 6 Incomplete ICE Table for the Reaction of $H_{2(g)}$ and $F_{2(g)}$

	$H_{2(g)}$	+	$F_{2(g)}$	$\rightleftharpoons$	$2\ HF_{(g)}$
Initial concentration (mol/L)	1.00		1.00		0.00
Change in concentration (mol/L)					
Equilibrium concentration (mol/L)					

By the time equilibrium is reached, a certain amount of $H_{2(g)}$ and $F_{2(g)}$ will have changed into $HF_{(g)}$. The balanced equation indicates that the change occurs in a 1:1:2 molar ratio, i.e. for every mole of $H_{2(g)}$ and $F_{2(g)}$ that react, two moles of $HF_{(g)}$ are formed, but we don't know what amount of the reactants is converted into product. We choose the variable x to represent changes in the concentrations of reactants and products, with the coefficients of x corresponding to the coefficients in the balanced equation. Since $H_{2(g)}$ and $F_{2(g)}$ are consumed in a 1:1 molar ratio, the change in the concentrations of $H_{2(g)}$ and $F_{2(g)}$ are represented by $(-x\ \text{mol/L})$ (**Table 7**).

The balanced equation indicates that 2 mol of $HF_{(g)}$ are produced for every mole of $H_{2(g)}$ and $F_{2(g)}$ that reacts. The change in concentration of $HF_{(g)}$ is therefore represented by $(+2x)$. The final concentrations of $H_{2(g)}$ and $F_{2(g)}$ will be their initial concentrations, 1.00 mol/L, minus x mol/L. Therefore, we place $1.00 - x$ in the Equilibrium concentration row under both $H_{2(g)}$ and $F_{2(g)}$. As there was no hydrogen fluoride at the beginning of the reaction, the final (equilibrium) concentration of this gas is $0 + 2x$, or simply $2x$. Therefore, place $2x$ in the Equilibrium row under $HF_{(g)}$.

The ICE table now looks like this:

LEARNING TIP

Remember that the multiples of x always correspond to the coefficients in the balanced equation.

Table 7 Completed ICE Table for the Reaction of $H_{2(g)}$ and $F_{2(g)}$

	$H_{2(g)}$	+	$F_{2(g)}$	$\rightleftharpoons$	$2\ HF_{(g)}$
Initial concentration (mol/L)	1.00		1.00		0.00
Change in concentration (mol/L)	$-x$		$-x$		$+2x$
Equilibrium concentration (mol/L)	$1.00 - x$		$1.00 - x$		$2x$

Knowing that the equilibrium concentration of $F_{2(g)}$ is 0.24 mol/L, you can determine the value of x.

$$1.00\ \text{mol/L} - x = 0.24\ \text{mol/L}$$
$$-x = 0.24\ \text{mol/L} - 1.00\ \text{mol/L}$$
$$-x = -0.76\ \text{mol/L}$$
$$x = 0.76\ \text{mol/L}$$

Now use the value of x to calculate the equilibrium concentrations of the other two entities.

$$[H_{2(g)}] = 1.00\ \text{mol/L} - x$$
$$= 1.00\ \text{mol/L} - 0.76\ \text{mol/L}$$
$$[H_{2(g)}] = 0.24\ \text{mol/L}$$

$$[HF_{(g)}] = 2x$$
$$= 2(0.76\ \text{mol/L})$$
$$[HF_{(g)}] = 1.52\ \text{mol/L}$$

The equilibrium concentrations of $H_{2(g)}$ and $HF_{(g)}$ are 0.24 mol/L and 1.52 mol/L, respectively.

The hydrogen fluoride formation example is relatively simple, because the mole ratio is 1:1:2. In the following problem, the ratio is not so simple.

2. **When ammonia is heated, it decomposes into nitrogen gas and hydrogen gas according to the following equation.**

$$2 NH_{3(g)} \rightleftharpoons N_{2(g)} + 3 H_{2(g)}$$

When 4.0 mol of $NH_{3(g)}$ is introduced into a 2.0-L rigid container and heated to a particular temperature, the amount of ammonia changes as shown in Figure 9. Determine the equilibrium concentrations of the other two entities.

First, read the amounts of $NH_{3(g)}$ from the graph, and convert these amounts to concentrations, using the volume of the container.

Initial amount of $NH_{3(aq)}$ = 4.0 mol

$$[NH_{3(g)}]_{initial} = \frac{4.0 \text{ mol}}{2.0 \text{ L}} = 2.0 \text{ mol/L}$$

Final (equilibrium) amount of $NH_{3(g)}$ = 2.0 mol

$$[NH_{3(g)}]_{equilibrium} = \frac{2.0 \text{ mol}}{2.0 \text{ L}} = 1.0 \text{ mol/L}$$

Next, set up an ICE table (**Table 8**). Notice that the multiples of x correspond to the coefficients in the balanced equation. The change is $-2x$ for $NH_{3(g)}$ since it is consumed in the reaction. The change values are $+x$ and $+3x$ for $N_{2(g)}$ and $H_{2(g)}$, respectively—the signs are positive as they are formed in the reaction.

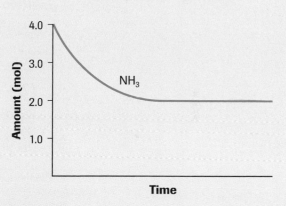

Decomposition of Ammonia

Figure 9

Table 8 ICE Table for the Decomposition of Ammonia

	$2 NH_{3(g)}$	$\rightleftharpoons$	$N_{2(g)}$	$+$	$3 H_{2(g)}$
Initial concentration (mol/L)	2.0		0.0		0.0
Change in concentration (mol/L)	$-2x$		$+x$		$+3x$
Equilibrium concentration (mol/L)	$2.0 - 2x$		x		$3x$

Now, determine the value of x using the initial, change, and equilibrium values of $NH_{3(g)}$ using the calculated equilibrium concentration for $NH_{3(g)}$.

$$[NH_{3(g)}] = 2.0 \text{ mol/L} - 2x$$
$$= 1.0 \text{ mol/L (from the earlier calculation)}$$

$$2.0 \text{ mol/L} - 2x = 1.0 \text{ mol/L}$$
$$-2x = -1.0 \text{ mol/L}$$
$$-x = -0.5 \text{ mol/L}$$
$$x = 0.5 \text{ mol/L}$$

Use the value of x to calculate the equilibrium concentrations of the other two entities.

$$[N_{2(g)}] = x$$
$$[N_{2(g)}] = 0.5 \text{ mol/L}$$

$$[H_{2(g)}] = 3x$$
$$= 3(0.5 \text{ mol/L})$$
$$[H_{2(g)}] = 1.5 \text{ mol/L}$$

The equilibrium concentrations of $N_{2(g)}$ and $H_{2(g)}$ are 0.5 mol/L and 1.5 mol/L, respectively.

Example

In a gaseous reaction system, 0.200 mol of hydrogen gas, $H_{2(g)}$, is added to 0.200 mol of iodine vapour, $I_{2(g)}$, in a rigid 2.00-L container at 448°C. At equilibrium the system contains 0.040 mol of hydrogen gas, $H_{2(g)}$. Determine the equilibrium concentrations of $H_{2(g)}$ and $HI_{(g)}$.

Solution

$$n_{H_{2(g)initial}} = 0.200 \text{ mol}$$
$$n_{I_{2(g)initial}} = 0.200 \text{ mol}$$
$$n_{H_{2(g)equilibrium}} = 0.040 \text{ mol}$$
$$\text{volume} = 2.00 \text{ L}$$

$$[H_{2(g)}]_{initial} = \frac{0.200 \text{ mol}}{2.00 \text{ L}} = 0.100 \text{ mol/L}$$

$$[I_{2(g)}]_{initial} = \frac{0.200 \text{ mol}}{2.00 \text{ L}} = 0.100 \text{ mol/L}$$

$$[H_{2(g)}]_{equilibrium} = \frac{0.040 \text{ mol}}{2.00 \text{ L}} = 0.020 \text{ mol/L}$$

Table 9 ICE Table for the Reaction of $H_{2(g)}$ and $I_{2(g)}$

	$H_{2(g)}$ +	$I_{2(g)}$ ⇌	$2 HI_{(g)}$
Initial concentration (mol/L)	0.100	0.100	0.000
Change in concentration (mol/L)	$-x$	$-x$	$+2x$
Equilibrium concentration (mol/L)	0.020	$0.100 - x$	$2x$

At equilibrium:

$$[H_{2(g)}] = 0.100 \text{ mol/L} - x$$
$$= 0.020 \text{ mol/L}$$

$$0.100 \text{ mol/L} - x = 0.020 \text{ mol/L}$$
$$-x = -0.080 \text{ mol/L}$$
$$x = 0.080 \text{ mol/L}$$

$$[I_{2(g)}] = 0.100 \text{ mol/L} - x$$
$$= 0.100 \text{ mol/L} - 0.080 \text{ mol/L}$$
$$[I_{2(g)}] = 0.020 \text{ mol/L}$$

$$[HI_{(g)}] = 2x$$
$$= 2(0.080 \text{ mol/L})$$
$$[HI_{(g)}] = 0.160 \text{ mol/L}$$

The equilibrium concentrations of $I_{2(g)}$ and $HI_{(g)}$ are 0.020 mol/L and 0.160 mol/L, respectively.

▶ Practice

Understanding Concepts

6. When carbon dioxide is heated in a closed container, it decomposes into carbon monoxide and oxygen according to the following equilibrium equation:

$$2\,CO_{2(g)} \rightleftharpoons 2\,CO_{(g)} + O_{2(g)}$$

When 2.0 mol of $CO_{2(g)}$ is placed in a 5.0-L closed container and heated to a particular temperature, the equilibrium concentration of $CO_{2(g)}$ is measured to be 0.39 mol/L. Use an ICE table to determine the equilibrium concentrations of $CO_{(g)}$ and $O_{2(g)}$.

7. At 35°C, 2.0 mol of pure $NOCl_{(g)}$ is introduced into a 2.0-L flask. The $NOCl_{(g)}$ partially decomposes according to the following equilibrium equation:

$$2\,NOCl_{(g)} \rightleftharpoons 2\,NO_{(g)} + Cl_{2(g)}$$

At equilibrium, the concentration of $NO_{(g)}$ is 0.032 mol/L. Use an ICE table to determine equilibrium concentrations of $NOCl_{(g)}$ and $Cl_{2(g)}$ at this temperature.

Answers

6. $[CO_{(g)}] = 0.01$ mol/L;
   $[O_{2(g)}] = 0.005$ mol/L

7. $[NOCl_{(g)}] = 0.968$ mol/L;
   $[Cl_{2(g)}] = 0.016$ mol/L

▶ Section 7.1 Questions

Understanding Concepts

1. For a chemical system at equilibrium,
   (a) describe the observable characteristics.
   (b) why is the equilibrium considered "dynamic"?
   (c) what is "equal" about the system?

2. Describe three equilibrium situations (not necessarily chemical equilibria) that you might encounter in your life. For each situation, state whether the system is in dynamic equilibrium or not, and explain your reasoning.

3. In a gaseous reaction system, 2.00 mol of methane, $CH_{4(g)}$, is added to 10.00 mol of chlorine, $Cl_{2(g)}$. At equilibrium, the system contains 1.40 mol of chloromethane, $CH_3Cl_{(g)}$, and some hydrogen chloride, $HCl_{(g)}$.
   (a) Write a balanced chemical equation for this equilibrium and determine the maximum possible yield of chloromethane product.
   (b) Calculate the percent reaction of this equilibrium and state whether products or reactants are favoured.

4. After 4.00 mol of $C_2H_{4(g)}$ and 2.50 mol of $Br_{2(g)}$ are placed in a sealed 1.0-L container, the reaction

$$C_2H_{4(g)} + Br_{2(g)} \rightleftharpoons C_2H_4Br_{2(g)}$$

reaches equilibrium. **Figure 10** shows the concentration of $C_2H_{4(g)}$ as it changes over time at a fixed temperature until equilibrium is reached.
   (a) Create an ICE table for all three substances and calculate equilibrium concentrations.

(b) Copy **Figure 10** and add lines to show how the concentration of each of the other two substances changes.
(c) Calculate the percent reaction at equilibrium.

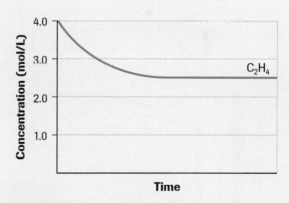

Reaction of Ethene

C_2H_4

Figure 10

5. For each of the following, write the chemical equation with appropriate equilibrium arrows, as shown in **Table 5**.
   (a) The Haber process is used to manufacture ammonia fertilizer from hydrogen and nitrogen gases. Under less than desirable conditions, only an 11% yield of ammonia is obtained at equilibrium.

(b) A mixture of carbon monoxide and hydrogen, known as coal gas, is used as a supplementary fuel in many large industries. At high temperatures, the reaction of coke and steam forms an equilibrium mixture in which the products (carbon monoxide and hydrogen gases) are favoured. (Assume that coke is pure carbon.)

(c) Because of the cost of silver, many high-school science departments recover silver metal from waste solutions containing silver compounds or silver ions. A quantitative reaction of waste silver ion solutions with copper metal results in the production of silver metal and copper(II) ions.

(d) One step in the industrial process used to manufacture sulfuric acid is the production of sulfur trioxide from sulfur dioxide and oxygen gases. Under certain conditions, the reaction produces a 65% yield of products.

6. The concept of equilibrium can be applied to many different chemical reaction systems. Use the generalizations from your study of organic chemistry to predict the position of equilibrium for bromine placed in a reaction container with ethylene at a high temperature.

7. Interpret **Figure 11** to answer the following questions about the reaction.
(a) All three substances are gases. If the container has a volume of 2.00 L, what amount (in moles) of each substance was present initially?
(b) What amount (in moles) of hydrogen iodide had formed at equilibrium? (Create an ICE table, then convert concentration to amount.)
(c) What is the percent reaction at equilibrium?
(d) In Chapter 6, the reaction rate was described as the slope of the concentration curve of a given reactant. Does this mean that the rate of reaction of all three substances becomes zero when their concentrations become constant at equilibrium? Explain.

Reaction of Hydrogen and Iodine

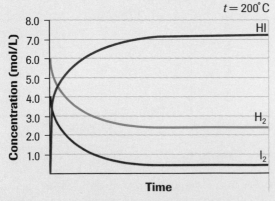

Figure 11

8. A 2.00-mol sample of phosphorus pentachloride, $PCl_{5(g)}$, is placed into a 2.00-L flask at 160°C. The reaction produces 0.200 mol of phosphorus trichloride, $PCl_{3(g)}$, and some chlorine, $Cl_{2(g)}$, at equilibrium.

$$PCl_{5(g)} \rightleftharpoons PCl_{3(g)} + Cl_{2(g)}$$

Calculate the concentration of $PCl_{5(g)}$ and $Cl_{2(g)}$ at equilibrium.

9. Methanol, $CH_3OH_{(g)}$, is manufactured from carbon monoxide, $CO_{(g)}$, and hydrogen, $H_{2(g)}$, according to the following equation:

$$CO_{(g)} + 2 H_{2(g)} \rightleftharpoons CH_3OH_{(g)}$$

A 1.00-L container is filled with 0.100 mol $CO_{(g)}$ and 0.200 mol $H_{2(g)}$. The reaction is allowed to proceed at 200°C. At equilibrium, there is 0.120 mol $H_{2(g)}$.
(a) What are the equilibrium concentrations of $CO_{(g)}$ and $CH_3OH_{(g)}$?
(b) Calculate the percent reaction.

Inquiry Skills

10. Two graduate students in a university research laboratory conducted an experiment to determine whether a saturated solution forms a dynamic equilibrium with excess solid. Given the following question, experimental design, and evidence, complete the analysis section of the report.

Question
Is a saturated solution in a state of dynamic equilibrium?

Experimental Design
A sample of solid iodine was placed in a flask containing a mixture of water and alcohol. The flask was sealed and allowed to stand undisturbed for a week. Then a sample of solid radioactive iodine, I-131, was added to the flask beside the original iodine. The flask was again sealed and left alone for a month. A Geiger counter was used to measure the radioactivity in the system after a week and after a month.

Evidence
One week after addition of I-131:
    Solution colour: red
    Amount of solid I_2 at the bottom: unchanged
One month after addition of I-131:
    Solution colour: red
    Size of both samples of iodine: unchanged
Amount of radioactivity after one month:
    the I-131 sample had lost some of its radioactivity; the solution and the original piece of iodine had become slightly radioactive.

Analysis
(a) Use the Evidence to answer the Question.

Making Connections

11. Many of the nutrients in digested food, such as the carbohydrate fructose, are absorbed into the cells of the small intestine by diffusion. This occurs because, after a meal rich in carbohydrates, the concentration of fructose in solution in the intestinal tract is higher than the concentration inside the surrounding cells.
(a) Will diffusion result in the absorption of all fructose molecules from the digesting food travelling through the intestine? Explain.
(b) Consult a general biology textbook or conduct Internet research to determine how cells ensure that they absorb the maximum possible amount of nutrients.

 www.science.nelson.com

Equilibrium Law in Chemical Reactions 7.2

When chemical reactions take place in closed systems, the forward and reverse reactions occur continuously, and the reaction mixture always contains reactants and products. In your previous studies in chemistry, you used stoichiometry to calculate amounts of reactants or products in reactions that proceeded to completion. However, in a system at equilibrium, the reaction never proceeds to completion: There are always both reactants and products in the reaction mixture. Is there a reliable quantitative measure that allows us to determine the amounts of reactants and products at equilibrium?

LAB EXERCISE 7.2.1

Develop an Equilibrium Law (p. 514)

Analyze experimental observations to discover a simple mathematical relationship among equilibrium concentrations .

The Equilibrium Constant, K

Detailed empirical studies of many equilibrium systems were conducted by two Norwegian chemists, Cato Maximilian Guldberg and Peter Waage, in the mid-1800s (**Figure 1**). By 1864 they had proposed a mathematical description of the equilibrium condition that they called the "law of mass action." Analyzing the results of their experiments, Guldberg and Waage noticed that, when they arranged the equilibrium concentrations into the following ratio, the resulting value was the same no matter what combinations of initial concentrations were mixed. They called this relationship the **equilibrium law** expression:

For the general chemical reaction

$$a\text{A} + b\text{B} \rightleftharpoons c\text{C} + d\text{D}$$

$$K = \frac{[\text{C}]^c[\text{D}]^d}{[\text{A}]^a[\text{B}]^b}$$

where
A, B, C, and D are chemical entities in gas or aqueous phases,
a, b, c, and d are the coefficients in the balanced chemical equation, and
K is a constant called the **equilibrium constant**.

Figure 1
Cato Maximilian Guldberg (1836–1902) and Peter Waage (1833–1900) were related by more than their interest in chemistry: They were brothers-in-law!

Let us follow their reasoning by taking, as an example, the formation of hydrogen iodide from its elements, according to the following reaction equation:

$$\text{H}_{2(g)} + \text{I}_{2(g)} \rightleftharpoons 2\,\text{HI}_{(g)}$$

Now consider three experiments in which we mix different concentrations of $\text{H}_{2(g)}$ and $\text{I}_{2(g)}$ in a 2.000-L closed container and allow the mixtures to react at 485°C until they achieve equilibrium. **Table 1** lists the intitial concentrations and equilibrium concentrations of hydrogen, iodine, and hydrogen iodide for each of the three experiments.

equilibrium constant, K
the value obtained from the mathematical combination of equilibrium concentrations using the equilibrium law expression

Table 1 Three Experiments with the $\text{H}_{2(g)}$-$\text{I}_{2(g)}$-$\text{HI}_{(g)}$ Equilibrium

Experiment	Initial concentration (mol/L)			Equilibrium concentration (mol/L)		
	$[\text{H}_{2(g)}]$	$[\text{I}_{2(g)}]$	$[\text{HI}_{(g)}]$	$[\text{H}_{2(g)}]$	$[\text{I}_{2(g)}]$	$[\text{HI}_{(g)}]$
1	2.000	2.000	0	0.442	0.442	3.116
2	0	0	2.000	0.221	0.221	1.560
3	0	0.010	0.350	0.035	0.045	0.280

For the hydrogen iodide reaction, the equilibrium law expression is

$$\frac{[HI_{(g)}]^2}{[H_{2(g)}][I_{2(g)}]}$$

Solving the expression for our three experiments results in the values of K shown in **Table 2**:

Table 2 Ratios of Equilibrium Concentrations

Experiment	Ratio of equilibrium concentrations	Value of K
1	$\dfrac{[HI_{(g)}]^2}{[H_{2(g)}][I_{2(g)}]} = \dfrac{(3.116)^2}{(0.442)(0.442)}$	49.8
2	$\dfrac{[HI_{(g)}]^2}{[H_{2(g)}][I_{2(g)}]} = \dfrac{(1.56)^2}{(0.221)(0.221)}$	49.8
3	$\dfrac{[HI_{(g)}]^2}{[H_{2(g)}][I_{2(g)}]} = \dfrac{(0.280)^2}{(0.035)(0.045)}$	49.8

Notice that the ratio yields a constant value of K for all three sets of equilibrium concentrations. If we carried out more experiments at the same temperature (using different initial concentrations), we would find that the ratio of equilibrium concentrations gives the same constant. Guldberg and Waage applied this analysis to a large number of different equilibrium systems and found that it produced a constant that was characteristic of that system (at a particular temperature) each time.

The equilibrium law expression and its associated constant, K, describe the behaviour of almost all gaseous and aqueous chemical equilibria. This expression can be used to accurately predict the amounts of reactants and products at equilibrium, given the amounts of the starting materials.

Note the following characteristics of the equilibrium law (equilibrium constant) expression:

- The molar concentrations of the *products* (right hand side of chemical equilibrium equation) are always multiplied by one another and written in the numerator, and the molar concentrations of the *reactants* (left hand side of chemical equilibrium equation) are always multiplied by one another and written in the denominator.
- The coefficients in the balanced chemical equation are equal to the exponents of the equilibrium law expression.
- The concentrations in the equilibrium law expression are the molar concentrations of the entities *at equilibrium*.

As usual, the concentrations of aqueous entities are given in mol/L, which means moles of aqueous entity per litre of solution. The concentrations of gaseous entities are also given in mol/L. This may seem strange, since pure gases are not in a "dissolved" state. Remember that for gases, the mol/L unit means moles of gaseous entity per litre occupied.

The Equilibrium Constant and Reaction Kinetics

Our definition of dynamic equilibrium as a balance between opposing processes occurring at the same rate indicates that reversible chemical reactions achieve equilibrium when the rate of the forward reaction equals the rate of the reverse reaction. We can use this relationship to generate a mathematical expression that helps explain the equilibrium constant.

In Chapter 6, you learned that the rate of a chemical reaction in a dilute solution is proportional to the molar concentrations of the reacting substances. For a reversible chemical reactions in which reactant A and B form products C and D,

$$A + B \overset{v_f}{\underset{v_r}{\rightleftharpoons}} C + D$$

v_f represents the rate of the forward reaction, and v_r represents the rate of the reverse reaction.

For the forward reaction,

$$A + B \overset{v_f}{\rightarrow} C + D,$$
$$v_f = k_f[A][B],$$

where k_f is the rate constant of the forward reaction.

For the reverse reaction,

$$C + D \overset{v_r}{\rightarrow} A + B,$$
$$v_r = k_r[C][D],$$

where k_r is the rate constant of the reverse reaction.

At equilibrium, the rate of the forward reaction is equal to the rate of the reverse reaction, $v_f = v_r$, and

$$k_f[A][B] = k_r[C][D],$$

where [A], [B], [C], and [D] are molar concentrations at equilibrium.

Solving for k_f/k_r, we get

$$\frac{k_f}{k_r} = \frac{[C][D]}{[A][B]}$$

The ratio of the rate constants of the forward and reverse reactions, k_f/k_r, is also a constant. This constant is the equilibrium constant, K, and (like the rate constants from which it can be derived) it varies with the temperature of the chemical system.

$$K = \frac{k_f}{k_r} = \frac{[C][D]}{[A][B]}$$

which, for an elementary process, where the equation coefficients are all 1, is equivalent to the equilibrium law expression proposed by Guldberg and Waage:

$$K = \frac{[C]^c[D]^d}{[A]^a[B]^b}$$

Writing Equilibrium Law Expressions for Equilibrium Reactions

Write the equilibrium law expression for the reaction in which nitrogen gas reacts with hydrogen gas in a closed system to produce gaseous ammonia as the only product.

Begin by writing a chemical equation for the equilibrium reaction.

$$N_{2(g)} + 3H_{2(g)} \rightleftharpoons 2NH_{3(g)}$$

Next, write the equilibrium law equation with product concentrations (multiplied by each other) in the numerator and reactant concentrations (multiplied by each other) in the denominator; and each concentration raised to exponents equal to their respective coefficients in the balanced equation.

$$K = \frac{[NH_{3(g)}]^2}{[N_{2(g)}][H_{2(g)}]^3}$$

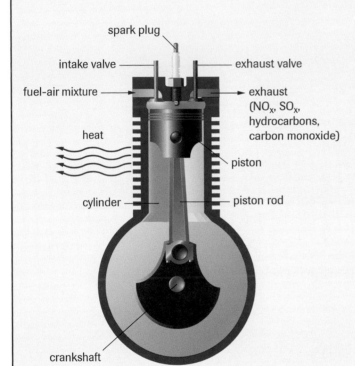

spark plug

intake valve

exhaust valve

fuel-air mixture

exhaust
(NO$_x$, SO$_x$,
hydrocarbons,
carbon monoxide)

heat

piston

cylinder

piston rod

crankshaft

Figure 2
A cylinder in an internal-combustion engine.

Example
The combustion of nitrogen, $N_{2(g)}$, in the cylinder of an internal-combustion engine produces several oxides of nitrogen (**Figure 2**). Some of these are released into the atmosphere, where they act as a source of acid rain. Write the equilibrium law equation for the reaction in which nitrogen monoxide reacts with oxygen to form nitrogen dioxide at SATP. (Assume a closed system.)

$$2NO_{(g)} + O_{2(g)} \rightleftharpoons 2NO_{2(g)}$$

Solution

$$K = \frac{[NO_{2(g)}]^2}{[NO_{(g)}]^2[O_{2(g)}]}$$

Understanding Concepts

1. Write the equilibrium law equation for each of the following reactions:
   (a) $2SO_{2(g)} + O_{2(g)} \rightleftharpoons 2SO_{3(g)}$
   (b) ammonia reacts with oxygen to form nitrogen and water vapour at SATP
   (c) $2NOBr_{(g)} \rightleftharpoons 2NO_{(g)} + Br_{2(g)}$

Calculating K

Clearly, you can calculate the value of an equilibrium constant if you know the concentrations of all entities in a chemical system at equilibrium.

Calculating K, Given Equilibrium Concentrations

Nitrogen and hydrogen combine to form ammonia (Figure 3), according to the following balanced equation:

$$N_{2(g)} + 3\,H_{2(g)} \rightleftharpoons 2\,NH_{3(g)}$$

Calculate the value of the equilibrium constant for this reaction if the following concentrations were measured at equilibrium, at 500°C:

$$[N_{2(g)}] = 1.50 \times 10^{-5}\ mol/L$$
$$[H_{2(g)}] = 3.45 \times 10^{-1}\ mol/L$$
$$[NH_{3(g)}] = 2.00 \times 10^{-4}\ mol/L$$

Start by writing the equilibrium law equation, making sure you place products in the numerator and reactants in the denominator.

$$K = \frac{[NH_{3(g)}]^2}{[N_{2(g)}][H_{2(g)}]^3}$$

Now, substitute the equilibrium concentrations into the equilibrium law equation and solve.

$$K = \frac{(2.00 \times 10^{-4}\ mol/L)^2}{(1.50 \times 10^{-5}\ mol/L)(3.45 \times 10^{-1}\ mol/L)^3}$$

$$K = 6.49 \times 10^{-2}$$

The equilibrium constant is 6.49×10^{-2} for this reaction at 500°C.

Note that we did not cancel units, or include units in the final statement of the value of K. It is common to ignore units and write only the numerical value of an equilibrium constant. The units of equilibrium constants vary, since they depend on the coefficients of the balanced chemical equations. For some equilibrium constants, all the units cancel. In this case, they do not; the units are $(mol^{-2} \cdot L^{-2})$. In subsequent examples we will not show units in the equilibrium law expression or with equilibrium constants. However, when doing equilibrium law calculations, you must ensure that all concentrations are expressed in mol/L before you substitute the values into the expression.

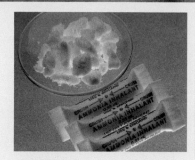

Figure 3
Ammonia is the gas, released from smelling salts, that irritates the nasal passages and stimulates a sharp intake of breath, rousing a woozy athlete.

Example

Calculate the value of the equilibrium constant for the decomposition of ammonia at 500°C into its elements—the reverse of the reaction used in the above sample problem. Use the same equilibrium concentrations.

Solution

$$2\,NH_{3(g)} \rightleftharpoons N_{2(g)} + 3\,H_{2(g)}$$

$$K = \frac{[N_{2(g)}][H_{2(g)}]^3}{[NH_{3(g)}]^2}$$

$$= \frac{[1.50 \times 10^{-5}][3.45 \times 10^{-1}]^3}{[2.00 \times 10^{-4}]^2}$$

$$K = 15.4$$

The equilibrium constant for the decomposition of ammonia at 500°C is 15.4.

▸

Answers

2. 10.6

3. 0.020

> ▶ **Practice**

Understanding Concepts

2. Carbon monoxide reacts with hydrogen to form methanol, according to the following equation:

$$CO_{(g)} + 2H_{2(g)} \rightleftharpoons CH_3OH_{(g)}$$

Calculate the value of the equilibrium constant at 327°C if an equilibrium mixture contains the following concentrations of reactants and products:

$[CO_{(g)}] = 0.079$ mol/L; $[H_{2(g)}] = 0.158$ mol/L; $[CH_3OH_{(g)}] = 0.021$ mol/L

3. For the reaction

$$2HI_{(g)} \rightleftharpoons H_{2(g)} + I_{2(g)}$$

the equilibrium concentrations at 440°C are
$[HI_{(g)}] = 1.870$ mol/L, $[H_{2(g)}] = 1.065$ mol/L, $[I_{2(g)}] = 0.065$ mol/L.
Calculate the value of K for this equilibrium.

If you look closely at the Sample Problem and Example above, you will see that the equilibrium constant for the formation of ammonia (6.49×10^{-2}) and the equilibrium constant for the decomposition of ammonia (15.4) are reciprocal values. For the purpose of comparison, we will let K represent the equilibrium constant of the forward reaction and K' represent the equilibrium constant of the reverse reaction.

Forward reaction: $N_{2(g)} + 3H_{2(g)} \rightleftharpoons 2NH_{3(g)}$

$$K = \frac{[NH_{3(g)}]^2}{[N_{2(g)}][H_{2(g)}]^3} = 6.49 \times 10^{-2}$$

Reverse reaction: $2NH_{3(g)} \rightleftharpoons N_{2(g)} + 3H_{2(g)}$

$$K' = \frac{[N_{2(g)}][H_{2(g)}]^3}{[NH_{3(g)}]^2} = 15.4$$

$$K = \frac{1}{K'}$$

$$= \frac{1}{15.4}$$

$$K = 6.49 \times 10^{-2}$$

In general, the equilibrium constant of a forward reaction and the equilibrium constant of the reverse reaction are reciprocal quantities.

> ▶ **Practice**

Understanding Concepts

4. The equilibrium constant equation for a reaction is

$$K = \frac{[H_{2(g)}]^4[CS_{2(g)}]}{[H_2S_{(g)}]^2[CH_{4(g)}]}$$

Write a balanced equation for this reaction.

5. Write a balanced equation for the reaction with the following equilibrium law equation expression:

$$K = \frac{[SO_{2(g)}]^2[H_2O_{(g)}]^2}{[O_{2(g)}]^3[H_2S_{(g)}]^2}$$

6. For the following equilibrium expression and its value, write a balanced equation for the reverse reaction and calculate the value of the equilibrium constant for the reverse reaction, K':

$$\frac{[C_2H_{6(g)}][H_{2(g)}]}{[CH_{4(g)}]^2} = 9.5 \times 10^{-13}$$

Answer

6. $K' = 1.1 \times 10^{12}$

Limitations of Equilibrium Constants and Percent Reaction Values

The *position* of equilibrium is a measure of the extent to which reactants become products in a closed system. Both methods of expressing the position of an equilibrium—the equilibrium constant and the percent reaction—have limited application. The value of the equilibrium constant, K, depends on temperature. Any stated numerical value for an equilibrium constant, or any calculation using an equilibrium law equation, must specify a temperature.

Notice how K values change with temperature for the reaction that produces ammonia.

$$N_{2(g)} + 3\,H_{2(g)} \rightleftharpoons 2\,NH_{3(g)}$$

$K = 4.26 \times 10^8$ at 25°C

$K = 1.02 \times 10^{-5}$ at 300°C

$K = 8.00 \times 10^{-7}$ at 400°C

Percent reaction values are dependent on both temperature and concentration.

Equilibrium constants and percent reaction values give no information about the rate of reaction; they provide only a measure of the equilibrium position of the reaction.

Heterogeneous Equilibria

Equilibrium systems can involve solids, liquids, gases, and aqueous solutions. Most of the systems you will study in this chapter are **homogeneous equilibria**. This means that the reactants and products are all in the same phase: all gases or all aqueous solutions. However, in some systems, the reactants and products are in different phases. These are called **heterogeneous equilibria**. For example, the electrolysis of water in a closed container involves liquid and gas phases:

$$2\,H_2O_{(l)} \rightleftharpoons 2\,H_{2(g)} + O_{2(g)}$$

If we follow the conventions for writing equilibrium law equations, the equation for this equilibrium is

$$K = \frac{[H_{2(g)}]^2[O_{2(g)}]}{[H_2O_{(l)}]^2}$$

However, the concentration of liquid water written in the denominator of this expression is problematic. Earlier you learned that, at a particular temperature, the value of K remains constant for all combinations of reactant and product concentrations at equilibrium. The concentration of a pure liquid cannot change: It is fixed and equal to the substance's

homogeneous equilibria equilibria in which all entities are in the same phase

heterogeneous equilibria equilibria in which reactants and products are in more than one phase

density. For example, a litre of liquid water at SATP has a mass of 1.00 kg. This is equal to 55.5 moles. Therefore, water has a "concentration" of 55.5 mol/L at SATP. In fact, this is simply water's density expressed in mol/L instead of the more common g/L or kg/m^3 units. Adding or removing water from the electrolysis reaction vessel does not change its concentration (density). However, adding or removing $H_{2(g)}$ or $O_{2(g)}$ does change their concentrations.

Since the concentration of water is itself a constant, its value is incorporated into the K value shown above, yielding the equilibrium constant that is reported for this reaction in standard reference tables.

$$K[H_2O_{(l)}]^2 = [H_{2(g)}]^2[O_{2(g)}]$$
$$K(55.5)^2 = [H_{2(g)}]^2[O_{2(g)}]$$

Since K is a constant, and $(55.5)^2$ is also constant, the equation can be rewritten as

$$K = [H_{2(g)}]^2[O_{2(g)}]$$

In general, the concentrations of entities in a condensed state (solids and liquids) are not included as variables in the equilibrium law expression, but rather are incorporated into the value of the equilibrium constant.

Note that if the electrolysis of water is carried out with water vapour instead of liquid water, the equilibrium law equation would be

$$K = \frac{[H_{2(g)}]^2[O_{2(g)}]}{[H_2O_{(g)}]^2}$$

Water vapour is a gas, just like hydrogen and oxygen, and its concentration varies.

▶ **SAMPLE** problem | **Writing Equations for Reactions Involving Condensed States**

Write the equilibrium law equation for the decomposition of solid ammonium chloride to gaseous ammonia and hydrogen chloride gas.

$$NH_4Cl_{(s)} \rightleftharpoons NH_{3(g)} + HCl_{(g)}$$

The concentration of the solid, $NH_4Cl_{(s)}$, is omitted from the equilibrium expression. The value of K includes the density of $NH_4Cl_{(s)}$.

$$K = [NH_{3(g)}][HCl_{(g)}]$$

Example
Write the equilibrium law equation for the preparation of quicklime, $CaO_{(s)}$, by heating calcium carbonate, $CaCO_{3(s)}$, from seashells.

Solution

$$CaCO_{3(s)} \rightleftharpoons CaO_{(s)} + CO_{2(g)}$$
$$K = [CO_{2(g)}]$$

Since equilibrium depends on the concentrations of reacting substances, substances must be represented in the equilibrium expression as they actually exist—meaning that ions in solution must be represented as individual entities. Equilibrium law expressions are always written from the *net ionic* form of reaction equations, balanced with simplest whole number (integral) coefficients.

Writing Equations for Reactions Involving Ions

Write the equilibrium law equation for the reaction of zinc in copper(II) chloride solution.

First, write a complete ionic equation to represent the reaction.

$$Zn_{(s)} + Cu^{2+}_{(aq)} + 2Cl^-_{(aq)} \rightleftharpoons Cu_{(s)} + Zn^{2+}_{(aq)} + 2Cl^-_{(aq)}$$

The factors for spectator ions simply cancel out, resulting in the following net ionic equation:

$$Zn_{(s)} + Cu^{2+}_{(aq)} \rightleftharpoons Cu_{(s)} + Zn^{2+}_{(aq)}$$

Next, write the equilibrium law equation.

$$K = \frac{[Cu_{(s)}][Zn^{2+}_{(aq)}]}{[Zn_{(s)}][Cu^{2+}_{(aq)}]}$$

We can rearrange this equation to separate the variables from the constants:

$$K\frac{[Zn_{(s)}]}{[Cu_{(s)}]} = \frac{[Zn^{2+}_{(aq)}]}{[Cu^{2+}_{(aq)}]}$$

$[Zn_{(s)}]/[Cu_{(s)}]$ is a constant, so is incorporated into the value of K, leaving

$$K = \frac{[Zn^{2+}_{(aq)}]}{[Cu^{2+}_{(aq)}]}$$

Note that the constant concentrations of the solids, as well as the concentration of spectator ions (the chloride ions in this example), don't appear in the equilibrium expression.

▶ Practice

Understanding Concepts

7. Write the equilibrium law equations for the following reactions.
   (a) iron in nickel(II) chloride solution
   (b) $3\,Zn_{(s)} + 2\,CrBr_{3(aq)} \rightleftharpoons 2\,Cr_{(s)} + 3\,ZnBr_{2(aq)}$
   (c) $2\,NaHCO_{3(s)} \rightleftharpoons Na_2CO_{3(s)} + H_2O_{(g)} + CO_{2(g)}$

The Magnitude of *K*

The magnitude of the equilibrium constant provides a measure of the extent to which the reaction has gone to completion when equilibrium is reached. Consider the reaction of carbon monoxide with oxygen to produce carbon dioxide.

$$2\,CO_{(g)} + O_{2(g)} \rightleftharpoons 2\,CO_{2(g)}$$

The value of *K* for this reaction is 3.3×10^{91} at 25°C. Therefore, at this temperature,

$$\frac{[CO_{2(g)}]^2}{[CO_{(g)}]^2[O_{2(g)}]} = 3.3 \times 10^{91}$$

The very large value of *K* tells us that the concentration of $CO_{2(g)}$ will be very large compared to the concentrations of $CO_{(g)}$ and $O_{2(g)}$ at equilibrium. We can assume that this reaction goes essentially to completion as it does in open systems (**Figure 4**).

For the equilibrium reaction

$$NO_{2(g)} + NO_{(g)} \rightleftharpoons N_2O_{(g)} + O_{2(g)} \qquad K = 0.914 \text{ at } 500°C$$

$$\frac{[N_2O_{(g)}][O_{2(g)}]}{[NO_{2(g)}][NO_{(g)}]} = 0.914$$

Figure 4
Fortunately, in an open system, poisonous carbon monoxide reacts almost completely with oxygen in the air to become relatively harmless carbon dioxide.

In this case, the value of K is very close to 1. This means that the concentrations of the products will be approximately equal to the concentrations of the reactants at equilibrium.

Now, consider the equilibrium constant for the thermal decomposition of water.

$$2\,H_2O_{(g)} \rightleftharpoons 2\,H_{2(g)} + O_{2(g)} \qquad\qquad K = 7.3 \times 10^{-10} \text{ at } 1000°C$$

$$\frac{[H_{2(g)}]^2[O_{2(g)}]}{[H_2O_{(g)}]^2} = 7.3 \times 10^{-10}$$

Since the value of K is so small, the concentration of water vapour (the reactant) is very much larger than the concentrations of hydrogen gas and oxygen gas (the products) at equilibrium. At this temperature, the reaction hardly proceeds at all to the right.

Table 3 summarizes the relationship between the equilibrium constant and the position of the reactants and products at equilibrium.

Table 3 The Magnitude of the Equilibrium Constant

$K \gg 1$ The reaction proceeds toward completion. The concentrations of products are much greater than the concentrations of reactants at equilibrium.	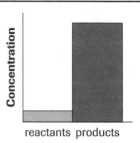
$K \doteq 1$ The concentrations of reactants and products are approximately equal at equilibrium.	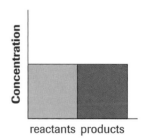
$K \ll 1$ Very small amounts of products formed. The concentrations of reactants are much greater than the concentrations of products at equilibrium.	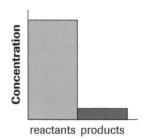

▶ *Section 7.2 Questions*

Understanding Concepts

1. Write a balanced equation with integer coefficients and the expression for the equilibrium constant for each of the following reaction systems.
   (a) Hydrogen gas reacts with chlorine gas to produce hydrogen chloride gas.
   (b) In the Haber process, nitrogen reacts with hydrogen to produce ammonia gas.

(c) At some time in the future, industry and consumers may make more extensive use of the combustion of hydrogen as an energy source.
(d) When aqueous ammonia is added to an aqueous nickel(II) ion solution, the $Ni(NH_3)_6^{2+}_{(aq)}$ complex ion is formed (**Figure 5**).
(e) In the Solvay process for making washing soda, one reaction involves heating solid calcium carbonate

(limestone) to produce solid calcium oxide (quicklime) and carbon dioxide.

(f) In a sealed can of soda, carbonic acid, $H_2CO_{3(aq)}$, decomposes to liquid water and carbon dioxide gas.

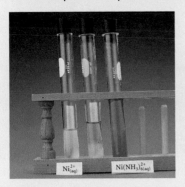

Figure 5
A $Ni^{2+}_{(aq)}$ solution is green. Ammonia reacts with the nickel(II) ion to form the intensely blue hexaaminenickel(II) ion, $Ni(NH_3)^{2+}_{6(aq)}$.

2. Write the expression of the equilibrium constant for the hydrogen–iodine–hydrogen iodide system at 485°C. Using the values in **Table 1** on page 439, calculate the value of the equilibrium constant.

3. What is the value of the equilibrium constant at 200°C for the decomposition of phosphorus pentachloride gas to phosphorus trichloride gas and chlorine gas? At equilibrium,

$[PCl_{3(g)}] = [Cl_{2(g)}]$
$\qquad = 0.014$ mol/L

$[PCl_{5(g)}] = 4.3 \times 10^{-4}$ mol/L

4. In the Haber process for synthesizing ammonia gas (**Figure 6**), the value of K is 8.00×10^{-7} for the reaction at 400°C. In a sealed container at equilibrium at 400°C, the concentrations of $H_{2(g)}$ and of $N_{2(g)}$ are measured to be 0.50 mol/L and 1.50 mol/L, respectively. Write the equilibrium constant equation and calculate the equilibrium concentration of $NH_{3(g)}$.

5. Liquid butane in lighters escapes in the gas phase when the lighter is opened. When closed, a small amount of the gas occupies the space above the liquid.
   (a) Write a chemical equation and an equilibrium law equation for this phase-change reaction.
   (b) Explain why this equation yields a constant value at a particular temperature.

recycling

impure
N_2, H_2

unwanted trace
gases removed

pure
N_2, H_2

catalytic
reactor

NH_3
$+N_2, H_2$

cooling
chamber

liquid NH_3
(yield 20%
on each
cycle)

Figure 6
The production of ammonia by the Haber process

(c) Is the concentration of butane gas in the lighter related to the quantity of liquid butane present? Explain.

6. At a certain constant (very high) temperature, 1.00 mol of HBr(g) is introduced into a 2.00-L container. Decomposition of this gas to hydrogen and bromine gases quickly establishes an equilibrium, at which point the molar concentration of HBr(g) is measured to be 0.100 mol/L.
   (a) Write a balanced equation for the reaction.
   (b) Write the equilibrium constant expression.
   (c) Calculate the amount (in moles) of $HBr_{(g)}$ present at equilibrium.
   (d) Calculate the amount of $HBr_{(g)}$ that reacted.
   (e) Calculate the amounts of $H_{2(g)}$ and $Br_{2(g)}$ produced.
   (f) Calculate the concentration of all substances present at equilibrium.
   (g) Calculate K for this reaction at this temperature.

7. (a) Describe two methods for stating how far a reaction proceeds.
   (b) How are these two methods similar, and how are they different?
   (c) What are the advantages and restrictions of each method?

Applying Inquiry Skills

8. A student carried out an experiment to verify that the equilibrium law expression for a chemical system at equilibrium yields a constant.

Question
Is the expression

$$\frac{[SO_{3(g)}]^2}{[SO_{2(g)}]^2[O_{2(g)}]}$$

a constant?

Experimental Design
Three trials were conducted in which different concentrations of gaseous sulfur dioxide, oxygen, and sulfur trioxide were combined and allowed to reach equilibrium in a closed container at 600°C.

Evidence
Table 4 Equilibrium Concentrations (mol/L)

	$[SO_{2(g)}]$	$[O_{2(g)}]$	$[SO_{3\,(g)}]$
Trial 1	1.50×10^{-1}	1.26×10^{-2}	3.50×10^{-2}
Trial 2	5.90×10^{-3}	4.50×10^{-4}	2.60×10^{-3}
Trial 3	1.00×10^{-2}	3.0×10^{-2}	3.6×10^{-3}

Analysis
(a) Use the Evidence to answer the Question.

Making Connections

9. Ozone is a poisonous form of oxygen that in small concentrations is sometimes used to disinfect drinking water. Given the following equilibrium constant, explain why the formation of ozone in the air of your classroom does not pose a health risk.

$$3\,O_{2(g)} \rightleftharpoons 2\,O_{3(g)} \qquad K = 1.6 \times 10^{-56} \text{ at } 25°$$

Consider again the student who is attempting to maintain her position while walking up the "down" escalator. What happens if the escalator suddenly starts moving faster? She notices that her steps are no longer keeping up with the escalator's new speed. She is drifting downward. To reestablish equilibrium she must increase her stepping rate to match the movement of the escalator. After doing so, she is again in dynamic equilibrium, but at a lower level on the escalator. The student, wanting to remain at rest, counteracts changes in the system in order to maintain her state of dynamic equilibrium. In doing so, she establishes a new equilibrium at a different level.

Similar adjustments occur when chemical systems at equilibrium are disturbed.

Le Châtelier's Principle

In 1884, the brilliant French chemist Henry Louis Le Châtelier (**Figure 1**) made observations of chemical systems at equilibrium. As a result of his studies, he developed a generalization that has become one of the most powerful laws of science:

> **Le Châtelier's Principle**
> When a chemical system at equilibrium is disturbed by a change in a property, the system adjusts in a way that opposes the change.

The application of Le Châtelier's principle involves an initial equilibrium state, a shifting "non-equilibrium" state, and a new equilibrium state.

Le Châtelier's principle provides a method of predicting the response of a chemical system to a change of conditions. Using this simple approach, chemical engineers can produce more of the desired product, making technological processes more efficient and more economical. For example, Fritz Haber used Le Châtelier's principle to devise a process for the economical production of ammonia from atmospheric nitrogen. (See the Haber process, Figure 6, Section 7.2.)

Le Châtelier's Principle and Concentration Changes

Le Châtelier's principle predicts that, if more of a reactant is added to a system at equilibrium, then that system will undergo an **equilibrium shift**.

The effect of adding more of a reactant is that we first observe the reactant concentration decreasing, as some of the added reactant changes to products. This period of change ends with the establishment of a new equilibrium state, in which concentrations are usually different from their original values. The system changes in a way that opposes the change. For example, the production of Freon-12, a chlorofluorocarbon (CFC), involves the following equilibrium reaction:

$$CCl_{4(l)} + 2\,HF_{(g)} \rightleftharpoons CCl_2F_{2(g)} + 2\,HCl_{(g)}$$
$$\text{Freon-12}$$

To improve the yield of Freon-12, more hydrogen fluoride is added to the equilibrium system. The additional reactant disturbs the equilibrium state and the system shifts to the right, consuming some of the added hydrogen fluoride by reaction with carbon tetrachloride. As a result, more Freon-12 is produced and a new equilibrium state is obtained.

Figure 1
Henry-Louis Le Châtelier, 1850–1936, a French chemist and engineer, worked in chemical industries. To maximize the yield of products, Le Châtelier used systematic trial and error. After measuring properties of many equilibrium states in chemical systems, he discovered a pattern and stated it as a generalization. This generalization has been supported extensively by evidence and is now considered an extremely useful tool in chemistry. It has become known as Le Châtelier's principle.

equilibrium shift movement of a system at equilibrium, resulting in a change in the concentrations of reactants and products

In chemical equilibrium shifts, the imposed concentration change is normally only *partially* counteracted, and the concentrations of the reactants and products in the final equilibrium are usually different from the values in the original equilibrium state. This can be seen in the concentration–time graph in **Figure 2**. Notice that the graph has three lines representing changes in the concentrations of $CO_{2(g)}$, $CO_{(g)}$ and $O_{2(g)}$. The amount of $CO_{2(g)}$ in the vessel is reduced in the shift.

The removal of a product (if the removal decreases concentration) will also shift an equilibrium forward, to the right. The carbon dioxide reaction can be shifted forward by removing either gaseous product (**Figure 3**), since decreasing the amount of a gas lowers its concentration in the reaction container.

The effects of forward and reverse shifts in equilibrium are shown in the iron (III) thiocyanate system in **Figure 4** (p. 452).

Adjusting an equilibrium state by adding and/or removing a substance is a common application of Le Châtelier's Principle. For industrial chemical reactions, engineers strive to design processes where reactants are added continuously and products are continuously removed, so that an equilibrium is never allowed to establish. If the reaction is always shifting forward, the process is always making product (and presumably, the industry is always making money).

Consider an industrial example, the final step in the production of nitric acid, which is represented by the reaction

$$3\,NO_{2(g)} + H_2O_{(l)} \rightleftharpoons 2\,HNO_{3(aq)} + NO_{(g)}$$

In this process, nitrogen monoxide gas is removed from the chemical system by a further reaction with oxygen gas. The removal of the nitrogen monoxide causes the system to shift to the right—more nitrogen dioxide and water react, replacing some of the removed nitrogen monoxide. As the system shifts, more of the desired product, nitric acid, is produced.

A vitally important biological equilibrium is that of hemoglobin, $Hb_{(aq)}$, (a protein complex in red blood cells that transports oxygen around the body), oxygen, and oxygenated hemoglobin, $HbO_{2(aq)}$.

$$Hb_{(aq)} + O_{2(aq)} \rightleftharpoons HbO_{2(aq)}$$

As blood circulates to the lungs, the high concentration of dissolved oxygen shifts the equilibrium to the right and the hemoglobin becomes oxygenated (**Figure 5**). As the blood circulates throughout the body, cellular respiration consumes oxygen. This removal of oxygen shifts the equilibrium to the left, and more oxygen is released.

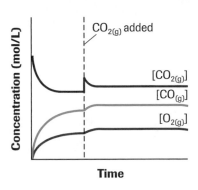

Figure 2
The reaction establishes an equilibrium that is then disturbed (at the time indicated by the vertical dotted line) by the addition of $CO_{2(g)}$. Some of the added $CO_{2(g)}$ reacts, decreasing its concentration, while the concentration of both products increases until a new equilibrium is established. The concentrations eventually become constant again, at a new level. However, the initial K value and the final K value are the same.

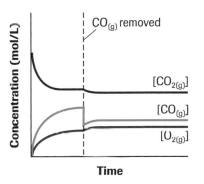

Figure 3
The reaction establishes an equilibrium that is disturbed (at the time indicated by the vertical dotted line) by the removal of $CO_{(g)}$. The equilibrium shifts forward; the concentration of $O_{2(g)}$ increases while the concentration of $CO_{2(g)}$ decreases, until a new equilibrium is established. The initial K value and the final K value are the same.

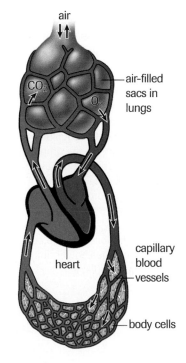

Figure 5
Oxygenated blood from the lungs is pumped by the heart to body tissues. The deoxygenated blood returns to the heart and is pumped to the lungs. Shifts in equilibrium in this system occur continuously.

Figure 4
Disturbing the iron(III) thiocyanate equilibrium

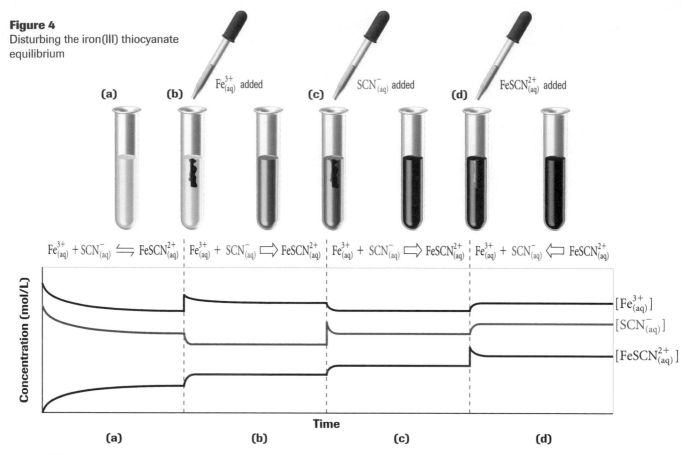

When solutions containing $Fe^{3+}_{(aq)}$ (colourless) and $SCN^-_{(aq)}$ (brown) are mixed, an equilibrium is reached with the product, $FeSCN^{2+}_{(aq)}$ (deep red), as shown by the constant, uniform light brown colour of the equilibrium solution. On the graph, notice that the concentrations of $Fe^{3+}_{(aq)}$ and $SCN^-_{(aq)}$ drop after mixing as they react to form $FeSCN^{2+}_{(aq)}$. All three concentrations become constant when equilibrium is reached (flat lines).

$Fe^{3+}_{(aq)}$ is added. In response the system shifts to the right, producing more red $FeSCN^{2+}_{(aq)}$. Notice the spike in the graph of $Fe^{3+}_{(aq)}$ when more is added, and that the concentration of $Fe^{3+}_{(aq)}$ subsequently drops. The concentration of $FeSCN^{2+}_{(aq)}$ rises as more is produced. As $SCN^-_{(aq)}$ ions are used up, the concentration drops. Equilibrium is reestablished at a new level (flat lines).

The addition of more solution containing $SCN^-_{(aq)}$ shifts the equilibrium to the right, producing more of the dark red $FeSCN^{2+}_{(aq)}$ ions. Note the corresponding changes in the graph.

Adding $FeSCN^{2+}_{(aq)}$ ions to the mixture forces the equilibrium to shift toward the reactants, giving the solution a slightly darker colour due to an overall increase in concentrations. Note the corresponding changes in the graph

Rate Theory and Concentration Changes

Kinetic theory provides a simple explanation of the equilibrium shift that occurs when a reactant concentration is increased. We assume that when reactant is added, with more reactant particles present per unit volume, collisions are suddenly much more frequent for the forward reaction. This *increases* the forward rate significantly. Since the reverse reaction rate is not immediately changed, the rates are no longer equal, and for a time the difference in rates results in an observed increase of products.

Of course, as the concentration of products increases, so does the reverse reaction rate. The forward rate decreases as reactant is consumed, until eventually the two rates (forward and reverse) become equal to each other again. The rates at the new equilibrium state are faster than those at the original state, because the system now contains a larger number of particles (and therefore, there are more collisions) in dynamic equilibrium.

If a substance is removed, causing an equilibrium shift, the explanation is similar except that the initial effect is to suddenly *decrease* one of the equilibrium reaction rates.

Remember that the addition or removal of a substance in solid or liquid state does not change the *concentration* of that substance. The reaction of condensed phases (solids and liquids) takes place only at an exposed surface—and if the surface area exposed is changed it is always exactly the same change in available area for both forward and reverse reaction collisions. The forward and reverse rates will change by exactly the same amount if they change at all, so equilibrium is not disturbed and no shift occurs.

Le Châtelier's Principle and Temperature Changes

The energy in a chemical equilibrium equation can be treated as though it were a reactant or a product.

Endothermic reaction: reactants + energy $\rightleftharpoons$ products

Exothermic reaction: reactants $\rightleftharpoons$ products + energy

Energy can be added to or removed from a system by heating or cooling the container. In either situation, the equilibrium shifts to minimize the change. If the system is cooled, the system tries to "warm" itself and the equilibrium shifts in the direction that produces heat. If heat is added, the equilibrium shifts in the direction that absorbs heat.

The equilibrium between two oxides of nitrogen illustrates the effect of temperature on a system at equilibrium. The equation for this reaction is

$N_2O_{4(g)}$ + energy $\rightleftharpoons$ 2 $NO_{2(g)}$
colourless   reddish-brown

When the system at equilibrium is heated, the reaction shifts to the right, increasing the concentration of nitrogen dioxide. This is made visible by the intensification of the reddish-brown colour of the reaction mixture (**Figure 6**). The energy that is added causes the system to shift to the right, absorbing some of the added energy.

In the exothermic production of sulfur trioxide, as part of the contact process for making sulfuric acid, the product is favoured if the temperature of the system is kept low.

2 $SO_{2(g)}$ + $O_{2(g)}$ $\rightleftharpoons$ 2 $SO_{3(g)}$ + energy

Removing energy (cooling) causes the system to shift to the right. This shift yields more sulfur trioxide while at the same time partially replacing the energy that was removed.

Rate Theory and Energy Changes

Kinetic theory explains the equilibrium shift that occurs when the energy of a system at equilibrium is increased or decreased. Consider the contact process reaction equation mentioned above—a typical exothermic reaction.

2 $SO_{2(g)}$ + $O_{2(g)}$ $\rightleftharpoons$ 2 $SO_{3(g)}$ + energy

Rate theory explains the result of cooling this exothermic system by assuming that both forward and reverse reaction rates are slower at the lower temperature, but that the reverse rate decreases *more* than the forward rate. While the rates remain unequal, the observed result is the production of more product and more energy. The shift causes concentration changes that will increase the reverse rate and decrease the forward rate until they become equal again, at the new, lower temperature (**Figure 7**).

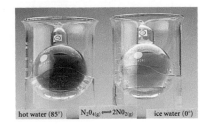

hot water (85°) $N_2O_{4(g)} \rightleftharpoons 2NO_{2(g)}$ ice water (0°)

Figure 6
Each of these flasks contains an equilibrium mixture of dinitrogen tetroxide and nitrogen dioxide. Shifts in equilibrium can be seen when one of the flasks is heated or cooled.

2 $SO_{2(g)}$ + $O_{2(g)}$ $\rightleftharpoons$ 2 $SO_{3(g)}$ + energy

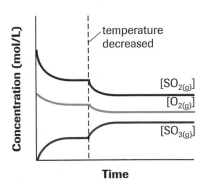

Figure 7
The reaction establishes an equilibrium that is disturbed (at the time indicated by the vertical dotted line) by a decrease in temperature. The equilibrium shifts forward, increasing the concentration of $SO_{3(g)}$ product while decreasing the concentration of both reactants, until a new equilibrium is established.

$$CO_{2(g)} + energy \rightleftharpoons 2\,CO_{(g)} + O_{2(g)}$$

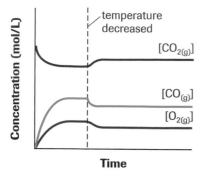

Figure 8
The reaction establishes an equilibrium that is then disturbed (at the time indicated by the vertical dotted line) by a decrease in temperature. The equilibrium shifts in the reverse direction, increasing the concentration of $CO_{2(g)}$ while decreasing the concentrations of $CO_{(g)}$ and $O_{2(g)}$, until a new equilibrium is established.

For endothermic reactions, the situation is reversed. Consider an endothermic reaction such as the decomposition of carbon dioxide into carbon monoxide and oxygen.

$$CO_{2(g)} + 566\,kJ \rightleftharpoons 2\,CO_{(g)} + O_{2(g)}$$

Cooling the system, and maintaining it at a lower temperature, causes the rate of the forward and reverse reactions to decrease, but the rate of the forward reaction decreases *more* than the reverse reaction. While the rates remain unequal, the observed result is the production of more reactant and more energy. The shift causes concentration changes that will increase the reverse rate and decrease the forward rate until they become equal again, at the new, lower temperature (**Figure 8**).

Le Châtelier's Principle and Gas Volume Changes

According to Boyle's law, the concentration of a gas in a container is directly related to the pressure of the gas.

Decreasing the volume to half its original value doubles the concentration of every gas in the container. Changing the volume of any equilibrium system involving gases *may* cause a shift in the equilibrium. To predict whether a change in pressure will affect the equilibrium state, you must consider the total number of moles of gas reactants and the total number of moles of gas products. For example, in the equilibrium reaction where nitrogen and hydrogen form ammonia, four moles of gaseous reactants produce two moles of gaseous product.

$$N_{2(g)} + 3\,H_{2(g)} \rightleftharpoons 2\,NH_{3(g)}$$
$$\underbrace{1\ mol + 3\ mol}_{4\ mol} \qquad \underbrace{2\ mol}$$

If the volume of the vessel containing this reaction mixture is decreased, the overall pressure is increased. Le Châtelier's principle suggests that the system will react in a way that resists the change—i.e., in a way that reduces the pressure. In this case, the equilibrium will shift to the right, which decreases the number of gas molecules in the container and reduces the pressure (**Figure 9**).

Key:

$N_{2(g)}$

$H_{2(g)}$

$NH_{3(g)}$

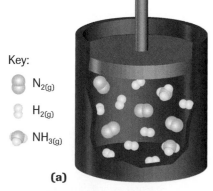

(a)

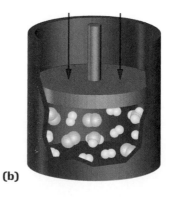

(b)

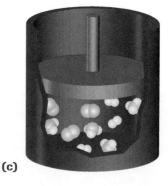

(c)

Figure 9
(a) An equilibrium mixture containing $N_{2(g)}$, $H_{2(g)}$, and $NH_{3(g)}$
(b) The volume is decreased, increasing the pressure.
(c) The reaction shifts to the right, toward the side with fewer molecules, to relieve pressure.

If the volume of the vessel is increased, the pressure is decreased, and the shift is in the opposite direction, to the left, which counteracts the change by producing more gas molecules.

A system with equal numbers of gas molecules on each side of the equation will not shift after a change in volume, since no shift can change the pressure in the vessel. Consider the equilibrium reaction between hydrogen and iodine to produce hydrogen iodide.

$$H_{2(g)} + I_{2(g)} \rightleftharpoons 2\,HI_{(g)}$$
1 mol + 1 mol
2 mol   2 mol

Systems involving only liquids or solids are not affected appreciably by changes in pressure. Substances in these condensed states are virtually incompressible, and so reactions involving them cannot counteract pressure changes.

Rate Theory and Gas Volume Changes

We can again explain the equilibrium shift observed when the volume of a system involving gaseous reactants and products is changed, as an imbalance of reaction rates.

$$2\,SO_{2(g)} + O_{2(g)} \rightleftharpoons 2\,SO_{3(g)} + 198\;kJ$$
2 mol + 1 mol
3 mol   2 mol

Kinetic theory explains the effect of a decrease in volume by assuming that both the forward and reverse reaction rates increase because the concentrations (partial pressures) of reactants and products increase. However, the forward rate increases more than the reverse rate because there are more particles involved in the forward reaction. This means that the increase in the total number of effective collisions is greater for the forward reaction. Again, while the rates remain unequal, the observed result is the production of more product. The shift causes concentration changes that eventually increase the reverse rate and decrease the forward rate until they become equal again (**Figure 10**).

Changes That Do Not Affect the Position of Equilibrium Systems

We have looked at three changes that have an effect on the equilibrium of a chemical system—concentration changes, energy changes, and pressure (volume) changes. There are other changes that have no effect whatever on the equilibrium position of a chemical system. Here is a brief look at these changes, and why they have no effect.

Adding Catalysts

Catalysts are used in most industrial chemical reactions. (Note: enzymes differ from inorganic catalysts when applied to equilibria. One enzyme speeds up the forward reaction but a different enzyme speeds up the reverse reaction.) A catalyst decreases the time required to reach the equilibrium position, but does not affect the final position of equilibrium. The presence of a catalyst in a chemical reaction system lowers the activation energy for both forward and reverse reactions by an equal amount, so the equilibrium establishes much more rapidly but at the *same position* as it would without the catalyst (**Figure 11**). Forward and reverse rates are increased equally. The value of catalysts in industrial processes is to decrease the time required for equilibrium shifts to occur, allowing a more rapid overall production of the desired product.

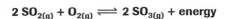

$$2\,SO_{2(g)} + O_{2(g)} \rightleftharpoons 2\,SO_{3(g)} + energy$$

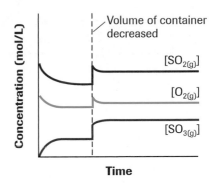

Figure 10
The equilibrium is disturbed by a decrease in the volume of the container (at the time indicated by the vertical dotted line). The equilibrium shifts forward (i.e., toward products). The concentration of $SO_{3(g)}$ increases while the concentration of reactants decreases, until a new equilibrium state is established.

(a)

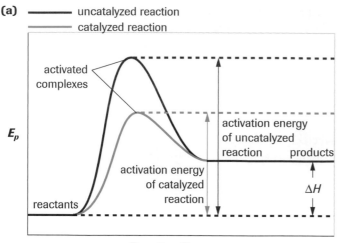

Figure 11
(a) A catalyst reduces the activation energy by the same amount whether the reaction proceeds to the right or to the left. **(b)** It does not affect the relative concentrations of entities.

(b)

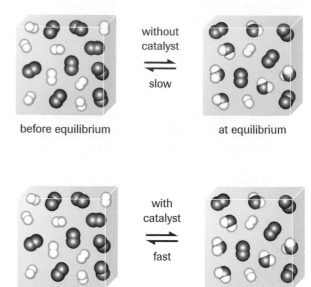

INVESTIGATION 7.3.1

Testing Le Châtelier's Principle (p. 514)
Le Châtelier's Principle provides guidance when predicting how a chemical system at equilibrium will respond to disturbance.

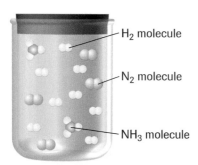

H₂ molecule

N₂ molecule

NH₃ molecule

He atom

Figure 12
Adding an inert gas such as helium to a system at equilibrium increases the pressure of the system (at constant volume) but does not cause a shift in the equilibrium position.

Adding Inert Gases

The pressure of a gaseous system at equilibrium can be changed by adding a gas while keeping the volume constant. If the gas is inert in the system, for example, if it is a noble gas or if it cannot react with the entities in the system, the equilibrium position of the system will not change. We can explain this by turning to rate theory: The presence of the inert gas changes the probability of successful collisions for both the reactants and the products equally, resulting in no shift in the equilibrium system (**Figure 12**).

SUMMARY *Variables Affecting Chemical Equilibria*

Table 1

Variable	Type of Change	Response of System
concentration	increase	shifts to consume some of the added reactant or product
	decrease	shifts to replace some of the removed reactant or product
temperature	increase	shifts to consume some of the added thermal energy
	decrease	shifts to replace some of the removed thermal energy
volume	increase (decrease in pressure)	shifts toward the side with the larger total amount of gaseous entities
	decrease (increase in pressure)	shifts toward the side with the smaller total amount of gaseous entities
Variables That Do Not Affect Chemical Equilibria		
catalysts	—	no effect
inert gases	—	no effect

▶ *Practice*

Understanding Concepts

1. How will the following system at equilibrium shift in each of the following cases?

   heat

   $$2 SO_{3(g)} \rightleftharpoons 2 SO_{2(g)} + O_{2(g)} \qquad \Delta H° = 197 \text{ kJ}$$

   (a) $SO_{2(g)}$ is added ←
   (b) the pressure is decreased by increasing the volume of the container →
   (c) the pressure is increased by adding $Ne_{(g)}$ ✗
   (d) the temperature is decreased ←
   (e) $O_{2(g)}$ is removed →

2. (a) Draw a concentration–time graph (similar to **Figure 10**) that illustrates the changes in concentration that occur when $F_{2(g)}$ is added to a sealed vessel containing the following equilibrium:

   $$Br_{2(g)} + 5 F_{2(g)} \rightleftharpoons 2 BrF_{5(g)}$$

   (b) Draw a concentration–time graph for the removal of some $HOCl_{(g)}$ from the following equilibrium:

   $$H_2O_{(g)} + Cl_2O_{(g)} \rightleftharpoons HOCl_{(g)}$$

3. The following reaction is used in the commercial production of hydrogen gas.

   $$CH_{4(g)} + 2 H_2O_{(g)} \rightleftharpoons CO_{2(g)} + 4 H_{2(g)}$$

   (a) In a closed system, how would a catalyst affect the establishment of equilibrium in the system?
   (b) How would the concentration of $H_{2(g)}$ at equilibrium be affected by the use of a $Ni_{(s)}$ catalyst?

4. Much of the brown haze hanging over large cities is nitrogen dioxide, $NO_{2(g)}$. Nitrogen dioxide reacts to form dinitrogen tetroxide, $N_2O_{4(g)}$, according to the equilibrium:

   $$2 NO_{2(g)} \rightleftharpoons N_2O_{4(g)} + 57.2 \text{ kJ}$$
   (brown) \qquad (colourless)

   Use this equilibrium to explain why the brownish haze over a large city disappears in the winter, only to reappear again in the spring. (Assume that the atmosphere over the city constitutes a closed system.)

5. A land-based vehicle will be an important part of any future exploration of the planet Mars. One proposed design for a Mars rover uses a methane gas fuel cell as its power supply. The methane fuel can be made on Mars using a chemical reaction that has been known for over 100 a—the Sabatier methanation reaction:

   $$CO_{2(g)} + 4 H_{2(g)} \rightleftharpoons CH_{4(g)} + 2 H_2O_{(g)}$$
   $$\Delta H = -165 \text{ kJ at } 250°C$$

   Predict, using Le Châtelier's principle, the conditions required in a closed Sabatier reactor to produce the maximum amount of methane.

6. In caves we sometimes see large structures known as stalactites and stalagmites (**Figure 13**).

   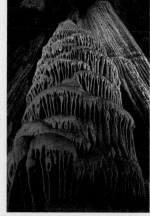

   Figure 13

   These formations are made of crystals of the insoluble compound calcium carbonate, $CaCO_{3(s)}$, also known as limestone. Calcium carbonate forms when dissolved

carbon dioxide and calcium ions combine, according to the following equilibrium:

$$CaCO_{3(s)} + 2 H^+_{(aq)} \rightleftharpoons Ca^{2+}_{(aq)} + H_2O_{(l)} + CO_{2(g)}$$

Use Le Châtelier's principle to answer the following:
(a) How would the stalagmites and stalactites be affected if the water in the cave became more acidic?
(b) How would the hardness of the water $[Ca^{2+}_{(aq)}]$ affect the growth of stalagmites and stalactites?

Applying Inquiry Skills

7. A student designed an experiment to measure the effect of the addition of chloride ions on the equilibrium point of the following system.

$$Cu^{2+}_{(aq)} + 4 Cl^-_{(aq)} \rightleftharpoons CuCl^{2-}_{4(aq)}$$
(blue)   (green)

Question

What effect does the addition of chloride ions have on the system at equilibrium?

Prediction

(a) Predict whether an equilibrium shift will occur and the effect of the shift on the system.

Experimental Design

Test tubes containing 10 mL of a copper(II) chloride solution are combined with 5-mL, 10-mL, and 15-mL samples of 1 mol/L hydrochloric acid, $HCl_{(aq)}$. All test tubes were stoppered, shaken, and allowed to reach equilibrium.

Evidence

Table 2 Observations on $Cu^{2+}_{(aq)}$ Equilibrium Investigation

Test tube (10 mL of $CuCl_{2(aq)}$ in each)	Vol. $HCl_{(aq)}$ added (mL)	Total vol. after mixing (mL)	Colour at equilibrium
1	5 mL	15 mL	blue
2	10 mL	20 mL	blue-green
3	15 mL	25 mL	green

Analysis

(b) Use the Evidence to answer the Question.

Evaluation

(c) Critique the Experimental Design.

Making Connections

8. The digestion of some high-protein foods, such as red meat, beans, lentils, and shellfish, releases uric acid, $HC_5H_3N_4O_{3(aq)}$, which ionizes into hydrogen ions, $H^+_{(aq)}$ and urate ions, $C_5H_3N_4O^-_{3(aq)}$, in the bloodstream. People whose kidneys do not function properly cannot excrete urate ion sufficiently quickly, leading to an increased concentration of urate in the blood. This sometimes leads to a painful form of arthritis known as gout, characterized by the formation of tiny needle-like crystals of sodium urate, $NaC_5H_3N_4O_{3(s)}$, in joints and tissues, according to the equation

$$NaC_5H_3N_4O_{3(s)} \rightleftharpoons Na^+_{(aq)} + C_5H_3N_4O^-_{3(aq)}$$

(a) Suppose you were a nutritionist. What advice could you give to your patients who suffer from gout? Explain why following the advice would be effective.

(b) Many women take calcium supplements on a daily basis to prevent the loss of bone mass (a condition known as osteoporosis). If a woman suffering from osteoporosis has gout too, she may also develop kidney stones (which can consist of calcium urate). Write a chemical equilibrium equation for this reaction and explain why this happens.

(c) Research and report on other non-dietary treatments of gout.

 www.science.nelson.com

▶ Section 7.3 Questions

Understanding Concepts

1. (a) List three environmental factors that may affect the position of a chemical equilibrium. Briefly explain how rate theory explains the effect in each case.
   (b) List two factors that do not affect the position of equilibrium.

2. For each of the following chemical systems at equilibrium, use Le Châtelier's principle to predict the effect of the change imposed on the chemical system. Indicate the direction in which the equilibrium is expected to shift, if at all. For each example, sketch a graph of concentration versus time, plotted from just before the possible change to the established equilibrium.
   (a) $H_2O_{(l)}$ + energy $\rightleftharpoons$ $H_2O_{(g)}$
   The container is heated.
   (b) $H_2O_{(l)}$ $\rightleftharpoons$ $H^+_{(aq)}$ + $OH^-_{(aq)}$
   A few crystals of $NaOH_{(s)}$ are added to the container.
   (c) $CaCO_{3(s)}$ + energy $\rightleftharpoons$
   $\qquad\qquad CaO_{(s)}$ + $CO_{2(g)}$
   $CO_{2(g)}$ is removed from the container.
   (d) $CH_3COOH_{(aq)}$ $\rightleftharpoons$ $H^+_{(aq)}$ + $CH_3COO^-_{(aq)}$
   A few drops of pure $CH_3COOH_{(l)}$ are added to the system.

3. The following equation is important in the industrial production of nitric acid. Predict the direction of the equilibrium shift for each of the following changes in a closed vessel. Explain any shift in terms of changes in rates of the forward and reverse reactions.

   $NH_{3(g)}$ + 5 $O_{2(g)}$ $\rightleftharpoons$ 4 $NO_{(g)}$ + 6 $H_2O_{(g)}$ + energy

   (a) $O_{2(g)}$ is added to the system.
   (b) The temperature of the system is increased.
   (c) $NO_{(g)}$ is removed from the system.
   (d) The pressure of the system is increased by decreasing the volume of the reaction vessel.
   (e) Argon gas is added to the system without changing the volume.

4. In a solution of copper(II) chloride, the following equilibrium exists:

   $CuCl_4^{2-}{}_{(aq)}$ + 4 $H_2O_{(l)}$ $\rightleftharpoons$ $Cu(H_2O)_4^{2+}{}_{(aq)}$ + 4 $Cl^-_{(aq)}$
   dark green   blue

Predict the shift in the equilibrium and draw a graph of concentration versus time for relevant reactants to communicate the shift after the following stresses are applied to the system:
(a) Hydrochloric acid is added.
(b) Silver nitrate is added.

5. The two oxyanions of chromium(IV) are the orange dichromate ion, $Cr_2O_7^{2-}{}_{(aq)}$, and the yellow chromate ion, $CrO_4^{2-}{}_{(aq)}$. Explain why a solution containing the following equilibrium system turns yellow when sodium hydroxide is added.

   $Cr_2O_7^{2-}{}_{(aq)}$ + $H_2O_{(l)}$ $\rightleftharpoons$ 2 $CrO_4^{2-}{}_{(aq)}$ + 2 $H^+_{(aq)}$
   orange   yellow

6. Identify the nature of the change imposed on the equilibrium system (**Figure 14**) at each of the times indicated A, B, C, D, and E.

   $C_2H_{4(g)}$ + $H_{2(g)}$ $\rightleftharpoons$ $C_2H_{6(g)}$ + energy

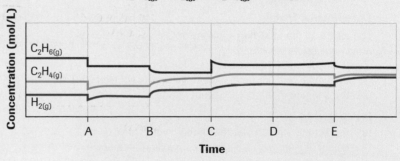

Figure 14

Applying Inquiry Skills

7. A student plans to test Le Châtelier's principle by increasing the pressure in a closed system containing a nitrogen dioxide–dinitrogen tetroxide equilibrium, $NO_{2(g)}$–$N_2O_{4(g)}$, and measuring changes in colour intensity.

   Question
   (a) State a Question for this experiment.

   Prediction
   (b) Predict an answer to the Question.

   Experimental Design
   (c) Propose an experimental design.

Evidence

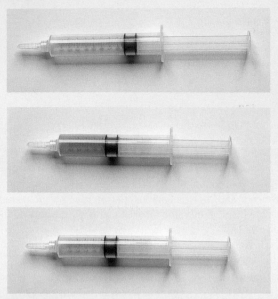

Evaluation

(d) Evaluate the experimental design.

8. A student plans an investigation into the effects of various changes on a chromate–dichromate equilibrium. The chromate ion, $CrO_4^{2-}{}_{(aq)}$, reacts with acid to form the following equilibrium with the dichromate ion, $Cr_2O_7^{2-}{}_{(aq)}$, and water.

$$2\,H^+_{(aq)} + 2\,CrO_4^{2-}{}_{(aq)} \rightleftharpoons Cr_2O_7^{2-}{}_{(aq)} + H_2O_{(l)}$$
$$\text{(yellow)} \qquad \text{(orange)}$$

Experimental Design

The effects of the following changes on a chromate–dichromate equilibrium are noted:

(i) the addition of 0.1 mol/L hydrochloric acid
(ii) the addition of 0.1 mol/L sodium hydroxide solution
(iii) the addition of 0.1 mol/L barium nitrate

Prediction

(a) Predict the effect that each change would have on the colour of a chromate-dichromate equilibrium mixture.

Synthesis

(b) After disposing of the contents of the test tube, a student discovers that the inside of the test tube is coated with a light yellow precipitate that cannot be easily washed off. What chemical could be added to the test tube to remove the precipitate?

Making Connections

9. Hydrogen sulfide is a foul-smelling and toxic byproduct of the processing of crude oil and natural gas. One method to recover H_2S so that it does not contaminate the environ-

ment is known as the Claus process, which involves the following reaction:

$$2\,H_2S_{(g)} + SO_{2(g)} \rightleftharpoons 3\,S_{(s)} + 2\,H_2O_{(g)} + \text{heat}$$

The Claus process is capable of removing up to 95% of the sulfur emissions from petroleum-processing plants.

(a) Research and report on the Claus process.
(b) Describe why it is advantageous to remove the sulfur from the process as quickly as it forms.

 www.science.nelson.com

10. An air purification system involving lithium hydroxide, LiOH, was used in NASA's Apollo missions to the moon. LiOH absorbs carbon dioxide.

$$2\,LiOH_{(s)} + CO_{2(g)} \rightleftharpoons Li_2CO_{3(s)} + H_2O_{(l)}$$

Use Le Châtelier's principle to explain why the amount of time astronauts can spend in a spacecraft is limited.

11. When the Olympic Games were held in Mexico in 1968, many athletes arrived early to train in the higher altitude (2.3 km above sea level) and lower atmospheric pressure of Mexico City. Exertion at high altitudes, for people who are not acclimatized, may make them "lightheaded" from lack of oxygen. A similar effect occurred at the 2002 Winter Olympics in Salt Lake City, Utah (1.3 km above sea level). Use the theory of dynamic equilibrium and Le Châtelier's principle to explain this observation. How are people who normally live at high altitudes physiologically adapted to their reduced-pressure environment?

 www.science.nelson.com

12. Hemoglobin, Hb, a protein molecule found in red blood cells, attracts and binds inhaled oxygen, which can then be transported throughout the body.

$$Hb_{(aq)} + O_{2(aq)} \rightleftharpoons HbO_{2(aq)}$$

Carbon monoxide, $CO_{(g)}$, binds more readily to hemoglobin than oxygen and can displace oxygen according to this equilibrium:

$$HbO_{2(aq)} + CO_{(g)} \rightleftharpoons HbCO_{(aq)} + O_{2\,(g)} \qquad K = 200 \text{ at } 37°C$$

Consider this scenario:

A patient, unconscious due to suspected carbon monoxide poisoning, has just been brought to the hospital emergency ward where you are the doctor in charge. Based on your knowledge of Le Châtelier's principle, what treatment would you recommend

 www.science.nelson.com

Case Study: The Haber Process: Ammonia for Food and Bombs

In the late nineteenth century, rapid population growth in Europe and North America began to outstrip the food supply. Research was showing that the addition of nitrogen-based fertilizers such as sodium nitrate, $NaNO_{3(s)}$, and ammonium nitrate, $NH_3NO_{3(s)}$, significantly increased crop yield to help ease the worldwide food shortage. However, there was initially not sufficient fertilizer to meet the growing demand. Guano (bird droppings), a traditional fertilizer from Peru, and sodium nitrate supplies from Chile would soon be exhausted, not least because those raw materials were also being used to produce explosives.

An alternative source of ammonia or nitrate had to be found.

In 1909, a leading German chemical company, Badische Anilin und Soda Fabrik (BASF), started to investigate the possibility of producing ammonia from atmospheric nitrogen, $N_{2(g)}$. Little did they know that one year earlier, Fritz Haber, a professor at a technical college in Karlsruhe, Germany, had discovered a method for doing just that (**Figure 1**).

Haber realized, after much experimentation, that nitrogen gas and hydrogen gas form an equilibrium mixture with ammonia. The optimum conditions include a closed container, a suitable catalyst (such as iron oxide), a temperature of 600°C, and a pressure of 30 MPa.

$$N_{2(g)} + 3 H_{2(g)} \overset{Fe_2O_{3(s)}}{\rightleftharpoons} 2 NH_{3(g)} \qquad \Delta H° = -92\,kJ$$

This method is now called the Haber process, in honour of its discoverer. BASF bought the rights to the Haber process and with the help of Carl Bosch, BASF's chief chemical engineer, built a giant industrial plant capable of producing 10 000 t of ammonia per year. Today, ammonia is in sixth position in a ranking of chemicals produced worldwide, with over 80 billion kilograms produced each year.

Figure 1
Fritz Haber (1868–1934) discovered a method for converting atmospheric nitrogen into ammonia at a technical college in Karlsruhe, Germany. He was awarded the Nobel Prize in Chemistry in 1918 for discovering the process that now bears his name.

> ▶ **Practice**

Understanding Concepts

1. Suggest five factors that could affect the production of ammonia in the Haber process. Explain the effect of each factor, using rate theory.

Making Connections

2. Create a concept map starting with "Haber process" and including at least two end uses of the product of this process.

The Temperature Puzzle

The reaction of nitrogen and hydrogen at low temperatures is so slow that the process becomes uneconomical. Adding heat increases the rate of the reaction (**Figure 2**), which is important in any continuous industrial process. However, in this reaction, the higher the temperature, the lower the yield of ammonia. The relationship between percentage yield and temperature is shown in **Figure 3**.

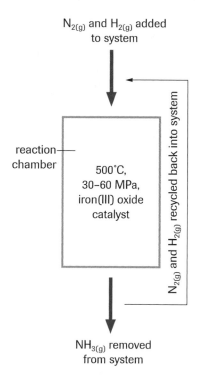

$N_{2(g)}$ and $H_{2(g)}$ added to system

reaction chamber

500°C, 30–60 MPa, iron(III) oxide catalyst

$N_{2(g)}$ and $H_{2(g)}$ recycled back into system

$NH_{3(g)}$ removed from system

Figure 2
Schematic of conditions for the Haber process

Ammonia Yield at Various Conditions of Temperature and Pressure

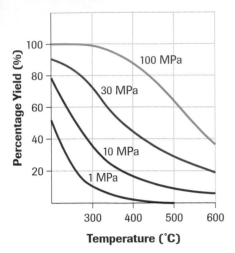

Figure 3

Figure 4
Ammonia fertilizer can be added directly to the soil.

Haber had to balance the rate of the reaction (increased by higher temperatures) against the equilibrium of the reaction (pushed to the right by lower temperatures). He discovered that using an iron oxide catalyst eliminates the need for excessively high temperatures, allowing the equilibrium position to move quickly to the right at lower temperatures. An industrial plant using a modification of the Haber process might operate at a temperature of about 500°C and a pressure of 50 MPa. After a suitable length of time under these conditions, the yield of ammonia is about 40%.

Today, the Haber process and modifications of it are used to produce large quantities of ammonia, which is used as a fertilizer (**Figure 4**). As a fertilizer, ammonia dissolves in moisture present in the soil and, if the soil is slightly acidic, it is converted into the ammonium ion. The ammonium ion enters the nitrogen cycle, where it is converted to nitrate ions by soil bacteria. Nitrate ions are absorbed by the roots of plants and used in the synthesis of proteins, chlorophyll, and nucleic acids. Without a source of nitrogen, plants do not grow, but produce yellow leaves and die prematurely.

Ammonia is also used to make explosives.

▶ **Practice**

Understanding Concepts

3. Why is a low temperature, which gives a higher percentage yield of ammonia, not used in the Haber process?

4. What role does iron oxide play in the Haber process?

Making Connections

5. (a) Ammonia, produced by the Haber process, can be oxidized to nitric acid, the raw material used in the manufacture of explosives. Perform library or Internet research to determine the most common types of explosives produced with nitric acid.
   (b) Draw structural formulas for the three most common nitrogen-based explosives. What are the specific uses of each?
   (c) Write the chemical equation that describes a nitroglycerine explosion. Why is this reaction explosive?
   (d) What is gun cotton? What are its uses? How is it made?

 www.science.nelson.com

▶ **Section 7.4** *Questions*

Making Connections

1. You have been hired as an efficiency consultant by a plant that produces ammonia by the Haber process.
   (a) Using equilibrium principles only, what advice would you give the company regarding the best environmental conditions for optimal ammonia production?
   (b) In what ways might the theoretically ideal conditions suggested in your answer to (a) be less than ideal, practically, for the company?
   (c) What additional advice could you give the company to help reduce the costs associated with your answer to (a)?

2. The Haber process requires nitrogen and hydrogen as reactants.
   (a) Suggest reasonable sources for each of these elements.

   (b) Conduct library and/or Internet research to learn how modern ammonia production facilities obtain pure hydrogen and nitrogen for the process.

    www.science.nelson.com

3. To be used by growing plants, elemental nitrogen must first be converted into another form (such as ammonia) in a natural process called nitrogen fixation. The Haber process is synthetic nitrogen fixation.
   (a) How do bacteria fix nitrogen naturally?
   (b) Currently, which of the two processes, synthetic and natural, fixes the most nitrogen?
   (c) What problems have arisen from the dramatic increase in nitrogen fixation in the last century? Pick one of these problems and suggest some remedies.

    www.science.nelson.com

Quantitative Changes in Equilibrium Systems 7.5

Now that we are familiar with the equilibrium law expression, can we predict, quantitatively, how much of the reactants will be consumed and how much product will be formed in any chemical system that reaches equilibrium?

Yes we can, if we combine our knowledge of equilibria with our understanding of stoichiometric techniques. We must remember, though, that at this stage we are only considering reactions taking place in closed systems. We will be able to calculate and predict the quantities involved in establishing an initial equilibrium (beginning with only reactants present), as well as those resulting from equilibrium shifts.

Recall the convention used for the equilibrium expression. For the general reaction

$$aA + bB \rightleftharpoons cC + dD$$

$$K = \frac{[C]^c[D]^d}{[A]^a[B]^b}$$

This tells us that, all things being equal, the larger the value of K, the more the reaction, as written, favours products. Consider the following two examples:

$$H_{2(g)} + I_{2(g)} \rightleftharpoons 2\,HI_{(g)} \qquad\qquad K = 50 \text{ at } 450°C$$
$$CO_{2(g)} + H_{2(g)} \rightleftharpoons CO_{(g)} + H_2O_{(g)} \qquad\qquad K = 1.1 \text{ at } 900°C$$

The larger value of K for the first reaction indicates that the first reaction proceeds farther to completion by the time equilibrium is established.

Generally then, we can say that a very large K value means a reaction favouring products, and a very small K value means a reaction favouring reactants.

We must also consider temperature in any quantitative calculation involving K. The only factor that changes the value of K for a given reaction is temperature: Changes in concentration do not have a significant effect on the numerical value of K, and neither does the presence or absence of a catalyst. A temperature change that increases the value of K for a reaction shifts equilibrium to the right (more complete), and a temperature change that decreases the value of K for a reaction shifts equilibrium to the left (less complete).

The Reaction Quotient, Q

If we know the concentrations of the substances in any closed chemical system, we may want to determine whether the system is at equilibrium—and, if not, in which direction the system will shift to reach equilibrium.

If a chemical system begins with reactants only, it is obvious that the reaction will initially proceed to the right, toward products. If, however, reactants and products are both present, the direction in which the reaction proceeds is usually less obvious. In such a case, we can substitute the concentrations into the equilibrium law expression to produce a trial value that is called a **reaction quotient, Q.** We can think of Q as being similar to K, with the difference being that K is calculated using concentrations at equilibrium, whereas Q may or may not be at equilibrium. The same mathematical equation is used for calculating both K and Q. The result of such a trial calculation must be one of three possible situations.

- Q is equal to K, and the system is at equilibrium.
- Q is greater than K, and the system must shift left (toward reactants) to reach equilibrium, because the product-to-reactant ratio is too high.
- Q is less than K, and the system must shift right (toward products) to reach equilibrium, because the product-to-reactant ratio is too low.

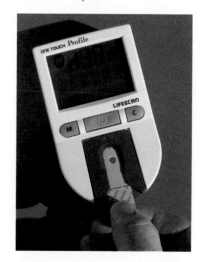

reaction quotient, Q a test calculation using measured concentration values of a system in the equilibrium expression

SAMPLE problem — Calculating Q and Comparing It to K

In a container at 450°C, nitrogen and hydrogen react to produce ammonia. The equilibrium constant, K, is 0.064. When the system is analyzed, the concentrations are found to be as follows: $[N_{2(g)}]$ is 4.0 mol/L, $[H_{2(g)}]$ is 2.0×10^{-2} mol/L, and $[NH_{3(g)}]$ is 2.2×10^{-4} mol/L. Determine whether the system is at equilibrium, and if it is not, predict the direction in which the reaction will proceed to reach equilibrium.

First, write a balanced chemical equation for the reaction.

$$N_{2(g)} + 3\,H_{2(g)} \rightleftharpoons 2\,NH_{3(g)} \qquad K = 0.064 \text{ at } 450°C$$

List the known values.

$$[N_{2(g)}] = 4.0 \text{ mol/L}$$
$$[H_{2(g)}] = 2.0 \times 10^{-2} \text{ mol/L}$$
$$[NH_{3(g)}] = 2.2 \times 10^{-4} \text{ mol/L}$$

Then calculate the value of the reaction quotient, Q.

$$Q = \frac{[NH_{3(g)}]^2}{[N_{2(g)}][H_{2(g)}]^3}$$
$$= \frac{(2.2 \times 10^{-4})^2}{(4.0)(2.0 \times 10^{-2})^3}$$
$$= 1.5 \times 10^{-3}$$
$$Q = 0.0015$$

Now compare the value of Q to the value of K, the equilibrium constant.

Since the value of Q (0.0015) is smaller than the value of K (0.064), the reaction is not at equilibrium. In order to attain equilibrium, the reaction will shift to the right (as written). The concentration of the reactants will decrease and the concentration of the products will increase.

Example

The following reaction occurs in a closed container at 445°C. The equilibrium constant, K, is 0.020.

$$2\,HI_{(g)} \rightleftharpoons H_{2(g)} + I_{2(g)}$$

Is the system at equilibrium in each of the following cases? If not, predict the direction in which the reaction will proceed to reach equilibrium.

(a) $[HI_{(g)}] = 0.14$ mol/L; $[H_{2(g)}] = 0.04$ mol/L; and $[I_{2(g)}] = 0.01$ mol/L
(b) $[HI_{(g)}] = 0.20$ mol/L; $[H_{2(g)}] = 0.15$ mol/L; and $[I_{2(g)}] = 0.09$ mol/L

Solution

(a) $$Q = \frac{[H_{2(g)}][I_{2(g)}]}{[HI_{(g)}]^2}$$
$$= \frac{(0.04)(0.01)}{(0.14)^2}$$
$$Q = 0.02$$

Since Q (0.02) is equal to K (0.02), the reaction is at equilibrium. No shift will occur.

(b) $$Q = \frac{[H_{2(g)}][I_{2(g)}]}{[HI_{(g)}]^2}$$
$$= \frac{(0.15)(0.09)}{(0.20)^2}$$
$$Q = 0.34$$

Since Q (0.34) is greater than K (0.02), the reaction is not at equilibrium. In order to attain equilibrium, the reaction will proceed to the left. The concentration of the reactants will increase and the concentration of the products will decrease.

▶ *Practice*

Understanding Concepts

1. Liquid dinitrogen tetroxide, $N_2O_{4(l)}$, was used as a fuel in Apollo missions to the moon (**Figure 1**). In a closed container the gas $N_2O_{4(g)}$ decomposes to nitrogen dioxide, $NO_{2(g)}$. The equilibrium constant, K, for this reaction is 0.87 at 55°C. A vessel filled with $N_2O_{4(g)}$ at this temperature is analyzed twice during the course of the reaction and found to contain the following concentrations:
   (a) $[N_2O_{4(g)}] = 5.30$ mol/L; $[NO_{2(g)}] = 2.15$ mol/L
   (b) $[N_2O_{4(g)}] = 0.80$ mol/L; $[NO_{2(g)}] = 1.55$ mol/L

   In each case, determine whether the system is in equilibrium, and if not, predict the direction in which the reaction will proceed to achieve equilibrium.

2. Given the equilibrium system

   $$PCl_{5(g)} \rightleftharpoons PCl_{3(g)} + Cl_{2(g)} \qquad K = 12.5 \text{ at } 60°C$$

   A 1.0-L reaction vessel is analyzed and found to contain 3.2 mol $Cl_{2(g)}$, 1.5 mol $PCl_{3(g)}$, and 2.0 mol $PCl_{5(g)}$. Show that the reaction mixture has not yet reached equilibrium.

Figure 1

Calculations Involving Equilibrium Systems

We can use stoichiometry to calculate concentrations of reactants and products for systems at equilibrium.

Calculating Equilibrium Concentrations from Knowns | **SAMPLE** problem ◀

Hydrogen iodide, $HI_{(g)}$, a compound used in the production of hydroiodic acid, $HI_{(aq)}$, is produced by reacting hydrogen gas and iodine vapour according to the following equation:

$$H_{2(g)} + I_{2(g)} \rightleftharpoons 2HI_{(g)}$$

The equilibrium constant, K, for this reaction is 49.70 at 458°C. Calculate $[HI_{(g)}]$ at equilibrium if the equilibrium concentration of the other two entities is 1.07 mol/L.

First, list the known values. Then, set the equilibrium constant equal to the equilibrium law expression for this reaction.

$$[H_{2(g)}]_{equilibrium} = 1.07 \text{ mol/L}$$
$$[I_{2(g)}]_{equilibrium} = 1.07 \text{ mol/L}$$
$$K = 49.70 \text{ at } 458°C$$
$$\frac{[HI_{(g)}]^2}{[H_{2(g)}][I_{2(g)}]} = K$$

▶

Since the equilibrium concentrations of $H_{2(g)}$ and $I_{2(g)}$ are given, substitute these into the equilibrium law expression and solve for $[HI_{(g)}]$.

$$\frac{[HI_{(g)}]^2}{(1.070)(1.070)} = 49.70$$

$$[HI_{(g)}]^2 = 49.70 \times 1.14$$

$$[HI_{(g)}]^2 = 56.9 \text{ mol/L}$$

The equilibrium concentration of $HI_{(g)}$ is 7.54 mol/L.

Example

Sulfur trioxide, $SO_{3(g)}$, a compound used in the production of sulfuric acid, is produced by reacting sulfur dioxide, $SO_{2(g)}$, with oxygen gas according to the following equation:

$$2\,SO_{2(g)} + O_{2(g)} \rightleftharpoons 2\,SO_{3(g)}$$

Calculate the equilibrium concentration of oxygen if 1.50 mol/L $SO_{2(g)}$ and 3.50 mol/L $SO_{3(g)}$ are found in an equilibrium mixture of this system at 600°C. K for the reaction at this temperature is 4.30.

Solution

$[SO_{2(g)}]_{equilibrium} = 1.50 \text{ mol/L}$

$[SO_{3(g)}]_{equilibrium} = 3.50 \text{ mol/L}$

$K = 4.30$ at 600°C

$$\frac{[SO_{3(g)}]^2}{[SO_{2(g)}]^2[O_{2(g)}]} = K$$

$$\frac{(3.50)^2}{(1.50)^2[O_{2(g)}]} = 4.30$$

$$[O_{2(g)}] = \frac{(3.50)^2}{(1.50)^2 \times 4.30}$$

$$[O_{2(g)}] = 1.27$$

The equilibrium concentration of oxygen is 1.27 mol/L.

▶ Practice

Understanding Concepts

Answers

3. 2.7×10^{-3} mol/L

4. 1.8×10^{-2} mol/L

3. For the following system at equilibrium

$$N_{2(g)} + 3\,H_{2(g)} \rightleftharpoons 2\,NH_{3(g)} \qquad\qquad K = 626 \text{ at } 200°C$$

the equilibrium concentrations of hydrogen and ammonia are 0.50 mol/L and 0.46 mol/L, respectively. Calculate the equilibrium concentration of nitrogen.

4. Consider the following equilibrium:

$$PCl_{5(g)} \rightleftharpoons PCl_{3(g)} + Cl_{2(g)} \qquad\qquad K = 32 \text{ at } 750°C$$

If the equilibrium concentrations of chlorine and phosphorus trichloride are 0.80 mol/L and 0.70 mol/L, respectively, find the equilibrium concentration of phosphorus pentachloride.

Calculating Equilibrium Concentrations from Initial Concentrations

Calculating equilibrium concentrations from initial concentrations is more complex than calculating an equilibrium concentration when other equilibrium concentrations are known. The following Sample shows some strategies you can use to deal with such problems.

Calculating Equilibrium Concentrations	*SAMPLE* problem ◀

Carbon monoxide reacts with water vapour to produce carbon dioxide and hydrogen. At 900°C, K is 4.200. Calculate the concentrations of all entities at equilibrium if 4.000 mol of each entity are initially placed in a 1.000-L closed container.

First, write the balanced equation for this reaction, including the value of K.

$$CO_{(g)} + H_2O_{(g)} \rightleftharpoons CO_{2(g)} + H_{2(g)} \qquad\qquad K = 4.200 \text{ at } 900°C$$

Now write the equilibrium law equation for this reaction.

$$K = \frac{[CO_{2(g)}][H_{2(g)}]}{[CO_{(g)}][H_2O_{(g)}]}$$

Next, calculate the initial concentrations of all entities. In this case,

$$[CO_{(g)}] - [H_2O_{(g)}] = [CO_{2(g)}] = [H_{2(g)}] = \frac{4.000 \text{ mol}}{1.000 \text{ L}}$$

$[CO_{(g)}] = 4.000 \text{ mol/L}$

$[H_2O_{(g)}] = 4.000 \text{ mol/L}$

$[CO_{2(g)}] = 4.000 \text{ mol/L}$

$[H_{2(g)}] = 4.000 \text{ mol/L}$

Calculate the value of Q, using the initial concentrations, and compare its value to K to determine whether the reaction is at equilibrium, and if not, the direction in which the reaction must proceed to reach equilibrium.

$$Q = \frac{[CO_{2(g)}][H_{2(g)}]}{[CO_{(g)}][H_2O_{(g)}]}$$

$$= \frac{(4.000)(4.000)}{(4.000)(4.000)}$$

$$Q = 1.000 \text{ mol/L}$$

Since Q (1.000 mol/L) is smaller than K (4.200), the reaction is not at equilibrium and will proceed to the right to achieve equilibrium.

We can set up the calculations needed to determine the equilibrium concentrations as an ICE table (**Table 1**). Begin by setting x as the change in concentration of carbon monoxide from its initial value to its (lower) equilibrium value.

The 1:1:1:1 stoichiometric ratio from the balanced equation indicates that the concentrations of both reactants will decrease by the same quantity, x, and the concentrations of both products will increase by that same amount.

Table 1 ICE Table for the Reaction of $CO_{(g)}$ and $H_2O_{(g)}$

	$CO_{(g)}$	+	$H_2O_{(g)}$	$\rightleftharpoons$	$CO_{2(g)}$	+	$H_{2(g)}$
Initial concentration (mol/L)	4.000		4.000		4.000		4.000
Change in concentration (mol/L)	$-x$		$-x$		$+x$		$+x$
Equilibrium concentration (mol/L)	$4.000 - x$		$4.000 - x$		$4.000 + x$		$4.000 + x$

Now substitute the equilibrium concentration values from the ICE table into the equilibrium law equation and solve for x.

$$\frac{[CO_{2(g)}][H_{2(g)}]}{[CO_{(g)}][H_2O_{(g)}]} = K$$

$$\frac{(4.000 + x)(4.000 + x)}{(4.000 - x)(4.000 - x)} = 4.200$$

$$\frac{(4.000 + x)^2}{(4.000 - x)^2} = 4.200$$

Notice that the left side is a perfect square. Taking the square root of both sides gives

$$\frac{(4.000 + x)}{(4.000 - x)} = \sqrt{4.200}$$

Multiplying both sides by $(4.000 - x)$ gives

$$4.000 + x = \sqrt{4.200} \times (4.000 - x)$$

$$4.000 + x = (2.050)(4.000 - x)$$

$$4.000 + x = 8.200 - 2.050x$$

$$3.050x = 4.200$$

$$x = \frac{4.200}{3.050}$$

$$x = 1.377$$

Now calculate the equilibrium concentrations by substituting the value of x.

$$
\begin{aligned}
[CO_{(g)}] &= 4.000 - x \\
&= 4.000 - 1.3777 \\
&= 2.623 \text{ mol/L}
\end{aligned}
$$

$$
\begin{aligned}
[H_2O_{(g)}] &= 4.000 - x \\
&= 4.000 - 1.3777 \\
&= 2.623 \text{ mol/L}
\end{aligned}
$$

$$
\begin{aligned}
[CO_{2(g)}] &= 4.000 + x \\
&= 4.000 + 1.3777 \\
&= 5.377 \text{ mol/L}
\end{aligned}
$$

$$
\begin{aligned}
[H_{2(g)}] &= 4.000 + x \\
&= 4.000 + 1.3777 \\
&= 5.377 \text{ mol/L}
\end{aligned}
$$

To check your work, substitute the calculated equilibrium concentrations into the reaction quotient expression, Q, and compare the result to the value of K.

$$Q = \frac{[CO_{2(g)}][H_{2(g)}]}{[CO_{(g)}][H_2O_{(g)}]}$$

$$= \frac{(5.377)^2}{(2.623)^2}$$

$$Q = 4.20 \text{ mol/L}$$

Since Q (4.20) is equal to K (4.20), the calculations were correct. The equilibrium concentrations are therefore

$$
\begin{aligned}
[CO_{(g)}] &= 2.623 \text{ mol/L} \\
[H_2O_{(g)}] &= 2.623 \text{ mol/L} \\
[CO_{2(g)}] &= 5.377 \text{ mol/L} \\
[H_{2(g)}] &= 5.377 \text{ mol/L}
\end{aligned}
$$

Example 1

Iodine, $I_{2(g)}$, and bromine, $Br_{2(g)}$, react in a closed 2.0-L container at 150°C to produce iodine monobromide, $IBr_{(g)}$, according to the following equation:

$$I_{2(g)} + Br_{2(g)} \rightleftharpoons 2\,IBr_{(g)}$$

The equilibrium constant, K, at this temperature is 1.2×10^2. What are the equilibrium concentrations of all entities in the mixture if the container initially contained 4.00 mol each of iodine and bromine?

Solution

$$\frac{[IBr_{(g)}]^2}{[I_{2(g)}][Br_{2(g)}]} = K$$

$$\frac{[IBr_{(g)}]^2}{[I_{2(g)}][Br_{2(g)}]} = 1.2 \times 10^2$$

Initial concentrations are

$$[I_{2(g)}] = [Br_{2(g)}] = \frac{4.00 \text{ mol}}{2.00 \text{ L}}$$
$$[I_{2(g)}] = [Br_{2(g)}] = 2.00 \text{ mol/L}$$
$$[IBr_{(g)}] = 0.00 \text{ mol/L}$$

Since the initial concentration of iodine monobromide is 0 mol/L, $Q = 0$. The reaction is not at equilibrium and must proceed to the right to achieve equilibrium (**Table 2**).

Table 2 ICE Table for the Formation of $IBr_{(g)}$

	$I_{2(g)}$	+	$Br_{2(g)}$	$\rightleftharpoons$	$2\,IBr_{(g)}$
Initial concentration (mol/L)	2.00		2.00		0.00
Change in concentration (mol/L)	$-x$		$-x$		$+2x$
Equilibrium concentration (mol/L)	$2.00 - x$		$2.00 - x$		$2x$

$$\frac{[IBr_{(g)}]^2}{[I_{2(g)}][Br_{2(g)}]} = K$$

$$\frac{(2x)^2}{(2.00 - x)^2} = 1.2 \times 10^2$$

$$\sqrt{\frac{(2x)^2}{(2.00 - x)^2}} = \sqrt{1.2 \times 10^2}$$

$$\frac{2x}{2.00 - x} = 10.95 \qquad \text{(extra digit carried)}$$

$$2x = 21.9 - 10.95x$$

$$x = 1.69$$

$$[I_{2(g)}] = 2.00 \text{ mol/L} - x$$
$$= 2.00 \text{ mol/L} - 1.69 \text{ mol/L}$$
$$[I_{2(g)}] = 0.31 \text{ mol/L}$$

$$[Br_{2(g)}] = 2.00 \text{ mol/L} - x$$
$$= 2.00 \text{ mol/L} - 1.69 \text{ mol/L}$$
$$[Br_{2(g)}] = 0.31 \text{ mol/L}$$

LEARNING TIP

Note that an extra digit is carried into the next calculation to avoid "rounding error." Never round intermediate values. Use all digits in your calculations. Round to the correct number of significant digits at the end of the calculation.

NEL

$$[IBr_{(g)}] = 2x$$
$$= 2(1.69 \text{ mol/L})$$
$$[IBr_{(g)}] = 3.38 \text{ mol/L}$$

Check:

$$Q = \frac{[IBr_{(g)}]^2}{[I_{2(g)}][Br_{2(g)}]}$$

$$= \frac{(3.38)^2}{(0.31)^2}$$

$$Q = 1.19 \times 10^2$$

The value of Q (1.19×10^2) and the value of K (1.20×10^2) are very close, so the calculation is correct. The equilibrium concentrations are:

$$[I_{2(g)}] = 0.31 \text{ mol/L}$$
$$[Br_{2(g)}] = 0.31 \text{ mol/L}$$
$$[IBr_{(g)}] = 3.38 \text{ mol/L}$$

Example 2

When hydrogen reacts with fluorine, hydrogen fluoride is formed according to the following equation:

$$H_{2(g)} + F_{2(g)} \rightleftharpoons 2\,HF_{(g)}$$

The equilibrium constant, K, is 1.15×10^2 at SATP. Calculate the concentratons of all entities at equilibrium if 4.00 mol of $H_{2(g)}$, 4.00 mol of $F_{2(g)}$, and 6.00 mol of $HF_{(g)}$ are initially placed into a 2.00-L reaction vessel.

Solution

Initial concentrations are

$$[H_{2(g)}] = [F_{2(g)}] = \frac{4.00 \text{ mol}}{2.00 \text{ L}}$$

$$[H_{2(g)}] = 2.00 \text{ mol/L}$$
$$[F_{2(g)}] = 2.00 \text{ mol/L}$$

$$[HF_{(g)}] = \frac{6.00 \text{ mol/L}}{2.00 \text{ L}}$$

$$[HF_{(g)}] = 3.00 \text{ mol/L}$$

$$Q = \frac{[HF_{(g)}]^2}{[H_{2(g)}][F_{2(g)}]}$$

$$= \frac{(3.00)^2}{(2.00)(2.00)}$$

$$Q = 2.25$$
$$K = 1.15 \times 10^2 \quad \text{(given)}$$

Since $Q < K$, the reaction is not at equilibrium and must proceed to the right to achieve equilibrium (**Table 3**).

Table 3 ICE Table for the Reaction of $H_{2(g)}$ and $F_{2(g)}$

	$H_{2(g)}$ +	$F_{2(g)}$ ⇌	$2\ HF_{(g)}$
Initial concentration (mol/L)	2.00	2.00	3.00
Change in concentration (mol/L)	$-x$	$-x$	$+2x$
Equilibrium concentration (mol/L)	$2.00 - x$	$2.00 - x$	$3.00 + 2x$

At equilibrium,

$$\frac{[HF_{(g)}]^2}{[H_{2(g)}][F_{2(g)}]} = K$$

$$\frac{(3.00 + 2x)^2}{(2.00 - x)(2.00 - x)} = 1.15 \times 10^2$$

$$\sqrt{\frac{(3.00 + 2x)^2}{(2.00 - x)(2.00 - x)}} = \sqrt{1.15 \times 10^2}$$

$$\frac{(3.00 + 2x)}{(2.00 - x)} = 10.724 \text{ (extra digit carried)}$$

$$3.00 + 2x = 10.724 (2.00 - x)$$

$$12.724\,x = 18.448$$

$$x = 1.450$$

$[H_{2(g)}]$ = 2.00 mol/L $- x$

$\phantom{[H_{2(g)}]}$ = 2.00 mol/L $-$ 1.450 mol/L

$[H_{2(g)}]$ = 0.550 mol/L

$[F_{2(g)}]$ = 2.00 mol/L $- x$

$\phantom{[F_{2(g)}]}$ = 2.00 mol/L $-$ 1.450 mol/L

$[F_{2(g)}]$ = 0.550 mol/L

$[HF_{(g)}]$ = 3.00 mol/L $+ 2x$

$\phantom{[HF_{(g)}]}$ = 3.00 mol/L $+$ 2(1.450 mol/L)

$\phantom{[HF_{(g)}]}$ = 3.00 mol/L $+$ 2.900 mol/L

$[HF_{(g)}]$ = 5.900 mol/L

Check:

$$Q = \frac{[HF_{(g)}]^2}{[H_{2(g)}][F_{2(g)}]}$$

$$= \frac{(5.900)^2}{(0.550)^2}$$

$$Q = 1.15 \times 10^2$$

Since the value of Q at equilibrium (1.15×10^2) and the value of K (1.15×10^2) are equal, our calculations are correct. The equilibrium concentrations are

$[H_{2(g)}]$ = 0.550 mol/L

$[F_{2(g)}]$ = 0.550 mol/L

$[HF_{(g)}]$ = 5.900 mol/L

Answers

5. $[CO_{2(g)}] = 0.0442$ mol/L
$[H_{2(g)}] = 0.0442$ mol/L
$[CO_{(g)}] = 0.0558$ mol/L
$[H_2O_{(g)}] = 0.0558$ mol/L
6. $[I_{2(g)}] = 0.045$ mol/L
$[Cl_{2(g)}] = 0.045$ mol/L
$[ICl_{2(g)}] = 0.41$ mol/L

Practice

Understanding Concepts

5. If 1.00 mol each of carbon dioxide and hydrogen is initially injected into a 10.0-L reaction chamber at 986°C, what would be the concentrations of each entity at equilibrium?

$CO_{2(g)} + H_{2(g)} \rightleftharpoons CO_{(g)} + H_2O_{(g)}$ $K = 1.60$ for 986°C

6. If 0.50 mol of iodine and 0.50 mol of chlorine are initially placed into a 2.00-L reaction vessel at 25°C, find the concentrations of all entities at equilibrium.

$I_{2(g)} + Cl_{2(g)} \rightleftharpoons 2\,ICl_{(g)}$ $K = 81.9$ at 25°C

Calculations With Imperfect Squares

Our ability to square both sides of the equilibrium law equation in the previous examples greatly simplified the calculation of equilibrium concentrations. In the absence of perfect squares, a different simplification technique helps us solve the problem.

SAMPLE problem — Calculating Equilibrium Concentrations When K Is Very Small

Carbon monoxide is a primary starting material in the synthesis of many organic compounds, including methanol, $CH_3OH_{(l)}$. At 2000°C, K is 6.40×10^{-7} for the decomposition of carbon dioxide into carbon monoxide and oxygen. Calculate the concentrations of all entities at equilibrium if 0.250 mol of $CO_{2(g)}$ is placed in a closed container and heated to 2000°C.

$2\,CO_{2(g)} \rightleftharpoons 2\,CO_{(g)} + O_{2(g)}$ $K = 6.40 \times 10^{-7}$ at 2000°C

$n = 0.250$ mol

$V = 1$ L

$$\frac{[CO_{(g)}]^2[O_{2(g)}]}{[CO_{2(g)}]^2} = K$$

Since there are no products in the initial condition, $Q = 0$ and the reaction will proceed to the right.

Table 4 ICE Table for the Decomposition of $CO_{2(g)}$

	$2\,CO_{2(g)}$	$\rightleftharpoons$	$2\,CO_{(g)}$	+	$O_{2(g)}$
Initial concentration (mol/L)	0.250		0.00		0.00
Change in concentration (mol/L)	$-2x$		$+2x$		$+x$
Equilibrium concentration (mol/L)	$0.250 - 2x$		$2x$		x

$$\frac{[CO_{(g)}]^2[O_{2(g)}]}{[CO_{2(g)}]^2} = 6.40 \times 10^{-7}$$

$$\frac{(2x)^2(x)}{(0.25 - 2x)^2} = 6.40 \times 10^{-7}$$

$$\frac{4x^3}{(0.25 - 2x)^2} = 6.40 \times 10^{-7}$$

This is a cubic equation that is difficult to solve directly. However, it may be simplified by recognizing that the equilibrium constant value is very small in comparison to the initial concentration of $CO_{2(g)}$. This means that very little $CO_{2(g)}$ decomposes into carbon monoxide and oxygen at this temperature. We can expect that the value of x will be exceedingly small, and that the value of $2x$ won't be much bigger. When this very small value is subtracted from 0.250 (a much larger value), the result will essentially remain 0.250. In other words, since the initial concentration is a quantity with uncertainty, adding or subtracting any number smaller than the uncertainty will not change the value.

We assume that

$$0.250 - 2x \doteq 0.250$$

At equilibrium,

$$\frac{4x^3}{(0.250)^2} \doteq 6.40 \times 10^{-7}$$

$$4x^3 \doteq 4.00 \times 10^{-8}$$

$$x^3 \doteq \frac{4.00 \times 10^{-8}}{4}$$

$$x \doteq \sqrt[3]{1.00 \times 10^{-8}}$$

$$x \doteq 2.15 \times 10^{-3}$$

We can now use the value of x we have calculated to test the validity of our earlier assumption. Does the result we obtained contradict or support the assumption?

$$0.250 - 2x \doteq 0.250 - 2(2.15 \times 10^{-3})$$

$$\doteq 0.250 - 0.0043$$

$$\doteq 0.247$$

The difference between 0.250 and 0.247 is 0.003, or 1.2%. This is a very small discrepancy that will have little effect on calculations of the equilibrium concentrations. In general, a difference of less than 5% justifies the simplifying assumption. It can be shown that the simplifying assumption will give an error of less than 5% if the concentration to which x is added or from which x is subtracted is at least 100 times greater than the value of K.

In this case,

$$\frac{[CO_{2(g)}]_{initial}}{K} = \frac{0.250}{6.40 \times 10^{-7}}$$

$$= 3.91 \times 10^5$$

Since $3.91 \times 10^5 > 100$, the assumption that $0.250 - 2x \doteq 0.250$ is warranted. We will call this the "hundred rule." You should apply the rule to determine whether a simplifying assumption is warranted *before the assumption is made*. If it is warranted, you may proceed with the calculation using the simplification. If not, you must solve the problem without the simplification. This is modelled in the next Sample Problem.

With our simplifying assumption validated, we can continue with the calculation.

$$[CO_{2(g)}] = 0.250 - 2x$$

$$\doteq 0.250$$

$$[CO_{2(g)}] \doteq 0.250 \text{ mol/L}$$

$$[CO_{(g)}] = 2x$$

$$\doteq 2(2.15 \times 10^{-3})$$

$$[CO_{(g)}] \doteq 4.30 \times 10^{-3} \text{ mol/L}$$

$$[O_{2(g)}] = x$$

$$[O_{2(g)}] \doteq 2.15 \times 10^{-3} \text{ mol/L}$$

LEARNING TIP

When making simplifying assumptions, you can only neglect an x that is added or subtracted in the equilibrium expression. You **cannot** neglect x values that are multiplied or divided.

You may ignore these values of x.

$x\,(0.50 - x)$ $\qquad\qquad$ $\frac{(0.50 - x)}{x}$

You are never justified in ignoring these values of x.

We can check the results by using the calculated equilibrium concentration values to generate Q and then compare it to K. If they are the same, we can assume that our calculations are correct.

$$Q = \frac{[CO_{(g)}]^2[O_{2(g)}]}{[CO_{2(g)}]^2}$$

$$Q = \frac{(4.30 \times 10^{-3})^2(2.15 \times 10^{-3})}{(0.250)^2}$$

$$Q = 6.36 \times 10^{-7}$$

The value of Q (6.36×10^{-7}) is sufficiently close to the value of K (6.40×10^{-7}). We can assume that our calculations are correct.

Example

Nitrosyl chloride, $NOCl_{(g)}$, decomposes to form nitrogen monoxide, $NO_{(g)}$, and chlorine gas, $Cl_{2(g)}$, according to the following equation:

$$2\,NOCl_{(g)} \rightleftharpoons 2\,NO_{(g)} + Cl_{2(g)}$$

At 35°C the equilibrium constant, K, is 1.60×10^{-5}. Calculate the concentration of all entities at equilibrium if 0.80 mol $NOCl_{(g)}$ is placed in an evacuated 2.00-L container at 35°C and allowed to reach equilibrium.

Solution

$n_{NOCl_{(g)}} = 0.80$ mol

$V = 2.00$ L

$t = 35°C$

$K = 1.60 \times 10^{-5}$ at 35°C

$$[NOCl_{(g)}]_{initial} = \frac{0.80 \text{ mol}}{2.00 \text{ L}}$$

$$[NOCl_{(g)}]_{initial} = 0.40 \text{ mol/L}$$

$$\frac{[NO_{(g)}]^2[Cl_{2(g)}]}{[NOCl_{(g)}]^2} = K$$

Since there is no product, $Q = 0$, and the reaction will proceed to the right.

Table 5 ICE Table for the Decomposition of $NOCl_{(g)}$

	$2\,NOCl_{(g)}$	$\rightleftharpoons$	$2\,NO_{(g)}$	+	$Cl_{2(g)}$
Initial concentration (mol/L)	0.40		0.00		0.00
Change in concentration (mol/L)	$-2x$		$+2x$		$+x$
Equilibrium concentration (mol/L)	$0.40 - 2x$		$2x$		x

$$\frac{[NO_{(g)}]^2[Cl_{2(g)}]}{[NOCl_{(g)}]^2} = K$$

$$\frac{(2x)^2(x)}{(0.40 - 2x)^2} = 1.60 \times 10^{-5}$$

$$\frac{4x^3}{(0.40 - 2x)^2} = 1.60 \times 10^{-5}$$

Is a simplification warranted? Use the hundred rule.

$$\frac{[NOCl_{(aq)}]_{\text{initial}}}{K} = \frac{0.40}{1.60 \times 10^{-5}}$$

$$= 2.4 \times 10^4$$

Since $2.4 \times 10^4 > 100$, we may assume that $0.40 - 2x \doteq 0.40$

At equilibrium,

$$\frac{4x^3}{(0.40)^2} \doteq 1.60 \times 10^{-5}$$

$$\frac{4x^3}{(0.16)} \doteq 1.60 \times 10^{-5}$$

$$4x^3 \doteq 2.56 \times 10^{-6}$$
$$x^3 \doteq 6.40 \times 10^{-7}$$

$$x \doteq \sqrt[3]{6.40 \times 10^{-7}}$$
$$x \doteq 8.62 \times 10^{-3}$$

Validate the assumption:

$$0.40 - 2x = 0.40 - 2(8.62 \times 10^{-3})$$
$$= 0.40 - 0.017$$
$$= 0.38$$

Difference = 5%, which is (just barely) acceptable. The assumption was valid.

$$[NOCl_{(g)}] = 0.40 - 2x$$
$$\doteq 0.40$$
$$[NOCl_{(g)}] \doteq 0.40 \text{ mol/L}$$

$$[NO_{(g)}] = 2x$$
$$\doteq 2(8.62 \times 10^{-3})$$
$$[NO_{(g)}] \doteq 1.72 \times 10^{-2} \text{ mol/L}$$

$$[Cl_{2(g)}] = x$$
$$[Cl_{2(g)}] \doteq 8.62 \times 10^{-3} \text{ mol/L}$$

Check:

$$Q = \frac{[NO_{(g)}]^2 [Cl_{2(g)}]}{[NOCl_{(g)}]^2}$$

$$Q = \frac{(1.72 \times 10^{-2})^2 (8.62 \times 10^{-3})}{(0.40)^2}$$

$$Q = 1.59 \times 10^{-5}$$

The value of Q (1.59×10^{-5}) is sufficiently close to the value of K (1.60×10^{-5}). We can assume that our calculations are correct.

At equilibrium,
$$[NOCl_{(g)}] \doteq 0.40 \text{ mol/L}$$
$$[NO_{(g)}] \doteq 1.72 \times 10^{-2} \text{ mol/L}$$
$$[Cl_{2(g)}] \doteq 8.62 \times 10^{-3} \text{ mol/L}$$

▶

▶ **Practice**

Understanding Concepts

7. The equilibrium constant, K, is 4.20×10^{-6} at a temperature of 1100 K for the reaction,

$$2\,H_2S_{(g)} \rightleftharpoons 2\,H_{2(g)} + S_{2(g)}$$

What concentration of $S_{2(g)}$ can be expected when 0.200 mol of $H_2S_{(g)}$ comes to equilibrium at 1100 K in a 1.00-L vessel?

8. Hydrogen chloride, $HCl_{(g)}$, decomposes into its elements according to the following equation:

$$HCl_{(g)} \rightleftharpoons H_{2(g)} + Cl_{2(g)}$$

The equilibrium constant, K, is 3.2×10^{-34} at 25°C. Calculate the equilibrium concentrations of all entities if 2.00 mol $HCl_{(g)}$ is initially placed in a closed 1.00-L vessel.

Answers

7. $[S_{2(g)}] = 3.48 \times 10^{-3}$ mol/L

8. $[H_{2(g)}] = 2.53 \times 10^{-17}$ mol/L

   $[Cl_{2(g)}] = 2.53 \times 10^{-17}$ mol/L

   $[HCl_{(g)}] = 2.00$ mol/L

In some cases, the value of K is too large to ignore. In these situations, the calculation may involve the need to solve a quadratic equation in the form

$$ax^2 + bx + c = 0$$

by using the quadratic formula

$$x = \frac{-b \pm \sqrt{b^2 - 4ac}}{2a}$$

as in the next sample problem.

▶ **SAMPLE** problem

Calculating Equilibrium Concentrations Involving a Quadratic Equation

If 0.50 mol of $N_2O_{4(g)}$ is placed in a 1.0-L closed container at 150°C, what will be the concentrations of $N_2O_{4(g)}$ and $NO_{2(g)}$ at equilibrium? ($K = 4.50$)

$$N_2O_{4(g)} \rightleftharpoons 2\,NO_{2(g)}$$

Rewrite the equation, including known values.

$$N_2O_{4(g)} \rightleftharpoons 2\,NO_{2(g)} \qquad K = 4.50 \text{ at } 150°C$$

$$n_{N_2O_{4(g)}} = 0.50 \text{ mol}$$

$$V = 1.0 \text{ L}$$

Write the equation for K.

$$\frac{[NO_{2(g)}]^2}{[N_2O_{4(g)}]} = K$$

Initial concentrations are

$$[N_2O_{4(g)}] = 0.50 \text{ mol/L}$$

$$[NO_{2(g)}] = 0.00 \text{ mol/L}$$

$$Q = \frac{[NO_{2(g)}]^2}{[N_2O_{4(g)}]}$$

$$= \frac{0.00}{0.50}$$

$$Q = 0$$

Since Q is less than K, the reaction will proceed to the right (**Table 6**).

Table 6 ICE Table for the Decomposition of $N_2O_{4(g)}$

	$N_2O_{4(g)}$	$\rightleftharpoons$	$2\ NO_{2(g)}$
Initial concentration (mol/L)	0.50		0.00
Change in concentration (mol/L)	$-x$		$+2x$
Equilibrium concentration (mol/L)	$0.50 - x$		$2x$

At equilibrium,

$$\frac{[NO_{2(g)}]^2}{[N_2O_{4(g)}]} = K$$

$$\frac{(2x)^2}{(0.50 - x)} = 4.50$$

This equation cannot be simplified by taking the square root of both sides, nor can it be simplified by ignoring the x in the denominator, as we can demonstrate using the hundred rule:

$$\frac{[N_2O_{4(g)}]_{initial}}{K} = \frac{0.50}{4.50}$$

$$= 0.11$$

Since $0.11 \ll 100$, the assumption that $0.50 - x \doteq 0.50$ is *not* warranted.
We must solve the equation using the quadratic formula. First, multiply both sides by $(0.50 - x)$:

$$4x^2 = 4.50\,(0.50 - x)$$

Next, collect like terms:

$$4x^2 + 4.50x - 2.25 = 0$$

This is a quadratic equation in the form

$$ax^2 + bx + c = 0$$

which can be solved for x with the quadratic formula:

$$x = \frac{-b \pm \sqrt{b^2 - 4ac}}{2a}$$

In the quadratic equation, $a = 4$, $b = 4.5$, and $c = -2.25$. Substituting these values into the quadratic formula gives:

$$x = \frac{-4.50 \pm \sqrt{(4.50)^2 - 4(4)(-2.25)}}{2(4)}$$

$$= \frac{-4.50 \pm 7.5}{8}$$

$$x = -1.50 \quad \text{or} \quad x = 0.375$$

The negative root would result in a negative concentration for $NO_{2(g)}$. Since negative concentrations are impossible, $x = 0.375$ is the only acceptable solution to the quadratic equation.

Substitute 0.38 (rounded to the correct certainty), but retaining the extra digit in calculations for x in the equilibrium concentration expressions in the ICE table for $NO_{2(g)}$ and $N_2O_{4(g)}$:

$$[N_2O_{4(g)}] = 0.50 \text{ mol/L} - x$$
$$= 0.50 \text{ mol/L} - 0.38 \text{ mol/L}$$
$$[N_2O_{4(g)}] = 0.12 \text{ mol/L}$$

$$[NO_{2(g)}] = 2x$$
$$= 2 (0.38 \text{ mol/L})$$
$$[NO_{2(g)}] = 0.75 \text{ mol/L}$$

At this point, it is a good idea to check your results by calculating Q again to see if it equals K. If so, the results are acceptable.

$$Q = \frac{[NO_{2(g)}]^2}{[N_2O_{4(g)}]}$$

$$= \frac{(0.75)^2}{0.125}$$

$$Q = 4.5$$

Since Q equals K, the solution is at equilibrium when $[NO_{2(g)}] = 0.12 \text{ mol/L}$, and $[N_2O_{4(g)}] = 0.75 \text{ mol/L}$.

Example

When hydrogen and iodine are placed in a closed container at 440°C, they react to form hydrogen iodide. At this temperature, the equilibrium constant, K, is 49.7. Determine the concentrations of all entities at equilibrium if 4.00 mol of hydrogen and 2.00 mol of iodine are placed in a 2.00-L reaction vessel.

Solution

$$H_{2(g)} + I_{2(g)} \rightleftharpoons 2 HI_{(g)} \qquad\qquad K = 49.7 \text{ at } 440°C$$

$n = 4.00 \text{ mol} \quad n = 2.00 \text{ mol}$
$V = 2.00 \text{ L}$

$$\frac{[HI_{(g)}]^2}{[H_{2(g)}][I_{2(g)}]} = K$$

Initial concentrations are

$$[H_{2(g)}] = \frac{4.00 \text{ mol}}{2.00 \text{ L}}$$
$$[H_{2(g)}] = 2.00 \text{ mol/L}$$

$$[I_{2(g)}] = \frac{2.00 \text{ mol}}{2.00 \text{ L}}$$
$$[I_{2(g)}] = 1.00 \text{ mol/L}$$

$$[HI_{(g)}] = 0.00 \text{ mol/L}$$

Since the initial $[HI_{(g)}]$ is 0.00 mol/L, $Q = 0$. The reaction is not at equilibrium and must proceed to the right to achieve equilibrium (**Table 8**).

Table 7   ICE Table for the Formation of $HI_{(g)}$

	$H_{2(g)}$	+	$I_{2(g)}$	$\rightleftharpoons$	$2\ HI_{(g)}$
Initial concentration (mol/L)	2.00		1.00		0.00
Change in concentration (mol/L)	$-x$		$-x$		$+2x$
Equilibrium concentration (mol/L)	$2.00 - x$		$1.00 - x$		$2x$

At equilibrium,

$$\frac{[HI_{(g)}]^2}{[H_{2(g)}][I_{2(g)}]} = K$$

$$\frac{(2x)^2}{(2.00 - x)(1.00 - x)} = 49.7$$

[Using the hundred rule reveals that a simplifying assumption is *not* warranted. Check for yourself.]

$$4x^2 = 49.7\,(2.00 - x)(1.00 - x)$$

$$0.92x^2 - 3.00x + 2.00 = 0$$

$$x = \frac{-b \pm \sqrt{b^2 - 4ac}}{2a}$$

$$x = \frac{3.00 \pm \sqrt{9.00 - 7.36}}{1.84}$$

$$x = 1.63 \pm 0.70$$

$$x = 2.33 \quad \text{or} \quad x = 0.93$$

The root $x = 2.33$ is rejected as the concentrations cannot have a negative value (i.e., $2.00 - 2.33$). Therefore, $x = 0.93$.

$$[H_{2(g)}] = 2.00\ \text{mol/L} - x$$
$$= 2.00\ \text{mol/L} - 0.93\ \text{mol/L}$$
$$[H_{2(g)}] = 1.07\ \text{mol/L}$$

$$[I_{2(g)}] = 1.00\ \text{mol/L} - x$$
$$= 1.00\ \text{mol/L} - 0.93\ \text{mol/L}$$
$$[I_{2(g)}] = 0.07\ \text{mol/L}$$

$$[HI_{(g)}] = 2x$$
$$= 2(0.93)\ \text{mol/L}$$
$$[HI_{(g)}] = 1.87\ \text{mol/L}$$

Check:

$$Q = \frac{[HI_{(g)}]^2}{[H_{2(g)}][I_{2(g)}]}$$

$$= \frac{(1.87)^2}{(1.07)(0.07)}$$

$$Q = 49.7$$

The calculated value of Q, 49.7, is equal to the given value of K, 49.7. Therefore, we assume that our calculations are correct.

The equilibrium concentrations are

$$[H_{2(g)}] = 1.07 \text{ mol/L}$$
$$[I_{2(g)}] = 0.07 \text{ mol/L}$$
$$[HI_{(g)}] = 1.87 \text{ mol/L}$$

▶ Practice

Answers

9. $[NO_{2(g)}] = 0.357$ mol/L
   $[N_2O_{4(g)}] = 0.147$ mol/L
10. $[H_{2(g)}] = 0.18$ mol/L
    $[I_{2(g)}] = 0.18$ mol/L
    $[HI_{(g)}] = 1.2$ mol/L

9. In a sealed container, nitrogen dioxide is in equilibrium with dinitrogen tetroxide.

$$2\,NO_{2(g)} \rightleftharpoons N_2O_{4(g)} \qquad\qquad K = 1.15 \text{ at } 55°C$$

Find the equilibrium concentration of nitrogen dioxide and dinitrogen tetroxide if the initial concentration of nitrogen dioxide is 0.650 mol/L.

10. The following equation describes the formation of $HI_{(g)}$

$$H_{2(g)} + I_{2(g)} \rightleftharpoons 2\,HI_{(g)} \qquad\qquad K = 46.0 \text{ at } 490°C$$

Initially, 0.40 mol of hydrogen and 0.40 mol of iodine is injected into a 500-mL electrically heated reaction vessel whose temperature is raised to 490°C. Find the concentrations of all entities at equilibrium.

SUMMARY *Solving Equilibrium Problems*

To solve equilibrium problems that require you to determine equilibrium concentrations when given the value of K, follow these steps:

1. Write a balanced equation for the reaction and list the known values.

2. Write the equilibrium constant equation.

3. Determine and list the initial concentrations.

4. Calculate Q and compare it to the value of K. Determine whether the system is at equilibrium, and if not, determine the direction in which it must proceed to attain equilibrium.

5. Construct an ICE table and input initial concentrations.

6. Let x represent the changes in concentration. When entering change information in the "change" row of the ICE table, make sure you multiply x by the appropriate coefficient in the balanced equation, and ensure that reactant concentrations all change in the same way (if one decreases, they all decrease), and that product concentrations all change in the opposite direction.

7. Substitute equilibrium concentrations (from the "equilibrium" row in the ICE table) into the equilibrium constant equation.

8. Apply appropriate simplifying assumptions, if possible (use the hundred rule).

9. Solve for x.

10. Justify any simplifying assumptions you have made (use the 5% rule).

11. Calculate equilibrium concentrations by substituting x into equilibrium concentration expressions from the "equilibrium" row of the ICE table.

12. Check your results by calculating Q using the calculated equilibrium concentration values and comparing the values of Q and K.

► **Section 7.5** *Questions*

Understanding Concepts

1. In a closed container, nitrogen and hydrogen react to produce ammonia. The equilibrium constant is 0.050. At a specific time in the reaction process $[N_{2(g)}]$ is 2.0×10^{-4} mol/L, $[H_{2(g)}]$ is 4.0×10^{-3} mol/L, and $[NH_{3(g)}]$ is 2.2×10^{-4} mol/L.

$$N_{2(g)} + 3H_{2(g)} \rightleftharpoons 2NH_{3(g)}$$

In which direction must the reaction proceed to establish equilibrium?

2. Consider the system

$$CO_{2(g)} + H_{2(g)} \rightleftharpoons CO_{(g)} + H_2O_{(g)}$$

Initially, 0.25 mol of water vapour and 0.20 mol of carbon monoxide are placed in a 1.00 L reaction vessel. At equilibrium, spectroscopic evidence shows that 0.10 mol of carbon dioxide is present. Calculate K for the reaction.

3. Consider the system

$$2HBr_{(g)} \rightleftharpoons H_{2(g)} + Br_{2(g)}$$

Initially, 0.25 mol of hydrogen and 0.25 mol of bromine are placed in a 500-mL reaction vessel that is heated electrically. K for the reaction at the temperature used is 0.020.
   (a) Calculate the concentrations at equilibrium.
   (b) Calculate the amount (in moles) of each substance present at equilibrium.
   (c) Calculate the extent of reaction as a percent reaction.

4. If 0.20 mol of hydrogen and 0.50 mol of iodine are initially introduced into a 0.500-L reaction chamber, calculate the concentrations of all entities at equilibrium.

$$H_{2(g)} + I_{2(g)} \rightleftharpoons 2HI_{(g)} \qquad K = 46.0 \text{ at } 490°C$$

5. Suppose that 4.00 mol of $N_2O_{4(g)}$ is injected into a 1.00-L container at 55°C and that the following reaction proceeds toward equilibrium:

$$2NO_{2(g)} \rightleftharpoons N_2O_{4(g)} \qquad K = 1.15 \text{ at } 55°C$$

What is the equilibrium concentration of nitrogen dioxide?

6. Consider the following equilibrium for the production of hydrogen chloride from its elements

$$H_{2(g)} + Cl_{2(g)} \rightleftharpoons 2HCl_{(g)} \qquad K = 4.4 \times 10^{-2} \text{ at } 0°C$$

Initially, 1.50 mol of hydrogen and 1.50 mol of chlorine are injected into a 750-mL refrigerated reaction vessel cooled to 0°C.
   (a) Find the concentrations of all entities at equilibrium.
   (b) Calculate the amount (in moles) of each entity present at equilibrium.
   (c) Calculate the reaction extent as a percent reaction.

7. Given that 2.5 mol of carbonyl chloride gas, $COCl_{2(g)}$, is initially introduced into a 10.00-L rigid container and the following reaction is allowed to reach equilibrium:

$$CO_{(g)} + Cl_{2(g)} \rightleftharpoons COCl_{2(g)} \qquad K = 8.2 \times 10^{-2} \text{ at } 626°C$$

Find the equilibrium concentrations of carbon monoxide and chlorine.

8. If 0.500 mol each of phosphorus trichloride and chlorine are injected into a 1.00-L container at 60°C, find the equilibrium concentrations of all three species in the equilibrium mixture.

$$PCl_{5(g)} \rightleftharpoons PCl_{3(g)} + Cl_{2(g)} \qquad K = 12.5 \text{ at } 60°C$$

Chemical Systems in Equilibrium **481**

Salt solutions are very common in nature. Seawater, lakewater, tapwater, tree sap, saliva, and blood plasma all contain a mixture of dissolved salts including $NaCl_{(aq)}$, $MgSO_{4(aq)}$, and $NaHCO_{3(aq)}$. These compounds are all highly soluble electrolytes. However, many useful salts are only slightly soluble in water. These include compounds such as magnesium hydroxide, $Mg(OH)_{2(s)}$, (milk of magnesia, an antacid), calcium sulfate, $CaSO_{4(s)}$ (gypsum, used to make wallboard), and barium sulfate, $BaSO_{4(s)}$ (used in X rays of the gastrointestinal tract).

Sparingly Soluble Solutes in Animals

When an animal digests the proteins and nucleic acids in its food, it produces toxic nitrogenous wastes that must be eliminated from its body. The source of the nitrogen in these waste products is the amino group ($-NH_2$) that is removed from the amino acids of proteins and the nitrogenous bases of nucleic acids when these macromolecules are metabolized for energy or converted into other useful molecules. Different classes of animals process the amino groups in different ways.

Most aquatic animals like fish convert the amino groups into ammonia, $NH_{3(aq)}$. Ammonia is highly toxic, but also highly soluble in water. Ammonia never builds up in the tissues of such organisms because of its solubility and is easily excreted through the gills into the surrounding water as dissolved ammonium ions, $NH_{4(aq)}^+$.

Mammals, amphibians, and sharks, on the other hand, convert amino groups into urea, $H_2NCONH_{2(aq)}$, a compound ten thousand times less toxic than ammonia. Urea may accumulate to relatively high concentrations in the circulatory system of an organism without ill effect. Eventually, it is filtered by the kidneys and excreted in liquid urine.

However, birds, insects, and reptiles do not convert the amino groups into urea. Instead, they form uric acid, $H_2C_5N_4O_{3(s)}$, a substance that is thousands of times less soluble in water than ammonia or urea. (The solubility of uric acid is 4.2×10^{-8} mol/L at 25°C.) The white portion of bird droppings is composed of precipitated uric acid crystals (**Figure 2**).

Production of uric acid and urea are the only two methods available to terrestrial animals for excreting nitrogen-containing waste products. Ammonia is far too toxic for terrestrial organisms—they cannot excrete nitrogenous wastes by letting them diffuse out of membranes into an external aqueous solution, the way fish do. In land animals, ammonia would accumulate to toxic levels very quickly. Biologists have discovered that the method of reproduction of an organism seems to play a role in determining which method a particular group of animals uses to eliminate nitrogenous waste.

Animals that excrete solid uric acid lay eggs with hard shells (birds and reptiles). The shells of these eggs are permeable to gases, but not liquids. As the embryo develops inside the egg it produces nitrogenous waste that must be removed, or the growing organism will be poisoned. Since the shell is impermeable to liquids, dissolved ammonia or urea would accumulate within the egg to toxic concentrations, killing the developing embryo. Uric acid, with its extremely low solubility in water, poses less of a threat. It precipitates out of solution even in low concentrations, and can be stored within the egg as solid waste until the bird hatches.

Another slightly soluble salt has a noticeable effect on the human body. When the concentrations of calcium ions, $Ca_{(aq)}^{2+}$, and oxalate ions, $C_2O_{4(aq)}^{2-}$, become sufficiently high in the bloodstream, calcium oxalate, $CaC_2O_{4(s)}$, "stones" precipitate in the kidneys.

Figure 1
Over thousands of years, the action of water on limestone (calcium carbonate) rock formations has created crevices and caves.

Figure 2
Uric acid (white) is present in bird droppings at concentrations far beyond its solubility.

What amount of a slightly soluble salt will dissolve in a given volume of water to form a saturated solution? What minimum concentration of aqueous calcium and oxalate ions causes the precipitation of the salts that form kidney stones?

These questions involve equilibria between the dissolved ions and solids of slightly soluble salts. Whether we want to determine the degree of dissolution (dissociation), or the conditions that result in precipitation of slightly soluble salts, equilibrium constants and equilibrium law equations are involved.

In previous chemistry courses you learned about **solubility**. You may have done investigations to find out how much of a solute would dissolve in a solvent under certain conditions. Using solutes with high solubility, you were able to add a visible amount of the solute and make it dissolve, forming an unsaturated solution. When you could make no more solute dissolve, you had prepared a saturated solution, one where there is a dynamic equilibrium between undissolved solute and dissolved solute. If the container were sealed, and there were no temperature changes, no further changes occurred in the concentration of the solution or in the quantity of undissolved solute. You had made a system in which the rate of dissolution was equal to the rate of crystallization— a dynamic equilibrium.

Solubility Product

A special case of equilibrium involves any situation where excess solute is in equilibrium with its aqueous solution. You can establish such an equilibrium either by starting with excess salt and dissolving it until the solution is saturated and excess solute remains (**Figure 3(a)**), or by mixing solutions of two salts that results in a product precipitating out of solution (**Figure 3(b)**). Once precipitation ends, the remaining solution is saturated, and the dissolved ions form a dynamic equilibrium with the precipitated crystals.

Consider the chemical equation and equilibrium law equation for a weak electrolyte such as copper(I) chloride. Notice that since copper(I) chloride dissolves very little in water, almost any amount dropped into water will result in the formation of a saturated solution, and there will be a heterogeneous equilibrium between the solid and dissolved $Cu^+_{(aq)}$ and $Cl^-_{(aq)}$ ions.

$$CuCl_{(s)} \rightleftharpoons Cu^+_{(aq)} + Cl^-_{(aq)}$$

The solubility equilibrium law equation is

$$K = \frac{[Cu^+_{(aq)}][Cl^-_{(aq)}]}{[CuCl_{(s)}]}$$

which simplifies to

$$K = [Cu^+_{(aq)}][Cl^-_{(aq)}]$$

as the concentration (density) of the $CuCl_{(s)}$ is constant and so is incorporated into the value of the equilibrium constant. The value of this equilibrium constant is

$$K = 1.7 \times 10^{-7} \text{ at } 25°C.$$

For any solute that forms ions in solution, the solubility equilibrium constant is the product of the concentrations of the ions in solution raised to the power equal to the coefficient of each in the balanced equation, and for that reason is often called the **solubility product constant** of the substance, symbolized as K_{sp}. The equilibrium law equation for the copper(I) chloride equilibrium is written

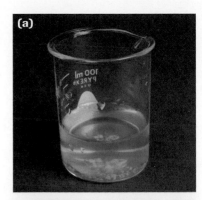

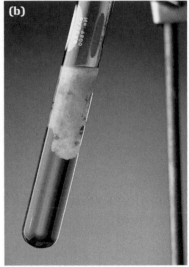

Figure 3
Saturated solutions.
(a) This saturated solution was formed by adding excess copper(II) sulfate to water. The solid is in equilibrium with the ions in solution.
(b) This saturated solution was formed by adding a solution of iron(III) nitrate to a solution of sodium phosphate. The result is a saturated solution of iron(III) phosphate, in which the ions in solution are in equilibrium with the solid precipitate.

solubility the concentration of a saturated solution of a solute in a particular solvent at a particular temperature; solubility is a specific maximum concentration

solubility product constant (K_{sp}) the value obtained from the equilibrium law applied to a saturated solution

$$K_{sp} = [Cu^+_{(aq)}][Cl^-_{(aq)}]$$
$$K_{sp} = 1.7 \times 10^{-7} \text{ at } 25°C$$

For ionic substances with a more complex formula, like calcium phosphate, the K_{sp} expression is also more complex. The balanced equation is

$$Ca_3(PO_4)_{2(s)} \rightleftharpoons 3\,Ca^{2+}_{(aq)} + 2\,PO^{3-}_{4(aq)}$$

The equilibrium expression is

$$K_{sp} = [Ca^{2+}_{(aq)}]^3[PO^{3-}_{4(aq)}]^2$$

with the value of the constant

$$K_{sp} = 2.1 \times 10^{-33} \text{ at } 25°C$$

In general, for the dissociation equilibrium equation

$$BC_{(s)} \rightleftharpoons b\,B^+_{(aq)} + c\,C^-_{(aq)}$$

where $BC_{(s)}$ is a slightly soluble salt, and $B^+_{(aq)}$ and $C^-_{(aq)}$ are aqueous ions,

$$K_{sp} = [B^+_{(aq)}]^b[C^-_{(aq)}]^c$$

K_{sp} values are listed in many chemistry reference sources (see **Table 1** and Appendix C6).

LEARNING TIP

As the units for K_{sp} vary from mol/L to (mol/L)^2 to (mol/L)^3, the convention in the scientific community is to omit units when writing K_{sp} values, as is the case for other K values.

Table 1 Some Solubility Product Constants at 25°C

Name	Formula	K_{sp}
cobalt(II) hydroxide	$Co(OH)_{2(s)}$	1.1×10^{-15}
lithium carbonate	$Li_2CO_{3(s)}$	8.2×10^{-4}
mercury(I) chloride	$Hg_2Cl_{2(s)}$	1.5×10^{-18}
nickel(II) carbonate	$NiCO_{3(s)}$	1.4×10^{-7}
tin(II) sulfide	$SnS_{(s)}$	3.2×10^{-28}
zinc hydroxide	$Zn(OH)_{2(s)}$	7.7×10^{-17}
calcium phosphate	$Ca_3(PO_4)_{2(s)}$	2.1×10^{-33}
magnesium fluoride	$MgF_{2(s)}$	$7.4. \times 10^{-11}$
lead(II) chloride	$PbCl_{2(s)}$	1.2×10^{-5}

Reference – *CRC Handbook of Chemistry and Physics* (76th Ed.)

LEARNING TIP

Do not confuse *solubility* with *solubility product*. The solubility of a salt is the amount of salt that dissolves in a given amount of solvent to give a saturated solution. Solubility product is the product of the molar concentrations of the ions in the saturated solution.

References typically list K_{sp} values only for ionic compounds with low solubility, because under ordinary laboratory conditions highly soluble ionic compounds do not form precipitates. Their solutions, as used, are not saturated—no solubility equilibrium is established. Solubilities of highly soluble substances are listed in mol/L or g/100 mL values rather than as K_{sp} values.

Calculating Solubility Using K_{sp} Values

A straightforward calculation will convert a solubility value to (or from) a K_{sp} value, as the following examples show.

Calculating the Solubility Product Constant, Given the Solubility | SAMPLE problem ◀

Magnesium fluoride is a hard, slightly soluble salt that is used to make spectral lenses for technical instruments. Calculate K_{sp} for magnesium fluoride at 25°C, given a solubility of 0.001 72 g/100 mL.

$$MgF_{2(s)} \rightleftharpoons Mg^{2+}_{(aq)} + 2 F^-_{(aq)}$$
$$K_{sp} = [Mg^{2+}_{(aq)}][F^-_{(aq)}]^2$$

From the balanced equation, we know that

$$[Mg^{2+}_{(aq)}] = [MgF_{2(aq)}]$$

The $[MgF_{2(aq)}]$ in the above equation refers to the concentration of MgF_2 formula units that produce the aqueous ions, $Mg^{2+}_{(aq)}$ and $F^-_{(aq)}$. This is a *very* small proportion of the $MgF_{2(s)}$ crystal.

Convert the solubility values (g/100mL) to concentration values (mol/L).

$$[Mg^{2+}_{(aq)}] = [MgF_{2(aq)}] = \frac{0.001\ 72\ g}{100\ mL} \times \frac{1\ mol}{62.31\ g} \times \frac{1000\ mL}{1\ L}$$

$$= 2.8 \times 10^{-4}\ mol/L$$

From the balanced reaction equation above,

$$[F^-_{(aq)}] = 2\,[Mg^{2+}_{(aq)}]$$
$$= 2 \times 2.8 \times 10^{-4}\ mol/L$$
$$= 5.5 \times 10^{-4}\ mol/L$$

$$K_{sp} = [Mg^{2+}_{(aq)}][F^-_{(aq)}]^2$$
$$= (2.8 \times 10^{-4})(5.5 \times 10^{-4})^2$$
$$K_{sp} = 8.4 \times 10^{-11}$$

OR
Once the concentration of the fraction of $MgF_{2(s)}$ that dissolves is calculated (2.8×10^{-4} mol/L), it is also possible to solve the problem using an ICE table (**Table 2**).

Table 2 ICE Table for Calculating K_{sp} from Solubility

	$MgF_{2(s)}$	$\rightleftharpoons$	$Mg^{2+}_{(aq)}$	+	$2\ F^-_{(aq)}$
Initial concentration (mol/L)	–		0		0
Change in concentration (mol/L)	–		$+x$		$+2x$
Equilibrium concentration (mol/L)	–		x		$2x$

$$x = 2.8 \times 10^{-4}$$
$$2x = 2(2.8 \times 10^{-4})$$
$$2x = 5.5 \times 10^{-4}$$

$$K_{sp} = [Mg^{2+}_{(aq)}][F^-_{(aq)}]^2$$
$$= (2.8 \times 10^{-4})(5.5 \times 10^{-4})^2$$
$$K_{sp} = 8.4 \times 10^{-11}$$

Example
Calculate the molar solubility of zinc hydroxide at 25°C, where K_{sp} is 7.7×10^{-17}.

$$Zn(OH)_{2(s)} \rightleftharpoons Zn^{2+}_{(aq)} + 2\,OH^-_{(aq)}$$

▶

Solution

$$K_{sp} = [Zn^{2+}_{(aq)}][OH^-_{(aq)}]^2$$
$$= 7.7 \times 10^{-17}$$

$$[OH^-_{(aq)}] = 2\,[Zn^{2+}_{(aq)}]$$

$$K_{sp} = [Zn^{2+}_{(aq)}](2\,[Zn^{2+}_{(aq)}])^2$$
$$7.7 \times 10^{-17} = 4[Zn^{2+}_{(aq)}]^3$$
$$[Zn^{2+}_{(aq)}] = \sqrt[3]{\dfrac{7.7 \times 10^{-17}}{4}}$$
$$= 2.7 \times 10^{-6} \text{ mol/L}$$

$$[Zn(OH)_{2(aq)}] = [Zn^{2+}_{(aq)}]$$
$$= 2.7 \times 10^{-6} \text{ mol/L}$$

The molar solubility of zinc hydroxide is 2.7×10^{-6} mol/L.

Or, using an ICE table (**Table 3**):

Table 3 ICE Table for Calculating K_{sp} from Solubility

	$Zn(OH)_{2(s)}$	$\rightleftharpoons$	$Zn^{2+}_{(aq)}$	+	$2\,OH^-_{(aq)}$
Initial concentration (mol/L)	—		0		0
Change in concentration (mol/L)	—		$+x$		$+2x$
Equilibrium concentration (mol/L)	—		x		$2x$

$$K_{sp} = [Zn^{2+}_{(aq)}][OH^-_{(aq)}]^2 = 7.7 \times 10^{-17}$$
$$K_{sp} = (x)(2x)^2$$
$$7.7 \times 10^{-17} = 4x^3$$
$$x = \sqrt[3]{\dfrac{7.7 \times 10^{-17}}{4}}$$
$$x = 2.7 \times 10^{-6} \text{ mol/L}$$

The molar solubility of zinc hydroxide is 2.7×10^{-6} mol/L

▶ Practice

Understanding Concepts

Answers

1. 1.2×10^{-8} mol/L
2. 5.9×10^{-6} mol/L
3. 2.3×10^{-6} mol/L
4. (a) 7.90×10^{-37}
   (b) 1.50×10^{-11}

1. Calculate the solubility of silver iodide at 25°C. The K_{sp} of $AgI_{(s)}$ is 1.5×10^{-16} at 25°C.

2. Calculate the solubility of iron(II) carbonate at 25°C. The K_{sp} of $FeCO_{3(s)}$ is 3.5×10^{-11} at 25°C.

3. Calculate the solubility of zinc hydroxide at 25°C. The K_{sp} of $Zn(OH)_{2(s)}$ is 4.5×10^{-17} at 25°C.

4. Given the following solubilities, calculate the value of the solubility product for each compound:
   (a) copper (II) sulfide, 8.89×10^{-19} mol/L
   (b) zinc carbonate, 3.87×10^{-6} mol/L

Table 4 Solubility of Ionic Compounds at SATP

		Anions						
		Cl^-, Br^-, I^-	S^{2-}	OH^-	SO_4^{2-}	CO_3^{2-}, PO_4^{3-}, SO_3^{2-}	$C_2H_3O_2^-$	NO_3^-
Cations	high solubility (aq) ≥0.1 mol/L (at SATP)	most	Group 1, NH_4^+ Group 2	Group 1, NH_4^+ Sr^{2+}, Ba^{2+}, Tl^+	most	Group 1, NH_4^+	most	all
			All Group 1 compounds, acids, and all ammonium compounds are assumed to have high solubility in water.					
	low Solubility (s) <0.1 mol/L (at SATP)	Ag^+, Pb^{2+}, Tl^+, Hg_2^{2+} (Hg^+), Cu^+	most	most	Ag^+, Pb^{2+}, Ca^{2+}, Ba^{2+}, Sr^{2+}, Ra^{2+}	most	Ag^+	none

Predicting Precipitation

In Section 7.5, you learned that the reaction quotient, Q, is a value that can be used to calculate whether or not a system is at equilibrium. We can also use the reaction quotient in the context of solubility—to predict whether a precipitate will form when we mix solutions of metal cations and nonmetal anions. (In previous chemistry courses, you used solubility tables like **Table 4** to determine qualitatively whether an ionic compound had high or low solubility.) We can now calculate Q to determine whether, after mixing, the ions are present in too high a concentration, in which case a precipitate will form. In this situation, the reaction quotient, Q, is sometimes called the **trial ion product**.

To predict whether a precipitate will form when solutions containing anions and cations are mixed, we compare the K_{sp} value for the salt of these ions (from a table like **Table 1**, page 484) to Q (as a trial ion product).

As you know, the value of K_{sp} equals the ion product of a saturated solution in which dissolved and undissolved solutes are in dynamic equilibrium. For example,

$$CuCl_{(s)} \rightleftharpoons Cu^+_{(aq)} + Cl^-_{(aq)} \qquad K_{sp} = 7.1 \times 10^{-7}$$

Consider a solution containing $Cu^+_{(aq)}$ and $Cl^-_{(aq)}$ ions, each at a concentration of 4.1×10^{-4} mol/L. Since the trial ion product, Q, is less than K_{sp},

$$Q = [Cu^+_{(aq)}][Cl^-_{(aq)}]$$
$$= (4.1 \times 10^{-4})(4.1 \times 10^{-4})$$
$$Q = 1.7 \times 10^{-7}$$
$$K_{sp} = 7.1 \times 10^{-7}$$
$$Q < K_{sp}$$

trial ion product the reaction quotient applied to the ion concentrations of a slightly soluble salt

no precipitation occurs—all aqueous ions are able to remain in the dissolved state. The dynamic equilibrium that exists between dissolved ions and any undissolved solute ensures that there will be no net crystallization. The solution is unsaturated.

Now consider a solution that is saturated. It is not capable of dissolving more ions. In this case, the trial ion product is equal to the value of K_{sp}. A saturated solution will not form a precipitate.

A third possibility is the case where the trial ion product is greater than the K_{sp} value. In this case, there are more ions in solution than are necessary for saturation. This is a **supersaturated solution**. A supersaturated solution is unstable; there is a tendency for the extra solute to precipitate.

We can also use the value of K_{sp} to determine whether a precipitate will form when two solutions are mixed. See the Sample Problem on page 488.

supersaturated solution a solution whose solute concentration exceeds the equilibrium concentration

Determining Whether a Precipitate Will Form

If 100 mL of 0.100 mol/L CaCl$_{2\,(aq)}$ and 100 mL of 0.0400 mol/L Na$_2$SO$_{4(aq)}$ are mixed at 20°C, determine whether a precipitate will form. For CaSO$_{4(aq)}$ at 20°C, K$_{sp}$ is 3.6 × 10^{-5}.

This problem essentially asks whether a double displacement reaction will occur when a solution of CaCl$_{2(aq)}$ is mixed with a solution of Na$_2$SO$_{4(aq)}$, and, if so, whether CaSO$_{4(s)}$ will precipitate.

We begin by writing the potential double displacement reaction.

$$CaCl_{2(aq)} + Na_2SO_{4(aq)} \rightarrow CaSO_{4(s)} + NaCl_{(aq)}$$

According to **Table 4**, CaSO$_{4(s)}$ is relatively insoluble. But are the concentrations of Ca$^{2+}_{(aq)}$ and SO$^{2-}_{4(aq)}$ high enough for precipitation to occur?

To determine the concentration of Ca$^{2+}_{(aq)}$ before mixing, we analyze the equation that describes the calcium chloride solution.

$$CaCl_{2(s)} \rightleftharpoons Ca^{2+}_{(aq)} + 2\,Cl^-_{(aq)}$$
$$[Ca^{2+}_{(aq)}] = [CaCl_{2(aq)}] = 0.100 \text{ mol/L} \qquad \text{(before mixing)}$$

Similarly, to determine the concentration of SO$^{2-}_{4(aq)}$ before mixing, we analyze the equation that describes the sodium sulfate solution.

$$[SO^{2-}_{4(aq)}] = [Na_2SO_{4(aq)}] = 0.0400 \text{ mol/L} \quad \text{(before mixing)}$$

Note that mixing two solutions always increases the overall volume, so the initial concentration of ions in both solutions is always *decreased* by the act of mixing them.

In this instance, after mixing,

$$[Ca^{2+}_{(aq)}] = 0.100 \text{ mol/L} \times \frac{100 \text{ mL}}{200 \text{ mL}} = 0.0500 \text{ mol/L}$$

Similarly, after mixing,

$$[SO^{2-}_{4(aq)}] = 0.0400 \text{ mol/L} \times \frac{100 \text{ mL}}{200 \text{ mL}} = 0.0200 \text{ mol/L}$$

Now we use these two concentrations to calculate the ion product, Q, for CaSO$_{4(s)}$, which we can write from the dissociation reaction of the salt.

$$CaSO_{4(s)} \rightleftharpoons Ca^{2+}_{(aq)} + SO^{2-}_{4(aq)}$$

$$Q = [Ca^{2+}_{(aq)}][SO^{2-}_{4(aq)}]$$

$$= (0.0500)(0.0200)$$
$$Q = 1.00 \times 10^{-3}$$

Q gives the reaction quotient for the component ions in CaSO$_{4(s)}$. Q is much greater than the K$_{sp}$ value (3.6 × 10^{-5}), indicating that more ions are present than would be present at equilibrium, so the reaction must shift toward the solid to establish equilibrium. Therefore, a precipitate forms.

(Note that, although in this case we are starting with ions and producing a precipitate, the equation is, by convention, written in the form above, with the solid salt on the left. Remember that the reaction under consideration is the backward, or reverse, reaction.)

Example

Would a precipitate of lead(II) sulfate, PbSO$_{4(s)}$, (K$_{sp}$ = 1.8 × 10^{-8}) form if 255 mL of 0.000 16 mol/L lead(II) nitrate, Pb(NO$_3$)$_{2(aq)}$, is poured into 456 mL of 0.000 23 mol/L sodium sulfate, Na$_2$SO$_{4(aq)}$?

Solution

$$Pb(NO_3)_{2(aq)} + Na_2SO_{4(aq)} \rightarrow PbSO_{4(s)} + 2\,NaNO_{3(aq)}$$

Before mixing:

$$Pb(NO_3)_{2(aq)} \rightleftharpoons Pb^{2+}_{(aq)} + 2\,NO^-_{3(aq)}$$
$$[Pb^{2+}_{(aq)}] = [Pb(NO_3)_{2(aq)}] = 0.000\ 16\ mol/L$$
$$Na_2SO_{4(aq)} \rightleftharpoons 2\,Na^+_{(aq)} + SO^{2-}_{4(aq)}$$
$$[SO^{2-}_{4(aq)}] = [Na_2SO_{4(aq)}] = 0.000\ 23\ mol/L$$

After mixing, there is

$$255\ mL + 456\ mL = 711\ mL$$

of the solution. Therefore, the concentrations of the lead(II) and sulfate ions in the mixed solution are calculated as

$$[Pb^{2+}_{(aq)}] = 0.000\ 16\ mol/L \times \frac{255\ mL}{711\ mL} = 5.74 \times 10^{-5}\ mol/L$$

$$[SO^{2-}_{4(aq)}] = 0.000\ 23\ mol/L \times \frac{456\ mL}{711\ mL} = 1.48 \times 10^{-4}\ mol/L$$

$$PbSO_{4(s)} \rightleftharpoons Pb^{2+}_{(aq)} + SO^{2-}_{4(aq)}$$

$$Q = [Pb^{2+}_{(aq)}][SO^{2-}_{4(aq)}]$$
$$= (5.74 \times 10^{-5})(1.48 \times 10^{-4})$$

$$Q = 8.46 \times 10^{-9}$$
$$K_{sp} = 1.8 \times 10^{-8}$$

Q is smaller than K_{sp}. Therefore, a precipitate does not form.

▶ Practice

Understanding Concepts

5. Refer to Appendix C8 to predict whether a precipitate forms if
   (a) 25.0 mL of 0.010 mol/L silver nitrate is mixed with 25.0 mL of 0.0050 mol/L potassium chloride.
   (b) equal volumes of 0.0010 mol/L calcium nitrate and 0.0020 mol/L potassium hydroxide are combined.
   (c) equal volumes of 0.010 mol/L lead(II) nitrate and 0.10 mol/L sodium chloride are combined.

Answers

5. (a) $Q = 1.25 \times 10^{-5}$
   (b) $Q = 5.0 \times 10^{-10}$
   (c) $Q = 1.2 \times 10^{-5}$
6. (a) $[Ba^{2+}_{(aq)}] = 1.0 \times 10^{-5}\ mol/L$

Making Connections

6. Barium sulfate is a white, insoluble ionic compound that is opaque to X rays. Prior to going for a gastrointestinal X ray, patients are sometimes given a chalky-white suspension of barium sulfate to drink. Because barium is opaque to X rays, the patient's gastrointestinal tract is clearly visible in the X ray. Barium sulfate has a K_{sp} of 1.1×10^{-10}.

$$BaSO_{4(s)} \rightleftharpoons Ba^{2+}_{(aq)} + SO^{2-}_{4(aq)}$$

Why is it safe for patients to consume barium sulfate even though barium ions are extremely toxic?

INVESTIGATION 7.6.1

Determining the K_{sp} of Calcium Oxalate (p. 517)
Calcium oxalate may crystallize out of solution in the kidneys and other parts of the human urinary tract, forming kidney stones. What is the K_{sp} of this low-solubility salt?

INVESTIGATION 7.6.2

Determining K_{sp} for Calcium Hydroxide (p. 519)
Design and carry out your own version of this classic experiment to find K_{sp}.

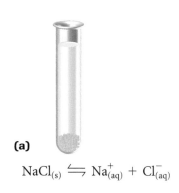

(a)

$$NaCl_{(s)} \rightleftharpoons Na^+_{(aq)} + Cl^-_{(aq)}$$

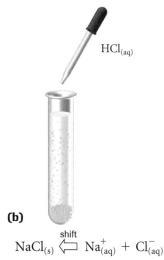

$HCl_{(aq)}$

(b)

$$NaCl_{(s)} \overset{shift}{\rightleftharpoons} Na^+_{(aq)} + Cl^-_{(aq)}$$

Figure 4
The common ion effect.
(a) A saturated solution of sodium chloride is in equilibrium with excess solid.
(b) Adding a common ion, $Cl^-_{(aq)}$, from hydrochloric acid, $HCl_{(aq)}$, shifts the equilibrium, causing sodium chloride to precipitate out of solution.

Using Trial Ion Product

When determining whether a precipitate will form in a particular solution, remember that K_{sp} is the ion product for a saturated solution—a solution in which there is a dynamic equilibrium between dissolved and undissolved solute. However, the reaction quotient Q (trial ion product) may be calculated for any solution, regardless of the extent to which the solute is dissolved.

For example, the K_{sp} of AgCl is 1.8×10^{-10} at 25°C. In a saturated solution of AgCl,

$$[Ag^+_{(aq)}] = [Cl^-_{(aq)}]$$

and

$$K_{sp} = [Ag^+_{(aq)}][Cl^-_{(aq)}]$$

Effectively then, the concentration of either of the two ions at equilibrium is the square root of the ion product, which (to two significant digits) is 1.3×10^{-5} mol/L. Logically then, any $AgCl_{(aq)}$ solution in which the concentration of both ions is less than 1.3×10^{-5} must be unsaturated, and no precipitate will form. The trial ion product for such a solution will be less than the K_{sp} for AgCl.

In an AgCl solution where the $[Ag^+_{(aq)}]$ and $[Cl^-_{(aq)}]$ are both greater than 1.3×10^{-5} mol/L, the trial ion product will be greater than the K_{sp}. This indicates that there is more solute dissolved in the solution than can be maintained in the dissolved state (a supersaturated solution). The excess solute will precipitate out of solution until the concentrations of $Ag^+_{(aq)}$ and $Cl^-_{(aq)}$ equal 1.3×10^{-5} mol/L, at which point the solution is saturated.

By comparing the trial ion product to the K_{sp} for a particular salt solution, we can predict whether a precipitate will form.

SUMMARY   Using Q to Predict Solubility

Ion product, $Q > K_{sp}$	(supersaturated solution)	Precipitate will form.
Ion product, $Q = K_{sp}$	(saturated solution)	Precipitate will not form.
Ion product, $Q < K_{sp}$	(unsaturated solution)	Precipitate will not form.

The Common Ion Effect

When equilibrium exists in a solution involving ions, the equilibrium can be shifted by dissolving into the solution any other compound that adds a common ion, or any compound that reacts with one of the ions already in solution.

Consider a saturated solution of sodium chloride, in equilibrium with a small amount of undissolved sodium chloride (**Figure 4(a)**).

$$NaCl_{(s)} \rightleftharpoons Na^+_{(aq)} + Cl^-_{(aq)}$$

If a few drops of concentrated hydrochloric acid is added to the equilibrium mixture, additional crystals of sodium chloride will form (**Figure 4(b)**). How can we explain this? With Le Châtelier's principle. The hydrochloric acid releases large numbers of chloride ions into the solution.

$$HCl_{(aq)} \rightleftharpoons H^+_{(aq)} + Cl^-_{(aq)}$$

These additional ions increase the concentration of chloride ions in the mixture, shifting the sodium chloride equilibrium to the left—i.e., causing $NaCl_{(s)}$ to precipitate out of the solution. In this example, the chloride ions were common to both solutions that were mixed. We could expect a similar result if we had added a solution containing $Na^+_{(aq)}$ ions instead. In that case the common ion would be the sodium ion. The lowering of the solubility of an ionic compound by the addition of a common ion is called the **common ion effect**.

common ion effect a reduction in the solubility of a salt caused by the presence of another salt having a common ion

SAMPLE problem ◀

Solubility in Solutions With Common Ions

What is the molar solubility of $PbCl_{2(s)}$ in a 0.2 mol/L $NaCl_{(aq)}$ solution at SATP?

In this problem, you are asked to determine the amount of $PbCl_{2(s)}$ that will dissolve into a solution that already contains $Cl^-_{(aq)}$ ions. The two salts have a common ion, $Cl^-_{(aq)}$.

$NaCl_{(s)}$ (which has high solubility) dissolves completely in water to form $Na^+_{(aq)}$ and $Cl^-_{(aq)}$ ions.

$$NaCl_{(s)} \rightarrow Na^+_{(aq)} + Cl^-_{(aq)}$$

Therefore, before the addition of the lead(II) chloride,

$$[Cl^-_{(aq)}] = [NaCl_{(aq)}] = 0.2 \text{ mol/L}$$

We can look up the solubility product constant of $PbCl_{2(s)}$ in a solubility table, for example, the table in Appendix C8. The low K_{sp} value shows that it is only slightly soluble in water. We can therefore say that $PbCl_{2(s)}$ establishes a dynamic equilibrium in solution according to the following equation:

$$PbCl_{2(s)} \rightleftharpoons Pb^{2+}_{(aq)} + 2Cl^-_{(aq)} \qquad K_{sp} = 1.7 \times 10^{-5} \text{ at } 25°C$$

In this problem, a $PbCl_2$ equilibrium is being established within a solution that already contains $Cl^-_{(aq)}$ ions. The result should be that the position of the $PbCl_2$ equilibrium is "left-shifted" from its normal point. In other words, the dissolution should not proceed to the right as much as it would in pure water. More of the lead chloride should stay in solid form (**Figure 5**).

We can use the $PbCl_{2(s)}$ equilibrium equation and an ICE table to determine the solubility of $PbCl_{2(s)}$, noting that an initial $[Cl^-_{(aq)}]$ of 0.2 mol/L already exists in the solution before any $PbCl_{2(s)}$ dissolves. When setting up the ICE table (**Table 5**) in a common ion problem, treat the overall process as two separate steps. The first step involves the initial solution into which a salt is going to be dissolved (if this is pure water, then the initial concentrations of all ions are assumed to be 0 mol/L). In this problem, the first step is determining $[Cl^-_{(aq)}]$ in the $NaCl_{(aq)}$ solution, which can be entered on the Initial concentration line in the table.

The second step in setting up the ICE table is to use the coefficients in the equilibrium equation to give the multiples for x in the Change in concentration line of the ICE table. As a result, the multiples of x will correspond to the ion coefficients in the balanced equilibrium equation.

Figure 5
Lead(II) chloride is a low solubility salt, but its solubility is lower still in a solution of sodium chloride.

LEARNING TIP

Note that the coefficients of the equilibrium chemical equation for the added salt do NOT apply to the original solute. Do NOT multiply original concentrations by the coefficients of the equilibrium chemical equation.

Table 5 ICE Table to Predict the Solubility of $PbCl_{2(s)}$ in a Solution Containing $NaCl_{(aq)}$

	$PbCl_{2(s)}$	$\rightleftharpoons$	$Pb^{2+}_{(aq)}$	+	$2Cl^-_{(aq)}$
Initial concentration (mol/L)	—		0		0.2
Change in concentration (mol/L)	—		$+x$		$+2x$
Equilibrium concentration (mol/L)	—		x		$0.2 + 2x$

$$K_{sp} = [Pb^{2+}_{(aq)}][Cl^-_{(aq)}]^2 = 1.7 \times 10^{-5}$$

$$K_{sp} = (x)(0.2 + 2x)^2 = 1.7 \times 10^{-5}$$

We can simplify the math in this question by noting that the K_{sp} is very small ($PbCl_2$ has a very low solubility in water). It follows that the value of x and therefore, $2x$ will be exceedingly small. When this very small value is added to 0.2 (a much larger value), the result will essentially remain 0.2. We make the simplifying assumption that

$$0.2 + 2x \doteq 0.2$$

Therefore,

$$K_{sp} \doteq (x)(0.2)^2 = 1.7 \times 10^{-5}$$

$$x \doteq \frac{1.7 \times 10^{-5}}{(0.2)^2}$$

$$x \doteq 4.2 \times 10^{-4}$$

To determine whether the assumption that $0.2 + 2x \doteq 0.2$ is appropriate, we notice that

$$2x = 2(4.2 \times 10^{-4})$$

$$= 8.4 \times 10^{-4}$$

which is indeed much smaller than 0.2. We accept the assumption as valid.

The molar solubility of lead(II) chloride in 0.2 mol/L $NaCl_{(aq)}$ solution is 4.2×10^{-4} mol/L.

▶ Practice

Understanding Concepts

7. Calculate the solubility of silver chloride in a 0.10 mol/L solution of sodium chloride at 25°C. At SATP, K_{sp} $AgCl_{(s)}$ = 1.8×10^{-10}.

8. Calculate the solubility of calcium sulfate in 0.010 mol/L calcium nitrate at SATP.

9. The K_{sp} of $Ag_2CrO_{4(s)}$ is 1.12×10^{-12}. Calculate the molar solubility of $Ag_2CrO_{4(s)}$
   (a) in pure water, and
   (b) in a solution of 0.10 mol/L sodium chromate, $Na_2CrO_{4(s)}$.
   (c) Compare your answers in (a) and (b). Is the difference reasonable? Explain.

10. Name two compounds that will decrease the solubility of barium sulfate, $BaSO_{4(s)}$.

11. Name two compounds that will decrease the solubility of copper(II) carbonate, $CuCO_{3(s)}$.

12. Consider the equilibrium:

$$AgCl_{(s)} \rightleftharpoons Ag^+_{(aq)} + Cl^-_{(aq)}$$

During the processing of photographic film, unreacted silver compounds are removed from the film by the addition of thiosulfate ions, $S_2O_3^{2-}_{(aq)}$, to produce the ion $Ag(S_2O_3)_2^{3-}_{(aq)}$.

Using Le Châtelier's principle, explain why the addition of a sodium thiosulfate solution makes the solid silver compounds dissolve.

Answers

7. 1.8×10^{-9} mol/L

8. 4.8×10^{-3} mol/L

9. (a) 6.54×10^{-5} mol/L
   (b) 1.7×10^{-6} mol/L

Understanding Concepts

1. Distinguish between *solubility* and *solubility product constant*.

2. (a) Define the common ion effect.
   (b) How does Le Châtelier's principle explain the common ion effect?

3. How do the concepts of a saturated solution and a super-saturated solution help explain the formation of the precipitate of a slightly soluble salt when certain soluble salt solutions are mixed?

4. Calculate the molar solubility of barium sulfate at 25°C.

5. Calculate the solubility at 25°C of silver bromide, in g/100 mL.

6. Calculate the molar concentration of fluoride ions in a saturated solution of strontium fluoride at 25°C.

7. Calculate the K_{sp} of thallium(I) chloride at 100°C. The concentration of a saturated solution of the salt at this temperature is 2.4 g/100 mL.

8. The concentration of a saturated solution of calcium fluoride is determined by evaporating a 100-mL sample of saturated solution to dryness. A mass of 0.0016 g of solid remains after evaporation. The original solution was formed at 20°C. Calculate the K_{sp} of the salt at this temperature.

9. Mercury(I) chloride dissolves as shown by the equation

$$Hg_2Cl_{2(s)} \rightleftharpoons Hg_2^{2+}_{(aq)} + 2\,Cl^-_{(aq)}$$

Calculate the mass of compound required to make 500 mL of a saturated solution of mercury(I) chloride at 25°C.

10. For each of the following mixtures, calculate a trial ion product, Q, to predict whether a precipitate forms. All mixtures are made at SATP.
    (a) 50 mL of 0.040 mol/L $Ca(NO_3)_{2(aq)}$ plus 150 mL of 0.080 mol/L $(NH_4)_2SO_{4(aq)}$
    (b) 50 mL of 2.2×10^{-9} mol/L $AgNO_{3(aq)}$ plus 50 mL of 0.050 mol/L $NH_4Cl_{(aq)}$
    (c) 100 mL of 2.1×10^{-3} mol/L $Pb(NO_3)_{2(aq)}$ plus 50 mL of 0.0060 mol/L $NaI_{(aq)}$

11. Calculate the molar solubility of $PbI_{2(s)}$ in a 0.10 mol/L solution of $NaI_{(aq)}$ at SATP.

Applying Inquiry Skills

12. Lead(II) chloride is a low-solubility salt that dissociates according to this equilibrium:

$$PbCl_{2(s)} \rightleftharpoons Pb^{2+}_{(aq)} + 2\,Cl^-_{(aq)}$$

Lead ions undergo a single displacement reaction with zinc:

$$Pb^{2+}_{(aq)} + Zn_{(s)} \rightleftharpoons Pb_{(s)} + Zn^{2+}_{(aq)}$$

Question

What is the K_{sp} of lead(II) chloride?

Experimental Design

A strip of zinc of known mass is placed into a saturated solution of lead(II) chloride overnight. The next day, the strip is removed from the solution, cleaned, and its new mass measured to determine how much mass was lost.

Evidence

The zinc strip was coated with a black layer of metallic lead. Volume of saturated lead(II) chloride solution: 100 mL
Mass loss of zinc strip: 0.094 g

Analysis

(a) Calculate the amount (in moles) of zinc reacted.
(b) Calculate the molar concentration of the lead ions in the lead(II) chloride solution.
(c) Calculate the solubility product for lead(II) chloride.

13. Silver acetate is a low-solubility salt that dissociates according to this equilibrium:

$$AgCH_3COO_{(s)} \rightleftharpoons Ag^+_{(aq)} + CH_3COO^-_{(aq)}$$

Silver ions undergo a single displacement reaction with copper:

$$2\,Ag^+_{(aq)} + Cu_{(s)} \rightleftharpoons 2\,Ag_{(s)} + Cu^{2+}_{(aq)}$$

Question

What is the K_{sp} of silver acetate?

Experimental Design

A coil of copper wire is placed in a saturated solution of silver acetate for two days. The copper is removed from the solution, cleaned, and its mass measured to determine how much mass it lost.

Evidence

The copper wire was coated with a silvery-grey layer.

Volume of the saturated silver acetate solution: 100 mL

Mass loss of copper wire: 0.16 g

Analysis

(a) Calculate the moles of copper reacted.
(b) Calculate the moles of silver that reacted.
(c) Calculate the molar concentration of the silver ions in the silver acetate solution.
(d) Calculate the solubility product for silver acetate.

Evaluation

(e) In this experiment, it is very difficult to scrape all the silver off the copper wire. Suppose some silver residue remained on the wire. What effect would this have on the calculation of the silver acetate K_{sp} value?

(a)

(b)

Figure 1
Spontaneous exothermic reactions

spontaneous reaction one that, given the necessary activation energy, proceeds without continuous outside assistance

In Chapter 5, you learned about the energy changes that take place during physical and chemical changes. You know that when different substances come in contact with one another, there is a possibility that they will react. A small piece of sodium placed in water reacts quickly (**Figure 1(a)**); the activation energy is provided by the thermal energy of the surroundings. Lighting a sparkler, however, requires an intense source of heat like a flame or a spark for activation (**Figure 1(b)**). Once lit, the available fuel combusts quickly and completely, releasing large amounts of energy as heat and light.

These reactions are exothermic: They release more energy than they absorb. They are also **spontaneous reactions**, meaning that, given the necessary activation energy to begin, the reaction occurs without continuous outside assistance, proceeding to completion in open systems. Spontaneous exothermic reactions in closed systems, such as the oxidation of iron to iron oxide, establish a state of dynamic equilibrium.

The decomposition of water into its elements, hydrogen and oxygen, is endothermic. A continuous supply of energy is needed to sustain the reaction. This is usually supplied in the form of electricity in an electrolysis apparatus. The reaction is nonspontaneous and will stop when energy is no longer supplied.

Unlike the electrolysis of water, the dissolution of ammonium nitrate in water is an endothermic and spontaneous process. Heat is transferred from the environment as the salt dissolves. This is the reaction that produces the characteristic cooling effect of some commercially available cold packs. The reaction occurs spontaneously (without continuous outside assistance).

It is evident that some reactions occur spontaneously, but others proceed only with continuous outside assistance. Is there a theory that helps predict whether materials will react spontaneously when they come in contact under certain environmental conditions?

Thermodynamics, the study of energy transformations, has led to three empirically derived, fundamental laws of nature that allow us to understand why certain changes occur but others do not. These laws have become known as the laws of thermodynamics. We can use the predictive powers of these laws for chemical and physical changes. The laws of thermodynamics explain why ammonium nitrate spontaneously dissolves in water at SATP, absorbing energy from its surroundings in the process. They can also accurately predict that ice will melt at temperatures above 0°C at 101.3 kPa atmospheric pressure. The first, second, and third laws of thermodynamics, as they are called, predict whether changes are spontaneous.

Back in Chapter 5, you learned to calculate the enthalpy changes, ΔH, associated with a chemical reaction. This aspect of thermodynamics, called thermochemistry, primarily deals with the exchange of energy between a chemical system and its surroundings.

The first law of thermodynamics, also known as the law of conservation of energy, essentially states that if an object or process gains an amount of energy, it does so at the expense of a loss in energy somewhere else in the universe. The total energy in the universe is constant.

First Law of Thermodynamics

The total amount of energy in the universe is constant. Energy can be neither created nor destroyed, but can be transferred from one object or place to another, or transformed from one form to another.

In a chemical reaction, changes in the potential energy of reactants and products result in the transfer of energy, as heat, from the surroundings to the chemical system (endothermic change), or from the system to the surroundings (exothermic change). In both cases, the total energy of the universe (system plus surroundings) remains constant.

When a chemical change occurs in a series of steps, the overall enthalpy change is equal to the algebraic sum of the enthalpy changes of the steps. This is Hess's Law. The first law of thermodynamics serves as a foundation for Hess's Law: The value of ΔH for any reaction that can be written in steps equals the sum of the ΔH values for each of the individual steps (see Chapter 5).

Enthalpy Changes and Spontaneity

Molecules possess stability because of the chemical bonds between their atoms. When atoms form covalent bonds they generally achieve a greater stability by attaining a stable configuration of valence electrons. However, some chemical bonds are more stable than others. **Bond energy** is a measure of the stability of a covalent bond. It is measured in kilojoules (kJ) and is equal to the minimum energy required to break the intramolecular bonds between one mole of molecules of a pure substance. It is also equal to the amount of energy released when a mole of a particular bond is formed. **Table 1** lists the average bond energies of a variety of common chemical bonds.

We assume that the energy needed to break a bond is equivalent to the relative stability of that bond. A bond that requires 436 kJ/mol to break, such as the H−H bond, is almost twice as stable as one that requires only 222 kJ/mol, such as the N−O bond. In a chemical reaction, the bonds between reactant molecules must be broken and the bonds between product molecules must form. Energy is absorbed when reactant bonds break and energy is released when product bonds form. The enthalpy of reaction is a measure of the difference between the energy used to break reactant bonds and the energy released when product bonds form. These energy changes are illustrated in the potential energy diagrams you constructed in Chapter 5.

Enthalpy and Entropy Changes
Together Determine Spontaneity

In Chapter 5 you learned that reactions with negative enthalpy changes ($\Delta H < 0$) are exothermic, and reactions with positive enthalpy changes ($\Delta H > 0$) are endothermic. In general, exothermic reactions tend to proceed spontaneously.

Endothermic reactions such as the electrolysis of water are nonspontaneous, occurring only when a continuous supply of energy is available. However, some endothermic reactions like the double displacement reaction between ammonium nitrate and barium hydroxide, and the dissolution of ammonium nitrate in water, are spontaneous even though the products are less energetically stable than the reactants.

bond energy the minimum energy required to break one mole of bonds between two particular atoms; a measure of the stability of a chemical bond

Table 1 Average Bond Energies

Bond type	Average bond energy (kJ/mol)
H−H	436
C−H	413
N−H	391
C−C	346
C=C	615
C−N	305
O−H	436
C−O	358
C=O	749
N−O	222

$$2\,NH_4NO_{3(aq)} + Ba(OH)_{2(aq)} + energy \xrightarrow{\text{spontaneous}} 2\,NH_4OH_{(aq)} + Ba(NO_3)_{2(aq)}$$

$$NH_4NO_{3(s)} + energy \xrightarrow[\quad H_2O_{(l)} \quad]{\text{spontaneous}} NH_{4(aq)}^+ + NO_{3(aq)}^-$$

Why do reactions yielding less stable products occur spontaneously? Studies in thermodynamics have determined that enthalpy is not the only factor that determines whether a chemical or physical change occurs spontaneously. A physical property called entropy must also be taken into consideration.

Entropy, S, is a measure of disorder or randomness. It may apply to a system, S_{system}, the surroundings, $S_{surroundings}$, or the universe as a whole, $S_{universe}$. Entropy increases when disorder increases. A change in entropy is calculated by subtracting the entropy of the reactants, $S_{reactants}$, from the entropy of the products, $S_{products}$.

$$\Delta S = S_{products} - S_{reactants}$$

When entropy increases in a reaction, the entropy of the products, $S_{products}$, is greater than the entropy of the reactants, $S_{reactants}$, yielding an overall positive change in entropy, ΔS.

$$\Delta S > 0 \quad \text{when}$$
$$S_{products} > S_{reactants}$$

When entropy decreases in a reaction, the entropy of the products is less than the entropy of reactants, yielding an overall negative change in entropy.

$$\Delta S < 0 \quad \text{when}$$
$$S_{products} < S_{reactants}$$

Entropy increases when a solute such as a salt crystal dissolves into a solvent. In this process, the distribution of the particles of the crystal become more disorganized as randomly moving aqueous ions. Entropy also increases when a liquid evaporates into a vapour and when a solid melts into a liquid. These are both spontaneous processes. It seems to be a universal phenomenon that a change resulting in an increase in entropy ($\Delta S > 0$) is more likely to occur spontaneously than a change in which entropy decreases ($\Delta S < 0$). There is a close relationship between entropy and statistical probability. Allowing a package of playing cards to fall through the air and flutter to the ground models an increase in entropy (**Figure 2**).

Figure 2
(a) A new package of playing cards is highly ordered.
(b) Cards fluttering through the air increases entropy.
(c) The entropy of the cards on the floor is higher than the entropy of the cards in the package.

The cards are more randomly assorted by the time they reach the ground than they were in the package. This occurs because when the cards flutter freely through the air they assort randomly. There is only one way for them to come to rest on the ground in perfect numerical sequence and organized into suits, but many millions of ways to be disorganized. The tendency for systems to achieve greater disorder is simply a result of the laws of probability. The same laws of chance that explain the randomization of the fluttering cards explain the tendency of the components of chemical and physical systems to become more disordered as they change. Water falling over a waterfall and rain falling to the ground are examples of physical changes involving an increase in entropy.

SUMMARY   *Predicting the Sign of ΔS*

In general, a system will experience an increase in entropy ($\Delta S > 0$) if
- the volume of a gaseous system increases (**Figure 3(a)**),
- the temperature of a system increases, (**Figure 3(b)**), or
- the physical state of a system changes from solid to liquid or gas, or liquid to gas ($S_{gas} > S_{liquid} > S_{solid}$) (**Figure 3(c)**).

Figure 3

(a) low entropy   higher entropy

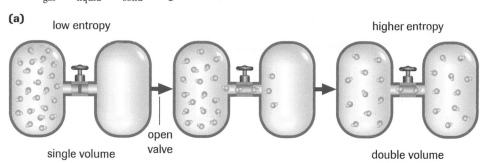

single volume   open valve   double volume

(a) When a gas expands into a larger volume, the particles move randomly to achieve a more probable, less ordered (higher entropy) state.

(b) low entropy   higher entropy

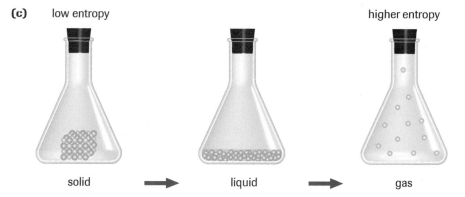

low temperature   higher temperature

(b) As temperature increases, the random movement of particles increases, and the entropy of the system increases.

(c) low entropy   higher entropy

solid   liquid   gas

(c) As a substance changes from a solid to a liquid to a gas, the entropy of the system increases because the particles become more randomly distributed.

In chemical reactions, entropy increases ($\Delta S > 0$) when
- fewer moles of reactant molecules form a greater number of moles of product molecules;

$$2\,NH_{3(g)} \rightarrow N_{2(g)} + 3\,H_{2(g)}$$

- complex molecules are broken down into simpler subunits (e.g., combustion of organic fuels into carbon dioxide and water);

$$C_6H_{12}O_{6(s)} + 6\,O_{2(g)} \rightarrow 6\,CO_{2(g)} + 6\,H_2O_{(g)}$$

- solid reactants become liquid or gaseous products (or liquids become gases).

$$2\,H_2O_{(l)} \rightarrow 2\,H_{2(g)} + O_{2(g)}$$

| **Predicting the Sign of ΔS** |

Predict whether the change in entropy is positive or negative for each of the following changes and predict their spontaneity. Explain your answers.
(a) Solid carbon dioxide sublimes into gaseous carbon dioxide.
(b) $N_2O_{4(g)} \rightarrow 2\,NO_{2(g)}$
(c) The synthesis reaction between oxygen and hydrogen forms liquid water.

(a) The entropy change is positive ($\Delta S > 0$) for this change since the particles of a gas are more randomly distributed. The change will tend to occur spontaneously.

(b) The entropy change is positive ($\Delta S > 0$) for this change since one mole of $N_2O_{4(g)}$ yields two moles of $NO_{2(g)}$ and the state of reactant and product remains the same. The change will tend to occur spontaneously.

(c) The balanced equation for the reaction is

$$2\,H_{2(g)} + O_{2(g)} \rightarrow 2\,H_2O_{(l)}$$

The entropy change is negative ($\Delta S < 0$) for this change since three moles of reactants yields two moles of product and the reactants are gases while the product is a liquid. This change will tend to be nonspontaneous.

▶ **Practice**

Understanding Concepts

1. Predict whether the entropy will be positive or negative for each of the following changes and determine whether the change will tend to be spontaneous as written.
   (a) $Ag^+_{(aq)} + Cl^-_{(aq)} \rightarrow AgCl_{(s)}$
   (b) Polyethene is produced from ethene gas.
   (c) $I_{2(s)} \rightarrow I_{2(g)}$
   (d) $NaCl_{(s)} \rightarrow Na^+_{(aq)} + Cl^-_{(aq)}$

Enthalpy, Entropy, and Spontaneous Change

Changes in the enthalpy, ΔH, and entropy, ΔS, of a system help us to predict whether a change will occur spontaneously. In some cases, these two factors work together in determining the spontaneity of a change while in other cases, they work against each other. Exothermic reactions ($\Delta H < 0$) involving an increase in entropy ($\Delta S > 0$) occur spontaneously, because both changes are favoured. Endothermic reactions ($\Delta H > 0$) involving a decrease in entropy ($\Delta S < 0$) are not spontaneous because neither change is favoured. But what happens in cases where the energy change is exothermic (favoured) and the entropy decreases (not favoured), and others where the energy change is endothermic (not favoured) but entropy increases (favoured)? In these situations, the temperature at which the change occurs becomes an important consideration. Predicting the spontaneity of these changes requires an understanding of the other two laws of thermodynamics and the concept of free energy.

Free Energy

In the late 1800s Josiah Willard Gibbs, an American physicist, discovered a relationship among the energy change, entropy change, and temperature of a reaction that predicts whether the reaction, carried out at constant temperature and pressure, will proceed spontaneously. Gibbs distinguished between energy and **free energy**. **Figure 4** illustrates the difference between total energy and free energy.

free energy (or Gibbs free energy) energy that is available to do useful work

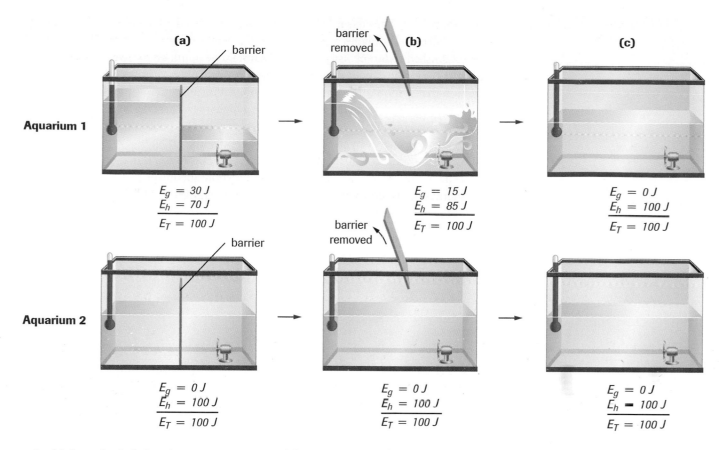

(a) barrier / barrier

Aquarium 1

$E_g = 30\ J$
$E_h = 70\ J$
$E_T = 100\ J$

barrier removed **(b)**

$E_g = 15\ J$
$E_h = 85\ J$
$E_T = 100\ J$

(c)

$E_g = 0\ J$
$E_h = 100\ J$
$E_T = 100\ J$

barrier removed

Aquarium 2

barrier

$E_g = 0\ J$
$E_h = 100\ J$
$E_T = 100\ J$

$E_g = 0\ J$
$E_h = 100\ J$
$E_T = 100\ J$

$E_g = 0\ J$
$L_h = 100\ J$
$E_T = 100\ J$

In this hypothetical situation, two aquariums of the same size contain the same amount of water, and both contain 100 J of total energy. A fan is placed in both aquariums as shown in **Figure 4**. The aquariums differ in two ways:

- the water temperature in Aquarium 1 is lower than that of Aquarium 2, so Aquarium 1 contains 30 J of thermal energy (E_h) *less* than Aquarium 2; and

- Aquarium 1 possesses 30 J of gravitational potential energy (E_g) *more* than Aquarium 2 because of a reservoir of water that is raised to a higher level and held there by a removable barrier (**Figure 4(a)**).

When the barrier is removed in Aquarium 1, water falls onto the fan blades and does 8 J of useful mechanical work as it turns the fan blades. The system transforms 15 J of gravitational potential energy into 10 J of mechanical work and 5 J of thermal energy (temperature increases).

The molecules of water in Aquarium 2 are in constant, random motion, on average moving faster than those in Aquarium 1 (before the barrier was removed). Like the water molecules in Aquarium 1, the molecules in Aquarium 2 also hit the blades of the fan. However, they hit the blades in every direction with approximately equal frequency and strength. Thus, the blades do not move in any one direction and useful work cannot be done.

When the barrier was removed in Aquarium 1, water moved from a more ordered state to a less ordered state, resulting in both a decrease of free energy and work being done on the blades of the fan. This change occurs spontaneously, without external input to the system.

In general, a change at constant temperature and pressure will occur spontaneously if it is accompanied by a decrease in Gibbs free energy, G. Changes are spontaneous if the change in G, ΔG, is negative, $\Delta G < 0$ (G_{final} is less than $G_{initial}$).

Figure 4
(a) Although the total energy, E_T content of aquarium 1 is equal to that of aquarium 2, aquarium 1 has 30 J of gravitational potential energy (E_g) available to do useful work (turning the blades of the fan).
(b) When the barriers are removed, water will fall and turn the fan's blades in aquarium 1 but not in aquarium 2. The conversion of gravitational potential energy to mechanical energy cannot be 100%. Some of the useful energy is lost as heat (increases in temperature).
(c) Aquariums 1 and 2 ultimately achieve the same outcome. All energy is in the form of thermal energy (E_h) (the random movement of particles).

In Aquarium 1, the water molecules are more disordered after the barrier is removed and they fall onto the blades of the fan. The high water reservoir in Aquarium 1 cannot be re-established spontaneously (water does not move uphill on its own). A change with a negative ΔG in one direction has an equivalent positive ΔG in the reverse direction. Thus, a reaction that is spontaneous in one direction is nonspontaneous in the reverse direction. Work must be done to move the water to a higher level (greater order and more free energy). The energy needed to do this work can only be obtained from an outside source that releases free energy by itself becoming more disordered.

These results are not restricted to this one hypothetical case, but apply to all changes that occur in the universe. All changes, whether spontaneous or not, are accompanied by an increase in the entropy (overall disorder) of the universe. This is called the **second law of thermodynamics** (also known as the law of entropy).

> **Second Law of Thermodynamics**
>
> All changes either directly or indirectly increase the entropy of the universe. Mathematically,
>
> $$\Delta S_{universe} > 0$$

The entropy of the universe is equal to the sum of the entropy of the system and the entropy of the surroundings.

$$S_{universe} = S_{system} + S_{surroundings}$$

A change in the entropy of the universe may occur as a result of a change in the entropy of the system or of the surroundings, or of both.

$$\Delta S_{universe} = \Delta S_{system} + \Delta S_{surroundings}$$

This means that a system's entropy, ΔS_{system}, can decrease (the system becomes more ordered), so long as there is a larger increase in the entropy of the surroundings, $\Delta S_{surroundings}$, so that the overall entropy change, $\Delta S_{universe}$, is positive.

The second law has far-reaching implications for us and other living things. Living organisms seem to violate the second law of thermodynamics. They build highly ordered molecules such as proteins and DNA from a random assortment of amino acids and nucleotides dissolved in cell fluids. They assemble highly organized membranes and other cell organelles as they grow. On a larger scale, living things organize the world around them by building highly ordered structures such as nests, webs, and space shuttles. These are all changes involving an increase in Gibbs free energy at the local level. Each of these activities seems to violate the second law of thermodynamics by causing the universe to become more ordered. However, studies show that, in each and every case, the apparent order created by nonspontaneous processes that increase the order (entropy) of the system (like growth, reproduction, the movement of materials that result in the building of a house) is accompanied by an *even greater* disorder caused by the processes used to harness the energy needed to make these nonspontaneous activities occur. This includes energy-yielding processes like the oxidation of nutrients by cellular respiration (a spontaneous process that yields free energy), and the burning of fossil fuels to power machines.

Living organisms obey the second law of thermodynamics because they create order out of chaos in a local area of the universe while creating a greater amount of disorder in the universe as a whole. Tracing the source of free energy, we notice that nutrient molecules like glucose used as a source of free energy were originally produced in a plant cell that used the free energy of the Sun (in the form of light energy) to assemble carbon dioxide and water molecules into glucose. The apparent increases in order generated by the photosynthetic process in the Earth system are accompanied by an even greater disorder of the surroundings (the universe). Since every unit of order created by photosynthesis on Earth is offset by an even larger disorder created in the universe, the net change in the entropy of the universe is positive, and the second law of thermodynamics is satisfied.

The same accounting may be done for every change, including all chemical changes. Over time, all sources of free energy in the universe will become completely disordered and useless (work will not be possible). The second law of thermodynamics predicts that the universe will eventually experience a final "thermal death" in which all particles and energy move randomly about. Life will come to an end because there won't be any sources of free energy to exploit; stars will stop shining. Waterfalls will stop falling. All radioactive nuclei disintegrate and become completely random. All energy will have become randomized (just like the cards that fell to the floor in our analogy near the beginning of this section). All of the energy that there ever was will still be there, except that it will be uniformly distributed throughout the universe, unable to apply an effective push or a pull on anything. According to the second law, a state of perfect equilibrium is the ultimate fate of the universe.

▶ EXPLORE an issue

Take a Stand: Can We Do Anything About Pollution?

Decision-Making Skills

○ Define the Issue ● Analyze the Issue ● Research
● Defend the Position ● Identify Alternatives ● Evaluate

Living organisms, especially humans, have a tendency to build things. Spiders build webs, birds build nests, and humans build skyscrapers, space shuttles, and Ferris wheels. All of this activity produces pollution. In fact, we produce pollution even as we sleep. Respiration breaks ordered glucose molecules down into carbon dioxide and water. The $CO_{2(g)}$ we breathe out and the other gases and liquids we excrete add to environmental pollution.

Many believe that pollution is inevitable because of the second law of thermodynamics. As we create order at a local level (building a car), we inadvertently create a greater amount of disorder in our surroundings (fumes, solvents, and rubbish, **Figure 5**). The problem is the result of the second law of thermodynamics and there's nothing we can do about it.

Others vehemently disagree. They contend that disorder is not necessarily the problem. The sand along a beach is, in fact, the disordered state of a mountain, but it is not pollution because it is nontoxic, even desirable. Pollution is the scattering of the unwanted or toxic byproducts of human activity. We must do everything we can to find ways of producing "cleaner" pollution and to eliminate the toxic and undesirable pollution in our environment.

Proposition: The second law of thermodynamics makes pollution an inevitable result of human activity. There's nothing we can do about it.

- In small groups, discuss the issue. Record arguments that support both sides.

- Conduct independent library or Internet research on both sides of the issue.

GO www.science.nelson.com

(a) Prepare an individual position paper on the issue.

Figure 5
Is garbage inevitable?

Predicting Spontaneity

The second law of thermodynamics helps us to predict whether a change will be spontaneous. If a change increases the entropy of the universe, the change is spontaneous as written. However, it is not always easy to determine whether a particular change in a system increases the entropy of the universe as a whole.

Is there a way of determining whether a change is spontaneous by considering the system's change in enthalpy, ΔH, change in entropy, ΔS, and absolute temperature, T? Indeed there is: There is an equation that incorporates all these factors to give a value for the change in Gibbs free energy, ΔG. The *sign* of ΔG accurately predicts whether the reaction is spontaneous. The spontaneity of any reaction carried out at constant temperature and pressure can be predicted by calculating the value of ΔG using the following equation, called the Gibbs-Helmholtz equation:

$$\Delta G = \Delta H - T\Delta S$$
where
ΔG is the Gibbs free energy change (kJ);
ΔH is the enthalpy change (kJ);
T is the absolute (kelvin) temperature of the system (K);
and
ΔS is the entropy change (kJ/K).

Exothermic reactions with an increase in entropy are spontaneous at all temperatures. Endothermic reactions with an accompanying decrease in entropy are nonspontaneous at all temperatures. (Nonspontaneous reactions are impossible as written, but can be forced to occur with a continuous input of free energy. They can never be made spontaneous, regardless of the temperature.) However, the spontaneity of the other two types of change (endothermic reactions with increase in entropy and exothermic reactions with decrease in entropy) depends on the system's absolute temperature, measured in kelvins.

ΔG, Spontaneity, and Free Energy

If ΔG for a reaction is negative, the change is spontaneous as written. If ΔG for a reaction is positive, the change is nonspontaneous as written.

$\Delta G < 0$ change is spontaneous
$\Delta G > 0$ change is nonspontaneous

The sign of ΔG predicts whether a given change is spontaneous at constant temperature and pressure, and the absolute value of ΔG is a measure of the total amount of free energy that is released in the change. However, the value of ΔG provides no information regarding the rate of reaction. The following equation describes the combustion of ethanol at 25°C (298 K) (**Figure 6**).

$$C_2H_5OH_{(l)} + 3\,O_{2(g)} \rightarrow 2\,CO_{2(g)} + 3\,H_2O_{(g)} \qquad \Delta G = -1299.8 \text{ kJ}$$

Since ΔG is negative, the combustion of ethanol is spontaneous. Given the necessary activation energy, the reaction will occur spontaneously. The absolute value of the Gibbs free energy change tells us that 1299.8 kJ of free energy will be released (either to do useful work, or as "waste" thermal energy) for every mole of ethanol burned.

In general, a reaction with a negative ΔG value will be spontaneous (the reactants will react to form product spontaneously) and will yield an amount of free energy given by the absolute value of ΔG.

Figure 6
The combustion of ethanol is a spontaneous process.

It may at first seem confusing that an endothermic reaction can proceed spontaneously, and do useful work. Although an endothermic reaction always absorbs thermal energy from its surroundings, it can simultaneously release free energy if its ΔS is positive and sufficiently large to make $\Delta H - T\Delta S$ a negative value. Such a reaction is spontaneous. It is not only the enthalpy change that determines whether a change is spontaneous and if it can do useful work. Entropy change plays an equally important role.

This new concept changes the way we interpret change, and it applies to *all* changes that occur in the universe, not just chemical change! Have you ever wondered why a cold pack really works? The change is endothermic. Why does ammonium chloride absorb thermal energy from its environment, even in a relatively cool environment, to dissolve into the water? It must be because the increase in entropy is sufficiently great to "overwhelm" the increase in enthalpy, and make ΔG negative. This is an "entropy-driven" reaction.

However, the magnitude of ΔG gives us no information whatsoever about the rate of reaction. Just because a candle has more total free energy available (large negative ΔG) than a single firecracker (smaller negative ΔG), and will release it spontaneously after you provide a bit of activation energy (e.g., with the heat from the flame of a match), doesn't mean that the candle releases its energy as quickly, more quickly, or more slowly than the firecracker. The fact that a reaction occurs spontaneously, and has lots of free energy available for release reveals nothing of its rate. Information about the rate of a reaction can be obtained only by performing kinetics experiments similar to those you learned about in Chapter 6. In fact, the reaction may occur so slowly that its spontaneity may not be readily evident. The rusting of iron is an example of a slow spontaneous reaction.

Let's look in detail at all the possible values of ΔH and ΔS (summarized in **Table 2**), to see whether a variety of reactions will occur spontaneously.

Reactions with a negative ΔH and positive ΔS all have a negative ΔG.

$$\Delta G = \Delta H - T\Delta S$$
$$= (-) - T(+)$$

Since T is always positive, ΔG will be negative at all temperatures. These reactions are spontaneous at all temperatures. An example of this is the combination of carbon and oxygen to produce carbon dioxide,

$$C_{(s)} + O_{2(g)} \rightarrow CO_{2(g)}$$

TRY THIS activity   *Stretching a Point*

In this activity you will analyze the thermodynamic properties of a rubber band.

Materials: a wide rubber band

- Stretch a rubber band between your index fingers and hold it close to your nose without touching for approximately 10 s.
- Gently touch the band to the tip of your nose and feel its temperature.
- Relax the band and quickly touch it to your nose to feel its temperature again.

(a) Determine the algebraic signs of ΔS, ΔH, and ΔG of the forward reaction (stretching).

(b) Determine the algebraic signs of ΔS, ΔH, and ΔG of the reverse reaction (relaxing).

(c) Are the molecules in the band becoming more ordered or disordered during the forward reaction? During the reverse reaction? Explain.

(d) Are bonds breaking or forming during the forward reaction? During the reverse reaction? Explain.

(e) If we assume that the molecules in the rubber band are all hydrocarbons, what type of bonds are being affected during the changes?

(f) Is the spontaneous contraction of the rubber band an entropy-driven or enthalpy-driven process? Explain.

(g) Predict whether the contraction process is spontaneous at all temperatures.

(h) Design an investigation to test your prediction. Carry out your experiment, with your teacher's approval. Write a brief report summarizing the thermodynamics of a rubber band.

Reactions with a positive ΔH and negative ΔS all have a positive ΔG.

$$\Delta G = \Delta H - T\Delta S$$
$$\quad = (+) - T(-)$$

Since T is always positive, ΔG will be positive at all temperatures. These reactions are nonspontaneous at all temperatures and will occur only with the continuous input of energy. An example of this type of reaction is the production of ozone, $O_{3(g)}$, from oxygen, $O_{2(g)}$.

$$3\,O_{2(g)} \rightarrow 2\,O_{3(g)}$$

The spontaneity of reactions with negative ΔH and negative ΔS or positive ΔH and positive ΔS depend on the value of T.
When ΔH is negative and ΔS is negative,

$$\Delta G = \Delta H - T\Delta S$$
$$\quad = (-) - T(-)$$

ΔG will be negative in temperature conditions where the value of the $T\Delta S$ factor is lower than the value of ΔH. An example of this type of reaction is the formation of sulfur trioxide from sulfur dioxide and oxygen,

$$2\,SO_{2(g)} + O_{2(g)} \rightarrow 2\,SO_{3(g)}$$

This is an exothermic change that is accompanied by a decrease in entropy. It is spontaneous at temperatures below 786°C (1059 K) and nonspontaneous above this temperature.
When ΔH is positive and ΔS is positive,

$$\Delta G = \Delta H - T\Delta S$$
$$\quad = (+) - T(+)$$

ΔG will be negative in temperature conditions where the value of the $T\Delta S$ factor is higher than the value of ΔH. An example of this type of change is the melting of ice,

$$H_2O_{(s)} \rightarrow H_2O_{(l)}$$

This is an endothermic change that is accompanied by an increase in entropy. As you know, the change is spontaneous above 0°C (273 K) and nonspontaneous below this temperature.

SUMMARY *Spontaneous and Nonspontaneous Reactions*

Table 2 Classification of Spontaneous and Nonspontaneous Reactions

	Endothermic ($\Delta H > 0$) (not favoured)	Exothermic ($\Delta H < 0$) (favoured)
Entropy increases ($\Delta S > 0$) **(favoured)** Example	spontaneity depends on T (spontaneous at higher temps) $H_2O_{(s)} \rightarrow H_2O_{(l)}$	spontaneous (at all temperatures) $C_{(s)} + O_{2(g)} \rightarrow CO_{2(g)}$
Entropy decreases ($\Delta S < 0$) **(not favoured)** Example	nonspontaneous (proceeds only with continuous input of energy) $3\,O_{2(g)} \rightarrow 2\,O_{3(g)}$	spontaneity depends on T (spontaneous at lower temperatures) $2\,SO_{2(g)} + O_{2(g)} \rightarrow 2\,SO_{3(g)}$

The Third Law and Standard Free Energy Changes

As you have just learned, we can use the Gibbs-Helmholtz equation

$$\Delta G = \Delta H - T\Delta S$$

to predict the spontaneity of a reaction, if the enthalpy change, ΔH, entropy change, ΔS, and absolute temperature, T, of the reaction are known.

In Chapter 5, you learned how to calculate the standard enthalpy change of reaction, $\Delta H°$, by subtracting the sum of the standard enthalpies of formation of the reactants (each multiplied by the respective coefficient in the balanced chemical equation) from the sum of the standard enthalpies of formation of the products (each multiplied by the respective coefficient in the balanced chemical equation). You used the equation

$$\Delta H° = \sum n \, \Delta H°_{f(products)} - \sum n \, \Delta H°_{f(reactants)}$$

Entropy, like enthalpy, can be measured. Our ability to measure the total amount of entropy possessed by a substance is based on the **third law of thermodynamics** (**Figure 7**).

Figure 7
The entropy of any subtance at absolute zero (0 K) is, according to the third law of thermodynamics, zero. In normal life you are unlikely to encounter any substance colder than the magnets in the interior of an MRI scanner, cooled by liquid helium at 4 K.

> **Third Law of Thermodynamics**
>
> The entropy of a perfectly ordered pure crystalline substance is zero at absolute zero.
>
> Mathematically,
> $$S = 0 \quad \text{at} \quad T = 0 \, K$$

It follows that the entropy of a pure substance is greater than zero at temperatures above absolute zero. The entropy of a substance is measured in units of joules per kelvin, J/K, and, if determined for one mole of substance at SATP, is called the **standard entropy** of that substance, $S°$ (**Table 3**).

Now you can calculate standard entropy changes, $\Delta S°$, for a given chemical reaction using standard entropy values, $S°$ (**Table 3**; a more complete table is found in Appendix C6), in an equation that is very similar to the enthalpy equation.

standard entropy the entropy of one mole of a substance at SATP; units (J/mol·K)

$$\Delta S° = \sum n_p \, S°_{products} - \sum n_r \, S°_{reactants}$$

where
$\Delta S°$ is the standard entropy change for a chemical reaction;
n_p is the number of moles of a product, as specified by the equation for the reaction;
$S°_{product}$ is the standard entropy of a product;
n_r is the number of moles of a reactant, as specified by the equation for the reaction; and
$S°_{reactant}$ is the standard entropy of a reactant.

Table 3 Standard Entropy Values

	S° (J/mol·K)
ethene, $C_2H_{4(g)}$	219.3
ethyne, $C_2H_{2(g)}$	201.0
water, $H_2O_{(l)}$	69.95

Note the following differences between tabulated $\Delta H°_f$ values and tabulated $S°$ values:

- $\Delta H°_f$ values are measures of the enthalpy of formation of one mole of compound from its elements under standard conditions, whereas tabulated $S°$ values are not measures of the entropy of formation, but of the total entropy possessed by a given substance in standard state (relative to $S° = 0$ at 0 K).

- Unlike standard enthalpy change ($\Delta H°$) values for elements, which are defined as zero, the standard entropy ($S°$) values of elements are *always greater than zero*, because all matter at temperatures above absolute zero has some entropy.

- The entropy term is given in standard entropy units (J/mol·K), whereas the units for standard enthalpy are kJ/mol. (Remember that "standard" values are always "per mole.")

Equipped with any balanced chemical equation, we can now calculate $\Delta H°$, $\Delta S°$, and $\Delta G°$, for substances in their standard state (the state they are in at 298 K (25°C)).

$$\Delta G° = \Delta H° - T\Delta S°$$
where
$\Delta G°$ is the *standard* Gibbs free energy change;
$\Delta H°$ is the *standard* enthalpy change;
$T = 298$ K (*standard* ambient temperature); and
$\Delta S°$ is the *standard* entropy change.

The sign of the calculated $\Delta G°$ value predicts whether the reaction is spontaneous at 298 K, and its absolute value gives a measure of the magnitude of the free energy change associated with the reaction.

If the value of $\Delta G°$ is negative, the reaction is spontaneous and the absolute value is a measure of the maximum amount of free energy that can be theoretically harnessed from the reaction to do useful work. If the value of $\Delta G°$ is positive, the reaction is non-spontaneous and the absolute value is a measure of the amount of work that theoretically needs to be done on the system in order for the reaction to occur as written. However, a positive result for $\Delta G°$ also indicates that the reverse reaction *is* spontaneous, and provides a measure of the maximum theoretical amount of energy available to do work if the reaction proceeds in the reverse direction.

When calculating $\Delta G°$ according to the methods outlined in the following samples and examples, you can find the values of $\Delta H°_f$ and $S°$ from the chart of thermodynamic data in Appendix C6.

SAMPLE problem

Calculating ΔG for a Reaction

Calculate the standard Gibbs free energy change asociated with the hydrogenation of ethene (Figure 8), and interpret the result.

$$C_2H_{4(g)} + H_{2(g)} \rightarrow C_2H_{6(g)}$$

First, find the standard enthalpies of formation, $\Delta H°_f$, from a reference source, then use the enthalpy change equation to calculate $\Delta H°$.

$$\Delta H°_{f(C_2H_{6(g)})} = -84.5 \text{ kJ/mol}$$
$$\Delta H°_{f(C_2H_{4(g)})} = 51.9 \text{ kJ/mol}$$
$$\Delta H°_{f(H_{2(g)})} = 0 \text{ kJ/mol}$$

$$\Delta H° = [\Delta H°_{f(C_2H_{6(g)})}] - [\Delta H°_{f(C_2H_{4(g)})} + \Delta H°_{f(H_{2(g)})}]$$
$$= [1 \text{ mol} \times (-84.5 \text{ kJ/mol})] - [(1 \text{ mol} \times (+51.9 \text{ kJ/mol})) + (1 \text{ mol} \times (0 \text{ kJ/mol})]$$
$$= (-84.5 \text{ kJ}) - (+51.9 \text{ kJ})$$
$$\Delta H° = -136.4 \text{ kJ}$$

Next, find the standard entropy, $S°$, of each substance from a reference source, then use the entropy change equation to calculate the value of $\Delta S°$.

$$S°_{C_2H_{6(g)}} = 229.5 \text{ J/mol·K}$$
$$S°_{C_2H_{4(g)}} = 219.8 \text{ J/mol·K}$$
$$S°_{H_{2(g)}} = 130.6 \text{ J/mol·K}$$

Figure 8
The hydrogenation of ethene is one of the methods of synthesizing ethyne, or acetylene, for use in cutting torches.

$$\Delta S^\circ = [S^\circ_{C_2H_{6(g)}}] - [S^\circ_{C_2H_{4(g)}} + S^\circ_{H_{2(g)}}]$$
$$= [1 \text{ mol} \times (+229.5 \text{ J/mol·K})] - [(1 \text{ mol} \times (+219.8 \text{ J/mol·K})) +$$
$$(1 \text{ mol} \times (+130.6 \text{ J/mol·K})]$$
$$= [+229.5 \text{ J/K}] - [(+219.8 \text{ J/K}) + (+130.6 \text{ J/K})]$$
$$= (+229.5 \text{ J/K}) - (+350.4 \text{ J/K})$$
$$\Delta S^\circ = -120.9 \text{ J/K}$$

Now, convert the calculated value of ΔS° to units of kJ/K.

$$\Delta S^\circ = -120.9 \text{ J/K} \times \frac{1 \text{ kJ}}{1000 \text{ J}}$$

$$\Delta S^\circ = -0.1209 \text{ kJ/K}$$

Now, substitute the calculated values of ΔH° and ΔS° into the Gibbs-Helmholtz equation:

$$\Delta G^\circ = \Delta H^\circ - T\Delta S^\circ$$
$$T = 298 \text{ K (standard ambient temperature)}$$

$$\Delta G^\circ = (-136.4 \text{ kJ}) - (+298 \text{ K})(-0.1209 \text{ kJ/K})$$
$$= (-136.4 \text{ kJ}) - (-36.03 \text{ kJ})$$
$$= -136.4 \text{ kJ} + 36.03 \text{ kJ}$$
$$\Delta G^\circ = -100.4 \text{ kJ}$$

Since ΔG° is negative, the hydrogenation of ethene is spontaneous under standard conditions and will proceed to the right as written. The value of ΔG° provides no information regarding the rate of this reaction. A total of 100.4 kJ of free energy is made available to do useful work for each mole of ethene that reacts.

Example

Calculate the standard Gibbs free energy change asociated with the reaction of urea (**Figure 9**) with water, and interpret the result.

$$CO(NH_2)_{2(aq)} + H_2O_{(l)} \rightarrow CO_{2(g)} + 2\,NH_{3(g)}$$

Solution

$$\Delta H^\circ_{fCO_{2(g)}} = -394 \text{ kJ/mol}$$

$$\Delta H^\circ_{fNH_{3(g)}} = -460 \text{ kJ/mol}$$

$$\Delta H^\circ_{fCO(NH_2)_{2(g)}} = -319.2 \text{ kJ/mol}$$

$$\Delta H^\circ_{fH_2O_{(l)}} = -285.8 \text{ kJ/mol}$$

$$\Delta H^\circ = [\Delta H^\circ_{f(CO_{2(g)})} + 2\,\Delta H^\circ_{f(NH_{3(g)})}] - [\Delta H^\circ_{f(CO(NH_2)_{2(aq)})} + \Delta H^\circ_{f(H_2O_{(l)})}]$$
$$= [(1 \text{ mol} \times (-394 \text{ kJ/mol}) + 2 \text{ mol} \times (-46.0 \text{ kJ/mol})] - [(1 \text{ mol} \times$$
$$(-319.2 \text{ kJ/mol}) + (1 \text{ mol} \times (-285.8 \text{ kJ/mol})]$$
$$= (-486 \text{ kJ}) - (-605.0 \text{ kJ})$$
$$\Delta H^\circ = +119 \text{ kJ}$$

$$S^\circ_{CO_{2(g)}} = +213.6 \text{ J/mol·K}$$

$$S^\circ_{NH_{3(g)}} = +192.5 \text{ J/mol·K}$$

$$S^\circ_{CO(NH_2)_{2(aq)}} = +173.8 \text{ J/mol·K}$$

$$S^\circ_{H_2O_{(l)}} = +69.96 \text{ J/mol·K}$$

Figure 9
Urea is a critical component in the indigo dyeing process. Before the 20th century, stale urine was the most convenient available source of urea for the dye industry.

$$\Delta S° = [S°_{CO_{2(g)}} + 2S°_{NH_{3(g)}}] - [S°_{CO(NH_2)_{2(aq)}} + S°_{H_2O_{(l)}}]$$

$$= [(1 \text{ mol} \times (+213.6 \text{ J/mol·K}) + 2 \text{ mol} \times (+192.5 \text{ J/mol·K}))] -$$
$$[(1 \text{ mol} \times (+173.8 \text{ J/mol·K})] + (1 \text{ mol} \times (+69.96 \text{ J/mol·K})]$$

$$= (+598.6 \text{ J/K}) - (+243.8 \text{ J/K})$$

$$\Delta S° = +354.8 \text{ J/K}$$

$$\Delta S° = +354.8 \text{ J/K} \times \frac{1 \text{ kJ}}{1000 \text{ J}}$$

$$\Delta S° = +0.3548 \text{ kJ/K}$$

$$\Delta G° = \Delta H° - (298 \text{ K}) \Delta S°$$

$$= (+119 \text{ kJ}) - (+298 \text{ K})(+0.3548 \text{ kJ/K})$$

$$= (+119 \text{ kJ}) - (+105.7 \text{ kJ})$$

$$\Delta G° = +13 \text{ kJ}$$

Since $\Delta G°$ is positive, the reaction is not spontaneous under standard conditions as written (but *is* spontaneous in reverse). A total of 13.2 kJ of free energy must be provided per mole of $CO(NH_2)_{2(aq)}$ reacted to make the reaction proceed according to the balanced equation.

> ▶ **Practice**

Understanding Concepts

2. Calculate the standard Gibbs free energy change associated with each of the following reactions and interpret the results in each case.
(a) $NH_{3(g)} + HCl_{(g)} \rightarrow NH_4Cl_{(s)}$
(b) the combustion of ethanol, producing carbon dioxide and water
(c) $2 NH_{3(g)} + 3 O_{2(g)} + 2 CH_{4(g)} \rightarrow 2 HCN_{(g)} + 6 H_2O_{(g)}$

Equilibrium and ΔG

When the value of ΔG is negative for a particular reaction, the change occurs spontaneously as written. When ΔG is positive, the change is nonspontaneous in the forward direction, but spontaneous in the reverse direction. What does a value of zero for ΔG mean? When ΔG is zero, the forward and reverse reactions are equally spontaneous. The system is in a state of dynamic equilibrium. When a reversible reaction reaches equilibrium in a closed system, ΔG equals zero.

In general,

- if $\Delta G < 0$, the reaction is spontaneous in the forward direction;

- if $\Delta G > 0$, the reaction is nonspontaneous in the forward direction but is spontaneous in the reverse direction;

- if $\Delta G = 0$, there is no "preferred" direction. The system is at equilibrium.

Since $\Delta G = 0$ for a system at equilibrium, such a system can do no useful work. Consider an electric cell (battery): When fully charged, the unequal distribution of charge on the two sides of an insulator is maximal and the value of ΔG for the system is large and negative (**Figure 10(a)**). When a conductor is connected between the terminals, charges spontaneously flow from the negatively charged compartment to the positively charged compartment, reducing the value of ΔG and releasing free energy that is able to perform useful work, such as turning a motor or lighting a light bulb (**Figure 10(b)**). When the electrical charges become equally distributed between the two compartments, dynamic equilibrium is reached and the value of ΔG reduces to zero (**Figure 10(c)**). At this stage, the probability of charges moving either way is equal. No net movement of charge is

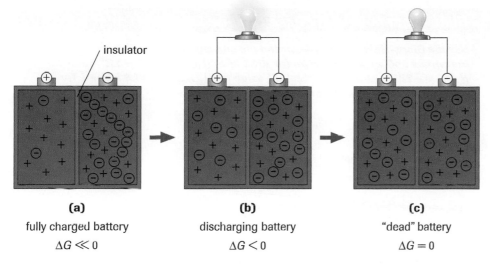

Figure 10
(a) A charged battery contains an unequal distribution of charges across the internal insulator. The value of ΔG is large and negative.
(b) A discharging battery releases free energy as the value of ΔG decreases. The free energy released is used to perform useful work.
(c) A "dead" battery has a ΔG value of zero. The system is in a state of dynamic equilibrium. The system has no more free energy to release and can do no work.

(a)
fully charged battery
$\Delta G \ll 0$

(b)
discharging battery
$\Delta G < 0$

(c)
"dead" battery
$\Delta G = 0$

possible, and no more work can be extracted from the system. We say that the cell is "dead," but it is just in a state of equilibrium.

Consider a sample of liquid water being cooled to a temperature below its freezing point at 101.3 kPa of atmospheric pressure. The equation for this process is

$$H_2O_{(l)} \rightleftharpoons H_2O_{(s)}$$

Above 0°C, ΔG is positive and freezing is nonspontaneous—the water remains liquid. At exactly 0°C, the system exists in a state of dynamic equilibrium as an ice–water mixture: slush. If the temperature remains at 0°C, the system will remain as slush indefinitely. Below 0°C, ΔG is negative and freezing occurs spontaneously.

Temperature and Equilibrium

Since the value of ΔG is equal to zero for any system at equilibrium, we can put the Gibbs-Helmholtz equation to practical use. We can use it to predict the temperature at which a phase changes occurs. For example, the normal boiling point of a liquid is defined as the temperature at which the liquid form of the material is in equilibrium with its gas phase. Since $\Delta G = 0$ for a system at equilibrium,

$$\Delta G = \Delta H - T\Delta S$$
$$0 = \Delta H - T\Delta S$$
$$T = \frac{\Delta H}{\Delta S}$$

We can use standard enthalpy and entropy values in calculations such as this if we assume that these values are independent of the temperature. This is not strictly true, of course, since the values of ΔH and ΔS change with changes in temperature. (Remember the third law of thermodynamics regarding entropy values and how they decrease with decreasing temperature, and discussions in Chapter 5 regarding enthalpy values.) As a result, these calculations yield only an estimate of phase change temperatures. Using standard values for enthalpy and entropy at equilibrium yields the equation

$$T = \frac{\Delta H°}{\Delta S°}$$

where
T is the absolute temperature at which a phase change occurs;
$\Delta H°$ is the standard enthalpy change for the process; and
$\Delta S°$ is the standard entropy change for the process.

Gibbs-Helmholtz and Phase Change Equilibria

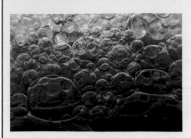

Figure 11
We can find the boiling point of water using the Gibbs-Helmholtz equation.

Use the Gibbs-Helmholtz equation and the concept of equilibrium to calculate the normal boiling point of water (at 101.3 kPa) (Figure 11). The $\Delta H°_f$ value for $H_2O_{(l)}$ is -285.9 kJ/mol, and the $\Delta H°_f$ value for $H_2O_{(g)}$ is -241.8 kJ/mol. The $S°$ value for $H_2O_{(l)}$ is $+69.96$ J/mol·K, and the $S°$ value for $H_2O_{(g)}$ is $+188.7$ J/mol·K.

$$\Delta H°_{f(H_2O_{(l)})} = -285.8 \text{ kJ/mol}$$
$$\Delta H°_{f(H_2O_{(g)})} = -241.8 \text{ kJ/mol}$$
$$S°_{H_2O_{(l)}} = +69.96 \text{ J/mol·K}$$
$$S°_{H_2O_{(g)}} = +188.7 \text{ J/mol·K}$$

First, write the chemical equation for this process.

$$H_2O_{(l)} \rightleftharpoons H_2O_{(g)}$$

The normal boiling point of water occurs when $\Delta G° = 0$.
Therefore,

$$T = \frac{\Delta H°}{\Delta S°}$$

First, calculate the value of $\Delta H°$.

$$\begin{aligned} \Delta H° &= [\Delta H°_{f(H_2O_{(g)})}] - [\Delta H°_{f(H_2O_{(l)})}] \\ &= (1 \text{ mol} \times (-241.8 \text{ kJ/mol})) - (1 \text{ mol} \times (-285.8 \text{ kJ/mol})) \\ &= (-241.8 \text{ kJ}) - (-285.8 \text{ kJ}) \\ \Delta H° &= +44.0 \text{ kJ} \end{aligned}$$

Then calculate the value of $\Delta S°$.

$$\begin{aligned} \Delta S° &= [S°_{H_2O_{(g)}}] - [S°_{H_2O_{(l)}}] \\ &= (1 \text{ mol} \times (+188.7 \text{ J/mol·K})) - (1 \text{ mol} \times (+69.96 \text{ J/mol·K})) \\ &= (+188.7 \text{ J/K}) - (+69.96 \text{ J/K}) \\ \Delta S° &= +118.7 \text{ J/K} \end{aligned}$$

Convert the value of $\Delta S°$ to kJ/K units.

$$\Delta S° = +118.7 \text{ J/K} \times \frac{1 \text{ kJ}}{1000 \text{ J}}$$
$$\Delta S° = +0.1187 \text{ kJ/K}$$

Now substitute the calculated values of $\Delta H°$ and $\Delta S°$ into the equation

$$T = \frac{\Delta H°}{\Delta S°}$$
$$= \frac{(+44.0 \text{ kJ})}{(+0.119 \text{ kJ/K})}$$
$$T = 370 \text{ K}$$

Now convert the absolute temperature value to a Celsius value.

$$\begin{aligned} t &= (T - 273)°C \\ &= (370 - 273)°C \\ t &= 97°C \end{aligned}$$

The normal boiling point of water is predicted to be 97°C.

Example

Use the Gibbs-Helmholtz equation and the concept of equilibrium to calculate the normal condensation point of hydrogen peroxide, $H_2O_{2(l)}$ (at 101.3 kPa). The $\Delta H°_f$ value for $H_2O_{2(l)}$ is -187.8 kJ/mol, and the $\Delta H°_f$ value for $H_2O_{2(g)}$ is -136.3 kJ/mol. The $S°$ value for $H_2O_{2(l)}$ is $+109.6$ J/mol·K, and the $S°$ value for $H_2O_{2(g)}$ is $+233$ J/mol·K.

Solution

$$H_2O_{2(g)} \rightleftharpoons H_2O_{2(l)}$$

$$\Delta H°_{fH_2O_{2(l)}} = 1.87.8 \text{ kJ/mol}$$

$$\Delta H°_{fH_2O_{2(g)}} = -136.3 \text{ kJ/mol}$$

$$\Delta S°_{H_2O_{2(l)}} = +109.6 \text{ J/mol·K}$$

$$\Delta S°_{H_2O_{2(g)}} = +233 \text{ J/mol·K}$$

$$\Delta H° = [\Delta H°_{f(H_2O_{2(l)})}] - [\Delta H°_{f(H_2O_{2(g)})}]$$

$$= (1 \text{ mol} \times (-187.8 \text{ kJ/mol})) - (1 \text{ mol} \times (-136.3 \text{ kJ/mol}))$$

$$= (-187.8 \text{ kJ}) - (-136.3 \text{ kJ})$$

$$\Delta H° = -51.5 \text{ kJ}$$

$$\Delta S° = [S°_{H_2O_{2(l)}}] - [S°_{H_2O_{2(g)}}]$$

$$= (1 \text{ mol} \times (+109.6 \text{ J/mol·K})) - (1 \text{ mol} \times (+233 \text{ J/mol·K}))$$

$$= (+109.6 \text{ J/K}) - (+233 \text{ J/K})$$

$$\Delta S° = -123 \text{ J/K}$$

$$\Delta S° = -123 \text{ J/K} \times \frac{1 \text{ kJ}}{1000 \text{ J}}$$

$$\Delta S° = -0.123 \text{ kJ/K}$$

$$T = \frac{\Delta H°}{\Delta S°}$$

$$= \frac{(-51.5 \text{ kJ})}{(-0.123 \text{ kJ/K})}$$

$$T = 417 \text{ K}$$

$$t = (T - 273)°C$$

$$= (417 - 273)°C$$

$$t = 144°C$$

The normal condensation point of hydrogen peroxide is predicted to be 144°C.

▶ Practice

Understanding Concepts

3. Use the Gibbs-Helmholtz equation and the concept of equilibrium to calculate the approximate Celsius temperature for the boiling point of bromine.

$$Br_{2(l)} \rightleftharpoons Br_{2(g)}$$

4. Estimate the normal boiling point of liquid mercury, $Hg_{(l)}$.

Answers

3. 58°C

4. 341°C

Understanding Concepts

1. What is meant by the term *spontaneous change*?

2. Distinguish between change in enthalpy and change in entropy.

3. Determine whether the entropy change will be positive or negative for each of the following changes:
   (a) sugar dissolves in a cup of tea
   (b) water condenses on a mirror
   (c) a balloon is broken open and a gas escapes

4. (a) State the second law of thermodynamics.
   (b) Briefly explain, using the analogy of a student doing a jigsaw puzzle, how the second law of thermodynamics is related to probability.
   (c) Does the existence of your school violate the second law of thermodynamics? Explain your answer.

5. What are the algebraic signs of ΔH and ΔS for changes that are
   (a) spontaneous at all temperatures?
   (b) nonspontaneous at all temperatures?

6. Potassium reacts violently with water according to the equation

   $$2\,K_{(s)} + H_2O_{(l)} \rightarrow H_{2(g)} + 2\,KOH_{(aq)}$$

   (a) Predict the signs of ΔH and ΔS for this reaction.
   (b) Use your prediction to explain why this reaction is spontaneous at room temperature.

7. Like all nitrates, ammonium nitrate is soluble in water.

   $$NH_4NO_{3(s)} \rightarrow NH_{4(aq)}^+ + NO_{3(aq)}^-$$

   Use the Gibbs-Helmholtz free energy equation to explain why this reaction occurs spontaneously at room temperature.

8. State the relationship between the value of ΔG and the rate of the chemical reaction.

9. The formation of a covalent bond between two atoms is an endothermic process that is accompanied by a decrease in entropy. Use the Gibbs-Helmholtz equation to explain why all compounds decompose into their individual elements if heated to high enough temperatures.

10. Predict the algebraic sign of the entropy change for each of the following reactions:
    (a) $N_{2(g)} + 3\,H_{2(g)} \rightarrow 2\,NH_{3(g)}$
    (b) $SO_{2(g)} + CaO_{(s)} \rightarrow CaSO_{3(s)}$
    (c) $3\,PbO_{(s)} + 2\,NH_{3(g)} \rightarrow 3\,Pb_{(s)} + N_{2(g)} + 3\,H_2O_{(g)}$

11. Calculate the value of $\Delta G°$ for the following reactions and interpret the results:
    (a) $Ca(OH)_{2(s)} + H_2SO_{4(l)} \rightarrow CaSO_{4(s)} + 2\,H_2O_{(l)}$
    (b) $2\,NH_4Cl_{(s)} + CaO_{(s)} \rightarrow CaCl_{2(s)} + H_2O_{(l)} + 2\,NH_{3(g)}$

12. Use Appendix C6, the Gibbs-Helmholtz equation and the concept of equilibrium to calculate the temperature (in °C) at which potassium melts.

13. When a bottle of ammonia, $NH_{3(g)}$, is opened near a bottle of hydrogen chloride, $HCl_{(g)}$, a white "smoke" containing microscopic ammonium chloride crystals, $NH_4Cl_{(s)}$, is visible in the space between the two bottles (**Figure 12**).

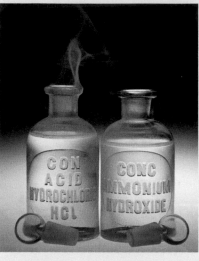

Figure 12

The equation for this reaction is

$$NH_{3(g)} + HCl_{(g)} \rightarrow NH_4Cl_{(s)}$$

   (a) Predict the signs of $\Delta G°$, $\Delta S°$, and $\Delta H°$ for this observed reaction. Provide theoretical support for your predictions.
   (b) Use values obtained from the table of thermodynamic data in Appendix C6 to calculate $\Delta H°$, $\Delta S°$, and $\Delta G°$ for this change.
   (c) How did your predicted values compare with the calculated values?

14. (a) Calculate the value of $\Delta G°$ for the combustion of ethene, $C_2H_{4(g)}$, to produce only gaseous products.
    (b) What does the value of $\Delta G°$ for this reaction tell you?

15. How are free energy and equilibrium related?

Making Connections

16. (a) List three changes you have witnessed recently that were spontaneous.
    (b) List three changes that were nonspontaneous, but occurred as a result of the application of continuous outside assistance.

17. How does the second law of thermodynamics affect the operation of an automobile?

18. Describe two chemical reactions you use in your home to obtain free energy to do useful work. Describe, using a word equation, the work done with the energy.

19. (a) Use the Gibbs-Helmholtz equation to calculate the normal boiling point of hydrazine, $N_2H_{4(l)}$. Conduct library or Internet research to identify commercial uses of hydrazine.
    (b) Make some suggestions about the handling and storage of hydrazine

 www.science.nelson.com

🧪 INVESTIGATION 7.1.1

Inquiry Skills

○ Questioning	● Planning	● Analyzing
○ Hypothesizing	● Conducting	● Evaluating
● Predicting	● Recording	● Communicating

Discovering the Extent of a Chemical Reaction

In a quantitative reaction—a reaction that goes to completion—the limiting reagent is completely consumed. We can test the final reaction mixture for the presence of the original reactants to determine whether a reaction has completely consumed at least one of the initial substances.

Purpose

The purpose of this investigation is to test the validity of the assumption that chemical reactions go to completion.

Question

What are the limiting and excess reagents in the chemical reaction of various quantities of aqueous sodium sulfate and aqueous calcium chloride?

Prediction

(a) Using your current state of knowledge, predict whether the reaction between aqueous sodium sulfate and aqueous calcium chloride is quantitative.

Experimental Design

Samples of sodium sulfate solution and calcium chloride solution are mixed in different proportions and the final mixture is filtered (**Figure 1**).

(b) Write the balanced equation for the overall reaction.

(c) Write the net ionic equation for the reaction that produces a precipitate.

If the reaction proceeds to completion, at least one of the reagents will be completely consumed, and will not appear in the filtrate.

Samples of the filtrate are tested for the presence of excess reagents, using the following diagnostic tests.

- If a few drops of $Ba(NO_3)_{2(aq)}$ are added to the filtrate and a precipitate forms, then excess sulfate ions are present.

$$Ba^{2+}_{(aq)} + SO^{2-}_{4(aq)} \rightarrow BaSO_{4(s)}$$

Figure 1
Filtering the final mixture

- If a few drops of $Na_2CO_{3(aq)}$ are added to the filtrate and a precipitate forms, then excess calcium ions are present.

$$Ca^{2+}_{(aq)} + CO^{2-}_{3(aq)} \rightarrow CaCO_{3(s)}$$

Materials

lab apron	2 50-mL or 100-mL beakers
eye protection	2 small test tubes
25 mL of 0.50 mol/L $CaCl_{2(aq)}$	10-mL or 25-mL
25 mL of 0.50 mol/L $Na_2SO_{4(aq)}$	graduated cylinder
1.0 mol/L $Na_2CO_{3(aq)}$	filtration apparatus
in dropper bottle	filter paper
saturated $Ba(NO_3)_{2(aq)}$	wash bottle
in dropper bottle	stirring rod

 Barium compounds are toxic. Solutions containing barium should be collected in a marked disposal container at the end of the lab.

Remember to wash your hands before leaving the laboratory.

Wear eye protection and a laboratory apron.

Procedure

(d) Write a procedure for testing whether the reaction between aqueous sodium sulfate and aqueous calcium chloride is quantitative by reacting different proportions of the two reactants in a closed system.

1. Obtain your teacher's approval, then carry out the experiments.

Analysis

(e) Write a statement describing what you observed, using the chemical names from the equation.

(f) Referring to the net ionic equation, write a statement about the anomaly that you observed.

Evaluation

(g) Evaluate your Prediction in (a), and the Experimental Design.

(h) Suggest improvements to the Experimental Design.

(i) Evaluate the wording of the Question.

LAB EXERCISE 7.2.1

Develop an Equilibrium Law

The following chemical equation represents a chemical equilibrium.

$$Fe^{3+}_{(aq)} + SCN^-_{(aq)} \rightleftharpoons FeSCN^{2+}_{(aq)}$$

This equilibrium is convenient to study because the reaction is clearly indicated by a colour change (**Figure 2**).

Question

What mathematical formula, using equilibrium concentrations of reactants and products, gives a constant for the iron(III) thiocyanate reaction system?

Experimental Design

Reactions are performed using various initial concentrations of iron(III) nitrate and potassium thiocyanate solutions. The equilibrium concentrations of the reactants and the product are determined from the measurement and analysis of the intensity of the colour. Possible mathematical relationships among the concentrations are tried, and then are analyzed to determine if the mathematical formula gives a constant value.

Figure 2
The two reactants combine to form a dark-red equilibrium mixture. The red colour of the solution is the colour of the aqueous thiocyanate iron(III) product, $FeSCN^{2+}_{(aq)}$.

Observations

Table 1 Iron(III)–Thiocyanate Equilibrium at SATP

Trial	$[Fe^{3+}_{(aq)}]$ (mol/L)	$[SCN^-_{(aq)}]$ (mol/L)	$[FeSCN^{2+}_{(aq)}]$ (mol/L)
1	3.91×10^{-2}	8.02×10^{-5}	9.22×10^{-4}
2	1.48×10^{-2}	1.91×10^{-4}	8.28×10^{-4}
3	6.27×10^{-3}	3.65×10^{-4}	6.58×10^{-4}
4	2.14×10^{-3}	5.41×10^{-4}	3.55×10^{-4}
5	1.78×10^{-3}	6.13×10^{-4}	3.23×10^{-4}

Analysis

(a) Test the following mathematical relationships to see which gives a constant value:

1. $[Fe^{3+}_{(aq)}][SCN^-_{(aq)}][FeSCN^{2+}_{(aq)}]$

2. $[Fe^{3+}_{(aq)}] + [SCN^-_{(aq)}] + [FeSCN^{2+}_{(aq)}]$

3. $\dfrac{[FeSCN^{2+}_{(aq)}]}{[Fe^{3+}_{(aq)}][SCN^-_{(aq)}]}$

4. $\dfrac{[Fe^{3+}_{(aq)}]}{[FeSCN^{2+}_{(aq)}]}$

5. $\dfrac{[SCN^-_{(aq)}]}{[FeSCN^{2+}_{(aq)}]}$

(b) Describe this relationship in words.

INVESTIGATION 7.3.1

Testing Le Châtelier's Principle

Le Châtelier's principle helps us to predict how reactions will respond under certain conditions, such as changes in temperature or pressure. In this investigation, you will make and test your own predictions on a variety of chemical systems.

Purpose

The purpose of this investigation is to test Le Châtelier's principle by applying stresses to seven different chemical equilibria.

Question

How does applying stresses to particular chemical equilibria affect the systems?

Prediction

(a) Read the Experimental Design, Materials, and Procedure, and use Le Châtelier's principle to predict

⚗ INVESTIGATION 7.3.1 *continued*

the change(s) that will occur when each equilibrium mixture is subjected to the stated stress.

Experimental Design

Stresses are applied to the following seven chemical equilibrium systems and evidence is gathered to test predictions made using Le Châtelier's principle. Control samples are used in all cases.

Part I Dinitrogen tetroxide–nitrogen dioxide equilibrium (demonstration)

$$N_2O_{4(g)} + energy \rightleftharpoons 2 NO_{2(g)}$$

Flasks containing an $N_2O_{4(g)}$–$NO_{2(g)}$ equilibrium mixture are placed into cold- and hot-water baths.

Part II Carbon dioxide–carbonic acid equilibrium

$$CO_{2(g)} + H_2O_{(l)} \rightleftharpoons H^+_{(aq)} + HCO^-_{3(aq)}$$

A $CO_{2(g)}$–$HCO^-_{3(aq)}$ equilibrium mixture is placed in a syringe and subjected to increased pressure.

Part III Cobalt(II) complexes

$$CoCl_4{}^{2-}_{(alcohol)} + 6 H_2O_{(l)} \rightleftharpoons Co(H_2O)_6{}^{2+}_{(aq)} + 4 Cl^-_{(aq)} + energy$$

Water, a solution of silver nitrate, and heat are added to, and heat is removed from, samples of the equilibrium mixture, which are provided for you.

Part IV Thymol blue indicator

$$H_2Tb_{(aq)} \rightleftharpoons H^+_{(aq)} + HTb^-_{(aq)}$$
$$HTb^-_{(aq)} \rightleftharpoons H^+_{(aq)} + Tb^{2-}_{(aq)} \quad \text{(HTb is short for thymol blue)}$$

Hydrochloric acid and sodium hydroxide are added to different samples of the equilibrium mixture, which are provided for you.

Part V Iron(III)–thiocyanate equilibrium

$$Fe^{3+}_{(aq)} + SCN^-_{(aq)} \rightleftharpoons FeSCN^{2+}_{(aq)}$$

Iron(III) nitrate, potassium thiocyanate, and sodium hydroxide are added to samples of the equilibrium mixture, which are provided for you.

Part VI Copper(II) complexes

$$Cu(H_2O)_4{}^{2+}_{(aq)} + 4 NH_{3(aq)} \rightleftharpoons Cu(NH_3)_4{}^{2+}_{(aq)} + 4 H_2O_{(l)}$$

Aqueous ammonia and hydrochloric acid are added to samples of the equilibrium mixture, which are provided for you.

Part VII Chromate–dichromate equilibrium

$$2 CrO_4{}^{2-}_{(aq)} + 2 H^+_{(aq)} \rightleftharpoons Cr_2O_7{}^{2-}_{(aq)} + H_2O_{(l)}$$

Aqueous sodium hydroxide, hydrochloric acid, and then aqueous barium nitrate are all added to a sample of the equilibrium mixture, which is provided for you.

Materials

lab apron
eye protection
100-mL beaker
large waste beaker
6 to 12 small test tubes
test-tube rack
small syringe with needle removed (5 to 50 mL)
solid rubber stopper to seal end of syringe
distilled water
crushed ice
hot-water bath
2 flasks or tubes containing an $N_2O_{4(g)}$–$NO_{2(g)}$ mixture
6.0 mol/L $NaOH_{(aq)}$
dropper bottles containing:
    carbon dioxide–bicarbonate equilibrium mixture (pH = 7)
    cobalt(II) chloride equilibrium mixture in ethanol
    0.2 mol/L $AgNO_{3(aq)}$
    thymol blue indicator
    bromothymol blue indicator
    0.1 mol/L $HCl_{(aq)}$
    0.1 mol/L $NaOH_{(aq)}$
    iron(III)–thiocyanate equilibrium mixture
    0.2 mol/L $Fe(NO_3)_{3(aq)}$
    0.2 mol/L $KSCN_{(aq)}$
    0.1 mol/L $CuSO_{4(aq)}$
    1.0 mol/L $NH_{3(aq)}$
    1.0 mol/L $HCl_{(aq)}$
    chromate–dichromate equilibrium mixture
    0.1 mol/L $Ba(NO_3)_{2(aq)}$

 Be careful with the flasks containing nitrogen dioxide: This gas is highly toxic. Use in a fume hood in case of breakage.

 The chemicals may be corrosive, irritating, and/or toxic. Exercise great care when using the chemicals and avoid skin and eye contact. Immediately rinse

INVESTIGATION 7.3.1 *continued*

the skin with cold water if there is any contact and flush the eyes for a minimum of 15 min and inform the teacher.

Ethanol is flammable. Make sure there are no open flames in the laboratory when using the ethanol solution of cobalt(II) chloride.

Solutions containing heavy metal ions must not be flushed down the sink. Collect them in a marked container.

Wear a lab apron and eye protection.

Procedure

As you finish each part, dispose of the chemicals as directed by your teacher. Most may be washed down the drain with large amounts of water. Those containing heavy metals must be collected in a marked container for separate disposal.

Part I Dinitrogen tetroxide–nitrogen dioxide equilibrium (demonstration)

1. Place the sealed $N_2O_{4(g)}$–$NO_{2(g)}$ flasks in hot- and cold-water baths and record your observations.

Be careful with the flasks containing nitrogen dioxide: This gas is highly toxic. Use in a fume hood.

Part II Carbon dioxide–carbonic acid equilibrium

2. Place two or three drops of bromothymol blue indicator in the carbon dioxide–bicarbonate equilibrium mixture.

3. Draw some of the carbon dioxide–bicarbonate equilibrium mixture into the syringe, then block the end with a rubber stopper.

4. Slowly depress the syringe plunger and record your observations.

Part III Cobalt(II) complexes

5. Obtain 25 mL of the equilibrium mixture with the cobalt(II) chloride complex ions.

Remember that the solute for this solution is alcohol, so keep it away from open flames.

6. Place a small amount of the mixture into each of five small test tubes. Use the fifth test tube as a control for comparison purposes.

7. Add drops of water to one test tube until a change is evident. Record your observations.

8. Add drops of 0.2 mol/L silver nitrate to another test tube and record your observations.

9. Heat a third sample of the equilibrium mixture in a water bath and record your observations.

10. Cool a fourth sample of the equilibrium mixture in an ice bath and record your observations.

Part IV Thymol blue indicator

11. Add about 5 mL of distilled water to each of two small test tubes.

12. Add 1 to 3 drops of thymol blue indicator to the water in each test tube to obtain a noticeable colour.

13. Use one test tube of solution as a control.

14. Add drops of 0.1 mol/L $HCl_{(aq)}$ to the experimental test tube to test for the predicted colour changes.

15. Add drops of 0.1 mol/L $NaOH_{(aq)}$ to the same test tube to test for the predicted colour changes.

Part V Iron(III)–thiocyanate equilibrium

16. Obtain about 20 mL of the iron(III)–thiocyanate equilibrium solution.

17. Place about 5 mL of the equilibrium solution in each of three test tubes.

18. Use one test tube as a control.

19. Add drops of $Fe(NO_3)_{3(aq)}$ to an equilibrium mixture until a change is evident.

20. Add drops of 6.0 mol/L $NaOH_{(aq)}$ to this new equilibrium mixture until a change occurs. (Remember that iron(III) hydroxide has a low solubility.)

6.0 mol/L $NaOH_{(aq)}$ is extremely corrosive. Use extreme care.

21. Add drops of $KSCN_{(aq)}$ to another equilibrium mixture until a change is evident.

Part VI Copper(II) complexes

22. Obtain 2 mL of 0.1 mol/L $CuSO_{4(aq)}$ in a small test tube.

23. Add three drops of 1.0 mol/L $NH_{3(aq)}$ to establish the equilibrium mixture.

24. Add more 1.0 mol/L $NH_{3(aq)}$ to the above equilibrium mixture and record the results.

25. Add 1.0 mol/L $HCl_{(aq)}$ to the equilibrium mixture from step 24 and record the results.

⚗ INVESTIGATION 7.3.1 *continued*

Part VII Chromate–dichromate equilibrium

26. Obtain 15 mL of the chromate–dichromate equilibrium mixture.

27. Place 5-mL samples of the equilibrium mixture into each of three small test tubes.

28. Add 0.1 mol/L $HCl_{(aq)}$ drop by drop to one sample or to 0.1 mol/L $K_2Cr_2O_{7(aq)}$ and record your observations.

29. Add 0.1 mol/L $NaOH_{(aq)}$ drop by drop to another sample (or, if you choose, the previous $HCl_{(aq)}$ sample or 0.1 mol/L $K_2Cr_2O_{7(aq)}$) and record the results.

30. Add 0.1 mol/L $Ba(NO_3)_{2(aq)}$ drop by-drop to a third sample and record your observations. (Remember that barium chromate has a low solubility.) Compare to **Table 2**.

31. Ensure that all equipment and surfaces are left clean and that your hands are washed thoroughly before leaving the laboratory.

Table 2   Diagnostic Test Colours

Ion or Compound	Colour
$CoCl_4^{2-}{}_{(aq)}$	blue
$Co(H_2O)_6^{2+}{}_{(aq)}$	pink
$H_2Tb_{(aq)}$	red
$HTb^-_{(aq)}$	yellow
$Tb^{2-}_{(aq)}$	blue
$Fe^{3+}_{(aq)}$	yellow
$SCN^{2-}_{(aq)}$	colourless
$FeSCN^{2+}_{(aq)}$	red
$Cu(H_2O)_4^{2+}{}_{(aq)}$	pale blue
$Cu(NH_3)_4^{2+}{}_{(aq)}$	deep blue
$CrO_4^{2-}{}_{(aq)}$	yellow
$Cr_2O_7^{2-}{}_{(aq)}$	orange
$N_2O_{4(g)}$	colourless
$NO_{2(g)}$	brown

Analysis

(b) Answer the Question.

(c) Summarize your observations in the form of a chart.

⚗ INVESTIGATION 7.6.1

Determining the K_{sp} of Calcium Oxalate

Calcium oxalate, $CaC_2O_{4(s)}$, is a slightly soluble salt that dissolves according to the chemical reaction

$$CaC_2O_{4(s)} \rightleftharpoons Ca^{2+}_{(aq)} + C_2O_4^{2-}{}_{(aq)}$$

Question

What is the K_{sp} of calcium oxalate?

Experimental Design

In this investigation, the solubility product constant of calcium oxalate, $CaC_2O_{4(s)}$, is determined by mixing a fixed volume of 0.1 mol/L sodium oxalate with a serial dilution of aqueous calcium nitrate in a series of spotplate wells. The K_{sp} of calcium oxalate is determined by calculating the product of the concentrations of the ions in the well containing the highest concentration of ions with no visible precipitate.

Materials

lab apron
eye protection
0.1 mol/L calcium nitrate, $Ca(NO_3)_{2(aq)}$
0.1 mol/L sodium oxalate, $Na_2C_2O_{4(aq)}$
minimum 24-well spotplate (12 × 2)
4 pipets
distilled water
dark coloured paper

 Sodium oxalate is toxic if ingested.

Evidence

(a) Prepare a 12-row, 5-column table in which to record your observations and calculations (**Table 3**).

Table 3 Observations and Calculations for Investigation 7.6.1

Well #	A initial $[Ca^{2+}_{(aq)}]$	B initial $[C_2O_4^{2-}_{(aq)}]$	C final $[Ca^{2+}_{(aq)}]$	D final $[C_2O_4^{2-}_{(aq)}]$	E (C × D) ion product
1					
2					
3					
4					
5					
6					
7					
8					
9					
10					
11					
12					

Procedure

1. Place a spotplate on a dark sheet of paper.

2. Add 5 drops of distilled water to each of 11 consecutive wells in row A of the spotplate (wells A2 to A12), leaving the first well, A1, empty (**Figure 3**).

Figure 3
The dark paper should make it easier to detect the formation of a light-coloured precipitate.

3. Add 10 drops of 0.1 mol/L calcium nitrate to well A1.

4. Draw the solution from A1 into an empty pipet, and place 5 drops of the solution into A2. (Return any excess solution to the first well.)

5. Using the same pipet, transfer 5 drops of the solution in A2 to the water in A3.

6. Repeat step 5 for each of the remaining wells. Discard the 5 drops from well A12 into a sink with lots of running water.

7. Using a new pipet, place 5 drops of 0.1 mol/L sodium oxalate solution into wells B1 through B12.

8. Use a clean pipet to transfer the entire contents of well A1 into well B1. Mix well with the tip of the pipet. Continue this process for wells A2 into B2, A3 into B3, etc., until A12 and B12 have been mixed.

9. Examine all of the wells. Identify the first well that appears to have no precipitate.

10. Dispose of chemical wastes as instructed by your teacher.

Analysis

(b) Calculate the initial $[Ca^{2+}_{(aq)}]$ in each well of row A. Record these in column A of your table.

(c) Record the initial $[C_2O_4^{2-}_{(aq)}]$ in each well of row B in column B of your table.

(d) Determine the $[Ca^{2+}_{(aq)}]$ and $[C_2O_4^{2-}_{(aq)}]$ in wells B1 to B12, and record these values in columns C and D, respectively.

(e) Calculate the ion product for the contents of each well. Record these in column E of the table.

(f) Write the K_{sp} expression for calcium oxalate and the calculation that determined its value in this experiment.

(g) Answer the Question by deciding which ion product in column E of the table corresponds to the K_{sp} of calcium oxalate. Give reasons for your decision.

(h) Use your experimental value of K_{sp} to determine the solubility of calcium oxalate.

Evaluation

(i) Identify sources of error and uncertainty in the Experimental Design. Provide suggestions for improvement.

(j) Compare your experimentally derived value for the K_{sp} of calcium oxalate to the accepted value contained in a reliable source such as the *CRC Handbook of Chemistry and Physics*. Calculate the percentage difference. Comment on the validity of your experimental result.

INVESTIGATION 7.6.2

Determining *K*_{sp} for Calcium Hydroxide

Read Investigation 7.6.1 and design your own investigation to determine the solubility product constant for calcium hydroxide.

Purpose

The purpose of this investigation is to use solubility equilibrium theory and experimental evidence to determine the solubility product constant for calcium hydroxide, and then to compare this value with the accepted reference value.

Question

(a) Write a question that you will attempt to answer.

Prediction

(b) Look up the accepted reference value for the solubility product constant for calcium hydroxide.

Experimental Design

(c) Write your design.

Materials

(d) List the materials you will need. Be sure to include any necessary safety precautions, particularly in the handling and disposal of calcium hydroxide.

Procedure

(e) Write a detailed series of steps for your Procedure.

1. With your teacher's approval, carry out your Procedure.

Analysis

(f) Analyze your Evidence to answer the Question.

Evaluation

(g) Compare your experimental value to the reference value, and determine the percentage difference.

(h) How much confidence do you have in your experimental value? Account for any uncertainty.

Key Expectations

- Illustrate the concept of dynamic equilibrium with reference to systems such as liquid-vapour equilibrium, weak electrolytes in solution, and chemical reactions. (7.1)

- Demonstrate an understanding of the law of chemical equilibrium as it applies to the concentrations of the reactants and products at equilibrium. (7.2)

- Identify effects of solubility on biological systems. (7.2, 7.6)

- Demonstrate an understanding of Le Châtelier's Principle; and apply the principle to predict how factors (such as changes in volume, pressure, concentration, or temperature) affect a chemical system at equilibrium, and confirm your predictions through experimentation. (7.3)

- Explain how equilibrium principles may be applied to optimize the production of industrial chemicals. (7.4)

- Solve equilibrium problems involving concentrations of reactants and products, and K (7.2) and K_{sp} (7.5).

- Describe, using the concept of equilibrium, the behaviour of ionic solutes in solutions that are unsaturated, saturated, and supersaturated. (7.6)

- Define constant expressions, such as K_{sp}. (7.6)

- Carry out experiments and calculations to determine equilibrium constants. (7.6)

- Predict the formation of precipitates by using the solubility product constant, K_{sp}. (7.6)

- Identify, in qualitative terms, entropy changes associated with chemical and physical processes. (7.7)

- Describe the tendency of reactions to achieve minimum energy and maximum entropy. (7.7)

- Use appropriate vocabulary to communicate ideas, procedures, and results related to chemical systems and equilibrium. (all sections)

Key Terms

bond energy

chemical reaction equilibrium

closed system

common ion effect

dissolution

dynamic equilibrium

entropy, S

equilibrium constant, K

equilibrium law expression

equilibrium shift

first law of thermodynamics

forward reaction

free energy (Gibbs free energy)

heterogeneous equilibria

homogeneous equilibria

Le Châtelier's principle

percent reaction

phase equilibrium

quantitative reaction

reaction quotient, Q

reverse reaction

reversible reaction

second law of thermodynamics

solubility

solubility equilibrium

solubility product constant, K_{sp}

spontaneous reaction

standard entropy

supersaturated solution

third law of thermodynamics

trial ion product

weak electrolytes

Key Symbols and Equations

- $K = \dfrac{[C]^c [D]^d}{[A]^a [B]^b}$ at equilibrium $\qquad Q = \dfrac{[C]^c [D]^d}{[A]^a [B]^b}$ at any time

- $K_{sp} = [B^+_{(aq)}]^b [C^-_{(aq)}]^c$

Extension

- $\Delta G = \Delta H - T\Delta S$ $\qquad T = \dfrac{\Delta H°}{\Delta S°}$

- $\Delta G° = \Delta H° - T\Delta S°, \ \ T = 298 \ K$

Problems You Can Solve

- What is the concentration of one entity at equilibrium, given the concentrations of the other entities? (7.1)

- What is the equilibrium law expression for a reaction? (7.2)

- What is the equilibrium constant for a reaction, given the equilibrium concentrations? (7.2)

- What is the equilibrium law expression for a reaction involving condensed states? (7.2)

- What is the equilibrium law expression for a reaction involving ions? (7.2)

- Is a system at equilibrium? Compare Q to K. (7.5)

- What is the concentration of one entity at equilibrium, given K and the equilibrium concentrations of the other entities, or K and the initial concentrations? (7.5)

- What are the equilibrium concentrations of entities, given initial concentrations and a very small value for K? (7.5)

- What are the equilibrium concentrations of entities, given K and the initial concentrations? (7.5)

- What is K_{sp}, given the solubility of a salt (or vice versa)? (7.6)

- Will a precipitate form when two ionic compounds of known concentration are mixed? (7.6)

- What is the molar solubility of an ionic compound in a solution containing a common ion? (7.6)

- Is the entropy change for a chemical system positive or negative? (7.7)

- What is the Gibbs free energy change for a reaction? (7.7)

- At what temperature does a given phase change occur, given values for $\Delta H°_f$ and $S°$? (7.7)

▶ *MAKE* a summary

You have learned several fundamental and useful scientific laws in this chapter. Create a concept map beginning with the term "Laws learned in Chapter 7" and building the concept map through levels describing concepts, equations, and special considerations.

Identify each of the following statements as true, false, or incomplete. If the statement is false or incomplete, rewrite it as a true statement.

1. Chemical equilibrium means that all chemical reactions have stopped.

2. Equilibrium can only occur in a closed system.

3. A catalyst shifts the position of equilibrium toward the products.

4. Reactions with a lower activation energy achieve equilibrium faster.

5. According to Le Châtelier's principle, a reaction at equilibrium whose ΔH is -512 kJ will shift to the right if heated.

6. If the trial ion product is less than K_{sp}, a precipitate will form.

7. If ΔH is negative and ΔS is negative, the reaction will be spontaneous only at high temperatures.

8. A reaction whose ΔG is zero is at equilibrium.

9. All chemical reactions are, in principle, reversible.

10. If $K = 1$, the concentrations of the reactants and products are approximately equal.

Identify the letter that corresponds to the best answer to each of the following questions.

11. In a dynamic equilibrium
    (a) macroscopic properties are constant.
    (b) reactants are converted to products.
    (c) products are converted to reactants.
    (d) rates of forward and reverse reactions are equal.
    (e) all of the above.

12. Which of the following affect the value of the equilibrium constant?
    (a) temperature
    (b) small changes in concentration
    (c) the density of a slightly soluble salt
    (d) small changes in the pressure of a gas
    (e) (a) and (b)

13. Which is the correct form of the equilibrium law expression for the following reaction?

    $2 SO_{3(g)} \rightleftharpoons 2 SO_{2(g)} + O_{2(g)}$

    (a) $K = \dfrac{[SO_{2(g)}][O_{2(g)}]}{[SO_{3(g)}]}$
    (c) $K = \dfrac{[SO_{2(g)}]^2[O_{2(g)}]}{[SO_{3(g)}]^2}$

    (b) $K = \dfrac{[SO_{3(g)}]^2}{[SO_{2(g)}]^2[O_{2(g)}]}$
    (d) $K = [SO_{2(g)}][O_{2(g)}]^2$

    (e) none of the above

14. If the equilibrium constant, K, for the conversion of isobutane to n-butane is 2.5, what is the value of the equilibrium constant for the reverse reaction?
    (a) 2.5   (c) 6.25   (e) 1.0
    (b) 0.4   (d) 1.3

15. What is the K_{sp} of $PbCl_2$ if, in a saturated solution of this salt, $[Cl^-_{(aq)}] = 0.032$ mol/L ?
    (a) 5.1×10^{-4} (d) 1.6×10^{-5}
    (b) 4.8×10^{-3} (e) 6.2×10^{-2}
    (c) 3.9×10^{-5}

16. Given the equilibrium

    $AgBr_{(s)} \rightleftharpoons Ag^+_{(aq)} + Br^-_{(aq)}$ $K_{sp} = 3.3 \times 10^{-13}$

    If $[Br^-_{(aq)}] = 0.50$ mol/L, what is $[Ag^+_{(aq)}]$?
    (a) 6.6×10^{-13} (d) 3.3×10^{-13}
    (b) 1.7×10^{-13} (e) 7.5×10^{12}
    (c) 1.5×10^{12}

17. What will happen to $[Pb^{2+}_{(aq)}]$ in the following equilibrium if $NaCl_{(s)}$ is added to the container?

    $PbCl_{2(s)} \rightleftharpoons Pb^{2+}_{(aq)} + 2Cl^-_{(aq)}$

    (a) It will decrease.
    (b) It will increase.
    (c) It will remain the same.
    (d) It will increase at first, then decrease.
    (e) The answer cannot be determined.

18. Predict the effect on the following equilibrium if the volume of the flask is decreased.

    $Cl_{2(g)} + 3 F_{2(g)} \rightleftharpoons 2 ClF_{3(g)} + heat$

    (a) The reaction shifts right.
    (b) The reaction shifts left.
    (c) There is no change in equilibrium position.
    (d) The reaction first shifts right, then left.
    (e) none of the above

19. For which of the following is the entropy change negative?
    (a) a skier skiing down a slope
    (b) paper burning
    (c) $H_2O_{(g)} \rightarrow H_2O_{(l)}$
    (d) $NaCl_{(s)} \rightarrow NaCl_{(aq)}$
    (e) none of the above

20. In which of the following cases does the tendency to maximum entropy favour the forward reaction?
    (a) $Br_{2(g)} \rightleftharpoons Br_{2(l)}$
    (b) $N_{2(g)} + 3 H_{2(g)} \rightleftharpoons 2 NH_{3(g)}$
    (c) $Li^+_{(aq)} + Cl^-_{(aq)} \rightleftharpoons LiCl_{(s)}$
    (d) $6 C_{(s)} + 3 H_{2(g)} \rightleftharpoons C_6H_{6(l)}$
    (e) $CaO_{(s)} + CO_{2(g)} \rightleftharpoons CaCO_{3(s)}$

An interactive version of the quiz is available online.   NEL

GO   www.science.nelson.com

Understanding Concepts

1. (a) What is chemical equilibrium?
   (b) On what concept does the idea of chemical equilibrium depend? (7.1)

2. What are two ways to describe the relative amounts of reactants and products present in a chemical reaction at equilibrium? (7.1)

3. Describe and explain a situation in which a soft drink is in
   (a) a non-equilibrium state.
   (b) an equilibrium state. (7.1)

4. Write a statement of Le Châtelier's principle. (7.2)

5. Scientists and technologists are particularly interested in the use of hydrogen as a fuel.

   $$2\,H_{2(g)} + O_{2(g)} \rightleftharpoons 2\,H_2O_{(g)} \qquad K = 1 \times 10^{80} \text{ at SATP}$$

   What interpretation can be made about the relative proportions of reactants and products in this system at equilibrium? (7.2)

6. What variables are commonly manipulated to shift the position of a system at equilibrium? (7.3)

7. How does a change in volume of a closed system containing gases affect the pressure of the system? (7.3)

8. In many processes in industry, engineers try to maximize the yield of a product. How can the concentrations of reactants or products be manipulated to increase the yield of product? (7.3)

9. For each of the following descriptions, write a chemical equation for the system at equilibrium. Communicate the position of the equilibrium with equilibrium arrows, then write a mathematical expression of the equilibrium law for each chemical system.
   (a) At high temperatures, the formation of water vapour from hydrogen and oxygen is quantitative.
   (b) The reaction of carbon monoxide with water vapour to produce carbon dioxide and hydrogen has a percentage yield of 67% at 500°C.
   (c) A combination of low pressure and high temperature provides a percentage yield of less than 10% for the formation of ammonia in the Haber process. (7.4)

10. In a sealed container, nitrogen dioxide is in equilibrium with dinitrogen tetroxide.

    $$2\,NO_{2(g)} \rightleftharpoons N_2O_{4(g)} \qquad K = 1.15,\, t = 55°C$$

    (a) Write the equilibrium law expression for this chemical system.
    (b) If the equilibrium concentration of nitrogen dioxide is 0.050 mol/L, predict the concentration of dinitrogen tetroxide.
    (c) Write a prediction for the shift in equilibrium that occurs when the concentration of nitrogen dioxide is increased. (7.3)

11. Predict the shift in the following equilibrium system resulting from each of the following changes.

    $$4\,HCl_{(g)} + O_{2(g)} \rightleftharpoons 2\,H_2O_{(g)} + 2\,Cl_{2(g)} + 113\text{ kJ}$$

    (a) an increase in the temperature of the system
    (b) an increase in the volume of the container
    (c) an increase in the concentration of oxygen
    (d) the addition of a catalyst
    (e) addition of $Ne_{(g)}$ at constant volume (7.3)

12. Chemical engineers use Le Châtelier's principle to predict shifts in chemical systems at equilibrium resulting from changes in the reaction conditions. Predict the changes necessary to maximize the yield of product for each of the following important industrial reactions.
    (a) the production of ethene (ethylene)

    $$C_2H_{6(g)} + \text{energy} \rightleftharpoons C_2H_{4(g)} + H_{2(g)}$$

    (b) the production of methanol

    $$CO_{(g)} + 2\,H_{2(g)} \rightleftharpoons CH_3OH_{(g)} + \text{heat} \qquad (7.3)$$

13. For each example, predict whether, and in which direction, the equilibrium is shifted by the change imposed. Explain any shift in terms of changes in forward and reverse reaction rates.
    (a) $Cu^{2+}_{(aq)} + 4\,NH_{3(g)} \rightleftharpoons Cu(NH_3)_4^{2+}_{(aq)}$
        $CuSO_{4(s)}$ is added
    (b) $CaCO_{3(s)} + \text{energy} \rightleftharpoons CaO_{(s)} + CO_{2(g)}$
        temperature is decreased
    (c) $Na_2CO_{3(s)} + \text{energy} \rightleftharpoons Na_2O_{(s)} + CO_{2(g)}$
        sodium carbonate is added
    (d) $H_2CO_{3(aq)} + \text{energy} \rightleftharpoons CO_{2(g)} + H_2O_{(l)}$
        volume is decreased
    (e) $KCl_{(s)} \rightleftharpoons K^+_{(aq)} + Cl^-_{(aq)}$ $AgNO_{3(s)}$ is added
    (f) $CO_{2(g)} + NO_{(g)} \rightleftharpoons CO_{(g)} + NO_{2(g)}$
        volume is increased
    (g) $Fe^{3+}_{(aq)} + SCN^-_{(aq)} \rightleftharpoons FeSCN^{2+}_{(aq)}$
        $Fe(NO_3)_{3(s)}$ is added (7.3)

14. In a container at high temperature, ethyne (acetylene) and hydrogen react to produce ethene (ethylene).

$$C_2H_{2(g)} + H_{2(g)} \rightleftharpoons C_2H_{4(g)}$$

The equilibrium constant is 0.072. At a specific time, the substance concentrations are $[C_2H_{2(g)}] = 0.40$ mol/L, $[H_{2(g)}] = 0.020$ mol/L, and $[C_2H_{4(g)}] = 3.2 \times 10^{-4}$ mol/L. Predict the direction of the reaction shift. (7.5)

15. For the reaction

$$H_{2(g)} + Br_{2(g)} \rightleftharpoons 2\,HBr_{(g)} \qquad K = 12.0 \text{ at } t°C$$

calculate the concentrations of all three substances at equilibrium, if the following amounts of reactants are mixed in a 2.00-L reaction container.
(a) 8.00 mol of hydrogen and 8.00 mol of bromine
(b) 12.0 mol of hydrogen and 12.0 mol of bromine
(c) 12.0 mol of hydrogen and 8.00 mol of bromine
(7.5)

16. $CO_{(g)} + H_2O_{(g)} \rightleftharpoons CO_{2(g)} + H_{2(g)}$
$K = 4.00$ at 900°C

In a container, carbon monoxide and water vapour are in the process of reacting to produce carbon dioxide and hydrogen. The concentrations are $[CO_{(g)}] = 4.00$ mol/L, $[H_2O_{(g)}] = 2.00$ mol/L, $[CO_{2(g)}] = 4.00$ mol/L, and $[H_{2(g)}] = 2.00$ mol/L. Determine the direction in which the reaction proceeds to establish equilibrium. (7.5)

17. Find K_{sp} for calcium fluoride at 25°C, given that a 1.00-L sample of the saturated solution, evaporated to dryness, produced 26.76 mg of solid CaF_2. (7.6)

18. Find the molar solubility of calcium oxalate at 25°C, where K_{sp} is 2.3×10^{-9}. (7.6)

Applying Inquiry Skills

19. Consider this system at equilibrium.

$$PCl_{5(g)} \rightleftharpoons PCl_{3(g)} + Cl_{2(g)} \qquad K = 0.40 \text{ at } 170°C$$

(a) One mole of phosphorus pentachloride was initially placed into a 1.0-L container. Once equilibrium had been reached, it was found that the equilibrium concentration of PCl$_5$ was 0.54 mol/L. **Figure 1** describes the change in $[PCl_5]$ over time.

Copy the graph. Sketch lines to indicate the changing concentrations of $[PCl_3]$ and $[Cl_2]$ over the same time period.

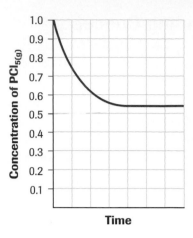

Figure 1

(b) 0.500 mol of PCl$_3$ and 0.300 mol of Cl$_2$ were placed into a 1.00-L container. Once equilibrium had been reached it was found that the equilibrium concentration of PCl$_3$ was 0.360 mol/L. **Figure 2** describes the change in $[PCl_3]$ over time.

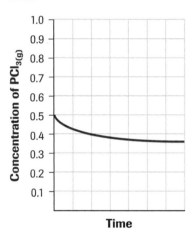

Figure 2

Copy the graph. Sketch how the concentrations of PCl$_5$ and Cl$_2$ change over the same time period. (7.1)

20. A student designed an experiment to test the prediction that the addition of a common ion, chloride, can alter the following equilibrium system.

$$Cu^{2+}_{(aq)} + 4\,Cl^-_{(aq)} \rightleftharpoons CuCl_4^{2-}_{(aq)}$$
(blue)   (green)

Question

What effect does the addition of a common ion (chloride) have on a system at equilibrium?

Experimental Design

Samples (5-mL, 10-mL, and 15-mL) of 1 mol/L hydrochloric acid are added to test tubes containing 10 mL of a copper(II) chloride solution. Similarly, samples (0.5-g, 1.0-g, and 1.5-g) of sodium chloride are added to test tubes containing 10 mL of the $CuCl_2$ solution. All test tubes were stoppered, shaken, and allowed to reach equilibrium.

Evidence

Table 1 Evidence for Copper Chloride

Test tube (10 mL of $CuCl_{2(aq)}$ solution in each)	Volume HCl added (mL)	Total volume after mixing (mL)	Colour of system at equilibrium
1	5	15	blue
2	10	20	blue-green
3	15	25	green
(10 mL of $CuCl_{2(aq)}$ solution in each)	Mass of $NaCl_{(s)}$ added (g)	Total volume after mixing (mL)	Colour of system at equilibrium
4	0.5	10	green
5	1.0	10	green
6	1.5	10	green

Analysis

(a) Use the Evidence to answer the Question.

Evaluation

(b) Critique the Experimental Design. What flaws can you see in the student's plan? Suggest improvements. (7.6)

Making Connections

21. The operation of a halogen lamp depends, in part, on the equilibrium system,

$$W_{(s)} + I_{2(g)} \rightleftharpoons WI_{2(g)}$$

Research to find the role of temperature in the operation of a halogen lamp. For example, how is it possible for a halogen lamp to operate with the filament at 2700°C when the tungsten would normally decompose/oxidize at this high temperature? Is such a high operating temperature desirable? (7.3)

 www.science.nelson.com

22. If the stoichiometric percent yield in the Haber process is 15%, suggest what should be done with the unreacted reagents after ammonia has been separated from the mixture. (7.4)

23. Explain how equilibrium principles are applied to optimize the industrial production of sulfuric acid. (7.5)

GO www.science.nelson.com

24. As a scuba diver descends, the surrounding water pressure increases. This forces more nitrogen in the compressed air the diver is breathing to dissolve into the diver's bloodstream, according to the equilibrium

$$N_{2(g)} \rightleftharpoons N_{2(aq)}$$

(a) Use Le Châtelier's principle to explain why the concentration of dissolved nitrogen increases.

(b) "The bends" or decompression sickness can occur if the diver ascends to the surface too quickly. This condition is caused by nitrogen coming out of solution too quickly, forming bubbles in blood vessels. The symptoms of the bends are dizziness, blindness, paralysis, or severe pain. Use Le Châtelier's principle to explain why "the bends" can occur during a diver's ascent. (7.5)

Extension

25. Temperature can have a significant effect on the value of the equilibrium constant. Consider this equilibrium

$$2\,NO_{2(g)} \rightleftharpoons N_2O_{4(g)} + 57.2\ kJ$$

Two values of K are given in **Table 2**. Match the equilibrium constant with the temperature at which it was determined. (7.5)

Table 2 K for NO_2–N_2O_4 Equilibrium

K	Temperature (°C)
1250	0 or 25?
200	0 or 25?

chapter

Acid–Base Equilibrium

▶ **In this chapter, you will be able to**

- compare strong and weak acids and bases using the concept of equilibrium;

- define equilibrium constant expressions, such as K_w, K_a, and K_b;

- use appropriate vocabulary to communicate ideas, procedures, and results related to acid–base systems;

- solve equilibrium problems involving K_a, K_b, pH, and pOH;

- predict, in qualitative terms, whether a solution of a specific salt will be acidic, basic, or neutral;

- solve problems involving acid–base titration data and the pH at the equivalence point;

- explain how buffering action affects our daily lives;

- describe the characteristics and components of a buffer solution.

Water is the most common liquid on Earth. It is found on the surface in lakes, rivers, and oceans; beneath the surface as ground water; and in the atmosphere as a vapour. The bodies of all living organisms are at least 66% water by mass. Water dissolves a wide range of ionic and polar substances, including table salt and table sugar, allowing for a huge variety of aqueous solutions. The most familiar of these solutions in the laboratory, workplace, and home are those of acids and bases. Vinegar, lemon juice, vitamin C, and battery fluid are common acidic solutions; drain cleaner, milk of magnesia, and household ammonia are common bases.

Living organisms are sensitive to the acidity of aqueous solutions in their internal and external environments. The pH of human blood must be kept at precisely 7.4. A sustained increase or decrease of only 0.2 pH units could mean death. How does the body maintain such a narrow range of pH when we consume so many acidic foods and beverages? We shall explore such concepts later in this chapter.

If the water in rivers and streams becomes even slightly acidic, trout and salmon will not survive. Soil pH is a prime determinant of the type of vegetation an area can support. Below a pH of 6, most plants absorb essential nutrients so poorly that growth is stunted and leaves turn yellow. Of course, there are exceptions — rhododendrons fail to thrive in soils with pH levels above 5.5.

Acids and bases are extremely useful materials and in some cases essential to the proper functioning of natural and synthetic processes. Acids are used to etch glass and digest food. Bases are used as cleaning agents, rocket fuel, and dyes. Both are active ingredients in a host of pharmaceutical drugs. Aspirin, the world's most popular analgesic, is an acid. Morphine, a powerful painkiller, is a base.

♀ REFLECT on your learning

1. Why is the pH of pure water at SATP equal to 7?

2. Solutions of acetic acid and hydrochloric acid of the same concentration are not equally acidic. Which of the two solutions has a lower pH? Why?

3. When equal amounts of the same concentrations of hydrochloric acid and sodium hydroxide are mixed, the resulting solution is neutral (pH = 7). When equal amounts of equal concentrations of acetic acid and sodium hydroxide are mixed, the resulting solution is basic (pH>7). Explain.

4. Most soft drinks are acidic solutions. When you consume a soft drink, acids are absorbed into the bloodstream. However, the pH of blood remains virtually unchanged. Why?

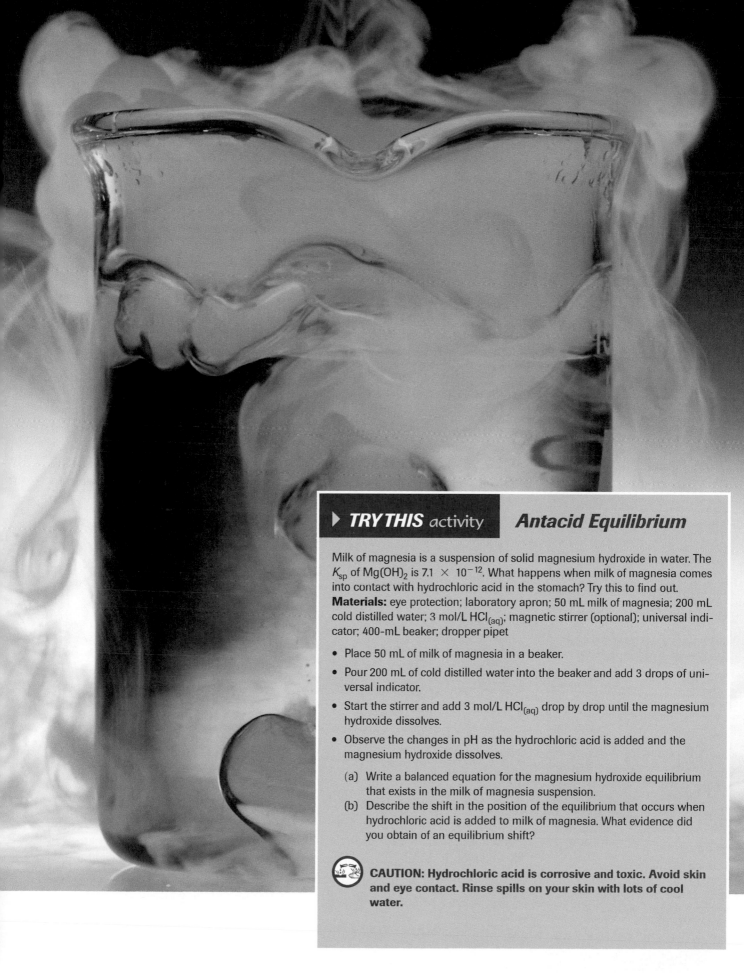

▶ TRY THIS activity　　*Antacid Equilibrium*

Milk of magnesia is a suspension of solid magnesium hydroxide in water. The K_{sp} of $Mg(OH)_2$ is 7.1×10^{-12}. What happens when milk of magnesia comes into contact with hydrochloric acid in the stomach? Try this to find out.

Materials: eye protection; laboratory apron; 50 mL milk of magnesia; 200 mL cold distilled water; 3 mol/L $HCl_{(aq)}$; magnetic stirrer (optional); universal indicator; 400-mL beaker; dropper pipet

- Place 50 mL of milk of magnesia in a beaker.
- Pour 200 mL of cold distilled water into the beaker and add 3 drops of universal indicator.
- Start the stirrer and add 3 mol/L $HCl_{(aq)}$ drop by drop until the magnesium hydroxide dissolves.
- Observe the changes in pH as the hydrochloric acid is added and the magnesium hydroxide dissolves.

(a) Write a balanced equation for the magnesium hydroxide equilibrium that exists in the milk of magnesia suspension.

(b) Describe the shift in the position of the equilibrium that occurs when hydrochloric acid is added to milk of magnesia. What evidence did you obtain of an equilibrium shift?

CAUTION: Hydrochloric acid is corrosive and toxic. Avoid skin and eye contact. Rinse spills on your skin with lots of cool water.

Acids and bases are electrolytes that form aqueous solutions with unique properties. Acidic solutions like vinegar are sour tasting, conduct electricity, and turn blue litmus red. Basic solutions, like aqueous ammonia, also conduct electricity, are generally bitter tasting, feel slippery, and turn red litmus blue.

Svante Arrhenius (**Figure 1**), a Swedish chemist, was the first to characterize acids and bases in terms of their chemical properties. According to Arrhenius, acids are solutes that produce hydrogen ions, $H^+_{(aq)}$, in aqueous solutions, while bases produce hydroxide ions, $OH^-_{(aq)}$, when dissolved in water. This model adequately explains the properties of most acids and ionic hydroxide bases, but fails to satisfactorily account for the basic properties of compounds that do not contain the hydroxide ion, such as ammonia ($NH_{3(aq)}$).

In 1923, Johannes Brønsted of Denmark (**Figure 2(a)**) and Thomas Lowry of England (**Figure 2(b)**) recognized that, in most acid–base interactions, a proton (H^+ ion) is transferred from one reactant to another.

Figure 1
Svante Arrhenius (1859–1927)

Brønsted-Lowry Theory

According to Brønsted and Lowry, when hydrogen chloride reacts with water, a proton is transferred from a hydrogen chloride molecule to a water molecule, forming a hydronium ion and a chloride ion (**Figure 3**). Hydrogen chloride acts as a Brønsted–Lowry acid; water acts as a Brønsted–Lowry base. Notice the single arrow in the equation, indicating that hydrogen chloride is a strong acid, ionizing quantitatively (completely) when it reacts with water.

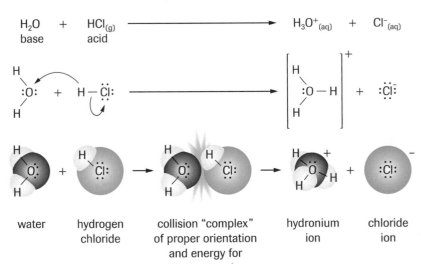

Figure 3

Figure 2
J. Brønsted (1879–1947) and T. Lowry (1874–1936) independently created new theoretical definitions for acids and bases based upon proton transfer.

When ammonia reacts with water, a water molecule acts as a Brønsted–Lowry acid, donating a proton to ammonia, the Brønsted–Lowry base. Notice the double arrow in the equation, indicating that ammonia is a weak base, ionizing incompletely and forming a dynamic equilibrium with the products of the reaction.

$$NH_{3(g)} + H_2O_{(l)} \rightleftharpoons NH^+_{4(aq)} + OH^-_{(aq)}$$
$$\text{base} \qquad \text{acid}$$

According to the Brønsted–Lowry concept,

> a Brønsted-Lowry acid is a proton donor, and
> a Brønsted-Lowry base is a proton acceptor.

A substance can be classified as a Brønsted–Lowry acid or base only for a specific reaction. This point is important—protons may be gained in a reaction with one substance, but lost in a reaction with another substance. (For example, in the reaction of HCl with water, water acts as a base, whereas in the reaction of NH_3 with water, water acts as an acid.) A substance that appears to act as a Brønsted–Lowry acid in some reactions and as a Brønsted–Lowry base in other reactions is called **amphoteric (amphiprotic)**.

In baking soda, the hydrogen carbonate ion, HCO_3^-, is amphoteric, as shown by the following reactions.

$$\underset{\text{base}}{HCO_{3(aq)}^-} + \underset{\text{acid}}{H_2O_{(l)}} \rightleftharpoons H_2CO_{3(aq)} + OH_{(aq)}^-$$

$$\underset{\text{acid}}{HCO_{3(aq)}^-} + \underset{\text{base}}{H_2O_{(l)}} \rightleftharpoons CO_{3(aq)}^{2-} + H_3O_{(aq)}^+$$

The advantage of the Brønsted–Lowry definitions over the Arrhenius definitions is that they enable us to define acids and bases in terms of chemical reactions rather than simply as substances that form acidic and basic aqueous solutions. A definition of acids and bases in terms of chemical reactions allows us to describe, explain, and predict many reactions in aqueous solution, non-aqueous solution, or pure states. For example, according to Arrhenius, an acid–base neutralization produces water and a salt as in the reaction between sodium hydroxide and hydrochloric acid,

$$NaOH_{(aq)} + HCl_{(aq)} \rightarrow H_2O_{(l)} + NaCl_{(aq)}$$

However, when ammonia, a basic substance, reacts with hydrogen chloride, an acidic substance, a neutralization occurs (in the gas state) that does not involve hydronium ions, hydroxide ions, or water,

$$NH_{3(g)} + HCl_{(g)} \rightarrow NH_4Cl_{(s)}$$

In this reaction, an H^+ ion (a proton) transfers from the Cl atom in the HCl molecule to the N atom of the NH_3 molecule.

According to the Brønsted–Lowry concept, acid–base reactions involve the transfer of a proton. These reactions are universally reversible and result in an acid–base equilibrium.

Reversible Acid–Base Reactions

In a proton transfer reaction at equilibrium, both forward and reverse reactions involve Brønsted–Lowry acids and bases. For example, in an acetic acid solution, we can describe the forward reaction as a proton transfer from acetic acid to water molecules and the reverse reaction as a proton transfer from hydronium to acetate ions.

$$\underset{\text{acid}\qquad\text{base}}{HC_2H_3O_{2(aq)} + H_2O_{(l)}} \rightleftharpoons \underset{\text{base}\qquad\text{acid}}{C_2H_3O_{2(aq)}^- + H_3O_{(aq)}^+}$$

amphoteric (amphiprotic)
in the Brønsted–Lowry model, a substance capable of acting as an acid or a base in different chemical reactions; a substance that may donate or accept a proton.

DID YOU KNOW?

Debatable Synonyms
The terms amphoteric and amphiprotic are commonly used as synonyms. However, there may be a difference. Amphoteric means "may act as an acid or a base" and amphiprotic means "may accept or donate protons." Since Brønsted-Lowry acids and bases are defined in terms of an entity's ability to accept or donate protons, an amphiprotic substance will always be amphoteric. However, in a more general model of acids and bases called the Lewis model (Section 8.3), acids and bases are defined according to an entity's ability to accept or donate a pair of electrons. In this case, an amphoteric substance may or may not be amphiprotic.

conjugate acid–base pair two substances whose formulas differ only by one H^+ unit

This equilibrium is typical of all Brønsted-Lowry acid–base reactions. There will always be two acids (in the above example, $HC_2H_3O_{2(aq)}$ and $H_3O_{(aq)}^+$) and two bases (in the above example, $H_2O_{(l)}$ and $C_2H_3O_{2(aq)}^-$) in any acid–base equilibrium. Furthermore, the base on the right ($C_2H_3O_{2(aq)}^-$) is formed by removal of a proton ($H_{(aq)}^+$ ion) from the acid on the left ($HC_2H_3O_{2(aq)}$). The acid on the right ($H_3O_{(aq)}^+$) is formed by the addition of a proton to the base on the left ($H_2O_{(l)}$). A pair of substances whose molecular formulas differ by a single H^+ ion (a proton) is called a **conjugate acid–base pair**. An acetic acid molecule and an acetate ion are a conjugate acid–base pair. Acetic acid is the conjugate acid of the acetate ion and the acetate ion is the conjugate base of acetic acid. The hydronium ion and water are the second conjugate acid–base pair in this equilibrium.

$$\text{conjugate pair}$$
$$HC_2H_3O_{2(aq)} + H_2O_{(l)} \rightleftharpoons C_2H_3O_{2(aq)}^- + H_3O^+{}_{(aq)}$$
$$\text{conjugate pair}$$

A Competition for Protons

The Brønsted–Lowry model of acids and bases allows us to view acid–base reactions as a competition for protons between two bases. For example, in the acetic acid equilibrium, the competition is between the acetate ion, $C_2H_3O_{2(aq)}^-$ and water, $H_2O_{(aq)}$.

$$H_2O_{(l)} \overset{\leftarrow}{\dots} H^+ \overset{\rightarrow}{\dots} C_2H_3O_{2(aq)}$$

In a 0.1 mol/L solution of acetic acid, $HC_2H_3O_{2(aq)}$, at SATP, electrical conductivity measurements show that, at equilibrium, only about 1.3% of the $HC_2H_3O_{2(aq)}$ molecules have reacted with water to produce acetate ions and hydronium ions. It appears that the ability of the $C_2H_3O_2^-$ part of the $HC_2H_3O_2$ molecule to hold on to its proton (H^+ ion) is much greater than the ability of H_2O to pull the proton away. Therefore, the percent ionization is low. This may help explain why the following equilibrium lies far to the left, and why we call acetic acid a *weak* acid.

$$\overset{1.3\%}{HC_2H_3O_{2(aq)} + H_2O_{(l)} \rightleftharpoons H_3O_{(aq)}^+ + C_2H_3O_{2(aq)}^-}$$

However, when HCl molecules react with H_2O, the chlorine atom in HCl has a much weaker affinity for the proton of the hydrogen atom it is bonded to than H_2O does. Thus, the water molecule wins the competition for HCl's proton, causing the H^+ ion to be completely transferred to H_2O, which becomes H_3O^+. Virtually every HCl molecule loses this competition, which helps explain why the following equilibrium lies very far to the right (thus, the one-way arrow), and why we call HCl a *strong* acid.

$$\overset{99\%}{HCl_{(aq)} + H_2O_{(l)} \rightarrow H_3O_{(aq)}^+ + Cl_{(aq)}^-}$$

In general, the terms *strong acid* and *weak acid* can be explained by the Brønsted–Lowry concept. Using $HA_{(aq)}$ as the general symbol for an acid and $A_{(aq)}^-$ as its conjugate base, we represent an acid ionization reaction as follows:

$$HA_{(aq)} + H_2O_{(l)} \rightleftharpoons H_3O_{(aq)}^+ + A_{(aq)}^-$$

The extent of the proton transfer between HA and H_2O determines the strength of $HA_{(aq)}$. In Brønsted–Lowry terms, when a strong acid reacts with water, the transfer of protons is virtually complete in the forward direction and almost no transfer of protons occurs via the reverse reaction. In other words, the reaction of a strong acid with water is essentially 100% complete. The strongest acids have the highest percent reaction (or percent ionization). The forward reactions of these acids are strongly favoured. Weaker acids have lower percent reactions, so their equilibrium position is farther to the left, favouring the reactants rather than the products.

In terms of a competition for protons, a strong acid has a very low attraction for its proton and easily donates it to a base, even a relatively weak base like water. This leads to the interpretation that the conjugate base, A^-, of a strong acid has a very weak attraction for protons: It is a very weak base. A useful generalization regarding the relative strengths of a conjugate acid–base pair is:

> The stronger an acid, the weaker its conjugate base, and conversely, the weaker an acid, the stronger its conjugate base.

It is common to represent the equation for the ionization of an acid in water by abbreviating the ionization equation,

$$HA_{(aq)} + H_2O_{(l)} \rightleftharpoons H_3O^+_{(aq)} + A^-_{(aq)}$$

to

$$HA_{(aq)} \rightleftharpoons H^+_{(aq)} + A^-_{(aq)}$$

For example,

$$HC_2H_3O_{2(aq)} + H_2O_{(l)} \rightleftharpoons H_3O^+_{(aq)} + C_2H_3O_2^-{}_{(aq)}$$

reduces to

$$HC_2H_3O_{2(aq)} \rightleftharpoons H^+_{(aq)} + C_2H_3O_2^-{}_{(aq)}$$

The abbreviated equation (bottom) is produced by removing the water molecule that is common to both sides of the complete equation (top). We will use the abbreviated form throughout this chapter. However, always keep in mind that, while it communicates the essential change that takes place in the reaction (the ionization of HA), it does not communicate the important role played by water in causing the acid to ionize, or the fact that the proton ($H^+_{(aq)}$) most probably exists in solution as a hydronium ion, $H_3O^+_{(aq)}$.

SUMMARY *Brønsted–Lowry Definitions*

- An acid is a proton donor.
- A base is a proton acceptor.
- An amphoteric substance is one that appears to act as a Brønsted–Lowry acid (a proton donor) in some reactions and as a Brønsted–Lowry base (a proton acceptor) in other reactions.
- A conjugate acid–base pair consists of two substances that differ only by a proton—the acid has one more proton than its conjugate base.

- A strong acid has a very weak attraction for protons. A strong base has a very strong attraction for protons.
- The stronger an acid, the weaker its conjugate base, and conversely, the weaker an acid, the stronger its conjugate base.

> ### ▶ Practice

Understanding Concepts

1. Use the Brønsted-Lowry definitions to identify the two conjugate acid–base pairs in each of the following acid–base reactions.

   (a) $HCO_{3(aq)}^- + S_{(aq)}^{2-} \rightleftharpoons HS_{(aq)}^- + CO_{3(aq)}^{2-}$

   (b) $H_2CO_{3(aq)} + OH_{(aq)}^- \rightleftharpoons HCO_{3(aq)}^- + H_2O_{(l)}$

   (c) $HSO_{4(aq)}^- + HPO_{4(aq)}^{2-} \rightleftharpoons H_2PO_{4(aq)}^- + SO_{4(aq)}^{2-}$

   (d) $H_2O_{(l)} + H_2O_{(l)} \rightleftharpoons H_3O_{(aq)}^+ + OH_{(aq)}^-$

2. Identify all the amphoteric entities in question 1.

3. Some ions can form more than one conjugate acid–base pair. List the two conjugate acid–base pairs involving a hydrogen carbonate ion in the reactions in question 1.

The Autoionization of Water

Water is never just a collection of H_2O molecules. Even a sample of contaminant-free water has a very slight conductivity that is observable if measured with very sensitive instruments. According to Arrhenius's theory, conductivity is due to the presence of ions. So, there must be ions in "pure" water. What is the nature of these ions, and what is their source? How many are there, and do their numbers change? Experiments have revealed that some water molecules react with each other to produce hydronium, $H_3O_{(aq)}^+$, and hydroxide, $OH_{(aq)}^-$, ions according to the following equation.

$$H_2O_{(l)} + H_2O_{(l)} \rightleftharpoons H_3O_{(aq)}^+ + OH_{(aq)}^-$$

Because the conductivity is so slight, there must be considerably more water molecules than ions in the equilibrium mixture at SATP. In every sample of water, an equilibrium is formed between hydronium ions, hydroxide ions, and water molecules that greatly favours the water molecules.

Of the billions of random collisions occurring among water molecules, a few are at the right energy and orientation to cause a reaction. This results in the transfer of a proton (H^+ ion) from one molecule of water to the other, producing a hydronium ion, $H_3O_{(aq)}^+$, and a hydroxide ion, $OH_{(aq)}^-$ (**Figure 4**).

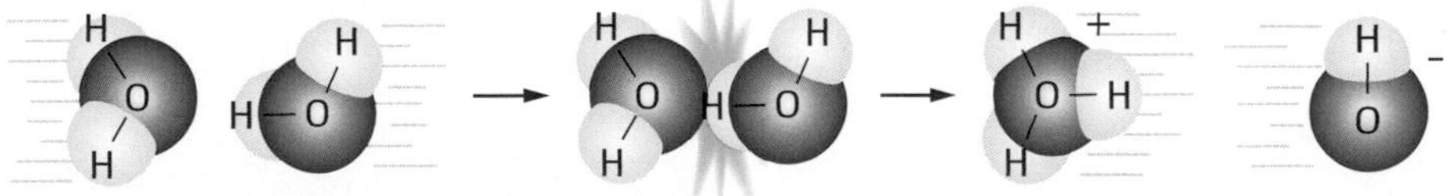

Figure 4
The collision that forms hydronium and hydroxide ions is very rare.

autoionization of water the reaction between two water molecules producing a hydronium ion and a hydroxide ion

This process is called the **autoionization of water**, since water molecules ionize one another. The H_3O^+ ion produced may be viewed as a water molecule, H_2O, with an H^+ ion (a proton) attached by a coordinate covalent bond to the oxygen atom. Chemists often omit the water molecule that carries the H^+ ion for convenience. In this way, the equilibrium may be written

$$H_2O_{(l)} \rightleftharpoons H_{(aq)}^+ + OH_{(aq)}^-$$

It is important to remember, however, that water molecules don't simply (or spontaneously) dissociate into H^+ and OH^- ions, but rather, that the production of ions occurs as the result of an ionization process in which a proton is transferred from one molecule to another. Evidence indicates that fewer than two water molecules in one billion ionize at SATP.

The water equilibrium, like all chemical equilibria, obeys the law of mass action (equilibrium law). Therefore, we can construct an equilibrium law equation:

$$\frac{[H^+_{(aq)}][OH^-_{(aq)}]}{[H_2O_{(l)}]} = K$$

The concentration of water molecules in pure water and in dilute aqueous solutions is essentially constant and equal to 55.6 mol/L. This value is derived from the density and molar mass of water:

$$[H_2O_{(l)}] = \frac{1.00 \times 10^3 \text{ g/L}}{18 \text{ g/mol}}$$

$$= 55.6 \text{ mol/L}$$

Therefore, a new constant, which incorporates both the constant value of $[H_2O_{(l)}]$ and the equilibrium constant, can be calculated. This new constant is called the **ion product constant for water, K_w**.

$$\frac{[H^+_{(aq)}][OH^-_{(aq)}]}{[H_2O_{(l)}]} = K$$

$$[H^+_{(aq)}][OH^-_{(aq)}] = \underbrace{K[H_2O_{(l)}]}_{\text{constant}}$$

$$[H^+_{(aq)}][OH^-_{(aq)}] = K_w$$

ion product constant for water, K_w equilibrium constant for the ionization of water; 1.0×10^{-14}

The equilibrium equation for the autoionization of water shows that hydrogen ions and hydroxide ions are formed in a 1:1 ratio. Therefore, the concentrations of hydrogen ions and hydroxide ions in pure water must be equal. Precise measurements of pure water at 25°C show that the concentrations of $H^+_{(aq)}$ and $OH^-_{(aq)}$ are, in fact, the same: 1.0×10^{-7} mol/L.

$$[H^+_{(aq)}] = [OH^-_{(aq)}] = 1.0 \times 10^{-7} \text{ mol/L}$$

Therefore, at SATP

$$K_w = [H^+_{(aq)}][OH^-_{(aq)}]$$
$$K_w = (1.0 \times 10^{-7} \text{ mol/L})(1.0 \times 10^{-7} \text{ mol/L})$$
$$K_w = 1.0 \times 10^{-14}$$

> **LEARNING TIP**
>
> It is customary to write the value of K_w without units. However, note that it could be written as 1.0×10^{-14} (mol/L)^2, allowing you to write concentrations calculated using K_w as mol/L.

As usual, we do not include units with the value of K_w. This will be the case with all equilibrium constants encountered in this chapter.

The autoionization of water takes place in *all* aqueous solutions. However, since acids and bases may be dissolved in water, the concentrations of $H^+_{(aq)}$ and $OH^-_{(aq)}$ ions may not be equal in all solutions. Dissolving acids increases $[H^+_{(aq)}]$ and dissolving bases increases $[OH^-_{(aq)}]$. Nevertheless, in all aqueous solutions at SATP, the product of the concentrations of $H^+_{(aq)}$ and $OH^-_{(aq)}$ is constant and equal to 1.0×10^{-14}, K_w.

Table 1 Values of K_w at Selected Temperatures

Temperature (°C)	K_w
0	1.5×10^{-15}
10	3.0×10^{-15}
20	6.8×10^{-15}
25	1.0×10^{-14}
30	1.5×10^{-14}
40	3.0×10^{-14}
50	5.5×10^{-14}
60	9.5×10^{-14}

In all aqueous solutions at SATP,
$$[H^+_{(aq)}][OH^-_{(aq)}] = K_w = 1.0 \times 10^{-14}$$

According to the Arrhenius theory, an acid is a substance that ionizes in water to produce hydrogen ions. The hydrogen ions provided by the acid increase the hydrogen ion concentration in the solution; the $[H^+_{(aq)}]$ will be greater than 10^{-7} mol/L at SATP, so the solution is *acidic*. A *basic* (or *alkaline*) solution is one in which the hydroxide ion concentration is greater than 10^{-7} mol/L at SATP, and a *neutral* solution is one where the hydrogen ion and hydroxide ion concentrations are the same and each equal to 10^{-7} mol/L at SATP.

A basic solution is produced, for example, by the dissociation, in water, of an ionic hydroxide such as sodium hydroxide.

In neutral solutions	$[H^+_{(aq)}] = [OH^-_{(aq)}]$
In acidic solutions	$[H^+_{(aq)}] > [OH^-_{(aq)}]$
In basic solutions	$[H^+_{(aq)}] < [OH^-_{(aq)}]$

strong acid an acid that is assumed to ionize quantitatively (completely) in aqueous solution (percent ionization is > .99%)

Another important point is that the numerical value given above for K_w is valid at SATP, but *not* at temperatures that are much higher or lower (**Table 1**). Recall that the value of any equilibrium constant depends on temperature. For higher temperatures, K_w has a greater value, so products are more favoured. This means that more water molecules become ionized in aqueous systems when the molecular collisions are more energetic and more frequent.

We can use the ion product constant for water, K_w, to calculate either the hydrogen ion concentration or the hydroxide ion concentration in an aqueous solution of a strong or weak acid or base at SATP, if the other concentration is known.

Since	$K_w = [H^+_{(aq)}][OH^-_{(aq)}]$
then	$[H^+_{(aq)}] = \dfrac{K_w}{[OH^-_{(aq)}]}$
and	$[OH^-_{(aq)}] = \dfrac{K_w}{[H^+_{(aq)}]}$

Strong Acids

A **strong acid** is an acid that ionizes quantitatively (completely) in water to form hydrogen ions. The percent ionization of strong acids is greater than 99%. However, we will assume that it is 100% in calculations. For example, we will assume that every molecule of $HCl_{(g)}$ that dissolves in water ionizes into $H^+_{(aq)}$ and $Cl^-_{(aq)}$.

$$HCl_{(g)} \xrightarrow[H_2O_{(l)}]{100\%} H^+_{(aq)} + Cl^-_{(aq)}$$

This means that, although the label on a bottle of hydrochloric acid may say 1.0 mol/L $HCl_{(aq)}$, we assume that the solution contains virtually no HCl molecules (**Figure 5**). Instead, we assume that it contains 1.0 mol/L $H^+_{(aq)}$ and 1.0 mol/L $Cl^-_{(aq)}$ ions only. Container labels usually indicate the concentration of the substance(s) used to make the solution and do not usually describe the concentration(s) of the substances that form from the ionization (or dissociation) of the starting material.

○ H^+　● A^-

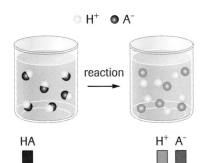

Figure 5
Strong acids ionize greater than 99% in aqueous solution.

There are relatively few strong acids: hydrochloric acid, $HCl_{(aq)}$, hydrobromic acid, $HBr_{(aq)}$, sulfuric acid, $H_2SO_{4(aq)}$, nitric acid, $HNO_{3(aq)}$, and phosphoric acid, $H_3PO_{4(aq)}$, are the most familiar. **Monoprotic acids** like $HCl_{(aq)}$ contain only one ionizable "acidic" hydrogen atom. Diprotic acids like $H_2SO_{4(aq)}$ have two ionizable hydrogen atoms and triprotic acids like $H_3PO_{4(aq)}$ have three. As with almost all acids, the ionizable hydrogen atoms of oxyacids such as sulfuric acid, nitric acid, and phosphoric acid are written first in the molecular formulas, but this does not necessarily give a clue to the structure of the molecule. In these molecules, the hydrogens are not attached to the sulfur, nitrogen, or phosphorous atoms but to oxygen atoms (**Figure 6**).

monoprotic acid an acid that possesses only one ionizable (acidic) proton

sulfuric acid,
$H_2SO_{4(aq)}$

nitric acid,
$HNO_{3(aq)}$

phosphoric acid,
$H_3PO_{4(aq)}$

Figure 6
The ionizable "acidic" hydrogen atoms of common oxyacids are covalently bonded to oxygen atoms.

We can now use two concepts—the assumption that strong acids ionize quantitatively in solution, and the value of K_w—to calculate the hydrogen ion or hydroxide ion concentrations of strong acid solutions.

Calculating [$OH^-_{(aq)}$] or [$H^+_{(aq)}$] of a Strong Acid   SAMPLE problem

A 0.15 mol/L solution of hydrochloric acid at SATP is found to have a hydrogen ion concentration of 0.15 mol/L. Calculate the concentration of the hydroxide ions.

When analyzing problems dealing with equilibrium solutions of acids and bases, we must account for all of the entities that may contribute to the solution's hydrogen ion and hydroxide ion concentrations, since these give rise to the acid/base properties of the solution. After identifying all of the entities that may affect the acid–base balance, we must determine the *major* entities (solution components that are present in relatively large amounts). In most cases, a major entity determines the acid–base properties of the solution, and the minor entities may be ignored.

Since hydrochloric acid is a strong acid, we assume that it undergoes 100% ionization in aqueous solution.

$$HCl_{(g)} \xrightarrow[\text{H}_2\text{O}_{(l)}]{100\%} H^+_{(aq)} + Cl^-_{(aq)}$$

Therefore, a 0.15 mol/L $HCl_{(aq)}$ solution will have 0.15 mol/L $H^+_{(aq)}$, and 0.15 mol/L $Cl^-_{(aq)}$. Water is a very weak electrolyte. It ionizes very little, according to the following equilibrium.

$$H_2O_{(l)} \rightleftharpoons H^+_{(aq)} + OH^-_{(aq)}$$

In pure water,

$[H^+_{(aq)}] = 1.0 \times 10^{-7}$ mol/L and
$[OH^-_{(aq)}] = 1.0 \times 10^{-7}$ mol/L

However, in an $HCl_{(aq)}$ solution, the [$H^+_{(aq)}$] and [$OH^-_{(aq)}$] contributed by the autoionization of water will be less than 1.0×10^{-7} mol/L because the $H^+_{(aq)}$ ions produced by the ionization of HCl cause the water equilibrium to shift to the left (by Le Châtelier's principle), reducing the contribution of $H^+_{(aq)}$ ions (and $OH^-_{(aq)}$) from the autoionization of water to less

LEARNING TIP

Unless stated otherwise, conditions in problems in this chapter are SATP.

than 1.0×10^{-7} mol/L. Compared to the 0.15 mol/L $H^+_{(aq)}$ contributed by HCl, the tiny contribution made by the autoionization of water may be safely ignored. In a similar way, the miniscule contribution of $OH^-_{(aq)}$ made by the autoionization of water may also be ignored.

Therefore, the major entities in solution are

$$H^+_{(aq)} \quad \text{and} \quad Cl^-_{(aq)}$$
$$\text{(from HCl)} \qquad \text{(from HCl)}$$

As indicated earlier, the concentration of chloride ions, $Cl^-_{(aq)}$, in this solution is 0.15 mol/L. This is a significant concentration. However, we may assume that this ion does not contribute to $[H^+_{(aq)}]$ or $[OH^-_{(aq)}]$ in the solution because it is the conjugate base of a strong acid. (Remember that the conjugate base of a strong acid is sufficiently weak for us to ignore its presence. We may assume this for the conjugate bases of all strong acids.) Therefore, in this problem, the major entity affecting the acid–base characteristics of the solution is the $H^+_{(aq)}$ ion produced by the ionization of HCl.

Since $HCl_{(aq)}$ ionizes quantitatively,

$$[H^+_{(aq)}] = 0.15 \text{ mol/L}$$

Now, use the K_w expression to calculate the concentration of hydroxide ions. (Remember that $K_w = 1.0 \times 10^{-14}$ at SATP.)

$$K_w = [H^+_{(aq)}][OH^-_{(aq)}]$$
$$[OH^-_{(aq)}] = \frac{K_w}{[H^+_{(aq)}]}$$
$$= \frac{1.0 \times 10^{-14}}{0.15}$$

$$[OH^-_{(aq)}] = 6.7 \times 10^{-14} \text{ mol/L}$$

Always remember that, while we omit units in the K_w expression, we must *always* write units with concentration values in the solution.

Example
Calculate the hydroxide ion concentration in a 0.25 mol/L $HBr_{(aq)}$ solution.

Solution
Hydrobromic acid is a strong acid.

$$HBr_{(aq)} \xrightarrow{100\%} H^+_{(aq)} + Br^-_{(aq)}$$

The major entities in solution are $H^+_{(aq)}$, $Br^-_{(aq)}$, and $H_2O_{(l)}$.

We can ignore the insignificant amount of $OH^-_{(aq)}$ produced by the autoionization of water and the presence of $Br^-_{(aq)}$ (a very weak base).

$$[H^+_{(aq)}] = 0.25 \text{ mol/L}$$
$$K_w = [H^+_{(aq)}][OH^-_{(aq)}]$$

$$[OH^-_{(aq)}] = \frac{K_w}{[H^+_{(aq)}]}$$

$$= \frac{1.00 \times 10^{-14}}{0.25}$$

$$[OH^-_{(aq)}] = 4.0 \times 10^{-14} \text{ mol/L}$$

Understanding Concepts

4. In a 0.30 mol/L $HNO_{3(aq)}$ solution,
   (a) what is the concentration of nitric acid molecules?
   (b) what is the hydroxide ion concentration?

5. Calculate the hydroxide ion concentration in a solution prepared by dissolving 0.37 g of hydrogen chloride in water to form 250 mL of solution.

6. The hydrogen ion concentration in an industrial effluent is 4.40 mmol/L (4.40×10^{-3} mol/L). Calculate the concentration of hydroxide ions in the effluent.

Applying Inquiry Skills

7. In a particular solution, chromate ions are in equilibrium with dichromate ions (**Figure 7**).

$$2\,CrO_{4(aq)}^{2-} + 2\,H_{(aq)}^{+} \rightleftharpoons Cr_2O_{7(aq)}^{2-} + H_2O_{(l)}$$

   The equilibrium concentration of $CrO_{4(aq)}^{2-}$ depends on the acidity of the solution. Complete the Prediction and Experimental Design (including diagnostic tests) of the investigation report.

   Question

   How does a change in the hydrogen ion concentration affect the chromate-dichromate equilibrium?

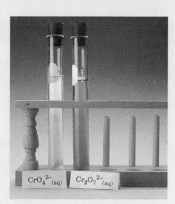

Figure 7
In aqueous solution, chromate ions, $CrO_{4(aq)}^{2-}$, produce a yellow colour; dichromate ions, $Cr_2O_{7(aq)}^{2-}$, produce an orange colour.

Answers

4. (a) 0.30 mol/L
   (b) 3.3×10^{-14} mol/L

5. 2.5×10^{-13} mol/L

6. 2.3×10^{-12} mol/L

Strong Bases

According to Arrhenius, a *base* is a substance that dissociates to increase the hydroxide ion concentration of a solution. Ionic hydroxides have varying solubility in water, but all are **strong bases** that dissociate quantitatively (completely) when they dissolve in water.

All of the hydroxides of Group 1 elements ($LiOH_{(s)}$, $NaOH_{(s)}$, $KOH_{(s)}$, $RbOH_{(s)}$, and $CsOH_{(s)}$) are strong bases. When these bases dissolve in water, one mole of hydroxide ion is produced for every mole of metal hydroxide that dissolves. For example,

$$NaOH_{(s)} \xrightarrow[H_2O_{(l)}]{100\%} Na_{(aq)}^{+} + OH_{(aq)}^{-}$$

These metal hydroxides are all highly soluble in water.

Group 2 elements form the strong hydroxides $Mg(OH)_{2(s)}$, $Ca(OH)_{2(s)}$, $Ba(OH)_{2(s)}$, and $Sr(OH)_{2(s)}$. When these bases dissolve in water, two moles of hydroxide ion are formed for every mole of metal hydroxide that dissolves in solution.

$$Ba(OH)_{2(s)} \xrightarrow[H_2O_{(l)}]{100\%} Ba_{(aq)}^{2+} + 2\,OH_{(aq)}^{-}$$

You may recall from Chapter 7 that these hydroxides are only slightly soluble in water. Their low solubility makes them useful in medical applications. Many antacids are suspensions of metal hydroxides in water, such as magnesium hydroxide in milk of magnesia.

strong base an ionic substance that (according to the Arrhenius definition) dissociates completely in water to release hydroxide ions

Their low solubility prevents large hydroxide ion concentrations that could damage the tissues of the mouth and esophagus as the suspension is being ingested. Once in the stomach, the hydroxide ions react with the hydrogen ions in stomach acid, shifting the equilibrium to the right, and causing the undissolved salts to dissolve and produce higher (more effective) $OH^-_{(aq)}$ ion concentrations.

Just as we did with acids, we can now use two concepts — the assumption that strong bases dissociate quantitatively in solution and the value of K_w — to calculate the hydrogen ion or hydroxide ion concentrations of solutions of strong bases.

▶ **SAMPLE** problem

Calculating $[H^+_{(aq)}]$ and $[OH^-_{(aq)}]$ of a Strong Base

Figure 8
Barium hydroxide has many uses, including in the refining of sugar.

1. **Calculate the hydrogen ion concentration in a 0.025 mol/L solution of barium hydroxide, a strong base (Figure 8).**

We analyze problems involving strong bases in the same way we analyzed problems associated with strong acids.

First, we must identify the major entities in solution.

Begin by writing the dissociation equation for barium hydroxide in water:

$$Ba(OH)_{2(aq)} \xrightarrow{100\%} Ba^{2+}_{(aq)} + \underbrace{2\,OH^-_{(aq)}}_{(from\ Ba(OH)_{2(aq)})}$$

The major entities are $Ba^{2+}_{(aq)}$, $OH^-_{(aq)}$, and $H_2O_{(l)}$.

Since $Ba(OH)_{2(aq)}$ is a strong base, we assume that it dissociates quantitatively (completely) into ions. From the balanced equation, we note that every mole of $Ba(OH)_2$ that dissociates produces one mole of $Ba^{2+}_{(aq)}$ and *two* moles of $OH^-_{(aq)}$.

Therefore,

$$[OH^-_{(aq)}] = 2(0.025\ mol/L)$$
$$[OH^-_{(aq)}] = 0.050\ mol/L$$

The autoionization of water also produces $OH^-_{(aq)}$ ions. However, since the concentration of $OH^-_{(aq)}$ contributed by this process (1.0×10^{-7} mol/L) is insignificant when compared to the 0.050 mol/L produced by the dissociation of $Ba(OH)_2$, it may be ignored. Also, since $Ba(OH)_2$ is a strong base, you may assume that the $Ba^{2+}_{(aq)}$ ions do not affect the acid–base properties of the solution. The $Ba^{2+}_{(aq)}$ ions have no affinity for $H^+_{(aq)}$ ions, nor can they produce $H^+_{(aq)}$ ions; they do not attract $OH^-_{(aq)}$ ions, nor can they produce $OH^-_{(aq)}$ ions. In general, you may ignore the presence of the cations of all ionic hydroxides when determining the acid–base properties of their aqueous solutions. This includes all Group 1 and 2 cations. However, in Section 8.3 you will learn that this assumption cannot be made for *all* dissolved metal cations.

Now use the K_w expression to calculate $[H^+_{(aq)}]$:

$$K_w = [H^+_{(aq)}][OH^-_{(aq)}]$$

$$[H^+_{(aq)}] = \frac{K_w}{[OH^-_{(aq)}]}$$

$$[H^+_{(aq)}] = \frac{1.00 \times 10^{-14}}{0.050}$$

$$[H^+_{(aq)}] = 2.00 \times 10^{-13}\ mol/L$$

The concentration of hydrogen ions in the barium hydroxide solutions is 2.00×10^{-14} mol/L.

2. *Determine the hydrogen ion and hydroxide ion concentrations in 500 mL of an aqueous solution containing 2.6 g of dissolved sodium hydroxide.*

As usual, we begin by writing a balanced equation for the dissolution of the strong base and use it to identify the major entities in solution.

$$NaOH_{(aq)} \rightarrow Na^+_{(aq)} + OH^-_{(aq)}$$

The major entities in solution are $Na^+_{(aq)}$, $OH^-_{(aq)}$, and $H_2O_{(l)}$.

In this case, we are not given $[NaOH_{(aq)}]$. However, we are given the mass of $NaOH_{(s)}$ and the volume of the solution. We can use these quantities to calculate the concentration of the base immediately before dissociation.

$$NaOH_{(aq)} \rightarrow Na^+_{(aq)} + OH^-_{(aq)}$$

$[NaOH_{(aq)}] = 2.6 \text{ g}/500 \text{ mL}$

$m_{NaOH} = 2.6 \text{ g}$

$M_{NaOH} = 40.00 \text{ g/mol}$

$n_{NaOH} = ?$

First, convert the $[NaOH_{(aq)}]$ from units of g/mL to units of mol/L.

$$n_{NaOH} = 2.6 \text{ g} \times \frac{1 \text{ mol}}{40.00 \text{ g}}$$

$$n_{NaOH} = 0.065 \text{ mol}$$

$$[NaOH] = \frac{0.065 \text{ mol}}{0.500 \text{ L}}$$

$$[NaOH] = 0.13 \text{ mol/L}$$

From the balanced equation, we note that every mole of NaOH dissociates into one mole of $Na^+_{(aq)}$ and one mole of $OH^-_{(aq)}$.

Therefore, after dissolution,

$$[OH^-_{(aq)}] = 0.13 \text{ mol/L}$$

We note that the concentration of $OH^-_{(aq)}$ produced by the autoionization of water is insignificant when compared to the 0.13 mol/L produced by the dissociation of NaOH. We also assume that since NaOH is a strong base, $Na^+_{(aq)}$ does not affect the acid–base properties of the solution.

Now we can use the K_w expression to calculate the $[H^+_{(aq)}]$, assuming that $[OH^-_{(aq)}] = 0.13$ mol/L.

$$K_w = [H^+_{(aq)}][OH^-_{(aq)}]$$

$$[H^+_{(aq)}] = \frac{K_w}{[OH^-_{(aq)}]}$$

$$= \frac{1.0 \times 10^{-14}}{0.13}$$

$$[H^+_{(aq)}] = 7.7 \times 10^{-14} \text{ mol/L}$$

Therefore, the $[OH^-_{(aq)}]$ is 0.13 mol/L, and the $[H^+_{(aq)}]$ is 7.7×10^{-14} mol/L.

Example

A cleaning solution contains 5.00 g of $KOH_{(aq)}$ in 2.00 L of solution. Calculate the $[H^+_{(aq)}]$ of the cleaning solution.

Solution

$$KOH_{(aq)} \xrightarrow{100\%} K^+_{(aq)} + OH^-_{(aq)}$$

The major entities are $K^+_{(aq)}$, $OH^-_{(aq)}$ and $H_2O_{(l)}$.

$$n_{KOH} = 5.00\ \cancel{g} \times \frac{1.00\ mol}{56.11\ \cancel{g}}$$

$$n_{KOH} = 0.089\ mol$$

$$[KOH_{(aq)}] = \frac{0.089\ mol}{2.00\ L}$$

$$[KOH_{(aq)}] = 4.46 \times 10^{-2}\ mol/L$$

Since one mole of KOH dissociates into one mole of $K^+_{(aq)}$ and one mole of $OH^-_{(aq)}$,

$$[OH^-_{(aq)}] = 4.46 \times 10^{-2}\ mol/L$$

Ignoring the $[OH^-_{(aq)}]$ produced by the autoionization of water, and $K^+_{(aq)}$,

$$K_w = [H^+_{(aq)}][OH^-_{(aq)}]$$

$$[H^+_{(aq)}] = \frac{K_w}{[OH^-_{(aq)}]}$$

$$= \frac{1.00 \times 10^{-14}}{4.46 \times 10^{-2}}$$

$$[H^+_{(aq)}] = 2.24 \times 10^{-13}\ mol/L$$

The concentration of hydrogen ions in the solution is 2.24×10^{-13} mol/L.

> ## ▶ Practice

Understanding Concepts

Answers

8. 7.2×10^{-13} mol/L

9. 3.34×10^{-11} mol/L

10. 1.40×10^{-14} mol/L

11. $10^{-5}\%$

8. Calculate the hydrogen ion concentration in a saturated solution of calcium hydroxide (limewater). Calcium hydroxide has a solubility of 6.9 mmol per litre of solution.

9. The hydroxide ion concentration in a household cleaning solution is 0.299 mmol/L. Calculate the hydrogen ion concentration in the cleaning solution.

10. What is the hydrogen ion concentration in a solution made by dissolving 20.0 g of potassium hydroxide in water to form 500 mL of solution?

11. Calculate the percent ionization of water at SATP. Recall that 1.000 L of water has a mass of 1000 g.

Hydrogen Ion Concentration and pH

A concentrated acid solution may have a hydrogen ion concentration exceeding 10 mol/L. A concentrated base solution may have a hydrogen ion concentration of 10^{-15} mol/L, or less. Similarly, the hydroxide ion concentration can vary widely. Because of the tremendous range of hydrogen ion and hydroxide ion concentrations, scientists rely on a simple system for communicating concentrations. This system, called the pH scale, was developed in 1909 by Danish chemist Sören Sörenson. Expressed as a numerical value without units, the **pH** of a solution is the negative of the logarithm to the base ten of the hydrogen ion concentration.

$$pH = -\log[H^+_{(aq)}]$$

pH values can be calculated from the hydrogen ion concentration. As shown in the following example, the digits preceding the decimal point in a pH value are determined by the digits in the exponent of the given hydrogen ion concentration. These digits serve to locate the position of the decimal point in the concentration value and have no connection with the certainty of the value. However, *the number of digits following the decimal point in the pH value is equal to the number of significant digits in the hydrogen ion concentration.* For example, a hydrogen ion concentration of 2.7×10^{-3} mol/L corresponds to a pH of 2.57. (Two significant digits in the value for $[H^+_{(aq)}]$ means we should give the pH value to two decimal places.)

Example

Calculate the pH of a solution with a hydrogen ion concentration of 4.7×10^{-11} mol/L.

Solution

$$pH = -\log[H^+_{(aq)}]$$
$$= -\log(4.7 \times 10^{-11}) \qquad \text{(two significant digits)}$$
$$= 10.33 \quad \text{(two digits following the decimal point)}$$

The solution has a pH of 10.33.

In pure water, and in any neutral solution at SATP, the concentrations of hydrogen and hydroxide ions are equal, and therefore the pH is 7.00:

$$[H^+_{(aq)}][OH^-_{(aq)}] = 1.0 \times 10^{-14} \text{ mol/L}$$
$$[H^+_{(aq)}] = [OH^-_{(aq)}]$$
$$[H^+_{(aq)}]^2 = 1.0 \times 10^{-14}$$
$$[H^+_{(aq)}] = \sqrt{1.0 \times 10^{-14} \text{ mol/L}}$$
$$[H^+_{(aq)}] = 1.0 \times 10^{-7} \text{ mol/L}$$
$$pH = -\log(1.0 \times 10^{-7})$$
$$pH = 7.00$$

At SATP, an acidic solution is one in which the $[H^+_{(aq)}]$ is greater than 10^{-7} mol/L, a basic solution is one where $[H^+_{(aq)}]$ is less than 10^{-7} mol/L, and a neutral solution is one where $[H^+_{(aq)}]$ is equal to 10^{-7} mol/L. At SATP:

neutral solution	pH $= 7.00$
acidic solution	pH < 7.00
basic solution	pH > 7.00

Note that the hydrogen ion concentration changes by a multiple of 10 for every increase or decrease of one pH unit. For example, at pH 4.0, $[H^+_{(aq)}]$ is 1×10^{-4} mol/L; at pH 3.0, $[H^+_{(aq)}]$ is 1×10^{-3} mol/L. At pH 3, the $H^+_{(aq)}$ concentration is ten times higher.

If pH is measured in an acid–base experiment, a conversion from pH to the molar concentration of hydrogen ions may be necessary. This conversion is based on the mathematical concept that a base ten logarithm represents an exponent.

$$[H^+_{(aq)}] = 10^{-pH}$$

The method of calculating the hydrogen ion concentration from the pH value is shown in the following example.

LEARNING TIP

On many calculators, $-\log(4.7 \times 10^{-11})$ may be entered by pushing the following sequence of keys.

Note that, on some calculators, the EXP button may be labelled EE. Nevertheless, the sequence of keys remains the same. If your calculator lacks either of these keys, consult the user's manual for instructions.

LEARNING TIP

Notice that the negative sign in the definition of pH establishes an inverse relationship between the magnitude of the hydrogen ion concentration and the magnitude of the pH value.

small pH value $=$ large value of $[H^+_{(aq)}]$

large pH value $=$ small value of $[H^+_{(aq)}]$

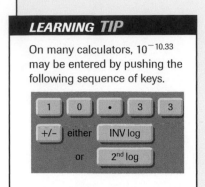
Example
Convert a pH of 10.33 to a hydrogen ion concentration.

Solution

$$[H^+_{(aq)}] = 10^{-pH}$$
$$= 10^{-10.33} \text{ mol/L} \qquad \text{(two digits following the decimal point)}$$
$$[H^+_{(aq)}] = 4.7 \times 10^{-11} \text{ mol/L} \qquad \text{(two significant digits)}$$

pOH and pK_w

The concentration of hydroxide ions is very small in dilute basic solutions. Therefore, it is convenient to describe hydroxide ion concentrations in a similar way as is done for $H^+_{(aq)}$ concentrations, by calculating **pOH**.

$$pOH = -\log[OH^-_{(aq)}]$$

A solution's pOH may be used to calculate the hydroxide ion concentration:

$$[OH^-_{(aq)}] = 10^{-pOH}$$

The following example shows how the pOH of a solution is calculated from the $[OH^-_{(aq)}]$.

Example
Calculate the pOH of a solution with a hydroxide ion concentration of 3.0×10^{-6} mol/L.

Solution

$$pOH = -\log[OH^-_{(aq)}]$$
$$= -\log(3.0 \times 10^{-6})$$
$$pOH = 5.52$$

The pOH of the solution is 5.52.

The mathematics of logarithms allows us to derive a simple relationship between pH and pOH. We derive this relationship below. However, before we do, we need to define a quantity, **pK_w**.

$$pK_w = -\log K_w$$

The numerical value of pK_w follows from the value of K_w, which is always 1×10^{-14} at SATP.

$$pK_w = -\log(1 \times 10^{-14})$$
$$= -(-14.00)$$
$$pK_w = 14.00 \qquad \text{(at SATP)}$$

According to the rules of logarithms,

$$\log(ab) = \log(a) + \log(b)$$

Using the equilibrium law expression for the autoionization of water,

$$[H^+_{(aq)}][OH^-_{(aq)}] = K_w$$
$$\log([H^+_{(aq)}][OH^-_{(aq)}]) = \log(K_w)$$
$$\log[H^+_{(aq)}] + \log[OH^-_{(aq)}] = \log(K_w)$$
$$(-\log[H^+_{(aq)}]) + (-\log[OH^-_{(aq)}]) = -\log(K_w)$$

or $\qquad$ pH + pOH = pK_w
and therefore $\qquad$ pH + pOH = 14.00 $\qquad$ (at SATP)

This relationship enables a quick conversion between pH and pOH.

Figure 9
Purple cabbage boiled in water produces an extract that changes colour in different solutions. The test tubes show the colour of the cabbage juice in solutions of (from left) a strong acid (pH 1), a weak acid (pH 4), a neutral solution (pH 7), a weak base (pH 9), and a strong base (pH 14). All concentrations are 0.10 mol/L.

Example
What is the pOH of a solution whose pH is measured to be 6.4?

Solution
$$pH + pOH = 14$$
$$pOH = 14 - pH$$
$$= 14 - 6.4$$
$$pOH = 7.6$$

The pOH of the solution is 7.6.

Measuring pH
There are several different ways of measuring the pH of a solution, some more precise than others. Many plant compounds and synthetic dyes change colour when mixed with an acid or a base (**Figure 9**).

Substances that change colour when they react with acids or bases are known as **acid–base indicators**. A common indicator used in school laboratories is litmus, a dye obtained from a lichen (**Figure 10**). It is prepared by soaking absorbent paper in litmus solution and then drying it. As you know, red and blue are the two colours of the litmus dye. Litmus dye is red below pH 4.7 and blue above pH 8.3. The colour change occurs over this pH range. Litmus remains brown at approximately pH 6.5, which is very close to neutral pH. A solution is acidic if it causes blue litmus to turn pink and (basic) alkaline if it causes red litmus to turn blue.

acid–base indicator a chemical substance that changes colour when the pH of the system changes

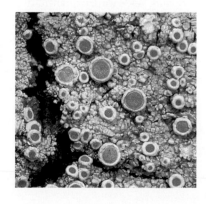

Figure 10
Lichen like this are used to make litmus.

pH meter a device used to measure pH; based on the electric potential of a silver–silver chloride glass electrode and a saturated calomel (dimercury(I) chloride) electrode

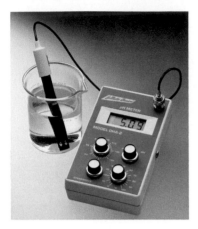

Figure 11
A pH meter measures the voltage generated by a pH-dependent voltaic cell and the scale converts the millivolt reading into a pH reading.

ACTIVITY 8.1.1

Determining the pH of Common Substances (p. 626)
Most common household cleaning agents, foods, and beverages are acidic or basic solutions. Which methods are most suitable for measuring the pH of common household materials?

A **pH meter** is an electronic instrument that measures the voltage between electrodes in a solution and displays this measurement as a pH value (**Figure 11**).

▶ **TRY THIS** activity *Magic Markers*

Colour-changing markers have become a popular toy. The ink in these markers changes colour when drawn over with a special white "magic" marker, creating a stunning visual effect. Why does the coloured ink change colour when mixed with the substance in the white marker?
Materials: lab apron; set of colour-changing markers (including the "magic" marker); 3 small plastic cups; 3 cotton swabs; sheet of blank white paper; white vinegar (5% acetic acid); dilute baking soda solution; distilled water; red cabbage leaf

- Draw horizontal parallel lines on the sheet of white paper, using all of the different coloured markers. The lines should be at least 0.5 cm wide, 6 cm long, and 1 cm apart.

- Draw another horizontal line equally spaced from the first three using the cut edge of a folded red cabbage leaf.

- Using the "magic" marker, draw a single vertical line, perpendicular to the coloured lines, that crosses each of the coloured lines. Label this line "magic marker."

- Pour small amounts of distilled water, vinegar, and baking soda solution into three different cups.

- Use litmus paper to determine whether the liquid in each cup is acidic, neutral, or basic.

- Dip the tip of a cotton swab into the distilled water and draw a vertical line across all of the coloured lines, as you did earlier with the magic marker. Label this line "water." Also indicate whether the liquid is acidic, neutral, or basic.

- Using a new cotton swab each time, draw new vertical lines with vinegar and baking soda solution. Label these lines accordingly.

- Record your observations in a suitable table.
  (a) Provide a hypothesis to explain the effect of the "magic" marker on the coloured ink and cabbage juice mark.
  (b) Describe a procedure to test your hypothesis. If possible, carry out the test.

The pH of Strong Acids

Strong acids ionize quantitatively in aqueous solution. As you already know, a 0.1 mol/L $HCl_{(aq)}$ solution has virtually no HCl molecules in it. We assume that it contains 0.1 mol/L $H^+_{(aq)}$ and 0.1 mol/L $Cl^-_{(aq)}$. The pH of this solution is 1.0, since the negative logarithm of 0.1 is 1. In general, the pH of solutions of strong monoprotic acids is calculated from the concentration of $H^+_{(aq)}$ ions, which is assumed to be equal to the molar concentration of the solute molecules (before ionization).

Calculating the pH, pOH, and [OH⁻$_{(aq)}$] of a Strong Acid

Calculate the pH, pOH, and [OH⁻$_{(aq)}$] of a 0.042 mol/L HNO$_{3(aq)}$ solution (Figure 12).

This problem is similar to an earlier sample problem in which we calculated the [OH⁻$_{(aq)}$] of a solution of a strong acid. In this case, we take the problem one step further and calculate the pH and pOH of the solution.

Appendix C7 lists HNO_3 as a strong acid. Therefore, we assume that it ionizes 100% when it dissolves in water at SATP:

$$HNO_{3(aq)} \xrightarrow{100\%} H^+_{(aq)} + NO^-_{3(aq)}$$

The major entities in solution are $H^+_{(aq)}$, $NO^-_{3(aq)}$, and $H_2O_{(l)}$.
In 0.04 mol/L $HNO_{3(aq)}$,

$$[H^+_{(aq)}] = 0.040 \text{ mol/L}$$

We can ignore the miniscule contributions to $[H^+_{(aq)}]$ made by the autoionization of water and the presence of $NO^-_{3(aq)}$ (the weak conjugate base of HNO_3).
First, calculate pH.

$$pH = -\log[H^+_{(aq)}]$$
$$= -\log(0.040)$$
$$pH = 1.40$$

Then calculate pOH.

$$pH + pOH = 14.00$$
$$pOH = 14.00 - pH$$
$$= 14.00 - 1.40$$
$$pOH = 12.60$$

Using the value of pOH, we can now calculate the concentration of hydroxide ions in the solution:

$$[OH^-_{(aq)}] = 10^{-pOH}$$
$$= 10^{-12.60} \text{ mol/L}$$
$$[OH^-_{(aq)}] = 2.5 \times 10^{-13} \text{ mol/L}$$

The pH of the solution is 1.40; the pOH is 12.60; and the $[OH^-_{(aq)}]$ is 2.5×10^{-13} mol/L.

Figure 12
Nitric acid is produced commercially using a process invented by Wilhelm Ostwald (1853–1932).

Example
Calculate the pH, pOH, and [OH⁻$_{(aq)}$] of a 0.0020 mol/L HBr$_{(aq)}$ solution. (HBr is a strong acid used mostly in the halogenation of organic chemicals.)

Solution

$$HBr_{(l)} \xrightarrow{100\%} H^+_{(aq)} + Br^-_{(aq)}$$

The major entities are $H^+_{(aq)}$, $Br^-_{(aq)}$, and $H_2O_{(l)}$.

$$[H^+_{(aq)}] = 0.0020 \text{ mol/L}$$

We can ignore the small $[H^+_{(aq)}]$ produced by the autoionization of water, and the presence of $Br^-_{(aq)}$.

$$pH = -\log[H^+_{(aq)}]$$
$$= -\log(0.0020)$$
$$pH = 2.70$$

$$pOH = 14.0 - pH$$
$$= 14.0 - 2.7$$
$$pOH = 11.30$$

$$[OH^-_{(aq)}] = 10^{-pOH}$$
$$= 10^{-11.30} \text{ mol/L}$$
$$[OH^-_{(aq)}] = 5.0 \times 10^{-12} \text{ mol/L}$$

▶ *Practice*

Understanding Concepts

12. Calculate the pH, pOH, and $[OH^-_{(aq)}]$ of each of the following solutions.
  (a) 0.006 mol/L $HI_{(aq)}$
  (b) 0.025 mol/L $HNO_{3(aq)}$
  (c) 0.010 mol/L $HCl_{(aq)}$

13. To clean a clogged drain, 26 g of sodium hydroxide is added to water to make 150 mL of solution. What are the pH and pOH values for the solution?

14. What mass of potassium hydroxide is contained in 500 mL of solution that has a pH of 11.5?

Making Connections

15. Food scientists and dietitians measure the pH of foods when they devise recipes and special diets. The juices of various fruits and vegetables are extracted. Various measurements related to their acidity are made and recorded in **Table 2** below.

Table 2 Acidity of Foods

Food	$[H^+_{(aq)}]$ (mol/L)	$[OH^-_{(aq)}]$ (mol/L)	pH	pOH
oranges	5.5×10^{-3}			
asparagus				5.6
olives		2.0×10^{-11}		
blackberries				10.60

  (a) Copy and complete **Table 2**.
  (b) Based on pH only, predict which of the foods would taste most sour, assuming that sour taste is directly proportional to pH.
  (c) Which of these foods might dietitians recommend to their patients to help relieve heartburn? Why?
  (d) There is some research that suggests that women's diets may affect the likelihood of their getting pregnant, as sperm are sensitive to pH. Research this topic, and make some diet suggestions for a woman who is trying to get pregnant.

 www.science.nelson.com

16. When food enters the stomach, it stimulates the production and secretion of hydrochloric acid for digestion, reducing the pH of the stomach contents from 4 to 2.
  (a) Compare the $[H^+_{(aq)}]$ before and after the change in pH.
  (b) Conduct library or Internet reseach to find out how the stomach protects itself from the corrosive effects of this low pH level.

 www.science.nelson.com

Answers

12. (a) pH = 2.2
    pOH = 11.8
    $[OH^-_{(aq)}] = 2 \times 10^{-12}$ mol/L

  (b) pH = 1.60
    pOH = 12.40
    $[OH^-_{(aq)}] = 4.0 \times 10^{-13}$ mol/L

  (c) pH = 2.00
    pOH = 12.00
    $[OH^-_{(aq)}] = 1.0 \times 10^{-12}$ mol/L

13. pH = 14.64
    pOH = −0.64

14. $m_{KOH} = 0.09$ g

15. (a) $[OH^-_{(aq)}]_{oranges} =$
    1.8×10^{-12} mol/L
    $pH_{oranges} = 2.26$
    $pOH_{oranges} = 11.74$

    $[H^+_{(aq)}]_{asparagus} =$
    4×10^{-9} mol/L
    $[OH^-_{(aq)}]_{asparagus} =$
    3×10^{-6} mol/L
    $pH_{asparagus} = 8.4$

    $[H^+_{(aq)}]_{olives} =$
    5.0×10^{-4} mol/L
    $pH_{olives} = 3.30$
    $pOH_{olives} = 10.70$

    $[H^+_{(aq)}]_{blackberries} =$
    4.0×10^{-4} mol/L
    $[OH^-_{(aq)}]_{blackberries} =$
    2.5×10^{-11} mol/L
    $pH_{blackberries} = 3.40$

The pH of Strong Bases

The pH and the conductivity of a $Ba(OH)_{2(aq)}$ solution are found to be higher than those of an $NaOH_{(aq)}$ solution of equal concentration. The barium hydroxide solution is more basic because barium hydroxide dissociates to yield two hydroxide ions per formula unit.

As with strong acids, the pOH and pH of strong bases are determined entirely by the $[OH^-_{(aq)}]$ contributed by the dissociation of the ionic hydroxide solute. The contribution made by the autoionization of water is so small, in comparison, as to be negligible. Also, we assume that the metal cation produced by the dissociation of a strong base has no effect on the pH of the solution. The pOH of a basic solution may be calculated from the solution's pH by applying the equation pH + pOH = 14.0. For example, if the measured pH of a basic solution is 10.2, then

$$pH + pOH = 14.0$$
$$pOH = 14.0 - 10.2$$
$$pOH = 3.8$$

Calculating the pH of a Strong Base

1. **Calculate the pH of a 0.02 mol/L $NaOH_{(aq)}$ solution (Figure 13).**

As usual, begin by writing the dissociation equation

$$NaOH_{(aq)} \xrightarrow[H_2O_{(l)}]{100\%} Na^+_{(aq)} + OH^-_{(aq)}$$

The major entities are $Na^+_{(aq)}$, $OH^-_{(aq)}$ and $H_2O_{(l)}$.
Since NaOH is a strong base, we assume that it dissociates completely in solution. Thus, the hydroxide ion concentration will equal the concentration of NaOH given (0.02 mol/L).

$$[OH^-_{(aq)}] = 0.02 \text{ mol/L}$$

We can ignore the very small $[OH^-_{(aq)}]$ produced by the autoionization of water, and the presence of $Na^+_{(aq)}$.

$$pOH = -\log[OH^-_{(aq)}]$$
$$= -\log(0.02)$$
$$pOH = 1.7$$

Finally, calculate pH.

$$pH + pOH = 14.00$$
$$pH = 14.00 - pOH$$
$$= 14.00 - 1.7$$
$$pH = 12.3$$

The pH of the sodium hydroxide solution is 12.3.

Figure 13
One of the many uses of sodium hydroxide is as a bleaching agent, to whiten wood pulp before it is made into paper or cardboard.

2. **Calculate the pH and pOH of a solution prepared by dissolving 4.3 g of $Ba(OH)_{2(s)}$ in water to form 1.5 L of solution.**

This problem is similar to an earlier sample problem in which we calculated the $[H^+_{(aq)}]$ of a $Ba(OH)_{2(aq)}$ solution. In this case, we take the problem one step farther and calculate the pH and pOH of the solution.

▶

First, we use the balanced equation for the dissociation of $Ba(OH)_{2(s)}$ in water to identify the major entities in solution.

$$Ba(OH)_{2(aq)} \xrightarrow[H_2O_{(l)}]{100\%} Ba^{2+}_{(aq)} + 2\,OH^-_{(aq)}$$

The major entities are $Ba^{2+}_{(aq)}$, $OH^-_{(aq)}$, and $H_2O_{(l)}$.
Now calculate the amount of $Ba(OH)_{2(aq)}$ that dissociates in solution.

$$n_{Ba(OH)_2} = 4.3\,\cancel{g} \times \frac{1\ mol}{171.3\,\cancel{g}}$$

$$n_{Ba(OH)_2} = 2.5 \times 10^{-2}\ mol$$

Use this value to calculate the $[Ba(OH)_{2(aq)}]$.

$$[Ba(OH)_{2(aq)}] = \frac{2.5 \times 10^{-2}\ mol}{1.5\ L}$$

$$[Ba(OH)_{2(aq)}] = 1.7 \times 10^{-2}\ mol/L$$

Every mole of $Ba(OH)_{2(aq)}$ that dissociates, forms two moles of $OH^-_{(aq)}$. Therefore,

$$[OH^-_{(aq)}] = 2(1.7 \times 10^{-2}\ mol/L)$$

$$[OH^-_{(aq)}] = 3.3 \times 10^{-2}\ mol/L$$

We can ignore the very small $[OH^-_{(aq)}]$ produced by the autoionization of water, and the presence of $Ba^{2+}_{(aq)}$. Therefore we can use the $[OH^-_{(aq)}]$ from the dissociation of $Ba(OH)_2$ to calculate the pOH of the solution.

$$pOH = -\log[OH^-_{(aq)}]$$

$$= -\log(3.3 \times 10^{-2})$$

$$pOH = 1.47$$

Finally, calculate the pH.

$$pH + pOH = 14.00$$

$$pH = 14.00 - pOH$$

$$= 14.00 - 1.47$$

$$pH = 12.53$$

The pH of the solution is 12.53, and the pOH is 1.47.

Example
Calculate the pH of a 0.002 mol/L $Ca(OH)_{2(aq)}$ solution (**Figure 14**).

Solution

$$Ca(OH)_{2(aq)} \xrightarrow{100\%} Ca^{2+}_{(aq)} + 2\,OH^-_{(aq)}$$

The major entities are $Ca^{2+}_{(aq)}$, $OH^-_{(aq)}$, and $H_2O_{(aq)}$.
In 0.002 mol/L $Ca(OH)_{2(aq)}$,

$$[OH^-_{(aq)}] = 2(0.002\ mol/L)$$

$$[OH^-_{(aq)}] = 0.004\ mol/L$$

Figure 14
Effluent at a sewage treatment plant tends to be acidic. The sewage can be neutralized by adding calcium hydroxide, also known as hydrated lime. In this automated plant, workers can control pH from terminals.

Ignore the very small $[OH^-_{(aq)}]$ produced by the autoionization of water, and the presence of $Ca^{2+}_{(aq)}$.

$$pOH = -\log[OH^-_{(aq)}]$$
$$= -\log(0.004)$$
$$pOH = 2.4$$

$$pH + pOH = 14.00$$
$$pH = 14.00 - pOH$$
$$= 14.00 - 2.4$$
$$pH = 11.6$$

The pH of the calcium hydroxide solution is 11.6.

▶ **Practice**

Understanding Basic Concepts

17. Calculate the pH of a 0.15 mol/L sodium hydroxide solution.

18. Calculate the pH of a 0.032 mol/L $Ba(OH)_{2(aq)}$ solution.

19. A solution is made by dissolving 0.078 g $Ca(OH)_{2(s)}$ in water to make 100 mL of final solution. Calculate the pH of the solution.

Answers

17. 13.18

18. 12.81

19. 12.3

SUMMARY **Summary pH and pOH**

$$K_w = [H^+_{(aq)}][OH^-_{(aq)}]$$

$$pK_w = -\log K_w$$

$$pH = -\log[H^+_{(aq)}] \qquad pOH = -\log[OH^-_{(aq)}]$$

$$[H^+_{(aq)}] = 10^{-pH} \qquad [OH^-_{(aq)}] = 10^{-pOH}$$

$$pH + pOH = 14.00 \quad \text{(at SATP)}$$

▶ **Section 8.1** *Questions*

Understanding Concepts

1. How does the hydrogen ion concentration compare with the hydroxide ion concentration if a solution is
   (a) neutral?
   (b) acidic?
   (c) basic?

2. What two diagnostic tests can distinguish a weak acid from a strong acid?

3. According to Arrhenius's theory, what do all bases have in common?

4. Calculate the mass of sodium hydroxide that must be dissolved to make 2.00 L of a solution with a pH of 10.35, at SATP.

5. According to the table of acid–base indicators (Appendix C10), what is the colour of each of the following indicators in the solutions of given pH?
   (a) Litmus in a solution with a pH of 8.2
   (b) Methyl orange in a solution with a pH of 3.9

6. Complete the analysis for each of the following diagnostic tests. *If* [the specified indicator] is added to a solution, *and* the colour of the solution turns [the given colour], *then* the solution pH is ___.

(a) methyl red; red
(b) alizarin yellow; red
(c) bromocresol green; blue
(d) bromothymol blue; green

7. Separate samples of an unknown solution turned both methyl orange and bromothymol blue to yellow, and turned bromocresol green to blue.
(a) Estimate the pH of the unknown solution.
(b) Calculate the approximate hydrogen ion concentration.

Applying Inquiry Skills

8. Create an experimental design, using a flow chart or a table, that will identify each of four colourless solutions as one of: a strong acid solution; a weak acid solution; a neutral molecular solution; a neutral ionic solution.

Making Connections

9. Conduct library or Internet research to obtain information to answer the following questions about gastroesophageal reflux disease, GERD.

(a) What is GERD?
(b) Who is usually affected by this condition?
(c) Describe the apparatus used to diagnose this condition.
(d) What treatments are currently available for this disease? Include examples of chemotherapeutic and surgical interventions.

 www.science.nelson.com

10. Conduct library or Internet research to answer the following questions regarding acid-free paper.
(a) What is acid-free paper? Why is it called "acid-free"?
(b) What are the primary uses of acid-free paper?
(c) List advantages and disadvantages of its use.

 www.science.nelson.com

Many common substances are weak acids or weak bases. The acid in vinegar, acetic acid, is a weak acid, as is Aspirin (acetylsalicylic acid, a pain reliever), and vitamin C (ascorbic acid, an essential part of your diet). Codeine, a cough suppressant and powerful painkiller, is a weak base. Like other weak solutes, weak acids and bases ionize incompletely when they react with water.

When dissolved in water, they form solutions containing unreacted molecules in dynamic equilibrium with the ions formed by the reaction of some molecules with water. We can describe the equilibria of weak acids and bases in both qualitative and quantitative terms. Like all equilibria, the reactions of weak acids and bases with water can be shifted to the right or left by the addition or removal of reactants or products. We can predict an equilibrium's behaviour when disturbed, and can calculate useful values such as ion concentrations and pH.

Weak Acids

A **weak acid** is a weak electrolyte that does not ionize quantitatively (completely) in water to form hydrogen ions. Most common acids are weak. Many naturally occurring acids are carboxylic acids, which are weak acids (**Figure 1**).

Common inorganic weak acids include hydrofluoric acid (glass etching), $HF_{(aq)}$, carbonic acid (in soft drinks), $H_2CO_{3(aq)}$, hydrosulfuric acid, $H_2S_{(aq)}$, and boric acid (eyewash solutions), $H_3BO_{3(aq)}$.

Weak Bases

According to Arrhenius's theory, bases are soluble ionic hydroxides that dissociate in water into positive metal ions and negative hydroxide ions. According to Le Châtelier's principle, the hydroxide ions added to water cause a shift to the left in the autoionization of water equilibrium, decreasing the hydrogen ion concentration and producing a pH greater than 7 at SATP. However, you are probably aware that some molecular and ionic compounds, other than hydroxides, also dissolve in water to produce basic solutions. These solutions are not as basic as ionic hydroxide solutions of the same concentration, and so the compounds are called **weak bases**.

The Arrhenius definition is too restricted to explain the characteristics of bases that are not hydroxide compounds. However, we can extend the definition of a base. Brønsted and Lowry defined bases as proton acceptors. A weak base is a compound that reacts non-quantitatively (incompletely) with water to form an equilibrium that includes $OH^-_{(aq)}$ ions according to the following general equation,

$$B_{(aq)} + H_2O_{(l)} \rightleftharpoons OH^-_{(aq)} + HB^+_{(aq)}$$

To act as a weak Brønsted-Lowry base, a compound must possess an atom with a lone pair of valence electrons. The lone pair of electrons accepts an H^+ ion from water, forming $OH^-_{(aq)}$. For this reason, the base is usually represented by the symbol B: (the two dots represent the lone pair of valence electrons),

$$B: + H_2O_{(l)} \rightleftharpoons OH^-_{(aq)} + HB^+_{(aq)}$$

benzoic acid,
$HC_6H_5O_2$ or C_6H_5COOH

Figure 1
Acetic acid, a monoprotic acid found in vinegar, occurs naturally from the oxidation of ethanol. Benzoic acid is a monoprotic acid used as a food preservative. Like the majority of carboxylic acids, these are weak acids.

weak acid an acid that partially ionizes in solution but exists primarily in the form of molecules

weak base a base that has a weak attraction for protons

For example, solutions such as ammonia and sodium phosphate are generally basic. The basic properties of these solutions are explained by the transfer of a hydrogen ion from a water molecule to an atom of the base that has a lone pair of valence electrons:

$$\overset{H^+}{\underset{\downarrow \quad\quad}{NH_{3(aq)} + H_2O_{(l)}}} \rightleftharpoons OH^-_{(aq)} + NH^+_{4(aq)}$$

Sodium phosphate undergoes two chemical changes to produce its alkaline (basic) characteristics.

$$Na_3PO_{4(aq)} \overset{100\%}{\longrightarrow} 3\,Na^+_{(aq)} + PO^{3-}_{4(aq)} \qquad \text{dissociation (strong)}$$
$$PO^{3-}_{4(aq)} + H_2O_{(l)} \rightleftharpoons OH^-_{(aq)} + HPO^{2-}_{4(aq)} \qquad \text{ionization (weak)}$$

Percent Ionization of Weak Acids

If you were to measure the pH of two solutions of equal concentrations—one a weak acid and one a strong acid—you would find that the pH of the weak acid is closer to 7. Measurements of pH indicate that most weak acids ionize less than 50% (**Figure 2**). Acetic acid, $HC_2H_3O_{2(aq)}$, a common weak acid, ionizes only 1.3% in solution at 25°C and 0.10 mol/L concentration.

$$HC_2H_3O_{2(aq)} \overset{1.3\%}{\rightleftharpoons} H^+_{(aq)} + C_2H_3O^-_{2(aq)}$$

Since one mole of $H^+_{(aq)}$ and one mole of $C_2H_3O^-_{2(aq)}$ are produced for every mole of $HC_2H_3O_{2(aq)}$ ionized, the concentration of acid ionized is equal to $[H^+_{(aq)}]$ and $[C_2H_3O^-_{2(aq)}]$.

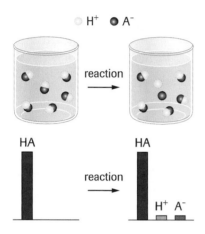

H$^+$ ● A$^-$

HA   HA

reaction

reaction

H$^+$ A$^-$

Figure 2
Weak acids ionize very little in aqueous solution.

Percent ionization, p, is defined as follows:

$$p = \frac{\text{concentration of acid ionized}}{\text{concentration of acid solute}} \times 100\%$$

For the general weak acid ionization reaction,

$$HA_{(aq)} \rightleftharpoons H^+_{(aq)} + A^-_{(aq)}$$

$$\frac{p}{100} = \frac{[H^+_{(aq)}]}{[HA_{(aq)}]}$$

and therefore

$$[H^+_{(aq)}] = \frac{p}{100} \times [HA_{(aq)}]$$

where p is percent ionization, and

$[HA_{(aq)}]$ is the concentration of the acid

As you learned earlier, in a 0.10 mol/L $HCl_{(aq)}$ solution, virtually all of the HCl molecules ionize. The concentration of hydrogen ions is therefore equal to the initial concentration of solute, and $p \doteq 100\%$.

$$HCl_{(aq)} \overset{100\%}{\longrightarrow} H^+_{(aq)} + Cl^-_{(aq)}$$
$$[H^+_{(aq)}] = \frac{100}{100} \times 0.10 \text{ mol/L}$$
$$[H^+_{(aq)}] = 0.10 \text{ mol/L}$$

In a 0.10 mol/L solution of acetic acid, only 1.3% of the $HC_2H_3O_{2(aq)}$ molecules ionize to form hydrogen ions.

$$\overset{1.3\%}{HC_2H_3O_{2(aq)} \rightleftharpoons H^+_{(aq)} + C_2H_3O^-_{2(aq)}}$$

$$[H^+_{(aq)}] = \frac{1.3}{100} \times 0.10 \text{ mol/L}$$

$$[H^+_{(aq)}] = 1.3 \times 10^{-3} \text{ mol/L}$$

If we know the pH of a weak acid solution, we can calculate the percent ionization of the acid

Calculating Percent Ionization	SAMPLE problem ◀

The pH of a 0.10 mol/L methanoic acid solution is 2.38. Calculate the percent ionization of methanoic acid (Figure 3).

First, write the ionization equation and list the known values:

$$HCO_2H_{(aq)} \rightleftharpoons H^+_{(aq)} + HCO^-_{2(aq)}$$

$$[HCO_2H_{(aq)}] = 0.10 \text{ mol/L}$$

$$pH = 2.38$$

Calculate the $[H^+_{(aq)}]$, using the solution's pH.

$$[H^+_{(aq)}] = 10^{-pH} \text{ mol/L}$$

$$= 10^{-2.30} \text{ mol/L}$$

$$[H^+_{(aq)}] = 4.2 \times 10^{-3} \text{ mol/L}$$

Figure 3
Methanoic, or formic, acid is used by ants to immobilize their prey.

Now, use the percent ionization equation to calculate percent ionization:

$$[H^+_{(aq)}] = \frac{p}{100} \times [HA_{(aq)}]$$

$$p = \frac{[H^+_{(aq)}]}{[HA_{(aq)}]} \times 100\%$$

$$p = \frac{[H^+_{(aq)}]}{[HCO_2H_{(aq)}]} \times 100\%$$

$$= \frac{4.2 \times 10^{-3} \text{ mol/L}}{0.10 \text{ mol/L}} \times 100\%$$

$$p = 4.2\%$$

Methanoic acid ionizes 4.2% in a 0.10 mol/L solution.

Example
Calculate the percent ionization of propanoic acid, $HC_3H_5O_{2(aq)}$, if a 0.050 mol/L solution has a pH of 2.78.

Solution
$$HC_2H_5CO_{2(aq)} \rightleftharpoons H^+_{(aq)} + C_2H_5CO^-_{2(aq)}$$

$$[HC_2H_5CO_{2(aq)}] = 0.050 \text{ mol/L}$$

$$pH = 2.78$$

$$[H^+_{(aq)}] = 10^{-pH} \text{ mol/L}$$

$$= 10^{-2.78} \text{ mol/L}$$

$$[H^+_{(aq)}] = 1.7 \times 10^{-3} \text{ mol/L}$$

▶

$$[H^+_{(aq)}] = \frac{p}{100} \times [HA_{(aq)}]$$

$$p = \frac{[H^+_{(aq)}]}{[HC_2H_5CO_{2(aq)}]} \times 100\%$$

$$= \frac{1.7 \times 10^{-3} \text{ mol/L}}{0.05 \text{ mol/L}} \times 100\%$$

$$p = 3.3\%$$

Propanoic acid ionizes 3.3% in a 0.050 mol/L solution.

> ▶ **Practice**

Understanding Concepts

Answers

1. 0.63%

2. 6.7%

1. The pH of a 0.46 mol/L acetic acid, $HC_2H_3O_{2(aq)}$, solution is 2.54. Calculate the percent ionization of acetic acid.

2. What is the percent ionization of a 0.15 mol/L $HF_{(aq)}$ solution whose pH is measured to be 2.00?

Ionization Constants for Weak Acids

In quantitative terms, we can consider an equilibrium solution of a weak acid dissolved in water to be just like the equilibrium systems we studied in Chapter 7. We can represent the system with an equilibrium law expression and calculate an equilibrium constant, as before. The only difference is that the equilibrium constant for a weak acid is known as the **acid ionization constant**, and is symbolized, K_a. The subscript "a" simply means that the constant applies to the equilibrium of an acid. For acetic acid,

acid ionization constant, K_a equilibrium constant for the ionization of an acid

$$HC_2H_3O_{2(aq)} \rightleftharpoons H^+_{(aq)} + C_2H_3O^-_{2(aq)}$$

$$K_a = \frac{[H^+_{(aq)}] [C_2H_3O^-_{2(aq)}]}{[HC_2H_3O_{2(aq)}]}$$

The value of K_a is usually determined experimentally by measuring the electrical conductivity of a weak acid solution of known concentration at equilibrium at SATP. The results of this test yield a value for the degree of ionization (the fraction of acid molecules that react with water to give ions) which can be expressed as a percent ionization. The percent ionization can then be used to calculate the K_a value, as in the following Sample Problem.

▶ **SAMPLE** problem | *Calculating K_a from Percent Ionization*

Calculate the acid ionization constant, K_a, of acetic acid if a 0.1000 mol/L solution at equilibrium at SATP has a percent ionization of 1.3%.

Begin by writing an equation and an acid ionization constant expression for the reaction

$$HC_2H_3O_{2(aq)} \rightleftharpoons H^+_{(aq)} + C_2H_3O^-_{2(aq)}$$

$$K_a = \frac{[H^+_{(aq)}] [C_2H_3O^-_{2(aq)}]}{[HC_2H_3O_{2(aq)}]}$$

A percent ionization of 1.3% means that the initial $[HC_2H_3O_{2(aq)}]$ is diminished by 1.3% by the time equilibrium is reached. Since the reaction occurs in a 1:1:1 molar ratio, $[H^+_{(aq)}]$ and $[C_2H_3O^-_{2(aq)}]$ increase by the same amount that $[HC_2H_3O_{2(aq)}]$ decreases.

We can solve this problem using an ICE table. First, we need to calculate the "change" value, x, for the ICE table. Since the percent ionization is given at 1.3%, we can obtain the change value by simply multiplying the initial concentration of acetic acid by 0.013.

$$x = 0.1 \text{ mol/L} \times 0.013$$
$$x = 0.0013 \text{ mol/L}$$

In this reaction, the $[HC_2H_3O_{2(aq)}]$ will be reduced by 0.0013 mol/L, and the $[H^+_{(aq)}]$ and $[C_2H_3O^-_{2(aq)}]$ will increase by 0.0013 mol/L.

Now, we can set up an ICE table (**Table 1**) to determine the concentrations of all entities at equilibrium.

Table 1   ICE Table for the Ionization of $HC_2H_3O_{2(aq)}$

	$HC_2H_3O_{2(aq)} \rightleftharpoons$	$H^+_{(aq)}$ +	$C_2H_3O^-_{2(aq)}$
Initial concentration (mol/L)	0.1000	0.0000	0.0000
Change in concentration (mol/L)	−0.0013	+0.0013	+0.0013
Equilibrium concentration (mol/L)	0.0987	0.0013	0.0013

Now, we can substitute the equilibrium values from the ICE table into the K_a expression and solve for K_a.

$$K_a = \frac{[H^+_{(aq)}][C_2H_3O^-_{2(aq)}]}{[HC_2H_3O_{2(aq)}]}$$

$$= \frac{(0.0013)^2}{0.0987}$$

$$K_a = 1.7 \times 10^{-5}$$

The acid ionization constant of acetic acid at SATP is calculated to be 1.7×10^{-5}. Notice that K_a values are written without units (just like K_w, and other K values you used in Chapter 7).

We can now compare the calculated value with the value listed in a standard table of acid ionization constants such as Appendix C7.

The K_a of acetic acid listed in the acid–base table in the Appendix is 1.8×10^{-5}. The calculated value (1.7×10^{-5} mol/L) differs from the listed value by only 5.5%, an acceptable error value.

The relatively small percent ionization and K_a values for acetic acid indicate that, at equilibrium, the concentration of (non-ionized) acetic acid molecules, $HC_2H_3O_{2(aq)}$, is much higher than the concentrations of $H^+_{(aq)}$ and $C_2H_3O^-_{2(aq)}$ ions. (Remember that $H^+_{(aq)}$ is an abbreviation for $H_3O^+_{(aq)}$.) This is evident from the Equilibrium values in **Table 1**.

Example

Calculate the K_a of hydrofluoric acid, $HF_{(aq)}$, if a 0.100 mol/L solution at equilibrium at SATP has a percent ionization of 7.8% (**Table 2**).

Solution

$$\overset{7.8\%}{HF_{(aq)} \rightleftharpoons H^+_{(aq)} + F^-_{(aq)}}$$

$$K_a = \frac{[H^+_{(aq)}][F^-_{(aq)}]}{[HF_{(aq)}]}$$

$$x = 0.100 \text{ mol/L} \times 0.078$$
$$x = 0.0078 \text{ mol/L}$$

Table 2 ICE Table for the Ionization of $HF_{(aq)}$

	$HF_{(aq)}$	$\rightleftharpoons$	$H^+_{(aq)}$	+	$F^-_{(aq)}$
Initial concentration (mol/L)	0.100		0.000		0.000
Change in concentration (mol/L)	−0.0078		+0.0078		+0.0078
Equilibrium concentration (mol/L)	0.0922		0.0078		0.0078

$$K_a = \frac{[H^+_{(aq)}][F^-_{(aq)}]}{[HF_{(aq)}]}$$

$$= \frac{0.0078^2}{0.0922}$$

$$K_a = 6.6 \times 10^{-4}$$

This value agrees with the value listed in Appendix C9: 6.6×10^{-4}.
The K_a of hydrofluoric acid at SATP is 6.6×10^{-4}.

> ▶ **Practice**

Understanding Concepts

Answers

3. 7.1×10^{-4}

4. (a) 7.8×10^{-3} mol/L
   (b) 7.8×10^{-6} mol/L

5. 1.16%

3. Calculate the K_a of nitrous acid if a 0.200 mol/L solution at equilibrium at SATP has a percent ionization of 5.8%.

4. Refer to the percent ionization values given in the Table of Acids and Bases in Appendix C9.
   (a) What is the hydrogen ion concentration of a 0.10 mol/L solution of hydrofluoric acid?
   (b) What is the hydrogen ion concentration of a 0.10 mol/L solution of hydrocyanic acid?
   (c) Which of the above solutions is most acidic?

5. A lab technician tests a 0.100 mol/L solution of propanoic acid and finds that its hydrogen ion concentration is 1.16×10^{-3} mol/L. Calculate the percent ionization of propanoic acid in water.

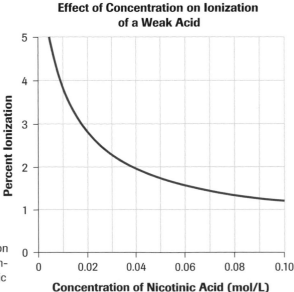

Effect of Concentration on Ionization of a Weak Acid

Figure 4
The percent ionization decreases as the concentration of nicotinic acid increases.

Percent Ionization and Concentration

The values of K_a for a series of acids provide a way of comparing the relative strengths of weak acids. Percent ionization values may also be used for this purpose, but only when solutions of equal initial concentration are compared. Percent ionization values vary with the concentration of the acid. For example, at SATP, 0.10 mol/L acetic acid has a much greater percent ionization (1.3%) than 1.0 mol/L acetic acid (0.42%). **Figure 4** shows how the percent ionization varies with the concentration of a solution (in this case, nicotinic acid).

Notice that the more dilute the solution, the greater the degree of ionization. We can explain this phenomenon using Le Châtelier's principle. Consider a solution of a weak acid, $HA_{(aq)}$, at equilibrium with its ions $H^+_{(aq)}$ and $A^-_{(aq)}$ according to the equation

$$HA_{(aq)} \rightleftharpoons H^+_{(aq)} + A^-_{(aq)}$$

If we dilute the solution by adding water to it, the concentration of all three entities is suddenly decreased. According to Le Châtelier's principle, the system adjusts by shifting the reaction in the direction that produces a greater number of entities (in this case, to the right). A shift to the right ionizes more of the acid. Note that the concentrations of all entities have been reduced beause water has been added to the system; however, since the degree of ionization is greater in the dilute solution, the percent ionization is greater too. In general, *the more dilute a weak acid solution, the greater the percent ionization*, and vice versa.

Ionization Constants for Weak Bases

Weak bases like ammonia, NH_3 (a cleaning agent), and hydrazine, N_2H_4 (a jet fuel), form dynamic equilibria in aqueous solutions. As such, the reaction of a weak base with water may be defined according to the equilibrium law, resulting in an equilibrium law expression and an equilibrium constant. As bases, these compounds produce $OH^-_{(aq)}$ ions that affect the acid–base characteristics of the solution. We will first develop an equilibrium law expression for a particular weak base, then derive an expression for weak bases in general.

Consider the equilibrium reaction of ammonia with water.

$$NH_{3(aq)} + H_2O_{(l)} \rightleftharpoons OH^-_{(aq)} + NH^+_{4(aq)}$$

The equilibrium law equation for this reaction is written:

$$K = \frac{[OH^-_{(aq)}][NH^+_{4(aq)}]}{[NH_{3(aq)}][H_2O_{(l)}]}$$

Since $[H_2O_{(l)}]$ is a constant, its concentration value (its density) is incorporated into the value of K (just as it was in the development of the equilibrium law expressions for K_a and K_w). This yields a new constant, K_b, called the **base ionization constant.**

$$K[H_2O_{(l)}] = \frac{[OH^-_{(aq)}][NH^+_{4(aq)}]}{[NH_{3(aq)}]}$$

$$K_b = \frac{[OH^-_{(aq)}][NH^+_{4(aq)}]}{[NH_{3(aq)}]}$$

Notice the similarities and differences between the mathematical expression for the base ionization constant, K_b, and that for the acid ionization constant, K_a, earlier in this section. For the general base equilibrium reaction

$$\underset{\text{base}}{B_{(aq)}} + \underset{\text{acid}}{H_2O_{(l)}} \rightleftharpoons \underset{\text{conjugate acid}}{HB^+_{(aq)}} + \underset{\text{conjugate base}}{OH^-_{(aq)}}$$

where $B_{(aq)}$ is a weak Brønsted-Lowry base (note that the two dots representing the weak base's unshared pair of electrons are omitted for convenience),
$H_2O_{(l)}$ is a Brønsted-Lowry acid,
$HB^+_{(aq)}$ is the conjugate acid of the weak base, B, and
$OH^-_{(aq)}$ is the conjugate base of the acid, $H_2O_{(l)}$.

$$K_b = \frac{[HB^+_{(aq)}][OH^-_{(aq)}]}{[B_{(aq)}]}$$

base ionization constant, K_b equilibrium constant for the ionization of a base

Table 3 lists the K_b and pK_b values of some common weak bases. Notice that many of the bases listed contain one or more nitrogen atoms. Others—such as the acetate ion, $C_2H_3O^-_{2(aq)}$, and the hypochlorite ion, $ClO^-_{(aq)}$—are conjugate bases of weak acids like acetic acid, $HC_2H_3O_{2(aq)}$, and hypochlorous acid, $HClO_{(aq)}$.

Table 3 K_b Values of Selected Weak Bases at 25°C

Name of Base	Formula	K_b
dimethylamine	$(CH_3)_2NH_{(aq)}$	9.6×10^{-4}
butylamine	$C_4H_9NH_{2(aq)}$	5.9×10^{-4}
methylamine	$CH_3NH_{2(aq)}$	4.4×10^{-4}
ammonia	$NH_{3(aq)}$	1.8×10^{-5}
hydrazine	$N_2H_{4(aq)}$	1.7×10^{-6}
morphine	$C_{17}H_{19}NO_{3(aq)}$	7.5×10^{-7}
hypochlorite ion	$ClO^-_{(aq)}$	3.45×10^{-7}
pyridine	$C_5H_5N_{(aq)}$	1.7×10^{-9}
acetate ion	$C_2H_3O^-_{2(aq)}$	5.6×10^{-10}
urea	$NH_2CONH_{2(aq)}$	1.5×10^{-14}

Organic Bases

Bases such as methylamine, $CH_3NH_{2(aq)}$, and urea, $NH_2CONH_{2(aq)}$, are organic compounds with one or more amine functional groups, $(-NH_2)$. As you learned when studying organic chemistry, these bases may be viewed as substituted ammonia molecules, where one of the hydrogen atoms of ammonia is replaced with an alkyl group such as the methyl group $(-CH_3)$ in methylamine. Pyridine, $C_6H_5N_{(l)}$, is similar to benzene, $C_6H_{6(l)}$, with one of the carbon atoms of the ring replaced with a nitrogen atom.

Amines (bases containing the amine group) are important compounds in living organisms. Many of them are used as chemical messengers that stimulate or inhibit specific cellular processes in animals. Adrenaline and dopamine (a neurotransmitter) are two such examples of human nervous-system stimulants.

Alkaloids are a group of nitrogen-containing organic bases, largely obtained from plants, that may also have powerful stimulating effects on animal nervous systems. Some plant alkaloids are used as pharmaceutical drugs such as ephedrine (a decongestant), codeine, novocaine, and morphine (painkillers), and quinine (used for treating malaria). Mescaline and cocaine, illicit drugs, have hallucinogenic effects, and caffeine and nicotine, licit drugs, are the active ingredients in coffee and tobacco.

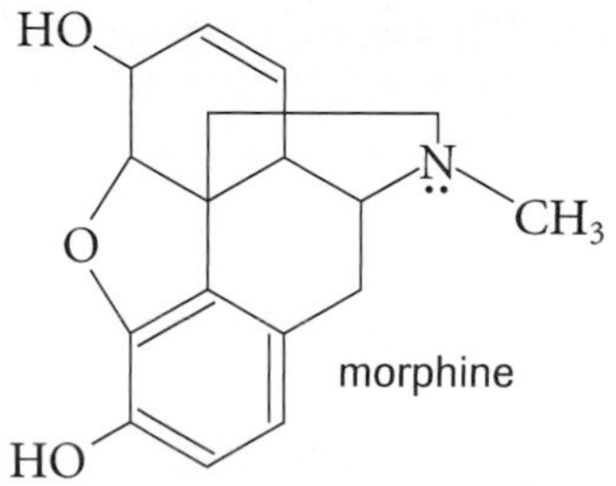

morphine

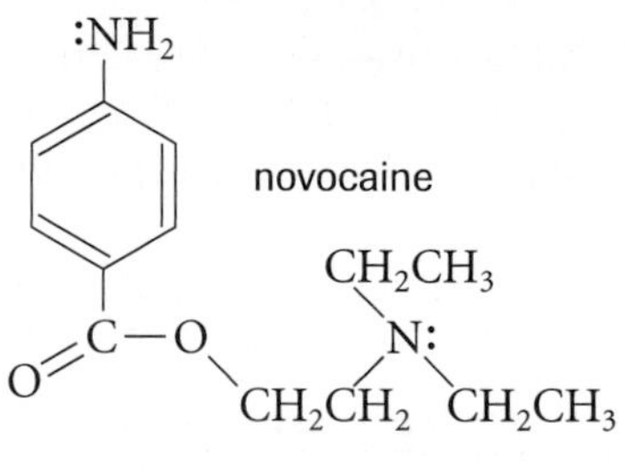

novocaine

Nitrogenous bases (bases containing nitrogen) are generally weak Brønsted–Lowry bases that undergo the same type of ionization reaction with water as ammonia. In each case, a nitrogen atom with an unshared pair of valence electrons acts as a proton acceptor (a Brønsted–Lowry base).

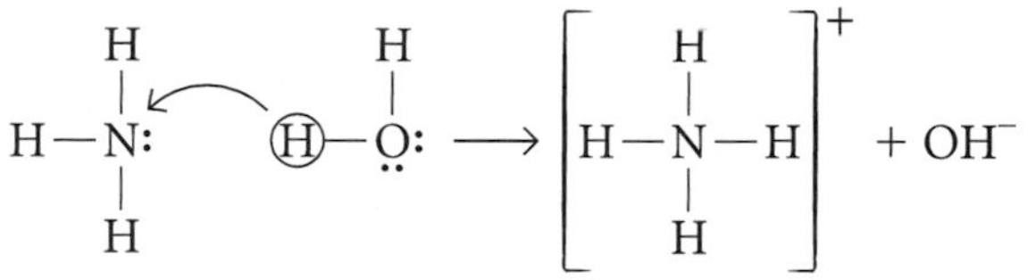

NH_4^+

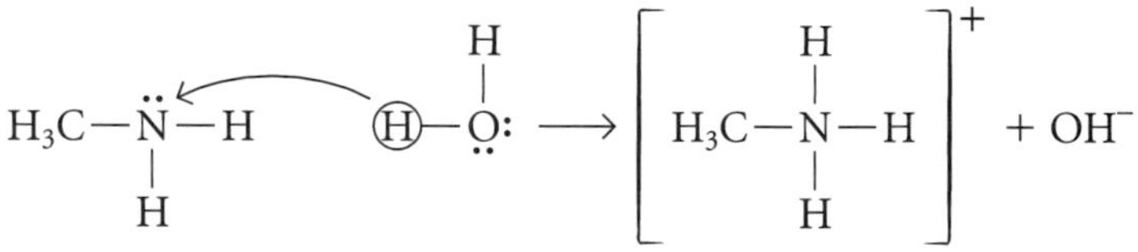

$CH_3NH_3^+$

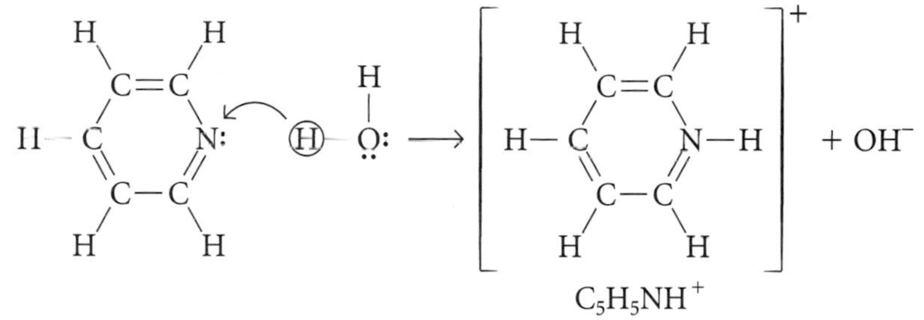

$C_5H_5NH^+$

The Relationship Between K_a and K_b

Acetic acid is a common weak acid that in aqueous solution forms a dynamic equilibrium with $H_{(aq)}^+$ and its conjugate base, the acetate ion, $C_2H_3O_{2(aq)}^-$. The chemical equation for this reaction and its associated acid equilibrium law expression follow.

$$HC_2H_3O_{2(aq)} \rightleftharpoons H_{(aq)}^+ + C_2H_3O_{2(aq)}^-$$

$$K_a = \frac{[H_{(aq)}^+][C_2H_3O_{2(aq)}^-]}{[HC_2H_3O_{2(aq)}]}$$

As a base, the acetate ion, $C_2H_3O_2^-{}_{(aq)}$, may also react with water according to the following equation,

$$C_2H_3O_2^-{}_{(aq)} + H_2O_{(l)} \rightleftharpoons OH^-{}_{(aq)} + HC_2H_3O_{2(aq)}$$

The base equilibrium law expression associated with this reaction is

$$K_b = \frac{[OH^-{}_{(aq)}][HC_2H_3O_{2(aq)}]}{[C_2H_3O_2^-{}_{(aq)}]}$$

Is there a relationship between the K_a of acetic acid, $HC_2H_3O_{2(aq)}$, and the K_b of its conjugate base, $C_2H_3O_2^-{}_{(aq)}$? Notice that most of the terms in the two equilibrium constant expressions are similar. If we add the two equations and cancel common entities that appear as reactants and products, we get

$$
\begin{array}{ll}
HC_2H_3O_{2(aq)} \rightleftharpoons H^+{}_{(aq)} + C_2H_3O_2^-{}_{(aq)} & K_a \\
C_2H_3O_2^-{}_{(aq)} + H_2O_{(l)} \rightleftharpoons OH^-{}_{(aq)} + HC_2H_3O_{2(aq)} & K_b \\
\hline
H_2O_{(l)} \rightleftharpoons H^+{}_{(aq)} + OH^-{}_{(aq)} & K_w
\end{array}
$$

It can be shown that, when reactions are summed like this, the resulting equilibrium constant, in this case, K_w, is equivalent to the product of the equilibrium constants of the constituent reactions. Thus,

$$K_w = K_a K_b$$

We can see that this is true by multiplying the K_a and K_b expressions and cancelling like terms:

$$K_a \times K_b = \frac{[H^+{}_{(aq)}][C_2H_3O_2^-{}_{(aq)}]}{[HC_2H_3O_{2(aq)}]} \times \frac{[OH^-{}_{(aq)}][HC_2H_3O_{2(aq)}]}{[C_2H_3O_2^-{}_{(aq)}]}$$

$$K_a \times K_b = [H^+{}_{(aq)}][OH^-{}_{(aq)}]$$

Since, as you know, $[H^+{}_{(aq)}][OH^-{}_{(aq)}] = K_w$, then $K_a \times K_b = K_w$.

For acids and bases whose chemical formulas differ only by a hydrogen, i.e., conjugate acid–base pairs,

$$K_a K_b = K_w \quad \textbf{or} \quad K_b = \frac{K_w}{K_a} \quad \textbf{or} \quad K_a = \frac{K_w}{K_b}$$

Notice that this equation allows us to convert the K_a values of acids into the K_b values of their conjugate bases and vice versa, given the value of K_w, which is a constant (1.0×10^{-14}) at SATP. Standard acid–base tables rarely list the K_a and K_b values of conjugate acid–base pairs. Usually, only the K_a of the acid is listed. If you need to know the K_b of a base, first look up the K_a value of its conjugate acid, then use the equation $K_a K_b = K_w$ to calculate the value of K_b. (Appendix C9 does give K_b values for some nitrogenous weak bases.)

The following sample problem shows how to convert between K_a and K_b values of conjugate acid–base pairs.

Calculating K_b or K_a

What is the value of the base ionization constant, K_b, for the acetate ion, $C_2H_3O_2{}^-_{(aq)}$, at SATP?

In order to calculate the K_b for this base, we must be given the K_a of its conjugate acid, $HC_2H_3O_{2(aq)}$. We know that K_w at SATP is 1.0×10^{-14}. The K_a of $HC_2H_3O_{2(aq)}$ may be obtained from a reference table such as Appendix C9.

$$K_w = 1.0 \times 10^{-14}$$

$$K_a = 1.8 \times 10^{-5} \text{ (from Appendix C9)}$$

$$K_a K_b = K_w$$

$$K_b = \frac{K_w}{K_a}$$

$$= \frac{1.0 \times 10^{-14}}{1.8 \times 10^{-5}}$$

$$K_b = 5.6 \times 10^{-10} \text{ mol/L}$$

The base ionization constant for the acetate ion is 5.6×10^{-10} mol/L.

Example

The K_b for hydrazine, $N_2H_{4(g)}$, a rocket fuel, is 1.7×10^{-6}. What is the K_a of its conjugate acid, $N_2H_5{}^+_{(aq)}$?

Solution

$$K_w = 1.0 \times 10^{-14}$$

$$K_b = 1.7 \times 10^{-6}$$

$$K_a K_b = K_w$$

$$K_a = \frac{K_w}{K_b}$$

$$= \frac{1.0 \times 10^{-14}}{1.7 \times 10^{-6}}$$

$$K_a = 5.9 \times 10^{-9}$$

The K_a for $N_2H_5{}^+_{(aq)}$ is 5.9×10^{-9}.

As mentioned earlier, the equation $K_a K_b = K_w$ is particularly useful. An important outcome of this mathematical relationship is that, for any conjugate acid–base pair, the product $K_a K_b$ is a constant. Thus, the larger the value of K_a, the smaller the value of K_b. In general, the weaker the acid, the stronger its conjugate base (**Figure 5**) — a relationship developed earlier in this chapter.

Note that, while a strong acid like $HCl_{(aq)}$ always has a very weak conjugate base, it is not necessarily true that a weak acid has a strong conjugate base, as can be seen in **Table 4**.

Table 4 Ionization Constants of Conjugate Pairs

Acid	K_a (tabulated)		Conjugate base	K_b (calculated from K_a)	
$HS^-_{(aq)}$	1.3×10^{-13}	very weak acids	$S^{2-}_{(aq)}$	7.7×10^{-2}	strong conjugate bases
$HPO_4{}^{2-}_{(aq)}$	4.2×10^{-13}		$PO_4{}^{3-}_{(aq)}$	2.4×10^{-2}	
$HC_2H_3O_{2(aq)}$	1.8×10^{-5}	weak acids	$C_2H_3O_2{}^-_{(aq)}$	5.6×10^{-10}	weak conjugate bases
$HCO_2H_{(aq)}$	1.8×10^{-4}		$CO_2H^-_{(aq)}$	5.6×10^{-11}	
$HCl_{(aq)}$	$\doteq 10^7$ (not tabulated)	strong	$Cl^-_{(aq)}$	$\doteq 10^{-21}$	very weak conjugate bases
$HI_{(aq)}$	$\doteq 10^{11}$ (not tabulated)		$I^-_{(aq)}$	$\doteq 10^{-25}$	

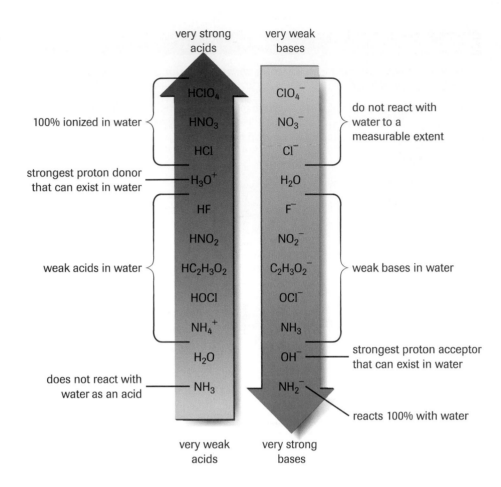

very strong acids — very weak bases

HClO₄ — ClO₄⁻
100% ionized in water { HNO₃ — NO₃⁻ } do not react with water to a measurable extent
HCl — Cl⁻

strongest proton donor that can exist in water —— H₃O⁺ — H₂O

HF — F⁻
HNO₂ — NO₂⁻
weak acids in water { HC₂H₃O₂ — C₂H₃O₂⁻ } weak bases in water
HOCl — OCl⁻
NH₄⁺ — NH₃

H₂O — OH⁻ —— strongest proton acceptor that can exist in water

does not react with water as an acid —— NH₃ — NH₂⁻ —— reacts 100% with water

very weak acids — very strong bases

Figure 5
The relative strengths of various conjugate acid–base pairs

Acetic acid is a weak acid, $K_a = 1.8 \times 10^{-5}$, but its conjugate base, the acetate ion, $C_2H_3O_{2\,(aq)}^-$, is an even weaker base, $K_b = 5.6 \times 10^{-10}$.

This analysis of K_a values for acids of different strengths and the strengths of their respective conjugate bases yields the following important generalizations:

1. The conjugate base of a strong acid is a very weak base.
2. The conjugate base of a weak acid is a weak base.
3. The conjugate base of a very weak acid is a strong base.
4. The conjugate acid of a very weak base is a strong acid.
5. The conjugate acid of a weak base is a weak acid.
6. The conjugate acid of a strong base is a very weak acid.

In this chapter, we primarily analyze solutions of strong and weak acids and bases. You will not have to deal with cases relating to the third and fourth generalizations above. In the following exercises, you will analyze the solutions of weak acids and bases — the other generalizations will apply.

Practice

Understanding Concepts

6. Use information from Appendix C9 and the value of K_w to calculate the base ionization constant, K_b, of the following bases:
   (a) hypochlorite ion, $ClO^-_{(aq)}$
   (b) nitrite ion, $NO^-_{2(aq)}$
   (c) benzoate ion, $C_7H_5O^-_{2(aq)}$

Answers

6. (a) 3.4×10^{-7}
   (b) 1.4×10^{-11}
   (c) 1.6×10^{-10}

The pH of Weak Acid Solutions

Since the value of K_a is constant over a range of acid concentrations, it can be used to calculate the hydrogen ion concentration and pH of weak acid solutions. We will assume SATP conditions for all problems in this chapter, unless otherwise stated.

Calculating $[H^+_{(aq)}]$ and pH of a Weak Acid, Given K_a

SAMPLE problem

Calculate the hydrogen ion concentration and the pH of a 0.10 mol/L acetic acid solution. K_a for acetic acid is 1.8×10^{-5}.

The solution of a weak acid like 0.10 mol/l $HC_2H_3O_{2(aq)}$ is an aqueous system in equilibrium, similar to the equilibrium solutions encountered in Chapter 7. Some of the problem-solving techniques used here will be similar.

Since we are being asked to calculate the hydrogen ion concentration and pH of an acid solution, we must first identify the major entities that may ionize to produce $H^+_{(aq)}$ ions. In this case, both $HC_2H_3O_{2(aq)}$ and $H_2O_{(l)}$ may produce $H^+_{(aq)}$ ions. We start by writing balanced equations for both ionization reactions, with their ionization constants.

$$HC_2H_3O_{2(aq)} \rightleftharpoons H^+_{(aq)} + C_2H_3O^-_{2(aq)} \qquad K_a = 1.8 \times 10^{-5}$$
$$H_2O_{(l)} \rightleftharpoons H^+_{(aq)} + OH^-_{(aq)} \qquad K_w = 1.0 \times 10^{-14}$$

Comparing the K_a value of $HC_2H_3O_{2(aq)}$ with the K_w value of $H_2O_{(l)}$, we notice that $HC_2H_3O_{2(aq)}$ is a much stronger acid than $H_2O_{(l)}$. We may assume that virtually all of the $H^+_{(aq)}$ ions are produced by the ionization of $HC_2H_3O_{2(aq)}$, and that $H_2O_{(l)}$ contributes an insignificant amount of $H^+_{(aq)}$ to the solution. We may also disregard the presence of the acetate ion, $C_2H_3O^-_{2(aq)}$, since it is a very weak conjugate base ($K_b = 5.7 \times 10^{-10}$) and will have no appreciable effect on the pH of the solution.

Therefore, the ionization of $HC_2H_3O_{2(aq)}$ will determine the $[H^+_{(aq)}]$ and the pH of this solution.

For the equilibrium,

$$HC_2H_3O_{2(aq)} \rightleftharpoons H^+_{(aq)} + C_2H_3O^-_{2(aq)}$$

the equilibrium law equation is

$$\frac{[H^+_{(aq)}][C_2H_3O^-_{2(aq)}]}{[HC_2H_3O_{2(aq)}]} = K_a$$

We now use the techniques developed in Chapter 7 for solving equilibrium problems to calculate the $[H^+_{(aq)}]$ at equilibrium.

An ICE table is a useful tool for organizing information. We will incorporate the following initial concentrations:

$[HC_2H_3O_{2(aq)}] = 0.10$ mol/L
$[H^+_{(aq)}] = 0.00$ mol/L

Note that this zero value for the initial $[H^+_{(aq)}]$ is an approximation since we are assuming that the 10^{-7} mol/L formed by the autoionization of water is insignificant.

$$[C_2H_3O^-_{2(aq)}] = 0.00 \text{ mol/L}$$

As usual, we let x represent the change in the $[HC_2H_3O_{2(aq)}]$ that is required for the reaction to reach equilibrium. Since the reaction proceeds in a 1:1:1 molar ratio, the value of x will be subtracted from the initial concentration of $HC_2H_3O_{2(aq)}$, and added to the initial concentrations of $H^+_{(aq)}$ and $C_2H_3O^-_{2(aq)}$.

The ICE table looks like this:

Table 5 ICE Table for the Ionization of $HC_2H_3O_{2(aq)}$

	$HC_2H_3O_{2(aq)} \rightleftharpoons$	$H^+_{(aq)} +$	$C_2H_3O^-_{2(aq)}$
Initial concentration (mol/L)	0.10	0.00	0.00
Change in concentration (mol/L)	$-x$	$+x$	$+x$
Equilibrium concentration (mol/L)	$0.10 - x$	$+x$	$+x$

Now, substitute the equilibrium concentration values into the K_a equation and solve for x.

$$\frac{[H^+_{(aq)}][C_2H_3O^-_{2(aq)}]}{[HC_2H_3O_{2(aq)}]} = K_a$$

$$\frac{(x)(x)}{(0.10 - x)} = 1.8 \times 10^{-5}$$

$$\frac{(x)^2}{(0.10 - x)} = 1.8 \times 10^{-5}$$

It is evident that this is a quadratic equation. The calculation may be simplified by assuming that, since K_a is so small, $HC_2H_3O_{2(aq)}$ will ionize very little and the value of x is expected to be very small. We will use the hundred rule to determine whether this assumption is warranted.

$$\frac{[HA]_{initial}}{K_a} = \frac{0.10}{1.8 \times 10^{-5}}$$

$$= \frac{0.10 \text{ mol/L}}{1.8 \times 10^{-5}}$$

$$\frac{[HA]_{initial}}{K_a} = 5.6 \times 10^3$$

Since $5.6 \times 10^3 > 100$, we will assume that $0.10 - x \doteq 0.10$.

The equilibrium equation becomes

$$\frac{(x)^2}{0.10} \doteq 1.8 \times 10^{-5}$$

which yields

$$x^2 \doteq 1.8 \times 10^{-6}$$
$$x \doteq 1.3 \times 10^{-3}$$

We must now validate the approximation that $0.10 - x \doteq 0.1$. Since K_a values are typically known to an accuracy of $\pm 5\%$, we will use this figure to judge the validity of the assumption. In general, the approximation will be considered valid if, for the acid HA,

$$\frac{x}{[HA]_{initial}} \times 100\% \leq 5\%$$

where x is the $[H^+_{(aq)}]$ calculated by using the simplifying assumption, and $[HA]_{initial}$ is the concentration of the acid before it ionizes in solution.

In this sample problem,

$$x = 1.3 \times 10^{-3} \text{ mol/L}$$

$$[HA]_{initial} = 0.10 \text{ mol/L}$$

$$\frac{x}{[HA]_{initial}} \times 100\% = \frac{1.3 \times 10^{-3} \text{ mol/L}}{0.10 \text{ mol/L}} \times 100\%$$

$$= 1.3\%$$

Since $1.3\% < 5\%$, we can consider the assumption we made valid, and the calculated value of x to be acceptable. (When a simplifying assumption is unjustified, the equilibrium expression will yield a quadratic equation that may be solved by applying the *quadratic formula*, as you did in similar problems in Chapter 7.)

Therefore,

$$[H^+_{(aq)}] = x$$

$$[H^+_{(aq)}] = 1.3 \times 10^{-3} \text{ mol/L}$$

and

$$pH \doteq -\log[H^+_{(aq)}]$$

$$\doteq -\log(1.3 \times 10^{-3})$$

$$pH \doteq 2.89$$

In some cases, a simplifying assumption is not justified because the initial concentration of the acid and the acid ionization constant are too close in value. When this is so, the unsimplified ionization constant expression will yield a quadratic equation that must be solved using the quadratic formula. The following sample problem shows how such a problem may be solved. This strategy is similar to the one used in Chapter 7 to solve equilibrium problems where a simplifying assumption was not justified.

2. **Chloracetic acid, $HC_2H_2O_2Cl_{(aq)}$ is a weak acid ($K_a = 1.36 \times 10^{-3}$ mol/L) that is used as a herbicide (Figure 6). Determine the pH of a 0.0100 mol/L solution of chloracetic acid.**

The major entities of the solution are: $HC_2H_2O_2Cl_{(aq)}$ and $H_2O_{(l)}$.

$$HC_2H_2O_2Cl_{(aq)} \rightleftharpoons H^+_{(aq)} + C_2H_2O_2Cl^-_{(aq)} \qquad K_a = 1.36 \times 10^{-3}$$
$$H_2O_{(l)} \rightleftharpoons H^+_{(aq)} + OH^-_{(aq)} \qquad K_w = 1.0 \times 10^{-14}$$

Since $HC_2H_2O_2Cl_{(aq)}$ is a much stronger acid than $H_2O_{(l)}$, it will dominate in the production of $H^+_{(aq)}$. We may ignore the miniscule contribution of $H^+_{(aq)}$ by the autoionization of water, and the presence of the chloracetate ion, $C_2H_2O_2Cl^-_{(aq)}$, a very weak conjugate base ($K_b = 7.4 \times 10^{-12}$).

$$\frac{[H^+_{(aq)}][C_2H_2O_2Cl^-_{(aq)}]}{[HC_2H_2O_2Cl_{(aq)}]} = K_a$$

Figure 6

Predict whether a simplifying assumption is warranted, using the hundred rule.

$$\frac{[\text{HA}]_{initial}}{K_a} = \frac{0.0100 \text{ mol/L}}{1.36 \times 10^{-3}}$$

$$= 7.35$$

Since $7.35 < 100$, we may *not* assume that $0.0100 - x \doteq 0.0100$. Therefore, the problem will have to be solved without making simplifying assumptions.

Construct an ICE table to track the changes in concentration that occur as the solution reacts to form an equilibrium mixture.

Table 6 ICE Table for the Ionization of $HC_2H_2O_2Cl_{(aq)}$

	$HC_2H_2O_2Cl_{(aq)} \rightleftharpoons$	$H^+_{(aq)} +$	$C_2H_2O_2Cl^-_{(aq)}$
Initial concentration (mol/L)	0.01	0	0
Change in concentration (mol/L)	$-x$	$+x$	$+x$
Equilibrium concentration (mol/L)	$0.0100 - x$	$+x$	$+x$

Substitute the equilibrium concentration values into the equilibrium law expression for this equilibrium.

$$\frac{[H^+_{(aq)}][C_2H_2O_2Cl^-_{(aq)}]}{[HC_2H_2O_2Cl_{(aq)}]} = K_a$$

$$\frac{x^2}{(0.01 - x)} = 1.36 \times 10^{-3}$$

$$x^2 = 1.36 \times 10^{-3}(0.0100 - x)$$

$$x^2 = (1.36 \times 10^{-5}) - (1.36 \times 10^{-3})x$$

$$x^2 + (1.36 \times 10^{-3})x - 1.36 \times 10^{-5} = 0$$

Solve for x using the quadratic formula.

For this problem, $a = 1, b = 1.36 \times 10^{-3}, c = -1.36 \times 10^{-5}$

$$x = \frac{-b \pm \sqrt{b^2 - 4ac}}{2a}$$

$$x = \frac{-(1.36 \times 10^{-3}) \pm \sqrt{(1.36 \times 10^{-3})^2 - 4(1)(-1.36 \times 10^{-5})}}{2(1)}$$

We obtain the following two possible solutions to the quadratic equation:
$x = 4.43 \times 10^{-3}$ mol/L and $x = -3.07 \times 10^{-3}$ mol/L.

Since a negative concentration is impossible, $x = 4.43 \times 10^{-3}$ mol/L.

Notice that the value of x (4.43×10^{-3} mol/L) is not negligible compared to the initial concentration of $HC_2H_2O_2Cl_{(aq)}$ (1.00×10^{-2} mol/L), confirming that a simplifying assumption could not be made.

The equilibrium concentrations are

$$[H^+_{(aq)}] = 4.43 \times 10^{-3} \text{ mol/L}$$

$$[C_2H_2O_2Cl^-_{(aq)}] = 4.43 \times 10^{-3} \text{ mol/L}$$

$$[HC_2H_2O_2Cl_{(aq)}] = 0.0100 \text{ mol/L} - 0.00443 \text{ mol/L}$$

$$= 5.57 \times 10^{-3} \text{ mol/L}$$

Calculate the pH:

$$\begin{aligned} pH &= -\log[H^+_{(aq)}] \\ &= -\log(4.43 \times 10^{-3}) \\ pH &= 2.35 \end{aligned}$$

The pH of a 0.0100 mol/L solution of chloracetic acid is 2.35.

Example

Barbituric acid, $HC_4H_3N_2O_{3(aq)}$, an organic acid used to manufacture hypnotic drugs and some plastics, is a weak acid with a K_a of 9.8×10^{-5}. An industrial process requires a 0.25 mol/L solution of barbituric acid. Calculate the $[H^+_{(aq)}]$ and pH of this solution.

Solution

The major entities are $HC_4H_3N_2O_{3(aq)}$ and $H_2O_{(l)}$.

$$HC_4H_3N_2O_{3(aq)} \rightleftharpoons H^+_{(aq)} + C_4H_3N_2O^-_{3(aq)} \qquad K_a = 9.8 \times 10^{-5}$$
$$H_2O_{(l)} \rightleftharpoons H^+_{(aq)} + OH^-_{(aq)} \qquad K_w = 1.0 \times 10^{-14}$$

Since $HC_4H_3N_2O_{3(aq)}$ is a much stronger acid than $H_2O_{(l)}$, it will dominate in the production of $H^+_{(aq)}$. We may disregard the presence of the barbiturate ion, $C_4H_3N_2O^-_{3(aq)}$, since it is a very weak conjugate base ($K_b = 1.0 \times 10^{-10}$).

$$\frac{[H^+_{(aq)}][C_4H_3N_2O^-_{3(aq)}]}{[HC_4H_3N_2O_{3(aq)}]} = K_a$$

Table 7 ICE Table for the Ionization of $HC_4H_3N_2O_{3(aq)}$

	$HC_4H_3N_2O_{3(aq)} \rightleftharpoons H^+_{(aq)} + C_4H_3N_2O^-_{3(aq)}$		
Initial concentration (mol/L)	0.25	0.00	0.00
Change in concentration (mol/L)	$-x$	$+x$	$+x$
Equilibrium concentration (mol/L)	$0.25 - x$	$+x$	$+x$

$$\frac{x^2}{0.25 - x} = K_a$$

Predict whether a simplifying assumption is warranted, using the hundred rule:

$$\frac{[HA]_{initial}}{K_a} = \frac{0.25 \text{ mol/L}}{9.8 \times 10^{-5}}$$

$$\frac{[HA]_{initial}}{K_a} = 2.55 \times 10^3$$

Since $2.55 \times 10^3 > 100$, we will assume that

$$0.25 - x \doteq 0.25.$$

Therefore,

$$\frac{x^2}{0.25} \doteq 9.8 \times 10^{-5}$$
$$x^2 \doteq 2.4 \times 10^{-5}$$
$$x \doteq 4.9 \times 10^{-3}$$

▶

Use the 5% rule to validate the assumption:

$$\frac{x}{[HA_{(aq)}]_{initial}} \times 100\% = \frac{4.9 \times 10^{-3}}{0.25} \times 100\%$$

$$= 2.0\%$$

Since 2.0 % < 5.0 %, the approximation is considered valid.

$$x \doteq 4.9 \times 10^{-3} \text{ mol/L}$$
$$[H_{(aq)}^+] \doteq 4.9 \times 10^{-3} \text{ mol/L}$$
$$pH \doteq -\log(4.9 \times 10^{-3})$$
$$pH \doteq 2.31$$

The $[H_{(aq)}^+]$ of a 0.25 mol/L solution of barbituric acid is 4.9×10^{-3} mol/L, and the pH is 2.31.

SUMMARY | *Calculating the pH of a Solution of a Weak Monoprotic Acid, HA$_{(aq)}$, Given the Value of K$_a$*

Step 1 List the major entities in solution.

Step 2 Write balanced equations for all entities that may produce $H_{(aq)}^+$.

Step 3 Identify the dominant equilibrium, and write the equilibrium constant equation for the dominant equilibrium.

Step 4 Use the coefficients in the balanced equation of the dominant equilibrium to determine the changes in initial concentrations that will occur as the dominant reaction proceeds to equilibrium. Record all values in an ICE table.

Step 5 Substitute the equilibrium concentrations of all entities (from the "Equilibrium" line in the ICE table) into the acid ionization constant equation.

Step 6 Assume that $[HA]_{initial} - x \doteq [HA]_{initial}$, but only if $\frac{[HA]_{initial}}{K_a} \geq 100$.

Step 7 Solve for x (which, for a monoprotic acid, equals $[H_{(aq)}^+]$).

Step 8 Use the 5% rule to validate any assumption made in step 6.

Step 9 Calculate pH from $[H_{(aq)}^+]$ (the value of x).

▶ **Practice**

Understanding Concepts

7. Lactic acid, $HC_3H_5O_{3(aq)}$, is a weak acid that gives yogurt its sour taste. Calculate the pH of a 0.0010 mol/L solution of lactic acid. The K_a for lactic acid is 1.4×10^{-4}.

8. Methanoic acid, $HCO_2H_{(aq)}$, also known as formic acid, is partly responsible for the characteristic itchy rash produced by the leaves of the stinging nettle plant. Calculate the pH of 0.150 mol/L methanoic acid. The K_a for methanoic acid is 1.8×10^{-4}.

Answers

7. 3.43

8. 2.28

Starting with pH

The above calculations can be reversed to calculate a K_a value from the pH of an acidic solution. The K_a value obtained can then be used to calculate the hydrogen ion concentration or pH over a range of concentrations. Acid ionization constants are useful for comparing the relative strengths of weak acids, as we did in the sample problem to determine whether the chloracetic acid equilibrium or the water equilibrium dominated the production of $H^+_{(aq)}$.

Finding K_a, Given Concentration and pH　　　　*SAMPLE* problem ◄

You measure the pH of a 0.100 mol/L hypochlorous acid, $HOCl_{(aq)}$, solution, and find it to be 4.23. What is the K_a for hypochlorous acid?

Since the pH is given, you can calculate the $[H^+_{(aq)}]$.

$$pH = 4.23$$
$$[H^+_{(aq)}] = 10^{-pH}$$
$$= 10^{-4.23}$$
$$[H^+_{(aq)}] = 5.9 \times 10^{-5} \text{ mol/L}$$

Now, write the equilibrium reaction equation and the equilibrium law equation for the ionization of $HOCl_{(aq)}$.

$$HOCl_{(aq)} \rightleftharpoons H^+_{(aq)} + OCl^-_{(aq)}$$

$$K_a = \frac{[H^+_{(aq)}][OCl^-_{(aq)}]}{[HOCl_{(aq)}]}$$

We assume that HOCl is the major entity producing $H^+_{(aq)}$ since the autoionization of water contributes an extremely small amount of $H^+_{(aq)}$. We may also ignore the basic effects of $OCl^-_{(aq)}$ since it is a very weak conjugate base ($K_b = 1.7 \times 10^{-10}$).
According to the balanced equilibrium reaction equation, one mole of $HOCl_{(aq)}$ ionizes to form one mole of $H^+_{(aq)}$ and one mole of $OCl^-_{(aq)}$. Therefore, at equilibrium,

$$[H^+_{(aq)}] = [OCl^-_{(aq)}] = 5.9 \times 10^{-5} \text{ mol/L}$$

$$K_a = \frac{(5.9 \times 10^{-5})^2}{0.100}$$

$$K_a = 3.5 \times 10^{-8}$$

The K_a of hypochlorous acid is 3.5×10^{-8}.

Example

Calculate the K_a of a 0.050 mol/L solution of nicotinic acid, $HC_2H_6NO_{2(aq)}$, with a pH of 3.08. Nicotinic acid is one of the B vitamins, a dietary requirement.

Solution

$$pH = 3.08$$

$$[H^+_{(aq)}] = 10^{-pH}$$
$$= 10^{-3.08}$$
$$[H^+_{(aq)}] = 8.3 \times 10^{-4} \text{ mol/L}$$

$$HC_2H_6NO_{2(aq)} \rightleftharpoons H^+_{(aq)} + C_2H_6NO_2^-_{(aq)}$$

▶

$$K_a = \frac{[H^+_{(aq)}][C_2H_6NO_{2^-(aq)}]}{[HC_2H_6NO_{2(aq)}]}$$

The $HC_2H_6NO_{2(aq)}$ equilibrium is the major source of $H^+_{(aq)}$. The contribution to $[H^+_{(aq)}]$ by the autoionization of water is insignificant, and so, at equilibrium,

$$[H^+_{(aq)}] = [C_2H_6NO_{2^-(aq)}]$$

Therefore,

$$K_a = \frac{(8.3 \times 10^{-4})^2}{0.050}$$

$$K_a = 1.4 \times 10^{-5}$$

The K_a of nicotinic acid is 1.4×10^{-5}.

The pH of Weak Base Solutions

Just as we used the acid ionization constant to convert between pH and $[H^+_{(aq)}]$, we can use the base ionization constant, K_b, to determine the pH, pOH, $[H^+_{(aq)}]$, and $[OH^-_{(aq)}]$ of solutions of weak bases. In general, the same restrictions and approximations apply. As usual, we need to determine whether or not we can use a simplifying assumption in a calculation by applying the hundred rule, and then justify any assumptions by an application of the 5% rule. If simplifications are not appropriate, we may have to solve quadratic equations using the quadratic formula, as we did in problems involving weak acids.

When solving problems involving weak bases, you will need to know the value of the base ionization constant, K_b, for the base in question. In some cases, the necessary value will be given. In other cases, you may find the value in a chart such as Appendix C7.

In other cases, you will have to calculate the K_b value from the K_a value of the conjugate acid. To calculate the K_b value for a base, you will have to

- determine the formula of the base's conjugate acid (add an H^+ to the formula of the base),
- locate the formula of the conjugate acid in the chart,
- note the conjugate acid's K_a value,
- substitute the K_a value into the equation $K_aK_b = K_w$, and solve for the K_b of the base ($K_w = 1.0 \times 10^{-14}$).

For example, if you had to find the K_b value for the fluoride ion, $F_{(aq)}^-$, look up the K_a value for its conjugate acid, hydrofluoric acid, $HF_{(aq)}$, in Appendix C7, and then use this value to calculate K_b for $F_{(aq)}^-$ as follows.

The value of K_a for $HF_{(aq)}$ reported in the chart is 6.6×10^{-4}.

$$K_a K_b = K_w$$

$$K_b = \frac{K_w}{K_a}$$

$$= \frac{1.0 \times 10^{-14}}{6.6 \times 10^{-4}}$$

$$K_b = 1.5 \times 10^{-11}$$

The K_b value of $F_{(aq)}^-$ is 1.5×10^{-11}.

Finding K_b, pK_b, or pH for Weak Bases
SAMPLE problem

Calculate the pH of a 0.100 mol/L aqueous solution of hydrazine, $N_2H_{4(aq)}$, a weak base. The K_b for hydrazine is 1.7×10^{-6}.

Our strategy for solving questions involving weak bases will be similar to the one we used in calculations with weak acids. First, identify the major entities in solution and determine which of them is able to furnish $OH_{(aq)}^-$ ions.

Since $N_2H_{4(aq)}$ is a weak base ($K_b = 1.7 \times 10^{-6}$), we may assume that most of the dissolved $N_2H_{4(aq)}$ molecules will remain intact. Therefore, the major entities in solution are $N_2H_{4(aq)}$ and $H_2O_{(l)}$.

Both of these substances may produce $OH_{(aq)}^-$ ions according to the following equilibrium reactions:

$$N_2H_{4(aq)} + H_2O_{(l)} \rightleftharpoons N_2H_{5(aq)}^+ + OH_{(aq)}^- \qquad K_b = 1.7 \times 10^{-6}$$

$$H_2O_{(l)} \rightleftharpoons H_{(aq)}^+ + OH_{(aq)}^- \qquad K_w = 1.0 \times 10^{-14}$$

We can neglect the contribution to $[OH_{(aq)}^-]$ made by the autoionization of water, since $K_w \ll K_b$. Therefore, the $N_2H_{4(aq)}$ equilibrium will predominate in this solution, and we will assume that all of the $OH_{(aq)}^-$ is produced by the $N_2H_{4(aq)}$ equilibrium. We may also neglect the presence of the conjugate acid $N_2H_{5(aq)}^+$ since it is a very weak acid ($K_a = 5.9 \times 10^{-9}$). The equilibrium expression for this reaction is

$$\frac{[N_2H_{5(aq)}^+][OH_{(aq)}^-]}{[N_2H_{4(aq)}]} = K_b$$

We begin by constructing an ICE table (**Table 8**) to show the changes in concentration that occur when hydrazine ionizes to form an equilibrium mixture. As usual, we let x represent the changes in concentration that occur as the reaction proceeds to equilibrium.

Notice that no values are written for water, since its concentration remains constant as the reaction proceeds to equilibrium.

Table 8 ICE Table for the Ionization of $N_2H_{4(aq)}$

	$N_2H_{4(aq)}$	$+ \quad H_2O_{(l)}$	$\rightleftharpoons \quad N_2H_{5(aq)}^+$	$+ \quad OH_{(aq)}^-$
Initial concentration (mol/L)	0.100	—	0.00	0.00
Change in concentration (mol/L)	$-x$	—	$+x$	$+x$
Equilibrium concentration (mol/L)	$0.100 - x$	—	$+x$	$+x$

Now, we substitute the equilibrium concentrations in the ICE table into the equilibrium expression for this reaction, and solve for $[OH^-_{(aq)}]$.

$$\frac{[N_2H^+_{5(aq)}][OH^-_{(aq)}]}{[N_2H_{4(aq)}]} = K_b$$

$$\frac{x^2}{(0.100 - x)} = 1.7 \times 10^{-6}$$

We will now use the hundred rule to predict whether a simplifying assumption may be used.

$$\frac{0.100}{1.7 \times 10^{-6}} = 5.9 \times 10^4$$

Since $5.9 \times 10^4 >> 100$, we will assume that $0.100 - x \doteq 0.100$.

$$\frac{x^2}{0.100} \doteq 1.7 \times 10^{-6}$$

$$x^2 \doteq 1.7 \times 10^{-7}$$

$$x \doteq 4.12 \times 10^{-4}$$

We must now justify the simplifying assumption that $0.100 - x \doteq 0.100$.

$$\frac{4.12 \times 10^{-4}}{0.100} \times 100\% = 0.41\%$$

Since $0.41\% << 5\%$, our assumption is justified.

From the ICE table we see that the value of x equals the $[OH^-_{(aq)}]$. We may now substitute the $[OH^-_{(aq)}]$ into the K_w equation and solve for $[H^+_{(aq)}]$.

$$K_w = [H^+_{(aq)}][OH^-_{(aq)}] = 1.00 \times 10^{-14}$$

(Remember that the K_w equation can be used for *all* aqueous solutions.) We can then take the negative logarithm of $[H^+_{(aq)}]$ to obtain the pH. Alternatively, we may calculate pOH from $[OH^-_{(aq)}]$ and substitute this value into the equation, pH + pOH = 14, to calculate pH directly. We will use the second method in this sample problem.

$$x \doteq 4.12 \times 10^{-4} \text{ mol/L}$$

$$[OH^-_{(aq)}] \doteq 4.12 \times 10^{-4} \text{ mol/L}$$

$$pOH = -\log[OH^-_{(aq)}]$$

$$\doteq -\log(4.12 \times 10^{-4})$$

$$pOH \doteq 3.38$$

$$pH + pOH = 14.00$$

$$pH = 14.00 - pOH$$

$$\doteq 14.00 - 3.38$$

$$pH \doteq 10.62$$

The pH of a 0.100 mol/L hydrazine solution is 10.62.

Example

Morphine, $C_{17}H_{19}NO_3$, is a weak base and a powerful painkiller. A solution of morphine has a concentration of 0.01 mol/L. Determine the pH of this solution. The K_b for morphine is 7.5×10^{-7}.

Solution

$$C_{17}H_{19}NO_{3(aq)} + H_2O_{(l)} \rightleftharpoons HC_{17}H_{19}NO_{3(aq)}^+ + OH_{(aq)}^- \qquad K_b = 7.5 \times 10^{-7}$$
$$H_2O_{(l)} \rightleftharpoons H_{(aq)}^+ + OH_{(aq)}^- \qquad\qquad\qquad\qquad K_w = 1.0 \times 10^{-14}$$

The major entities are $C_{17}H_{19}NO_{3(aq)}$ and $H_2O_{(l)}$ (**Table 9**).

We can neglect the contribution to $[OH_{(aq)}^-]$ made by the autoionization of water since $K_w \ll K_b$. We also neglect the presence of the very weak conjugate acid $HC_{17}H_{19}NO_{3(aq)}^+$ ($K_a = 1.3 \times 10^{-8}$).

$$\frac{[HC_{17}H_{19}NO_{3(aq)}^+][OH_{(aq)}^-]}{[C_{17}H_{19}NO_{3(aq)}]} = K_b$$

Table 9 ICE Table for the Ionization of $C_{17}H_{19}NO_{3(aq)}$

	$C_{17}H_{19}NO_{3(aq)}$	$+ H_2O_{(l)}$	$\rightleftharpoons HC_{17}H_{19}NO_{3(aq)}^+$	$+ OH_{(aq)}^-$
Initial concentration (mol/L)	0.010	—	0.00	0.00
Change in concentration (mol/L)	$-x$	—	$+x$	$+x$
Equilibrium concentration (mol/L)	$0.010 - x$	—	x	x

$$\frac{[HC_{17}H_{19}NO_{3(aq)}^+][OH_{(aq)}^-]}{[C_{17}H_{19}NO_{3(aq)}]} = K_b$$

$$\frac{x^2}{0.010 - x} = 7.5 \times 10^{-7}$$

Predicting the validity of a simplification assumption …

$$\frac{0.010}{7.5 \times 10^{-7}} = 1.3 \times 10^4$$

Since $1.3 \times 10^4 \gg 100$, we assume that $0.010 - x \doteq 0.010$.

$$\frac{x^2}{0.010} \doteq 7.5 \times 10^{-7}$$
$$x \doteq 8.7 \times 10^{-5}$$

Justifying the simplification assumption …

$$\frac{8.7 \times 10^{-5}}{0.010} \times 100\% = 0.87\%$$

Since $0.87\% \ll 5\%$, our assumption is justified.

$$x \doteq 8.7 \times 10^{-5} \text{ mol/L}$$
$$[OH_{(aq)}^-] \doteq 8.7 \times 10^{-5} \text{ mol/L}$$
$$pOH \doteq -\log(8.7 \times 10^{-5})$$
$$\doteq 4.06$$
$$pH + pOH = 14.00$$
$$pH = 14.00 - pOH$$
$$\doteq 14.00 - 4.06$$
$$pH \doteq 9.94$$

The pH of a 0.010 mol/L morphine solution is 9.94.

> ▶ **Practice**

Understanding Concepts

12. Calculate the pH and $[H^+_{(aq)}]$ of a 0.30 mol/L solution of butanoic acid, $HC_4H_7O_{2(aq)}$. The K_a of butanoic acid is 1.52×10^{-5}.

13. Strychnine, $C_{21}H_{22}N_2O_{2(aq)}$, is a weak base but a powerful poison. Calculate the pH of a 0.001 mol/L solution of strychnine. The K_b of strychnine is 1.0×10^{-6}.

SUMMARY ⬛ *Calculating the pH of the Solution of a Weak Base, Given the Value of K_b*

Step 1 List the major entities in solution.

Step 2 Write balanced equations for all entities that may produce $OH^-_{(aq)}$.

Step 3 Identify the dominant equilibrium, and write the equilibrium constant equation for the dominant equilibrium.

Step 4 Use the coefficients in the balanced equation of the dominant equilibrium to determine the changes in initial concentrations that will occur as the dominant reaction proceeds to equilibrium. Record all values in an ICE table.

Step 5 Substitute the equilibrium concentrations of all entities (from the "Equilibrium" line in the ICE table) into the base ionization constant equation.

Step 6 Assume that $[B]_{initial} - x \doteq [B]_{initial}$, but only if $\dfrac{[B]_{initial}}{K_b} \geq 100$, where $[B]_{initial}$ represents the concentration of the base before the ionization process begins.

Step 7 Solve for x (which equals $[OH^-_{(aq)}]$).

Step 8 Use the 5% rule to validate assumptions made in step 6.

Step 9 Calculate pOH from $[OH^-_{(aq)}]$ (the value of x).

Step 10 Calculate the pH by substituting the value of pOH into the equation pH + pOH = 14.

Polyprotic Acids

polyprotic acid an acid with more than one ionizable (acidic) proton

Some acids, like sulfuric acid, $H_2SO_{4(aq)}$, and boric acid, $H_3BO_{3(aq)}$, possess more than one ionizable proton. Because of this, they are called **polyprotic acids**. More specifically, sulfuric acid is a *diprotic* acid because it can donate two $H^+_{(aq)}$ ions per molecule, and boric acid is a *triprotic* acid because it can produce three $H^+_{(aq)}$ ions. Polyprotic acids do not donate all of their protons simultaneously when they react. They always ionize in a stepwise fashion, releasing one proton at a time. The following reaction sequence illustrates the ionization of oxalic acid, $H_2C_2O_{4(aq)}$. Note that each ionization reaction possesses its own acid ionization constant, designated K_{a1} for the first ionization reaction and K_{a2} for the second reaction. Note the values of the acid ionization constants in each case.

$$H_2C_2O_{4(aq)} \rightleftharpoons H^+_{(aq)} + HC_2O^-_{4(aq)}$$

$$K_{a1} = \frac{[H^+_{(aq)}][HC_2O^-_{4(aq)}]}{[H_2C_2O_{4(aq)}]}$$

$$K_{a1} = 5.4 \times 10^{-2}$$

$$HC_2O^-_{4(aq)} \rightleftharpoons H^+_{(aq)} + AsO^{2-}_{4(aq)}$$

$$K_{a2} = \frac{[H^+_{(aq)}][C_2O^{2-}_{4(aq)}]}{[HC_2O^-_{4(aq)}]}$$

$$K_{a2} = 5.4 \times 10^{-5}$$

Notice that the value of K_{a1} is much greater than the value of K_{a2}. This is true for most polyprotic acids. That is, the acid involved in each step of the multi-step ionization process is successively weaker. (See **Table 10**.)

Table 10 Acid Ionization Constants for Polyprotic Acids at SATP

Acid Name	Formula	K_{a1}	K_{a2}	K_{a3}
oxalic acid	$H_2C_2O_{4(aq)}$	5.4×10^{-2}	5.4×10^{-5}	–
ascorbic acid	$H_2C_6H_6O_{6(aq)}$	7.9×10^{-5}	1.6×10^{-12}	–
sulfuric acid	$H_2SO_{4(aq)}$	very large	1.0×10^{-2}	–
hydrosulfuric acid	$H_2S_{(aq)}$	1.1×10^{-7}	1.3×10^{-13}	–
phosphoric acid	$H_3PO_{4(aq)}$	7.1×10^{-3}	6.3×10^{-8}	4.2×10^{-13}
arsenic acid	$H_3AsO_{4(aq)}$	5×10^{-3}	8×10^{-8}	4.0×10^{-12}
carbonic acid	$H_2CO_{3(aq)}$	4.4×10^{-7}	4.7×10^{-11}	–

In general, for a polyprotic acid,
$$K_{a1} > K_{a2} > K_{a3} \ ...$$

The multiple ionizations and the different values of the ionization constants seem to indicate that calculating the pH of a polyprotic acid is a difficult problem. However, the fact that the first ionization constant is usually much larger than the subsequent K_a values makes such calculations relatively straightforward. In general, for a typical polyprotic acid in aqueous solution, only the first ionization step is used in determining the pH of the solution. The following sample problem shows a suitable procedure.

Calculating the pH of a Polyprotic Acid　　SAMPLE problem ◀

Calculate the pH of a 0.10 mol/L solution of ascorbic acid, $H_2C_6H_6O_{6(aq)}$, and the equilibrium concentrations of $H_2C_6H_6O_{6(aq)}$, $HC_6H_6O^-_{6(aq)}$, and $C_6H_6O^{2-}_{6(aq)}$.

First, we need to find the K_a values for the two equilibriums. **Table 10** lists the K_a values for the ionization of ascorbic acid as

$$K_{a1} = 7.9 \times 10^{-5}$$
$$K_{a2} = 1.6 \times 10^{-12}$$

Since K_{a2} is much smaller than K_{a1}, we will assume that the first ionization reaction forms the major equilibrium and thus determines the pH of the solution. Note that ascorbic acid is a weak acid since its K_{a1} is merely 7.9×10^{-5}.

The major entities in solution are $H_2C_6H_6O_{6(aq)}$ and $H_2O_{(l)}$.

$$H_2C_6H_6O_{6(aq)} \rightleftharpoons H^+_{(aq)} + HC_6H_6O^-_{6(aq)} \qquad K_{a1} = 7.9 \times 10^{-5}$$
$$H_2O_{(l)} \rightleftharpoons H^+_{(aq)} + OH^-_{(aq)} \qquad K_w = 1.0 \times 10^{-14}$$

Since $H_2C_6H_6O_{6(aq)}$ is a much stronger acid than $H_2O_{(l)}$, it will dominate in the production of $H^+_{(aq)}$. We can disregard the presence of the ascorbate ion, $HC_6H_6O^-_{6(aq)}$, since it is a very weak conjugate base ($K_b = 1.3 \times 10^{-10}$).

$$\frac{[H^+_{(aq)}][HC_6H_6O^-_{6(aq)}]}{[H_2C_6H_6O_{6(aq)}]} = K_{a1}$$

Construct an ICE table as follows.

Table 11 ICE Table for the Ionization of $H_2C_6H_6O_{6(aq)}$

	$H_2C_6H_6O_{6(aq)}$ $\rightleftharpoons$	$HC_6H_6O^-_{6(aq)}$ +	$H^+_{(aq)}$
Initial concentration (mol/L)	0.10	0.00	0.00
Change in concentration (mol/L)	$-x$	$+x$	$+x$
Equilibrium concentration (mol/L)	$0.10 - x$	x	x

$$\frac{[H^+_{(aq)}][HC_6H_6O^-_{6(aq)}]}{[H_2C_6H_6O_{6(aq)}]} = 7.9 \times 10^{-5}$$

$$\frac{x^2}{0.10 - x} = 7.9 \times 10^{-5}$$

Predicting whether $0.10 - x \doteq 0.10$ ...

$$\frac{[HA]_{initial}}{K_{a1}} = \frac{0.10}{7.9 \times 10^{-5}}$$

$$\frac{[HA_{initial}]}{K_{a1}} = 1.3 \times 10^3$$

Since $1.3 \times 10^3 > 100$, we assume that $0.10 - x \doteq 0.10$.

$$\frac{x^2}{0.10} \doteq 7.9 \times 10^{-5}$$
$$x^2 \doteq 7.9 \times 10^{-6}$$
$$x \doteq 2.8 \times 10^{-3}$$

Validating the assumption ...

$$\frac{2.8 \times 10^{-3}}{0.10} \times 100\% = 2.8\%$$

Since $2.8\% < 5.0\%$, the approximation is valid.

$$x \doteq 2.8 \times 10^{-3} \text{ mol/L}$$
$$[H^+_{(aq)}] \doteq 2.8 \times 10^{-3} \text{ mol/L}$$
$$pH \doteq -\log(2.8 \times 10^{-3})$$
$$pH \doteq 2.55$$

We can calculate the concentrations of $H_2C_6H_6O_{6(aq)}$ and $HC_6H_6O^-_{6(aq)}$ by substituting the value of x into the equilibrium concentration expressions for these entities in the ICE table.

$[H_2C_6H_6O_{6(aq)}]$ = 0.10 mol/L − x

$\qquad\qquad\quad$ = 0.10 mol/L − 2.8 × 10^{-3} mol/L

$[H_2C_6H_6O_{6(aq)}]$ = 0.097 mol/L

$[HC_6H_6O_{6(aq)}^-]$ = x

$[HC_6H_6O_{6(aq)}^-]$ = 2.8 × 10^{-3} mol/L

The concentration of $C_6H_6O_{6(aq)}^{2-}$ cannot be calculated using expressions derived from the first ionization reaction because $C_6H_6O_{6(aq)}^{2-}$ is not produced by this ionization reaction: It is produced in the second ionization reaction. We must use the second ionization constant expression and the value of the second ionization constant, K_{a2}, to calculate this value.

$$HC_6H_6O_{6(aq)}^- \rightleftharpoons H_{(aq)}^+ + C_6H_6O_{6(aq)}^{2-}$$

$$\frac{[H_{(aq)}^+][C_6H_6O_{6(aq)}^{2-}]}{[HC_6H_6O_{6(aq)}^-]} = K_{a2}$$

Rearrange this equation to isolate $[C_6H_6O_{6(aq)}^{2-}]$ on the left:

$$[C_6H_6O_{6(aq)}^{2-}] = \frac{[HC_6H_6O_{6(aq)}^-]K_{a2}}{[H_{(aq)}^+]}$$

Find the value of K_{a2} from **Table 10**.

K_{a2} − 1.6 × 10^{-12}

We also $^+$substitute the values of $[H_{(aq)}^+]$ and $[HC_6H_6O_{6(aq)}^-]$, calculated earlier, into this equation. Remember our assumption that the second ionization reaction is insignificant and so does not change these concentration values appreciably.

$$[C_6H_6O_{6(aq)}^{2-}] = \frac{(2.8 \times 10^{-3})(1.6 \times 10^{-12})}{2.8 \times 10^{-3}}$$

$$[C_6H_6O_{6(aq)}^{2-}] = 1.6 \times 10^{-12} \text{ mol/L}$$

Notice that $[C_6H_6O_{6(aq)}^{2-}] = K_{a2}$

Example
Calculate the pH of 1.00 mol/L phosphoric acid, $H_3PO_{4(aq)}$.

Solution
Table 10 lists the K_a values for the ionization of phosphoric acid as

K_{a1} = 7.1 × 10^{-3}

K_{a2} = 6.3 × 10^{-8}

K_{a3} = 4.2 × 10^{-13}

Since K_{a2} and K_{a3} are much smaller than K_{a1}, we will assume that the first ionization reaction forms the major equilibrium and thus determines the pH of the solution.

The major entities in solution are $H_3PO_{4(aq)}$ and $H_2O_{(l)}$.

$H_3PO_{4(aq)} \rightleftharpoons H_{(aq)}^+ + H_2PO_{4(aq)}^-$ $\qquad K_{a1}$ = 7.1 × 10^{-3}

$H_2O_{(l)} \rightleftharpoons H_{(aq)}^+ + OH_{(aq)}^-$ $\qquad\quad K_w$ = 1.0 × 10^{-14}

Since $H_3PO_{4(aq)}$ is a much stronger acid than $H_2O_{(l)}$, it will dominate in the production of $H^+_{(aq)}$. We can disregard the presence of the dihydrogen phosphate ion, $H_2PO^-_{4(aq)}$, since it is a very weak conjugate base ($K_b = 1.4 \times 10^{-12}$).

Construct an ICE table as follows.

$$\frac{[H^+_{(aq)}][H_2PO^-_{4(aq)}]}{[H_3PO_{4(aq)}]} = K_{a1}$$

Table 12 ICE Table for the Ionization of $H_3PO_{4(aq)}$

	$H_3PO_{4(aq)}$ $\rightleftharpoons$	$H^+_{(aq)}$ +	$H_2PO^-_{4(aq)}$
Initial concentration (mol/L)	1.00	0.00	0.00
Change in concentration (mol/L)	$-x$	$+x$	$+x$
Equilibrium concentration (mol/L)	$1.00 - x$	x	x

$$\frac{[H^+_{(aq)}][H_2PO^-_{4(aq)}]}{[H_3PO_{4(aq)}]} = 7.1 \times 10^{-3}$$

$$\frac{x^2}{1.00 - x} = 7.1 \times 10^{-3}$$

Predicting whether $1.00 - x \doteq 1.00$ ...

$$\frac{[HA]_{initial}}{K_{a1}} = \frac{1.00}{7.1 \times 10^{-3}}$$

$$= 141$$

Since $141 > 100$, we assume that $1.00 - x \doteq 1.00$.

$$\frac{x^2}{1.00} \doteq 7.1 \times 10^{-3}$$

$$x^2 \doteq 7.1 \times 10^{-4}$$

$$x \doteq 2.7 \times 10^{-2}$$

Validating the assumption ...

$$\frac{2.7 \times 10^{-2}}{1.00} \times 100\% = 2.7\%$$

Since $2.7\% < 5.0\%$, the approximation is valid.

$$x \doteq 2.7 \times 10^{-2} \text{ mol/L}$$

$$[H^+_{(aq)}] \doteq 2.7 \times 10^{-2} \text{ mol/L}$$

$$pH = -\log[H^+_{(aq)}]$$

$$\doteq -\log(2.7 \times 10^{-2})$$

$$pH \doteq 1.6$$

The pH of a 1.00 mol/L $H_3PO_{4(aq)}$ solution is 1.6

> ## Practice

Understanding Concepts

14. Calculate the pH of the following aqueous solutions:
  - (a) 1.00 mol/L sulfuric acid, $H_2SO_{4(aq)}$
  - (b) 0.001 mol/L sulfuric acid, $H_2SO_{4(aq)}$
  - (c) 0.010 mol/L hydrosulfuric acid, $H_2S_{(aq)}$

Answers

14. (a) 0.000

    (b) 3

    (c) 4.48

▶ **Section 8.2** *Questions*

Understanding Concepts

1. What is the difference between a weak acid and a dilute solution of a strong acid?

2. Calculate the pH of a 0.18 mol/L solution of cyanide ion, $CN^-_{(aq)}$.

3. Phenol, C_6H_5OH, a powerful disinfectant, ionizes according to the equation:

   $$C_6H_5OH_{(aq)} \rightleftharpoons H^+_{(aq)} + C_6H_5O^-_{(aq)}$$

   If the pH of a 0.0200 mol/L phenol solution is 5.9, calculate the percent ionization of phenol.

4. Benzoic acid is a weak monoprotic acid that is often used as a preservative in foods. The benzoic acid in a 0.100 mol/L solution is found to be 2.5% ionized. Calculate the K_a for benzoic acid.

5. The pH of a 0.100 mol/L solution of nitrous acid, HNO_2, is 2.1. Calculate the value for the acid ionization constant for nitrous acid.

6. Hypobromous acid, HBrO, is a weak acid with $K_a = 2.5 \times 10^{-9}$. What is the pH of a 0.200 mol/L solution of HBrO?

7. (a) Use Appendix C7 to rank the following acids in order of increasing percent ionization:

   $HCN_{(aq)}$, $HNO_{3(aq)}$, $HCO_2H_{(aq)}$, $HF_{(aq)}$

   (b) Estimate the approximate pH of a 1.0 mol/L solution of each acid.

8. (a) List the following acids in order of decreasing acid strength.
   (b) List the acids in order of decreasing pH.
   (c) List the conjugate bases of the following acids in order of decreasing base strength.

hydrosulfuric acid	$H_2S_{(aq)} \rightleftharpoons H^+_{(aq)} + HS^-_{(aq)}$	$K_a = 1.0 \times 10^{-7}$
acetic acid	$HC_2H_3O_{2(aq)} \rightleftharpoons H^+_{(aq)} + C_2H_3O^-_{2(aq)}$	$K_a = 1.8 \times 10^{-5}$
ammonium ion	$NH^+_{4(aq)} \rightleftharpoons H^+_{(aq)} + NH_{3(aq)}$	$K_a = 5.6 \times 10^{-10}$
nitrous acid	$HNO_{2(aq)} \rightleftharpoons H^+_{(aq)} + NO^-_{2(aq)}$	$K_a = 4.5 \times 10^{-4}$
phosphoric acid	$H_3PO_{4(aq)} \rightleftharpoons H^+_{(aq)} + H_2PO^-_{4(aq)}$	$K_a = 7.5 \times 10^{-3}$

9. For each of the following weak bases, write the chemical equilibrium equation and the mathematical expression for K_b.

   (a) $CN^-_{(aq)}$

   (b) $SO^{2-}_{4(aq)}$

   (c) $NO^-_{2(aq)}$

   (d) $F^-_{(aq)}$

10. The medicines in **Table 13** are all weak bases.
    (a) Rank the medicines in order of increasing base strength.
    (b) Calculate the pH of a 0.1-mol/L solution of each base.

Table 13 Base Ionization Constants for Some Medicines

Medicine	K_b
atropine (a treatment for Parkinson's disease)	3.2×10^{-5}
morphine (a pain reliever)	7.9×10^{-7}
erythromycin (an antibiotic)	6.3×10^{-6}

11. The hydroxide ion concentration in a 0.157 mol/L solution of sodium propanoate, $NaC_3H_5O_{2(aq)}$, is found to be 1.1×10^{-5} mol/L. Calculate the base ionization constant for the propanoate ion.

12. Using the acid–base table, determine the K_b for the nitrite ion.

13. Codeine (use Cod as an abbreviated chemical symbol) has a K_b of 1.73×10^{-6} mol/L. Calculate the pH of a 0.020 mol/L codeine solution.

14. Ammonia, NH_3, is the conjugate base of the ammonium ion, $NH^+_{4(aq)}$.
    (a) Given that the K_b for ammonia is 1.72×10^{-5}, write the K_b expression for the ionization of ammonia.
    (b) Given that the K_a for the ammonium ion is 5.80×10^{-10}, write the K_a expression for the ammonium ion.
    (c) Use the equations in (a) and (b) to show that $K_aK_b = K_w$.

15. (a) Write the chemical formula for the conjugate acid for each of the following bases. (See Appendix C7.)

    $NH_{3(aq)}$, $HS^-_{(aq)}$, $SO^{2-}_{4(aq)}$

    (b) Use Appendix C7 to determine the K_b for each of the bases given in part (a).

16. Morphine is one of the most effective pain relievers. Given that a 0.0100 mol/L solution of morphine has a pH of 10.10, calculate the K_b of morphine.

17. An aqueous solution of ammonia is an effective glass cleaner. What is the pH of a 0.100 mol/L solution of ammonia? (K_b for ammonia is 1.77×10^{-5}.)

18. Calculate the pH of a 0.0500 mol/L sodium oxalate solution, given that the K_b for the oxalate ion is 1.7×10^{-10}.

Applying Inquiry Skills

19. List some empirical properties that would be useful when distinguishing strong bases from weak bases.

20. Complete the Analysis section of the following report.

    Question

    Which of the unknown solutions provided is $HBr_{(aq)}$, $HC_2H_3O_{2(aq)}$, $NaCl_{(aq)}$, and $C_{12}H_{22}O_{11(aq)}$?

    Experimental Design

    The evidence in **Table 14** was obtained by testing a 0.1 mol/L sample of each of the above solutions with a conductivity apparatus and red and blue litmus.

Evidence

Table 14 Litmus and Conductivity Tests on Unknown Solutions

Solution	Red litmus	Blue litmus	Conductivity
1	no change	red	very high
2	no change	no change	none
3	no change	red	high
4	no change	no change	high

Making Connections

21. Aniline, $C_6H_5NH_2$, is a colourful weak base closely related to ammonia, NH_3.
    (a) If the pH of a 0.10 mol/L aniline solution is 8.81, calculate the K_b for aniline.
    (b) Aniline has been used as a pigment for centuries. What colours are aniline dyes? Research and write a short report on aniline dyes.

 www.science.nelson.com

22. The stomach wall has a protective mucous lining that prevents stomach acid from attacking the underlying tissues. Frequent use of Aspirin (acetylsalicylic acid), $HC_8H_7O_2CO_{2(aq)}$, can damage the stomach wall. Aspirin is a weak carboxylic acid with a K_a of 3.2×10^{-4}.

$$HC_8H_7O_2CO_{2(aq)} \rightleftharpoons H^+_{(aq)} + C_8H_7O_2CO^-_{2(aq)}$$

(a) Stomach acid has a pH of about 1.5. Given the acidity of the stomach, would Aspirin dissolved in stomach fluid be mostly in its ionized or un-ionized form?

(b) Un-ionized Aspirin molecules can readily penetrate the stomach lining into a region of less acidity. This is where the stomach irritation associated with Aspirin occurs. Use Le Châtelier's principle to explain why irritation occurs in this area.

(c) Conduct library or Internet research to obtain information on enteric-coated tablets or capsules. How do these help alleviate the problems described in (a) and (b)?

 www.science.nelson.com

23. (a) During a cross-country race, the concentration of lactic acid in the fluid surrounding muscles can be 5.6 mmol/L. Given that the K_a for lactic acid is 7.94×10^{-5}, calculate the pH of the fluid around the muscles of a runner.

(b) What are the symptoms of lactic acid buildup in the muscles?

(c) Long-distance runners make use of a quantity called the lactic acid threshold (or lactate threshold) to help them avoid lactic acid buildup. Describe the lactic acid threshold and explain how runners make use of this value in training and competition.

 www.science.nelson.com

Acid–Base Properties of Salt Solutions 8.3

Salts are solids (at SATP) composed of cations and anions arranged in a crystalline lattice. When they dissolve in water, they dissociate into individual hydrated ions that may or may not affect the pH of the solution. ▣▯

Sodium chloride, $NaCl_{(s)}$, like all sodium salts, is a strong electrolyte. When dissolved in water it dissociates quantitatively into aqueous sodium and chloride ions.

$$NaCl_{(s)} \rightarrow Na^+_{(aq)} + Cl^-_{(aq)}$$

When the pH of a sodium chloride solution is tested with a pH meter or acid–base indicator, it shows a pH of 7 at SATP: neutral. Sodium chloride does not produce hydronium or hydroxide ions in solution, so it is classed as a neutral electrolyte.

Sodium carbonate, $Na_2CO_{3(s)}$, is another highly soluble ionic compound. It dissociates quantitatively in water to produce aqueous sodium and carbonate ions.

$$Na_2CO_{3(s)} \rightarrow 2\,Na^+_{(aq)} + CO_3^{2-}{}_{(aq)}$$

When tested with a pH meter or acid–base indicator, the sodium carbonate solution proves to have a pH greater than 7: basic. However, sodium carbonate cannot contribute hydronium or hydroxide ions to the solution directly. So, why is a sodium carbonate solution basic while a sodium chloride solution is neutral? The reason must lie in the action of aqueous carbonate ions, not sodium ions, because both salts release sodium ions in solution.

The basic character of carbonate solutions can be explained by the Brønsted–Lowry theory. As in the case of aqueous ammonia, a Brønsted–Lowry base acts as a proton acceptor, and may remove a proton from water to form hydroxide ions in solution. The following equation describes how the carbonate ion acts as a Brønsted–Lowry base.

$$\overset{\displaystyle H^+}{\overbrace{CO_3^{2+}{}_{(aq)} + H_2O_{(l)}}} \rightleftharpoons OH^-_{(aq)} + HCO_3^-{}_{(aq)}$$

   base   acid

Empirical studies have shown that the carbonate ion is a relatively weak base with a base ionization constant, K_b, of 2.1×10^{-4}.

Ammonium chloride, $NH_4Cl_{(s)}$, is a soluble salt that forms an acidic solution when dissolved in water. The dissociation of the ions in aqueous solution is described by the equation

$$NH_4Cl_{(s)} \rightarrow NH_4^+{}_{(aq)} + Cl^-_{(aq)}$$

The pH of an $NaCl_{(aq)}$ solution is neutral, so pH is unaffected by the presence of $Cl^-_{(aq)}$ ions. Is it possible that the acidity of ammonium chloride is caused by the $NH_4^+{}_{(aq)}$ ion? We can hypothesize that a Brønsted-Lowry proton-transfer reaction occurs between ammonium ions and water molecules:

$$NH_4^+{}_{(aq)} + H_2O_{(l)} \rightleftharpoons H_3O^+_{(aq)} + NH_{3(aq)}$$

Laboratory tests show that the ammonium ion is indeed a weak acid with an acid ionization constant, K_a, of 5.8×10^{-10}.

In general, since salts contain two different ions, the pH of an aqueous salt solution may be affected by the anion, the cation, or both.

▣ **INVESTIGATION 8.3.1**

The pH of Salt Solutions (p. 627)
Do soluble salts affect the pH of aqueous solutions?

The reaction of an ion with water to produce an acidic or basic solution is called **hydrolysis** (from the Greek *hydro*, meaning "water," and *lysis*, meaning "splitting").

In the case of sodium chloride, neither sodium nor chloride ions hydrolyze, so neither ion affects the acid–base properties of an aqueous solution; the solution remains neutral. Conversely, when ammonium carbonate dissolves, carbonate ions hydrolyze to produce hydroxide ions (which may produce a basic solution) *and* ammonium ions hydrolyze to produce hydronium ions (which may produce an acidic solution). You must always consider both ions when assessing the effect of a salt on the pH of an aqueous solution.

So, can we accurately predict if a salt will produce an acidic or a basic solution? In the following analysis, we will develop models that will help you make such predictions.

Salts That Form Neutral Solutions

In general, salts that consist of the cations of strong bases (like $Na^+_{(aq)}$ and $K^+_{(aq)}$) and the anions of strong acids (like $Cl^-_{(aq)}$ and $NO^-_{3(aq)}$) have no effect on the pH of an aqueous solution. Examples include $NaCl_{(aq)}$, $KCl_{(aq)}$, $NaI_{(aq)}$, and $NaNO_{3(aq)}$ (**Table 1**).

Table 1 Composition of "Neutral" Salts

Salt	Cation from strong base	Anion from strong acid
NaCl	NaOH	HCl
KCl	KOH	HCl
NaI	NaOH	HI
NaNO$_3$	NaOH	HNO$_3$

Salts That Form Acidic Solutions

Cations such as the ammonium ion, $NH^+_{4(aq)}$, and hydrazinium ion, $N_2H^+_{5(aq)}$, act as Brønsted–Lowry acids. They hydrolyze to form hydronium ions in solution. Consequently, solutions of $NH_4Cl_{(aq)}$ and $N_2H_5Cl_{(aq)}$ are acidic.

In general, cations that are conjugate acids of weak molecular bases act as weak acids that have a tendency to lower the pH of a solution (**Table 2**).

Table 2 Composition and Hydrolysis of "Acidic" Salts

Salt	Cation (acid) of weak base		Hydrolysis reaction
NH$_4$Cl	$NH^+_{4(aq)}$	NH_3	$NH^+_{4(aq)} + H_2O_{(l)} \rightleftharpoons H_3O^+_{(aq)} + NH_{3(aq)}$
N$_2$H$_5$Br	$N_2H^+_{5(aq)}$	N_2H_4	$N_2H^+_{5(aq)} + H_2O_{(l)} \rightleftharpoons H_3O^+_{(aq)} + N_2H_{4(aq)}$

Certain metal cations also hydrolyze and act as acids. An aluminum nitrate solution, $Al(NO_3)_{3(aq)}$, is acidic. However, other solutions, such as sodium nitrate, $NaNO_{3(aq)}$, and calcium nitrate, $Ca(NO_3)_{2(aq)}$, are neutral. Why are some metal salts acidic and some neutral? We could propose a hypothesis that the aluminum ion is responsible for the acidity of the solution. Further tests with aluminum solutions indeed establish that Al^{3+} behaves as an acid in water (increases $[H^+_{(aq)}]$). The research indicates that Group 1 and 2 metal ions (except for Be^{2+}) do not produce acidic solutions, but that highly charged small ions *do* form acidic solutions. (See **Table 3** and Appendix C7.)

Ions such as $Al^{3+}_{(aq)}$, $Fe^{3+}_{(aq)}$, and $Sn^{2+}_{(aq)}$ have large (positive) charge densities — a large amount of charge in a small volume. These cations produce hydronium ions indirectly by a slightly different reaction than the ammonium ion example above. When a highly charged metal ion such as Al^{3+} dissolves in water, it becomes hydrated with six water molecules (waters of hydration) according to the following equation:

$$Al^{3+}_{(aq)} + 6\,H_2O_{(l)} \rightleftharpoons Al(H_2O)^{3+}_{6(aq)}$$

Table 3 K_a for Some Metal Ions at SATP

Metal cation*	K_a
$Zr^{4+}_{(aq)}$	2.1
$Sn^{2+}_{(aq)}$	2.0×10^{-2}
$Fe^{3+}_{(aq)}$	1.5×10^{-3}
$Cr^{3+}_{(aq)}$	1.0×10^{-4}
$Al^{3+}_{(aq)}$	9.8×10^{-6}
$Be^{2+}_{(aq)}$	3.2×10^{-7}
$Fe^{2+}_{(aq)}$	1.8×10^{-7}
$Pb^{2+}_{(aq)}$	1.6×10^{-8}
$Cu^{2+}_{(aq)}$	1.0×10^{-8}

* The aqueous metal ion is a hydrated complex ion (e.g., $Cu(H_2O)^{2+}_{4(aq)}$). Aqueous ions of transition metals are usually written in a simplified form, without showing the number of water molecules present in the actual hydrated complex ion, as shown in **Table 3**.

This means that six water molecules bond to the ion with relatively weak electrostatic forces of attraction. The high-charge density of the $Al^{3+}_{(aq)}$ ion increases the polarity of the —OH bonds in the H_2O molecules. That is, the H_2O molecules in $Al(H_2O)_6^{3+}{}_{(aq)}$ are more likely to transfer a proton to the solvent ($H_2O_{(l)}$) than are $H_2O_{(l)}$ molecules in pure water. Although we might imagine that several (or all) of the water molecules in $Al(H_2O)_6^{3+}{}_{(aq)}$ could transfer a proton, experiments show that only one H_2O will. Thus, $Al(H_2O)_6^{3+}{}_{(aq)}$ behaves as a weak monoprotic acid in aqueous solution, according to the following equation:

$$\underbrace{Al(H_2O)_6^{3+}{}_{(aq)}}_{\text{acid}} + \underbrace{H_2O_{(l)}}_{\text{base}} \rightleftharpoons \underbrace{H_3O^+_{(aq)}}_{\substack{\text{conjugate}\\\text{base}}} + \underbrace{Al(H_2O)_5(OH)^{2+}_{(aq)}}_{\substack{\text{conjugate}\\\text{acid}}}$$

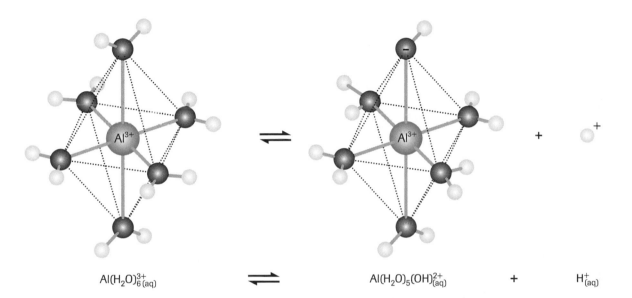

$$Al(H_2O)_6^{3+}{}_{(aq)} \rightleftharpoons Al(H_2O)_5(OH)^{2+}_{(aq)} + H^+_{(aq)}$$

Notice that this weak acid and its acid ionization constant appear in Appendix C9.

No cation with low charge density acts as an acid in this way. This includes all of the singly charged ions of Group 1 metals, Li^+, Na^+, K^+, Rb^+, and Cs^+. Since they cannot hydrolyze, they do not affect the pH of aqueous solutions. With the notable exception of Be^{2+}, none of the Group 2 cations acts as an acid either.

Salts That Form Basic Solutions

A sodium chloride solution, $NaCl_{(aq)}$, is neutral, but a sodium acetate solution, $NaC_2H_3O_{2(aq)}$, is basic. Why? Both solutions contain aqueous sodium ions that do not hydrolyze in solution, so the change in pH must be due to the effect of the acetate ion, $C_2H_3O_2^-{}_{(aq)}$. It is reasonable to suspect hydrolysis to be the cause. In this case, hydrolysis caused by the acetate ion produces hydroxide ions and increases the pH of the solution. The following two equations show the dissociation of sodium acetate followed by the hydrolysis of the resulting acetate ion:

$$NaC_2H_3O_{2(s)} \rightarrow Na^+_{(aq)} + C_2H_3O_2^-{}_{(aq)} \qquad \text{(dissociation)}$$

$$C_2H_3O_2^-{}_{(aq)} + H_2O_{(l)} \rightleftharpoons HC_2H_3O_{2(aq)} + OH^-_{(aq)} \qquad \text{(hydrolysis)}$$

Earlier, we explained that Group 1 anions like $Cl^-_{(aq)}$ cannot hydrolyze. So, why is it that the acetate ion is able to hydrolyze, but the chloride ion cannot? Earlier in this chapter, you learned that the stronger the acid, the weaker its conjugate base. Extremely strong acids such as $HCl_{(aq)}$ and $HNO_{3(aq)}$ possess extremely weak conjugate bases, $Cl^-_{(aq)}$ and $NO^-_{3(aq)}$. These bases attract protons (H^+ ions) so poorly that they are unable to hydrolyze to any significant degree. If they did attract protons strongly, HCl and HNO_3 would not be strong acids and would not ionize so readily in aqueous solution. In general, the conjugate bases of strong acids (e.g., $Cl^-_{(aq)}$, $Br^-_{(aq)}$, $I^-_{(aq)}$, $NO^-_{3(aq)}$) do not affect the pH of an aqueous solution.

Conversely, the acetate ion, $C_2H_3O^-_{2(aq)}$, is the conjugate base of acetic acid, $HC_2H_3O_{2(aq)}$. Acetic acid is a weak acid because the acetate ion portion of the molecule possesses a relatively high affinity for the ionizable proton (H^+ ion). This relatively strong attraction for protons makes the acetate ion able to hydrolyze to form $OH^-_{(aq)}$ ions in aqueous solution. A solution of sodium acetate is basic because the sodium ions cannot act as acids, but the acetate ions do act as weak bases according to the hydrolysis equation above. Remember that, in general, the conjugate base of a weak acid is also weak. The K_a of acetic acid is 1.8×10^{-5} (weak), and the K_b of its conjugate base, the acetate ion, is 5.6×10^{-10} (weak). However, when the acetate ion is introduced as the anion of a salt containing a nonhydrolyzing cation like $Na^+_{(aq)}$ or $K^+_{(aq)}$, the acetate ion forms the major acid–base equilibrium in solution. (The only other reaction is the autoionization of water.) This results in a net production of $OH^-_{(aq)}$ and a basic solution.

When determining whether an anion will affect the pH of a solution, it helps to remember that

- the anion of a strong acid is too weak a base to affect the pH of an aqueous solution (e.g., $Cl^-_{(aq)}$ and $NO^-_{3(aq)}$), and
- the anion of a weak acid, although it is also weak, is a strong enough base (when acting alone) to affect the pH of an aqueous solution. It will tend to increase the pH of the solution and make it more basic.

In general, a salt made of a nonhydrolyzing cation (like $Na^+_{(aq)}$ or $K^+_{(aq)}$), and an anion that is the conjugate base of a weak acid, will form a basic aqueous solution (pH > 7).

Salts That Act as Acids and Bases

Some salts contain the cation of a weak base and the anion of a weak acid — both ions can hydrolyze. To obtain a precise estimate of the pH of such a solution, we would have to solve two hydrolysis equilibria simultaneously. This is a difficult mathematical problem. However, we can predict whether the solution will be acidic, basic, or (approximately) neutral without performing any calculations.

Consider the ions produced when ammonium cyanide, $NH_4CN_{(s)}$, dissolves in water.

$$NH_4CN_{(s)} \rightarrow NH^+_{4(aq)} + CN^-_{(aq)} \quad \text{(dissociation)}$$

The cation, $NH^+_{4(aq)}$, is the conjugate acid of the weak molecular base, $NH_{3(aq)}$, so it will hydrolyze to produce hydronium ions in solution. Its tendency is to lower the pH. The anion, $CN^-_{(aq)}$, is the conjugate base of the weak acid, $HCN_{(aq)}$, so will hydrolyze to produce hydroxide ions in solution. Its tendency is to increase the pH. What is the net effect? That depends on the relative strengths of the two ions, as judged by the K_a value for the acid and the K_b value for the base. If these values are equal, then the salt will have no net effect on the pH of the solution. If the K_a of the acid is larger than the K_b of the base, then the solution will be more acidic. Conversely, a higher K_b would result in a basic solution. In the case of ammonium cyanide,

$$NH_{4(aq)}^+ + H_2O_{(l)} \rightarrow NH_{3(aq)} + H_3O_{(aq)}^+ \qquad K_a = 5.8 \times 10^{-10} \text{ mol/L}$$
$$CN_{(aq)}^- + H_2O_{(l)} \rightarrow HCN_{(aq)} + OH_{(aq)}^- \qquad K_b = 1.6 \times 10^{-5} \text{ mol/L}$$

Since the K_b of the cyanide ion (1.6×10^{-5} mol/L) is much larger than the K_a of the ammonium ion (5.8×10^{-10} mol/L), an aqueous solution of ammonium cyanide will be basic.

SUMMARY	**Predicting the Acid–Base Behaviour of a Salt**

The following steps will help you predict whether the ions of a salt have an effect on the pH of an aqueous solution.

Step 1 Determine if the cation is the conjugate acid of a weak base or a cation with high charge density. If so, it will make the solution more acidic. If not, it will not affect the pH of the solution.

Step 2 Determine if the anion is the conjugate base of a weak acid. If so, it will make the solution more basic. If not, it will not affect the pH of the solution.

Step 3 If the salt has a cation and an anion that can both hydrolyze, compare the K_a and K_b values of the cation and anion. If $K_a > K_b$, then the solution will become more acidic. If $K_a < K_b$, the solution will become more basic. If $K_a = K_b$, the solution will be neutral.

Predicting the Acidic or Basic Nature of Solutions   **SAMPLE** problem

(a) **Predict whether a 0.10 mol/L solution of $NaNO_{2(aq)}$ will be acidic, basic, or neutral.**

(b) **Calculate the pH of a 0.10 mol/L solution of $NaNO_{2(aq)}$.**

(a) First, write the dissociation equation for $NaNO_{2(s)}$.

$$NaNO_{2(s)} \rightarrow Na_{(aq)}^+ + NO_{2(aq)}^-$$

Next, examine the cation. $Na_{(aq)}^+$ is a Group 1 metal cation so will have no effect on pH. Now, examine the anion, $NO_{2(aq)}^-$. It is the conjugate base of the weak acid, $HNO_{2(aq)}$, so will act as a base and increase the pH.

   A 0.1 mol/L solution of $NaNO_{2(aq)}$ will be basic.

(b) Since $Na_{(aq)}^+$ cannot hydrolyze, the pH of a 0.10 mol/L $NaNO_{2(aq)}$ solution is determined solely by the reaction of $NO_{2(aq)}^-$ with water.

   Write the chemical equation for the hydrolysis reaction between the base, $NO_{2(aq)}^-$, and water, then write the corresponding K_b expression.

$$NO_{2(aq)}^- + H_2O_{(l)} \rightleftharpoons OH_{(aq)}^- + HNO_{2(aq)}$$

$$\frac{[OH_{(aq)}^-][HNO_{2(aq)}]}{[NO_{2(aq)}^-]} = K_b$$

Calculate the value of K_b for $NO_{2(aq)}^-$ from the value of K_a for $HNO_{2(aq)}$ (which you can find in Appendix C7).

$$K_a K_b = K_w$$

$$K_b = \frac{K_w}{K_a}$$

$$= \frac{1.0 \times 10^{-14}}{7.2 \times 10^{-4}}$$

$$K_b = 1.4 \times 10^{-11}$$

Now, construct an ICE table to show the changes in concentration that occur as the reaction reaches equilibrium. Note that, in the reaction, one mole of $HNO_{2(aq)}$ and one mole of $OH_{(aq)}^-$ are produced for every mole of $NO_{2(aq)}^-$ that reacts. Let x represent the changes in concentration that occur as the reaction establishes equilibrium.

Table 4 ICE Table for the Hydrolysis of $NO_{2(aq)}^-$

	$NO_{2(aq)}^-$	$+$ $H_2O_{(l)}$	$\rightleftharpoons$ $OH_{(aq)}^-$	$+$ $HNO_{2(aq)}$
Initial concentration (mol/L)	0.10	—	0	0
Changes in concentration (mol/L)	$-x$	—	$+x$	$+x$
Equilibrium concentration (mol/L)	$0.10 - x$	—	x	x

$$\frac{[OH_{(aq)}^-][HNO_{2(aq)}]}{[NO_{2(aq)}^-]} = K_b$$

$$[OH_{(aq)}^-] = [HNO_{2(aq)}] = x$$

$$[NO_{2(aq)}^-] = 0.10 - x = 0.10$$

(Since the value of K_b is so small, we will make the assumption that $(0.10 - x) \doteq 0.10$. This simplification is warranted by the hundred rule. Check it yourself.)

$$\frac{x^2}{0.10} \doteq 1.4 \times 10^{-11}$$

$$x \doteq \sqrt{1.4 \times 10^{-12}}$$

$$x \doteq 1.2 \times 10^{-6}$$

As usual, we now justify our simplifying assumption, using the 5% rule.

$$\frac{1.2 \times 10^{-6}}{0.10} \times 100\% = 1.2 \times 10^{-3}\%$$

Since $1.2 \times 10^{-3}\% < 5\%$, our assumption is justified. Therefore,

$$x \doteq 1.2 \times 10^{-6}$$

$$[OH_{(aq)}^-] \doteq 1.2 \times 10^{-6} \text{ mol/L}$$

Now, we can calculate pOH and pH.

$$pOH = -\log [OH_{(aq)}^-]$$

$$\doteq -\log(1.2 \times 10^{-6} \text{ mol/L})$$

$$pOH \doteq 5.92$$

$$pH + pOH = 14.00$$

$$= 14.00 - pOH$$

$$\doteq 14.00 - 5.92$$

$$pH \doteq 8.08$$

The pH of a 0.10 mol/L $NaNO_{2(aq)}$ solution is 8.08.

Example

(a) Predict whether a 0.20 mol/L solution of ammonium chloride, $NH_4Cl_{(aq)}$, will be acidic, basic, or neutral.

(b) Calculate the pH of a 0.20 mol/L solution of $NH_4Cl_{(aq)}$.

Solution

(a) First, write the dissociation equation for $NH_4Cl_{(aq)}$.

$$NH_4Cl_{(aq)} \rightarrow NH_{4(aq)}^+ + Cl_{(aq)}^-$$

Since $Cl_{(aq)}^-$ is the conjugate base of a strong acid, it will not affect the pH of the solution. $NH_{4(aq)}^+$ is the conjugate acid of the weak base, $NH_{3(aq)}$, so it will hydrolyze according to the following equation:

$$NH_{4(aq)}^+ + H_2O_{(l)} \rightleftharpoons H_3O_{(aq)}^+ + NH_{3(aq)}$$

A solution of $NH_4Cl_{(aq)}$ will be acidic.

(b) $NH_{4(aq)}^+ + H_2O_{(l)} \rightleftharpoons H_3O_{(aq)}^+ + NH_{3(aq)}$

$$K_a = 5.8 \times 10^{-10} \text{ (from Appendix C7)}$$

$$\frac{[H_3O_{(aq)}^+][NH_{3(aq)}]}{[NH_{4(aq)}^+]} = K_a$$

Table 5 ICE Table for the Hydrolysis of $NH_{4(aq)}^+$

	$NH_{4(aq)}^+$ + $H_2O_{(l)}$ $\rightleftharpoons$ $H_3O_{(aq)}^+$ + $NH_{3(aq)}$			
Initial concentration (mol/L)	0.20	—	0	0
Changes in concentration (mol/L)	$-x$	—	$+x$	$+x$
Equilibrium concentration (mol/L)	$0.20 - x$	—	x	x

$$\frac{[H_3O^+{}_{(aq)}][NH_{3(aq)}]}{[NH_4{}^+{}_{(aq)}]} = K_a$$

$$\frac{x^2}{0.20 - x} = 5.8 \times 10^{-10}$$

Predicting whether $0.20 - x \doteq 0.20$...

$$\frac{[HA]_{initial}}{K_a} = \frac{0.20}{5.8 \times 10^{-10}}$$

$$= 3.4 \times 10^8$$

Since $3.4 \times 10^8 >> 100$, we assume that $0.20 - x \doteq 0.20$.

$$\frac{x^2}{0.20} \doteq 5.8 \times 10^{-10}$$

$$x^2 \doteq 2.9 \times 10^{-9}$$

$$x \doteq 5.4 \times 10^{-5}$$

Since $x = [H_3O_{(aq)}^+]$,

$$[H_3O_{(aq)}^+] \doteq 5.4 \times 10^{-5}$$

$$pH = -\log [H_3O_{(aq)}^+]$$

$$\doteq -\log(5.4 \times 10^{-5})$$

$$pH \doteq 4.$$

The pH of a 0.20 mol/L $NH_4Cl_{(aq)}$ solution is 4.

▶ **Practice**

Understanding Concepts

1. Predict whether the following solutions are acidic, basic, or neutral. Provide explanations to support your predictions.

   (a) ammonium phosphate, $(NH_4)_3PO_{4(aq)}$ (fertilizer)

   (b) ammonium sulfate, $(NH_4)_2SO_{4(aq)}$ (fertilizer)

   (c) magnesium oxide, $MgO_{(aq)}$ (milk of magnesia)

2. Predict whether a solution of sodium sulfite, $Na_2SO_{3(aq)}$ (photographic developer) will be acidic, basic, or neutral.

3. Calculate the pH of a 0.30 mol/L ammonium nitrate (fertilizer) solution.

4. Calculate the pH of 0.25 mol/L $NH_4Br_{(aq)}$.

5. Predict whether a solution of $NH_4C_2H_3O_{2(aq)}$ is acidic, basic, or neutral. Explain.

Making Connections

6. What kind of fertilizers would be appropriate for acid-loving plants like evergreens?

Hydrolysis of Amphoteric Ions

Remember that all polyatomic ions whose chemical formulas begin with hydrogen, H (e.g., $HCO_{3(aq)}^-$ and $HSO_{4(aq)}^-$) are amphoteric. As mentioned in Section 8.1, the term amphiprotic may also be used to describe such entities because they can either donate or accept a hydrogen ion (proton).

$NaHCO_{3(s)}$ dissolves in water, forming a conducting solution containing sodium and hydrogen carbonate ions.

$$NaHCO_{3(s)} \rightarrow Na_{(aq)}^+ + HCO_{3(aq)}^-$$

Because it is amphoteric, the $HCO_{3(aq)}^-$ ion may hydrolyze as an acid or a base, according to the following equilibrium equations:

$$HCO_{3(aq)}^- + H_2O_{(l)} \rightleftharpoons H_3O_{(aq)}^+ + CO_{3(aq)}^{2-} \qquad \text{(acid hydrolysis)}$$
$$HCO_{3(aq)}^- + H_2O_{(l)} \rightleftharpoons OH_{(aq)}^- + H_2CO_{3(aq)} \qquad \text{(base hydrolysis)}$$

Testing a sodium hydrogen carbonate solution with litmus paper reveals that it is basic. Therefore, we assume that the base hydrolysis equilibrium dominates in solution.

Similarly, the acidic character of a sodium hydrogen sulfate, $NaHSO_{4(s)}$ solution is explained by the dissociation and subsequent acid hydrolysis of the hydrogen sulfate ion.

$$NaHSO_{4(s)} \rightarrow Na_{(aq)}^+ + HSO_{4(aq)}^- \qquad \text{(dissociation)}$$
$$HSO_{4(aq)}^- + H_2O_{(l)} \rightleftharpoons H_3O_{(aq)}^+ + SO_{4(aq)}^{2-} \qquad \text{(acid hydrolysis)}$$

In both cases, the hydrolyzing ion may act as an acid or a base, but one equilibrium predominates and gives rise to the overall acidic or basic character of the aqueous solution. How do we predict which one will win out? The following discussion describes the theory.

Picture a laboratory setting in which a student tests the pH of a sodium dihydrogen borate, $NaH_2BO_{3(aq)}$, solution. The solution has a pH greater than 7. What reaction occurs to make the solution basic?

The following equations describe the dissociation of $NaH_2BO_{3(s)}$ in water, and the hydrolysis of the $H_2BO_{3(aq)}^-$ ions in solution.

$$NaH_2BO_{3(s)} \rightarrow Na_{(aq)}^+ + H_2BO_{3(aq)}^- \qquad \text{(dissociation)}$$
$$H_2BO_{3(aq)}^- + H_2O_{(l)} \rightleftharpoons H_3O_{(aq)}^+ + HBO_{3(aq)}^{2-} \qquad \text{(acid hydrolysis)}$$
$$H_2BO_{3(aq)}^- + H_2O_{(l)} \rightleftharpoons OH_{(aq)}^- + H_3BO_{3(aq)} \qquad \text{(base hydrolysis)}$$

Since the solution is basic when tested with an acid–base indicator, $H_2BO_{3(aq)}^-$ must hydrolyze primarily as a base. Is there a way to predict the acid–base character of a solution without testing it directly in the laboratory? Remember that when a salt containing both a hydrolyzing anion and a hydrolyzing cation, such as $NH_4CN_{(aq)}$, is dissolved in water, we predict the acid–base characteristics of the solution by comparing K_a and K_b values. The following sample problem will provide you with a similar model for predicting hydrolysis in solutions of amphoteric ions.

Predicting the Acidity of Amphoteric Ions
SAMPLE problem

Predict whether an aqueous solution of baking soda, $NaHCO_{3(s)}$, is acidic, basic, or neutral.

First, write the dissociation equation.

$$NaHCO_{3(s)} \rightarrow Na_{(aq)}^+ + HCO_{3(aq)}^- \qquad \text{(dissociation)}$$

Next, find the acid ion dissociation constant (from a reference table), as if the hydrogen carbonate ion were to act as an acid:

$$HCO_{3(aq)}^- + H_2O_{(l)} \rightleftharpoons CO_{3(aq)}^{2-} + H_3O_{(aq)}^+ \qquad K_a = 4.7 \times 10^{-11} \text{ (hydrolysis as an acid)}$$

Next, find the base ion dissociation constant, as if the hydrogen carbonate ion were to act as a base (use either a table of K_b values, or calculate from K_a):

$$HCO_{3(aq)}^- + H_2O_{(l)} \rightleftharpoons H_2CO_{3(aq)} + OH_{(aq)}^- \qquad K_b = 2.7 \times 10^{-8} \text{ (hydrolysis as a base)}$$

According to the relative values of K_a and K_b, the baking soda solution should be basic because the K_b is larger than the K_a. The hydrogen carbonate ion is a stronger base than it is an acid.

Example
Predict whether a solution of $NaHBO_{3(aq)}$ is acidic, basic, or neutral.

Solution

$$NaH_2BO_{3(s)} \rightarrow Na_{(aq)}^+ + H_2BO_{3(aq)}^- \qquad \text{(dissociation)}$$

$$H_2BO_{3(aq)}^- + H_2O_{(l)} \rightleftharpoons HBO_{3(aq)}^{2-} + H_3O_{(aq)}^+ \qquad K_a = 5.8 \times 10^{-10} \text{ (hydrolysis as an acid)}$$

$$H_2BO_{3(aq)}^- + H_2O_{(l)} \rightleftharpoons H_3BO_{3(aq)} + OH_{(aq)}^- \qquad K_b = 1.7 \times 10^{-5} \text{ (hydrolysis as a base)}$$

Since $K_b > K_a$, a solution of $NaH_2BO_{3(aq)}$ is basic.

▶ Practice

Understanding Concepts

7. Make a list of all amphoteric ions from your acid–base table, Appendix C7.

8. Predict whether the following solutions are acidic, basic, or neutral.
   (a) $NaHSO_{4(aq)}$
   (b) $Na_2HPO_{4(aq)}$

Hydrolysis of Metal and Nonmetal Oxides

When calcium oxide, $CaO_{(s)}$, is dissolved in water it produces a basic solution. However, an aqueous carbon dioxide solution, $CO_{2(aq)}$, is acidic. Can we explain this evidence using the hydrolysis theory? First, we must realize that calcium oxide and carbon dioxide both have low solubility in water. Therefore, we show the pure substance *reacting* with water rather than dissolving in water.

$$CaO_{(s)} + H_2O_{(l)} \rightarrow Ca^{2+}_{(aq)} + 2\,OH^-_{(aq)}$$

As you know, most metal oxides have low solubility in water, but the accepted theory is that the solid state oxide ions are converted quantitatively into aqueous hydroxide ions by the reaction with water to form a basic solution, according to the following equation.

$$O^{2-}_{(s)} + H_2O_{(l)} \rightarrow 2\,OH^-_{(aq)}$$

The $O^{2-}_{(s)}$ ions do not exist in aqueous solution; they are quantitatively converted to $OH^-_{(aq)}$ ions from the solid state (in $CaO_{(s)}$). However, the equation may be used to explain the basic nature of the solutions of metal oxides.

Now, let us try to explain the acidic character of carbon dioxide in water. A possible explanation is the two-step process presented below.

$$CO_{2(g)} + H_2O_{(l)} \rightleftharpoons H_2CO_{3(aq)}$$
$$\underline{H_2CO_{3(aq)} + H_2O_{(l)} \rightleftharpoons H_3O^+_{(aq)} + HCO^-_{3(aq)}}$$
$$CO_{2(g)} + 2\,H_2O_{(l)} \rightleftharpoons H_3O^+_{(aq)} + HCO^-_{3(aq)} \quad \text{(net equation)}$$

Chemists have done numerous tests on metal oxides and nonmetal oxides to determine the acidic and basic character of the solutions formed. Their evidence led to the following generalizations.

- Metal oxides react with water to produce basic solutions.
- Nonmetal oxides react with water to produce acidic solutions.

▶ **SAMPLE** problem | **Predicting the Acidity of Solutions of Oxides**

Predict, using the hydrolysis concept, whether solutions of the following oxides will be acidic, neutral, or basic. Write an appropriate hydrolysis equation in each case.
(a) magnesium oxide, $MgO_{(s)}$
(b) sulfur dioxide, $SO_{2(g)}$

(a) Magnesium oxide, $MgO_{(s)}$, is a metal oxide, so will react with water to form a basic solution, according to the following hydrolysis equation:

$$MgO_{(s)} + H_2O_{(l)} \rightleftharpoons Mg^{2+}_{(aq)} + 2\,OH^-_{(aq)}$$

The hydroxide ions are produced according to the following equation:

$$O^{2-}_{(s)} + H_2O_{(l)} \rightleftharpoons 2\,OH^-_{(aq)}$$

(b) Sulfur dioxide, $SO_{2(g)}$, is a nonmetal oxide, so will react with water to form an acidic solution, according to the following hydrolysis equation:

$$SO_{2(g)} + 2\,H_2O_{(l)} \rightleftharpoons H_3O^+_{(aq)} + HSO^-_{3(aq)}$$

Example

Use hydrolysis concepts to predict whether an aqueous solution of copper(II) oxide, $CuO_{(s)}$, is acidic, basic, or neutral.

Solution

Copper (II) oxide, $CuO_{(s)}$, is a metal oxide, so will react with water to form a basic solution, according to the following equation:

$$CuO_{(s)} + H_2O_{(l)} \rightleftharpoons Cu^{2+}_{(aq)} + 2\,OH^-_{(aq)}$$
$$O^{2-}_{(s)} + H_2O_{(l)} \rightleftharpoons 2\,OH^-_{(aq)}$$

The acidic properties of nonmetal oxides are responsible for many natural processes such as the the weathering of minerals, the absorption of nutrients by the roots of plants, and the chemistry of tooth decay. For example, limestone, $CaCO_{3(s)}$, dissolves in water that is made acidic by the dissolution of $CO_{2(g)}$ from the atmosphere, according to the following two-step process:

$$CO_{2(g)} + H_2O_{(l)} \rightleftharpoons H^+_{(aq)} + HCO^-_{3(aq)} \qquad \text{(acidification of rain or ground water)}$$
$$H^+_{(aq)} + CaCO_{3(s)} \rightleftharpoons Ca^{2+}_{(aq)} + HCO^-_{3(aq)} \qquad \text{(dissolution of } CaCO_{3(s)} \text{ in acidic solution)}$$

Limestone caves and the stalagmites and stalactites they contain are formed by the action of acidic ground water on limestone deposits in Earth's crust.

A cave is formed when the above reactions proceed to the right and calcium carbonate in underground limestone deposits is dissolved. Stalactites are formed in the cave when the aqueous solution on the ceiling evaporates. As it evaporates, carbon dioxide escapes. This shifts the equilibrium to the left, causing solid calcium carbonate to precipitate. This precipitation causes stalactites to form on the ceilings and stalagmites to form where drops of the solution drip onto the cave floor. Stalactite and stalagmite formation is very slow; the growth rate is, on average, about one millimetre per century.

Tooth decay is also caused by the dissolution of minerals in acidic solutions. Tooth enamel is composed of the mineral hydroxyapatite, $Ca_5(PO_4)_3OH$ ($K_{sp} = 6.8 \times 10^{-37}$). Acids (in saliva, from fruits and fruit juices, or formed when sugars are metabolized by bacteria in the mouth) react with hydroxyapatite, leading to erosion of tooth enamel and eventually tooth decay (cavities). Fluoride salts (as a source of fluoride ions) are often added to toothpaste and to drinking water in treatment plants in some communities to help prevent tooth decay. The fluoride ions react with the $Ca_5(PO_4)_3OH$ in tooth enamel to form more decay-resistant fluorapatite $Ca_5(PO_4)_3F$ ($K_{sp} = 1.0 \times 10^{-60}$).

DID YOU *KNOW* ?

Growing Up or Growing Down?
When viewing photos of stalactites and stalagmites in a cave, you may not be able to tell which way is up. Look at the tips of the formations. Stalactites almost always have pointed tips, whereas stalagmites are usually rounded or flat. Is the photo right-side up or upside down?

▶ Practice

Understanding Concepts

9. For each of the following solutions of compounds, write an ionization or dissociation equation where appropriate, and then write a net equation showing reactions with water to produce either hydronium or hydroxide ions (consistent with the evidence).
   (a) $Na_2O_{(s)}$ in solution turns red litmus blue.
   (b) $SO_{3(g)}$ in solution turns blue litmus red.

SUMMARY *Acid–Base Characteristics of Salts*

Table 6

Type of Salt	Examples	Description	pH
cation of strong base and anion of strong acid	$NaCl_{(aq)}$, $KNO_{3(aq)}$ $NaI_{(aq)}$	does not hydrolyze as an acid or as a base	neutral
cation of strong base and anion of weak acid	$NaC_2H_3O_{2(aq)}$ $KF_{(aq)}$	anion hydrolyzes as a base; cation does not hydrolyze	basic
cation is conjugate acid of a weak base; anion of a strong acid	$NH_4NO_{3(aq)}$ $NH_4Cl_{(aq)}$	cation hydrolyzes as an acid; anion does not hydrolyze	acidic
cation is conjugate acid of a weak base; anion is conjugate base of a weak acid	$NH_4C_2H_3O_{2(aq)}$ $NH_4F_{(aq)}$	cation hydrolyzes as an acid; anion hydrolyzes as a base	acidic if $K_a > K_b$ basic if $K_a < K_b$ neutral if $K_a = K_b$
cation is highly charged metal ion; anion of a strong acid	$AlCl_{3(aq)}$ $FeI_{3(aq)}$	hydrated cation hydrolyzes as an acid; anion does not hydrolyze	acidic
metal oxides	$CuO_{(s)}$	solid state oxide ion reacts with water to form $OH^-_{(aq)}$	basic
nonmetal oxides	$CO_{2(g)}$	compound reacts with water to form $H_3O^+_{(aq)}$	acidic

The Lewis Model of Acids and Bases

Lewis acid an electron-pair acceptor

Lewis base an electron-pair donor

The Arrhenius and Brønsted–Lowry models successfully explain much of the behaviour of acids and bases. Nevertheless, both of these models contain limitations. Remember that the Arrhenius model could not adequately explain the basic properties of an aqueous ammonia solution. In the early 1920s, G. N. Lewis expanded the Brønsted–Lowry model to encompass a number of substances that would not normally be classified as Brønsted–Lowry acids or bases. According to the Lewis model, a **Lewis acid** is an electron-pair acceptor and a **Lewis base** is an electron-pair donor. In order to act as a Lewis base, a substance must possess a non-bonded pair of electrons in one of its orbitals. Conversely, in order to act as a Lewis acid, a substance must possess an empty valence orbital that may accept (share) a pair of non-bonding valence electrons from a Lewis base. The following structural formula equation illustrates the reaction between a Lewis acid, $H^+_{(aq)}$, and a Lewis base, $H_2O_{(l)}$, to form $H_3O^+_{(aq)}$.

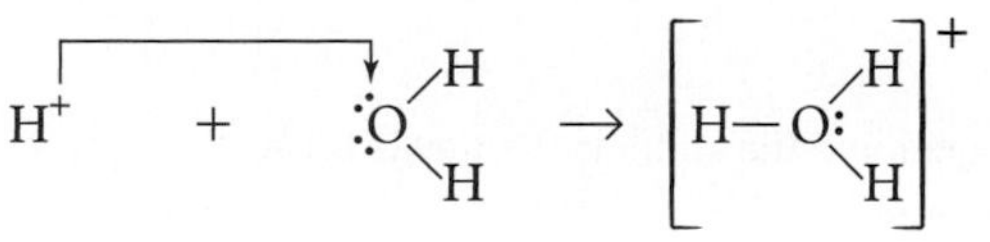

H⁺ ion · water · hydronium ion
(Lewis acid) · (Lewis base)

In the above reaction, the $H^+_{(aq)}$ ion (proton) acts as the Lewis acid (electron pair acceptor) and the water molecule acts as a Lewis base (electron-pair donor).

In this case, the Lewis base is also a Brønsted–Lowry base, and the Lewis acid is also a Brønsted–Lowry acid. However, in the following example, this is not the case.

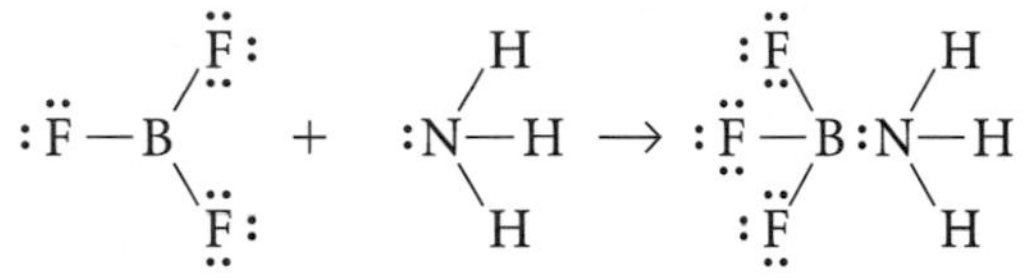

boron trifluoride · ammonia · boron trifluoride
(Lewis acid) · (Lewis base) · ammonia complex

Note that the Brønsted–Lowry model also accounts for ammonia as a base, but does not characterize boron trifluoride as an acid.

The Lewis acid–base theory explains the reaction between BF_3 (boron trifluoride) and NH_3 (ammonia). Boron trifluoride is a trigonal planar molecule with sp^2 hybrid orbitals. This arrangement leaves an empty $2p$ orbital on the boron atom that is able to accommodate the pair of nonbonding electrons in the sp^3 hybrid orbital of NH_3. A covalent bond forms between the boron and the nitrogen, forming the compound BF_3NH_3.

The Lewis acid–base theory can also explain why small, highly positive ions such as Al^{3+} form complex ions in water:

$$Al^{3+}_{(aq)} + 6\,H_2O_{(l)} \rightleftharpoons Al(H_2O)_6^{3+}_{(aq)}$$

This is a Lewis acid–base reaction. The water molecules each possess nonbonding pairs of electrons and so act as Lewis bases, and the $Al^{3+}_{(aq)}$ ion possesses empty $3s$, $3p$, and $3d$ orbitals that may accommodate electron pairs. The electron configuration of the Al^{3+} ion can be represented as Al^{3+}: [Ne] $3s^0\,3p^0$. The complex $Al(H_2O)_6^{3+}_{(aq)}$ is formed when an Al^{3+} ion, acting as a Lewis acid, bonds with six water molecules, each acting as a Lewis base.

The Lewis model is a more general model of acids and bases that not only encompasses the Brønsted–Lowry model, but extends it. Brønsted–Lowry acids and bases are acids and bases in the Lewis model, but the reverse is not always true.

Example

Identify the Lewis acid and the Lewis base in the following reaction.

$$SO_{3(aq)} + H_2O_{(l)} \rightleftharpoons H_2SO_{4(aq)}$$

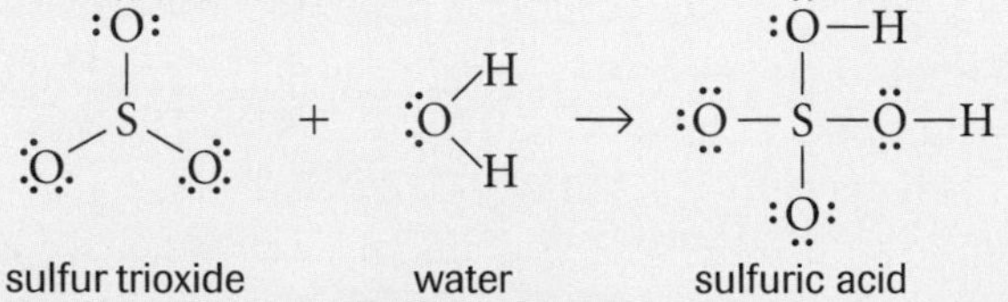

sulfur trioxide · water · sulfuric acid
(Lewis acid) · (Lewis base)

Solution

$SO_{3(aq)}$ is the Lewis acid and the water is the Lewis base.

> ### Practice

Understanding Concepts

12. Identify the Lewis acid and the Lewis base in each of the following reactions.
    (a) $H^+_{(aq)} + OH^-_{(aq)} \rightarrow H_2O_{(l)}$
    (b) $H^+_{(aq)} + NH_{3(aq)} \rightleftharpoons NH^+_{4(aq)}$

> ## Section 8.3 Questions

Understanding Concepts

1. Predict whether the following solutions are acidic, basic, or neutral. Provide explanations to support your predictions.
   (a) table salt, $NaCl_{(aq)}$ (saline or brine) solution
   (b) aluminum chloride, $AlCl_{3(aq)}$ (antiperspirant)
   (c) $Na_2CO_{3(aq)}$ (washing soda)

2. Predict whether a solution of ammonium carbonate (a component of baking powder) will be acidic, basic, or neutral.

3. What is the strongest possible acid in an aqueous solution, and what is the strongest possible base in an aqueous solution?

4. Will an aqueous solution of $BeCl_{2(aq)}$ turn litmus red or blue? Explain.

5. Predict whether the following solutions are acidic, basic, or neutral.
   (a) a carbonated beverage containing $CO_{2(aq)}$ (pop)
   (b) strontium oxide, $SrO_{(s)}$

Applying Inquiry Skills

6. Analyze the Evidence (**Table 7**) to determine the kind of solutions (acidic, basic, or neutral) formed when Period 3 oxides are placed in water.

Experimental Design

Oxides of elements in Period 3 are tested in water, using litmus paper. To all the oxides, a strong acid (hydrochloric acid) and a strong base (sodium hydroxide) are added to determine if a neutralization reaction occurs.

Evidence

Table 7 Litmus and Neutralization Tests on Oxides

Oxide	Litmus test	$HCl_{(aq)}$ test	$NaOH_{(aq)}$ test
$Na_2O_{(s)}$	red to blue	neutralizes	no reaction
$MgO_{(s)}$	red to blue	neutralizes	no reaction
$Al_2O_{3(s)}$	(insoluble)	neutralizes	neutralizes
$SiO_{2(s)}$	(insoluble)	no reaction	neutralizes
$P_2O_{3(s)}$	blue to red	no reaction	neutralizes
$SO_{3(g)}$	blue to red	no reaction	neutralizes
$Cl_2O_{(g)}$	blue to red	no reaction	neutralizes

Extension

7. Nitrogen oxides such as nitrogen dioxide, $NO_{2(g)}$, are produced in automobile engines and released into the atmosphere via exhaust fumes. In the atmosphere, nitrogen dioxide will dissolve in droplets of rain. Use hydrolysis concepts to predict what will happen to the pH of rain when $NO_{2(g)}$ dissolves.

In your previous chemistry course, you became familiar with acid–base titrations. A **titration** (**Figure 1**) is a technique involving the progressive addition of a solution (called the **titrant**) from a buret into a measured volume of a solution (called the **sample**) in an Erlenmeyer flask. The purpose is to determine the amount of a specified chemical in the sample, from which the molar mass and the concentration of the chemical may be determined. This is possible because the titrant and the sample contain substances that react according to known stoichiometry. In general, the sample is placed in a receiving flask, and the titrant is dispensed from a buret.

titration the precise addition of a solution in a buret into a measured volume of a sample solution

titrant the solution in a buret during a titration

sample the solution being analyzed in a titration

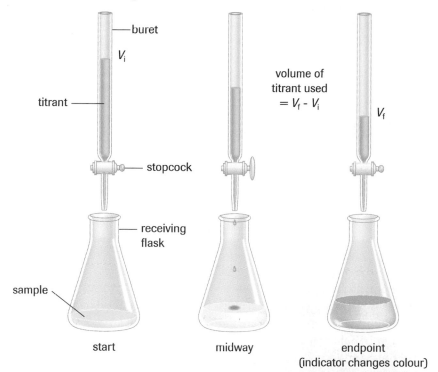

Figure 1
In an acid–base titration, the concentration of an acid or base solution of unknown concentration is determined by the delivery (from a buret) of a measured volume of a solution of known concentration (the titrant). If the sample in the flask is an acid, the titrant used is a base, and vice versa.

Before the sample can be analyzed, it is important that we know, to a considerable degree of accuracy, the concentration of the titrant, because this concentration is used to calculate the concentration of the sample. Measuring the titrant's concentration is called "standardizing" the titrant, and is often the first stage of a titration. Common **primary standards** are sodium carbonate, $Na_2CO_{3(s)}$ (a base used to standardize an acid titrant), and potassium hydrogen phthalate, $KHC_7H_4O_{4(s)}$ (an acid used to standardize a basic titrant). These are appropriate choices because, due to their purity, we can be confident that their stated concentrations are accurate. The standardization process is itself a titration.

Hydrochloric acid and sodium hydroxide are not used as primary standards because hydrogen chloride gas vaporizes, especially from concentrated hydrochloric acid solutions, and solid sodium hydroxide is hygroscopic—it gains mass by absorbing water from the air.

Notice that the primary standards are solids at SATP; they are not hygroscopic like sodium hydroxide, and they do not vaporize like hydrochloric acid. They are available in very pure form, and produce colourless aqueous solutions.

primary standard a chemical, available in a pure and stable form, for which an accurate concentration can be prepared; the solution is then used to determine precisely, by means of titrating, the concentration of a titrant

An acid–base titration involves the reaction between an acid and a base. In a typical titration, a measured volume of standardized titrant is added to a known volume of the sample. The addition continues until the amount of reactant in the sample is just consumed by the reactant in the titrant. This is called the **equivalence point** or the stoichiometric point.

Before beginning an acid–base titration, a drop or two of an acid–base indicator is added to the sample. The acid–base indicator signals the end of the titration by sharply and permanently changing colour when the equivalence point is reached. At this point, the volume of titrant added is recorded, and the number of moles of titrant used to reach the equivalence point is calculated. Ideally, an acid–base indicator is chosen such that the **endpoint** occurs precisely at the equivalence point (so the colour change occurs sharply at the point where a complete reaction is attained). However, it is virtually impossible to achieve such precision in the laboratory. Consequently, titrations are subject to considerable experimental error. Bromothymol blue and phenolphthalein are common indicators used in acid–base titrations. Bromothymol blue changes from yellow to blue in the range pH 6.0 to 7.6. Phenolphthalein changes from colourless to pink in the range pH 8.2 to 10.0.

Titrating a Strong Acid with a Strong Base

No doubt you are familiar with the titration of a strong acid with a strong base from previous chemistry courses. Consider the reaction between hydrochloric acid, $HCl_{(aq)}$, a strong acid, with sodium hydroxide, $NaOH_{(aq)}$, a strong base:

$HCl_{(aq)} + NaOH_{(aq)} \rightarrow H_2O_{(l)} + NaCl_{(aq)}$ (molecular equation)

$H_3O^+_{(aq)} + OH^-_{(aq)} \rightarrow 2\,H_2O_{(l)}$ (net ionic equation)

$H^+_{(aq)} + OH^-_{(aq)} \rightarrow H_2O_{(l)}$ (abbreviated net ionic equation)

What chemical changes occur during a reaction such as this?

The stoichiometric analysis that follows a titration usually involves the average of at least three consistent titration trials. Chemists demand high reproducibility from titration results. Equivalence points that are more than ±0.2 mL from a set of consistent results are recorded but not included in the average volume of titrant used. Titrations must involve reactions that obey the assumptions required of stoichiometric calculations—the reactions must be stoichiometric (reactants react according to the ratio of coefficients in the reaction equation), spontaneous, fast, and quantitative (the reaction proceeds until all reacting entities are consumed).

▶ **SAMPLE** problem | **Chemical Changes During a Strong Acid/Strong Base Titration**

In a titration, 20.00 mL of 0.300 mol/L $HCl_{(aq)}$ is titrated with standardized 0.300 mol/L $NaOH_{(aq)}$. What is the amount of unreacted $HCl_{(aq)}$ and the pH of the solution after the following volumes of $NaOH_{(aq)}$ have been added?

(a) 0 mL

(b) 10.0 mL

(c) 20.0 mL

Since $HCl_{(aq)}$ is a strong acid, the major entities in the sample solution are $H^+_{(aq)}$, $Cl^-_{(aq)}$, and $H_2O_{(l)}$.

(a) Before adding titrant ($NaOH_{(aq)}$), the pH of the solution is equal to the pH of 0.300 mol/L $HCl_{(aq)}$.

Since $HCl_{(aq)}$ ionizes completely,

$[H^+_{(aq)}] = 0.300$ mol/L

$pH = -\log(0.300)$

$pH = 0.5$

Now, we calculate the amount of unreacted $HCl_{(aq)}$, n_{HCl}.

$V_{HCl} = 20.00$ mL

$C_{HCl} = 0.300$ mol/L

$n_{HCl} = V_{HCl} \times C_{HCl}$

$= 20.00$ mL̸ $\times 0.300$ mol/L̸

$n_{HCl} = 6.00$ mmol

Since HCl is a strong acid, 6.00 mmol $HCl_{(aq)}$ contains 6.00 mmol $H^+_{(aq)}$ and 6.00 mmol $Cl^-_{(aq)}$.

(b) Adding 10.00 mL of 0.300 mol/L $NaOH_{(aq)}$ introduces the following amount of $NaOH_{(aq)}$ to the solution.

$V_{NaOH} = 10.00$ mL

$C_{NaOH} = 0.300$ mol/L $NaOH_{(aq)}$

$n_{NaOH} = V_{NaOH} \times C_{NaOH}$

$= 10.00$ mL̸ $\times 0.300$ mol/L̸

$n_{NaOH} = 3.00$ mmol

Since NaOH is a strong base, 3.00 mmol $NaOH_{(aq)}$ introduces 3.00 mmol $Na^+_{(aq)}$ and 3.00 mmol $OH^-_{(aq)}$ into the solution.

Before reaction, the solution contains the following amounts of the major entities (in addition to $H_2O_{(l)}$):

From HCl:

$H^+_{(aq)}$	$Cl^-_{(aq)}$
6.00 mmol	6.00 mmol

From NaOH:

$OH^-_{(aq)}$	$Na^+_{(aq)}$
3.00 mmol	3.00 mmol

Hydrogen ions react with hydroxide ions, according to the neutralization reaction:

	$H^+_{(aq)}$	$+$	$OH^-_{(aq)}$	$\rightarrow$	$H_2O_{(l)}$
Before reaction:	6.00 mmol		3.00 mmol		
After reaction:	6.00 mmol $-$ 3.00 mmol		3.00 mmol $-$ 3.00 mmol		
	3.00 mmol		0 mmol		
	(excess)				

Sodium ions and chloride ions remain in solution. As you learned in Section 8.3, $Na^+_{(aq)}$ does not hydrolyze and thus cannot affect the pH of an aqueous solution. Chloride ions also do not affect the pH because as the conjugate bases of the strong acid, HCl, they do not hydrolyze.

Notice that the total volume of the solution after the addition of 10 mL $NaOH_{(aq)}$ is 20.00 mL $+$ 10.00 mL $=$ 30.00 mL.

The pH of the sample is calculated using the amount of excess $H^+_{(aq)}$ and the total volume of the sample:

$$[H^+_{(aq)}] = \frac{3.00 \text{ mmol}}{30.00 \text{ mL}}$$

$$[H^+_{(aq)}] = 0.100 \text{ mol/L}$$

We can now calculate the pH:

$$pH = -\log[H^+_{(aq)}]$$
$$= -\log(0.100 \text{ mol/L})$$
$$pH = 1.000$$

The pH rises as more and more titrant is added to the receiving flask. **Table 1** records the results of calculations like those above for successive additions of base to the mixture.

Table 1 Titration of 20.00 mL of 0.300 mol/L $HCl_{(aq)}$ with 0.300 mol/L $NaOH_{(aq)}$

Initial Volume of $HCl_{(aq)}$ (mL)	Initial Amount of $HCl_{(aq)}$ (mol)	Volume of $NaOH_{(aq)}$ Added (mL)	Amount of $NaOH_{(aq)}$ (mol)	Amount of Excess Reagent (mol)	Total Volume of Solution (mL)	Molar Concentration of Ion in Excess (mol/L)	pH
20.00	6.000×10^{-3}	0	0	6.000×10^{-3} (H^+)	20.00	0.3000 (H^+)	0.52
20.00	6.000×10^{-3}	5.00	1.500×10^{-3}	4.500×10^{-3} (H^+)	25.00	0.1400 (H^+)	0.85
20.00	6.000×10^{-3}	15.00	4.500×10^{-3}	1.500×10^{-3} (H^+)	35.00	4.286×10^{-2} (H^+)	1.36
20.00	6.000×10^{-3}	19.00	5.700×10^{-3}	3.000×10^{-4} (H^+)	39.00	7.692×10^{-3} (H^+)	2.11
20.00	6.000×10^{-3}	19.90	5.970×10^{-3}	3.000×10^{-5} (H^+)	39.90	7.519×10^{-4} (H^+)	3.12
20.00	6.000×10^{-3}	19.99	5.997×10^{-3}	3.000×10^{-6} (H^+)	39.99	7.502×10^{-5} (H^+)	4.12
20.00	6.000×10^{-3}	20.00	6.000×10^{-3}	0	40.00	0	7.00
20.00	6.000×10^{-3}	20.01	6.003×10^{-3}	3.000×10^{-6} (OH^-)	40.01	7.498×10^{-5} (OH^-)	9.87
20.00	6.000×10^{-3}	20.10	6.030×10^{-3}	3.000×10^{-5} (OH^-)	40.10	7.481×10^{-4} (OH^-)	10.87
20.00	6.000×10^{-3}	21.00	6.300×10^{-3}	3.000×10^{-4} (OH^-)	41.00	7.317×10^{-3} (OH^-)	11.86
20.00	6.000×10^{-3}	40.00	1.200×10^{-2}	6.000×10^{-3} (OH^-)	80.00	7.500×10^{-2} (OH^-)	12.88

(c) From **Table 1** you can see that, when 20.00 mL of $NaOH_{(aq)}$ has been added, all of the $HCl_{(aq)}$ is neutralized and the equivalence point is reached. The resulting solution contains only H_2O, $Na^+_{(aq)}$, and $Cl^-_{(aq)}$ ions. Earlier, you learned that neither of these ions is able to hydrolyze. Therefore, the concentration of hydrogen ions is equal to that in pure water, 1.0×10^{-7} mol/L, and the pH is 7.00.

If we were to add more $NaOH_{(aq)}$ to the receiving flask, we would simply be adding more $Na^+_{(aq)}$ and $OH^-_{(aq)}$ ions. The pH of the resulting solution is determined by the amount of $OH^-_{(aq)}$ added, taking into account the increase in volume of the solution.

Consider the acid–base titration in the above Sample Problem. When the pH of the solution in the receiving flask is plotted against the volume of 0.300 mol/L $NaOH_{(aq)}$ added, the result is a titration curve (**Figure 2**). The curve for the titration of 0.300 mol/L $HCl_{(aq)}$ with 0.300 mol/L $NaOH_{(aq)}$ is typical of that for the titration of any strong acid with any strong base.

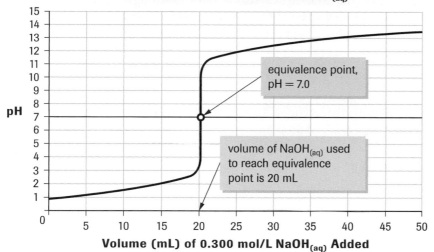

Titration Curve for Titration of 20 mL of 0.300 mol/L HCl$_{(aq)}$ with 0.300 mol/L Standardized NaOH$_{(aq)}$

equivalence point, pH = 7.0

volume of NaOH$_{(aq)}$ used to reach equivalence point is 20 mL

pH

Volume (mL) of 0.300 mol/L NaOH$_{(aq)}$ Added

Figure 2
This curve is typical of curves depicting the titration of a strong acid with a strong base. Notice that the curve sweeps up and to the right as NaOH$_{(aq)}$ is added, beginning at a pH below 7 and ending at a pH above 7. The equivalence point is reached at pH 7.

Notice the shape of the titration curve. At the beginning, the pH rises gradually as base is added. In this section, the pH remains relatively constant even though small amounts of base are being added. This occurs because the first amount of titrant is immediately consumed, leaving an excess of strong acid, and the pH is changed very little. Following this flat region there is a very rapid increase in pH for a very small additional volume of the titrant. The *midpoint* of the sharp increase in pH occurs at a pH of 7.00. This is the equivalence point of the reaction and corresponds (with an appropriate indicator) to the endpoint of the titration. Theoretically, the equivalence point represents the stoichiometric quantity of titrant required by the balanced chemical equation. This is true of all titrations of a strong monoprotic acid with any strong base. Since the conjugates of a strong acid and a strong base cannot hydrolyze (being a weak conjugate base and weak conjugate acid, respectively), the pH is determined by the auto-ionization of water only, and is thus 7.00. The curve then bends to reflect a more gradual increase in pH as excess base is added.

▶ *Practice*

Understanding Concepts

1. (a) When 25 mL of 0.10 mol/L HBr$_{(aq)}$ is titrated with 0.10 mol/L NaOH$_{(aq)}$, what is the pH at the equivalence point?
   (b) Select an appropriate indicator for this titration from Appendix C10.

2. In a titration, how many millilitres of 0.23 mol/L NaOH$_{(aq)}$ must be added to 11 mL of 0.18 mol/L HI$_{(aq)}$ to reach the equivalence point?

Applying Inquiry Skills

3. (a) When a titration is being performed, it is common to wash a clinging drop of titrant into the receiving flask with a stream of distilled water from a wash bottle. Why is it important that a drop of titrant clinging to the tip of the buret be forced into the sample solution?
   (b) Will the added water affect the results of the titration? If so, why? If not, why not?

Answers

1. (a) 7

2. 8.6 mL

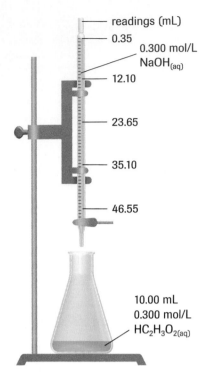

readings (mL)

0.35

0.300 mol/L NaOH$_{(aq)}$

12.10

23.65

35.10

46.55

10.00 mL
0.300 mol/L
HC$_2$H$_3$O$_{2(aq)}$

Figure 3
Even though the base is strong and the acid is weak, if they are in the same concentrations, an equivalent amount of base will be contained in the same volume as the sample of acid.

Titrating a Weak Acid with a Strong Base

Now we will consider the titration of a weak acid with a strong base. In this titration we place 20.00 mL of 0.300 mol/L HC$_2$H$_3$O$_{2(aq)}$ in the receiving flask and a standardized solution of 0.300 mol/L NaOH$_{(aq)}$ in the buret (**Figure 3**). Acetic acid ionizes very little in aqueous solution, forming the following equilibrium:

$$HC_2H_3O_{2(aq)} \rightleftharpoons H^+_{(aq)} + C_2H_3O^-_{2(aq)} \qquad K_a = 1.8 \times 10^{-5}$$

The low K_a indicates that acetic acid exists primarily as HC$_2$H$_3$O$_2$ molecules in solution. When NaOH$_{(aq)}$ is added drop by drop to the acetic acid solution in the receiving flask, the OH$^-_{(aq)}$ ions react with the HC$_2$H$_3$O$_{2(aq)}$ molecules according to the following neutralization equation.

$$HC_2H_3O_{2(aq)} + OH^-_{(aq)} \rightarrow C_2H_3O^-_{2(aq)} + H_2O_{(l)}$$

As OH$^-_{(aq)}$ from NaOH are slowly added to the acetic acid solution, more and more acetic acid molecules are consumed in the neutralization reaction. It is important to remember that although acetic acid is weak, it reacts *quantitatively* with the hydroxide ions until essentially all of the molecules are consumed.

▶ **SAMPLE** problem | **Chemical Changes During a Weak Acid/Strong Base Titration**

In a titration of 20.00 mL of 0.300 mol/L HC$_2$H$_3$O$_{2(aq)}$ with standardized 0.300 mol/L NaOH$_{(aq)}$, what is the amount of unreacted HC$_2$H$_3$O$_{2(aq)}$ and the pH of the solution:

(a) before titration begins;

(b) during titration but before the equivalence point (10.00 mL of 0.300 mol/L NaOH$_{(aq)}$ added);

(c) at the equivalence point (20.00 mL of 0.300 mol/L NaOH$_{(aq)}$ added); and

(d) beyond the equivalence point.

This reaction consumes HC$_2$H$_3$O$_{2(aq)}$ and continually shifts the equilibrium to the right until all HC$_2$H$_3$O$_{2(aq)}$ molecules have been consumed and the reaction reaches the equivalence point. Beginning with 20.00 mL of 0.300 mol/L HC$_2$H$_3$O$_{2(aq)}$ in the flask, we know that an equivalent amount of OH$^-_{(aq)}$ will be contained in 20.00 mL of 0.300 mol/L NaOH$_{(aq)}$.

Calculating the pH of the acetic acid sample before adding sodium hydroxide is staightforward and similar to pH calculations you performed earlier in this chapter. However, when sodium hydroxide is added to the solution, it reacts with acetic acid and causes the acetic acid equilibrium to shift to the right. The extent of this shift determines the pH of the solution. The calculation of the solution's pH will be simplified if you deal with the stoichiometry of the acid–base (acetic acid–sodium hydroxide) reaction *separately* from the shift in the acetic acid equilibrium. Therefore, we will carry out two separate calculations:

1. A *stoichiometry calculation* to determine the concentration of weak acid (acetic acid) remaining after the acid–base reaction, and

2. An *equilibrium calculation* to determine the new position of the weak acid (acetic acid) equilibrium and the solution's pH.

Remember to always carry out these two calculations separately.

(a) **Before titration**, the pH of the solution in the receiving flask is the pH of a 0.300 mol/L $HC_2H_3O_{2(aq)}$ solution, so we do not need to perform a stoichiometric calculation, only the equilibrium calculation familiar from Section 8.2.

Equilibrium Calculation

The major entities in solution (before ionization occurs) are

$HC_2H_3O_{2(aq)}$ and $H_2O_{(l)}$.

We will begin by constructing an ICE table for the ionization process, letting x represent the changes in concentration that occur as equilibrium is established.

Table 2 ICE Table for the Ionization of $HC_2H_3O_{2(aq)}$

	$HC_2H_3O_{2(aq)}$	$\rightleftharpoons$	$H^+_{(aq)}$	+	$C_2H_3O^-_{2(aq)}$
Initial concentration (mol/L)	0.300		0.00		0.00
Change in concentration (mol/L)	$-x$		$+x$		$+x$
Equilibrium concentration (mol/L)	$0.300 - x$		$+x$		$+x$

Now, substitute the equilibrium concentration values into the K_a expression for this equilibrium, and solve for x.

$$\frac{[H^+_{(aq)}][C_2H_3O^-_{2(aq)}]}{[HC_2H_3O_{2(aq)}]} = K_a$$

(From Appendix C7, $K_a = 1.8 \times 10^{-5}$)

$$\frac{x^2}{0.300 - x} = 1.8 \times 10^{-5}$$

If we assume that $0.300 - x \doteq 0.300$ (use the hundred rule), the equilibrium expression becomes

$$\frac{x^2}{0.300} \doteq 1.8 \times 10^{-5}$$

$$x^2 \doteq 5.4 \times 10^{-6}$$

$$x \doteq 2.3 \times 10^{-3}$$

Validating the assumption ...

$$\frac{2.3 \times 10^{-3}}{0.300} \times 100\% = 0.8\%$$

Since 0.8% < 5%, the assumption is valid, and

$$x \doteq 2.3 \times 10^{-3} \text{ mol/L}$$
$$[H^+_{(aq)}] \doteq 2.3 \times 10^{-3} \text{ mol/L}$$

$$pH = -\log[H^+_{(aq)}]$$
$$\doteq -\log(2.3 \times 10^{-3})$$
$$pH \doteq 2.62$$

Before titration begins, the pH of the sample is 2.62.

Calculating the amount of $HC_2H_3O_{2(aq)}$ in the initial sample is also familiar.

$$V_{HC_2H_3O_2} = 20.00 \text{ mL}$$
$$C_{HC_2H_3O_2} = 0.300 \text{ mol/L}$$

$$n_{HC_2H_3O_2} = V_{HC_2H_3O_2} \times C_{HC_2H_3O_2}$$
$$= 20.00 \text{ mL} \times 0.300 \text{ mol/L}$$
$$n_{HC_2H_3O_2} = 6.00 \text{ mmol}$$

Notice, we are assuming that none of the $HC_2H_3O_{2(aq)}$ is in ionized form (owing to the small K_a value for acetic acid).

(b) During titration, remember to perform stoichiometry calculations separately from equilibrium calculations.

Stoichiometry Calculations
The major entities in the solution (before reaction) are:
$HC_2H_3O_{2(aq)}$, $OH^-_{(aq)}$, $Na^+_{(aq)}$, and $H_2O_{(l)}$

The $OH^-_{(aq)}$ ions from the added $NaOH_{(aq)}$ will react with the strongest proton donor (acid) in solution. Although water may act as an acid, it is a much weaker acid ($K_w = 1.0 \times 10^{-14}$) than acetic acid ($K_a = 1.8 \times 10^{-5}$). Therefore, $OH^-_{(aq)}$ ions react with $HC_2H_3O_{2(aq)}$, according to the following *neutralization reaction* equation:

$$HC_2H_3O_{2(aq)} + OH^-_{(aq)} \rightarrow C_2H_3O^-_{2(aq)} + H_2O_{(l)}$$

Adding 10.00 mL of 0.300 mol/L $NaOH_{(aq)}$ introduces the following amount of $NaOH_{(aq)}$ to the solution.

$$V_{NaOH} = 10.00 \text{ mL}$$
$$C_{NaOH} = 0.300 \text{ mol/L}$$

$$n_{NaOH} = V_{NaOH} \times C_{NaOH}$$
$$= 10.00 \text{ mL} \times 0.300 \text{ mol/L}$$
$$n_{NaOH} = 3.00 \text{ mmol}$$

Since NaOH is a strong base, 3.00 mmol $NaOH_{(aq)}$ introduces 3.00 mmol $Na^+_{(aq)}$ and 3.00 mmol $OH^-_{(aq)}$ into the solution. This amount of $NaOH_{(aq)}$ reacts with an equal amount (3.00 mmol) of $HC_2H_3O_{2(aq)}$, according to the neutralization reaction, leaving 3.00 mmol of acetic acid unreacted. (Remember that $Na^+_{(aq)}$ does not affect the acid–base characteristics of an aqueous solution, and water does not react with $OH^-_{(aq)}$ because it is a much weaker acid than acetic acid.)
The volume of the solution is now

20.00 mL	+	10.00 mL	=	30.00 mL
(volume of original solution)		(volume of $NaOH_{(aq)}$ added)		(total volume)

Equilibrium Calculations
First, we list the major entities in the solution *after the neutralization reaction has taken place*. The major entities are:
$HC_2H_3O_2$, $C_2H_3O^-_{2(aq)}$, $Na^+_{(aq)}$, and $H_2O_{(l)}$.

Notice that there are no $OH^-_{(aq)}$ ions in the solution; they were all consumed in the neutralization.
The acetic acid and acetate ions are components of the following equilibrium:
$$HC_2H_3O_{2(aq)} \rightleftharpoons H^+_{(aq)} + C_2H_3O^-_{2(aq)}$$

We can now construct an ICE table to monitor the changes that occur as the equilibrium shifts in response to the neutralization we analyzed above. However, before we construct the ICE table, we must calculate the $[HC_2H_3O_{2(aq)}]$ and $[C_2H_3O_{2(aq)}^-]$ in the new volume *after the neutralization reaction has taken place, but before a shift in equilibrium occurs*. (This is a purely theoretical condition, as these two changes actually occur at the same time. However, it simplifies the calculations considerably.)

$$[HC_2H_3O_{2(aq)}] = \frac{3.00 \text{ mmol}}{30.0 \text{ mL}}$$

$$[HC_2H_3O_{2(aq)}] = 0.100 \text{ mol/L}$$

and

$$[C_2H_3O_{2(aq)}^-] = \frac{3.00 \text{ mmol}}{30.0 \text{ mL}}$$

$$[C_2H_3O_{2(aq)}^-] = 0.100 \text{ mol/L}$$

Let x represent the changes in concentration that occur as the system re-establishes equilibrium. The ICE table looks like this:

Table 3   ICE Table for the Ionization of $HC_2H_3O_{2(aq)}$

	$HC_2H_3O_{2(aq)}$ $\rightleftharpoons$	$H_{(aq)}^+$ +	$C_2H_3O_{2(aq)}^-$
Initial concentration (mol/L)	0.100	0.00	0.100
Change in concentration (mol/L)	$-x$	$+x$	$0.100 + x$
Equilibrium concentration (mol/L)	$0.100 - x$	$+x$	$0.100 + x$

Substituting the equilibrium values into the following K_a expression and solving for x, we get:

$$\frac{[H_{(aq)}^+][C_2H_3O_{2(aq)}^-]}{[HC_2H_3O_{2(aq)}]} = K_a$$

$$\frac{x(0.100 + x)}{0.100 - x} = 1.8 \times 10^{-5}$$

Apply the hundred rule to show that a simplifying assumption is warranted:

$$\frac{[HC_2H_3O_{2(aq)}]}{K_a} = \frac{0.100}{1.8 \times 10^{-5}}$$

$$\frac{[HC_2H_3O_{2(aq)}]}{K_a} = 5.6 \times 10^3$$

Since $5.6 \times 10^3 > 100$, we can assume that
$0.100 - x \doteq 0.100$, and
$0.100 + x \doteq 0.100$.

The equilibrium equation simplifies to:

$$\frac{(x)(0.100)}{0.100} \doteq 1.8 \times 10^{-5}$$

$$x \doteq 1.8 \times 10^{-5}$$

Now we validate the simplifying assumption with the 5% rule:

$$\frac{1.8 \times 10^{-5}}{0.100} \times 100\% = 0.018\%$$

Since $0.018\% < 5\%$, the simplifying assumption is justified.

$$x \doteq 1.8 \times 10^{-5}$$
$$[H^+_{(aq)}] \doteq 1.8 \times 10^{-5}$$

$$pH \doteq -\log(1.8 \times 10^{-5})$$
$$pH \doteq 4.74$$

The pH of the solution after adding 10.00 mL of 0.300 mol/L $NaOH_{(aq)}$ is 4.74.

(c) At the equivalence point, 20.00 mL of 0.300 mol/L $NaOH_{(aq)}$ has been added.

The H^+ ions of all $HC_2H_3O_{2(aq)}$ molecules in the flask have combined with the $OH^-_{(aq)}$ ions of all the added $NaOH_{(aq)}$ to form $H_2O_{(l)}$. What remains is a solution containing the following major entities:
$Na^+_{(aq)}$, $C_2H_3O^-_{2(aq)}$, and $H_2O_{(l)}$.

Being a Group 1 metal ion, $Na^+_{(aq)}$ does not hydrolyze, so does not affect the pH of the solution. However, the $C_2H_3O^-_{2(aq)}$ ion is the conjugate base of a weak acid (acetic acid), and *does* hydrolyze. The pH of the solution will therefore be determined by the extent of this hydrolysis reaction. As in part (b), we begin with stoichiometry.

Stoichiometry Calculations
The reaction of acetate ions with water is used to determine the pH of the solution:

$$C_2H_3O^-_{2(aq)} + H_2O_{(l)} \rightleftharpoons HC_2H_3O_{2(aq)} + OH^-_{(aq)}$$

This is a typical weak base equilibrium characterized by the following base ionization constant equation:

$$\frac{[HC_2H_3O_{2(aq)}][OH^-_{(aq)}]}{[C_2H_3O^-_{2(aq)}]} = K_b$$

The value of K_b for $C_2H_3O^-_{2(aq)}$ can be determined from the K_a of its conjugate acid, $HC_2H_3O_{2(aq)}$, and K_w, as follows.

$$K_a = 1.8 \times 10^{-5}$$
$$K_w = 1.0 \times 10^{-14}$$
$$K_aK_b = K_w$$

$$K_b = \frac{K_w}{K_a}$$

$$= \frac{1.0 \times 10^{-14}}{1.8 \times 10^{-5}}$$
$$K_b = 5.6 \times 10^{-10}$$

Therefore,

$$\frac{[HC_2H_3O_{2(aq)}][OH^-_{(aq)}]}{[C_2H_3O^-_{2(aq)}]} = 5.6 \times 10^{-10}$$

To reach the equivalence point, we added 20.00 mL of $NaOH_{(aq)}$ to the original 20.00 mL of solution. At the equivalence point, the total volume of the solution is 40.00 mL.

Since we began with 6.00 mmol $HC_2H_3O_{2(aq)}$, we will end up with 6.00 mmol of $C_2H_3O^-_{2(aq)}$ at the equivalence point. This amount of $C_2H_3O^-_{2(aq)}$ is dissolved in 40.00 mL of solution:

$$[C_2H_3O^-_{2(aq)}] = \frac{6.00 \text{ mmol}}{40.00 \text{ mL}}$$

$$[C_2H_3O^-_{2(aq)}] = 0.150 \text{ mol/L}$$

Notice that this is the acetate ion concentration *before* equilibrium is established.

Equilibrium Calculations

We construct an ICE table to monitor the changes in concentrations as equilibrium is established.

Table 4   ICE Table for the Hydrolysis Reaction of $C_2H_3O_2^-{}_{(aq)}$

	$C_2H_3O_2^-{}_{(aq)}$ + $H_2O_{(l)}$ $\rightleftharpoons$ $HC_2H_3O_{2(aq)}$ + $OH^-{}_{(aq)}$			
Initial concentration (mol/L)	0.150	—	0.00	0.000
Change in concentration (mol/L)	$-x$	—	$+x$	$+x$
Equilibrium concentration (mol/L)	$0.150 - x$	—	x	x

$$C_2H_3O_2^-{}_{(aq)} + H_2O_{(l)} \rightleftharpoons HC_2H_3O_{2(aq)} + OH^-{}_{(aq)} \quad K_b = 5.6 \times 10^{-10} \text{ (from an earlier calculation)}$$

$$\frac{[HC_2H_3O_{2(aq)}][OH^-{}_{(aq)}]}{[C_2H_3O_2^-{}_{(aq)}]} = K_b$$

Substituting the equilibrium values into the ionization constant equation, we get:

$$\frac{x^2}{(0.150 - x)} = 5.6 \times 10^{-10}$$

If we assume that $0.150 - x \doteq 0.150$ (after using the hundred rule)

$$\frac{x^2}{0.150} \doteq 5.6 \times 10^{-10}$$
$$x^2 \doteq 8.4 \times 10^{-11}$$
$$x \doteq 9.2 \times 10^{-6}$$

(The 5% rule justifies the assumption that $0.150 - x \doteq 0.150$.)

$$[OH^-{}_{(aq)}] \doteq 9.2 \times 10^{-6} \text{ mol/L}$$
$$pOH \doteq -\log(9.2 \times 10^{-6})$$
$$pOH \doteq 5.03$$

Since

$$pH + pOH = 14.00$$
$$pH = 14.00 - pOH$$
$$\doteq 14.00 - 5.03$$
$$pH \doteq 8.97$$

The pH of the solution at the equivalence point is 8.97, and there are no $HC_2H_3O_{2(aq)}$ molecules left in solution.

It is important to realize that the equivalence point does not necessarily mean the point at which the sample is neutral (with a pH of 7). While this is true of all strong acid–strong base titrations, it may not be true of other titrations. Rather, the equivalence point is the point at which equivalent amounts of reactants have reacted, according to the balanced chemical equation. At the equivalence point, there may still be ions in solution that affect the pH.

Not surprisingly, the titration of any weak acid with a strong base generally produces a basic solution with a pH greater than 7.

Since the reactants and products of this titration are all clear, colourless solutions, the analyst would select an indicator that changes colour at or near a pH of 8.97, to show that the equivalence point has been reached. Phenolphthalein is an ideal choice since it changes from colourless to pink between pH 8.2 and 10.0. In order to select an appropriate indicator, the pH at the equivalence point must be calculated *before* the titration.

Table 5 records more of the results of the titration in the Sample Problem above, and
Figure 4 illustrates the titration curve. Notice the similarities and the differences between
this titration curve and that of the strong acid–strong base titration (**Figure 2**). At the
beginning, the pH rises gradually as base is added. This relatively flat region of the
curve is where a *buffering action* occurs, which means that the pH remains relatively
constant even though small amounts of strong base are being added. This occurs
because the first amount of titrant is immediately consumed, leaving an excess of acid,
and the pH is changed very little. Following this buffering region is a very rapid increase
in pH for a very small additional volume of the titrant. You will learn more about
buffering action in Section 8.5.

Table 5 Titration of 20.00 mL of 0.3000 mol/L HC$_2$H$_3$O$_{2(aq)}$ with 0.3000 mol/L NaOH$_{(aq)}$

Volume of base added (mL)	Molar concentration of entity in parentheses (mol/L)	pH
None	2.3×10^{-3} (H$^+$)	2.6
5.00	5.5×10^{-5} (H$^+$)	4.26
19.90	9.1×10^{-8} (H$^+$)	7.04
19.99	9.1×10^{-9} (H$^+$)	8.04
20.00	9.3×10^{-6} (OH$^-$)	8.97
20.01	7.6×10^{-5} (OH$^-$)	9.88
20.10	7.4×10^{-4} (OH$^-$)	10.87
21.00	7.3×10^{-3} (OH$^-$)	11.86
30.00	6.0×10^{-2} (OH$^-$)	12.78

Titration Curve for Titrating 0.300 mol/L HC$_2$H$_3$O$_{2(aq)}$ with 0.300 mol/L NaOH$_{(aq)}$

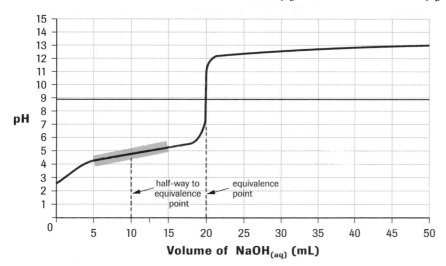

Figure 4
This curve is typical of curves depicting the titration of a weak acid with a strong base. Notice that the curve sweeps up and to the right as NaOH$_{(aq)}$ is added, beginning at a pH below 7 and ending at a pH above 7. The equivalence point is reached at a pH greater than 7.

▶ *Practice*

Understanding Concepts

4. If 25.00 mL of 0.20 mol/L HCO$_2$H$_{(aq)}$ is titrated with 0.20 mol/L NaOH$_{(aq)}$ (the titrant), determine the pH
  (a) before titration begins;
  (b) after 10.00 mL of NaOH$_{(aq)}$ has been added;
  (c) at the equivalence point.

5. If 10.0 mL of 0.250 mol/L NaOH$_{(aq)}$ is added to 30.0 mL of 0.17 mol/L HOCN$_{(aq)}$, what is the pH of the resulting solution?

Answers

4. (a) 2.22
   (b) 3.57
   (c) 8.37

5. 3.74

Titrating a Weak Base with a Strong Acid

The titration of a weak base with a strong acid can be analyzed the same way we analyzed the titration of a weak acid and a strong base. Consider the titration of 20.0 mL of 0.100 mol/L NH$_{3(aq)}$ with 0.100 mol/L HCl$_{(aq)}$. The relevant chemical equations are:

$$NH_{3(aq)} + HCl_{(aq)} \rightarrow NH_4Cl_{(aq)} \qquad \text{(molecular)}$$
$$NH_{3(aq)} + H^+_{(aq)} \rightarrow NH^+_{4(aq)} \qquad \text{(net ionic)}$$

Like the other titrations we have studied in this section, the titration of a weak base with a strong acid can be analyzed at the following four points in the titration:

(a) Before titration begins, when the receiving flask contains a dilute solution of NH$_{3(aq)}$. (We can find the pH by using the K_b of NH$_{3(aq)}$.)

(b) During titration, but before the equivalence point, when the solution contains significant amounts of unreacted NH$_{3(aq)}$ and NH$^+_{4(aq)}$ ions. At this stage, we can find the pH by using the K_b for NH$_{3(aq)}$ (or the K_a of NH$^+_{4(aq)}$) and by performing separate stoichiometry and equilibrium calculations. Remember to use the *total* volume of the solution in the flask.

(c) At the equivalence point, where the solution contains H$_2$O$_{(l)}$, NH$^+_{4(aq)}$, and Cl$^-_{(aq)}$ ions only. Since Cl$^-_{(aq)}$ ions do not hydrolyze, we can calculate the pH of the solution by using the K_a of NH$^+_{4(aq)}$. Again, keep stoichiometric and equilibrium calculations separate and remember to use the total volume of the solution in the receiving flask when calculating final concentrations.

(d) Beyond the equivalence point, where the pH decreases quantitatively. We find the pH by calculating the $[H^+_{(aq)}]$ produced by the ionization of the excess $HCl_{(aq)}$ added.

In general, the pH at the equivalence point, for a titration of a weak base with a strong acid, will be lower than 7 (**Figure 5**).

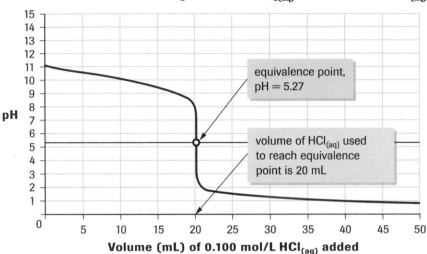

Titration Curve for Titrating 0.100 mol/L NH$_{3(aq)}$ with 0.100 mol/L HCl$_{(aq)}$

equivalence point, pH = 5.27

volume of HCl$_{(aq)}$ used to reach equivalence point is 20 mL

pH

Volume (mL) of 0.100 mol/L HCl$_{(aq)}$ added

Figure 5
This curve is typical of curves depicting the titration of a weak base with a strong acid. Notice that the curve sweeps down and to the right as HCl$_{(aq)}$ is added, beginning at a pH higher than 7 and ending at a pH below 7. The equivalence point is reached at a pH lower than 7.

Answer

6. (a) 11.21
   (b) 5.18

SUMMARY *Titration Characteristics*

Table 6

Type of titration	pH at equivalence point	Entity determining pH at equivalence point
strong acid and weak base	< 7	conjugate acid of weak base
strong base and strong acid	7	autoionization of water
strong base and weak acid	> 7	conjugate base of weak acid

▶ **Practice**

Understanding Concepts

6. For the titration of 20.0 mL of 0.1500 mol/L NH$_{3(aq)}$ with 0.1500 mol/L HI$_{(aq)}$ (the titrant), calculate
   (a) the pH before any HI$_{(aq)}$ is added.
   (b) the pH at the equivalence point.

Acid–Base Indicators

The behaviour of acid–base indicators depends, in part, on both the Brønsted-Lowry concept and the equilibrium concept. An indicator is a conjugate weak acid–base pair formed when an indicator dye dissolves in water. If we use HIn$_{(aq)}$ to represent the acid form and In$^-_{(aq)}$ to represent the base form of any indicator, the following equilibrium can be written. (The colours of litmus are given below the equation as an example.)

conjugate pair

$$HIn_{(aq)} + H_2O_{(l)} \rightleftharpoons In^-_{(aq)} + H_3O^+_{(aq)}$$

acid   base

red   (litmus colour)   blue

According to Le Châtelier's principle, an increase in the hydronium ion concentration shifts the above equilibrium to the left. In acidic solutions, the primary form of the indicator is its un-ionized (acid) form. This happens, for example, when litmus is added to an acidic solution. Similarly, in basic solutions the hydroxide ions remove hydronium ions with the result that the equilibrium shifts to the right. Then the base colour of the indicator (In^-) predominates. Since different indicators have different acid strengths, the acidity or pH of the solution at which an indicator changes colour varies (**Figure 6**). These pH values have been measured and are reported in **Table 7** and in Appendix C9.

We can use the indicator equilibrium equation above to derive the following acid (indicator) ionization constant equation:

$$K_{In} = \frac{[In^-_{(aq)}][H_3O^+_{(aq)}]}{[HIn_{(aq)}]}$$

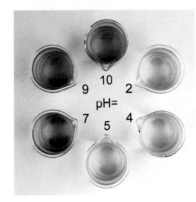

(a) bromothymol blue

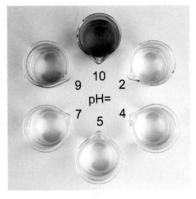

(b) phenolphthalein

Figure 6
Colour changes of common acid–base indicators

Table 7   Acid–Base Indicators

Common name of indicator	Suggested symbol	Colour of $HIn_{(aq)}$	Approximate pH range	Colour of $In^-_{(aq)}$	pK_{In}
methyl violet	HMv	yellow	0.0–1.6	blue	0.8
thymol blue*	H_2Tb	red	1.2–2.8	yellow	1.6
methyl yellow	HMy	red	2.9–4.0	yellow	3.3
congo red	HCr	blue	3.0–5.0	red	4.0
methyl orange	HMo	red	3.2–4.4	yellow	4.2
bromocresol green	HBg	yellow	3.8–5.4	blue	4.7
methyl red	HMr	red	4.8–6.0	yellow	5.0
chlorophenol red	HCh	yellow	5.2–6.8	red	6.0
bromothymol blue	HBb	yellow	6.0–7.6	blue	7.1
litmus	HLt	red	6.0–8.0	blue	7.2
phenol red	HPr	yellow	6.6–8.0	red	7.4
metacresol purple	HMp	yellow	7.4–9.0	purple	8.3
thymol blue*	HTb^-	yellow	8.0–9.6	blue	8.9
phenolphthalein	HPh	colourless	8.2–10.0	red	9.4
thymolphthalein	HTh	colourless	9.4–10.6	blue	9.9
alizarin yellow r	HAy	yellow	10.1–12.0	red	11.0
indigo carmine	HIc	blue	11.4–13.0	yellow	12.2
Clayton yellow	HCy	yellow	12.0–13.2	amber	12.7

*Thymol blue is a diprotic indicator that changes colour twice

In an acid–base titration, the pH changes sharply near the equivalence point. This large change in pH shifts the indicator's equilibrium from one colour state to another. The change in colour actually occurs over a small range in pH but, in a titration, the change in pH occurs so quickly that we see it as a sudden colour change occurring at the indicator's **transition point**. Note that, because a small amount of indicator is used in a titration (a few drops at most), it does not contribute to (or affect) the pH of the solution. Instead, the pH of the solution determines the position of the indicator equilibrium. The indicator "responds" to the pH conditions of the solution.

When selecting an indicator for a particular acid–base titration, the pH at the equivalence point must be known. Ideally, the pH of the titration's equivalence point should be reached at the point where half of all indicator molecules have changed colour so that at the equivalence point there will be equal concentrations of both forms of the indicator.

The equilibrium constant equation for the indicator equilibrium is

$$K_{In} = \frac{[In^-_{(aq)}][H_3O^+_{(aq)}]}{[HIn_{(aq)}]}$$

If, at the equivalence point,

$$[In^-_{(aq)}] = [HIn_{(aq)}]$$

then, $K_{In} = \dfrac{[\cancel{In^-_{(aq)}}][H_3O^+_{(aq)}]}{[\cancel{HIn_{(aq)}}]}$

and

$$K_{In} = [H_3O^+_{(aq)}]$$

Since K_{In} and $[H_3O^+_{(aq)}]$ (or $[H^+_{(aq)}]$) values are very small, we can conveniently convert them into pK_{In} and pH by taking the negative logarithm of each value as follows:

$$pK_{In} = -\log K_{In} \quad \text{and, of course,}$$
$$pH = -\log[H^+_{(aq)}]$$

Thus, for an ideal indicator,

$$pK_{In} = pH \quad \text{(at the equivalence point)}$$

This means that an indicator will be ideally suited to mark the endpoint of a titration if its pK_{In} equals the pH at the equivalence point of that particular titration (**Figure 7**).

Figure 7
Thymol blue is an unsuitable indicator for this titration because it changes colour before the equivalence point (pH 7). Alizarin yellow is also unsuitable because it changes colour after the equivalence point. Bromothymol blue is suitable because its endpoint of pH 6.8 (assume the middle of its pH range) closely matches the equivalence point of pH 7, and the colour change is completely on the vertical portion of the pH curve, within the range where there is rapid change in pH.

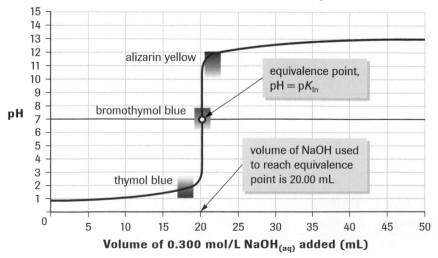

Titration Curve for Titrating 20.00 mL of 0.300 mol/L HCl$_{(aq)}$ with 0.300 mol/L NaOH$_{(aq)}$

alizarin yellow

equivalence point, pH = pK_{In}

bromothymol blue

volume of NaOH used to reach equivalence point is 20.00 mL

thymol blue

pH / Volume of 0.300 mol/L NaOH$_{(aq)}$ added (mL)

> ▶ **Practice**

Understanding Concepts

For the following questions, use Appendix C10.

7. Explain why bromocresol green is a better indicator than alizarin yellow in the titration of dilute ammonia with dilute hydrochloric acid.

8. Why must a very small amount of indicator be used in a titration?

9. If methyl red is used in the titration of dilute benzoic acid, $HC_7H_5O_{2(aq)}$, with dilute sodium hydroxide,$NaOH_{(aq)}$, will the endpoint of the titration correspond to the equivalence point? Explain.

Polyprotic Acid Titrations

The pH curve for the titration of hydrochloric acid with sodium hydroxide has only one observable endpoint (**Figure 7**), but the pH curve for the addition of $HCl_{(aq)}$ (titrant) to $Na_2CO_{3(aq)}$ (**Figure 8**) displays two equivalence points—two rapid changes in pH. pH curves such as this are typical of the titration of polyprotic acids or bases. Here, for example, two successive reactions occur. The two endpoints in **Figure 9** can be explained by two different proton transfer equations.

Remember that sodium carbonate is a strong electrolyte and so fully dissociates into $Na^+_{(aq)}$ and $CO_3^{2-}_{(aq)}$ ions.

$$Na_2CO_{3(aq)} \rightarrow 2\,Na^+_{(aq)} + CO_3^{2-}_{(aq)}$$

Therefore, the major entities in the receiving flask are $Na^+_{(aq)}$, $CO_3^{2-}_{(aq)}$, and $H_2O_{(l)}$. At the beginning of the titration, $H^+_{(aq)}$ ions from $HCl_{(aq)}$ react with $CO_3^{2-}_{(aq)}$ ions, since carbonate ions are the strongest base present in the initial mixture.

$$\underset{\text{from } HCl_{(aq)}}{H^+_{(aq)}} \quad + \quad \underset{\text{from } Na_2CO_{3(aq)}}{CO_3^{2-}_{(aq)}} \quad \rightarrow \quad HCO_3^-_{(aq)}$$

Then, in a second reaction, protons from $HCl_{(aq)}$ react with the hydrogen carbonate ions formed in the first reaction.

$$\underset{\text{from } HCl_{(aq)}}{H^+_{(aq)}} \quad + \quad \underset{\text{from first reaction}}{HCO_3^-_{(aq)}} \quad \rightarrow \quad H_2CO_{3(aq)}$$

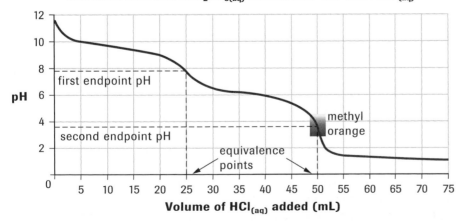

25.0 mL of 0.50 mol/L $Na_2CO_{3(aq)}$ Titrated with 0.50 mol/L $HCl_{(aq)}$

Figure 8

A pH curve for the addition of 0.50 mol/L $HCl_{(aq)}$ to a 25.0 mL sample of 0.50 mol/L $Na_2CO_{3(aq)}$.

Acid–Base Equilibrium **611**

Notice from observing the pH curve that each reaction requires about 25 mL of hydrochloric acid to reach the equivalence point and that the methyl orange colour change marks the second endpoint of the titration.

As you know, acids that can donate more than one proton are called *polyprotic* acids. This term applies to bases as well. The carbonate ion is a polyprotic base, called diprotic because it can accept two protons. Other polyprotic bases include sulfide ions and phosphate ions. Sulfide ions are diprotic; phosphate ions are triprotic. Their reactions with a strong acid are shown below.

$$S^{2-}_{(aq)} + H^+_{(aq)} \rightarrow HS^-_{(aq)}$$
$$HS^-_{(aq)} + H^+_{(aq)} \rightarrow H_2S_{(aq)}$$

$$PO_4^{3-}{}_{(aq)} + H^+_{(aq)} \rightarrow HPO_4^{2-}{}_{(aq)}$$
$$HPO_4^{2-}{}_{(aq)} + H^+_{(aq)} \rightarrow H_2PO_4^-{}_{(aq)}$$
$$H_2PO_4^-{}_{(aq)} + H^+_{(aq)} \rightarrow H_3PO_4{}_{(aq)}$$

Polyprotic acids, such as oxalic acid and phosphoric acid, can donate more than one proton. Oxalic acid is a diprotic acid; phosphoric acid is a triprotic acid. Their reactions with strong bases are shown below.

$$H_2C_2O_4{}_{(aq)} + OH^-_{(aq)} \rightarrow HC_2O_4^-{}_{(aq)}$$
$$HC_2O_4^-{}_{(aq)} + OH^-_{(aq)} \rightarrow C_2O_4^{2-}{}_{(aq)}$$

$$H_3PO_4{}_{(aq)} + OH^-_{(aq)} \rightarrow H_2PO_4^-{}_{(aq)}$$
$$H_2PO_4^-{}_{(aq)} + OH^-_{(aq)} \rightarrow HPO_4^{2-}{}_{(aq)}$$
$$HPO_4^{2-}{}_{(aq)} + OH^-_{(aq)} \rightarrow PO_4^{3-}{}_{(aq)}$$

Evidence from pH measurements indicates that polyprotic substances become weaker acids or bases with every proton donated or accepted. This occurs because it is easier to remove a H^+ ion (a proton) from neutral $H_3PO_4{}_{(aq)}$ than from the negatively charged $H_2PO_4^-{}_{(aq)}$ ion or the even more negatively charged $HPO_4^{2-}{}_{(aq)}$ ion. According to Le Châtelier's principle, with each successive proton removal, there are more $H^+_{(aq)}$ ions in solution pushing the reaction back toward the reactants.

Figure 9 shows the pH curve for phosphoric acid titrated with sodium hydroxide. Only two endpoints are present, corresponding to equivalence points of 25 mL and 50 mL of $NaOH_{(aq)}$ titrant added. At the first equivalence point, equal amounts of $H_3PO_4{}_{(aq)}$ and $OH^-_{(aq)}$ have been added.

$$H_3PO_4{}_{(aq)} + OH^-_{(aq)} \rightarrow H_2PO_4^-{}_{(aq)}$$

Since all the $H_3PO_4{}_{(aq)}$ has reacted, the second plateau (30 mL to 50 mL $NaOH_{(aq)}$ added) must represent the reaction of $OH^-_{(aq)}$ with $H_2PO_4^-{}_{(aq)}$. The second equivalence point (50 mL $NaOH_{(aq)}$ added) corresponds to the completion of the reaction of $H_2PO_4^-{}_{(aq)}$ with an additional 25 mL of $NaOH_{(aq)}$ solution added. No $H_2PO_4^-{}_{(aq)}$ remains.

$$H_2PO_4^-{}_{(aq)} + OH^-_{(aq)} \rightarrow HPO_4^{2-}{}_{(aq)} + H_2O_{(l)}$$

Notice that there is no apparent endpoint at 75 mL for the possible reaction of $HPO_4^{2-}{}_{(aq)}$ with $OH^-_{(aq)}$. A clue to this missing third endpoint can be obtained from Appendix C9. The hydrogen phosphate ion is an extremely weak acid ($K_a = 4.2 \times 10^{-13}$) and apparently does not quantitatively lose its proton to $OH^-_{(aq)}$. As a general rule, *only quantitative reactions produce detectable endpoints in an acid–base titration.*

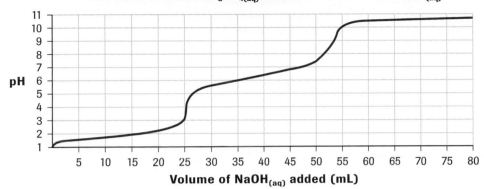

Figure 9
A pH curve for the addition of 0.50 mol/L NaOH$_{(aq)}$ to a 25.0-mL sample of 0.50 mol/L H$_3$PO$_{4(aq)}$ displays only two rapid changes in pH. This means that there are only two quantitative reactions for phosphoric acid with sodium hydroxide.

▶ *Practice*

Understanding Concepts

10. In an acid–base titration, 25.0 mL of 0.50 mol/L Na$_3$PO$_{4(aq)}$ was titrated with 0.50 mol/L HCl$_{(aq)}$ (**Figure 10**).
    (a) Write three Brønsted-Lowry equations that describe the reactions that may take place during the titration.
    (b) At what volumes of HCl$_{(aq)}$ added do the equivalence points occur?
    (c) Why do only two equivalence points show in Figure 10?

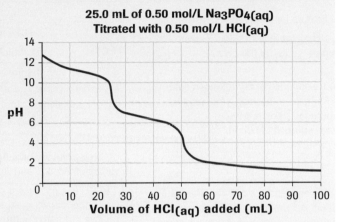

Figure 10

Answer

10 (b) 25 mL, 50 mL

▶ *Section 8.4 Questions*

Understanding Concepts

1. An acetic acid sample is titrated with sodium hydroxide.
   (a) Based on **Figure 11**, estimate the endpoint and the equivalence point.
   (b) Choose an appropriate indicator for this titration.
   (c) Write a Brønsted-Lowry equation for this reaction.

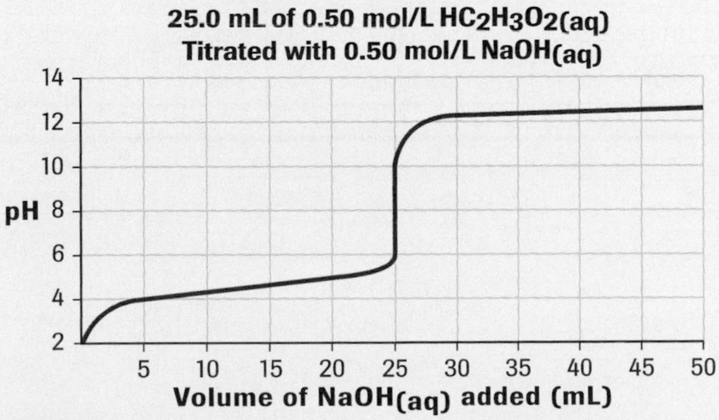

25.0 mL of 0.50 mol/L HC$_2$H$_3$O$_{2(aq)}$ Titrated with 0.50 mol/L NaOH$_{(aq)}$

Figure 11

2. Predict whether the pH endpoint is ≐ 7, > 7, or < 7 for each of the following acid–base titrations. Justify your predictions.
   (a) hydroiodic acid with sodium hydroxide
   (b) boric acid with sodium hydroxide
   (c) hydrochloric acid with magnesium hydroxide
   (d) hydrochloric acid with aqueous ammonia

3. Predict the pH of the following solutions. Justify your predictions.
   (a) NH$_4$Cl$_{(aq)}$ (b) Na$_2$S$_{(aq)}$ (c) KNO$_{3(aq)}$

4. How is a pH curve used to choose an indicator for a titration?

5. According to the table of acid–base indicators in Appendix C10, what is the colour of each of the following indicators in the solutions of given pH?
   (a) phenolphthalein in a solution with a pH of 11.7
   (b) bromothymol blue in a solution with a pH of 2.8
   (c) litmus in a solution with a pH of 8.2
   (d) methyl orange in a solution with a pH of 3.9

6. For the titration of 25.00 mL of 0.100 mol/L benzoic acid, $HC_7H_5O_{2(aq)}$, with 0.100 mol/L of sodium hydroxide, $NaOH_{(aq)}$, calculate the pH
   (a) before adding any $NaOH_{(aq)}$;
   (b) after 10 mL of $NaOH_{(aq)}$ has been added;
   (c) at the equivalence point.

7. Select an appropriate indicator for the titration in question 6.

8. (a) What is the pH at the equivalence point for each of the titrations in **Table 8**?

Table 8

Titrant	Sample
(i) 0.200 mol/L $HCl_{(aq)}$	20.0 mL of 0.100 mol/L $NH_{3(aq)}$
(ii) 0.150 mol/L $NaOH_{(aq)}$	10.0 mL of 0.350 mol/L $HC_2H_3O_{2(aq)}$
(iii) 0.250 mol/L $HBr_{(aq)}$	15.0 mL of 0.150 mol/L $N_2H_{4(aq)}$

(b) Select appropriate indicators for the titrations in **Table 8**.

9. If 25 mL of 0.23 mol/L $NaOH_{(aq)}$ were added to 45 mL of 0.10 mol/L $HC_2H_3O_{2(aq)}$, what would be the pH of the resulting solution?

10. In an investigation, separate samples of an unknown solution turned both methyl orange and bromothymol blue to yellow, and turned bromocresol green to blue.
    (a) Estimate the pH of the unknown solution.
    (b) Calculate the approximate hydronium ion concentration.

Applying Inquiry Skills

11. Oxalic acid reacts quantitatively in a two-step reaction with a sodium hydroxide solution. Assuming that an excess of sodium hydroxide is added drop by drop, sketch a pH curve (without any numbers) for the titration.

12. Given the following experimental design and evidence, determine the approximate pH of three unknown solutions.

Experimental Design

The unknown solutions were labelled A, B, and C. Each solution was tested by adding each of several indicators to samples.

Evidence

Table 9

Solution	Indicator	Indicator colour
A	methyl violet	blue
	methyl orange	yellow
	methyl red	red
	phenolphthalein	colourless
B	indigo carmine	blue
	phenol red	yellow
	bromocresol green	blue
	methyl red	yellow
C	phenolphthalein	colourless
	thymol blue	yellow
	bromocresol green	yellow
	methyl orange	red

13. Design an experiment that uses indicators to identify which of three unknown solutions labelled X, Y, and Z have pH values of 3.5, 5.8, and 7.8. There are several acceptable designs!

14. Given the following experimental evidence, determine the relative strengths of the acids in **Table 10**. All acid solutions were of equal molar concentration

Table 10 Indicator Colours with Various Acids

Acid	Methyl violet	Thymol blue	Benzopurpurine-48	Congo red	Chlorophenol red
hydrofluoric, $HF_{(aq)}$	blue	orange	violet	blue	yellow
acetic, $HC_2H_3O_{2(aq)}$	blue	yellow	purple	blue	yellow
nitric, $HNO_{3(aq)}$	green	red	violet	blue	yellow
hydrocyanic, $HCN_{(aq)}$	blue	yellow	red	red	yellow
methanoic, $HCHO_{2(aq)}$	blue	orange	purple	blue	yellow
hydrochloric $HCl_{(aq)}$	green	red	violet	blue	yellow

All pH curves involving a weak acid or weak base have at least one region where a buffering action occurs—where the pH changes very little, despite the addition of an appreciable amount of acid or base. The curves in these relatively constant pH regions are most nearly horizontal at a volume of titrant that is one-half the volume at the equivalence point (halfway between successive equivalence points for polyprotic substances). The solution near these points has a special significance and is known as a *buffer solution* or simply a **buffer**. For example, in the titration of acetic acid with sodium hydroxide (**Figure 1**), the pH is approximately 4.7 at a volume of 12.5 mL of sodium hydroxide. Since one-half of the equivalence volume has been added, one-half of the original acetic acid has reacted.

$$HC_2H_3O_{2(aq)} + OH^-_{(aq)} \rightarrow C_2H_3O_2^-_{(aq)} + H_2O_{(l)}$$

buffer a mixture of a conjugate acid–base pair that maintains a nearly constant pH when diluted or when a strong acid or base is added; an equal mixture of a weak acid and its conjugate base

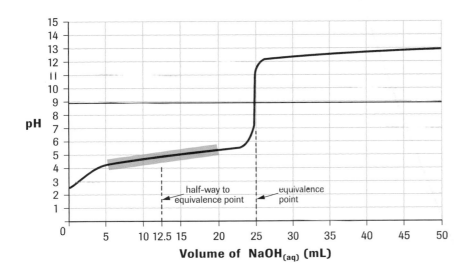

Figure 1
When a volume of titrant that is close to one-half the volume at the equivalence point is added in the titration of a weak acid and a strong base (or a strong acid and a weak base), the pH of the sample solution changes very little despite the addition of more base. The solution is called a buffer. The highlighted plateau near the halfway mark to the equivalence point shows the buffering region of the titration.

The mixture in this buffering region contains approximately equal amounts of the unreacted weak acid, $HC_2H_3O_{2(aq)}$, and its conjugate base, $C_2H_3O_2^-_{(aq)}$, produced in the reaction. A buffer has the ability to maintain a nearly constant pH when small amounts of a strong acid, $H^+_{(aq)}$, or base, $OH^-_{(aq)}$, are added.

Although a titration produces a solution with buffering action, if a buffer is desired it is usually prepared on its own, by mixing a weak acid such as acetic acid with a soluble salt of its conjugate base, such as sodium acetate, $NaC_2H_3O_{2(s)}$. Alternatively, a buffer may be formed by mixing a weak base such as ammonia, $NH_{3(aq)}$, with a soluble salt of its conjugate acid, such as ammonium chloride, $NH_4Cl_{(s)}$. The two components of a buffer are mixed to produce approximately equal molar concentrations of the conjugate acid–base pair.

Explaining Buffers

Buffering action can be explained using Brønsted–Lowry equations. When a small amount of $NaOH_{(aq)}$ is added to the acetic acid–acetate ion buffer, the following reaction occurs.

$$HC_2H_3O_{2(aq)} + OH^-_{(aq)} \rightarrow C_2H_3O_2^-_{(aq)} + H_2O_{(l)}$$

The small amount of $OH^-_{(aq)}$ ions added would convert a small amount of acetic acid to acetate ions. The overall effect is a small decrease in the ratio of acetic acid to acetate ions in the buffer and a slight increase in the pH. This small change and the consumption of some of the added hydroxide ions in the process explains why the pH change is small. This buffer would work equally well if a small amount of a strong acid, such as $HCl_{(aq)}$, were added to the buffer:

$$C_2H_3O^-_{2(aq)} + H^+_{(aq)} \rightarrow HC_2H_3O_{2(aq)}$$

The hydrogen ions are consumed and the mixture now has a slightly higher ratio of acetic acid to acetate ions and a slightly lower pH than it would have had if there were no buffer present.

The CRC Handbook of Chemistry and Physics provides recipes for preparing buffer solutions. In general, buffers are synthesized or exist naturally (e.g., blood) for a purpose, often to counteract small amounts of an acid or base that may be inadvertently added to the mixture. The compounds that form the buffer are mixed to produce a buffer that possesses a particular pH that can be tolerated by the system or process that uses the buffer.

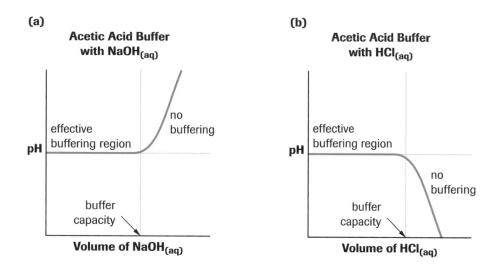

Figure 2
A batch of buffer is eventually depleted and then the pH changes quickly.

The Capacity of a Buffer

As you can see in **Figure 2**, there is a limit to the amount of strong acid or base that a buffer can neutralize before its pH begins to rise rapidly. A buffer's capacity is determined by the concentrations of its conjugate acid–base pair. The following sample problem demonstrates the effectiveness of a buffer.

▶ *SAMPLE* problem	*Calculating pH in a Buffer*

A 1.0-L buffer is prepared that contains 0.20 mol/L acetic acid and 0.20 mol/L sodium acetate at equilibrium.
(a) Calculate the pH of the buffer.
(b) If 0.10 mol of $H^+_{(aq)}$ is added to the buffer without changing its volume, calculate the pH. (It is possible to add H^+ ions without changing volume, if they are generated by some internal process in the buffer mixture.)
(c) Compare the change in pH to the change expected if the same amount of $H^+_{(aq)}$ is added to water to make a 1.0-L solution.

(a) We can calculate the pH of the buffer before the addition using the K_a for acetic acid, since the buffer contains the equilibrium

$$HC_2H_3O_{2(aq)} \rightleftharpoons H^+_{(aq)} + C_2H_3O_{2(aq)}^- \qquad\qquad K_a = 1.8 \times 10^{-5}$$

$$\frac{[H^+_{(aq)}]\ [C_2H_3O_{2(aq)}^-]}{[HC_2H_3O_{2(aq)}]} = K_a$$

Solving for $[H^+_{(aq)}]$, we get

$$[H^+_{(aq)}] = K_a \frac{[HC_2H_3O_{2(aq)}]}{[C_2H_3O_{2(aq)}^-]}$$

Substituting, we get

$$[H^+_{(aq)}] = \frac{1.8 \times 10^{-5}\ (0.20)}{(0.20)}$$

$$[H^+_{(aq)}] = 1.8 \times 10^{-5}\ \text{mol/L}$$

$$pH = -\log[H^+_{(aq)}]$$

$$= -\log(1.8 \times 10^{-5})$$

$$pH = 4.74$$

The pH of the initial buffer solution is 4.74.
(Notice that this will *always* be the pH of an acetic acid/sodium acetate buffer if the buffer contains equal concentrations of acetic acid and acetate ions.)

(b) When $H^+_{(aq)}$ ions are produced in the buffer, they react with acetate ions:

$$H^+_{(aq)} + C_2H_3O_{2(aq)}^- \rightarrow HC_2H_3O_{2(aq)}$$

As this reaction is quantitative, adding 0.1 mol of $H^+_{(aq)}$ causes the amount of acetate ions to decrease by 0.1 mol and the amount of acetic acid to increase by 0.1 mol. Since the volume of the mixture is 1.0 L,

$$[C_2H_3O_{2(aq)}^-]_{final} = 0.2\ \text{mol/L} - 0.1\ \text{mol/L}$$

$$= 0.1\ \text{mol/L}$$

$$[HC_2H_3O_{2(aq)}]_{final} = 0.2\ \text{mol/L} + 0.1\ \text{mol/L}$$

$$= 0.3\ \text{mol/L}$$

Rearranging and substituting into the equilibrium constant equation we get

$$\frac{[H^+_{(aq)}]\ [C_2H_3O_{2(aq)}^-]}{[HC_2H_3O_{2(aq)}]} = K_a$$

$$[H^+_{(aq)}] = K_a \frac{[HC_2H_3O_{2(aq)}]}{[C_2H_3O_{2(aq)}^-]}$$

$$= 1.8 \times 10^{-5} \frac{(0.10)}{(0.30)}$$

$$[H^+_{(aq)}] = 5.4 \times 10^{-5}\ \text{mol/L}$$

$$pH = -\log[H^+_{(aq)}]$$

$$= -\log(5.4 \times 10^{-5})$$

$$pH = 4.27$$

The pH of the mixture dropped from 4.74 to 4.27, a rather small decrease.

(c) In the water case,

$$[H^+_{(aq)}] = 0.10 \text{ mol/L}$$

$$\begin{aligned} pH &= -\log[H^+_{(aq)}] \\ &= -\log(0.10) \\ pH &= 1.00 \end{aligned}$$

The pH of the water solution drops from 7.00 (pure water) to 1.00, a substantial difference.

You may want to determine yourself what happens if a large amount of $H^+_{(aq)}$ is added to the buffer.

Example

Calculate the change in pH that occurs when 0.10 mol $HCl_{(g)}$ is added to 1.0 L of an ammonia-ammonium chloride buffer containing 0.33 mol/L $NH_{3(aq)}$ and 0.33 mol/L $NH^+_{4(aq)}$ at equilibrium. Assume no change in the voume of the buffer.

0.10 mol $HCl_{(g)}$ added

$[NH_{3(aq)}] = 0.33$ mol/L

$[NH^+_{4(aq)}] = 0.33$ mol/L

$$NH_{3(aq)} + H_2O_{(l)} \rightleftharpoons NH^+_{4(aq)} + OH^-_{(aq)}$$

$$\frac{[NH^+_{4(aq)}][OH^-_{(aq)}]}{[NH_{3(aq)}]} = K_b$$

$$K_{a_{NH^+_{4(aq)}}} = 5.6 \times 10^{-10} \text{ (from Appendix C7)}$$

$$\begin{aligned} K_a K_b &= K_w \\ K_b &= \frac{K_w}{K_a} \\ &= \frac{1.0 \times 10^{-14}}{5.6 \times 10^{-10}} \\ K_b &= 1.8 \times 10^{-5} \end{aligned}$$

$$\begin{aligned} [OH^-_{(aq)}] &= K_b \frac{[NH_{3(aq)}]}{[NH^+_{4(aq)}]} \\ &= 1.8 \times 10^{-5} \frac{(0.33 \cancel{\text{mol/L}})}{(0.33 \cancel{\text{mol/L}})} \\ [OH^-_{(aq)}] &= 1.8 \times 10^{-5} \\ pOH &= -\log[OH^-_{(aq)}] \\ &= -\log(1.8 \times 10^{-5}) \\ pOH &= 4.74 \end{aligned}$$

$$\begin{aligned} pH + pOH &= 14.00 \\ pH &= 14.00 - pOH \\ &= 14.00 - 4.74 \\ pH &= 9.26 \end{aligned}$$

Since $HCl_{(aq)}$ is a strong acid, it ionizes completely into $H^+_{(aq)}$ and $Cl^-_{(aq)}$ ions.

Therefore, $[H^+_{(aq)}]$ added $= \dfrac{0.10 \text{ mol}}{1.0 \text{ L}}$

$\qquad\qquad [H^+_{(aq)}]$ added $= 0.10 \text{ mol/L}$

The added $H^+_{(aq)}$ is neutralized in a 1:1 reaction shown by the following equation:

$$NH_{3(aq)} + H^+_{(aq)} \rightarrow NH^+_{4(aq)}$$

$[NH_{3(aq)}]_{final} = (0.33 - 0.10) \text{ mol/L}$

$[NH_{3(aq)}]_{final} = 0.23 \text{ mol/L}$

$[NH^+_{4(aq)}]_{final} = (0.33 + 0.10) \text{ mol/L}$

$[NH^+_{4(aq)}]_{final} = 0.43 \text{ mol/L}$

$$\frac{[NH^+_{4(aq)}][OH^-_{(aq)}]}{[NH_{3(aq)}]} = K_b$$

$$[OH^-_{(aq)}] = K_b \frac{[NH_{3(aq)}]}{[NH^+_{4(aq)}]}$$

$$= 1.8 \times 10^{-5} \frac{(0.23 \text{ mol/L})}{(0.43 \text{ mol/L})}$$

$[OH^-_{(aq)}] = 9.6 \times 10^{-6} \text{ mol/L}$

$\qquad pOH = -\log[OH^-_{(aq)}]$

$\qquad\qquad = -\log(9.6 \times 10^{-6})$

$\qquad pOH = 5.02$

$pH + pOH = 14.00$

$\qquad\qquad = 14.00 - pOH$

$\qquad\qquad = 14.00 - 5.02$

$\qquad pH = 8.98$

The pH of the ammonia-ammonium chloride buffer decreased from 9.26 to 8.98.

Buffers in Action

The ability of buffers to maintain a relatively constant pH is important in many biological processes where certain chemical reactions occur at a specific pH. Many aspects of cell function and metabolism in living organisms are very sensitive to pH changes. For example, each enzyme carries out its function optimally over a small pH range. One important buffer within living cells is the conjugate acid–base pair, $H_2PO^-_{4(aq)}/HPO^{2-}_{4(aq)}$. The major buffer system in blood and other body fluids is the conjugate acid–base pair $H_2CO_{3(aq)}/HCO^-_{3(aq)}$. Blood plasma has a remarkable buffering capacity, as shown by **Table 1**.

Table 1 Buffering Action of Neutral Saline ($NaCl_{(aq)}$) Solution and of Blood Plasma

Solution (1.0 L)	Initial pH of mixture	Final pH after adding 1 mL of 10 mol/L $HCl_{(aq)}$
neutral saline	7.0	2.0
blood plasma	7.4	7.2

Human blood plasma normally has a pH of about 7.4. Any change in pH of more than 0.2, induced by poisoning or disease, is life-threatening. If the blood were not buffered, the acid absorbed from a glass of orange juice would probably be fatal.

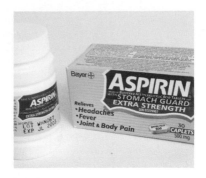

Figure 3
Many consumer and commercial products contain buffers. Buffered Aspirin is a well-known example. Blood plasma and capsules for making buffer solutions (for example, to calibrate pH meters) are commercial examples of buffers.

INVESTIGATION 8.5.1

Buffer Action (p. 629)
Test the effectiveness of a phosphate buffer system.

Buffers are also important in many consumer, commercial, and industrial applications (**Figure 3**).

Fermentation and the manufacture of antibiotics require buffering to optimize yields and to avoid undesirable side reactions. The production of various cheeses, yogurt, and sour cream are dependent on buffers to control pH levels, since an optimum pH is needed to manage the growth of microorganisms and to allow enzymes to catalyze fermentation processes efficiently. Sodium nitrite and vinegar are widely used to preserve food: Part of their function is to prevent the fermentation that takes place only at certain pH values.

▶ *Practice*

Understanding Concepts

1. (a) Describe the empirical characteristics of a good buffer solution.
   (b) What two entities must a buffer contain, and in what relative amounts?

2. (a) Write an equilibrium reaction equation for each of the following buffer mixtures:
   (i) $NH_{3(aq)}$ and $NH_4Cl_{(aq)}$
   (ii) $HC_7H_5O_{2(aq)}$ and $NaC_7H_5O_{2(aq)}$
   (b) In what direction will each equilibrium shift if a small amount of $HCl_{(aq)}$ is added to the buffer solution? Write a reaction equation that illustrates the cause of each shift.

3. A 1.0-L buffer is prepared from 1.5 mol of formic acid ($HCO_2H_{(aq)}$) and 1.5 mol of formate. Calculate the change in pH if 0.13 mol of $H^+_{(aq)}$ is formed in the buffer.

▶ *Section 8.5 Questions*

Understanding Concepts

1. What is a buffer?

2. List two buffers that help maintain a normal pH level in your body.

3. (a) Write an equilibrium reaction equation for a carbonic acid–hydrogen carbonate ion buffer.
   (b) Write a reaction equation that illustrates what happens when a small amount of $HCl_{(aq)}$ is added to the buffer.
   (c) Write a reaction equation that illustrates what happens when a small amount of $NaOH_{(aq)}$ is added to the buffer.

4. What happens if a large amount of a strong acid or base is added to a buffer? Explain, using an example.

5. Use Le Châtelier's principle to predict what reaction occurs and how the pH of the solution changes in an acetic acid–acetate ion buffer when
   (a) a small amount of $HCl_{(aq)}$ is added.
   (b) a small amount of $NaOH_{(aq)}$ is added.

6. Predict whether each of the following buffers is acidic, basic, or near neutral, then rank them in order from lowest to highest pH.

   (a) hydrogen phosphate ion–phosphate ion
   (b) $HCO_2H_{(aq)}$–$CO_2H^-_{(aq)}$
   (c) $H_2CO_{3(aq)}$–$HCO_3^-_{(aq)}$
   (d) hydrogen sulfite ion–sulfite ion

7. Does each of the following mixtures form an effective buffer? Explain.
   (a) $HNO_{3(aq)}$ and $NaNO_{3(aq)}$
   (b) $NH_{3(aq)}$ and $NH_4Cl_{(aq)}$
   (c) $HCl_{(aq)}$ and $NaOH_{(aq)}$
   (d) $HC_7H_5O_2$ and $NaC_7H_5O_2$

8. Aqueous solutions of nitric acid and nitrous acid of the same concentration are prepared.
   (a) How do their pH values compare?
   (b) Explain your answer, using the Brønsted–Lowry concept.

9. A 1.0-L buffer is prepared from 0.25 mol of acetic acid and 0.25 mol of sodium acetate. Calculate the change in pH if 0.15 mol of $OH^-_{(aq)}$ is formed in the buffer.

The term *acid rain* is a familiar one in our society. No doubt you have been introduced to the idea that atmospheric pollutants dissolve and ionize in rain to have serious consequences once they reach the ground. However, solutes can also be carried by snow, fog, and other forms of precipitation. A general term to cover all these situations is **acid deposition**.

Although natural emissions (from volcanoes, lightning, and microbial action) contribute to acid deposition, it seems clear that its primary source is human activity. Empirical work indicates that the main causes of acid deposition in North America are sulfur dioxide, $SO_{2(g)}$, and nitrogen oxides, NO_x. The major sources of $SO_{2(g)}$ emissions in North America are coal-fired power generating stations (**Figure 1**) and non-ferrous ore smelters. When coal is burned in power stations, sulfur in the coal is oxidized to $SO_{2(g)}$. The roasting of sulfide ores in smelters also produces $SO_{2(g)}$. In the atmosphere, $SO_{2(g)}$ reacts with water to produce sulfurous acid, $H_2SO_{3(aq)}$, or is further oxidized to sulfuric acid, $H_2SO_{4(aq)}$. Because nitrogen oxides are produced whenever fuel is burned at high temperatures, the main source of NO_x is motor vehicle emissions. At the high temperatures of combustion reactions, the nitrogen and oxygen present in the air combine to form a variety of nitrogen oxides, which produce nitrous and nitric acid when they react with atmospheric water and oxygen.

Sulfuric and sulfurous acids cause considerable environmental damage when they fall in the form of acid deposition. Experiments indicate that virtually anything that the acids contact (soil, water, plants, and structural materials) is affected to some degree (**Figure 2**). Scientists have repeatedly shown that acid deposition has increased the acidity of some lakes and streams to the point where aquatic life is depleted and waterfowl populations are threatened. Some environmental groups claim that 14 000 Canadian lakes have been damaged by acid deposition. Apparently, the greatest damage is done to lakes that are poorly buffered. When natural alkaline buffers such as limestone are present, they neutralize the acidic compounds from acid deposition. However, lakes lying on granite rock are susceptible to immediate damage because acids cause metal ions to go into solution in a process called leaching. Especially harmful are cadmium, mercury, lead, arsenic, aluminum, and chromium ions, because they are toxic to living organisms. For example, aluminum is highly toxic to fish because it impairs gill function. Acid rain may also increase human exposure—through food and drinking water—to these dangerous metals.

Acid deposition is also suspected as one of the causes of forest decline, particularly in forests at high altitudes and colder latitudes. Evidence continues to accumulate that acid deposition is causing serious harm to forests throughout the Northern Hemisphere. The Black Forest of Germany has been particularly hard hit. Some observers contend that many forests receive as much as 30 times more acid than they would if rain fell through clean air. Damage to trees includes yellowing, premature loss of needles, and eventual death. Studies of tree rings reveal that trees grow more slowly in regions that are prone to acid deposition. Some research indicates that, as the concentration of trace metal ions increases, ring growth decreases. Research also suggests that acid deposition damages the needles and leaves of trees, cutting down carbohydrate production.

Disputes over the effect of acid rain on forests highlight the wide range of interpretation of the empirical data. A small minority of scientists insists that there is currently no direct evidence linking acid deposition to elevated tree mortality rates and to decreases in the ring widths of tree trunks. They cite evidence suggesting that the reduction in

Figure 1
One of Ontario's coal-powered generating plants.

acid deposition acidic rain, snow, fog, dust, etc.

Figure 2
Acid rain modifies the work of human artists. This figure's cheek was smooth fewer than 100 years ago.

the growth rate of trees at high altitudes and latitudes may be more directly related to a reduction in mean annual temperatures in those regions. These researchers point out that growth ring data correspond to fluctuations in mean annual temperatures over the last century. Some researchers report that acid painted on seedlings in soil with inadequate nutrients actually has a beneficial effect on growth. Other research indicates that ground-level ozone, rather than acid deposition, is implicated in the extensive damage to Germany's Black Forest. For example, lichens, which are adversely affected by $SO_{2(g)}$, have been found growing abundantly on dying trees, which would not be an expected finding if sulfur dioxide emissions were the cause of the trees' death.

The complex nature of acid deposition makes disentangling the separate effects of pollution and climate change difficult. The effects of both these factors become more severe with increasing altitude, where trees are growing at their limits—winter temperatures are the coldest they can normally tolerate. Under such conditions, any decrease in temperature moves the trees into a lethal climatic range. Scientists have also found that the moisture in clouds tends to be much more acidic than rainfall. The reasonable conclusion, supported by research, is that the acid deposition phenomenon elevates tree mortality most seriously at high altitudes, where trees are in frequent contact with clouds and droplets of water from clouds are the main source of moisture (**Figure 4**).

Figure 4
Damage to trees caused by acid rain

Technology of Acid Deposition

Both technological attempts to deal with the causes and effects of acid rain and the instruments used in scientific research reflect some important aspects of the nature of technology.

Taller smokestacks were an early, and relatively inexpensive, technological fix for local air pollution problems. The thinking was that, if the pollutants were released at higher altitudes, they would be diluted by air and would not reach harmful concentrations at ground level. The rationale for tall stacks was soon shown to be flawed. Local air quality did improve, but the taller stacks spewed pollutants higher into the air than shorter ones could. High-tech instruments provide evidence that bands of smoke from tall stacks sometimes travelled hundreds of kilometres. Even after the visible smoke has dispersed, the invisible pollutants continue to travel thousands of kilometres from their source, often crossing international boundaries. Monitoring air chemistry with special instruments indicates that more than half the acid deposition in Eastern Canada originates as emissions from industries in the United States. Canadian emissions also contribute to acid deposition in the United States; between 10% and 25% of the deposition in the northeastern states apparently originates in Canada.

One technology for reducing acid deposition is the chemical scrubber, a device that processes the gases emitted by smelters and power plants, dissolving or precipitating the pollutants. Catalysts that reduce the nitrogen oxides produced by combustion reactions represent another technological response to the problem. For example, new automobiles are now outfitted with catalytic converters.

Technology also counterbalances some of the effects of acid rain. Adding basic materials (for example, lime and limestone) to lakes to neutralize the acid has had some success. Other research has found that certain types of bacteria can oxidize sulfur compounds, while other types can reduce sulfur. This finding suggests that microorganisms might play a beneficial role in the control of lake acidification, particularly when the water remains in the lake for a long time. This research may lead to new technologies for using microorganisms.

The various strategies for reducing acid rain involve annual investments of billions of dollars. Because the costs are so high, it is essential that the atmospheric conditions involved in producing and transporting acidic precipitation be well understood. This explains a scientific focus on development of computer models to identify the source of the acid and the physical and chemical mechanisms by which it is transported to other locations.

Development of sophisticated technology has made possible the tracking of airborne acidic material from smokestacks. After a tracer compound is released in different parts of Canada and the United States, a sensitive detector samples the air downwind from the source. Science and technology were partners in causing the problem of acid deposition, and they are now partners in the search for solutions.

Social Aspects of Acid Deposition

Acid deposition is a societal issue that reveals some important aspects of the interaction among science, technology, and society.

Acid deposition is causing serious environmental, economic, and social problems in Eastern Canada. The environmental problems described above generate enormous economic problems. Acid deposition is endangering fishing, tourism, agriculture, and forestry. The resource base at risk sustains approximately 8% of Canada's gross national product (GNP). It is estimated that acid deposition causes about one billion dollars' worth of damage in Canada annually.

Besides the social costs of economic losses, there are other human costs as well. Many respiratory problems are associated with exposure to air pollution. These problems range from aggravation of asthma cases and a consequent increase in hospital admissions to eventual chronic lung disease. The acute effects of sulfuric acid on humans are particularly pronounced in asthmatics.

The economic and social problems caused by acid rain must be dealt with in the political arena. This entails mediating between the pro-development and pro-environment lobby groups. In addition, the fact that acid deposition crosses borders makes it a contentious political issue between Canada and the United States.

Although the acid deposition problem is not completely understood, both the Canadian and the American governments have passed legislation intended to reduce the discharge of sulfur and nitrogen oxides. Opponents of these measures argue that the regulations could severely hinder economic growth, because the cleanup effort will affect the operation of coal-burning electric power plants that emit large amounts of SO_2. These power plants are perceived to be essential to the industrial growth, economic well-being, and social fabric of Ontario and the Northeastern United States, the very regions believed to be responsible for the acid deposition that is ravaging Eastern Canada.

Understanding Concepts

1. Create a table with two columns, headed "Evidence of Acid Rain Damage" and "Alternative Interpretations." In the table, show how the same information can lead researchers to different conclusions.

2. Within the context of the acid-deposition issue, provide an example of:
    (a) science assisting technology
    (b) technology assisting science
    (c) technology affecting society
    (d) society affecting science
    (e) society affecting technology

Making Connections

3. The ability of soils and bedrock to neutralize acid deposits depends on their capacity to release bases when they weather. Limestone-based soils weather rapidly and contain the highest concentration of basic minerals like calcite, $CaCO_{3(s)}$, and dolomite, $CaMg(CO_3)_{2(s)}$, two forms of limestone.

    $$CaMg(CO_3)_{2(s)} \rightleftharpoons Ca^{2+}_{(aq)} + Mg^{2+}_{(aq)} + 2\,CO^{2-}_{3(aq)}$$

    Areas of Ontario that are near the Precambrian Shield have quartzite- or granite-based rock primarily composed of insoluble $SiO_{2(s)}$ and little topsoil. These areas do not carry enough buffering capacity to neutralize even small amounts of acid falling on the soil and the lakes.

    (a) Calculate the pH of natural rain, that is, rainwater saturated with carbon dioxide ($[CO_{2(aq)}] = 1.2 \times 10^{-5}$ mol/L), given the following equilibrium reaction equation,

    $$CO_{2(aq)} + H_2O_{(l)} \rightleftharpoons H_2CO_{3(aq)} \qquad\qquad K = 3.5 \times 10^{-2}$$

    (b) Explain how the following equilibrium reaction equations may act as buffers that help neutralize the $H^+_{(aq)}$ ions in acid rain.

    $$CO^{2-}_{3(aq)} + H^+_{(aq)} \rightleftharpoons HCO^-_{3(aq)}$$

    $$HCO^-_{3(aq)} + H^+_{(aq)} \rightleftharpoons H_2CO_{3(aq)} \rightleftharpoons H_2O_{(l)} + CO_{2(g)}$$

    (c) Conduct library or Internet research to learn about techniques (in addition to those described here) being currently used to increase the buffering capacity (buffering ability) of granite-based lakes.

     www.science.nelson.com

▸ **EXPLORE** an issue

Take a Stand:
Acting to Reduce the Effects

Decision-Making Skills

| ○ Define the Issue | ● Analyze the Issue | ● Research |
| ● Defend the Position | ● Identify Alternatives | ● Evaluate |

(a) Knowing some of the likely causes of acid deposition, research and suggest at least one action that North Americans could take to reduce the negative effects of acid deposition.

 www.science.nelson.com

(b) Now, consider your suggestion from several (about five) different perspectives. Think of examples of people who might have such a perspective. Imagine how each person might respond to your suggestion.

(c) Write a newspaper article as if you were summarizing the presentations made at a town-hall meeting to discuss ways to reduce the effects of acid deposition.

(d) Exchange articles with a classmate. Read your classmate's article. If you had to make a decision on whether or not to implement the suggested course of action, what would your decision be? (Remember that your classmate's suggested action is likely to be different from your own.)

Hospital Pharmacist

Hospital pharmacists oversee the dispensing and storage of all medicines given to all patients in the hospital. Pharmacists are often involved in deciding (with the physician) which medication is best for each patient. Pharmacists need to understand how various drugs affect the body, and how they might interact with each other. They are also on the front line, counselling and advising patients under stress, so need to have good people skills.

Quality-Control Chemist

A quality control chemist may work for a pharmaceutical company or any other company that produces chemicals that must be pure. The job involves testing samples for the purity of both the reactants and the products of a process, and investigating any returned materials. Two or more testing techniques might be used simultaneously, in an effort to find out which one is better for a particular purpose. Quality-control chemists perform statistical analyses and present formal reports. In a large company, a quality-control chemist is likely to report to a quality-control lab supervisor. A B.S. in chemistry is a minimum requirement.

Inorganic Laboratory Analyst

An inorganic laboratory analyst is likely to be part of a team that analyzes environmental samples (such as groundwater, drinking water, and soil from industrial sites) for trace heavy metals, such as mercury, and other potentially dangerous substances. The analyst must provide the lab's clients with high-quality analytical data in a timely, cost-effective manner. Laboratory analysts may be required to design tests for specific contaminants, or tests that are quicker, more effective, or less expensive.

Environmental Chemist

Environmental protection and monitoring of industries and facilities is a booming growth area. An environmental chemist may spend much time in the field, using portable analytical equipment and computers. Environmental chemists may be required to respond to environmental emergencies, or investigate reports of pollution or contamination, or advise companies about likely environmental consequences of processes they plan to use. A senior environmental chemist would likely require a Ph.D. in chemistry, plus several years of experience.

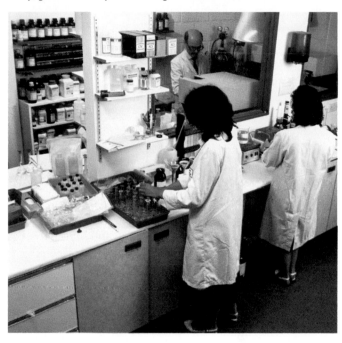

▸ **Practice**

Making Connections

1. For each of the careers mentioned here, suggest how acid–base equilibrium reactions might be important.
2. Select one of the careers mentioned (or a similar career of your choice) and find
   (a) at least three universities that provide the necessary degree(s);
   (b) what the tuition fees are at each of the universities;
   (c) what scholarships or bursaries are available to students taking these degrees.

 www.science.nelson.com

ACTIVITY 8.1.1

Determining the pH of Common Substances

One reason for the wide acceptance of the pH scale is the availability of convenient, rapid, and precise methods for measuring pH. The purpose of this activity is to measure the pH of common household liquids and to determine the suitability of the various methods to measure the pH of the different substances.

Materials
lab apron
eye protection
red or blue litmus paper
wide-range pH paper (pH 1–14) and colour chart
pH meter and pH 7 buffer solution
wash bottle of distilled water
400-mL waste beaker
several 100-mL beakers and an equal number of watch
    glasses
several droppers
various household cleaning agents, such as "liquid"
    ammonia, glass cleaner, drain cleaner, and shampoo
various foods and beverages such as tap water, mineral
    water, juices, pop, vinegar, and milk

 Many household cleaning products are corrosive or toxic. Handle them with care and do not mix them together.

Wash your hands upon completion of the investigation.

• Place a small amount of each substance to be measured in a different 100-mL beaker. The amount depends on the type of pH electrode supplied with the pH meter you are using. Your teacher will tell you what volume you should use.

• Using a different dropper for each liquid, place a drop of the substance onto a small piece of litmus paper and lay the paper onto a watch glass. Record your observations.

• Repeat your pH measurements using wide-range pH paper instead of litmus paper.

• Measure the pH of each sample using a pH meter. For each sample, rinse the electrode of the pH meter with distilled water. Place the pH meter electrode in a standard pH 7 buffer solution and calibrate the instrument by adjusting the meter to read the pH of the buffer. Using a wash bottle and a 400-mL waste beaker, rinse the pH meter electrode with distilled water. Place the electrode in a beaker containing the sample and record the pH reading.

• Dispose of samples in the sink with lots of running water. Discard the used pieces of pH paper and litmus paper in the wastepaper basket. Return the pH meter to storage, according to your teacher's instructions.

• Wash your hands with soap and water.

(a) Classify the samples as acidic, neutral, or basic.

(b) Provide a generalization regarding the acid–base properties of the various materials tested and their household uses.

(c) Provide advantages and disadvantages of using litmus paper, wide-range pH paper, and the pH meter in determining the pH of each of the samples tested.

(d) Why was the pH meter electrode calibrated with the buffer solution before use?

INVESTIGATION 8.3.1

The pH of Salt Solutions

Inquiry Skills

○ Questioning	● Planning	● Analyzing
○ Hypothesizing	● Conducting	● Evaluating
● Predicting	● Recording	● Communicating

When some salts are dissolved in water they form neutral solutions; others form acidic or basic solutions. The theory that a variety of ions may affect the acid–base characteristics of an aqueous solution can be tested in the laboraory.

Purpose

The purpose of this investigation is to test the theory that ions of soluble salts may affect the pH of aqueous solutions.

Question

Which salt solutions are acidic, which are basic, and which neutral?

Prediction

(a) Predict the pH of a 0.1 mol/L aqueous solution of each of the salts in the Materials list.

Experimental Design

The pH of a variety of aqueous salt solutions is measured using a suitable pH measuring system.

Materials

lab apron
eye protection
pure water
0.10 mol/L aqueous solutions of:
  sodium carbonate
  sodium phosphate
  aluminum sulfate
  aluminum chloride
  sodium chloride
  ammonium chloride

ammonium oxalate
ammonium acetate
ammonium carbonate
ammonium sulfate
potassium sulfate
copper(II) sulfate
iron(III) sulfate
iron(III) chloride
sodium hydrogen carbonate
sodium hydrogen sulfate
pH paper, pH meter, and/or universal indicator
containers (small beakers, test tubes, or spot plates)
waste beakers

Procedure

(b) Write a Procedure for your investigation. If you use a pH meter to measure pH, include a description of the setup procedure and any precautions you must consider.

1. With your teacher's approval, carry out your Procedure.

Analysis

(c) Analyze your observations and use them to answer the Question. Display your answer in an appropriate table of evidence.

Evaluation

(d) Compare your answers for (c) to your Predictions. Does the evidence support your predictions? Suggest an explanation for any discrepancies.

ACTIVITY 8.4.1

Quantitative Titration

In this activity, you will standardize a sodium hydroxide solution, then determine the concentration of an unknown acid by titration with the standardized base.

Experimental Design

This is a two-part activity. In the first part, you will standardize a sodium hydroxide solution by titrating it with the primary standard, potassium hydrogen phthalate, $KHC_8H_4O4_{(aq)}$. The concentration of the unknown acid solution is then determined by titrating it with the standardized sodium hydroxide solution.

Materials

lab apron
eye protection
sheet of blank white paper
electronic balance
pH meter
distilled water
wash bottle with distilled water
125-mL Erlenmeyer flask
1000-mL glass or plastic bottle
rubber stoppers
$NaOH_{(s)}$
$KHC_8H_4O_{4(s)}$
1% phenolphthalein
vinegar, lemon juice, or other acid solution
   of unknown concentration
dropper
stirring rod
buret
buret stand
100-mL graduated cylinder
10-mL graduated cylinder
weighing paper

 Acids are corrosive and may also be toxic. Solid sodium hydroxide is extremely corrosive, and especially dangerous to eyes and skin. Do not touch your eyes. If sodium hydroxide enters your eye, flush continuously with cold water for 10 min and get immediate medical attention. Wear a lab apron and eye protection.

Procedure

Part I Standardization of NaOH_{(aq)}

1. In a 125-mL Erlenmeyer flask, dissolve approximately 10 g of $NaOH_{(s)}$ in 50 mL of previously boiled, distilled water.

2. Transfer 10.0 mL of the solution to a 1000-mL bottle and dilute with 500 mL previously boiled, distilled water. Stir the solution but do not shake it.

3. Weigh approximately 0.4 g $KHC_8H_4O_{4(s)}$ and record the mass to three significant digits.

4. Place the $KHC_8H_4O_{4(s)}$ into a clean, dry Erlenmeyer flask. Add 50.0 mL distilled water and two to three drops of phenolphthalein. Swirl to mix.

5. Allow several millilitres of the $NaOH_{(aq)}$ solution to flow through a buret, making sure that the solution wets all of the inside surfaces.

6. Fill the wetted buret with the $NaOH_{(aq)}$ solution and record the volume in a suitable chart.

7. Place the Erlenmeyer flask containing $KHC_8H_4O_{4(aq)}$ over a sheet of white paper and titrate with the $NaOH_{(aq)}$ solution in the buret until the endpoint is reached.

8. Repeat steps 3 to 7 two more times and calculate the mean of the three volumes of $NaOH_{(aq)}$ used to reach the endpoint. Rinse out the flask.

Part II Determining the [H⁺_{(aq)}] in a solution of unknown concentration

9. Refill the buret you used in Part A with standardized $NaOH_{(aq)}$ solution.

10. Place 25.00 mL of an unknown acidic solution into a clean, dry Erlenmeyer flask. Add 2 to 3 drops of phenolphthalein. Swirl to mix.

11. Place the Erlenmeyer flask containing the acid solution over a sheet of white paper and titrate with standardized $NaOH_{(aq)}$ solution in the buret until the endpoint is reached.

12. Repeat steps 10 to 11 two more times, and calculate the mean of the three volumes of $NaOH_{(aq)}$ used to reach the endpoint. Use this average value to calculate the concentration of $H^+_{(aq)}$ in the acidic solution.

13. Calibrate a pH meter and use it to measure the pH of a sample of the unknown acid solution. Record the value.

14. Discard all solutions in the sink with lots of running water. Return materials and equipment to their proper location. Wash your hands with soap and water.

Analysis

Part 1

(a) Use the average of your titration volumes to calculate the concentration of the $NaOH_{(aq)}$ solution.

(b) Why was $KHC_8H_4O_{4(s)}$ used as the acid in the standardization titration?

(c) Why should you not shake the $KHC_8H_4O_{4(aq)}$ solution before titrating it with $NaOH_{(aq)}$?

(d) Why was boiled distilled water used to prepare the solutions?

Evaluation

(f) Evaluate the Procedure and suggest changes that might correct any sources of error.

ACTIVITY 8.4.1 *continued*

Part II

(e) Use the evidence from your titration to calculate $[H^+_{(aq)}]$ of the unknown solution.

Evaluation

(f) Evaluate the Procedure and suggest changes that might correct any sources of error.

INVESTIGATION 8.5.1

Buffer Action

Inquiry Skills

○ Questioning	● Planning	● Analyzing
○ Hypothesizing	● Conducting	● Evaluating
● Predicting	● Recording	● Communicating

You will investigate the buffering capacity of a $H_2PO_4^-{}_{(aq)}$/$HPO_4^{2-}{}_{(aq)}$ buffer.

$$H_2PO_4^-{}_{(aq)} + OH^-{}_{(aq)} \rightarrow HPO_4^{2-}{}_{(aq)} + H_2O_{(l)}$$

excess   limiting   (base part
(acid part   reagent   of buffer)
of buffer)

Question

How does the pH change when a strong acid and a strong base are slowly added to a $H_2PO_4^-{}_{(aq)}$/$HPO_4^{2-}{}_{(aq)}$ buffer?

Prediction

(a) Predict how the pH of the $H_2PO_4^-{}_{(aq)}$/$HPO_4^{2-}{}_{(aq)}$ buffer will change when $HCl_{(aq)}$ and $NaOH_{(aq)}$ are added to different samples of the buffer, as described in the Procedure.

Experimental Design

(b) Read the Procedure and describe the design in a brief paragraph.

Materials

(c) Create a suitable list of materials. Have your list approved by your teacher before continuing. Include relevant safety precautions and disposal procedures beside each material in your list.

Procedure

1. Obtain 50 mL of 0.10 mol/L $KH_2PO_4{}_{(aq)}$ and 29 mL of 0.10 mol/L $NaOH_{(aq)}$ in separate graduated cylinders.

2. Pour the $KH_2PO_4{}_{(aq)}$ and then the $NaOH_{(aq)}$ into a beaker to prepare a buffer with a pH of 7.

3. Pour an equal amount of the buffer into test tubes 1 and 2.

4. Add 0.10 mol/L $NaCl_{(aq)}$ as a control into Tubes 3 and 4.

5. Add two drops of bromocresol green to Tubes 1 and 3.

6. Add and count drops of 0.10 mol/L $HCl_{(aq)}$ to Tubes 1 and 3 until the colour changes.

7. Repeat steps 5 and 6 with phenolphthalein and 0.10 mol/L $NaOH_{(aq)}$ in Tubes 2 and 4.

8. Dispose of all solutions down the drain with lots of running water.

Evidence

(d) After reading the Procedure, create a table in which to record observations.

Analysis

(e) Answer the Question.

Evaluation

(f) Evaluate the Experimental Design.

(g) Evaluate your Prediction based on the evidence gathered and your confidence in the design.

Key Expectations

- Define constant expressions, such as K_w (8.1), K_a (8.2), and K_b (8.2).
- Compare strong and weak acids and bases using the concept of equilibrium. (8.1, 8.2)
- Use appropriate vocabulary to communicate ideas, procedures, and results related to acid–base systems and equilibria. (all sections)
- Solve equilibrium problems involving concentrations of reactants and products and the following quantities: K_a, K_b, pH, pOH. (8.2, 8.3, 8.4)
- Predict, in qualitative terms, whether a solution of a specific salt will be acidic, basic, or neutral. (8.3, 8.4)
- Solve problems involving acid–base titration data and the pH at the equivalence point. (8.4)
- Describe the characteristics and components of a buffer solution. (8.5)
- Explain how buffering action affects our daily lives, using examples. (8.5, 8.6)

Key Terms

acid–base indicator

acid deposition

acid ionization
   constant, K_a

amphoteric

autoionization of water

base ionization
   constant, K_b

Brønsted-Lowry acid

Brønsted-Lowry base

buffer

conjugate acid–base
   pair

endpoint

equivalence point

hydrolysis

ion product constant
   for water, K_w

Lewis acid

Lewis base

monoprotic acid

pH

pH meter

pK_w

pOH

polyprotic acid

primary standard

sample

strong acid

strong base

titrant

titration

transition point

weak acid

weak base

Key Symbols and Equations

$$[H^+_{(aq)}][OH^-_{(aq)}] = K_w$$

$$pH = -\log[H^+_{(aq)}] \qquad [H^+_{(aq)}] = 10^{-pH}$$

$$pOH = -\log[OH^-_{(aq)}] \qquad [OH^-_{(aq)}] = 10^{-pOH}$$

$$pH + pOH = pK_w \qquad pH + pOH = 14.00 \text{ (at SATP)}$$

$$\frac{p}{100} = \frac{[H^+_{(aq)}]}{[HA_{(aq)}]} \qquad [H^+_{(aq)}] = \frac{p}{100} \times [HA_{(aq)}]$$

(where p = percent ionization and
$[HA_{(aq)}]$ = concentration of the acid)

$$K_a = \frac{[H^+_{(aq)}][A^-_{(aq)}]}{[HA_{(aq)}]} \qquad K_b = \frac{[HB^+_{(aq)}][OH^-_{(aq)}]}{[B_{(aq)}]}$$

For conjugate acid–base pairs,

$$K_a K_b = K_w \qquad K_b = \frac{K_w}{K_a} \qquad K_a = \frac{K_w}{K_b}$$

Problems You Can Solve

- What is the $[OH^-_{(aq)}]$ and pOH or $[H^+_{(aq)}]$ and pH of a solution of a strong acid or a strong base, given the concentration? (8.1)
- What is the percent ionization of a weak acid in solution, given its pH and concentration? (8.2)
- What is the acid ionization constant, K_a, of an acid, given the percent ionization and concentration? (8.2)
- What is the K_b of the conjugate base of a weak acid? (8.2)
- What is the $[H^+_{(aq)}]$, pH, or K_a, given two of those quantities? (8.2)
- What is the K_b, pK_b, or pH for a weak base, given its concentration? (8.2)
- What is the pH of a polyprotic acid, given its concentration? (8.2)
- Predict if the solution of a salt, or oxide, will be acidic, basic, or neutral.
- Predict whether amphoteric ions in solution are likely to be acidic or basic. (8.3)
- Analyze a sample using titration of strong acid/strong base, weak acid/strong base, or weak base/strong acid. (8.4)
- Predict the pH of a buffer after a given amount of strong acid or base is added. (8.5)

▶ **MAKE** *a summary*

Select a favourite storybook. Write a six-page storybook in the author's style—one page for each section of this chapter—summarizing the key concepts and problem-solving skills you learned. Use illustrations as much as possible.

Identify each of the following statements as true, false, or incomplete. If the statement is false or incomplete, rewrite it as a true statement.

1. The stronger a Brønsted–Lowry acid is, the stronger is its conjugate base.

2. Group 1 metal ions produce basic aqueous solutions.

3. Cations that are conjugate acids of weak bases lower the pH of an aqueous solution.

4. In the titration of a weak acid with a strong base, the pH at the equivalence point is greater than 7.

5. Given that the bicarbonate ion, $HCO_{3(aq)}^-$, is amphoteric with $K_a = 5.6 \times 10^{-11}$ and $K_b = 2.4 \times 10^{-8}$, a solution of sodium bicarbonate is neutral.

6. Given that $K_w = 3.0 \times 10^{-14}$ at 40°C, the pH of pure water is 7.00 at this temperature.

7. A buffer may be composed of an acid and the salt of its conjugate base in unequal concentrations.

8. Most dyes that act as acid–base indicators are strong bases.

9. An acid–base indicator should be selected for a particular titration if its pK_{In} equals the pH at the equivalence point of the titration.

Identify the letter that corresponds to the best answer to each of the following questions.

10. A Brønsted–Lowry base is defined as
   (a) a proton donor.
   (b) a proton acceptor.
   (c) a hydroxide donor.
   (d) a hydroxide acceptor.
   (e) an electron pair acceptor.

11. The pH of a 1.25×10^{-3} mol/L $NaOH_{(aq)}$ solution is
   (a) 7.00.   (d) 3.90.
   (b) 11.10.   (e) 4.90.
   (c) 12.00.

12. Arsenic acid, $H_3AsO_{4(aq)}$, reacts with water as follows:

   $$H_3AsO_{4(aq)} + H_2O_{(l)} \rightleftharpoons H_2AsO_{4(aq)}^- + H_3O_{(aq)}^+$$

   The conjugate base of $H_3AsO_{4(aq)}$ in this reaction is:
   (a) $H_3AsO_{4(aq)}$ (d) $OH_{(aq)}^-$
   (b) $H_2O_{(l)}$ (e) $H_2AsO_{4(aq)}^-$
   (c) $H_3O_{(aq)}^+$

13. What is the value of K_a for the reaction

   $$NH_{4(aq)}^+ + H_2O_{(l)} \rightleftharpoons NH_{3(aq)} + H_3O_{(aq)}^+$$

at SATP, given that K_b for ammonia is 1.74×10^{-5} at SATP?
   (a) 5.75×10^{-10} (d) 2.98×10^{-2}
   (b) 5.75×10^4 (e) 112
   (c) 1.74×10^{-5}

14. Which of the following 0.1 mol/L aqueous solutions would be basic?
   (a) $NH_4(ClO_4)_{(aq)}$ (d) $H_2C_2O_{4(aq)}$
   (b) $NaCN_{(aq)}$ (e) $NaCl_{(aq)}$
   (c) $Ca(NO_3)_{2(aq)}$

15. The pH of a 0.10 mol/L aqueous solution of $Fe(NO_3)_3$ is not 7.00. The equation that best accounts for this observation is:
   (a) $Fe_{(aq)}^{3+} + 3H_2O_{(l)} \rightleftharpoons Fe(OH)_{3(aq)} + 3\,H_{(aq)}^+$
   (b) $NO_{3(aq)}^- + H_2O_{(l)} \rightleftharpoons HNO_{3(aq)} + OH_{(aq)}^-$
   (c) $Fe(H_2O)_{6(aq)}^{3+} + H_2O_{(l)} \rightleftharpoons$
        $Fe(H_2O)_5(OH)_{(aq)}^{2+} + H_3O_{(aq)}^+$
   (d) $Fe(H_2O)_{6(aq)}^{3+} + H_2O_{(l)} \rightleftharpoons$
        $Fe(OH_2)_5(H_3O)_{4(aq)}^+ + OH_{(aq)}^-$
   (e) $HNO_{3(aq)} + H_2O_{(l)} \rightleftharpoons H_3O_{(aq)}^+ + NO_{3(aq)}^-$

16. The percentage ionization in a 0.05 mol/L $NH_{3(aq)}$ solution whose pH is 11.00 is
   (a) 11%.   (d) 5%.
   (b) 0.02%.   (e) 2%.
   (c) 3%.

17. In a titration, 50.0 mL of an acetic acid solution required 20.0 mL of a standard solution of 0.200 mol/L $NaOH_{(aq)}$. The concentration of the acetic acid solution is
   (a) 0.08 mol/L.   (d) 1.0 mol/L.
   (b) 0.50 mol/L.   (e) 0.30 mol/L.
   (c) 0.05 mol/L.

18. Which of the following titrations might begin at pH = 11 and end with pH = 1?
   (a) strong acid with strong base
   (b) weak base with strong acid
   (c) weak acid with strong base
   (d) weak acid with weak base
   (e) none of the above

19. Which of the following titrations has pH = 7 at the equivalence point?
   (a) hydrochloric acid and sodium hydroxide
   (b) hydrochloric acid and ammonia
   (c) nitric acid and ammonia
   (d) acetic acid and sodium hydroxide
   (e) acetic acid and ammonia

Understanding Concepts

1. If 8.50 g of sodium hydroxide is dissolved to make 500 mL of cleaning solution, calculate the pOH of the solution. (8.1)

2. (a) Calculate the pH and pOH of a hydrochloric acid solution prepared by dissolving 30.5 kg of hydrogen chloride gas in 806 L of water at SATP.
   (b) What assumptions are made when doing this calculation? (8.1)

3. Sketch a flow chart or concept map that summarizes the conversion of $[H^+_{(aq)}]$ to and from $[OH^-_{(aq)}]$, pH, and concentration of solute. Make your flow chart large enough that you can write the procedure between the quantity symbols in the diagram. (8.1)

4. Unlike the rest of the hydrogen halides, hydrogen fluoride is a weak acid. It is used to etch glass and to produce frosted effects on glass (**Figure 1**). Write the K_a expression for hydrofluoric acid and calculate the hydrogen and fluoride ion concentrations in a 2.0 mol/L solution of this acid at 25°C. (8.2)

Figure 1

5. At 25°C, the hydroxide ion concentration in normal human blood is 2.5×10^{-7} mol/L. Calculate the hydrogen ion concentration and the pH of blood. (8.2)

6. Acetic acid is the most common weak acid used in industry. Determine the pH and pOH of an acetic acid solution prepared by dissolving 60.0 kg of pure, liquid acetic acid to make 1.25 kL of solution. (8.2)

7. Acetylsalicylic acid (ASA) is a painkiller used in many headache tablets. This drug forms an acidic solution that can sometimes damage the lining of the digestive system. The *Merck Index* lists its K_a at 25°C as 3.27×10^{-4}.
   (a) Calculate the pH of a saturated 0.018 mol/L solution of acetylsalicylic acid, $HC_{10}H_7CO_{4(aq)}$.
   (b) How might the pH change as the temperature changes to 37°C? (8.2)

8. Hydrocyanic acid is a very weak acid.
   (a) Write an equilibrium expression for the ionization of 0.10 mol/L $HCN_{(aq)}$. Include the percent ionization at SATP.
   (b) Calculate $[H^+_{(aq)}]$ and the pH of a 0.10 mol/L solution of $HCN_{(aq)}$ at SATP. (8.2)

9. Sodium ascorbate, the sodium salt of ascorbic acid, is used as an antioxidant in food products. A 0.15 mol/L solution of the ascorbate ion, $HC_6H_6O_6^-{}_{(aq)}$, has a pH of 8.65. Calculate K_b for the ascorbate ion. (8.2)

10. A series of experiments with a non-aqueous solvent determined that the products are highly favoured in each of the following acid–base equilibria.

$$(C_6H_5)_3C^- + C_4H_4NH \rightleftharpoons (C_6H_5)_3CH + C_4H_4N^-$$
$$HC_2H_3O_2 + HS^- \rightleftharpoons H_2S + C_2H_3O_2^-$$
$$O^{2-} + (C_6H_5)_3CH \rightleftharpoons (C_6H_5)_3C^- + OH^-$$
$$C_4H_4N^- + H_2S \rightleftharpoons C_4H_4NH + HS^-$$

   (a) Identify the Brønsted-Lowry acids, bases, and conjugate acid–base pairs in these chemical reactions.
   (b) Arrange the acids in the four chemical reactions in order of decreasing acid strength; that is, prepare a table of acids and bases. (8.2)

11. Sodium methoxide, $NaCH_3O_{(s)}$, is dissolved in water. Will the final solution be acidic, basic, or neutral? Explain your answer, using a net ionic equation. (8.3)

12. Compounds may be classified as ionic or molecular. Each of these classes can be subdivided into neutral substances, acids, and bases. Construct a flow chart that includes two examples for each of the six categories under the headings "Ionic" and "Molecular." (8.3)

13. If the pH of a solution is 6.8, what is the colour of each of the following indicators in this solution?
   (a) methyl red   (d) phenolphthalein
   (b) chlorophenol red   (e) methyl orange
   (c) bromothymol blue   (8.4)

14. Three separate 25.0-mL samples of a diluted rust-removing solution containing phosphoric acid were each titrated to the second endpoint using 1.50 mol/L sodium hydroxide. The average volume of $NaOH_{(aq)}$ required to reach the equivalence point was 17.9 mL. What is the concentration of phosphoric acid in the rust-removing solution? (8.4)

15. Sketch the pH (titration) curve for 15.0 mL of 0.10 mol/L $HCl_{(aq)}$ being added to 10.0 mL of 0.10 mol/L $NH_{3(aq)}$. Include the following information in your sketch. Show all calculations.
   (a) the equivalence point of the reaction
   (b) the initial pH of the ammonia solution
   (c) the pH after adding 5.0 mL of $HCl_{(aq)}$
   (d) the entities present at the equivalence point
   (e) the pH after adding 10.0 mL of $HCl_{(aq)}$
   (f) the pH after adding 15.0 mL of $HCl_{(aq)}$
   (g) an indicator for an endpoint (8.4)

16. Liquid ammonia can be used as a solvent for acid–base reactions.
    (a) What is the strongest acid species that could exist in this solvent?
    (b) What is the strongest base that exists in pure liquid ammonia?
    (c) The autoionization equilibrium of ammonia as a solvent is similar to that of water as a solvent. Write the equilibrium equation for the autoionization of the ammonia.
    (d) Sketch a titration curve for the addition of the strongest acid in ammonia to the strongest base. Instead of pH, what do you think would be used on the vertical axis of your graph?   (8.4)

17. Samples of an unknown solution were tested with indicators. Congo red was red and chlorophenol red was yellow in the solution. Estimate the pH and hydronium ion concentration of the solution.   (8.4)

18. Baking soda was added to a solution of sodium hydroxide and to a solution of hydrochloric acid. The pH of the sodium hydroxide changed from 13.0 to 9.5. The pH of the hydrochloric acid changed from 1.0 to 4.5. Provide an explanation of these results, with chemical equations to describe the reactions.   (8.5)

Applying Inquiry Skills

19. Write two experimental designs to rank a group of bases in order of strength.   (8.2)

20. (a) Predict whether each of the following chemical systems will be acidic, basic, or neutral. Explain your reasoning.
       (i)   aqueous hydrogen bromide
       (ii) aqueous potassium nitrite
       (iii) aqueous ammonia
       (iv) aqueous sodium hydrogen sulfate
       (v)   carbonated beverages
       (vi) limewater
       (vii) vinegar
    (b) Write an experimental design to test the predictions.   (8.3)

21. Each of seven unlabelled beakers was known to contain one of the following 0.10 mol/L solutions: $HC_2H_3O_{2(aq)}$, $Ba(OH)_{2(aq)}$, $NH_{3(aq)}$, $C_2H_4(OH)_{2(aq)}$, $H_2SO_{4(aq)}$, $HCl_{(aq)}$, and $NaOH_{(aq)}$. Describe diagnostic test(s) required to distinguish the solutions. Use a table to communicate your answer.   (8.3)

22. Critique the following experimental designs.

    (a) Sodium hydroxide is titrated against a phosphoric acid solution to the third equivalence point, using the bromothymol blue colour change as the endpoint.
    (b) The concentration of hydroxide ions in an ammonia solution is determined by precipitating the ions with a silver nitrate solution.
    (c) Hydrochloric acid is used as a primary standard to determine the concentration of sodium sulfide solution.
    (d) Litmus is used as a diagnostic test of the reaction between sodium hydrogen carbonate and sodium hydroxide.
    (e) Cobalt chloride paper is used in a diagnostic test for the production of water in a strong acid–strong base reaction.
    (f) The strength of an acid is determined by titration.   (8.4)

Making Connections

23. Many antacids contain carbonates such as $CaCO_3$, $MgCO_3$, and $NaHCO_3$ as their active ingredients. Other antacids are based on hydroxides such as $Al(OH)_3$ and $Mg(OH)_2$.
    (a) Write a chemical reaction to show how hydroxide antacids neutralize excess stomach acid.
    (b) Bicarbonate-based antacids, which contain HCO_3 as the active ingredient, are very common. Consuming excess bicarbonate can lead to a medical condition known as alkalosis. Conduct library or Internet research to determine the chemical equilibrium that is involved in alkalosis.
    (c) Visit a drugstore and examine antacid labels. Which active ingredient is most common?   (8.5)

 www.science.nelson.com

Extension

24. Many properties can be used to construct a titration curve. A pH curve is the most familiar example. Write an experimental design to determine the concentration of a barium hydroxide solution, using a conductivity titration with sulfuric acid. Explain your method with appropriate equations and predict the titration curve.

25. Calculate the pH of a 1.0×10^{-2} mol/L solution of $H_2SO_{4(aq)}$.

> **Criteria**

Process

- Make and evaluate predictions.
- Design an appropriate experimental procedure.
- Develop appropriate quantitative methods for calculating concentrations.
- Choose and safely use laboratory materials.
- Accurately carry out a series of titrations and record observations with appropriate precision.
- Analyze evidence.
- Evaluate experimental procedures and suggest improvements.
- Conduct research and analyze findings.

Product

- Clearly show your use of quantitative methods for calculating concentrations.
- Prepare a suitable lab report.

Chemical Analyst for a Day

Chemical analysts conduct quantitative tests to identify unknown chemicals, determine their composition, and measure their physical and chemical properties. You are already familiar with acid–base titration techniques, and how they may be used to determine concentrations, but you may not know about EDTA titrations. EDTA (ethylenediaminetetraacetic acid) is a "chelating agent" that can bond with entities like $Ca^{2+}_{(aq)}$ and $Mg^{2+}_{(aq)}$, forming complex ions (**Figure 1**).

Calcium and magnesium ions are the chief causes of water hardness. EDTA molecules bind hard-water ions, removing them from solution. A common indicator used in EDTA titrations is eriochrome black T (EBT), which changes from pink to blue at a single endpoint, when all calcium (and magnesium) ions have reacted with EDTA. EDTA titrations must be performed at a pH of approximately 10. An ammonia/ammonium chloride buffer is used to maintain the correct pH throughout the titration.

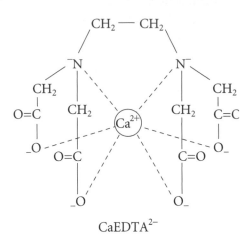

Figure 1
A calcium ion bound by EDTA in an ion complex.

Task

You will be a chemical analyst. You will apply concepts, lab skills, and problem-solving skills learned in this unit to determine the concentrations of inorganic solutes in a variety of aqueous solutions (e.g., drinking water, milk, simulated body fluids). Calcium ions are a beneficial constituent of many aqueous mixtures, including milk and mineral water.

Investigation: Calcium Content of Common Liquids

Questions

What is the $[Ca^{2+}_{(aq)}]$ in the control and experimental samples?
How do the experimental conditions affect the $[Ca^{2+}_{(aq)}]$ in the experimental sample?

Prediction

(a) Predict how the $[Ca^{2+}_{(aq)}]$ will differ after the experimental sample has been altered.

Experimental Design

The $[Ca^{2+}_{(aq)}]$ of two identical aqueous samples is determined by performing EDTA titrations in a pH 10 ammonia/ammonium chloride buffer using EBT indicator. One sample is titrated "as is": the control. The other sample is first subjected to conditions (of your design) that change the $[Ca^{2+}_{(aq)}]$ in a predictable way. This is the experimental sample.

Materials

120 mL EDTA solution
2 60-mL samples of a test solution
30 mL pH 10 ammonia/ammonium
   chloride buffer
eriochrome black T indicator in
   dropper bottle

distilled water in wash bottle
125-mL Erlenmeyer flask
50-mL buret
ring stand with clamp
20-mL volumetric pipet
10-mL graduated cylinder

100-mL beaker   sheet of white paper
safety pipet bulb

(b) List the materials that you will require to modify your experimental sample.

Procedure

Part I Modifying the Experimental Sample

(c) Plan and write a Procedure for modifying the $[Ca^{2+}_{(aq)}]$ of the sample.

1. With your teacher's permission, carry out your Procedure.

Part II EDTA Titration

2. Set up a clean buret for a titration using 60 mL of EDTA solution.

3. Obtain a clean 125-mL Erlenmeyer flask, and carefully transfer 20 mL of the control sample to be analyzed into the flask. Using a graduated cylinder, add 4 mL of the pH 10 buffer and 3 drops of EBT indicator solution to the flask. Swirl to mix.

4. Place a sheet of unlined white paper under the receiving flask. Slowly titrate by adding EDTA solution to the sample in the receiving flask. As the endpoint approaches, the colour begins to shift from pink to violet to sky blue. When the violet colour is reached, slowly add EDTA drop-wise until the sky blue colour persists. Record a final reading on your buret.

5. Discard the contents of the receiving flask in the sink with lots of running water. Repeat the titration two more times.

6. Repeat the titration with the experimental sample.

Analysis

(d) Research to find the reaction mechanism and ratios of entities in EDTA's reaction with calcium.

(e) Use your evidence to calculate $[Ca^{2+}_{(aq)}]$ of your two samples to answer the Questions.

Evaluation

(f) Describe sources of experimental error and suggest improvements that would help reduce or remove error.

(g) Evaluate your predictions on the basis of the evidence gathered.

Synthesis

(h) Compare an EDTA titration and an acid–base titration.

(i) Describe the equilibria and the changes they undergo as the EDTA titration progresses. (Use chemical equations where possible.)

(j) Explain why an EDTA titration has to occur in pH 10 buffer. How does the pH of the solution affect EDTA's reaction mechanism? Describe the buffer used in the EDTA titration using Brønsted–Lowry equations.

(k) What are some commercial and pharmaceutical uses for EDTA? Name three products you can purchase at a grocery store or pharmacy that contain EDTA and explain why EDTA is added to the product.

(l) Is EDTA safe? Describe safety considerations associated with the use of EDTA as a laboratory reagent and as a consumable product.

 www.science.nelson.com

Identify each of the following statements as true, false, or incomplete. If the statement is false or incomplete, rewrite it as a true statement.

1. At equilibrium, the concentrations of all reactants and products are equal.

2. Condensed phases, such as liquids, are omitted when writing equilibrium constant expressions.

3. For the following equilibrium,

$$2\,NOCl_{(g)} \rightleftharpoons 2\,NO_{(g)} + Cl_{2(g)} \qquad \Delta H° = +77.1\ kJ$$

the equilibrium concentration of chlorine can be increased by raising the temperature.

4. Catalysts generally favour the forward reaction in equilibrium.

5. Adding an inert gas, such as helium, to the following equilibrium system will increase the equilibrium concentration of carbon dioxide.

$$2\,CO_{2(g)} \rightleftharpoons 2\,CO_{(g)} + O_{2(g)}$$

6. Consider the reaction:

$$2\,HI_{(g)} \rightleftharpoons H_{2(g)} + I_{2(g)} + 52.8\ kJ \qquad K = 50\ at\ 450°C$$

The value of the equilibrium constant will increase as the temperature increases.

7. Consider these solubility products:

Compound	K_{sp}
CaF_2	3.9×10^{-11}
$AgCl$	1.8×10^{-10}

Based on this information, silver chloride is more soluble than calcium fluoride.

8. Silver chloride is less soluble in a solution of sodium chloride than in distilled water.

9. The activation energy for the forward and reverse reaction in an equilibrium system is always the same.

10. Exothermic reactions always occur spontaneously.

11. The pH of 0.1 mol/L acetic acid is 1.

12. Hypochlorite ion, $ClO^-_{(aq)}$ ($K_a = 3.45 \times 10^{-7}$) is a stronger base than ammonia, $NH_{3(aq)}$ ($K_b = 1.8 \times 10^{-5}$).

13. Metal oxides form acidic solutions while nonmetal oxides form basic solutions.

14. Potassium sulfate, $K_2SO_{4(s)}$, (a fertilizer) dissolves in water to form an acidic solution.

15. Ammonia, $NH_{3(aq)}$, may act as a Lewis base and a Brønsted–Lowry base.

16. The pH of a sulfurous acid solution, $H_2SO_{3(aq)}$ ($K_{a1} = 1.2 \times 10^{-2}$; $K_{a2} = 6.6 \times 10^{-8}$), may be calculated from its K_{a1} value only.

17. At the equivalence point of any titration, pH $= 7$.

18. Methyl orange ($pK_{In} = 4$) is a good indicator for strong acid–strong base titrations.

19. The highlighted portion of **Figure 1** indicates the part of a titration where a buffering action occurs.

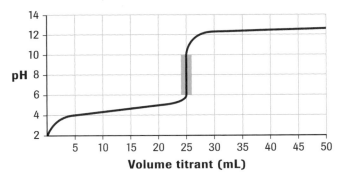

Figure 1

20. **Figure 2** is a curve for the titration of a weak base (sample) with a strong acid (titrant).

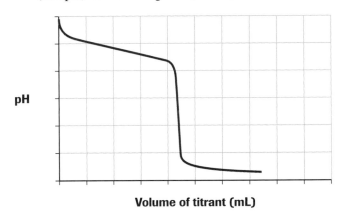

Figure 2

21. An effective acid–base buffer contains approximately equal concentrations of a strong acid and its conjugate base.

22. An $H_2CO_{3(aq)}/HCO^-_{3(aq)}$ buffer is one of several buffer systems used to control pH in blood.

Identify the letter that corresponds to the best answer to each of the following questions.

23. In the following reaction, the two Brønsted–Lowry acids are

$$HSO^-_{4(aq)} + HSO^-_{3(aq)} \rightleftharpoons H_2SO_{3(aq)} + SO^{2-}_{4(aq)}$$

636 Unit 4

An interactive version of the quiz is available online. NEL

GO www.science.nelson.com

(a) $HSO_{4(aq)}^{-}$ and $HSO_{3(aq)}^{-}$

(b) $HSO_{4(aq)}^{-}$ and $H_2SO_{3(aq)}$

(c) $SO_{4(aq)}^{2-}$ and $H_2SO_{3(aq)}$

(d) $HSO_{3(aq)}^{-}$ and $H_2SO_{3(aq)}$

(e) $HSO_{3(aq)}^{-}$ and $SO_{4(aq)}^{2-}$

24. Which of the following can act as a Brønsted–Lowry acid and a Brønsted–Lowry base in water.

(a) $NH_{3(aq)}$

(b) $HPO_{4(aq)}^{2-}$

(c) $OH_{(aq)}^{-}$

(d) $HI_{(aq)}$

(e) $NH_{2(aq)}^{-}$

25. If the $[H_{(aq)}^{+}]$ in a solution is 2.0×10^{-4} mol/L, the hydroxide ion concentration is

(a) 5.0×10^{-10} mol/L

(b) 2.0×10^{-4} mol/L

(c) 5.0×10^{-4} mol/L

(d) 2.0×10^{-11} mol/L

(e) 5.0×10^{-11} mol/L

26. The acid ionization constant for acetic acid, $HC_2H_3O_{2(aq)}$, is 1.8×10^{-5}. Which of the following is true of a 0.1 mol/L acetic acid solution?

(a) $[H_{(aq)}^{+}][C_2H_3O_{2(aq)}^{-}] > [HC_2H_3O_{2(aq)}]$

(b) $[H_{(aq)}^{+}][C_2H_3O_{2(aq)}^{-}] < [HC_2H_3O_{2(aq)}]$

(c) $[H_{(aq)}^{+}][C_2H_3O_{2(aq)}^{-}] = [HC_2H_3O_{2(aq)}]$

(d) $[H_{(aq)}^{+}] > [C_2H_3O_{2(aq)}^{-}]$

(e) $[H_{(aq)}^{+}] < [C_2H_3O_{2(aq)}^{-}]$

27. Consider the acid ionization constants of the acids in **Table 1**.

Table 1 Acid Ionization Constants

Acid	K_a
$HF_{(aq)}$	6.6×10^{-4}
$HCO_2H_{(aq)}$	1.8×10^{-4}
$HC_2H_3O_{2(aq)}$	1.8×10^{-5}
$HClO_{(aq)}$	2.9×10^{-8}
$HCN_{(aq)}$	6.2×10^{-10}

The correct order of increasing pH of a 0.1 mol/L solution of these acids is

(a) $HCN_{(aq)}$, $HClO_{(aq)}$, $HC_2H_3O_{2(aq)}$, $HCO_2H_{(aq)}$, $HF_{(aq)}$

(b) $HF_{(aq)}$, $HCO_2H_{(aq)}$, $HC_2H_3O_{2(aq)}$, $HClO_{(aq)}$, $HCN_{(aq)}$

(c) $HCO_2H_{(aq)}$, $HC_2H_3O_{2(aq)}$, $HClO_{(aq)}$, $HCN_{(aq)}$, $HF_{(aq)}$

(d) $HF_{(aq)}$, $HCN_{(aq)}$, $HClO_{(aq)}$, $HCO_2H_{(aq)}$, $HC_2H_3O_{2(aq)}$

(e) $HCN_{(aq)}$, $HF_{(aq)}$, $HC_2H_3O_{2(aq)}$, $HCO_2H_{(aq)}$, $HClO_{(aq)}$

28. Which of the following salts hydrolyzes to give a basic solution?

(a) $KCl_{(s)}$

(b) $AlCl_{3(s)}$

(c) $Na_2SO_{3(s)}$

(d) $Na_2SO_{4(s)}$

(e) $CaCl_{2(s)}$

29. Which of the following salts hydrolyzes to give an acidic solution?

(a) $KCl_{(s)}$

(b) $AlCl_{3(s)}$

(c) $Na_2SO_{3(s)}$

(d) $Na_2SO_{4(s)}$

(e) $CaCl_{2(s)}$

30. Which of the following salts produces a neutral solution when dissolved in water?

(a) $Na_2CO_{3(s)}$

(b) $NH_4Cl_{(s)}$

(c) $Ba(OH)_{2(s)}$

(d) $KHSO_{4(s)}$

(e) $Ca(NO_3)_{2(s)}$

31. Consider 0.1 mol/L aqueous solutions of the following compounds:

I $NaNO_{3(aq)}$

II $Cr(NO_3)_{3(aq)}$

III $Al(NO_3)_{3(aq)}$

IV $Fe(NO_3)_{3(aq)}$

V $Na_3PO_{4(aq)}$

In order of increasing pH, the solutions are:

(a) V, I, III, II, IV

(b) I, II, III, IV, V

(c) IV, II, III, I, V

(d) IV, II, III, V, I

(e) IV, III, II, I, V

32. The hydroxide ion concentration in a 1.0 mol/L solution of acetic acid at 25°C is

(a) 4.2×10^{-3} mol/L

(b) 2.4×10^{-11} mol/L

(c) 2.4×10^{-12} mol/L

(d) 2.4×10^{-3} mol/L

(e) 1.8×10^{-5} mol/L

33. Barium hydroxide completely dissociates in aqueous solution. What is the $[OH_{(aq)}^{-}]$ in a 0.05 mol/L solution of barium hydroxide?

(a) 0.05 mol/L

(b) 0.025 mol/L

(c) 0.075 mol/L

(d) 0.1 mol/L

(e) 0.01 mol/L

34. Which of the following titrations has a pH of 9 at the equivalence point?

(a) hydrochloric acid and sodium hydroxide

(b) hydrochloric acid and ammonia

(c) acetic acid and sodium hydroxide

(d) nitric acid and ammonia

(e) acetic acid and ammonia

35. The graph in **Figure 3** is the titration curve of

(a) a strong monoprotic acid with a strong base

(b) a strong monoprotic acid with a weak base

(c) a weak monoprotic acid with a strong base
(d) a weak monoprotic acid with a weak base
(e) none of the above

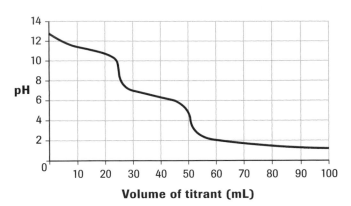

Figure 3

36. The titration of a weak base with a strong acid results in a solution with pH
    (a) equal to 7   (d) equal to 14
    (b) greater than 7   (e) equal to 1
    (c) less than 7

37. Phenolphthalein (abbrev. HPh) is an acid–base indicator commonly used during titrations. The colour changes observed in the titration are based on the following equilibrium

    $HPh_{(aq)} \rightleftharpoons H^+_{(aq)} + Ph^-_{(aq)}$
    (colourless)   (pink)

    During the titration of a strong acid sample with a strong base titrant, the colour of the sample
    (a) changes from pink to colourless.
    (b) changes from colourless to pink.

(c) remains colourless throughout.
(d) remains pink throughout.
(e) changes from pink to colourless then back to pink.

38. If the $[H^+_{(aq)}]$ at the equivalence point is calculated to be 1.0×10^{-4} mol/L for a particular titration, which of the following acid–base indicators should be used to mark the endpoint?
    (a) methyl orange
    (b) bromocresol green
    (c) bromothymol blue
    (d) phenol red
    (e) phenolphthalein

39. Which of the following combinations of chemicals could form an effective buffer solution?
    (a) $HCl_{(aq)}$ and $KOH_{(aq)}$
    (b) $HF_{(aq)}$ and $HNO_{3(aq)}$
    (c) $NH_{3(aq)}$ and $NaOH_{(aq)}$
    (d) $HC_2H_3O_{2(aq)}$ and $C_2H_3O^-_{2(aq)}$
    (e) $NH^+_{4(aq)}$ and $Br^-_{(aq)}$

40. A Lewis acid is
    (a) a proton donor.
    (b) a proton acceptor.
    (c) an electron pair donor.
    (d) an electron pair acceptor.
    (e) both (a) and (d).

41. Identify the Lewis base
    (a) $H^+_{(aq)}$ (d) BH_3
    (b) SO_3 (e) $O^{2-}_{(aq)}$
    (c) BeH_2

An interactive version of the quiz is available online.   NEL

GO   www.science.nelson.com

Understanding Concepts

1. The equilibrium constant for the following reaction has a value of 279 at 1000 K.

$$2 SO_{2(g)} + O_{2(g)} \rightleftharpoons 2 SO_{3(g)}$$

Use this information to calculate the value of the equilibrium constant for the reaction

$$2 SO_{3(g)} \rightleftharpoons 2 SO_{2(g)} + O_{2(g)} \qquad (7.2)$$

2. Nitrogen monoxide is a pollutant produced during the combustion of gasoline in cars. The reaction involved is

$$N_{2(g)} + O_{2(g)} \rightleftharpoons 2 NO_{(g)}$$

An equilibrium mixture at 2000°C contains these concentrations:

$[N_{2(g)}] = 0.63$ mol/L; $[O_{2(g)}] = 0.21$ mol/L; $[NO_{(g)}] = 0.015$ mol/l

Calculate the value of the equilibrium constant for this reaction. (7.2)

3. Consider this reaction:

$$C_{(s)} + H_2O_{(g)} \rightleftharpoons CO_{(g)} + H_{2(g)} \qquad \Delta H° = 131.3 \text{ kJ}$$

Predict the effect, if any, of the following changes on the equilibrium concentration of carbon monoxide, CO.
(a) increasing the concentration of hydrogen gas
(b) cooling the reaction vessel
(c) increasing the volume of the reaction vessel
(d) adding a catalyst
(e) grinding the carbon into a fine powder
(f) removing hydrogen
(g) adding inert argon gas to the reaction vessel (7.3)

4. Methanol can be produced using the following reaction:

$$CO_{(g)} + 2 H_{2(g)} \rightleftharpoons CH_3OH_{(g)} \qquad \Delta H° = -90.1 \text{ kJ}$$

Suggest three ways in which to increase the methanol concentration at equilibrium. (7.3)

5. Urea is an important nitrogen-based fertilizer.
(a) The first step in the production of urea fertilizer is

$$2 NH_{3(g)} + CO_{2(g)} \rightleftharpoons NH_2CO_2NH_{4(aq)} + \text{energy}$$

(ammonium carbamate)

Explain why running the reaction at high pressure increases the yield of this reaction to almost 100%.

(b) The second step in making urea involves the decomposition of ammonium carbamate:

$$NH_2CO_2NH_{4(aq)} + \text{energy} \rightleftharpoons CO(NH_2)_{2(s)} + H_2O_{(g)}$$

(urea)

How could the temperature and pressure be altered to maximize the yield of urea? (7.3)

6. (a) Calculate the solubility of silver chloride at 25°C.
(b) Calculate the solubility of silver chloride in 0.015 mol/L sodium chloride.
(c) Use Le Châtelier's principle to explain why the answers in (a) and (b) differ. (7.3)

7. Ammonia, an important chemical used in the production of chemical fertilizers, is synthesized from its elements according to the reaction

$$N_{2(g)} + 3 H_{2(g)} \rightleftharpoons 2 NH_{3(g)}$$

Consider the following equilibrium constants at different temperatures for this reaction.

Table 1 K at Different Temperatures

Temp (°C)	K
473	612
873	4.2
1073	3.9×10^{-2}

Is the synthesis of ammonia endothermic or exothermic? Explain your answer. (7.4)

8. Nitrosyl chloride, $NOCl_{(g)}$, decomposes to form nitrogen monoxide, $NO_{(g)}$, and chlorine gas, $Cl_{2(g)}$, according to the following chemical equation:

$$2 NOCl_{(g)} \rightleftharpoons 2 NO_{(g)} + Cl_{2(g)}$$

At 35°C the equilibrium constant, K, is 1.60×10^{-5}.
(a) At this temperature, are the products or reactants favoured?
(b) Calculate the equilibrium concentration of $NOCl_{(g)}$, if the equilibrium concentrations of $NO_{(g)}$ and $Cl_{2(g)}$ are both 0.10 mol/L. (7.5)

9. Consider the following equilibrium system:

$$N_2O_{4(g)} \rightleftharpoons 2 NO_{2(g)} \qquad K = 6.13 \times 10^{-3} \text{ at 25°C}$$

(colourless) (reddish-brown)

If the molar concentration of $N_2O_{4(g)}$ and $NO_{2(g)}$ are 8.00×10^{-4} mol/L and 4.00×10^{-3} mol/L respectively, is the system at equilibrium? If not, what colour change would be observed as the system shifts to approach equilibrium? (7.5)

10. Consider this equilibrium:

$$H_{2(g)} + CO_{2(g)} \rightleftharpoons H_2O_{(g)} + CO_{(g)}$$

When 0.100 mol each of hydrogen and carbon dioxide are introduced into a 1.00-L container at 986°C, the equilibrium concentration of carbon monoxide is found to be 0.056 mol/L.
  (a) Calculate the equilibrium concentrations of all substances.
  (b) Calculate the value of the equilibrium constant. (7.5)

11. Phosphorus pentachloride decomposes according to the reaction

$$PCl_{5(g)} \rightleftharpoons PCl_{3(g)} + Cl_{2(g)}$$

When 0.60 mol of $PCl_{5(g)}$ are initially put into a 2.00-L container, the equilibrium concentration of chlorine is found to be 0.26 mol/L.
  (a) Calculate the equilibrium concentration of all substances.
  (b) Calculate the value of the equilibrium constant. (7.5)

12. Hydrogen and iodine combine to form hydrogen iodide according to the equation,

$$H_{2(g)} + I_{2(g)} \rightleftharpoons 2\,HI_{(g)} \qquad K = 49.7 \text{ at } 458°C$$

Calculate the equilibrium concentrations of all substances if 0.50 mol $H_{2(g)}$ and 0.50 mol $I_{2(g)}$ are initially placed into a 5.00-L reaction vessel at 458°C. (7.5)

13. Ammonia decomposes into its elements according to the reaction

$$2\,NH_{3(g)} \rightleftharpoons N_{2(g)} + 3\,H_{2(g)} \qquad K = 1.60 \times 10^{-3} \text{ at } 200°C$$

If the initial concentration of ammonia is 0.20 mol/L, what are the equilibrium concentrations of all substances? (7.5)

14. Consider the formation of hydrogen iodide from its elements:

$$H_{2(g)} + I_{2(g)} \rightleftharpoons 2\,HI_{(g)} \qquad K = 49.7 \text{ at } 458°C$$

If 2.00 mol of hydrogen and 1.00 mol of iodine are initially placed into a 5.00-L reaction vessel, what is the equilibrium concentration of hydrogen iodide? (7.5)

15. Sulfur trioxide decomposes according to the following equation:

$$2\,SO_{3(g)} \rightleftharpoons 2\,SO_{2(g)} + O_{2(g)} \qquad K = 6.9 \times 10^{-7} \text{ at } 1500°C$$

What concentration of oxygen can be expected when 0.400 mol of $SO_{3(g)}$ comes to equilibrium at 1500°C in a 2.00-L vessel? (7.5)

16. Calculate the molar solubility of calcium sulfate at 25°C. (7.6)

17. Calculate the molar concentration of chloride ions in a saturated solution of lead(II) chloride, $PbCl_2$, at 25°C. (7.6)

18. The chloride ion concentration in tap water can be 2.2×10^{-4} mol/L. Will a precipitate form if 250.0 mL of water is mixed with 250.0 mL of 0.010 mol/L silver nitrate at SATP? (7.6)

19. A solution is prepared by mixing 100.0 mL of 0.015 mol/L magnesium nitrate with 300 mL of 0.10 mol/L potassium fluoride at SATP. Will a precipitate form? (7.6)

20. Distinguish between the terms solubility and solubility product. (7.6)

21. Name two substances that will decrease the solubility of calcium sulfate, $CaSO_{4(s)}$ in water. (7.6)

22. Predict whether the change in entropy is positive or negative for each of the following changes. Explain your answers.
  (a) $Na_{(s)} \rightarrow Na_{(l)}$
  (b) $Pb(NO_3)_{2(aq)} + 2\,KI_{(aq)} \rightarrow$ $PbI_{2(s)} + 2\,KNO_{3(aq)}$
  (c) $H_2O_{(g)} + Cl_2O_{(g)} \rightarrow 2\,HOCl_{(g)}$
  (d) $NH_4Cl_{(s)} \rightarrow NH_{3(g)} + HCl_{(g)}$ (7.7)

23. Use the Gibbs-Helmholtz equation to explain qualitatively why table salt, $NaCl_{(s)}$, does not spontaneously decompose into its elements at room temperature. (7.7)

24. Cold packs contain solid ammonium chloride, which absorbs thermal energy from the water. Calculate the standard Gibbs free energy change associated with the dissolving of ammonium chloride and interpret the results.

$$NH_4Cl_{(s)} \rightarrow NH_4Cl_{(aq)} \qquad (7.7)$$

25. Methane, $CH_{4(g)}$, is a major component of natural gas. Calculate the value of $\Delta G°$ for the combustion of methane to produce only gaseous products. What does the value of $\Delta G°$ for this reaction tell you? (7.7)

26. Use the Gibbs-Helmholtz equation and the concept of equilibrium to calculate the normal condensation point of ethanol, $C_2H_5OH_{(l)}$. (7.7)

27. Complete the following Brønsted–Lowry acid–base equations.
    (a) $HNO_{2(aq)} + SO_{4(aq)}^{2-} \rightleftharpoons$
    (b) $HCO_{3(aq)}^{-} + NH_{4(aq)}^{+} \rightleftharpoons$
    (c) $NH_{3(aq)} + HS_{(aq)}^{-} \rightleftharpoons$ (8.1)

28. Consider the autoionization of water:

$$H_2O_{(l)} \rightleftharpoons H_{(aq)}^{+} + OH_{(aq)}^{-}$$

The value for the ion product constant, K_w, like all equilibrium constants, depends on temperature. Consider **Table 2**.

Table 2 K_w at Different Temperatures

Temperature (°C)	K_w
25	1.0×10^{-14}
37	2.7×10^{-14}
60	9.6×10^{-14}

(a) Is the autoionization of water exothermic or endothermic?
(b) How does the pH of water change as its temperature increases? (8.1)

29. Normal rain has a pH of 5.6 whereas the pH of acidic rain can be as low as 4. Calculate the ratio of $[H_{(aq)}^{+}]$ in acidic rain to $[H_{(aq)}^{+}]$ in normal rain. (8.1)

30. Cola soft drinks are acidic because of carbonation and the addition of phosphoric acid. The pH of a cola can be as low as 2.7. Calculate the hydrogen and hydroxide ion concentrations of an acidic cola. (8.1)

31. Cyanic acid, $HOCN_{(aq)}$, is a weak acid used in the manufacture of certain pesticides. A 0.100 mol/L solution of cyanic acid is found to be 5.9% ionized. Calculate the K_a for cyanic acid. (8.2)

32. Saccharin, $HC_7H_4NO_3S_{(aq)}$, the oldest artificial sweetener (currently banned in Canada as a suspected carcinogen), is a weak monoprotic acid. Given that the K_a of saccharin is 2.1×10^{-12}, calculate the pH of a 0.500 mol/L solution of saccharin. (8.2)

33. The formate ion, $HCO_{2(aq)}^{-}$ is the conjugate base of formic acid, $HCO_2H_{(aq)}$.
    (a) Write the K_b expression for the ionization of the formate ion.
    (b) Write the K_a expression for the ionization of formic acid.
    (c) Use the equations in (a) and (b) to show that $K_a K_b = K_w$. (8.2)

34. Controlling the concentration of hypochlorous acid, $HClO_{(aq)}$, is important to maintaining a clean swimming pool. $HClO_{(aq)}$ is a weak acid with $K_a = 2.9 \times 10^{-8}$. Calculate the pH of a 0.100 mol/L solution of hypochlorous acid. (8.2)

35. Write a Brønsted–Lowry reaction in which the hydrogen sulfite ion, $HSO_{3(aq)}^{-}$, acts as an acid and another equation in which it acts as a base. (8.3)

36. Why do 0.1 mol/L solutions of $NaHSO_{4(aq)}$ and $NaHSO_{3(aq)}$ have different pH values? (8.3)

37. Most crops grow well in soil within a pH range of 6 to 8. Minerals such as limestone (calcium carbonate, $CaCO_{3(s)}$) and alum (aluminum sulfate, $Al_2(SO_4)_{3(s)}$) can be added to raise or lower soil pH as needed.
    (a) What effect would the addition of limestone have on soil pH?
    (b) What effect would the addition of alum have on soil pH? (8.3)

38. Sodium phosphate, $Na_3PO_{4(aq)}$, is an excellent cleaner of stubborn grease stains.
    (a) Predict whether a 0.10 mol/L solution of sodium phosphate is acidic, basic, or neutral.
    (b) Calculate the pH of a 0.10 mol/L solution of sodium phosphate. (8.3)

39. (a) Complete the following chemical equations:
       (i) $HPO_{4(aq)}^{2-} + H_2O_{(l)} \rightleftharpoons$
           (hydrolyzing as an acid)
       (ii) $HPO_{4(aq)}^{2-} + H_2O_{(l)} \rightleftharpoons$
           (hydrolyzing as a base)
    (b) Compare the K_a and K_b values for the monohydrogen phosphate ion.
    (c) Is a solution of Na_2HPO_4 acidic, basic, or neutral? (8.3)

40. Predict whether the pH of each of the following solutions is 7; > 7; or < 7.
    (a) $NH_4NO_{3(aq)}$ (d) $KHCO_{3(aq)}$
    (b) $Na_2S_{(aq)}$ (e) $NaHSO_{4(aq)}$
    (c) $CuCl_{2(aq)}$ (f) $Na_2SO_{4(aq)}$ (8.3)

41. Identify the Lewis acid and Lewis base in these reactions:
    (a) $HCl_{(g)} + NH_{3(g)} \rightarrow NH_4Cl_{(s)}$
    (b) $Cu_{(aq)}^{2+} + 6 H_2O_{(l)} \rightarrow Cu[H_2O]_{6(aq)}^{2+}$
    (c) $Al(OH)_{3(aq)} + OH_{(aq)}^{-} \rightarrow Al(OH)_{4(aq)}^{-}$ (8.3)

42. Match the pH at the equivalence point with the type of titration for the two columns in **Table 3**. (8.4)

Table 3 Matching pH

Type of titration	pH at equivalence point
strong acid/strong base	9
strong acid/weak base	7
weak acid/strong base	6

43. Sketch the pH (titration) curve for 15.00 mL of 0.100 mol/L sodium hydroxide (the titrant) being added to 10.00 mL of 0.100 mol/L hydrochloric acid (the sample). Include the following information on your sketch.
    (a) the equivalence point for the titration
    (b) the initial pH of the hydrochloric acid solution
    (c) the pH after adding 5.00 mL of base
    (d) the entities present at the equivalence point
    (e) the pH after adding 10.00 mL of base
    (f) the pH after adding 15.00 mL of base
    (g) the name of a suitable indicator for this reaction (8.4)

44. In a titration, 20.00 mL of 0.100 mol/L hydrochloric acid is titrated with standardized 0.100 mol/L sodium hydroxide (the titrant). Calculate the pH after the following volumes of NaOH have been added, and sketch the titration curve. Label important regions of the curve and identify key points.
    (a) 0.00 mL   (d) 19.99 mL
    (b) 10.00 mL   (e) 20.01 mL
    (c) 19.90 mL   (f) 25.00 mL   (8.4)

45. Consider the titration of hydrofluoric acid, $HF_{(aq)}$, with sodium hydroxide, $NaOH_{(aq)}$.
    (a) Write the chemical equation for this reaction.
    (b) Sketch the titration curve for this reaction.
    (c) Write the chemical equation for the equilibrium that occurs during the first buffering region. Use the equation to explain why the pH remains relatively constant even though small amounts of strong base are being added.
    (d) Predict whether the pH at the equivalence point is equal to 7, is greater than 7, or is less than 7. (8.4)

46. **Figure 1** shows curves for the titration of four different acids with 0.10 mol/L $NaOH_{(aq)}$ as the titrant.

Titration of Four Acids with NaOH$_{(aq)}$

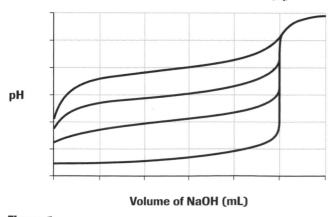

Volume of NaOH (mL)

Figure 1

(a) Which titration curve in **Figure 1** represents the titration of a strong acid with $NaOH_{(aq)}$? Explain.
(b) Identify the titration curve of the acid with the smallest K_a value.
(c) The vertical red bar on the graph is meant to draw your attention to an important generalization about acid–base titrations. Describe the generalization. (8.4)

47. According to the table of acid–base indicators on page 609, what is the colour of each of the indicators in the solutions of given pH (**Table 4**)? (8.4)

Table 4 pH of Solutions

Indicator	pH
methyl red	5
thymolphthalein	6
bromothymol blue	7
indigo carmine	13

48. Consider the pH curve for the titration of acetic acid with sodium hydroxide (**Figure 2**). Select the most suitable indicator for this titration from **Table 5**. (8.4)

HC$_2$H$_3$O$_{2(aq)}$ Titrated with NaOH$_{(aq)}$

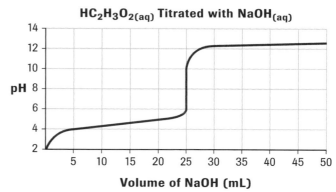

Volume of NaOH (mL)

Figure 2

Table 5 pH of Indicators

Indicator	pH colour change range
bromothymol blue	6.0 – 7.6
methyl orange	3.1 – 4.4
thymolphthalein	9.3 – 10.5

49. Describe how to prepare 100 mL of acetic acid–acetate buffer using 0.5 mol/L acetic acid and solid sodium acetate, $NaC_2H_3O_{2(s)}$. The concentration of acetic acid and acetate should be the same in the buffer. (8.4)

50. Consider the titration curve of hydrochloric acid with sodium hydroxide (**Figure 3**).

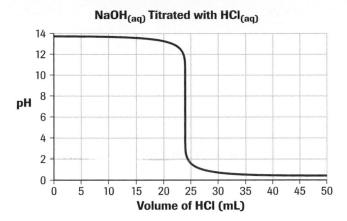

NaOH$_{(aq)}$ Titrated with HCl$_{(aq)}$

Figure 3

(a) Explain why bromothymol blue is an ideal indicator for this titration.

(b) Explain why phenolphthalein is sometimes used even though it has a pK_{In} of 9.4. (8.4)

51. Use Le Châtelier's principle to predict the effect of the following stresses on the given equilibrium.
(a) the addition of a small volume of HCl$_{(aq)}$ to an NH$_4^+{}_{(aq)}$/NH$_{3(aq)}$ buffer:

$$NH_4^+{}_{(aq)} \rightleftharpoons H^+{}_{(aq)} + NH_{3(aq)}$$

(b) the addition of NaOH$_{(aq)}$ to a phenolphthalein indicator solution (HPh$_{(aq)}$):

$$HPh_{(aq)} \rightleftharpoons H^+{}_{(aq)} + Ph^-{}_{(aq)}$$

(c) the addition of NaC$_2$H$_3$O$_{2(s)}$ to vinegar:

$$HC_2H_3O_{2(aq)} \rightleftharpoons H^+{}_{(aq)} + C_2H_3O_2^-{}_{(aq)} \qquad (8.5)$$

52. Calculate the change in pH that would occur if 0.10 mol HCl$_{(g)}$ is added to 1.0 L of a formic acid/sodium formate buffer containing 0.25 mol/L HCO$_2$H$_{(aq)}$ and 0.15 mol/L CO$_2$H$^-{}_{(aq)}$ at equilibrium. Assume that the volume of the buffer stays constant. (8.5)

Applying Inquiry Skills

53. Silver sulfate is a common laboratory reagent.

Question

What is the K_{sp} of silver sulfate?

Experimental Design

A saturated solution of silver sulfate is prepared by adding about 5 g of silver sulfate to 200 mL of distilled water. The next day, all the undissolved silver sulfate settles to the bottom of the container. A 50.00-mL sample of the saturated solution above the white solid is carefully removed and transferred to an evaporating dish. The dish is heated to remove all the water, leaving behind a white crust of silver sulfate. The mass of the remaining silver sulfate is determined.

Evidence

volume of saturated silver sulfate solution: 50.00 mL
volume of silver sulfate collected: 0.25 g

Analysis

(a) What amount of silver sulfate was in the 50.00-mL sample?

(b) Calculate the molar solubility of silver sulfate.

(c) What is the molar concentration of the silver ion in the saturated solution?

(d) What is the molar concentration of the sulfate ion in the saturated solution?

(e) Write the K_{sp} expression for silver sulfate.

(f) Calculate the K_{sp} for silver sulfate.

Evaluation

(g) Look up the K_{sp} for silver sulfate in a standard table of K_{sp} values. Compare your calculated value with the accepted value by calculating a percentage difference. Is the calculated value acceptable? (7.6)

54. Extensive use of the pesticide DDT in North America in the 1960s and 1970s resulted in the near extinction of the bald eagle. The accumulation of DDT in their tissue caused eagles to produce eggs with so little calcium that they would break before hatching. Today's modern chicken farmer must also be aware of the calcium content of eggshells. The amount of calcium in the egg is an indicator of the health of the bird.

An acid–base titration can be used to determine the calcium content of eggshells.

Question

What is the percent by mass of calcium carbonate in an eggshell?

Experimental Procedure

An eggshell is carefully cleaned and then dried in an oven for 24 h. The eggshell is ground into a fine powder, its mass measured, and then added to a titration flask. 10.00 mL of 1.00 mol/L hydrochloric acid and 25 mL of water are added to the flask. The flask is heated to boiling and then allowed to cool. The reaction occurring in the flask is

$$2\,HCl_{(aq)} + CaCO_{3(s)} \rightarrow Ca^{2+}{}_{(aq)} + CO_{2(g)} + H_2O_{(l)} + 2\,Cl^-{}_{(aq)}$$

The acid not consumed by this reaction is titrated with a standardized sodium hydroxide solution to a phenolphthalein endpoint.

Evidence

mass of eggshell:	0.45 g
volume of acid used:	25.00 mL of 1.00 mol/L $HCl_{(aq)}$
volume of base required:	17.40 mL of 1.00 mol/L $NaOH_{(aq)}$

Analysis

(a) Calculate the amount of unreacted acid.
(b) Calculate the amount of calcium carbonate in the eggshell.
(c) Answer the Question.

Evaluation

(d) Why was it necessary to grind the eggshell into a fine powder?
(e) Why was it necessary to heat the contents of the titration flask?
(f) Explain why the addition of water to the titration flask does not affect the outcome of the experiment. (8.4)

55. In an experiment, a solution of sodium hydroxide is standardized by titrating it with the primary standard, potassium hydrogen iodate, $KH(IO_3)_{2(aq)}$. The chemical equation for this reaction is

$$KH(IO_3)_{2(aq)} + NaOH_{(aq)} \rightarrow H_2O_{(l)} + KIO_{3(aq)} + NaIO_{3(aq)}$$

Question

What is the molar concentration of a sodium hydroxide solution?

Experimental Design

A known mass of potassium hydrogen iodate is titrated with a sodium hydroxide solution of unknown concentration to a bromothymol blue endpoint. The titration is repeated two more times.

Evidence

mass of $KH(IO_3)_{2(s)}$ used in each titration: 1.00 g
average volume of sodium hydroxide: 20.54 mL

Analysis

(a) Calculate the concentration of the $NaOH_{(aq)}$ solution.
(b) Potassium hydrogen iodate is a strong acid. Explain why either bromothymol blue or phenolphthalein would be suitable indicators for this titration.
(c) The following equation describes the reaction of dissolved carbon dioxide with hydroxide ions:

$$CO_{2(g)} + 2 OH^-_{(aq)} \rightarrow CO_3^{2-}_{(aq)} + H_2O_{(l)}$$

Why is it necessary to boil the water used to prepare the sodium hydroxide solution?
(d) Why is it sometimes recommended to place an inverted test tube over the top of a buret containing sodium hydroxide? (8.4)

Making Connections

56. Liquid carbon dioxide may soon replace the organic solvents currently being used in dry cleaning. Liquid carbon dioxide is just as effective in dissolving grease and far more environmentally friendly. To prevent liquid carbon dioxide from evaporating, the washing machines must be pressurized to a minimum of 517.6 kPa at 56.6°C.
(a) Use Le Châtelier's principle and the following equilibrium to explain why high pressure is required to keep carbon dioxide in the liquid state.

$$CO_{2(g)} \rightleftharpoons CO_{2(l)} + heat$$

(b) High levels of carbon dioxide can be dangerous. What safety precautions should be observed in the building where the dry cleaning is done?
(c) Conduct library or Internet research to determine the advantages of using liquid carbon dioxide as a dry-cleaning solvent. (7.3)

 www.science.nelson.com

57. It has been said that seawater contains more gold than all of the world's banks. One estimate of the molar concentration of dissolved gold in seawater is 5.6×10^{-11} mol/L.
(a) Would a precipitate form if equal volumes of seawater and 0.100 mol/L sodium chloride are combined?

$$K_{sp(AuCl_{(aq)})} = 2.0 \times 10^{-13}$$
$$K_{sp(AuCl_{3(aq)})} = 3.2 \times 10^{-25}$$

(b) Explain whether this is a feasible way to extract gold from seawater. (7.6)

58. Lye, or sodium hydroxide, is a common ingredient of oven cleaners.
(a) Calculate the pH and pOH of the resulting solution when 24.00 g of sodium hydroxide are dissolved into enough water to make 750 mL of cleaning solution.

(b) What are some safety precautions that should be taken when using this solution for cleaning purposes? (8.1)

59. The minerals in your teeth become more soluble as pH drops below 5.5.
    (a) Why is the consumption of sweets bad for your teeth?
    (b) Suggest ways to prevent tooth decay caused by acid degradation of tooth enamel. (8.1)

60. Chlorine-containing compounds such as calcium hypochlorite are added to backyard swimming pools to kill the microorganisms in pool water and oxidize organic matter in the pool. Once dissolved in water, the hypochlorite forms the equilibrium

$$HClO_{(aq)} \rightleftharpoons H^+_{(aq)} + ClO^-_{(aq)}$$

Maintaining the backyard pool at pH 7.4 —the pH of human tears—is ideal. At this level, the concentration of hypochlorous acid and hypochlorite are approximately equal.
    (a) Predict the effect of decreasing pH has on the concentration of $HClO_{(aq)}$ in the pool. (This effect makes pool water irritating to the swimmers' eyes.)
    (b) Predict the effect of increasing pH on the concentration of $ClO^-_{(aq)}$. (This is not desirable because hypochlorite decomposes in sunlight.)
    (c) Superchlorination or "shock treatment" is sometimes necessary. This involves adding a large excess of chlorine-containing compounds to the water. Consult a pool supply store or a pool owner in the neighbourhood to determine when "shocking" the pool is necessary.
    (d) Use Le Châtelier's principle to predict the effect of a shock treatment on pool pH. (8.5)

61. Calcium hydroxide (or slaked lime), $Ca(OH)_{2(aq)}$, is sometimes used to neutralize lakes damaged by acid rain. In many cases, solid calcium hydroxide is dumped directly into an acid lake in an effort to neutralize it. Describe advantages and disadvantages of this practice. Suggest an alternative strategy. (8.6)

Extension/Challenge

62. The following reaction may be used to remove sulfur compounds from power plant smokestack emissions.

$$2 H_2S_{(g)} + SO_{2(g)} \rightleftharpoons 3 S_{(s)} + 2 H_2O_{(g)} \quad \Delta H° = -145 \text{ kJ}$$

The value of the equilibrium constant for this reaction is 8.0×10^{15} at 25°C.
    (a) Given the value of K, is this reaction an effective way to clean smokestack emissions of sulfur compounds?
    (b) Would the efficiency of this reaction increase or decrease with an increase in pressure?
    (c) How does the efficiency of this reaction in removing sulfur dioxide change as the temperature is increased? (7.3)

63. More sulfuric acid is produced each year in North America than any other industrial chemical. One step in the manufacture of sulphuric acid is

$$2 SO_{2(g)} + O_{2(g)} \overset{V_2O_5 \text{ catalyst}}{\rightleftharpoons} 2 SO_{3(g)} \quad \begin{array}{l} \Delta H° = -197.8 \text{ kJ} \\ K = 0.0114 \text{ at } 400°C \end{array}$$

    (a) Explain why each of the following factors would help to increase the yield of sulfur trioxide.
       (i)   increasing the pressure
       (ii) removing $SO_{3(g)}$ as quickly as it forms
       (iii) continually adding $SO_{2(g)}$ and $O_{2(g)}$
    (b) Why does a catalyst increase the rate of production of $SO_{3(g)}$?
    (c) Engineers have found that $SO_{3(g)}$ can be produced faster by increasing the temperature to 400°C even though this change favours the reverse reaction. Suggest an explanation. (7.3)

64. (a) Calculate the solubility of calcium carbonate in water at 25°C.
    (b) A 2.0-L kettle has 5.00 g of calcium carbonate scale on the bottom. How many times would the kettle have to be filled and emptied with water at 25°C in order to remove all the scale?
    (c) The K_{sp} of $CaCO_{3(s)}$ is lower at 100°C than it is at 25°C. How does this explain the formation of solid calcium carbonate on the bottom of the kettle?
    (d) Where else in a typical home would you expect to find $CaCO_{3(s)}$ precipitate? (8.3)

Electrochemistry

Gillian Goward
Assistant Professor of Chemistry
McMaster University

"Research in my field is driven by the desire to develop environmentally friendly power sources. For example, polymer-electrolyte fuel cells (PEM-FCs) offer the huge environmental advantage of reducing to zero emissions from vehicles. In cities where smog is a problem, such as Los Angeles, or Hamilton, the need for such vehicles is crucial. However, many challenges still need to be overcome before this technology will be widely available. I chose to become a research scientist because I can use my imagination to unravel problems at the microscopic level and understand materials from the building-blocks vantage point. The focus of my research with PEM-FCs is on the mechanism of proton transport within the polymer membrane, which I examine with a microscope using a technique called solid-state nuclear magnetic resonance. By acquiring NMR spectra of membrane materials under a variety of conditions, I hope to ascertain the details of proton transport, which will aid in designing new and improved membranes."

▶ Overall Expectations

In this unit, you will be able to

- demonstrate an understanding of fundamental concepts related to oxidation–reduction and the interconversion of chemical and electrical energy;

- build and explain the functioning of simple galvanic and electrolytic cells; use equations to describe these cells; solve quantitative problems related to electrolysis; and

- describe some uses of batteries and fuel cells; explain the importance of electrochemical technology to the production and protection of metals; and assess environmental and safety issues associated with these technologies.

▶ **Prerequisites**

Concepts

- Identify the major components of atoms, and distinguish between atoms and ions.
- Explain the formation of ionic and covalent bonds in terms of electrons.
- Recognize that the type of chemical reaction depends on the nature of the reactants.
- Relate the reactivity of a series of elements to their position on the periodic table.
- Demonstrate an understanding of the Arrhenius theory of substances in solution.
- Demonstrate an understanding of the mole concept and the quantitative relationships expressed in a chemical equation.

Skills

- Demonstrate the skills required to carry out laboratory studies of chemical reactions.
- Represent chemical reactions using balanced chemical equations.
- Conduct appropriate chemical (diagnostic) tests to identify common substances.
- Use appropriate scientific vocabulary to communicate ideas related to chemical reactions and chemical calculations.
- Solve stoichiometry problems involving solutions.

Safety and Technical Skills

1. What should you check before plugging in electrical equipment?
2. List some safety precautions for operating electrical equipment.

Knowledge and Understanding

3. When a metal atom forms an ion, the atom _____ electrons to form a _____ charged ion.

4. When a nonmetal atom forms an ion, the atom _____ electrons to form a _____ charged ion.

5. According to trends in the periodic table and your general chemistry knowledge, copy and complete **Table 1**.

Table 1 Reactivity of Elements

Category	Groups or examples
most reactive metals	
least reactive metals	
most reactive nonmetals	
least reactive nonmetals	

6. Complete the following chemical equations using Lewis symbols or structures for the products.

   (a) $K \cdot + \cdot \overset{..}{\underset{..}{Cl}} : \longrightarrow$

   potassium + chlorine →

   (b) $\cdot \overset{..}{\underset{.}{P}} \cdot + \cdot \overset{..}{\underset{..}{Cl}} : \longrightarrow$

   phosphorus + chlorine →

   (c) Compare electron rearrangement in (a) to electron rearrangement in (b).
   (d) According to the bonding theory you have studied, what is believed to determine whether two atoms transfer or share electrons?

7. (a) Write the generalized chemical equation for a single displacement reaction.
   (b) How do you know what class of element, metal or nonmetal, forms in a single displacement reaction?

8. Complete and balance the chemical equation for each of the following reactions.
   (a) __ $Zn_{(s)}$ + __ $AgNO_{3(aq)}$ →
   (b) __ $Cl_{2(aq)}$ + __ $KBr_{(aq)}$ →
   (c) __ $Al_{(s)}$ + __ $HCl_{(aq)}$ →
   (d) __ $C_3H_{8(g)}$ + __ $O_{2(g)}$ → __ $CO_{2(g)}$ + __

Inquiry and Communication

9. When studying chemical reactions, diagnostic tests are important to identify products (see Appendix A6). Interpret the evidence in **Table 2** to identify the product or type of product formed in a chemical reaction.

Table 2 Diagnostic Tests of Unknown Products

Diagnostic test	Evidence obtained	Analysis
halogen		(a)
		(b)
acid, base, neutral	bromothymol blue added	(c)
		(d)
ions		(e)
		(f)

chapter

Electric Cells

Since their invention in 1888, vehicles powered entirely by electricity have drifted in and out of fashion. Many experts predict that will end in the next decade, when electric vehicles finally make a breakthrough. A combination of political, economic, and environmental factors are slowly making electric power a viable alternative to gasoline power. One advantage of electric cars over gasoline-fuelled cars is environmental. There is real promise of a significant reduction in pollution from the vehicles we drive. Also, cars powered by gasoline engines are about 15% efficient, but many electric cars are 90% efficient. (Of course, overall efficiency depends on how the electricity and gasoline are produced in the first place.) Other attractive features of electric vehicles are that they are nearly silent and require minimal maintenance.

The most serious obstacle to the widespread use of electric cars is the lack of a powerful, lightweight, inexpensive battery to increase range and usefulness. Scientists and engineers are researching alternatives to the common lead-acid battery. Perhaps the most promising alternative is not a battery that needs to be periodically recharged, but one that runs continuously as fuel is supplied. One such alternative is the aluminum-air fuel cell, which uses aluminum metal as the fuel and oxygen from the air to produce electricity. Another possibility is the solid polymer hydrogen (or methanol) fuel cell, in which hydrogen (or a hydrogen-rich fuel) and oxygen from the air produce electricity. The discovery of the hydrogen fuel cell led to a four-fold improvement in power, so that liquid and gas fuel cells are now feasible batteries for electric cars.

In this chapter, you will study the technological development and scientific understanding of cells and batteries that produce electricity for many uses, including, potentially, electric cars.

REFLECT on your learning

1. How do cells and batteries work?
2. What are the key scientific concepts or principles that can be used to understand and explain the internal components and processes of a cell that produces electricity?
3. List the types and uses of a variety of common electric cells. Include an assessment of the impact of each one on our lives.

▶ *TRY THIS* activity ― *A Simple Electric Cell*

A cell that produces electricity can be amazingly simple because it uses very common materials and requires no technical expertise to construct. Anyone can make one and then improve its efficiency without much understanding of the scientific principles involved. That is why the electric cell was used for more than 100 years before scientists understood how it worked.

Materials: copper and zinc metal strips (or any two different metals); steel wool, orange, apple, potato (and other fruits or vegetables); LCD clock; voltmeter (or multimeter) with leads

* Clean the metals with steel wool to remove any coating or oxides.
* Stick both metal strips into the orange. Make sure that the metal strips are not in contact inside the orange.
* Momentarily touch the leads (red—positive; black—negative) from the voltmeter, one to each metal strip. Now reverse the leads and test again.
  (a) Record and describe what happened in each case.
* Connect the leads to the LCD clock, paying attention to positive and negative connections.
  (b) Does the clock work? If it does not, suggest a solution to make it work. Try it.
  (c) Explain, in your own words, what you think happened in (b).
* Repeat the above process using other fruits and vegetables.
  (d) Did you produce electricity in each case?
  (e) What do all fruits and vegetables have in common?
  (f) How could you improve upon your electric cell?

 Wash your hands after completing this activity.

In prehistoric times, people learned to extract metals from rocks and minerals (**Figure 1**). This discovery initiated both the technology of metallurgy and humanity's progression from the Stone Age, through the Bronze Age and the Iron Age, to our increasingly technological modern age. Only a few metals, such as gold and silver, exist naturally in the form of a pure element. Most metals exist in a variety of compounds mixed with other substances in rocks called ores. The pure metals must be extracted, or refined, from the ores. For some metals, the basic procedures are quite simple and were developed early in human history; for others, more complex procedures have been developed more recently.

Figure 1
The technology of metallurgy has a long history, preceding by thousands of years the scientific understanding of the processes.

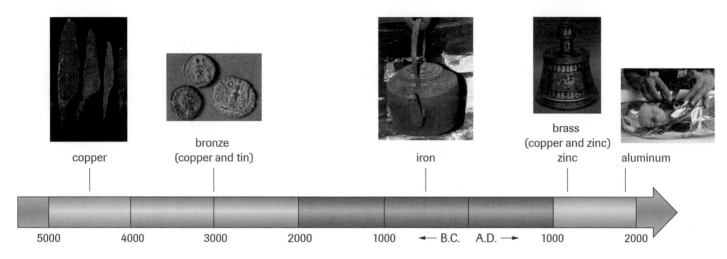

copper

bronze
(copper and tin)

iron

brass
(copper and zinc)
zinc

aluminum

5000   4000   3000   2000   1000   ← B.C.   A.D. →   1000   2000

From metallurgy, the term *reduction* came to be associated with producing metals from their compounds. Red iron(III) oxide "reduced" by carbon monoxide gas to iron metal is a typical example of this process. The production of tin and copper metals are other examples where a metal compound is reduced to the metal.

$$Fe_2O_{3(s)} + 3\,CO_{(g)} \rightarrow 2\,Fe_{(s)} + 3\,CO_{2(g)}$$
$$SnO_{2(s)} + C_{(s)} \rightarrow Sn_{(s)} + CO_{2(g)}$$
$$CuS_{(s)} + H_{2(g)} \rightarrow Cu_{(s)} + H_2S_{(g)}$$

As you can see from these chemical equations, another substance, called a *reducing agent*, causes or promotes the reduction of a metal compound to an elemental metal. In the preceding examples, carbon monoxide is the reducing agent for the production of iron from iron(III) oxide. Carbon (charcoal) is the reducing agent for the production of tin from tin(IV) oxide, and hydrogen is the reducing agent for the production of copper from copper(II) sulfide. These are three of the most common reducing agents used in metallurgical processes.

Although the discovery of fire occurred much earlier than that of metal refining, both discoveries advanced the development of civilization significantly. There are also important similarities in the chemistry behind these technological developments. Of course, it does not require a detailed scientific understanding of the processes to use fire (**Figure 2**). Only relatively recently, in the 18th century, have we come to realize the role of oxygen in burning. Understanding the connection between corrosion and burning is an even more recent development. Corrosion, including the rusting of metals, is now understood to

Figure 2
Making steel requires higher temperatures than those provided by a simple wood fire. Only a few cultures developed the technology to make steel early in their history. In Japan, steel was used in the crafting of samurai swords. At a time when there was no written language, the process of sword-making was made into a ritual so that it could be more accurately passed on from one generation to the next.

be similar to combustion, although corrosion reactions occur more slowly. Reactions of substances with oxygen, whether they were the explosive combustion of gunpowder, the burning of wood, or the slow corrosion of iron, came to be called *oxidation*. As the study of chemistry developed, it became apparent that oxygen was not the only substance that could cause reactions similar to oxidation reactions. For example, metals can be converted to compounds by most nonmetals and by some other substances as well. The rapid reaction process we call burning may even take place with gases other than oxygen, such as chlorine or bromine (**Figure 3**). The term "oxidation" has been extended beyond reactions with oxygen to include a wide range of combustion and corrosion reactions, such as the following:

$$2\,Mg_{(s)} \ + \ O_{2(g)} \rightarrow 2\,MgO_{(s)}$$
$$2\,Al_{(s)} \ + \ 3\,Cl_{2(g)} \rightarrow 2\,AlCl_{3(s)}$$
$$Cu_{(s)} \ + \ Br_{2(g)} \rightarrow CuBr_{2(s)}$$

A substance that causes or promotes the oxidation of a metal to produce a metal compound is called an *oxidizing agent*. In the reactions shown above, the oxidizing agents are oxygen, chlorine, and bromine. As you will see, an understanding of reduction and oxidation is necessary to explain how electric cells (batteries) work.

Figure 3
Copper metal is oxidized by reactive nonmetals such as bromine.

▶ Practice

Understanding Concepts

1. Write an empirical definition for each of the following terms:
   (a) reduction
   (b) oxidation
   (c) oxidizing agent
   (d) reducing agent
   (e) metallurgy
   (f) corrosion

2. For each of the following, classify the reaction of the metal or metal compound as reduction or oxidation, and identify the oxidizing agent or the reducing agent.
   (a) $4\,Fe_{(s)} \ + \ 3\,O_{2(g)} \rightarrow 2\,Fe_2O_{3(s)}$
   (b) $2\,PbO_{(s)} \ + \ C_{(s)} \rightarrow 2\,Pb_{(s)} \ + \ CO_{2(g)}$
   (c) $NiO_{(s)} \ + \ H_{2(g)} \rightarrow Ni_{(s)} \ + \ H_2O_{(l)}$
   (d) $Sn_{(s)} \ + \ Br_{2(l)} \rightarrow SnBr_{2(s)}$
   (e) $Fe_2O_{3(s)} \ + \ 3\,CO_{(g)} \rightarrow 2\,Fe_{(s)} \ + \ 3\,CO_{2(g)}$
   (f) $Cu_{(s)} \ + \ 4\,HNO_{3(aq)} \rightarrow Cu(NO_3)_{2(aq)} \ + \ 2\,H_2O_{(l)} \ + \ 2\,NO_{2(g)}$

3. List three reducing agents used in metallurgy.

4. What class of elements behaves as oxidizing agents for metals?

Making Connections

5. In the history of metallurgy, which came first, technological applications or scientific understanding? Elaborate on your answer.

Extension

6. Archaeometallurgy is the study of ancient metallurgy using modern analytical techniques (**Figure 4**). Give some examples of research in the field. What metals and time periods have been studied? Can a metal from one mine be distinguished from the same metal from another mine? How is this information used in archaeological studies?

 www.science.nelson.com

Figure 4
Aslihan Yener pioneered the use of modern X-ray techniques to identify metals from as early as 8000 B.C.E.

Figure 5
A common single displacement reaction is the reaction of active metals with an acid, such as zinc with hydrochloric acid.

🧪 **INVESTIGATION 9.1.1**

Single Displacement Reactions (p. 715)
Some familiar single displacement reactions provide evidence for reduction and oxidation processes.

Electron Transfer Theory 🧪

Single displacement reactions are familiar reactions in which one element replaces another similar element in a compound. These reactions are useful to investigate first because they provide a relatively simple introduction to the modern theoretical definitions of oxidation and reduction. A useful idea is to imagine that a reaction is a combination of two parts, called *half-reactions*. A half-reaction represents what is happening to only one reactant in an overall reaction. It tells only part of the story. Another half-reaction is required to complete the description of the reaction. Splitting a chemical reaction equation into two parts not only makes the explanations simpler but also leads to some important applications that are discussed later in this unit.

For example, when zinc metal is placed into a hydrochloric acid solution, gas bubbles form as the zinc slowly dissolves (**Figure 5**). Diagnostic tests show that the gas is hydrogen and that zinc ions are present in the solution. Notice that zinc metal is oxidized to zinc chloride. This is a corrosion of zinc caused by the hydrochloric acid.

$$Zn_{(s)} \ + \ 2\,HCl_{(aq)} \ \rightarrow \ ZnCl_{2(aq)} \ + \ H_{2(g)}$$

What happens to the zinc and what happens to the hydrochloric acid? The half-reactions help to answer these questions. The zinc atoms in the solid, $Zn_{(s)}$, are converted to zinc ions in solution, $Zn^{2+}_{(aq)}$. Atomic theory requires that the zinc atoms lose two electrons, as shown by the following half-reaction equation:

$$Zn_{(s)} \ \rightarrow \ Zn^{2+}_{(aq)} \ + \ 2\,e^-$$

Simultaneously, hydrogen ions in the solution gain electrons and are converted into hydrogen gas, as shown below.

$$2\,H^+_{(aq)} \ + \ 2\,e^- \ \rightarrow \ H_{2(g)}$$

Notice that both of these half-reaction equations, or half-reactions, are balanced by mass (same number of element symbols on both sides) and by charge (same total charge on both sides).

Figure 6
A piece of copper before it is placed into a beaker of silver nitrate solution (left). Note the changes after the reaction has occurred (right). The blue colour of the solution indicates $Cu^{2+}_{(aq)}$ ions are present.

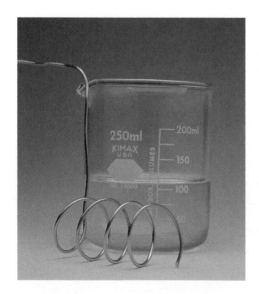
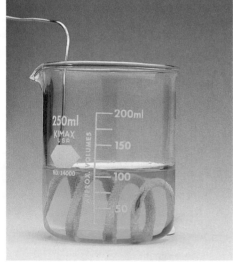

In a laboratory, single displacement reactions in aqueous solution are easier to study than the metallurgy or corrosion reactions discussed earlier in this chapter. However, all of these reactions share a common feature—ions are converted to atoms and atoms are converted to ions. For example, consider the reduction of aqueous silver nitrate to silver metal in the presence of solid copper (**Figure 6**). According to atomic theory, silver atoms are electrically neutral particles ($47p^+, 47e^-$) and silver ions are charged particles ($47p^+, 46e^-$). In this reaction, an electron is required to convert a silver ion into a silver atom. The following half-reaction equation explains the reduction of silver ions using the theoretical rules for atoms and ions.

$$Ag^+_{(aq)} + e^- \rightarrow Ag^0_{(s)} \quad \text{(reduction)}$$

> According to modern theory, the gain of electrons is called **reduction**.

Although this theoretical definition of reduction is in agreement with current atomic theory, it does not explain where the electrons come from. As crystals of silver metal are produced, the solution becomes blue, indicating that copper atoms are being converted to copper(II) ions. According to atomic theory, copper atoms ($29p^+, 29e^-$) must be losing electrons as they form copper(II) ions ($29p^+, 27e^-$).

$$Cu^0_{(s)} \rightarrow Cu^{2+}_{(aq)} + 2e^- \quad \text{(oxidation)}$$

> According to modern theory, the loss of electrons is called **oxidation**.

Evidence shows that the silver-coloured solid and the blue colour of the solution are simultaneously formed near the surface of the copper metal. Therefore, scientists believe that the electrons required by the silver ions are supplied when silver ions collide with copper atoms on the metal surface.

Theory suggests that the *total number of electrons gained in a reaction must equal the total number of electrons lost*. Also, oxidation and reduction are separate processes. This theoretical description requires oxidation and reduction to occur simultaneously rather than sequentially. Oxidation–reduction reactions are often simply called "redox" reactions.

The equations for reduction and oxidation half-reaction and the overall (net) ionic equation summarize the electron transfer that is believed to take place during a redox reaction. As in other chemical reactions, the net equation must be balanced.

reduction a process in which electrons are gained

oxidation a process in which electrons are lost

Writing and Balancing Half-Reaction Equations

SAMPLE problem ◀

1. **Write a balanced net equation for the reaction of copper metal with aqueous silver nitrate.**

To show that the number of electrons gained equals the number lost in two half-reaction equations, it may be necessary to multiply one or both half-reaction equations by an integer to balance the electrons. In this example, the silver half-reaction equation must be multiplied by 2.

$$Cu_{(s)} \rightarrow Cu^{2+}_{(aq)} + 2e^- \quad \text{(two electrons lost by one atom)}$$

$$2\,[Ag^+_{(aq)} + e^- \rightarrow Ag_{(s)}] \quad \text{(two electrons gained by two ions)}$$

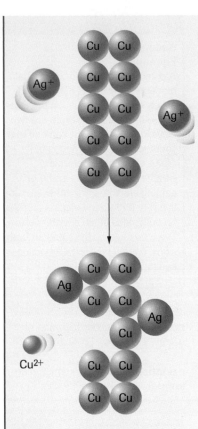

Figure 7
A model of the reaction of copper metal and silver nitrate solution illustrates aqueous silver ions reacting at the surface of a copper strip.

LEARNING TIP

Cancelling Terms for a Net Ionic Equation
The terms you can cancel must be identical, including their states of matter. Electrons must always cancel completely. This is because the electrons that appear in each half-reaction equation are the same electrons. They are the electrons that transfer from one particle to another.

Now, add the half-reaction equations and cancel terms that appear on both sides of the equation to obtain the net ionic equation.

$$2\,Ag^+_{(aq)} + 2\,e^- + Cu_{(s)} \rightarrow 2\,Ag_{(s)} + Cu^{2+}_{(aq)} + 2\,e^-$$
$$2\,Ag^+_{(aq)} + Cu_{(s)} \rightarrow 2\,Ag_{(s)} + Cu^{2+}_{(aq)}$$

Silver ions are reduced to silver metal by reaction with copper metal. Simultaneously, copper metal is oxidized to copper(II) ions by reaction with silver ions (**Figure 7**).

oxidized to metal ion

$$2\,Ag^+_{(aq)} + Cu_{(s)} \rightarrow 2\,Ag_{(s)} + Cu^{2+}_{(aq)}$$

reduced to metal

To evaluate this theory of oxidation and reduction you should look to see if it is consistent with other accepted theories and definitions. The theoretical definitions of oxidation and reduction are consistent with the historical, empirical definitions presented earlier in this chapter; for example, a metal-containing compound is reduced to a metal and a metal is oxidized to form a compound. Redox theory is also consistent with accepted atomic theory and the collision–reaction theory. Most importantly, redox theory explains the observations made by scientists.

Example 1
Write and label two balanced half-reaction equations to describe the reaction of zinc metal with aqueous lead(II) nitrate, as given by the following chemical equation.

$$Zn_{(s)} + Pb(NO_3)_{2(aq)} \rightarrow Pb_{(s)} + Zn(NO_3)_{2(aq)}$$

Solution

$$Zn_{(s)} \rightarrow Zn^{2+}_{(aq)} + 2\,e^- \qquad\qquad \text{(oxidation)}$$
$$Pb^{2+}_{(aq)} + 2\,e^- \rightarrow Pb_{(s)} \qquad\qquad \text{(reduction)}$$

SUMMARY *Electron Transfer Theory*

- A *redox reaction* is a chemical reaction in which electrons are transferred between particles.
- The total number of electrons gained in the reduction equals the total number of electrons lost in the oxidation.
- *Reduction* is a process in which electrons are gained.
- *Oxidation* is a process in which electrons are lost.

▶ Practice

Understanding Concepts

7. Write a theoretical definition for each of the following terms:
   (a) redox reaction (b) reduction (c) oxidation
8. Write a pair of balanced half-reaction equations—one showing a gain of electrons and one showing a loss of electrons—for each of the following reactions:
   (a) $Zn_{(s)} + Cu^{2+}_{(aq)} \rightarrow Zn^{2+}_{(aq)} + Cu_{(s)}$
   (b) $Mg_{(s)} + 2\,H^+_{(aq)} \rightarrow Mg^{2+}_{(aq)} + H_{2(g)}$

9. For each of the following, write and label the oxidation and reduction half-reaction equations. Ignore spectator ions.
   (a) $Ni_{(s)} + Cu(NO_3)_{2(aq)} \rightarrow Cu_{(s)} + Ni(NO_3)_{2(aq)}$
   (b) $Pb_{(s)} + Cu(NO_3)_{2(aq)} \rightarrow Cu_{(s)} + Pb(NO_3)_{2(aq)}$
   (c) $Ca_{(s)} + 2\,HNO_{3(aq)} \rightarrow H_{2(g)} + Ca(NO_3)_{2(aq)}$
   (d) $2\,Al_{(s)} + Fe_2O_{3(s)} \rightarrow 2\,Fe_{(l)} + Al_2O_{3(s)}$ **(Figure 8)**

10. We have only looked at one type of single displacement reaction—a metal displacing another metal from an ionic compound. A nonmetal can also displace another nonmetal from an ionic compound. For example,

    $$Cl_{2(aq)} + 2\,NaI_{(aq)} \rightarrow I_{2(s)} + 2\,NaCl_{(aq)}$$

    Using your knowledge of atoms and ions and the ideas presented in this chapter, write a pair of balanced half-reaction equations for this reaction—one showing a gain of electrons and one showing a loss of electrons

11. Ionic compounds can react in double displacement reactions. For example,

    $$FeCl_{3(aq)} + 3\,NaOH_{(aq)} \rightarrow Fe(OH)_{3(s)} + 3\,NaCl_{(aq)}$$

    According to ideas discussed in this chapter, has a redox reaction taken place in the reaction above? Explain your answer.

Figure 8
The reduction of iron(III) oxide by aluminum is called the "thermite" reaction. Because this reaction is rapid and very exothermic, molten white-hot iron is produced. Here a falling aluminum wrench momentarily sparks a thermite reaction when it strikes a rusted iron block.

Oxidation States

Historically, oxidation and reduction were considered separate processes, more of interest for technology than for science. With modern atomic theory came the idea of an electron transfer involving both a gain of electrons by one particle and a loss of electrons by another particle. This theory of redox reactions is most easily understood for atoms or monatomic ions. Metals and monatomic anions tend to lose electrons (become oxidized), whereas nonmetals and monatomic cations tend to gain electrons (become reduced).

More complex redox reactions, such as the reduction of iron(III) oxide by carbon monoxide in the process of iron production, the oxidation of glucose in the biological process of respiration, and the use of dichromate ions as a strong oxidizing agent in chemical analysis are not adequately described or explained with simple redox theory.

In order to describe oxidation and reduction of molecules and polyatomic ions, chemists have developed a method of "electron bookkeeping" to keep track of the loss and gain of electrons. The method is arbitrary but it works well. In this system, the *oxidation state* of an atom in an entity is defined as the apparent net electric charge that an atom would have if electron pairs in covalent bonds belonged entirely to the more electronegative atom. An oxidation state is a useful idea for keeping track of electrons but it does not usually represent an actual charge on an atom—oxidation states are arbitrary charges.

An **oxidation number** is a positive or negative number corresponding to the oxidation state assigned to an atom. In a covalently bonded molecule or polyatomic ion, the more electronegative atoms are considered to be negative and the less electronegative atoms are considered to be positive. For example, in a water molecule the oxygen atom (electronegativity 3.5) is assigned the bonding electron from each hydrogen atom (electronegativity 2.1); that is, the oxidation number of the oxygen atom is -2 and the oxidation number of each hydrogen atom is $+1$ (**Figure 9**). In order to distinguish these numbers from actual electrical charges, oxidation numbers are written in this book as positive or negative numbers, that is, with the sign preceding the number. Oxidation

Figure 9
An oxygen atom has $8p^+$ and $8e^-$. If the oxygen atom gets to count the two hydrogen electrons in the two shared pairs of electrons, then $8p^+$ and $10e^-$ produces an apparent net charge of $2-$. Each hydrogen atom with $1p^+$ has no additional electrons. Its one electron has already been counted by the oxygen atom. Therefore, the hydrogen has an apparent net charge of $1+$.

oxidation number a positive or negative number corresponding to the apparent charge that an atom in a molecule or ion would have if the electron pairs in covalent bonds belonged entirely to the more electronegative atom

Oxidation Number Format
Oxidation numbers are simply positive or negative numbers assigned on the basis of a set of arbitrary rules. It is important for you to realize that these are not electric charges. For this reason, we write an oxidation with the sign preceding the number, as in +2 or "positive two." This differs from an ion charge of 2+ or "two positive" units of electric charge.

numbers can be assigned to many common atoms and ions (**Table 1**) and they can then be used to determine the oxidation numbers of other atoms.

Oxidation numbers are simply a systematic way of counting electrons. Therefore, the sum of the oxidation numbers in a compound or ion must equal the total charge—zero for neutral compounds, and the ion charge for ions.

Table 1 Common Oxidation Numbers

Atom or ion	Oxidation number	Examples
all atoms in elements	0	Na is 0, Cl in Cl_2 is 0
hydrogen in all compounds	+1	H in HCl is +1
except hydrogen in hydrides	−1	H in LiH is −1
oxygen in all compounds	−2	O in H_2O is −2
except oxygen in peroxides	−1	O in H_2O_2 is −1
all monatomic ions	charge on ion	Na^+ is +1, S^{2-} is −2

▶ **SAMPLE** problem | **Oxidation Number in a Molecular Compound**

1. What is the oxidation number of carbon in methane, CH_4?

This is determined by assigning an oxidation number of +1 to hydrogen (**Table 1**).

$$x \quad +1$$
$$CH_4$$

Now solve for x. Since a methane molecule is electrically neutral, the oxidation numbers of the one carbon atom and the four hydrogen atoms (4 times +1) must equal zero.

$$x + 4(+1) = 0$$
$$x = -4$$

$$x \quad +1 \qquad\qquad -4 \quad +1$$
$$CH_4 \quad becomes \quad CH_4$$

Carbon in methane has an oxidation number of −4.

2. What is the oxidation number of manganese in a permanganate ion, MnO_4^-.

In a polyatomic ion, like a neutral compound, the total of the oxidation numbers of all atoms must equal the charge. The oxidation number of manganese in the permanganate ion, MnO_4^-, is determined using the oxidation number of oxygen as −2 (**Table 1**) and the knowledge that the charge on the ion is 1−. The total of the oxidation numbers of the one manganese atom (x) and the four oxygen atoms (4 times −2) must equal the charge on the ion (1−).

$$x + 4(-2) = -1$$
$$x = +7$$

$$x \quad -2 \qquad\qquad +7 \quad -2$$
$$MnO_4^- \quad becomes \quad MnO_4^-$$

The oxidation number of manganese in MnO_4^- is +7.

Example

What is the oxidation number of sulfur in sodium sulfate?

Solution

$$+1 \quad x \quad -2 \qquad 2(+1) + x + 4(-2) = 0$$
$$Na_2SO_4 \qquad\qquad\qquad x = +6$$

The oxidation number of sulfur in sodium sulfate is +6.

SUMMARY | Determining Oxidation Numbers

Step 1 Assign common oxidation numbers (**Table 1**).

Step 2 The total of the oxidation numbers of atoms in a molecule or ion equals the value of the net electric charge on the molecule or ion.

(a) The sum of the oxidation numbers for a compound is zero.

(b) The sum of the oxidation numbers for a polyatomic ion equals the charge on the ion.

Step 3 Any unknown oxidation number is determined algebraically from the sum of the known oxidation numbers and the net charge on the entity.

▶ Practice

Understanding Concepts

12. Determine the oxidation number of
    (a) S in SO_2
    (b) Cl in $HClO_4$
    (c) S in SO_4^{2-}
    (d) Cr in $Cr_2O_7^{2-}$
    (e) I in MgI_2
    (f) H in CaH_2

13. Determine the oxidation number of nitrogen in
    (a) $N_2O_{(g)}$
    (b) $NO_{(g)}$
    (c) $NO_{2(g)}$
    (d) $NH_{3(g)}$
    (e) $N_2H_{4(g)}$
    (f) $NaNO_{3(s)}$
    (g) $N_{2(g)}$
    (h) $NH_4Cl_{(s)}$

14. Determine the oxidation number of carbon in
    (a) graphite (elemental carbon)
    (b) glucose
    (c) sodium carbonate
    (d) carbon monoxide

15. In a breathalyzer (**Figure 10**), ethanol is oxidized by an acidic dichromate solution.
    (a) Determine the oxidation number of every atom or ion in the following breatha-lyzer reaction:

    $$16\,H^+_{(aq)} + 2\,Cr_2O_7^{2-}{}_{(aq)} + 3\,C_2H_5OH_{(aq)} \rightarrow 4\,Cr^{3+}_{(aq)} + 3\,CH_3COOH_{(aq)} + H_2O_{(l)}$$

    (b) What colour change would you expect in this reaction?

16. Carbon can be progressively oxidized in a series of organic reactions. Determine the oxidation number of carbon in each of the compounds in the following series of oxidations:

    methane→ methanol→ methanal→ methanoic acid→ carbon dioxide

Extension

17. Assigning oxidation numbers using the rules we have established may occasion-ally produce some unusual results. For example, consider Fe_3O_4.
    (a) Determine the oxidation number of iron in Fe_3O_4.
    (b) What is unusual about your answer?
    (c) Suggest a reason why this might occur.

Figure 10
It is a criminal offence to drive a vehicle when you have a blood alcohol content greater than 0.08 g/100 mL of blood. The breath-alyzer measures the alcohol content of exhaled breath, which is assumed to be proportional to the blood alcohol content. Inside the device, the alcohol in the breath sample is oxidized by acidic potassium dichromate in an acidic solution.

Answers

12. (a) +4   (d) +6
    (b) +7   (e) −1
    (c) +6   (f) −1

13. (a) +1   (e) −2
    (b) +2   (f) +5
    (c) +4   (g) 0
    (d) −3   (h) −3

14. (a) 0   (c) +4
    (b) 0   (d) +2

15. +1; +6,−2; −2, +1, −2, +1; +3; 0,+1, 0, −2, +1; +1; −2

16. −4; −2; 0; +2; +4

oxidation an increase in oxidation number

reduction a decrease in oxidation number

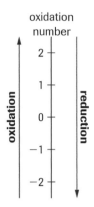

Figure 11
In a redox reaction, both oxidation and reduction occur.

![lab icon] **LAB EXERCISE 9.1.1**

Oxidation States of Vanadium (p. 716)
Vanadium is an interesting element because it forms many different ions with different colours.

Oxidation Numbers and Redox Reactions

Although the concept of oxidation states is somewhat arbitrary, because it is based on assigned charges, it is self-consistent and allows predictions of electron transfer. If the oxidation number of an atom or ion changes during a chemical reaction, then an electron transfer (that is, an oxidation–reduction reaction) has occurred. An increase in the oxidation number is defined as an **oxidation** and a decrease in the oxidation number is a **reduction**. If oxidation numbers are listed as positive and negative numbers on a number line as they are in **Figure 11**, then the process of oxidation involves a change to a more positive value ("up" on the number line) and reduction is a change to a more negative value ("down" on the number line). If the oxidation numbers do not change, this is interpreted as no transfer of electrons. A reaction in which all oxidation numbers remain the same is not a redox reaction.

> • An oxidation is an increase in oxidation number.
> • A reduction is a decrease in oxidation number.
> • In a redox reaction, oxidation numbers change.

Coal is a fossil fuel that is burned in huge quantities in some electrical power generating stations. If we assume pure carbon and a complete combustion, carbon is converted to carbon dioxide. In this reaction, the oxidation number of carbon changes from 0 in $C_{(s)}$ to +4 in $CO_{2(g)}$. Simultaneously, oxygen is reduced from 0 in $O_{2(g)}$ to -2 in $CO_{2(g)}$.

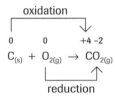

The main purpose of assigning oxidation numbers is to see how these numbers change as a result of a chemical reaction. In any redox reaction, like the combustion of carbon, there will always be both an oxidation and a reduction. We will use these changes in the next section to balance redox equations, but first we will look at some additional examples. ![icon]

▶ **SAMPLE** problem | **Oxidation Number Changes in a Reaction**

1. **You have seen the reaction of active metals like zinc with an acid (Figure 5). Identify the oxidation and reduction in the reaction of zinc metal with hydrochloric acid.**

First, you need to write the chemical equation, as it is not provided. Net ionic equations are best, but the procedure will still work if you write a non-ionic equation.

$$Zn_{(s)} + 2\,H^+_{(aq)} \rightarrow Zn^{2+}_{(aq)} + H_{2(g)}$$

After writing the equation, determine all oxidation numbers.

$$\overset{0}{Zn_{(s)}} + 2\,\overset{+1}{H^+_{(aq)}} \rightarrow \overset{+2}{Zn^{2+}_{(aq)}} + \overset{0}{H_{2(g)}}$$

Now look for the oxidation number of an atom/ion that increases as a result of the reaction and label the change as oxidation. There must also be an atom/ion whose oxidation number decreases. Label this change as reduction.

oxidation

$$\overset{0}{Zn_{(s)}} + 2\overset{+1}{H^+_{(aq)}} \rightarrow \overset{+2}{Zn^{2+}_{(aq)}} + \overset{0}{H_{2(g)}}$$

reduction

2. **When natural gas burns in a furnace, carbon dioxide and water form. Identify oxidation and reduction in this reaction.**

First, write the equation.

$$CH_{4(g)} + 2O_{2(g)} \rightarrow CO_{2(g)} + 2H_2O_{(g)}$$

Now we can insert the oxidation numbers and arrows.

oxidation

$$\overset{-4\ +1}{CH_{4(g)}} + 2\overset{0}{O_{2(g)}} \rightarrow \overset{+4\ -2}{CO_{2(g)}} + 2\overset{+1\ -2}{H_2O_{(g)}}$$

reduction

Carbon is oxidized from -4 in methane to $+4$ in carbon dioxide as it reacts with oxygen. Simultaneously, oxygen is reduced from 0 in oxygen gas to -2 in both products.

Notice that the oxygen atoms in the reactant are distributed between the two products. This does not change our procedure because we are only looking for the change from reactant to product. We say that "oxygen is reduced" in this reaction and it does not matter where the reduced oxygen appears in the products.

Example

The determination of blood alcohol content from a sample of breath or blood involves the reaction of the sample with acidic potassium dichromate solution. If ethanol is present, chromium(III) ions, water, and acetic acid are produced. Identify the oxidation and reduction in the following chemical reaction:

$$2Cr_2O_7^{2-}{}_{(aq)} + 16H^+_{(aq)} + 3C_2H_5OH_{(aq)} \rightarrow 4Cr^{3+}_{(aq)} + 11H_2O_{(g)} + 3CH_3COOH_{(aq)}$$

Solution

oxidation

$$2\overset{+6\ -2}{Cr_2O_7^{2-}{}_{(aq)}} + 16\overset{+1}{H^+_{(aq)}} + 3\overset{-2\,+1\,-2\,+1}{C_2H_5OH_{(aq)}} \rightarrow 4\overset{+3}{Cr^{3+}_{(aq)}} + 11\overset{+1\ -2}{H_2O_{(g)}} + 3\overset{0\,+1\ 0\,-2\,-2\,+1}{CH_3COOH_{(aq)}}$$

reduction

Chromium atoms in $Cr_2O_7^{2-}$ are reduced ($+6$ to $+3$). Carbon atoms in C_2H_5OH are oxidized (-2 to 0).

SUMMARY *Oxidation States*

- Oxidation is an increase in oxidation number.
- Reduction is a decrease in oxidation number.
- A redox reaction involves both an oxidation and a reduction.

DID YOU KNOW ?

Redox in Biological Systems
Biologists often classify oxidation and reduction in terms of the addition or removal of oxygen or hydrogen.

- Removal of oxygen decreases the oxidation number of carbon (i.e., a reduction),

  e.g., $HCOOH \rightarrow HCHO$

  C is $+2$ C is 0

- Removal of hydrogen increases the oxidation number of carbon (i.e., an oxidation),

  e.g., $C_2H_6 \rightarrow C_2H_4$

  C is -3 C is -2

DID YOU KNOW ?

Redox Reactions in Living Organisms
The ability of carbon to take on different oxidation states is essential to life on Earth. Photosynthesis involves a series of reduction reactions in which the oxidation number of carbon changes from $+4$ in carbon dioxide to an average of 0 in sugars such as glucose. In cellular respiration, carbon undergoes a series of oxidations, after which the oxidation number of carbon is again $+4$ in carbon dioxide.

Understanding Concepts

Answers

18. (a) $-2, +1, -2, +1; +7,$
$-2; +1; 0, +1, -2; +2;$
$+1, -2$

19. (a) $0; +1, +5, -2; 0; +2,$
$+5, -2$

(b) $+2, +5, -2; +1, -1; +2,$
$-1; +1, +5, -2$

(c) $0; +1, -1; 0; +1, -1$

(d) $+1, -1; 0; 0$

(e) $+1, -1; +1, -2, +1; +1,$
$-2, +1; +1, -1$

(f) $0; 0; +3, -1$

(g) $-2\frac{1}{2}, +1; 0; +4, -2; +1,$
-2

(h) $+1, -1; +1, -2; 0$

18. Methanol reacts with acidic permanganate ions as shown below:

$$5\,CH_3OH_{(l)} + 2\,MnO_{4(aq)}^- + 6\,H_{(aq)}^+ \rightarrow 5\,CH_2O_{(l)} + 2\,Mn_{(aq)}^{2+} + 8\,H_2O_{(l)}$$

(a) Assign oxidation numbers to all atoms/ions.
(b) Which atom/ion is oxidized? Label the oxidation above the equation.
(c) Which atom/ion is reduced? Label the reduction below the equation.

19. For each of the following chemical reactions, assign oxidation numbers to each atom/ion and indicate whether the equation represents a redox reaction. If it does, identify the oxidation and reduction.

(a) $Cu_{(s)} + 2\,AgNO_{3(aq)} \rightarrow 2\,Ag_{(s)} + Cu(NO_3)_{2(aq)}$
(b) $Pb(NO_3)_{2(aq)} + 2\,KI_{(aq)} \rightarrow PbI_{2(s)} + 2\,KNO_{3(aq)}$
(c) $Cl_{2(aq)} + 2\,KI_{(aq)} \rightarrow I_{2(s)} + 2\,KCl_{(aq)}$
(d) $2\,NaCl_{(l)} \rightarrow 2\,Na_{(l)} + Cl_{2(g)}$
(e) $HCl_{(aq)} + NaOH_{(aq)} \rightarrow HOH_{(l)} + NaCl_{(aq)}$
(f) $2\,Al_{(s)} + 3\,Cl_{2(g)} \rightarrow 2\,AlCl_{3(s)}$
(g) $2\,C_4H_{10(g)} + 13\,O_{2(g)} \rightarrow 8\,CO_{2(g)} + 10\,H_2O_{(l)}$
(h) $2\,H_2O_{2(l)} \rightarrow 2\,H_2O_{(l)} + O_{2(g)}$

20. Classify the chemical equations in question 19 (a−h) using the five reaction types—formation, decomposition, single displacement, double displacement, and combustion. Which reaction type does not appear to be a redox reaction?

21. Hydrogen peroxide, $H_2O_{2(l)}$, can either be oxidized or reduced depending on the substance with which it reacts. Use oxidation numbers to explain why this is possible.

Making Connections

22. Earth has an oxidizing atmosphere of oxygen. The planet Saturn has a reducing atmosphere of hydrogen and methane. Describe the two types of atmospheres in terms of changes in oxidation numbers of carbon and of the likely reactions.

SUMMARY **Electron Transfer and Oxidation States**

- According to current theory, a redox reaction is a chemical reaction in which electrons are transferred and the oxidation numbers change.

- Oxidation is the increase in oxidation number and corresponds to a loss of electrons.

- Reduction is the decrease in oxidation number and corresponds to a gain of electrons.

> ▶ **Section 9.1 Questions**

Understanding Concepts

1. Copy and complete the following table to distinguish between oxidation and reduction:

	Electron transfer	**Oxidation states**
oxidation		
reduction		

2. Write and label a pair of balanced half-reaction equations for each of the following reactions.
   (a) $Pb_{(s)} + Cu^{2+}_{(aq)} \rightarrow Pb^{2+}_{(aq)} + Cu_{(s)}$
   (b) $Cl_{2(aq)} + 2 Br^-_{(aq)} \rightarrow 2 Cl^-_{(aq)} + Br_{2(l)}$

3. What is an oxidation number?

4. State two ways you can recognize a redox reaction, using a chemical reaction equation.

5. In a redox reaction, what happens to the oxidation number of
   (a) an atom that is oxidized?
   (b) an atom that is reduced?

6. Write the oxidation number of each atom/ion in the following substances:
   (a) carbon monoxide, $CO_{(g)}$, a toxic gas
   (b) ozone, $O_{3(g)}$, ozone layer
   (c) ammonium chloride, $NH_4Cl_{(s)}$, used in dry cells (batteries)
   (d) phosphoric acid, $H_3PO_{4(aq)}$, in cola soft drinks
   (e) sodium thiosulfate, $Na_2S_2O_{3(s)}$, antidote for cyanide poisoning
   (f) sodium tripolyphosphate, $Na_5P_3O_{10(s)}$, in laundry detergents

7. Redox reactions are common in organic chemistry. For example, carboxyl groups can be oxidized to form carbon dioxide. In the following chemical equation, permanganate ions convert oxalate ions to carbon dioxide in an acidic solution.

$$2 MnO^-_{4(aq)} + 5 C_2O^{2-}_{4(aq)} + 16 H^+_{(aq)} \rightarrow$$
$$2 Mn^{2+}_{(aq)} + 8 H_2O_{(l)} + 10 CO_{2(g)}$$

   (a) Assign oxidation numbers to all atoms/ions.
   (b) Which atom is oxidized? State the change.
   (c) Which atom is reduced? State the change.

8. When carbon dioxide is released into the atmosphere from natural or human activities, some of it reacts with water to form carbonic acid. This accounts for the natural acidity of rainwater and may also contribute to acid rain.
   (a) Write the balanced chemical equation for the reaction of carbon dioxide with water to form carbonic acid.
   (b) Is this a redox reaction? Justify your answer.

Applying Inquiry Skills

9. In the Try This Activity at the beginning of this chapter, you put copper and zinc strips into an orange (or other fruit) and a reaction occurred to produce electricity. Write a brief design (plan) to determine if either of the copper or zinc metals is oxidized or reduced.

Making Connections

10. The dark tarnish that forms on silver objects is silver sulfide. A common home remedy is to restore the silver with baking soda and aluminum foil (**Figure 12**).
    (a) Write the single displacement reaction of silver sulfide and aluminum metal.
    (b) Identify what is oxidized and what is reduced.
    (c) Do you think this is a better method of cleaning than polishing or scrubbing the silver? Why or why not?

Figure 12
If the tarnished silver is placed in a hot solution of baking soda on aluminum foil in a nonmetallic dish (e.g., glass), a redox reaction converts the tarnish back into silver metal.

11. Police forces use various instruments and processes to determine whether a person is legally impaired. What is the difference in operation between the Breathalyzer and the Intoxilyzer? Describe briefly how redox reactions are involved.

 www.science.nelson.com

Knowing the ratio of reacting chemicals is necessary in many applications. Chemists use the mole ratio from a balanced chemical equation to study the nature of the reaction. The stoichiometry of a reaction is essential in many types of chemical analysis such as the breathalyzer. Finally, chemical industries need to know the quantities of reactants to mix and the yield of a desired product. In this and previous courses, you have already seen many examples of the use of balanced chemical equations.

Simple redox reaction equations can be balanced by inspection or by a trial-and-error method. You have done this often in previous courses for reactions such as single displacement and combustion. More complex redox reactions may be very difficult to balance this way due to the number and complexity of the reactants and products. As you will see in this section, oxidation numbers and half-reaction equations can be used to balance any redox equation.

Oxidation Number Method

One way of recognizing a redox reaction is to assign oxidation numbers to each atom or ion and then look for any changes in the numbers. Any change in the oxidation number of a particular atom or ion is believed to be related to a change in the number of electrons. Because electrons are transferred in a redox reaction, the total number of electrons lost by one atom/ion must equal the total number of electrons gained by another atom/ion. In terms of oxidation numbers, this means that the changes in oxidation numbers must also be balanced.

> The total increase in oxidation number for a particular atom/ion must equal the total decrease in oxidation number of another atom/ion.

Let's look at a simple example first. You could easily balance this equation by inspection, but we will use it to illustrate the main points of the oxidation number method.

SAMPLE problem | **Balancing Redox Equations**

Hydrogen sulfide is an unpleasant constituent of "sour" natural gas. Hydrogen sulfide is not only very toxic but it also smells terrible, like rotten eggs. It is common practice to "flare" or burn small quantities of sour natural gas that occur with oil deposits (Figure 1). The gas is burned because it is not worth recovering and treating a small quantity of gas. When this gas is burned, hydrogen sulfide is converted to sulfur dioxide.

$$H_2S_{(g)} + O_{2(g)} \rightarrow SO_{2(g)} + H_2O_{(g)}$$

Balance this equation.

If you are using oxidation numbers to balance a redox equation, the first step is to assign oxidation numbers to all atoms/ions and look for the numbers that change. Circle or highlight the oxidation numbers that change.

LEARNING TIP

Electron Transfer
All redox reactions are electron transfer reactions. This means that electrons that are lost by one particle are the same electrons that are gained by another. This is like you giving five dollars to a friend. You lose five dollars and your friend gains five dollars—five dollars has been exchanged. Obviously, the money lost must equal the money gained. The same is true for electrons in a redox reaction.

oxidation

$$\overset{+1\ -2}{H_2S_{(g)}} + \overset{0}{O_{2(g)}} \longrightarrow \overset{+4\ -2}{SO_{2(g)}} + \overset{+1\ -2}{H_2O_{(g)}}$$

reduction

Notice that a sulfur atom is oxidized from -2 to $+4$. This is a change of 6 and means $6e^-$ have been transferred. An oxygen atom is reduced from 0 to -2, a change of 2 or $2e^-$ transferred. Because the substances in the equation are molecules, not atoms, we need to specify the change in the number of electrons per molecule.

$$\overset{+1\ -2}{H_2S_{(g)}} \quad + \quad \overset{0}{O_{2(g)}} \quad \rightarrow \quad \overset{+4\ -2}{SO_{2(g)}} \quad + \quad \overset{+1\ -2}{H_2O_{(g}}$$

$6e^-/$S atom $\qquad 2e^-/$O atom

$6e^-/H_2S \qquad 4e^-/O_2$

An H_2S molecule contains one sulfur atom. Therefore, the number of electrons transferred per sulfur atom is the same number per H_2S molecule. An O_2 molecule contains two O atoms. Therefore, when one O_2 molecule reacts, two oxygen atoms transfer $2e^-$ each for a total of $4e^-$.

The next step is to determine the simplest whole numbers that will balance the number of electrons transferred for each reactant. The numbers become the coefficients for the reactants.

$$\overset{+1\ -2}{H_2S_{(g)}} \quad + \quad \overset{0}{O_{2(g)}} \quad \rightarrow \quad \overset{+4\ -2}{SO_{2(g)}} \quad + \quad \overset{+1\ -2}{H_2O_{(g)}}$$

$6e^-/$S atom $\qquad 2e^-/$O atom

$6e^-/H_2S \qquad 4e^-/O_2$

$$\frac{\times 2}{12} \quad = \quad \frac{\times 3}{12}$$

Now you have the coefficients for the reactants.

$$\mathbf{2}\,H_2S_{(g)} + \mathbf{3}\,O_{2(g)} \rightarrow SO_{2(g)} + H_2O_{(g)}$$

The coefficients of the products can easily be obtained by balancing the atoms whose oxidation numbers have changed and then any other atoms. The final balanced equation is shown below.

$$\mathbf{2}\,H_2S_{(g)} \quad + \quad \mathbf{3}\,O_{2(g)} \quad \rightarrow \quad \mathbf{2}\,SO_{2(g)} \quad + \quad \mathbf{2}\,H_2O_{(g)}$$

Figure 1
Oil deposits often contain small quantities of natural gas. The gas is simply burned ("flared"). Since the gas often contains hydrogen sulfide, this practice can be a significant source of pollutants. It is also a waste of energy when many of these flares operate in a large oil field.

LEARNING TIP

You can adjust the number of electrons per atom to the number per molecule by multiplying the number per atom by the subscript of the atom in the chemical formula.

Sometimes you may not know all of the reactants and products of a redox reaction. The main reactants and oxidized/reduced products will always be given and you will know if the reaction took place in an acidic or basic solution. Experimental evidence shows that water molecules, hydrogen ions, and hydroxide ions play important roles in reactions in such solutions. The procedure for balancing such equations is initially the same as the one we have discussed, but you will need to add water molecules, hydrogen ions, and/or hydroxide ions to finish the balancing of the overall equation. The following two sample problems illustrate this procedure.

1. **Chlorate ions and iodine react in an acidic solution to produce chloride ions and iodate ions.**

$$ClO_{3(aq)}^- + I_{2(aq)} \rightarrow Cl_{(aq)}^- + IO_{3(aq)}^-$$

Balance the equation for this reaction.

Assign oxidation numbers to each atom/ion and note which numbers change.

$$\overset{+5\ -2}{ClO_{3(aq)}^-} + \overset{0}{I_{2(aq)}} \rightarrow \overset{-1}{Cl_{(aq)}^-} + \overset{+5\ -2}{IO_{3(aq)}^-}$$

A chlorine atom is reduced from +5 to −1, a change of 6. Simultaneously, an iodine atom is oxidized from 0 to +5, a change of 5. Record the change in the number of electrons per atom and per molecule or polyatomic ion.

$$\overset{+5\ -2}{ClO_{3(aq)}^-} + \overset{0}{I_{2(aq)}} \rightarrow \overset{-1}{Cl_{(aq)}^-} + \overset{+5\ -2}{IO_{3(aq)}^-}$$

$6e^-/Cl \qquad 5e^-/I$

$6e^-/ClO_3^- \qquad 10e^-/I_2$

The total number of electrons transferred by each reactant must be the same. Multiply the numbers of electrons by the simplest whole numbers to make the totals equal, in this case, $30e^-$. You can now write the coefficients for the reactants and the products.

$$\overset{+5\ -2}{\mathbf{5}\ ClO_{3(aq)}^-} + \overset{0}{\mathbf{3}\ I_{2(aq)}} \rightarrow \overset{-1}{\mathbf{5}\ Cl_{(aq)}^-} + \overset{+5\ -2}{\mathbf{6}\ IO_{3(aq)}^-}$$

$6e^-/Cl \qquad\qquad 5e^-/I$

$6e^-/ClO_3^- \qquad\quad 10e^-/I_2$

$\times\ \mathbf{5} \qquad\qquad\quad \times\ \mathbf{3}$

Although the chlorine and iodine atoms are now balanced, notice that the oxygen atoms are not; 15 on the left versus 18 on the right. Because this reaction occurs in an aqueous solution, we can add H_2O molecules to balance the O atoms. The reactant side requires 3 oxygen atoms (from 3 water molecules) to equal the total of 18 oxygen atoms on the product side.

$$\mathbf{3}\ H_2O_{(l)} + \mathbf{5}\ ClO_{3(aq)}^- + \mathbf{3}\ I_{2(aq)} \rightarrow \mathbf{5}\ Cl_{(aq)}^- + \mathbf{6}\ IO_{3(aq)}^-$$

In adding water molecules, we are also adding H atoms. Because this reaction occurs in an acidic solution, we will add $H_{(aq)}^+$ to balance the hydrogen.

$$\mathbf{3}\ H_2O_{(l)} + \mathbf{5}\ ClO_{3(aq)}^- + \mathbf{3}\ I_{2(aq)} \rightarrow \mathbf{5}\ Cl_{(aq)}^- + \mathbf{6}\ IO_{3(aq)}^- + \mathbf{6}\ H_{(aq)}^+$$

The redox equation should now be completely balanced. Check your work by checking the total number of each atom/ion on each side and checking the total electric charge, which should also be balanced.

2. **Methanol reacts with permanganate ions in a basic solution. The main reactants and products are shown below.**

$$CH_3OH_{(aq)} + MnO_{4(aq)}^- \rightarrow CO_{3(aq)}^{2-} + MnO_{4(aq)}^{2-}$$

Balance the equation for this reaction.

LEARNING TIP

A balanced chemical reaction equation includes both a mass and charge balance. Mass is balanced using the atomic symbols. *If the symbols balance but not the charge, the equation is not balanced.* Be sure to check both symbol and charge.

We will follow the same procedure as in the previous problem, adjusting for a basic solution at the end: assign oxidation numbers; note which ones change and by how much per reactant; and then balance the total number of electrons to obtain the coefficients for the main reactants and products.

$$\overset{-2\ +1\ -2\ +1}{\textbf{1 }CH_3OH_{(aq)}} \quad + \quad \overset{+7\ -2}{\textbf{6 }MnO_{4\ (aq)}^-} \quad \rightarrow \quad \overset{+4\ -2}{\textbf{1 }CO_{3(aq)}^{2-}} + \quad \overset{+6\ -2}{\textbf{6 }MnO_{4(aq)}^{2-}}$$

$6e^-/C$ $1e^-/Mn$
$6e^-/CH_3OH$ $1e^-/MnO_4^-$
$\times$ **1** $\times$ **6**

Just as before, add $H_2O_{(l)}$ to balance the O atoms. The reactant side requires 2 oxygen atoms (from 2 water molecules) to equal the 27 oxygen atoms on the product side. Next, balance the H atoms using $H_{(aq)}^+$. The product side requires 8 hydrogens to balance the 8 on the reactant side (4 in water and 4 in methanol).

$$\textbf{2 }H_2O_{(l)} + CH_3OH_{(aq)} + 6\,MnO_{4\,(aq)}^- \rightarrow CO_{3(aq)}^{2-} + 6\,MnO_{4(aq)}^{2-} + \textbf{8 }H_{(aq)}^+$$

If this reaction occurred in an acidic solution, you would now be finished. For a basic solution, however, we add enough $OH_{(aq)}^-$ to both sides to equal the number of $H_{(aq)}^+$ present. The hydrogen and hydroxide ions on the same side of the equation are then combined to form water.

$$\textbf{8 }OH_{(aq)}^- + 2\,H_2O_{(l)} + CH_3OH_{(aq)} + 6\,MnO_{4\,(aq)}^- \rightarrow CO_{3(aq)}^{2-} + 6\,MnO_{4(aq)}^{2-} + \underbrace{8\,H_{(aq)}^+ + \textbf{8 }OH_{(aq)}^-}_{\textbf{8 }H_2O}$$

Finally, cancel the same number of H_2O molecules on both sides. In this case, the H_2O on the reactant side can be cancelled by also removing 2 H_2O from the product side, leaving the extra 6 H_2O in the final equation.

$$8\,OH_{(aq)}^- + CH_3OH_{(aq)} + 6\,MnO_{4\,(aq)}^- \rightarrow CO_{3(aq)}^{2-} + 6\,MnO_{4(aq)}^{2-} + 6\,H_2O_{(l)}$$

SUMMARY

Procedure for Balancing Redox Equations Using Oxidation Numbers

Step 1 Assign oxidation numbers and identify the atoms/ions whose oxidation numbers change.

Step 2 Using the change in oxidation numbers, write the number of electrons transferred per atom.

Step 3 Using the chemical formulas, determine the number of electrons transferred per reactant. (Use the formula subscripts to do this.)

Step 4 Calculate the simplest whole number coefficients for the reactants that will balance the total number of electrons transferred. Balance the reactants and products.

Step 5 Balance the O atoms using $H_2O_{(l)}$, and then balance the H atoms using $H_{(aq)}^+$.

For basic solutions only,

Step 6 Add $OH_{(aq)}^-$ to both sides equal in number to the number of $H_{(aq)}^+$ present.

Step 7 Combine $H_{(aq)}^+$ and $OH_{(aq)}^-$ on the same side to form $H_2O_{(l)}$, and cancel the same number of $H_2O_{(l)}$ on both sides.

Check the balancing of the final equation. Make sure that both symbols and charge are balanced.

Example 1
Balance the chemical equation for the oxidation of ethanol by dichromate ions in a breathalyzer to form chromium(III) ions and acetic acid in an acidic solution.

Solution

$$\overset{}{16\,H^+_{(aq)}} + \overset{+6\ -2}{2\,Cr_2O^{2-}_{7(aq)}} + \overset{-2+1\ -2+1}{3\,C_2H_5OH_{(aq)}} \rightarrow \overset{+3}{4\,Cr^{3+}_{(aq)}} + \overset{0\ +1\ 0\ -2\ +1}{3\,CH_3COOH_{(aq)}} + 11\,H_2O_{(l)}$$

$$3e^-/Cr \qquad\qquad 2e^-/C$$
$$6e^-/Cr_2O_7^{2-} \qquad 4e^-/C_2H_5OH$$
$$\times\,2 \qquad\qquad\qquad \times\,3$$

Example 2
Balance the following chemical equation, assuming the reaction occurs in a basic solution.

$$NO_{2(aq)}^- + I_{2(aq)} \rightarrow NO_{3(aq)}^- + I_{(aq)}^-$$

Solution

$$2\,OH^-_{(aq)} + \overset{+3\ -2}{NO_2^-_{(aq)}} + \overset{0}{I_{2(aq)}} \rightarrow \overset{+5\ -2}{NO_3^-_{(aq)}} + \overset{-1}{2\,I^-_{(aq)}} + H_2O_{(l)}$$

$$2e^-/N \qquad 1e^-/I$$

▶ **Practice**

Understanding Concepts

1. Why is the change in oxidation number of an atom the same as the number of electrons transferred?

2. Balance the following chemical equations for reactions in an acidic solution:
   (a) $Cr_2O^{2-}_{7(aq)} + Cl^-_{(aq)} \rightarrow Cr^{3+}_{(aq)} + Cl_{2(aq)}$
   (b) $IO^-_{3(aq)} + HSO^-_{3(aq)} \rightarrow SO^{2-}_{4(aq)} + I_{2(s)}$
   (c) $HBr_{(aq)} + H_2SO_{4(aq)} \rightarrow SO_{2(g)} + Br_{2(l)}$

3. Balance the following chemical equations for reactions in a basic solution:
   (a) $MnO^-_{4(aq)} + SO^{2-}_{3(aq)} \rightarrow SO^{2-}_{4(aq)} + MnO_{2(s)}$
   (b) $ClO^-_{3(aq)} + N_2H_{4(aq)} \rightarrow NO_{(g)} + Cl^-_{(aq)}$

4. Ammonia gas undergoes a combustion to produce nitrogen dioxide gas and water vapour. Write and balance the reaction equation.

Answers

2. (a) $1\,Cr_2O_7^{2-}$, $6\,Cl^-$, $14\,H^+$; $2\,Cr^{3+}$, $3\,Cl_2$, $7\,H_2O$

   (b) $2\,IO_3^-$, $5\,HSO_3^-$; $5\,SO_4^{2-}$, $1\,I_2$, $3\,H^+$, $1\,H_2O$

   (c) $2\,HBr$, $1\,H_2SO_4$; $1\,Br_2$, $2\,H_2O$

3. (a) $2\,MnO_4^-$, $3\,SO_3^{2-}$, $1\,H_2O$; $3\,SO_4^{2-}$, $2\,MnO_2$, $2\,OH^-$

   (b) $4\,ClO_3^-$, $3\,N_2H_4$; $6\,NO$, $4\,Cl^-$, $6\,H_2O$

Half-Reaction Method

An alternative to the oxidation number method for balancing redox equations is to write balanced oxidation and reduction half-reaction equations. Once these half-reaction equations are obtained, it is a simple matter to balance electrons and obtain the final balanced redox equation. We will first address the writing of an individual half-reaction equation.

Although most metals and nonmetals have relatively simple half-reaction equations, polyatomic ions and molecular compounds undergo more complicated oxidation and reduction processes. In most of these processes, the reaction takes place in an aqueous solution that is very often acidic or basic. As before, we must consider the important role that water molecules, hydrogen ions, and hydroxide ions play an important role in these half-reactions. A method of writing half-reactions for polyatomic ions and molecular compounds requires that water molecules and hydrogen or hydroxide ions be included. This method, illustrated in the following sample problem, is sometimes called the "half-reaction" or "ion-electron" method.

Writing Half-Reaction Equations

1. **Nitrous acid can be reduced in an acidic solution to form nitrogen monoxide gas. What is the reduction half-reaction for nitrous acid?**

The first step is to write the reactants and products.

$$HNO_{2(aq)} \rightarrow NO_{(g)}$$

If necessary, you should balance all atoms other than oxygen and hydrogen in this partial equation. In this example, there is only one nitrogen atom on each side.

Next, add water molecules, present in an aqueous solution, to balance the oxygen atoms, just as we did in the oxidation number method.

$$HNO_{2(aq)} \rightarrow NO_{(g)} + H_2O_{(l)}$$

Because the reaction takes place in an acidic solution, hydrogen ions are present, and these are used to balance the hydrogen on both sides of the equation.

$$H^+_{(aq)} + HNO_{2(aq)} \rightarrow NO_{(g)} + H_2O_{(l)}$$

At this stage, all of the atoms should be balanced, but the charge on both sides will not be balanced. Add an appropriate number of electrons to balance the charge. Because electrons carry a negative charge, they are always added to the less negative, or more positive, side of the half-reaction.

$$e^- + H^+_{(aq)} + HNO_{2(aq)} \rightarrow NO_{(g)} + H_2O_{(l)}$$

This balanced half-reaction equation represents a gain of electrons—in other words, a reduction of the nitrous acid. Check to make sure that both the atom symbols and the charge are balanced.

In a basic solution, the concentration of hydroxide ions greatly exceeds that of hydrogen ions. For basic solutions, we will develop the half-reaction as if it occurred in an acidic solution and then convert the hydrogen ions into water molecules using hydroxide ions. This trick works because a hydrogen ion and a hydroxide ion react in a 1:1 ratio to form a water molecule. The following problem illustrates the procedure for writing half-reaction equations that occur in basic solutions.

> ### LEARNING TIP
>
> Notice the similarity in balancing O atoms and H atoms with what you did in the oxidation number method. Reactions in a basic solution are also treated in the same way as you did for the oxidation number method.

2. **Copper metal can be oxidized in a basic solution to form copper(I) oxide. What is the half-reaction for this process?**

Following the same steps as before, we write the formula and balance the atoms, other than oxygen and hydrogen. Here the copper atoms must be balanced.

$$2\,Cu_{(s)} \rightarrow Cu_2O_{(s)}$$

Next, balance the oxygen using water molecules and balance the hydrogen using hydrogen ions, assuming, for the moment, an acidic solution. The charge is balanced using electrons.

$$H_2O_{(l)} + 2\,Cu_{(s)} \rightarrow Cu_2O_{(s)} + 2\,H^+_{(aq)} + 2\,e^-$$

Because the half-reaction occurs in a basic solution, add the same number of hydroxide ions as there are hydrogen ions, to both sides of the equation. This is done to maintain the balance of mass and charge.

$$2\,OH^-_{(aq)} + H_2O_{(l)} + 2\,Cu_{(s)} \rightarrow Cu_2O_{(s)} + 2\,H^+_{(aq)} + 2\,e^- + 2\,OH^-_{(aq)}$$

Combine equal numbers of hydrogen ions and hydroxide ions to form water molecules.

$$2\,OH^-_{(aq)} + H_2O_{(l)} + 2\,Cu_{(s)} \rightarrow Cu_2O_{(s)} + 2\,H_2O_{(l)} + 2\,e^-$$

Finally, cancel H_2O and anything else that is the same from both sides of the equation. Check that the atom symbols and charge are balanced.

$$2\,OH^-_{(aq)} + \cancel{H_2O}_{(l)} + 2\,Cu_{(s)} \rightarrow Cu_2O_{(s)} + 2\,\cancel{H_2O}_{(l)} + 2\,e^-$$
$$2\,OH^-_{(aq)} + 2\,Cu_{(s)} \rightarrow Cu_2O_{(s)} + H_2O_{(l)} + 2\,e^-$$

Example

Chlorine is converted to perchlorate ions in an acidic solution. Write the half-reaction equation. Is this half-reaction an oxidation or a reduction?

Solution

$$4\,H_2O_{(l)} + Cl_{2(aq)} \rightarrow 2\,ClO_4^-{}_{(aq)} + 8\,H^+_{(aq)} + 8\,e^-$$

This half-reaction is an oxidation.

Example

Aqueous permanganate ions are reduced to solid manganese(IV) oxide in a basic solution. Write the half-reaction equation. Is the half-reaction an oxidation or a reduction?

Solution

$$4\,OH^-_{(aq)} + 3\,e^- + 4\,H^+_{(aq)} + MnO_4^-{}_{(aq)} \rightarrow MnO_{2(s)} + 2\,H_2O_{(l)} + 4\,OH^-_{(aq)}$$
$$4\,H_2O_{(l)} + 3\,e^- + MnO_4^-{}_{(aq)} \rightarrow MnO_{2(s)} + 2\,H_2O_{(l)} + 4\,OH^-_{(aq)}$$
$$MnO_4^-{}_{(aq)} + 2\,H_2O_{(l)} + 3\,e^- \rightarrow MnO_{2(s)} + 4\,OH^-_{(aq)}$$

This half-reaction is a reduction.

LEARNING TIP

LEO says GER
Recall that loss of electrons is oxidation (LEO) and gain of electrons is reduction (GER)

SUMMARY   *Writing Half-Reaction Equations*

Step 1   Write the chemical formulas for the reactants and products.

Step 2   Balance all atoms, other than O and H.

Step 3   Balance O by adding $H_2O_{(l)}$.

Step 4   Balance H by adding $H^+_{(aq)}$.

Step 5   Balance the charge on each side by adding e^- and cancel anything that is the same on both sides.

For basic solutions only,

Step 6   Add $OH^-_{(aq)}$ to both sides to equal the number of $H^+_{(aq)}$ present.

Step 7   Combine $H^+_{(aq)}$ and $OH^-_{(aq)}$ on the same side to form $H_2O_{(l)}$. Cancel equal amounts of $H_2O_{(l)}$ from both sides.

> ▶ *Practice*

Understanding Concepts

5. For each of the following, complete the half-reaction equation and classify it as an oxidation or a reduction.
  (a) dinitrogen oxide to nitrogen gas in an acidic solution
  (b) nitrite ions to nitrate ions in a basic solution
  (c) silver(I) oxide to silver metal in a basic solution
  (d) nitrate ions to nitrous acid in an acidic solution
  (e) hydrogen gas to water in a basic solution

Balancing Redox Equations Using Half-Reaction Equations

A redox reaction includes both an oxidation and a reduction. In other words, one substance has to lose electrons as another substance gains electrons. Now that you know how to write half-reaction equations, we can combine an oxidation half-reaction equation and a reduction half-reaction equation to obtain the overall balanced redox equation.

For a particular reaction, chemists know the main starting materials and the reaction conditions (e.g., acidic or basic). A chemical analysis of the products determines the oxidized and reduced species produced in the reaction. This provides a skeleton equation showing the main reactants and products. The details of the final redox equation will be provided by the individual balanced half-reaction equations.

Balancing Redox Equations Using Half-Reactions

***SAMPLE* problem** ◄

In a chemical analysis, a solution of dichromate ions is reacted with an acidic solution of iron(II) ions (Figure 2). The products formed are iron(III) and chromium(III) ions as shown by the following skeleton equation.

$$Fe^{2+}_{(aq)} + Cr_2O_7^{2-}_{(aq)} \rightarrow Fe^{3+}_{(aq)} + Cr^{3+}_{(aq)}$$

Balance the equation.

The first step is to separate the equation into two skeleton half-reaction equations, keeping related entities together.

$$Fe^{2+}_{(aq)} \rightarrow Fe^{3+}_{(aq)}$$
$$Cr_2O_7^{2-}_{(aq)} \rightarrow Cr^{3+}_{(aq)}$$

Now you can treat each half-reaction equation separately to obtain a balanced equation. The iron(II) half-reaction requires only the addition of an electron to balance the charge.

$$Fe^{2+}_{(aq)} \rightarrow Fe^{3+}_{(aq)} + e^-$$

For the dichromate half-reaction, you need to follow the same procedure as you did before: balance atoms other than O and H atoms; balance O atoms by adding $H_2O_{(l)}$; balance H atoms by adding $H^+_{(aq)}$; and finally, balance the charge by adding electrons.

$$6\,e^- + 14\,H^+_{(aq)} + Cr_2O_7^{2-}_{(aq)} \rightarrow 2\,Cr^{3+}_{(aq)} + 7\,H_2O_{(l)}$$

Figure 2
The concentration of dichromate ions can be determined by a reaction of a standard iron(II) solution. A redox indicator is usually added to produce a sharp colour change at the completion of the reaction.

▶

Recall that the total number of electrons lost must equal the total number of electrons gained. Using simple whole numbers, multiply one or both half-reaction equations so that the electrons will be balanced. In this example, the iron(II) half-reaction equation must be multiplied by a factor of 6 to balance $6e^-$ in the dichromate half-reaction equation.

$$6\,[Fe^{2+}_{(aq)} \rightarrow Fe^{3+}_{(aq)} + e^-]$$
$$6\,e^- + 14\,H^+_{(aq)} + Cr_2O_7^{2-}{}_{(aq)} \rightarrow 2\,Cr^{3+}_{(aq)} + 7\,H_2O_{(l)}$$

Add the two half-reaction equations.

$$6\,Fe^{2+}_{(aq)} + 6\,e^- + 14\,H^+_{(aq)} + Cr_2O_7^{2-}{}_{(aq)} \rightarrow 2\,Cr^{3+}_{(aq)} + 7\,H_2O_{(l)} + 6\,Fe^{3+}_{(aq)} + 6\,e^-$$

Cancel the electrons and anything else that is the same on both sides of the equation.

$$6\,Fe^{2+}_{(aq)} + 6\,\cancel{e^-} + 14\,H^+_{(aq)} + Cr_2O_7^{2-}{}_{(aq)} \rightarrow 2\,Cr^{3+}_{(aq)} + 7\,H_2O_{(l)} + 6\,Fe^{3+}_{(aq)} + 6\,\cancel{e^-}$$
$$6\,Fe^{2+}_{(aq)} + 14\,H^+_{(aq)} + Cr_2O_7^{2-}{}_{(aq)} \rightarrow 2\,Cr^{3+}_{(aq)} + 7\,H_2O_{(l)} + 6\,Fe^{3+}_{(aq)}$$

Check the final redox equation to make sure that both the atom symbols and the charge are balanced.

For reactions that occur in basic solutions, it is easier to follow the same procedure outlined above and then convert to a basic solution. In other words, create the balanced redox equation for an acidic solution, then add $OH^-_{(aq)}$ to convert the $H^+_{(aq)}$ to water molecules. This is the same procedure you used for obtaining a redox equation using the oxidation number method. An example for a basic solution is shown below.

Example

Permanganate ions and oxalate ions react in a basic solution to produce carbon dioxide and manganese(IV) oxide.

$$MnO_4^-{}_{(aq)} + C_2O_4^{2-}{}_{(aq)} \rightarrow CO_{2(g)} + MnO_{2(s)}$$

Write the balanced redox equation for this reaction.

Solution

$$2\,[3\,e^- + 4\,H^+_{(aq)} + MnO_4^-{}_{(aq)} \rightarrow MnO_{2(s)} + 2\,H_2O_{(l)}]$$
$$3\,[C_2O_4^{2-}{}_{(aq)} \rightarrow 2\,CO_{2(g)} + 2\,e^-]$$
$$8\,H^+_{(aq)} + 2\,MnO_4^-{}_{(aq)} + 3\,C_2O_4^{2-}{}_{(aq)} \rightarrow 2\,MnO_{2(s)} + 4\,H_2O_{(l)} + 6\,CO_{2(g)}$$
$$8\,OH^-_{(aq)} + 8\,H^+_{(aq)} + 2\,MnO_4^-{}_{(aq)} + 3\,C_2O_4^{2-}{}_{(aq)} \rightarrow 2\,MnO_{2(s)} + 4\,H_2O_{(l)} + 6\,CO_{2(g)} + 8\,OH^-_{(aq)}$$
$$4\,H_2O_{(l)} + 2\,MnO_4^-{}_{(aq)} + 3\,C_2O_4^{2-}{}_{(aq)} \rightarrow 2\,MnO_{2(s)} + 6\,CO_{2(g)} + 8\,OH^-_{(aq)}$$

SUMMARY — Balancing Redox Equations Using Half-Reaction Equations

Step 1 Separate the skeleton equation into the start of two half-reaction equations.

Step 2 Balance each half-reaction equation.

Step 3 Multiply each half-reaction equation by simple whole numbers to balance the electrons lost and gained.

Step 4 Add the two half-reaction equations, cancelling the electrons and anything else that is exactly the same on both sides of the equation.

For basic solutions only,

Step 5   Add $OH^-_{(aq)}$ to both sides equal in number to the number of $H^+_{(aq)}$ present.

Step 6   Combine $H^+_{(aq)}$ and $OH^-_{(aq)}$ on the same side to form $H_2O_{(l)}$, and cancel the same number of $H_2O_{(l)}$ on both sides.

▶ **Practice**

Understanding Concepts

6. Balance the following skeleton redox equations using the half-reaction method. All reactions occur in an acidic solution.
   (a) $Zn_{(s)} + NO^-_{3(aq)} \rightarrow NH^+_{4(aq)} + Zn^{2+}_{(aq)}$
   (b) $Cl_{2(aq)} + SO_{2(g)} \rightarrow Cl^-_{(aq)} + SO^{2-}_{4(aq)}$

7. Balance the following skeleton redox equations using the half-reaction method. All reactions occur in a basic solution.
   (a) $MnO^-_{4(aq)} + I^-_{(aq)} \rightarrow MnO_{2(s)} + I_{2(s)}$
   (b) $CN^-_{(aq)} + IO^-_{3(aq)} \rightarrow CNO^-_{(aq)} + I^-_{(aq)}$
   (c) $OCl^-_{(aq)} \rightarrow Cl^-_{(aq)} + ClO^-_{3(aq)}$

Extension

8. Balance the following redox equation.
   $KMnO_{4(aq)} + H_2S_{(aq)} + H_2SO_{4(aq)} \rightarrow K_2SO_{4(aq)} + MnSO_{4(aq)} + S_{(s)}$

Answers

6. (a) $4\ Zn$, $1\ NO_3^-$, $10\ H^+$; $1\ NH_4^+$, $4\ Zn^{2+}$, $3\ H_2O$

   (b) $1\ Cl_2$, $1\ SO_2$, $2\ H_2O$; $2\ Cl^-$, $1\ SO_4^{2-}$, $4\ H^+$

7. (a) $2\ MnO_4^-$, $6\ I^-$, $4\ H_2O$; $2\ MnO_2$, $3\ I_2$, $8\ OH^-$

   (b) $3\ CN^-$, $1\ IO_3^-$; $1\ I^-$, $3\ CNO^-$

   (c) $3\ OCl^-$; $2\ Cl^-$, $1\ ClO_3^-$

8. $2\ KMnO_4$, $5\ H_2S$, $3\ H_2SO_4$; $1\ K_2SO_4$, $2\ MnSO_4$, $5\ S$, $8\ H_2O$

▶ **Section 9.2 Questions**

Understanding Concepts

1. In what way are the two methods of balancing redox equations similar?

2. Compare oxidation and reduction in terms of oxidation numbers and electrons transferred.

3. Balance the following equations representing reactions that occur in an acidic solution:
   (a) $Cu_{(s)} + NO^-_{3(aq)} \rightarrow Cu^{2+}_{(aq)} + NO_{2(g)}$
   (b) $Mn^{2+}_{(aq)} + HBiO_{3(aq)} \rightarrow Bi^{3+}_{(aq)} + MnO^-_{4(aq)}$
   (c) $H_2O_{2(aq)} + Cr_2O^{2-}_{7(aq)} \rightarrow Cr^{3+}_{(aq)} + O_{2(g)} + H_2O_{(l)}$

4. Balance the following equations representing reactions that occur in a basic solution:
   (a) $Cr(OH)_{3(s)} + IO^-_{3(aq)} \rightarrow CrO^{2-}_{4(aq)} + I^-_{(aq)}$
   (b) $Ag_2O_{(s)} + CH_2O_{(aq)} \rightarrow Ag_{(s)} + CHO^-_{2(aq)}$
   (c) $S_2O^{2-}_{4(aq)} + O_{2(g)} \rightarrow SO^{2-}_{4(aq)}$

Applying Inquiry Skills

5. State two general experimental designs that could help determine the balancing of the main species in a redox reaction.

Making Connections

6. Many commercially available drain cleaners contain a basic solution of sodium hydroxide, which helps to remove any grease in the drains. Some solid drain cleaners contain solid sodium hydroxide and finely divided aluminum metal. When mixed with water this produces a very vigorous, exothermic reaction shown by the following skeleton equation:

   $Al_{(s)} + H_2O_{(l)} \rightarrow Al(OH)^-_{4(aq)} + H_{2(g)}$

   (a) Complete the balanced redox equation for this reaction.
   (b) Describe and discuss some possible health and safety issues associated with the use of solid drain cleaners.

Extension

7. The analysis of iron by an oxidation–reduction titration is a common analytical method. A common titrant used in this analyis is a solution of the cerium(IV) ion, which is reduced to cerium(III) in the analysis. In one chemical analysis of some iron ore, the sample is treated to convert all of the iron to iron(II) ions. A 25.0-mL sample of iron(II) is titrated with 0.125 mol/L cerium(IV) solution using a redox indicator. The average volume of cerium(IV) required to reach the endpoint was 15.1 mL. Calculate the concentration of the iron(II) ions in the sample.

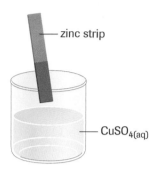

Figure 1
Copper(II) ions react spontaneously with zinc metal. A copper(II) ion has a stronger attraction for the valence electrons of a zinc atom than zinc does.

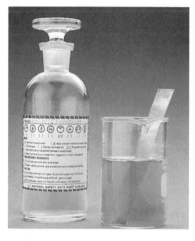

Figure 2
The green nickel(II) ion colour remains and the copper metal does not react. Collisions between copper atoms and nickel(II) ions apparently do not result in the transfer of electrons.

reducing agent a substance that loses or gives up electrons to another substance in a redox reaction

oxidizing agent a substance that gains or removes electrons from another substance in a redox reaction

A redox reaction may be explained as a transfer of valence electrons from one substance to another. Evidence indicates that the majority of atoms, molecules, and ions are stable and do not readily release electrons. Since two particles must be involved in an electron transfer, this transfer can be explained as a competition for electrons. Using a tug-of-war analogy, each particle pulls on the same electrons. If one particle is able to pull electrons away from the other, a spontaneous reaction occurs (**Figure 1**). Otherwise, no reaction occurs (**Figure 2**). In the spontaneous reaction of copper(II) ions and zinc metal, the Cu^{2+} ion is electron deficient and pulls electrons from a Zn atom. The reaction occurs because Cu^{2+} pulls harder on Zn's electrons than Zn does. Cu^{2+} wins the two valence electrons from a Zn atom. A successful electron transfer has occurred.

Without mixing all possible reactants and observing any evidence of reaction, how can we predict if a reaction will occur? If a reaction occurs, what will be the products? The answers to these questions cannot be obtained easily using redox theory. By observing many successful and unsuccesful reactions, patterns emerge and empirical generalizations can be made.

Oxidizing and Reducing Agents

Before we look at these patterns, some terms commonly used by chemists need to be defined. When discussing possible reactants and comparing their reactivities, it is customary and convenient to classify the reactants in a redox reaction. This classification originated historically but is now defined in terms of an ability to lose or gain electrons. In any redox reaction an electron transfer occurs, which means that one reactant is oxidized and one reactant is reduced.

	X	+	Y	→	products
electron change	loses		gains		
oxidation number	increases		decreases		
	is oxidized		is reduced		

Examples:

$$Zn_{(s)} + Cu^{2+}_{(aq)} \rightarrow Zn^{2+}_{(aq)} + Cu_{(s)}$$
$$2\,Br^-_{(aq)} + Cl_{2(g)} \rightarrow Br_{2(l)} + 2\,Cl^-_{(aq)}$$
$$3\,CO_{(g)} + Fe_2O_{3(s)} \rightarrow 3\,CO_{2(g)} + 2\,Fe_{(s)}$$

Rather than saying "the reactant that is oxidized" and "the reactant that is reduced," chemists use the terms **reducing agent** and **oxidizing agent**. These terms originated in the early history of metallurgy and corrosion. For example, to "reduce" a larger volume of iron(III) oxide to a smaller volume of pure iron, a substance called a reducing agent was required, e.g., $CO_{(g)}$.

$$\text{reducing agent} + \overset{+3}{Fe_2}O_{3(s)} \rightarrow \overset{0}{Fe} + \text{other products}$$

reduction

Similarly, oxidation was originally associated with an oxidizing agent. For example, a metal could be oxidized by certain substances called oxidizing agents. At first, oxygen was the only known oxidizing agent, but others (e.g., halogens) can also oxidize or corrode metals.

oxidation

0 → 2+

oxidizing agent + $MgO_{(s)} \rightarrow Mg^{2+}$ + other products

The terms oxidizing and reducing agents developed separately, long before any redox theory of electron transfer emerged. Today, chemists routinely think in terms of electron transfer to explain redox reactions. A redox reaction is recognized as an electron transfer between an oxidizing agent and a reducing agent (**Figure 3**).

Development of a Redox Table

Some redox reactions such as single displacement reactions are easy to study experimentally. The evidence of a reaction is immediately obvious and the interpretation of an electron transfer is relatively simple. In the past, you have generally assumed that all single displacement reactions are spontaneous. However, by testing several combinations of metals and metal ions, it can easily be shown that some combinations react immediately, but many do not react at all. The question that arises is, "How do you know when a chemical reaction will occur spontaneously without actually doing the reaction?" 🧪

Let's look at some examples of combinations of metals and metal ions. Suppose copper, lead, silver, and zinc metals were combined one at a time with each of copper(II), lead(II), silver, and zinc ion solutions. We can rank the ability of the metal ions to react with the metals (**Table 1**).

Table 1 Reactivities of Metal Ions with Metals

Ions	$Ag^+_{(aq)}$	$Cu^{2+}_{(aq)}$	$Pb^{2+}_{(aq)}$	$Zn^{2+}_{(aq)}$
Reacted with	$Cu_{(s)}, Pb_{(s)}, Zn_{(s)}$	$Pb_{(s)}, Zn_{(s)}$	$Zn_{(s)}$	none
Number of reactions	3	2	1	0
Reactivity order	most ──────────────────────────────→			least

The most reactive metal ion, $Ag^+_{(aq)}$, has the greatest tendency to gain electrons. On the other hand, $Zn^{2+}_{(aq)}$ shows no tendency to gain electrons in the combinations tested. Therefore, the order of reactivity is also the order of strengths as oxidizing agents.

strongest oxidizing agent

weakest oxidizing agent

$$Ag^+_{(aq)} + e^- \rightarrow Ag_{(s)}$$
$$Cu^{2+}_{(aq)} + 2e^- \rightarrow Cu_{(s)}$$
$$Pb^{2+}_{(aq)} + 2e^- \rightarrow Pb_{(s)}$$
$$Zn^{2+}_{(aq)} + 2e^- \rightarrow Zn_{(s)}$$

The order of reactivity of the four metals can be obtained in a similar way (**Table 2**).

Table 2 Reactivities of Metals with Metal Ions

Metals	$Zn_{(s)}$	$Pb_{(s)}$	$Cu_{(s)}$	$Ag_{(s)}$
Reacted with	$Ag^+_{(aq)}, Cu^{2+}_{(aq)}, Pb^{2+}_{(aq)}$	$Ag^+_{(aq)}, Cu^{2+}_{(aq)}$	$Ag^+_{(aq)}$	none
Number of reactions	3	2	1	0
Reactivity order	most ──────────────────────────────→			least

e^-

OA + RA ⟶

Figure 3
In all redox reactions, electrons are transferred from a reducing agent to an oxidizing agent.

LEARNING TIP

Oxidizing and Reducing Agents
If a positively charged metal ion reacts, then it is usually converted to a metal atom. According to redox theory, this requires a gain of electrons and hence the metal ion is behaving as an oxidizing agent. Similarly, if a metal atom reacts, then it is always converted to a positively charged ion by losing electrons. Metals always behave as reducing agents.

INVESTIGATION 9.3.1

Spontaneity of Redox Reactions (p. 716)
How many reactions will occur?

The most reactive metal, $Zn_{(s)}$, has the greatest tendency to lose electrons and $Ag_{(s)}$ shows no tendency to lose electrons in the combinations tested. Metals behave as reducing agents and so $Zn_{(s)}$ is the strongest reducing agent among those tested.

strongest reducing agent

$$Zn_{(s)} \rightarrow Zn^{2+}_{(aq)} + 2\,e^-$$
$$Pb_{(s)} \rightarrow Pb^{2+}_{(aq)} + 2\,e^-$$
$$Cu_{(s)} \rightarrow Cu^{2+}_{(aq)} + 2\,e^-$$
$$Ag_{(s)} \rightarrow Ag^+_{(aq)} + e^-$$

weakest reducing agent

In these reactions, the metal ions are the oxidizing agents and the silver ion is the strongest oxidizing agent (SOA) of the four ions because it is the most reactive in our group. The metals are the reducing agents and the zinc metal is the strongest reducing agent (SRA). The two lists of reactivity can be summarized using a single set of half-reactions as shown in **Table 3**.

Table 3 Relative Strengths of Oxidizing and Reducing Agents

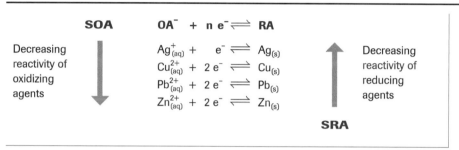

In **Table 3**, the metal ions are on the left side of the equations and the metal atoms are on the right side. For metal ions (the oxidizing agents), the half-reaction equations are read from left to right in the table. For metal atoms (the reducing agents), the half-reaction equations are read from right to left.

▶ **Practice**

Understanding Concepts

1. Oxidation and reduction are processes, and oxidizing agents and reducing agents are substances. Explain this statement, using definitions of the terms.

2. If a substance is a very strong oxidizing agent, what does this mean in terms of electrons?

3. If a substance is a very strong reducing agent, what does this mean in terms of electrons?

Refer to **Tables 1, 2,** and **3** to answer questions 4 to 8.

4. List the metal(s) that react spontaneously with a copper(II) ion solution.

5. Which metal(s) did not appear to react with a copper(II) ion solution?

6. Start with the position of $Cu^{2+}_{(aq)}$ in **Table 3** and note the position of the metal(s) that reacted and the metal(s) that did not react. For a metal that reacts spontaneously with $Cu^{2+}_{(aq)}$, where does the metal appear on a table of reduction half-reactions (**Table 3**)?

7. Repeat questions 4 to 6 for the $Pb^{2+}_{(aq)}$ ion.

Applying Inquiry Skills

8. Your answers to 6 and 7 form an empirical hypothesis that can be tested by making predictions for the other metal ions. Use **Table 3** to predict which of the reactions should be spontaneous. Are the predictions correct? Is your hypothesis verified?

9. An experiment similar to the example of metals and metal ions was conducted using halogens and halide ions.

Question

What is the table of relative strengths of oxidizing and reducing agents for the halogens?

Evidence

Only three combinations produced evidence of a reaction (**Figure 4**, **Table 4**).

(a)

(b)

(c)

Analysis

(a) Prepare a table of half-reaction equations like **Table 3** for the halogens.

Table 4 Reactions of Halogens with Solutions of Halides

	$Br_{2(aq)}$	$Cl_{2(aq)}$	$I_{2(aq)}$
$Br^-_{(aq)}$	no reaction	yellow-brown	no reaction
$Cl^-_{(aq)}$	no reaction	no reaction	no reaction
$I^-_{(aq)}$	yellow-brown	yellow-brown	no reaction

Figure 4
None of the combinations of aqueous solutions of chlorine, bromine, and iodine with their corresponding halides show any evidence of reaction except for the reaction between **(a)** bromine and iodide ions, **(b)** chlorine and bromide ions, and **(c)** chlorine and iodide ions.

The Spontaneity Rule

Evidence obtained from the study of many redox reactions has been used to establish a generalization, called the **redox spontaneity rule**. **Figure 5** illustrates how you can use the rule, along with a table of oxidizing and reducing agents, to predict whether or not a reaction is spontaneous.

redox spontaneity rule a spontaneous redox reaction occurs only if the oxidizing agent (OA) is above the reducing agent (RA) in a table of relative strengths of oxidizing and reducing agents.

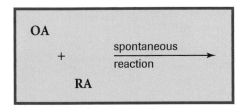

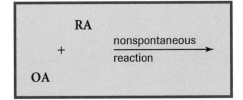

Figure 5
The redox spontaneity rule

Another Method for Building Redox Tables

Once a spontaneity rule is developed from experimental evidence, the rule may be used to generate half-reaction tables. The evidence to be analyzed in this case is a net ionic equation, accompanied by observations of spontaneity. In the following method, the spontaneity rule, rather than the number of reactions observed, is used to order the oxidizing and reducing agents to produce a redox table. The procedure for this type of analysis and synthesis is illustrated by the following example.

 **LAB EXERCISE 9.3.1**

Building a Redox Table (p. 717)
Several groups of experimental evidence are combined to make one larger table.

Creating a Redox Table

Three reactions among indium, cobalt, palladium, and copper were investigated. The reaction equations below indicate that two spontaneous reactions occurred and only one combination did not react. Using these equations, construct a redox table of half-reaction equations showing the relative strengths of the oxidizing and reducing agents.

$$3\ Co^{2+}_{(aq)}\ +\ 2\ In_{(s)}\ \rightarrow 2\ In^{3+}_{(aq)}\ +\ 3\ Co_{(s)}$$

$$Cu^{2+}_{(aq)} + Co_{(s)} \rightarrow Co^{2+}_{(aq)}\ +\ Cu_{(s)}$$

$$Cu^{2+}_{(aq)}\ +\ Pd_{(s)} \rightarrow \text{no evidence of reaction}$$

To construct a table from this information, work with one equation at a time. Identify the oxidizing and reducing agents for the first reaction, and arrange them in two columns using the spontaneity rule. For the first reaction, this step is shown in **Figure 6(a)**. $Co^{2+}_{(aq)}$ is the oxidizing agent and $In_{(s)}$ is the reducing agent. Since the reaction is spontaneous, the oxidizing agent is above the reducing agent in the list.

In the second reaction, $Cu^{2+}_{(aq)}$ is the oxidizing agent and $Co_{(s)}$ is the reducing agent. This reaction is also spontaneous; therefore, $Cu^{2+}_{(aq)}$ is above $Co_{(s)}$ in the list. Since a metal appears on the same line as its ion in a half-reaction table, add $Co_{(s)}$ and extend the list as shown in **Figure 6(b)**.

No reaction occurs for the third pair of reactants. If a reaction had occurred, $Cu^{2+}_{(aq)}$ would be the oxidizing agent and $Pd_{(s)}$ would be the reducing agent. As this reaction is not spontaneous, the oxidizing agent appears below the reducing agent. **Figure 6(c)** shows the list extended to include $Pd_{(s)}$. To complete the table, write balanced half-reaction equations for each oxidizing/reducing agent pair.

SOA $\quad Pd^{2+}_{(aq)} + 2\ e^- \rightleftharpoons Pd_{(s)}$

$\qquad\quad Cu^{2+}_{(aq)} + 2\ e^- \rightleftharpoons Cu_{(s)}$

$\qquad\quad Co^{2+}_{(aq)} + 2\ e^- \rightleftharpoons Co_{(s)}$

$\qquad\quad In^{2+}_{(aq)}\ + 3\ e^- \rightleftharpoons In_{(s)}\quad$ **SRA**

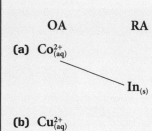
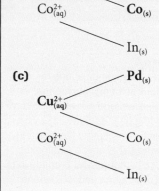

Figure 6
The relative position of a pair of oxidizing and reducing agents indicates whether a reaction will be spontaneous.

▶ **Practice**

Understanding Concepts

10. The following reactions were performed. Construct a table of relative strengths of oxidizing and reducing agents.

$$Co^{2+}_{(aq)}\ +\ Zn_{(s)} \rightarrow Co_{(s)}\ +\ Zn^{2+}_{(aq)}$$

$$Mg^{2+}_{(aq)}\ +\ Zn_{(s)} \rightarrow \text{no evidence of reaction}$$

11. In a school laboratory four metals were combined with each of four solutions. Construct a table of relative strengths of oxidizing and reducing agents.

$$Be_{(s)}\ +\ Cd^{2+}_{(aq)} \rightarrow Be^{2+}_{(aq)}\ +\ Cd_{(s)}$$

$$Cd_{(s)}\ +\ 2\ H^+_{(aq)} \rightarrow Cd^{2+}_{(aq)}\ +\ H_{2(g)}$$

$$Ca^{2+}_{(aq)}\ +\ Be_{(s)} \rightarrow \text{no evidence of reaction}$$

$$Cu_{(s)}\ +\ H^+_{(aq)}\ \rightarrow \text{no evidence of reaction}$$

12. Is the redox spontaneity rule empirical or theoretical? Justify your answer.

13. Use the relative strengths of nonmetals and metals as oxidizing and reducing agents, as indicated in the following unbalanced equations, to construct a table of half-reactions.

$$Ag_{(s)}\ +\ Br_{2(l)}\ \rightarrow AgBr_{(s)}$$

$$Ag_{(s)}\ +\ I_{2(s)}\ \rightarrow \text{no evidence of reaction}$$

$$Cu^{2+}_{(aq)}\ +\ I^-_{(aq)} \rightarrow \text{no redox reaction}$$

$$Br_{2(l)}\ +\ Cl^-_{(aq)} \rightarrow \text{no evidence of reaction}$$

An Extended Redox Table

Evidence collected in many experiments has been analyzed to produce an extended redox table of oxidizing and reducing agents such as the one found in Appendix C11. A table such as this represents the combined efforts of many people over many years. A redox table is an important reference for chemists. You can use this table to compare oxidizing and reducing agents, and to predict spontaneous redox reactions.

▶ *Practice*

Understanding Concepts

Use the redox table in Appendix C11 or the *CRC Handbook of Chemistry and Physics* to answer the following questions.

14. Arrange the following metal ions in order of decreasing strength as oxidizing agents: lead(II) ions, silver ions, zinc ions, and copper(II) ions. How does this order compare with the one in **Table 3**?

15. What classes of substances (e.g., metals, nonmetals, acidic, basic) usually behave as
    (a) oxidizing agents?
    (b) reducing agents?

16. Use atomic theory to explain why nonmetals behave as oxidizing agents and metals behave as reducing agents. Is there logical consistency between atomic theory and the empirically determined table of oxidizing and reducing agents?

17. Trends in the reactivity of elements show that fluorine is the most reactive nonmetal. How does this relate to the position of fluorine in the redox table of oxidizing and reducing agents? State one reason why this element is the most reactive nonmetal. Why is your reason an explanation? (Keep asking a series of "why" questions until your theoretical knowledge is expended. Does your theory pass the test of being able to explain the empirically determined table?)

18. Identify three oxidizing agents (other than $Fe^{2+}_{(aq)}$, shown in **Figure 7**) from the table that can also act as reducing agents. Try to explain this unique behaviour.

19. Use the redox spontaneity rule to predict whether the following mixtures will show evidence of a reaction; that is, predict whether the reactions are spontaneous. (Do not write the equations for the reaction.)
    (a) nickel metal in a solution of silver ions
    (b) zinc metal in a solution of aluminum ions
    (c) an aqueous mixture of copper(II) ions and iodide ions
    (d) chlorine gas bubbled into a bromide ion solution
    (e) an aqueous mixture of copper(II) ions and tin(II) ions
    (f) copper metal in nitric acid

Applying Inquiry Skills

20. Describe two experimental designs or methods to collect evidence from which half-reaction tables can be built.

Making Connections

21. From your own knowledge, list two metals that are found as elements and two that are never found as elements in nature. Test your answer by referring to the position of these metals in the table of oxidizing and reducing agents.

22. Of the two parallel ways of knowing, empirical and theoretical, which, to this point, has been the most useful to you in predicting the spontaneity of redox reactions? Explain.

Fe
↑   GER/OA
Fe^{2+}
↓   LEO/RA
Fe^{3+}

Figure 7
Iron(II) ions can either lose or gain electrons and, therefore, can act as either reducing agents or oxidizing agents.

DID YOU *KNOW* ❓

Getting Rid of Skunk Odour
The smell of a skunk is caused by a thiol compound (R—SH). To deodorize a pet sprayed by a skunk, you need to convert the smelly thiol to an odourless compound. Hydrogen peroxide in a basic solution (usually from sodium bicarbonate) acts as an oxidizing agent to change the thiol to a disulfide compound (RS—SR), which is odourless.

Table 5 Hints for Listing and Labelling Entities

- Aqueous solutions contain $H_2O_{(l)}$ molecules.
- Acidic solutions contain $H^+_{(aq)}$ ions.
- Basic solutions contain $OH^-_{(aq)}$ ions.
- Some oxidizing and reducing agents are combinations, for example, $MnO^-_{4(aq)}$ and $H^+_{(aq)}$.
- $H_2O_{(l)}$, $Fe^{2+}_{(aq)}$, $Cu^+_{(aq)}$, $Sn^{2+}_{(aq)}$ and $Cr^{2+}_{(aq)}$ may act as either oxidizing or reducing agents. Label both possibilities in your list.

Predicting Redox Reactions in Solution

Arrhenius's ideas about solutions provide an important starting point for predicting redox reactions. In solutions, molecules and ions act approximately independently of each other. A first step in predicting redox reactions is to list all entities that are present. (Some helpful reminders are listed in **Table 5**.) For example, when copper metal is placed into an acidic potassium permanganate solution, copper atoms, potassium ions, permanganate ions, hydrogen ions, and water molecules are all present. Next, using your knowledge of oxidizing and reducing agents and Appendix C11, label all possible oxidizing and reducing agents in the starting mixture. The permanganate ion is listed as an oxidizing agent only in an acidic solution. To indicate this combination, draw an arc between the permanganate and hydrogen ions as shown, and label the pair as an oxidizing agent. This procedure of listing and identifying entities present is a crucial step in predicting redox reactions.

$$
\begin{array}{ccccc}
& \textbf{OA} & \overset{\textbf{OA}}{\frown} & \textbf{OA} & \textbf{OA} \\
Cu_{(s)} & K^+_{(aq)} & MnO^-_{4(aq)} & H^+_{(aq)} & H_2O_{(l)} \\
\textbf{RA} & & & & \textbf{RA}
\end{array}
$$

> ▶ *Practice*

Understanding Concepts

23. List all entities initially present in the following mixtures and identify all possible oxidizing and reducing agents.
    (a) A lead strip is placed in a copper(II) sulfate solution.
    (b) A gold coin is placed in a nitric acid solution.
    (c) A potassium dichromate solution is added to an acidic iron(II) nitrate solution.
    (d) An aqueous chlorine solution is added to a phosphorous acid solution.
    (e) A potassium permanganate solution is mixed with an acidified tin(II) chloride solution.
    (f) Iodine solution is added to a basic mixture containing manganese(IV) oxide.

We can use a redox table to identify the strongest oxidizing and reducing agents in a mixture and then predict which reactions will occur. If we assume that collisions are completely random, the strongest oxidizing agent and the strongest reducing agent will react. (In some cases, further reactions may occur as well, but we will consider only the initial reaction, unless otherwise specified.) When using the redox table in Appendix C11 to predict redox reactions,

---

INVESTIGATION 9.3.2

The Reaction of Sodium with Water (p. 718)

Test a prediction of a redox reaction.

- Choose the strongest oxidizing agent present in your mixture by starting at the top left corner of the redox table and going down the list until you find the oxidizing agent that is in your mixture.

- Choose the strongest reducing agent in your mixture by starting at the bottom right corner of the redox table and going up the list until you find the reducing agent that is in your mixture.

- Reduction half-reaction equations are read from left to right (following the forward arrow).

- Oxidation half-reaction equations are read from right to left (following the reverse arrow).

- Any substances not present in the redox table will be assumed to be spectator ions. You do not need to label or consider these substances.

Using SOA and SRA to Predict Reactions

Suppose a solution of potassium permanganate is slowly poured into an acidified iron(II) sulfate solution. Does a redox reaction occur and, if it does, what is the reaction equation?

To make a prediction, the entities initially present are identified as oxidizing agents, reducing agents, or both, as shown below.

$$
\begin{array}{cccccc}
\text{OA} & \text{OA} \ \text{OA} & \text{OA} & \text{OA} & \text{OA} \ \text{OA} \\
K^+_{(aq)} & MnO^-_{4(aq)} \ \ H^+_{(aq)} & Fe^{2+}_{(aq)} & SO^{2-}_{4(aq)} & H_2O_{(l)} \\
 & \text{RA} & & \text{RA} &
\end{array}
$$

Use the table in Appendix C9 to choose the strongest oxidizing agent and the strongest reducing agent from your list and indicate them with SOA and SRA.

$$
\begin{array}{cccccc}
\text{OA} & \text{SOA} \ \text{OA} & \text{OA} & \text{OA} \ \text{OA} \\
K^+_{(aq)} & MnO^-_{4(aq)} \ \ H^+_{(aq)} & Fe^{2+}_{(aq)} & SO^{2-}_{4(aq)} & H_2O_{(l)} \\
 & \text{SRA} & & \text{RA} &
\end{array}
$$

Now, write the half-reaction equation for the reduction of the SOA.

$$MnO^-_{4(aq)} + 8\,H^+_{(aq)} + 5\,e^- \rightarrow Mn^{2+}_{(aq)} + 4\,H_2O_{(l)}$$

Write the half-reaction equation for the oxidation of the SRA. Remember you are reading from right to left on the table.

$$Fe^{2+}_{(aq)} \rightarrow Fe^{3+}_{(aq)} + e^-$$

Before combining the half-reaction equations, balance the number of electrons transferred by multiplying one or both half-reaction equations by an integer so that the number of electrons gained by the oxidizing agent equals the number of electrons lost by the reducing agent.

In this case, the iron half-reaction must be multiplied by 5. Add the two equations, but remember to cancel any common terms. You can cancel terms as you add (e.g., $5\,e^-$) or after you add the two half-reactions.

$$
\begin{aligned}
MnO^-_{4(aq)} + 8\,H^+_{(aq)} + 5\,e^- &\rightarrow Mn^{2+}_{(aq)} + 4\,H_2O_{(l)} \\
5\,[Fe^{2+}_{(aq)} &\rightarrow Fe^{3+}_{(aq)} + e^-] \\
\hline
MnO^-_{4(aq)} + 8\,H^+_{(aq)} + 5\,Fe^{2+}_{(aq)} &\rightarrow 5\,Fe^{3+}_{(aq)} + Mn^{2+}_{(aq)} + 4\,H_2O_{(l)}
\end{aligned}
$$

Finally, use the spontaneity rule to predict whether the net ionic equation represents a spontaneous redox reaction. Indicate this by writing "spont." or "non-spont." over the equation arrow.

$$MnO^-_{4(aq)} + 8\,H^+_{(aq)} + 5\,Fe^{2+}_{(aq)} \xrightarrow{\text{spont.}} 5\,Fe^{3+}_{(aq)} + Mn^{2+}_{(aq)} + 4\,H_2O_{(l)}$$

This prediction may be tested by mixing the solutions (**Figure 8**) and performing some diagnostic tests. If the solutions are mixed and the purple colour of the permanganate ion disappears, then it is likely that the permanganate ion reacted. If the pH of the solution is tested before and after reaction, and the pH has increased, then the hydrogen ions likely reacted.

Figure 8
A solution of potassium permanganate is being added to an acidic solution of iron(II) ions. The dark purple colour of $MnO^-_{4(aq)}$ ions instantly disappears. The interpretation is that $MnO^-_{4(aq)}$ ions react with $Fe^{2+}_{(aq)}$ ions to produce the yellow-brown $Fe^{3+}_{(aq)}$ and $Mn^{2+}_{(aq)}$ ions.

Figure 9
Copper in hydrochloric acid does not appear to react.

Example 1

In a chemical industry, could copper pipe be used to transport a hydrochloric acid solution? To answer this question,
(a) predict the redox reaction and its spontaneity, and
(b) describe two diagnostic tests that could be done to test your prediction.

Solution

(a)

$$
\begin{array}{cccc}
 & \textbf{SOA} & & \textbf{OA} \\
Cu_{(s)} & H^+_{(aq)} & Cl^-_{(aq)} & H_2O_{(l)} \\
\textbf{SRA} & & \textbf{RA} \quad \textbf{RA} & \textbf{RA}
\end{array}
$$

$$2\,H^+_{(aq)} + 2\,e^- \rightarrow H_{2(g)}$$

$$\underline{Cu_{(s)} \rightarrow Cu^{2+}_{(aq)} + 2\,e^-}$$

$$2\,H^+_{(aq)} + Cu_{(s)} \xrightarrow{\text{nonspont.}} H_{2(g)} + Cu^{2+}_{(aq)}$$

Since the reaction is nonspontaneous, it should be possible to use a copper pipe to carry hydrochloric acid.

(b) If no gas is produced when the mixture is observed, then it is likely that no hydrogen gas was produced (**Figure 9**). If the colour of the solution did not change to blue, then copper probably did not react to produce copper(II) ions. (If the solution is tested for pH before and after adding the copper, and the pH did not increase, then the hydrogen ions probably did not react.)

SUMMARY   *Predicting Redox Reactions*

Step 1 List all entities present and classify each as an oxidizing agent, reducing agent, or both. Do not label spectator ions.

Step 2 Choose the strongest oxidizing agent as indicated in the table of relative strengths of oxidizing and reducing agents, and write the equation for its reduction.

Step 3 Choose the strongest reducing agent as indicated in the table, and write the equation for its oxidation.

Step 4 Balance the number of electrons lost and gained in the half-reaction equations by multiplying one or both equations by a number. Then add the two balanced half-reaction equations to obtain a net ionic equation.

Step 5 Using the spontaneity rule, predict whether the net ionic equation represents a spontaneous or nonspontaneous redox reaction.

▶ **Practice**

Understanding Concepts

24. Predict the most likely redox reaction in each of the following situations. For any spontaneous reaction, describe one diagnostic test to identify a primary product.
  (a) During a demonstration, zinc metal is placed in a hydrochloric acid solution.
  (b) A gold ring is placed into a hydrochloric acid solution.
  (c) Nitric acid is painted onto a copper sheet to etch a design.

25. In your previous chemistry course, predictions of reactions were made according to the single displacement generalization assuming the formation of the most common ion.
    (a) Use the generalization about single displacement reactions to predict the reaction of iron metal with a copper(II) sulfate solution.
    (b) Use redox theory and a table showing half-reactions to predict the most likely redox reaction of iron metal with a copper(II) sulfate solution.
    (c) Can both predictions be correct? Which do you think is likely correct and why?
26. Oxygen gas is bubbled into an aqueous solution of iron(II) iodide containing excess hydrochloric acid. Predict all spontaneous reactions, in the order in which they will occur.

Applying Inquiry Skills

27. Write one qualitative and one quantitative experimental design to test the two different predictions made for the reaction between iron metal and the copper(II) sulfate solution in question 25.
28. Write a Prediction, with your reasoning, and an Experimental Design (including diagnostic tests) to complete the investigation report.

    Question
    What are the products of the reaction of tin(II) chloride with an ammonium dichromate solution acidified with hydrochloric acid?

Making Connections

29. When aluminum pots are used for cooking, small pits often develop in the metal. Use your knowledge of redox reactions to explain the formation of these pits. Suggest why this might be a slow process.

Extensions

30. Fluoride treatments of children's teeth have been found to significantly reduce tooth decay. When this was first discovered, toothpastes were produced containing tin(II) fluoride.

    Problem
    What is the concentration of tin(II) ions in a solution prepared for research on toothpaste?

    Experimental Design
    An acidified tin(II) solution was titrated with a standardized potassium permanganate solution.

    Evidence
    volume of tin(II) solution = 10.00 mL
    concentration of permanganate solution = 0.0832 mol/L
    average volume of permanganate reacted = 12.4 mL

    Analysis
    (a) Calculate the molar concentration of the tin(II) solution.

DID YOU KNOW?

Aluminum Oxide Clouds

The solid rocket boosters of the space shuttle contain the main reactants ammonium perchlorate and aluminum powder. Ammonium perchlorate is a powerful oxidizing agent and aluminum is a relatively strong reducing agent. Their very exothermic reaction produces finely divided aluminum oxide, which forms the billows of white smoke you see in the photograph.

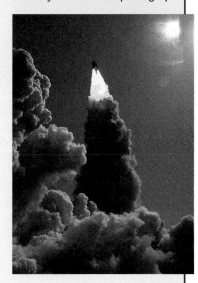

Answer

30. (a) 0.258 mol/L

Section 9.3 Questions

Understanding Concepts

1. What is the key idea used to explain a redox reaction?
2. Write a theoretical definition of oxidation and reduction.
3. Distinguish between oxidation and oxidizing agent.
4. Distinguish between reduction and reducing agent.
5. Write and label two half-reaction equations to describe each of the following reactions:
   (a) $Co_{(s)} + Cu(NO_3)_{2(aq)} \rightarrow Cu_{(s)} + Co(NO_3)_{2(aq)}$
   (b) $Cd_{(s)} + Zn(NO_3)_{2(aq)} \rightarrow Zn_{(s)} + Cd(NO_3)_{2(aq)}$
   (c) $Br_{2(l)} + 2\,KI_{(aq)} \rightarrow I_{2(s)} + 2\,KBr_{(aq)}$

6. Using the redox table in Appendix C9, predict the spontaneity of each of the reactions shown in 5(a) to (c).

7. What is the relative strength of oxidizing and reducing agents for strontium, cerium, nickel, hydrogen, platinum, and their aqueous ions? Use the following information to construct a table of relative strengths of oxidizing and reducing agents.

$$3 Sr_{(s)} + 2 Ce^{3+}_{(aq)} \rightarrow 3 Sr^{2+}_{(aq)} + 2 Ce_{(s)}$$

$$Ni_{(s)} + 2 H^+_{(aq)} \rightarrow Ni^{2+}_{(aq)} + H_{2(g)}$$

$$2 Ce^{3+}_{(aq)} + 3 Ni_{(s)} \rightarrow \text{no evidence of reaction}$$

$$Pt_{(s)} + 2 H^+_{(aq)} \rightarrow \text{no evidence of reaction (assume } Pt^{4+}_{(aq)})$$

8. In the industrial production of iodine, chlorine gas is bubbled into seawater. Using only water and iodide ions in seawater as the possible reactants, predict the most likely redox reaction, including appropriate equations for the half-reactions.

9. The steel of an automobile fender is exposed to acidic rain. (Assume that steel is made mainly of iron.) Predict the most likely redox reactions, including the equations for the relevant half-reactions.

10. A chemical technician prepares several solutions for use in a chemical analysis. Will each of the solutions listed below be stable if stored for a long period of time? Justify your answer.
    (a) acidic tin(II) chloride in an inert glass container
    (b) copper(II) nitrate in a tin can

11. An excess of cobalt metal was left in an aqueous mixture containing silver ions, iron(III) ions, and copper(II) ions for an extended period of time. Write a balanced redox equation for every reaction that occurs.

Applying Inquiry Skills

12. Prepare a redox table of half-reactions showing the relative strengths of oxidizing and reducing agents in **Table 6**.

Table 6 Reactions of Group 13 Elements and Ions

	$Al^{3+}_{(aq)}$	$Tl^+_{(aq)}$	$Ga^{3+}_{(aq)}$	$In^{3+}_{(aq)}$
Al	X	√	√	√
Tl	X	X	X	X
Ga	X	√	X	√
In	X	√	X	X

X no evidence of a redox reaction
√ a spontaneous reaction occurred

Making Connections

13. Ursula Franklin (**Figure 10**) is an internationally recognized scientist in her field. She has many interests and is outspoken on many topics. Briefly describe her pioneering scientific work and outline her views about science and technology and funding for scientific research. What other causes does she support?

 www.science.nelson.com

Figure 10
Ursula Franklin was born in Germany in 1921 and received her Ph.D. in 1948 from the Technical University in Berlin. She continued her studies at the University of Toronto, where she became a professor in 1973. In 1984 she was appointed a University Professor at the University of Toronto, an honour acknowledging that her academic and scientific interests go far beyond a single discipline.

Before 1800, scientists knew that static electricity was produced by the friction created by two moving objects in contact. They discovered ways of storing the charges temporarily, but when the energy was released in the form of an electrical spark, it could not be put to practical use. Practical applications of electricity were developed only after 1800, the year in which Alessandro Volta announced his invention of the **electric cell**.

Volta invented the first electric cell but he got his inspiration from the work, almost 30 years earlier, of the Italian physician Luigi Galvani. Galvani noticed that the muscles in a frog's leg would twitch when a spark hit the leg. Galvani's crucial observation was that two different metals could make the muscle twitch. Unfortunately, Galvani thought his discovery was due to some mysterious "animal electricity." It was Volta who recognized that this effect had nothing to do with animals or muscle tissue, and everything to do with conductors and electrolytes, as you have already observed in the activity at the beginning of this chapter.

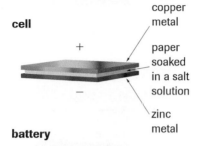

Cells and Batteries

The electric cells Volta invented produced very little electricity. Eventually, he came up with a better design by joining several cells together. A **battery** is a group of two or more electric cells connected to each other, in series, like railway cars in a train. Volta's first battery consisted of several bowls of brine (aqueous sodium chloride) connected by metals that dipped from one bowl into the next (**Figure 1**). This arrangement of metal strips and electrolytes produced a steady flow of electric current. Volta improved the design of this battery by replacing the strips of metal with flat sheets, and replacing the bowls with paper or leather soaked in brine. This produced more electric current for a longer period of time. As shown in **Figure 2**, Volta stacked cells on top of each other to form a battery, known as a voltaic pile. When a loop of wire was attached to the top and bottom of this voltaic pile, a steady electric current flowed. Volta assembled voltaic piles containing more than 100 cells.

Volta's invention was an immediate success because it produced an electric current more simply and more reliably than methods that depended on static charges. It also produced a steady electric current—something no other device could do. The development of this technology led to many advances in physics (for example, the theory and description of current electricity), in chemistry (for example, the discovery of Groups 1 and 2 metals), and in electrical and chemical engineering.

Figure 2
Volta's revised cell design, simpler than the first, consisted of a sandwich of two metals separated by paper soaked in salt water (the electrolyte). A cell consisted of a layer of zinc metal separated from a layer of copper metal by the brine-soaked paper. A large pile of cells could be constructed to give more electrical energy.

electric cell a device that continuously converts chemical energy into electrical energy.

battery a group of two or more electric cells connected in series

Figure 1
A version of Volta's first battery. Each bowl contains two different metals, copper and zinc, in an electrolyte, salt water. A series of bowls forms a series of cells (battery) whose total voltage is the sum of the individual voltages of all cells.

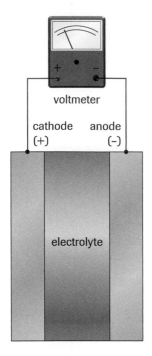

voltmeter

cathode (+)   anode (−)

electrolyte

Figure 3
A cell always contains two electrodes—an anode and a cathode—and an electrolyte. When testing the voltage of a cell or battery, the red (+) lead of the voltmeter is connected to the positive electrode (cathode), and the black (−) lead is connected to the negative electrode (anode).

electrode a solid electrical conductor

electrolyte an aqueous electrical conductor

electric potential difference (voltage) the potential energy difference per unit charge

volt (V) the SI unit for electric potential difference; 1 V = 1 J/C

electric current the rate of flow of charge past a point

ampere (A) the SI unit for electric current; 1 A = 1 C/s

coulomb (C) the SI unit for electric charge

Basic Cell Design and Properties

Each electric cell is composed of two **electrodes** and one **electrolyte** (**Figure 3**). In the cells we buy for home use, the electrolyte is usually a moist paste, containing only enough conducting solution to make the cell function. The electrodes are usually two metals, or graphite and a metal. In some designs, one of the electrodes is the container of the cell. One of the electrodes is marked positive (+) and the other is marked negative (−).

> In an electric cell or battery, the cathode is the positive electrode and the anode is the negative electrode.

According to the theory that electricity is the flow of electrons, electrons move from the anode of a battery through some conducting materials to the cathode. A battery produces electricity only when there is an external conducting path, such as a wire, through which electrons can move. Disconnecting the wire from the battery immediately stops the electric current.

A voltmeter is a device that can be used to measure the energy difference, per unit electric charge, between any two points in an electric circuit. The energy difference per unit charge is called the **electric potential difference** or the **voltage**, and is measured in **volts** (V). For example, the electrons transferred via a 1.5-V cell release only one-sixth as much energy as the electrons from a 9-V battery.

Since the voltage is a ratio of energy to charge, it is not dependent on the size of the cell. You may have noticed that you can buy the same type and brand of 1.5-V cells in a variety of sizes, such as AA, B, C, and D. All are rated at 1.5 V. The larger cells can produce more energy at the same time as transferring more charge, but the ratio of energy to charge is the same as the smaller cells. *The voltage of a cell depends mainly on the chemical composition of the reactants in the cell.*

Electric current, measured by an ammeter in **amperes** (A), is a measure of the rate of flow of charge past a point in an electrical circuit (**Figure 4**). The larger the electric cell of a particular kind, the greater the current that can be produced by the cell. The charge transferred by a cell or battery is measured in **coulombs** (C) and expresses the total charge transferred by the movement of charged particles. The *power* of a cell or battery is the rate at which it produces electrical energy. Power is measured in watts (W), and is calculated as the product of the current and the voltage of the battery. The *energy density*, or *specific energy* of a battery, is a measure of the quantity of energy stored or supplied per unit mass. Energy density may be measured in joules per kilogram (J/kg). **Table 1** summarizes electrical quantities and their units of measurement.

Table 1 Electrical Quantities and SI Units

Quantity	Symbol	Meter	Unit	Unit symbol
charge	q	−	coulomb	C
current	I	ammeter	ampere	A (1 A = 1 C/s)
potential difference	V	voltmeter	volt	V (1 V = 1 J/C)
power	P	−	watt	W (1 W = 1 J/s)
energy density	−	−	joules per kilogram	J/kg

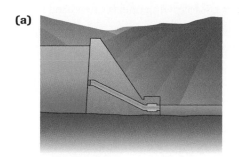

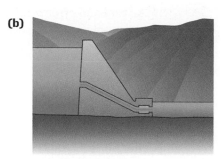

(a)   **(b)**

Figure 4
(a) A dam built across a stream or river stops the flow of water. Each kilogram of water that backs up behind the dam has a certain quantity of potential energy relative to the bottom of the dam. In other words, there is a potential energy difference between a kilogram of water at the top of the dam and a kilogram of water at the bottom of the dam, A voltmeter can be used to measure the height of the "dam" inside a battery, that is, the potential energy difference between a unit number of electrons at the cathode and a unit number of electrons at the anode.
(b) If water is released from behind the dam, it naturally flows from the region of higher potential energy (behind the dam) to a lower potential energy below the dam. Similarly, when the circuit is connected to a cell or battery, the electrons naturally flow because there is a difference in potential energy.

SUMMARY	*Components of an Electric Cell*

- An electric cell must have two electrodes and an electrolyte.
- An electrode is a solid conductor.
- An electrolyte is an aqueous conductor.
- The cathode is the electrode labelled positive.
- The anode is the electrode labelled negative.
- The electron flow is from the anode to the cathode.

▶ **Practice**

Understanding Concepts

1. What are the parts of a simple electric cell?
2. Write an empirical definition of electrode and electrolyte, and a conventional definition of anode and cathode.
3. If a cassette player requires 6 V to operate, how many 1.5-V "dry" cells corrected in series would it need?
4. Differentiate between electric current and voltage.

Making Connections

5. Why do manufacturers of battery-operated devices print a diagram showing the correct orientation of the batteries? (Supply two answers to this question—one from a scientific perspective and one from a technological perspective.)

Technological Problem Solving

The initial development of cells and batteries preceded much of the current scientific understanding of these devices. Cells and batteries existed almost 100 years before the electron was discovered. The study of electric cells is a good illustration of tremendous advances in technology based on very limited scientific knowledge. Technological development or problem solving is similar in some ways to scientific problem solving, but its purpose differs. The purpose of technological problem solving is to find a realistic way around a practical difficulty—to make something work—while the purpose of scientific problem-solving is to describe, explain, and ultimately understand natural and technological phenomena. Technology and science are dependent on each other. Although scientific knowledge can be used to guide the creation of a technology, the technology may create new scientific understanding.

DID YOU KNOW ?

A "Not Quite Dry" Cell
The electrolyte in the "dry cell" is
actually a moist paste. If the cell
were completely dry it would not
work because the ions in the electrolyte must be able to carry the
electric current to complete the circuit. Just enough water is added so
that the ions can move, but not
enough to make the mixture liquid.

primary cell an electric cell that
cannot be recharged

Figure 5
Like a flashlight D cell, the dry cell
on the left has a voltage of 1.5 V.
The 9 V battery on the right is made
up of six 1.5-V dry cells in series.

secondary cell an electric cell that
can be recharged

A systematic trial-and-error process, such as the following one, is often used in technological problem solving (Appendix A3):

- Develop a general design for problem-solving trials; for example, select which
  variables to manipulate and which to control.

- Follow several prediction–procedure–evidence–analysis cycles, manipulating
  and systematically studying one variable at a time.

- Complete an evaluation based on criteria such as efficiency, reliability, cost, and
  simplicity.

This technological problem-solving model was important in the early development of
practical electric cells.

Consumer, Commercial, and Industrial Cells

Since Volta's invention of the electric cell and battery, there have been many advances in
electrochemistry and technology. Invented in 1865, the zinc chloride cell is commonly
referred to as a dry cell because this design was the first to use a sealed container. These
1.5-V dry cells were used to make the first 9-V battery (**Figure 5**). Both the 1.5-V dry cell
and the 9-V battery are simple, reliable, and relatively inexpensive. Other cells, such as
the alkaline dry cell and the mercury cell (**Table 2**), were developed to improve the performance of the original dry cell. One problem with all of these cells is that the chemicals are eventually depleted and irreversible reactions prevent these cells from being
recharged. Cells that cannot be recharged are called **primary cells**. Later, we will discuss
two other types of cells that do not have this disadvantage.

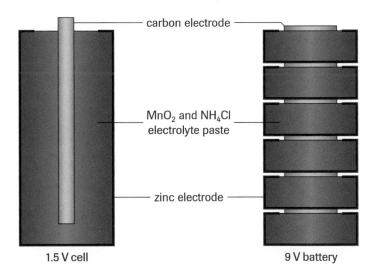

Zinc Chloride Dry Cells

carbon electrode

MnO_2 and NH_4Cl
electrolyte paste

zinc electrode

1.5 V cell

9 V battery

Secondary Cells

Secondary cells can be recharged by using electricity to reverse the chemical reaction
that occurs when electricity is produced by the cell. Secondary cells and batteries include
the nickel-cadmium (Ni-Cad) cell and the lead-acid battery (**Table 2** and **Figure 6**). A
relatively recently developed secondary cell with a unique design is the lithium-ion cell,
or Molicel (**Figure 7,** page 690).

Table 2 Primary, Secondary, and Fuel Cells

Type	Name of Cell	Half-Reactions	Characteristics and Uses
primary cells	dry cell (1.5 V)	$2\,MnO_{2(s)} + 2\,NH_{4(aq)}^{+} + 2\,e^{-} \rightarrow Mn_2O_{3(s)} + 2\,NH_{3(aq)} + H_2O_{(l)}$ $Zn_{(s)} \rightarrow Zn_{(aq)}^{2+} + 2\,e^{-}$	• inexpensive, portable, many sizes • flashlights, radios, many other consumer items
	alkaline dry cell (1.5 V)	$2\,MnO_{2(s)} + H_2O_{(l)} + 2\,e^{-} \rightarrow Mn_2O_{3(s)} + 2\,OH_{(aq)}^{-}$ $Zn_{(s)} + 2\,OH_{(aq)}^{-} \rightarrow ZnO_{(s)} + H_2O_{(l)} + 2\,e^{-}$	• longer shelf life; higher currents for longer periods compared with dry cell • same uses as dry cell
	mercury cell (1.35 V)	$HgO_{(s)} + H_2O_{(l)} + 2\,e^{-} \rightarrow Hg_{(l)} + 2\,OH^{-}$ $Zn_{(s)} + 2\,OH_{(aq)}^{-} \rightarrow ZnO_{(s)} + H_2O_{(l)} + 2\,e^{-}$	• small cell; constant voltage during its active life • hearing aids, watches
secondary cells	Ni-Cad cell (1.25 V)	$2\,NiO(OH)_{(s)} + 2\,H_2O_{(l)} + 2\,e^{-} \rightarrow 2\,Ni(OH)_{2(s)} + 2\,OH^{-}$ $Cd_{(s)} + 2\,OH_{(aq)}^{-} \rightarrow Cd(OH)_{2(s)} + 2\,e^{-}$	• can be completely sealed; lightweight but expensive • all normal dry cell uses, as well as power tools, shavers, portable computers
	lead-acid cell (2.0 V)	$PbO_{2(s)} + 4\,H_{(aq)}^{+} + SO_{4(aq)}^{2-} + 2\,e^{-} \rightarrow PbSO_{4(s)} + 2\,H_2O_{(l)}$ $Pb_{(s)} + SO_{4(aq)}^{2-} \rightarrow PbSO_{4(s)} + 2\,e^{-}$	• very large currents; reliable for many recharges • all vehicles
fuel cells	aluminum-air cell (2 V)	$3\,O_{2(g)} + 6\,H_2O_{(l)} + 12\,e^{-} \rightarrow 12\,OH_{(aq)}^{-}$ $4\,Al_{(s)} \rightarrow 4\,Al_{(aq)}^{3+} + 12\,e^{-}$	• very high energy density; made from readily available aluminum alloys • designed for electric cars
	hydrogen-oxygen cell (1.2 V)	$O_{2(g)} + 2\,H_2O_{(l)} + 4\,e^{-} \rightarrow 4\,OH_{(aq)}^{-}$ $2\,H_{2(g)} + 4\,OH_{(aq)}^{-} \rightarrow 4\,H_2O_{(l)} + 4\,e^{-}$	• lightweight; high efficiency; can be adapted to use hydrogen-rich fuels • vehicles and space shuttle

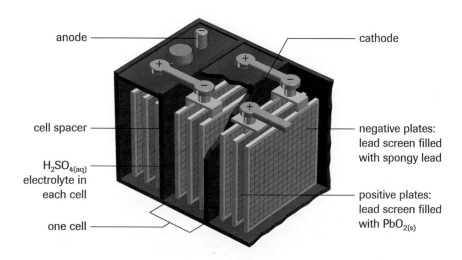

anode

cathode

cell spacer

$H_2SO_{4(aq)}$ electrolyte in each cell

one cell

negative plates: lead screen filled with spongy lead

positive plates: lead screen filled with $PbO_{2(s)}$

Figure 6
The anodes of a lead-acid car battery are composed of spongy lead and the cathodes are composed of lead(IV) oxide on a metal screen. The large electrode surface area is designed to deliver sufficient current to start a car engine.

LEARNING TIP

Discharging and Charging
Discharging a cell or battery is like letting the water spontaneously run out from the higher level behind a dam. Charging (or recharging) is like pumping the water up behind the dam. This is not a spontaneous process and requires energy.

One of the most common and reliable secondary cells is the lead-acid cell in a typical car battery. The discharging of this cell (see lead-acid cell, **Table 2**) produces approximately 2.0 V for the following net equation.

$$\text{Pb}_{(s)} + \text{PbO}_{2(s)} + 2\,\text{H}_2\text{SO}_{4(aq)} \xrightarrow{\text{discharging}} 2\,\text{PbSO}_{4(s)} + 2\,\text{H}_2\text{O}_{(l)}$$

To charge (or recharge) this cell requires the input (from the car's alternator) of at least 2.0 V to force the products to change back to the reactants. The half-reactions for the lead-acid cell listed in **Table 2** both need to be reversed to obtain the following net equation.

$$2\,\text{PbSO}_{4(s)} + 2\,\text{H}_2\text{O}_{(l)} \xrightarrow{\text{charging}} \text{Pb}_{(s)} + \text{PbO}_{2(s)} + 2\,\text{H}_2\text{SO}_{4(aq)}$$

A battery can be recharged if the products are stable with no further reactions occurring and if the products are able to travel through the electrolyte toward the appropriate electrode.

▶ *Practice*

Understanding Concepts

6. What is the relationship between scientific knowledge and technological problem solving?

7. What steps are involved in technological problem solving?

8. Suppose you decided to develop and market an aluminum-can cell. (See Activity 9.4.1.) How and why would you alter the electrolyte?

9. Distinguish between primary and secondary cells, including a common example of each.

10. What are some advantages and disadvantages of the zinc chloride dry cell?

Making Connections

11. Find out how commercially available AA, C, and D cells differ. How do these differences affect their performance?

12. What do the designs of the dry-cell container and the ice-cream cone have in common?

13. Portable electronic devices can be found everywhere. Laptop computers, cellular telephones, mobile radios, cordless phones, portable disc and MP3 players, and digital cameras all require an electric cell.
    (a) What are some of the requirements for cells used in these applications?
    (b) Why are some rechargeable batteries used in various portable devices supposed to be totally "drained" (discharged) before recharging?

 www.science.nelson.com

14. Moli Energy of Maple Ridge, BC was the first company in the world to develop a commercial, rechargeable lithium cell, called a Molicel (**Figure 7**). Research the characteristics and advantages of Molicels compared with other secondary cells.

 www.science.nelson.com

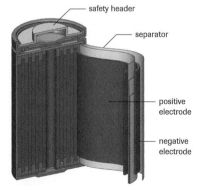

safety header
separator
positive electrode
negative electrode

Figure 7
Invented and manufactured in British Columbia, the Molicel is a high-energy, rechargeable cell in a unique, jellyroll design.

Fuel Cells

A **fuel cell** is a different solution to the problem of the limited life of a primary cell. Fuel cells produce electricity by the reaction of a fuel that is continuously supplied to keep the cell operating. In principle, the fuel cell could be used forever, provided the fuel is continuously supplied. The fuel cell offers several advantages over methods that produce electricity by the combustion of fossil fuels. For example, fuel cells generate electricity more efficiently (**Table 3**), without the production of greenhouse gases or substances that contribute to acid rain. The development of a cost-effective fuel cell is currently the focus of much scientific study and technological research and development.

The first fuel cell was invented accidentally by William Grove in 1839 using platinum electrodes, hydrogen and oxygen as fuels, and sulfuric acid as the electrolyte. Grove was actually studying the reverse process—using electricity to convert water into hydrogen and oxygen. After one experiment, he reconnected the two electrodes, without a power supply attached, and found that a small current was produced spontaneously as hydrogen and oxygen combined to form water. Grove continued to work on this cell but eventually decided it was not a practical device because the electric characteristics, such as voltage, current, and energy capacity, were very low. Although many attempts to improve this cell were made, including those by Nobel Prize winners Fritz Haber and Walther Hermann Nernst, no significant progress was made. Many variables were manipulated, such as different electrodes and electrolytes, but the reaction rates were too low and corrosion of the electrodes was often a serious problem.

Finally, in 1955, Francis Bacon succeeded where many others had failed. He produced a practical hydrogen-oxygen fuel cell using an alkaline electrolyte and electrodes constructed of porous nickel (**Figure 8**). Although the idea had been around for a long time, Bacon's cell was really the first practical fuel cell. NASA quickly adopted the hydrogen-oxygen fuel cell as an electrical power source for space flights, because hydrogen and oxygen are already available for propulsion systems and the product, water, can be purified for drinking. NASA's fuel cell, a modification of the original Bacon cell, is an alkaline cell using potassium hydroxide as the electrolyte (**Table 2**). It produces 12 kW of electricity and operates at 70% efficiency. Unfortunately, NASA's fuel cell is expensive and has a relatively short working life, primarily due to the corrosive electrolyte. As a result, NASA's cell is not economically viable for general or commercial applications.

The Ballard Fuel Cell

A variation of a hydrogen-oxygen fuel cell, also known more simply as the hydrogen fuel cell, was developed for commercial applications by Ballard Power Systems in Vancouver, BC. The Ballard fuel cell employs a proton exchange membrane (PEM) in place of a liquid electrolyte. Normal electric cells use the ions in the liquid electrolyte to transfer electric charge within the cell. In a hydrogen fuel cell, the PEM is a membrane made from a solid proton-conducting polymer that transfers charge within the cell (**Figure 9**). The PEM is simple, robust, eliminates corrosive liquids, and permits a high energy density.

The Ballard fuel cell consists of an anode and a cathode separated by a polymer membrane electrolyte. Hydrogen fuel admitted through a porous anode is then converted into hydrogen ions (protons) and free electrons in the presence of a catalyst at the anode. An external circuit conducts the free electrons and produces the desired electrical current. Water and heat are produced when the protons, after migrating through the polymer membrane to the cathode, react both with oxygen molecules from the air and with the free electrons from the external circuit. Fuel cells can be connected in series (stacked) to increase the voltage and power output (**Figure 10**). For example, an experimental

fuel cell an electric cell that produces electricity by a continually supplied fuel

Table 3 Efficiencies of Different Technologies*

Technology	Efficiency*
Fuel Cells	40–70%
Electric power plants	30–40%
Automobile engines	17–23%
Gasoline lawn mower	about 12%

*Efficiency is the fraction of the maximum available energy that is actually usable.

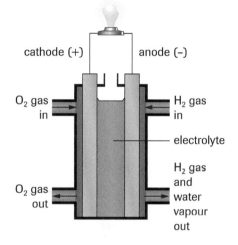

cathode (+)   anode (−)

O_2 gas in H_2 gas in

electrolyte

O_2 gas out H_2 gas and water vapour out

Figure 8
Hydrogen and oxygen gases are continuously pumped into the cell, and each reacts at a different electrode. Unused gases are removed, filtered, and then recycled.

LEARNING *TIP*

Inventions Require Patience and Insight
Notice that the work done to try and improve the original Grove cell was very much a trial-and-error development. Patience is required. It took over 90 years to turn the Grove cell into a technologically useful cell.

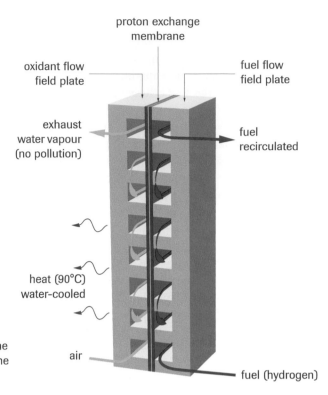

proton exchange
membrane

oxidant flow
field plate

fuel flow
field plate

exhaust
water vapour
(no pollution)

fuel
recirculated

heat (90°C)
water-cooled

air

fuel (hydrogen)

Figure 9
The hydrogen fuel cell has the same design as Volta's original cell but the electrolyte is a conducting solid.

LAB EXERCISE 9.4.1

Characteristics of a Hydrogen Fuel Cell (p. 720)
Compare the electrical characteristics of a hydrogen fuel cell and a dry cell.

Vancouver transit bus uses an electric motor powered by a Ballard fuel cell that is capable of 205 kW (or 250 hp).

Ballard has development agreements with most major car manufacturers to use its cells in future electric cars. The zero-emission engines convert hydrogen, or hydrogen-rich fuels such as natural gas and even methanol, into electricity, producing water and heat as the main byproducts.

Although the Ballard hydrogen fuel cell looks very promising, there are several problems yet to be solved. Cost is a major factor, which may be partially solved by mass production. The fuel is also under debate. If hydrogen gas is used, where does it come from? Electrolysis of water uses a lot of electrical energy and is an expensive way to obtain hydrogen. How would the hydrogen gas be distributed and stored on board the vehicle? There are important safety concerns associated with the handling and storage of hydrogen, which is flammable. Many scientists and engineers believe that the solution is not to use hydrogen gas directly but to use hydrogen-rich fuels. We have a lot of knowledge of reforming hydrocarbons to produce hydrogen. If natural gas or even gasoline were reformed as needed on board the vehicle, then we would have a familiar fuel source and an infrastructure in place to supply this fuel. Not everyone agrees that this is a good solution.

Figure 10
Vehicles require high-power-density fuel cells, i.e., ones that can produce large quantities of energy per second for every kilogram of fuel cell. Fuel cell stacks developed by Ballard have continued to improve. The stack on the left (2001) has a power density about 16 times greater than the one on the rght (1989), yet the size is similar.

Aluminum–Air Cell

Another type of fuel cell that has not received the media attention of hydrogen fuel cells is the metal-air fuel cell, the most common of which is the aluminum-air cell (**Table 2**). This is actually an aluminum-oxygen cell and has been developed for possible use in electric cars. Air is pumped into the cell and oxygen reacts at the cathode while a replaceable mass of aluminum reacts at the anode (**Figure 11**). The fuel is solid aluminum metal and the product, aluminum hydroxide, can be recycled back to aluminum metal. The simple design means that this cell can be assembled in almost any size. The high energy density of these cells results from the fact that three moles of electrons are released from every mole of aluminum, and aluminum is a very lightweight metal. Unlike hydrogen, storage and transportation of a solid fuel do not pose a problem. Estimates from prototypes suggest that the aluminum anode will need replacement every 2500 km in an electric car.

Large-Scale Commercial and Industrial Fuel Cells

The requirements for electrical power fuel cells for large-scale use in businesses and industry are similar with regard to the fuel, but there is less concern about volume, weight, or energy density. However, there is a need for cells with much longer lifetimes. Fuel cells for large-scale commercial and industrial use are almost always co-generation units. This means that they produce electricity as well as heat for space heating. Co-generation means that the overall efficiencies can be as high as 90%. Commercially viable fuel cells today are usually acid electrolyte cells such as the phosphoric acid fuel cell, which can produce 400 MW of power, sufficient for the electrical energy needs of a small city (**Figure 12**). These cells usually use natural gas as a source of hydrogen for the fuel cell and operate at temperatures of 200°C.

▶ Practice

Understanding Concepts

15. Using several perspectives, state some advantages and disadvantages of a fuel cell.

16. List some potential uses of fuel cells.

17. For both the hydrogen-oxygen fuel cell and the aluminum-air fuel cell,
    (a) write the two half-reaction equations (**Table 2**).
    (b) label each equation from (a) as an oxidation or a reduction.
    (c) write the net ionic equation for each cell.

18. List some problems that must be solved before the Ballard cell sees widespread use.

19. Another Ballard-type fuel cell uses methanol as a fuel. What are some advantages of methanol over hydrogen or natural gas?

Making Connections

20. One of the most successful batteries has been the lead-acid car battery.
    (a) Identify the anode, cathode, and electrolyte.
    (b) How are the large currents produced that are necessary to start a car?
    (c) What has been the social impact of this battery?
    (d) What are some possible environmental impacts of this battery?

21. Plastic batteries were the dream of the 1980s, the disappointment of the 1990s, and the subject of the 2000 Nobel Prize for Chemistry. Now it appears that some commercial products will eventually result from the research and development invested in plastic batteries. Briefly describe the electrodes and electrolyte for a plastic battery. How is this battery similar to and different from an ordinary battery? What are some advantages and disadvantages?

 www.science.nelson.com

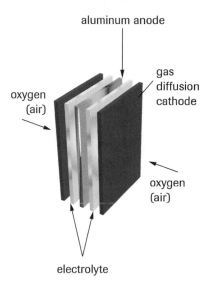

Figure 11
Aluminum Power Inc., based in Toronto, ON, has done extensive development of the aluminum–air solid fuel cell.

Figure 12
The world's first commercial phosphoric acid fuel cell, produced by ONSI/International Fuel Cells. It has been available since 1992 and uses natural gas, waste methane, propane, or hydrogen as fuels. This unit produces 200 kW electricity and 200 kW heat at a total system efficiency of 80%.

Debate: Hydrogen Fuel Cells

No one doubts that internal-combustion vehicles are a major source of pollution and environmental damage and all major automobile manufacturers are racing to develop viable alternatives (e.g., an electric car). Judging from media reports and automobile advertisements, fuel cells are seen as the "pollution-free" alternative to the internal-combustion engine. Specifically, hydrogen fuel cells are being widely promoted as the "green alternative." Often mentioned in the media, the product of a hydrogen fuel cell is pure water. What could be better than that?

(a) In small groups, prepare for a debate on the proposition, "Hydrogen fuel cells are the ideal 'green' solution to the internal-combustion engine."

○ Define the Issue	● Identify Alternatives	● Research
● Analyze the Issue	● Defend the Position	○ Evaluate

In your research, consider:
- where does the hydrogen come from?
- from source to final end product, are hydrogen fuel cells nonpolluting?
- other perspectives, such as economic, social, political, and technological.
- alternatives to the hydrogen fuel cell.

(b) Develop your opinion, defending or opposing the proposition. Brainstorm and research arguments in support of your position.

 GO   www.science.nelson.com

▶ *Section 9.4 Questions*

Understanding Concepts

1. Draw a simple diagram of an electric cell and label: electrodes, electrolyte, cathode, anode, signs for cathode and anode, and direction of electron flow through an external wire.

2. What is the evidence that an electric cell involves a redox reaction?

3. What are the three types of electric cells used in consumer and commercial operations? Briefly describe the main feature of each cell.

4. State two common examples of consumer cells and where they may be used.

5. A silver oxide cell is often used when a miniature cell or battery is required, as in watches, calculators, and cameras. The following half-reaction equations occur in the cell:

$$Ag_2O_{(s)} + H_2O_{(l)} + 2e^- \rightarrow 2Ag_{(s)} + 2OH^-_{(aq)}$$
$$Zn_{(s)} + 2OH^-_{(aq)} \rightarrow Zn(OH)_{2(s)} + 2e^-$$

   (a) In which direction does the electric current flow—silver to zinc or zinc to silver?
   (b) Which is the anode and which is the cathode?
   (c) Write the net redox equation for the discharging of the silver oxide cell.

Making Connections

6. Suppose cells and batteries did not exist. What impact would that have on your life?

7. (a) Why is there a great deal of interest in electric cars?
   (b) Suggest some reasons why we don't use lead-acid batteries as the only power source for electric cars.
   (c) How have advances in hydrogen fuel cells facilitated the development of electric cars?

8. Criteria used to evaluate a battery include its reliability, cost, simplicity of use, safety (leakage), size (volume), shelf life, active life, energy density, power capacity, maintenance, disposal, environmental impact, and ability to be recharged. Gather some information and analyze it to determine what is the best cell or battery for a portable radio, CD, or MP3 player.

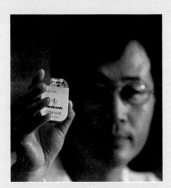

Figure 13
A pacemaker includes the electronics and a built-in battery. The whole unit is only a few centimetres in size and is implanted under the skin near the collarbone.

9. People whose heart occasionally beats too slowly or too quickly often have pacemakers to keep the heart beating regularly (**Figure 15**). Pacemakers use a battery for electric power. What kind of battery is commonly used today? How long does it last? How does the doctor know when the battery is nearing the end of its life and needs to be replaced? Why are rechargeable batteries generally not used?

 GO   www.science.nelson.com

Electric cells were invented about 1800 and were developed to serve practical purposes. They were not explained scientifically until about 100 years after their invention. Their use, however, contributed to scientific understanding of redox reactions and, later, this knowledge helped explain reactions inside the cell itself. Electric cells adapted for scientific study are often called *galvanic cells* (in recognition of Luigi Galvani) or *voltaic cells* (in recognition of Alessandro Volta).

From a scientific perspective, the design of a cell "plays a trick" on oxidizing and reducing agents, resulting in electrons passing through an external circuit rather than directly from one substance to another. You have seen that the individual components of a cell—electrodes and electrolytes—determine electrical characteristics such as voltage and current. Why is this so? What happens in different parts of a cell? To answer these questions, chemists use a cell with a different design, with the parts of the cell separated so they can be studied more easily. This is not a very practical arrangement but it greatly facilitates the study of cells. Each electrode is in contact with an electrolyte, but the electrolytes surrounding each electrode are separated. This is accomplished by a porous boundary, a barrier that separates electrolytes at least over a short time while still permitting ions to move through tiny openings between the two solutions. Two common examples of porous boundaries are the salt bridge and the porous cup, shown in **Figure 1**.

Figure 1
(a) A salt bridge is a U-shaped tube containing an inert (unreactive) aqueous electrolyte such as sodium sulfate. The cotton plug allows ions to move into or out of the ends of the tube when the ends are immersed in electrolytes.

(b) An unglazed porcelain (porous) cup containing one electrolyte sits in a container of a second electrolyte. The two solutions are separated, but ions can move in and out of the cup through the pores in the porcelain.

(a)

electrolyte

electrolyte

electrolyte

ions

ions

cotton plugs

(b)

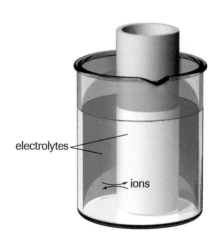

electrolytes

ions

With this design modification, a cell can be split into two parts connected by a porous boundary. Each part, called a **half-cell**, consists of one electrode and one electrolyte. For example, the copper-zinc cell shown in **Figure 2** has two half-cells, copper metal in a solution of copper ions, and zinc metal in a solution of zinc ions. It can be represented using the following abbreviated ("shorthand") notation, called a cell notation:

$Cu_{(s)} \mid Cu(NO_3)_{2(aq)} \parallel Zn(NO_3)_{2(aq)} \mid Zn_{(s)}$

In this notation, a single line ($\mid$) indicates a phase boundary such as the interface of an electrode and an electrolyte in a half-cell. A double line ($\parallel$) represents a physical boundary

half-cell an electrode and an electrolyte forming half of a complete cell

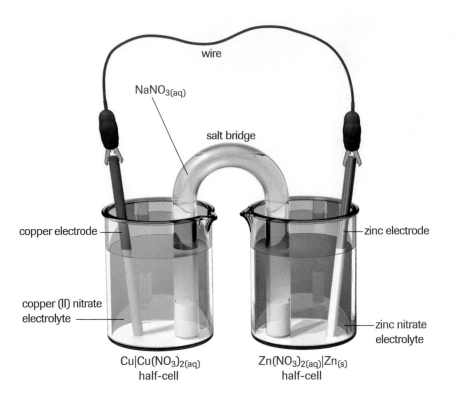

Figure 2
The essential parts of a cell are two electrodes and an electrolyte. In this design each electrode is in its own electrolyte, forming a half-cell. The two half-cells are connected by a salt bridge (containing $NaNO_{3(aq)}$) and by an external conductor to make a complete circuit.

wire

$NaNO_{3(aq)}$

salt bridge

copper electrode

zinc electrode

copper (II) nitrate electrolyte

zinc nitrate electrolyte

$Cu|Cu(NO_3)_{2(aq)}$
half-cell

$Zn(NO_3)_{2(aq)}|Zn_{(s)}$
half-cell

galvanic cell an arrangement of two half-cells that can produce electricity spontaneously

 **ACTIVITY 9.5.1**

Galvanic Cell Design (p. 721)
Examine the design and operation of a galvanic cell.

such as a porous boundary between half-cells. A **galvanic cell** is an arrangement of two half-cells that can produce electricity spontaneously. Cells such as the one in **Figure 2** are especially suitable for scientific study.

A Theoretical Description of a Galvanic Cell

Observation of a galvanic cell as it operates provides evidence that explains what is happening inside the cell. For example, the study of a silver-copper cell in Activity 9.5.1 provides the evidence listed in **Table 1**. A theoretical interpretation of each point is included in the table and is shown in **Figure 3**.

Table 1 Evidence and Interpretations of the Silver-Copper Cell

Evidence	Interpretation
The copper electrode decreases in mass and the intensity of the blue colour of the electrolyte increases.	Oxidation of copper metal is occurring: $$Cu_{(s)} \rightarrow Cu^{2+}_{(aq)} + 2\,e^-$$ blue
The silver electrode increases in mass as long, silver-coloured crystals grow.	Reduction of silver ions is occurring: $$Ag^+_{(aq)} + e^- \rightarrow Ag_{(s)}$$
A blue colour slowly moves up the U-tube from the copper half-cell to the silver half-cell and the solution remains electrically neutral.	Copper(II) ions move toward the cathode. Negative ions (anions) move toward the anode.
A voltmeter indicates that the silver electrode is the cathode (positive) and the copper electrode is the anode (negative).	Electrons move from the copper electrode to the silver electrode.
An ammeter shows that the electric current flows between the copper electrode and the silver electrode.	Electrons leave the copper half-cell and enter the silver half-cell.

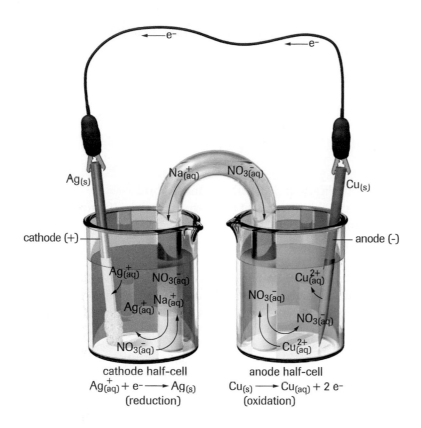

cathode half-cell
$Ag^+_{(aq)} + e^- \longrightarrow Ag_{(s)}$
(reduction)

anode half-cell
$Cu_{(s)} \longrightarrow Cu_{(aq)} + 2\ e^-$
(oxidation)

Figure 3
A theoretical interpretation of the silver-copper cell

According to the electron transfer theory and the concept of relative strengths of oxidizing and reducing agents, silver ions are the strongest oxidizing agents in the cell; they undergo a reduction half-reaction at the cathode. The strongest oxidizing agent in the cell always undergoes a reduction at the cathode. Copper atoms, which are the strongest reducing agents in the cell, give up electrons in an oxidation half-reaction and enter the solution at the anode. The strongest reducing agent in the cell always undergoes an oxidation at the anode. Therefore, the **cathode** is the electrode where reduction occurs and the **anode** is the electrode where oxidation occurs.

cathode the electrode where reduction occurs

anode the electrode where oxidation occurs

- The strongest oxidizing agent present in the cell always undergoes a reduction at the cathode.
- The strongest reducing agent present in the cell always undergoes an oxidation at the anode.

Electrons released by the oxidation of copper atoms at the anode travel through the connecting wire to the silver cathode. The direction of electron flow can be explained in terms of competition for electrons. According to the table of relative strengths of oxidizing and reducing agents in Appendix C11, silver ions are stronger oxidizing agents than copper(II) ions. Silver ions win the tug of war for the electrons available from the conducting wire.

To write the net equation for the silver-copper galvanic cell, identify the strongest oxidizing and reducing agents present in the mixture. (This is the same procedure you followed when predicting redox reactions in Section 9.3.) Then follow the same procedure for predicting half-reactions in which the two materials are in contact with each other.

$$\begin{array}{cc} \textbf{SOA} & \textbf{OA} \\ Ag_{(s)} \mid Ag^+_{(aq)} \parallel Cu^{2+}_{(aq)} \mid Cu_{(s)} \\ \textbf{RA} & \textbf{SRA} \end{array}$$

reduction at the cathode	$2\,[Ag^+_{(aq)} + e^- \rightarrow Ag_{(s)}]$
oxidation at the anode	$Cu_{(s)} \rightarrow Cu^{2+}_{(aq)} + 2\,e^-$
net	$Cu_{(s)} + 2\,Ag^+_{(aq)} \rightarrow Cu^{2+}_{(aq)} + 2\,Ag_{(s)}$

The electrical neutrality in the half-cells and the salt bridge can be explained in terms of the half-reactions and the movement of electrons and ions (**Figure 4**). If cations did not move to the cathode, the removal of silver ions from the solution near the cathode would create a net negative charge around the cathode and the buildup of negative charge would prevent electrons from being transferred. Migration of cations toward the cathode solution ensures that electrical neutrality is maintained. Likewise, the formation of copper(II) ions at the anode would create a net positive charge, but this is balanced by the movement of negative ions to the anode compartment through the salt bridge or porous cup. The salt bridge permits the redistribution of charge that is needed to maintain electrical neutrality in the electrolyte solutions of the half-cells.

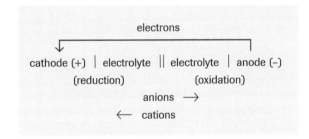

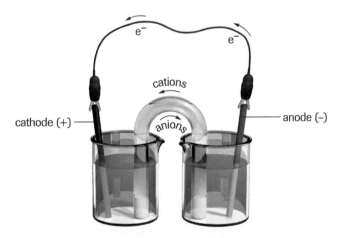

Figure 4
In any operating cell, the electrical circuit is completed by the electron flow in the external part (wires) of the cell and the ion flow in the internal part (solutions) of the cell.

Galvanic Cells with Inert Electrodes

For cells containing metals and metal ions, the electrodes are usually the metals, and half-reactions take place on the surface of the metals. What happens if an oxidizing or a reducing agent other than these is used? For example, an acidic dichromate solution is a strong oxidizing agent that reacts spontaneously with copper metal. To construct this cell you can use a copper half-cell, as in **Figure 5**, but an electrode is required for the dichromate half-cell. You cannot use solid sodium dichromate as an electrode because solid ionic compounds do not conduct electricity and solid sodium dichromate would also dissolve in the solution. You need a solid conductor that will not react in the cell or interfere with the desired cell reaction. In other words, you need an unreactive or **inert electrode**. Inert electrodes provide a location to connect a wire and a surface on which a half-reaction can occur. A carbon (graphite) rod (**Figure 5**) or platinum metal foil are two inert electrodes that are commonly used.

inert electrode a solid conductor that will not react with any substances present in a cell (usually carbon or platinum)

Example

(a) Write equations for the half-reactions and the overall reaction that occur in the following cell:

$$C_{(s)} \mid Cr_2O_7{}^{2-}{}_{(aq)}, H^+{}_{(aq)} \parallel Cu^{2+}{}_{(aq)} \mid Cu_{(s)}$$

(b) Draw a diagram of the cell, labelling electrodes, electrolytes, the direction of electron flow, and the direction of ion movement.

Solution

$$\overset{\text{SOA OA}}{\overset{\displaystyle\frown}{\quad}} \quad \overset{\text{OA}}{\quad}$$

$$C_{(s)} \mid Cr_2O_7{}^{2-}{}_{(aq)}, H^+{}_{(aq)} \parallel Cu^{2+}{}_{(aq)} \mid Cu_{(s)}$$

$$\underset{\text{SRA}}{\underset{\displaystyle\smile}{\quad}}$$

cathode	$Cr_2O_7{}^{2-}{}_{(aq)} + 14\ H^+{}_{(aq)} + 6\ e^- \rightarrow 2\ Cr^{3+}{}_{(aq)} + 7\ H_2O_{(l)}$
anode	$3\ [\ Cu_{(s)} \rightarrow Cu^{2+}{}_{(aq)} + 2\ e^-\ ]$

net $Cr_2O_7{}^{2-}{}_{(aq)} + 14\ H^+{}_{(aq)} + 3\ Cu_{(s)} \rightarrow 3\ Cu^{2+}{}_{(aq)} + 2\ Cr^{3+}{}_{(aq)} + 7\ H_2O_{(l)}$

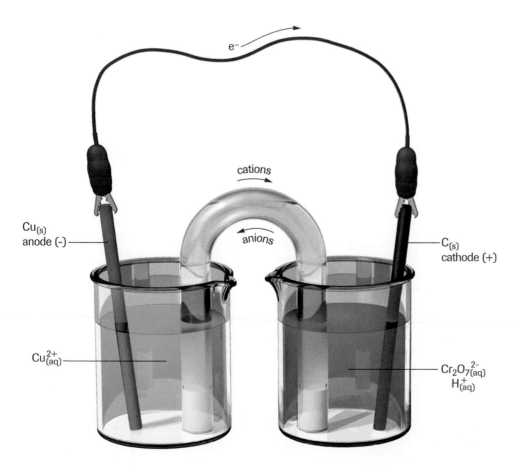

Cu$_{(s)}$
anode (−)

Cu$^{2+}_{(aq)}$

cations

anions

e$^-$

C$_{(s)}$
cathode (+)

Cr$_2$O$_7{}^{2-}_{(aq)}$
H$^+_{(aq)}$

LEARNING TIP

Cell Names
There are a variety of names used for cells based upon spontaneous redox reactions—electric, voltaic, galvanic, and electrochemical. In this book, electric cell is used for consumer cells and galvanic cell for scientific research cells.

Figure 5
The copper electrode decreases in mass, and the blue colour of the electrolyte increases, indicating oxidation at the anode. The carbon electrode remains unchanged, but the orange colour of the dichromate solution becomes less intense and greenish yellow in colour, evidence that reduction is occurring in this half cell.

Galvanic Cells

- A galvanic cell consists of two half-cells separated by a porous boundary with solid electrodes connected by an external circuit.
- The cathode is the positive electrode. Reduction of the strongest oxidizing agent present in the cell occurs at the cathode.
- The anode is the negative electrode. Oxidation of the strongest reducing agent present in the cell occurs at the anode.
- Electrons travel in the external circuit from the anode to the cathode.
- Internally, anions move toward the anode and cations move toward the cathode as the cell operates. The solution remains electrically neutral.

▶ *Practice*

Understanding Concepts

1. Write an empirical description of each of the following terms: galvanic cell, half-cell, porous boundary, and inert electrode.

2. Write a theoretical definition of a cathode and an anode.

3. Indicate whether the following processes occur at the cathode or at the anode of a galvanic cell.
   (a) reduction half-reaction
   (b) oxidation half-reaction
   (c) reaction of the strongest reducing agent
   (d) reaction of the strongest oxidizing agent

4. When is an inert electrode used?

5. What are the characteristics of the solution in a salt bridge? Provide an example.

6. For each of the following cells, use the given cell notation to identify the strongest oxidizing and reducing agents. Write chemical equations to represent the cathode, anode, and net cell reactions. Draw a diagram of each cell, labelling the electrodes, electrolytes, direction of electron flow, and direction of ion movement.
   (a) $Ag_{(s)} \mid Ag^+_{(aq)} \parallel Zn^{2+}_{(aq)} \mid Zn_{(s)}$
   (b) $Pt_{(s)} \mid Na^+_{(aq)}, Cl^-_{(aq)}, O_{2(g)}, H_2O_{(l)} \parallel Al^{3+}_{(aq)} \mid Al_{(s)}$

7. Ions move through a porous boundary between the two half-cells of a voltaic cell.
   (a) Why do the ions move? Take your answer and convert it into another "why" question. Now answer this question.
   (b) In what direction do the cations and anions move?

8. Draw and label a diagram for a galvanic cell constructed from some (not all) of the following materials:

strip of cadmium metal	voltmeter
strip of nickel metal	connecting wires
solid cadmium sulfate	glass U-tube
solid nickel(II) sulfate	cotton
solid potassium sulfate	various beakers
distilled water	porous porcelain cup

9. Redesign the galvanic cell in question 8 by changing at least one electrode and one electrolyte. The net reaction should remain the same for the redesigned cell.

Standard Cells and Cell Potentials

The investigations and activities you have completed show that the design and composition of a cell affect its operation. To make comparisons and scientific study easier, chemists specify the composition of a cell and the conditions under which the cell operates. A **standard cell** is a galvanic cell in which each half-cell contains all entities shown in the half-reaction equation at SATP conditions, with a concentration of 1.0 mol/L for the aqueous solutions. If a metal is not part of a half-cell, then an inert electrode is used to construct the standard cell. For example, for a standard dichromate-zinc cell, the cell description is

$$C_{(s)} \mid Cr_2O_{7(aq)}^{2-}, H_{(aq)}^+, Cr_{(aq)}^{3+} \parallel Zn_{(aq)}^{2+} \mid Zn_{(s)} \quad \text{at SATP}$$
$$\qquad \text{1.0 mol/L} \qquad\qquad\qquad \text{1.0 mol/L}$$

The **standard cell potential** $\Delta E°$ is the maximum electric potential difference (voltage) of the cell operating under standard conditions; $\Delta E°$ represents the energy difference (per unit of charge) between the cathode and the anode. The degree sign (°) indicates that standard 1.0 mol/L and SATP conditions apply. Based on the idea of competition for electrons, a **standard reduction potential** $E_r°$ represents the ability of a standard half-cell to attract electrons, thus undergoing a reduction. The half-cell with the greater attraction for electrons—that is, the one with the more positive reduction potential—gains electrons from the half-cell with the lower reduction potential. The standard cell potential is the difference between the reduction potentials of the two standard half-cells.

$$\Delta E° = E_r° - E_r°$$
$$\text{cell} \qquad \text{cathode} \qquad \text{anode}$$

It is impossible to determine experimentally the reduction potential of a single half-cell because electron transfer requires both an oxidizing agent and a reducing agent. Note that a voltmeter can only measure a potential difference, $\Delta E°$. In order to assign values for standard reduction potentials, we measure the "reducing" strength of all possible half-cells relative to an accepted, standard half-cell. The half-cell used for this purpose is the standard hydrogen half-cell. A half-cell such as this, that is chosen as a reference and arbitrarily assigned an electrode potential of *exactly zero volts*, is called a **reference half-cell**.

Standard Hydrogen Half-Cell

The standard hydrogen half-cell (**Figure 6**) consists of an inert platinum electrode immersed in a 1.00 mol/L solution of hydrogen ions, with hydrogen gas at a pressure of 100 kPa bubbling over the electrode. The pressure and temperature of the cell are kept at SATP conditions. Standard reduction potentials for all other half-cells are measured relative to that of the standard hydrogen half-cell. The reduction potential of the hydrogen ion reduction half-reaction is defined to be exactly zero volts.

$$2\,H_{(aq)}^+ + 2\,e^- \rightleftharpoons H_{2(g)} \qquad E_r° = 0.00\ V$$

As a result, a numerical value can be assigned to the reduction potential associated with every other other reaction. When a half-reaction, written as a reduction, has a positive reduction potential, we conclude that the oxidizing agent in that half-reaction is a stronger oxidizing agent than hydrogen ions. Therefore, if a half-cell based on that half reaction were connected to a standard hydrogen half-cell, electrons would be drawn away from the standard hydrogen electrode. A negative reduction potential means that the oxidizing agent in the half-cell connected to the hydrogen half-cell attracts electrons less strongly

standard cell a galvanic cell in which each half-cell contains all entities shown in the half-reaction equation at SATP conditions, with a concentration of 1.0 mol/L for the aqueous entities

standard cell potential $\Delta E°$ is the maximum electric potential difference (voltage) of a cell operating under standard conditions

standard reduction potential $E_r°$ represents the ability of a standard half-cell to attract electrons in a reduction half-reaction

reference half-cell a half-cell arbitrarily assigned an electrode potential of exactly zero volts; the standard hydrogen half-cell

---

LEARNING TIP

Think of the standard cell potential $\Delta E°$ as representing the *difference* in ability of two half-cells to gain electrons. This potential difference can only be measured accurately if no current is allowed to flow. A good-quality voltmeter has a large internal resistance to prevent current flow.

IUPAC now recommends the use of SATP as standard conditions for reduction potentials. The change from 101.325 kPa to 100 kPa will not noticeably affect measuring and reporting values previously determined at 101.325 kPa.

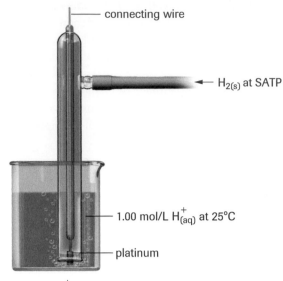

connecting wire

$H_{2(s)}$ at SATP

1.00 mol/L $H^+_{(aq)}$ at 25°C

platinum

$$Pt_{(s)} \mid H_{2(g)}, H^+_{(aq)} \quad E^\circ_r = 0.00 \text{ V}$$

Figure 6
The standard hydrogen half-cell is used internationally as the reference half-cell in electrochemical research.

than hydrogen ions do. The choice of the standard hydrogen half-cell as a reference is the accepted convention. If a different half-cell had been chosen as the reference, individual reduction potentials would be different, but their relative values would remain the same.

Measuring Standard Reduction Potentials

The standard reduction potential of a half-cell can be measured by constructing a standard cell using a hydrogen reference half-cell and the half-cell whose reduction potential you want to measure. There are two things you need to know—the voltage and the direction of the current. The magnitude of the voltage determines the numerical value of the half-cell potential and the direction of the current determines the sign of the half-cell potential. The cell potential is measured with a voltmeter, which will also show the direction that the electrons tend to flow from the sign of the voltage. If ΔE° is positive, then the positive terminal on the voltmeter is connected to the cathode and the oxidizing agent at the cathode is stronger than hydrogen ions.

The cell shown in **Figure 7** can be represented as follows:

$$\underset{\text{cathode}}{Cu_{(s)} \mid Cu^{2+}_{(aq)}} \| \underset{\text{anode}}{H_{2(g)}, H^+_{(aq)} \mid Pt_{(s)}} \qquad \Delta E^\circ = +0.34 \text{ V}$$

The voltmeter shows that the copper electrode is the cathode and is 0.34 V higher in potential than the platinum anode (**Figure 7**). If the voltmeter is replaced by a connecting wire so that the current is allowed to flow, the blue colour of the copper(II) ion disappears and the pH of the hydrogen half-cell decreases as more hydrogen ions are produced and the solution becomes more acidic. Based on this evidence, copper(II) ions are being reduced to copper metal and hydrogen molecules are being oxidized to hydrogen ions. Since this redox reaction is spontaneous, copper(II) ions are stronger oxidizing agents than hydrogen ions.

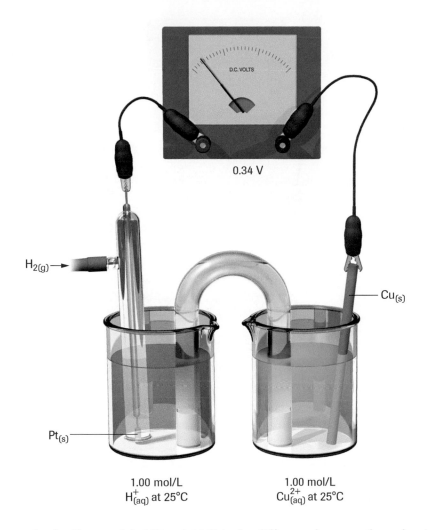

1.00 mol/L
$H^+_{(aq)}$ at 25°C

1.00 mol/L
$Cu^{2+}_{(aq)}$ at 25°C

Figure 7
A copper-hydrogen standard cell.

The standard cell potential, $\Delta E° = 0.34$ V, is the difference between the reduction potentials of these two half-cells;

cathode	$Cu^{2+}_{(aq)} + 2\,e^- \rightarrow Cu_{(s)}$
anode	$H_{2(g)} \rightarrow 2\,H^+_{(aq)} + 2\,e^-$
net	$Cu^{2+}_{(aq)} + H_{2(g)} \rightarrow Cu_{(s)} + 2\,H^+_{(aq)}$

$\Delta E° = 0.34$ V $-$ 0.00 V

$\Delta E° = +0.34$ V

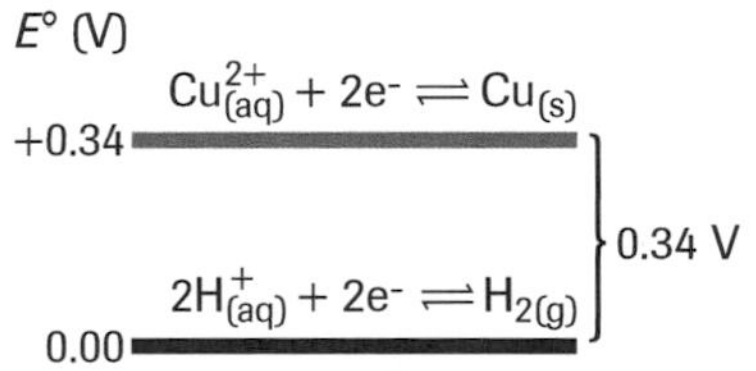

Suppose a standard aluminum half-cell is set up with a standard hydrogen half-cell (**Figure 8**).

$Pt_{(s)} \mid H^+_{(aq)}\ H_{2(g)} \parallel Al^{3+}_{(aq)} \mid Al_{(s)}$ $\qquad \Delta E° = +1.66$ V

cathode $\qquad\qquad\qquad$ anode

$E°$ (V)

$Cu^{2+}_{(aq)} + 2e^- \rightleftharpoons Cu_{(s)}$

$+0.34$

$\left. \begin{array}{c} \\ \\ \end{array} \right\}$ 0.34 V

$2H^+_{(aq)} + 2e^- \rightleftharpoons H_{2(g)}$

0.00

Figure 8
As you already know from the redox table (Appendix C11), copper(II) ions are stronger oxidizing agents than hydrogen ions. The cell potential provides a quantitative measurement of how much stronger

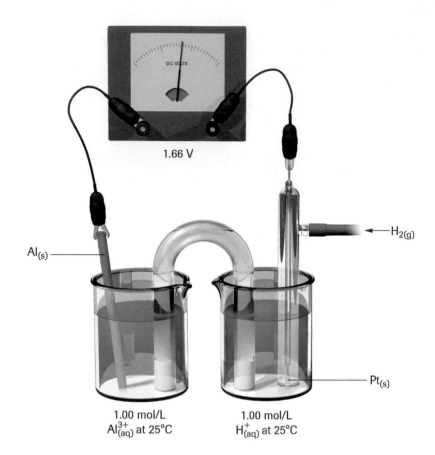

1.66 V

Al$_{(s)}$

H$_{2(g)}$

Pt$_{(s)}$

1.00 mol/L
Al$^{3+}_{(aq)}$ at 25°C

1.00 mol/L
H$^+_{(aq)}$ at 25°C

Figure 9
An aluminum-hydrogen standard cell

According to the voltmeter, the platinum electrode is the cathode and the aluminum electrode is the anode. This indicates that hydrogen ions are stronger oxidizing agents than aluminum ions, by 1.66 V. Since the reduction potential of hydrogen ions is defined as 0.00 V, the reduction potential of the aluminum ions must be 1.66 V below that of hydrogen, or −1.66 V (**Figure 10**).

$$2 \, H^+_{(aq)} + 2 \, e^- \rightleftharpoons H_{2(g)} \qquad E^\circ_r = 0.00 \text{ V}$$
$$Al^{3+}_{(aq)} + 3 \, e^- \rightleftharpoons Al_{(s)} \qquad E^\circ_r = -1.66 \text{ V}$$

The standard cell potential, $\Delta E^\circ = 1.66$ V, is the difference between the reduction potentials of these two half-cells. To obtain the net or overall cell reaction, add the reduction and oxidation half-reactions, but remember to balance and cancel the electrons.

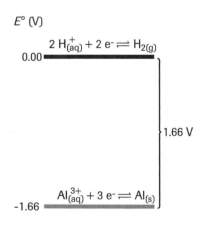

E° (V)

$2 \, H^+_{(aq)} + 2 \, e^- \rightleftharpoons H_{2(g)}$

0.00

1.66 V

$Al^{3+}_{(aq)} + 3 \, e^- \rightleftharpoons Al_{(s)}$

−1.66

cathode	$3 \, [2 \, H^+_{(aq)} + 2 \, e^- \rightarrow H_{2(g)}]$
anode	$2 \, [Al_{(s)} \rightarrow Al^{3+}_{(aq)} + 3 \, e^-]$
net	$6 \, H^+_{(aq)} + 2 \, Al_{(s)} \rightarrow 3 \, H_{2(g)} + 2 \, Al^{3+}_{(aq)}$

$$\Delta E^\circ = 0.00 \text{ V} - (-1.66 \text{ V}) = 1.66 \text{ V}$$

Figure 10
On a redox table, hydrogen ions are stronger oxidizing agents than aluminum ions. The cell potential tells us the hydrogen ions are 1.66 V above aluminum ions.

Notice that the half-reaction equations were multiplied by appropriate factors to balance the electrons, but the *reduction potentials are not altered by the factors used to balance the electrons*. Electric potential represents energy per coulomb of charge (1 V = 1 J/C). Multiplying the aluminum half-reaction by a factor of 2 doubles both the energy and the charge transferred, so that the ratio of energy (J) to charge (C), that is the voltage, is unaffected.

In both of these examples, the strongest oxidizing agent reacts at the cathode and the strongest reducing agent reacts at the anode. The measured cell potential is the difference between the reduction potentials at the cathode and at the anode.

> A positive cell potential ($\Delta E > 0$) indicates that the net reaction is spontaneous—a requirement for all galvanic cells.

In **Figure 11**, the results from the copper-hydrogen and aluminum-hydrogen standard cells are combined. This process of measuring standard cell potentials can quickly be extended to more and more oxidizing agents. Notice that this process, although started with the hydrogen reference cell, does not require that it be used for all cell measurements. For example, knowing that the reduction potential of copper(II) ions is 0.34 V, we can now set up many cells that include a standard copper half-cell. A more extensive list of reduction potentials is found in the table of relative strengths of oxidizing and reducing agents in Appendix C11.

Using the table in Appendix C11, you can predict the reaction that occurs spontaneously in any galvanic cell operating under standard conditions. The standard cell potential is predicted as follows:

$$\Delta E^\circ = \underset{\text{cathode}}{E_r^\circ} - \underset{\text{anode}}{E_r^\circ}$$

This order of subtraction is necessary to confirm the spontaneity from the sign of ΔE°. If ΔE° is positive, the reaction is spontaneous.

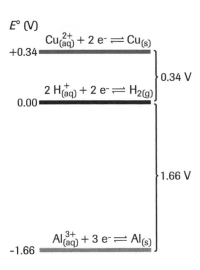

Figure 11
Measurements of standard cell potentials show that the reduction potential of $Cu^{2+}_{(aq)}$ is 0.34 V greater than that of $H^+_{(aq)}$, which is 1.66 V greater than that of $Al^{3+}_{(aq)}$. If you set up a standard cell using copper and aluminum, what would be the cell potential, ΔE? (Answer: 2.00 V)

SUMMARY   *Rules for Analyzing Standard Cells*

Given the contents of the cell, you will need to do one or more of the following steps:

- The cathode is the electrode where the strongest oxidizing agent present in the cell reacts, i.e., the oxidizing agent in the cell that is closest to the top on the left side of the redox table.
  If required, copy the reduction half-reaction for the strongest oxidizing agent and its reduction potential.
- The anode is the electrode where the strongest reducing agent present in the cell reacts, i.e., the reducing agent in the cell that is closest to the bottom on the right side of the redox table.
  If required, copy the oxidation half-reaction (reverse the half-reaction by reading from right to left) for the strongest reducing agent and the reduction potential listed in the table.
- Balance the electrons for the two half-reaction equations (but do not change the E_rs) and add the half-reaction equations to obtain the overall or net cell reaction.
- Calculate the standard cell potential, ΔE°. 🧪▐

LEARNING *TIP*

To ensure a correct interpretation, always write the cathode half-reaction of the SOA first. This will help you to remember to subtract the reduction potentials in the correct order.

🧪 **INVESTIGATION 9.5.1**

Investigating Galvanic Cells (p. 722)
Construct galvanic cells and evaluate cell potentials.

Example 1

A standard dichromate-lead cell is constructed. Write the cell notation, label the electrodes, and calculate the standard cell potential.

Solution

$$C_{(s)} \mid Cr_2O_{7(aq)}^{2-}, H_{(aq)}^+, Cr_{(aq)}^{3+}, \parallel Pb_{(aq)}^{2+}, \mid Pb_{(s)}$$

cathode   anode

$$\Delta E° = 1.23\text{ V} - (-0.13\text{ V}) = 1.36\text{ V}$$

Example 2

A standard copper-scandium cell is constructed and the cell potential measured. The voltmeter indicates that copper metal is the cathode.

$$Cu_{(s)} \mid Cu_{(aq)}^{2+}, \parallel Sc_{(aq)}^{3+}, \mid Sc_{(s)} \qquad \Delta E° = 2.36\text{ V}$$

Write and label the half-reaction and net equations and calculate the standard reduction potential of the scandium ion.

Solution

cathode	$3\,[Cu_{(aq)}^{2+} + 2\,e^- \rightarrow Cu_{(s)}]$	$E_r° = 0.34\text{ V}$
anode	$2\,[Sc_{(s)} \rightarrow Sc_{(aq)}^{3+} + 3\,e^-]$	$E_r° = ?$
net	$3\,Cu_{(aq)}^{2+} + 2\,Sc_{(s)} \rightarrow 3\,Cu_{(s)} + 2\,Sc_{(aq)}^{3+}$	$\Delta E° = 2.36\text{ V}$

$$\Delta E° = E°_{Cu^{2+}} - E°_{Sc^{3+}}$$
$$2.36\text{ V} = 0.34\text{ V} - E°_{Sc^{3+}}$$
$$E°_{Sc^{3+}} = -2.02\text{ V}$$

Cell Potentials Under Nonstandard Conditions

The electric potential difference or voltage of a cell decreases slowly as the cell operates. Simultaneously, colour changes, and precipitate formation occurs. If the cell is left for a very long time, the voltage would eventually become zero and no further changes would be observed in the cell. When people refer to a "dead" cell or battery, this is often what is meant.

The electric potential difference of a cell is a measure of the tendency for electrons to flow. Ideally, during a measurement of the cell potential, a voltmeter should not allow any electrons to flow. If electrons flow, oxidation and reduction reactions occur which, in turn, change the concentrations from the standard 1.0 mol/L value. The value that is measured by a voltmeter represents an electric potential or stored energy just as the water behind a lock in a canal has gravitational potential energy (**Figure 12(a)**). Connecting

Figure 12

(a) The water behind the gates in a lock has a certain potential energy, ΔE, relative to the bottom of the closed outlet.

(b) When the outlet is opened, water spontaneously flows to the lower level on the other side of the gates. Potential energy, ΔE, is converted to kinetic energy of the flowing water. The water flowing through the outlet is analogous to electron flow.

(c) The flow of water ceases when the levels on both sides of the gates become equal. The gates open, and the ship can then exit to the next lock.

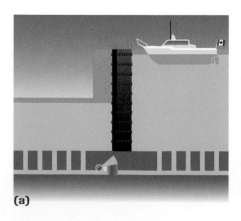

(a)

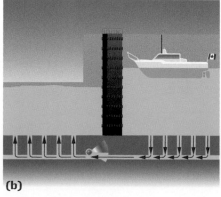

(b)

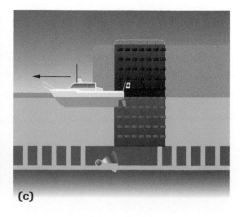

(c)

the electrodes of a cell in a circuit allows the electrons to flow from the anode to the cathode. This is analogous to opening the valve or sluice and allowing the water to flow from behind the gates to a lower point in front of the gates (**Figure 12(b)**). In both cases, stored potential energy is converted to kinetic energy of electrons or water. If the water available behind a lock is allowed to flow out, then eventually no more water will flow. The level (potential energy) of the water on the two sides of the gate is equalized. An equilibrium is reached with no potential energy difference (**Figure 12(c)**). A similar situation occurs with an operating cell. If electrons are allowed to flow, eventually an equilibrium will be reached when the flow ceases. The rate of the forward reaction, which predominates initially, decreases as the rate of the reverse reaction increases, until the two rates become equal. This is the equilibrium condition and no net flow of electrons will occur. At this time, the electric potential difference as measured by a voltmeter becomes zero.

Standard cell potentials can be determined readily using standard reduction potentials. If the concentrations are not standard, the cell potential can be predicted using a relationship discovered by Walther Hermann Nernst. A simplified version of the Nernst equation is shown below:

$$\Delta E = \Delta E° - \frac{0.0592\,\text{V}}{n} \log Q \qquad \text{(at 25°C only)}$$

where
ΔE is the cell potential at 25°C and non-standard concentrations
$\Delta E°$ is the cell potential at 25°C and standard concentrations (1 mol/L)
n is the amount, in moles, of electrons transferred according to the cell reaction
Q is the reaction quotient

SUMMARY   *Standard Cell Potential*

- A standard cell is one in which all entities shown in the half-reaction equation are present and at SATP. The concentration of aqueous entities is 1.0 mol/L.

- The standard cell potential $\Delta E°$ is the maximum electric potential difference between the cathode and anode of a galvanic cell at standard conditions.

$$\Delta E°_{cell} = E°_{r_{cathode}} - E°_{r_{anode}}$$

- A positive standard cell potential ($\Delta E° > 0$) indicates that the overall cell reaction is spontaneous.

- The standard reduction potential $E°_r$ represents the ability of a standard half-cell to attract electrons, relative to the reference half-cell.

- The reference half-cell is $Pt_{(s)} | H_{2(g)}, H^+_{(aq)}$, which has, by definition, a standard reduction potential of exactly zero volts.

Understanding Concepts

10. For each of the following cells, write the equations for the reactions occurring at the cathode and at the anode, and an equation for the overall or net cell reaction. Calculate the standard cell potential. (Use the redox table in Appendix C11.)

    (a) $Sn_{(s)} \mid Sn^{2+}_{(aq)} \parallel Cr^{2+}_{(aq)} \mid Cr_{(s)}$

    (b) $C_{(s)} \mid SO^{2-}_{4(aq)}, H^{+}_{(aq)}, H_2SO_{3(aq)} \parallel Co^{2+}_{(aq)} \mid Co_{(s)}$

    (c) $Pt_{(s)} \mid OH^{-}_{(aq)}, O_{2(g)} \parallel H_{2(g)}, OH^{-}_{(aq)} \mid Pt_{(s)}$

11. For each of the following standard cells, refer to the redox table of relative strengths of oxidizing and reducing agents in Appendix C11, to represent the cell using the standard cell notation. Identify the cathode and anode and calculate the standard cell potential without writing half-reaction equations.

    (a) copper-lead standard cell

    (b) nickel-zinc standard cell

    (c) iron(III)-hydrogen standard cell

12. One experimental design for determining the position of a half-cell reaction that is not included in a table of oxidizing and reducing agents is shown below. Use the following standard cell, refer to the standard reduction potential of gold in Appendix C11, and calculate the reduction potential for the indium(III) ion.

$$Au_{(s)} \mid Au^{3+}_{(aq)} \parallel In^{3+}_{(aq)} \mid In_{(s)} \qquad \Delta E° = 1.84 \text{ V}$$
$$\text{cathode} \qquad\qquad\qquad \text{anode}$$

13. Any standard half-cell could have been chosen as the reference half-cell—the zero point of the reduction potential scale. What would be the standard reduction potentials for copper and zinc half-cells, assuming that the standard lithium cell were chosen as the reference half-cell, with its reduction potential defined as 0.00 V?

14. A zinc-iron cell is constructed and allowed to operate until the measured potential difference becomes zero. What interpretation can be made about the chemical system at this point?

Applying Inquiry Skills

15. Develop a table of oxidizing agents and reduction potentials from experimental evidence.

    Experimental Design

    Several cells are investigated; each cell has at least one half-cell in common with one of the other cells. The cell potentials are measured and the positive and negative electrodes of each cell are identified.

    Evidence

Positive electrode	Negative electrode	
$C_{(s)} \mid Cr_2O^{2-}_{7(aq)}, H^{+}_{(aq)}$	$\parallel Pd^{2+}_{(aq)} \mid Pd_{(s)}$	$\Delta E° = +\ 0.28$ V
$Tl_{(s)} \mid Tl^{+}_{(aq)}$	$\parallel Ti^{2+}_{(aq)} \mid Ti_{(s)}$	$\Delta E° = +\ 1.29$ V
$Pd_{(s)} \mid Pd^{2+}_{(aq)}$	$\parallel Tl^{+}_{(aq)} \mid Tl_{(s)}$	$\Delta E° = +\ 1.29$ V

    Analysis

    (a) Using the given evidence, complete a table of relative strengths of oxidizing and reducing agents, including reduction potentials.

16. Complete the Prediction for the following investigation. Include your reasoning.

    Question

    What is the total electric potential difference of two cells connected in series?

    Experimental Design

    Copper-silver and copper-zinc standard cells are connected as shown in **Figure 13**. The total electric potential difference of the two cells is measured with a voltmeter connected to the silver and zinc electrodes.

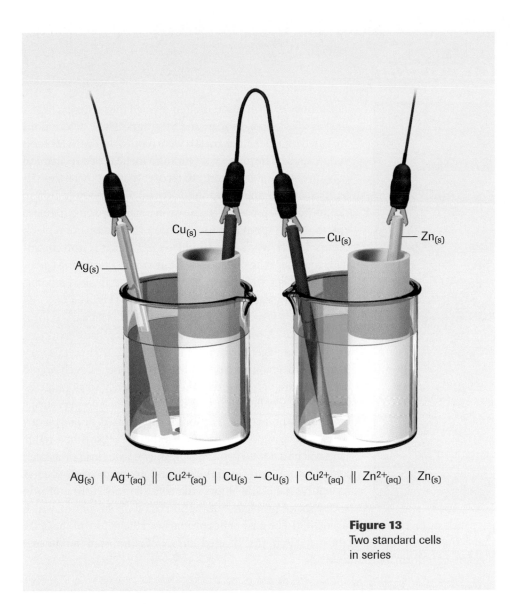

$$Ag_{(s)} \mid Ag^+_{(aq)} \parallel Cu^{2+}_{(aq)} \mid Cu_{(s)} - Cu_{(s)} \mid Cu^{2+}_{(aq)} \parallel Zn^{2+}_{(aq)} \mid Zn_{(s)}$$

Figure 13
Two standard cells
in series

▶ *Section 9.5* *Questions*

Understanding Concepts

1. (a) For a given cell, how is the cell potential predicted?
   (b) What are the restrictions on this prediction?

2. How does the cell potential indicate spontaneity of the reaction?

3. Why are the reactions in galvanic cells always spontaneous?

4. Define the hydrogen reference cell, including contents and conditions.

5. Why is a reference half-cell necessary?

6. (a) What is the cell potential of a standard cobalt-zinc cell?
   (b) What is the theoretical interpretation of this cell potential?

7. For each of the following cells,
   • use the given cell notation to identify the strongest oxidizing and reducing agents;
   • write chemical equations to represent the cathode, anode, and overall (net) cell reactions (include the half-cell and cell potentials); and
   • draw a diagram of each cell, labelling the electrodes, polarity (signs) of electrodes, electrolytes, direction of electron flow, and direction of ion movement.

   (a) $Cu_{(s)} \mid Cu^{2+}_{(aq)} \parallel Zn^{2+}_{(aq)} \mid Zn_{(s)}$
   (b) $C_{(s)} \mid Cr_2O_7^{2-}{}_{(aq)}, H^+_{(aq)} \parallel Sn^{2+}_{(aq)} \mid Sn_{(s)}$

8. You can determine a possible identity of an unknown half-cell from the cell potential involving a known half-cell. Use the following evidence and the table of reduction potentials in Appendix C9 to determine the reduction potential and possible identity of the unknown $X^{2+}_{(aq)} \mid X_{(s)}$ redox pair.

   $$2\,Ag^+_{(aq)} + X_{(s)} \rightarrow 2\,Ag_{(s)} + X^{2+}_{(aq)} \qquad \Delta E^\circ = +1.08\,V$$

9.6 Corrosion

Figure 1
Large ships have steel hulls. The rusting of steel involves the oxidation of iron in the steel and is a constant headache for shipping companies.

corrosion an electrochemical process in which a metal reacts with substances in the environment, returning the metal to an ore-like state

INVESTIGATION 9.6.1

The Corrosion of Iron (p. 723)
Study the factors that affect the corrosion of iron.

DID YOU KNOW?

Rates of Corrosion
A tin can (tin on steel) will corrode completely in about 100 a; an aluminum can in about 400 a; and a glass bottle in about 100 ka.

The history of civilization is often divided into different "ages" such as the Copper, Bronze, and Steel ages. These descriptions are based on when these metals were refined and used for tools and weapons. The process of refining a metal is electrochemical in nature and requires energy to recover the pure metal from its naturally occurring compounds (ores). **Corrosion** is also an electrochemical process. Because we live in an oxidizing (oxygen) environment, spontaneous oxidation (corrosion) of a metal occurs. In fact, we need to produce metals such as iron continually to replace the metals lost to corrosion. Preventing corrosion and dealing with the effects of corrosion are major economic and technological problems for our society (**Figure 1**).

As a metal is oxidized, metal atoms lose electrons to form positive ions. A redox table of relative strengths of oxidizing and reducing agents provides the evidence that metals vary greatly in their ability to be oxidized. Some metals, such as gold and silver, are noble because they are relatively weak reducing agents. On the other hand, Group 1 and 2 metals are very strong reducing agents and are, therefore, easily oxidized. In general, any metal appearing below the oxygen half-reactions in a redox table will be oxidized in our environment. Iron (including steel) and aluminum are such metals, and are extensively used as structural materials. Why is the corrosion or rusting of iron such a major problem, but the corrosion of aluminum, which is a much stronger reducing agent, not? The answer lies primarily in the nature of the oxide that forms on the surface of the metal. A freshly cleaned surface of aluminum rapidly oxidizes in air to form aluminum oxide.

$$4\,Al_{(s)} + 3\,O_{2(g)} \rightarrow 2\,Al_2O_{3(s)}$$

The aluminum oxide adheres tightly to the surface of the metal. This prevents further corrosion by effectively sealing any exposed surfaces.

Unfortunately, the iron compounds that form on the surface of exposed iron do not adhere very well. They flake off, exposing new iron to be corroded. In addition, the corrosion of iron is a complex process that is significantly affected by the presence of substances other than oxygen.

Rusting of Iron

Studies of the corrosion of iron have shown that the presence of both oxygen and water is required and the iron is converted into iron hydroxides and oxides. The first step of the mechanism is thought to be the oxidation of iron at a wet exposed surface (**Figure 2**).

$$Fe_{(s)} \rightarrow Fe^{2+}_{(aq)} + 2\,e^-$$

Iron(II) ions diffuse through the water on the iron surface while the electrons easily travel through the iron metal, which is an electrical conductor. The electrons are picked up by oxygen molecules dissolved in water on the surface at a point away from the original oxidation site (**Figure 2**).

$$\frac{1}{2}O_{2(g)} + H_2O_{(l)} + 2\,e^- \rightarrow 2\,OH^-_{(aq)}$$

The combination of iron(II) ions and hydroxide ions forms a low-solubility precipitate of iron(II) hydroxide, which is further oxidized by oxygen and water to form iron(III) hydroxide, a yellow-brown solid. The familiar red-brown rust is formed by the dehydration of iron(III) hydroxide to form a mixture of iron(III) hydroxide and hydrated iron(III) oxide. The amount of the hydroxide and the oxide varies, so rust is referred to as a hydrated oxide of indeterminate formula, $Fe_2O_3 \cdot xH_2O_{(s)}$.

Hydrated oxide
Iron(III) hydroxide can be converted to iron(III) oxide trihydrate as shown below.

$$2\,Fe(OH)_{3(s)} \rightarrow Fe_2O_3 \cdot 3H_2O_{(s)}$$

In fact, it is difficult to determine how much of the iron(III) exists in rust as the hydroxide or hydrated oxide. Warming this mixture can drive off some of the waters of hydration.

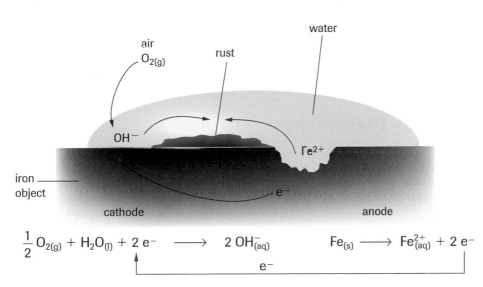

$$\frac{1}{2}\,O_{2(g)} + H_2O_{(l)} + 2\,e^- \longrightarrow 2\,OH^-_{(aq)} \qquad\qquad Fe_{(s)} \longrightarrow Fe^{2+}_{(aq)} + 2\,e^-$$

Figure 2
The corrosion of iron is a small electrochemical cell with iron oxidation at one location (the anode) and oxygen reduction at another location (the cathode).

This simplified mechanism for the rusting of iron can be used to explain why certain conditions promote rusting. If the iron is kept in a dry environment (low humidity) or if air has been removed from the water, little or no corrosion occurs (**Figure 3**). Eliminating either water or the oxygen in the water makes the reduction of aqueous oxygen impossible. Iron cannot be oxidized unless a suitable oxidizing agent is present. If oxidizing agents other than oxygen are present, such as certain metal ions, nonmetals, or hydrogen ions,

Figure 3
Rusting of exposed iron is almost negligible when the relative humidity is less than 50%. This iron pillar in Delhi, India, has existed for about 1500 years because of the very dry and unpolluted environment.

▶ **TRY THIS** activity   *Home Corrosion Experiment*

Soft drinks are acidic and contain electrolytes. Would different brands of pop corrode iron at different rates?

Materials: Coca-Cola, 7-Up, 2 identical steel nails, 2 plastic glassses
   (a) Predict which drink will cause faster corrosion.
 • Test your prediction, using a clean steel nail placed in a fresh sample of each soft drink.
   (b) Explain the results.

Fighting Corrosion

Concrete structures are reinforced with steel bars (sometimes referred to as rebar). These bars are made from scrap steel that is melted, reshaped, and then air-cooled. This cooling step allows the carbon in the steel to precipitate, forming microscopic carbide "fingers" that strengthen the steel. Unfortunately, they also act as little batteries. Electrons tend to flow from the iron toward the carbide fingers. As the iron loses electrons, it corrodes. If a concrete structure is in an area prone to corrosion, such as at the water line or even near ocean spray, this electron loss and corrosion occur even faster, resulting in a general weakening of the steel, and therefore, the concrete. The concrete structure may look just as solid, but it is no longer as strong as when it was first built. One solution being researched is to cool the steel rebar with water instead of air. The steel cools more quickly, which seems to reduce the growth of carbide fingers. The result is a less "electrically active" steel that may last longer.

cathodic protection a method of corrosion prevention in which the metal being protected is forced to become the cathode of a cell, using either an impressed current or a sacrificial anode

the iron can still be corroded through spontaneous redox reactions. This helps to explain the corrosion of iron in acidic environments, for example, why acid rain corrodes iron more than natural rain does.

In general, electrolytes accelerate rusting. Ships rust more rapidly in seawater than in fresh water and cars rust more rapidly in places where salt is used on roads. Chloride ions from salt are known to inhibit the adherence of protective oxide coatings on many metals, thus exposing more metal to be corroded. Electrolytes like sodium chloride conduct electricity and improve charge transfer, accelerating the rusting process. It is well known to plumbers that you cannot use steel straps or nails to hold copper pipes in place, because corrosion of the iron will be accelerated. Any moisture that is present sets up an electric cell similar in principle to Volta's original discovery of electricity from dissimilar metals (section 9.4). As the cell operates, the iron corrodes to form rust.

> The rusting of iron requires the presence of oxygen and water and is accelerated by the presence of acidic solutions, electrolytes, mechanical stresses, and contact with less active metals.

Corrosion Prevention

Methods used for preventing or minimizing the corrosion of iron can be divided into two categories: barrier methods that employ *protective coatings* and the method of *cathodic protection*. In some critical situations, such as a large fuel tank, both methods may be used.

Paint and other similar coatings are a simple method of corrosion prevention. This method works well as long as the surface is completely covered and the coating remains intact. Unfortunately, a scratch or chip in the surface can easily expose a small surface of iron and corrosion begins.

Both tin and zinc are used as metallic coatings. Tin, as in the familiar tin can, adheres well to the iron and provides a strong, shiny coating. The outer surface of the tin coating has a thin, strongly adhering layer of tin oxide that protects the tin. If a crack or break occurs in the tin layer, moisture can collect in the crack and an electric cell with tin and iron electrodes is established. Since iron is more easily oxidized than tin, iron becomes the anode in this cell. The electrons released by the oxidation of iron flow to the tin and corrosion is accelerated. Evidence of this is the typical iron rust on tin cans that have been crushed and left outside.

A spontaneous electric cell also arises when a zinc coating on an iron object is broken. However, in this case, the zinc is more easily oxidized than the iron. The zinc is preferentially oxidized, preventing corrosion of the iron. Zinc plating (galvanizing) of steel or iron provides double protection—a protective layer and preferential corrosion of the zinc.

Cathodic Protection

According to the redox theory of a cell, oxidation is the loss of electrons and occurs at the anode of a cell. Therefore, an effective method of preventing corrosion of iron is **cathodic protection**, forcing the iron to become the cathode by supplying the iron with electrons.

For a battery or DC generator connected in a circuit, electrons flow out of the negative terminal and into the positive terminal. If the negative terminal is connected to the iron object and the positive terminal to an inert carbon electrode, an electric current is forced to flow to the iron, through an electrolyte such as ground water, from the carbon electrode. The iron is forced to become the cathode and is prevented from corroding. An *impressed current* is an electric current forced to flow toward an iron object by an external potential difference. This method of corrosion prevention requires a constant electric power supply (typically 8 mV) and is used as cathodic protection for pipelines and culverts.

A less common but simpler method of cathodic protection is the use of a sacrificial anode. A *sacrificial anode* is a metal more easily oxidized than iron and connected to the iron object to be protected. The practice of zinc plating (galvanizing) iron objects is a common example of this method. Sacrificial zinc anodes are also connected to the exposed underwater metal surfaces of ships and boats to prevent the corrosion of the iron in the steel. Blocks of magnesium can also be used as sacrificial anodes (**Figure 4**). In all cases, the more active metal (appearing below iron in a half-reaction table) is slowly consumed or sacrificed at the anode, forcing the iron object to be the cathode of the cell.

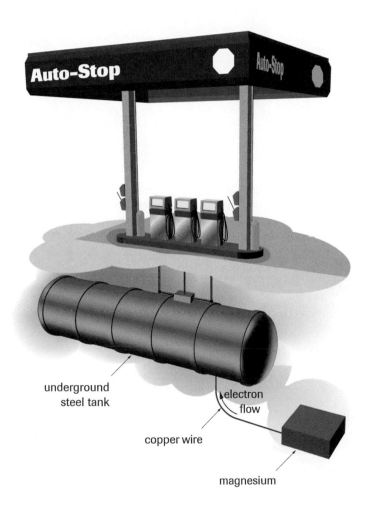

underground
steel tank

electron
flow

copper wire

magnesium

Figure 4
Corrosion of iron involves the oxidation of iron at the anode of a cell. If the iron is attached directly or connected electrically to a metal that is more easily oxidized (a sacrificial anode), then a spontaneous cell develops in which iron is the cathode. The electrolyte of the cell is the moisture in the ground.

Understanding Concepts

1. What are the minimum requirements for the corrosion of iron?

2. List some factors that accelerate or promote the corrosion of iron.

3. Write the balanced net ionic equation for the corrosion of iron to iron(II) ions in the presence of oxygen and water.

4. Although the corrosion of iron is a serious problem, other metals are also corroded in air or other environments. For each of the following situations, use your knowledge of writing and balancing redox equations to write and label the half-reaction and net ionic equations:
   (a) Zinc is an active metal that oxidizes quickly when exposed to air and water.
   (b) A lead pipe corrodes if it is used to transport acidic solutions that also contain dissolved oxygen.
   (c) In dry air, minute quantities of hydrogen sulfide gas can slowly react with silver objects to produce hydrogen gas and silver sulfide, recognized by the dark tarnish on the surface of the silver.

5. You may have noticed that when rusting appears on a car body, the rust appears around the break or chip in the paint but the damage may extend under the painted surface for some distance.
   (a) What is the evidence for damage extending well beyond the break in the paint?
   (b) Suggest an explanation why the damage may extend far from the break in the paint.

6. Would a basic solution prevent or slow down the corrosion of iron? Provide your reasoning.

7. Why is a zinc coating on iron better than a tin coating?

8. What are the two methods of cathodic protection and how are they similar?

Applying Inquiry Skills

9. The following investigation looks at the reactivity of oxygen and various acids with a metal. Evidence of a spontaneous reaction would be the corrosion of the metal.

Question
What effect do oxygen and various acids have on the corrosion of copper metal?

Experimental Design
Several test tubes are set up with a clean piece of copper metal in each. Various possible oxidizing agents will be tested for reaction wth copper: oxygen gas only, oxgyen bubbled into water, oxygen bubbled into each of dilute hydrochloric acid, nitric acid, sulfuric acid, and phosphoric acid. (If oxygen is not available, use air.)
   (a) Identify the independent, dependent, and controlled variables.
   (b) Prepare a list of materials and write a procedure including safety and disposal instructions. When your teacher has approved your work, conduct and report on this experiment.

Figure 5
When this pipeline was being constructed, a zinc wire was attached to and buried with the pipe.

Making Connections

10. A zinc wire is connected to and buried with a pipeline when it is built (**Figure 5**).
    (a) Why is this done? Include a brief description of the principles involved.
    (b) Is this the only type of corrosion protection used with major pipelines?
    (c) Discuss the environmental and safety issues associated with protecting and also not protecting pipelines.

 www.science.nelson.com

11. State several examples of metal corrosion of manufactured materials. Which examples involve environmental, health, or safety issues? Are there examples of corrosion that are desirable? Discuss briefly.

12. Search the Internet, using the key words "iron corrosion." How many different titles did the search find? What is the implication of this number? Find a general site and list the different classes of iron corrosion.

INVESTIGATION 9.1.1

Single Displacement Reactions

Inquiry Skills

○ Questioning	● Planning	● Analyzing
○ Hypothesizing	● Conducting	● Evaluating
● Predicting	● Recording	● Communicating

In this investigation you will observe some common single displacement reactions and then interpret the changes in terms of electron transfer.

Purpose
The purpose of this investigation is to gather some evidence about single displacement reactions.

Question
What are the products of the single displacement reactions for the following sets of reactants?

- copper and aqueous silver nitrate
- aqueous chlorine and aqueous sodium bromide
- magnesium and hydrochloric acid
- zinc and aqueous copper(II) sulfate
- aqueous chlorine and aqueous potassium iodide

Prediction

(a) According to the single displacement generalization, predict the balanced chemical equations for each set of reactants listed above.

Experimental Design
Small quantities of reactants are mixed and diagnostic tests (Appendix A6) are used to determine the products of each reaction.

(b) For each chemical equation listed in your prediction, record diagnostic tests that you will use. (Some diagnostic tests will be very specific and some will be a general observation you expect.)

Materials
lab apron	dropper bottles of 0.1 mol/L
eye protection	—aqueous silver nitrate
five small test tubes	—aqueous sodium bromide
two test-tube stoppers	—aqueous copper(II) sulfate
test-tube rack	—aqueous potassium iodide
emery paper or steel wool	—hydrochloric acid
wash bottle	chlorine water (or bleach)
matches	cyclohexane
copper wire or strip	
magnesium metal ribbon	
zinc strip	

The substances used in this experiment are toxic and corrosive. Avoid skin contact. Wash any splashes on the skin or clothing with plenty of water. If any chemical is splashed in your eyes, rinse for at least 15 min and inform your teacher.

Magnesium and cyclohexane are highly flammable. Keep away from open flame.

Keep the cyclohexane sealed to avoid evaporation and inhalation of the vapours. Dispose of the hydrocarbon mixtures as directed by your teacher.

Make sure matches are extinguished by dipping in water.

Procedure

1. Set up five test tubes, each filled to a depth of 2–3 cm with one of the five aqueous solutions.
2. Add the appropriate element to each solution (see Question) in each test tube.
3. Perform diagnostic tests on each of the five mixtures. Record your observations.
4. Dispose of the solutions as directed by your teacher.

Evidence

(c) Design a table to record your observations.

Analysis

(d) Interpret your evidence and record the products that you can reasonably conclude you obtained in each reaction.

Evaluation

(e) Evaluate the experimental design, materials, and procedure by considering any possible flaws and improvements.

(f) Use your answer to (a). What is your judgment of the quality and quantity of evidence obtained?

(g) How confident are you in the answer obtained? Provide your reasons.

(h) Evaluate the Prediction and provide your reasons.

Oxidation States of Vanadium

Vanadium is a transition metal that forms many different ions with different oxidation states (**Table 1**). Vanadium and its compounds have many different uses, including colouring for glass, ceramics, and plastics.

Table 1 Colours of Vanadium Ions

Ion name	Ion formula	Colour
vanadate(V)	$VO_{3(aq)}^{-}$	
vanadate (IV)	$VO_{(aq)}^{2+}$	
vanadium(III)	$V_{(aq)}^{3+}$	
vanadium(II)	$V_{(aq)}^{2+}$	

Purpose

The purpose of this lab exercise is to investigate some redox chemistry of vanadium compounds.

Question

What are the oxidation states and changes in oxidation number for vanadium ions?

Evidence

Table 2 Reactions of Vanadium Ions

Procedure	Final solution colours
(1) ammonium vanadate(V) dissolved in sulfuric acid	yellow
(2) yellow solution with three subsequent additions of small quantities of zinc dust	yellow turned blue, then green, then violet
(3) violet solution left sitting in an open container	slowly turned green
(4) yellow solution mixed with potassium iodide solution	very dark, almost black
(5) blue solution mixed with potassium iodide solution	stayed blue; no change
(6) violet solution slowly mixed with acidic potassium permanganate	violet to green to blue to yellow

Analysis

(a) Using **Table 1**, identify the vanadium ions in the sequence of reactions in **Table 2**.

(b) In each case, is the vanadium being oxidized or reduced? Justify your answer, using oxidation numbers.

(c) Explain the observations made in (3) to (6) in **Table 2** above. Suggest what is causing these changes.

(d) Vanadium ion chemistry can be quite complicated. For example, the initial yellow solution is likely an equilibrium between $VO_{3(aq)}^{-}$ and $VO_{2(aq)}^{+}$. Does this alter your analysis in the previous questions? Explain briefly.

INVESTIGATION 9.3.1

Inquiry Skills

Spontaneity of Redox Reactions

○ Questioning	● Planning	● Analyzing
○ Hypothesizing	● Conducting	● Evaluating
● Predicting	● Recording	● Communicating

In the past, we have usually assumed that all chemical reactions are spontaneous; that is, they occur of their own accord once reactants are placed in contact, without a continuous addition of energy to the system. Spontaneous redox reactions in solution generally provide visible evidence of a reaction within a few minutes.

Purpose

The scientific purpose of this investigation is to test the assumption that all single displacement reactions are spontaneous.

INVESTIGATION 9.3.1 *continued*

Question

Which combinations of copper, lead, silver, and zinc metals and their aqueous metal ion solutions produce spontaneous reactions?

Prediction

(a) State and justify your answer to the question.

Experimental Design

A drop of each metal ion solution is placed in separate locations on a clean area of each of the four metal strips.

Materials

lab apron
eye protection
reusable strips of copper, lead, silver, and zinc metals
0.10 mol/L solutions of copper(II) nitrate, lead(II) nitrate, silver nitrate, and zinc nitrate in dropper bottles
steel wool or emery paper

LEARNING *TIP*

Distinguishing Lead and Zinc
When cleaned, lead and zinc metals look very similar. However, lead is much softer. You can distinguish between them because lead strips bend and scratch much more easily than zinc strips.

The solutions used are toxic—especially the lead solution—and irritants. Avoid skin contact. Remember to wash your hands before leaving the laboratory.

Dispose of reaction products according to your teacher's instructions.

Rinse all of the metal strips thoroughly and return them so they can be used again.

Procedure

(b) Write a brief procedure for this investigation. Have your teacher approve your procedure before you start.

Evidence

(c) Design a table to record your observations.

Analysis

(d) Based on your evidence, which combinations of reactants produced spontaneous reactions?

Evaluation

(e) Were the experimental design, materials, and procedure sufficient to answer the question? Justify your answer.

(f) Suggest an improvement to increase the certainty of the evidence.

(g) Evaluate the prediction, including your reasons.

(h) What does your answer to (g) tell you about the assumption on which the prediction was based? What should be done with this assumption?

LAB EXERCISE 9.3.1

Building a Redox Table

Suppose that a research team is developing a table of relative strengths of oxidizing and reducing agents. One team member had completed an investigation summarized in Table 3, Section 9.3, and another had completed the investigation reported in Practice question 9, Section 9.3. A third member used the combination of metals, nonmetals, and solutions shown below. By completing this exercise, you will see how scientists have developed more extensive tables of relative strengths of oxidizing and reducing agents.

Inquiry Skills

○ Questioning	○ Planning	● Analyzing
○ Hypothesizing	○ Conducting	○ Evaluating
○ Predicting	○ Recording	● Communicating

Purpose

The purpose of this lab exercise is to construct a table of relative strengths of oxidizing and reducing agents.

Question

What is the table of relative strengths of oxidizing and reducing agents for copper, silver, bromine, and iodine?

Evidence

Table 3 Reactions of Metals and Nonmetals with Solutions of Ions

	$I_{2(aq)}$	$Cu^{2+}_{(aq)}$	$Ag^+_{(aq)}$	$Br_{2(aq)}$
$I^-_{(aq)}$	X	X	√	√
$Cu_{(s)}$	√	X	√	√
$Ag_{(s)}$	X	X	X	√
$Br^-_{(aq)}$	X	X	X	X

X no evidence of a redox reaction
√ evidence redox reaction occurred

Analysis

(a) Using the results from **Table 3**, prepare a table of relative strengths of oxidizing and reducing agents.

Synthesis

(b) Compare **Table 3** in Section 9.3, your analysis table from Practice question 9 in section 9.3, and your table from (a) above. Note that there are several substances that appear in two of these tables. Combine all three tables in one larger table showing the order of oxidizing and reducing agents.

> **LEARNING TIP**
>
> **Tables of Oxidizing and Reducing Agents**
> All tables of oxidizing and reducing agents follow the same general format.
> $OA + n\,e^- \rightleftharpoons RA$

INVESTIGATION 9.3.2

The Reaction of Sodium with Water

This demonstration will provide practice in both predicting and testing a chemical reaction.

Purpose

The purpose of this demonstration is to test the five-step method for predicting redox reactions.

Question

What are the products of the reaction of sodium metal with water?

Prediction

(a) Answer the question and provide your reasoning.

Experimental Design

(b) Write a general plan for this reaction.

(c) Suggest one diagnostic test for each predicted product, using the "If [procedure], and [evidence], then [analysis]" format for every product predicted. (This format is described in Appendix A6.)

(d) What control(s) should be used with these tests?

Inquiry Skills

○ Questioning	● Planning	● Analyzing
○ Hypothesizing	○ Conducting	● Evaluating
● Predicting	● Recording	● Communicating

 This reaction of sodium metal with water must be demonstrated with great care, because a great deal of heat is produced.

Evidence

(e) Record all observations, including the controls, in a suitable table.

Analysis

(f) According to the evidence collected, what products were obtained?

Evaluation

(g) Assuming the evidence is of suitable quality, evaluate your prediction.

(h) Evaluate the method of writing redox reactions used to make your prediction.

(i) This experiment is not sufficient to provide a judgment of the method of predicting redox reactions. What should be done next?

ACTIVITY 9.4.1

Developing an Electric Cell

In this activity, an aluminum soft-drink can is both the container and one of the electrodes (**Figure 1**). The other electrode is a solid conductor such as graphite from a pencil, an iron nail, or a piece of copper wire or pipe. The electrolyte may be a salt solution or an acidic or basic solution. Although the overall performance of a cell depends on many factors, only the voltage is investigated here.

The purpose of this activity is to use a technological problem-solving (trial-and-error) approach to construct a working electric cell with the highest possible voltage. You will need to control variables and work in a systematic way. Be prepared to alter your materials if your results are not promising and to maximize your results.

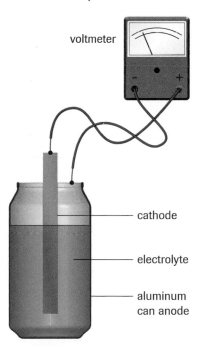

Figure 1
An aluminum can cell is an efficient design, since one of the electrodes also serves as the container.

Materials

lab apron
eye protection
various electrodes
acidic, basic, and neutral
   ionic solutions
bottle of distilled water
steel wool or emery paper

voltmeter
2 plug–and–clip wires
an aluminum can with the
   top removed

Be careful when handling acidic and basic solutions because they are corrosive. Wear eye protection and work near a source of water. Electrolytes may be toxic or irritants; follow all safety precautions. Avoid eye and skin contact.

Dispose of solutions according to your teacher's instructions.

Design

- Using the same electrolyte and the aluminum can as the control variables, test two or three different materials as the second electrode. Measure the voltage of each cell. (Scrape the paint from the can where the wire is attached.)

- Using the same two electrodes as the control variables, test two or three possible electrolytes. Measure the voltage of each cell.

- Test additional combinations, based on the analysis of the initial trials.

Evidence

(a) Keep careful records of all observations, including what worked and did not work.

(b) Set up a table or organized list to record your observations and analysis of each trial in your problem-solving cycle.

Analysis

(c) What is the best result you obtained? Report the details of the design and voltage.

Evaluation

(d) Evaluate the quality of your evidence, including any sources of experimental error or uncertainty.

(e) If you were to continue this process, what changes or improvements would you make?

(f) Evaluate the suitability of your final electric cell for potential commercial development. Consider a variety of factors.

Characteristics of a Hydrogen Fuel Cell

Inquiry Skills

○ Questioning	○ Planning	● Analyzing
○ Hypothesizing	○ Conducting	● Evaluating
○ Predicting	○ Recording	○ Communicating

A fuel cell supplies electrical energy in the same way as any electric cell. Ions transfer the charge within the cell, and electrons flow through an external circuit between the electrodes. In a hydrogen fuel cell, the electrodes are not consumed, and there is no need to reverse the cell to recharge it. The reactants, hydrogen and oxygen, are continuously supplied and consumed to produce water and electricity.

Commercial hydrogen fuel cells, such as the Ballard cell, utilize a solid electrolyte called a proton exchange membrane (PEM). This polymer is bonded to two porous carbon cloth electrodes (**Figure 2**).

Purpose

The purpose of this lab exercise is to compare the electrical characteristics of a hydrogen fuel cell and a dry cell.

Question

How do the trends of the voltage-current and power-current graphs compare for a hydrogen fuel cell and a typical dry cell?

Experimental Design

Hydrogen and oxygen gases, produced from the electrolysis of water, are passed into a hydrogen fuel cell. The voltage of the cell is measured with an open circuit. Various resistances are added to the circuit, and the voltage and current are measured for each resistance (load). The fuel cell is disconnected and replaced by a dry cell. The same procedure is followed to collect voltage and current measurements for the same set of resistances in the circuit.

Evidence

Fuel cell		Dry cell	
Voltage (V)	**Current (mA)**	**Voltage (V)**	**Current (mA)**
0.81	0	1.44	0
0.80	5	1.42	5
0.79	6	1.41	10
0.78	16	1.40	22
0.73	66	1.34	90
0.70	115	1.30	146
0.67	175	1.25	212
0.61	315	1.14	335

Analysis

(a) Construct and label a graph with voltage on the vertical axis and current on the horizontal axis. Plot and label the line for the fuel cell and another line for the dry cell.

(b) Describe and compare the trend shown by the voltage-current line for the fuel cell with that of the dry cell. Does the hydrogen fuel cell behave like a regular (dry) cell?

(c) The power output of a cell is a measure of the quantity of energy per second delivered by the cell. For each voltage and current, calculate the power supplied by using the formula $P = VI$. (Recall that power is measured in watts (W) and that $1 \text{ V·A} = 1 \text{ W}$.) Summarize your results in a table for both the fuel cell and the dry cell.

(d) Repeat (a) and (b) to construct a power curve for each cell. Plot power on the vertical axis and current on the horizontal axis.

(e) Use the results of your analysis to answer the Question.

Synthesis

(f) Compare the designs of fuel cells and dry cells.

(g) What makes fuel cells more practical for powering electric cars compared to other types of cells?

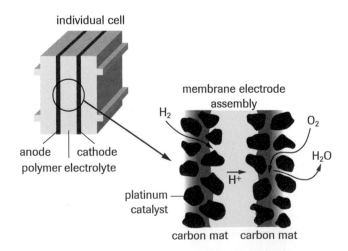

Figure 2
The membrane electrode assembly includes the electrodes (carbon mats) and the proton exchange polymer. Each carbon particle has tiny platinum particles on its surface.

ACTIVITY 9.5.1

Galvanic Cell Design

The purpose of this activity is to demonstrate the design and operation of a galvanic cell used in scientific research. A cell with only one electrolyte is compared with similar cells containing the same electrodes but two electrolytes (**Figure 3**).

Materials

lab apron
eye protection
silver and copper electrodes
steel wool (for cleaning electrodes)
voltmeter with leads
4 medium-sized beakers
U-tube
cotton plugs
porous cup
bottle of distilled water
solutions of sodium nitrate, silver nitrate,
   copper(II) nitrate

 Solutions used are irritants and are toxic if ingested. Avoid contact with skin and eyes. Silver nitrate will temporarily blacken your skin.

Dispose of solutions according to your teacher's instructions.

• Construct the three cells shown in **Figure 3**.

(a) Which design is most similar to Volta's invention? Compare the three cell designs.

• Use a voltmeter to determine which electrode is positive and which is negative, and measure the electric potential difference (voltage) of each cell.

(b) According to the voltmeter test, which electrode is the cathode and which is the anode?

(c) Why is your answer to (b) the same for all three cells?

(d) Suggest a reason why two of the voltages measured are very similar and the third is very different.

• With the voltmeter connected, remove and then replace the various parts of each cell.

(e) Why does the voltmeter reading go to zero when one of the parts of the cell is removed?

(f) What common device in your home and school also "breaks" the circuit?

• For each cell, connect the two electrodes directly with a wire. Record any evidence of a reaction after several minutes, and after one or two days. Measure the electric potential difference after several days.

(g) What is the design of a control that can be used to compare changes with each of the three cells?

(h) State some diagnostic tests that could be done to obtain more specific evidence for the operation of each cell.

(i) Suggest a reason why all solutions were nitrates.

salt bridge:
$Ag_{(s)} | AgNO_{3(aq)} \| Cu(NO_3)_{2(aq)} | Cu_{(s)}$

no porous boundary:
$Ag_{(s)} | NaNO_{3(aq)} | Cu_{(s)}$

porous cup:
$Ag_{(s)} | AgNO_{3(aq)} \| Cu(NO_3)_{2(aq)} | Cu_{(s)}$

Figure 3
Three different cell designs

Investigating Galvanic Cells

Inquiry Skills

○ Questioning	● Planning	● Analyzing
○ Hypothesizing	● Conducting	● Evaluating
● Predicting	● Recording	● Communicating

In this investigation, you are given the opportunity to construct galvanic cells and compare your observations with the rules and concepts you have learned.

Purpose
The purpose of this investigation is to compare the predictions of cell potentials and electrodes of various cells with those measured in the lab.

Question
In cells constructed from various combinations of copper, lead, silver, and zinc half-cells, what are the standard cell potentials, and which is the anode and cathode in each case?

Prediction
(a) According to redox concepts and a redox table (Appendix C11), prepare a table of all possible combinations of half-cells and answer the question.

Experimental Design
(b) Write a brief general plan to answer the question, including the identification of variables.

Materials
lab apron
safety glasses
voltmeter and connecting wires
U-tube with cotton plugs, porous cups, or filter-paper
   strips
four 100-mL beakers or well plate
distilled water
steel wool or emery paper
$Cu_{(s)}$, $Pb_{(s)}$, $Ag_{(s)}$, and $Zn_{(s)}$ strips
1.0 mol/L $CuSO_{4(aq)}$, $Pb(NO_3)_{2(aq)}$, $AgNO_{3(aq)}$, $NaNO_{3(aq)}$,
   and $ZnSO_{4(aq)}$

 Some of the solutions used are toxic and/or irritants. Avoid skin and eye contact.

Procedure
(c) Based on the equipment supplied, write a specific procedure to collect the evidence to answer the question. Be sure to include safety and disposal instructions. Have your teacher approve your procedure before starting.

Evidence/Analysis
(d) Prepare a table to record your observations and the predicted cell potentials. Include a column for expressing the accuracy of each result (in terms of a percentage difference).

(e) Note and record any unexpected observations.

Evaluation
(f) List all sources of experimental error or uncertainty. Considering this list, state your judgment of the overall quality of the evidence obtained.

(g) For each cell, compare the electrodes you predicted to be the cathodes and the anodes with your evidence. How well do these agree?

(h) Limitations of the equipment and materials mean that some experimental uncertainties are unavoidable. Assuming that about 5% difference is unavoidable, is the agreement between your predicted and measured cell potentials acceptable? Justify your answer.

(i) Is there any pattern to the accuracy of your measured cell potentials? Suggest some reasons to explain this.

(j) Evaluate the design, procedure, and materials. Note any flaws or possible improvements.

🔬 INVESTIGATION 9.6.1

The Corrosion of Iron

Inquiry Skills		
○ Questioning	● Planning	● Analyzing
○ Hypothesizing	● Conducting	● Evaluating
● Predicting	● Recording	● Communicating

The knowledge gained from this experiment is used to help explain corrosion and to develop methods of corrosion prevention.

 The solvent used is very flammable and may be toxic.

Purpose

The purpose of this investigation is to test your predictions of factors affecting the rate of corrosion of iron.

Question

What factors, chemical and electrical, affect the rate of corrosion of iron?

Prediction

(a) Based on your experience and knowledge of electro-chemistry, predict some factors that may affect the rate of corrosion of iron. Provide your reasoning.

Experimental Design

Several iron nails or pieces of iron wire are thoroughly cleaned with steel wool, rinsed with water, and dried with a solvent (alcohol or acetone). In part 1, the iron is exposed to different conditions in separate test tubes. A clean piece of iron in a dry empty test tube is the control. All test tubes are observed immediately and after one day. In part 2, two iron-carbon cells are connected to 9-V batteries, with the electrodes attached oppositely for the two cells.

Materials

(b) Prepare a list of materials. Check with your teacher to make sure your materials are suitable and available.

Procedure

(c) Using the experimental design and your list of materials, write a procedure to answer the question. Include safety and disposal instructions. Obtain approval from your teacher before proceeding.

Evidence

(d) Prepare a table to record your observations.

Analysis

(e) Compare the evidence from the factors you tested with the control. Which factors appear to accelerate the rate of corrosion of iron? To the extent possible, arrange your factors from least to greatest apparent effect.

(f) List any factors that did not seem to affect the rate of corrosion of iron compared to the control.

Evaluation

(g) Evaluate the design, noting any flaws or improvements that could be made. Check the variables and control to see if these are all appropriate. Suggest changes if necessary.

(h) Evaluate the procedure and materials. Were you able to gather sufficient and suitable evidence? Suggest improvements where required.

(i) What are the main sources of experimental error or uncertainty? How serious would you judge these to be?

(j) Using your answers to (g) to (i), state your judgment of the quality of the evidence. How confident are you about the evidence?

(k) Assuming your evidence is of reasonable quality, state your judgment of the prediction.

(l) Judge the reasoning you used to make your prediction. How useful was it? Why is there a need for more empirical and theoretical knowledge about corrosion?

Key Expectations

- Demonstrate an understanding of oxidation and reduction in terms of the transfer of electrons or change in oxidation number. (9.1)
- Identify and describe the functioning of the components in electric cells. (9.1, 9.2, 9.3, 9.4, 9.5)
- Demonstrate an understanding of oxidation–reduction reactions through experiments and analysis of these reactions. (9.1, 9.3, 9.4, 9.5)
- Write balanced chemical equations for redox reactions using half-reaction equations. (9.1, 9.2, 9.3)
- Predict the spontaneity of redox reactions and cell potentials using a table of half-cell reduction potentials. (9.3)
- Describe examples of common cells and evaluate their environmental and social impact. (9.4)
- Research and assess environmental, health, and safety issues involving electrochemistry. (9.4, 9.6)
- Describe galvanic cells in terms of oxidation and reduction half-cells and electric potential differences. (9.5)
- Describe the function of the hydrogen reference half-cell in assigning reduction potential values. (9.5)
- Determine oxidation and reduction half-cell reactions, current and ion flow, electrode polarity and cell potentials of typical galvanic cells. (9.5)
- Explain corrosion as an electrochemical process, and describe corrosion-inhibiting techniques. (9.6)
- Use appropriate scientific and technological vocabulary related to electrochemistry. (all sections)

Key Terms

ampere

anode

battery

cathode

cathodic protection

corrosion

coulomb

electric cell

electric current

electric potential difference (voltage)

electrode

electrolyte

fuel cell

galvanic cell

half-cell

inert electrode

oxidation

oxidation number

oxidizing agent

primary cell

redox spontaneity rule

reducing agent

reduction

reference half-cell

secondary cell

standard cell

standard cell potential

standard reduction potential

volt

Key Symbols and Equations

- $E_r°$, $\Delta E°$

- $\Delta E°_{cell} = E°_{r\,cathode} - E°_{r\,anode}$

Problems You Can Solve

- Assign oxidation numbers and use them to identify oxidation and reduction processes, and to balance redox reactions.
- Write and balance half-reaction equations to obtain a balanced redox equation.
- Construct a table of relative strengths of oxidizing and reducing agents.
- Predict spontaneity of redox reactions using a redox table.
- Analyze components and processes occurring in electric and galvanic cells.

▶ **MAKE** *a summary*

Start with "Cell" and make a flow chart that connects as many terms, concepts, and symbols as possible from those listed above.

Identify each of the following statements as true, false, or incomplete. If the statement is false or incomplete, rewrite it as a true statement.

1. Oxidation corresponds to an increase in oxidation number.

2. Reduction is a process in which electrons are lost or donated by an atom or ion in a redox reaction.

3. An oxidizing agent gains electrons and one of its atoms decreases in oxidation number.

4. The strongest oxidizing agent in a galvanic cell is above the strongest reducing agent in the redox table, producing a cell potential that is negative.

5. The cathode of a cell is the electrode where electrons are lost or given up by the reducing agent.

6. Only the cell potential can be experimentally measured and a reference half-cell, assigned a zero value, is necessary to calculate reduction potentials.

7. The cell potential of a standard lead-nickel cell is -0.39 V.

8. Corrosion is an electrochemical process in which electrons are transferred.

9. The development of electric cars is closely related to advances in the hydrogen fuel cell.

Identify the letter that corresponds to the best answer to each of the following questions.

10. A reducing agent can be described as a substance that
    (a) loses electrons and causes reduction.
    (b) loses electrons and becomes reduced.
    (c) gains electrons and causes oxidation.
    (d) gains electrons and becomes reduced.
    (e) gains electrons and becomes oxidized.

11. Which of the following solutions should **not** be stored in a tin-plated container?

    I $NaNO_{3(aq)}$ III $SnBr_{2(aq)}$
    II $AgNO_{3(aq)}$ IV $Cl_{2(aq)}$

    (a) I only
    (b) II, III, IV
    (c) II and III
    (d) II and IV
    (e) III and IV

12. The oxidation number of the carbon atom in the carbonate ion is
    (a) $+8$
    (b) $+6$
    (c) $+4$
    (d) $+2$
    (e) 0

13. In a galvanic cell,
    (a) electrons are provided by the reducing agent at the negative electrode.
    (b) electrons are gained by the oxidizing agent at the negative electrode.
    (c) electrons flow through the solution from the anode to the cathode.
    (d) electrons flow through the solution from the cathode to the anode.
    (e) electrons flow through the porous barrier from the cathode to the anode.

14. A porous boundary, or a salt bridge, is required in a standard cell to
    (a) conduct electrons from the anode to the cathode.
    (b) transfer ions between the half-cells.
    (c) keep the electrodes from contacting.
    (d) maintain standard conditions in both half cells.
    (e) stop current flow when electrodes are connected.

15. If the electrodes of a standard copper-silver cell are connected with a wire,
    (a) silver is plated at the anode.
    (b) a voltmeter would show a reading of 1.14 V.
    (c) electrons flow from the silver to copper electrodes.
    (d) the solution at the anode becomes darker blue.
    (e) the silver ion concentration increases.

16. The standard hydrogen half-cell can be represented as
    (a) $Pt_{(s)} \mid H_{2(g)}$ (d) $Pt_{(s)} \mid H_{2(g)} \parallel H^+_{(aq)}$
    (b) $Pt_{(s)} \mid H^+_{(aq)}$ (e) $Pt_{(s)} \mid H_{2(g)} \parallel H^+_{(aq)}, OH^-_{(aq)},$
    (c) $Pt_{(s)} \mid H_{2(g)}, H^+_{(aq)}$

17. The corrosion of iron is accelerated by
    (a) low humidity.   (d) nonelectrolytes.
    (b) lack of oxygen.   (e) low pH.
    (c) low temperature.

18. Cathodic protection of iron is effective because the iron
    (a) is forced to become the anode of a cell.
    (b) is forced to become the cathode of a cell.
    (c) is no longer a part of any electrochemical cell.
    (d) is attached to the positive terminal of a battery.
    (e) is electrically prevented from gaining electons.

Understanding Concepts

1. Write theoretical definitions for each of the following words in terms of both electrons and oxidation states:
   (a) oxidation
   (b) reduction
   (c) redox reaction (9.1)

2. Write and label balanced half-reaction equations for each of the following redox reactions:
   (a) $2\,Fe^{3+}_{(aq)} + Ni_{(s)} \rightarrow 2\,Fe^{2+}_{(aq)} + Ni^{2+}_{(aq)}$
   (b) $Br_{2(aq)} + 2\,I^-_{(aq)} \rightarrow 2Br^-_{(aq)} + I_{2(s)}$
   (c) $Pd^{2+}_{(aq)} + Sn^{2+}_{(aq)} \rightarrow Pd_{(s)} + Sn^{4+}_{(aq)}$ (9.1)

3. Assign an oxidation number to
   (a) I in $I_{2(s)}$ (d) H in $NH_{3(g)}$
   (b) I in $CaI_{2(s)}$ (e) H in $AlH_{3(s)}$ (9.1)
   (c) I in $HIO_{(aq)}$

4. Assign oxidation numbers to all atoms/ions and indicate which atom/ion is oxidized and which is reduced.
   (a) $2\,Al_{(s)} + Fe_2O_{3(s)} \rightarrow 2\,Fe_{(s)} + Al_2O_{3(s)}$
   (b) $In_{(s)} + 3\,Tl^+_{(aq)} \rightarrow In^{3+}_{(aq)} + 3\,Tl_{(s)}$
   (c) $2\,Cr^{3+}_{(aq)} + Sn^{2+}_{(aq)} \rightarrow 2\,Cr^{2+}_{(aq)} + Sn^{4+}_{(aq)}$
   (d) $Cl_{2(aq)} + 2\,I^-_{(aq)} \rightarrow 2Cl^-_{(aq)} + I_{2(aq)}$
   (e) $UCl_{4(s)} + 2\,Ca_{(s)} \rightarrow 2CaCl_{2(s)} + U_{(s)}$ (9.1)

5. Make a list of everything that must be balanced in a net ionic equation representing a redox reaction. (9.2)

6. The silver(II) ion, used in chemical analysis, reacts spontaneously with water according to the following (unbalanced) equation:

   $$Ag^{2+}_{(aq)} + H_2O_{(l)} \rightarrow Ag^+_{(aq)} + O_{2(g)}$$

   (a) Assign an oxidation number to each atom or ion.
   (b) Balance the equation, assuming an acid solution. (9.2)

7. Use the oxidation number method to balance the reaction equations for the following redox reactions in acid solutions:
   (a) $Cu_{(s)} + HNO_{3(aq)} \rightarrow$
   $\quad Cu(NO_3)_{2(aq)} + NO_{(g)} + H_2O_{(l)}$
   (b) $MnO^-_{4(aq)} + H_2C_2O_{4(aq)} \rightarrow$
   $\quad Mn^{2+}_{(aq)} + CO_{2(g)} + H_2O_{(l)}$
   (c) $KIO_{3(aq)} + KI_{(aq)} + HCl_{(aq)} \rightarrow$
   $\quad KCl_{(aq)} + I_{2(s)} + H_2O_{(l)}$ (9.2)

8. Write equations for the reduction and oxidation half-reactions, and balanced net redox equation.
   (a) $O_{3(g)} + I^-_{(aq)} \rightarrow IO^-_{3(aq)} + O_{2(g)}$ (acidic)
   (b) $Pt_{(s)} + NO^-_{3(aq)} + Cl^-_{(aq)} \rightarrow$
   $\quad PtCl^{2-}_{6(aq)} + NO_{2(g)}$ (acidic)

(c) $CN^-_{(aq)} + ClO^-_{2(aq)} \rightarrow$
$\quad CNO^-_{(aq)} + Cl^-_{(aq)}$ (basic)
(d) $PH_{3(g)} + CrO^{2-}_{4(aq)} \rightarrow$
$\quad Cr(OH)^-_{4(aq)} + P_{4(s)}$ (basic)
(e) $MnO^{2-}_{4(aq)} \rightarrow$
$\quad Mn^{2+}_{(aq)} + MnO^-_{4(aq)}$ (acidic) (9.2)

9. While working on the development of a new electrochemical cell, a research chemist places selected Period 4 transition metal strips into aqueous solutions of their ionic compounds. She observes that the following combinations of metal and cation react spontaneously:

   $$V_{(s)} + Mn^{2+}_{(aq)} \rightarrow V^{2+}_{(aq)} + Mn_{(s)}$$
   $$V^{2+}_{(aq)} + Ti_{(s)} \rightarrow V_{(s)} + Ti^{2+}_{(aq)}$$
   $$Co^{2+}_{(aq)} + Mn_{(s)} \rightarrow Co_{(s)} + Mn^{2+}_{(aq)}$$

   (a) Use this information to develop a table of oxidizing and reducing agents for these metals and their ions.
   (b) Identify the strongest oxidizing and the strongest reducing agent in your table. (9.3)

10. For each of the following situations, list and classify the entities present, write the equations for the half-reactions and the overall equation, and then predict whether a spontaneous reaction will be observed.
    (a) Nitric acid is added to aqueous potassium bromide.
    (b) Aqueous potassium permanganate is used to titrate an acidic solution of iron(II) sulfate.
    (c) A strip of copper is placed in a beaker of hydrochloric acid.
    (d) An iron pipe is exposed to the wind and the rain.
    (e) Aqueous cobalt(II) sulfate is mixed with a basic solution of sodium sulfite.
    (f) Aqueous solutions of chromium(II) nitrate and tin(II) nitrate are mixed together. (9.3)

11. Calcium metal spontaneously reacts with water.
    (a) Write the half-reaction and net ionic reaction equation for this reaction.
    (b) Describe diagnostic tests that could be done to test for the predicted products. (9.3)

12. State and describe the three main components of a simple electric cell. (9.4)

13. The mercury cell is a special cell for products such as watches and hearing aids. The equations for the half-reactions are

    $$HgO_{(s)} + H_2O_{(l)} + 2\,e^- \rightarrow Hg_{(l)} + 2\,OH^-_{(aq)}$$
    $$Zn_{(s)} + 2\,OH^-_{(aq)} \rightarrow ZnO_{(s)} + H_2O_{(l)} + 2\,e^-$$

(a) Write the equation for the overall or net cell reaction.

(b) In which direction do the electrons flow—mercury to zinc or zinc to mercury? Explain briefly.

(c) Identify the anode and cathode in this cell. (9.5)

14. What is the predicted cell potential of each of the following standard cells? Include half-cell reaction equations.
(a) permanganate-silver cell
(b) tin-zinc cell (9.5)

15. Predict the potential of the following standard cells:
(a) $Co_{(s)} | Co^{2+}_{(aq)} \| Zn^{2+}_{(aq)} | Zn_{(s)}$
(b) $Cu_{(s)} | Cu^{2+}_{(aq)} \| Sn^{2+}_{(aq)} | Sn_{(s)}$
(c) $C_{(s)} \| MnO^-_{4(aq)}, H^+_{(aq)}, Mn^{2+}_{(aq)} \| Ni^{2+}_{(aq)} | Ni_{(s)}$ (9.5)

16. A standard nickel-cadmium cell is constructed and tested.
(a) Predict which electrode will be the cathode and which one will be the anode.
(b) List all entities present, write the half-cell and net cell reaction equations, and calculate the cell potential.
(c) Sketch and label a cell diagram for a standard nickel-cadmium cell. Specify all substances, label important cell components, and show the direction of electron and ion movement. (9.5)

17. Identify and describe the components of the standard half-cell used as a reference for reduction potential. (9.5)

18. Given the potential of the following standard cell with a cadmium anode, predict the reduction potential of the cerium(III) ion half-cell.

$Ce_{(s)} | Ce^{3+}_{(aq)} \| Cd^{2+}_{(aq)} | Cd_{(s)} \quad \Delta E° = 1.94 \ V$ (9.5)

19. What is the voltage and chemical state of a "dead" galvanic cell? (9.5)

20. One method of reducing the corrosion of an iron object is by applying a protective coating to its surface. Describe the chemical processes involved in protecting iron with the following metals:
(a) tin   (b) zinc (9.6)

21. Describe the following methods of cathodic protection and use redox theory to explain how each method can prevent the corrosion of iron pipes and tanks.
(a) impressed current
(b) sacrificial anode (9.6)

Applying Inquiry Skills

22. A group of students are given the following materials and told to construct an operating galvanic cell.

a strip of copper	1.0 mol/L copper(II) sulfate
a strip of lead	1.0 mol/L lead(II) nitrate
connecting wire	1.0 mol/L sodium nitrate
a U-tube and cotton	

(a) Draw a diagram to show how these materials can be used to construct an operating galvanic cell:
(b) Write the equations for the anode half-reaction; cathode half-reaction; overall reaction.
(c) Predict the cell potential. (9.5)

23. Your challenge is to identify three unknown solutions using only the materials listed: 0.25 mol/L solutions of unknowns A, B, and C; silver, zinc, and magnesium strips; dropper bottles of 0.25 mol/L aqueous solutions of sodium sulfate, sodium carbonate, and sodium hydroxide; steel wool; test tubes and test-tube rack; 50-mL beakers; 400-mL waste beaker.
(a) Assuming all possible spontaneous reactions are rapid and that the nitrate ion is a spectator ion, write a procedure to identify which solution is sodium nitrate, which is lead(II) nitrate, and which is calcium nitrate.
(b) Describe the expected results. (9.5)

Making Connections

24. For decades, the use of electric cars has been impeded by the lack of a powerful, lightweight, inexpensive battery. Recent advances in the hydrogen fuel cell are facilitating the introduction of electric cars.
(a) Describe the operation of the Ballard fuel cell.
(b) Speculate on the environmental and social impacts of the widespread use of electric cars. (9.4)

25. In a methane fuel cell, the chemical energy of this compound is converted into electrical energy instead of the heat that would flow during the combustion of methane.
(a) Using only the following half-reactions and reduction potentials, write a net reaction equation and determine the approximate potential for the methane fuel cell:

$$CO^{2-}_{3(aq)} + 7 \ H_2O_{(g)} + 8 \ e^- \rightarrow CH_{4(g)} + 10 \ OH^-_{(aq)}$$
$$E_r = +0.17 \ V$$
$$O_{2(g)} + 2 \ H_2O_{(g)} + 4 \ e^- \rightarrow 4 \ OH^-_{(l)} \quad E_r = +0.40 \ V$$

(b) Discuss some of the advantages and disadvantages of this technology. (9.5)

10

Electrolytic Cells

In this chapter, you will be able to

- identify and describe the functioning of the components in electrolytic cells;

- describe electrolytic cells in terms of oxidation and reduction half-cells and electric potential differences;

- demonstrate an understanding of the interrelationships of time, current, and the quantity of substance produced or consumed in an electrolytic process;

- use appropriate scientific and technological vocabulary related to electrochemistry;

- write balanced chemical equations for redox reactions using half-reaction equations;

- determine oxidation and reduction half-cell reactions, current and ion flow, electrode polarity, and cell potentials of typical electrolytic cells;

- predict the spontaneity of redox reactions and cell potentials using a table of half-cell reduction potentials;

- solve problems based on Faraday's law;

- explain how electrolytic processes are involved in industrial processes;

- research and assess environmental, health, and safety issues involving electrochemistry.

All living things and many industries depend on spontaneous redox reactions. These reactions can also produce electricity, as you have seen in the previous chapter.

What about redox reactions that are not spontaneous? You may be surprised to learn that common reactions essential for producing many substances are often nonspontaneous. The loonie, all aluminum products, including pop cans, some jewellery, and cutlery, the chlorine used in water-treatment plants, household bleach, copper electrical wiring, and magnesium alloys, as in "mag" wheels, are all produced through redox reactions that must be forced.

In this chapter, you will learn how electricity can be used to produce these and other products and how the technology has developed to do this on a large scale.

REFLECT on your learning

1. What does it mean if the cell potential is a negative value?

2. How are elements produced from naturally occurring substances?

3. What is the relationship between the amount, in moles, of electrons transferred in a cell and the amount of product produced at an electrode?

4. What has been the relationship of science, technology, and society in the evolution of electrolytic (nonspontaneous) cells?

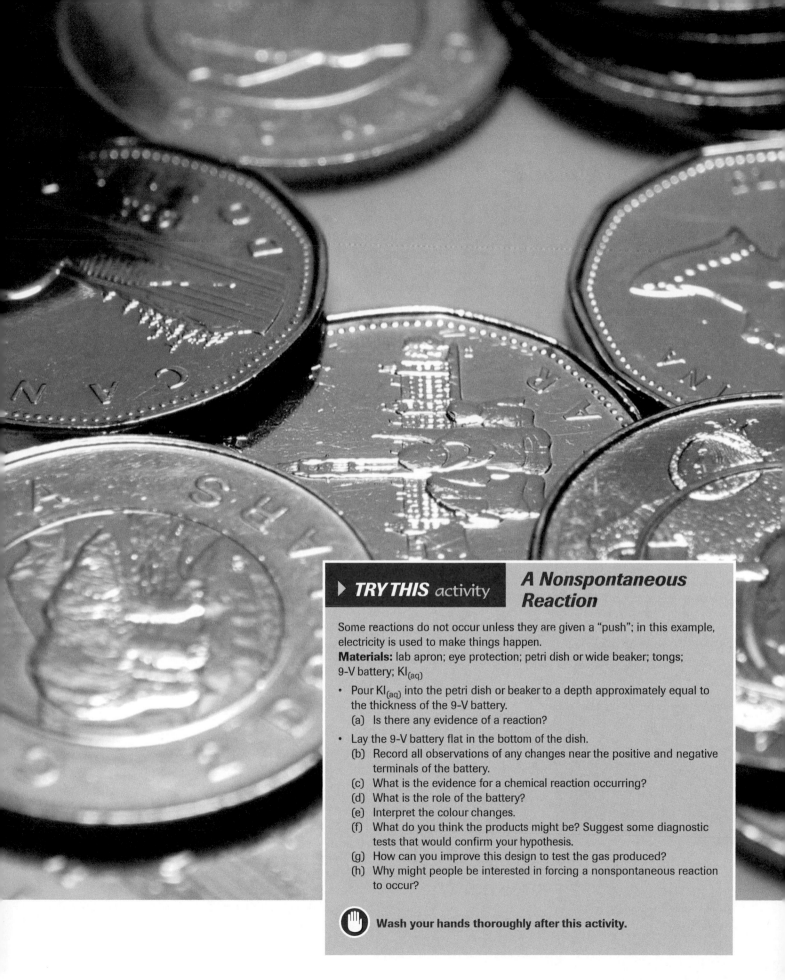

▶ *TRY THIS* activity

A Nonspontaneous Reaction

Some reactions do not occur unless they are given a "push"; in this example, electricity is used to make things happen.

Materials: lab apron; eye protection; petri dish or wide beaker; tongs; 9-V battery; $KI_{(aq)}$

- Pour $KI_{(aq)}$ into the petri dish or beaker to a depth approximately equal to the thickness of the 9-V battery.
  - (a) Is there any evidence of a reaction?

- Lay the 9-V battery flat in the bottom of the dish.
  - (b) Record all observations of any changes near the positive and negative terminals of the battery.
  - (c) What is the evidence for a chemical reaction occurring?
  - (d) What is the role of the battery?
  - (e) Interpret the colour changes.
  - (f) What do you think the products might be? Suggest some diagnostic tests that would confirm your hypothesis.
  - (g) How can you improve this design to test the gas produced?
  - (h) Why might people be interested in forcing a nonspontaneous reaction to occur?

🤚 **Wash your hands thoroughly after this activity.**

10.1 Electrolysis

An electric cell contains reactants chosen to react spontaneously to convert their chemical energy into electrical energy. These cells or batteries can be used to power a portable MP3 player, start a car, or plate silver metal on jewellery.

A scientific research cell or galvanic cell produces electricity spontaneously because each half-cell contains both oxidized and reduced entities. The cell potential ΔE is always greater than zero. In a redox table, the stronger oxidizing agent present in the cell will always be above the stronger reducing agent present.

If a cell does not contain all oxidized and reduced species shown in the half-reaction equation, it is possible that the reactants (electrodes and electrolyte) present will not react spontaneously. For example, if lead electrodes are placed in a solution of zinc sulfate and the electrodes are connected with a wire, there is no evidence of any reaction.

$Pb_{(s)} \mid ZnSO_{4(aq)} \mid Pb_{(s)}$

	$Pb^{2+}_{(aq)}$	+	$2\,e^-$	$\rightleftharpoons$	$Pb_{(s)}$	**SRA**	
	$Ni^{2+}_{(aq)}$	+	$2\,e^-$	$\rightleftharpoons$	$Ni_{(s)}$		$\Delta E<0$
	$Fe^{2+}_{(aq)}$	+	$2\,e^-$	$\rightleftharpoons$	$Fe_{(s)}$		
SOA	$Zn^{2+}_{(aq)}$	+	$2\,e^-$	$\rightleftharpoons$	$Zn_{(s)}$		

Figure 1
According to the redox spontaneity rule, if the strongest oxidizing agent present is below the strongest reducing agent present, no spontaneous reaction will occur.

> **LEARNING TIP**
>
> **Standard Cells**
> Recall from the previous chapter that a standard cell contains all entities listed in the equation for the half-reaction. In addition, the concentration of aqueous entities is 1.0 mol/L and the conditions are SATP.

The strongest oxidizing agent present in this cell is $Zn^{2+}_{(aq)}$ and the strongest reducing agent present is $Pb_{(s)}$. A quick check in the redox table shows that the oxidizing agent, $Zn^{2+}_{(aq)}$, is well below the position of the reducing agent, $Pb_{(s)}$, and the ΔE is negative (**Figure 1**). Let us calculate the $\Delta E°$ for the only reaction that could occur.

$$Zn^{2+}_{(aq)} + 2\,e^- \rightarrow Zn_{(s)} \qquad E_r° = -0.76\text{ V}$$

$$Pb_{(s)} \rightarrow Pb^{2+}_{(aq)} + 2\,e^- \qquad E_r° = -0.13\text{ V}$$

$$Zn^{2+}_{(aq)} + Pb_{(s)} \rightarrow Zn_{(s)} + Pb^{2+}_{(aq)}$$

$$\Delta E°_{\text{cell}} = E_r°_{\text{cathode}} - E_r°_{\text{anode}}$$

$$= -0.76\text{ V} - (-0.13\text{ V})$$

$$= -0.89\text{ V}$$

Since the $\Delta E°$ for the reaction is negative, we conclude that the lead will not be oxidized spontaneously in the zinc sulfate solution. (Note that the reverse reaction would be spontaneous but cannot occur because neither $Pb^{2+}_{(aq)}$ nor $Zn_{(s)}$ is present initially.)

Strictly speaking, the zinc sulfate cell is not a standard cell, even if the concentration of the zinc sulfate were 1.0 mol/L. Therefore, the cell potential that is calculated is not accurate, but will be close enough for our purposes. This cell would not produce electricity because the reaction is nonspontaneous. Why would anyone be interested in a cell like this? Certainly not to use in a battery. However, by supplying electrical energy to a nonspontaneous cell, we can force the reaction to occur. As you will see later in this chapter, this is especially useful for producing substances, particularly elements. For example, the zinc sulfate cell discussed above is similar to the cell used in the industrial production of zinc metal (**Figure 2**).

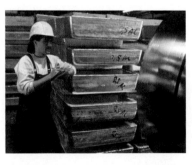

Figure 2
Cominco in Trail, B.C., operates the world's largest zinc and lead smelter, producing almost 300 kt of zinc annually.

The term *electrochemical cell* is often used in chemistry to refer to either a cell with a spontaneous reaction, such as the electric or galvanic cell, or a cell with a nonspontaneous reaction, which we will call an **electrolytic cell** (**Figure 3**), which uses a process called **electrolysis.** The external power supply acts as an "electron pump"; the electric energy is used to do work on the electrons to cause an electron transfer inside the electrolytic cell. In an electrolytic cell, the chemical reaction is the reverse of that of a spontaneous cell. However, most of the scientific principles you've already studied also apply to electrolytic cells (**Table 1**).

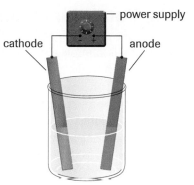

Figure 3
Electrons are pulled from the anode and pushed to the cathode by the battery or power supply.

electrolytic cell a cell that consists of a combination of two electrodes, an electrolyte, and an external battery or power source

electrolysis the process of supplying electrical energy to force a nonspontaneous redox reaction to occur

reactants $\underset{\text{electrolytic cell}}{\overset{\text{electric/galvanic cell}}{\rightleftarrows}}$ products + electrical energy

Table 1 Electrochemical Cells: Galvanic and Electrolytic

	Galvanic cell	Electrolytic cell
Spontaneity	spontaneous reaction	nonspontaneous reaction
Standard cell potential, $\Delta E°$	positive	negative
Cathode	• strongest oxidizing agent present undergoes a *reduction* • positive electrode	• strongest oxidizing agent present undergoes a *reduction* • negative electrode
Anode	• strongest reducing agent present undergoes an *oxidation* • negative electrode	• strongest reducing agent present undergoes an *oxidation* • positive electrode
Direction of electron movement	anode → cathode	anode → cathode
Direction of ion movement	anions → anode cations → cathode	anions → anode cations → cathode

A secondary cell is a rechargeable cell such as a nickel-cadmium (Ni-Cad) cell. A secondary cell can be used to illustrate the difference between an electric or galvanic cell and an electrolytic cell. As the cell discharges, electrical energy is spontaneously produced and the cell functions as an electric cell. When the cell is recharged, the electrical energy forces the products to react and re-form the original reactants. During recharging, the secondary cell is functioning as an electrolytic cell (**Figure 4**).

Figure 4
Recall the analogy of the can for electric potential difference. Water spontaneously flows from a region of higher gravitational potential energy to a region of lower gravitational potential energy. In a similar analogy for an electrolytic cell, water from a river (A) is pumped (B) up into a water tower (C). Work must be done (energy is consumed) to raise the water to the higher *gravitational* potential energy. In an electrolytic cell, electrical work is done to increase the *chemical* potential energy.

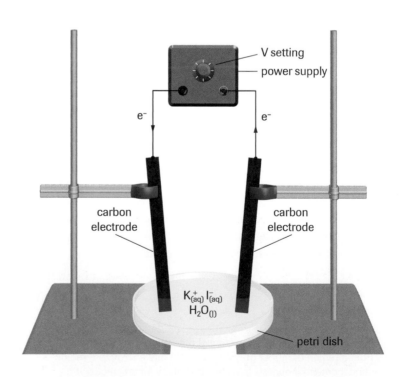

A Potassium Iodide Electrolytic Cell (p. 724)
Construct a simple electrolytic cell and observe it in action.

LEARNING TIP

Don't Forget the Water!
At first glance, the cell notation, $C_{(s)} \mid KI_{(aq)} \mid C_{(s)}$, does not show that water is involved in the reaction. Of course, the subscript (aq) means dissolved in water. Don't forget to consider the presence of water, because it is often a reactant in aqueous electrolytic cells.

The Potassium Iodide Electrolytic Cell: A Synthesis

In the potassium iodide electrolytic cell (**Figure 5**), litmus paper does not change colour in the initial solution and turns blue only near the electrode from which gas bubbles. At the other electrode, a yellow-brown colour and a dark precipitate forms. The yellow-brown substance produces a violet colour in a halogen test. This chemical evidence agrees with the interpretation supplied by the following half-reaction equations. According to the redox table of relative strengths of oxidizing and reducing agents, water is the stronger oxidizing agent present and iodide ions are the stronger reducing agents present in a potassium iodide solution. ▨▮

$$
\begin{array}{ccc}
\text{OA} & & \text{SOA} \\
K^+_{(aq)} & I^-_{(aq)} & H_2O_{(l)} \\
& \text{SRA} &
\end{array}
$$

cathode: $2\,H_2O_{(l)} + 2e^- \rightarrow H_{2(g)} + 2\,OH^-_{(aq)}$
gas bubbles   blue litmus

anode: $2I^-_{(aq)} \rightarrow I_{2(s)} + 2\,e^-$
purple in cyclohexane

Evidence from the study of this and many other aqueous electrolytic cells suggests that the generalizations for electric or galvanic cells also apply to electrolytic cells. From a theoretical perspective, the strongest oxidizing agent present in a particular mixture has the greatest attraction for electrons and gains electrons at the cathode. Notice that it does not matter where the electrons originate, from a power supply or directly from another electrode. The strongest reducing agent present in the mixture has the least attraction for electrons and loses electrons at the anode.

Figure 5
A power supply provides the energy for the chemical reactions at the two electrodes.

V setting
power supply
e^- e^-
carbon electrode   carbon electrode
$K^+_{(aq)}\ I^-_{(aq)}$
$H_2O_{(l)}$
petri dish

The theoretical definitions of cathode and anode are the same for both galvanic and electrolytic cells (**Table 1**).

Observation of a potassium iodide cell indicates that the transfer of electrons is not spontaneous. When a voltage is supplied to the cell, electrons that are supplied from the negative terminal of the battery flow toward the cathode of the electrolytic cell and are consumed (gained) by water molecules, which have the more positive reduction potential. Simultaneously, electrons flow from iodide ions on the surface of the anode to the positive terminal of the battery. This explanation agrees with previous redox concepts and agrees with the observations, so we can judge the explanation acceptable. Predictions of cathode, anode, and overall cell reactions for electrolytic cells follow the same steps outlined for galvanic cells in Section 9.5.

Analyzing Electrolytic Cells   SAMPLE problem ◀

What is the cell potential of the potassium iodide electrolytic cell?

First, we identify the major entities in the solution and use the table of Relative Strengths of Oxidizing and Reducing Agents (Appendix C9) to identify the strongest oxidizing and reducing agents, as in Chapter 9.

OA   SOA

$K^+_{(aq)}$, $I^-_{(aq)}$, $H_2O_{(l)}$

          SRA

Now we can calculate the cell potential. The potassium iodide cell is not a standard cell because the products of the reactions are not present initially. Therefore, the reduction potentials given in the table of half-reactions are not strictly applicable, but we will use them to approximate the cell potential.

cathode: $2\,H_2O_{(l)} + 2\,e^- \rightarrow H_{2(g)} + 2\,OH^-_{(aq)}$ $E^\circ_r = -0.83\,V$

anode: $2\,I^-_{(aq)} \rightarrow I_{2(s)} + 2\,e^-$ $E^\circ_r = +0.54\,V$

net $2\,H_2O_{(l)} + 2\,I^-_{(aq)} \rightarrow H_{2(g)} + 2\,OH^-_{(aq)} + I_{2(s)}$

ΔE° = E°_r – E°_r
cell   cathode   anode

    = $-0.83\,V$ – $(+0.54\,V)$

    = $-1.37\,V$

A negative sign for a cell potential indicates that the chemical process is nonspontaneous. The more negative the cell potential, the more energy is required. In this case, to force the cell reactions, electrons must be supplied with a minimum of $+1.37\,V$ from an external battery or other power supply. In practice, however, a greater voltage is required.

SUMMARY   *Procedure for Analyzing Electrolytic Cells*

Step 1   Use the cell notation as a list or make a list of all substances present. Label all possible oxidizing and reducing agents present. (Do not forget to include water for aqueous electrolytes.)

Step 2   Use the redox table (Appendix C9) to identify the strongest oxidizing agent present in the cell. Write the equation for its reduction half-reaction, including the reduction potential. (This is the reaction at the cathode.)

▶

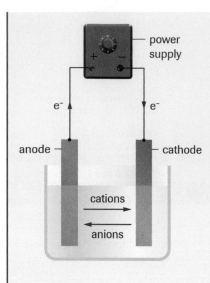

anode — | — cathode

cations →

anions ←

power supply

e⁻ | e⁻

Step 3 Use the redox table (Appendix C11) to identify the strongest reducing agent present in the cell. Write the equation for its oxidation half-reaction (by reversing the reduction equation) and write the reduction potential. (This is the reaction at the anode.)

Step 4 Balance electrons and write the equation for the overall or net cell reaction.

Step 5 Calculate the cell potential. (If this is negative, then the cell reaction is nonspontaneous.)

$$\Delta E° = E°_r - E°_r$$
$$ \text{cathode} \quad \text{anode}$$

Step 6 If required, state the minimum electric potential (voltage) to force the reaction to occur. (The minimum voltage is the absolute value of $\Delta E°$.)

Example 1

An electrolytic cell containing cobalt(II) chloride solution and lead electrodes is assembled.

$$Pb_{(s)} \mid Co^{2+}_{(aq)}, Cl^-_{(aq)} \mid Pb_{(s)}$$

(a) Predict the reactions at the cathode and anode, and in the overall cell.
(b) Draw and label a cell diagram for this electrolytic cell, including the power supply.
(c) What minimum voltage must be applied to make this cell work?

Solution

(a) SRA SOA

$$Pb_{(s)} \mid Co^{2+}_{(aq)} \ Cl^-_{(aq)} \mid Pb_{(s)}$$

cathode $\quad Co^{2+}_{(aq)} + 2\,e^- \rightarrow Co_{(s)}$

anode $\qquad\qquad\quad Pb_{(s)} \rightarrow Pb^{2+}_{(aq)} + 2\,e^-$

net $\quad Co^{2+}_{(aq)} + Pb_{(s)} \rightarrow Co_{(s)} + Pb^{2+}_{(aq)}$

(b)

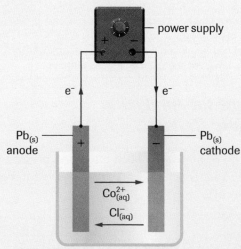

power supply

e⁻ | e⁻

Pb_{(s)} anode

Pb_{(s)} cathode

$Co^{2+}_{(aq)}$

$Cl^-_{(aq)}$

(c) $\Delta E° = E_r°_{\text{cathode}} - E_r°_{\text{anode}}$

$= -0.28\ V - (-0.13\ V)$

$= -0.15\ V$

A minimum applied voltage of +0.15 V is required.

Example 2

An electrolytic cell is set up with a power supply connected to two nickel electrodes immersed in an aqueous solution containing cadmium nitrate and zinc nitrate. Predict the equations for the initial reaction at each electrode and the net cell reaction. Calculate the minimum voltage that must be applied to make the reactions occur.

Solution

SRA SOA

$Ni_{(s)}$, $H_2O_{(l)}$, $Cd^{2+}_{(aq)}$, $NO^-_{3(aq)}$, $Zn^{2+}_{(aq)}$

$$Cd^{2+}_{(aq)} + 2\,e^- \rightarrow Cd(s) \qquad E_r° = -0.40\ V$$

$$Ni_{(s)} \rightarrow Ni^{2+}_{(aq)} + 2\,e^- \qquad E_r° = -0.26\ V$$

$$Cd^{2+}_{(aq)} + Ni_{(s)} \rightarrow Cd_{(s)} + Ni^{2+}_{(aq)}$$

$\Delta E° = E_r°_{\text{cathode}} - E_r°_{\text{anode}}$

$= -0.40\ V - (-0.26\ V)$

$= -0.14\ V$

A minimum applied voltage of +0.14 V is required.

Practice

1. Predict the cathode, anode, and net cell reactions for each of the following electrolytic cells. Calculate the minimum potential difference that must be applied to force the cell reaction to occur.
   (a) $C_{(s)} \mid Ni^{2+}_{(aq)}, I^-_{(aq)} \mid C_{(s)}$ (b) $Pt_{(s)} \mid Na^+_{(aq)}, OH^-_{(aq)} \mid Pt_{(s)}$
2. What is the minimum electric potential difference of an external power supply that produces chemical changes in the following electrolytic cells?
   (a) $C_{(s)} \mid Cr^{3+}_{(aq)}, Br^-_{(aq)} \mid C_{(s)}$ (b) $Cu_{(s)} \mid Cu^{2+}_{(aq)}, SO^{2-}_{4(aq)} \mid Cu_{(s)}$
3. Two tin electrodes are placed in an aqueous solution containing potassium nitrate and magnesium iodide.
   (a) If a power supply is connected to force any reactions to occur, what would be the reactions at the cathode, anode, and in the overall cell?
   (b) Draw and label a cell diagram, including electrodes, electrolyte, power supply, and the direction of movement of electrons and ions.
   (c) Would a 1.5-V cell be suitable as a power supply? Justify your answer.
4. Would the following cell have a spontaneous reaction? Explain.
   $C_{(s)} \mid Na_2SO_{4(aq)} \mid Pb_{(s)}$

Making Connections

5. Describe a specific consumer product that you use sometimes as an electric cell and sometimes as an electrolytic cell.

Answers

1. (a) 0.80 V
   (b) 1.23 V
2. (a) 1.48 V
   (b) 0 V

DID YOU KNOW?

Cold Fusion

Fusion is the type of nuclear reaction that takes place in the Sun at extremely high temperatures. "Cold fusion" is the combining of the nuclei of hydrogen isotopes to yield larger nuclei at room temperature rather than at solar temperatures. Researchers claimed in 1989 that cold fusion had been achieved during the electrolysis of heavy water, $^2_1H_2O$, with palladium electrodes. Some experimental results indicated a possible net gain in energy. An unusual aspect of this story is that the report of these results was made through the media before the results were published in a professional journal. The hypothesis proposed was that hydrogen-2 molecules that had dissolved in the palladium crystal lattice had undergone fusion to produce helium-3, a neutron, and some energy. There has been little supporting evidence from other laboratories and the hypothesis of cold fusion has not been supported by theoreticians.

SUMMARY Electrolytic Cells

- An electrolytic cell is based upon a reaction that is nonspontaneous; the $\Delta E°$ for the reaction is negative. An applied voltage of at least the absolute value of ΔE is required to force the reactions to occur.

- The strongest oxidizing agent present in the cell undergoes reduction at the cathode (negative electrode).

- The strongest reducing agent present in the cell undergoes oxidation at the anode (positive electrode).

- Electrons are forced by a power supply to travel from the anode to the cathode through the external circuit.

- Internally, anions in the cell move toward the anode and cations move toward the cathode. ▨▮

▨ INVESTIGATION 10.1.2

Investigating Several Electrolytic Cells (p. 755)
The products of several electrolytic cells with common substances and inert electrodes are predicted and then tested.

▶ *Section 10.1 Questions*

Understanding Concepts

1. Describe the type of agent reacting and the process occurring at the cathode and anode of an electrolytic cell.

2. What is different about the cathode and anode of an electrolytic cell versus a galvanic cell?

3. Describe the direction of movement of electrons and ions within an electrolytic cell.

4. Explain why a power supply is necessary for an electrolytic cell.

5. Which of the following cells would produce a spontaneous reaction? Justify each answer, using the cell potential.
   (a) $C_{(s)} \mid Cr(NO_3)_{2(aq)} \mid C_{(s)}$
   (b) $Ag_{(s)} \mid FeCl_{3(aq)} \mid Ag_{(s)}$
   (c) $Cu_{(s)} \mid Pb(NO_3)_{2(aq)} \mid Cu_{(s)}$

6. For each of the following electrolytic cells, write equations for the cathode and anode half-reactions and the net reaction. Determine the minimum potential difference that must be applied to make the cell operate.
   (a) $C_{(s)} \mid K_2SO_{4(aq)} \mid Cd_{(s)}$
   (b) $Pt_{(s)} \mid SnBr_{2(aq)} \mid Pt_{(s)}$

7. Draw a diagram of an electrolytic cell containing a zinc iodide solution and inert carbon electrodes.
   - Label the power supply and electrodes, including signs, the electrolyte, as well as the directions of electron and ion movement.
   - Write half-reaction and net equations.
   - Calculate the cell potential, using standard values.

8. German silver is an alloy containing copper, zinc, and nickel. A piece of German silver is used as the anode in an electrolytic cell containing aqueous sodium sulfate. The other electrode is platinum metal.
   (a) As the applied voltage is slowly increased, in what order will the half-reactions occur at the anode? Write an equation for each half-reaction.
   (b) Describe what happens at the cathode.

9. Discuss the spontaneity of the reaction in the following cell when no potential difference is applied.

$$Co_{(s)} \mid CoSO_{4(aq)} \mid Co_{(s)}$$

The invention of the electric cell by Volta in 1800 immediately resulted in many discoveries in chemistry. One of these discoveries was that electric cells could be used as an electric power source for electrolytic cells. Many natural substances, such as soda (sodium carbonate) and potash (potassium carbonate) that were thought to be elements, were shown to be composed of the previously unknown elements sodium and potassium (**Figure 1**). Industrial applications of electrolytic cells include the production of elements, the refining of metals, and the plating of metals onto the surface of an object. The study of electrolysis in industry reveals the strong relationship between science and technology.

Production of Elements

Most elements occur naturally combined with other elements in compounds. For example, ionic compounds of sodium, potassium, lithium, magnesium, calcium, and aluminum are abundant, but these reactive metals are not found uncombined in nature. The explanation is that the reduction potentials for these metals are very negative. Consequently, the metals are easily oxidized by practically all other substances. Even water has a more positive reduction potential than any of these metal ions, so if the metals did exist naturally, a spontaneous reaction would convert them into their ions (**Figure 2**).

$$
\begin{array}{c}
\text{SOA} \\
\downarrow
\end{array}
\quad
\begin{aligned}
Zn^{2+}_{(aq)} + 2\,e^- &\rightleftharpoons Zn_{(s)} \\
2H_2O_{(l)} + 2\,e^- &\rightleftharpoons H_{2(g)} + 2OH^-_{(aq)} \\
Mg^{2+}_{(aq)} + 2\,e^- &\rightleftharpoons Mg_{(s)} \\
Na^+_{(aq)} + e^- &\rightleftharpoons Na_{(s)}
\end{aligned}
$$

Figure 2
In an aqueous cell, metal cations can undergo a reduction to the metal as long as the metal cation is above water; i.e., when the metal cation is a stronger oxidizing agent. If you try to reduce active metal cations such as sodium and magnesium ions, water will react instead.

Many metals can be produced by electrolysis of solutions of their ionic compounds, but two difficulties arise. First, many naturally occurring ionic compounds have low solubility in water and second, water is a stronger oxidizing agent than active metal cations. To overcome these difficulties, a technological design in which water is not present can be used. Fortunately, ionic compounds can be melted. These molten ionic compounds are good electrical conductors and can function as the electrolyte in a cell. In the electrolysis of molten binary ionic compounds, only one oxidizing agent and one reducing agent are present. The production of active metals (strong reducing agents) from their minerals typically involves the electrolysis of molten compounds of the metal, a technology first used in the scientific work of Humphry Davy.

Strontium metal was one of many active metals discovered by Davy using the electrolysis of molten salts. Strontium chloride was first melted in an electrolytic cell with inert electrodes. In this cell, there are only two kinds of ions present, $Sr^{2+}_{(l)}$ and $Cl^-_{(l)}$. You may recall from the previous chapter that metal cations generally tend to undergo a reduction and nonmetal anions tend to undergo an oxidation. In this cell, there are no other competing substances. Therefore, the strontium ions will consume (gain) electrons at the cathode to form strontium metal.

Figure 1
In his youth, Sir Humphry Davy (1778–1829) worked as an assistant to a physician who was interested in the therapeutic properties of gases. Davy studied nitrous oxide (laughing gas) by conducting experiments on himself. He was eventually fired from his job, supposedly because of his liking for explosive chemical reactions. Davy's main fame came from his experiments with electricity. He constructed a voltaic pile with over 250 metal plates. He used this powerful cell to decompose stable compounds and discover the elements sodium, potassium, barium, strontium, calcium, and magnesium. Davy had many other scientific accomplishments; for example, he was the first to show that chlorine is an element and will support combustion, and that hydrogen is the key component of acids. Given his habit of tasting, inhaling, and exploding new chemicals, it is not surprising to learn that he was an invalid in his early thirties and died in middle age, probably of chemical poisoning.

> **LEARNING TIP**
>
> No reduction potentials can be listed for the electrolysis of a molten salt. The table of oxidizing and reducing agents in Appendix C11 lists only electric potentials for half-reactions in 1.0 mol/L aqueous solutions at SATP.

$$Sr^{2+}_{(l)} + 2\,e^- \rightarrow Sr_{(s)} \qquad \text{(reduction at the cathode)}$$

At the anode, chloride ions will give up (lose) electrons to form chlorine gas.

$$2\,Cl^-_{(l)} \rightarrow Cl_{2(g)} + 2\,e^- \qquad \text{(oxidation at the anode)}$$

Electrons are balanced and adding the two equations gives the overall reaction in the cell.

$$Sr^{2+}_{(l)} + 2\,Cl^-_{(l)} \rightarrow Sr_{(s)} + Cl_{2(g)}$$

This reaction would not be possible in an aqueous solution because water is a stronger oxidizing agent (i.e., has a more positive reduction potential) than aqueous strontium ions.

> In molten-salt electrolysis, metal cations move to the cathode and are reduced to metals, and nonmetal anions move to the anode and are oxidized to nonmetals.

Production of Sodium

Electrolysis of molten ionic compounds is expensive; a significant quantity of energy must be used and the electrolysis cell must be specially designed to withstand the high temperatures involved. One common method to reduce the temperature is to add an inert compound to form a mixture that melts at a lower temperature. In general, the melting point of any substance is lowered by adding an impurity. (This is the reason people sprinkle salt on roads or sidewalks in winter to melt ice—adding salt lowers the melting point of ice.) Pure sodium chloride has a melting point of about 800°C, but when mixed with calcium chloride, the melting point is about 600°C. In such a cell, the potential difference that is applied to the mixture must be controlled to reduce sodium ions but not calcium ions (**Figure 3**). The electrolysis of molten sodium chloride in a Down's cell is the main source of sodium metal (**Figure 4**).

Figure 4
One of the uses of sodium metal is the production of sodium vapour lamps used for street lighting. Sodium lamps produce a yellow light that penetrates farther than white light and allows better vision in fog. Sodium lamps are also fifteen times more efficient than regular incandescent lamps.

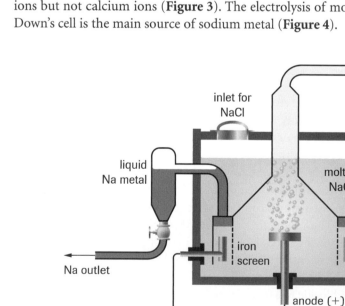

Figure 3
At the operating temperature of the cell (600°C), sodium is liquid because its melting point is only 98°C. Liquid sodium metal is formed at the cylindrical cathode and then floats to the top of the molten sodium chloride. Chlorine gas forms on the carbon anode and rises out of the cell.

Example 1

Lithium is the least dense of all metals and is a very strong reducing agent; both qualities make it an excellent anode for batteries. Lithium can be produced by electrolysis of molten lithium chloride at a temperature greater than 605°C, the melting point of lithium chloride. Write the equations for the cathode and anode half-reactions, and the net cell reaction.

Solution

$$\text{cathode:} \quad 2[Li^+_{(l)} + e^- \rightarrow Li_{(s)}]$$
$$\text{anode:} \quad 2\,Cl^-_{(l)} \rightarrow Cl_{2(g)} + 2\,e^-$$
$$\overline{2\,Li_{(l)} + 2\,Cl^-_{(l)} \rightarrow 2\,Li_{(s)} + Cl_{2(g)}}$$

Production of Aluminum

Aluminum is the third most abundant element on Earth. It was discovered in France in the early 1800s. At the time, aluminum was more expensive than gold. The wonderful properties of aluminum—shiny, light, strong, and corrosion resistant—made it ideal for jewellery and cutlery, so there was a high demand for the metal, especially among the aristocracy. However, the supply of aluminum was limited because the technology for producing aluminum was not yet practical or economically viable for mass production.

Initial efforts to produce aluminum by electrolysis were unproductive because its common ore, $Al_2O_{3(s)}$, has a high melting point, 2072°C. No material could be found to hold the molten compound. In 1886 two scientists, working independently and knowing nothing of each other's work, made the same discovery. Charles Martin Hall in the United States and Paul-Louis-Toussaint Héroult in France discovered that $Al_2O_{3(s)}$ dissolves in a molten mineral called cryolite, Na_3AlF_6. In this design, the cryolite acts as an inert solvent for the electrolysis of aluminum oxide and forms a molten conducting mixture with a melting point around 1000°C. Aluminum (m.p. 660°C) can be produced electrolytically from this molten mixture (**Figure 5**). This discovery had an immediate effect on the supply and cost of aluminum. Around 1855, aluminum was sold for $45,000 per kilogram; a few years after the Hall-Héroult invention, the cost was about 90 cents.

Aluminum Production in Canada
The production of aluminum is important to Canada's economy, although Canada does not have large deposits of aluminum ore. Hydroelectric power is used to produce aluminum metal from concentrated, imported bauxite in an electrolytic cell. Recycling aluminum from soft drink and beer cans requires only 5% of the energy required to produce aluminum by electrolysis.

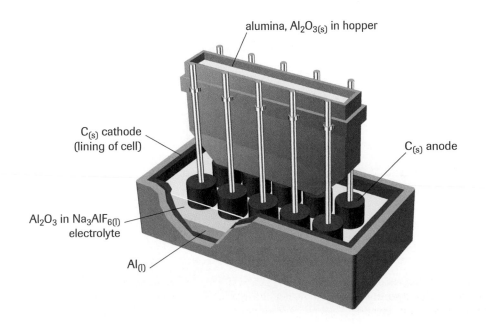

alumina, $Al_2O_{3(s)}$ in hopper

$C_{(s)}$ cathode (lining of cell)

$C_{(s)}$ anode

Al_2O_3 in $Na_3AlF_{6(l)}$ electrolyte

$Al_{(l)}$

Figure 5
The Hall-Héroult cell for the production of aluminum. The cathode is the carbon lining of the steel cell. At the cathode, aluminum ions are reduced to produce liquid aluminum, which collects at the bottom of the cell and is periodically drained away. At the carbon anode, oxide ions are oxidized to produce oxygen gas. The oxygen produced at the anode reacts with the carbon electrodes, producing carbon dioxide, so these electrodes must be replaced frequently.

Aluminum oxide is obtained from bauxite, an aluminum ore. Once the ore is purified, the aluminum oxide is dissolved in molten cryolite and it dissociates into individual ions. The reactions occurring at the electrodes in a Hall-Héroult cell are summarized below.

cathode: $\qquad$ $4[Al^{3+}_{(cryolite)} + 3\,e^- \rightarrow Al_{(l)}]$

anode: $\qquad$ $3[2\,O^{2-}_{(cryolite)} \rightarrow O_{2(g)} + 4\,e^-]$

$$4\,Al^{3+}_{(cryolite)} + 6\,O^{2-}_{(cryolite)} \rightarrow 4\,Al_{(l)} + 3\,O_{2(g)}$$

The overall cell reaction is a decomposition of aluminum oxide.

$$2\,Al_2O_{3(s)} \rightarrow 4\,Al_{(s)} + 3\,O_{2(g)}$$

The Chlor-Alkali Process

Instead of eliminating or replacing water as a solvent in the electrolytic production of elements, another design overcomes the difficulty of producing active metals by simply "overpowering" the reduction of the water. A high voltage leads to the reduction of metal ions rather than water because the reduction of water is a relatively slow reaction. Aqueous sodium chloride can be electrolyzed in this manner to produce chlorine, hydrogen, and sodium hydroxide, all very important industrial chemicals.

$$2\,NaCl_{(aq)} + 2\,H_2O_{(l)} \rightarrow H_{2(g)} + Cl_{2(g)} + 2\,NaOH_{(aq)}$$

This process, called the *chlor-alkali process*, is electrolysis of a concentrated sodium chloride solution. One common design (**Figure 6**) uses high voltages to force the reduction of aqueous sodium ions to sodium metal. The sodium metal is then reacted with water to produce hydrogen gas and sodium hydroxide. Chlorine is preferentially produced at the anode instead of oxygen, in spite of its more favourable position in the redox table. There are several factors that contribute to this effect.

Figure 6

Design of a chlor-alkali cell. The sodium metal forms rapidly at the cathode and is dissolved and carried away by a liquid mercury cathode as soon as it forms. Water is later added to the sodium-mercury solution to form hydrogen gas and a sodium hydroxide solution. Chlorine gas is formed and collected at the anodes.

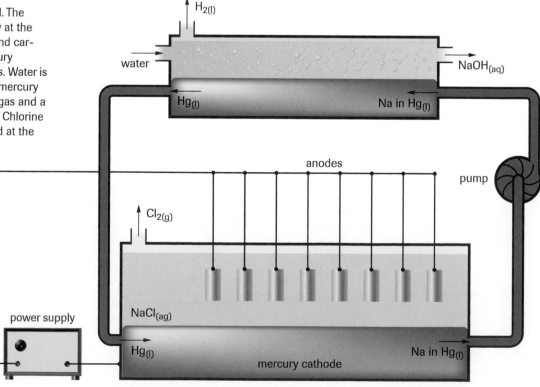

The chlor-alkali technology requires large quantities of electrical energy and, in the past, employed mercury as the cathode. The extreme toxicity of mercury endangered the safety of workers and the environment. Newer chlor-alkali plants now use a process that relies on an ion-exchange membrane to separate the sodium and chloride ions during electrolysis. This new technology not only eliminates mercury but is also less expensive. For all chlor-alkali processes, the overall reaction is still the same.

Hydrogen gas is used to make ammonia, hydrogen peroxide, and margarine, and to crack petroleum. It may also be used on site as a fuel to produce electricity. Chlorine is used as a disinfectant for drinking water and to manufacture bleach (sodium hypochlorite), plastics, pesticides, and solvents. Sodium hydroxide is used on a large scale in industry to make cellophane, pulp and paper, aluminum, and detergents.

▶ *Practice*

Understanding Concepts

1. (a) Describe two difficulties associated with the electrolysis of aqueous ionic compounds in the production of active metals.
   (b) What two designs can be used to offset these difficulties?

2. Scandium is a metal with a low density and a melting point that is higher than that of aluminum. These properties are of interest to engineers who design space vehicles. Scandium metal is produced by electrolysis of molten scandium chloride. List all ions present in the electrolysis cell, and write the equations for the reactions that occur at the cathode and anode and the net cell reaction.

3. The following statements summarize the steps in the chemical technology of obtaining magnesium from seawater. Write a balanced equation to represent each reaction.
   (a) Slaked lime (solid calcium hydroxide) is added to seawater (ignore all solutes except $MgCl_{2(aq)}$) in a double displacement reaction to precipitate magnesium hydroxide.
   (b) Hydrochloric acid is added to the magnesium hydroxide precipitate.
   (c) After the magnesium chloride product is separated and dried, it is melted in preparation for electrolysis. List all ions present in the electrolysis, and write the equations for the reactions that occur at the cathode and anode and the net cell reaction.
   (d) An alternative process produces magnesium from dolomite, a mineral containing $CaCO_3$ and $MgCO_3$. Suggest some technological advantages and disadvantages of the dolomite process compared with the seawater process.

Making Connections

4. What products in your home may have originated from substances produced in the chlor-alkali process?

5. Why should we recycle metals such as aluminum? State several arguments that you might use in a debate.

6. Research and describe some of the variety of uses of aluminum summarized in the exhibition, "Aluminum by Design: Jewellery to Jets."

    www.science.nelson.com

Extension

7. Research and describe the newer, ion-exchange membrane cell design for the chlor-alkali process. Include a labelled cell diagram and the function of the membrane. Why is it superior to other designs?

    www.science.nelson.com

Take a Stand: The Case For and Against Chlorine

Chlorine is a controversial chemical. There is no doubt that chlorine and products made from chlorine have been very beneficial to society, but there are concerns. Should the production or use of chlorine be limited to certain essential uses?

(a) Research the production, storage, transportation, and uses of chlorine and assess one environmental, health, or safety issue.

(b) What is the best resolution of the issue you assessed? Present your solution in a way designed to influence decision makers. Include an outline of your findings in your presentation.

 www.science.nelson.com

---

electrorefining production of a pure metal at the cathode of an electrolytic cell using impure metal at the anode

Refining of Metals

In the production of metals, the initial product is usually an impure metal. Impurities are often other metals that come from various compounds in the original ore. To purify or refine a metal, a variety of methods are used. However, a common method, known as **electrorefining**, uses an electrolytic cell to obtain high-grade metals at the cathode from an impure metal at the anode.

Figure 7
When the electrolytic cell is operated at a carefully controlled voltage, only copper and metals more easily oxidized than copper, such as iron and zinc, are oxidized to ions and dissolve at the anode. Only copper is reduced at the cathode. Other impurities in the anode, such as silver, gold, and platinum, do not react; these fall to the bottom of the cell as a sludge called anode mud. Removed from the cell periodically, the anode mud undergoes further processing to extract valuable metals.

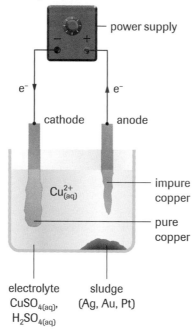

electrolyte
CuSO$_{4(aq)}$,
H$_2$SO$_{4(aq)}$

sludge
(Ag, Au, Pt)

A good example is the electrorefining of copper. The presence of impurities in copper lowers its electrical conductivity, not a desirable property considering that one of the most common uses of copper is in electrical wiring. The initial smelting process produces copper that is about 99% pure, containing some silver, gold, platinum, iron, and zinc. These valuable impurities can be recovered and sold to help pay for the process. As shown in **Figure 7**, a slab of impure copper is the anode of an electrolytic cell that contains copper(II) sulfate dissolved in sulfuric acid. The cathode is a thin sheet of very pure copper. As the cell operates, copper and some of the other metals in the anode are oxidized, but only copper is reduced at the cathode. An understanding of oxidation, reduction, and reduction potentials allows precise control over what is oxidized and what is reduced, so that after electrorefining, the copper is about 99.98% pure (**Figure 8**). The half-reactions are:

cathode:	reduction of copper	$Cu^{2+}_{(aq)} + 2e^- \rightarrow Cu_{(s)}$
anode:	oxidation of copper	$Cu_{(s)} \rightarrow Cu^{2+}_{(aq)} + 2e^-$
	oxidation of zinc	$Zn_{(s)} \rightarrow Zn^{2+}_{(aq)} + 2e^-$
	oxidation of iron	$Fe_{(s)} \rightarrow Fe^{2+}_{(aq)} + 2e^-$

Figure 8
A bundle of cathodes shows the corrugation of alternating cathodes that helps ensure more efficient subsequent melting.

Another related method of purifying metals is to reduce metal cations from a molten or aqueous electrolyte at the cathode of an electrolytic cell, much like the production of elements discussed previously. This method, which uses a molten salt, is known as *electrowinning*. It is the only way to obtain some active metals, such as those in Group 1. Many other metals, such as zinc, can be produced by electrowinning an aqueous solution. For example, Cominco's operation at Trail, BC uses the electrolysis of an acidic zinc sulfate solution with a specially treated lead anode to deposit very pure zinc metal at the cathode.

cathode	$2[Zn^{2+}_{(aq)} + 2e^- \rightarrow Zn_{(s)}]$
anode	$2H_2O_{(l)} \rightarrow O_{2(g)} + 4H^+_{(aq)} + 4e^-$
net	$2Zn^{2+}_{(aq)} + 2H_2O_{(l)} \rightarrow 2Zn_{(s)} + O_{2(g)} + 4H^+_{(aq)}$

Electroplating

Several metals, such as silver, gold, zinc, and chromium, are valuable because of their resistance to corrosion. However, products made from these metals in their pure form are either too expensive or they lack suitable mechanical properties, such as strength and hardness. To achieve the best compromise among price, mechanical properties, appearance, and corrosion resistance, utensils or jewellery may be made of a relatively inexpensive, yet strong, alloy such as steel, and then coated (plated) with another metal or alloy to improve appearance or corrosion resistance. Plating of a metal at the cathode of an electrolytic cell is called **electroplating** and is a common technology that is used to cover the surface of an object with a thin layer of metal. The design of a process for plating metals is obtained by systematic trial and error, involving the careful manipulation of one possible variable at a time. In this situation, a scientific perspective helps identify variables but cannot usually provide successful predictions.

As mentioned earlier, the development and use of electric cells preceded scientific understanding of the processes involved. Today, we still have examples of technological processses that are not fully understood, such as chromium plating (**Figure 9**) and silver plating. For example, there is no satisfactory explanation for why silver deposited in an electrolysis of a silver nitrate solution does not adhere well to any surface, whereas silver plated from silver cyanide solution does.

Figure 9
Chromium is best plated from a solution of chromic acid. A thin layer of chromium metal is very shiny and, like aluminum, protects itself from corrosion by forming a tough oxide layer.

SUMMARY **Applications of Electrolytic Cells**

- In molten-salt electrolysis, metal cations are reduced to metal atoms at the cathode and nonmetal anions are oxidized at the anode.
- Mixtures of salts are used to lower the melting point for a more practical and economical molten-salt electrolysis.
- The use of high voltages favours the reduction of active metal cations over the reduction of water.
- Electrorefining is a process used to obtain high-grade metals at the cathode from an impure metal at the anode.
- Electroplating is a process in which a metal is deposited on the surface of an object placed at the cathode of an electrolytic cell.

DID YOU KNOW ?

Dow Chemical Company
Herbert Dow (1866–1930) was born in Belleville, Ontario, but grew up in Cleveland, Ohio. The first of his 107 patents was the electrolytic production of bromine. Based on this process, he formed the Midland Chemical Co. When his financial backers refused to fund his invention of the chlor-alkali process, he formed the Dow Chemical Company, now one of the world's largest chemical producers.

▶ **Practice**

Understanding Concepts

8. When refining metals in an electrolytic cell, why must the metal product form at the cathode?

9. High-purity copper metal is produced using electrorefining.
   (a) At which electrode is the impure copper placed? Why?
   (b) What is the minimum electric potential difference required for this cell?
   (c) Why is it unlikely that your answer to (b) is what is used? Discuss briefly.

10. How can you predict which metals might be refined from an aqueous solution?

11. List some reasons for, and examples of, electroplating.

Applying Inquiry Skills

12. Suppose you want to set up an electrolytic cell to electroplate some metal spoons with a thin layer of silver.
   (a) As part of the experimental design, draw the cell and label the electrodes, power supply, electrolyte, and the directions of electron and ion movement.
   (b) What variables do you need to consider when planning the electrolysis?

Making Connections

13. There are companies that specialize in bronzing baby booties, sports equipment, and keepsakes. Find one and research the service it offers.
   (a) How does it make a nonconductor like a shoe into an electrode? Which electrode?
   (b) Briefly describe the process and the approximate costs for typical items.

 www.science.nelson.com

14. Electroplating industries produce considerable waste that is expensive to manage and an environmental hazard if not treated properly. List four different types of electroplating waste, including potential hazards. Describe some ways companies reduce, recover, and treat electroplating wastes.

 www.science.nelson.com

15. Aluminum cans are widely used for beverages. Write a short report about the production of aluminum cans, including how the can is made, how the top is attached to the can, how the construction of the can has changed since the first model, and the advantages of using recycled aluminum instead of new aluminum.

 www.science.nelson.com

Technicians, engineers, and research scientists all play roles in the production of the metals and metal objects that are so important in our society.

Electroplating Technician (Electroplater)

When you look around, you will find many items that are metal-plated: faucets, jewellery, trophies, coins, etc. Electroplating technicians produce these items manually or by using controlled automatic equipment. There are also people who specialize in anodizing, metal preparation, powder coatings, and in other types of surfaces.

▸ Practice

Making Connections

16. Research one of the careers listed, or another related career that may interest you. Write a report that
    (a) provides a general description of the work and how it is related to metals;
    (b) describes the education and training required;
    (c) outlines current opportunities in this area, including typical salaries.

     www.science.nelson.com

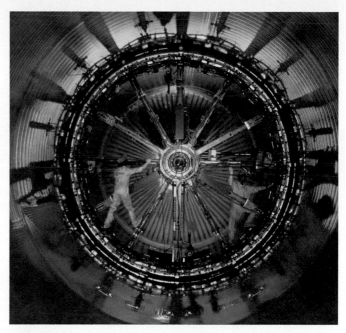

Welder

Whether it is bridges, large buildings, manufacturing equipment, or pipelines, welders are required to weld or join metals in a safe and reliable way. There are many different types of welders with different skills for specific materials and welding processes.

Electrochemist

An electrochemist is a highly trained scientist who may research fundamental aspects of electrochemical cells, specialize in fuel cell development, or develop sophisticated analytical methods based on electrochemistry.

Metallurgical Engineer

Metallurgists research, control, and develop processes for extracting metals, refining metals, and solving problems associated with metals, alloys, and plating. This occupation covers a wide area and may involve work closely associated with mining operations or the study and applications of metals.

Understanding Concepts

1. List three uses of electrolytic cells in industry.

2. Why were many metals discovered only after the invention of the electric cell?

3. How does the occurrence of metals in nature relate to the redox table of relative strengths of oxidizing and reducing agents?

4. How is the problem of solids with high melting points solved in industrial electrolysis? Provide some examples.

5. If ionic compounds can be electrolyzed in the aqueous and molten states, why can this not be done in their solid state?

6. Draw and label a simple cell for the electrolysis of molten potassium iodide (m.p. 682°C). Label electrodes and power supply, directions of electron and ion flow, and write half-reaction and net equations.

7. An electrolytic cell is set up to produce pure tin using tin(II) sulfate solution as the electrolyte. One electrode is a thin strip of pure tin and the other electrode is a large piece of impure tin containing silver and copper.
   (a) Which metal piece should be the cathode and which the anode? Explain briefly.
   (b) Draw a diagram of the cell and label the electrodes, electrolyte, power supply, including polarity, and the directions of ion and electron movement.
   (c) What will be reduced and at which electrode? Write equations for the half-reaction(s).
   (d) What will be oxidized and at which electrode? Write equations for the half-reaction(s).
   (e) What is the range of potential difference that can be applied to a cell to obtain pure tin at the cathode? Justify your answer.
   (f) Assuming that the cell operates as planned, what happens to the impurities?

Applying Inquiry Skills

8. Design a cell to electroplate zinc onto an iron spoon. In your cell diagram, include:
   —ions in the solution
   —substances used for the electrodes
   —anode and cathode labels
   —power supply, showing signs and connections
   —direction of ion and electron flow

9. Suppose you work for a mining company and you are given a job to design a process that will recover nickel metal from a waste solution containing nickel(II) ions.
   (a) Propose an experimental design involving electrolysis that could be tested in the laboratory on a small scale.
   (b) What are some possible complications or factors that need to be considered? List these as questions to be answered.

Making Connections

10. Describe how electrolytic processes are involved in the production of zinc metal and in the production of galvanized (zinc-plated) objects.

11. The loonie (**Figure 10**) replaced the one-dollar bill, which typically wore out in the space of a few months. The Royal Canadian Mint wanted to produce a dollar coin with a richer sheen than the shiny metals used in coins of lower value. Sherritt Gordon of Fort Saskatchewan, AB developed a

Figure 10
The production line for the Canadian loonie

unique process for plating the loonie coin.
   (a) Research the production and composition of the loonie.
   (b) What is the golden "aureate" finish on the loonie? Describe the materials and process for producing this finish.
   (c) Why did the coin end up with a loon stamped on it?

 www.science.nelson.com

Stoichiometry of Cell Reactions 10.3

In the production of elements, the refining of metals, and electroplating, the quantity of electricity that passes through a cell determines the masses of substances that react or are produced at the electrodes. As you know from oxidation and reduction half-reactions, a specific number of electrons are lost or gained. For example, when zinc is plated onto a steel pipe to galvanize it, two moles of electrons must be gained by one mole of zinc ions to deposit one mole of zinc atoms as metal.

$$Zn^{2+}_{(aq)} + 2\,e^- \rightarrow Zn_{(s)}$$

As in all stoichiometry, this relationship establishes a mole ratio of electrons to some other substance in the half-reaction equation. Unfortunately, there is no meter or instrument for measuring directly (or counting directly) the number of electrons. The number of electrons (as moles of electrons) is determined indirectly. In the past, you have measured mass and then converted to an amount in moles of a substance; a similar procedure is necessary for amounts of electrons.

Before we can look at the amount of electrons, it is necessary to see how the charge is determined. Charge, q, in coulombs, is determined from the electric current, I, in amperes (coulombs per second), and the time, t, in seconds, according to the following definition:

$$q = It$$

One coulomb (C) is the quantity of charge transferred by a current of one ampere (A) during a time of one second.

Calculating Electric Charge

SAMPLE problem

The technology of the Hall-Héroult cell for producing aluminum has improved considerably since the first industrial factory. Modern electrolytic cells may use up to 300 kA of current. What is the charge that passes through one of these cells in a 24-h period?

By definition, a current in amperes (A) is the number of coulombs per second, 1 A = 1 C/s. You always need to convert the time into seconds before time can be used in the ulation of charge.

$$I = 300\ \text{kA} = 300 \times 10^3\ \frac{\text{C}}{\text{s}}$$

$$t = 24\ \cancel{\text{h}} \times \frac{3600\ \text{s}}{1\ \cancel{\text{h}}} = 8.6 \times 10^4\ \text{s}$$

Now the charge in coulombs can be calculated as follows:

$$\begin{aligned} q &= It \\ &= 300 \times 10^3\ \frac{\text{C}}{\cancel{\text{s}}} \times 8.6 \times 10^4\ \cancel{\text{s}} \\ &= 2.6 \times 10^{10}\ \text{C} \end{aligned}$$

Therefore, a current of 300 kA for 24 h transfers 2.6×10^{10} C of charge. This is a huge quantity of charge. For comparison, the charge passing through a 100-W light bulb in 24 h is about 3200 C.

Understanding Concepts

Answers

1. 45 C

2. 3.89 A

3. 7.13 C

4. 3.91 min

1. Calculate the charge transferred by a current of 1.5 A flowing for 30 s.

2. In an electrolytic cell, 87.6 C of charge is transferred in 22.5 s. Determine the electric current.

3. Calculate the charge transferred by a current of 250 mA in a time of 28.5 s.

4. How long, in minutes, does it take a current of 1.60 A to transfer a charge of 375 C?

Faraday's Law

Faraday's law the mass of a substance formed or consumed at an electrode is directly related to the charge transferred

Faraday constant the charge of one mole of electrons; $F = 9.65 \times 10^4$ C/mol

The relationship between electricity and electrochemical changes was first investigated by Michael Faraday in the 1830s. Faraday continued Humphry Davy's work in electrochemistry, coining the terms electrolysis, electrolyte, electrode, anode, cathode, and ion. His quantitative study of electrolysis identified the factors that determine the mass of an element produced or consumed at an electrode. He discovered that this mass was directly proportional to the time the cell operated, as long as the current was constant (**Faraday's law**). Furthermore, he found that *9.65 × 10⁴ C of charge is transferred for every mole of electrons that flows in the cell.* In modern terms, this value is the molar charge of electrons, also called the **Faraday constant**, F.

$$F = 9.65 \times 10^4 \, \frac{C}{mol}$$

LEARNING TIP

Molar Quantities
Note the similarity between various calculations using molar quantities:

$$n_{substance} = \frac{m}{M}$$

$$n_{gas} = \frac{v}{V}$$

$$n_{e^-} = \frac{q}{F}$$

This constant can be used as a conversion factor in converting electric charge to an amount in moles—in the same way that molar mass is used to convert mass to an amount in moles.

$$n_{e^-} = \frac{q}{F}$$

and since $q = It$, the amount of electrons can now be written as

$$n_{e^-} = \frac{It}{F}$$

Calculations Using the Faraday Constant

What amount of electrons is transferred in a cell that operates for 1.25 h at a current of 0.150 A?

Recall from the calculation of electric charge that the time must always be in seconds, because the ampere is defined as coulombs per second (1 A = 1 C/s).

$$t = 1.25 \, h \times \frac{3600 \, s}{1 \, h} = 4.50 \times 10^3 \, s$$

Now you can calculate the amount, in moles, of electrons using the Faraday:

$$n_{e^-} = \frac{It}{F}$$

$$= \frac{0.150 \, \frac{C}{s} \times 4.50 \times 10^3 \, s}{9.65 \times 10^4 \, \frac{C}{mol}}$$

$$n_{e^-} = 6.99 \times 10^{-3} \, mol$$

The amount of electrons transferred is 6.99 mmol.

The same formula can be used to calculate electric current or time if the other variables are known, as shown in Examples 1 and 2.

Example 1

Convert a current of 1.74 A for 10.0 min into an amount in moles of electrons.

Solution

$I = 1.74 \text{ A}$

$t = 10.0 \text{ min} \times \dfrac{60 \text{ s}}{1 \text{ min}} = 600 \text{ s}$

$F = 9.65 \times 10^4 \dfrac{C}{\text{mol}}$

$n_{e^-} = \dfrac{It}{F}$

$\quad = \dfrac{1.74 \dfrac{C}{s} \times 600 \text{ s}}{9.65 \times 10^4 \dfrac{C}{\text{mol}}}$

$n_{e^-} = 0.0108 \text{ mol}$

The amount of electrons transferred is 0.0108 mol or 10.8 mmol.

Example 2

How long, in minutes, will it take a current of 3.50 A to transfer 0.100 mol of electrons?

Solution

$I = 3.50 \text{ A}$

$n_{e^-} = 0.100 \text{ mol}$

$F = 9.65 \times 10^4 \dfrac{C}{\text{mol}}$

$n_{e^-} = \dfrac{It}{F}$

$t = \dfrac{n_{e^-}F}{I}$

$\quad = \dfrac{0.100 \text{ mol} \times 9.65 \times 10^4 \dfrac{C}{\text{mol}}}{3.50 \dfrac{C}{s}}$

$\quad = 2.76 \times 10^3 \text{ s} \times \dfrac{1 \text{ min}}{60 \text{ s}}$

$t = 46.0 \text{ min}$

It would take 46.0 min to transfer 0.100 mol of electrons at a current of 3.50 A.

▶ Practice

Understanding Concepts

5. An electroplating cell operates for 35 min with a current of 1.9 A. Calculate the amount, in moles, of electrons transferred.

6. A cell transferred 0.146 mol of electrons with a constant current of 1.24 A. How long, in hours, did this take?

7. Calculate the current required to transfer 0.015 mol of electrons in a time of 20 min.

Answers

5. 0.041 mol

6. 3.16 h

7. 1.2 A

Half-Cell Calculations

Since the mass of an element produced at an electrode depends on the amount in moles of transferred electrons, a half-reaction equation showing the number of electrons involved is necessary to do stoichiometric calculations. This applies to all electrochemical cells, whether galvanic or electrolytic. Separate calculations are carried out for each electrode, although the same charge and therefore the same amount in moles of electrons passes through each electrode in a cell or a group of cells in series. As the following examples show, concepts of stoichiometry used in other calculations also apply to half-cell calculations. The only new part of the stoichiometry is the calculation of the amount of electrons based on the Faraday constant. 🧪▮

🧪 **INVESTIGATION 10.3.1**

Investigating an Electrolytic Cell (p. 756)
Determine the value of the Faraday constant.

▶ **SAMPLE** problem

Half-Cell Stoichiometry

What is the mass of copper deposited at the cathode of a copper electrorefining cell operated at 12.0 A for 40.0 min?

Your first step should be to identify and write the appropriate half-cell equation. Because copper is being deposited at the cathode, copper(II) ions must be gaining electrons to form copper metal. Write the equation for this reduction and list all information given, including constants such as molar mass and Faraday.

$$Cu^{2+}_{(aq)} + 2e^- \rightarrow Cu_{(s)}$$

40.0 min	m
12.0 A	63.55 g/mol
9.65×10^4 C/mol	

Notice that we have all of the information necessary to calculate the amount of electrons. Don't forget to make sure the time is converted to units of seconds, if necessary.

$$n_{e^-} = \frac{It}{F}$$

$$= \frac{12.0\frac{C}{s} \times 40.0 \text{ min} \times \frac{60 \text{ s}}{1 \text{ min}}}{\frac{9.65 \times 10^4 \text{ C}}{\text{mol}}}$$

$$n_{e^-} = 0.298 \text{ mol}$$

The procedure that is common to all stoichiometry is the use of the mole ratio from a balanced equation. The mole ratio is what allows us to convert from an amount in moles of one substance to another. In the reduction half-reaction given, notice that 1 mol of copper metal is formed when 2 mol of electrons are transferred.

$$n_{Cu} = 0.298 \text{ mol } e^- \times \frac{1 \text{ mol Cu}}{2 \text{ mol } e^-}$$

$$n_{Cu} = 0.149 \text{ mol}$$

The final step is to convert to the quantity requested in the question, in this case, the mass of copper metal.

$$m_{Cu} = 0.149 \text{ mol} \times \frac{63.55 \text{ g}}{1 \text{ mol}}$$

$$m_{Cu} = 9.48 \text{ g}$$

The mass of copper metal deposited is 9.48 g.

DID YOU KNOW ?

Copper Extraction
A solvent extraction process is being used to recover formerly "unrecoverable" copper at the Gibraltar Mine near McLeese Lake, B.C. Past mining operations at the site produced broken rock containing the copper sulfide mineral chalcopyrite ($CuFeS_2$), at concentrations too low to process economically by conventional methods. A unique electrolytic refining technique has changed this. The rock waste is sprinkled with a weak solution of sulfuric acid, which percolates through the rocks, dissolving copper and forming a copper sulfate solution. Naturally occurring bacteria act as a catalyst to speed up the oxidation. A concentrated copper electrolyte solution is created and a DC current passed through it between an anode and a cathode plate. Pure copper forms on the cathode.

LEARNING TIP

Stoichiometry Procedure
Note the similarity of the procedure for stoichiometry calculations of half-cells to all other stoichiometry calculations you have done in the past. Essentially, the only difference is a new relationship (formula based on Faraday's law) to convert to and from amount (in moles).

SUMMARY   *Procedure for Half-Cell Stoichiometry*

Step 1   Write the balanced equation for the half-cell reaction of the substance produced or consumed. List the measurements and conversion factors for the given and required entities.

Step 2   Convert the given measurements to an amount in moles by using the appropriate conversion factor (M, C, F).

Step 3   Calculate the amount of the required substance by using the mole ratio from the half-reaction equation.

Step 4   Convert the calculated amount to the final quantity by using the appropriate conversion factor (M, C, F).

Example

In a silver electroplating cell, 0.175 g of silver is to be deposited from a silver cyanide solution in a time of 10.0 min. Predict the current required.

Solution

$$Ag^+_{(aq)} \quad + \quad e^- \quad\quad \rightarrow \quad Ag_{(s)}$$

$$\begin{array}{ll} \text{10.0 min} & \text{0.175 g} \\ I & \text{107.87 g/mol} \\ 9.65 \times 10^4 \text{ C/mol} & \end{array}$$

$$n_{Ag} = 0.175 \ g \times \frac{1 \text{ mol}}{107.87 \ g}$$

$$n_{Ag} = 1.62 \times 10^{-3} \text{ mol}$$

$$n_{e^-} = 1.62 \times 10^{-3} \text{ mol } Ag \times \frac{1 \text{ mol } e^-}{1 \text{ mol } Ag}$$

$$= 1.62 \times 10^{-3} \text{ mol}$$

$$n_{e^-} = \frac{It}{F}$$

$$I = \frac{n_{e^-}F}{t}$$

$$= \frac{1.62 \times 10^{-3} \ mol \times 9.65 \times 10^4 \ \frac{C}{mol}}{10.0 \ min \times \frac{60 \text{ s}}{min}}$$

$$I = 0.261 \frac{C}{s}$$

The current required to plate 0.175 g of silver in 10.0 min is 0.261 A.

▶ Practice

Understanding Concepts

8. A student reconstructs Volta's electric battery using sheets of copper and zinc, and a current of 0.500 A is produced for 10.0 min. Calculate the mass of zinc oxidized to aqueous zinc ions.

Answers

8. 0.102 g

9. Electroplating is a common technological process for coating objects with a metal to enhance the appearance of the object or its resistance to corrosion.
   (a) A car bumper is plated with chromium using chromium(III) ions in solution. If a current of 54 A flows in the cell for 45 min 30 s, determine the mass of chromium deposited on the bumper.
   (b) For corrosion resistance, a steel bolt is plated with nickel from a solution of nickel(II) sulfate. If 0.250 g of nickel produces a plating of the required thickness and a current of 0.540 A is used, predict how long in minutes the process will take.

10. During the electrolysis of molten aluminum chloride in an electrolytic cell, 5.40 g of aluminum is produced at the cathode. Predict the mass of chlorine produced at the anode.

11. Chromium metal can be plated onto an object from an acidic solution of dichromate ions. What average current is required to plate 17.8 g of chromium metal in a time of 2.20 h? (You will need to construct your own equation for the half-reaction.)

Applying Inquiry Skills

12. The purpose of this experiment is to test the method of stoichiometry in cells.

    Question
    What is the mass of tin electroplated at the cathode of a tin-plating cell by a current of 3.46 A for 6.00 min?

    Prediction
    (a) Predict the mass of tin that should form. Show your reasoning.

    Experimental Design
    A steel can is placed in an electroplating cell as the cathode. An electric current of 3.46 A flows through the cell, which contains a 3.25 mol/L solution of tin(II) chloride, for 6.00 min.

    Evidence
    initial mass of can = 117.34 g
    final mass of can = 118.05 g

    Analysis
    (b) Based on the evidence, what is the mass of tin produced?

    Evaluation
    (c) Calculate the accuracy of the result, using a percentage difference.
    (d) Evaluate the prediction and method used to make the prediction.

Making Connections

13. A rapidly developing technology is the production of less expensive, more durable, and more energy-dense electrochemical cells, that is, cells with a high energy-to-mass ratio.
    (a) A car battery has a rating of 125 A·h (ampere-hours). What does this tell you about the electrical capacity of this battery?
    (b) Why is this a useful way to rate batteries?
    (c) What mass of lead is oxidized as this battery discharges?
    (d) If an aluminum-oxygen fuel cell has the same rating as the car battery in (a), what mass of aluminum metal would be oxidized?
    (e) Comment on the implications of your answers to (c) and (d).

Understanding Concepts

1. A battery delivers 0.300 A for 15.0 min. What amount of electrons, in moles, is transferred?

2. A current of 55 kA passes through a chlor-alkali cell. What mass of chlorine is formed during an 8.0-h time period?

3. A family wishes to plate an antique teapot with 10.00 g of silver. If the current to be used is 1.80 A, what length of time, in minutes, is required?

4. A typical Hall-Héroult cell produces 425 kg of molten aluminum in 24.0 h. Calculate the current used.

5. Magnesium metal is produced in an electrolytic cell containing molten magnesium chloride. A current of 2.0×10^5 A is passed through the cell for 18.0 h.
   (a) Determine the mass of magnesium produced.
   (b) What mass of chlorine is produced at the same time?

6. Cobalt metal is plated from 250.0 mL of cobalt(II) sulfate solution. What is the minimum concentration of cobalt(II) sulfate required for this cell to operate for 2.05 h with a current of 1.14 A?

7. A 25.72-g piece of copper metal is the anode in a cell in which a current of 0.876 A flows for 75.0 min. Determine the final mass of the copper electrode.

Making Connections

8. Using a specific example of an electrolytic cell, describe how Faraday's law is useful in designing and controlling the process.

9. A chemical technician's job involves solving problems and monitoring processes as part of a team. You don't have to be at the top of your class to become a well-trained, well-paid technician.
   (a) List some specific jobs that technicians do.
   (b) Identify some educational requirements to become a chemical technician.
   (c) List at least two educational facilities where training can be received.
   (d) Check the job section of a newspaper and list any chemical jobs available.

⚗ INVESTIGATION 10.1.1

A Potassium Iodide Electrolytic Cell

In this investigation, you will first observe a simple cell without an external battery or power supply and then compare this with the observations when a battery is connected. In this way, you will see firsthand the operation of an electrolytic cell.

Purpose

The purpose of this investigation is to observe the operation of an electrolytic cell and to determine its reaction products.

Question

What are the products of the reaction during the operation of an aqueous potassium iodide electrolytic cell?

Experimental Design

Inert electrodes are placed in a 0.50 mol/L solution of potassium iodide and connected directly with a wire. Then a battery or power supply is added to the circuit to provide a direct current of electricity to the cell. The litmus and halogen diagnostic tests (Appendix A6) are conducted to test the solution near each electrode. The litmus and halogen tests before the battery or power supply is added serve as a control for the same tests done after electric power is supplied.

Materials

lab apron	ring stand and utility clamp
eye protection	small test tube with stopper
U-tube	plastic pipet with long tip
2 clean carbon electrodes	dropper bottle of cyclo-
two connecting wires	hexane
3-V to 9-V battery or power	bottle of distilled water
supply	0.50 mol/L $KI_{(aq)}$
red and blue litmus paper	

Cyclohexane is highly flammable. Do not use near an open flame. Avoid inhaling fumes of cyclohexane.

Dispose of the solutions as directed by your teacher.

Procedure

1. Set up the $KI_{(aq)}$ cell as shown in **Figure 1**, but with a single wire connecting the electrodes (i.e., no power supply).

2. Record observations of the cell.

3. Use the medicine dropper to remove some solution near each electrode in the cell.

4. Test each solution using the litmus and halogen tests.

5. Using two connecting wires, hook up the power supply (**Figure 1**).

6. Turn on the power supply and record observations made while the cell is operating.

7. Perform both diagnostic tests at each electrode by repeating steps 3 and 4.

8. Deposit any hexane mixtures into the labelled disposal container.

Evidence

(a) Create a convenient table to compare diagnostic test results. In your table, identify the electrodes according to the terminal (red–positive, black–negative) on the power supply. Be sure to record any other general observations of the cell.

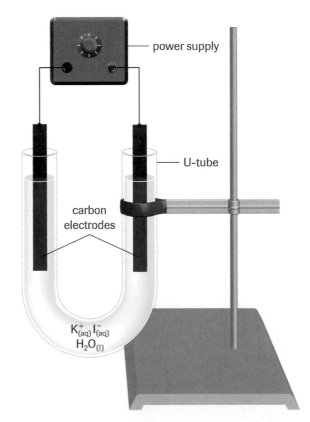

Figure 1
A U-tube is a convenient container for the aqueous potassium iodide solution because the inert, carbon electrodes can be separated by a reasonable distance.

INVESTIGATION 10.1.1 *continued*

Analysis

(b) Interpret your evidence and answer the question as well as you can with the evidence collected.

Evaluation

(c) Evaluate the design. For example, were the control tests sufficient to show clearly whether changes occurred?

(d) Which product were you not able to identify? Why not?

(e) Suggest some improvements to the materials and to the procedure used to collect and identify the gas produced.

(f) How certain are you about the other two products? Justify your answer.

(g) Why is it necessary to set up the apparatus so that the electrodes are not touching?

(h) Explain the observations at the bottom of the U-tube.

(i) Overall, how would you judge the quality of the evidence obtained? Provide reasons.

INVESTIGATION 10.1.2

Investigating Several Electrolytic Cells

Inquiry Skills		
○ Questioning	○ Planning	● Analyzing
○ Hypothesizing	○ Conducting	● Evaluating
● Predicting	● Recording	● Communicating

The purpose of this demonstration is to evaluate the method of predicting the products of a reaction occurring in an electrolytic cell.

Questions

What are the products obtained when an electrolytic cell is made by immersing inert electrodes in

- aqueous copper(II) sulfate?
- aqueous sodium sulfate?
- aqueous sodium chloride?

Prediction

(a) According to redox concepts and the table of reduction potentials, predict the products of the reaction occurring in each electrolytic cell. For your reasoning, show the equations for the cathode and anode half-reactions and the net cell reaction. Calculate the minimum potential difference that must be applied in each case.

Experimental Design

Electrolysis of aqueous copper(II) sulfate is carried out in a U-tube with carbon electrodes. Electrolysis of aqueous sodium sulfate and sodium chloride is carried out with platinum electrodes in a Hoffman apparatus (**Figure 2**) so the gases can be collected. Diagnostic tests, with any necessary control tests, are conducted to establish whether any of the predicted products are obtained.

 Copper(II) sulfate is toxic and and an irritant. Avoid skin and eye contact. If you spill copper(II) sulfate solution on your skin, wash the affected area with lots of cool water.

During electrolysis, corrosive substances are produced; avoid skin and eye contact.

Remember to wash your hands when you have finished this investigation.

Evidence

(b) Set up a table that includes the titles: Cell, Cathode, Anode. Record the cell notation and applied voltage under Cell. Record the observations for cathode and anode half-cells, including observations made during the control tests.

Analysis

(c) Based on your interpretation of the evidence collected, prepare a table or list of the products at each electrode for each of the three cells.

Evaluation

(d) Evaluate the experimental design. Consider whether the question was answered, any flaws were present, and the controls were adequate. Suggest some improvements.

(e) What is your judgment of the overall quality of the evidence? How certain are you about this? Provide reasons.

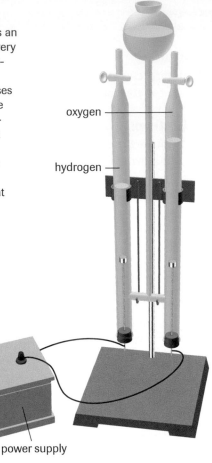

INVESTIGATION 10.1.2 *continued*

(f) Evaluate each of the three predictions, considering both cathode and anode products.

(g) Overall, how would you judge the redox concepts and the table of reduction potentials used to make these predictions? Provide reasons.

Synthesis

(h) Which one of the half-reactions that was expected to occur did not occur? Using the table of reduction potentials, write the equations for the cathode and anode half-reactions and the net reaction to obtain the observed products. Calculate the minimum potential difference required to force this reaction to occur.

(i) Compare the minimum potential differences for oxygen as a product (see Prediction) and chlorine as a product.

(j) Suggest one reason why it was possible to produce chlorine.

(k) Considering that one half-reaction out of six did not agree with the prediction, should our rules be restricted, revised, or replaced? Discuss briefly.

Figure 2
A Hoffman apparatus is an electrolytic cell that is very useful when doing electrolysis reactions that produce gases. The gases rise in a graduated tube (like a burette) and displace the solution back into the bulb (in the middle). The gases can easily be removed by opening the stopcock at the top of each tube.

oxygen

hydrogen

power supply

INVESTIGATION 10.3.1

Investigating an Electrolytic Cell

Inquiry Skills

○ Questioning	○ Planning	● Analyzing
○ Hypothesizing	● Conducting	● Evaluating
○ Predicting	● Recording	● Communicating

Suppose we design an experiment to determine the value of a scientific constant such as the Faraday. Since this constant is well known and reliable, we can use the accuracy of our experimentally determined constant to evaluate the design of our experiment or our understanding of the processes occurring.

Purpose

To test the reactions in an electrolytic cell by determining the value of a known constant.

Question

What is the value of the Faraday in the electrolysis of copper(II) sulfate solution?

Prediction

(a) According to a modern reference, what is the value of the Faraday?

Experimental Design

An electrolytic cell with copper electrodes and copper(II) sulfate solution is set up as shown in **Figure 3**. Measurements of the change in mass of the electrodes, current, and time are used to determine the Faraday constant. An important controlled variable is the electric current.

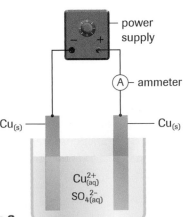

power supply

ammeter

$Cu_{(s)}$

$Cu_{(s)}$

$Cu^{2+}_{(aq)}$

$SO^{2-}_{4(aq)}$

Figure 3

INVESTIGATION 10.3.1 *continued*

Materials
lab apron
eye protection
two 250-mL beakers
2 strips of copper foil (8 cm by 3 cm)
steel wool or emery paper
connecting wires
ammeter (0–5 A)
variable power supply
stopwatch
cardboard holder
centigram balance
bottle of distilled water
0.5 mol/L copper(II) sulfate

(T) Copper(II) sulfate is toxic and and an irritant. Avoid skin and eye contact. If you spill copper(II) sulfate solution on your skin, wash the affected area with lots of cool water.

Procedure
1. Carefully clean the copper strips with steel wool, rinse with water, and dry. Once they are cleaned and dry, do not touch the surface of the strips with your fingers, except at the top.
2. Label each copper strip and then record the mass of each strip.
3. Fill a beaker close to the top with copper(II) sulfate solution.
4. Place the copper strips into the cardboard holder and set this into the beaker of solution.
5. Assemble the rest of the apparatus as shown in **Figure 3**. Note which copper strip is the cathode and which is the anode.
6. Set the variable power supply voltage to about the halfway position.
7. Plug in the power supply and reset the stopwatch to zero.
8. Turn on the power supply and start the stopwatch at the same time.
9. Immediately after the power is on, check the ammeter and adjust the power supply voltage to set the current between one half and two amperes. If necessary, adjust the power supply voltage to keep the current constant.

10. Record the current used.
11. Record observations of the contents of the cell.
12. Let the cell run for at least 25 min, longer if possible.
13. Turn off the power supply and stopwatch at the same time. Record the time.
14. Fill the other beaker with distilled water.
15. Carefully remove the cardboard holder with the copper electrodes and dip the electrodes into the distilled water to remove any $CuSO_{4(aq)}$. Be careful not to lose any deposit.
16. Let the electrodes dry completely. (An alcohol or acetone rinse may be available.)
17. Measure the mass of each copper electrode. Wait 5 min and measure the mass again.
18. Recycle the copper(II) sulfate solution to the appropriate container.

Analysis
(b) Calculate the change in mass of each electrode.
(c) Using the average of the two masses determined in (b), determine the amount, in moles, of electrons.
(d) Calculate the charge in coulombs that passed through the cell, using the current and the time.
(e) Using your results from (c) and (d), answer the Question.

Evaluation
(f) Evaluate the evidence collected in this experiment. Consider the design, materials, procedure, and skills. Note any flaws and improvements.
(g) What is your judgment of the quality of the evidence? How certain are you about this?
(h) Calculate the accuracy of your experimental result.
(i) Assume that the evidence was of reasonable quality. What does the accuracy tell you about the processes occurring within the cell? Justify your answer, using equations for the half-reactions and any other observations you have.

Synthesis
(j) Describe a technological application of the cell used in this investigation.

Key Expectations

Throughout this chapter, you have had the opportunity to do the following:

- Identify and describe the functioning of the components in electrolytic cells. (10.1)
- Describe electrolytic cells in terms of oxidation and reduction half-cells and electric potential differences. (10.1)
- Predict the spontaneity of redox reactions and cell potentials using a table of half-cell reduction potentials. (10.1)
- Determine oxidation and reduction half-cell reactions, current and ion flow, electrode polarity, and cell potentials of typical electrolytic cells. (10.1, 10.2)
- Explain how electrolytic processes are involved in industrial processes. (10.2)
- Research and assess environmental, health, and safety issues involving electrochemistry. (10.2)
- Demonstrate an understanding of the interrelationships of time, current, and the quantity of substance produced or consumed in an electrolytic process. (10.3)
- Solve problems based on Faraday's law. (10.3)
- Use appropriate scientific and technological vocabulary related to electrochemistry. (all sections)
- Write balanced chemical equations for redox reactions using half-reaction equations. (all sections)

Key Terms

electrolysis

electrolytic cell

electroplating

electrorefining

Faraday

Faraday's law

Key Symbols and Equations

- $\Delta E°$, $E_r°$, q, I, F

- $\Delta E° = E_r°_{\text{cathode}} - E_r°_{\text{anode}}$

- $q = It \qquad n_{e^-} = \dfrac{q}{F}$

- $n_{e^-} = \dfrac{It}{F}$

Problems You Can Solve

- Use the redox table and a cell notation or contents to predict the half-reaction equations and cell potential. (10.1)
- Predict the spontaneity of the reaction in a cell and determine the minimum potential difference. (10.1)
- Use Faraday's law and a half-reaction equation to calculate the mass of a substance produced or consumed at the electrode of a cell (10.3)

▶ **MAKE** a summary

- Draw a diagram of a simple, general electrolytic cell showing all components. Label all components, including names and signs where appropriate. Show the directions of electron and ion movement. State the process occurring at each electrode and how the cell potential is determined.
- List and describe four general categories of electrolytic products or processes. Include at least one consumer or industrial example with each category.

Identify each of the following statements as true, false, or incomplete. If the statement is false or incomplete, rewrite it as a true statement.

1. Electrolytic cells are based on nonspontaneous reactions and have a negative cell potential.

2. Reduction occurs at the anode and oxidation occurs at the cathode in an electrolytic cell.

3. Electrolytic cells generally have a single electrolyte so they are really half-cells with the power supply serving as the other half-cell.

4. The minimum potential difference for an aqueous cadmium sulfate cell with inert electrodes is 1.63 V.

5. The charge transferred in a cell is directly proportional to both the current and the time.

6. The mass of a nonmetal formed at the anode in an electrolytic cell is directly related to the amount of electrons transferred at the electrode.

7. If you want to deposit twice a given mass of silver in an electrolytic cell, then you must use twice the current for double the time.

Identify the letter that corresponds to the best answer to each of the following questions.

8. In an electrolytic cell, electrons are transferred through the
   (a) solution from the anode to the cathode.
   (b) solution from the cathode to the anode.
   (c) porous barrier from the cathode to the anode.
   (d) external wire from the cathode to the anode.
   (e) external wire from the anode to the cathode.

9. Oxidation and reduction half-reactions occur
   (a) at the surface of an electrode.
   (b) in the power supply.
   (c) in the salt bridge.
   (d) where the wire connects to the electrode.
   (e) at the porous barrier.

10. In an electrolytic cell, the strongest oxidizing agent reacts at the
    (a) anode and undergoes an oxidation.
    (b) cathode and undergoes a reduction.
    (c) anode and gains electrons.
    (d) cathode and loses electrons.
    (e) porous barrier and transfers electrons.

11. If we assume standard conditions, what is the cell potential for the cell

    $Pt_{(s)} | ZnBr_{2(aq)} | Pt_{(s)}$?

    (a) $+1.83$ V   (d) -1.83 V
    (b) $+0.31$ V   (e) -2.06 V
    (c) -0.07 V

12. What are the products obtained at the electrodes in the following cell with a small potential difference applied?

    $Sn_{(s)} | Sn(NO_3)_{2(aq)} | Sn_{(s)}$

	cathode	anode
(a)	$Sn^{2+}_{(aq)}$	$Sn^{4+}_{(aq)}$
(b)	$Sn_{(s)}$	$Sn^{2+}_{(aq)}$
(c)	$H_{2(g)}, OH^-_{(aq)}$	$O_{2(g)}, H^+_{(aq)}$
(d)	$NO_{2(g)}, H_2O_{(l)}$	$Sn^{2+}_{(aq)}$
(e)	$Sn^{4+}_{(aq)}$	$Sn^{2+}_{(aq)}$

13. The Faraday constant relates
    (a) charge and current.
    (b) current and time.
    (c) time and mass.
    (d) mass and charge of electrons.
    (e) charge and amount of electrons.

14. In the electrolysis of aqueous potassium hydroxide with inert electrodes, the product(s) at the anode will be
    (a) $O_{2(g)}, H^+_{(aq)}$
    (b) $K_{(s)}$
    (c) $O_{2(g)}, H_2O_{(l)}$
    (d) $H_{2(g)}, OH^-_{(aq)}$
    (e) $KOH_{(s)}$

15. The mass of the gas produced at the anode in the electrolysis of aqueous potassium hydroxide using 5.9 A of current for 22 min is
    (a) 0.65 g.   (d) 0.020 g.
    (b) 0.32 g.   (e) 0.017 g.
    (c) 0.081 g.

16. Electrolytic cells used in industrial processes may do all of the following *except*
    (a) produce elements.
    (b) refine metals.
    (c) plate metals.
    (d) produce electricity.
    (e) act as cathodic protectors.

Understanding Concepts

1. Electrolysis is used in the industrial production of several important elements and compounds.
   (a) Describe, in your own words, the meaning of electrolysis.
   (b) In an electrolytic cell, what type of half-reaction occurs at the anode? at the cathode?
   (c) Compare the electrolysis of molten compounds with the electrolysis of aqueous solutions. State some similarities and differences. (10.1)

2. While galvanic cells and electrolytic cells share many similarities, they differ in some important ways. Compare galvanic cells and electrolytic cells by listing the similarities and differences of the components and processes. (10.1)

3. In which of the following mixtures must an external voltage be applied to inert electrodes to observe evidence of a spontaneous redox reaction?
   (a) a solution of cadmium nitrate
   (b) a solution of iron(III) iodide
   (c) solutions of iron(III) chloride and tin(II) sulfate in separate half-cells connected by a salt bridge
   (d) solutions of potassium iodide and zinc nitrate in separate half-cells (10.1)

4. Determine the minimum potential difference that must be applied to the following electrolytic cells to cause a chemical reaction. (You do not need to write the half-cell reaction equations.)
   (a) iron(II) sulfate electrolyte with inert electrodes
   (b) hydrochloric acid electrolyte with silver electrodes
   (c) tin(II) chloride electrolyte with tin electrodes (10.1)

5. Write the equations for reactions at the cathode and anode, and the net cell reaction. Calculate the minimum potential difference that must be applied to each of the following electrolytic cells to cause a reaction.
   (a) $C_{(s)} | NaBr_{(aq)} | C_{(s)}$ (c) $C_{(s)} | CoCl_{2(aq)} | C_{(s)}$
   (b) $Pt_{(s)} | KOH_{(aq)} | Pt_{(s)}$ (10.1)

6. Volta's invention of the electric battery in 1800 led to a flurry of scientific research using this new technology. A few weeks after he heard about Volta's battery, William Nicholson, an English chemist, built his own battery and passed a current through slightly acidified water. With the current flowing, bubbles of colourless gases formed at each electrode. This was the first demonstration that an electric current could bring about a chemical reaction.
   (a) Write equations for the cathode, anode, and net reactions occurring in Nicholson's demonstration.

(b) Determine the minimum potential difference needed for the reaction. (10.1)

7. An important industrial use of electrolysis is the production of elements. Write equations for the anode, cathode, and net reactions for the
   (a) Hall-Héroult cell for the production of aluminum.
   (b) chlor-alkali process for the production of chlorine. (10.2)

8. Metals produced by the initial refining process contain traces of other elements. Electrolytic cells are used to further refine metals to a high degree of purity.
   (a) Describe the composition of the anode, cathode, and electrolyte used in the electrorefining of copper.
   (b) Explain how electrorefining separates traces of silver, gold, platinum, iron, and zinc from the copper.
   (c) What technological applications require copper of high purity? (10.2)

9. Electroplating is used to apply corrosion-resistant metals to the surface of more reactive metals. Write equations for the anode, cathode and net reactions in a nickel-plating cell using nickel(II) sulfate electrolyte and inert electrodes. (10.2)

10. Potassium hydroxide is obtained commercially by electrolysis of aqueous potassium chloride.
    (a) Sketch a diagram of a cell that could be used to electrolyze an aqueous solution of potassium chloride. Label electrodes, electrolyte, power supply, and the directions of the electron and ion flow.
    (b) Write equations for the cathode, anode, and net reactions, and calculate the minimum potential difference for the electrolysis of aqueous potassium chloride. (10.2)

11. In a school chemistry experiment, an electrolytic cell containing aqueous copper(II) sulfate is constructed to electroplate an iron bolt with copper metal.
    (a) At which electrode should the bolt be attached? Why?
    (b) What are the reaction and product(s) at the other electrode?
    (c) Determine the mass of copper plated out by a 1.5-A current running for 30 min. (10.3)

12. In a chemistry demonstration, three electrolytic cells, A, B, and C, are joined in series (i.e., the same current flows through each one). Cell A contains $Al^{3+}_{(aq)}$, Cell B

contains $Ni^{2+}_{(aq)}$, and Cell C contains $Ag^+_{(aq)}$. If a 2.50-A current flows for 45.0 min, calculate the mass of metal produced at the cathode of each cell. (10.3)

13. A commercial operation uses a 75.0-A current in one of its electrolysis cells. Predict the length of time it would take this cell to plate out 100 g of the following metals:

    (a) $Cr_{(s)}$ from $Cr^{3+}_{(aq)}$ (c) $Sn_{(s)}$ from $Sn^{4+}_{(aq)}$

    (b) $Cu_{(s)}$ from $Cu^{2+}_{(aq)}$ (10.3)

14. One technological process for refining zinc metal involves the electrolysis of a zinc sulfate solution.

    (a) Write equations for the cathode, anode, and net reactions, and calculate the minimum potential difference for the electrolysis of a zinc sulfate solution.

    (b) Calculate the time required to produce 1.00 kg of zinc using a 5.00 kA current. (10.3)

15. Determine the current required to produce 1.00 kg of aluminum per hour in a single Hall-Héroult cell for the production of aluminum. (10.3)

16. The electrolysis of copper(II) sulfate is demonstrated to a chemistry class by using carbon electrodes in an electrolysis cell. In the demonstration, a 1.50-A current passes through 75.0 mL of 0.125 mol/L copper(II) sulfate solution. How long, in minutes, would it take to plate all of the copper from the solution? (10.3)

17. Given that the typical current used in the chlor-alkali plant is 55 kA, predict the rate that chlorine gas is produced (in moles per hour). (10.3)

Applying Inquiry Skills

18. The purpose of the following experiment is to calibrate an ammeter.

    Question

    What is the average current flowing through an electroplating cell?

    Prediction

    According to the ammeter connected in series with the electroplating cell, the current is 1.85 A.

    Experimental Design

    An electroplating cell is set up using an aqueous copper(II) sulfate electrolyte, a copper strip at the anode, and a stainless steel strip at the cathode. Measurement of the change in mass of the electrodes and the elapsed time are used to determine the current flowing through the cell.

Evidence

initial mass of anode = 53.14 g
final mass of anode = 52.96 g
initial mass of cathode = 20.85 g
final mass of cathode = 21.03 g
elapsed time = 300 s

Analysis

(a) Use the change in mass of the electrodes to calculate the average current in the cell.

Evaluation

(b) List some sources of experimental error.

(c) Calculate the percentage difference between experimental current and the ammeter reading. Comment on the accuracy of the ammeter. (10.3)

Making Connections

19. Plastic components used in the automotive and electronic industries are often electroplated with chromium, nickel, or copper to give them a metallic appearance (**Figure 2**). The two major challenges for electroplating plastic are to make the surface of the material electrically conductive and to effectively bond the metal to the plastic.

    (a) How is plastic made electrically conductive?

    (b) How is the metal bonded to the plastic?

    (c) What are some of the limitations/problems associated with electroplating plastics?

    Figure 2

    (10.3)

     www.science.nelson.com

20. Sludge and wastewater from electroplating processes contain chromium, copper, gold, nickel, and silver, as well as cyanide ions. Sending these metals to landfill or discharging them in waste-water, instead of recycling them, is detrimental to the environment and wasteful of energy. Technologies are being developed to separate and purify these metals, including precipitation, electrolysis, and reverse osmosis. Research one technology designed to reduce the environmental impact of the electroplating industry, and write a short account of how it works. (10.3)

     www.science.nelson.com

PERFORMANCE TASK

▶ Criteria

Process

- Create and justify a prediction.
- Develop an appropriate experimental design.
- Choose and safely use suitable materials.
- Carry out the approved investigation.
- Record observations with appropriate precision and units.
- Analyze the results.
- Evaluate the experiment and discuss improvements.

Product

- Prepare a lab report, including a discussion of the method used.
- Electroplate a spoon with copper.
- Demonstrate an understanding of the relevant redox concepts and skills.
- Use terms, symbols, equations, and SI units correctly.

Electroplating

Most pure metals do not have the desired physical properties or corrosion resistance necessary for the many applications of metals in our society. Alloys provide one answer, but metal plating is also a common process (**Figure 1**). Knowledge of modern electrochemistry provides some understanding of the plating process and product. This scientific knowledge helps with the design of many aspects of the process but it cannot completely predict all of the factors involved in successful metal plating. Technological problem-solving methods are also an important component in producing the desired result.

Many technological processes are developed through trial and error. This involves a systematic investigation of variables to achieve some final goal, which may be a specific product or process. Thomas Edison (1847–1931), generally considered the greatest inventor ever, patented about thirteen hundred different inventions. This is a record that no one else has even approached. His successes are well known; his difficulties and failures are less well known. For example, he spent more than a year on a thousand attempts to invent the incandescent light bulb. He made eight thousand attempts to devise a new storage battery, failing every time, but he managed to retain a philosophical outlook, commenting, "Well, at least we know eight thousand things that don't work." One of his more famous quotes is, "Genius is one percent inspiration and ninety-nine percent perspiration."

Figure 1
Tools and utensils are often electroplated with a corrosion-resistant metal.

Investigation: Electroplating Copper

In the early stages of any research and development, it is important to clearly identify all potential variables. This involves some brainstorming and may also involve some related research. Once variables are identified, one variable is chosen and a plan is developed to investigate how this variable will be manipulated and how other variables will be controlled. After completing an experimental cycle from prediction to analysis, the results are evaluated to plan further refinements, or perhaps a fresh start if a "dead end" is reached. It is important to realize that something is learned from every attempt.

Purpose

The purpose is to create a process for plating copper onto a metal object.

Question

What design produces a smooth layer of copper metal that adheres to a metal object?

Prediction

(a) Based on your knowledge of the concepts developed in this unit and your research, predict the general plan and key independent variable to answer the question. Provide your reasoning.

Experimental Design

A small metal object, such as a spoon, a key, or a piece of metal, is carefully cleaned and plated with copper. The dependent variable is the quality of the copper plating, as determined by its thickness, appearance, and adherence to the object.

(b) Complete the design, including a labelled diagram.

(c) List all potential variables and identify your independent variable.

Materials

(d) List all materials. Check the availability of each material, and use MSDS information to check any safety precautions needed in their handling and disposal.

Procedure

(e) Write a complete and specific procedure, including safety and disposal instructions. Have your procedure approved by your teacher before performing this investigation.

Analysis

(f) Interpret your evidence to establish the quality of the copper plating.

(g) Calculate the mass of copper plated onto the object.

Evaluation

(h) Evaluate your experiment by carefully considering the design, materials, and procedure.

(i) Does it appear that you have a promising method? Discuss the merits of your method and the problems that remain to be solved.

(j) What minor or major adjustments should be made before starting the next experimental cycle? (If time allows, complete another cycle.)

(k) What did you learn from your experiences?

Identify each of the following statements as true, false, or incomplete. If the statement is false or incomplete, rewrite it as a true statement.

1. In a redox reaction, electrons are transferred from the reducing agent to the oxidizing agent.

2. Reduction is the gain of electrons and occurs at the anode of any cell.

3. Oxidation is the decrease in oxidation number and reduction is the increase in oxidation number.

4. For both electric and electrolytic cells, electrons flow from the anode to the cathode.

5. Inert electrodes are required for all electrolytic cells.

6. Galvanic cells are based on spontaneous redox reactions; electrolytic cells are based on nonspontaneous redox reactions.

7. The hydrogen half-cell at standard conditions is defined as the reference half-cell for assigning reduction potentials.

8. The cell potential is determined by adding the reduction potentials for the two half-cell reactions.

9. In a standard cell, a porous boundary allows ions to pass through while preventing immediate mixing of the solutions in each half-cell.

10. A standard hydrogen-cobalt cell has a cell potential of -0.28 V.

11. In a standard copper-lead cell, lead is the cathode and copper is the anode.

12. The power supply in an electrolytic cell must supply a potential difference at least equal to the absolute value of the calculated cell potential.

13. The charge transferred by a 1.5-A current in a time of 2.0 min is 3.0 C.

14. Metals are always plated at the cathode of a cell.

15. If we assume a constant current, twice the mass of a metal can be refined in twice the time.

16. Corrosion of a metal can be described as an electrochemical cell in which the metal is the anode.

17. Both tin and zinc plating work equally well in inhibiting the corrosion of iron.

18. Large galvanic cells are used to refine metals and to produce nonmetals like chlorine.

Identify the letter that corresponds to the best answer to each of the following questions.

19. A redox reaction involves
    (a) a transfer of electrons from the oxidizing agent to the reducing agent.
    (b) a transfer of electrons from the reducing agent to the oxidizing agent.
    (c) either a reduction or an oxidation.
    (d) a transfer of a proton between two agents.
    (e) a transfer of electrons through a porous barrier.

20. The metal molybdenum, $Mo_{(s)}$, reacts to form $MoO_{2(s)}$. The half-reaction equation that explains the change in oxidation state of molybdenum can be written as
    (a) $Mo_{(s)} + 2\,e^- \rightarrow Mo^{2+}_{(s)}$
    (b) $Mo_{(s)} \rightarrow Mo^{2+}_{(s)} + 2\,e^-$
    (c) $Mo^{4+}_{(s)} + 4\,e^- \rightarrow Mo_{(s)}$
    (d) $Mo^{2+}_{(s)} \rightarrow Mo^{4+}_{(s)} + 2\,e^-$
    (e) $Mo_{(s)} \rightarrow Mo^{4+}_{(s)} + 4\,e^-$

21. During the process of photosynthesis,

    $$6\,CO_{2(g)} + 6\,H_2O_{(g)} \rightarrow C_6H_{12}O_{6(aq)} + 6\,O_{2(g)}$$

    (a) carbon in carbon dioxide is oxidized.
    (b) hydrogen in water is reduced.
    (c) oxygen in carbon dioxide and/or water is oxidized.
    (d) oxygen in glucose is oxidized.
    (e) carbon in glucose is oxidized.

22. When copper metal is immersed in aqueous silver nitrate, a spontaneous reaction is observed. This reaction is best explained by stating that
    (a) copper(II) ions have a greater attraction for electrons than do silver ions.
    (b) copper(II) ions have a lesser attraction for electrons than do copper atoms.
    (c) silver ions have a greater attraction for electrons than do copper(II) ions.
    (d) silver ions have a lesser attraction for electrons than do silver atoms.
    (e) silver atoms have a lesser attraction for electrons than do copper atoms.

23. Rank the following solutions in order of strongest oxidizing agent to weakest oxidizing agent.

    1 sulfuric acid
    2 lithium hydroxide
    3 gold(III) fluoride
    4 chromium(II) nitrate

(a) 2 3 1 4   (c) 2 4 1 3   (e) 1 2 3 4

(b) 3 4 1 2   (d) 3 1 4 2

24. Which of the following equations describes a redox reaction?

(a) $HCOOH_{(aq)} \rightarrow CO_{(g)} + H_2O_{(l)}$

(b) $H^+_{(aq)} + OH^-_{(aq)} \rightarrow H_2O_{(l)}$

(c) $Ag^+_{(aq)} + Cl^-_{(aq)} \rightarrow AgCl_{(s)}$

(d) $HMnO_{4(aq)} \rightarrow H^+_{(aq)} + MnO^-_{4(aq)}$

(e) $C_2H_{4(g)} + 3 O_{2(g)} \rightarrow 2 CO_{2(g)} + 2 H_2O_{(g)}$

25. A high school laboratory's waste container is used to dispose of aqueous solutions of sodium nitrate, potassium sulfate, hydrochloric acid, and tin(II) chloride. The most likely net redox reaction predicted to occur inside the waste container is represented by the equation:

(a) $2 H^+_{(aq)} + 2 K^+_{(aq)} \rightarrow H_{2(g)} + K_{(s)}$

(b) $Sn^{2+}_{(aq)} + 2 NO^-_{3(aq)} + 4 H^+_{(aq)} \rightarrow$
$\qquad\qquad 2 NO_{2(g)} + 2 H_2O_{(l)} + Sn^{4+}_{(aq)}$

(c) $SO^{2-}_{4(aq)} + 4 H^+_{(aq)} + 2 Cl^-_{(aq)} \rightarrow$
$\qquad\qquad H_2SO_{3(aq)} + H_2O_{(l)} + Cl_{2(g)}$

(d) $Cl_{2(g)} + Sn^{2+}_{(aq)} \rightarrow Cl^-_{(aq)} + Sn_{(s)}$

(e) $SnSO_{4(s)} \rightarrow Sn^{2+}_{(aq)} + SO^{2-}_{4(aq)}$

26. All galvanic and electrolytic cells require

(a) an external power supply.

(b) a voltmeter.

(c) one electrode and two electrolytes.

(d) two electrodes and one electrolyte.

(e) a porous barrier.

27. In a galvanic cell, the reduction potentials of two standard half-cells are +0.35 V and −1.13 V. The predicted cell potential of the galvanic cell constructed from these two half-cells is

(a) 1.48 V.   (c) 0.78 V.   (e) 0.13 V.

(b) 1.13 V.   (d) 0.35 V.

28. If we assume standard conditions, the minimum potential difference required to electrolyze a solution of nickel(II) sulfate is

(a) 0.17 V.   (c) 0.97 V.   (e) 2.06 V.

(b) 0.43 V.   (d) 1.49 V.

29. When molten aluminum bromide is electrolyzed, the products are

	at the cathode	at the anode
(a)	$Al^{3+}_{(l)}$	$Br^-_{(l)}$
(b)	$H_{2(g)}$	$Br_{2(g)}$
(c)	$Al_{(s)}$	$O_{2(g)}, H^+_{(aq)}$
(d)	$H_{2(g)}, OH^-_{(aq)}$	$O_{2(g)}, H^+_{(aq)}$
(e)	$Al_{(l)}$	$Br_{2(g)}$

30. Standard reduction potentials for half-cells are based on the strengths of

(a) oxidizing agents relative to hydrogen gas.

(b) oxidizing agents relative to hydrogen ions.

(c) reducing agents relative to hydrogen ions.

(d) reducing agents relative to a standard acidic solution.

(e) reducing agents relative to an inert electrode.

31. In the plating of nickel from a nickel(II) ion solution, the mass of nickel obtained from the transfer of 0.250 mol of electrons is

(a) 0.125 g.   (c) 7.34 g.   (e) 21.9 g.

(b) 0.250 g.   (d) 14.7 g.

32. How long does it take to produce 4.50 g of scandium metal in the electrolysis of molten scandium chloride using a current of 8.5 A?

(a) 57 min   (c) 6.3 min   (e) 0.54 min

(b) 19 min   (d) 1.1 min

33. The process of corrosion is *most* similar to the principle behind

(a) a simple decomposition reaction.

(b) a combustion reaction.

(c) an electric cell.

(d) an electrolytic cell.

(e) a metal plating circuit.

34. Why does the use of salt on the roads in the winter promote the rusting of objects containing iron?

(a) Salt lowers the freezing point of water.

(b) Salt bonds to the iron objects.

(c) Salt contains sodium which is an active metal.

(d) Salt is an electrolyte which improves the charge transfer.

(e) Salt contains chlorine, which is a corrosive element.

35. Which one of the following metals would be *most* likely to oxidize if a clean surface of the metal were exposed to the atmosphere?

(a) aluminum   (c) silver   (e) gold

(b) iron   (d) zinc

36. A sacrificial anode for the protection of iron is

(a) a metal less easily oxidized than iron.

(b) a metal more easily oxidized than iron.

(c) any substance that is connected to an anode of a battery.

(d) an inert electrode.

(e) a metal that does not corrode.

Understanding Concepts

1. Write theoretical definitions for the following terms, using both oxidation states and electron transfer.
   (a) reduction
   (b) oxidation
   (c) redox reaction (9.1)

2. Are there chemical reactions that are not redox reactions? How can you recognize these? Provide some examples. (9.1)

3. What is the oxidation number of sulfur in each of the following substances?
   (a) $H_2S_{(g)}$
   (b) $H_2SO_{3(aq)}$
   (c) $H_2SO_{4(aq)}$
   (d) $SO_{2(g)}$
   (e) $S_{8(s)}$ (9.1)

4. Identify the oxidation number for each atom/ion and indicate which is oxidized and which is reduced.
   (a) $Sn^{4+}_{(aq)} + Co_{(s)} \rightarrow Sn^{2+}_{(aq)} + Co^{2+}_{(aq)}$
   (b) $Fe^{3+}_{(aq)} + Zn_{(s)} \rightarrow Fe^{2+}_{(aq)} + Zn^{2+}_{(aq)}$
   (c) $Cl_{2(aq)} + I^-_{(aq)} \rightarrow Cl^-_{(aq)} + I_{2(s)}$
   (d) $C_2O_4^{2-}{}_{(aq)} + MnO_4^-{}_{(aq)} + H^+_{(aq)} \rightarrow$
       $CO_{2(g)} + Mn^{2+}_{(aq)} + H_2O_{(l)}$
   (e) $Cl_{2(g)} + SO_3^{2-}{}_{(aq)} + OH^-_{(aq)} \rightarrow$
       $Cl^-_{(aq)} + SO_4^{2-}{}_{(aq)} + H_2O_{(l)}$ (9.1)

5. Balance the equation for each of the following reactions by constructing and labelling equations for the oxidation and reduction half-reactions.
   (a) $Au^{3+}_{(aq)} + SO_{2(aq)} \rightarrow SO_4^{2-}{}_{(aq)} + Au_{(s)}$ (acidic)
   (b) $Ag_{(s)} + NO_3^-{}_{(aq)} \rightarrow Ag^+_{(aq)} + NO_{(g)}$ (acidic)
   (c) $Zn_{(s)} + BrO_4^-{}_{(aq)} \rightarrow Zn(OH)_4^{2-}{}_{(aq)} + Br^-_{(aq)}$ (basic)
   (d) $ClO^-_{(aq)} \rightarrow ClO_2^-{}_{(aq)} + Cl_{2(g)}$ (basic)
   (e) $S_2O_3^{2-}{}_{(aq)} + OCl^-_{(aq)} \rightarrow S_4O_6^{2-}{}_{(aq)} + Cl^-_{(aq)}$ (acidic)
   (9.2)

6. The copper(I) ion undergoes the following reaction in aqueous solution...

   $$Cu^+_{(aq)} \rightarrow Cu_{(s)} + Cu^{2+}_{(aq)}$$

   (a) State the oxidation number of each species in the equation.
   (b) Write a balanced equation for the oxidation process.
   (c) Write a balanced equation for the reduction process.
   (d) Complete the balancing of the net equation.

(e) List three other examples of ions that can behave in the same way as the copper(I) ion. (9.2)

7. The gold(I) ion is unstable in aqueous solution, reacting as shown in the following unbalanced equation. Use oxidation numbers to balance the equation.

   $$Au^+_{(aq)} + H_2O_{(l)} \rightarrow Au_2O_{3(s)} + Au_{(s)} + H^+_{(aq)}$$ (9.2)

8. Iodide ion can be oxidized to iodate ion by the reaction with elemental chlorine in an acidic solution. Write balanced equations for the
   (a) oxidation reaction.
   (b) reduction reaction.
   (c) overall reaction. (9.2)

9. Balance the following equations:
   (a) $Ni_{(s)} + H_2SO_{4(aq)} \rightarrow$
       $NiSO_{4(aq)} + H_2O_{(l)} + SO_{2(g)}$ (acidic)
   (b) $I_{2(s)} + NO_3^-{}_{(aq)} \rightarrow IO_3^-{}_{(aq)} + NO_{2(g)}$ (acidic)
   (c) $Cr_2O_7^{2-}{}_{(aq)} + Cl^-_{(aq)} \rightarrow Cr^{3+}_{(aq)} + Cl_{2(g)}$ (acidic)
   (d) $Zn_{(s)} + H_2SO_{4(aq)} \rightarrow$
       $ZnSO_{4(aq)} + H_2S_{(g)} + H_2O_{(l)}$ (acidic)
   (e) $I_{2(s)} \rightarrow I^-_{(aq)} + IO_3^-{}_{(aq)}$ (basic) (9.2)

10. Chromium steel alloys are analyzed using a series of redox reactions. The alloy is initially reacted with perchloric acid that converts the chromium metal into dichromate ions while the perchloric acid is reduced to chlorine gas. The dichromate ions are then reduced to chromium(III) ions by adding an excess of iron(II) solution. The unreacted iron(II) is then titrated with a solution of cerium(IV) ions, which reduces them to cerium(III) ions. Write equations for the half-reactions and the balanced redox equation for each step. (9.2)

11. When a spontaneous redox reaction occurs, what kinds of evidence might be observed? (9.3)

12. Predict whether a spontaneous redox reaction will occur in the following situations:
    (a) A copper penny is dropped into hydrochloric acid.
    (b) A nickel is dropped into nitric acid.
    (c) A silver earring is dropped into sulfuric acid. (9.3)

13. For each of the following mixtures, list and classify the entities present, predict the half-reaction and net ionic reaction equations, and predict whether or not a spontaneous reaction will be observed.
    (a) Chlorine gas is bubbled into an iron(II) sulfate solution.

(b) Nickel(II) nitrate solution is mixed with a tin(II) sulfate solution.

(c) A zinc coating on a drainpipe is exposed to air and water.

(d) An acidic solution of sodium sulfate is spilled on a steel laboratory stand. (Consider only the iron in the steel.)

(e) For use in a titration, a sodium hydroxide solution is added to a potassium sulfite solution to make it basic. (9.3)

14. What are two technological solutions to the problem of batteries "going dead"? (9.4)

15. From the information in this unit, list two or three examples of situations in which technology preceded scientific explanations. (9.4)

16. Rechargeable nickel-metal hydride (NiMH) batteries have twice the energy density of Ni-Cd batteries and a similar operating voltage (**Figure 1**). The NiMH battery makes use of alloys that are capable of absorbing hydrogen equivalent to a thousand times their own volume and then releasing the absorbed hydrogen as the battery operates. The cells in NiMH batteries use $NiO(OH)_{(s)}$ as one electrode, a hydrogen-absorbing alloy as the other, and an alkaline electrolyte. In the following reduction half-equations, M indicates a hydrogen-absorbing alloy and H_{ab} indicates absorbed hydrogen.

Figure 1

$$NiO(OH)_{(s)} + H_2O_{(l)} + e^- \rightarrow$$
$$Ni(OH)_{2(s)} + OH^-_{(aq)} \quad E_r° = +0.49\ V$$
$$M_{(s)} + H_2O_{(l)} + e^- \rightarrow MH_{ab} + OH^-_{(aq)} \quad E_r° = -0.71\ V$$

(a) Write balanced equations for the anode, cathode, and net reactions occurring during the operation of an NiMH cell.

(b) Calculate the cell potential.

(c) List some of the technological, economic, and environmental considerations involved in evaluating the NiMH battery. (9.4)

17. A lead-cobalt standard cell is constructed and tested.

(a) Predict which electrode will be the cathode and which one will be the anode.

(b) List all entities present, write the half-cell and net cell reaction equations, and calculate the cell potential.

(c) Sketch and label a cell diagram. Specify all substances, label important cell components, and show the directions of electron and ion movement. (9.5)

18. Predict the cell potential of the following cells at standard conditions.

(a) $Cd_{(s)} \mid Cd^{2+}_{(aq)} \parallel Cr^{2+}_{(aq)} \mid Cr_{(s)}$

(b) $Pb_{(s)} \mid Pb^{2+}_{(aq)} \parallel Zn^{2+}_{(aq)} \mid Zn_{(s)}$

(c) $C_{(s)} \mid Cr_2O_7^{2-}_{(aq)}, H^+_{(aq)} \mid Cr^{3+}_{(aq)} \parallel Co^{2+}_{(aq)} \mid Co_{(s)}$ (9.5)

19. An experiment is designed to determine the identity of a half-cell by using known half-cells and measuring the potential difference.

(a) Use the evidence gathered to determine the reduction potential and the identity of the unknown $X^{2+}_{(aq)} \mid X_{(s)}$ redox pair.

$$Cu^{2+}_{(aq)} + X_{(s)} \rightarrow Cu_{(s)} + X^{2+}_{(aq)} \quad \Delta E° = +0.48\ V$$

(b) What is the significance of a negative value for the reduction potential obtained? (9.5)

20. Given the potential of the following standard cell, predict the standard reduction potential of the neodymium(III) ion. (9.5)

$$Cd_{(s)} \mid Cd^{2+}_{(aq)} \parallel Nd^{3+}_{(aq)} \mid Nd_{(s)} \quad \Delta E° = +1.85\ V$$

21. Silver, gold, and platinum are referred to as precious metals. What chemical properties do these metals have that contribute to making them precious? (9.6)

22. Pure silver is too soft for most jewellery but has the advantage of being relatively corrosion-resistant. Sterling silver contains 92.5% silver and 7.5% copper. Sterling silver is much more durable than pure silver, but tarnishes more easily. Using your knowledge of electrochemistry, suggest a reason why sterling silver is more easily corroded than pure silver. (9.6)

23. Compare chrome-plated steel, tin-plated steel, and galvanized steel in terms of appearance, oxidation of the plated metal, and protection of the steel from corrosion. (9.6)

24. Explain how an impressed current can be used to prevent the corrosion of a buried steel pipe. (9.6)

25. Explain why corrosion often occurs in places where two different metals (e.g., copper and iron) are joined together. (9.6)

26. Electrochemical cells are very important technological devices in our society. Discuss the main differences between galvanic and electrolytic cells in terms of their purpose and the chemical reactions that occur in them. (10.1)

27. In 1807 Humphry Davy used over 250 metal plates to construct the most powerful battery ever built at that time (**Figure 2**). When Davy used his battery to run electric current through molten potash (K_2CO_3), a globule of silvery metal formed at the cathode. Davy dropped the newly formed metal into water and witnessed a vigorous reaction that released a colourless gas that ignited and burned with a violet flame.
    (a) Write balanced equations for the half-reaction occurring at the cathode, the reaction of the metal with water, and the combustion of the colourless gas.
    (b) Why did the gas burn with a violet flame? (10.1)

Figure 2

28. Davy was the first person to isolate the metals potassium, sodium, barium, strontium, calcium, and magnesium from their molten compounds.
    (a) Why was it necessary to melt the compounds?

(b) Why couldn't Davy produce these metals by electrolyzing aqueous solutions of the compounds? (10.1)

29. After successfully electrolyzing molten potash to produce potassium, Davy used the same experimental design to produce several other metals from their compounds. Predict the reactions at the cathode and anode and the net cell reactions for the electrolysis of the following molten compounds (all electrolyzed for the first time by Davy):
    (a) lime, CaO
    (b) caustic soda, NaOH
    (c) magnesia, MgO
    (d) barium hydroxide, $Ba(OH)_2$ (10.1)

30. Potassium metal is produced by the electrolysis of the mineral sylvite, $KCl_{(s)}$ (**Figure 3**).
    (a) Sketch a diagram of a cell that could be used to electrolyze molten potassium chloride. Label electrodes, electrolyte, power supply, and the directions of the electron and ion flow.

Figure 3
Sylvite makes up about one-half of potash ore. Most of the ore is used to make fertilizer.

    (b) Write equations for the cathode, anode, and net reactions for the electrolysis of molten potassium chloride.
    (c) Why can't the table of relative strengths of oxidizing and reducing agents be used to calculate the minimum potential difference for this process? (10.2)

31. Assuming that Davy's battery produced 1.5 A for 30 min during his experiments, predict the mass of metal he produced by electrolysis of the following molten compounds:
    (a) strontium oxide
    (b) potassium hydroxide (10.3)

32. A student electroplates onto carbon electrodes using a power supply set at 2.0 A. Predict how long it will take to produce 1.0 g of metal from each of the following electrolytes:
    (a) $CuSO_{4(aq)}$
    (b) $AgNO_{3(aq)}$ (10.3)

33. Predict the current required to produce 1.00 kg of pure copper per hour in an electrorefining process. (10.3)

Applying Inquiry Skills

34. While investigating the oxidizing strength of Period 5 metal ions, a research chemist places selected metal strips into aqueous solutions of their ionic compounds. He observes that the following combinations of metal and cations react spontaneously:

$$In_{(s)} + Pd^{2+}_{(aq)} \rightarrow In^{3+}_{(aq)} + Pd_{(s)}$$
$$Cd^{2+}_{(aq)} + Y_{(s)} \rightarrow Cd_{(s)} + Y^{3+}_{(aq)}$$
$$Cd_{(s)} + In^{3+}_{(aq)} \rightarrow Cd^{2+}_{(aq)} + In_{(s)}$$

   (a) Use the evidence above to develop a table of oxidizing and reducing agents for these metals and their ions.
   (b) Which is the strongest oxidizing agent in the experiment? Why?
   (c) Which is the strongest reducing agent in the experiment? Why?
   (d) Write a balanced net ionic equation for the reaction between a strip of yttrium and an aqueous solution of palladium(II) nitrate, and predict the spontaneity of the reaction.

35. Electroplating finishes are often done in layers. For example, chromium plating does not work well on a zinc base, so a layer of copper is applied to the zinc and then a layer of nickel is added before the top chromium layer is plated on.
   (a) Propose a general design of an experiment to place a final chromium layer onto a galvanized metal. Include a labelled diagram and general plan.
   (b) In any electroplating, especially layers of metals, a particular thickness of metal is desired. Outline the experimental variables and the type of calculations that need to be done to plan a particular thickness of a metal plating.

Making Connections

36. Battery technology is a very active area of research. One proposal that shows some promise is a vanadium redox flow cell, also known as the All Vanadium Redox Battery. Describe the general construction of this battery, including electrodes, electrolytes, porous boundary, and external tanks. What redox reactions occur at the electrodes within this cell? List some unique aspects of this technology, as well as some advantages and proposed uses.

 www.science.nelson.com

37. The current technology for the manufacture of computer chips uses aluminum interconnects ("wires" or paths connecting electrical components) on the silicon surface. The next generation of Ultra Large-Scale Integration (ULSI) chips proposes to use copper in place of aluminum. Why is copper better than aluminum for interconnects? Briefly describe two electrochemical techniques used to create the copper interconnects. Include a brief description of the redox concepts involved in each technique.

38. Road salt is commonly used on Canadian roads, mostly during the winter months. Recently, this use has become an issue and Environment Canada is assessing whether road salt should be classified as a "toxic substance." What is "road salt"? Compare Ontario's use of road salt with that of other provinces. Describe some of the benefits of road salt related to its use, and some environmental and safety issues, including specific examples related to electrochemistry and other areas. What are some alternatives to the current use of road salt?

 www.science.nelson.com

Extensions

39. Describe how to connect car batteries to give someone a "boost." Why should the final connection be made to ground at a distance from both batteries?

40. Most chemical reactions are explained as being either electron transfer reactions or proton transfer reactions.
   (a) What are the similarities and differences between electron and proton transfer reactions?
   (b) State some evidence for energy changes in both electron and proton transfer reactions.
   (c) Identify a combination of chemicals that might produce either an electron or a proton transfer reaction, and describe some diagnostic tests that could be used to determine which reaction predominates.

 www.science.nelson.com

Appendixes

A1 **Planning an Investigation**

In our attempts to further our understanding of the natural world, we encounter questions, mysteries, or events that are not readily explainable. To develop explanations, we investigate using scientific inquiry. The methods used in scientific inquiry depend, to a large degree, on the purpose of the inquiry.

Controlled Experiments

A controlled experiment is an example of scientific inquiry in which an independent variable is purposefully and steadily changed to determine its effect on a second (dependent) variable. All other variables are controlled or kept constant. Controlled experiments are performed when the purpose of the inquiry is to create, test, or use a scientific concept.

The common components of controlled experiments are outlined below. *Even though the presentation is linear, there are normally many cycles through the steps during an actual experiment.*

Stating the Purpose

Every investigation in science has a purpose; for example,

- to develop a scientific concept (a theory, law, generalization, or definition);
- to test a scientific concept;
- to use a scientific concept to perform a chemical analysis;
- to determine a scientific constant; or
- to test an experimental design, an apparatus, a procedure, or a skill.

Determine which of these is the purpose of your investigation. Indicate your decision in a statement of the purpose.

Asking the Question

Your question forms the basis for your investigation: the investigation is designed to answer the question. Controlled experiments are about relationships, so the question could be about the effects on variable A when variable B is changed. The question can be general or specific.

Hypothesizing/Predicting

A hypothesis is a tentative explanation—an answer to a general question. To be scientific, a hypothesis must be testable. Hypotheses can range in certainty from an educated guess to a concept that is widely accepted in the scientific community.

A prediction is based upon a hypothesis or a more established scientific explanation, such as a theory or a law. A prediction is a tentative answer to a specific question. In the prediction you state what outcome you expect from your experiment.

Designing the Investigation

The design of a controlled experiment identifies how you plan to manipulate the independent variable, measure the response of the dependent variable, and control all the other variables in pursuit of an answer to your question. It is a summary of your plan for the experiment.

Gathering, Recording, and Organizing Observations

There are many ways to gather and record observations during your investigation. It is helpful to plan ahead and think about what you will need to answer the question and how best to record it. This helps to clarify your thinking about the question posed at the beginning, the variables, the number of trials, the procedure, the materials, and your skills. It will also help you organize your evidence for easier analysis.

Analyzing the Observations

After thoroughly analyzing your observations, you may have sufficient and appropriate evidence to enable you to answer the question posed at the beginning of the investigation.

Evaluating the Evidence and the Hypothesis/Prediction

At this stage of the investigation, you evaluate the processes that you followed to plan and perform the investigation.

You will also evaluate the outcome of the investigation, which involves evaluating any prediction you made, and the hypothesis or more established concept ("authority") the prediction was based on. You must identify and take into account any sources of error and uncertainty in your measurements.

Finally, compare the answer you hypotherized or predicted with the answer generated by analyzing the evidence. Is your hypothesis, or the authority, acceptable or not?

Reporting on the Investigation

In preparing your report, your objectives should be to describe your planning process and procedure clearly and in sufficient detail that the reader could repeat (replicate) the experiment as you performed it, and to report your evidence, your analysis, and your evaluation of your experiment accurately and honestly.

A2 Decision Making

Modern life is filled with environmental and social issues that have scientific and technological dimensions. An issue is defined as a problem that has at least two possible solutions rather than a single answer. There can be many positions, generally determined by the values that an individual or a society holds, on a single issue. Which solution is "best" is a matter of opinion; ideally, the solution that is implemented is the one that is most appropriate for society as a whole.

The common processes involved in the decision-making process are outlined below. *Even though the sequence is presented as linear, you may go through several cycles before deciding you are ready to defend a decision.*

Defining the Issue

The first step in understanding an issue is to explain why it is an issue, describe the problems associated with the issue, and identify the individuals or groups, called stakeholders, involved in the issue. You could brainstorm the following questions to research the issue: Who? What? Where? When? Why? How? Develop background information on the issue by clarifying information and concepts, and identifying relevant attributes, features, or characteristics of the problem.

Identifying Alternatives/Positions

Examine the issue and think of as many alternative solutions as you can. At this point it does not matter if the solutions seem unrealistic. To analyze the alternatives, you should examine the issue from a variety of perspectives. Stakeholders may bring different viewpoints to an issue and these may influence their position on the issue. Brainstorm or hypothesize how different stakeholders would feel about your alternatives. Perspectives that stakeholders may adopt while approaching an issue are listed in **Table 1**.

Researching the Issue

Formulate a research question that helps to limit, narrow, or define the issue. Then develop a plan to identify and find reliable and relevant sources of information. Outline the stages of your information search: gathering, sorting, evaluating, selecting, and integrating relevant information. You may consider using a flow chart, concept map, or other graphic organizer to outline the stages of your information search. Gather information from many sources, including newspapers, magazines, scientific journals, the Internet, and the library.

Table 1 Some Possible Perspectives on an Issue

cultural	focused on customs and practices of a particular group
environmental	focused on effects on natural processes and living things
economic	focused on the production, distribution, and consumption of wealth
educational	focused on the effects on learning
emotional	focused on feelings and emotions
aesthetic	focused on what is artistic, tasteful, beautiful
moral/ethical	focused on what is good/bad, right/wrong
legal	focused on rights and responsibilities
spiritual	focused on the effects on personal beliefs
political	focused on the aims of an identifiable group or party
scientific	focused on logic or the results of relevant inquiry
social	focused on effects on human relationships, the community
technological	focused on the use of machines and processes

Analyzing the Issue

In this stage, you will analyze the issue in an attempt to clarify where you stand. First, you should establish criteria for evaluating your information to determine its relevance and significance. You can then evaluate your sources, determine what assumptions may have been made, and assess whether you have enough information to make your decision.

There are five steps that must be completed to effectively analyze the issue:

1. Establish criteria for determining the relevance and significance of the information you have gathered.

2. Evaluate the sources of information.

3. Identify and determine what assumptions have been made. Challenge unsupported evidence.

4. Determine any causal, sequential, or structural relationships associated with the issue.

5. Evaluate the alternative solutions, possibly by conducting a risk-benefit analysis.

Defending the Decision

After analyzing your information, you can answer your research question and take an informed position on the issue. You should be able to defend your preferred solution in an appropriate format — debate, class discussion, speech, position paper, multimedia presentation (e.g., computer slide show), brochure, poster, video...

Your position on the issue must be justified using the supporting information that you have discovered in your research and tested in your analysis. You should be able to defend your position to people with different perspectives. In preparing for your defence, ask yourself the following questions:

- Do I have supporting evidence from a variety of sources?
- Can I state my position clearly?
- Do I have solid arguments (with solid evidence) supporting my position?
- Have I considered arguments against my position, and identified their faults?
- Have I analyzed the strong and weak points of each perspective?

Evaluating the Process

The final phase of decision making includes evaluating the decision the group reached, the process used to reach the decision, and the part you played in decision making. After a decision has been reached, carefully examine the thinking that led to the decision. Some questions to guide your evaluation follow:

- What was my initial stand on the issue? How has my position changed since I first began to explore the issue?
- How did we make our decision? What process did we use? What steps did we follow?
- In what ways does our decision resolve the issue?
- What are the likely short- and long-term effects of our decision?

- To what extent am I satisfied with our decision?
- What reasons would I give to explain our decision?
- If we had to make this decision again, what would I do differently?

A Risk–Benefit Analysis Model

Risk–benefit analysis is a tool used to organize and analyze information gathered in research. A thorough analysis of the risks and benefits associated with each alternative solution can help you decide on the best alternative.

- Research as many aspects of the proposal as possible. Look at it from different perspectives.
- Collect as much evidence as you can, including reasonable projections of likely outcomes if the proposal is adopted.
- Classify every individual potential result as being either a benefit or a risk.
- Quantify the size of the potential benefit or risk (perhaps as a dollar figure, or a number of lives affected, or in severity on a scale of 1 to 5).
- Estimate the probability (percentage) of that event occurring.
- By multiplying the size of a benefit (or risk) by the probability of its happening, you can assign a significance value for each potential result.
- Total the significance values of all the potential risks, and all the potential benefits and compare the sums to help you decide whether to accept the proposed action.

Note that although you should try to be objective in your assessment, your beliefs will have an effect on the outcome—two people, even if using the same information and the same tools, could come to a different conclusion about the balance of risk and benefit for any proposed solution to an issue.

A3 **Technological Problem Solving**

There is a difference between scientific and technological processes. The goal of science is to understand the natural world. The goal of technological problem solving is to develop or revise a product or a process in response to a human need. The product or process must fulfill its function but, in contrast with scientific problem solving, it is not essential to understand why or how it works. Technological solutions are evaluated based on such criteria as simplicity, reliability, efficiency, cost, and ecological and political ramifications.

Even though the sequence presented below is linear, there are normally many cycles through the steps in any problem-solving attempt.

Defining the Problem

This process involves recognizing and identifying the need for a technological solution. You need to clearly state the question(s) that you want to investigate to solve the problem and the criteria you will use as guidelines and to evaluate your solution. In any design, some criteria may be more important than others. For example, if a product is economical, but is not safe, then it is clearly unacceptable.

Identifying Possible Solutions

Use your knowledge and experience to propose possible solutions. Creativity is also important in suggesting novel solutions.

You should generate as many ideas as possible about the functioning of your solution and about potential designs. During brainstorming, the goal is to generate many ideas without judging them. They can be evaluated and accepted or rejected later.

To visualize the possible solutions it is helpful to draw sketches. Sketches are often better than verbal descriptions to communicate an idea.

Planning

Planning is the heart of the entire process. Your plan will outline your processes, identify potential sources of information and materials, define your resource parameters, and establish evaluation criteria.

Seven types of resources are generally used in developing technological solutions to problems — people, information, materials, tools, energy, capital, and time.

Constructing/Testing Solutions

In this phase, you will construct and test your prototype using systematic trial and error. Try to manipulate only one variable at a time. Use failures to inform decisions for your next trial. You may also do a cost-benefit analysis on the prototype.

To help you decide on the best solution, you can rate each potential solution on each of the design criteria using a five-point rating scale, with 1 being poor, 2 fair, 3 good, 4 very good, and 5 excellent. You can then compare your proposed solutions by totalling the scores.

Once you have made the choice among the possible solutions, you need to produce and test a prototype. While making the prototype you may need to experiment with the characteristics of different components. A model, on a smaller scale, might help you decide whether the product will be functional. The test of your prototype should answer three basic questions:

- Does the prototype solve the problem?
- Does it satisfy the design criteria?
- Are there any unanticipated problems with the design?

If these questions cannot be answered satisfactorily, you may have to modify the design or select another potential solution.

Presenting the Preferred Solution

You now need to communicate your solution, identify potential applications, and put your solution to use.

Once the prototype has been produced and tested, the best presentation of the solution is a demonstration of its use— a test under actual conditions. This demonstration can also serve as a further test of the design. Any feedback should be considered for future redesign. Remember that no solution should be considered the final solution.

Evaluating the Solution and Process

The technological problem-solving process is cyclical. At this stage, evaluating your solution and the process you used to arrive at your solution may lead to a revision of the solution.

Evaluation is not restricted to the final step; however, it is important to evaluate the final product using the criteria established earlier and to evaluate the processes used while arriving at the solution. Consider the following questions:

- To what degree does the final product meet the design criteria?
- Did you have to make any compromises in the design? If so, are there ways to minimize the effects of the compromises?
- Are there other possible solutions that deserve future consideration?
- Did you exceed any of the resource parameters?
- How did your group work as a team?

A4 Lab Reports

When carrying out investigations, it is important that scientists keep records of their plans and results, and share their findings. In order to have their investigations repeated (replicated) and accepted by the scientific community, scientists generally share their work by publishing papers in which details of their design, materials, procedure, evidence, analysis, and evaluation are given.

Lab reports are prepared after an investigation is completed. To ensure that you can accurately describe the investigation, it is important to keep thorough and accurate records of your activities as you carry out the investigation.

Investigators use a similar format in their final reports or lab books, although the headings and order may vary. Your lab book or report should reflect the type of scientific inquiry that you used in the investigation and should be based on the following headings, as appropriate.

Title

At the beginning of your report, write the number and title of your investigation. In this course the title is usually given, but if you are designing your own investigation, create a title that suggests what the investigation is about. Include the date the investigation was conducted and the names of all lab partners (if you worked as a team).

Purpose

State the purpose of the investigation. Why are you doing this investigation?

Question

This is the question that you attempted to answer in the investigation. If it is appropriate to do so, state the question in terms of independent and dependent variables.

Hypothesis/Prediction

Based on your reasoning or on a concept that you have studied, formulate a tentative explanation for what should happen (a hypothesis). From your hypothesis or an accepted concept you may make a prediction, a statement of what you expect to observe, before carrying out the investigation. Depending on the nature of your investigation, you may or may not have a hypothesis or a prediction.

Experimental Design

This is a brief general overview (one to three sentences) of what was done. If your investigation involved independent, dependent, and controlled variables, list them. Identify any control or control group that was used in the investigation.

Materials

This is a detailed list of all materials used, including sizes and quantities where appropriate. Be sure to include safety equipment such as eye protection, lab apron, latex gloves, and tongs, where needed. Draw a diagram to show any complicated setup of apparatus.

Procedure

Describe, in detailed, numbered steps, the procedure you followed in carrying out your investigation. Include steps to clean up and dispose of waste.

Observations

This includes all qualitative and quantitative observations that you made. Be as precise as appropriate when describing quantitative observations; include any unexpected observations; and present your information in a form that is easily understood. If you have only a few observations, this could be a list; for controlled experiments and for many observations, a table is more appropriate.

Analysis

Interpret your observations and present the evidence in the form of tables, graphs, or illustrations, each with a title. Include any calculations, the results of which can be shown in a table. Make statements about any patterns or trends you observed. Conclude the analysis with a statement, based only on the evidence you have gathered, answering the question that initiated the investigation.

Evaluation

The evaluation is your judgment about the quality of evidence obtained and about the validity of the prediction and hypothesis (if present). This section can be divided into two parts:

- Did your observations provide reliable and valid evidence to enable you to answer the question? Are you confident enough in the evidence to use it to evaluate any prediction and/or hypothesis you made?

- Was the prediction you made before the investigation supported or falsified by the evidence? Based on your evaluation of the evidence and prediction, is your hypothesis or the authority you used to make your prediction supported, or should it be rejected?

The leading questions that follow should help you through the process of evaluation.

Evaluation of the Evidence

1. Were you able to answer the question using the chosen experimental design? Are there any obvious flaws in the design? What alternative designs (better or worse) are available? As far as you know, is this design the best available in terms of controls, efficiency, and cost? How great is your confidence in the chosen design?

   You may sum up your conclusions about the design in a statement like: "The experimental design [name or describe in a few words] is judged to be adequate/ inadequate because …"

2. Were the materials you used adequate to gather reliable evidence? Sum up your conclusions about the materials in a statement like: "The materials are judged to be adequate/inadequate because …."

3. Were the steps that you used in the laboratory correctly sequenced, and adequate to gather sufficient evidence? What improvements could be made to the procedure? What steps, if not done correctly, would have significantly affected the results?

   Sum up your conclusions about the procedure in a statement like: "The procedure is judged to be adequate/inadequate because …"

4. Which specialized skills, if any, might have the greatest effect on the experimental results? Was the evidence from repeated trials reasonably similar? Can the measurements be made more precise?

   Sum up your conclusions: "The technological skills are judged to be adequate/inadequate because …"

5. You should now be ready to sum up your evaluation of the experiment. Do you have enough confidence in your experimental results to proceed with your evaluation of the authority being tested? Based on uncertainties and errors you have identified in the course of your evaluation, what would be an acceptable percent difference for this experiment (1%, 5%, or 10%)?

State your confidence level in a summary statement: "Based upon my evaluation of the experiment, I am not certain/I am moderately certain/I am very certain of my experimental results. The major sources of uncertainty or error are …"

Evaluation of the Prediction and Authority

1. Calculate the percent difference for your experiment.

$$\% \text{ difference} = \frac{|\text{experimental value} - \text{predicted value}|}{|\text{predicted value}|} \times 100\%$$

2. Judge the prediction based on the percent difference. How does the percent difference compare with your estimated total uncertainty (i.e. is the percent difference greater or smaller than the difference you've judged acceptable for this experiment)? Does the predicted answer clearly agree with the experimental answer in your analysis? Can the percent difference be accounted for by the sources of uncertainty listed earlier in the evaluation?

   Sum up your evalution of the prediction: "The prediction is judged to be verified/inconclusive/falsified because …"

3. If the prediction was verified, the hypothesis or the authority behind it is supported by the experiment. If the results of the experiment were inconclusive or the prediction was falsified, then doubt is cast upon the hypothesis or authority. How confident do you feel about any judgment you can make based on the experiment? Is there a need for a new or revised hypothesis, or to restrict, revise, or replace the authority being tested?

   Sum up your evaluation of the authority: "[The hypothesis or authority] being tested is judged to be acceptable/unacceptable because …."

A5 Math Skills

Scientific Notation

It is difficult to work with very large or very small numbers when they are written in common decimal notation. Usually it is possible to accommodate such numbers by changing the SI prefix so that the number falls between 0.1 and 1000; for example, 237 000 000 mm can be expressed as 237 km and 0.000 000 895 kg can be expressed as 0.895 mg. However, this prefix change is not always possible, either because an appropriate prefix does not exist or because it is essential to use a particular unit of measurement. In these cases, the best method of dealing with very large and very small numbers is to write them using scientific notation. Scientific notation expresses a number by writing it in the form $a \times 10^n$, where $1 < |a| < 10$ and the digits in the coefficient a are all significant. **Table 2** shows situations where scientific notation would be used.

Table 2: Examples of Scientific Notation

Expression	Common decimal notation	Scientific notation
124.5 million kilometres	124 500 000 km	1.245×10^8 km
154 thousand picometres	154 000 pm	1.54×10^{-5} pm
602 sextillion /mol	602 000 000 000 000 000 000 000 /mol	6.02×10^{23}/mol

To multiply numbers in scientific notation, multiply the coefficients and add the exponents; the answer is expressed in scientific notation. Note that when writing a number in scientific notation, the coefficient should be between 1 and 10 and should be rounded to the same certainty (number of significant digits) as the measurement with the least certainty (fewest number of significant digits). Look at the following examples:

$$(4.73 \times 10^5 \text{ m})(5.82 \times 10^7 \text{ m}) = 27.5 \times 10^{12} \text{ m}^2 = 2.75 \times 10^{13} \text{ m}^2$$
$$(3.9 \times 10^4 \text{ N})(5.3 \times 10^{-3} \text{ m}) = 0.74 \times 10^7 \text{ N·m} = 7.4 \times 10^6 \text{ N·m}$$

On many calculators, scientific notation is entered using a special key, labelled EXP or EE. This key includes "× 10" from the scientific notation; you need to enter only the exponent. For example, to enter

7.5×10^4	press	7.5 EXP 4
3.6×10^{-3}	press	3.6 EXP +/−3

Uncertainty in Measurements

There are two types of quantities that are used in science: exact values and measurements. Exact values include defined quantities (1 m = 100 cm) and counted values (5 cars in a parking lot). Measurements, however, are not exact because there is some uncertainty or error associated with every measurement.

There are two types of measurement error. **Random error** results when an estimate is made to obtain the last significant figure for any measurement. The size of the random error is determined by the precision of the measuring instrument. For example, when measuring length, it is necessary to estimate between the marks on the measuring tape. If these marks are 1 cm apart, the random error will be greater and the precision will be less than if the marks are 1 mm apart.

Systematic error is associated with an inherent problem with the measuring system, such as the presence of an interfering substance, incorrect calibration, or room conditions. For example, if the balance is not zeroed at the beginning, all measurements will have a systematic error; if using a metre stick that has been worn slightly at the ends, all measurements will contain an error.

The precision of measurements depends upon the gradations of the measuring device. **Precision** is the place value of the last measurable digit. For example, a measurement of 12.74 cm is more precise than a measurement of 127.4 cm because the first value was measured to hundredths of a centimetre whereas the latter was measured only to tenths of a centimetre.

When adding or subtracting measurements of different precision, the answer is rounded to the same precision as the least precise measurement. For example, using a calculator, add

$$11.7 \text{ cm} + 3.29 \text{ cm} + 0.542 \text{ cm} = 15.532 \text{ cm}$$

The answer must be rounded to 15.5 cm because the first measurement limits the precision to a tenth of a centimetre.

No matter how precise a measurement is, it still may not be accurate. **Accuracy** refers to how close a value is to its true value. The comparison of the two values can be expressed as a percentage difference. The percentage difference is calculated as:

$$\% \text{ difference} = \frac{|\text{experimental value} - \text{predicted value}|}{\text{predicted value}} \times 100$$

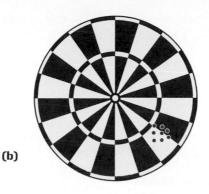
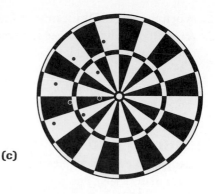

(a) **(b)** **(c)**

Figure 1
The positions of the darts in each of these figures are analogous to measured or calculated results in a laboratory setting.
The results in **(a)** are precise and accurate, in **(b)** they are precise but not accurate, and in **(c)** they are neither precise nor accurate.

Figure 1 shows an analogy between precision and accuracy, and the positions of darts thrown at a dartboard.

How certain you are about a measurement depends on two factors: the precision of the instrument used and the size of the measured quantity. More precise instruments give more certain values. For example, a mass measurement of 13 g is less precise than a measurement of 12.76 g; you are more certain about the second measurement than the first. Certainty also depends on the measurement. For example, consider the measurements 0.4 cm and 15.9 cm; both have the same precision. However, if the measuring instrument is precise to $\pm$ 0.1 cm, the first measurement is 0.4 $\pm$ 0.1 cm (0.3 cm or 0.5 cm) or an error of 25%, whereas the second measurement could be 15.9 $\pm$ 0.1 cm (15.8 cm or 16.0 cm) for an error of 0.6%. For both factors—the precision of the instrument used and the value of the measured quantity—the more digits there are in a measurement, the more certain you are about the measurement.

Significant Digits

The **certainty** of any measurement is communicated by the number of significant digits in the measurement. In a measured or calculated value, significant digits are the digits that are certain plus one estimated (uncertain) digit. Significant digits include all digits correctly reported from a measurement.

Follow these rules to decide if a digit is significant:

1. If a decimal point is present, zeros to the left of the first non-zero digit (leading zeros) are not significant.

2. If a decimal point is not present, zeros to the right of the last non-zero digit (trailing zeros) are not significant.

3. All other digits are significant.

4. When a measurement is written in scientific notation, all digits in the coefficient are significant.

5. Counted and defined values have infinite significant digits.

Table 3 shows some examples of significant digits.

Table 3: Certainty in Significant Digits

Measurement	Number of significant digits
32.07 m	4
0.0041 g	2
5×10^5 kg	1
6400 s	2
204.0 cm	4
10.0 kJ	3
100 people (counted)	infinite

An answer obtained by multiplying and/or dividing measurements is rounded to the same number of significant digits as the measurement with the fewest number of significant digits. For example, if we use a calculator to solve the following equation:

$$77.8 \text{ km/h} \times 0.8967 \text{ h} = 69.76326 \text{ km}$$

However, the certainty of the answer is limited to three significant digits, so the answer is rounded up to 69.8 km.

Rounding Off

The following rules should be used when rounding answers to calculations.

1. When the first digit discarded is less than five, the last digit retained should not be changed.
   3.141 326 rounded to 4 digits is 3.141

2. When the first digit discarded is greater than five, or if it is a five followed by at least one digit other than zero, the last digit retained is increased by 1 unit.
   2.221 672 rounded to 4 digits is 2.222
   4.168 501 rounded to 4 digits is 4.169

3. When the first digit discarded is five followed by only zeros, the last digit retained is increased by 1 if it is odd, but not changed if it is even.

2.35 rounded to 2 digits is 2.4
2.45 rounded to 2 digits is 2.4
-6.35 rounded to 2 digits is -6.4

Measuring and Estimating

Many people believe that all measurements are *reliable* (consistent over many trials), *precise* (to as many decimal places as possible), and *accurate* (representing the actual value). But there are many things that can go wrong when measuring.

- There may be limitations that make the instrument or its use unreliable (inconsistent).
- The investigator may make a mistake or fail to follow the correct techniques when reading the measurement to the available precision (number of decimal places).
- The instrument may be faulty or inaccurate; a similar instrument may give different readings.

For example, when measuring the temperature of a liquid, it is important to keep the thermometer at the correct depth and the bulb of the thermometer away from the bottom and sides of the container. If you set a thermometer with its bulb on the bottom of a liquid-filled container, you will be measuring the temperature of the bottom of the container, and not the temperature of the liquid. There are similar concerns with other measurements.

To be sure that you have measured correctly, you should repeat your measurements at least three times. If your measurements appear to be reliable, calculate the mean and use that value. To be more certain about the accuracy, repeat the measurements with a different instrument.

Logarithms

Any positive number N can be expressed as a power of some base b where $b > 1$. Some obvious examples are:

$16 = 2^4$ base 2, exponent 4
$25 = 5^2$ base 5, exponent 2
$27 = 3^3$ base 3, exponent 3
$0.001 = 10^{-3}$ base 10, exponent -3

In each of these examples, the exponent is an integer; however, exponents may be any real number, not just an integer. If you use the x^y button on your calculator, you can experiment to obtain a better understanding of this concept.

The most common base is base 10. Some examples for base 10 are

$10^{0.5} = 3.162$
$10^{1.3} = 19.95$
$10^{-2.7} = 0.001995$

By definition, the exponent to which a base b must be raised to produce a given number N is called the **logarithm** of N to base b (abbreviated as $\log_b$). When the value of the base is not written, it is assumed to be base 10. Logarithms to base 10 are called **common logarithms**. We can express the previous examples as logarithms:

$\log 3.162 = 0.5000$
$\log 19.95 = 1.299$
$\log 0.001995 = -2.700$

Most measurement scales you have encountered are linear in nature. For example, a speed of 80 km/h is twice as fast as a speed of 40 km/h and four times as fast as a speed of 20 km/h. However, there are several examples in science where the range of values of the variable being measured is so great that it is more convenient to use a logarithmic scale to base 10. One example of this is the scale for measuring the acidity of a solution (the pH scale). For example, a solution with a pH of 3 is 10 times more acidic than a solution with a pH of 4 and 100 times (10^2) more acidic than a solution with a pH of 5. Other situations that use logarithmic scales are sound intensity (the dB scale) and the intensity of earthquakes (the Richter scale).

Tables and Graphs

Both tables and graphs are used to summarize information and to illustrate patterns or relationships. Preparing tables and graphs requires some knowledge of accepted practice and some skill in designing the table or graph to best describe the information.

Tables

1. Write a title that describes the contents or the relationship among the entries in the table.

2. The rows or columns with the controlled variables and independent variable usually precedes the row or column with the dependent variable.

3. Give all rows and columns a heading, including unit symbols in parentheses where necessary. Units are not usually written in the main body of the table (**Table 4**).

Table 4 The Effect of Concentration on Reaction Time

Concentration of HCl$_{(aq)}$	Time for Reaction
(mol/L)	*(s)*
2.0	70
1.5	80
1.0	144
0.5	258

Graphs

1. Write a title and label the axes (**Figure 2**).
   (a) The title should be at the top of the graph. A statement of the two variables is often used as a title; for example, "Solubility versus Temperature for Sodium Chloride."
   (b) Label the horizontal (*x*) axis with the name of the independent variable and the vertical (*y*) axis with the name of the dependent variable.
   (c) Include the unit symbols in parentheses on each axis label, for example, "Time (s)."

2. Assign numbers to the scale on each axis.
   (a) As a general rule, the points should be spread out so that at least one-half of the graph paper is used.
   (b) Choose a scale that is easy to read and has equal divisions. Each division (or square) must represent a small simple number of units of the variable; for example, 0.1, 0.2, 0.5, or 1.0.

(c) It is not necessary to have the same scale on each axis or to start a scale at zero.
(d) Do not label every division line on the axis. Scales on graphs are labelled in a way similar to the way scales on rulers are labelled.

3. Plot the points.
   (a) Locate each point by making a small dot in pencil. When all points are drawn and checked, draw an X over each point or circle each point in ink. The size of the circle can be used to indicate the precision of the measurement.
   (b) Be suspicious of a point that is obviously not part of the pattern. Double check the location of such points, but do not eliminate the point from the graph just because it does not align with the rest.

4. Draw the best fitting curve.
   (a) Using a sharp pencil, draw a line that best represents the trend shown by the collection of points. Do not force the line to go through each point. Imprecision of experimental measurements may cause some of the points to be misaligned.
   (b) If the collection of points appears to fall in a straight line, use a ruler to draw the line. Otherwise draw a smooth curve that best represents the pattern of the points.
   (c) Since the points are ink and the line is pencil, it is easy to change the position of the line if your first curve does not fit the points to your satisfaction.

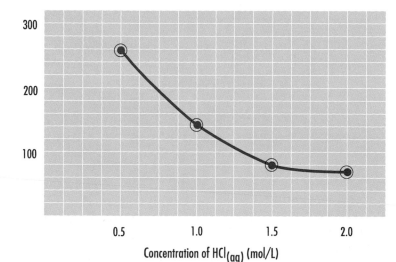

Figure 2

Using Graphs

A graph is constructed using a limited number of measured values, but the pattern may be used to extend the information gained from your experiment.

- *Interpolation* is used to find values between measured points on the graph.

- *Extrapolation* is used to find values beyond the measured points on a graph. A dotted line on a graph indicates an extrapolation.

- The scattering of points gives a visual indication of the imprecision in the experiment. A point that is obviously not part of the pattern may require a remeasurement to check for an error or may indicate the influence of an unexpected variable.

A6 Laboratory Skills and Techniques

Using a Laboratory Burner

The procedure outlined below should be practised and memorized. Note the safety caution. You are responsible for your safety and the safety of others near you.

1. Turn the air and gas adjustments to the off position (**Figure 3**).

2. Connect the burner hose to the gas outlet on the bench.

3. Turn the bench gas valve to the fully on position.

4. If you suspect that there may be a gas leak, replace the burner. (Give the leaky burner to your teacher.)

5. While holding a lit match above and to one side of the barrel, open the burner gas valve until a small yellow flame results (**Figure 4**). If a striker is used instead of matches, generate sparks over the top of the barrel (**Figure 5**).

6. Adjust the air flow and obtain a pale blue flame with a dual cone (**Figure 6**). For most laboratory burners, rotating the barrel adjusts the air intake. Rotate the barrel slowly. If too much air is added, the flame may go out. If this happens, immediately turn the gas flow off and relight the burner following the procedure outlined above.

7. Adjust the gas valve on the burner to increase or decrease the height of the blue flame. The hottest part of the flame is the tip of the inner blue cone. Usually a 5 to 10 cm flame, which just about touches the object heated, is used.

8. Laboratory burners, when lit, should not be left unattended. If the burner is on but not being used, adjust the air and gas intakes to obtain a small yellow flame. This flame is more visible and therefore less likely to cause problems.

Figure 4
A yellow flame is relatively cool and easier to obtain on lighting.

Figure 5
To generate a spark with a striker, pull the side of the handle containing the flint up and across.

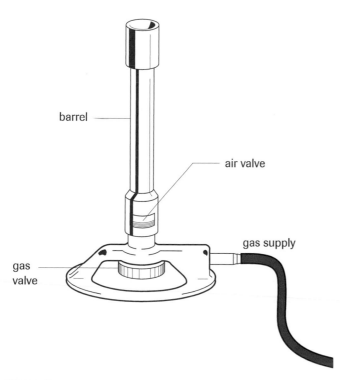

Figure 3
The parts of a common laboratory burner

barrel

air valve

gas supply

gas valve

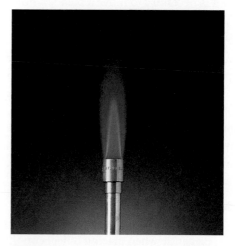

Figure 6
A pale blue/violet flame is much hotter than a yellow flame. The hottest point is at the tip of the inner blue cone.

Using a Laboratory Balance

A balance is a sensitive instrument used to measure the mass of an object. There are two types of balances — electronic (**Figure 7**) and mechanical (**Figure 8**). There are some general rules that you should follow when using a balance.

Figure 7
An electronic balance

- All balances must be handled carefully and kept clean.
- Always place chemicals into a container such as a beaker or plastic boat to avoid contamination and corrosion of the balance pan.
- To avoid error due to convection currents in the air, allow hot or cold samples to return to room temperature before placing them on the balance.
- Always record masses showing the correct precision. On a centigram balance, mass is measured to the nearest hundredth of a gram (0.01 g).
- When it is necessary to move a balance, hold the instrument by the base and steady the beam. Never lift a balance by the beams or pans.
- To avoid contaminating a whole bottle of reagent, a scoop should not be placed in the original container of a chemical. A quantity of the chemical should be poured out of the original reagent bottle into a clean, dry beaker or bottle, from which samples can be taken. Another acceptable technique for dispensing a small quantity of chemical is to rotate or tap the chemical bottle.

Using an Electronic Balance

Electronic balances are sensitive instruments requiring care in their use. Be gentle when placing objects on the pan, and remove the pan when cleaning it. Electronic balances are sensitive to small movements and changes in level; do not lean on the counter when using the balance.

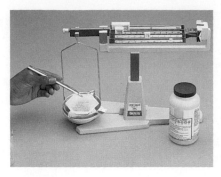

Figure 8
On this type of mechanical balance the sample is balanced by moving masses on several beams.

1. Place a container or weighing paper on the balance.
2. Reset (tare) the balance so the mass of the container registers as zero.
3. Add chemical until the desired mass of chemical is displayed. The last digit may not be constant, indicating uncertainty due to air currents or the high sensitivity of the balance.
4. Remove the container and sample.

Using a Mechanical Balance

There are different kinds of mechanical balance; however, a general procedure applies to most.

1. Clean and zero the balance. (Turn the zero adjustment screw so that the beam is balanced when the instrument is set to read 0 g and no load is on the pan.)
2. Place the container on the pan.
3. Move the largest beam mass one notch at a time until the beam drops, then move the mass back one notch.
4. Repeat this process with the next smaller mass and continue until all masses have been moved and the beam is balanced. (If you are using a dial type balance, the final step will be to turn the dial until the beam balances.)
5. Record the mass of the container.
6. If you need a specific mass of a substance, set the masses on the beams to correspond to the total mass of the container plus the desired sample.
7. Add the chemical until the beam is once again balanced.
8. Remove the sample from the pan and return all beam masses to the zero position. (For a dial type balance, return the dial to the zero position.)

Using a Pipet

A pipet is a specially designed glass tube used to measure precise volumes of liquids. There are two types of pipets and a variety of sizes for each type. A volumetric pipet (**Figure 9(a)**) transfers a fixed volume, such as 10.00 mL or 25.00 mL, accurate, for example, to within 0.04 mL. A graduated pipet (**Figure 9(b)**) measures a range of volumes within the limit of the scale, just as a graduated cylinder does. A 10-mL graduated pipet delivers volumes accurate to within 0.1 mL.

(a) 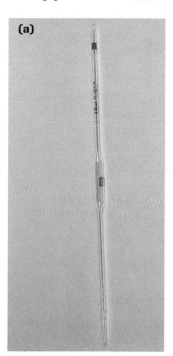 **(b)**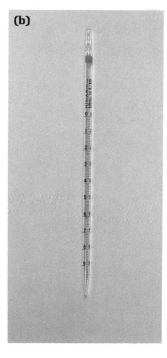

Figure 9
(a) A volumetric pipet delivers the volume printed on the label if the temperature is near room temperature.
(b) To use a graduated pipet you must be able to start and stop the flow of the liquid.

To use a pipet:

1. Rinse the pipet with small volumes of distilled water using a wash bottle, then with the sample solution. A clean pipet has no visible residue or liquid drops clinging to the inside wall. Rinsing with aqueous ammonia and scrubbing with a pipe cleaner might be necessary to clean the pipet.

2. Hold the pipet with your thumb and fingers near the top. Leave your index finger free.

3. Place the pipet in the sample solution, resting the tip on the bottom of the container if possible. Be careful that the tip does not hit the sides of the container.

4. Squeeze the bulb into the palm of your hand and place the bulb firmly and squarely on the end of the pipet (**Figure 10**) with your thumb across the top of the bulb.

Figure 10
Release the bulb slowly. Pressing down with your thumb placed across the top of the bulb maintains a good seal. Setting the pipet tip on the bottom slows the rise or fall of the liquid.

5. Release your grip on the bulb until the liquid has risen above the calibration line. (This may require bringing the level up in stages: remove the bulb, put your finger on the pipet, squeeze the air out of the bulb, replace the bulb, and continue the procedure.)

6. Remove the bulb, placing your index finger over the top.

7. Wipe all solution from the outside of the pipet using a paper towel.

8. While touching the tip of the pipet to the inside of a waste beaker, gently roll your index finger (or rotate the pipet between your thumb and fingers) to allow the liquid level to drop until the bottom of the meniscus reaches the calibration line (**Figure 11**). To

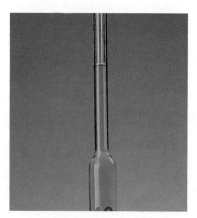

Figure 11
To allow the liquid to drop slowly to the calibration line, it is necessary for your finger and the pipet top to be dry. Also keep the tip on the bottom to slow down the flow.

avoid parallax errors, set the meniscus at eye level. Stop the flow when the bottom of the meniscus is on the calibration line. Use the bulb to raise the level of the liquid again if necessary.

9. While holding the pipet vertically, touch the pipet tip to the inside wall of a clean receiving container. Remove your finger or adjust the valve and allow the liquid to drain freely until the solution stops flowing.

10. Finish by touching the pipet tip to the inside of the container held at about a 45° angle (**Figure 12**). Do not shake the pipet. The delivery pipet is calibrated to leave a small volume in the tip.

Figure 12
A vertical volumetric pipet is drained by gravity and then the tip is placed against the inside wall of the container. A small volume is expected to remain in the tip.

Crystallization

Crystallization is used to separate a solid from a solution by evaporating the solvent or lowering the temperature. Evaporating the solvent is useful for quantitative analysis of a solution; lowering the temperature is commonly used to purify and separate a solid whose solubility is temperature-sensitive. Chemicals that have a low boiling point or decompose on heating cannot be separated by crystallization using a heat source. Fractional distillation is an alternative design for the separation of a mixture of liquids.

1. Measure the mass of a clean beaker or evaporating dish.

2. Place a precisely measured volume of the solution in the container.

3. Set the container aside to evaporate the solution slowly, or warm the container gently on a hot plate or with a laboratory burner.

4. When the contents appear dry, measure the mass of the container and solid.

5. Heat the solid with a hot plate or burner, cool it, and measure the mass again.

6. Repeat step 5 until the final mass remains constant. (Constant mass indicates that all of the solvent has evaporated.)

Filtration

In filtration, solid is separated from a mixture using a porous filter paper. The more porous papers are called qualitative filter papers. Quantitative filter papers allow only invisibly small particles through the pores of the paper.

1. Set up a filtration apparatus (**Figure 13**): stand, funnel holder, filter funnel, waste beaker, wash bottle, and a stirring rod with a flat plastic or rubber end for scraping.

Figure 13
The tip of the funnel should touch the inside wall of the collecting beaker.

2. Fold the filter paper along its diameter and then fold it again to form a cone. A better seal of the filter paper on the funnel is obtained if a small piece of the outside corner of the filter paper is torn off (**Figure 14**).

3. Measure and record the mass of the filter paper after removing the corner.

4. While holding the open filter paper in the funnel, wet the entire paper and seal the top edge firmly against the funnel with the tip of the cone centred in the bottom of the funnel.

5. With the stirring rod touching the spout of the beaker, decant most of the solution into the funnel (**Figure 15**). Transferring the solid too soon clogs the pores of the filter paper. Keep the level of liquid about two-thirds up the height of the filter paper. The stirring rod should be rinsed each time it is removed.

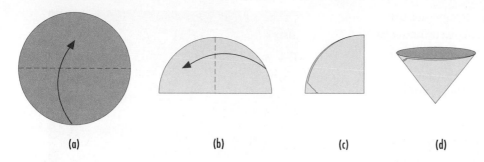

(a)   (b)   (c)   (d)

Figure 14
To prepare a filter paper, fold it in half twice and then remove the outside corner as shown.

6. When most of the solution has been filtered, pour the remaining solid and solution into the funnel. Use the wash bottle and the flat end of the stirring rod to clean any remaining solid from the beaker.

7. Use the wash bottle to rinse the stirring rod and the beaker.

8. Wash the solid two or three times to ensure that no solution is left in the filter paper. Direct a gentle stream of water around the top of the filter paper.

9. When the filtrate has stopped dripping from the funnel, remove the filter paper. Press your thumb against the thick (three-fold) side of the filter paper and slide the paper up the inside of the funnel.

10. Transfer the filter paper from the funnel onto a labelled watch glass and unfold the paper to let the precipitate dry.

11. Determine the mass of the filter paper and dry precipitate.

Figure 15
The separation technique of pouring off clear liquid is called decanting. Pouring along the stirring rod prevents drops of liquid from going down the outside of the beaker when you stop pouring.

Preparation of Standard Solutions

Laboratory procedures often call for the use of a solution of specific, precise concentration. The apparatus used to prepare such a solution is a volumetric flask. A meniscus finder is useful in setting the bottom of the meniscus on the calibration line (**Figure 16**).

Figure 16
Raise the meniscus finder along the back of the neck of the volumetric flask until the meniscus is outlined as a sharp, black line against a white background.

Preparing a Standard Solution from a Solid Reagent

1. Calculate the required mass of solute from the volume and concentration of the solution.

2. Obtain the required mass of solute in a clean, dry beaker or weighing boat. (Refer to Using a Laboratory Balance earlier in this section.)

3. Dissolve the solid in pure water using less than one-half of the final solution volume.

4. Transfer the solution and all water used to rinse the equipment into a clean volumetric flask. (The beaker and any other equipment should be rinsed two or three times with pure water.)

5. Add pure water, using a medicine dropper for the final few millilitres while using a meniscus finder to set the bottom of the meniscus on the calibration line.

6. Stopper the flask and mix the solution by slowly inverting the flask several times.

Preparing a Standard Solution by Dilution

1. Calculate the volume of concentrated reagent required.

2. Add approximately one-half of the final volume of pure water to the volumetric flask.

3. Measure the required volume of stock solution using a pipet. (Refer to Using a Pipet earlier in this section).

4. Transfer the stock solution slowly into the volumetric flask while mixing.

5. Add pure water and then use a medicine dropper and a meniscus finder to set the bottom of the meniscus on the calibration line.

6. Stopper the flask and mix the solution by slowly inverting the flask several times.

Titration

Titration is used in the volumetric analysis of an unknown concentration of a solution. Titration involves adding a solution (the titrant) from a buret to another solution (the sample) in an Erlenmeyer flask until a recognizable endpoint, such as a colour change, occurs.

1. Rinse the buret with small volumes of distilled water using a wash bottle. Using a buret funnel, rinse with small volumes of the titrant (**Figure 17**). (If liquid droplets remain on the sides of the buret after rinsing, scrub the buret with a buret brush. If the tip of the buret is chipped or broken, replace the tip or the whole buret.)

2. Using a small buret funnel, pour the solution into the buret until the level is near the top. Open the stopcock for maximum flow to clear any air bubbles from the tip and to bring the liquid level down to the scale.

3. Record the initial buret reading to the nearest 0.1 mL. Avoid parallax errors by reading volumes at eye level with the aid of a meniscus finder.

4. Pipet a sample of the solution of unknown concentration into a clean Erlenmeyer flask. Place a white piece of paper beneath the Erlenmeyer flask to make it easier to detect colour changes.

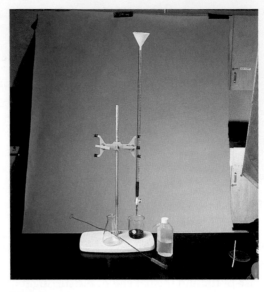

Figure 17
A buret should be rinsed with water and then the titrant before use. Use a buret brush only if necessary.

5. Add an indicator if one is required. Add the smallest quantity necessary (usually 1 to 2 drops) to produce a noticeable colour change in your sample.

6. Add the solution from the buret quickly at first, and then slowly, drop-by-drop, near the endpoint (**Figure 18**). Stop as soon as a drop of the titrant produces a permanent colour change in the sample solution. A permanent colour change is considered to be a noticeable change that lasts for 10 s after swirling.

7. Record the final buret reading to the nearest 0.1 mL.

8. The final buret reading for one trial becomes the initial buret reading for the next trial. Three trials with

Figure 18
Near the endpoint, continuous gentle swirling of the solution is particularly important.

results within 0.2 mL are normally required for a reliable analysis of an unknown solution.

9. Drain and rinse the buret with pure water. Store the buret upside down with the stopcock open.

Diagnostic Tests

The tests described in **Table 5** are commonly used to detect the presence of a specific substance. Thousands more are possible. All diagnostic tests include a brief procedure, some expected evidence, and an interpretation of the evidence obtained. This is conveniently communicated using the format — "If [procedure] and [evidence], then [analysis]."

Diagnostic tests can be constructed using any characteristic property of a substance. For example, diagnostic tests for acids, bases, and neutral substances can be specified in terms of the pH of the solutions. For specific chemical reactions, properties of the products that the reactants do not have, such as the insolubility of a precipitate, the production of a gas, or the colour of ions in aqueous solutions, can be used to construct diagnostic tests.

If possible, you should use a control to illustrate that the test does not give the same results with other substances. For example, in the test for oxygen, inserting a glowing splint into a test tube that contains only air is used to compare the effect of air on the splint with a test tube in which you expect oxygen has been collected.

For a test to be valid, it usually has to be conducted both before and after a chemical change. Consider this control when planning your designs and procedures.

Table 5 Some Standard Diagnostic Tests

Substance Tested	Diagnostic Test
water	If cobalt(II) chloride paper is exposed to a liquid or vapour, and the paper turns from blue to pink, then water is likely present.
oxygen	If a glowing splint is inserted into the test tube, and the splint glows brighter or relights, then oxygen gas is likely present.
hydrogen	If a flame is inserted into the test tube, and a squeal or pop is heard, then hydrogen is likely present.
carbon dioxide	If the unknown gas is bubbled into a limewater solution, and the limewater turns cloudy, then carbon dioxide is likely present.
halogens	If a few millilitres of chlorinated hydrocarbon solvent is added, with shaking, to a solution in a test tube, and the colour of the solvent appears to be: • light yellow-green, then chlorine is likely present; • orange, then bromine is likely present; • purple, then iodine is likely present.
acid	If strips of blue and red litmus paper are dipped into the solution, and the blue litmus turns red, then an acid is present.
base	If strips of blue and red litmus paper are dipped into the solution, and the red litmus turns blue, then a base is present.
neutral solution	If strips of blue and red litmus paper are dipped into the solution, and neither changes colour, then only neutral substances are likely present.
neutral ionic solution	If a neutral substance is tested for conductivity with a voltmeter or multimeter, and the solution conducts a current, then a neutral ionic substance is likely present.
neutral molecular solution	If a neutral solution is tested and does not conduct a current, then a neutral molecular substance is likely present.

B1 **Safety Conventions and Symbols**

Although every effort is undertaken to make the science experience a safe one, there are inherent risks associated with some scientific investigations. These risks are generally associated with the materials and equipment used and the disregard of safety instructions that accompany investigations. However, there may also be risks associated with the location of the investigation, whether in the science laboratory, at home, or outdoors. Most of these risks pose no more danger than one would normally experience in everyday life. With an awareness of the possible hazards, knowledge of the rules, appropriate behaviour, and a little common sense, these risks can be practically eliminated.

Remember, you share the responsibility not only for your own safety, but also for the safety of those around you. Always alert the teacher in case of an accident.

In this text, chemicals, equipment, and procedures that are hazardous are highlighted in red and are preceded by the appropriate Workplace Hazardous Materials Information System (WHMIS) symbol or by ✋ .

WHMIS Symbols and HHPS

The Workplace Hazardous Materials Information System (WHMIS) provides workers and students with complete and accurate information regarding hazardous products. All chemical products supplied to schools, businesses, and industries must contain standardized labels and be accompanied by Material Safety Data Sheets (MSDS) providing detailed information about the product. Clear and standardized labelling is an important component of WHMIS (**Table 1**). These labels must be present on the product's original container or be added to other containers if the product is transferred.

The Canadian Hazardous Products Act requires manufacturers of consumer products containing chemicals to include a symbol specifying both the nature of the primary hazard and the degree of this hazard. In addition, any secondary hazards, first aid treatment, storage, and disposal must be noted. Household Hazardous Product Symbols (HHPS) are used to show the hazard and the degree of the hazard by the type of border surrounding the illustration (**Figure 1**).

	Corrosive
	This material can burn your skin and eyes. If you swallow it, it will damage your throat and stomach.
	Flammable
	This product or the gas (or vapour) from it can catch fire quickly. Keep this product away from heat, flames, and sparks.
	Explosive
	Container will explode if it is heated or if a hole is punched in it. Metal or plastic can fly out and hurt your eyes and other parts of your body.
	Poison
	If you swallow or lick this product, you could become very sick or die. Some products with this symbol on the label can hurt you even if you breathe (or inhale) them.

Danger

Warning

Caution

Figure 1
Hazardous household product symbols

Table 1 The Workplace Hazardous Materials Information System (WHMIS)

Class and type of compounds	WHMIS symbol	Risks	Precautions
Class A: *Compressed Gas* Material that is normally gaseous and kept in a pressurized container		• could explode due to pressure • could explode if heated or dropped • possible hazard from both the force of explosion and the release of contents	• ensure container is always secured • store in designated areas • do not drop or allow to fall
Class B: *Flammable and Combustible Materials* Materials that will continue to burn after being exposed to a flame or other ignition source		• may ignite spontaneously • may release flammable products if allowed to degrade or when exposed to water	• store in designated areas • work in well-ventilated areas • avoid heating • avoid sparks and flames • ensure that electrical sources are safe
Class C: *Oxidizing Materials* Materials that can cause other materials to burn or support combustion		• can cause skin or eye burns • increase fire and explosion hazards • may cause combustibles to explode or react violently	• store away from combustibles • wear body, hand, face, and eye protection • store in container that will not rust or oxidize
Class D: *Toxic Materials Immediate and Severe* Poisons and potentially fatal materials that cause immediate and severe harm		• may be fatal if ingested or inhaled • may be absorbed through the skin • small volumes have a toxic effect	• avoid breathing dust or vapours • avoid contact with skin or eyes • wear protective clothing, and face and eye protection • work in well-ventilated areas and wear breathing protection
Class D: *Toxic Materials Long Term Concealed* Materials that have a harmful effect after repeated exposures or over a long period		• may cause death or permanent injury • may cause birth defects or sterility • may cause cancer • may be sensitizers causing allergies	• wear appropriate personal protection • work in a well-ventilated area • store in appropriate designated areas • avoid direct contact • use hand, body, face, and eye protection • ensure respiratory and body protection is appropriate for the specific hazard
Class D: *Biohazardous Infectious Materials* Infectious agents or a biological toxin causing a serious disease or death		• may cause anaphylactic shock • includes viruses, yeasts, moulds, bacteria, and parasites that affect humans • includes fluids containing toxic products • includes cellular components	• special training is required to handle materials • work in designated biological areas with appropriate engineering controls • avoid forming aerosols • avoid breathing vapours • avoid contamination of people and/or area • store in special designated areas
Class E: *Corrosive Materials* Materials that react with metals and living tissue		• eye and skin irritation on exposure • severe burns/tissue damage on longer exposure • lung damage if inhaled • may cause blindness if contacts eyes • environmental damage from fumes	• wear body, hand, face, and eye protection • use breathing apparatus • ensure protective equipment is appropriate • work in a well-ventilated area • avoid all direct body contact • use appropriate storage containers and ensure nonventing closures
Class F: *Dangerously Reactive Materials* Materials that may have unexpected reactions		• may react with water • may be chemically unstable • may explode if exposed to shock or heat • may release toxic or flammable vapours • may vigorously polymerize • may burn unexpectedly	• handle with care avoiding vibration, shocks, and sudden temperature changes • store in appropriate containers • ensure storage containers are sealed • store and work in designated areas

B2 Safety in the Laboratory

General Safety Rules

Safety in the laboratory is an attitude and a habit more than it is a set of rules. It is easier to prevent accidents than to deal with the consequences of an accident. Most of the following rules are common sense.

- Do not enter a laboratory or prep room unless a teacher or other supervisor is present, or you have permission to do so.
- Familiarize yourself with your school's safety regulations.
- Make your teacher aware of any allergies or other health problems you may have.
- Listen carefully to any instructions given by your teacher, and follow them closely.
- Wear eye protection, lab aprons or coats, and protective gloves when appropriate.
- Wear closed shoes (not sandals) when working in the laboratory.
- Place your books and bags away from the work area. Keep your work area clear of all materials except those that you will use in the investigation.
- Do not chew gum, eat, or drink in the laboratory. Food should not be stored in refrigerators in laboratories.
- Know the location of MSDS information, exits, and all safety equipment, such as the fire blanket, fire extinguisher, and eyewash station.
- Use stands, clamps, and holders to secure any potentially dangerous or fragile equipment that could be tipped over.
- Avoid sudden or rapid motion in the laboratory that may interfere with someone carrying or working with chemicals or using sharp instruments.
- Never engage in horseplay or practical jokes in the laboratory.
- Ask for assistance when you are not sure how to do a procedural step.
- Never attempt unauthorized experiments.
- Never work in a crowded area or alone in the laboratory.
- Report all accidents.
- Clean up all spills, even spills of water, immediately.
- Always wash your hands with soap and water before or immediately after you leave the laboratory. Wash your hands before you touch any food.

- Do not forget safety procedures when you leave the laboratory. Accidents can also occur outdoors, at home, or at work.

Eye and Face Safety

- Wear approved eye protection in a laboratory, no matter how simple or safe the task appears to be. Keep the eye protection over your eyes, not on top of your head. For certain experiments, full face protection (safety goggles or a face shield) may be necessary.
- Never look directly into the opening of flasks or test tubes.
- If, in spite of all precautions, you get a chemical in your eye, quickly use the eyewash or nearest cold running water. Continue to rinse the eye with water for at least 15 min. This is a very long time—have someone time you. Have another student inform your teacher of the accident. The injured eye should be examined by a doctor.
- If you must wear contact lenses in the laboratory, be extra careful; whether or not you wear contact lenses, do not touch your eyes without first washing your hands. If you do wear contact lenses, make sure that your teacher is aware of it. Carry your lens case and a pair of glasses with you.
- If a piece of glass or other foreign object enters your eye, seek immediate medical attention.
- Do not stare directly at any bright source of light (e.g., a piece of burning magnesium ribbon, lasers, or the Sun). You will not feel any pain if your retina is being damaged by intense radiation. You cannot rely on the sensation of pain to protect you.
- When working with lasers, be aware that a reflected laser beam can act like a direct beam on the eye.

Handling Glassware Safely

- Never use glassware that is cracked or chipped. Give such glassware to your teacher or dispose of it as directed. Do not put the item back into circulation.
- Never pick up broken glassware with your fingers. Use a broom and dustpan.
- Do not put broken glassware into garbage containers. Dispose of glass fragments in special containers marked "Broken Glass."

- Heat glassware only if it is approved for heating. Check with your teacher before heating any glassware.

- If you cut yourself, inform your teacher immediately. Embedded glass or continued bleeding requires medical attention.

- If you need to insert glass tubing or a thermometer into a rubber stopper, get a cork borer of a suitable size. Insert the borer in the hole of the rubber stopper, starting from the small end of the stopper. Once the borer is pushed all the way through the hole, insert the tubing or thermometer through the borer. Ease the borer out of the hole, leaving the tubing or thermometer inside. To remove the tubing or thermometer from the stopper, push the borer from the small end through the stopper until it shows from the other end. Ease the tubing or thermometer out of the borer.

- Protect your hands with heavy gloves or several layers of cloth before inserting glass into rubber stoppers.

- Be very careful while cleaning glassware. There is an increased risk of breakage from dropping when the glassware is wet and slippery.

Using Sharp Instruments Safely

- Make sure your instruments are sharp. Surprisingly, one of the main causes of accidents with cutting instruments is the use of a dull instrument. Dull cutting instruments require more pressure than sharp instruments and are therefore much more likely to slip.

- Select the appropriate instrument for the task. Never use a knife when scissors would work better.

- Always cut away from yourself and others.

- If you cut yourself, inform your teacher immediately and get appropriate first aid.

- Be careful when working with wire cutters or wood saws. Use a cutting board where needed.

Heat and Fire Safety

- In a laboratory where burners or hot plates are being used, never pick up a glass object without first checking the temperature by lightly and quickly touching the item, or by placing your hand near, but not on, the item. Glass items that have been heated stay hot for a long time, but do not appear to be hot. Metal items such as ring stands and hot plates can also cause burns; take care when touching them.

- Do not use a laboratory burner near wooden shelves, flammable liquids, or any other item that is combustible.

- Before using a laboratory burner, make sure that long hair is tied back. Do not wear loose clothing (wide long sleeves should be tied back or rolled up).

- Never look down the barrel of a laboratory burner.

- Always pick up a burner by the base, never by the barrel.

- Never leave a lighted laboratory burner unattended.

- If you burn yourself, *immediately* run cold water gently over the burned area or immerse the burned area in cold water and inform your teacher.

- Make sure that heating equipment, such as a burner, hot plate, or electrical equipment, is secure on the bench and clamped in place when necessary.

- Always assume that hot plates and electric heaters are hot and use protective gloves when handling.

- Keep a clear workplace when performing experiments with heat.

- When heating a test tube over a laboratory burner, use a test-tube holder and a spurt cap. Holding the test tube at an angle, facing away from you and others, gently move the test tube backwards and forwards through the flame.

- Remember to include a "cooling" time in your experiment plan; do not put away hot equipment.

- Very small fires in a container may be extinguished by covering the container with a wet paper towel or ceramic square.

- For larger fires, inform the teacher and follow the teacher's instructions for using fire extinguishers, blankets, and alarms, and for evacuation. Do not attempt to deal with a fire by yourself.

- If anyone's clothes or hair catch fire, tell the person to drop to the floor and roll. Then use a fire blanket to help smother the flames.

Electrical Safety

- Water or wet hands should never be used near electrical equipment.

- Do not operate electrical equipment near running water or any large containers of water.

- Check the condition of electrical equipment. Do not use if wires or plugs are damaged, or if the ground pin has been removed.

- Make sure that electrical cords are not placed where someone could trip over them.
- When unplugging equipment, remove the plug gently from the socket. Do not pull on the cord.
- When using variable power supplies, start at low voltage and increase slowly.

Handling Chemicals Safely

Many chemicals are hazardous to some degree. When using chemicals, operate under the following principles:

1. Never underestimate the risks associated with chemicals. Assume that any unknown chemicals are hazardous.
2. If you can substitute, use a less hazardous chemical wherever possible.
3. Reduce exposure to chemicals to the absolute minimum. Avoid direct skin contact if possible.
4. When using chemicals, ensure that there is adequate ventilation.

The following guidelines do not address every possible situation but, used with common sense, are appropriate for situations in the high-school laboratory.

- Consult the MSDS before you use a chemical.
- Wear appropriate eye protection at all times where chemicals are used or stored. Wear a lab coat and/or other protective clothing (e.g., aprons, gloves).
- When carrying chemicals, hold containers carefully using two hands, one around the container and one underneath.
- Read all labels to ensure that the chemicals you have selected are the intended ones. Never use the contents of a container that has no label or has an illegible label. Give any such containers to your teacher.
- Label all chemical containers correctly to avoid confusion about contents.
- Never pipet or start a siphon by mouth. Use a pipet bulb or equivalent device.
- Pour liquid chemicals carefully (down the side of the receiving container or down a stirring rod) to ensure that they do not splash. Always pour from the side opposite the label—if everyone follows this rule, drips will always form on the same side, away from your hand.
- Always pour volatile chemicals in a fume hood or in a well-ventilated area.

- Never smell or taste chemicals.
- Return chemicals to their correct storage place. Chemicals are stored by hazard class.
- If you spill a chemical, use a chemical spill kit to clean up.
- Do not return surplus chemicals to stock bottles. Dispose of excess chemicals in the appropriate manner.
- Clean up your work area, the fume hood, and any other area where chemicals were used.
- Wash hands immediately after handling chemicals and before leaving the lab, even if you wore gloves.

Waste Disposal

Waste disposal at school, at home, or at work is a social and environmental issue. To protect the environment, federal and provincial governments have regulations to control wastes, especially chemical wastes. For example, the WHMIS program applies to controlled products that are being handled. (When being transported, they are regulated under the *Transport of Dangerous Goods Act*, and for disposal they are subject to federal, provincial, and municipal regulations.) Most laboratory waste can be washed down the drain, or, if it is in solid form, placed in ordinary garbage containers. However, some waste must be treated more carefully. It is your responsibility to follow procedures and dispose of waste in the safest possible manner according to the teacher's instructions.

Flammable Substances

Flammable liquids should not be washed down the drain. Special fire-resistant containers are used to store flammable liquid waste. Waste solids that pose a fire hazard should be stored in fireproof containers. Care must be taken not to allow flammable waste to come into contact with any sparks, flames, other ignition sources, or oxidizing materials. The method of disposal depends on the nature of the substance.

Corrosive Solutions

Solutions that are corrosive but not toxic, such as acids and bases, can usually be washed down the drain, but care should be taken to ensure that they are first either neutralized or diluted to low concentration. While disposing of such substances, use large quantities of water and continue to pour water down the drain for a few minutes after all the substance has been washed away.

Heavy Metal Solutions

Heavy metal compounds (for example, lead, mercury, and cadmium compounds) should not be flushed down the drain. These substances are cumulative poisons and should be kept out of the environment. Pour any heavy metal waste into the special container marked "Heavy Metal Waste." Remember that paper towels used to wipe up solutions of heavy metals, as well as filter papers with heavy metal compounds embedded in them, should be treated as solid toxic waste.

Disposal of heavy metal solutions is usually accomplished by precipitating the metal ion (for example, as lead(II) silicate) and disposing of the solid. Heavy metal compounds should not be placed in school garbage containers. Usually, waste disposal companies collect materials that require special disposal and dispose of them as required by law.

Toxic Substances

Toxic chemicals and solutions of toxic substances should not be poured down the drain. They should be retained for disposal by a licensed waste disposal company.

Organic Material

Remains of plants and animals can generally be disposed of in the normal school garbage containers. Animal dissection specimens should be rinsed thoroughly to rid them of any excess preservative and sealed in plastic bags.

Fungi and bacterial cultures should be autoclaved or treated with a fungicide or antibacterial soap before disposal.

First Aid

The following guidelines apply if an injury, such as a burn, cut, chemical spill, ingestion, inhalation, or splash in eyes, happens to you or to one of your classmates.

- If an injury occurs, inform your teacher immediately. If the injury appears serious, call for emergency assistance immediately.

- Know the location of the first-aid kit, fire blanket, eyewash station, and shower, and be familiar with the contents/operation.

- If the injury is the result of chemicals, drench the affected area with a continuous flow of water for 30 min. Clothing should be removed as necessary. Inform your teacher. Retrieve the Material Safety Data Sheet (MSDS) for the chemical; this sheet provides information about the first-aid requirements for the chemical. If the chemicals are splashed in your eyes, have another student assist you in getting to the eyewash station immediately. Rinse with the eyes open for at least 15 min.

- If you have ingested or inhaled a hazardous substance, inform your teacher immediately. The MSDS will give information about the first-aid requirements for the substance in question. Contact the Poison Control Centre in your area.

- If the injury is from a burn, immediately immerse the affected area in cold water. This will reduce the temperature and prevent further tissue damage.

- In the event of electrical shock, do not touch the affected person or the equipment the person was using. Break contact by switching off the source of electricity or by removing the plug.

- If a classmate's injury has rendered him/her unconscious, notify the teacher immediately. The teacher will perform CPR if necessary. Do not administer CPR unless under specific instructions from the teacher. You can assist by keeping the person warm and reassured.

C1 Units, Symbols, and Prefixes

Throughout *Nelson Chemistry 12* and in this reference section, we have attempted to be consistent in the presentation and usage of quantities, units, and their symbols. As far as possible, the text uses the Système international d'unités (SI). However, some other units have been included because of their practical importance, wide usage, or use in specialized fields. In our interpretations and usage, *Nelson Chemistry 12* has followed the most recent *Canadian Metric Practice Guide* (CAN/CSA–Z234.1–89), published in 1989 and reaffirmed in 1995 by the Canadian Standards Association.

Numerical Prefixies

Prefix	Power	Symbol
deca-	10^1	da
hecto-	10^2	h
kilo-	10^3	k*
mega-	10^6	M*
giga-	10^9	G*
tera-	10^{12}	T
peta-	10^{15}	P
exa-	10^{18}	E
deci-	10^{-1}	d
centi-	10^{-2}	c*
milli-	10^{-3}	m*
micro-	10^{-6}	μ*
nano-	10^{-9}	n*
pico-	10^{-12}	p
femto-	10^{-15}	f
atto-	10^{-18}	a

* commonly used

Some Examples of Prefix Use

0.0034 mol = 3.4×10^{-3} mol =	3.4 **milli**moles or 3.4 mmol	
1530 L = 1.53×10^3 L = 1.53 **kilo**litres or 1.53 kL		

SI Base Units

Quantity	Symbol	Unit name	Symbol
amount of substance	n	mole	mol
electric current	I	ampere	A
length	L, l, h, d, w	metre	m
luminous intensity	I_v	candela	cd
mass	m	kilogram	kg
temperature	T	kelvin	K
time	t	second	s

Defined (Exact) Quantities

1 mL	=	$1 cm^3$
1 kL	=	$1 m^3$
1000 kg	=	1 t
1 Mg	=	1 t
1 atm	=	101.325 kPa
0°C	=	273.15 K
STP	=	0°C and 101.325 kPa
SATP	=	25°C and 100 kPa

Common Multiples

Multiple	Prefix
1	mono-
2	bi–, di–
3	tri–
4	tetra–
5	penta
6	hexa
7	hepta–
8	octa
9	nona–
10	deca–

C2 Common Chemicals

You live in a chemical world. As one bumper sticker asks, "What in the world isn't chemistry?" Every natural and technologically produced substance around you is composed of chemicals. Many of these chemicals are used to make your life easier or safer, and some of them have life-saving properties. Following is a list of selected common chemicals.

Common name	Recommended name	Formula	Common use/source
acetic acid	ethanoic acid	$HC_2H_3O_{2(aq)}$; $CH_3COOH_{(aq)}$	vinegar
acetone	propanone	$(CH_3)_2CO_{(l)}$	nail polish remover
acetylene	ethyne	$C_2H_{2(g)}$	cutting/welding torch
ASA (Aspirin®)	acetylsalicylic acid	$HC_9H_7O_{4(s)}$; $C_6H_4COOCH_3COOH_{(s)}$	for pain-relief medication
baking soda	sodium hydrogen carbonate	$NaHCO_{3(s)}$	leavening agent
battery acid	sulfuric acid	$H_2SO_{4(aq)}$	car batteries
bleach	sodium hypochlorite	$NaClO_{(s)}$	bleach for clothing
bluestone	copper(II) sulfate pentahydrate	$CuSO_4 \cdot 5\ H_2O_{(s)}$	algicide, fungicide
brine	aqueous sodium chloride	$NaCl_{(aq)}$	water-softening agent
CFC	chlorofluorocarbon	$C_xCl_yF_{z(l)}$; e.g., $C_2Cl_2F_{4(l)}$	refrigerant
charcoal/graphite	carbon	$C_{(s)}$	fuel, lead pencils
citric acid	2-hydroxy-1,2,3-propanetricarboxylic acid	$H_3C_8H_5O_{7(s)}C_3H_4OH(COOH)_{3(s)}$	in fruit and beverages
carbon dioxide	carbon dioxide	$CO_{2(g)}$	dry ice, carbonated beverages
ethylene	ethene	$C_2H_{4(g)}$	for polymerization
ethylene glycol	1,2-ethanediol	$C_2H_4(OH)_{2(l)}$	radiator antifreeze
freon-12	dichlorodifluoromethane	$CCl_2F_{2(l)}$	refrigerant
Glauber's salt	sodium sulfate decahydrate	$Na_2SO_4 \cdot 10\ H_2O_{(s)}$	solar heat storage
glucose	D-glucose; dextrose	$C_6H_{12}O_{6(s)}$	in plants and blood
grain alcohol	ethanol (ethyl alcohol)	$C_2H_5OH_{(l)}$	beverage alcohol
gypsum	calcium sulfate dihydrate	$CaSO_4 \cdot 2\ H_2O_{(s)}$	wallboard
lime (quicklime)	calcium oxide	$CaO_{(s)}$	masonry
limestone	calcium carbonate	$CaCO_{3(s)}$	chalk and building materials
lye (caustic soda)	sodium hydroxide	$NaOH_{(s)}$	oven/drain cleaner
malachite	copper(II) hydroxide carbonate	$Cu(OH)_2 \cdot CuCO_{3(s)}$	copper mineral
methyl hydrate	methanol (methyl alcohol)	$CH_3OH_{(l)}$	gas line antifreeze
milk of magnesia	magnesium hydroxide	$Mg(OH)_{2(s)}$	antacid (for indigestion)
MSG	monosodium glutamate	$NaC_5H_8NO_{4(s)}$	flavour enhancer
muriatic acid	hydrochloric acid	$HCl_{(aq)}$	concrete etching
natural gas	methane	$CH_{4(g)}$	fuel
PCBs	polychlorinated biphenyls	$(C_6H_xCl_y)_2$; e.g., $(C_6H_4Cl_2)_{2(l)}$	in transformers
potash	potassium chloride	$KCl_{(s)}$	fertilizer
road salt	calcium chloride or sodium chloride	$CaCl_{2(s)}$ or $NaCl_{(s)}$	melts ice
rotten-egg gas	hydrogen sulfide	$H_2S_{(g)}$	in natural gas
rubbing alcohol	2-propanol (also isopropanol)	$CH_3CHOHCH_{3(l)}$	for massage
sand (silica)	silicon dioxide	$SiO_{2(s)}$	in glassmaking
slaked lime	calcium hydroxide	$Ca(OH)_{2(s)}$	limewater
soda ash	sodium carbonate	$Na_2CO_{3(s)}$	in laundry detergents
sugar	sucrose	$C_{12}H_{22}O_{11(s)}$	sweetener
table salt	sodium chloride	$NaCl_{(s)}$	seasoning
vitamin C	ascorbic acid	$H_2C_6H_6O_{6(s)}$	vitamin supplement
washing soda	sodium carbonate decahydrate	$Na_2CO_3 \cdot 10\ H_2O_{(s)}$	water softener

C3 Using VSEPR Theory to Predict Molecular Shape

Note: This is an expansion of the table found on page 245 of the text.

Table 1 Using VSEPR Theory to Predict Molecular Shape

General formula*	Bond pairs	Lone pairs	Total pairs	Molecular shape Geometry**	Shape diagram	Examples
AX_2E	2	1	3	V-shaped (trigonal planar)		$SnCl_2$
AX_5	5	0	5	trigonal bipyramidal (trigonal bipyramidal)		$SbCl_5$
AX_4E	4	1	5	seesaw (trigonal bipyramidal)		SF_4
AX_3E_2	3	2	5	T-shaped (trigonal bipyramidal)		BrF_3
AX_2E_3	2	3	5	linear (trigonal bipyramidal)		XeF_2
AX_6	6	0	6	octahedral (octahedral)		SF_6
AX_5E	5	1	6	square pyramidal (octahedral)		BrF_{25}
AX_4E_2	4	2	6	square planar (octahedral)		XeF_4

* A is the central atom; X is another atom; E is a lone pair of electrons.
** Electron-pair arrangement is in parentheses.

C4 Specific Heat Capacities

Specific Heat Capacities of Pure Substances

Substance	Specific Heat Capacity* $(J/(g \cdot °C))$	Substance	Specific Heat Capacity* $(J/(g \cdot °C))$
aluminum	0.900	nickel	0.444
calcium	0.653	potassium	0.753
copper	0.385	silver	0.237
gold	0.129	sodium	1.226
hydrogen	14.267	sulfur	0.732
iron	0.444	tin	0.213
lead	0.159	zinc	0.388
lithium	3.556	ice, $H_2O_{(s)}$	2.01
magnesium	1.017	water, $H_2O_{(l)}$	4.18
mercury	0.138	steam, $H_2O_{(g)}$	2.01

*Elements at SATP state

C5 Molar Enthalpies of Combustion

Substance	Molar Enthalpy of Combustion (kJ/mol)
Methanel	−890
Ethane	−1560
Propane	−2220
Butane	−2871
Hexane	−4163
Octane	−5450
Methanol	−727
Ethanol	−1367
Propanol	−2020
Butanol	−2676

C6 Standard Molar Entropies and Enthalpies of Formation

Chemical Name	Formula	ΔH_f° (kJ/mol)	S° (J/(mol·K))	Chemical Name	Formula	ΔH_f° (kJ/mol)	S° (J/(mol·K))
acetone	$(CH_3)_2CO_{(l)}$	−248.1	198.8	carbon disulfide	$CS_{2(l)}$	+89.0	−
aluminum oxide	$Al_2O_{3(s)}$	−1675.7	50.92	carbon monoxide	$CO_{(g)}$	−110.5	197.66
ammonia	$NH_{3(g)}$	−45.9	192.78	chloroethene	$C_2H_3Cl_{(g)}$	+37.3	263.9
ammonium chloride	$NH_4Cl_{(s)}$	−314.4	94.6	chromium(III) oxide	$Cr_2O_{3(s)}$	−1139.7	81.2
ammonium chloride	$NH_4Cl_{(aq)}$	−299.7	169.9	copper(I) oxide	$Cu_2O_{(s)}$	−168.6	93.1
ammonium nitrate	$NH_4NO_{3(s)}$	−365.6	151.08	copper(II) oxide	$CuO_{(s)}$	−157.3	42.6
barium carbonate	$BaCO_{3(s)}$	−1216.3	112.1	copper(I) sulfide	$Cu_2S_{(s)}$	−79.5	120.9
barium hydroxide	$Ba(OH)_{2(s)}$	−944.7	107	copper(II) sulfide	$CuS_{(s)}$	−53.1	66.5
barium oxide	$BaO_{(s)}$	−553.5	72.07	cyclopropane	$C_3H_{6(g)}$	+17.8	−
barium sulfate	$BaSO_{4(s)}$	−1473.2	132.2	1,2-dichloroethane	$C_2H_4Cl_{2(l)}$	−126.9	−
benzene	$C_6H_{6(l)}$	+49.0	173.4	ethane	$C_2H_{6(g)}$	−83.8	229.1
bromine (vapour)	$Br_{2(g)}$	+30.9	245.47	1,2-ethanediol	$C_2H_4(OH)_{2(l)}$	−454.8	163.2
butane	$C_4H_{10(g)}$	−125.6	310.1	ethanoic (acetic) acid	$CH_3COOH_{(l)}$	−432.8	159.9
calcium carbonate	$CaCO_{3(s)}$	−1206.9	91.7	ethanol	$C_2H_5OH_{(l)}$	−235.2	161.0
calcium chloride	$CaCl_{2(s)}$	−795.8	104.6	ethanol	$C_2H_5OH_{(g)}$	−235.2	282.70
calcium hydroxide	$Ca(OH)_{2(s)}$	−986.1	83.4	ethene (ethylene)	$C_2H_{4(g)}$	+52.5	219.3
calcium oxide	$CaO_{(s)}$	−634.9	38.1	ethyne (acetylene)	$C_2H_{2(g)}$	+228.2	201.0
calcium sulphate	$CaSO_{4(s)}$	−1434.1	108.4	glucose	$C_6H_{12}O_{6(s)}$	−1273.1	212.1
carbon dioxide	$CO_{2(g)}$	−393.5	213.78				

Chemical Name	Formula	ΔH°_f (kJ/mol)	S° (J/(mol·K))	Chemical Name	Formula	ΔH°_f (kJ/mol)	S° (J/(mol·K))
hexane	$C_6H_{14(l)}$	−198.7	296.1	pentane	$C_5H_{12(l)}$	−173.5	262.7
hydrazine	$N_2H_{4(g)}$	+95.4	237.11	phenylethene (styrene)	$C_6H_5CHCH_{2(l)}$	+103.8	237.6
hydrazine	$N_2H_{4(l)}$	50.6	121.2	phosphorus pentachloride	$PCl_{5(g)}$	−443.5	364.6
hydrogen bromide	$HBr_{(g)}$	−36.3	198.70	phosphorus trichloride	$PCl_{3(l)}$	−319.7	217.2
hydrogen chloride	$HCl_{(g)}$	−92.3	186.90	phosphorus trichloride	$PCl_{3(g)}$	−287.0	311.8
hydrogen cyanide	$HCN_{(g)}$	+135.1	201.81	potassium	$K_{(s)}$	0.0	75.90
hydrogen iodide	$HI_{(g)}$	+26.5	206.59	potassium	$K_{(l)}$	2.3	71.46
hydrogen peroxide	$H_2O_{2(l)}$	−187.8	109.6	potassium chlorate	$KClO_{3(s)}$	−397.7	143.1
hydrogen sulfide	$H_2S_{(g)}$	−20.6	205.81	potassium chloride	$KCl_{(s)}$	−436.7	82.55
iodine (vapour)	$I_{2(g)}$	+62.4	180.79	potassium hydroxide	$KOH_{(s)}$	−424.8	78.9
iron(III) oxide	$Fe_2O_{3(s)}$	−824.2	87.40	propane	$C_3H_{8(g)}$	−104.7	270.2
iron(II, III) oxide	$Fe_3O_{4(s)}$	−1118.4	145.27	silicon dioxide	$SiO_{2(s)}$	−910.7	41.46
lead(II) oxide	$PbO_{(s)}$	−219.0	66.5	silver bromide	$AgBr_{(s)}$	−100.4	107.11
lead(IV) oxide	$PbO_{2(s)}$	−277.4	68.60	silver chloride	$AgCl_{(s)}$	−127.0	96.25
magnesium carbonate	$MgCO_{3(s)}$	−1095.8	65.7	silver iodide	$AgI_{(s)}$	−61.8	115.5
magnesium chloride	$MgCl_{2(s)}$	−641.3	89.63	sodium bromide	$NaBr_{(s)}$	−361.1	86.82
magnesium hydroxide	$Mg(OH)_{2(s)}$	−924.5	63.24	sodium chloride	$NaCl_{(s)}$	−411.2	115.5
magnesium oxide	$MgO_{(s)}$	−601.6	26.95	sodium hydroxide	$NaOH_{(s)}$	−425.6	64.4
manganese(II) oxide	$MnO_{(s)}$	−385.2	59.8	sodium iodide	$NaI_{(s)}$	−287.8	98.50
manganese(IV) oxide	$MnO_{2(s)}$	−520.0	53.1	sucrose	$C_{12}H_{22}O_{11(s)}$	−2225.5	360.2
mercury	$Hg_{(l)}$	0.0	75.90	sulfur dioxide	$SO_{2(g)}$	−296.8	248.22
mercury	$Hg_{(g)}$	61.4	174.97	sulfur trioxide (liquid)	$SO_{3(l)}$	−441.0	−
mercury(II) oxide	$HgO_{(s)}$	−90.8	70.25	sulfur trioxide (vapour)	$SO_{3(g)}$	−395.7	256.77
mercury(II) sulfide	$HgS_{(s)}$	−58.2	82.4	sulfuric acid	$H_2SO_{4(l)}$	−814.0	156.90
methanal (formaldehyde)	$CH_2O_{(g)}$	−108.6	218.8	tin(II) oxide	$SnO_{(s)}$	−280.7	57.17
methane	$CH_{4(g)}$	−74.4	186.3	tin(IV) oxide	$SnO_{2(s)}$	−577.6	49.04
methanoic (formic) acid	$HCOOH_{(l)}$	−425.1	129.0	2,2,4-trimethylpentane	$C_8H_{18(l)}$	−259.2	328.0
methanol	$CH_3OH_{(l)}$	−239.1	126.8	urea	$CO(NH_2)_{2(s)}$	−333.5	104.6
methylpropane	$C_4H_{10(g)}$	−134.2	294.6	water (liquid)	$H_2O_{(l)}$	−285.8	69.95
nickel(II) oxide	$NiO_{(s)}$	−239.7	38.00	water (vapour)	$H_2O_{(g)}$	−241.8	188.84
nitric acid	$HNO_{3(l)}$	−174.1	155.60	zinc oxide	$ZnO_{(s)}$	−350.5	43.65
nitrogen dioxide	$NO_{2(g)}$	+33.2	240.1	zinc sulfide	$ZnS_{(s)}$	−206.0	57.7
nitrogen monoxide	$NO_{(g)}$	+90.2	210.76				
nitromethane	$CH_3NO_{2(l)}$	−113.1	171.8				
octane	$C_8H_{18(l)}$	−250.1	−				
ozone	$O_{3(g)}$	+142.7	163.2				

- Standard molar enthalpies (heats) of formation are measured at SATP (25°C and 100 kPa). The values were obtained from *The CRC Handbook of Chemistry and Physics*, 71st Edition.

- The standard molar enthalpies of elements in their standard states are defined as zero.

C7 Cations and Anions

Common Cations

Ion	Name
H^+	hydrogen
Li^+	lithium
Na^+	sodium
K^+	potassium
Cs^+	cesium
Be^{2+}	beryllium
Mg^{2+}	magnesium
Ca^{2+}	calcium
Ba^{2+}	barium
Al^{3+}	aluminum
Ag^+	silver

Common Anions

Ion	Name
H^-	hydride
F^-	fluoride
Cl^-	chloride
Br^-	bromide
I^-	iodide
O^{2-}	oxide
S^{2-}	sulfide
N^{3-}	nitride
P^{3-}	phosphide

Ion Colours

Ion	Solution colour
Groups 1, 2, 17	colourless
$Cr^{2+}_{(aq)}$	blue
$Cr^{3+}_{(aq)}$	green
$Co^{2+}_{(aq)}$	pink
$Cu^+_{(aq)}$	green
$Cu^{2+}_{(aq)}$	blue
$Fe^{2+}_{(aq)}$	pale green
$Fe^{3+}_{(aq)}$	yellow-brown
$Mn^{2+}_{(aq)}$	pale pink
$Ni^{2+}_{(aq)}$	green
$CrO_4{}^{2-}_{(aq)}$	yellow
$Cr_2O_7{}^{2-}_{(aq)}$	orange
$MnO_4{}^-_{(aq)}$	purple

Ion	Flame
Li^+	bright red
Na^+	yellow
K^+	violet
Ca^{2+}	yellow-red
Sr^{2+}	bright red
Ba^{2+}	yellow-green
Cu^{2+}	blue (halides) green (others)
Pb^{2+}	light blue-grey
Zn^{2+}	whitish green

Common Polyatomic Ions

Ion	Name	Ion	Name
$C_2H_3O_2{}^-$	acetate	$CO_3{}^{2-}$	carbonate
$ClO_3{}^-$	chlorate*	$CrO_4{}^{2-}$	chromate
$ClO_2{}^-$	chlorite*	$Cr_2O_7{}^{2-}$	dichromate
CN^-	cyanide	$HPO_4{}^{2-}$	hydrogen phosphate
$H_2PO_4{}^-$	dihydrogen phosphate	$C_2O_4{}^{2-}$	oxalate
$HCO_3{}^-$	hydrogen carbonate (bicarbonate)	$O_2{}^{2-}$	peroxide
$HSO_4{}^-$	hydrogen sulfate (bisulfate)	$SiO_3{}^{2-}$	silicate
HS^-	hydrogen sulfide (bisulfide)	$SO_4{}^{2-}$	sulfate
$HSO_3{}^-$	hydrogen sulfite (bisulfite)	$SO_3{}^{2-}$	sulfite
ClO^-, OCl^-	hypochlorite*	$S_2O_3{}^{2-}$	thiosulfate
OH^-	hydroxide	$BO_3{}^{3-}$	borate
$NO_2{}^-$	nitrite	$PO_4{}^{3-}$	phosphate
$NO_3{}^-$	nitrate	$P_3O_{10}{}^{5-}$	tripolyphosphate
$ClO_4{}^-$	perchlorate*	$NH_4{}^+$	ammonium
$MnO_4{}^-$	permanganate	H_3O^+	hydronium
SCN^-	thiocyanate	$Hg_2{}^{2+}$	mercury(I)

*There are also corresponding ions containing Br and I instead of Cl.

Solubility of Ionic Compounds at SATP

		Anions						
		Cl^-, Br^-, I^-	S^{2-}	OH^-	$SO_4{}^{2-}$	$CO_3{}^{2-}$, $PO_4{}^{3-}$, $SO_3{}^{2-}$	$C_2H_3O_2{}^-$	$NO_3{}^-$
Cations	High solubility (aq) ≥0.1 mol/L (at SATP)	most	Group 1, $NH_4{}^+$ Group 2	Group 1, $NH_4{}^+$ Sr^{2+}, Ba^{2+}, Tl^+	most	Group 1, $NH_4{}^+$	most	all
		All Group 1 compounds, acids, and all ammonium compounds are assumed to have high solubility in water.						
	Low Solubility (s) <0.1 mol/L (at SATP)	Ag^+, Pb^{2+}, Tl^+, $Hg_2{}^{2+}$ (Hg^+), Cu^+	most	most	Ag^+, Pb^{2+}, Ca^{2+}, Ba^{2+}, Sr^{2+}, Ra^{2+}	most	Ag^+	none

C8 Solubility Product Constants (K_{sp})

Solubility Product Constants at 25°C

Name	Formula	K_{sp}
barium carbonate	$BaCO_{3(s)}$	2.6×10^{-9}
barium chromate	$BaCrO_{4(s)}$	1.2×10^{-10}
barium sulfate	$BaSO_{4(s)}$	1.1×10^{-10}
calcium carbonate	$CaCO_{3(s)}$	5.0×10^{-9}
calcium oxalate	$CaC_2O_{4(s)}$; $CaOOCCOO_{(s)}$	2.3×10^{-9}
calcium phosphate	$Ca_3(PO_4)_{2(s)}$	2.1×10^{-33}
calcium sulfate	$CaSO_{4(s)}$	7.1×10^{-5}
copper(I) chloride	$CuCl_{(s)}$	1.7×10^{-7}
copper(I) iodide	$CuI_{(s)}$	1.3×10^{-12}
copper(II) iodate	$Cu(IO_3)_{2(s)}$	6.9×10^{-8}
copper(II) sulfide	$CuS_{(s)}$	6.0×10^{-37}
iron(II) hydroxide	$Fe(OH)_{2(s)}$	4.9×10^{-17}
iron(II) sulfide	$FeS_{(s)}$	6.0×10^{-19}
iron(III) hydroxide	$Fe(OH)_{3(s)}$	2.6×10^{-39}
lead(II) bromide	$PbBr_{2(s)}$	6.6×10^{-6}
lead(II) chloride	$PbCl_{2(s)}$	1.2×10^{-5}
lead(II) iodate	$Pb(IO_3)_{2(s)}$	3.7×10^{-13}
lead(II) iodide	$PbI_{2(s)}$	8.5×10^{-9}
lead(II) sulfate	$PbSO_{4(s)}$	1.8×10^{-8}
magnesium carbonate	$MgCO_{3(s)}$	6.8×10^{-6}
magnesium fluoride	$MgF_{2(s)}$	6.4×10^{-9}
magnesium hydroxide	$Mg(OH)_{2(s)}$	5.6×10^{-12}
mercury(I) chloride	$Hg_2Cl_{2(s)}$	1.5×10^{-18}
silver bromate	$AgBrO_{3(s)}$	5.3×10^{-5}
silver bromide	$AgBr_{(s)}$	5.4×10^{-13}
silver carbonate	$Ag_2CO_{3(s)}$	8.5×10^{-12}
silver chloride	$AgCl_{(s)}$	1.8×10^{-10}
silver chromate	$Ag_2CrO_{4(s)}$	1.1×10^{-12}
silver iodate	$AgIO_{3(s)}$	3.2×10^{-8}
silver iodide	$AgI_{(s)}$	8.5×10^{-17}
strontium carbonate	$SrCO_{3(s)}$	5.6×10^{-10}
strontium fluoride	$SrF_{2(s)}$	4.3×10^{-9}
strontium sulfate	$SrSO_{4(s)}$	3.4×10^{-7}
zinc hydroxide	$Zn(OH)_{2(s)}$	7.7×10^{-17}
zinc sulfide	$ZnS_{(s)}$	2.0×10^{-25}

- Values in this table are taken from *The CRC Handbook of Chemistry and Physics,* 76th Edition.

C9 K_a and K_b for Common Acids and Weak Bases

Monoprotic Acids

Name	Formula of Acid	Formula of Conjugate Base	Equilibrium Constant, K_a
perchloric acid	$HClO_{4(aq)}$	$ClO_{4(aq)}^-$	very large
hydroiodic acid	$HI_{(aq)}$	$I_{(aq)}^-$	very large
hydrobromic acid	$HBr_{(aq)}$	$Br_{(aq)}^-$	very large
hydrochloric acid	$HCl_{(aq)}$	$Cl_{(aq)}^-$	very large
nitric acid	$HNO_{3(aq)}$	$NO_{3(aq)}^-$	very large
hydronium ion	$H_3O^+_{(aq)}$	$H_2O_{(l)}$	1.0
iron(III) ion	$Fe(H_2O)_6^{3+}{}_{(aq)}$	$Fe(H_2O)_5(OH)^{2+}_{(aq)}$	1.5×10^{-3}
citric acid	$H_3C_6H_5O_{7(aq)}$	$H_2C_6H_5O_{7(aq)}^-$	7.4×10^{-4}
nitrous acid	$HNO_{2(aq)}$	$NO_{2(aq)}^-$	7.2×10^{-4}
hydrofluoric acid	$HF_{(aq)}$	$F_{(aq)}^-$	6.6×10^{-4}
hydrogen cyanate	$HOCN$	$OCN_{(aq)}^-$	3.5×10^{-4}
methanoic acid	$HCHO_2$; $HCOOH_{(aq)}$	$CHO_{2(aq)}^-$	1.8×10^{-4}
chromium(III) ion	$Cr(H_2O)_6^{3+}{}_{(aq)}$	$Cr(H_2O)_5(OH)^{2+}_{(aq)}$	1.0×10^{-4}
methyl orange	$HMo_{(aq)}$	$Mo_{(aq)}^-$	$\sim 10^{-4}$
benzoic acid	$HC_7H_5O_{2(aq)}$; $C_6H_5COOH_{(aq)}$	$C_6H_5O_{2(aq)}^-$	6.3×10^{-5}
ethanoic (acetic) acid	$HC_2H_3O_{2(aq)}$; $CH_3COOH_{(aq)}$	$C_2H_3O_{2(aq)}^-$	1.8×10^{-5}
aluminum ion	$Al(H_2O)_6^{3+}{}_{(aq)}$	$Al(H_2O)_5(OH)^{2+}_{(aq)}$	9.8×10^{-6}
bromothymol blue	$HBb_{(aq)}$	$Bb_{(aq)}^-$	$\sim 10^{-7}$
hypochlorous acid	$HClO_{(aq)}$	$ClO_{(aq)}^-$	2.9×10^{-8}
phenolphthalein	$HPh_{(aq)}$	$Ph_{(aq)}^-$	$\sim 10^{-10}$
hydrocyanic acid	$HCN_{(aq)}$	$CN_{(aq)}^-$	6.2×10^{-10}
ammonium ion	$NH_4^+{}_{(aq)}$	$NH_{3(aq)}$	5.8×10^{-10}
boric acid	$H_3BO_{3(aq)}$	$H_2BO_{3(aq)}^-$	5.8×10^{-10}
phenol	$C_6H_5OH_{(aq)}$	$C_6H_5O_{(aq)}^-$	1.0×10^{-10}
hydrogen peroxide	$H_2O_{2(aq)}$	$HO_{2(aq)}^-$	2.2×10^{-12}
water	$H_2O_{(l)}$	$OH_{(aq)}^-$	1.0×10^{-14}
hydroxide ion	$OH_{(aq)}^-$	$O^{2-}_{(aq)}$	very small

- Values in this table are taken from *Lange's Handbook of Chemistry*, 13th Edition for 25°C.

Weak Bases

Name	Formula	Equilibrium Constant, K_b
dimethylamine	CH_3CH_3NH	9.6×10^{-4}
methylamine	CH_3NH_2	4.4×10^{-4}
ethylamine	$CH_3CH_2NH_2$	4.3×10^{-4}
trimethylamine	$CH_3CH_3CH_3N$	7.4×10^{-5}
ammonia	NH_3	1.8×10^{-5}
hydrazine	N_2H_4	9.6×10^{-7}
hydroxylamine	NH_2OH	6.6×10^{-9}
pyridine	C_5H_5N	1.5×10^{-9}
aniline	$C_5H_5NH_2$	4.1×10^{-10}

Polyprotic Acids

Name	Formula of Acid	Formula of Conjugate Base	K_{a_1}	K_{a_2}	K_{a_3}
sulfuric acid	$H_2SO_{4(aq)}$	$HSO_{4(aq)}^-$	very large	1.0×10^{-2}	
oxalic acid	$H_2C_2O_{4(aq)}$; $HOOCCOOH_{(aq)}$	$HC_2O_{4(aq)}^-$	5.4×10^{-2}	5.4×10^{-5}	
sulfurous acid ($SO_2 + H_2O$)	$H_2SO_{3(aq)}$	$HSO_{3(aq)}^-$	1.3×10^{-2}	6.2×10^{-8}	
phosphoric acid	$H_3PO_{4(aq)}$	$H_2PO_{4(aq)}^-$	7.1×10^{-3}	6.3×10^{-8}	4.2×10^{-13}
carbonic acid ($CO_2 + H_2O$)	$H_2CO_{3(aq)}$	$HCO_{3(aq)}^-$	4.4×10^{-7}	4.7×10^{-11}	
hydrosulfuric acid	$H_2S_{(aq)}$	$HS_{(aq)}^-$	1.1×10^{-7}	1.3×10^{-13}	

- Values in this table are taken from *Lange's Handbook of Chemistry*, 13th Edition for 25°C.

C10 Acids and Bases

Oxyacids

Acid	Name
$HNO_{3(aq)}$	nitric acid
$HNO_{2(aq)}$	nitrous acid
$H_2SO_{4(aq)}$	sulfuric acid
$H_2SO_{3(aq)}$	sulfurous acid
$H_3PO_{4(aq)}$	phosphoric acid
$HC_2H_3O_{2(aq)}$	acetic acid
$HClO_{4(aq)}$	perchloric acid
$HBrO_{4(aq)}$	perbromic acid
$HIO_{4(aq)}$	periodic acid
$HClO_{3(aq)}$	chloric acid
$HBrO_{3(aq)}$	bromic acid
$HIO_{3(aq)}$	iodic acid
$HClO_{2(aq)}$	chlorous acid
$HClO_{(aq)}$	hypochlorous acid
$HBrO_{(aq)}$	hypobromous acid
$HIO_{(aq)}$	hypoiodous acid
$HFO_{(aq)}$	hypofluorous acid

Concentrated Reagents•

Reagent	Formula	Molar mass (g/mol)	Concentration (mol/L)	Concentration (mass %)
acetic acid	$HC_2H_3O_{2(aq)}$	60.05	17.45	99.8
carbonic acid	$H_2CO_{3(aq)}$	62.03	0.039	0.17
formic acid	$HCOOH_{(aq)}$	46.03	23.6	90.5
hydrobromic acid	$HBr_{(aq)}$	80.91	8.84	48.0
hydrochloric acid	$HCl_{(aq)}$	36.46	12.1	37.2
hydrofluoric acid	$HFl_{(aq)}$	20.01	28.9	49.0
nitric acid	$HNO_{3(aq)}$	63.02	15.9	70.4
perchloric acid	$HClO_{4(aq)}$	100.46	11.7	70.5
phosphoric acid	$H_3PO_{4(aq)}$	98.00	14.8	85.5
sulfurous acid	$H_2SO_{3(aq)}$	82.08	0.73	6.0
sulfuric acid	$H_2SO_{4(aq)}$	98.08	18.0	96.0
ammonia	$NH_{3(aq)}$	17.04	14.8	28.0
potassium hydroxide	$KOH_{(aq)}$	56.11	11.7	45.0
sodium hydroxide	$NaOH_{(aq)}$	40.00	19.4	50.5

• Typical concentrations of commercial concentrated reagents

Acid–Base Indicators

Common Name	Colour of $HIn_{(aq)}$	pH range	Colour of $In^-_{(aq)}$	Common name	Colour of $HIn_{(aq)}$	pH range	Colour of $In^-_{(aq)}$
methyl violet	yellow	0.0 – 1.6	blue	p-nitrophenol	colourless	5.3 – 7.6	yellow
cresol red (acid range)	red	0.2 – 1.8	yellow	litmus	red	6.0 – 8.0	blue
cresol purple (acid range)	red	1.2 – 2.8	yellow	bromothymol blue	yellow	6.2 – 7.6	blue
thymol blue (acid range)	red	1.2 – 2.8	yellow	neutral red	red	6.8 – 8.0	yellow
tropeolin oo	red	1.3 – 3.2	yellow	phenol red	yellow	6.4 – 8.0	red
orange iv	red	1.4 – 2.8	yellow	m-nitrophenol	colourless	6.4 – 8.8	yellow
benzopurpurine-4B	violet	2.2 – 4.2	red	cresol red	yellow	7.2 – 8.8	red
2,6-dinotrophenol	colourless	2.4 – 4.0	yellow	m-cresol purple	yellow	7.6 – 9.2	purple
2,4-dinotrophenol	colourless	2.5 – 4.3	yellow	thymol blue	yellow	8.0 – 9.6	blue
methyl yellow	red	2.9 – 4.0	yellow	phenolphthalein	colourless	8.0 – 10.0	red
congo red	blue	3.0 – 5.0	red	α-naphtholbenzein	yellow	9.0 – 11.0	blue
methyl orange	red	3.1 – 4.4	yellow	thymolphthalein	colourless	9.4 – 10.6	blue
bromophenol blue	yellow	3.0 – 4.6	blue-violet	alizarin yellow r	yellow	10.0 – 12.0	violet
bromocresol green	yellow	4.0 – 5.6	blue	tropeolin o	yellow	11.0 – 13.0	orange-brown
methyl red	red	4.4 – 6.2	yellow	nitramine	colourless	10.8 – 13.0	orange-brown
chlorophenol red	yellow	5.4 – 6.8	red	indigo carmine	blue	11.4 – 13.0	yellow
bromocresol purple	yellow	5.2 – 6.8	purple	1,3,5-trinitrobenzene	colourless	12.0 – 14.0	orange
bromophenol red	yellow	5.2 – 6.8	red				

C11 Relative Strengths of Oxidizing and Reducing Agents

Oxidizing Agents	Reducing Agents	E°_r (V)
$F_{2(g)} + 2e^- \rightleftharpoons$	$2F^-_{(aq)}$	+2.87
$PbO_{2(s)} + SO^{2-}_{4(aq)} + 4H^+_{(aq)} + 2e^- \rightleftharpoons$	$PbSO_{4(s)} + 2H_2O_{(l)}$	+1.69
$MnO^-_{4(aq)} + 8H^+_{(aq)} + 5e^- \rightleftharpoons$	$Mn^{2+}_{(aq)} + 4H_2O_{(l)}$	+1.51
$Au^{3+}_{(aq)} + 3e^- \rightleftharpoons$	$Au_{(s)}$	+1.50
$ClO^-_{4(aq)} + 8H^+_{(aq)} + 8e^- \rightleftharpoons$	$Cl^-_{(aq)} + 4H_2O_{(l)}$	+1.39
$Cl_{2(g)} + 2e^- \rightleftharpoons$	$2Cl^-_{(aq)}$	+1.36
$2HNO_{2(aq)} + 4H^+_{(aq)} + 4e^- \rightleftharpoons$	$N_2O_{(g)} + 3H_2O_{(l)}$	+1.30
$Cr_2O^{2-}_{7(aq)} + 14H^+_{(aq)} + 6e^- \rightleftharpoons$	$2Cr^{3+}_{(aq)} + 7H_2O_{(l)}$	+1.23
$O_{2(g)} + 4H^+_{(aq)} + 4e^- \rightleftharpoons$	$2H_2O_{(l)}$	+1.23
$MnO_{2(s)} + 4H^+_{(aq)} + 2e^- \rightleftharpoons$	$Mn^{2+}_{(aq)} + 2H_2O_{(l)}$	+1.22
$2IO^-_{3(aq)} + 12H^+_{(aq)} + 10e^- \rightleftharpoons$	$I_{2(s)} + 6H_2O_{(l)}$	+1.20
$Br_{2(l)} + 2e^- \rightleftharpoons$	$2Br^-_{(aq)}$	+1.07
$Hg^{2+}_{(aq)} + 2e^- \rightleftharpoons$	$Hg_{(l)}$	+0.85
$ClO^-_{(aq)} + H_2O_{(l)} + 2e^- \rightleftharpoons$	$Cl^-_{(aq)} + 2OH^-_{(aq)}$	+0.84
$Ag^+_{(aq)} + e^- \rightleftharpoons$	$Ag_{(s)}$	+0.80
$NO^-_{3(aq)} + 2H^+_{(aq)} + e^- \rightleftharpoons$	$NO_{2(g)} + H_2O_{(l)}$	+0.80
$Fe^{3+}_{(aq)} + e^- \rightleftharpoons$	$Fe^{2+}_{(aq)}$	+0.77
$O_{2(g)} + 2H^+_{(aq)} + 2e^- \rightleftharpoons$	$H_2O_{2(l)}$	+0.70
$MnO^-_{4(aq)} + 2H_2O_{(l)} + 3e^- \rightleftharpoons$	$MnO_{2(s)} + 4OH^-_{(aq)}$	+0.60
$I_{2(s)} + 2e^- \rightleftharpoons$	$2I^-_{(aq)}$	+0.54
$Cu^+_{(aq)} + e^- \rightleftharpoons$	$Cu_{(s)}$	+0.52
$O_{2(g)} + 2H_2O_{(l)} + 4e^- \rightleftharpoons$	$4OH^-_{(aq)}$	+0.40
$Cu^{2+}_{(aq)} + 2e^- \rightleftharpoons$	$Cu_{(s)}$	+0.34
$SO^{2-}_{4(aq)} + 4H^+_{(aq)} + 2e^- \rightleftharpoons$	$H_2SO_{3(aq)} + H_2O_{(l)}$	+0.17
$Sn^{4+}_{(aq)} + 2e^- \rightleftharpoons$	$Sn^{2+}_{(aq)}$	+0.15
$Cu^{2+}_{(aq)} + e^- \rightleftharpoons$	$Cu^+_{(aq)}$	+0.15
$S_{(s)} + 2H^+_{(aq)} + 2e^- \rightleftharpoons$	$H_2S_{(aq)}$	+0.14
$AgBr_{(s)} + e^- \rightleftharpoons$	$Ag_{(s)} + Br^-_{(aq)}$	+0.07
$2H^+_{(aq)} + 2e^- \rightleftharpoons$	$H_{2(g)}$	0.00
$Pb^{2+}_{(aq)} + 2e^- \rightleftharpoons$	$Pb_{(s)}$	−0.13
$Sn^{2+}_{(aq)} + 2e^- \rightleftharpoons$	$Sn_{(s)}$	−0.14
$AgI_{(s)} + e^- \rightleftharpoons$	$Ag_{(s)} + I^-_{(aq)}$	−0.15
$Ni^{2+}_{(aq)} + 2e^- \rightleftharpoons$	$Ni_{(s)}$	−0.26
$Co^{2+}_{(aq)} + 2e^- \rightleftharpoons$	$Co_{(s)}$	−0.28
$H_3PO_{4(aq)} + 2H^+_{(l)} + 2e^- \rightleftharpoons$	$H_3PO_{3(aq)} + H_2O_{(l)}$	−0.28
$PbSO_{4(s)} + 2e^- \rightleftharpoons$	$Pb_{(s)} + SO^{2-}_{4(aq)}$	−0.36
$Se_{(s)} + 2H^+_{(aq)} + 2e^- \rightleftharpoons$	$H_2Se_{(aq)}$	−0.40
$Cd^{2+}_{(aq)} + 2e^- \rightleftharpoons$	$Cd_{(s)}$	−0.40
$Cr^{3+}_{(aq)} + e^- \rightleftharpoons$	$Cr^{2+}_{(aq)}$	−0.41
$Fe^{2+}_{(aq)} + 2e^- \rightleftharpoons$	$Fe_{(s)}$	−0.44
$Ag_2S_{(s)} + 2e^- \rightleftharpoons$	$2Ag_{(s)} + S^{2-}_{(aq)}$	−0.69
$Zn^{2+}_{(aq)} + 2e^- \rightleftharpoons$	$Zn_{(s)}$	−0.76
$Te_{(s)} + 2H^+_{(aq)} + 2e^- \rightleftharpoons$	$H_2Te_{(aq)}$	−0.79
$2H_2O_{(l)} + 2e^- \rightleftharpoons$	$H_{2(g)} + 2OH^-_{(aq)}$	−0.83
$Cr^{2+}_{(aq)} + 2e^- \rightleftharpoons$	$Cr_{(s)}$	−0.91
$SO^{2-}_{4(aq)} + H_2O_{(l)} + 2e^- \rightleftharpoons$	$SO^{2-}_{3(aq)} + 2OH^-_{(aq)}$	−0.93
$Al^{3+}_{(aq)} + 3e^- \rightleftharpoons$	$Al_{(s)}$	−1.66
$Mg^{2+}_{(aq)} + 2e^- \rightleftharpoons$	$Mg_{(s)}$	−2.37
$Na^+_{(aq)} + e^- \rightleftharpoons$	$Na_{(s)}$	−2.71
$Ca^{2+}_{(aq)} + 2e^- \rightleftharpoons$	$Ca_{(s)}$	−2.87
$Ba^{2+}_{(aq)} + 2e^- \rightleftharpoons$	$Ba_{(s)}$	−2.91
$K^+_{(aq)} + e^- \rightleftharpoons$	$K_{(s)}$	−2.93
$Li^+_{(aq)} + e^- \rightleftharpoons$	$Li_{(s)}$	−3.04

SOA Strongest Oxidizing Agent

Decreasing Strength of Oxidizing Agents

Decreasing Strength of Reducing Agents

SRA Strongest Reducing Agent

- All E° values are reduction potentials measured relative to the standard hydrogen electrode. E° values are measured using standard half-cells with both the oxidizing and reducing agents present at SATP using 1.0 mol/L solutions.
- Values in this table are taken from *The CRC Handbook of Chemistry and Physics*, 71st Edition.

D1 Matter and Chemical Bonding

SUMMARY

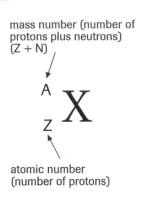

Figure 1
Symbolism representing an individual atom of an element

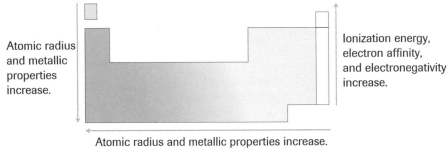

Figure 2
Trends in periodic properties

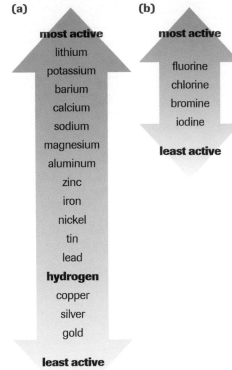

Figure 3
(a) In the activity series of metals, each metal will displace any metal listed below it. Hydrogen is usually included in the series, even though it is not a metal, because hydrogen can form positive ions, just like the metals.
(b) The halogens can also be ordered in an activity series.

Table 1 Summary of Bonding Characteristics

Intramolecular force	Bonding model
ionic bond	• involves an electron transfer, resulting in the formation of cations and anions • cations and anions attract each other
polar covalent bond	• involves unequal sharing of pairs of electrons by atoms of two different elements • bonds can involve 1, 2, or 3 pairs of electrons, i.e., single (weakest), double, or triple (strongest) bonds
nonpolar covalent bond	• involves equal sharing of pairs of electrons • bonds can involve 1, 2, or 3 pairs of electrons, i.e., single (weakest), double, or triple (strongest) bonds

Table 2 Summary of Reaction Type Generalizations

Reaction type	Reactants	Products
combustion	metal + oxygen	metal oxide
	nonmetal + oxygen	nonmetal oxide
	fossil fuel + oxygen	carbon dioxide + water
synthesis	element + element	compound
	element + compound	more complex compound
	compound + compound	more complex compound
decomposition	binary compound	element + element
	complex compound	simpler compound + simpler compound or simpler compound + element(s)
single displacement	A + BC	B + AC
double displacement	AB + CD	AD + CB

Table 3 Classical and IUPAC Names of Common Multivalent Metal Ions

Metal	Ion	Classical name	IUPAC name
iron	Fe^{2+}	ferrous	iron(II)
	Fe^{3+}	ferric	iron(III)
copper	Cu^{+}	cuprous	copper(I)
	Cu^{2+}	cupric	copper(II)
tin	Sn^{2+}	stannous	tin(II)
	Sn^{4+}	stannic	tin(IV)
lead	Pb^{2+}	plumbous	lead(II)
	Pb^{4+}	plumbic	lead(IV)
antimony	Sb^{3+}	stibnous	antimony(III)
	Sb^{5+}	stibnic	antimony(V)
cobalt	Co^{2+}	cobaltous	cobalt(II)
	Co^{3+}	cobaltic	cobalt(III)
gold	Au^{+}	aurous	gold(I)
	Au^{3+}	auric	gold(III)
mercury	Hg^{+}	mercurous	mercury(I)
	Hg^{2+}	mercuric	mercury(II)

Table 4 Prefixes Used When Naming Binary Molecular Compounds

Subscript in chemical formula	Prefix in chemical nomenclature
1	mono
2	di
3	tri
4	tetra
5	penta
6	hexa
7	hepta
8	octa
9	nona
10	deca

D

Table 6 Solubility of Ionic Compounds at SATP

		Anions						
		Cl^-, Br^-, I^-	S^{2-}	OH^-	SO_4^{2-}	$CO_3^{2-}, PO_4^{3-}, SO_3^{2-}$	$C_2H_3O_2^-$	NO_3^-
Cations	High solubility (aq) ≥ 0.1 mol/L (at SATP)	most	Group 1, NH_4^+ Group 2	Group 1, NH_4^+ Sr^{2+}, Ba^{2+}, Tl^+	most	Group 1, NH_4^+	most	all
		All Group 1 compounds, acids, and all ammonium compounds are assumed to have high solubility in water.						
	Low Solubility (s) <0.1 mol/L (at SATP)	$Ag^+, Pb^{2+}, Tl^+,$ $Hg_2^{2+} (Hg^+),$ Cu^+	most	most	$Ag^+, Pb^{2+}, Ca^{2+},$ $Ba^{2+}, Sr^{2+}, Ra^{2+}$	most	Ag^+	none

▶ Practice

1. Write the chemical name and symbol corresponding to each of the following theoretical descriptions:
   (a) 3 protons, 4 neutrons, and 3 electrons
   (b) 20 protons, mass number 40, and 18 electrons
   (c) 10 electrons, net charge of 2−
   (d) 6 protons, 8 neutrons, no charge

2. When a gas is heated, the gas will emit light. Use the Bohr model of the atom to explain why this phenomenon occurs.

3. Use the periodic table to predict the most common charges on ions of chlorine, potassium, and calcium. Provide a theoretical explanation of your answer.

4. Are the following pairs of atoms more likely to form ionic or covalent bonds? Give reasons for your answer.
   (a) chlorine and chlorine
   (b) potassium and iodine
   (c) carbon and oxygen
   (d) magnesium and fluorine

5. Draw a Lewis structure and a structural formula for each of the following:
   (a) O_2 (f) N_2H_4
   (b) CH_4 (g) HCN
   (c) NH_3 (h) H_2S
   (d) PF_3 (i) OH^-
   (e) CO_2 (j) H_3O^+

6. Identify the more polar bond in each of the following pairs:
   (a) C—H; O—H   (d) S—H; O—H
   (b) C—O; N—O   (e) H—Cl; H—I
   (c) C—C; C—H

7. Predict whether carbon tetrachloride, CCl_4, is a polar or nonpolar substance. Give reasons for your answer.

8. Write the formula, including state of matter, for each of the following compounds.
   (a) aluminum chloride
   (b) copper(II) sulfate
   (c) calcium hydroxide
   (d) lead(II) nitrate
   (e) sulfuric acid
   (f) ferrous iodide
   (g) ammonium nitrate
   (h) sodium phosphate
   (i) stannic bromide
   (j) iron(III) carbonate

   (k) potassium dichromate
   (l) cobalt(III) sulfate

9. Write the IUPAC name for each of the following:
   (a) $CuCl_{(s)}$
   (b) $Fe_2O_{3(s)}$
   (c) plumbic iodide
   (d) $SF_{6(l)}$
   (e) $NH_4ClO_{3(s)}$
   (f) $Cu(NO_3)_{2(s)}$
   (g) hydrochloric acid
   (h) pentaphosphorus decaoxide
   (i) $SnH_{4(g)}$
   (j) $Ca(HCO_3)_{2(s)}$
   (k) $KMnO_{4(s)}$
   (l) $CuSO_4 \cdot 5H_2O_{(s)}$

10. For each of the following reactions, write a balanced equation and classify the reaction as synthesis, decomposition, combustion, single displacement, or double displacement:
    (a) iron + copper(I) nitrate → iron(II) nitrate + copper
    (b) phosphorus + oxygen → diphosphorus pentoxide
    (c) calcium carbonate → calcium oxide + carbon dioxide
    (d) propane + oxygen → carbon dioxide + water
    (e) lead(II) hydroxide → lead(II) oxide + water
    (f) ammonia + sulfuric acid → ammonium sulfate
    (g) potassium phosphate + magnesium chloride → magnesium phosphate + potassium chloride

11. For each of the following, use an activity series to determine which single displacement reactions will proceed. For the reactions that do occur, predict the products and complete and balance the equation. Note reactions that do not occur with NR.
    (a) $Cu_{(s)} + HCl_{(aq)} \rightarrow$
    (b) $Au_{(s)} + ZnSO_{4(aq)} \rightarrow$
    (c) $Pb_{(s)} + CuSO_{4(aq)} \rightarrow$
    (d) $Cl_{2(g)} + NaBr_{(aq)} \rightarrow$
    (e) $Fe_{(s)} + AgNO_{3(aq)} \rightarrow$

12. Predict the products potentially formed by double displacement reactions in aqueous solutions of each of the following pairs of compounds. In each case, write a balanced chemical equation indicating the physical state of the products formed, and predict whether the reaction will proceed.
    (a) copper(II) chloride and magnesium nitrate
    (b) ammonium sulfate and silver nitrate
    (c) barium hydroxide and potassium sulfate

D2 Quantities in Chemical Reactions

SUMMARY

Table 7 Stoichiometry, Symbols and Units

Symbol	Quantity	Unit
n	amount	mol
m	mass	mg, y, kg
M	molar mass	g/mol
N	number of entities	atoms, ions, formula units, molecules
N_A	Avogadro's constant, 6.02×10^{23}/mol	–

Calculating Mass of Reactants and Products

Begin with a balanced chemical equation, with the measured mass of reactant or product written beneath the corresponding formula.

1. Convert the measured mass into an amount in moles.

2. Use the mole ratio in the balanced equation to predict the amount in moles of desired substance.

3. Convert the predicted amount in moles into mass (See example, **Figure 4**).

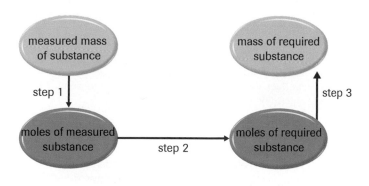

$$Fe_2O_{3(s)} + 3\,CO_{(g)} \rightarrow 2\,Fe_{(s)} + 3\,CO_{2(g)}$$

Figure 4
Steps showing calculations

Determining the Limiting Reactant

1. Write a balanced equation for the reaction.

2. Select one of the reactants and calculate the amount in moles *available*.

3. Use mole ratios in the balanced equation to calculate the amount in moles *needed* of the *other* reactants.

4. Calculate the *available* amount in moles of the *other* reactants. If the available amount of a reactant is more than sufficient, it is in excess. If the available amount is insufficient, it is limiting. (See example, **Figure 5**)]

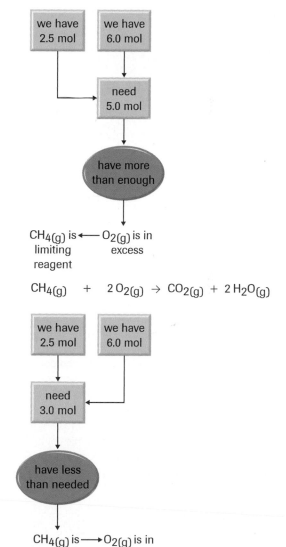

(a) $\quad CH_{4(g)} + 2\,O_{2(g)} \rightarrow CO_{2(g)} + 2\,H_2O_{(g)}$

(b) $\quad CH_{4(g)} + 2\,O_{2(g)} \rightarrow CO_{2(g)} + 2\,H_2O_{(g)}$

Figure 5
Steps showing limiting reagent

1. Calculate the molar mass of each of the following. Express your answers in g/mol.
   (a) nitrogen gas
   (b) $C_8H_{18(6)}$
   (c) oxygen gas
   (d) nickel(II) nitrate
   (e) zinc hydrogen carbonate
   (f) $CuSO_4 \cdot 5H_2O_{(s)}$
   (g) helium gas
   (h) sulfur trioxide liquid
   (i) ammonia gas
   (j) hydrochloric acid

2. What is the amount (in moles) of each type of atom in each of the following samples?
   (a) 3.0 mol of chlorine gas
   (b) 2.0 mol of iron(III) nitrate
   (c) 4.5 mol of potassium dichromate
   (d) 1.5 mol of liquid nitrogen
   (e) 5.0 mol of ammonium sulfate

3. Calculate the mass of each of the following:
   (a) 2.5 mol of $Mg(OH)_{2(s)}$
   (b) 0.25 mol of glucose, $C_6H_{12}O_{6(s)}$
   (c) 6.75 mmol of oxygen molecules
   (d) 1.20×10^{24} atoms of copper
   (e) 3.01×10^{22} molecules of methane, $CH_{4(g)}$.

4. Calculate the amount (in moles) of each of the following samples:
   (a) 10.00 g of $H_2O_{(l)}$
   (b) 1.50 kg of aluminum oxide
   (c) 2.35 mg of sodium phosphate
   (d) 1.20×10^{-5} g of hydrogen
   (e) 1.00×10^{25} molecules of $CO_{2(g)}$

5. Calculate the percentage composition of each of the following:
   (a) $H_2SO_{4(l)}$
   (b) 2.50 g of $AgNO_{3(s)}$
   (c) $NH_4NO_{3(s)}$

6. An oxide of nitrogen was found to contain 36.8% nitrogen by mass.
   (a) Find the empirical formula for this compound.
   (b) The molar mass of this compound was found to be 76.02 g/mol. What is the molecular formula of this compound?

7. A gaseous compound contains 16.0 g of hydrogen and 96.0 g of carbon. If the molar mass of this compound is 28.06 g/mol, what is its molecular formula?

8. Balance the following equations. (You can use whole or fractional coefficents.)
   (a) $NH_{3(g)} + O_{2(g)} \rightarrow NO_{(g)} + H_2O_{(l)}$
   (b) $NO_{2(g)} + H_2O_{(l)} \rightarrow HNO_{3(aq)} + NO_{(g)}$
   (c) $C_{12}H_{22}O_{11(s)} + O_{2(g)} \rightarrow CO_{2(g)} + H_2O_{(l)}$
   (d) $KClO_{3(s)} \rightarrow KCl_{(s)} + O_{2(g)}$
   (e) $MnO_{2(s)} + HCl_{(aq)} \rightarrow MnCl_{2(aq)} + Cl_{2(g)} + H_2O_{(l)}$
   (f) $Al_2O_{3(s)} \rightarrow Al_{(s)} + O_{2(g)}$
   (g) $Ni_{(s)} + AgNO_{3(aq)} \rightarrow Ag_{(s)} + Ni(NO_3)_{2(aq)}$
   (h) $KOH + H_3PO_4 \rightarrow K_3PO_4 + H_2O$

9. Write a balanced chemical equation for each of the following reactions:
   (a) phosphorus + oxygen → diphosphorus pentoxide
   (b) aluminum sulfate + calcium hydroxide →
       aluminum hydroxide + calcium sulfate
   (c) ammonia + oxygen → nitrogen + water
   (d) calcium chloride + nitric acid →
       calcium nitrate + hydrochloric acid
   (e) ammonium sulfide + lead(II) nitrate →
       ammonium nitrate + lead(II) sulfide
   (f) aluminum sulfate + ammonium bromide →
       aluminum bromide + ammonium sulfate
   (g) sodium nitrate → sodium nitrite + oxygen
   (h) potassium phosphate + magnesium chloride →
       magnesium phosphate + potassium chloride
   (i) ammonia + sulfuric acid → ammonium sulfate
   (j) mercury(II) hydroxide + phosphoric acid →
       mercury(II) phosphate + water

10. Methanol, $CH_3OH_{(l)}$, burns in excess oxygen to produce carbon dioxide and water, according to the following equation:

    $2 CH_3OH_{(l)} + 3 O_{2(g)} \rightarrow 2 CO_{2(g)} + 4 H_2O_{(g)}$
    (a) What amount of oxygen is required to completely burn 5 mol of methanol?
    (b) What amount of carbon dioxide is produced when 12.5 mol of methanol is completely burned?

11. Magnesium metal reacts with chlorine gas to produce magnesium chloride.
    (a) Write a balanced equation for the reaction.
    (b) What mass of magnesium metal is needed to completely react with 15.00 g of chlorine gas?
    (c) What mass of magnesium metal is required to produce, in excess chlorine, 8.00 g of magnesium chloride?

12. Calcium hydroxide reacts with aqueous sodium carbonate to produce sodium hydroxide and calcium carbonate.
    (a) Write a balanced equation for this reaction.
    (b) What mass of sodium carbonate is needed to completely react with 175.0 g of calcium hydroxide?
    (c) What mass of sodium hydroxide is produced when 175.0 g of calcium hydroxide is completely reacted in an excess of sodium carbonate?

13. A single displacement reaction occurs when zinc metal is immersed in lead(II) nitrate solution.
    (a) Predict the products of the reaction.
    (b) Write a balanced equation for the reaction.
    (c) Predict the mass of lead formed when 4.55 g of zinc is completely reacted in an excess of lead(II) nitrate.
    (d) What mass of zinc metal is required to produce 50.0 g of lead in this reaction, in an excess of lead(II) nitrate?

14. Propane, $C_3H_{8(g)}$, burns in oxygen to produce carbon dioxide and water, according to the following equation:

$$C_3H_{8(g)} + 5\ O_{2(g)} \rightarrow 3\ CO_{2(g)} + 4\ H_2O_{(g)}$$

Which is the limiting reagent if:
(a) 1 mol of propane and 1 mol of oxygen are available.
(b) 5 mol of propane and 5 mol of oxygen are available.
(c) 2 mol of propane and 5 mol of oxygen are available.
(d) 2 mol of propane and 12 mol of oxygen are available.
(e) 0.36 mol of propane and 1.60 mol of oxygen are available.

15. In a blast furnace, iron(III) oxide reacts with carbon monoxide to produce iron and carbon dioxide.
(a) Write a balanced equation for the reaction.
(b) Identify the limiting reagent if 2.50 mol of iron(III) oxide and 6.50 mol of carbon monoxide are available.
(c) Identify the limiting reagent if 200.0 g of iron(III) oxide and 100.0 g of carbon monoxide are available.
(d) Predict the mass of iron produced in the reaction when 200.0 g of iron(III) oxide and 100.0 g of carbon monoxide are available.

16. When a solution containing 15.0 g of aluminum chloride is mixed with a solution containing 15.0 g of sodium hydroxide, a double displacement reaction occurs.
(a) Predict the mass of aluminum hydroxide produced.
(b) What mass of the excess reagent remains unreacted?

17. Silicon tetrafluoride is produced from the reaction of silicon dioxide and hydrofluoric acid, with water as the other product.
(a) What mass of silicon tetrafluoride can be produced from 15.00 g of silicon dioxide in excess hydrofluoric acid?
(b) If the actual yield of silicon tetrafluoride is 17.92 g, what is the percentage yield?

18. When 8.40 g of zinc metal is placed in a solution in which 11.6 g of $HCl_{(g)}$ is dissolved, hydrogen gas and zinc chloride are produced.
(a) Calculate the expected yield of hydrogen gas.
(b) If 0.19 g of hydrogen gas is produced, what is the percentage yield?

D3 Solutions and Solubility

SUMMARY

Molar Concentration (mol/L)

$$\text{molar concentration} = \frac{\text{amount of solute (in moles)}}{\text{volume of solution (in litres)}}$$

$$C = \frac{n}{v}, \qquad n = vC, \qquad v = \frac{n}{C}$$

Preparing Standard Solution by Diluting Stock Solution

$$v_i C_i = v_f C_f$$

where
v_i = initial volume (volume of stock solution used)
C_i = initial concentration
  (concentration of stock solution used)
v_f = final volume (volume of dilute solution)
C_f = final concentration (concentration of dilute solution)

Hydrogen Ion Concentration and pH

pH is the negative power of ten of the hydrogen ion concentration.

$$pH = -\log[H^+_{(aq)}] \qquad \text{or} \qquad [H^+_{(aq)}] = 10^{-pH}$$

solution:	acidic	neutral	basic
$[H^+_{(aq)}]$:	$>10^{-7}$	10^{-7}	$<10^{-7}$
pH:	<7	7	>7

Note the inverse relationship between $[H^+_{(aq)}]$ and pH. The higher the hydrogen ion molar concentration, the lower the pH.

▶ *Practice*

1. Write equations to represent the dissociation of the following ionic compounds when they are placed in water:
(a) sodium chloride
(b) potassium sulfate
(c) ammonium nitrate

2. Calculate the molar concentration (mol/L) of each of the following solutions:

(a) 0.174 mol of sodium hydroxide dissolved in water to a final volume of 0.250 L of solution
(b) 60.0 g of $NaOH_{(s)}$ dissolved in water to a final volume of 750.0 mL of solution
(c) 15.0 g of glucose, $C_6H_{12}O_{6(s)}$, dissolved in water to a final volume of 125.0 mL of solution

3. What volume of a 0.36-mol/L solution of $KCl_{(aq)}$ contains 0.09 mol of the solute?

4. What mass of sodium carbonate is required to make 0.500 L of a 0.12 mol/L solution?

5. The solubility of NaCl in water at 0°C is 31.6 g/100 mL. What mass of $NaCl_{(s)}$ can be dissolved in 375 mL of solution at 0°C?

6. Calculate the molar concentration of a solution that contains 13.8 g of potassium bicarbonate in 354 mL of solution.

7. A 0.500-L sample of a sodium sulfate solution contains 0.320 mol of the solute. Calculate the molar concentration of
   (a) sodium sulfate
   (b) sodium ions
   (c) sulfate ions

8. Calculate the amount, in moles, of solute in 24.9 mL of a 0.200 mol/L solution of $NaOH_{(aq)}$.

9. What mass of copper(II) sulfate pentahydrate is needed to prepare 150.0 mL of a 0.125 mol/L solution?

10. A 15.0-mL sample of 11.6 mol/L $HCl_{(aq)}$ is added to water to make a final volume of 500.0 mL. Calculate the concentration of the final $HCl_{(aq)}$ solution.

11. What volume of concentrated 17.8 mol/L stock solution of sulfuric acid would you need in order to prepare 2.00 L of 0.215 mol/L sulfuric acid?

12. The density of water is 1.00 g/mL.
   (a) Calculate the mass of H_2O in 1.00 L of water.
   (b) Calculate the amount, in moles, of $H_2O_{(l)}$ in 1.00 L of water.
   (c) What is the molar concentration of water?
   (d) Does the concentration of water change?

13. Write the net ionic reaction for each of the following reactions:
   (a) aqueous barium chloride and aqueous sodium sulfate
   (b) aqueous copper(II) sulfate and aluminum
   (c) aqueous lead(II) nitrate and aqueous potassium iodide

14. A 27.5-mL sample of 0.112 mol/L $CuSO_{4(aq)}$ solution is added to 45.0 mL of 0.088 mol/L $Na_2CO_{3(aq)}$. A precipitate is formed.
   (a) Write a balanced equation for the reaction.
   (b) Identify the limiting reagent in the reaction.
   (c) Calculate the mass of $CuCO_3$ that is produced in the reaction.

15. When 5.00 mL of a solution of $KCl_{(aq)}$ is added to an excess of 1.00 mol/L $Pb(NO_3)_{2(aq)}$, a precipitate of $PbCl_{2(s)}$ is formed. The mass of the precipitate is found to be 0.075 g.
   (a) Write a balanced equation for the reaction.
   (b) Calculate the molar concentration of the $KCl_{(aq)}$ solution.

16. Write a sentence to distinguish between the terms in each of the following pairs:
   (a) dissociation and ionization
   (b) a strong acid and a weak acid

(c) a strong base and a weak base

17. Calculate the pH of each of the following solutions:
   (a) a vinegar solution with $[H^+_{(aq)}] = 1 \times 10^{-2}$ mol/L
   (b) an antacid solution with a hydrogen ion concentration of 4.5×10^{-11} mol/L
   (c) orange juice with $[H^+_{(aq)}] = 5.5 \times 10^{-3}$ mol/L
   (d) a household cleaner with $[H^+_{(aq)}] = 7.2 \times 10^{-10}$ mol/L

18. Calculate the concentration of hydrogen ions in solutions with the following pH values:
   (a) pH = 5.00
   (b) pH = 2.1
   (c) pH = 9.88
   (d) pH = 7.00

19. The pH of a hydrochloric acid solution was measured to be 1.1.
   (a) Write an ionization equation for hydrochloric acid.
   (b) What is the concentration of hydrogen ions in the solution?
   (c) What is the concentration of the $HCl_{(aq)}$ solution?

20. How do acids differ from bases
   (a) according to the Arrhenius definitions?
   (b) according to the Brønsted-Lowry definitions?

21. Identify the two acid–base conjugate pairs in each of the following reactions:
   (a) $H_3O^+_{(aq)} + NH_{3(aq)} \rightarrow H_2O_{(l)} + NH^+_{4(aq)}$
   (b) $OH^-_{(aq)} + HSO_{3(aq)}^- \rightarrow H_2O_{(l)} + SO^{2-}_{3(aq)}$
   (c) $HPO^{2-}_{4(aq)} + HSO^-_{4(aq)} \rightarrow H_2PO^-_{4(aq)} + SO^{2-}_{4(aq)}$
   (d) $HS^-_{(aq)} + HCO^-_{3(aq)} \rightarrow CO^{2-}_{3(aq)} + H_2S_{(aq)}$

22. A 25.0-mL portion of 0.125 mol/L hydrochloric acid requires 21.4 mL of potassium hydroxide solution for neutralization. Calculate the molar concentration of the potassium hydroxide solution.

23. A 20.0-mL portion of sulfuric acid solution requires 16.8 mL of 0.250 mol/L sodium hydroxide solution for neutralization. Calculate the molar concentration of the sulfuric acid solution.

24. A 10.0-mL portion of calcium hydroxide solution neutralizes 15.5 mL of 0.100 mol/L nitric acid. Calculate the molar concentration of the barium hydroxide solution.

25. Calculate the molar concentration of a solution of phosphoric acid if 17.8 mL of it neutralizes 20.0 mL of 0.050 mol/L calcium hydroxide.

26. A solution of KOH is prepared by dissolving 2.00 g of KOH in water to a final volume of 250 mL of solution. What volume of this solution will neutralize 20.0 mL of 0.115 mol/L sulfuric acid?

27. Oxalic acid dihydrate, $(COOH)_2 \cdot 2H_2O$, reacts with sodium hydroxide according to the following equation:

$(COOH)_2 \cdot 2H_2O_{(s)} + 2 NaOH_{(aq)} \rightarrow$
$(COONa)_{2(aq)} + 4 H_2O_{(l)}$

If a 0.118-g sample of oxalic acid dihydrate is dissolved in water and exactly neutralized with 10.4 mL of a NaOH solution, what is the molar concentration of the NaOH solution?

D4 Gases and Atmospheric Chemistry

SUMMARY

Gas Laws

STP: 0°C and 101.325 kPa (exact values)
SATP: 25°C and 100 kPa (exact values)

101.325 kPa − 1 atm = 760 mm Hg (exact values)
or 101 kPa (for calculation)

absolute zero = 0 K
or −273.15°C, or −273°C (for calculation)

T (K) = t (°C) + 273 (for calculation)

Boyle's law: $p_1v_1 = p_2v_2$

(for constant temperature and amount of gas)

Charles's law: $\dfrac{v_1}{T_1} = \dfrac{v_2}{T_2}$

(for constant pressure and amount of gas)

pressure–temperature law: $\dfrac{p_1}{T_1} = \dfrac{p_2}{T_2}$

(for constant volume and amount of gas)

combined gas law: $\dfrac{p_1v_1}{T_1} = \dfrac{p_2v_2}{T_2}$

(for constant amount of gas)

Ideal gas law:

$pv = nRT$

where n = amount (in moles)

R = 8.31 kPa·L/(mol·K)

Other Concepts

Dalton's law of partial pressures the total pressure of a mixture of nonreacting gases is equal to the sum of the partial pressures of the individual gases.

$p_{total} = p_1 + p_2 + p_3 + ...$

Avogadro's theory equal volumes of gases at the same temperature and pressure contain equal numbers of molecules

molar volume the volume that one mole of a gas occupies at a specified temperature and pressure

V_{STP} = 22.4 L/mol; V_{SATP} = 24.8 L/mol

▶ Practice

1. A balloon filled to 2.00 L at 98.0 kPa is taken to an altitude at which the pressure is 82.0 kPa, the temperature remaining the same. What is the new volume of the balloon?

2. What volume will a sample of gas occupy at 88°C if it occupies 1.50 L at 32°C?

3. A sample of gas in a metal cylinder has a pressure of 135.0 kPa at 298 K. What is the pressure in the cylinder if the gas is heated to a temperature of 398 K?

4. A balloon has a volume of 2.75 L at 22.0°C and 101.0 kPa. What is its volume at 37.0°C and 90.0 kPa?

5. A sample of gas occupies 1.00 L at 22°C and has a pressure of 700.0 kPa. What volume would this gas occupy at STP?

6. Calculate the volume occupied by 2.50 mol of nitrogen gas at 58.6 kPa and −40.0°C.

7. Calculate the pressure exerted by 6.60 g of carbon dioxide gas at 25°C in a 2.00-L container.

8. What amount of chlorine gas is present in a sample that has a volume of 500.0 mL at 20°C and exerts a pressure of 450.0 kPa?

9. Calculate the volume of 240.0 g of hydrogen gas when it is at STP.

10. 1.00 L of an unknown gas has a mass of 1.25 g at STP. Calculate the molar mass of the gas.

11. A sample of a mixture of gases contains 80.0% nitrogen gas and 20.0% oxygen gas by volume. Calculate the mass of 1.00 L of this mixture at STP.

12. Hydrogen gas reacts with nitrogen gas to produce ammonia gas. In an experiment, 75.0 L of hydrogen gas is reacted with an excess of nitrogen gas. All gases are at the same temperature and the pressure is kept constant.
    (a) What volume of nitrogen gas is required to react completely with the hydrogen gas?
    (b) What volume of ammonia gas is produced?

13. In a laboratory, hydrogen gas was collected by water displacement at an atmospheric pressure of 98.2 kPa and a temperature of 22.0°C. Calculate the partial pressure of the dry hydrogen gas. (The vapour pressure of water at 22.0°C is 2.64 kPa.)

14. Hydrogen gas is produced when zinc metal is added to hydrochloric acid. What mass of zinc is necessary to produce 250.0 mL of hydrogen at STP?

15. Ammonium nitrate, a solid, can decompose rapidly to produce nitrogen gas, oxygen gas, and water vapour.
    (a) Write a balanced equation for the decomposition of ammonium nitrate.
    (b) What is the total volume of the gases, measured at SATP, produced from the decomposition of 1.00 kg of ammonium nitrate?

D5 Hydrocarbons and Energy

SUMMARY

Hydrocarbons

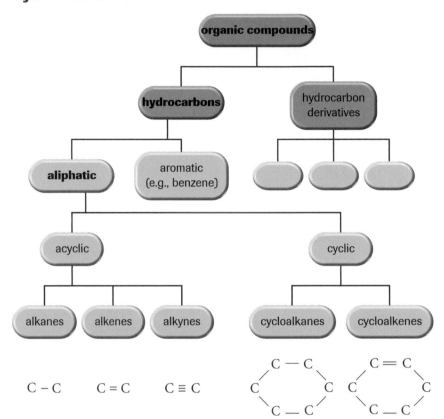

Table 8 Prefixes in Naming Alkanes, Alkenes, and Alkynes

Prefix	Number of carbon atoms
meth–	1
eth–	2
prop–	3
but–	4
pent–	5
hex–	6
hept–	7
oct–	8
non–	9
dec–	10

Figure 6
This classification system helps scientists organize their knowledge of organic compounds.

Isomers

Structural isomers: chemicals with the same molecular formula, but with different structures and different names.

Geometric (cis-trans) isomers: organic molecules that differ in structure only by the position of groups attached on either side of a carbon–carbon double bond. (A cis isomer has both groups on the same side of the molecular structure; a trans isomer has groups on opposite sides of the molecular structure.)

Specific Heat Capacity

A measure of the quantity of heat required to change the temperature of a unit mass of a substance by one degree Celsius (represented by c).

The specific heat capacity for water, $c = 4.18$ J/(g•°C)

The quantity of heat energy, q, transferred to or from a sample can be calculated:

$$q = mc\Delta T$$

Thermochemical Equations

endothermic reaction: reactants + energy (kJ) $\rightarrow$ products
exothermic reaction: reactants $\rightarrow$ products + energy (kJ)

▶ *Practice*

1. Draw a structural diagram for each of the following hydrocarbons:
   (a) 3-ethyl-2-methylhexane
   (b) 2,2,3-trimethyloctane
   (c) 1,3-dimethylcyclopentane
   (d) 4-ethyl-2-hexene
   (e) 3,4-dimethyl-2-pentene
   (f) 1-butyne

2. Write IUPAC names for the following hydrocarbons:
   (a) $CH{\equiv}C-CH_2-CH_2-CH_2-CH_3$

   (b) $CH_3-CH-CH{=}CH_2-CH_3$
       $\qquad\;\;|\quad\;\;|$
       $\qquad CH_3\; CH_3$

   (c) $CH_3-CH{=}CH_2-CH-CH-CH_3$
       $\qquad\qquad\qquad\quad\;\;|$
       $\qquad\qquad\qquad\quad CH_3$

   (d) CH_2-CH_2
       $\quad\;\;|\qquad|$
       $\quad CH_2-CH_2$

   (e) $CH_3(CH_2)_7CH_3$

   (f) $CH_2-CH_2-CH_3$
       $\;\;|$
       $CH_2-CH-CH_2-CH_2-CH_2-CH_3$
       $\qquad\quad\;|$
       $\qquad\quad CH_3$

3. Draw structural diagrams and write the IUPAC names for the five structural isomers of $C_4H_{8(g)}$.

4. Draw structural diagrams and write the IUPAC names for the geometric isomers of 2-pentene.

5. Write a balanced equation for the complete combustion of butane.

6. Classify each of the following hydrocarbons as saturated or unsaturated:
   (a) cyclohexane
   (b) ethyne
   (c) $C_3H_{8(g)}$
   (d) a compound containing only single covalent bonds
   (e) a hydrocarbon that reacts rapidly with bromine water or potassium permanganate solution

7. Calculate the quantity of heat required to raise the temperature of 1.50 L of water from 15.0°C to 75.0°C. The specific heat capacity of water is 4.18 J/(g•°C).

8. Calculate the quantity of heat required to raise the temperature of 500.0 g of water in a 325.0 g copper pot, from 12.0°C to 60.0°C. The specific heat capacity of copper is 0.385 J/(g•°C).

9. When 5.0 g of urea, $NH_2CONH_{2(s)}$, is completely dissolved in 150.0 mL of water, the temperature of the water changes from 22.0°C to 18.3°C.
   (a) Is the dissolving of urea in water endothermic or exothermic?
   (b) Calculate the specific heat of solution of urea (the energy change in dissolving 1.0 g of urea).
   (c) Calculate the molar heat of solution of urea (the energy change in dissolving 1.0 mol of urea).

10. When methanol, $CH_3OH_{(l)}$, burns in air, the products formed are carbon dioxide gas and water vapour. When 10.0 g of methanol is completely combusted, 227.0 kJ of heat is transferred.
   (a) Is the combustion of methanol endothermic or exothermic?
   (b) Calculate the molar heat of combustion of methanol.
   (c) Write a thermochemical equation for the combustion of 1.0 mol of methanol.
   (d) Write a thermochemical equation for the combustion of 3.0 mol of methanol.

This section includes answers to section questions and questions in Chapter and Unit Reviews that require calculation.

Unit 1 Organic Chemistry
Chapter 1 Organic Compounds

Section 1.3 Questions
5. 17% greater

Section 1.6 Questions
10. (b) 0.003 mol/L

Lab Exercise 1.3.1:
Preparation of Ethyne
(c) 0.050 mol $Ca(OH)_2$
(d) 1.30 g
(e) 47.2%

Chapter 1 Self-Quiz
1. False
2. False
3. True
4. False
5. True
6. (b)
7. (b)
8. (a)
9. (e)
10. (b)
11. (c)
12. (e)
13. (c)
14. (e)
15. (c)

Chapter 1 Review
11. (b) 87.0%

Chapter 2 Polymers – Plastics, Nylons, and Food

Chapter 2 Self-Quiz
1. False
2. True.
3. True.
4. False
5. False
6. (d)
7. (d)
8. (b)
9. (e)
10. (c)
11. (d)
12 (b)
13. (b)
14. (e)
15. (c)

Unit I Self-Quiz
1. False
2. True
3. False
4. False
5. False
6. True
7. False
8. False

9. False
10. True
11. (d)
12. (e)
13. (b)
14. (e)
15. (d)
16. (c)
17. (c)
18. (d)
19. (d)
20. (a)
21. (c)
22. (d)
23. (c)
24. (b)
25. (e)
26. (e)

Unit 1 Review
28. theoretical yield: 47.6 g; percent yield: 73.9%

Unit 2 Structure and Properties
Are You Ready?
6. hydrogen atom: 1,1,0
sodium atom: 11, 11, 0
chlorine atom: 17, 17, 0
hydrogen ion: 1, 0, 1+
sodium ion: 11, 10, 1+
chloride ion: 17, 18, 1–

Chapter 3 Atomic Theories
Section 3.3 Questions
6. (a) 3.6×10^{-19} J
(b) 3.6×10^{-19} J
7. (a) UV: 9.9×10^{-19} J; IR: 2.2×10^{-19} J
(b) 4.5:1

Section 3.4 Questions
13. (a) 485 nm.
(b) 6.19×10^{14} Hz
(c) 4.1×10^{-19} J
(d) 654 nm; 4.59×10^{14} Hz; 3.0×10^{-19} J
(e) 1.1×10^{-19} J

Section 3.6 Questions
1. (a) 2
(b) 8
(c) 18
(d) 32
2. (a) 1; 2
(b) 3; 6
(c) 5; 10
(d) 7; 14

Activity 3.4.2 The Hydrogen Line Spectrum and the Bohr Theory
(a) 410 nm, 434 nm, 486 nm, and 655 nm
(b) 656 nm
For H $n_i = 4$, $n_f = 2$, wavelength = 486 nm
For H $n_i = 5$, $n_f = 2$, wavelength = 434 nm
For H $n_i = 6$, $n_f = 2$, wavelength = 410 nm

Chapter 3 Self-Quiz
1. False
2. False
3. False
4. True
5. True
6. False
7. False
8. True
9. True
10. True
11. False
12. (b)
13. (d)
14. (a)
15. (c)
16. (c)
17. (b)
18. (e)
19. (d)

Chapter Review
16. (a) 2
(b) 8
(c) 18
(d) 32

Chapter 4 Chemical Bonding
Chapter 4 Self-Quiz
1. False
2. True
3. False
4. False
5. False
6. False
7. True
8. False
9. False
10. True
11. (e)
12. (b)
13. (d)
14. (a)
15. (c)
16. (e)
17. (c)
18. (a)

19. (b)
10. (d)

Unit 2 Self-Quiz
1. False
2. False
3. True
4. True
5. False
6. False
7. True
8. False
9. True
10. True
11. False
12. True
13. False
14. False
15. True
16. True
17. True
18. False
19. True
20. (e)
21. (b)
22. (c)
23. (a)
24. (a)
25. (a)
26. (c)
27. (d)
28. (b)
29. (d)
30. (b)
31. (e)
32. (a)
33. (b)
34. (c)
35. (e)
36. (c)
37. (b)
38. (c)
39. (a)
40. (d)

Unit 2 Review
34. (b) 7.8%
43. red – 4.29×10^{14} Hz; blue – 7.50×10^{14} Hz
44. highest – 4.97×10^{-19} J; lowest – 2.84×10^{-19} J
46. UV – 6.63×10^{-19} J; orange – 3.32×10^{-19} J

Unit 3 Energy Changes and Rates of Reaction
Are You Ready?
4. (c) 12540 J or 12 kJ
8. (b) 2.5 mol $NaHCO_3$/min
(c) 10 mol
(d) 2.5 mol CO_2/min

Chapter 5 Thermochemistry
Section 5.2 Questions
1. (a) 7.8 MJ
   (b) 2.08 MJ
2. 12°C
3. 1.50 g
4. 242 kJ

Section 5.3 Questions
4. (a) −11.0 MJ/mol
   (d) 17%

Section 5.4 Questions
1. (b) −247.5 kJ
2. −78.5 kJ
3. 492 kJ
4. (b) Experiment 1:−20.9 kJ;
   Experiment 2: −34.3 kJ;
   Experiment 3: −56.0 kJ/mol
   (c) 1.4 %

Section 5.5 Questions
2. (a) 100.7 kJ
   (b) −1411 kJ
   (c) −5640 kJ
3. (b) −96.6 kJ
4. (a) −1.79 MJ/mol acetone
   (h) −1.5 MJ/mol acetone
   (c) 16%

Lab Exercise 5.5.1 Testing Enthalpies of Formation
(a) − 726 kJ
(b) −597 kJ/mol
(c) 18%

Chapter 5 Self-Quiz
1. False
2. False
3. True
4. False
5. True
6. True
7. False
8. True
9. False
10. True
11. (c)
12. (b)
13. (e)
14. (c)
15. (c)
16. (c)
17. (d)
18. (a)

Chapter 5 Review
2. 1.10 J/(g•°C)
4. 170 kJ
5. 547 g
9. (c) −253.9 kJ
10. 206 kJ
11. 25.7 g
12. −117 kJ
13. (a) 382.8 kJ/mol NH_3
    (b) 2.25×10^4 kJ
    (c) 6.25 m^2

14. 129 kJ
15. −388.3 kJ/mol
16. (c) − 55 kJ
    (d) +19 kJ
18. (a) −44 kJ
    (b) −285.5 kJ/mol
    (c) − 1.7×10^9 kJ
    (d) $\Delta H_{condenstion}$: 0.4 cm;
        $\Delta H^{\circ}_{f(H_2O_{(l)})}$: 3 cm;
        ΔH_{fusion}: 1000 km

Chapter 6 Chemical Kinetics
Section 6.1 Questions
1. (a) 1.2 mol/(L•s)
   (b) 2.5 mol/(L•s)
   (c) 1.2 mol/(L•s)
   (d) 2.5 mol/(L•s)

Section 6.3 Questions
2. (a) 1 with respect to $Cl_{2(g)}$; 2
      with respect to $NO_{(g)}$
   (b) × 2
   (c) × 9
   (d) 3.0 L/(mol•s)
   (e) $8.2(5) \times 10^{-4}$ mol/(L•s)
3. (b) 0.495 a
   (c) 2.5 g
4. (a) 0.039 g

Section 6.4 Questions
3. (a) 60 kJ
   (b) −35 kJ

Lab Exercise 6.1.1
(c) (i) 0.4 mol/(Lmin)
    (ii) 0.075 mol/(Lmin)
(d) (i) 0.41 mol/(Lmin)
    (ii) 0.075 mol/(Lmin)

Chapter 6 Self-Quiz
1. False
2. True
3. False
4. True
5. True
6. False
7. True
8. True
9. False
10. False
11. (b)
12. (e)
13. (d)
14. (c)
15. (a)
16. (b)
17. (d)
18. (b)

Chapter 6 Review
3. 80 mL/s
5. (a) 1.47 mL/s
7. (c) 18 L^2/(mol²•s)
   (d) 0.65 mol/(L•s)
8. (b) 6.25%

Unit 3 Self-Quiz
1. False
2. True

3. False
4. True
5. False
6. True
7. True
8. False
9. False
10. True
11. (b)
12. (c)
13. (e)
14. (a)
15. (e)
16. (c)
17. (b)
18. (d)
19. (b)
20. (e)
21. (b)
22. (a)
23. (c)
24. (c)
25. (e)
26. (c)
27. (b)
28. (d)
29. (c)
30. (d)

Unit 3 Review
1. 340 kJ
2. 657 kJ
3. (a) −5.57 kJ/mol
5. (c) 44.6 kJ
6. −2572.4 kJ
7. (b) −3536.3 kJ
   (c) 982 kJ
9. (b) 0.130 mol/L
11. (b) 0.80; 1.30; 1.80; 2.20
    (c) (i) 0.092 mol/(L•h)
        (ii) 0.18 mol/(L•h)
        (iii) 0.046 mol/(L•h)
    (d) 0.14 mol/(L•h);
        0.058 mol/(L•h)
14. (b) 2.0×10^{-2} mol/(L•s) for
    $[O_{2(g)}]$; 1.2×10^{-2} mol/(L•s)
    for $[CO_{2(g)}]$
20. (a) −1.96 MJ/mol
21. (a) −104 kJ/mol
22. (a) −125.7 kJ/mol
27. (b) efficient −5470 kJ;
    non-efficient −3942 kJ
    (c) 28%
    (d) 3.7×10^2 g
    (e) 6.1×10^2 g

Unit 4 Chemical Systems and Equilibrium
Are You Ready?
(page 420)
2. (b) 0.1 mol $MgCl_2$
   (c) 0.4 mol/L
4. −1923.7 kJ/mol
5. (b) 0.027 mol/L
6. (g) 15.00 mL NaOH
   (h) 7
   (j) 1.0×10^{-7} mol/L

7. (a) 26
   (b) 0.28
   (c) 0.52 or −1.7

Chapter 7 Chemical Systems in Equilibrium
Section 7.1 Questions
3. (a) 2.00 mol
   (b) 70.0%
4. (a) $[C_2H_4]$ = 2.50 mol/L;
       $[Br_2]$ = 1.00 mol/L;
       $[C_2H_4Br_2]$ = 1.50 mol/L
   (c) 60.0%
7. (a) 0.0 mol HI; 8.0 mol I_2;
       12.0 mol H_2
   (b) 14 mol HII
   (c) 88%
8. $[PCl_5]$ = 0.90 mol/L;
   $[Cl_2]$ = 0.10 mol/L
9. (a) [CO] = 0.0600 mol/L;
       $[CH_3OH]$ = 0.0400 mol/L
   (b) 40.0%

Section 7.2 Questions
2. 49.70
3. 0.46
4. 3.9×10^{-4} mol/L
6. (c) 0.200 mol
   (d) 0.800 mol HBr
   (e) 0.400 mol H_2, 0.400 mol
       Br_2
   (f) 0.200 mol/L H_2,
       0.200 mol/L Br_2
   (g) 4.00

Section 7.5 Questions
2. 1.5
3. (a) [HBr] = 0.78 mol/L;
       $[H_2]$ = $[Br_2]$ = 0.011 mol/L
   (b) 0.39 mol HBr, 0.055 mol
       H_2, 0.055 mol Br_2
   (c) 78%
4. $[H_2]$ = 0.010 mol/L; $[I_2]$ =
   0.31 mol/L; [HI] = 0.38
   mol/L
5. $[NO_2]$ = 1.66 mol/L
6. (a) [HCl] = 0.38 mol/L;
       $[H_2]$ = $[Cl_2]$ = 1.81 mol/L
   (b) 0.285 mol HCl; 1.36 mol
       H_2; 1.36 mol Cl_2
   (c) 9.50%
7. [CO] = $[Cl_2]$ = 0.25 mol/L
8. $[PCl_5]$ = 0.199 mol/L; $[Cl_2]$
   = $[PCl_3]$ = 0.480 mol/L

Section 7.6 Questions
4. 1.0×10^{-5} mol/L
5. 1.4×10^{-5} g/100 mL
6. 2.0×10^{-3} mol/L
7. 1.0×10^{-2}
8. 3.4×10^{-11}
9. 1.7×10^{-4} g
10 (a) 6.0×10^{-4}
   (b) 2.8×10^{-11}
   (c) 5.6×10^{-9}
11. 8.5×10^{-7} mol/L

12. (a) 1.4×10^{-3} mol
    (b) 1.4×10^{-2} mol/L
    (c) 1.2×10^{-5}
13. (a) 2.5×10^{-3} mol
    (b) 5.0×10^{-3} mol
    (c) 5.0×10^{-2} mol/L
    (d) 2.5×10^{-3}

Section 7.7 Questions
11. (a) -207.5 kJ
    (b) $+803.8$ kJ
12. $300°C$
13. (b) $\Delta H° = -176.2$ kJ; $\Delta S° = -284.8$ J/K·mol ; $\Delta G° = -91.3$ kJ
14. (a) -1314.4 kJ
19. (a) 387 K

Chapter 7 Self-Quiz
1. False
2. True
3. False
4. True
5. False
6. False
7. False
8. True
9. True
10. True
11. (e)
12. (a)
13. (c)
14. (b)
15. (d)
16. (a)
17. (a)
18. (a)
19. (c)
20. (e)

Chapter 7 Review
10. (b) 2.9×10^{-3} mol/L
15. (a) $[H_2] = 1.46$ mol/L; $[Br_2] = 1.46$ mol/L; $[HBr] = 5.07$ mol/L
    (b) $[H_2] = 2.20$ mol/L; $[Br_2] = 2.20$ mol/L; $[HBr] = 2.61$ mol/L
    (c) $[H_2] = 3.00$ mol/L; $[Br_2] = 1.00$ mol/L; $[HBr] = 6.00$ mol/L
17. 1.61×10^{-10}
18. 4.8×10^{-5} mol/L

Chapter 8 Acid–Base Equilibrium

Section 8.1 Questions
4. 0.018 g

Section 8.2 Questions
2. 11.23
3. $6 \times 10^{-3}\%$
4. 6.3×10^{-5}
5. 7×10^{-4}
6. 4.65

10. (b) atropine 11.25; morphine 10.45; erythromycin 10.90
11. 7.7×10^{-10}
12. 1.4×10^{-11}
13. 10.27
15 (b) NH_3 1.7×10^{-5}; HS^- 9.1×10^{-8}; $SO_4{}^{2-}$ 1.0×10^{-12}
16. 1.6×10^{-6}
17. 11.124
18. 8.46
21. (a) 4.2×10^{-10}
23. (a) 3.20

Section 8.4 Questions
6. (a) 2.600
   (b) 4.025
   (c) 10.450
8. (a) (i) 5.206
       (ii) 8.883
       (iii) 4.283
9. 12.25

Section 8.5 Questions
9. 61 increase

Chapter 8 Self-Quiz
1. False
2. False
3. True
4. True
5. False
6. False
7. False
8. False
9. True
10. (b)
11. (b)
12. (e)
13. (a)
14. (b)
15. (c)
16. (e)
17. (a)
18. (b)
19. (a)

Chapter 8 Review
1. 0.372
2. (a) pH $= 0.0161$; pOH $= 13.984$
4. $[H^+_{(aq)}] = [F^-_{(aq)}] = 3.6 \times 10^{-2}$
5. $[H^+_{(aq)}] = 4.0 \times 10^{-8}$; pH $= 7.40$
6. pH $= 2.421$; pOH $= 11.579$
7. (a) 2.644
8. (b) $[H^+_{(aq)}] = 7.9 \times 10^{-6}$; pH $= 5.10$
9. 1.3×10^{-10}
14. 0.537 mol/L
15. (a) 5.27
    (b) 11.12
    (c) 9.26
    (e) 5.27
    (f) 1.70
25. 1.79

Unit 4 Self-Quiz
1. False
2. True
3. True
4. False
5. False
6. False
7. True
8. True
9. False
10. False
11. False
12. False
13. False
14. False
15. True
16. True
17. False
18. False
19. False
20. True
21. False
22. True
23. (b)
24. (b)
25. (e)
26. (b)
27. (b)
28. (c)
29. (b)
30. (e)
31. (c)
32. (c)
33. (d)
34. (c)
35. (e)
36. (c)
37. (b)
38. (b)
39. (d)
40. (d)
41. (e)

Unit 4 Review
1. 3.58×10^{-3}
2. 1.7×10^{-3}
6. (a) 1.3×10^{-5} mol/L
   (b) 1.2×10^{-8} mol/L
8. 7.91 mol/L
10. (a) $[H_2] = [CO_2] = 0.044$ mol/L; $[H_2O] = [CO] = 0.056$ mol/L
    (b) 1.6
11. (a) $[PCl_5] = 0.040$ mol/L; $[PCl_3] = [Cl_2] = 0.26$ mol/L
    (b) 1.7
12. $[H_2] = [I_2] = 0.0221$ mol/L; $[HI] = 0.156$ mol/L
13. $[NH3] = 0.14$ mol/L; $[N_2] = 0.032$ mol/L; $[H_2] = 0.097$ mol/L
14. 0.375 mol/L
15. 3.255×10^{-3} mol/L
16. 8.4×10^{-3} mol/L
17. 0.029 mol/L

24. -7.7 kJ
25. -801.2 kJ
26. 348 K
29. $40{:}1$
30. $[H^+_{(aq)}] = 2 \times 10^{-3}$ mol/L; $[OH^-_{(aq)}] = 5 \times 10^{-12}$ mol/L
31. 3.5×10^{-6}
34. 4.27
38. (b) 12.58
43. (a) 7.000
    (b) 1.000
    (c) 1.477
    (e) 7.000
    (f) 12.301
44. (a) 1.000
    (b) 1.477
    (c) 3.601
    (d) 4.602
    (e) 9.400
    (f) 12.046
52. 0.62 decrease
53. (a) 8.0×10^{-4} mol
    (b) 0.016 mol/L
    (c) 0.032 mol/L
    (d) 0.016 mol/L
    (f) 1.6×10^{-5}
54. (a) 1.740×10^{-4} mol
    (b) 0.38 g
    (c) 84 %
55. (a) 0.185 mol/L
58. (a) pOH $= 0.0969$; pH $= 13.903$
62. (a) 7.1×10^{-5} mol/L
    (b) 350
64. (a) 7.1×10^{-5} mol/L
    (b) 352

Unit 5 Electrochemistry
Chapter 9 Electric Cells

Section 9.2 Questions
7. 75.5 mmol/L

Section 9.5 Questions
6. $+0.48$ V
7. (a) $+1.10$ V
   (b) $+1.37$
8. -0.28 V

Chapter 9 Self-Quiz
1. True
2. False
3. True
4. False
5. False
6. True
7. False
8. True
9. True
10. (a)
11. (d)
12. (c)
13. (a)
14. (b)
15. (d)
16. (c)
17. (e)
18. (b)

Chapter 9 Review
14. (a) +0.71 V
    (b) +0.62 V
15. (a) +0.48 V
    (b) +0.48 V
    (c) +1.77 V
16. (b) +0.14 V
18. +1.54 V
22. (c) +0.47 V
25. (a) +0.23 V

Chapter 10 Electrolytic Cells
Section 10.1 Questions
5. (a) $\Delta E^\circ = -0.50$ V
    (b) $\Delta E^\circ = -0.03$ V
    (c) $\Delta E^\circ = -0.47$ V
6. (a) 0.43 V
    (b) 0.29 V
7. −1.30 V

Section 10.3 Questions
1. 2.80 mmol
2. 0.58 Cl_2 or 0.58 t
3. 82.8 min
4. 52.8 kA
5. (a) 1.63 Mg or 1.63 t
    (b) 4.76 Mg or 4.76 t
6. 0.174 mol/L
7. 24.42 g

Chapter 10 Self-Quiz
1. True
2. False
3. False
4. True
5. True
6. True
7. False
8. (e)
9. (a)
10. (b)
11. (d)
12. (b)
13. (e)
14. (c)
15. (a)
16. (d)

Chapter 10 Review
4. (a) 1.22 V
    (b) 0.80 V
    (c) 0.00 V
5. (a) 1.90 V
    (b) 1.23 V
    (c) 1.51 V
6. (b) 1.23 V
10. (b) 2.19 V
11. (c) 0.889 g
12. Al: 0.629 g; Ni: 2.05 g; Ag: 7.54 g
13. (a) 7.42×10^3 s
    (b) 4.05×10^3 s
    (c) 4.34×10^3 s
14. (a) 1.99 V
    (b) 590 s
15. 2.98 kA

16. 20.1 min
17. 1.03 kmol/h
18. (a) 1.8 A
    (b) 2%

Unit 5 Self-Quiz
1. True
2. False
3. False
4. True.
5. False
6. True
7. True
8. False
9. True
10. False
11. False
12. True
13. False
14. True
15. True
16. True
17. False
18. False
19. (b)
20. (e)
21. (c)
22. (c)
23. (d)
24. (e)
25. (b)
26. (d)
27. (a)
28. (d)
29. (e)
30. (b)
31. (c)
32. (a)
33. (c)
34. (d)
35. (a)
36. (b)

Unit 5 Review
3. (a) −2
    (b) +4
    (c) +6
    (d) +4
    (e) 0
4. (a) Sn +4; Co 0; Sn 2+; Co +2
    (b) Fe +3; Zn 0; Fe +2; Zn +2
    (c) Cl 0; I −1; Cl −1; I 0
    (d) C +3; O −2; Mn +7; O −2; H +1; C +4; O −2; Mn +2; H +1; O −2
    (e) Cl 0; S +4; O −2; O −2; H +1; Cl −1; S +6; O −2 H +1, O −2
6. (a) Cu +1; Cu O; Cu +2
16. (b) 1.20 V
18. (a) 0.51 V
    (b) 0.63 V
    (c) 1.51 V
19. (a) −0.14 V $Sn^{2+}|Sn$
20. −2.25 V

31. (a) 1.2 g
    (b) 1.1 g
32. (a) 1.5×10^3 s
    (b) 450 s
33. 0.844 kA

Appendix D
Chemistry 11 Review
Unit 2 Quantities in Chemical Reactions
1. (a) 28.02 g/mol
    (b) 114.26 g/mol
    (c) 32.00 g/mol
    (d) 182.71 g/mol
    (e) 187.42 g/mol
    (f) 249.71 g/mol
    (g) 4.00 g/mol
    (h) 80.06 g/mol
    (i) 17.04 g/mol
    (j) 36.46 g/mol
2. (a) 6 mol
    (b) Fe: 2 mol; N: 6 mol; O: 18 mol
    (c) K: 9 mol; Cr 9 mol; O: 31.5 mol
    (d) 3 mol
    (e) N: 10 mo; H: 40 mol; S: 5 mol; O: 20 mol
3. (a) 146 g
    (b) 45.0 g
    (c) 216 mg
    (d) 126 g
    (e) 0.803 g
4. (a) 0.555 mol
    (b) 14.7 mol
    (c) 1.43×10^{-5} mol
    (d) 5.94×10^{-6} mol
    (e) 16.6 mol
5. (a) H: 2.06%; S: 32.69%; O: 65.25%
    (b) Ag: 63.498%; N 8.247%; O: 28.26%
    (c) N: 35.00%; H: 5.05%; O: 59.96%
10. (a) 7.5 mol
    (b) 12.5 mol
11. (b) 5.144 g
    (c) 2.04 g
12. (b) 250.3 g.
    (c) 189.0 g
13. (c) 14.4 g
    (d) 15.8 g
15. (d) 132.9 g
16. (a) 8.82 g.
    (b) 1.44 g
17. (a) 25.98 g
    (b) 68.98%
18. (a) 0.259 g
    (b) 73%

Unit 3 Solutions and Solubility
2. (a) 0.696 mol/L
    (b) 2.00 mol/L
    (c) 0.664 mol/L

3. 0.25 L
4. 6.4 g
5. 119 g
6. 0.390 mol/L
7. (a) 0.640 mol/L
    (b) 1.28 mol/L
    (c) 0.640 mol/L
8. 4.98×10^{-3} mol
9. 4.69 g
10. 0.348 mol/L
11. 24.2 mL
12. (a) 1.00×10^3 g
    (b) 55.5 mol
    (c) 55.5 mol/L
14 (c) 0.381 g
15. (b) 0.11 mol/L
17. (a) 2
    (b) 10.35
    (c) 2.26
    (d) 9.14
18. (a) 1.0×10^{-5} mol/L
    (b) 8×10^{-3} mol/L
    (c) 1.6×10^{-10} mol/L
    (d) 1.0×10^{-7} mol/L
22. 0.146 mol/L
23. 0.0105 mol/L
24. 0.0775 mol/L
25. 0.112 mol/L
26. 32.4 mL
27. 0.180 mol/L

Unit 4 Gases and Atmospheric Chemistry
1. 2.39 L
2. 1.78 L
3. 180 kPa
4. 3.25 L
5. 6.98 L
6. 82.6 L
7. 186 kPa
8. 0.092 mol
9. 2660 L
10. 27.96 g/mol
11. 1.3 g
12. (a) 25.0 L
    (b) 50.0 L
13. 95.6 kPa
14. 0.732 g
15. (b) 981 L

Unit 5 Hydrocarbons and Energy
7. 376 kJ
8. 1.00×10^2 kJ
9. (b) 0.46 kJ/g
    (c) 28 kJ/mol
10. (b) 728 kJ/mol

Glossary

A

absorption spectrum a series of dark lines (i.e., missing parts) of a continuous spectrum; produced by placing a gas between the continuous spectrum source and the observer; also known as a dark-line spectrum

acid deposition acidic rain, snow, fog, dust

acid ionization constant, K_a equilibrium constant for the ionization of an acid

acid–base indicator a chemical substance that changes colour when the pH of the system changes

actinides the 14 metals in each of periods 6 and 7 that range in atomic number from 57–70 and 89–102, respectively; the elements filling the f block

activated complex an unstable chemical species containing partially broken and partially formed bonds representing the maximum potential energy point in the change; also known as *transition state*

activation energy the minimum increase in potential energy of a system required for molecules to react

addition polymer a polymer formed when monomer units are linked through addition reactions; all atoms present in the monomer are retained in the polymer

addition reaction a reaction of alkenes and alkynes in which a molecule, such as hydrogen or a halogen, is added to a double or triple bond

alcohol an organic compound characterized by the presence of a hydroxyl functional group; R–OH

aldehyde an organic compound characterized by a terminal carbonyl functional group; that is, a carbonyl group bonded to at least one H atom

aldose a sugar molecule with an aldehyde functional group at C 1

aliphatic hydrocarbon a compound that has a structure based on straight or branched chains or rings of carbon atoms; does not include aromatic compounds such as benzene

alkane a hydrocarbon with only single bonds between carbon atoms

alkene a hydrocarbon that contains at least one carbon–carbon double bond; general formula, C_nH_{2n}

alkyl group a hydrocarbon group derived from an alkane by the removal of a hydrogen atom; often a substitution group or branch on an organic molecule

alkyl halide an alkane in which one or more of the hydrogen atoms have been replaced with a halogen atom as a result of a substitution reaction

alkyne a hydrocarbon that contains at least one carbon2carbon triple bond; general formula, C_nH_{2n-2}

alpha–helix a right-handed spiralling structure held by intramolecular hydrogen bonding between groups along a polymer chain

amide an organic compound characterized by the presence of a carbonyl functional group ($C=O$) bonded to a nitrogen atom

amine an ammonia molecule in which one or more H atoms are substituted by alkyl or aromatic groups

amino acid a compound in which an amino group and a carboxyl group are attached to the same carbon atom

ampere (A) the SI unit for electric current; $1 A = 1 C/s$

amphoteric (amphiprotic) a substance capable of acting as an acid or a base in different chemical reactions

anode the electrode where oxidation occurs

aromatic alcohol an alcohol that contains a benzene ring

aromatic hydrocarbon a compound with a structure based on benzene: a ring of six carbon atoms

aufbau principle "aufbau" is German for building up; each electron is added to the lowest energy orbital available in an atom or ion

autoionization of water the reaction between two water molecules producing a hydronium ion and a hydroxide ion

average rate of reaction the speed at which a reaction proceeds over a period of time (often measured as change in concentration of a reactant or product over time)

B

base ionization constant, K_b equilibrium constant for the ionization of a base

battery a group of two or more electric cells connected in series

bond dipole the electronegativity difference of two bonded atoms represented by an arrow pointing from the lower ($\delta 1$) to the higher ($\delta 2$) electronegativity

bond energy the minimum energy required to break one mole of bonds between two particular atoms; a measure of the stability of a chemical bond

bright-line spectrum a series of bright lines of light produced or emitted by a gas excited by, for example, heat or electricity

Brønsted-Lowry acid a proton donor

Brønsted-Lowry base a proton acceptor

buffer a mixture of a conjugate acid–base pair that maintains a nearly constant pH when diluted or when a strong acid or base is added; an equal mixture of a weak acid and its conjugate base

C

calorimetry the technological process of measuring energy changes in a chemical system

carbohydrate a compound of carbon, hydrogen, and oxygen, with a general formula $C_x(H_2O)_y$

carbonyl group a functional group containing a carbon atom joined with a double covalent bond to an oxygen atom; $C=O$

carboxyl group a functional group consisting of a hydroxyl group attached to the C atom of a carbonyl group; $-COOH$

carboxylic acid one of a family of organic compounds that is characterized by the presence of a carboxyl group; $-COOH$

catalyst a substance that alters the rate of a chemical reaction without itself being permanently changed

cathode the electrode where reduction occurs

cathodic protection a method of corrosion prevention in which the metal being protected is forced to become the cathode of a cell, using either an impressed current or a sacrificial anode

cellulose a polysaccharide of glucose; produced by plants as a structural material

central atom the atom or atoms in a molecule that has or have the most bonding electrons; form the most bonds

chemical change a change in the chemical bonds between atoms, resulting in the rearrangement of atoms into new substances

chemical kinetics the area of chemistry that deals with rates of reactions

chemical reaction equilibrium a dynamic equilibrium between reactants and products of a chemical reaction in a closed system

chemical system a set of reactants and products under study, usually represented by a chemical equation

chiral able to exist in two forms that are mirror images of each other

closed system a system that may exchange energy but not matter with its surroundings

closed system one in which energy can move in or out, but not matter

collision theory the theory that a reaction occurs between two molecules if they collide at the correct orientation and if the energy of the collision is sufficient to break the chemical bonds within the molecules

combustion reaction the reaction of a substance with oxygen, producing oxides and energy

common ion effect a reduction in the solubility of a salt caused by the presence of another salt having a common ion

condensation polymer a polymer formed when monomer units are linked through condensation reactions; a small molecule is formed as a byproduct

condensation reaction a reaction in which two molecules combine to form a larger product, with the elimination of a small molecule such as water or an alcohol

conjugate acid–base pair two substances whose formulas differ only by one H+ unit

corrosion an electrochemical process in which a metal reacts with substances in the environment, returning the metal to an ore-like state

coulomb (C) the SI unit for electric charge

covalent bond or **nonpolar bond** a bond in which the bonding electrons are shared equally between atoms

covalent bonding the sharing of valence electrons between atomic nuclei within a molecule or complex ion

covalent network a 3-D arrangement of covalent bonds between atoms that extends throughout the crystal

crystal lattice a regular, repeating pattern of atoms, ions, or molecules in a crystal

cyclic alcohol an alcohol that contains an alicyclic ring

cyclic hydrocarbon a hydrocarbon whose molecules have a closed ring structure

D

dehydration reaction a reaction that results in the removal of water

deoxyribonucleic acid (DNA) a polynucleotide that carries genetic information; the cellular instructions for making proteins

dimer a molecule made up of two monomers

dipeptide two amino acids joined together with a peptide bond

dipole–dipole force a force of attraction between polar molecules

disaccharide a carbohydrate consisting of two monosaccharides

dissolution the process of dissolving

double helix the coiled structure of two complementary, antiparallel DNA chains

dynamic equilibrium a balance between forward and reverse processes occurring at the same rate

E

electric cell a device that continuously converts chemical energy into electrical energy

electric current the rate of flow of charge past a point

electric potential difference (voltage) the potential energy difference per unit charge

electrode a solid electrical conductor

electrolysis the process of supplying electrical energy to force a nonspontaneous redox reaction to occur

electrolyte an aqueous electrical conductor

electrolytic cell a cell that consists of a combination of two electrodes, an electrolyte, and an external battery or power source

electron configuration a method for communicating the location and number of electrons in electron energy levels; e.g., Mg: $1s^2 \, 2s^2 \, 2p^6 \, 3s^2$

electron probability density a mathematical or graphical representation of the chance of finding an electron in a given space

electroplating depositing a layer of metal onto another object at the cathode of an electrolytic cell

electrorefining production of a pure metal at the cathode of an electrolytic cell using impure metal at the anode

elementary step a step in a reaction mechanism that only involves one-, two-, or three-particle collisions

elimination reaction a type of organic reaction that results in the loss of a small molecule from a larger molecule; e.g., the removal of H_2 from an alkane

endothermic absorbing thermal energy as heat flows into the system

endpoint the point in a titration at which a sharp change in a measurable and characteristic property occurs; e.g, a colour change in an acid–base indicator

enthalpy change (ΔH) the difference in enthalpies of reactants and products during a change

entropy, S, a measure of the randomness or disorder of a system, or the surroundings

enzyme a molecular substance (protein) in living cells that controls the rate of a specific biochemical reaction

equilibrium constant, K the value obtained from the mathematical combination of equilibrium concentrations using the equilibrium law expression

equilibrium law for any equilibrium, the ratio of the product of the concentrations of the products, raised to the power of their coefficients in the equilibrium equation, to the product of the concentrations of the reactants, also raised to the power of their coefficients in the equilibrium equation, is a constant, K

equilibrium shift reaction of a system at equilibrium, resulting in a change in the concentrations of reactants and products

equivalence point the measured quantity of titrant recorded at the point at which chemically equivalent amounts have reacted

ester an organic compound characterized by the presence of a carbonyl group bonded to an oxygen atom

esterification a condensation reaction in which a carboxylic acid and an alcohol combine to produce an ester and water

ether an organic compound with two alkyl groups (the same or different) attached to an oxygen atom

exothermic releasing thermal energy as heat flows out of the system

F

Faraday Constant the charge of one mole of electrons; $F = 9.65 \times 10^4$ C/mol

Faraday's law the mass of a substance formed or consumed at an electrode is directly related to the charge transferred

fatty acid a long-chain carboxylic acid

first law of thermodynamics the total amount of energy in the universe is constant. Energy can be neither created nor destroyed, but can be transferred from one object or place to another, or transformed from one form to another

forward reaction in an equilibrium equation, the left-to-right reaction

fractional distillation the separation of components of petroleum by distillation, using differences in boiling points; also called fractionation

free energy (or Gibbs free energy) energy that is available to do useful work

fuel cell an electric cell that produces electricity by a continually supplied fuel

functional group a structural arrangement of atoms that, because of their electronegativity and bonding type, imparts particular characteristics to the molecule

G

galvanic cell an arrangement of two half-cells that can produce electricity spontaneously

glycogen a polysaccharide of glucose; produced by animals for energy storage

half-cell an electrode and an electrolyte forming half of a complete cell

H

half-life the time for half of the nuclei in a radioactive sample to decay, or for half the amount of a reactant to be used up (in a first-order reaction)

heat amount of energy transferred between substances

Heisenberg uncertainty principle it is impossible to simultaneously know exact position and speed of a particle

Hess's Law the value of the ΔH for any reaction that can be written in steps equals the sum of the values of ΔH for each of the individual steps

heterogeneous catalyst a catalyst in a reaction in which the reactants and the catalyst are in different physical states

heterogeneous equilibria equilibria in which reactants and products are in more than one phase

homogeneous catalyst a catalyst in a reaction in which the reactants and the catalyst are in the same physical state

homogeneous equilibria equilibria in which all entities are in the same phase

Hund's rule one electron occupies each of several orbitals at the same energy before a second electron can occupy the same orbital

hybrid orbital an atomic orbital obtained by combining at least two different orbitals

hybridization a theoretical process involving the combination of atomic orbitals to create a new set of orbitals that take part in covalent bonding

hydration reaction a reaction that results in the addition of a water molecule

hydrocarbon an organic compound that contains only carbon and hydrogen atoms in its molecular structure

hydrogen bonding the attraction of hydrogen atoms bonded to N, O, or F atoms to a lone pair of electrons of N, O, or F atoms in adjacent molecules

hydrolysis a reaction in which a bond is broken by the addition of the components of water, with the formation of two or more products

hydrolysis a reaction of an ion with water to produce an acidic or basic solution (hydronium or hydroxide ions)

hydroxyl group an –OH functional group characteristic of alcohols

I

inert electrode a solid conductor that will not react with any substances present in a cell (usually carbon or platinum)

instantaneous rate of reaction the speed at which a reaction is proceeding at a particular point in time

intermolecular force the force of attraction and repulsion between molecules

ion product constant for water, **Kw** equilibrium constant for the dissociation of water; 1.0×10^{-14}

ionic bond a bond in which the bonding pair of electrons is mostly with one atom/ion

ionic bonding the electrostatic attraction between positive and negative ions in the crystal lattice of a salt

isoelectronic having the same number of electrons per atom, ion, or molecule

isolated system an ideal system in which neither matter nor energy can move in or out

isomer a compound with the same molecular formula as another compound, but a different molecular structure

isotope ($_Z^A$X) a variety of atoms of an element; atoms of this variety have the same number of protons as all atoms of the element, but a different number of neutrons

IUPAC International Union of Pure and Applied Chemistry; the organization that establishes the conventions used by chemists

K

ketone an organic compound characterized by the presence of a carbonyl group bonded to two carbon atoms

ketose a sugar molecule with a ketone functional group, usually at C 2

L

lanthanides the 14 metals in each of periods 6 and 7 that range in atomic number from 57–70 and 89–102, respectively; the elements filling the f block

Le Châtelier's principle when a chemical system at equilibrium is disturbed by a change in a property, the system adjusts in a way that opposes the change

Lewis acid an electron-pair acceptor

Lewis base an electron-pair donor

London force the simultaneous attraction of an electron by nuclei within a molecule and by nuclei in adjacent molecules

M

macromolecule a large molecule composed of several subunits

magnetic quantum number, m_l relates primarily to the direction of the electron orbit. The number of values for ml is the number of independent orientations of orbits that are possible

Markovnikov's rule When a hydrogen halide or water is added to an alkene or alkyne, the hydrogen atom bonds to the carbon atom within the double bond that *already has more hydrogen atoms*. This rule may be remembered simply as "the rich get richer."

molar enthalpy of reaction, ΔH_x the energy change associated with the reaction of one mole of a substance (also called molar enthalpy change)

molar enthalpy, ΔH_x the enthalpy change associated with a physical, chemical, or nuclear change involving one mole of a substance

monomer a molecule of relatively low molar mass that is linked with other similar molecules to form a polymer

monoprotic acid an acid that possesses only one ionizable (acidic) proton

monosaccharide a carbohydrate consisting of a single sugar unit

N

neutron ($_0^1$n or n) a neutral (uncharged) subatomic particle present in the nucleus of atoms

nonpolar bond a nonpolar bond results from a zero difference in electronegativity between the bonded atoms; a covalent bond with equal sharing of bonding electrons

nonpolar molecule a molecule that has either nonpolar bonds or polar bonds whose bond dipoles cancel to zero

nuclear change a change in the protons or neutrons in an atom, resulting in the formation of new atoms

nucleotide a monomer of DNA, consisting of a ribose sugar, a phosphate group, and one of four possible nitrogenous bases

O

open system one in which both matter and energy can move in or out

orbital a region of space around the nucleus where an electron is likely to be found

order of reaction the exponent value that describes the initial concentration dependence of a particular reactant

organic family a group of organic compounds with common structural features that impart characteristic physical properties and reactivity

organic halide a compound of carbon and hydrogen in which one or more hydrogen atoms have been replaced by halogen atoms

overall order of reaction the sum of the exponents in the rate law equation

oxidation a process in which electrons are lost; an increase in oxidation number

oxidation number a positive or negative number corresponding to the apparent charge that an atom in a molecule or ion would have if the electron pairs in covalent bonds belonged entirely to the more electronegative atom

oxidation reaction a chemical transformation involving a loss of electrons; historically used in organic chemistry to describe any reaction involving the addition of oxygen atoms or the loss of hydrogen atoms

oxidizing agent a substance that gains or removes electrons from another substance in a redox reaction

P

Pauli exclusion principle no two electrons in an atom can have the same four quantum numbers; no two electrons in the same atomic orbital can have the same spin; only two electrons with opposite spins can occupy any one orbital

peptide bond the bond formed when the amine group of one amino acid reacts with the acid group of the next

percent reaction the yield of product measured at equilibrium compared with the maximum possible yield of product

pH meter a device used to measure pH; based on the electric potential of a silver–silver chloride glass electrode and a saturated calomel (dimercury(I) chloride) electrode

pH the negative of the logarithm to the base ten of the concentration of hydrogen (hydronium) ions in a solution

phase equilibrium a dynamic equilibrium between different physical states of a pure substance in a closed system

photoelectric effect the release of electrons from a substance due to light striking the surface of a metal

photon a quantum of light energy

physical change a change in the form of a substance, in which no chemical bonds are broken

pi (Π) bond a bond created by the side-by-side (or parallel) overlap of atomic orbitals, usually p orbitals

pK_w $pk_w = -\log Kw$

plastic a synthetic substance that can be moulded (often under heat and pressure) and that then retains its given shape

pleated-sheet conformation a folded sheetlike structure held by intramolecular or intermolecular hydrogen bonding between polymer chains

pOH a solution's pOH may be used to calculate the hydroxide ion concentration

polar bond a polar bond results from a difference in electronegativity between the bonding atoms; one end of the bond is, at least partially, positive and the other end is equally negative

polar covalent bond a bond in which electrons are shared somewhat unequally

polar molecule a molecule that has polar bonds with dipoles that do not cancel to zero

polyalcohol an alcohol that contains more than one hydroxyl functional group

polyamide a polymer formed by condensation reactions resulting in amide linkages between monomers

polyester a polymer formed by condensation reactions resulting in ester linkages between monomers

polymer a molecule of large molar mass that consists of many repeating subunits called monomers

polymerization the process of linking monomer units into a polymer

polypeptide a polymer made up of amino acids joined together with peptide bonds

polyprotic acid an acid with more than one ionizable (acidic) proton

polysaccharide a polymer composed of monosaccharide monomers

potential energy diagram a graphical representation of the energy transferred during a physical or chemical change

primary alcohol an alcohol in which the hydroxyl functional group is attached to a carbon which is itself attached to only one other carbon atom

primary cell an electric cell that cannot be recharged

primary standard a chemical, available in a pure and stable form, for which an accurate concentration can be prepared; the solution is then used to determine precisely, by the means of titrating, the concentration of a titrant

primary structure the sequence of the monomers in a polymer chain; in polypeptides and proteins, it is the sequence of amino acid subunits

principal quantum number n the principal quantum number relates primarily to the main energy of an electron, $n = 1, 2, 3, 4$

proton ($_0^1$p or p+) a positively charged subatomic particle found in the nucleus of atoms

Q

quantitative reaction a reaction in which virtually all of the limiting reagent is consumed

quantum a small discrete, indivisible quantity (plural, quanta); a quantum of light energy is called a photon

quantum mechanics the current theory of atomic structure based on wave properties of electrons; also known as wave mechanics

quaternary structure Some proteins are complexes formed from two or more protein subunits, joined by van der Waals forces and hydrogen bonding between protein subunits. For example, hemoglobin has four subunits held together in a roughly tetrahedral arrangement.

R

rate constant the proportionality constant in the rate law equation

rate law equation the relationship among rate, the rate constant, the initial concentrations of reactants, and the orders

of reaction with respect to the reactants; also called rate equation or rate law

rate of reaction the speed at which a chemical change occurs, generally expressed as change in concentration per unit time

rate-determining step the slowest step in a reaction mechanism

reaction intermediates molecules formed as short-lived products in reaction mechanisms

reaction mechanism a series of elementary steps that makes up an overall reaction

reaction quotient, Q a test calculation using measured concentration values of a system in the equilibrium expression

redox spontaneity rule a spontaneous redox reaction occurs only if the oxidizing agent (OA) is above the reducing agent (RA) in a table of relative strengths of oxidizing and reducing agents

reducing agent a substance that loses or gives up electrons to another substance in a redox reaction

reduction a process in which electrons are gained; a decrease in oxidation number

reference half-cell a half-cell arbitrarily assigned an electrode potential of exactly zero volts; the standard hydrogen half-cell

representative elements the metals and nonmetals in the main blocks, Groups 1–2, 13–18, in the periodic table; in other words, the s and p blocks

reverse reaction in an equilibrium equation, the right-to-left reaction

reversible reaction a reaction that can achieve equilibrium in the forward or reverse direction

ribonucleic acid (RNA) a polynucleotide involved as an intermediary in protein synthesis

S

sample the solution being analyzed in a titration

saponification a reaction in which an ester is hydrolyzed

saponification: the reaction in which a triglyceride is hydrolyzed by a strong base, forming a fatty acid salt; soap making

second law of thermodynamics all changes either directly or indirectly increase the entropy of the universe

secondary alcohol an alcohol in which the hydroxyl functional group is attached to a carbon which is itself attached to two other carbon atoms

secondary cell an electric cell that can be recharged

secondary quantum number relates primarily to the shape of the electron orbit. The number of values for l equals the volume of the principal quantum number.

secondary structure the three-dimensional organization of segments of a polymer chain, such as alpha-helices and pleated-sheet structures

shell main energy level; the shell number is given by the principal quantum number, n; for the representative elements, the shell number also corresponds to the period number on the periodic table for the s and p subshells

sigma (σ) bond a bond created by the end-to-end overlap of atomic orbitals

solubility equilibrium a dynamic equilibrium between a solute and a solvent in a saturated solution in a closed system

solubility product constant (K_{sp}) the value obtained from the equilibrium law applied to a saturated solution

solubility the concentration of a saturated solution of a solute in a particular solvent at a particular temperature; solubility is a specific maximum concentration

specific heat capacity quantity of heat required to raise the temperature of a unit mass of a substance 1°C or 1K

spectroscopy a technique for analyzing spectra; the spectra may be visible light, infrared, ultraviolet, X-ray, and other types

spin quantum number, m_s relates to a property of an electron that can best be described as its spin. The spin quantum number can only be $+1/2$ or $-1/2$ for any electron

spontaneous reaction one that, given the necessary activation energy, proceeds without continuous outside assistance

standard cell a galvanic cell in which each half-cell contains all entities shown in the half-reaction equation at SATP conditions, with a concentration of 1.0 mol/L for the aqueous entities

standard cell potential $\Delta E°$ is the maximum electric potential difference (voltage) of a cell operating under standard conditions

standard enthalpy of formation the quantity of energy associated with the formation of one mole of a substance from its elements in their standard states

standard entropy the entropy of one mole of a substance at SATP; units (J/mol·K)

standard molar enthalpy of reaction, $\Delta H°_x$ the energy change associated with the reaction of one mole of a substance at 100 kPa and a specified temperature (usually 25°C)

standard reduction potential $\Delta E_r°$ represents the ability of a standard half-cell to attract electrons in a reduction half-reaction

starch a polysaccharide of glucose; produced by plants for energy storage

stationary state a stable energy state of an atomic system that does not involve any emission of radiation

strong acid an acid that is assumed to ionize quantitatively (completely) in aqueous solution (percent ionization is +99%)

strong base an ionic substance that (according to the Arrhenius definition) dissociates completely in water to release hydroxide ions

subshell orbitals of different shapes and energies, as given by the secondary quantum number, l; the subshells are most often referred to as s, p, d, and f

substitution reaction a reaction in which a hydrogen atom is replaced by another atom or group of atoms; reaction of alkanes or aromatics with halogens to produce organic halides and hydrogen halides

supersaturated solution a solution whose solute concentration exceeds the equilibrium concentration

surroundings all matter around the system that is capable of absorbing or releasing thermal energy

T

temperature average kinetic energy of the particles in a sample of matter

tertiary alcohol an alcohol in which the hydroxyl functional group is attached to a carbon which is itself attached to three other carbon atoms

tertiary structure a description of the three-dimensional folding of the alpha-helices and pleated-sheet structures of polypeptide chains

thermal energy energy available from a substance as a result of the motion of its molecules

thermochemistry the study of the energy changes that accompany physical or chemical changes in matter

third law of thermodynamics the entropy of a perfectly ordered pure crystalline substance is zero at absolute zero

$$S = 0 \quad \text{at} \quad T = 0\,K$$

threshold energy the minimum kinetic energy required to convert kinetic energy to activation energy during the formation of the activated complex

titrant the solution in a buret during a titration

titration the precise addition of a solution in a buret into a measured volume of a sample solution

transition elements the metals in Groups 3–12; elements filling d orbitals with electrons

transition point the pH at which an indicator changes colour

transition the jump of an electron from one stationary state to another

trial ion product the reaction quotient applied to the ion concentrations of a slightly soluble salt

triglyceride an ester of three fatty acids and a glycerol molecule

V

valence bond theory atomic orbitals or hybrid orbitals overlap to form a new orbital containing a pair of electrons of opposite spin

volt (V) the SI unit for electric potential difference; $1 \text{ V} = 1 \text{ J/C}$

VSEPR **V**alence **S**hell **E**lectron **P**air **R**epulsion; pairs of electrons in the valence shell of an atom stay as far apart as possible to minimize the repulsion of their negative charges

W

weak acid an acid that partially ionizes in solution but exists primarily in the form of molecules

weak base a base that has a weak attraction for protons

weak electrolyte a substance which has a relatively poor electrical conductivity in water; a substance that only partially dissociates or ionizes into ions in water

Index

A

Abegg, Richard, 224
absorption spectrometer, 205
absorption spectrum, 175
accuracy, 778
acetaldehyde, 49
acetic acid, 59, 551
acetone, 50, 51, 124
acetylene, 506
acetylsalicylic acid (ASA), 59, 80
acid–base equilibria
    autoionization of water, 532–34
    Brønsted–Lowry theory, 528–29
    reversible acid–base reactions, 529–31
acid–base indicators, 543, 596, 608–10, 803
acid–base titration, 595–614
acid deposition, 621–23
acidic solutions, 534
    salts that form, 582–83
acid ionization constant, 554–56
acids. *See also* strong acids; weak acids
    Brønsted–Lowry, 528–29
    equilibrium constants for, 802
    Lewis, 592–94
    monoprotic, 535, 802
    polyprotic, 574–78, 611–13, 802
actinides, 195
activated complex, 386
activation energy, 384–87, 394
actone, 313
addition polymers, 100–7
addition reactions, 25
adipocytes, 133
air conditioning, 34–35
alcohols
    combustion, 43
    condensation reaction, 47
    hydrogenation reaction, 54
    naming, 38–41
    oxidation reaction, 52–53
    preparing, 42–43
    properties of, 42
    reactions, 42–44
    toxicity, 38
aldehydes
    defined, 49
    hydrogenation reactions, 54
    naming, 50
    oxidation reactions, 52–53
    properties, 51
aldose, 125
aliphatic hydrocarbons, 11
alkaline solutions, 534
alkaloids, 558
alkanes, 11
    naming, 12–16, 814
    reactions, 24–25
    related alkyl groups, 12
alkenes, 11
    and alcohols, 43
    naming, 16–19, 814

preparing, 36
    reactions, 25–27
alkyl group, 11
alkyl halides, 24, 32–34, 36
alkynes, 11
    naming, 16–19, 814
    reactions, 25–27
alpha-helix, 121, 122
aluminum
    corrosion of, 710
    production of, 739–40
aluminum-air cell, 693
aluminum oxide, 683, 710
amber, 128
amides
    defined, 69
    naming/preparing, 74–75
    properties, 76
    reactions, 77
amines
    defined, 69
    naming, 69–70
    as organic bases, 558
    preparing, 73
    properties, 72
amino acids, 117
    and DNA, 129
    polypeptides from, 118–19
amino groups, 70, 482
ammonia, 69, 443, 449, 450, 461–62, 482
ampere, 686
amphiprotic, 529, 588
amphoteric, 529
amphoteric ions, hydrolysis of, 588–89
amylase, 369
Ångström, Anders, 213
anions
    common, 800
    energy-level diagrams for, 189–90
anode, 686, 697, 702
    sacrificial, 713
antacids, 537
anthracene, 12
antibiotics, 130, 293, 620
antifreeze, 39
ants, 59
aromatic alcohols, 40
aromatic hydrocarbons, 12
    naming, 19–21
    reactions, 28–30
Arrhenius, Svante, 162, 528
Arrhenius equation, 399
artificial sweeteners, 120
ascorbic acid, 59
aspartame, 120
Aspirin, 620
Aston, Francis, 164
atmospheric chemistry, summary, 813
atomic-force microscope research
    technologist, 275
atomic spectra, 174–75
atomic structure, 185–98

atomic theories
    Bohr, 175–79, 213
    Dalton, 162, 224
    early history of, 162–66
    Rutherford, 163–64, 174
    Thomson, 162–63
atoms, energy-level diagrams for, 186–89
aufbau principle, 188
autoionization of water, 532–34
average rate of reaction, 362
Avogadro's theory, 813

B

Bacon, Francis, 691
Bader, Richard, 249
balances, 784
Ballard fuel cell, 395, 691–92
Balmer, Jacob, 176
bar code scanners, 204, 205
Bardeen, John, 201
base ionization constant, 557–58
bases. *See also* strong bases; weak bases
    Brønsted–Lowry, 528–29
    Lewis, 592–94
    organic, 558–59
    polyfunctional/polyprotic, 612
BASF, 461
basic solutions, 534
    salts that form, 583–84
batch processing, 429
batteries, 508–9, 685–94, 768
BCS theory, 201
bears, 135
beeswax, 135
benzaldehyde, 49, 51
benzene, 12, 29, 241
benzene rings, 12, 29
benzoic acid, 59, 551
Berthellot, J., 433
Berti, Paul, 293
beryllium, 234
beta-pleated sheet, 122
binary covalent compounds, prefixes for, 806
biochemist, 398
bioinformatics, 137
blackbody, 169
blood glucose, 463
blood plasma, 619
Bohr, Niels, 175–79, 200
boiling points, predicting, 258–59
bomb calorimeter, 299
bond dipole, 254
bonding. *See* chemical bonding
Bosch, Carl, 461
Boyle's law, 454, 813
breathalyzer test, 62, 659
bright-line spectrum, 174–75, 205, 213
Broglie, Louis de, 199, 203
bromothymol blue, 596
Brønsted, Johannes, 528
Brønsted–Lowry theory, 528–29, 551, 581, 593

Brooks, Harriet, 167–68
buckminsterfullerene, 238
buffering action, 606
buffers
    capacity of, 616–19
    defined, 615
    examples, 619–20
    explaining, 615–16
Bunsen, Robert, 174
buret, 788
butane lighters, 428
butanol, 39
butter, rancid, 134
butyl groups, 13

C

cabbage, 543
caffeine, 80
calcium hydroxide, 548
calorimeter, 295, 299, 309
calorimetry, 300–2
    of chemical changes, 311
    of physical changes, 308–9
camels, 135
CANDU reactors, 342
capillary action, 257, 262–63
car battery, 689, 690
carbohydrates, 125
carbon, covalent networks of, 272
carbonated beverages, 424
carbon atoms, 6
carbon–carbon multiple bonds, 9
carbon-14, 379
carbon monoxide, 447
carbonyl group, 49
carboxyl group, 59
carboxylic acids
    defined, 58
    esterification, 64
    naming, 59, 530
    preparing, 61–62
    properties, 60–61
    as weak acids, 551
Carothers, Wallace, 109
catalysts
    in equilibrium systems, 455–56
    in industry/biochemical systems, 405
    and reaction rates, 369–70
    theoretical effect of, 395
catalytic antibodies, 393
catalytic converters, 369, 623
cathode, 686, 697, 702
cathode rays, 209
cathode ray tube, 162
cathodic protection, 712–13
cations
    common, 800
    electron configurations for, 193
    energy-level diagrams for, 190
cat scan, 206
cell notation, 695, 698

cell potential. *See* standard cell potential
cell reactions, stoichiometry of, 747–51
cells, 685–94. *See also* electric cells
cellular respiration, 319, 661
cellulose, 111, 125, 126–27
central atom, 243
certainty, 779
Chadwick, James, 165
charge, 747
Charles's law, 813
chemical bond
    electronegativity/polarity of, 251–52
    nature of, 231–41
chemical bonding
    Lewis theory, 224–30
    summary, 805–7
    unsuccessful theories, 226
chemical changes, 304
    calorimetry of, 311
chemical kinetics, 360. *See also* reaction rates
    explaining and applying, 392–96
chemical reaction equilibrium, 424, 429–30
chemicals, common, list of, 797
chemical systems, 298–99
Chernobyl, 343–44
chickens, and sweat, 457
chiral molecules, 117
chlor-alkali process, 740–41
chlorine, 742
chlorofluorocarbons (CFCs), 34
chlorophyll, 369
chocolates, 127
cholesterol, 40
chromium, 743
cis isomer, 814
citric acid, 59
class D fires, 653
closed systems, 299, 424
coal, 326, 339
cockroaches, 61
cold fusion, 735
collision geometry, 393
collision theory, 383–90
combined gas law, 813
combustion reaction, 24, 314, 317, 806
Cominco, 730, 743
common ion effect, 490–91
computer analyst, in bioinformatics, 137
computerized axial tomography (CT), 206
Comte, Auguste, 175
concentrated reagents, 803
concentrated solutions, 557
concentration
    and percent ionization, 556–57
    and reaction rate, 368, 372–74
    theoretical effect of, 393
    of titrant, 595
concentration changes, and Le Châtelier's principle, 450–53
concrete, 303
condensation polymers, 108–13
condensation reaction, 47

condensed states, 446
conjugate acid–base pairs, 530, 562
contact lenses, 114–16
continuous processing, 429
Cooper, William, 201
copper
    electrorefining of, 742
    extraction, 750
corrosion, 652–53, 710–13
    preventing, 712
coulomb, 686
covalent bonding, 224
covalent bonds
    double and triple, 236–40
    defined, 252
covalent network crystals, 270–73
Crick, Francis, 131
Crookes, William, 162
crosslinking, 104–6
crude oil, 11
crystal lattice, 268
crystallization, 786
crystallography, 238
cubists, 226
cyclamates, 120
cyclic alcohols, 40
cyclic hydrocarbon, 11

D

Dacron, 108
Dalton, John, 162
Dalton's law of partial pressures, 813
dark-line spectrum, 175, 205
Davisson, Clinton, 199
Davy, Sir Humphry, 737, 748, 749, 768
DDT, 32
Debye, Peter, 182
decision making, 773–74
decomposition, 806
dehydration reactions, 43
Democritus, 162
denaturation
    of DNA, 131
    of proteins, 124
deoxyribonucleic acid. *See* DNA
diabetes, 463
diagnostic tests, 789
diamond, 271, 332
Diamond, Miriam, 419
diapers, disposable, 111–12
dicarboxylic acid, 108
dienes, 104–5
digesting the precipitate, 426
dilute solution, 557
dimers, 108
diol, 108
dipeptides, 118
dipole–dipole forces, 257–58
dipping, 743
disaccharides, 126
dissolution, 425

Index

Index

Index

UNIT 1 Opener, p.2, Imagestate/Firstlight.ca, inset, Courtesy of Eugenia Kumachev

CHAPTER 1 Opener, p.7, Reuters New Media/CORBIS/MAGMA; p.8, Paul A. Souders/CORBIS/MAGMA; p.11, Paul A. Souders/CORBIS/MAGMA; p.22, James L. Amos/CORBIS/MAGMA; p.24, Fig.1, top, Charles and Josette Lenars/CORBIS/MAGMA, below, Darius Koehli/Firstlight.ca; p.29, Fig.3, Alfred Pasieka/SPL; p.31, Fig.5, The Mariner's Museum/CORBIS/MAGMA; p.34, Fig.1, NASA/SPL; p.36, SPL; p.40, Fig.2, David M. Martin, M.D./SPL; p.45, Ron Watts/CORBIS/MAGMA; p.49, Fig.2, Joe MacDonald/CORBIS/MAGMA; p.50, Fig.3, Richard Seimens; p.51, Fig.4a, Michelle Garrett/CORBIS/MAGMA, Fig.4b, CORBIS/MAGMA; p.52, Gail Mooney/CORBIS/MAGMA; p.58, Fig.1, AFP/CORBIS/MAGMA; p.59, Fig.2, Michael Freeman/CORBIS/MAGMA; p.61, top, Catherine Karnow/CORBIS/MAGMA; p.64, Fig.4, S. Jezerinac/Custom Medical Stock Photo, Fig.5, CORBIS/MAGMA; p 67, Fig.6, Richard Seimens; p.68, Fig.8, Elio Ciol/CORBIS/MAGMA; p.69, Fig.1, Sinclair Stammers/SPL; p.72, Custom Medical Stock Photo; p.80, Fig.1b, Imperial Oil Ltd.; p.87, Fig.1, Michelle Garrett/CORBIS/MAGMA

CHAPTER 2 Opener, p.99, Zefa Visual Media, Germany/MaXx Images Inc., inset, Fig.1, EyeSquared; p.100, Fig.1, CORBIS/MAGMA; p.101, Fig.2, Greg Epperson/MaXx Images Inc.; p.106, Fig.8, Image Port/MaXx Images Inc., Fig.9, Foodpix; p.114, Fig.1, CORBIS/MAGMA; p.123, Fig.5, Jeffrey L. Rotman/CORBIS/MAGMA; p.124, Bruce Iverson; p.128, Dean Conger/CORBIS/MAGMA; p.130, Fig.3a, McCoy E. Langridge/Firstlight.ca, Fig.3b, Digital Art/Firstlight.ca; p.133, Fig.1, Ron Boardman/Frank Lane Agency/CORBIS/MAGMA; p.135, Minden Shots/Firstlight.ca; p.136, Amos Nachourn/Firstlight.ca; p.137, top left, Kitt Kittle/CORBIS/MAGMA, lower left, Mug Shots/Firstlight.ca, top right, Lester Lefkowitz/Firstlight.ca, lower right, James Holmes/Fulmer Research/SPL; p.141, Fig.2, Japack Company/CORBIS/MAGMA; p.148, Fig.1, EyeSquared

UNIT 2 Opener, p.156, David Taylor/SPL, inset, Courtesy of Geoffrey A. Ozin; p.159, Fig.2, Courtesy of Intel, Fig.3, Alfred Pasieka/SPL

CHAPTER 3 Opener, p.161, Keith Kent/SPL, inset, Lawrence Manning/CORBIS/MAGMA; p.162, Fig.1, D. Boone/CORBIS/MAGMA, lower, Jean-Loup Charmet/SPL; p.163, Fig.3, Bettman/CORBIS/MAGMA; p.167, Fig.1, SPL, Fig.2 and inset, Reproduced by permission of the McGill University Archives PR017759 and PR00678; p.169, Fig.1, Marko Modic/CORBIS/MAGMA; p.170, Fig.4, Bettman/CORBIS/MAGMA; p.179, CORBIS/MAGMA; p.184, NSO/SEL/Roger Ressmeyer/CORBIS/MAGMA; p.185, Fig.1, Dr. Fred Espenak/SPL; p.192, Fig.8, Novosti Press Agency/SPL; p.198, Fig.12, Ed Eckstein/CORBIS/MAGMA; p.199, Fig.1, SPL; p.201, Fig.5, Charles O'Rear/CORBIS/MAGMA; p.203, top, Charles O'Rear/CORBIS/MAGMA, lower, Klaus Guldbransen/SPL; p.204, left, MacDuff Everton/CORBIS/MAGMA, Fig.1, Sam Ogden/SPL, Fig.2, D. Dickenson/Custom Medical Stock Photo; p.206, Fig.4, AFP/CORBIS/MAGMA, p.207, Roger Ressmeyer/CORBIS/MAGMA; p.213, Figs.2,3, Physics Dept., Imperial College/SPL; p.221, Fig.1, Robert Estall/CORBIS/MAGMA

CHAPTER 4 Opener, p.223, W. Perry Conway/CORBIS/MAGMA; p.224, Fig.2, Science Pictures Ltd. /CORBIS/MAGMA; p.226, Edimedia/CORBIS/MAGMA; p.231, Fig.1, Roger

Ressmeyer/CORBIS/MAGMA; p.238, left, Sidney Moulds/SPL, Fig.14, CORBIS/MAGMA; p.242, top, Bettman/CORBIS/MAGMA, Fig.1, Courtesy of Dr. Ronald Gillespie; p.249, Fig.4,5, Courtesy of Dr. Richard Bader; p.257, Fig.2, Thomas Wiewand/MaXx Images Inc.; p.259, Duke University Archives; p.262, Fig.7, Robert Pickett/CORBIS/MAGMA, Fig.6, Will and Deni McIntrye/Photo Researchers Inc.; p.263, Fig.9, Sinclair Stammers/SPL; p.264, Fig.12, Picture Quest, Fig.13, Courtesy of Brian M. Goldstein; p.265, Fig.14, Courtesy of Dr. Robert LeRoy; p.268, Fig.2, David Spears/CORBIS/MAGMA; p.269, Fig.3, Bruce Iverson; p.270, Fig.7a,b, Jose Manuel Sanchi/CORBIS/MAGMA, Fig.7c, Biophoto Associates/SPL/Photo Researchers; p.272, Fig.11, Charles O'Rear/CORBIS/MAGMA; p.273, Fig.13, NASA/CORBIS/MAGMA; p.275, left, David Parker/SPL, top right, Yuri Gogotski/SPL, lower right, Charles O'Rear/CORBIS/MAGMA; p.283, Fig.1, Randy Jolly/CORBIS/MAGMA; p.284, Fig.1, top, Courtesy of Royal Canadian Mint, Fig.1, lower, Richard Megna/Fundamental Photographs, N.Y.C.; p.285, Ed Wheeler/Firstlight.ca; p.291, Fig.1, Courtesy of Indigo® Instruments (indigo.com)

UNIT 3 Opener, p.292, Superstock, inset, Courtesy of Paul Berti; p.294, Fig.1a, Jay Severson/CORBIS/MAGMA, Fig.1b, Duomo/CORBIS/MAGMA, Fig.1c, Eye Ubiquitous/CORBIS/MAGMA, Fig.1d, CORBIS/MAGMA;

CHAPTER 5 Opener, p.297, CORBIS/MAGMA; p.298, Fig.1a, AFP/CORBIS/MAGMA, Fig.1b, CORBIS/MAGMA, Fig.1c, European Space Agency/SPL; p.299, Fig.3, Richard Seimens, Fig.4, TJP Photos; p.305, Fig.8, Mehau Kulyk/SPL; p.312, Fig.2, Courtesy of allantimes.com; p.313, Fig.1, Jonathan Blair/CORBIS/MAGMA; p.314, Fig.3, CORBIS/MAGMA; p.315, Fig.4, Courtesy of Inco; p.317, Fig.5, Greg Locke/Firstlight.ca; p.322, Fig.1, CORBIS/MAGMA; p.338, Fig.2, CORBIS/MAGMA; p.339, Fig.3, Copyright 2001 Ontario Power Generation Inc.; p.341, CORBIS/MAGMA; p.343, Fig.5b, Ruet Stephane/Sygma/CORBIS/MAGMA; p.344, Fig.6, Bettman/CORBIS/MAGMA; p.345, Fig.7, Roger Ressmeyer/CORBIS/MAGMA

CHAPTER 6 Opener, p.359, CORBIS/MAGMA; p.360, Fig.1, Rosenfeld Images Ltd./SPL; p.365, Fig.6,7,8, Dave Starrett; p.367, Fig.1b,c, PhotoEdit; p.368, Fig.3, Frank Lane Picture Agency/CORBIS/MAGMA; p.369, Fig.4, Ontario Ministry of Natural Resources, Fig.5, EyeSquared; p.374, Fig.1, Duomo/CORBIS/MAGMA; p.379, Fig.5, James King Holmes/SPL; p.384, Fig.2, Nik Wheeler/CORBIS/MAGMA; p.385, Fig.4, Eye Ubiquitous/CORBIS/MAGMA; p. 392, Fig.1a,b, Dave Starrett; p.395, Fig.7, Courtesy of Ballard Power Systems; p.398, Bob Krist/CORBIS/MAGMA; p.401, Fig.1, TJP Photos; p.410, Fig.1, Bruce Iverson

UNIT 4: Opener, p.418, Alfred Pasieka/SPL, inset, Courtesy of Miriam Diamond;

CHAPTER 7 Opener, p.423, Richard Megna/Fundamental Photographs N.Y.C.; p.429, Fig.5, Courtesy of Dr. Malcolm Lefcort, President, Heuristic Engineering, Vancouver, B.C.; p.439, Fig.1, Reproduced courtesy of the Library and Information Centre, Royal Society of Chemistry; p.443, Fig.3, Richard Megna/Fundamental Photographs N.Y.C.; p.447, Fig.4, Martin Bough/Fundamental Photographs N.Y.C.; p.449, Fig.5, Richard Seimens; p.450, Fig.1, SPL; p.453, Fig.6, Richard Seimens; p.457, Fig.13, CORBIS/MAGMA, right, Lowell Georgia/CORBIS/MAGMA; p.460, Dave Starrett; p.461,

Fig.1, Austrian Archives/CORBIS/MAGMA; p.462, Fig.4, Potash and Phosphate Institute; p.463, Bruce Iverson; p.465, Fig.1, NASA/SPL; p.482, Fig.1, Pat Anderson/Visuals Unlimited, Fig.2, EyeSquared; p.494, Fig.1a, Richard Megna/Fundamental Photographs N.Y.C., Fig.1b, Philip James Corwin/CORBIS/MAGMA; p.501, Fig.5, CORBIS/MAGMA; p.502, Fig.6, Jeff Greenberg/Visuals Unlimited; p.505, Fig.7, Courtesy of Oxford Magnet Technology; p.506, Fig.8, Geoff Tompkinson/SPL; p.507, Fig.9, Michael Freeman/CORBIS/MAGMA; p.510, Fig.11, Alfred Pasieka/SPL; p.512, Fig.12, Richard Megna/Fundamental Photographs N.Y.C.

CHAPTER 8 Opener, p.527, Richard Megna/Fundamental Photographs N.Y.C.; p.528, Fig.1, Hulton-Deutsch Collection/CORBIS/MAGMA, Fig.2a,b, Reproduced courtesy of the Library and Information Centre, Royal Society of Chemistry; p.537, Fig.7, Richard Seimens; p.538, Fig.8, Philip Gould/CORBIS/MAGMA; p.543, Fig.9, Richard Seimens, Fig.10, Stephen Sharnoff; p.544, Fig.11, Richard Megna/Fundamental Photographs N.Y.C.; p.545, Fig.12, SPL/Photo Researchers; p.547, Fig.13, David Lees/CORBIS/MAGMA; p.548, Fig.14, Maximilian Stock Ltd./SPL; p.553, Fig.3, Gallo Images/CORBIS/MAGMA; p.565, Fig.6, Tom Pantages; p.591, Adam Woolfit/CORBIS/MAGMA; p.609, Fig.6a,b, Richard Megna/Fundamental Photographs N.Y.C.; p.621, Fig.1, Copyright 2001 Ontario Power Generation Inc., Fig.2, EyeSquared; p.622, Fig.4, Tom McHugh/Photo Researchers Inc.; p.625, left, CORBIS/MAGMA, right, James Holmes/Thomson Laboratories/SPL; p.632, Fig.1, Richard Megna/Fundamental Photographs N.Y.C.

UNIT 5: Opener, p.646, Robert Pickett/CORBIS/MAGMA, inset, Courtesy of Gillian Goward

CHAPTER 9 Opener, p. 651, top left, AFP/CORBIS/MAGMA, top right, Lake County Museum/CORBIS/MAGMA, lower, Pierre Paul Poulin/MAGMA; p. 652, Fig.1, from left, Photo courtesy of P.E.

Martin, S.R. Martin and Michigan Technological University Archaeology Lab, Jonathan Blair/CORBIS/MAGMA, Charles Philip/Visuals Unlimited, Angelo Hornak/CORBIS/MAGMA, Jim Amos/SPL/Photo Researchers, Fig.2, Bettman/CORBIS/MAGMA; p.653, Fig.3, From Chemistry: A Human Venture by Stan Percival and Ross Wilson, Toronto, ON: Irwin Publishing, 1988, Photo by Cary Smith. Used with permission of the publisher, Fig.4, Courtesy of Aslihan Yener, University of Chicago; p.654, Fig.6, Richard Seimens; p.657, Fig.8, Health and Safety Laboratory/SPL; p.659, Fig.10, Jim Varney/SPL; p.665, Fig.1, Sindo Farina/SPL; p.674, Fig.2, Richard Seimens; p.679, Tom Brakefield/CORBIS/MAGMA; p.683, NASA; p.684, Fig.10, CP (Chuck Mitchell); p.688, Lester V. Bergman/CORBIS/MAGMA; p.692, Fig.10, Courtesy of Ballard Power Systems; p.693, Fig.12, Courtesy of UTC Fuel Cells; p. 694, Fig.13, AFP/CORBIS/MAGMA; p.710, Fig.1, Craig Aurness/CORBIS/MAGMA; p.711, Fig.3, Eye Ubiquitous/CORBIS/MAGMA; p.712, EyeSquared; p.714, Fig.5, Science VU/NASA/Visuals Unlimited; p.716, Table 1, Tom Pantages

CHAPTER 10 Opener, p.729, EyeSquared; p.730, Fig.2, Maximilian Stock Ltd./SPL; p.737, Fig.1, Mary Evans Picture Library; p.738, Fig.4, Ben Johnson/SPL; p.739, Paul A. Souders/CORBIS/MAGMA; p.742, Fig.8, Kennecott; p.743, Fig.9, CORBIS/MAGMA; p.745, left, Sam Ogden/SPL, right, David Parker for ESA/CNES/Arianespace/SPL; p.746, Fig.10, Courtesy of Westaim Corporation; p.749, SPL; p.761, Fig.2, EyeSquared; p.767, Fig.1, Larry Stepanowicz/Visuals Unlimited; p.768, Fig.2, Mary Evans Picture Library, Fig.3, Paul Silverman/Fundamental Photographs N.Y.C.

APPENDIX: Opener, Phil Schermeister/CORBIS/MAGMA; APPENDIX A: Fig. 4,5,6, 7,8,10,11,12,13,15,16,17,18, Richard Seimens; Fig 9a,b, Dave Starrett

Setup Photographs Dave Starrett

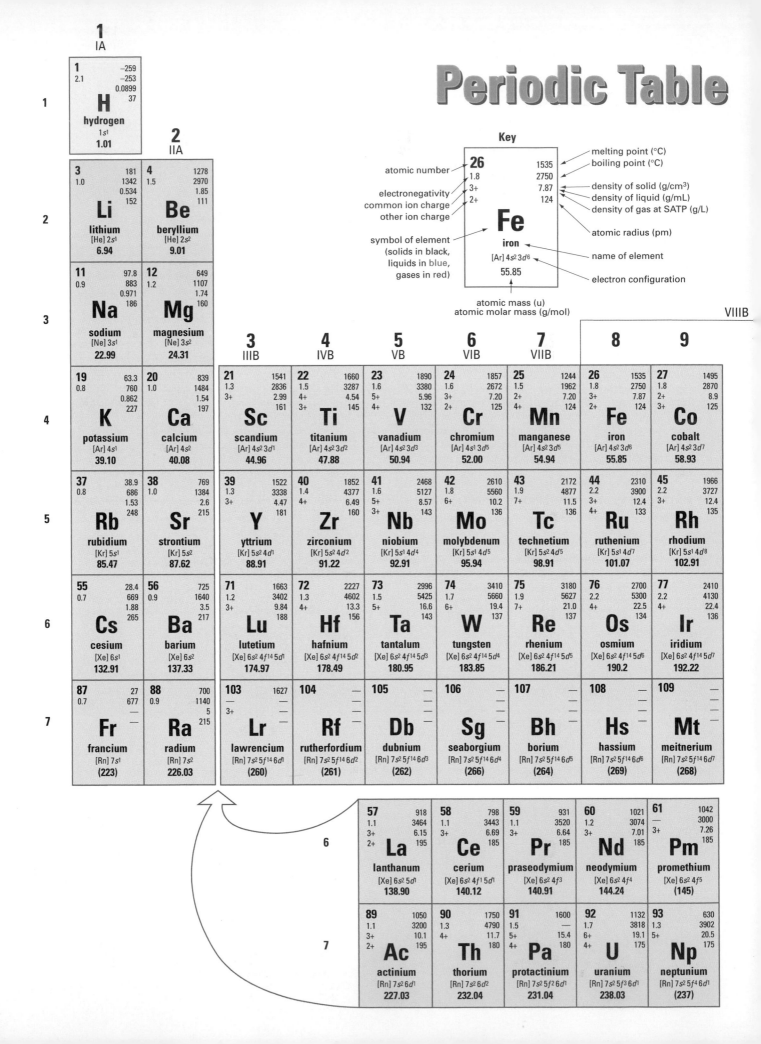

Periodic Table

of the Elements

18 VIIIA

	2	−272
	—	−269
	X	0.179
		50
	He	
	helium	
	1s2	
	4.00	

1

13 IIIA **14 IVA** **15 VA** **16 VIA** **17 VIIA**

Period 2

5 — 2300	6 — 3550	7 — −210	8 — −218	9 — −220	10 — −249
2.0 — 2550	2.5 — 4827	3.0 — −196	3.5 — −183	4.0 — −188	— −246
X — 2.34	X — 2.26	— 1.25	— 1.43	— 1.70	X — 0.900
88	77	70	66	64	62
B	**C**	**N**	**O**	**F**	**Ne**
boron	carbon	nitrogen	oxygen	fluorine	neon
[He] 2s2 2p1	[He] 2s2 2p2	[He] 2s2 2p3	[He] 2s2 2p4	[He] 2s2 2p5	[He] 2s2 2p6
10.81	12.01	14.01	16.00	19.00	20.18

Period 3

13 — 660	14 — 1410	15 — 44.1	16 — 113	17 — −101	18 — −189
1.5 — 2467	1.8 — 2355	2.1 — 280	2.5 — 445	3.0 — −34.6	— −186
— 2.70	X — 2.33	— 1.82	— 2.07	— 3.21	X — 1.78
143	117	110	104	99	95
Al	**Si**	**P**	**S**	**Cl**	**Ar**
aluminum	silicon	phosphorus	sulfur	chlorine	argon
[Ne] 3s2 3p1	[Ne] 3s2 3p2	[Ne] 3s2 3p3	[Ne] 3s2 3p4	[Ne] 3s2 3p5	[Ne] 3s2 3p6
26.98	28.09	30.97	32.06	35.45	39.95

10 **11 IB** **12 IIB**

Period 4

28 — 1455	29 — 1083	30 — 420	31 — 29.8	32 — 937	33 — 817	34 — 217	35 — −7.2	36 — −157
1.8 — 2730	1.9 — 2567	1.6 — 907	1.6 — 2403	1.8 — 2830	2.0 — 613	2.4 — 684	2.8 — 58.8	— −152
2+ — 8.90	2+ — 8.92	2+ — 7.14	3+ — 5.90	4+ — 5.35	— 5.73	— 4.81	— 3.12	X — 3.74
3+ 124	1+ 128	133	122	123	121	117	114	112
Ni	**Cu**	**Zn**	**Ga**	**Ge**	**As**	**Se**	**Br**	**Kr**
nickel	copper	zinc	gallium	germanium	arsenic	selenium	bromine	krypton
[Ar] 4s2 3d8	[Ar] 4s1 3d10	[Ar] 4s2 3d10	[Ar] 4s2 3d10 4p1	[Ar] 4s2 3d10 4p2	[Ar] 4s2 3d10 4p3	[Ar] 4s2 3d10 4p4	[Ar] 4s2 3d10 4p5	[Ar] 4s2 3d10 4p6
58.69	63.55	65.38	69.72	72.61	74.92	78.96	79.90	83.80

Period 5

46 — 1554	47 — 962	48 — 321	49 — 157	50 — 232	51 — 631	52 — 450	53 — 114	54 — −112
2.2 — 2970	1.9 — 2212	1.7 — 765	1.7 — 2080	1.8 — 2270	1.9 — 1750	2.1 — 990	2.5 — 184	— −107
2+ — 12.0	1+ — 10.5	2+ — 8.64	3+ — 7.30	4+ — 7.31	3+ — 6.68	— 6.2	— 4.93	X — 5.89
4+ 138	144	149	163	2+ 140	5+ 141	137	133	130
Pd	**Ag**	**Cd**	**In**	**Sn**	**Sb**	**Te**	**I**	**Xe**
palladium	silver	cadmium	indium	tin	antimony	tellurium	iodine	xenon
[Kr] 4d10	[Kr] 5s1 4d10	[Kr] 5s2 4d10	[Kr] 5s2 4d10 5p1	[Kr] 5s2 4d10 5p2	[Kr] 5s2 4d10 5p3	[Kr] 5s2 4d10 5p4	[Kr] 5s2 4d10 5p5	[Kr] 5s2 4d10 5p6
106.42	107.87	112.41	114.82	118.69	121.75	127.60	126.90	131.29

Period 6

78 — 1772	79 — 1064	80 — −39.0	81 — 304	82 — 328	83 — 271	84 — 254	85 — 302	86 — −71
2.2 — 3827	2.4 — 2808	1.9 — 357	1.8 — 1457	1.8 — 1740	1.9 — 1560	2.0 — 962	2.2 — 337	— −61.8
4+ — 21.5	3+ — 19.3	2+ — 13.5	1+ — 11.85	2+ — 11.3	3+ — 9.80	2+ — 9.40	— —	X — 9.73
2+ 138	1+ 144	1+ 160	3+ 170	4+ 175	5+ 155	4+ 167	142	140
Pt	**Au**	**Hg**	**Tl**	**Pb**	**Bi**	**Po**	**At**	**Rn**
platinum	gold	mercury	thallium	lead	bismuth	polonium	astatine	radon
[Xe] 6s1 4f14 5d9	[Xe] 6s1 4f14 5d10	[Xe] 6s2 4f14 5d10	[Xe] 6s2 4f14 5d10 6p1	[Xe] 6s2 4f14 5d10 6p2	[Xe] 6s2 4f14 5d10 6p3	[Xe] 6s2 4f14 5d10 6p4	[Xe] 6s2 4f14 5d10 6p5	[Xe] 6s2 4f14 5d10 6p6
195.08	196.97	200.59	204.38	207.20	209.98	(209)	(210)	(222)

Period 7

110 —	111 —	112 —	113	114 —	115	116 —	117	118
Uun	**Uuu**	**Uub**		**Uuq**		**Uuh**		
ununnilium	unununium	ununbium		ununquadium		ununhexium		
[Rn] 7s2 5f14 6d8	[Rn] 7s2 5f14 6d9	[Rn] 7s2 5f14 6d10		[Rn] 7s2 5f14 6d10 7p2		[Rn] 7s2 5f14 6d10 7p4		
(269, 271)	(272)	(277)		(285)		(289)		

Lanthanides

62 — 1074	63 — 822	64 — 1313	65 — 1356	66 — 1412	67 — 1474	68 — 1529	69 — 1545	70 — 819
1.2 — 1794	— 1527	1.1 — 3273	1.2 — 3230	1.2 — 2567	1.2 — 2700	1.2 — 2868	1.2 — 1950	1.1 — 1196
3+ — 7.52	3+ — 5.24	3+ — 7.90	3+ — 8.23	3+ — 8.55	3+ — 8.80	3+ — 9.07	3+ — 9.32	3+ — 6.97
2+ 185	185	180	175	175	175	175	175	2+ 175
Sm	**Eu**	**Gd**	**Tb**	**Dy**	**Ho**	**Er**	**Tm**	**Yb**
samarium	europium	gadolinium	terbium	dysprosium	holmium	erbium	thulium	ytterbium
[Xe] 6s2 4f6	[Xe] 6s2 4f7	[Xe] 6s2 4f7 5d1	[Xe] 6s2 4f9	[Xe] 6s2 4f10	[Xe] 6s2 4f11	[Xe] 6s2 4f12	[Xe] 6s2 4f13	[Xe] 6s2 4f14
150.36	151.96	157.25	158.92	162.50	164.93	167.26	168.93	173.04

Actinides

94 — 641	95 — 994	96 — 1340	97 — 986	98 — 900	99 — 860	100 — 1527	101 — 1021	102 — 863
1.3 — 3232	1.3 — 2607	— 3110	—	—	—	—	— 3074	—
4+ — 19.8	3+ — 13.7	3+ — 13.5	3+ — 14	3+ —	3+ —	3+ —	2+ —	2+ —
6+ 175	4+ 175		4+				3+	3+
Pu	**Am**	**Cm**	**Bk**	**Cf**	**Es**	**Fm**	**Md**	**No**
plutonium	americium	curium	berkelium	californium	einsteinium	fermium	mendelevium	nobelium
[Rn] 7s2 5f6	[Rn] 7s2 5f7	[Rn] 7s2 5f7 6d1	[Rn] 7s2 5f9	[Rn] 7s2 5f10	[Rn] 7s2 5f11	[Rn] 7s2 5f12	[Rn] 7s2 5f13	[Rn] 7s2 5f14
(244)	(243)	(247)	(247)	(251)	(252)	(257)	(258)	(259)